Digital Signal Processing

Digital Signal Processing
Fundamentals and Applications

Third Edition

Lizhe Tan

Jean Jiang

Academic Press is an imprint of Elsevier
125 London Wall, London EC2Y 5AS, United Kingdom
525 B Street, Suite 1650, San Diego, CA 92101, United States
50 Hampshire Street, 5th Floor, Cambridge, MA 02139, United States
The Boulevard, Langford Lane, Kidlington, Oxford OX5 1GB, United Kingdom

Notices

Knowledge and best practice in this field are constantly changing. As new research and experience broaden our understanding, changes in research methods, professional practices, or medical treatment may become necessary.

Practitioners and researchers must always rely on their own experience and knowledge in evaluating and using any information, methods, compounds, or experiments described herein. In using such information or methods they should be mindful of their own safety and the safety of others, including parties for whom they have a professional responsibility.

To the fullest extent of the law, neither the Publisher nor the authors, contributors, or editors, assume any liability for any injury and/or damage to persons or property as a matter of products liability, negligence or otherwise, or from any use or operation of any methods, products, instructions, or ideas contained in the material herein.

Library of Congress Cataloging-in-Publication Data
A catalog record for this book is available from the Library of Congress

British Library Cataloguing-in-Publication Data
A catalogue record for this book is available from the British Library

ISBN: 978-0-12-815071-9

Publisher: Katey Birtcher
Acquisition Editor: Steve Merken
Editorial Project Manager: Jennifer Pierce/Susan Ikeda
Production Project Manager: Sruthi Satheesh
Cover Designer: Mark Rogers

Typeset by SPi Global, India

Last digit is the print number: 9 8 7 6 5 4 3 2 1

Contents

Preface

Technology such as microprocessors, microcontrollers, and digital signal processors have become so advanced that they have had a dramatic impact on the disciplines of electronics engineering, computer engineering, and biomedical engineering. Engineers and technologists need to become familiar with digital signals and systems and basic digital signal processing (DSP) techniques. The objective of this book is to introduce students to the fundamental principles of these subjects and to provide a working knowledge such that they can apply DSP in their engineering careers.

This book can be used in an introductory DSP course at the junior or senior level in undergraduate electrical, computer, and biomedical engineering programs. The book is also useful as a reference to the undergraduate engineering students, science students, and practicing engineers.

The material has been tested for the DSP course in a signal processing sequence at Purdue University Northwest in Indiana. With the background established from this book, students will be well prepared to move forward to take other senior-level courses that deal with digital signals and systems for communications and control.

The textbook consists of 14 chapters, organized as follows:

- Chapter 1 introduces concepts of DSP and presents a general DSP block diagram. Application examples are included.
- Chapter 2 covers the sampling theorem described in time domain and frequency domain and also covers signal reconstruction. Some practical considerations for designing analog anti-aliasing lowpass filters and anti-image lowpass filters are included. The chapter ends with a section dealing with analog-to-digital conversion (ADC) and digital-to-analog conversion (DAC), as well as signal quantization and encoding.
- Chapter 3 introduces digital signals, linear time-invariant system concepts, difference equations, and digital convolutions.
- Chapter 4 introduces the discrete Fourier transform (DFT) and digital signal spectral calculations using the DFT. Methods for applying the DFT to estimate the spectra of various signals, including speech, seismic signals, electrocardiography data, and vibration signals, are demonstrated. The chapter ends with a section dedicated to illustrating fast Fourier transform (FFT) algorithms.
- Chapter 5 is devoted to the z-transform and difference equations.
- Chapter 6 covers digital filtering using difference equations, transfer functions, system stability, digital filter frequency response, and implementation methods such as direct-form I and direct-form II.
- Chapter 7 deals with various methods of finite impulse response (FIR) filter design, including the Fourier transform method for calculating FIR filter coefficients, window method, frequency sampling design, and optimal design. This chapter also includes applications using FIR filters for noise reduction and digital crossover system design.
- Chapter 8 covers various methods of infinite impulse response (IIR) filter design, including the bilinear transformation (BLT) design, impulse invariant design, and pole-zero placement design. Applications using IIR filters include the audio equalizer design, biomedical signal enhancement, dual-tone multifrequency (DTMF) tone generation, and detection with the Goertzel algorithm.

- Chapter 9 covers adaptive filters, least mean squares (LMS) algorithm, and recursive least squares (RLS) algorithm with applications such as noise cancellation, system modeling, line enhancement, cancellation of periodic interferences, echo cancellation, and 60-Hz interference cancellation in biomedical signals.
- Chapter 10 is devoted to speech quantization and compression, including pulse code modulation (PCM) coding, μ-law compression, adaptive differential pulse code modulation (ADPCM) coding, windowed modified discrete cosine transform (W-MDCT) coding, and MPEG audio format, specifically MP3 (MPEG-1, layer 3).
- Chapter 11 deals with topics pertaining to multirate DSP and applications, as well as principles of oversampling ADC, such as sigma-delta modulation. Undersampling for bandpass signals is also examined.
- Chapter 12 introduces a subband coding system and its implementation. Perfect reconstruction conditions for a two-band system are derived. Subband coding with an application of data compression is demonstrated. Furthermore, the chapter covers the discrete wavelet transform (DWT) with applications to signal coding and denoising.
- Chapter 13 covers image enhancement using histogram equalization and filtering methods, including edge detection. The chapter also explores pseudo-color image generation and detection, two-dimensional spectra, JPEG compression using DCT, image coding using the DWT, and the mixing of two images to create a video sequence. Finally, motion compensation of the video sequence is explored, which is a key element of video compression used in MPEG.
- Finally, Chapter 14 introduces DSP architectures, software and hardware, and fixed-point and floating-point implementations of digital filters. The advanced real-time implementation examples of adaptive filtering, signal quantization and coding, and sampling rate conversion are included.

MATLAB programs are listed whenever they are possible. Therefore, a MATLAB tutorial should be given to students who are new to the MATLAB environment.

- Appendix A serves as a MATLAB tutorial.
- Appendix B reviews key fundamentals of analog signal processing. Topics include Fourier series, Fourier transform, Laplace transform, and analog system basics.
- Appendixes C, D, and E overview Butterworth and Chebyshev filters, sinusoidal steady-state responses in digital filters, and derivation of the FIR filter design equation via the frequency sampling method, respectively.
- Appendix F details the derivations of wavelet analysis and synthesis equations.
- Appendix G briefly covers a review of the discrete-time random signals.
- Appendix H offers general useful mathematical formulas.

In this third edition, MATLAB projects dealing with the practical applications are included in Chapters 2, 4, 6–12. In addition, the advanced problems are added to Chapters 2–10.

Instructor support, including solutions, can be found at http://textbooks.elsevier.com. MATLAB programs and exercises for students, plus real-time C programs can be found at https://www.elsevier.com/books/digital-signal-processing/tan/978-0-12-815071-9.

Thanks to all the faculty and staff at Purdue University Northwest, Indiana for their encouragement and support. We are also indebted to all former students in our DSP classes at the Purdue University Northwest for their feedback over the years, which helped refine this edition.

Special thanks go to Steve Merken (Senior Acquisitions Editor), Jennifer Pierce and Susan Ikeda (Editorial Project Manager), Sruthi Satheesh (Project Manager) and team members at Elsevier Science Publishing for their encouragement and guidance in developing the third edition.

The book has benefited from many constructive comments and suggestions from the following reviewers and anonymous reviewers. The authors take this opportunity to thank them for their significant contributions. These include the reviewers for the third edition:

Professor Harold Broberg, Indiana University Purdue University-Fort Wayne, IN; Professor Mehmet Celenk, Ohio University; Professor James R. Marcus, University of New Haven, Connecticut; Professor Sudarshan R. Nelatury, Pennsylvania State University, Erie, PA; Professor Siripong Potisuk, The Citadel, the Military College of South Carolina, SC; the reviewers for the second edition: Professor Oktay Alkin, Southern Illinois University, Edwardsville; Professor Rabah Aoufi, DeVry University-Irving, TX; Dr. Janko Calic, University of Surrey, UK; Professor Erik Cheever, Swarthmore College; Professor Samir Chettri, University of Maryland Baltimore County; Professor Nurgun Erdol, Florida Atlantic University; Professor Richard L. Henderson, DeVry University, Kansas City, MO; Professor JeongHee Kim, San Jose State University; Professor Sudarshan R. Nelatury, Penn State University, Erie, PA; Professor Javad Shakib, DeVry University, Pomona, California; Dr.ir. Herbert Wormeerster, University of Twente, The Netherlands; Professor Yongpeng Zhang, Prairie View A&M University; and the reviewers for the first edition: Professor Mateo Aboy, Oregon Institute of Technology; Professor Jean Andrian, Florida International University; Professor Rabah Aoufi, DeVry University; Professor Larry Bland, John Brown University; Professor Phillip L. De Leon, New Mexico State University; Professor Mohammed Feknous, New Jersey Institute of Technology; Professor Richard L. Henderson, DeVry University; Professor Ling Hou, St. Cloud State University; Professor Robert C. (Rob) Maher, Montana State University; Professor Abdulmagid Omar, DeVry University; Professor Ravi P. Ramachandran, Rowan University; Professor William (Bill) Routt, Wake Technical Community College; Professor Samuel D. Stearns, University of New Mexico; Professor Les Thede, Ohio Northern University; Professor Igor Tsukerman, University of Akron; Professor Vijay Vaidyanathan, University of North Texas; Professor David Waldo, Oklahoma Christian University.

Finally, we thank readers who report corrections and provide feedback to us.

Lizhe Tan, Jean Jiang

CHAPTER

INTRODUCTION TO DIGITAL SIGNAL PROCESSING

1

CHAPTER OUTLINE

1.1 BASIC CONCEPTS OF DIGITAL SIGNAL PROCESSING

Digital signal processing (DSP) technology and its advancements have dramatically impacted our modern society everywhere. Without DSP, we would not have digital/Internet audio or video; digital recording; CD, DVD, MP3 players, iPhone, and iPad; digital cameras; digital and cellular telephones; digital satellite and TV; or wire and wireless networks. Medical instruments would be less efficient. It would be impossible to provide precise diagnoses if there were no digital electrocardiography (ECG), or digital radiography and other medical imaging modalities. We would also live in many different ways, since we would not be equipped with voice recognition systems, speech synthesis systems, and image and video editing systems. Without DSP, scientists, engineers, and technologists would have no powerful tools to analyze and visualize data and perform their design, and so on.

The concept of DSP is illustrated by the simplified block diagram in Fig. 1.1, which consists of an analog filter, an analog-to-digital conversion (ADC) unit, a digital signal (DS) processor, a digital-to-analog conversion (DAC) unit, and a reconstruction (anti-image) filter.

Digital Signal Processing. https://doi.org/10.1016/B978-0-12-815071-9.00001-4

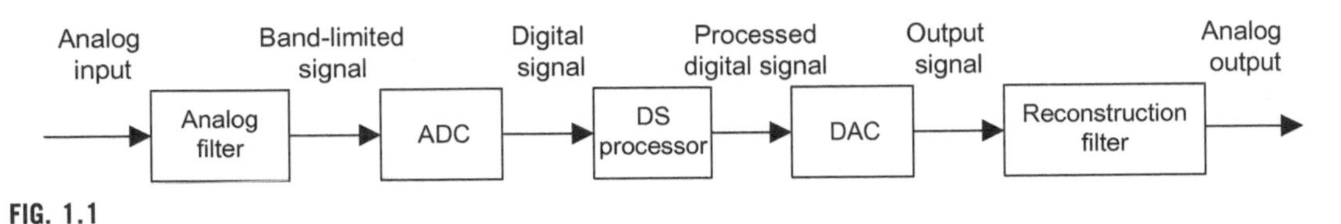

FIG. 1.1

A digital signal processing scheme.

As shown in the diagram, the analog input signal, which is continuous in time and amplitude, is generally encountered in our real life. Examples of such analog signals include current, voltage, temperature, pressure, and light intensity. Usually a transducer (sensor) is used to convert the nonelectrical signal to the analog electrical signal (voltage). This analog signal is fed to an analog filter, which is applied to limit the frequency range of analog signals prior to the sampling process. The purpose of filtering is to significantly attenuate *aliasing distortion*, which will be explained in Chapter 2. The band-limited signal at the output of the analog filter is then sampled and converted via the ADC unit into the digital signal, which is discrete both in time and in amplitude. The digital signal processor then accepts the digital signal and processes the digital data according to DSP rules such as lowpass, highpass, and bandpass digital filtering, or other algorithms for different applications. Note that the digital signal processor unit is a special type of a digital computer and can be a general-purpose digital computer, a microprocessor, or an advanced microcontroller; furthermore, DSP rules can be implemented using software in general.

With the digital signal processor and corresponding software, a processed digital output signal is generated. This signal behaves in a manner based on the specific algorithm used. The next block in Fig. 1.1, the DAC unit, converts the processed digital signal to an analog output signal. As shown, the signal is continuous in time and discrete in amplitude (usually a sample-and-hold signal, to be discussed in Chapter 2). The final block in Fig. 1.1 is designated as a function to smooth the DAC output voltage levels back to the analog signal via a reconstruction (anti-image) filter for the real-world applications.

In general, analog signal processing does not require software, algorithm, ADC, and DAC. The processing relies entirely on the electrical and electronic devices such as resistors, capacitors, transistors, operational amplifiers, and integrated circuits (ICs).

DSP systems, on the other hand, use software, digital processing, and algorithms; therefore, they have more flexibility, less noise interference, and no signal distortion in various applications. However, as shown in Fig. 1.1, DSP systems still require minimum analog processing such as the anti-aliasing and reconstruction filters, which are musts for converting real-world information to digital form and back again to real-world information.

Note that there are many real-world DSP applications that do not require DAC, such as the data acquisition and digital information display, speech recognition, data encoding, and so on. Similarly, DSP applications that need no ADC include CD players, text-to-speech synthesis, and digital tone generators, among others. We will review some of them in the following sections.

1.2 BASIC DIGITAL SIGNAL PROCESSING EXAMPLES IN BLOCK DIAGRAMS

We first look at digital noise filtering and signal frequency analysis, using block diagrams.

1.2.1 DIGITAL FILTERING

Let us consider the situation shown in Fig. 1.2, depicting a digitized noisy signal obtained from digitizing analog voltages (sensor output) containing useful low-frequency signal and noise that occupy all of the frequency range. After ADC, the digitized noisy signal $x(n)$, where n is the sample number, can be enhanced using digital filtering.

Since our useful signal contains low-frequency components, the high-frequency components above the cutoff frequency of our useful signal are considered as noise, which can be removed by using a digital lowpass filter. We set up the DSP block in Fig. 1.2 to operate as a simple digital lowpass filter. After processing the digitized noisy signal $x(n)$, the digital lowpass filter produces a clean digital signal $y(n)$. We can apply the cleaned signal $y(n)$ to another DSP algorithm for a different application or convert it to analog signal via DAC and the reconstruction filter.

The digitized noisy signal and clean digital signal, respectively, are plotted in Fig. 1.3, where the top plot shows the digitized noisy signal, while the bottom plot demonstrates the clean digital signal obtained by applying the digital lowpass filter. Typical applications of noise filtering include acquisition of clean digital audio and biomedical signal and enhancement of speech recording and others (Embree, 1995; Rabiner and Schafer, 1978; Webster, 2009).

FIG. 1.2

The simple digital filtering block.

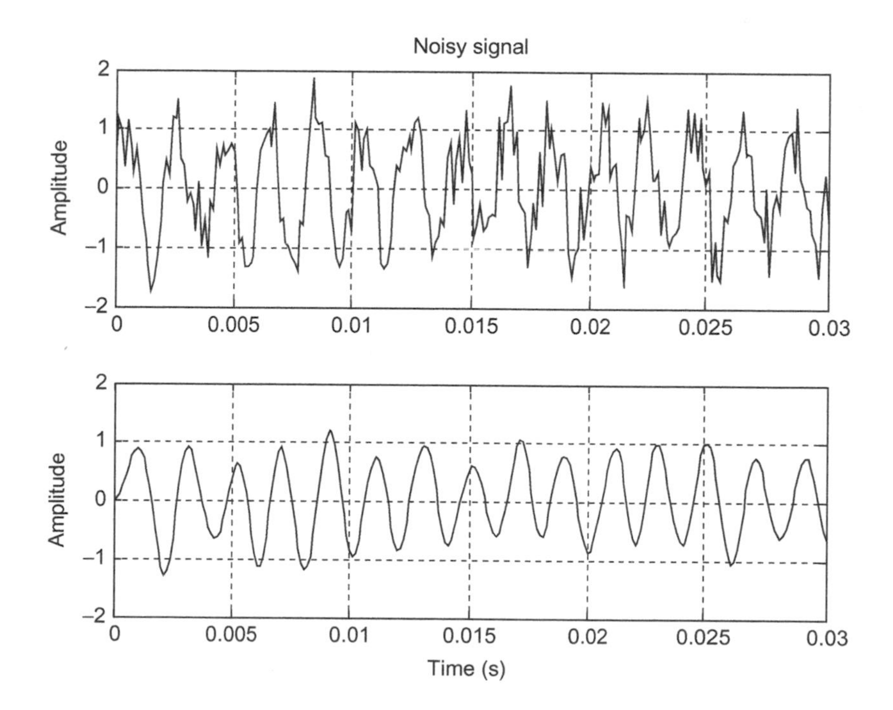

FIG. 1.3

(Top) Digitized noisy signal. (Bottom) Clean digital signal using the digital lowpass filter.

1.2.2 SIGNAL FREQUENCY (SPECTRUM) ANALYSIS

As shown in Fig. 1.4, certain DSP applications often require that time domain information and the frequency content of the signal be analyzed. Fig. 1.5 shows a digitized audio signal and its calculated signal spectrum (frequency content), defined as the signal amplitude vs. its corresponding frequency for the time being via a DSP algorithm, called *fast Fourier transform* (FFT), which will be studied in Chapter 4. The plot in Fig. 1.5A is the time domain display of a recorded audio signal with a frequency of 1000 Hz sampled at 16,000 samples per second, while the frequency content display of plot (B) displays the calculated signal spectrum vs. frequencies, in which the peak amplitude is clearly located at 1000 Hz. Plot (C) shows a time domain display of an audio signal consisting of one signal

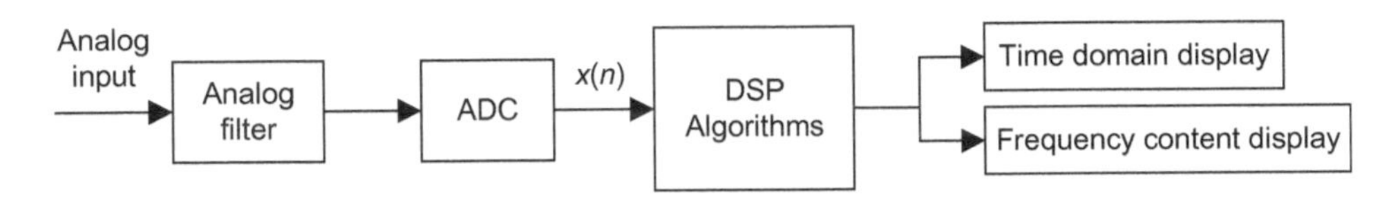

FIG. 1.4

Signal spectral analysis.

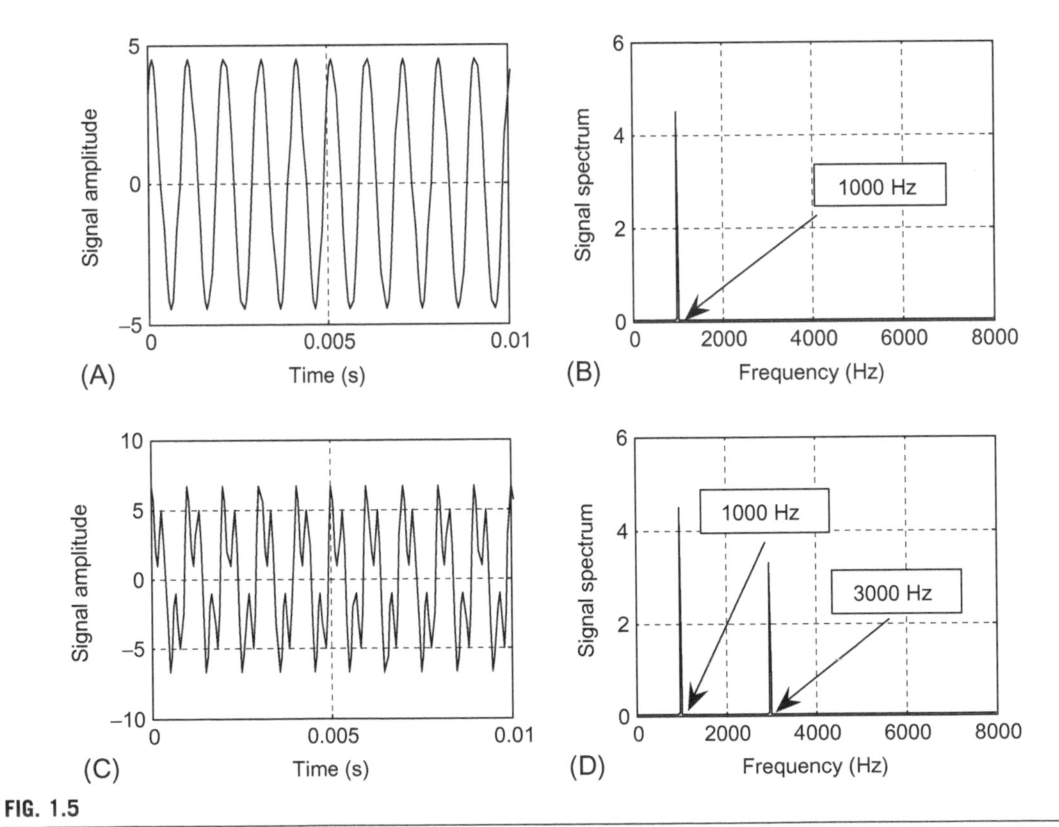

FIG. 1.5

Audio signals and their spectra. (A) 1000 Hz audio signal. (B) 1000 Hz audio signal spectrum. (C) Audio signal containing 1000 and 3000 Hz frequency components. (D) Audio signal spectrum containing 1000 and 3000 Hz frequency components.

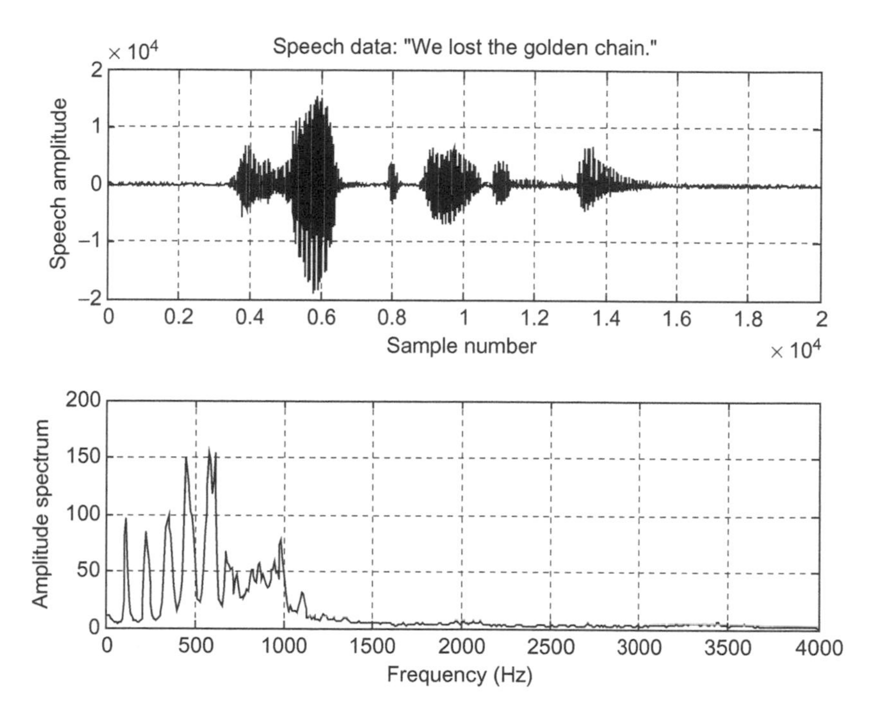

FIG. 1.6

Speech samples and speech spectrum.

of 1000 Hz and another of 3000 Hz sampled at 16,000 samples per second. The frequency content display shown in plot (D) gives two locations (1000 and 3000 Hz) where the peak amplitudes reside, hence the frequency content display presents clear frequency information of the recorded audio signal.

As another practical example, we often perform spectral estimation of a digitally recorded speech or audio (music) waveform using the FFT algorithm in order to investigate the spectral frequency details of speech information. Fig. 1.6 shows a speech signal produced by a human in the time domain and frequency content displays. The top plot shows the digital speech waveform vs. its digitized sample number, while the bottom plot shows the frequency content information of speech for a range from 0 to 4000 Hz. We can observe that there are about 10 spectral peaks, called *speech formants,* in the range between 0 and 1500 Hz. Those identified speech formants can be used for applications such as speech modeling, speech coding, speech feature extraction for speech synthesis and recognition, and so on (Deller et al., 1999).

1.3 OVERVIEW OF TYPICAL DIGITAL SIGNAL PROCESSING IN REAL-WORLD APPLICATIONS

1.3.1 DIGITAL CROSSOVER AUDIO SYSTEM

An audio system is required to operate in an entire audible range of frequencies, which may be beyond the capability of any single speaker driver. Several drivers, such as the speaker cones and horns, each covering a different frequency range, can be used to cover the full audio frequency range.

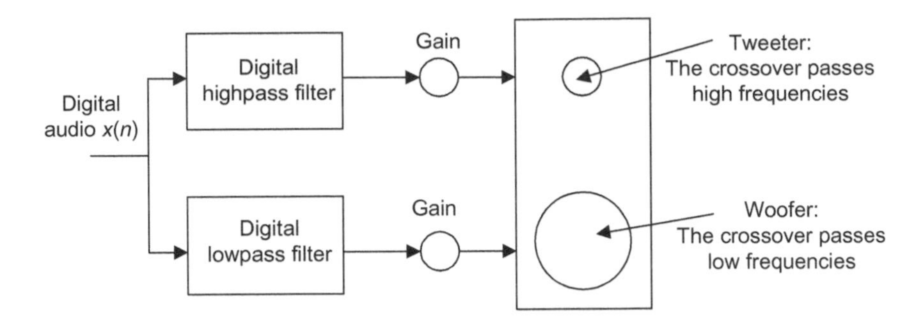

FIG. 1.7

Two-band digital crossover.

Fig. 1.7 shows a typical two-band digital crossover system consisting of two speaker drivers: a woofer and a tweeter. The woofer responds to low frequencies, while the tweeter responds to high frequencies. The incoming digital audio signal is split into two bands using a digital lowpass filter and a digital highpass filter in parallel. Then the separated audio signals are amplified. Finally, they are sent to their corresponding speaker drivers. Although the traditional crossover systems are designed using the analog circuits, the digital crossover system offers a cost-effective solution with programmable ability, flexibility, and high quality. This topic is taken up in Chapter 7.

1.3.2 INTERFERENCE CANCELLATION IN ELECTROCARDIOGRAPHY

In ECG recording, there often exists unwanted 60-Hz interference in the recorded data (Webster, 2009). The analysis shows that the interference comes from the power line and includes magnetic induction, displacement currents in leads or in the body of the patient, effects from equipment interconnections, and other imperfections. Although using proper grounding or twisted pairs minimizes such 60-Hz effects, another effective choice can be the use of a digital notch filter, which eliminates the 60-Hz interference while keeping all the other useful information. Fig. 1.8 illustrates a 60-Hz interference eliminator using a digital notch filter. As shown in Fig. 1.8, the acquired ECG signal containing the 60-Hz interference passes through the digital notch filter. The digital notch filter eliminates the 60-Hz interference and only outputs the clean ECG signal. With such enhanced ECG recording, doctors in clinics could provide accurate diagnoses for patients. This technique can also be used to remove 60-Hz interference in audio systems. This topic is explored in depth in Chapter 8.

1.3.3 SPEECH CODING AND COMPRESSION

One of the speech coding methods, called *waveform coding*, is depicted in Fig. 1.9 (top plot), describing the encoding process, while Fig. 1.9 (bottom plot) shows the decoding processing. As shown in Fig. 1.9 (top plot), the analog signal is first filtered by an analog lowpass filter to remove high-frequency noise components and is then passed through the ADC unit, where the digital values at sampling instants are captured by the digital signal processor. Next, the captured data are compressed using data compression rules to reduce the storage requirement. Finally, the compressed digital information is sent to storage

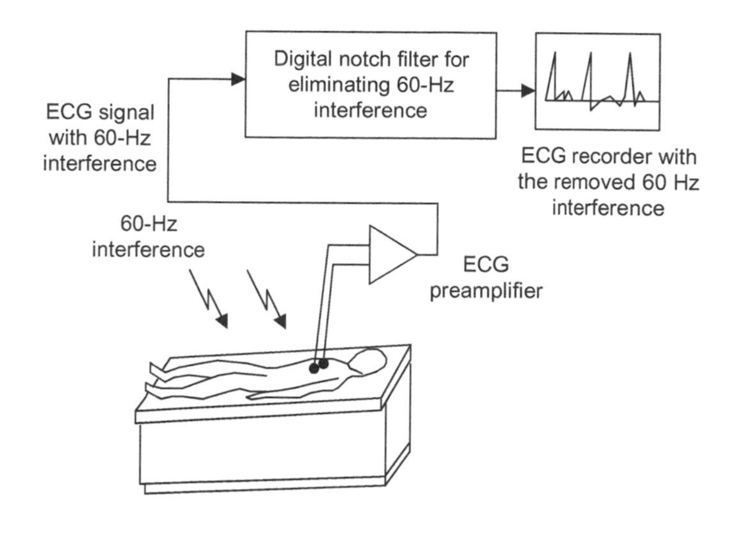

FIG. 1.8

Elimination of 60-Hz interference in electrocardiography (ECG).

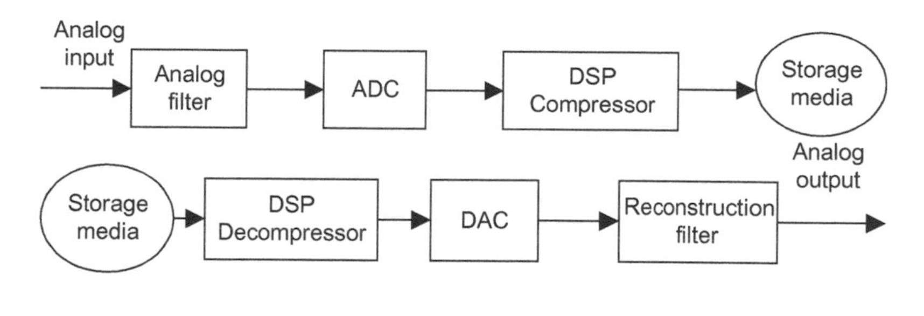

FIG. 1.9

(Top plot) Simplified data compressor. (Bottom plot) Simplified data expander (decompressor).

media. The compressed digital information can also be transmitted efficiently, since compression reduces the original data rate. Digital voice recorders, digital audio recorders, and MP3 players are products that use compression techniques (Deller et al., 1999; Li et al., 2014; Pan, 1995).

To retrieve the information, the reverse process is applied. As shown in Fig. 1.9 (bottom plot), the digital signal processor decompresses the data from the storage media and sends the recovered digital data to DAC. The analog output is acquired by filtering the DAC output via a reconstruction filter.

1.3.4 COMPACT-DISC RECORDING SYSTEM

A compact-disc (CD) recording system is described in Fig. 1.10 (top plot). The analog audio signal is sensed from each microphone and then fed to the anti-aliasing lowpass filter. Each filtered audio signal is sampled at the industry standard rate of 44.1 kilo-samples per second, quantized, and coded to 16 bits

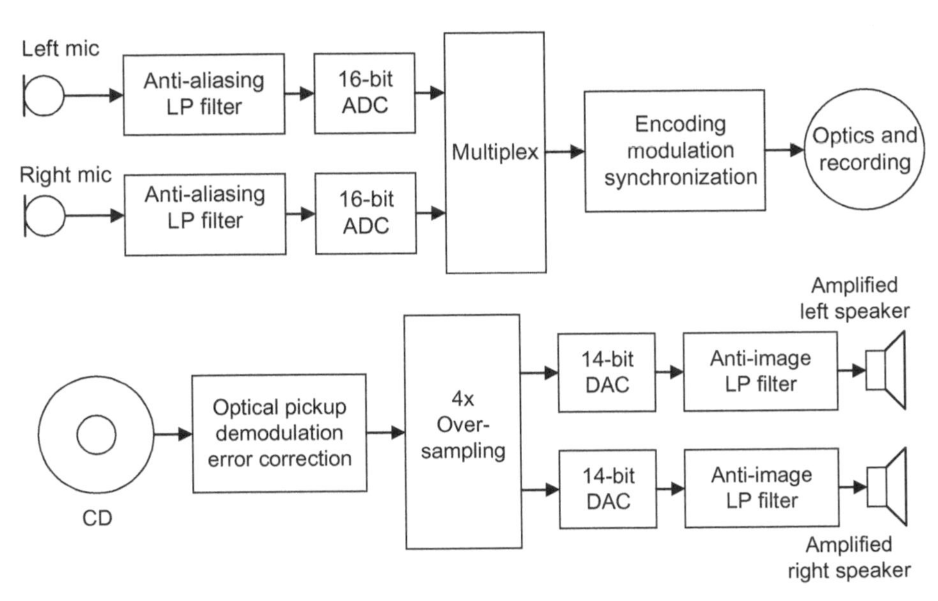

FIG. 1.10

(Top plot) Simplified encoder of the CD recording system. (Bottom plot) Simplified decoder of the CD recording system.

for each digital sample in each channel. The two channels are further multiplexed and encoded, and extra bits are added to provide information such as playing time and track number for the listener. The encoded data bits are modulated for storage, and more synchronized bits are added for subsequent recovery of sampling frequency. The modulated signal is then applied to control a laser beam that illuminates the photosensitive layer of a rotating glass disc. When the laser turns on and off, the digital information is etched on the photosensitive layer as a pattern of pits and lands in a spiral track. This master disc forms the basis for mass production of the commercial CD from the thermoplastic material.

During playback, as illustrated in Fig. 1.10 (bottom plot), a laser optically scans the tracks on a CD to produce digital signal. The digital signal is then demodulated. The demodulated signal is further oversampled by a factor of 4 to acquire a sampling rate of 176.4 kHz for each channel and is then passed to the 14-bit DAC unit. For the time being, we consider the oversampling process as interpolation, that is, adding three samples between every two original samples in this case, as we shall see in Chapter 11. After DAC, the analog signal is sent to the anti-image analog filter, which is a lowpass filter to smooth the voltage steps from the DAC unit. The output from each anti-image filter is fed to its amplifier and loudspeaker. The purpose of the oversampling is to relieve the higher-filter-order requirement for the anti-image lowpass filter, making the circuit design much easier and economical (Ambardar, 1999).

Software audio players that play music from CDs, such as Windows Media Player and RealPlayer, installed on computer systems, are examples of DSP applications. The audio player has many advanced features, such as a graphical equalizer, which allows users to change audio with sound effects including boosting low-frequency content or emphasizing high-frequency content to make music sound more entertaining (Ambardar, 1999; Embree, 1995; Ifeachor and Jervis, 2002).

1.3.5 VIBRATION SIGNATURE ANALYSIS FOR DEFECTED GEAR TOOTH

Gearboxes are widely used in industry and vehicles (Spectra Quest, Inc.). During the extended service lifetimes, the gear teeth will inevitably be worn, chipped, or missing. Hence, with DSP techniques, effective diagnostic methods can be developed to detect and monitor the defected gear teeth in order to enhance the reliability of the entire machine before any unexpected catastrophic events occur. Fig. 1.11A shows the gearbox, in which two straight bevel gears with a transmission ratio of 1.5:1 inside the gearbox are shown Fig. 1.11B. The number of teeth on the pinion is 18. The gearbox input shaft

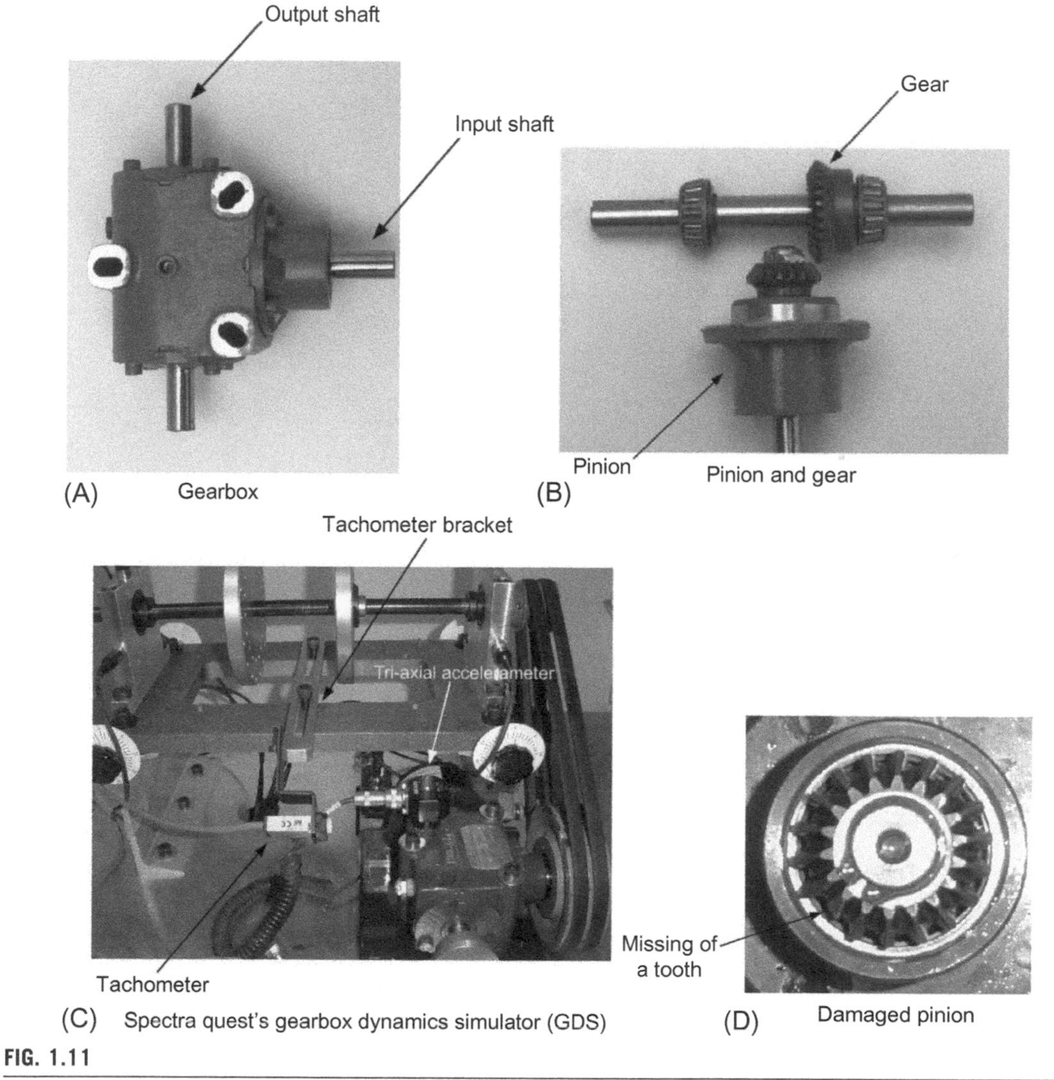

FIG. 1.11

Vibration signature analysis of the gearbox. (A) Gearbox, (B) Pinion and gear, (C) Spectra Quest's Gearbox Dynamics Simulator (GDS), (D) Damaged pinion.

Courtesy of SpectaQuest, Inc.

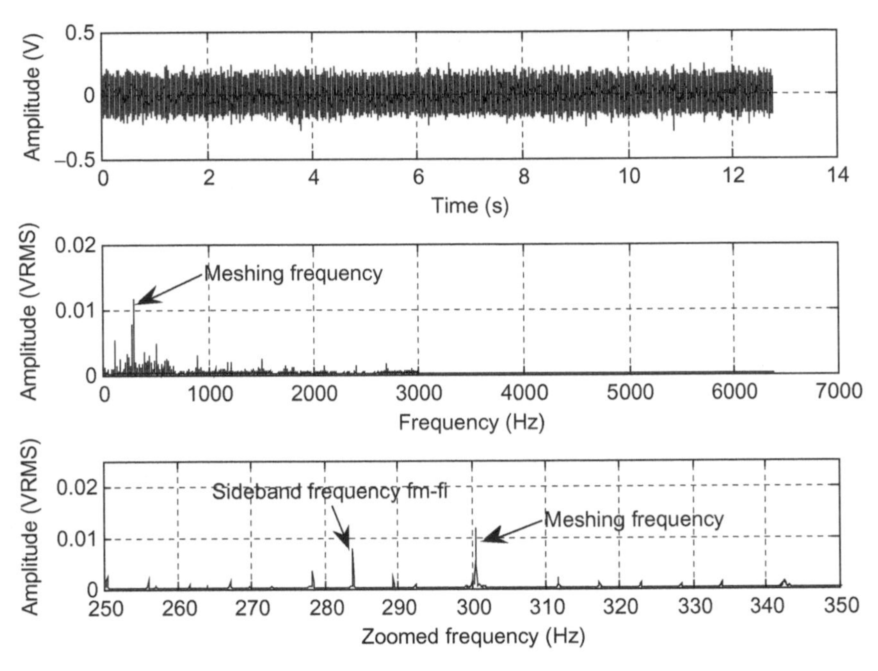

FIG. 1.12

Vibration signal and spectrum from the good condition gearbox.

Data provided by SpectaQuest, Inc.

is connected to a sheave and driven by a "V" belt drive. The vibration data can be collected by triaxial accelerometer installed on the top of the gearbox, as shown in Fig. 1.11C. The data acquisition system uses a sampling rate of 12.8 kHz. Fig. 1.11D shows that a pinion has a missing tooth. During the test, the motor speed is set to 1000 rpm (revolutions per minute) so the meshing frequency is determined as $f_m = 1000(\text{rpm}) \times 18/60 = 300\,\text{Hz}$ and input shaft frequency is $f_i = 1000(\text{rpm})/60 = 16.17\,\text{Hz}$. The baseline signal and spectrum (excellent condition) from x direction of the accelerometer are displayed in Fig. 1.12, where we can see that the spectrum contains the meshing frequency component of 300 Hz and a sideband frequency component of 283.33 (300–16.67) Hz. Fig. 1.13 shows the vibration signature for the damaged pinion in Fig. 1.11D. For the damaged pinion, the sidebands ($f_m \pm f_i$, $f_m \pm 2f_i \ldots$) become dominant. Hence, the vibration failure signature is identified. More details can be found in Robert Bond Randall (2011).

1.3.6 DIGITAL IMAGE ENHANCEMENT

We can look at another example of signal processing in two dimensions. Fig. 1.14A shows a picture of an outdoor scene taken by a digital camera on a cloudy day. Due to this weather condition, the image is improperly exposed in natural light and comes out dark. The image processing technique called *histogram equalization* (Gonzalez and Wintz, 1987) can stretch the light intensity of an image using the digital information (pixels) to increase the image contrast, therefore, detailed information can easily be seen in the image, as we can see in Fig. 1.14B. We will study this technique in Chapter 13.

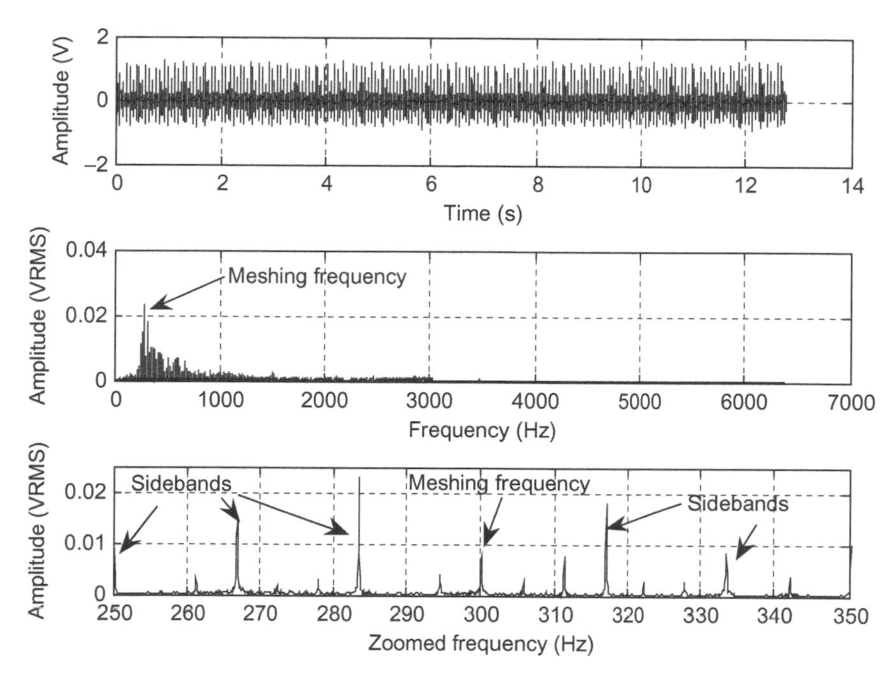

FIG. 1.13

Vibration signal and spectrum from the damaged gearbox.

Data provided by SpectaQuest, Inc.

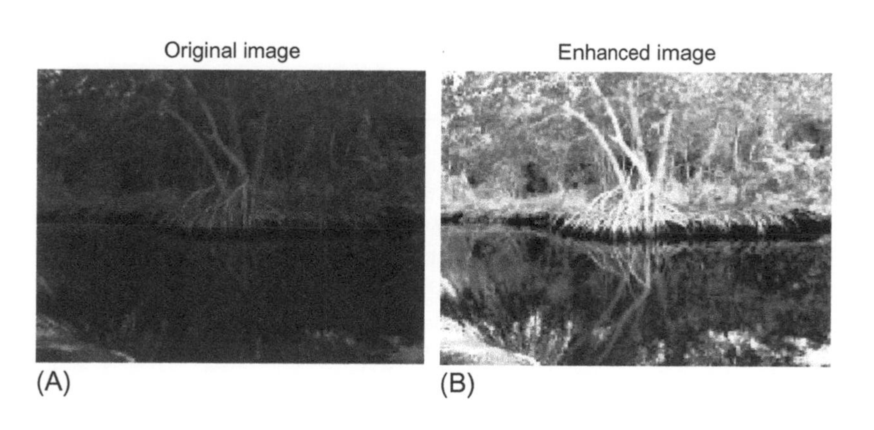

FIG. 1.14

Image enhancement. (A) Original image taken on a cloudy day. (B) Enhanced image using the histogram equalization technique.

Table 1.1 Applications of Digital Signal Processing

Digital Audio and Speech
Digital audio coding such as CD players, MP3 players, digital crossover, digital audio equalizers, digital stereo and surround sound, noise reduction systems, speech coding, data compression and encryption, speech synthesis and speech recognition
Digital telephone
Speech recognition, high-speed modems, echo cancellation, speech synthesizers, DTMF (dual-tone multifrequency) generation and detection, answering machines
Automobile Industry
Active noise control systems, active suspension systems, digital audio and radio, digital controls, vibration signal analysis
Electronic Communications
Cellular phones, digital telecommunications, wireless LAN (local area networking), satellite communications
Medical Imaging Equipment
ECG analyzers, cardiac monitoring, medical imaging and image recognition, digital X-rays and image processing
Multimedia
Internet phones, audio and video, hard disk drive electronics, iPhone, iPad, digital pictures, digital cameras, text-to-voice, and voice-to-text technologies

1.4 DIGITAL SIGNAL PROCESSING APPLICATIONS

Applications of DSP are increasing in many areas where analog electronics are being replaced by DSP chips, and new applications depend on DSP techniques. With the decrease in the cost of digital signal processors and increase in their performance, DSP will continue to affect engineering design in our modern daily life. Some application examples using DSP are listed in Table 1.1.

However, the list in the table by no means covers all the DSP applications. Many application areas are increasingly being explored by engineers and scientists. Applications of DSP techniques will continue to have profound impacts and improve our lives.

1.5 SUMMARY

1. An analog signal is continuous in both time and amplitude. Analog signals in the real world include current, voltage, temperature, pressure, light intensity, and so on. The digital signal contains the digital values converted from the analog signal at the specified time instants.
2. Analog-to-digital signal conversion requires an ADC unit (hardware) and a lowpass filter attached ahead of the ADC unit to block the high-frequency components that ADC cannot handle.
3. The digital signal can be manipulated using arithmetic. The manipulations may include digital filtering, calculation of signal frequency content, and so on.
4. The digital signal can be converted back to an analog signal by sending the digital values to DAC to produce the corresponding voltage levels and applying a smooth filter (reconstruction filter) to the DAC voltage steps.
5. DSP finds many applications in areas such as digital speech and audio, digital and cellular telephones, automobile controls, vibration signal analysis, communications, biomedical imaging, image/video processing, and multimedia.

CHAPTER

SIGNAL SAMPLING AND QUANTIZATION

2

CHAPTER OUTLINE

2.1 SAMPLING OF CONTINUOUS SIGNAL

As discussed in Chapter 1, Fig. 2.1 describes a simplified block diagram of a digital signal processing (DSP) system. The analog filter processes analog input to obtain the band-limited signal, which is sent to the analog-to-digital conversion (ADC) unit. The ADC unit samples the analog signal, quantizes the sampled signal, and encodes the quantized signal level to the digital signal.

Here we first develop concepts of sampling processing in time domain. Fig. 2.2 shows an analog (continuous-time) signal (solid line) defined at every point over the time axis (horizontal line) and amplitude axis (vertical line). Hence, the analog signal contains an infinite number of points.

It is impossible to digitize an infinite number of points. Furthermore, the infinite points are not appropriate to be processed by a digital signal processor or a computer, since they require infinite amount of memory and infinite amount of processing power for computations. Sampling can solve such a problem by taking samples at the fixed time interval as shown in Figs. 2.2 and 2.3, where the time T represents the sampling interval or sampling period in seconds.

As shown in Fig. 2.3, each sample maintains its voltage level during the sampling interval T to give the ADC enough time to convert it. This process is called *sample and hold*. Since there exits one amplitude level for each sampling interval, we can sketch each sample amplitude level at its corresponding sampling time instant shown in Fig. 2.2, where 14 sampled samples at their sampling time instants are plotted, each using a vertical bar with a solid circle at its top.

Digital Signal Processing. https://doi.org/10.1016/B978-0-12-815071-9.00002-6

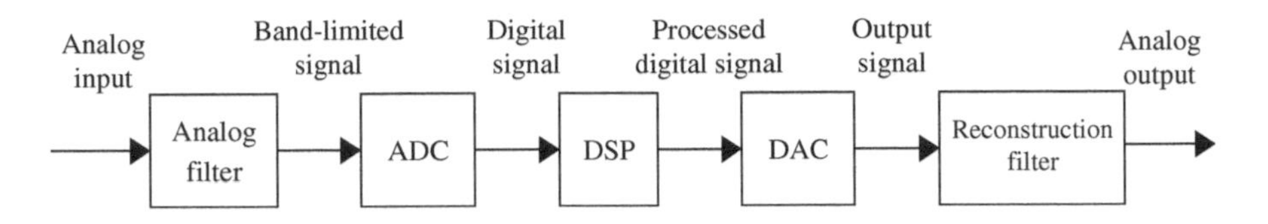

FIG. 2.1

A digital signal processing scheme.

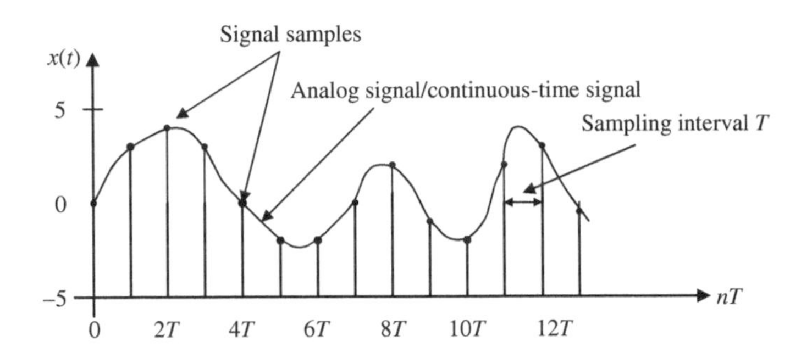

FIG. 2.2

Display of the analog (continuous) signal and display of digital samples vs. the sampling time instants.

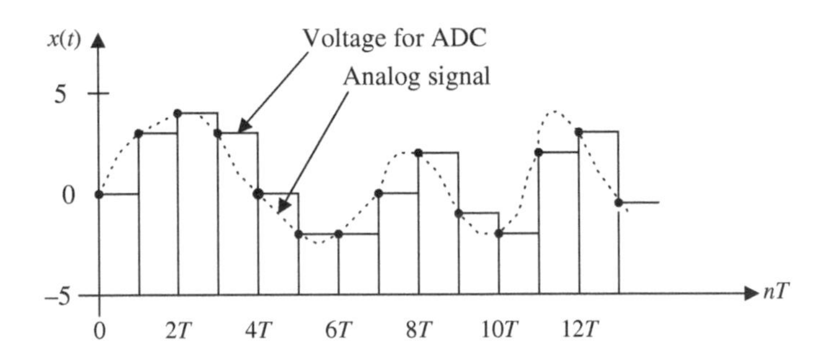

FIG. 2.3

Sample-and-hold analog voltage for ADC.

For a given sampling interval T, which is defined as the time span between two neighboring sample points, the sampling rate is therefore given by

$$f_s = \frac{1}{T} \text{ samples per second (Hz).}$$

For example, if a sampling period is $T = 125$ μs, the sampling rate is determined as $f_s = 1/125\ \mu s = 8000$ samples per second (Hz).

After the analog signal is sampled, we obtain the sampled signal whose amplitude values are taken at the sampling instants, thus the processor is able to handle the sample points. Next, we have to ensure that samples are collected at a rate high enough that the original analog signal can be reconstructed or recovered later. In other words, we are looking for a minimum sampling rate to acquire a complete reconstruction of the analog signal from its sampled version. If an analog signal is not appropriately sampled, *aliasing* will occur, which causes unwanted signals in the desired frequency band.

The sampling theorem guarantees that an analog signal can be in theory perfectly recovered as long as the sampling rate is at least twice of the highest-frequency component of the analog signal to be sampled. The condition is described as

$$f_s \geq 2f_{\max},$$

where $f_{\max}$ is the maximum-frequency component of the analog signal to be sampled.

For example, to sample a speech signal containing frequencies up to 4 kHz, the minimum sampling rate is chosen to be at least 8 kHz, or 8000 samples per second; to sample an audio signal with frequencies up to 20 kHz, a sampling rate with at least 40,000 samples per second or 40 kHz is required.

Fig. 2.4 illustrates sampling of two sinusoids, where the sampling interval between sample points is $T=0.01$ s, thus the sampling rate is $f_s=100$ Hz. The first plot in the figure displays a sine wave with a frequency of 40 Hz and its sampled amplitudes. The sampling theorem condition is satisfied since $2f_{\max}=80<f_s$. The sampled amplitudes are labeled using the circles shown in the first plot. We note that the 40-Hz signal is adequately sampled, since the sampled values clearly come from the analog version of the 40-Hz sine wave. However, as shown in the second plot, the sine wave with a frequency

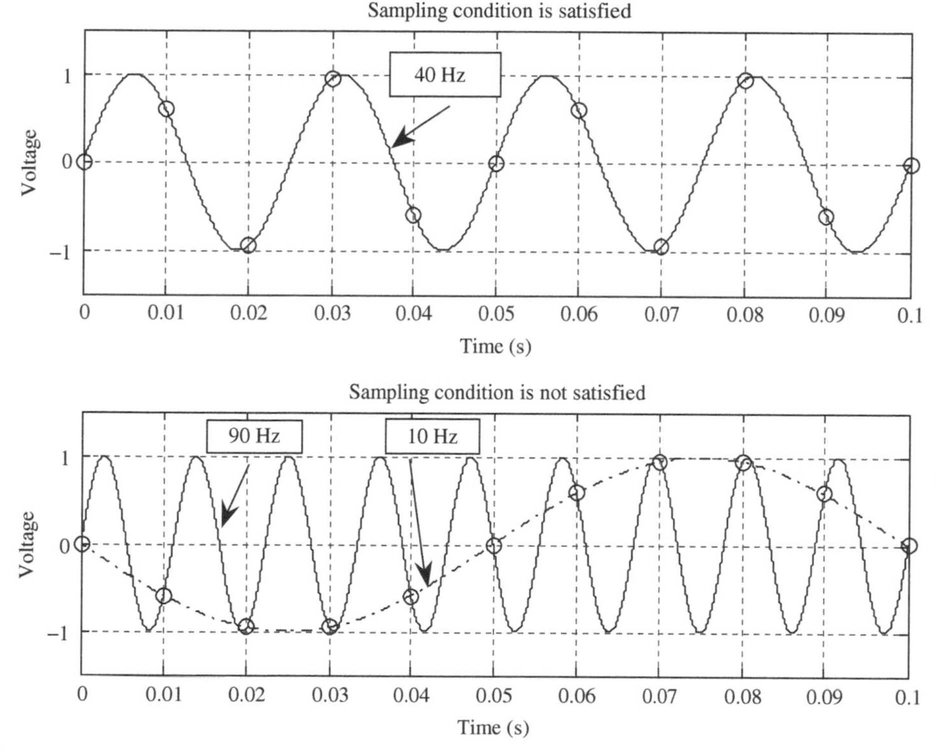

FIG. 2.4

Plots of the appropriately sampled signals and inappropriately sampled (aliased) signals.

of 90 Hz is sampled at 100 Hz. Since the sampling rate of 100 Hz is relatively low compared with the 90-Hz sine wave, the signal is undersampled due to $2f_{max}=180>f_s$. Hence, the condition of the sampling theorem is not satisfied. Based on the sample amplitudes labeled with the circles in the second plot, we cannot tell whether the sampled signal comes from sampling a 90-Hz sine wave (plotted using the solid line) or from sampling a 10-Hz sine wave (plotted using the dot-dash line). They are not distinguishable. Thus they are *aliases* of each other. We call the 10-Hz sine wave the aliasing noise in this case, since the sampled amplitudes actually come from sampling the 90-Hz sine wave.

Now let us develop the sampling theorem in the frequency domain, that is, the minimum sampling rate requirement for sampling an analog signal. As we shall see, in practice this can help us design the anti-aliasing filter (a lowpass filter that will reject high frequencies that cause aliasing) to be applied before sampling, and the anti-image filter (a reconstruction lowpass filter that will smooth the recovered sample-and-hold voltage levels to an analog signal) to be applied after the digital-to-analog conversion (DAC).

Fig. 2.5 depicts the sampled signal $x_s(t)$ obtained by sampling the continuous signal $x(t)$ at a sampling rate of f_s samples per second.

Mathematically, this process can be written as the product of the continuous signal and the sampling pulses (pulse train):

$$x_s(t)=x(t)p(t), \tag{2.1}$$

where $p(t)$ is the pulse train with a period $T=1/f_s$. The pulse train can be expressed as

$$p(t)=\sum_{n=-\infty}^{\infty}\delta(t-nT) \tag{2.2}$$

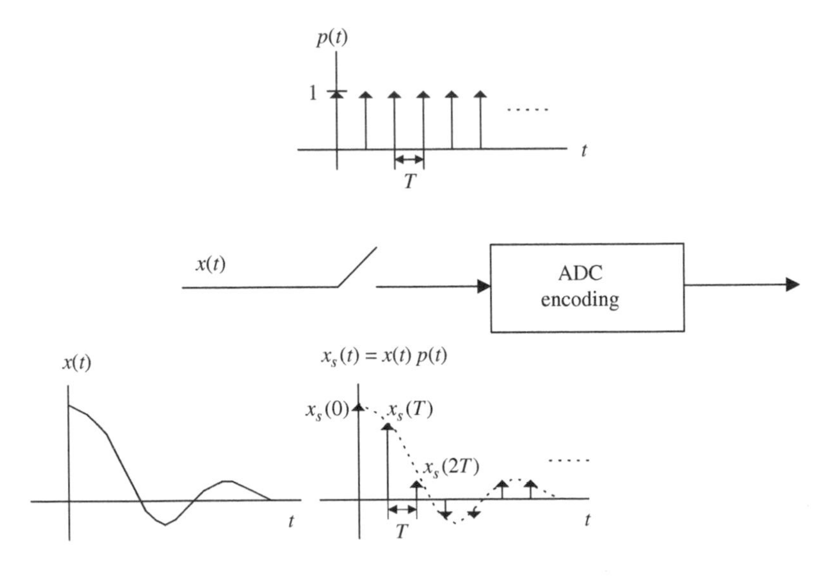

FIG. 2.5

The simplified sampling process.

with a fundamental frequency of $\omega_0 = 2\pi/T = 2\pi f_s$ rad/s. $p(t)$ can be expanded by the Fourier series (see Appendix B), that is,

$$p(t) = \sum_{k=-\infty}^{\infty} a_k e^{jk\omega_0 t}, \tag{2.3}$$

where a_k are the Fourier coefficients which can be determined by

$$a_k = \frac{1}{T}\int_{-\infty}^{\infty} \delta(t) e^{-jk\omega_0 t} dt = \frac{1}{T}. \tag{2.4}$$

Substituting Eq. (2.4) into Eq. (2.3), the pulse train is given by

$$p(t) = \sum_{k=-\infty}^{\infty} \frac{1}{T} e^{jk\omega_0 t}. \tag{2.5}$$

Again, substituting Eq. (2.5) into Eq. (2.1) leads to

$$x_s(t) = \sum_{k=-\infty}^{\infty} \frac{1}{T} x(t) e^{jk\omega_0 t}. \tag{2.6}$$

Applying Fourier transform (see Appendix B) on Eq. (2.6), we yield the following:

$$X_s(f) = FT\left\{\sum_{k=-\infty}^{\infty} \frac{1}{T} x(t) e^{jk\omega_0 t}\right\} = \sum_{k=-\infty}^{\infty} \frac{1}{T} FT\left\{x(t) e^{jk\omega_0 t}\right\} = \sum_{k=-\infty}^{\infty} \frac{1}{T}\int_{-\infty}^{\infty} \left\{x(t) e^{jk\omega_0 t}\right\} e^{-j\omega t} dt, \tag{2.7}$$

that is,

$$X_s(f) = \sum_{k=-\infty}^{\infty} \frac{1}{T}\int_{-\infty}^{\infty} x(t) e^{-j(\omega - k\omega_0)t} dt = \sum_{k=-\infty}^{\infty} \frac{1}{T}\int_{-\infty}^{\infty} x(t) e^{-j2\pi(f - kf_s)t} dt. \tag{2.8}$$

From the definition of Fourier transform, we note that

$$X(f) = \sum_{k=-\infty}^{\infty} \frac{1}{T}\int_{-\infty}^{\infty} x(t) e^{-j2\pi f t} dt. \tag{2.9}$$

Using Eq. (2.9), the original spectrum (frequency components) $X(f)$ and the sampled signal spectrum $X_s(f)$ in terms of Hz are related as

$$X_s(f) = \frac{1}{T}\sum_{k=-\infty}^{\infty} X(f - kf_s), \tag{2.10}$$

where $X(f)$ is assumed to be the original baseband spectrum while $X_s(f)$ is its sampled signal spectrum, consisting of the original baseband spectrum $X(f)$ and its replicas $X(f \pm kf_s)$. The derivation for Eq. (2.10) can also be found in well-known texts (Ahmed and Natarajan, 1983; Ambardar, 1999; Alkin, 1993; Oppenheim and Shafer, 1975; Proakis and Manolakis, 2007).

Expanding Eq. (2.10) leads to the sampled signal spectrum in Eq. (2.11)

$$X_s(f) = \cdots + \frac{1}{T}X(f + f_s) + \frac{1}{T}X(f) + \frac{1}{T}X(f - f_s) + \cdots. \tag{2.11}$$

Eq. (2.11) indicates that the sampled signal spectrum is the sum of the scaled original spectrum and copies of its shifted versions, called *replicas*. The sketch of Eq. (2.11) is given in Fig. 2.6, where three possible sketches are classified. Given the original signal spectrum $X(f)$ plotted in Fig. 2.6A, the sampled signal spectrum according to Eq. (2.11) is plotted in Fig. 2.6B, where the replicas, $\frac{1}{T}X(f)$, $\frac{1}{T}X(f-f_s)$, $\frac{1}{T}X(f+f_s)$,…, have separations between them. Fig. 2.6C shows that the baseband spectrum and its replicas, $\frac{1}{T}X(f)$, $\frac{1}{T}X(f-f_s)$, $\frac{1}{T}X(f+f_s)$,…, are just connected, and finally, in Fig. 2.6D, the original spectrum $\frac{1}{T}X(f)$, and its replicas $\frac{1}{T}X(f-f_s)$, $\frac{1}{T}X(f+f_s)$,…, are overlapped; that is, there are many overlapping portions in the sampled signal spectrum.

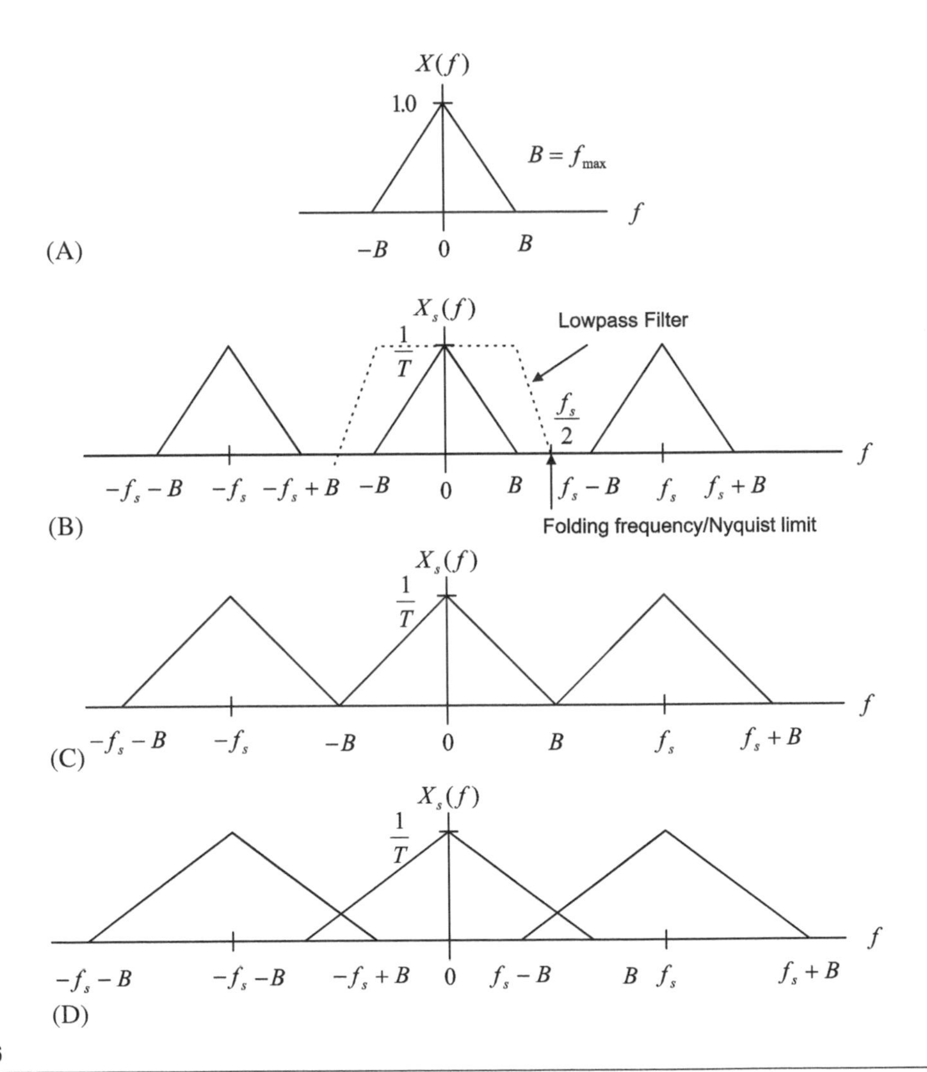

FIG. 2.6

Plots of the sampled signal spectrum. (A) Original signal spectrum. (B) Sampled signal spectrum for $f_s > 2B$. (C) Sampled signal spectrum for $f_s = 2B$. (D) Sampled signal spectrum for $f_s < 2B$.

From Fig. 2.6, it is clear that the sampled signal spectrum consists of the scaled baseband spectrum centered at the origin, and its replicas centered at the frequencies of $\pm kf_s$(multiples of the sampling rate) for each of $k=1, 2, 3, \ldots$.

If applying a lowpass reconstruction filter to obtain exact reconstruction of the original signal spectrum, the following condition must be satisfied:

$$f_s - f_{\max} \geq f_{\max}. \tag{2.12}$$

Solving Eq. (2.12) gives

$$f_s \geq 2f_{\max}. \tag{2.13}$$

In terms of frequency in radians per second, Eq. (2.13) is equivalent to

$$\omega_s \geq 2\omega_{\max}. \tag{2.14}$$

This fundamental conclusion is well known as the *Shannon sampling theorem*, which is formally described as below:

> For a uniformly sampled DSP system, an analog signal can be perfectly recovered as long as the sampling rate is at least twice as large as the highest-frequency component of the analog signal to be sampled.

We summarize two key points here:

(1) Sampling theorem establishes a minimum sampling rate for a given band-limited analog signal with the highest-frequency component $f_{\max}$. If the sampling rate satisfies Eq. (2.13), then the analog signal can be recovered via its sampled values using a lowpass filter, as described in Fig. 2.6B.

(2) Half of the sampling frequency $f_s/2$ is usually called the *Nyquist frequency* (Nyquist limit) or *folding frequency*. The sampling theorem indicates that a DSP system with a sampling rate of f_s can ideally sample an analog signal with its highest frequency up to half of the sampling rate without introducing spectral overlap (aliasing). Hence, the analog signal can be perfectly recovered from its sampled version.

Let us study the following example:

EXAMPLE 2.1

Suppose that an analog signal is given as

$x(t) = 5\cos(2\pi \times 1000t)$, for $t \geq 0$,

and is sampled at the rate 8000 Hz.

(a) Sketch the spectrum for the original signal.

(b) Sketch the spectrum for the sampled signal from 0 to 20 kHz.

Solution:

(a) Since the analog signal is sinusoid with a peak value of 5 and frequency of 1000 Hz, we can write the sine wave using Euler's identity:

$$5\cos(2\pi \times 1000t) = 5 \cdot \left(\frac{e^{j2\pi \times 1000t} + e^{-j2\pi \times 1000t}}{2}\right) = 2.5e^{j2\pi \times 1000t} + 2.5e^{-j2\pi \times 1000t},$$

Continued

EXAMPLE 2.1—CONT'D

which is a Fourier series expansion for a continuous periodic signal in terms of the exponential form (see Appendix B). We can identify the Fourier series coefficients as

$$c_1 = 2.5 \text{ and } c_{-1} = 2.5.$$

Using the magnitudes of the coefficients, we then plot the two-side spectrum as Fig. 2.7A.

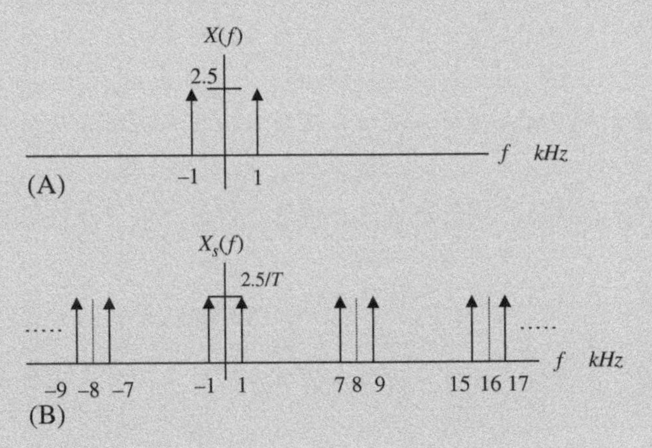

FIG. 2.7

Spectrum of the analog signal (A) and sampled signal (B) in Example 2.1.

(b) After the analog signal is sampled at the rate of 8000 Hz, the sampled signal spectrum and its replicas centered at the frequencies $\pm kf_s$, each with the scaled amplitude being $2.5/T$, are as shown in Fig. 2.7B:

Note that the spectrum of the sampled signal shown in Fig. 2.7B contains the images of the original spectrum shown in Fig. 2.7A; that the images repeat at multiples of the sampling frequency f_s (e.g., 8, 16, 24 kHz, etc.); and that all images must be removed, since they convey no additional information.

2.2 SIGNAL RECONSTRUCTION

In this section, we investigate the recovery of analog signal from its sampled signal version. Two simplified steps are involved, as described in Fig. 2.8. First, the digitally processed data $y(n)$ are converted to the ideal impulse train $y_s(t)$, in which each impulse has its amplitude proportional to digital output $y(n)$, and two consecutive impulses are separated by a sampling period of T; second, the analog reconstruction filter is applied to the ideally recovered sampled signal $y_s(t)$ to obtain the recovered analog signal.

To study the signal reconstruction, we let $y(n) = x(n)$ for the case of no DSP, so that the reconstructed sampled signal and the input sampled signal are ensured to be the same; that is, $y_s(t) = x_s(t)$. Hence, the

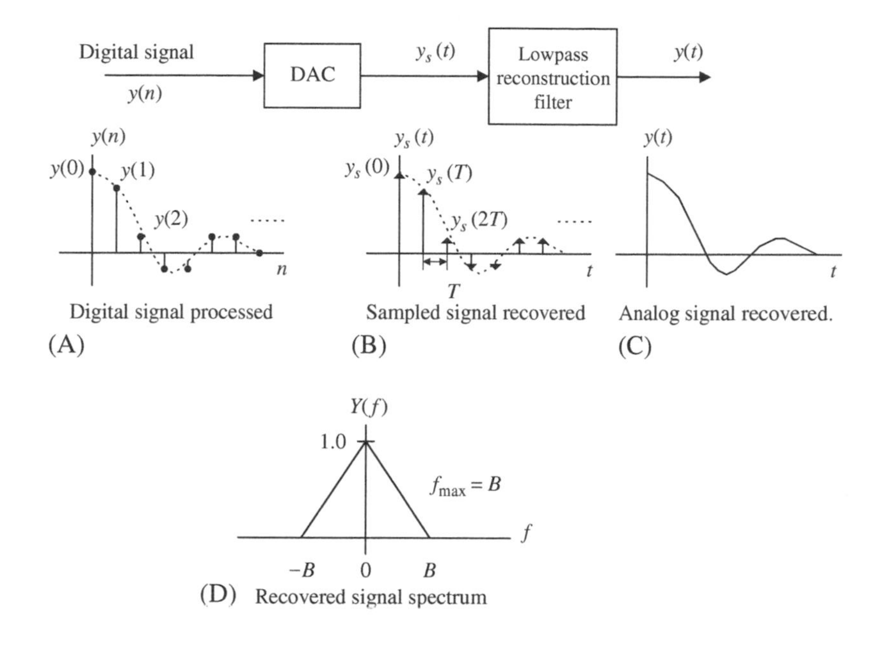

FIG. 2.8

Signal notations at its reconstruction stage: (A) Digital signal processed, (B) Sampled signal recovered, (C) Analog signal recovered, (D) Recovered signal spectrum.

spectrum of the sampled signal $y_s(t)$ contains the same spectral content of the original spectrum $X(f)$, that is, $Y(f)=X(f)$, with a bandwidth of $f_{max}=B$ Hz (described in Fig. 2.8D) and the images of the original spectrum (scaled and shifted versions). The following three cases are discussed for recovery of the original signal spectrum $X(f)$.

Case 1: $f_s=2f_{max}$

As shown in Fig. 2.9, where the Nyquist frequency is equal to the maximum frequency of the analog signal $x(t)$, an ideal lowpass reconstruction filter is required to recover the analog signal spectrum. This is an impractical case.

Case 2: $f_s>2f_{max}$

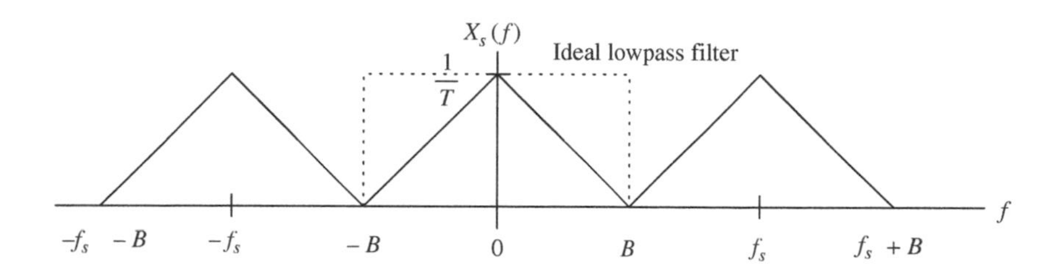

FIG. 2.9

Spectrum of the sampled signal when $f_s=2f_{max}$.

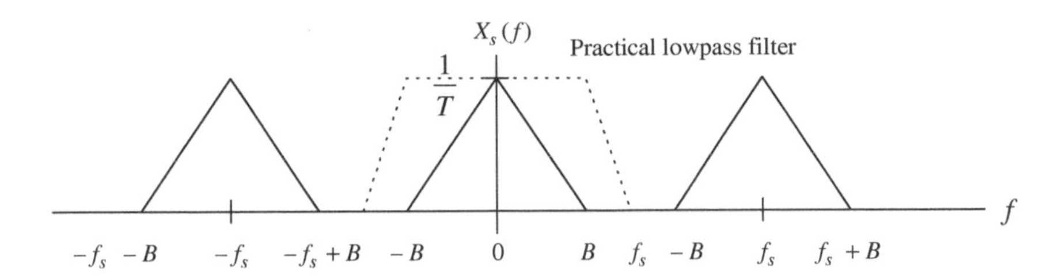

FIG. 2.10

Spectrum of the sampled signal when $f_s > 2f_{max}$.

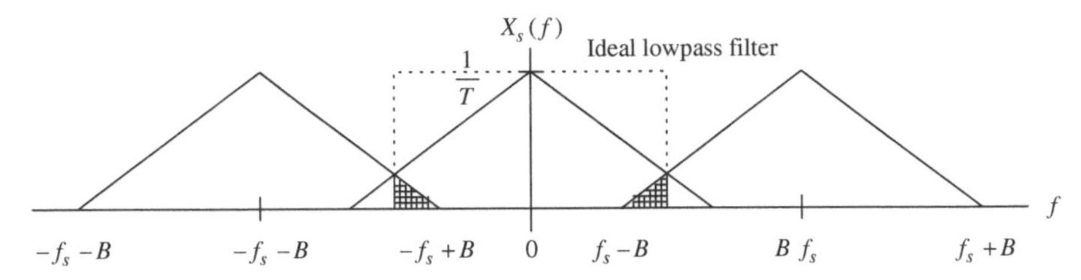

FIG. 2.11

Spectrum of the sampled signal when $f_s < 2f_{max}$.

In this case, as shown in Fig. 2.10, there is a separation between the highest-frequency edge of the baseband spectrum and the lower edge of the first replica. Therefore, a practical lowpass reconstruction (anti-image) filter can be designed to reject all the images and achieve the original signal spectrum.

Case 3: $f_s < 2f_{max}$

Case 3 violates the condition of the Shannon sampling theorem. As we can see, Fig. 2.11 depicts the spectral overlapping between the original baseband spectrum and the spectrum of the first replica and so on. Even we apply an ideal lowpass filter to remove these images, in the baseband there still some foldover frequency components from the adjacent replica. This is aliasing, where the recovered baseband spectrum suffers spectral distortion, that is, contains aliasing noise spectrum; in time domain, the recovered analog signal may consist of the aliasing noise frequency or frequencies. Hence, the recovered analog signal is incurably distorted.

Note that if an analog signal with a frequency f is undersampled, the aliasing frequency component f_{alias} in the baseband is simply given by the following expression:

$$f_{alias} = f_s - f.$$

The following examples give a spectrum analysis of the signal recovery.

EXAMPLE 2.2

Assuming that an analog signal is given by

$$x(t) = 5\cos(2\pi \times 2000t) + 3\cos(2\pi \times 3000t), \text{ for } t \geq 0$$

and it is sampled at the rate of 8000 Hz.

(a) Sketch the spectrum of the sampled signal up to 20 kHz.

(b) Sketch the recovered analog signal spectrum if an ideal lowpass filter with a cutoff frequency of 4 kHz is used to filter the sampled signal ($y(n)=x(n)$ in this case) to recover the original signal.

Solution:

(a) Using Euler's identity, we get

$$x(t)=\frac{3}{2}e^{-j2\pi\times3000t}+\frac{5}{2}e^{-j2\pi\times2000t}+\frac{5}{2}e^{j2\pi\times2000t}+\frac{3}{2}e^{j2\pi\times3000t}.$$

The two-sided amplitude spectrum for the sinusoid is displayed in Fig. 2.12.

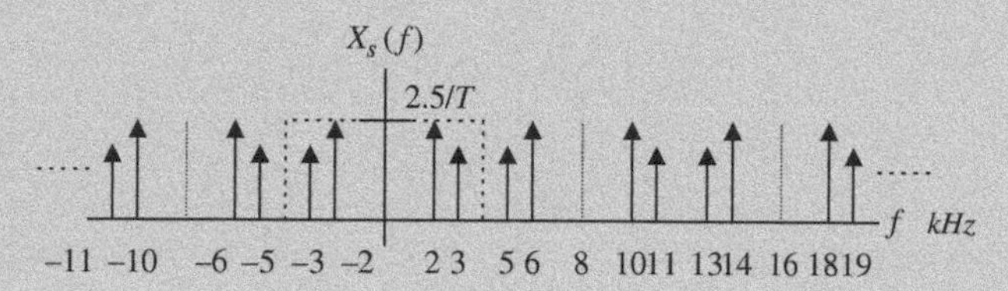

FIG. 2.12

Spectrum of the sampled signal in Example 2.2.

(b) Based on the spectrum in (a), the sampling theorem condition is satisfied; hence, we can recover the original spectrum using a reconstruction lowpass filter. The recovered spectrum is shown in Fig. 2.13.

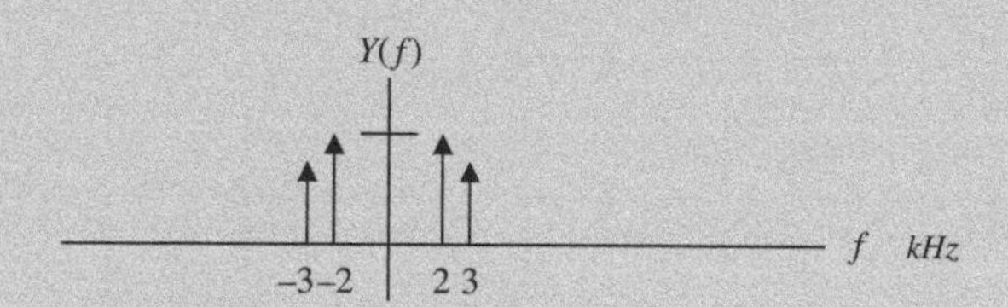

FIG. 2.13

Spectrum of the recovered signal in Example 2.2.

EXAMPLE 2.3

Given an analog signal

$x(t)=5\cos(2\pi\times2000t)+1\cos(2\pi\times5000t)$, for $t\geq0$,

which is sampled at a rate of 8000 Hz.

(a) Sketch the spectrum of the sampled signal up to 20 kHz.

(b) Sketch the recovered analog signal spectrum if an ideal lowpass filter with a cutoff frequency of 4 kHz is used to recover the original signal ($y(n)=x(n)$ in this case).

Solution:

(a) The spectrum for the sampled signal is sketched in Fig. 2.14:

Continued

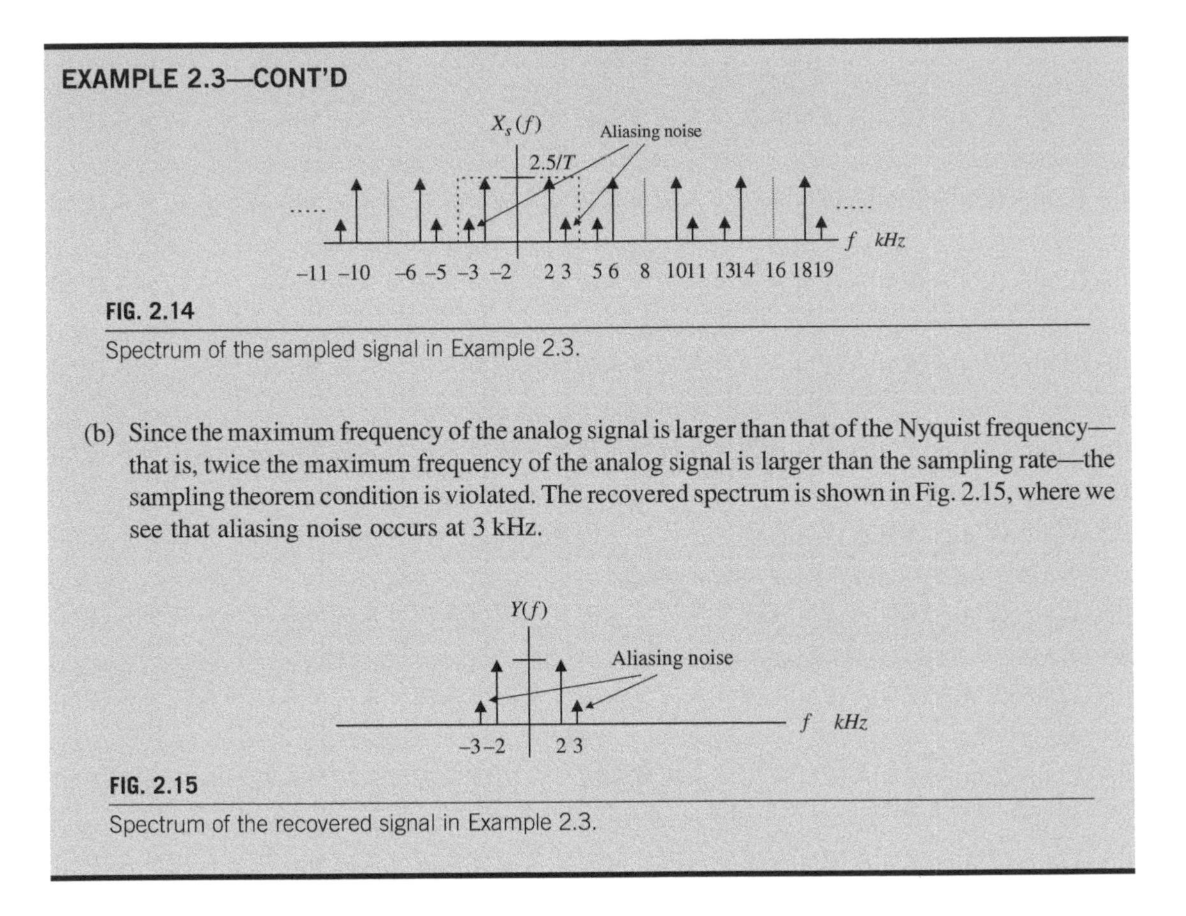

FIG. 2.14

Spectrum of the sampled signal in Example 2.3.

(b) Since the maximum frequency of the analog signal is larger than that of the Nyquist frequency—that is, twice the maximum frequency of the analog signal is larger than the sampling rate—the sampling theorem condition is violated. The recovered spectrum is shown in Fig. 2.15, where we see that aliasing noise occurs at 3 kHz.

FIG. 2.15

Spectrum of the recovered signal in Example 2.3.

2.2.1 PRACTICAL CONSIDERATIONS FOR SIGNAL SAMPLING: ANTI-ALIASING FILTERING

In practice, the analog signal to be digitized may contain the other frequency components whose frequencies are larger than the folding frequency, such as high-frequency noise. To satisfy the sampling theorem condition, we apply an anti-aliasing filter to limit the input analog signal, so that all the frequency components are less than the folding frequency (half of the sampling rate). Considering the worst case, where the analog signal to be sampled has a flat frequency spectrum, the band-limited spectrum $X(f)$ and sampled spectrum $X_s(f)$ are depicted in Fig. 2.16, where the shape of each replica in the sampled signal spectrum is the same as that of the anti-aliasing filter magnitude frequency response.

Due to nonzero attenuation of the magnitude frequency response of the anti-aliasing lowpass filter, the aliasing noise from the adjacent replica still appears in the baseband. However, the level of the aliasing noise is greatly reduced. We can also control the aliasing noise level by either using a higher-order lowpass filter or increasing the sampling rate. For illustrative purpose, we use a Butterworth filter. The method can also be extended to other filter types such as the Chebyshev filter. The Butterworth magnitude frequency response with an order of n is given by

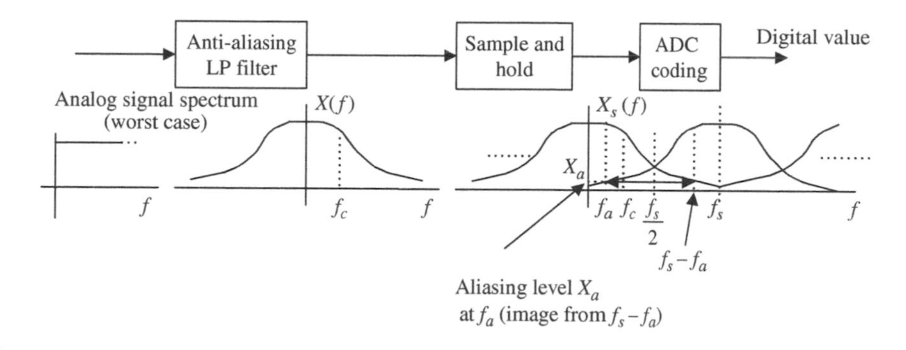

FIG. 2.16

Spectrum of the sampled analog signal with a practical anti-aliasing filter.

$$|H(f)| = \frac{1}{\sqrt{1 + (f/f_c)^{2n}}}. \tag{2.15}$$

For a second-order Butterworth lowpass filter with the unit gain, the transfer function (which will be discussed in Chapter 8) and its magnitude frequency response are given by

$$H(s) = \frac{(2\pi f_c)^2}{s^2 + 1.4141 \times (2\pi f_c)s + (2\pi f_c)^2}, \tag{2.16}$$

$$|H(f)| = \frac{1}{\sqrt{1 + (f/f_c)^4}}. \tag{2.17}$$

A unit gain second-order lowpass using a Sallen-Key topology is shown in Fig. 2.17. Matching the coefficients of the circuit transfer function to that of the second-order Butterworth lowpass transfer function in Eq. (2.16) as depicted in Eq. (2.18) gives the design formulas shown Fig. 2.17, where for a given cutoff frequency of f_c in Hz, and a capacitor value of C_2, we can determine the values for other elements using the formulas listed in the figure.

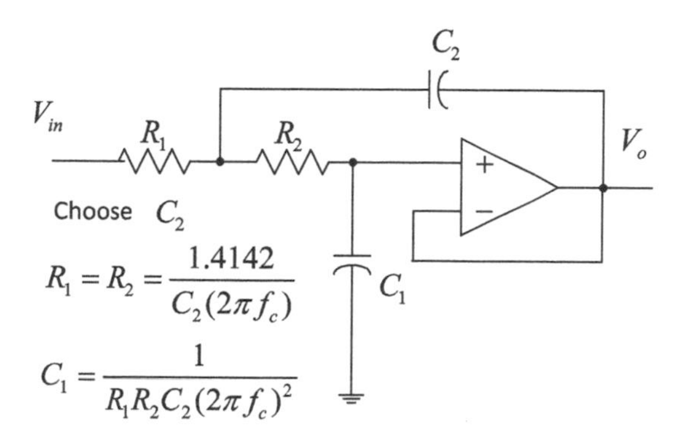

FIG. 2.17

Second-order unit-gain Sallen-Key lowpass filter.

$$\frac{\frac{1}{R_1R_2C_1C_2}}{s^2+\left(\frac{1}{R_1C_2}+\frac{1}{R_2C_2}\right)s+\frac{1}{R_1R_2C_1C_2}}=\frac{(2\pi f_c)^2}{s^2+1.4141\times(2\pi f_c)s+(2\pi f_c)^2}. \tag{2.18}$$

As an example, for a cutoff frequency of 3400 Hz, and by selecting $C_2=0.01$ microfarad (μF), we can get $R_1=R_2=6620\Omega$, and $C_1=0.005\mu F$.

Fig. 2.18 shows the magnitude frequency response, where the absolute gain of the filter is plotted. As we can see, the absolute attenuation begins at the level of 0.7 at 3400 Hz and reduces to 0.3 at 6000 Hz. Ideally, we want the gain attenuation to be zero after 4000 Hz if our sampling rate is 8000 Hz. Practically speaking, aliasing will occur anyway with some degree. We will study achieving the higher-order analog filter via Butterworth and Chebyshev prototype function tables in Chapter 8. More details of the circuit realization for the analog filter can be found in Chen (1986).

According to Fig. 2.16, we could derive the percentage of the aliasing noise level using the symmetry of the Butterworth magnitude function and its first replica. It follows that

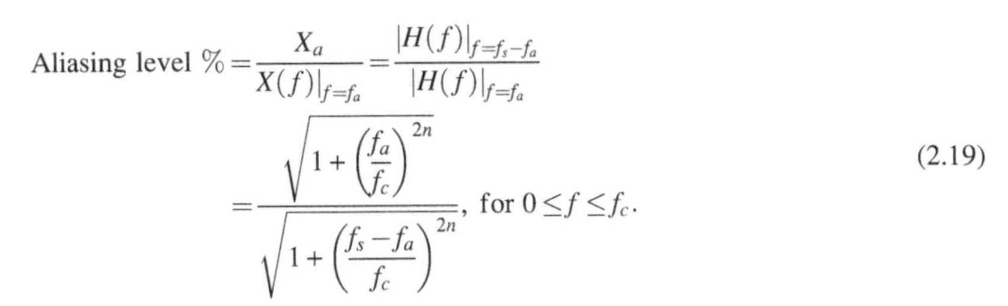

$$\text{Aliasing level }\%=\frac{X_a}{X(f)|_{f=f_a}}=\frac{|H(f)|_{f=f_s-f_a}}{|H(f)|_{f=f_a}}=\frac{\sqrt{1+\left(\frac{f_a}{f_c}\right)^{2n}}}{\sqrt{1+\left(\frac{f_s-f_a}{f_c}\right)^{2n}}},\text{ for }0\le f\le f_c. \tag{2.19}$$

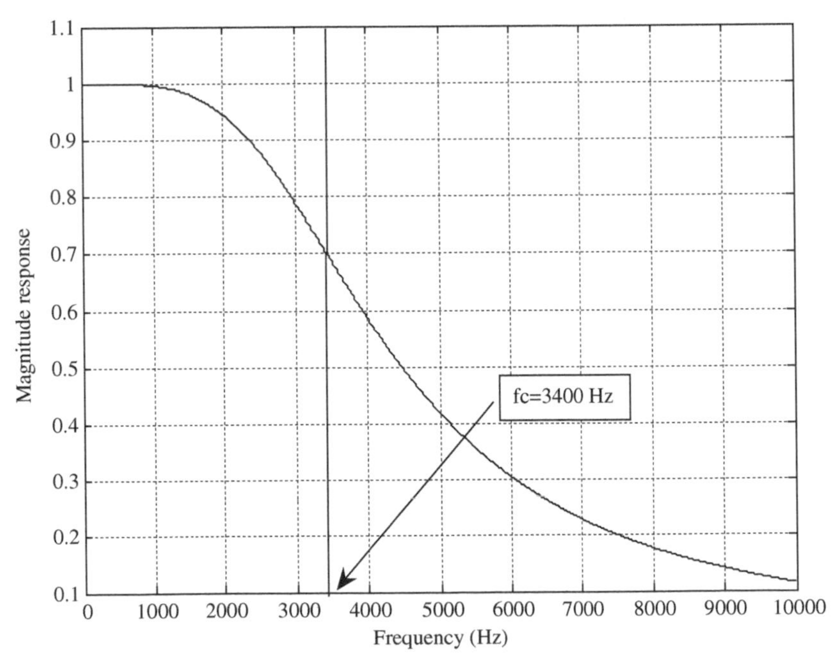

FIG. 2.18

Magnitude frequency response of the second-order Butterworth lowpass filter.

With the Eq. (2.19), we can estimate the aliasing level, or choose a higher-order anti-aliasing filter to satisfy requirement for the percentage of aliasing level.

EXAMPLE 2.4

Given the DSP system shown in Figs. 2.16–2.18, where a sampling rate of 8000 Hz is used and the anti-aliasing filter is a second-order Butterworth lowpass filter with a cutoff frequency of 3.4 kHz,

(a) Determine the percentage of aliasing level at the cutoff frequency.

(b) Determine the percentage of aliasing level at the frequency of 1000 Hz.

Solution:

$$f_s = 8000,\ f_c = 3400,\ \text{and}\ n = 2.$$

(a) Since $f_a = f_c = 3400$ Hz, we compute

$$\text{Aliasing level\%} = \frac{\sqrt{1+\left(\frac{3.4}{3.4}\right)^{2\times 2}}}{\sqrt{1+\left(\frac{8-3.4}{3.4}\right)^{2\times 2}}} = \frac{1.4142}{2.0858} = 67.8\%.$$

(b) With $f_a = 1000$ Hz, we have

$$\text{Aliasing level\%} = \frac{\sqrt{1+\left(\frac{1}{3.4}\right)^{2\times 2}}}{\sqrt{1+\left(\frac{8-1}{3.4}\right)^{2\times 2}}} = \frac{1.03007}{4.3551} = 23.05\%.$$

Let us examine another example with an increased sampling rate.

EXAMPLE 2.5

Given the DSP system shown in Figs. 2.16–2.18, where a sampling rate of 16,000 Hz is used and the anti-aliasing filter is a second-order Butterworth lowpass filter with a cutoff frequency of 3.4 kHz, determine the percentage of aliasing level at the cutoff frequency.

Solution:

$$f_s = 16000, f_c = 3400, \text{and}\, n = 2.$$

Since $f_a = f_c = 3400$ Hz, we have

$$\text{Aliasing level\%} = \frac{\sqrt{1+\left(\frac{3.4}{3.4}\right)^{2\times 2}}}{\sqrt{1+\left(\frac{16-3.4}{3.4}\right)^{2\times 2}}} = \frac{1.4142}{13.7699} = 10.26\%.$$

As a comparison with the result in Example 2.4, increasing the sampling rate can reduce the aliasing noise level.

The following example shows how to choose the order of an anti-aliasing filter.

EXAMPLE 2.6

Given the DSP system shown in Fig. 2.16, where a sampling rate of 40,000 Hz is used, the anti-aliasing filter is the Butterworth lowpass filter with a cutoff frequency 8 kHz, and the percentage of aliasing level at the cutoff frequency is required to be less than 1%, determine the order of the anti-aliasing lowpass filter.

Solution:

Using $f_s = 40,\ 000$, $f_c = 8000$, and $f_a = 8000$ Hz, we try each of the following filters with the increasing number of the filter order.

$$n = 1,\ \text{Aliasing level\%} = \frac{\sqrt{1 + \left(\frac{8}{8}\right)^{2\times 1}}}{\sqrt{1 + \left(\frac{40-8}{8}\right)^{2\times 1}}} = \frac{1.4142}{\sqrt{1+(4)^2}} = 34.30\%,$$

$$n = 2,\ \text{Aliasing level\%} == \frac{1.4142}{\sqrt{1+(4)^4}} = 8.82\%,$$

$$n = 3,\ \text{Aliasing level\%} = \frac{1.4142}{\sqrt{1+(4)^6}} = 2.21\%,$$

$$n = 4,\ \text{Aliasing level\%} = \frac{1.4142}{\sqrt{1+(4)^8}} = 0.55\% < 1\%.$$

To satisfy 1% aliasing noise level, we choose $n = 4$.

2.2.2 PRACTICAL CONSIDERATIONS FOR SIGNAL RECONSTRUCTION: ANTI-IMAGE FILTER AND EQUALIZER

Consider that a unit impulse function $\delta(t)$ is passed through the hold transfer function, $H_h(s)$. The impulse response $h(t)$ is expected in Fig. 2.19 and can be expressed as

$$h(t) = u(t) - u(t-T), \tag{2.20}$$

where T is the sampling period and $u(t)$ is the unit step function, that is,

$$u(t) = \begin{cases} 1 & t \geq 0 \\ 0 & t < 0 \end{cases}. \tag{2.21}$$

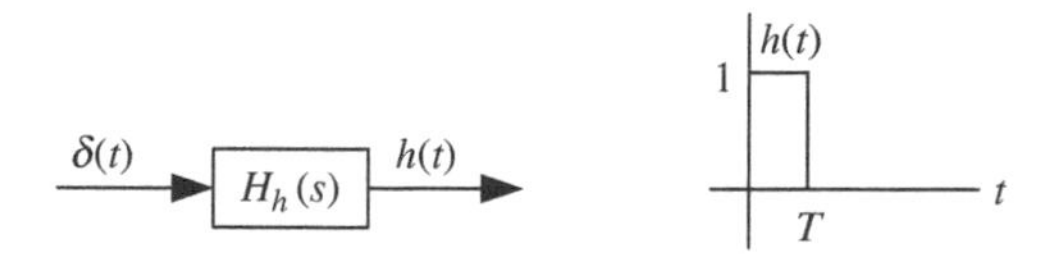

FIG. 2.19

Impulse response of the hold unit.

The transfer function $H_h(s)$ can be obtained by Laplace transform of the impulse response $h(t)$, that is,

$$H_h(s) = L\{h(t)\} = L\{u(t) - u(t-T)\} = \frac{1}{s} - \frac{1}{s}e^{-sT} = \frac{1-e^{-sT}}{s}. \tag{2.22}$$

The analog signal recovery for a practical DSP system is illustrated as shown in Fig. 2.20.

As shown in Fig. 2.20, the DAC unit converts the processed digital signal $y(n)$ to a sampled signal $y_s(t)$, and then the hold circuit produces the sample-and-hold voltage $y_H(t)$. The transfer function of the hold circuit is derived as

$$H_h(s) = \frac{1-e^{-sT}}{s}. \tag{2.23}$$

We can obtain the frequency response of the DAC with the hold circuit by substituting $s = j\omega$ to Eq. (2.23). It follows that

$$H_h(j\omega) = \frac{1-e^{-j\omega T}}{j\omega} = e^{-j\omega T/2}\frac{\left(e^{j\omega T/2} - e^{-j\omega T/2}\right)}{j\omega} = Te^{-j\omega T/2}\frac{\sin(\omega T/2)}{\omega T/2}. \tag{2.24}$$

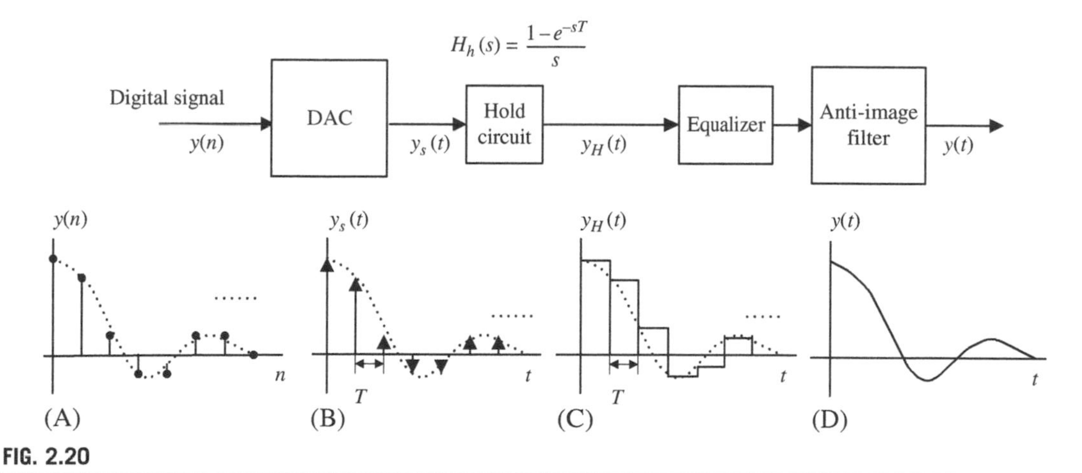

FIG. 2.20

Signal notations at the practical reconstruction stage: (A) Processed digital signal, (B) Recovered ideal sampled signal, (C) Recovered sample-and-hold voltage, and (D) Recovered analog signal.

The magnitude and phase responses are given by

$$|H_h(\omega)| = T\left|\frac{\sin(\omega T/2)}{\omega T/2}\right| = T\left|\frac{\sin(x)}{x}\right|, \tag{2.25}$$

$$\angle H_h(\omega) = -\omega T/2, \tag{2.26}$$

where $x = \omega T/2$. In terms of Hz, we have

$$|H_h(f)| = T\left|\frac{\sin(\pi f T)}{\pi f T}\right|, \tag{2.27}$$

$$\angle H_h(f) = -\pi f T. \tag{2.28}$$

The plot of the magnitude effect is shown in Fig. 2.21.

Fig. 2.22A describes the recovery stage.

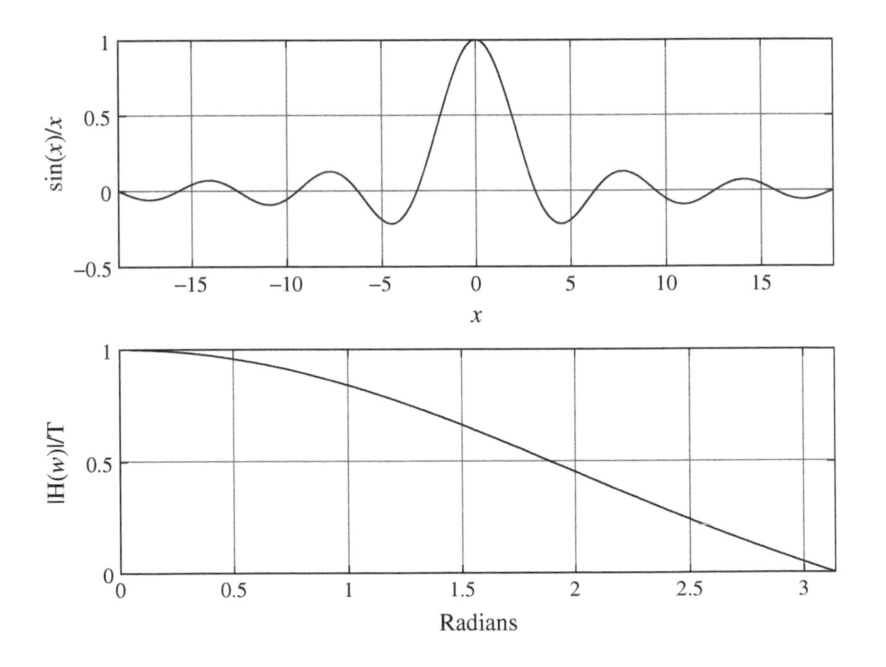

FIG. 2.21

Sample-and-hold lowpass filtering effect.

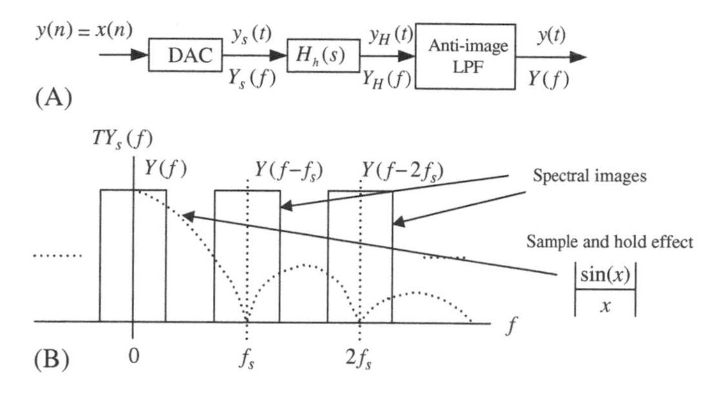

FIG. 2.22

(A) Signal recovery stage and (B) Sample-and-hold effect and distortion.

Assuming that there is no digital processing, that is, $y_s(t) = x_s(t)$, then

$$Y_s(f) = X_s(f) = \frac{1}{T}\sum_{k=-\infty}^{\infty} X(f - kf_s). \tag{2.29}$$

As shown in Fig. 2.22A, $y_H(t)$ is the sample-and-hold signal in time domain while $Y_H(f)$ is the spectrum of the sampled and hold signal, which is given by

$$Y_H(f) = H_h(f)Y_s(f) = e^{-j\pi fT}T\frac{\sin(\pi fT)}{\pi fT}Y_s(f). \tag{2.30}$$

The magnitude frequency response of the sampled and hold signal becomes

$$|Y_H(f)| = T\left|\frac{\sin(\pi fT)}{\pi fT}\right||Y_s(f)|. \tag{2.31}$$

The magnitude frequency response acts like lowpass filtering and shapes the sampled signal spectrum of $Y_s(f)$. This shaping effect distorts the sampled signal spectrum $Y_s(f)$ in the desired frequency band, as illustrated in Fig. 2.22B. On the other hand, the spectral images are attenuated due to the lowpass effect of $\sin(x)/x$. This sample-and-hold effect can help us design the anti-image filter.

Since the magnitude frequency response of the sampled signal using an ideal sampler is $T|Y_s(f)|$, therefore, the spectral distortion at the recovery stage can be derived as

$$\text{Distortion} = \frac{T|Y_s(f)| - |Y_H(f)|}{T|Y_s(f)|} = 1 - \frac{|Y_H(f)|}{T|Y_s(f)|} = 1 - \left|\frac{\sin(\pi fT)}{\pi fT}\right|. \tag{2.32}$$

The percentage of distortion in the desired frequency band is given by

$$\text{Distortion}\% = \left(1 - \left|\frac{\sin(\pi fT)}{\pi fT}\right|\right) \times 100\%. \tag{2.33}$$

Let us look at Example 2.7

EXAMPLE 9.1

Given a DSP system with a sampling rate of 8000 Hz and a hold circuit used after DAC,
(a) Determine the percentage of distortion at the frequency of 3400 Hz,
(b) Determine the percentage of distortion at the frequency of 1000 Hz.

Solution:

(a) Since $fT = 3400 \times 1/8000 = 0.425$,

$$\text{Distortion}\% = \left(1 - \left|\frac{\sin(0.425\pi)}{0.425\pi}\right|\right) \times 100\% = 27.17\%.$$

(b) Since $fT = 1000 \times 1/8000 = 0.125$,

$$\text{Distortion}\% = \left(1 - \left|\frac{\sin(0.125\pi)}{0.125\pi}\right|\right) \times 100\% = 2.55\%.$$

To overcome the sample-and-hold effect, the following methods can be applied:

(1) We can compensate the sample-and-hold shaping effect using an equalizer whose magnitude response is opposite to the shape of the hold circuit magnitude frequency response, which is shown as the solid line in Fig. 2.23.

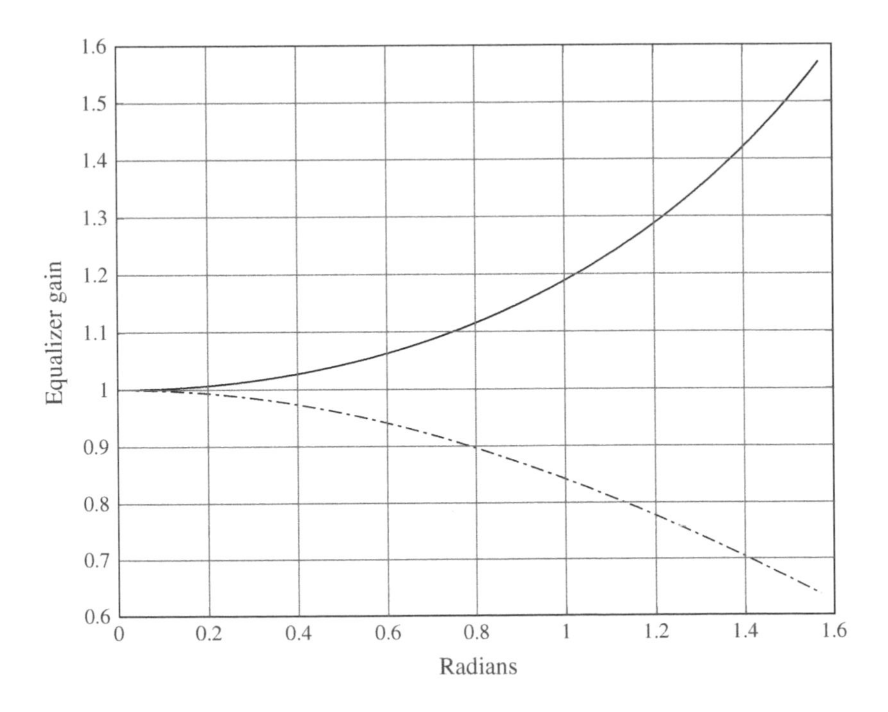

FIG. 2.23

Ideal equalizer magnitude frequency response to overcome the distortion introduced by the sample-and-hold process.

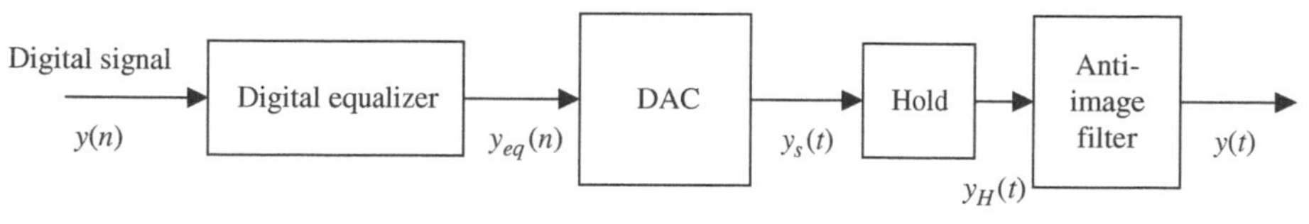

FIG. 2.24

Possible implementation using the digital equalizer.

(2) We can increase the sampling rate using oversampling and interpolation methods when a higher sampling rate is available at the DAC. Using the interpolation will increase the sampling rate without affecting the signal bandwidth, so that the baseband spectrum and its images are separated further apart and a lower-order anti-aliasing filter can be used. This subject will be discussed in Chapter 11.

(3) We can change the DAC configuration and perform digital pre-equalization using the flexible digital filter whose magnitude frequency response is against the spectral shape effect due to the hold circuit. Fig. 2.24 shows a possible implementation. In this way, the spectral shape effect can be balanced before the sampled signal passes through the hold circuit. Finally, the anti-image filter will remove the rest of images and recover the desired analog signal.

The following practical example illustrates the design of an anti-image filter using a higher sampling rate while making use of the sample-and-hold effect.

EXAMPLE 2.8

Determine the cutoff frequency and the order for the anti-image filter given a DSP system with a sampling rate of 16,000 Hz and specifications for the anti-image filter as shown in Fig. 2.25.

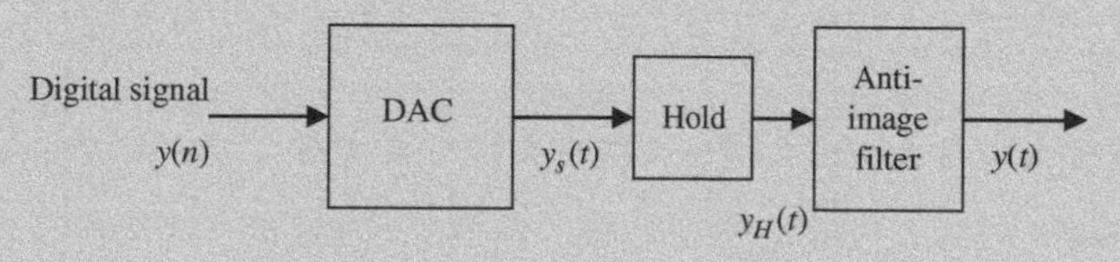

FIG. 2.25

DSP recovery system for Example 2.8.

Design requirements:

- Maximum allowable gain variation from 0 to 3000 Hz = 2 dB
- 33 dB rejection at the frequency of 13,000 Hz
- Butterworth filter is assumed for the anti-image filter

Solution:

We first determine the spectral shaping effects at $f = 3000$ Hz, and $f = 13,000$ Hz; that is,

$$f = 3000\text{Hz}, fT = 3000 \times 1/16{,}000 = 0.1875,$$

$$\text{Gain} = \left| \frac{\sin(0.1875\pi)}{0.1875\pi} \right| = 0.9484 = -0.46\text{dB},$$

and

Continued

EXAMPLE 2.8—CONT'D

$$f = 13{,}000\,\text{Hz}, fT = 13{,}000 \times 1/16{,}000 = 0.8125,$$

$$\text{Gain} = \left|\frac{\sin(0.8125\pi)}{0.8125\pi}\right| = 0.2177 \approx -13\,\text{dB}.$$

This gain would help the attenuation requirement as shown in Fig. 2.26.

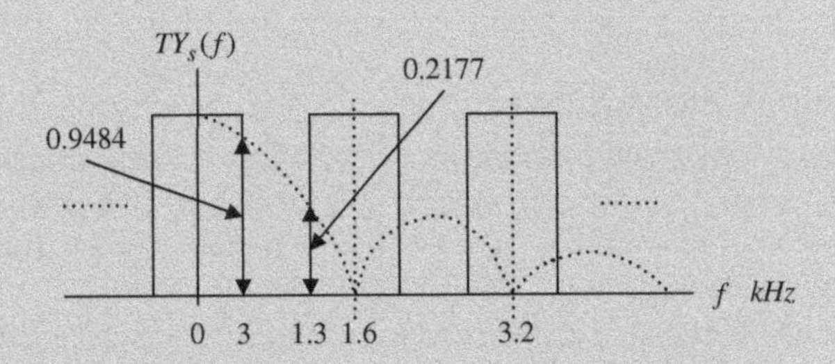

FIG. 2.26

Spectral shaping by the sample-and-hold effect in Example 2.8.

Hence, the design requirements for the anti-image filter are:

- Butterworth lowpass filter
- Maximum allowable gain variation from 0 to 3000 Hz $= 2 - 0.46 = 1.54\,\text{dB}$
- $33 - 13 = 20\,\text{dB}$ rejection at frequency 13,000 Hz.

We set up equations using log operations of the Butterworth magnitude function as

$$20\log\left(1 + (3000/f_c)^{2n}\right)^{1/2} \le 1.54$$

$$20\log\left(1 + (13{,}000/f_c)^{2n}\right)^{1/2} \ge 20.$$

From these two equations, we have to satisfy

$$(3000/f_c)^{2n} = 10^{0.154} - 1$$

$$(13{,}000/f_c)^{2n} = 10^2 - 1.$$

Taking the ratio of these two equations yields

$$\left(\frac{13{,}000}{3000}\right)^{2n} = \frac{10^2 - 1}{10^{0.154} - 1}.$$

Then

$$n = \frac{1}{2}\log\left((10^2 - 1)/(10^{0.154} - 1)\right)/\log(13{,}000/3000) = 1.86 \approx 2.$$

Finally, the cutoff frequency can be computed as

$$f_c = \frac{13{,}000}{\left(10^2 - 1\right)^{1/(2n)}} = \frac{13{,}000}{\left(10^2 - 1\right)^{1/4}} = 4121.30\,\text{Hz}$$

$$f_c = \frac{3000}{\left(10^{0.154} - 1\right)^{1/(2n)}} = \frac{3000}{\left(10^{0.154} - 1\right)^{1/4}} = 3714.23\,\text{Hz}.$$

We choose the smaller one, that is,

$$f_c = 3714.23\,\text{Hz}.$$

With the filter order and cutoff frequency, we can realize the anti-image (reconstruction) filter using a second-order unit-gain Sallen-Key lowpass filter described in Fig. 2.17.

Note that the specifications for anti-aliasing filter designs are similar to anti-image (reconstruction) filters, except for their stopband edges. The anti-aliasing filter is designed to block the frequency components beyond the folding frequency before the ADC operation, while the reconstruction filter is to block the frequency components beginning at the lower edge of the first image after the DAC.

2.3 ANALOG-TO-DIGITAL CONVERSION, DIGITAL-TO-ANALOG CONVERSION, AND QUANTIZATION

During the ADC process, amplitudes of the analog signal to be converted have infinite precision. The continuous amplitude must be converted to a digital data with finite precision, which is called the *quantization*. Fig. 2.27 shows that quantization is a part of ADC.

There are several ways to implement ADC. The most common ones are

- Flash ADC,
- Successive approximation ADC, and
- Sigma-delta ADC.

In this chapter, we will focus on a simple 2-bit flash ADC unit, described in Fig. 2.28, for illustrative purpose. Sigma-delta ADC is studied in Chapter 11.

As shown in Fig. 2.28, the 2-bit flash ADC unit consists of a serial reference voltage created by the equal value resistors, a set of comparators, and logic units.

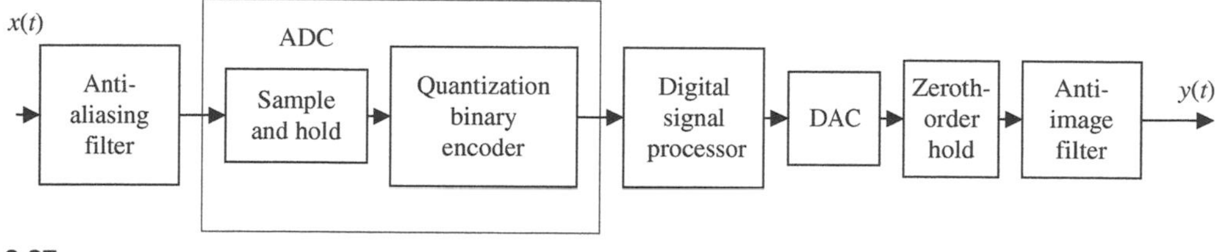

FIG. 2.27

A block diagram for a DSP system.

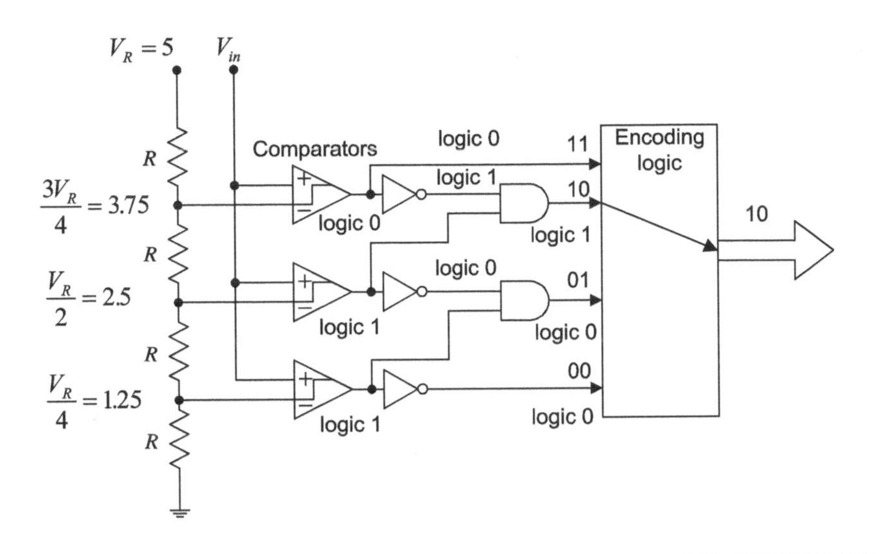

FIG. 2.28

An example of a 2-bit flash ADC.

As an example, the reference voltages in the figure are 1.25, 2.5, 3.75, and 5 V, respectively. If an analog sample-and-hold voltage is $V_{in} = 3$ V, then the lower two comparators will have output logic 1 of each. Through the logic units, only the line labeled 10 is actively high, and the rest of lines are active low. Hence, the encoding logic circuit outputs a 2-bit binary code of 10.

Flash ADC offers the advantage of high conversion speed, since all bits are acquired at the same time. Fig. 2.29 illustrates a simple 2-bit DAC unit using an R-2R ladder. The DAC contains the R-2R

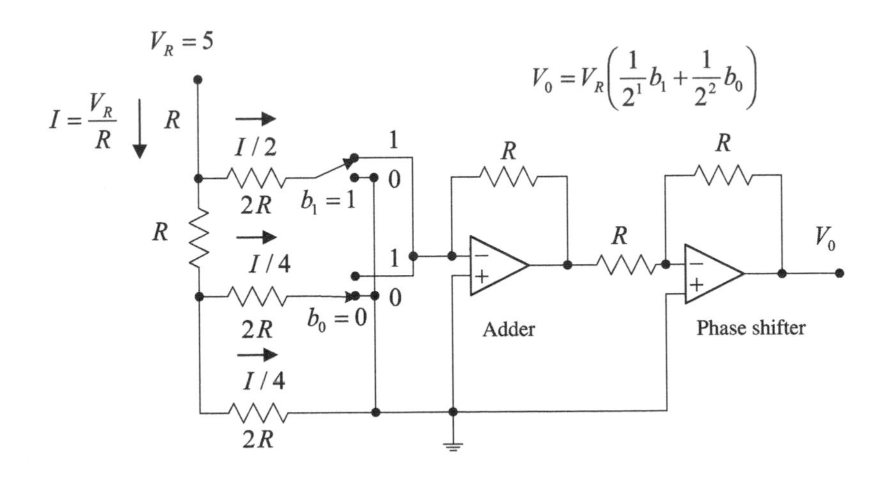

FIG. 2.29

R-2R ladder DAC.

ladder circuit, a set of single-throw switches, an adder, and a phase shifter. If a bit is logic 0, the switch connects a 2R resistor to ground. If a bit is logic 1, the corresponding 2R resistor is connected to the branch to the input of the operational amplifier (adder). When the operational amplifier operates in a linear range, the negative input is virtually equal to the positive input. The adder adds all the currents from all branches. The feedback resistor R in the adder provides overall amplification. The ladder network is equivalent to two 2R resistors in parallel. The entire network has a total current of $I = \frac{V_R}{R}$ using ohm's law, where V_R is the reference voltage, chosen to be 5 V, for example. Hence, half of the total current flows into the b_1 branch, while the other half flows into the rest of the network. The halving process repeats for each branch successively to the lower bit branches to get lower bit weights. The second operational amplifier acts like a phase shifter to cancel the negative sign of the adder output. Using the basic electric circuit principle, we can determine the DAC output voltage as

$$V_0 = V_R\left(\frac{1}{2^1}b_1 + \frac{1}{2^2}b_0\right),$$

where b_1 and b_0 are bits in the 2-bit binary code, with b_0 as the least significant bit (LSB).

As an example shown in Fig. 2.29, where we set $V_R = 5$ and $b_1b_0 = 10$, the ADC output is expected to be

$$V_0 = 5 \times \left(\frac{1}{2^1} \times 1 + \frac{1}{2^2} \times 0\right) = 2.5\text{ V}.$$

As we can see, the recovered voltage of $V_0 = 2.5$ V introduces voltage error as compared with $V_{in} = 3$ V, as discussed in the ADC stage. This is due to the fact that in the flash ADC unit, we use only four (i.e., finite) voltage levels to represent continuous (infinitely possible) analog voltage values. The voltage error is called *quantization error*, obtained by subtracting the original analog voltage from the recovered analog voltage. For example, we have the quantization error as

$$V_0 - V_{in} = 2.5 - 3 = -0.5\text{ V}.$$

Next, we focus on quantization development. The process of converting analog voltage with infinite precision to finite precision is called the *quantization process*. For example, if the digital processor has only a 3-bit word, the amplitudes can be converted into eight different levels.

A *unipolar quantizer* deals with analog signals ranging from 0 V to a positive reference voltage, and a *bipolar quantizer* has an analog signal range from a negative reference to a positive reference. The notations and general rules for quantization are:

$$\Delta = \frac{(x_{max} - x_{min})}{L} \tag{2.34}$$

$$L = 2^m \tag{2.35}$$

$$i = \text{round}\left(\frac{x - x_{min}}{\Delta}\right) \tag{2.36}$$

$$x_q = x_{min} + i\Delta \quad i = 0, 1, \ldots, L-1, \tag{2.37}$$

where x_{max} and x_{min} are the maximum value and minimum values, respectively, of the analog input signal x. The symbol L denotes the number of quantization levels, which is determined by Eq. (2.35), where m is the number of bits used in ADC. The symbol Δ is the step size of the quantizer or the ADC

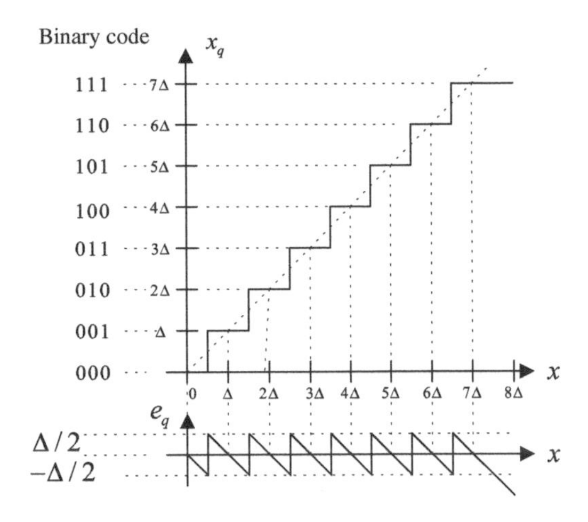

FIG. 2.30

Characteristics for a unipolar quantizer.

resolution. Finally, x_q indicates the quantization level, and i is an index corresponding to the binary code.

Fig. 2.30 depicts a 3-bit unipolar quantizer and corresponding binary codes. From Fig. 2.30, we see that $x_{\min}=0$, $x_{\max}=8\Delta$, and $m=3$. Applying Eq. (2.37) gives each quantization level as follows: $x_q=0+i\Delta$, $i=0, 1, \ldots, L-1$, where $L=2^3=8$ and i is the integer corresponding to the 3-bit binary code. Table 2.1 details quantization for each input signal subrange.

Similarly, a 3-bit bipolar quantizer and binary codes are shown in Fig. 2.31, where we have

$$x_{\min}=-4\Delta, x_{\max}=4\Delta, \text{and } m=3.$$

The corresponding quantization table is given in Table 2.2.

Table 2.1 Quantization Table for a 3-Bit Unipolar Quantizer (Step Size $=\Delta=(x_{\max}-x_{\min})/2^3$, $x_{\max}=$ maximum voltage, and $x_{\min}=0$)

Binary Code	Quantization Level x_q (V)	Input Signal Subrange (V)
0 0 0	0	$0 \le x < 0.5\Delta$
0 0 1	Δ	$0.5\Delta \le x < 1.5\Delta$
0 1 0	2Δ	$1.5\Delta \le x < 2.5\Delta$
0 1 1	3Δ	$2.5\Delta \le x < 3.5\Delta$
1 0 0	4Δ	$3.5\Delta \le x < 4.5\Delta$
1 0 1	5Δ	$4.5\Delta \le x < 5.5\Delta$
1 1 0	6Δ	$5.5\Delta \le x < 6.5\Delta$
1 1 1	7Δ	$6.5\Delta \le x < 7.5\Delta$

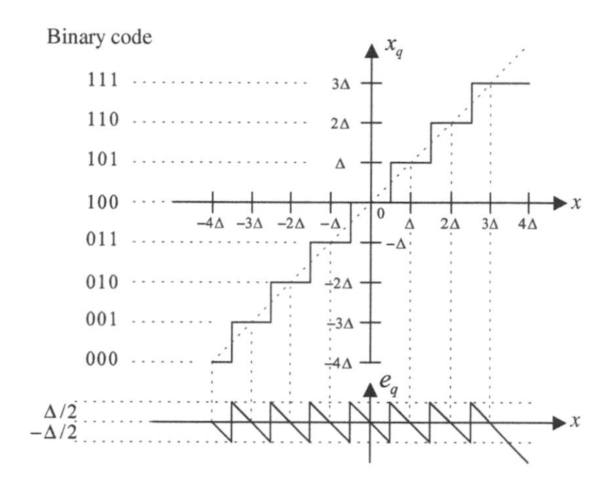

FIG. 2.31

Characteristics for a bipolar quantizer.

Table 2.2 Quantization Table for a 3-Bit Bipolar Quantizer (Step Size = $\Delta = (x_{max} - x_{min})/2^3$, x_{max} = maximum voltage, and $x_{min} = -x_{max}$)

Binary Code	Quantization Level x_q (V)	Input Signal Subrange (V)
0 0 0	-4Δ	$-4\Delta \le x < -3.5\Delta$
0 0 1	-3Δ	$-3.5\Delta \le x < -2.5\Delta$
0 1 0	-2Δ	$-2.5\Delta \le x < -1.5\Delta$
0 1 1	$-\Delta$	$-1.5\Delta \le x < -0.5\Delta$
1 0 0	0	$-0.5\Delta \le x < 0.5\Delta$
1 0 1	Δ	$0.5\Delta \le x < 1.5\Delta$
1 1 0	2Δ	$1.5\Delta \le x < 2.5\Delta$
1 1 1	3Δ	$2.5\Delta \le x < 3.5\Delta$

EXAMPLE 2.9

Assuming that a 3-bit ADC channel accepts analog input ranging from 0 to 5 V, determine the following:

(a) Number of quantization levels
(b) Step size of quantizer or resolution
(c) Quantization level when the analog voltage is 3.2 V
(d) Binary code produced by the ADC.

Solution:

Since the range is from 0 to 5 V and the 3-bit ADC is used, we have

$$x_{min} = 0\ \text{V}, x_{max} = 5\ \text{V}, \text{and}\ m = 3\ \text{bits}.$$

Continued

EXAMPLE 2.9—CONT'D

(a) Using Eq. (2.35), we get the number of quantization levels as

$$L = 2^m = 2^3 = 8.$$

(b) Applying Eq. (2.34) yields

$$\Delta = \frac{5-0}{8} = 0.625\,\text{V}.$$

(c) When $x = 3.2\frac{\Delta}{0.625} = 5.12\Delta$, from Eq. (2.36) we get

$$i = \text{round}\left(\frac{x - x_{\min}}{\Delta}\right) = \text{round}(5.12) = 5.$$

(d) From Eq. (2.37), we determine the quantization level as

$$x_q = 0 + 5\Delta = 5 \times 0.625 = 3.125\,\text{V}.$$

The binary code is determined as 101, either from Fig. 2.30 or Table 2.1.

After quantizing the input signal x, the ADC produces binary codes, as illustrated in Fig. 2.32.

The DAC process is shown in Fig. 2.33. As shown in the figure, the DAC unit takes the binary codes from the digital signal processor. Then it converts the binary code using the zero-order hold circuit to reproduce the sample-and-hold signal. Assuming that the spectrum distortion due to sample-and-hold effect can be ignored for our illustration, the recovered sample-and-hold signal is further processed using the anti-image filter. Finally, the analog signal is yielded.

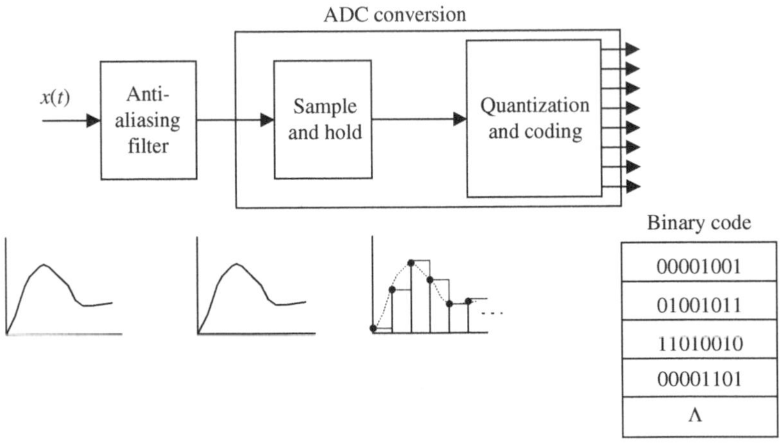

FIG. 2.32

Typical ADC process.

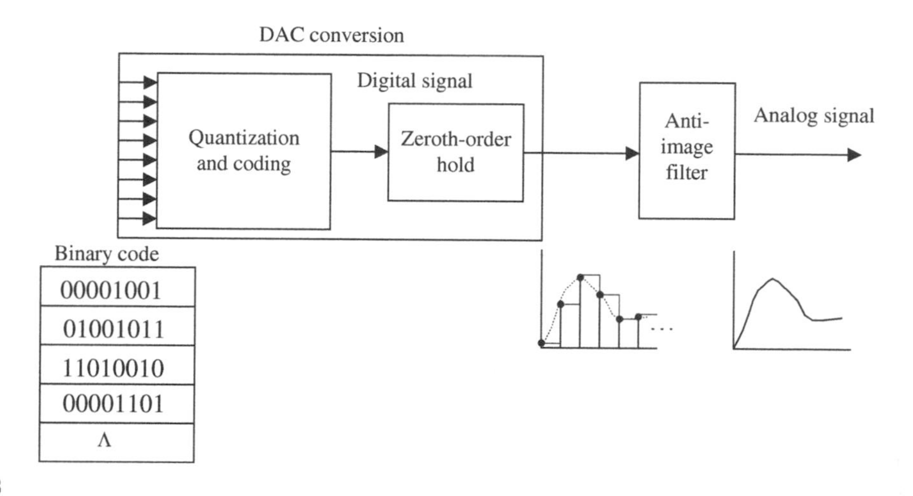

FIG. 2.33

Typical DAC process.

When the DAC outputs the analog amplitude x_q with finite precision, it introduces the quantization error defined as

$$e_q = x_q - x. \tag{2.38}$$

The quantization error as shown in Fig. 2.30 is bounded by half of the step size, that is,

$$-\frac{\Delta}{2} < e_q \le \frac{\Delta}{2}, \tag{2.39}$$

where Δ is the quantization step size, or the ADC resolution. We also refer to Δ as $V_{\min}$ (minimum detectable voltage) or the LSB value of the ADC.

EXAMPLE 2.10

Using Example 2.9, determine the quantization error when the analog input is 3.2 V.

Solution:

Using Eq. (2.38), we obtain

$$e_q = x_q - x = 3.125 - 3.2 = -0.075\,\text{V}.$$

Note that the quantization error is less than the half of the step size, that is,

$$|e_q| = 0.075 < \frac{\Delta}{2} = 0.3125\,\text{V}.$$

In practice, we can empirically confirm that the quantization error appears in uniform distribution when the step size is much smaller than the dynamic range of the signal samples and there are a sufficiently large number of samples. Assuming that e_q is a uniformly distributed random variable which has a range within a quantized interval Δ and has the following probability density function, $f(e_q)$, as shown in Fig. 2.34.

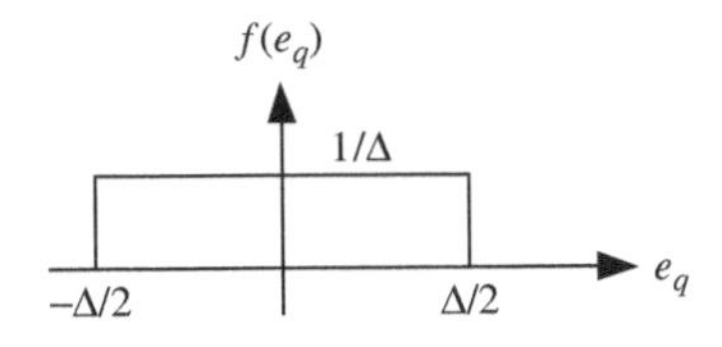

FIG. 2.34

Quantized noise distribution.

The quantized noise power can be derived as

$$E\left\{e_q^2\right\} = \int_{-\Delta/2}^{\Delta/2} e_q^2 f(e_q) de_q = \int_{-\Delta/2}^{\Delta/2} e_q^2 \frac{1}{\Delta} de_q = \left.\frac{e_q^3}{3\Delta}\right|_{-\Delta/2}^{\Delta/2} = \frac{\Delta^2}{12}. \tag{2.40}$$

Hence, the power of quantization noise is related to the quantization step and given by

$$E\left(e_q^2\right) = \frac{\Delta^2}{12}, \tag{2.41}$$

where $E()$ is the expectation operator, which actually averages the squared values of the quantization error [the reader can get more information from the texts by Roddy and Coolen (1997), Tomasi (2004), and Stearns and Hush (1990)]. The ratio of signal power to quantization noise power (SNR) due to quantization can be expressed as

$$SNR = \frac{E(x^2)}{E\left(e_q^2\right)}. \tag{2.42}$$

If we express the SNR in terms of decibels (dB), it follows that

$$SNR_{dB} = 10 \times \log_{10}(SNR)\,\text{dB}. \tag{2.43}$$

Substituting Eq. (2.41) and $E(x^2) = x^2{}_{rms}$ into Eq. (2.43), we achieve

$$SNR_{dB} = 10.79 + 20 \times \log_{10}\left(\frac{x_{rms}}{\Delta}\right), \tag{2.44}$$

where x_{rms} is the RMS (root-mean-squared) value of the signal to be quantized x.

Practically, the SNR can be calculated using the following formula:

$$SNR = \frac{\frac{1}{N}\sum_{n=0}^{N-1} x^2(n)}{\frac{1}{N}\sum_{n=0}^{N-1} e_q^2(n)} = \frac{\sum_{n=0}^{N-1} x^2(n)}{\sum_{n=0}^{N-1} e_q^2(n)}, \tag{2.45}$$

where $x(n)$ is the nth sample amplitude and $e_q(n)$ the quantization error from quantizing $x(n)$.

EXAMPLE 2.11

If the analog signal to be quantized is a sinusoidal waveform, that is,

$$x(t) = A\sin(2\pi \times 1000t),$$

and if the bipolar quantizer uses m bits, determine the SNR in terms of m bits.

Solution:

Since $x_{rms} = 0.707A$ and $\Delta = 2A/2^m$, substituting x_{rms} and Δ into Eq. (2.44) leads to

$$\begin{aligned} SNR_{dB} &= 10.79 + 20 \times \log_{10}\left(\frac{0.707A}{2A/2^m}\right) \\ &= 10.79 + 20 \times \log_{10}(0.707/2) + 20m \times \log_{10}2. \end{aligned}$$

After simplifying the numerical values, we get

$$SNR_{dB} = 1.76 + 6.02m\text{dB}. \tag{2.46}$$

EXAMPLE 2.12

For a speech signal, if a ratio of the RMS value over the absolute maximum value of the analog signal (Roddy and Coolen, 1997) is given, that is, $\left(\frac{x_{rms}}{|x|_{\max}}\right)$, and the ADC quantizer uses m bits, determine the SNR in terms of m bits.

Solution:

Since

$$\Delta = \frac{x_{\max} - x_{\min}}{L} = \frac{2|x|_{\max}}{2^m},$$

Substituting Δ in Eq. (2.44) achieves

$$\begin{aligned} SNR_{dB} &= 10.79 + 20 \times \log_{10}\left(\frac{x_{rms}}{2|x|_{\max}/2^m}\right) \\ &= 10.79 + 20 \times \log_{10}\left(\frac{x_{rms}}{|x|_{\max}}\right) + 20m \times \log_{10}2 - 20 \times \log_{10}2. \end{aligned}$$

Thus, after numerical simplification, we have

$$SNR_{dB} = 4.77 + 20 \times \log_{10}\left(\frac{x_{rms}}{|x|_{\max}}\right) + 6.02m. \tag{2.47}$$

From Examples 2.11 and 2.12, we observed that increasing 1 bit of the ADC quantizer can improve SNR due to quantization by 6dB.

EXAMPLE 2.13

Given a sinusoidal waveform with a frequency of 100 Hz,

$$x(t) = 4.5\sin(2\pi \times 100t),$$

sampled at 8000 Hz,

(a) Write a MATLAB program to quantize the $x(t)$ using 4 bits to obtain and plot the quantized signal x_q, assuming the signal range is between −5 and 5 V.

(b) Calculate the SNR due to quantization.

Solution:

(a) Program 2.1. MATLAB program for Example 2.13.

```
%Example 2.13
clear all; close all
disp('Generate 0.02-second sine wave of 100 Hz and Vp=5');
fs=8000;                                    % sampling rate
T=1/fs;                                     % sampling interval
t=0:T:0.02;                                 % duration of 0.02 second
sig = 4.5*sin(2*pi*100*t);                  % generate sinusoids
bits = input('input number of bits =>');
lg = length(sig);                           % length of signal vector sig
for x=1:lg
 [Index(x) pq]=biquant(bits, -5,5, sig(x)); % Output quantized index
end
% transmitted
% received
for x=1:lg
 qsig(x)=biqtdec(bits, -5,5, Index(x));    %Recover the quantized value
end
 qerr=qsig-sig;                             %Calculate quantized error
stairs(t,qsig); hold                        % plot signal in stair case style
plot(t,sig); grid;                          % plot signal
xlabel('Time (sec.)'); ylabel('Quantized x(n)')
disp('Signal to noise ratio due to quantization noise')
snr(sig,qsig)
```

Fig. 2.35 shows plots of the quantized signal and the original signal.

(b) Theoretically, applying Eq. (2.46) leads to

$$SNR_{dB} = 1.76 + 6.02 \times 4 = 25.84\text{dB}.$$

Practically, using Eq. (2.45), the simulated result is obtained as

$$SNR_{dB} = 25.78\,\text{dB}.$$

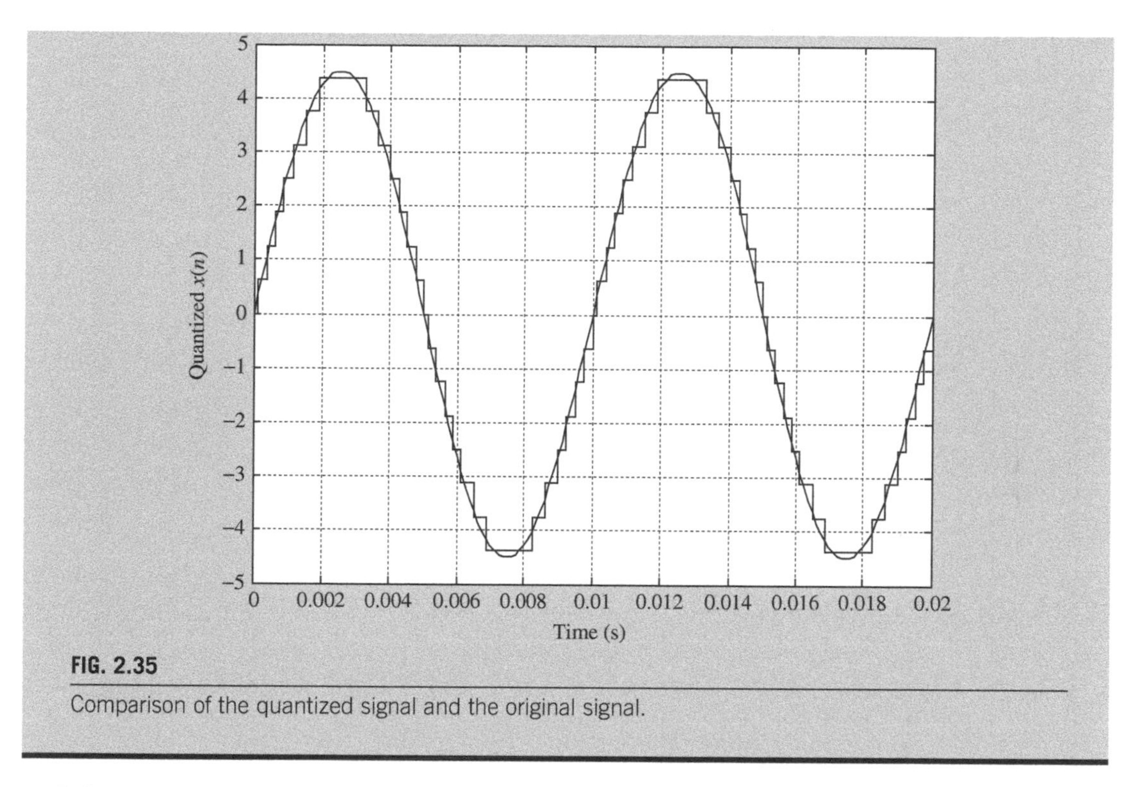

FIG. 2.35

Comparison of the quantized signal and the original signal.

It is clear from this example that the ratios of signal power to noise power due to quantization achieved from theory and from simulation are very close. Next, we look at an example for the quantizing a speech signal.

EXAMPLE 2.14

Given a speech signal sampled at 8000 Hz given in the file *we.dat*,

(a) Write a MATLAB program to quantize the $x(t)$ using 4-bits quantizers to obtain the quantized signal x_q, assuming the signal range is from −5 to 5 V.

(b) Plot the original speech, quantized speech, and quantization error, respectively.

(c) Calculate the SNR due to quantization using the MATLAB program.

Solution:

(a) Program 2.2 MATLAB program for Example 2.14.

```
%Example 2.14
clear all; close all
disp('load speech: We');
load we.dat                         % Load speech data at the current folder
sig = we;                           % Provided by the instructor
fs=8000;                            % Sampling rate
lg=length(sig);                     % Length of signal vector
T=1/fs;                             % Sampling period
```

Continued

EXAMPLE 2.14—CONT'D

```
t=[0:1:lg-1]*T;                          % Time instants in second
sig=4.5*sig/max(abs(sig));               % Normalizes speech in the range from −4.5 to 4.5
Xmax = max(abs(sig));                    % Maximum amplitude
Xrms = sqrt( sum(sig .* sig) / length(sig))   % RMS value
disp('Xrms/Xmax')
k=Xrms/Xmax
disp('20*log10(k)=>');
k = 20*log10(k)
bits = input('input number of bits =>');
lg = length(sig);
for x=1:lg
  [Index(x) pq]=biquant(bits, −5,5, sig(x));   %Output quantized index.
end
% transmitted
% received
for x=1:lg
  qsig(x) = biqtdec(bits, -5,5, Index(x));    %Recover the quantized value
end
  qerr = sig-qsig;                            %Calculate the quantized error
subplot(3,1,1);plot(t,sig);
ylabel('Original speech');title('we.dat: we');
subplot(3,1,2);stairs(t, qsig);grid
ylabel('Quantized speech')
subplot(3,1,3);stairs(t, qerr);grid
ylabel('Quantized error')
xlabel('Time (sec.)');axis([0 0.25 −1 1]);
disp('signal to noise ratio due to quantization noise')
snr(sig,qsig)                  % Signal to noise ratio in dB:
                               % sig = original signal vector,
                               % qsig =quantized signal vector
```

(b) In Fig. 2.36, the top plot shows the speech wave to be quantized, while the middle plot displays the quantized speech signal using 4 bits. The bottom plot shows the quantization error. It also shows that the absolute value of quantization error is uniformly distributed in a range between −0.3125 and 0.3125.

(c) From the MATLAB program, we have $\frac{x_{rms}}{|x|_{\max}} = 0.203$. Theoretically, from Eq. (2.47), it follows that

$$SNR_{dB} = 4.77 + 20\log_{10}\left(\frac{x_{rms}}{|x|_{\max}}\right) + 6.02 \times 4$$

$$= 4.77 + 20\log_{10}(0.203) + 6.02 \times 4 = 15\text{dB}.$$

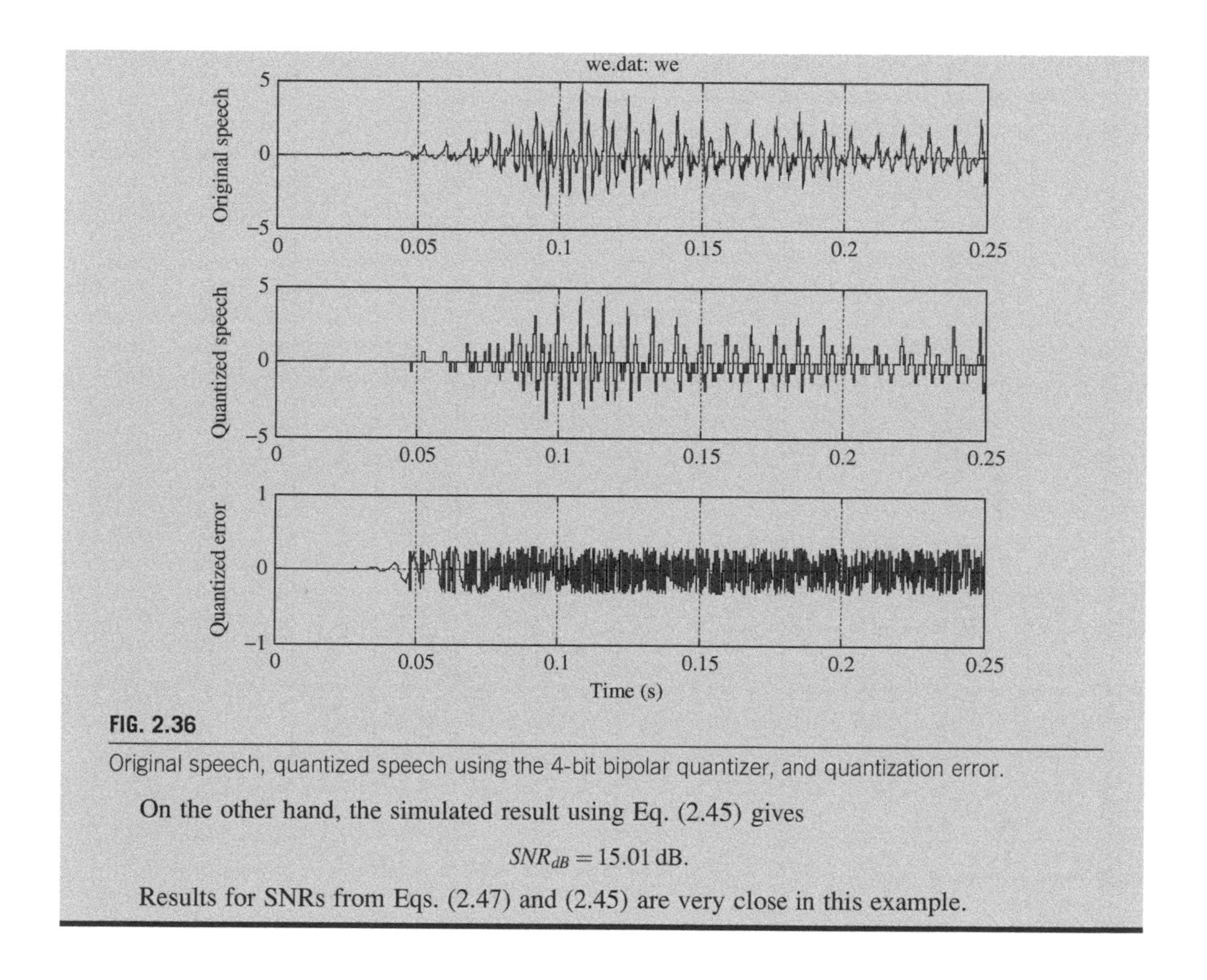

FIG. 2.36

Original speech, quantized speech using the 4-bit bipolar quantizer, and quantization error.

On the other hand, the simulated result using Eq. (2.45) gives

$$SNR_{dB} = 15.01 \text{ dB}.$$

Results for SNRs from Eqs. (2.47) and (2.45) are very close in this example.

2.4 SUMMARY

1. Analog signal is sampled at a fixed time interval so the ADC will convert the sampled voltage level to the digital value; this is called the sampling process.
2. The fixed time interval between two samples is the sampling period, and the reciprocal the sampling period is the sampling rate. The half of sampling rate is the folding frequency (Nyquist limit).
3. The sampling theorem condition that the sampling rate be larger than twice of the highest frequency of the analog signal to be sampled, must be met in order to have the analog signal be recovered.
4. The sampled spectrum is explained using the following well-known formula

$$X_s(f) = \cdots + \frac{1}{T}X(f+f_s) + \frac{1}{T}X(f) + \frac{1}{T}X(f-f_s) + \cdots.$$

That is, the sampled signal spectrum is a scaled and shifted version of its analog signal spectrum and its replicas centered at the frequencies that are multiples of the sampling rate.

5. The analog anti-aliasing lowpass filter is used before ADC to remove frequency components higher than the folding frequency to avoid aliasing.
6. The reconstruction (analog lowpass) filter is adopted after DAC to remove the spectral images that exist in the sampled-and-hold signal and obtain the smoothed analog signal. The sample-and-hold DAC effect may distort the baseband spectrum, but it also reduces image spectrum.
7. Quantization means the ADC unit converts the analog signal amplitude with infinite precision to digital data with finite precision (a finite number of codes).
8. When the DAC unit converts a digital code to a voltage level, quantization error occurs. The quantization error is bounded by half of the quantization step size (ADC resolution), which is a ratio of the full range of the signal over the number of the quantization levels (number of the codes).
9. The performance of the quantizer in terms of the signal-to-quantization noise ratio (SNR), in dB, is related to the number of bits in ADC. Increasing 1 bit used in each ADC code will improve 6-dB SNR due to quantization.

2.5 MATLAB PROGRAMS

Program 2.3. MATLAB function for uniform quantization encoding.

```
function [ I, pq]= biquant(NoBits,Xmin,Xmax,value)
% function pq = biquant(NoBits, Xmin, Xmax, value)
% This routine is created for simulation of uniform quantizer.
%
% NoBits: number of bits used in quantization.
% Xmax: overload value.
% Xmin: minimum value
% value: input to be quantized.
% pq: output of quantized value
% I: coded integer index
L=2^NoBits;
delta=(Xmax-Xmin)/L;
I=round((value-Xmin)/delta);
if ( I==L)
  I=I-1;
end
if I<0
  I=0;
end
pq=Xmin+I*delta;
```

Program 2.4. MATLAB function for uniform quantization decoding.

```
function pq = biqtdec(NoBits,Xmin,Xmax,I)
% function pq = biqtdec(NoBits,Xmin, Xmax, I)
% This routine recover the quantized value.
%
% NoBits: number of bits used in quantization.
% Xmax: overload value
% Xmin: minimum value
% pq: output of quantized value
% I: coded integer index
L=2^NoBits;
delta=(Xmax-Xmin)/L;
pq=Xmin+I*delta;
```

Program 2.5. MATLAB function for calculation of signal-to-quantization noise ratio.

```
function snr = calcsnr(speech, qspeech)
% function snr = calcsnr(speech, qspeech)
% this routine is created for calculation of SNR
%
% speech: original speech waveform.
% qspeech: quantized speech.
% snr: output SNR in dB.
%
qerr = speech-qspeech;
snr = 10*log10(sum(speech.*speech)/sum(qerr.*qerr))
```

2.6 PROBLEMS

2.1. Given an analog signal

$$x(t) = 5\cos(2\pi \times 1500t), \text{ for } t \geq 0,$$

sampled at a rate of 8000 Hz,

(a) sketch the spectrum of the original signal;
(b) sketch the spectrum of the sampled signal from 0 up to 20 kHz.

2.2. Given an analog signal

$$x(t) = 5\cos(2\pi \times 2500t) + 2\cos(2\pi \times 3200t), \text{ for } t \geq 0,$$

sampled at a rate of 8000 Hz,

(a) sketch the spectrum of the sampled signal up to 20 kHz;
(b) sketch the recovered analog signal spectrum if an ideal lowpass filter with a cutoff frequency of 4 kHz is used to filter the sampled signal in order to recover the original signal.

2.3. Given an analog signal

$$x(t) = 3\cos(2\pi \times 1500t) + 2\cos(2\pi \times 2200t), \text{ for } t \geq 0,$$

sampled at a rate of 8000 Hz,

(a) sketch the spectrum of the sampled signal up to 20 kHz;
(b) sketch the recovered analog signal spectrum if an ideal lowpass filter with a cutoff frequency of 4 kHz is used to filter the sampled signal in order to recover the original signal.

2.4. Given an analog signal

$$x(t) = 3\cos(2\pi \times 1500t) + 2\cos(2\pi \times 4200t), \text{ for } t \geq 0,$$

sampled at a rate of 8000 Hz,

(a) sketch the spectrum of the sampled signal up to 20 kHz;
(b) sketch the recovered analog signal spectrum if an ideal lowpass filter with a cutoff frequency of 4 kHz is used to filter the sampled signal in order to recover the original signal.

2.5. Given an analog signal

$$x(t) = 5\cos(2\pi \times 2500t) + 2\cos(2\pi \times 4500t), \text{ for } t \geq 0,$$

sampled at a rate of 8000 Hz,

(a) sketch the spectrum of the sampled signal up to 20 kHz;
(b) sketch the recovered analog signal spectrum if an ideal lowpass filter with a cutoff frequency of 4 kHz is used to filter the sampled signal in order to recover the original signal;
(c) determine the frequency/frequencies of aliasing noise.

2.6. Assuming a continuous signal is given as

$$x(t) = 10\cos(2\pi \times 5500t) + 5\sin(2\pi \times 7500t), \text{ for } t \geq 0,$$

sampled at a rate of 8000 Hz,

(a) sketch the spectrum of the sampled signal up to 20 kHz;
(b) sketch the recovered analog signal spectrum if an ideal lowpass filter with a cutoff frequency of 4 kHz is used to filter the sampled signal in order to recover the original signal;
(c) determine the frequency/frequencies of aliasing noise.

2.7. Assuming a continuous signal is given as

$$x(t) = 8\cos(2\pi \times 5000t) + 5\sin(2\pi \times 7000t), \text{ for } t \geq 0,$$

sampled at a rate of 8000 Hz,

(a) sketch the spectrum of the sampled signal up to 20 kHz;
(b) sketch the recovered analog signal spectrum if an ideal lowpass filter with a cutoff frequency of 4 kHz is used to filter the sampled signal in order to recover the original signal;
(c) determine the frequency/frequencies of aliasing noise.

2.8. Assuming a continuous signal is given as

$$x(t) = 10\cos(2\pi \times 5000t) + 5\sin(2\pi \times 7500t), \text{ for } t \geq 0,$$

sampled at a rate of 8000 Hz,

(a) sketch the spectrum of the sampled signal up to 20 kHz;
(b) sketch the recovered analog signal spectrum if an ideal lowpass filter with a cutoff frequency of 4 kHz is used to filter the sampled signal in order to recover the original signal;
(c) determine the frequency/frequencies of aliasing noise.

2.9. Given the following second-order anti-aliasing lowpass filter (Fig. 2.37) which is a Butterworth type, determine the values of circuit elements if we want the filter to have a cutoff frequency of 1000 Hz.
2.10. From Problem 2.9, determine the percentage of aliasing level at the frequency of 500 Hz, assuming that the sampling rate is 4000 Hz.
2.11. Given the following second-order anti-aliasing lowpass filter (Fig. 2.38) which is a Butterworth type, determine the values of circuit elements if we want the filter to have a cutoff frequency of 800 Hz.
2.12. From Problem 2.11, determine the percentage of aliasing level at the frequency of 400 Hz, assuming that the sampling rate is 4000 Hz.
2.13. Given a DSP system in which a sampling rate of 8000 Hz is used and the anti-aliasing filter is a second-order Butterworth lowpass filter with a cutoff frequency of 3.2 kHz, determine

(a) the percentage of aliasing level at the cutoff frequency;
(b) the percentage of aliasing level at the frequency of 1000 Hz.

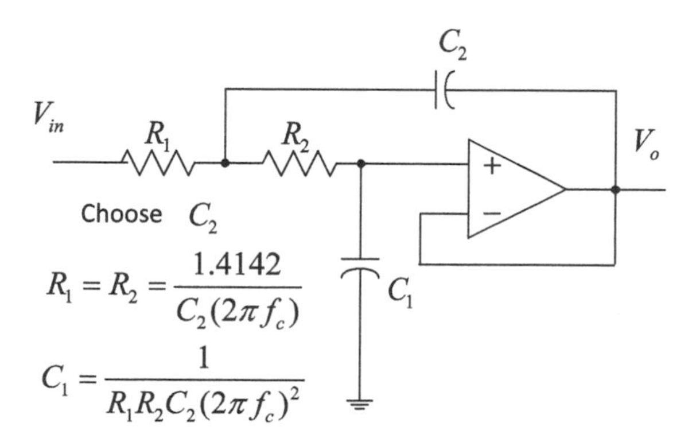

FIG. 2.37

Filter Circuit in Problem 2.9.

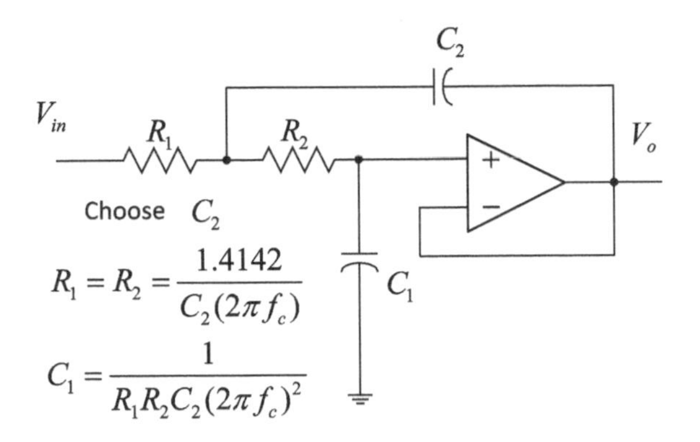

FIG. 2.38

Filter circuit in Problem 2.11.

2.14. Given a DSP system in which a sampling rate of 8000 Hz is used and the anti-aliasing filter is a Butterworth lowpass filter with a cutoff frequency 3.2 kHz, determine the order of the Butterworth lowpass filter for the percentage of aliasing level at the cutoff frequency required to be less than 10%.

2.15. Given a DSP system in which a sampling rate of 8000 Hz is used and the anti-aliasing filter is a second-order Butterworth lowpass filter with a cutoff frequency of 3.1 kHz, determine

(a) the percentage of aliasing level at the cutoff frequency;
(b) the percentage of aliasing level at the frequency of 900 Hz.

2.16. Given a DSP system in which a sampling rate of 8000 Hz is used and the anti-aliasing filter is a Butterworth lowpass filter with a cutoff frequency 3.1 kHz, determine the order of the Butterworth lowpass filter for the percentage of aliasing level at the cutoff frequency required to be less than 10%.

2.17. Given a DSP system (Fig. 2.39) with a sampling rate of 8000 Hz and assuming that the hold circuit is used after DAC, determine

(a) the percentage of distortion at the frequency of 3200 Hz;
(b) the percentage of distortion at the frequency of 1500 Hz.

2.18. A DSP system is given with the following specifications:
Design requirements:

Sampling rate 20,000 Hz;
Maximum allowable gain variation from 0 to 4000 Hz = 2 dB;

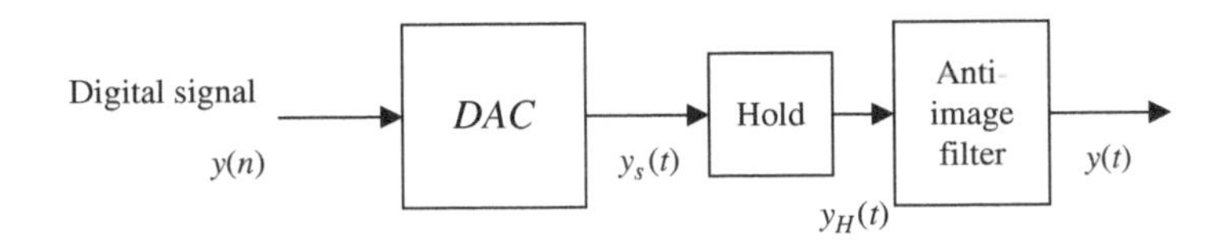

FIG. 2.39

Analog signal reconstruction in Problem 2.18.

40 dB rejection at the frequency of 16,000 Hz; and
Butterworth filter assumed.

Determine the cutoff frequency and order for the anti-image filter.

2.19. Given a DSP system with a sampling rate of 8000 Hz and assuming that the hold circuit is used after DAC, determine

(a) the percentage of distortion at the frequency of 3000 Hz;
(b) the percentage of distortion at the frequency of 1600 Hz.

2.20. A DSP system (Fig. 2.40) is given with the following specifications:
Design requirements:

Sampling rate 22,000 Hz;
Maximum allowable gain variation from 0 to 4000 Hz = 2 dB;
40 dB rejection at the frequency of 18,000 Hz; and
Butterworth filter assumed

Determine the cutoff frequency and order for the anti-image filter.

2.21. Given the 2-bit flash ADC unit with an analog sample-and-hold voltage of 2 V shown in Fig. 2.41, determine the output bits.

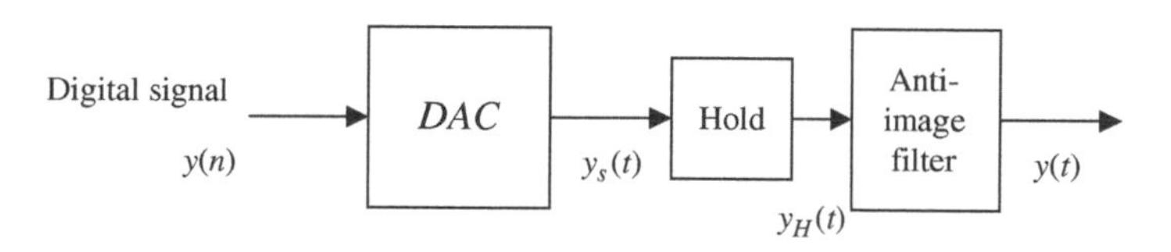

FIG. 2.40

Analog signal reconstruction in Problem 2.20.

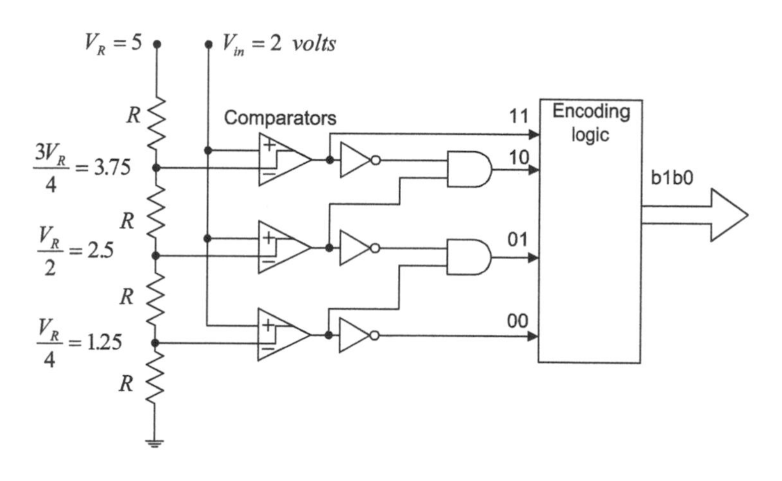

FIG. 2.41

A 2-Bit flash ADC in Problem 2.21.

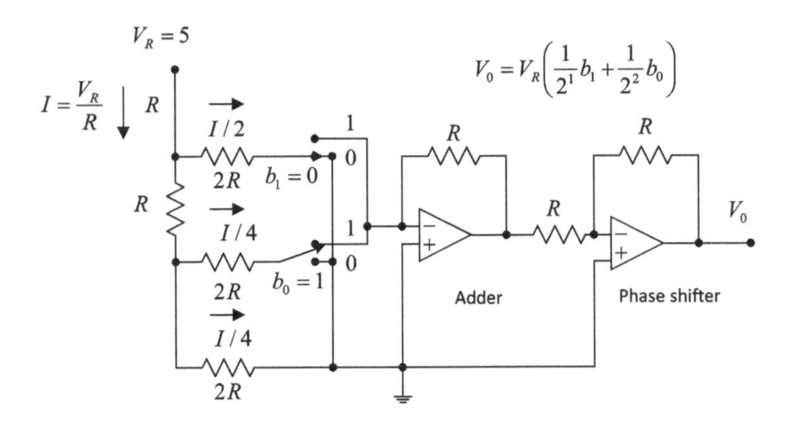

FIG. 2.42

A 2-Bit R-2R DAC in Problem 2.22.

2.22. Given the R-2R DAC unit with a 2-bit value as $b_1b_0 = 01$ shown in Fig. 2.42, determine the converted voltage.

2.23. Given the 2-bit flash ADC unit with an analog sample-and-hold voltage of 3.5 V shown in Fig. 2.41, determine the output bits.

2.24. Given the R-2R DAC unit with 2-bit values as $b_1b_0 = 11$ and $b_1b_0 = 10$, respectively, and shown in Fig. 2.42, determine the converted voltages.

2.25. Assuming that a 4-bit ADC channel accepts analog input ranging from 0 to 5 V, determine the following:

(a) Number of quantization levels;
(b) Step size of quantizer or resolution;
(c) Quantization level when the analog voltage is 3.2 V;
(d) Binary code produced by the ADC;
(e) Quantization error.

2.26. Assuming that a 5-bit ADC channel accepts analog input ranging from 0 to 4 V, determine the following:

(a) Number of quantization levels;
(b) Step size of quantizer or resolution;
(c) Quantization level when the analog voltage is 1.2 V;
(d) Binary code produced by the ADC;
(e) Quantization error.

2.27. Assuming that a 3-bit ADC channel accepts analog input ranging from −2.5 to 2.5 V, determine the following:

(a) Number of quantization levels;
(b) Step size of quantizer or resolution;

(c) Quantization level when the analog voltage is -1.2 V;
(d) Binary code produced by the ADC;
(e) Quantization error.

2.28. Assuming that an 8-bit ADC channel accepts analog input ranging from -2.5 to 2.5 V, determine the following:

(a) Number of quantization levels;
(b) Step size of quantizer or resolution;
(c) Quantization level when the analog voltage is 1.5 V;
(d) Binary code produced by the ADC;
(e) Quantization error.

2.29. If the analog signal to be quantized is a sinusoidal waveform, that is, $x(t) = 9.5\sin(2000 \times \pi t)$, and if a bipolar quantizer uses 6 bits, determine

(a) Number of quantization levels;
(b) Quantization step size or resolution, Δ, assuming the signal range is from -10 to 10 V;
(c) The signal power to quantization noise power ratio.

2.30. For a speech signal, if the ratio of the RMS value over the absolute maximum value of the signal is given, that is, $\left(\frac{x_{rms}}{|x|_{\max}}\right) = 0.25$ and the ADC bipolar quantizer uses 6 bits, determine

(a) Number of quantization levels;
(b) Quantization step size or resolution, Δ, if the signal range is 5 V;
(c) The signal power-to-quantization noise power ratio.

Computer problems with MATLAB: Use the MATLAB programs in Section 2.5 to solve the following problems.

2.31. Given a sinusoidal waveform of 100 Hz,

$$x(t) = 4.5\sin(2\pi \times 100t), \text{ for } t \geq 0,$$

sample it at 8000 samples per second and

(a) Write a MATLAB program to quantize $x(t)$ using a 6-bits bipolar quantizer to obtain the quantized signal x_q, assuming that the signal range to be from -5 to 5 V;
(b) Plot the original signal and quantized signal;
(c) Calculate the SNR due to quantization using the MATLAB program.

2.32. Given a signal waveform,

$$x(t) = 3.25\sin(2\pi \times 50t) + 1.25\cos(2\pi \times 100t + \pi/4), \text{ for } t \geq 0,$$

sample it at 8000 samples per second and

(a) Write a MATLAB program to quantize $x(t)$ using a 6-bits bipolar quantizer to obtain the quantized signal x_q, assuming that the signal range to be from -5 to 5 V;
(b) Plot the original signal and quantized signal;
(c) Calculate the SNR due to quantization using the MATLAB program.

2.33. Given a speech signal sampled at 8000 Hz, as shown in Example 2.14,

(a) Write a MATLAB program to quantize $x(t)$ using a 6-bits bipolar quantizer to obtain the quantized signal x_q, assuming that the signal range is from −5 to 5 V;
(b) Plot the original speech waveform, quantized speech, and quantization error;
(c) Calculate the SNR due to quantization using the MATLAB program.

MATLAB Projects

2.34. Performance evaluation of speech quantization:
Given an original speech segment "speech.dat" sampled at 8000 Hz with each sample encoded in 16 bits, use Programs 2.3–2.5 and modify Program 2.2 to quantize the speech segment using 3–15 bits, respectively. The SNR in dB must be measured for each quantization. MATLAB function: "sound(x/max(abs(x)),fs)" can be used to evaluate sound quality, where "x" is the speech segment while "fs" is the sampling rate of 8000 Hz. In this project, create a plot of the measured SNR (dB) versus the number of bits and discuss the effect of the sound quality. For comparisons, plot the original speech and the quantized one using 3, 8, and 15 bits.

2.35. Performance evaluation of seismic data quantization:
The seismic signal, a measurement of the acceleration of ground motion, is required for applications in the study of geophysics. The seismic signal ("seismic.dat" provided by the USGS Albuquerque Seismological Laboratory) has a sampling rate of 15 Hz with 6700 data samples, and each sample is encoded using 32 bits. Quantizing each 32-bit sample down to the lower number of bits per sample can reduce the memory storage requirement with the reduced signal quality. Use Programs 2.3–2.5 and modify Program 2.2 to quantize the seismic data using 13, 15, 17,..., 31 bits. The SNR in dB must be measured for each quantization. Create a plot of the measured SNR (dB) versus the number of bits. For comparison, plot the seismic data and the quantized one using 13, 18, 25, and 31 bits.

Advanced Problems

2.36–2.38. Given the following sampling system (see Fig. 2.43), $x_s(t)=x(t)p(t)$

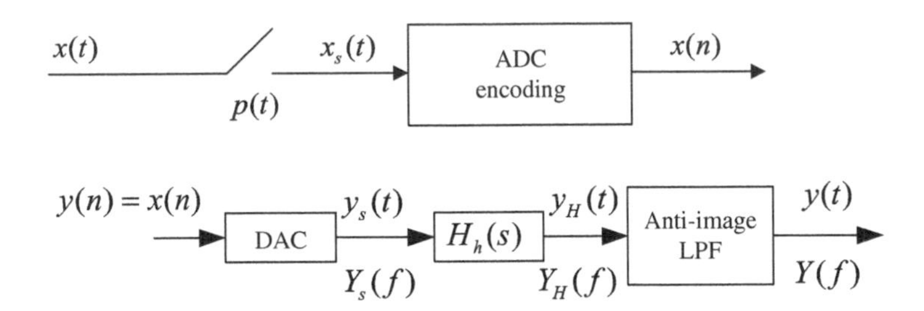

FIG. 2.43

Analog signal reconstruction in Problems 2.36–2.38.

2.36. If the pulse train used is depicted in Fig. 2.44

(a) Determine the Fourier series expansion for $p(t)=\sum_{k=-\infty}^{\infty} a_k e^{jkw_0 t}$;
(b) Determine $X_s(f)$ in terms of $X_s(f)$ using Fourier transform, that is,

$$X_s(f)=FT\{x_s(t)\}=FT\{x(t)p(t)\};$$

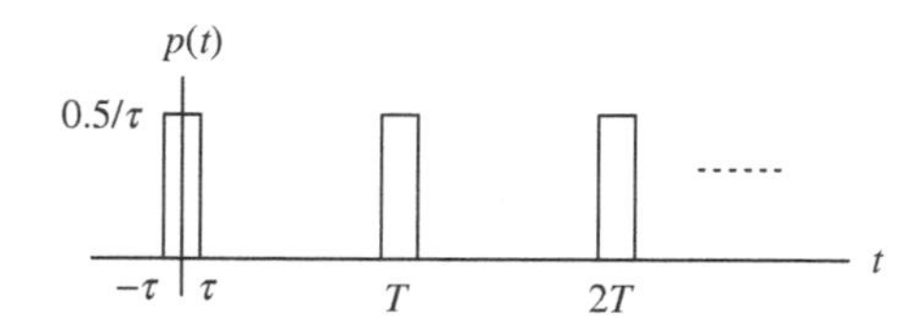

FIG. 2.44

Pulse train in Problem 2.36.

(c) Determine spectral distortion referring to $X_s(f)$ for $-f_s/2<f<f_s/2$.

2.37. If the pulse train used is depicted in Fig. 2.45

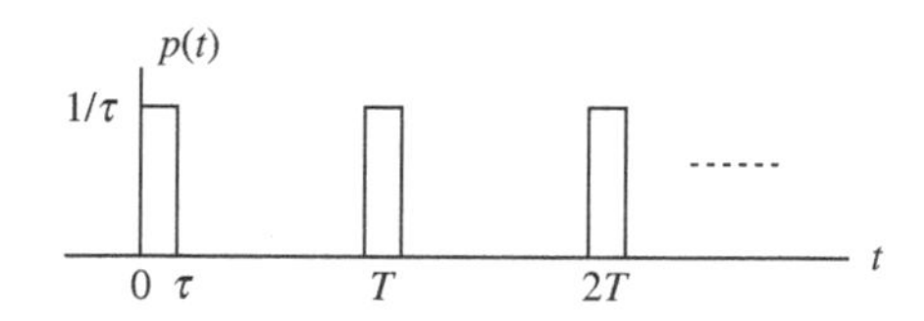

FIG. 2.45

Pulse train in Problem 2.37.

(a) Determine the Fourier series expansion for $p(t)=\sum_{k=-\infty}^{\infty} a_k e^{jkw_0t}$;
(b) Determine $X_s(f)$ in terms of $X(f)$ using Fourier transform, that is, $X_s(f)=FT\{x_s(t)\}=FT\{x(t)p(t)\}$;
(c) Determine spectral distortion referring to $X(f)$ for $-f/2<f<f_s/2$.

2.38. If the pulse train used is depicted in Fig. 2.46

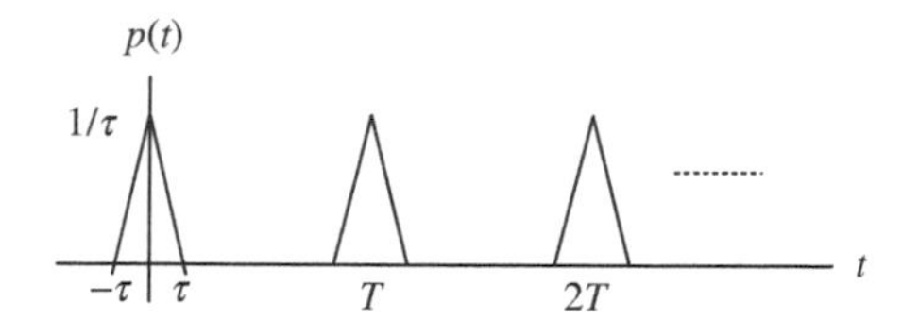

FIG. 2.46

Pulse train in Problem 2.38.

(a) Determine the Fourier series expansion for $p(t)=\sum_{k=-\infty}^{\infty} a_k e^{jkw_0t}$;
(b) Determine $X_s(f)$ in terms of $X(f)$ using Fourier transform, that $X_s(f)=FT\{x_s(t)\}=FT\{x(t)p(t)\}$;
(c) Determine spectral distortion referring to $X(f)$ for $-f_s/2<f<f_s/2$.

2.39. In Fig. 2.16, a Chebyshev lowpass filter is chosen to serve as anti-aliasing lowpass filter, where the magnitude frequency response of the Chebyshev filter with an order n is given by

$$|H(f)|=\frac{1}{\sqrt{1+\varepsilon^2 C_n^2(f/f_c)}},$$

where ε is the absolute ripple, and

$$C_n(x) = \begin{cases} \cos\left[n\cos^{-1}(x)\right] & x<1 \\ \cosh\left[n\cosh^{-1}(x)\right] & x>1 \end{cases} \text{ and } \cosh^{-1}(x) = \ln\left(x+\sqrt{x^2-1}\right).$$

(a) Derive the formula for the aliasing level; and

(b) When the sampling frequency is 8 kHz, a cutoff frequency is 3.4 kHz, ripple is 1 dB, and order equals $n=4$, determine the aliasing level at frequency of 1 kHz.

2.40. Given the following signal

$$x(t) = \sum_{i=1}^{N} A_i \cos(\omega_i t + \phi_i)$$

with the signals ranging from $-\sum_{i=1}^{N} A_i$ to $\sum_{i=1}^{N} A_i$, determine the signal to quantization noise power ratio using m bits.

2.41. Given the following modulated signal

$$x(t) = A_1 \cos(\omega_t t + \phi_1) \times A_2 \cos(\omega_2 t + \phi_2)$$

with the signal ranging from $-A_1A_2$ to A_1A_2, determine the signal to quantization noise power ratio using m bits.

2.42. Assume that truncation of the continuous signal $x(n)$ in Problem 2.40 is defined as:

$$x_q(n) = x(n) + e_q(n),$$

where $-\Delta < e_q(n) \le 0$ and $\Delta = 2\sum_{i=1}^{N} A_i/2^m$. The quantized noise has the distribution given in Fig. 2.47, determine the SNR using m bits.

2.43. Assume that truncation of the continuous signal $x(n)$ in Problem 2.41 is defined as

$$x_q(n) = x(n) + e_q(n),$$

where $0 \le e_q(n) < \Delta$ and $\Delta = 2A_1A_2/2^m$. The quantized noise has the distribution given in Fig. 2.48, determine the SNR using m bits.

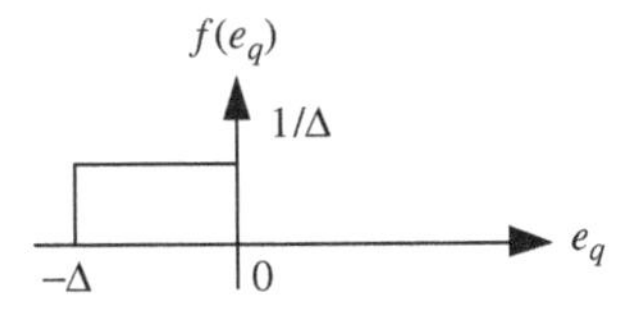

FIG. 2.47

The truncated error distribution in Problem 2.42.

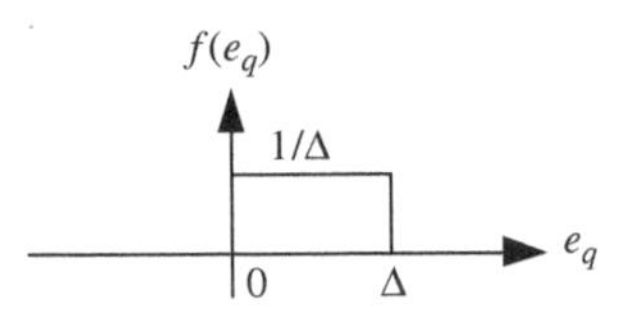

FIG. 2.48

The truncated error distribution in Problem 2.43.

CHAPTER

DIGITAL SIGNALS AND SYSTEMS 3

CHAPTER OUTLINE

3.1 DIGITAL SIGNALS

In our daily lives, analog signals appear in forms such as speech, audio, seismic, biomedical, and communications signals. To process an analog signal using a digital signal processor, the analog signal must be converted into a digital signal, that is, the analog-to-digital conversion (DAC) must take place, as discussed in Chapter 2. Then the digital signal is processed via digital signal processing (DSP) algorithm(s).

A typical digital signal $x(n)$ is shown in Fig. 3.1, where both the time and the amplitude of the digital signal are discrete. Note that the amplitudes of the digital signal samples are given and sketched only at their corresponding time indices, where $x(n)$ represents the amplitude of the nth sample and n is the time index or sample number. From Fig. 3.1, we learn that

$x(0)$: zeroth sample amplitude at the sample number $n=0$,
$x(1)$: first sample amplitude at the sample number $n=1$,
$x(2)$: second sample amplitude at the sample number $n=2$,
$x(3)$: third sample amplitude at the sample number $n=3$, and so on.

Digital Signal Processing. https://doi.org/10.1016/B978-0-12-815071-9.00003-8

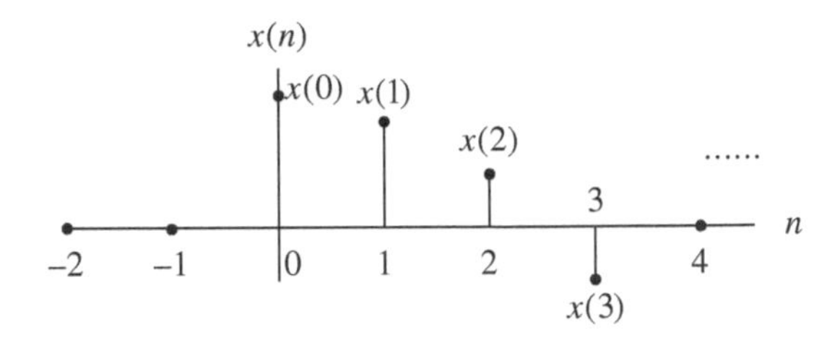

FIG. 3.1

Digital signal notation.

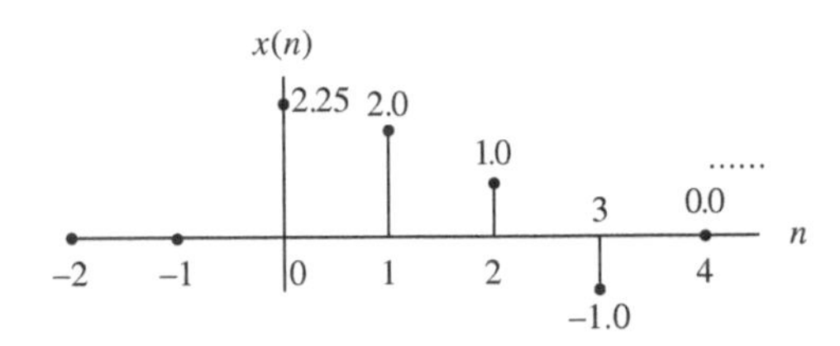

FIG. 3.2

Plot of the digital signal samples.

Furthermore, Fig. 3.2 illustrates the digital samples whose amplitudes are discrete encoded values represented in the digital signal processor. Precision of the data is based on the number of bits used in the DSP system. The encoded data format can be either an integer if a fixed-point digital signal processor is used or a floating-point number if a floating-point digital signal processor is used. As shown in Fig. 3.2 for the floating-point digital signal processor, we can identify the first five sample amplitudes at their time indices as follows:

$$x(0) = 2.25$$
$$x(1) = 2.0$$
$$x(2) = 1.0$$
$$x(3) = -1.0$$
$$x(4) = 0.0$$
$$\cdots$$

Again, note that each sample amplitude is plotted using a vertical bar with a solid dot. This notation is well accepted in the DSP literatures.

3.1.1 COMMON DIGITAL SEQUENCES

Let us study some special digital sequences that are widely used. We define and plot each of them as follows:

Unit-impulse sequence (digital unit-impulse function):

$$\delta(n)=\begin{cases}1 & n=0\\0 & n\neq 0\end{cases}. \tag{3.1}$$

The plot of the unit-impulse function is given in Fig. 3.3. The unit-impulse function has the unit amplitude at only $n=0$ and zero amplitudes at other time indices.

Unit-step sequence (digital unit-step function):

$$u(n)=\begin{cases}1 & n\geq 0\\0 & n<0\end{cases}. \tag{3.2}$$

The plot is given in Fig. 3.4. The unit-step function has the unit amplitude at $n=0$ and for all the positive time indices, and amplitudes of zero for all the negative time indices.

The shifted unit-impulse and unit-step sequences are displayed in Fig. 3.5.

As shown in Fig. 3.5, the shifted unit-impulse function $\delta(n-2)$ is obtained by shifting the unit-impulse function $\delta(n)$ to the right by two samples, and the shifted unit-step function $u(n-2)$ is achieved by shifting the unit-step function $u(n)$ to the right by two samples; similarly, $\delta(n+2)$ and $u(n+2)$ are acquired by shifting $\delta(n)$ and $u(n)$ via two samples to the left, respectively.

A sequence $x(n)$ is called a causal sequence if $x(n)=0$, for $n<0$. Otherwise, $x(n)$ is called noncausal sequence, that is, $x(n)$ has nonzero value (s) for $n<0$.

Sinusoidal and exponential sequences are depicted in Figs. 3.6 and 3.7, respectively.

For a sinusoidal sequence $x(n)=A\cos(0.125\pi n)u(n)$ and $A=10$, we can calculate the digital values for the first eight samples and list their values in Table 3.1. Note that $u(n)$ is used to ensure the sinusoidal sequence $x(n)$ is a causal sequence, and amplitudes of $x(n)$ are discrete-time values (encoded values in the floating format).

For the exponential sequence $x(n)=A(0.75)^n u(n)$, the calculated digital values for the first eight samples with $A=10$ are listed in Table 3.2.

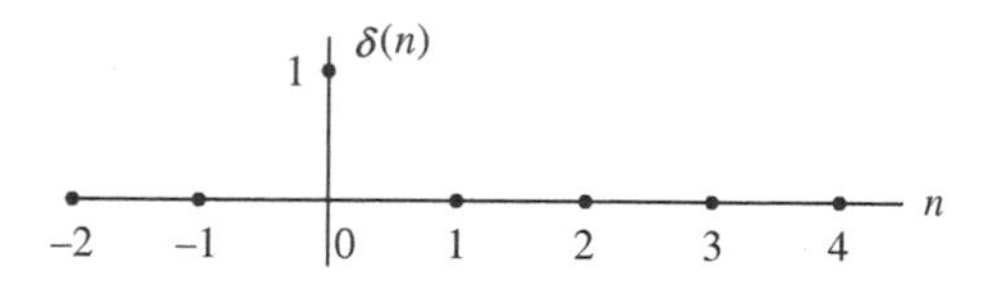

FIG. 3.3

Unit-impulse sequence.

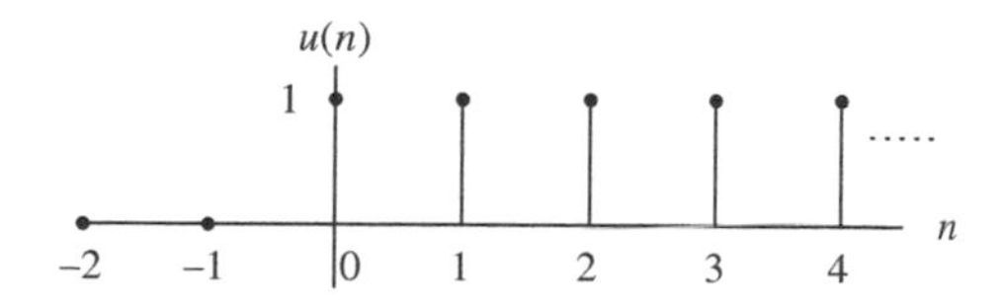

FIG. 3.4

Unit-step sequence.

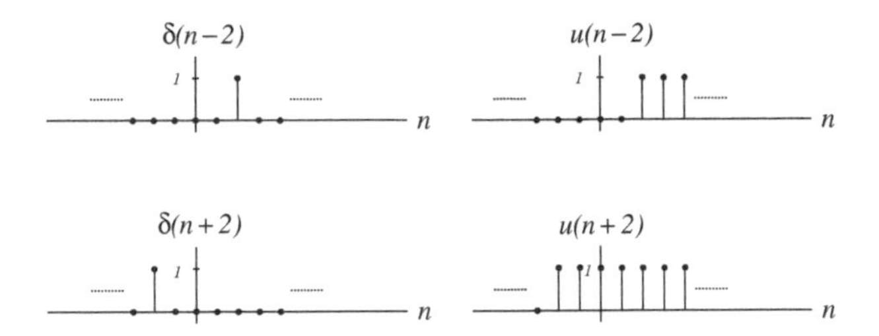

FIG. 3.5

Shifted unit-impulse and unit-step sequences.

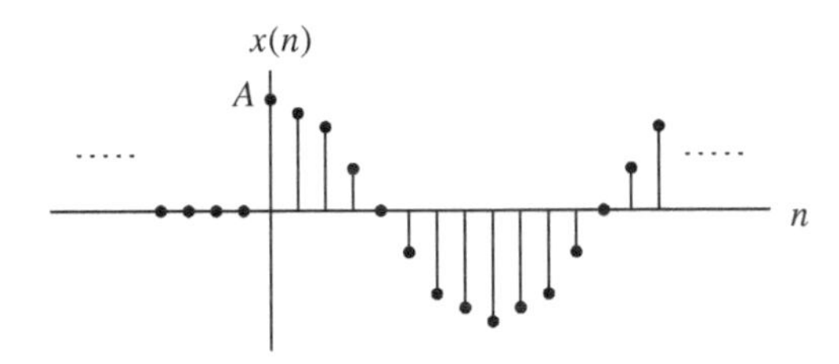

FIG. 3.6

Plot of samples of the sinusoidal function.

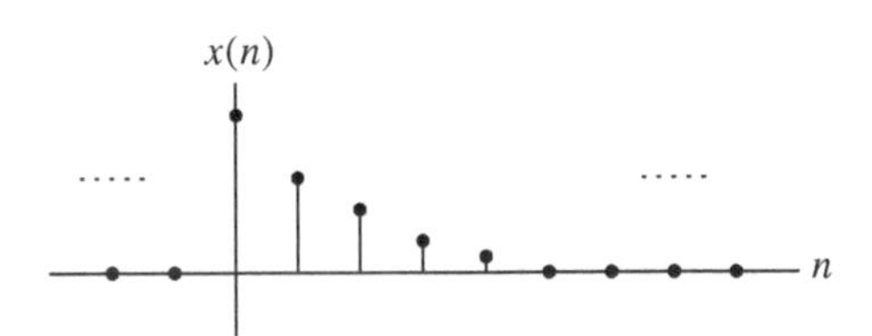

FIG. 3.7

Plot of samples of the exponential function.

Table 3.1 Sample Values Calculated from the Sinusoidal Function

n	$x(n) = 10\cos(0.125\pi n)u(n)$
0	10.0000
1	9.2388
2	7.0711
3	3.8628
4	0.0000
5	−3.8628
6	−7.0711
7	−9.2388

Table 3.2 Sample Values Calculated from the Exponential Function

n	$10(0.75)^n u(n)$
0	10.0000
1	7.5000
2	5.6250
3	4.2188
4	3.1641
5	2.3730
6	1.7798
7	1.3348

EXAMPLE 3.1

Given the following,

$$x(n) = \delta(n+1) + 0.5\delta(n-1) + 2\delta(n-2),$$

Sketch this sequence.

Solution:

According to the shift operation, $\delta(n+1)$ is obtained by shifting $\delta(n)$ to the left by one sample, while $\delta(n-1)$ and $\delta(n-2)$ are yielded by shifting $\delta(n)$ to right by one sample and two samples, respectively. Using the amplitude of each impulse function, we yield the sketch as shown in Fig. 3.8.

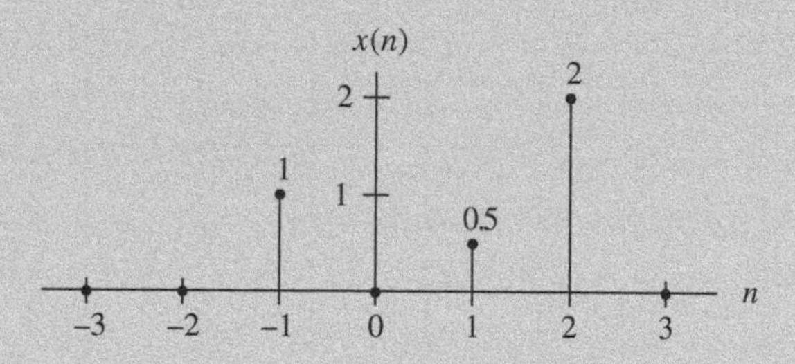

FIG. 3.8

Plot of digital sequence in Example 3.1.

3.1.2 GENERATION OF DIGITAL SIGNALS

Given the sampling rate of a DSP system to sample the analytical function of an analog signal, the corresponding digital function or digital sequence (assuming its sampled amplitudes are encoded to have finite precision) can be found. The digital sequence is often used to

1. Calculate the encoded sample amplitude for a given sample number n;
2. Generate the sampled sequence for simulation.

The procedure to develop a digital sequence from its analog signal function is as follows.

Assuming that an analog signal $x(t)$ is uniformly sampled at the time interval of $\Delta t = T$, where T is the sampling period, the corresponding digital function (sequence) $x(n)$ gives the *instant encoded values* of the analog signal $x(t)$ at all the time instants $t = n\Delta t = nT$ and can be achieved by substituting time $t = nT$ into the analog signal $x(t)$, that is,

$$x(n) = x(t)|_{t=nT} = x(nT). \tag{3.3}$$

Also note that for sampling the unit-step function $u(t)$, we have

$$u(t)|_{t=nT} = u(nT) = u(n). \tag{3.4}$$

The following example will demonstrate the use of Eqs. (3.3) and (3.4).

EXAMPLE 3.2

Assuming a DSP system with a sampling time interval of 125 μs,

(a) Convert each of following analog signal $x(t)$ to the digital signal $x(n)$.
1. $x(t) = 10e^{-5000t}u(t)$
2. $x(t) = 10\sin(2000\pi t)u(t)$

(b) Determine and plot the sample values from each obtained digital function.

Solution:

(a) Since $T = 0.000125$ s in Eq. (3.3), substituting $t = nT = n \times 0.000125 = 0.000125n$ into the analog signal $x(t)$ expressed in (1) leads to the digital sequence
1. $x(n) = x(nT) = 10e^{-5000 \times 0.000125n}u(nT) = 10e^{-0.625n}u(n)$.

Similarly, the digital sequence for (2) is achieved as follows:

2. $x(n) = x(nT) = 10\sin(2000\pi \times 0.000125n)u(nT) = 10\sin(0.25\pi n)u(n)$

(b) 1. The first five sample values are calculated and plotted in Fig. 3.9.

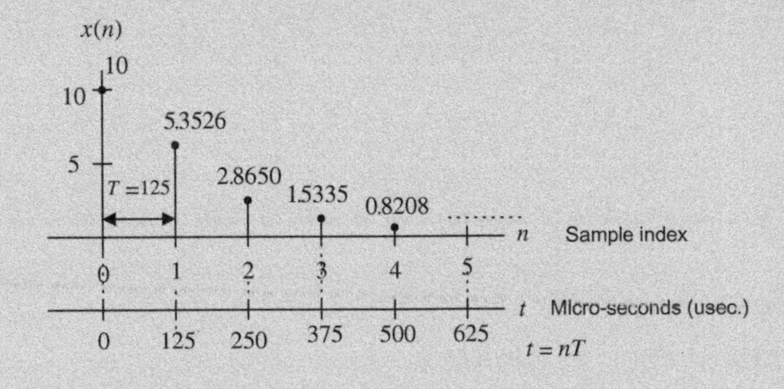

FIG. 3.9

Plot of the digital sequence for (1) in Example 3.2.

$$x(0) = 10e^{-0.625\times 0}u(0) = 10.0$$
$$x(1) = 10e^{-0.625\times 1}u(1) = 5.3526$$
$$x(2) = 10e^{-0.625\times 2}u(2) = 2.8650$$
$$x(3) = 10e^{-0.625\times 3}u(3) = 1.5335$$
$$x(4) = 10e^{-0.625\times 4}u(4) = 0.8208$$

2. The first eight amplitudes are computed and sketched in Fig. 3.10.

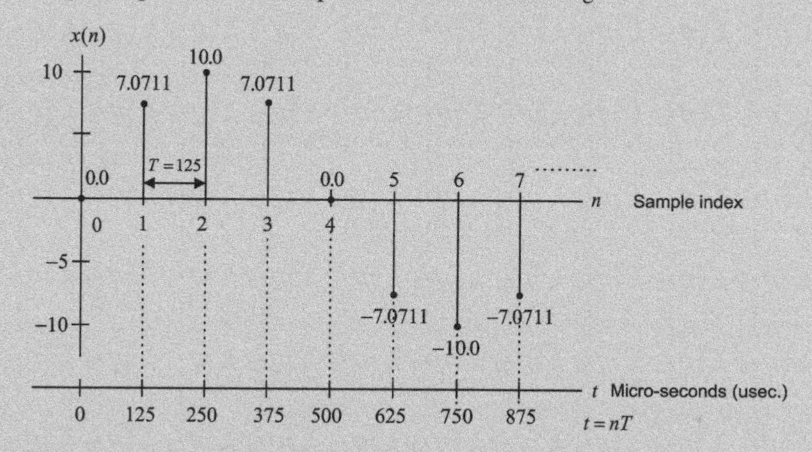

FIG. 3.10

Plot of the digital sequence for (2) in Example 3.2.

$$x(0) = 10\sin(0.25\pi \times 0)u(0) = 0$$
$$x(1) = 10\sin(0.25\pi \times 1)u(1) = 7.0711$$
$$x(2) = 10\sin(0.25\pi \times 2)u(2) = 10.0$$
$$x(3) = 10\sin(0.25\pi \times 3)u(3) = 7.0711$$
$$x(4) = 10\sin(0.25\pi \times 4)u(4) = 0.0$$
$$x(5) = 10\sin(0.25\pi \times 5)u(5) = -7.0711$$
$$x(6) = 10\sin(0.25\pi \times 6)u(6) = -10.0$$
$$x(7) = 10\sin(0.25\pi \times 7)u(7) = -7.0711$$

3.2 LINEAR TIME-INVARIANT, CAUSAL SYSTEMS

In this section, we study linear time-invariant causal systems and focus on properties such as linearity, time invariant, and causality.

3.2.1 LINEARITY

A linear system is illustrated in Fig. 3.11, where $y_1(n)$ is the system output using an input $x_1(n)$, and $y_2(n)$ the system output with an input $x_2(n)$.

Fig. 3.11 illustrates that the system output due to the weighted sum inputs $\alpha x_1(n)+\beta x_2(n)$ is equal to the same weighted sum of the individual outputs obtained from their corresponding inputs, that is,

$$y(n)=\alpha y_1(n)+\beta y_2(n), \tag{3.5}$$

where α and β are constants.

For example, considering a digital amplifier: $y(n)=10x(n)$, where the input is multiplied by 10 to generate the output. Then, the inputs $x_1(n)$ and $x_2(n)$ generate the outputs

$$y_1(n)=10x_1(n) \text{ and } y_2(n)=10x_2(n), \text{respectively.}$$

If, as described in Fig. 3.11, we apply to the system using the combined input $x(n)$, where the first input multiplied by a constant α while the second input multiplied by a constant β, that is,

$$x(n)=\alpha x_1(n)+\beta x_2(n),$$

then the system output due to the combined input is obtained as

$$y(n)=10x(n)=10(\alpha x_1(n)+\beta x_2(n))=10\alpha x_1(n)+10\beta x_2(n). \tag{3.6}$$

If we verify the weighted sum of the individual outputs, we see that

$$\alpha y_1(n)+\beta y_2(n)=\alpha[10x_1(n)]+\beta[10x_2(n)]. \tag{3.7}$$

Comparing Eqs. (3.6) and (3.7) verifies

$$y(n)=\alpha y_1(n)+\beta y_2(n). \tag{3.8}$$

Since this relationship holds for all inputs, system $y(n)=10x(n)$ is a linear system. The linearity means that the system obeys the superposition, as shown in Eq. (3.8). Let us verify a system whose output is a square of its input,

$$y(n)=x^2(n).$$

Applying to the system with the inputs $x_1(n)$ and $x_2(n)$ leads to

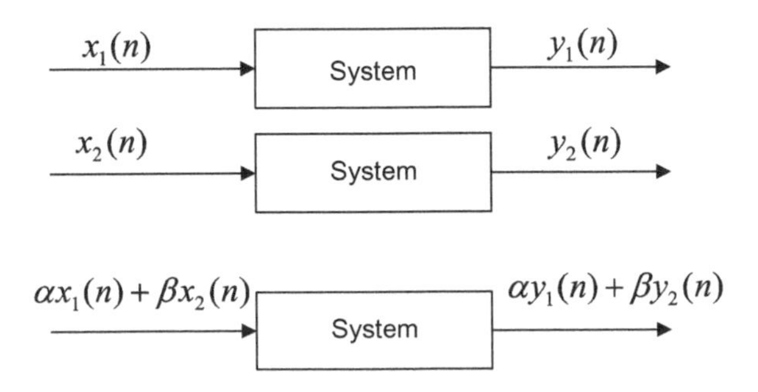

FIG. 3.11

Digital linear system.

$$y_1(n) = x_1^2(n) \text{ and } y_2(n) = x_2^2(n).$$

We can determine the system output using a combined input, which is the weighed sum of the individual inputs with constants α and β, respectively. Working on algebra, we see that

$$\begin{aligned} y(n) &= x^2(n) = (\alpha x_1(n) + \beta x_2(n))^2 \\ &= \alpha^2 x_1^2(n) + 2\alpha\beta x_1(n)x_2(n) + \beta^2 x_2^2(n). \end{aligned} \quad (3.9)$$

Again, we express the weighted sum of the two individual outputs with the same constants α and β as

$$\alpha y_1(n) + \beta y_2(n) = \alpha x_1^2(n) + \beta x_2^2(n). \quad (3.10)$$

It is obvious that

$$y(n) \neq \alpha y_1(n) + \beta y_2(n). \quad (3.11)$$

Hence, the system is a nonlinear system, since the linear property, superposition, does not hold, as shown in Eq. (3.11).

3.2.2 TIME INVARIANCE

The time-invariant system is illustrated in Fig. 3.12, where $y_1(n)$ is the system output for the input $x_1(n)$. Let $x_2(n) = x_1(n - n_0)$ be shifted version of $x_1(n)$ by n_0 samples. The output $y_2(n)$ obtained with the shifted input $x_2(n) = x_1(n - n_0)$ is equivalent to the output $y_2(n)$ acquired by shifting $y_1(n)$ by n_0 samples, $y_2(n) = y_1(n - n_0)$.

This can simply be viewed as the following:

If the system is time invariant and $y_1(n)$ is the system output due to the input $x_1(n)$, then the shifted system input $x_1(n - n_0)$ will produce a shifted system output $y_1(n - n_0)$ by the same amount of time n_0.

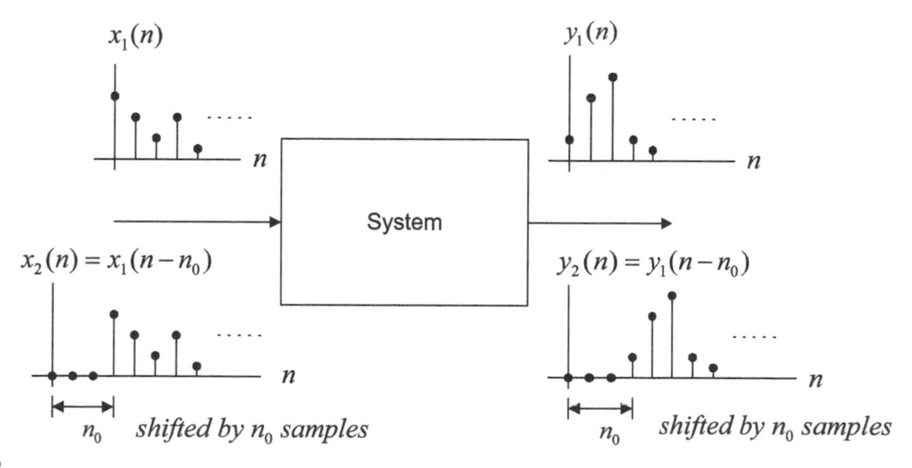

FIG. 3.12

Illustration of the linear time-invariant digital system.

EXAMPLE 3.3

Given the linear systems

(a) $y(n) = 2x(n-5)$

(b) $y(n) = 2x(3n)$,

determine whether each of the following systems is time invariant.

Solution:

(a) Let the input and output be $x_1(n)$ and $y_1(n)$, respectively; then the system output is $y_1(n) = 2x_1(n-5)$. Again, let $x_2(n) = x_1(n-n_0)$ be the shifted input and $y_2(n)$ be the output due to the shifted input. We determine the system output using the shifted input as

$$y_2(n) = 2x_2(n-5) = 2x_1(n-n_0-5).$$

Meanwhile, shifting by n_0 samples leads to

$$y_1(n-n_0) = 2x_1(n-5-n_0).$$

We can verify that $y_2(n) = y_1(n-n_0)$. Thus the shifted input of n_0 samples causes the system output to be shifted by the same n_0 samples, thus the system is time invariant.

(b) Let the input and output be $x_1(n)$ and $y_1(n)$, respectively; then the system output is $y_1(n) = 2x_1(3n)$. Again, let the input and output be $x_2(n)$ and $y_2(n)$, where $x_2(n) = x_1(n-n_0)$, a shifted version, and the corresponding output is $y_2(n) = 2x_2(3n)$. Since $x_2(n) = x_1(n-n_0)$, replacing n by $3n$ leads to $x_2(3n) = x_1(3n-n_0)$. We then have

$$y_2(n) = 2x_2(3n) = 2x_1(3n-n_0).$$

On the other hand, if we shift $y_1(n)$ by n_0 samples which replaces n in $y_1(n) = 2x_1(3n)$ by $n-n_0$, we yield

$$y_1(n-n_0) = 2x_1(3(n-n_0)) = 2x_1(3n-3n_0).$$

Clearly, we know that $y_2(n) \neq y_1(n-n_0)$. Since the system output $y_2(n)$ using the shifted input shifted by n_0 samples is not equal to the system output $y_1(n)$ shifted by the same n_0 samples, hence the system is not time invariant, that is, time variant.

3.2.3 CAUSALITY

A causal system is the one in which the output $y(n)$ at time n depends only on the current input $x(n)$ at time n, and its past input sample values such as $x(n-1)$, $x(n-2)$,.... Otherwise, if a system output depends on the future input values such as $x(n+1)$, $x(n+2)$,..., the system is noncausal. The noncausal system cannot be realized in real time.

EXAMPLE 3.4

Given the following linear systems

(a) $y(n) = 0.5x(n) + 2.5x(n-2)$, for $n \geq 0$,

(b) $y(n) = 0.25x(n-1) + 0.5x(n+1) - 0.4y(n-1)$, for $n \geq 0$,

determine whether each is causal.

Solution:

(a) Since for $n \geq 0$, the output $y(n)$ depends on the current input $x(n)$ and its past value $x(n-2)$, the system is causal.

(b) Since for $n \geq 0$, the output $y(n)$ depends on the current input $x(n)$ and its future value $x(n+1)$, the system is a noncausal.

3.3 DIFFERENCE EQUATIONS AND IMPULSE RESPONSES

Now we study the difference equation and its impulse response.

3.3.1 FORMAT OF DIFFERENCE EQUATION

A causal, linear, time-invariant system can be described by a difference equation having the following general form:

$$y(n) + a_1 y(n-1) + \cdots + a_N y(n-N) = b_0 x(n) + b_1 x(n-1) + \cdots + b_M x(n-M), \quad (3.12)$$

where $a_1, \ldots, a_N$, and $b_0, b_1, \ldots, b_M$ are the coefficients of the difference equation. M and N are the memory lengths for input $x(n)$ and output $y(n)$, respectively. Eq. (3.12) can further be written as

$$y(n) = -a_1 y(n-1) - \cdots - a_N y(n-N) + b_0 x(n) + b_1 x(n-1) + \cdots + b_M x(n-M), \quad (3.13)$$

or

$$y(n) = -\sum_{i=1}^{N} a_i y(n-i) + \sum_{j=0}^{M} b_j x(n-j). \quad (3.14)$$

Note that $y(n)$ is the current output which depends on the past output samples $y(n-1), \ldots, y(n-N)$, the current input sample $x(n)$, and the past input samples, $x(n-1), \ldots, x(n-N)$. We will examine the specific difference equations in the following examples.

EXAMPLE 3.5

Given the following difference equation:

$$y(n) = 0.25y(n-1) + x(n),$$

identify the nonzero system coefficients.

Solution:

Comparison with Eq. (3.13) leads to

$$b_0 = 1$$

$$-a_1 = 0.25$$

that is, $a_1 = -0.25$.

EXAMPLE 3.6

Given a linear system described by the difference equation

$$y(n) = x(n) + 0.5x(n-1),$$

determine the nonzero system coefficients.

Solution:

By comparing Eq. (3.13), we have

$$b_0 = 1 \text{ and } b_1 = 0.5$$

3.3.2 SYSTEM REPRESENTATION USING ITS IMPULSE RESPONSE

A linear time-invariant system can be completely described by its unit-impulse response, which is defined as the system response due to the impulse input $\delta(n)$ with zero-initial conditions, depicted in Fig. 3.13.

With the obtained unit-impulse response $h(n)$, we can represent the linear time-invariant system in Fig. 3.14.

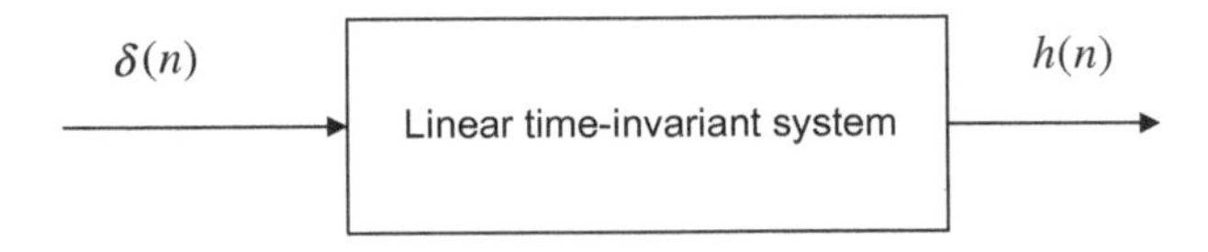

FIG. 3.13

Unit-impulse response of the linear time-invariant system.

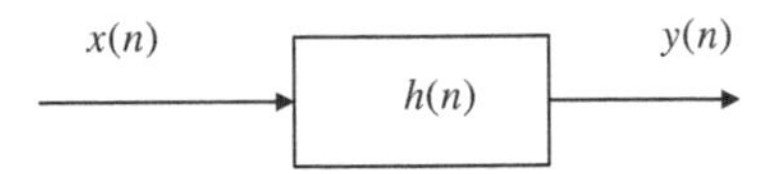

FIG. 3.14

Representation of linear time-invariant system using the impulse response.

EXAMPLE 3.7

Given the linear time-invariant system

$y(n) = 0.5x(n) + 0.25x(n-1)$ with an initial condition $x(-1) = 0$,

(a) Determine the unit-impulse response $h(n)$.

(b) Draw the system block diagram.

(c) Write the output using the obtained impulse response.

Solution:

(a) According to Fig. 3.13, let $x(n) = \delta(n)$, then

$$h(n) = y(n) = 0.5x(n) + 0.25x(n-1) = 0.5\delta(n) + 0.25\delta(n-1).$$

Thus, for this particular linear system, we have

$$h(n) = \begin{cases} 0.5 & n = 0 \\ 0.25 & n = 1 \\ 0 & \text{elsewhere} \end{cases}.$$

(b) The block diagram of the linear time-invariant system is shown in Fig. 3.15.

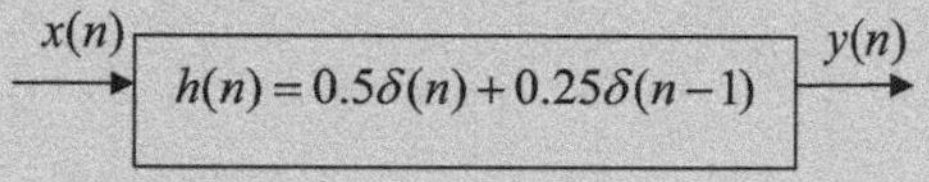

$x(n)$ → $h(n) = 0.5\delta(n) + 0.25\delta(n-1)$ → $y(n)$

FIG. 3.15

The system block diagram in Example 3.7.

(c) The system output can be rewritten as

$$y(n) = h(0)x(n) + h(1)x(n-1).$$

From the result in Example 3.7, it is noted that if the difference equation without the past output terms, $y(n-1), \ldots, y(n-N)$, that is, the corresponding coefficients $a_1,\ldots,a_N$, are zeros, and the impulse response $h(n)$ has a finite number of terms. We call this a *finite impulse response* (FIR) system. In general, Eq. (3.12) contains the past output terms and resulting impulse response $h(n)$ has an infinite number of terms. We can express the output sequence of a linear time-invariant system from its impulse response and inputs as

$$y(n) = \cdots + h(-1)x(n+1) + h(0)x(n) + h(1)x(n-1) + h(2)x(n-2) + \cdots = \sum_{k=-\infty}^{\infty} h(k)x(n-k). \quad (3.15)$$

Eq. (3.15) is called the *digital convolution sum*, which is explored in a later section.

We can verify Eq. (3.15) by substituting the impulse sequence $x(n) = \delta(n)$ to get the impulse response

$$h(n) = \cdots + h(-1)\delta(n+1) + h(0)\delta(n) + h(1)\delta(n-1) + h(2)\delta(n-2) + \cdots = \sum_{k=-\infty}^{\infty} h(k)\delta(n-k),$$

where $h(k)$ are the amplitudes of the impulse response at the corresponding time indices. Now let us look at another example.

EXAMPLE 3.8

Given the difference equation

$$y(n) = 0.25y(n-1) + x(n), \text{ for } n \geq 0 \text{ and } y(-1) = 0,$$

(a) Determine the unit-impulse response $h(n)$.
(b) Draw the system block diagram.
(c) Write the output using the obtained impulse response.
(d) For a step input $x(n) = u(n)$, verify and compare the output responses for the first three output samples using the difference equation and digital convolution sum (Eq. 3.15).

Solution:

(a) Let $x(n) = \delta(n)$, then

$$h(n) = 0.25h(n-1) + \delta(n).$$

To solve for $h(n)$, we evaluate

$$h(0) = 0.25h(-1) + \delta(0) = 0.25 \times 0 + 1 = 1$$

$$h(1) = 0.25h(0) + \delta(1) = 0.25 \times 1 + 0 = 0.25$$

$$h(2) = 0.25h(1) + \delta(2) = 0.25 \times 0.5 + 0 = 0.0625$$

With the calculated results, we can predict the impulse response as

$$h(n) = (0.25)^n u(n) = \delta(n) + 0.25\delta(n-1) + 0.0625\delta(n-2) + \cdots.$$

(b) The system block diagram is given in Fig. 3.16.

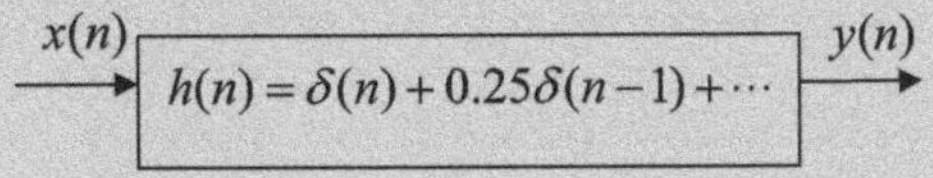

$x(n)$ → $h(n) = \delta(n) + 0.25\delta(n-1) + \cdots$ → $y(n)$

FIG. 3.16

The system block diagram in Example 3.8.

(c) The output sequence is a sum of infinite terms expressed as

$$\begin{aligned} y(n) &= h(0)x(n) + h(1)x(n-1) + h(2)x(n-2) + \cdots \\ &= x(n) + 0.25x(n-1) + 0.0625x(n-2) + \cdots \end{aligned}$$

(d) From the difference equation and using the zero-initial condition, we have

$$y(n) = 0.25y(n-1) + x(n) \text{ for } n \geq 0 \text{ and } y(-1) = 0$$

$$n = 0,\ y(0) = 0.25y(-1) + x(0) = u(0) = 1$$

$$n = 1,\ y(1) = 0.25y(0) + x(1) = 0.25 \times u(0) + u(1) = 1.25$$

$$n = 2,\ y(2) = 0.25y(1) + x(2) = 0.25 \times 1.25 + u(2) = 1.3125$$

Applying the convolution sum in Eq. (3.15) yields

$$y(n) = x(n) + 0.25x(n-1) + 0.0625x(n-2) + \cdots$$

$$n = 0,\ y(0) = x(0) + 0.25x(-1) + 0.0625x(-2) + \cdots$$
$$= u(0) + 0.25 \times u(-1) + 0.125 \times u(-2) + \cdots = 1$$

$$n = 1,\ y(1) = x(1) + 0.25x(0) + 0.0625x(-1) + \cdots$$
$$= u(1) + 0.25 \times u(0) + 0.125 \times u(-1) + \cdots = 1.25$$

$$n = 2,\ y(2) = x(2) + 0.25x(1) + 0.0625x(0) + \cdots$$
$$= u(2) + 0.25 \times u(1) + 0.0625 \times u(0) + \cdots = 1.3125$$
$$\cdots$$

Comparing the results, we verify that a linear time-invariant system can be represented by the convolution sum using its impulse response and input sequence. Note that we verify only the causal system for simplicity, and the principle works for both causal and noncausal systems.

Note that this impulse response $h(n)$ contains an infinite number of terms in its duration due to the past output term $y(n-1)$. Such a system as described in the preceding example is called an *infinite impulse response* (IIR) system, which is studied in the later chapters.

3.4 DIGITAL CONVOLUTION

Digital convolution plays an important role in digital filtering. As we verifies in the last section, a linear time-invariant system can be represented by a digital convolution sum. Given a linear time-invariant system, we can determine its unit-impulse response $h(n)$, which relates the system input and output. To find the output sequence $y(n)$ for any input sequence $x(n)$, we write the digital convolution as shown in Eq. (3.15) as:

$$\begin{aligned} y(n) &= \sum_{k=-\infty}^{\infty} h(k)x(n-k) \\ &= \cdots + h(-1)x(n+1) + h(0)x(n) + h(1)x(n-1) + h(2)x(n-2) + \cdots. \end{aligned} \tag{3.16}$$

The sequences $h(k)$ and $x(k)$ in Eq. (3.16) are interchangeable. In Eq. (3.16), let $m = n - k$, we have an alternative form as

$$\begin{aligned} y(n) &= \sum_{m=\infty}^{-\infty} h(n-m)x(m) = \sum_{k=-\infty}^{\infty} x(k)h(n-k) \\ &= \cdots + x(-1)h(n+1) + x(0)h(n) + x(1)h(n-1) + x(2)h(n-2) + \cdots. \end{aligned} \tag{3.17}$$

Using a conventional notation, we express the digital convolution as

$$y(n) = h(n)^* x(n). \tag{3.18}$$

Note that for a causal system, which implies its impulse response

$$h(n) = 0 \text{ for } n < 0.$$

The lower limit of the convolution sum begins at 0 instead of $-\infty$, that is

$$y(n) = \sum_{k=0}^{\infty} h(k)x(n-k). \tag{3.19}$$

The alternative for Eq. (3.19) can be expressed as

$$y(n) = \sum_{k=-\infty}^{n} x(k)h(n-k). \tag{3.20}$$

We will focus on evaluating the convolution sum based on Eq. (3.17). Let us examine first a few outputs from Eq. (3.17):

$$y(0) = \sum_{k=-\infty}^{\infty} x(k)h(-k) = \cdots + x(-1)h(1) + x(0)h(0) + x(1)h(-1) + x(2)h(-2) + \cdots$$

$$y(1) = \sum_{k=-\infty}^{\infty} x(k)h(1-k) = \cdots + x(-1)h(2) + x(0)h(1) + x(1)h(0) + x(2)h(-1) + \cdots$$

$$y(2) = \sum_{k=-\infty}^{\infty} x(k)h(2-k) = \cdots + x(-1)h(3) + x(0)h(2) + x(1)h(1) + x(2)h(0) + \cdots$$

We see that the convolution sum requires the sequence $h(n)$ to be reversed and shifted. The graphical, formula, and table methods are discussed for evaluating the digital convolution via the several examples. To begin with evaluating the convolution sum graphically, we need to apply the reversed sequence and shifted sequence. The reversed sequence is defined as follows: If $h(n)$ is the given sequence, $h(-n)$ is the reversed sequence. The reversed sequence is a mirror image of the original sequence, assuming the vertical axis as the mirror. Let us study the reversed sequence and shifted sequence via the following example.

EXAMPLE 3.9

Given a sequence,

$$h(k) = \begin{cases} 3, & k=0,1 \\ 1, & k=2,3 \\ 0 & \text{elsewhere} \end{cases}$$

where k is the time index or sample number,

(a) Sketch the sequence $h(k)$ and reversed sequence $h(-k)$.

(b) Sketch the shifted sequences $h(-k+3)$ and $h(-k-2)$.

Solution:

(a) Since $h(k)$ is defined, we plot it in Fig. 3.17.

Next, we need to find the reversed sequence $h(-k)$. We examine the following for

$$k>0, h(-k)=0$$

$$k=0, h(-0)=h(0)=3$$

$$k=-1, h(-k)=h(-(-1))=h(1)=3$$

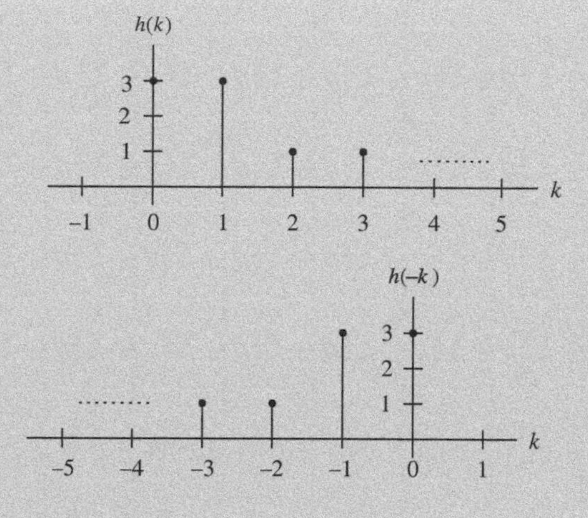

FIG. 3.17

Plots of the digital sequence and its reversed sequence in Example 3.9.

$$k=-2, h(-k)=h(-(-2))=h(2)=1$$

$$k=-3, h(-k)=h(-(-3))=h(3)=1$$

One can verify that $k \leq -4$, $h(-k)=0$. Then the reversed sequence $h(-k)$ is shown as the second plot in Fig. 3.17.

As shown in the sketches, $h(-k)$ is just a mirror image of the original sequence $h(k)$.

(b) Based on the definition of the original sequence, we know that

$h(0)=h(1)=3$, $h(2)=h(3)=1$, and the others are zeros. The time indices correspond to the following:

$$-k+3=0, k=3$$

$$-k+3=1, k=2$$

$$-k+3=2, k=1$$

$$-k+3=3, k=0$$

Thus we can sketch $h(-k+3)$ as shown in Fig. 3.18.

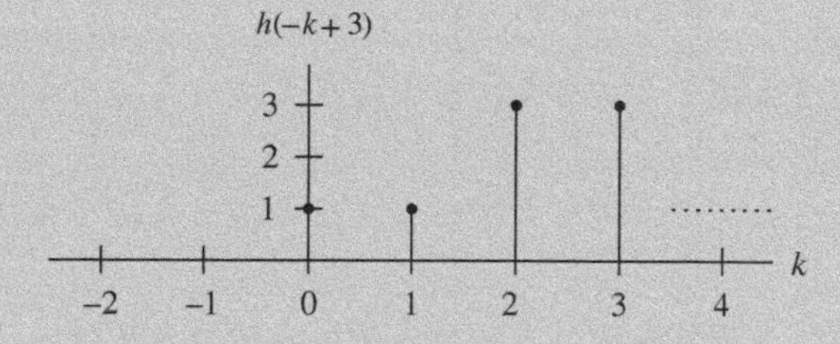

FIG. 3.18

Plot of the sequence $h(-k+3)$ in Example 3.9.

Continued

EXAMPLE 3.9—CONT'D

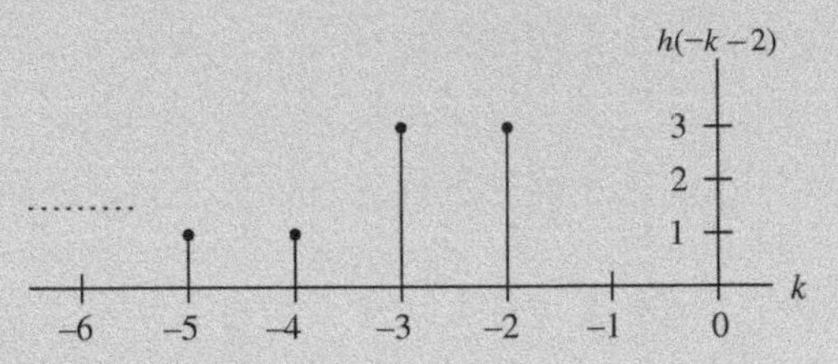

FIG. 3.19

Plot of the sequence $h(-k-2)$ in Example 3.9.

Similarly, $h(-k-2)$ is yielded in Fig. 3.19.
We can get $h(-k+3)$ by shifting $h(-k)$ to the right by three samples, and we can obtain $h(-k-2)$ by shifting $h(-k)$ to the left by two samples.
In summary, given $h(-k)$, we can obtain $h(n-k)$ by shifting $h(-k)$ n samples to the right or the left, depending on weather n is positive or negative.

Once we understand the shifted sequence and reversed sequence, we can perform digital convolution of two sequences $h(k)$ and $x(k)$, defined in Eq. (3.17) graphically. From that equation, we see that each convolution value $y(n)$ is the sum of the products of two sequences $x(k)$ and $h(n-k)$, the latter of which is the shifted version of the reversed sequence $h(-k)$ by $|n|$ samples. Hence, we can summarize the graphical convolution procedure in Table 3.3.

Table 3.3 Digital Convolution Using the Graphical Method

Step 1. Obtain the reversed sequence $h(-k)$.
Step 2. Shift $h(-k)$ by $\|n\|$ samples to get $h(n-k)$. If $n \geq 0$, $h(-k)$ will be shifted to right by n samples; but if $n<0$, $h(-k)$ will be shifted to the left by $\|n\|$ samples.
Step 3. Perform the convolution sum that is the sum of products of two sequences $x(k)$ and $h(n-k)$ to get $y(n)$.
Step 4. Repeat steps (1)–(3) for the next convolution value $y(n)$.

We illustrate digital convolution sum via the following example.

EXAMPLE 3.10

Using the following sequences defined in Fig. 3.20, evaluate the digital convolution

$$y(n) = \sum_{k=-\infty}^{\infty} x(k)h(n-k),$$

(a) By the graphical method.
(b) By applying the formula directly.

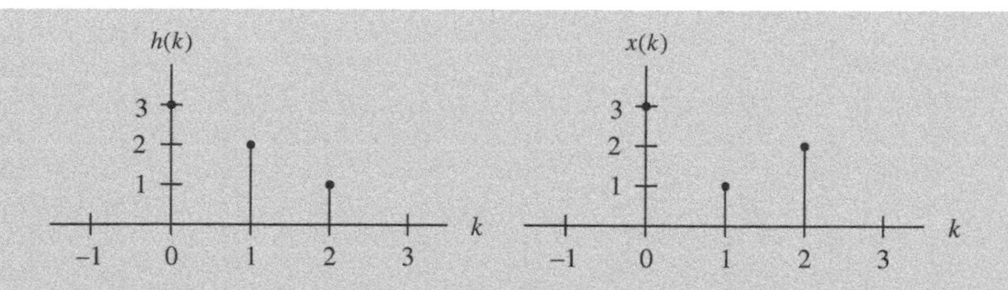

FIG. 3.20

Plots of digital input sequence and impulse sequence in Example 3.10.

Solution:

(a) To obtain $y(0)$, we need the reversed sequence $h(-k)$; and to obtain $y(1)$, we need the reversed sequence $h(1-k)$, and so on. Using the technique we have discussed, sequences $h(-k)$, $h(-k+1)$, $h(-k+2)$, $h(-k+3)$, and $h(-k+4)$ are achieved and plotted in Fig. 3.21, respectively.

Again, using the information in Figs. 3.20 and 3.21, we can compute the convolution sum as:

$$\text{Sum of product of } x(k) \text{ and } h(-k): y(0)=3\times3=9$$

$$\text{Sum of product of } x(k) \text{ and } h(1-k): y(1)=1\times3+3\times2=9$$

$$\text{Sum of product of } x(k) \text{ and } h(2-k): y(2)=2\times3+1\times2+3\times1=11$$

$$\text{Sum of product of } x(k) \text{ and } h(3-k): y(3)=2\times2+1\times1=5$$

$$\text{Sum of product of } x(k) \text{ and } h(4-k): y(4)=2\times1=2$$

Sum of product of $x(k)$ and $h(5-k): y(n)=0$ for $n>4$, since sequences $x(k)$ and $h(n-k)$ do not overlap.

Finally, we sketch the output sequence $y(n)$ in Fig. 3.22.

(b) Applying Eq. (3.20) with zero-initial conditions leads to

$$y(n)=x(0)h(n)+x(1)h(n-1)+x(2)h(n-2)$$

$$n=0,\ y(0)=x(0)h(0)+x(1)h(-1)+x(2)h(-2)=3\times3+1\times0+2\times0=9$$

$$n=1,\ y(1)=x(0)h(1)+x(1)h(0)+x(2)h(-1)=3\times2+1\times3+2\times0=9$$

$$n=2,\ y(2)=x(0)h(2)+x(1)h(1)+x(2)h(0)=3\times1+1\times2+2\times3=11$$

$$n=3,\ y(3)=x(0)h(3)+x(1)h(2)+x(2)h(1)=3\times0+1\times1+2\times2=5$$

$$n=4,\ y(4)=x(0)h(4)+x(1)h(3)+x(2)h(2)=3\times0+1\times0+2\times1=2$$

$$n\geq5,\ y(n)=x(0)h(n)+x(1)h(n-1)+x(2)h(n-2)=3\times0+1\times0+2\times0=0.$$

In simple cases such as Example 3.10, it is not necessary to use the graphical or formula methods. We can compute the convolution by treating the input sequence and impulse response as number sequences and sliding the reversed impulse response past the input sequence, cross-multiplying, and summing the nonzero overlap terms at each step. The procedure and calculated results are listed in Table 3.4.

Continued

EXAMPLE 3.10—CONT'D

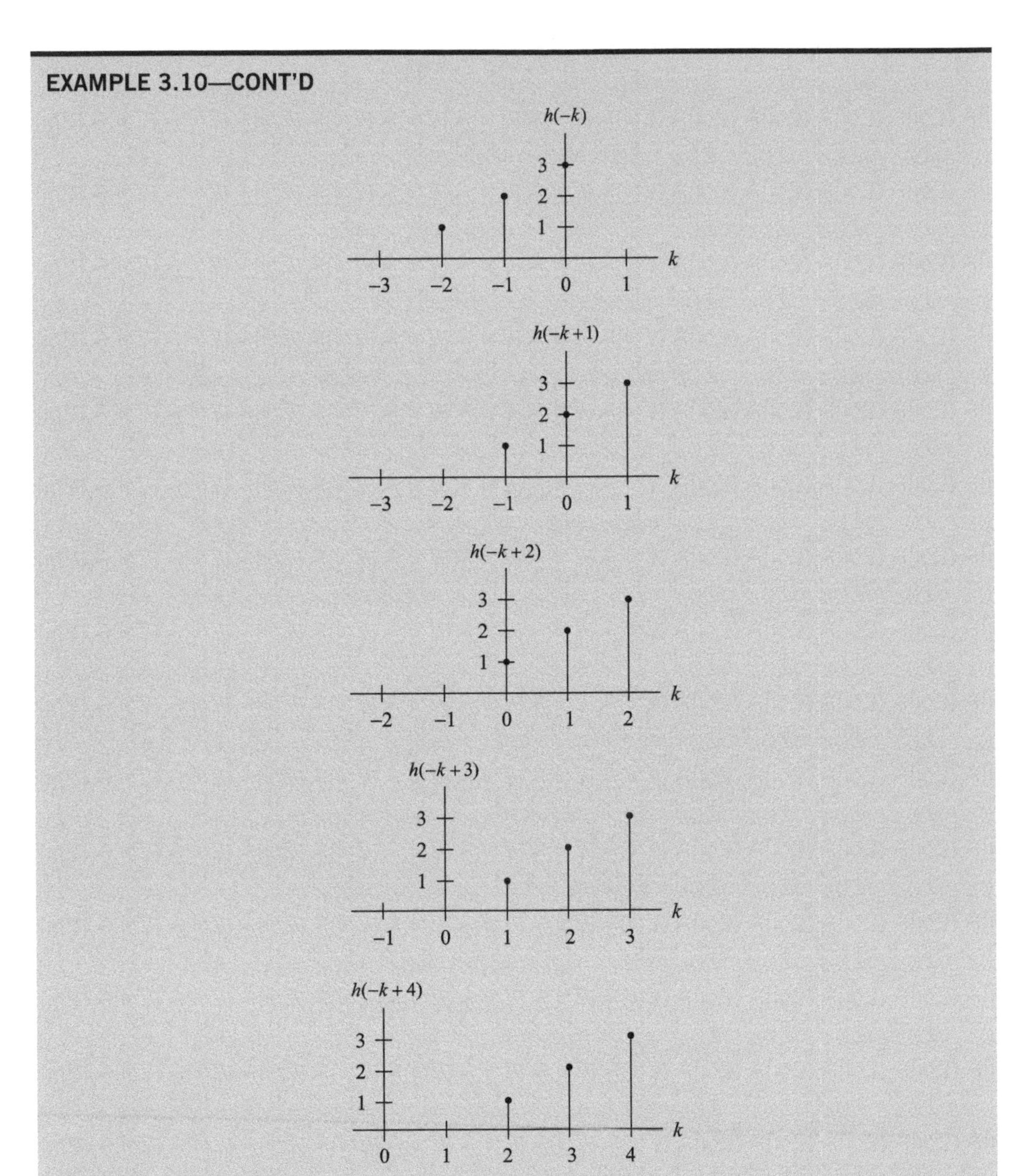

FIG. 3.21

Illustration of convolution of two sequences $x(k)$ and $h(k)$ in Example 3.10.

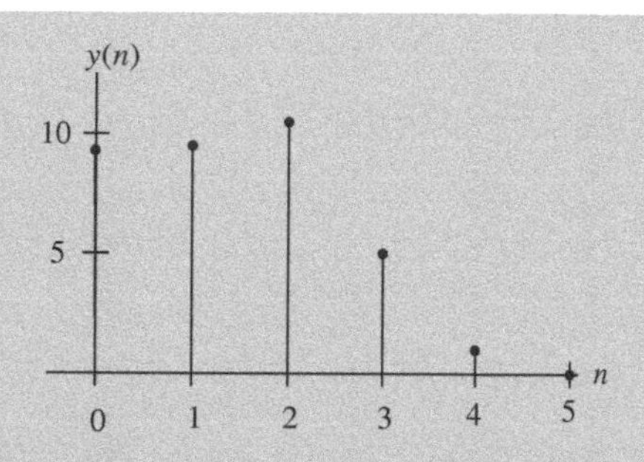

FIG. 3.22

Plot of the convolution sum in Example 3.10.

Table 3.4 Convolution Sum Using the Table Method

k:	−2	−1	0	1	2	3	4	5	
$x(k)$:			3	1	2				
$h(-k)$:	1	2	3						$y(0)=3\times3=9$
$h(1-k)$		1	2	3					$y(1)=3\times2+1\times3=9$
$h(2-k)$			1	2	3				$y(2)=3\times1+1\times2+2\times3=11$
$h(3-k)$				1	2	3			$y(3)=1\times1+2\times2=5$
$h(4-k)$					1	2	3		$y(4)=2\times1=2$
$h(5-k)$						1	2	3	$y(5)=0$ (no overlap)

We can see that the calculated results using all the methods are consistent. The steps using the table method are concluded in Table 3.5.

Table 3.5 Digital Convolution Steps via the Table

Step 1. List the index k covering a sufficient range.

Step 2. List the input $x(k)$.

Step 3. Obtain the reversed sequence $h(-k)$, and align the rightmost element of $h(n-k)$ to the leftmost element of $x(k)$.

Step 4. Cross-multiply and sum the nonzero overlap terms to produce $y(n)$.

Step 5. Slide $h(n-k)$ to the right by one position.

Step 6. Repeat Step 4; stop if all the output values are zero or if required.

EXAMPLE 3.11

Given the following two rectangular sequences,

$$x(n) = \begin{cases} 1 & n=0,1,2 \\ 0 & \text{otherwise} \end{cases} \quad \text{and} \quad h(n) = \begin{cases} 0 & n=0 \\ 1 & n=1,2 \\ 0 & \text{otherwise} \end{cases},$$

Convolve them using the table method.

Solution:

Using Table 3.5 as a guide, we list the operations and calculations in Table 3.6.

Table 3.6 Convolution Sum in Example 3.11

k:	−2	−1	0	1	2	3	4	5	…	
$x(k)$:			1	1	1				…	
$h(-k)$:	1	1	0							$y(0)=0$ (no overlap)
$h(1-k)$		1	1	0						$y(1)=1\times1=1$
$h(2-k)$			1	1	0					$y(2)=1\times1+1\times1=2$
$h(3-k)$				1	1	0				$y(3)=1\times1+1\times1=2$
$h(4-k)$					1	1	0			$y(4)=1\times1=1$
$h(n-k)$						1	1	0		$y(n)=0$, $n\geq5$ (no overlap)
										Stop

Note that the output should show the trapezoidal shape.

Let us examine convolving a finite long sequence with an infinite long sequence.

EXAMPLE 3.12

A system representation using the unit-impulse response for the linear system

$$y(n) = 0.25y(n-1) + x(n) \text{ for } n \geq 0 \text{ and } y(-1) = 0$$

is determined in Example 3.8 as

$$y(n) = \sum_{k=-\infty}^{\infty} x(k)h(n-k),$$

where $h(n) = (0.25)^n u(n)$. For a step input $x(n) = u(n)$, determine the output response for the first three output samples using the table method.

Solution:

Using Table 3.5 as a guide, we list the operations and calculations in Table 3.7.

Table 3.7 Convolution Sum in Example 3.13

k:	−2	−1	0	1	2	3	…	
$x(k)$:			1	1	1	1	…	
$h(-k)$:	0.0625	0.25	1					$y(0)=1\times 1=1$
$h(1-k)$		0.0625	0.25	1				$y(1)=1\times 0.25+1\times 1=1.25$
$h(2-k)$			0.0625	0.25	1			$y(2)=1\times 0.0625+1\times 0.25$ $+1\times 1=1.3125$
								Stop as required

As expected, the output values are the same as those obtained in Example 3.8.

3.5 BOUNDED-INPUT AND BOUNDED-OUTPUT STABILITY

We are interested in designing and implementing stable linear time-invariant systems. A stable system is one for which every bounded input produces a bounded output (BIBO). There are many other stability definitions. To find the stability criterion for the linear time-invariant system, consider the linear time-invariant representation with the bounded input as $|x(n)|<M$, where M is a positive finite number. Taking absolute value of Eq. (3.15) leads to the following inequality:

$$|y(n)|=\left|\sum_{k=-\infty}^{\infty} x(k)h(n-k)\right| < \sum_{k=-\infty}^{\infty} |x(k)||h(n-k)|. \tag{3.21}$$

Using the bounded input, we obtain

$$|y(n)|<M(\cdots+|h(-1)|+|h(0)|+|h(1)|+|h(2)|+\cdots). \tag{3.22}$$

If the absolute sum in Eq. (3.22) is a finite number, the product of the absolute sum and the maximum input value is therefore a finite number. Hence, we obtain a bounded output with a bounded input. This concludes that a linear time-invariant system is stable if only if the sum of its absolute impulse response coefficients is a finite positive number, that is,

$$S=\sum_{k=-\infty}^{\infty} |h(k)|=\cdots+|h(-1)|+|h(0)|+|h(1)|+\cdots<\infty. \tag{3.23}$$

Fig. 3.23 illustrates a linear time-invariant stable system, where the impulse response decreases to zero in finite amount of time so that the summation of its absolute impulse response coefficients is guaranteed to be finite.

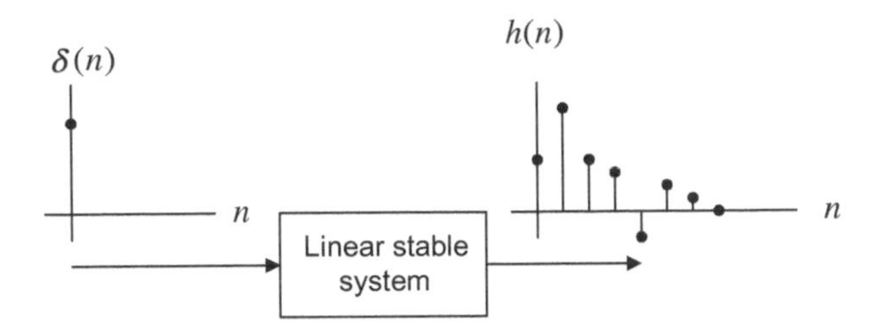

FIG. 3.23

Illustration of stability of the digital linear time-invariant system.

EXAMPLE 3.13

Given the linear time-invariant system in Example 3.8,

$$y(n) = 0.25y(n-1) + x(n) \text{ for } n \geq 0 \text{ and } y(-1) = 0,$$

which is described by the unit-impulse response

$$h(n) = (0.25)^n u(n),$$

determine whether this system is stable or not.

Solution:

Using Eq. (3.23), we have

$$S = \sum_{k=-\infty}^{\infty} |h(k)| = \sum_{k=-\infty}^{\infty} \left|(0.25)^k u(k)\right|.$$

Applying the definition of the unit-step function $u(k)$ for $u(k)$, we have

$$S = \sum_{k=0}^{\infty} (0.25)^k = 1 + 0.25 + 0.25^2 + \cdots$$

Using the formula for a sum of the geometric series (see Appendix H),

$$\sum_{k=0}^{\infty} a^k = \frac{1}{1-a},$$

where $a = 0.25 < 1$, we conclude

$$S = \sum_{k=0}^{\infty} (0.25)^k = 1 + 0.25 + 0.25^2 + \cdots = \frac{1}{1-0.25} = \frac{4}{3} < \infty.$$

Since the summation is a finite number, the linear system is stable.

3.6 SUMMARY

1. Concepts of digital signals are explained. Digital signal samples are sketched, using their encoded amplitude versus sample numbers with vertical bars topped by solid circles located at their

sampling instants, respectively. Impulse sequence, unit-step sequence, and their shifted versions are sketched in this notation.

2. An analog signal function can be sampled to its digital (discrete time) version by substituting time $t = nT$ into the analog function, that is,

$$x(n) = x(t)|_{t=nT} = x(nT).$$

The digital function values can be calculated for the given time index (sample number).

3. The DSP system we wish to design is typically a linear, time invariant, causal system. Linearity means that the superposition principle exists. Time-invariance requires that the shifted input generates the corresponding shifted output with the same amount of time. Causality indicates that the system output depends on only its current input sample and past input sample(s).
4. The difference equation describing a linear, time-invariant system has a format such that the current output depends on the current input, past input(s), and past output (s) in general.
5. The unit-impulse response can be used to fully describe a linear, time-invariant system. Given the impulse response, the system output is the sum of the products of the impulse response coefficients and corresponding input samples, called the digital convolution sum.
6. Digital convolution sum, which represents a DSP system, is evaluated in three ways: the graphical method, evaluation of the formula, and the table method. The table method is found to be most effective.
7. BIBO is a type of stability in which a bounded input will produce a bounded output. The condition for a BIBO linear time-invariant system requires that the sum of the absolute impulse response coefficients be a finite positive number.

3.7 PROBLEMS

3.1 Sketch each of the following special digital sequences:

(a) $5\delta(n)$

(b) $-2\delta(n-5)$

(c) $-5u(n)$

(d) $5u(n-2)$

3.2 Calculate the first eight sample values and sketch each of the following sequences:

(a) $x(n) = 0.5^n u(n)$

(b) $x(n) = 5\sin(0.2\pi n)u(n)$

(c) $x(n) = 5\cos(0.1\pi n + 30^0)u(n)$

(d) $x(n) = 5(0.75)^n \sin(0.1\pi n)u(n)$

3.3 Sketch each of the following special digital sequences:

(a) $8\delta(n)$

(b) $-3.5\delta(n-4)$

(c) $4.5u(n)$

(d) $-6u(n-3)$

3.4 Calculate the first eight sample values and sketch each of the following sequences:

(a) $x(n) = 0.25^n u(n)$

(b) $x(n) = 3\sin(0.4\pi n)u(n)$
(c) $x(n) = 6\cos(0.2\pi n + 30^0)u(n)$
(d) $x(n) = 4(0.5)^n \sin(0.1\pi n)u(n)$

3.5 Sketch the following sequence:
(a) $x(n) = 3\delta(n+2) - 0.5\delta(n) + 5\delta(n-1) - 4\delta(n-5)$
(b) $x(n) = \delta(n+1) - 2\delta(n-1) + 5\delta(n-4)$

3.6 Given the digital signals $x(n)$ in Figs. 3.24 and 3.25, write an expression for each digital signal using the unit-impulse sequence and its shifted sequences.

3.7 Sketch the following sequences:
(a) $x(n) = 2\delta(n+3) - 0.5\delta(n+1) - 5\delta(n-2) - 4\delta(n-5)$
(b) $x(n) = 2\delta(n+2) - 2\delta(n+1) + 5u(n-3)$

3.8 Given the digital signals $x(n)$ as shown in Figs. 3.26 and 3.27, write an expression for each digital signal using the unit-impulse sequence and its shifted sequences.

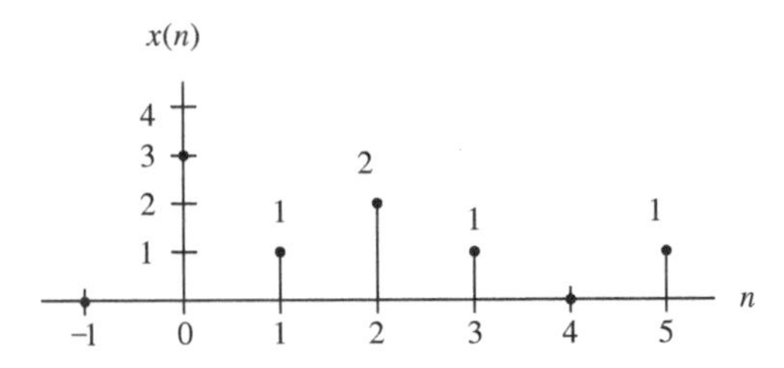

FIG. 3.24

The first digital signal in Problem 3.6.

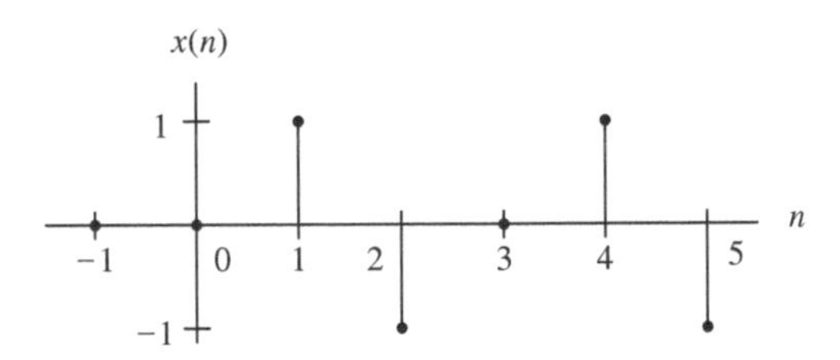

FIG. 3.25

The second digital signal in Problem 3.6.

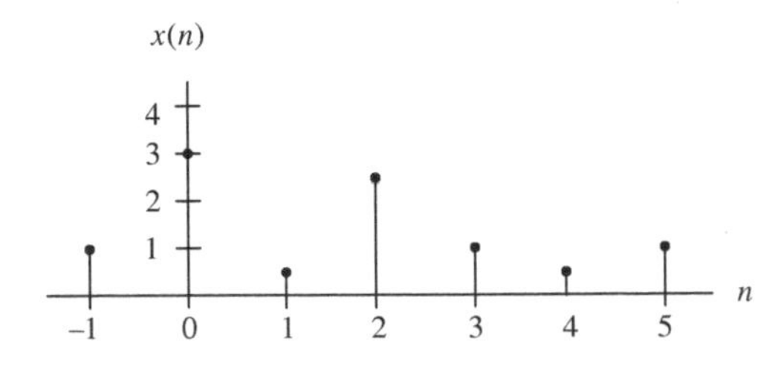

FIG. 3.26

The first digital signal in Problem 3.8.

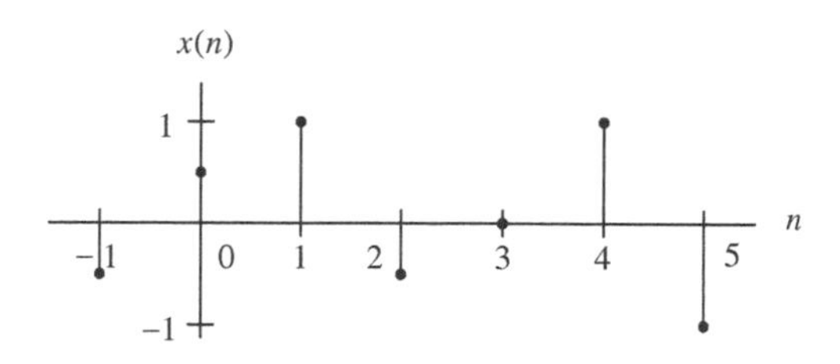

FIG. 3.27

The second digital signal in Problem 3.8.

3.9 Assume that a digital signal processor with a sampling time interval of 0.01 s converts each of the following analog signals $x(t)$ to a digital signal $x(n)$, determine the digital sequences for each of the following analog signals.

(a) $x(t)=e^{-50t}u(t)$

(b) $x(t)=5\sin(20\pi t)u(t)$

(c) $x(t)=10\cos(40\pi t+30^0)u(t)$

(d) $x(t)=10e^{-100t}\sin(15\pi t)u(t)$

3.10 Determine which of the following systems is a linear system.

(a) $y(n)=5x(n)+2x^2(n)$

(b) $y(n)=x(n-1)+4x(n)$

(c) $y(n)=4x^3(n-1)-2x(n)$

3.11 Assume that a digital signal processor with a sampling time interval of 0.005 s converts each of the following analog signals $x(t)$ to a digital signal $x(n)$, determine the digital sequences for each of the following analog signals.

(a) $x(t)=e^{-100t}u(t)$

(b) $x(t)=4\sin(60\pi t)u(t)$

(c) $x(t)=7.5\cos(20\pi t+60^0)u(t)$

(d) $x(t)=20e^{-200t}\sin(60\pi t)u(t)$

3.12 Determine which of the following systems is a linear system.

(a) $y(n)=4x(n)+8x^3(n)$

(b) $y(n)=x(n-3)+3x(n)$

(c) $y(n)=5x^2(n-1)-3x(n)$

3.13 Given the following linear systems, find which one is time invariant.

(a) $y(n)=-5x(n-10)$

(b) $y(n)=4x(n^2)$

3.14 Determine which of the following linear systems is causal.

(a) $y(n)=0.5x(n)+100x(n-2)-20x(n-10)$

(b) $y(n)=x(n+4)+0.5x(n)-2x(n-2)$

3.15 Determine the causality for each of the following linear systems.

(a) $y(n)=0.5x(n)+20x(n-2)-0.1y(n-1)$

(b) $y(n)=x(n+2)-0.4y(n-1)$

(c) $y(n)=x(n-1)+0.5y(n+2)$

3.16 Find the unit-impulse response for each of the following linear systems.

(a) $y(n)=0.5x(n)-0.5x(n-2)$; for $n\geq 0$, $x(-2)=0$, $x(-1)=0$

(b) $y(n)=0.75y(n-1)+x(n)$; for $n \geq 0$, $y(-1)=0$
(c) $y(n)=-0.8y(n-1)+x(n-1)$; for $n \geq 0$, $x(-1)=0$, $y(-1)=0$

3.17 Determine the causality for each of the following linear systems.
(a) $y(n)=5x(n)+10x(n-4)-0.1y(n-5)$
(b) $y(n)=2x(n+2)-0.2y(n-2)$
(c) $y(n)=0.1x(n+1)+0.5y(n+2)$

3.18 Find the unit-impulse response for each of the following linear systems.
(a) $y(n)=0.2x(n)-0.3x(n-2)$; for $n \geq 0$, $x(-2)=0$, $x(-1)=0$
(b) $y(n)=0.5y(n-1)+0.5x(n)$; for $n \geq 0$, $y(-1)=0$
(c) $y(n)=-0.6y(n-1)-x(n-1)$; for $n \geq 0$, $x(-1)=0$, $y(-1)=0$

3.19 For each of the following linear systems, find the unit-impulse response, and draw the block diagram.
(a) $y(n)=5x(n-10)$
(b) $y(n)=x(n)+0.5x(n-1)$

3.20 Given the sequence

$$h(k)=\begin{cases} 2, & k=0,1,2 \\ 1, & k=3,4 \\ 0 & \text{elsewhere} \end{cases}$$

where k is the time index or sample number,
(a) sketch the sequence $h(k)$ and the reverse sequence $h(-k)$;
(b) sketch the shifted sequences $h(-k+2)$ and $h(-k-3)$.

3.21 Given the sequence

$$h(k)=\begin{cases} -1 & k=0,1 \\ 2 & k=2,3 \\ -2 & k=4 \\ 0 & \text{elsewhere} \end{cases}$$

where k is the time index or sample number,
(a) sketch the sequence $h(k)$and the reverse sequence$h(-k)$;
(b) sketch the shifted sequences and $h(-k-2)$.

3.22 Using the following sequence definitions,

$$h(k)=\begin{cases} 2, & k=0,1,2 \\ 1, & k=3,4 \\ 0 & \text{elsewhere} \end{cases} \quad \text{and} \quad x(k)=\begin{cases} 2, & k=0 \\ 1, & k=1,2 \\ 0 & \text{elsewhere} \end{cases}$$

evaluate the digital convolution

$$y(n)=\sum_{k=-\infty}^{\infty} x(k)h(n-k)$$

(a) using the graphical method;
(b) using the table method;
(c) applying the convolution formula directly.

3.23 Using the sequence definitions

$$h(k) = \begin{cases} 2, & k=0,1,2 \\ 1, & k=3,4 \\ 0 & \text{elsewhere} \end{cases} \quad \text{and} \quad x(k) = \begin{cases} 2, & k=0 \\ 1, & k=1,2 \\ 0 & \text{elsewhere} \end{cases}$$

evaluate the digital convolution

$$y(n) = \sum_{k=-\infty}^{\infty} x(k)h(n-k)$$

(a) the graphical method;
(b) the table method;
(c) applying the convolution formula directly.

3.24 Convolve the following two rectangular sequences:

$$x(n) = \begin{cases} 1 & n=0,1 \\ 0 & \text{otherwise} \end{cases} \quad \text{and} \quad h(n) = \begin{cases} 0 & n=0 \\ 1 & n=1,2 \\ 0 & \text{otherwise} \end{cases}$$

using the table method.

3.25 Determine the stability for the following linear system.

$$y(n) = 0.5x(n) + 100x(n-2) - 20x(n-10)$$

3.26 For each of the following linear systems, find the unit-impulse response, and draw the block diagram.

(a) $y(n) = 2.5x(n-5)$
(b) $y(n) = 2x(n) + 1.2x(n-1)$

3.27 Determine the stability for the following linear system.

$$y(n) = 0.5x(n) + 100x(n-2) - 20x(n-10)$$

3.28 Determine the stability for each of the following linear systems.

(a) $y(n) = \sum_{k=0}^{\infty} 0.75^k x(n-k)$
(b) $y(n) = \sum_{k=0}^{\infty} 2^k x(n-k)$

3.29 Determine the stability for each of the following linear systems.

(a) $y(n) = \sum_{k=0}^{\infty} (-1.5)^k x(n-k)$
(b) $y(n) = \sum_{k=0}^{\infty} (-0.5)^k x(n-k)$

Advanced Problems

3.30 Given each of the following discrete-time systems,

(a) $y(n)=x(-n+3)$
(b) $y(n)=x(n-1)+0.5y(n-2)$
(c) $y(n)=nx(n-1)+x(n)$
(d) $y(n)=|x(n)|$

determine if the system is (1) linear or nonlinear; (2) time invariant or time varying; (3) causal or noncausal; and (4) stable or unstable.

3.31 Given each of the following discrete-time systems,

(a) $y(n)=\text{sign}[x(n)]$, where $\text{sign}(x)=\begin{cases} 1 & x>0 \\ 0 & x=0 \\ -1 & x<0 \end{cases}$
(b) $y(n)=\text{truncate}[x(n)]$
(c) $y(n)=\text{round}[x(n)]$

determine if the system is (1) linear or nonlinear; (2) time invariant or time varying; (3) causal or noncausal; and (4) stable or unstable.

3.32 Given each of the following discrete-time systems,

(a) $y(n)=x(n)x(n-1)$
(b) $y(n)=x(n)-0.2x(n-1)y(n-2)$

determine if the system is (1) linear or nonlinear; (2) time invariant or time varying; (3) causal or noncausal; and (4) stable or unstable.

3.33 For $N>M$, show that

(a) $\sum_{k=M}^{N} r^k = \begin{cases} \dfrac{r^M - r^{N+1}}{1-r} & \text{for } r \neq 1 \\ N-M+1 & \text{for } r=1 \end{cases}$
(b) For $|r|<1$, $\sum_{k=0}^{\infty} r^k = \frac{1}{1-r}$

3.34 Given a relaxed discrete-time system,

$$y(n)-ay(n-1)=x(n),$$

(a) Show that the impulse response is

$$h(n)=a^n u(n);$$

(b) If the impulse response of a relaxed discrete-time system is found as

$$h(n)=\begin{cases} a^n & n\geq 0, n=\text{even} \\ 0 & \text{otherwise} \end{cases},$$

determine the discrete-time system equation.

3.35 Given the following relaxed discrete-time system with a causal input $x(n)$, where $x(n)=0$, for $n<0$:

$$y(n)-\frac{n}{n+1}y(n-1)=\frac{1}{n+1}x(n),$$

show that the impulse response is $h(n)=1/(n+1)$.

3.36 Given $x(n)=a^n u(n)$, where $|a|<1$, and $h(n)=u(n)$

determine

(a) $y(n)=h(n)*x(n)$

(b) $y(n)=h(n)*x(-n)$.

3.37 A causal system output $y(n)$ is expressed as

$$y(n)=\sum_{k=-\infty}^{n} x(k)h(n-k),$$

where $h(n)$ and $x(n)$ are the impulse response and system input, respectively. Show that

$$y(n)=\sum_{k=0}^{\infty} h(k)x(n-k).$$

3.38 A system output $y(n)$ is expressed as

$$y(n)=\sum_{k=-\infty}^{\infty} x(k)h(n-k).$$

If both $h(n)$ and $x(n)$ are causal, show that

$$y(n)=\sum_{k=0}^{n} h(k)x(n-k)=\sum_{k=0}^{n} x(k)h(n-k).$$

CHAPTER

DISCRETE FOURIER TRANSFORM AND SIGNAL SPECTRUM

4

CHAPTER OUTLINE

4.1 DISCRETE FOURIER TRANSFORM

In time domain, representation of digital signals describes the signal amplitude vs. the sampling time instant or the sample number. However, in some applications, the signal frequency content is very useful than the digital signal samples. The representation of the digital signal in terms of its frequency component in a frequency domain, that is, the signal spectrum, needs to be developed. As an example shown in Fig. 4.1, the top plot displays 32 signal samples (vertical bars with circles) from sampling a continuous 1000-Hz sinusoid (dashed line) at a sampling rate of 8000 Hz in time domain representation; the bottom plot shows the magnitude of signal spectrum in frequency domain representation, where we can clearly observe that the amplitude peak is located at the frequency of 1000 Hz in the calculated spectrum. Hence, the spectral plot better displays frequency information of a digital signal.

The algorithm transforming the time domain signal samples to the frequency domain components is known as the *discrete Fourier transform*, or DFT. The DFT also establishes a relationship between the time domain representation and the frequency domain representation. Therefore, we can apply the DFT to perform frequency analysis of a time domain sequence. In addition, the DFT is widely used in many other areas, including spectral analysis, acoustics, imaging/video, audio, instrumentation, and communications systems.

Digital Signal Processing. https://doi.org/10.1016/B978-0-12-815071-9.00004-X

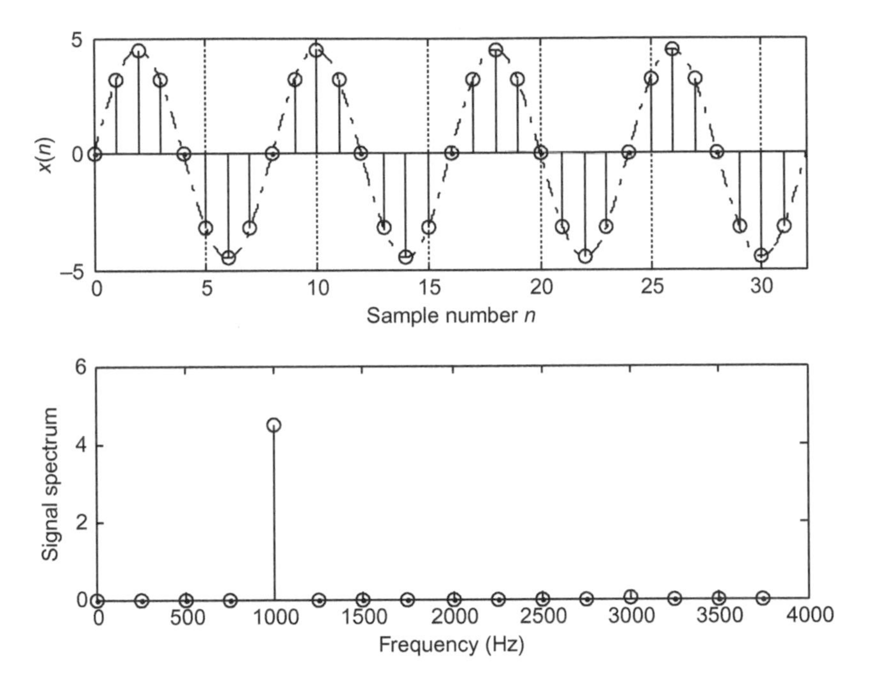

FIG. 4.1

Example of the digital signal and its amplitude spectrum.

To be able to develop the DFT and understand how to use it, we first study the spectrum of periodic digital signals using the Fourier series. (Detailed discussion on Fourier series is provided in Appendix B).

4.1.1 FOURIER SERIES COEFFICIENTS OF PERIODIC DIGITAL SIGNALS

Let us look at a process in which we want to estimate the spectrum of a periodic digital signal $x(n)$ sampled at a rate of f_s Hz with the fundamental period $T_0 = NT$, as shown in Fig. 4.2, where there are N samples within the duration of the fundamental period and $T = 1/f_s$ is the sampling period. For the time being, we assume that the periodic digital signal is band limited to have all harmonic frequencies less than the folding frequency $f_s/2$ so that aliasing does not occur.

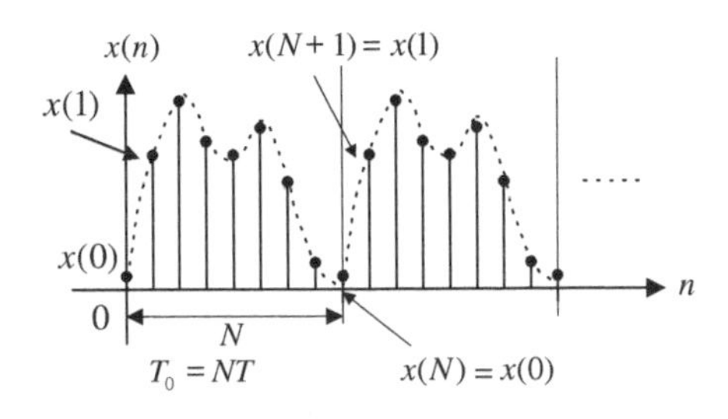

FIG. 4.2

Periodic digital signal.

According to Fourier series analysis (Appendix B), the coefficients of the Fourier series expansion of the periodic signal $x(t)$ in a complex form is

$$c_k = \frac{1}{T_0}\int_{T_0} x(t)e^{-jk\omega_0 t}dt, \text{ for } -\infty < k < \infty, \tag{4.1}$$

where k is the number of harmonics corresponding to the harmonic frequency of kf_0, and $\omega_0 = 2\pi/T_0$ and $f_0 = 1/T_0$ are the fundamental frequency in radians per second and the fundamental frequency in Hz, respectively. To apply Eq. (4.1), we substitute $T_0 = NT$, $\omega_0 = 2\pi/T_0$ and approximate the integration over one period using a summation by substituting $dt = T$ and $t = nT$. We obtain

$$c_k = \frac{1}{N}\sum_{n=0}^{N-1} x(n)e^{-j2\pi kn/N}, \text{ for } -\infty < k < \infty. \tag{4.2}$$

We refer c_k as discrete-time Fourier series coefficients. Since the coefficients c_k are obtained from the Fourier series expansion in the complex form, the resultant spectrum c_k will have two sides. There is an important feature of Eq. (4.2) in which the Fourier series coefficient c_k is periodic of N. We can verify this as follows:

$$c_{k+N} = \frac{1}{N}\sum_{n=0}^{N-1} x(n)e^{-j2\pi(k+N)n/N} = \frac{1}{N}\sum_{n=0}^{N-1} x(n)e^{-j2\pi kn/N}e^{-j2\pi n}. \tag{4.3}$$

Since $e^{-j2\pi n} = \cos(2\pi n) - j\sin(2\pi n) = 1$, it follows that

$$c_{k+N} = c_k. \tag{4.4}$$

Therefore, the two-sided line amplitude spectrum $|c_k|$ is periodic, as shown in Fig. 4.3.

We note the following points:

(a) As displayed in Fig. 4.3, only the line spectral portion between the frequencies $-f_s/2$ and $f_s/2$ (folding frequency) represents frequency information of the periodic signal.

(b) Note that the spectral portion from $f_s/2$ to f_s is a copy of the spectrum in the negative frequency range from $-f_s/2$ to 0 Hz due to the spectrum being periodic for every Nf_0 Hz. Again, the amplitude spectral components indexed from $f_s/2$ to f_s can be folded at the folding frequency $f_s/2$ to match the

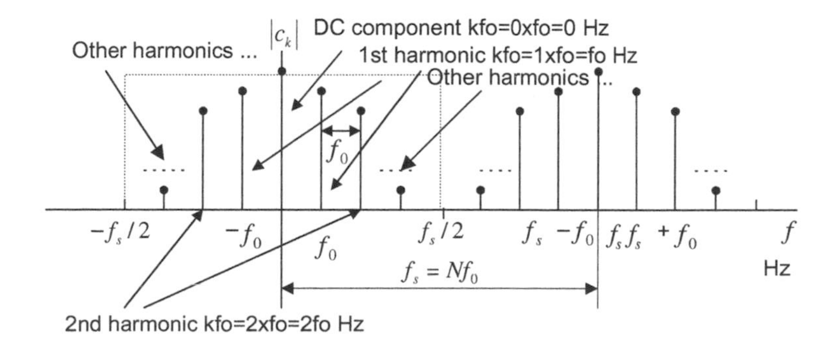

FIG. 4.3

Amplitude spectrum of the periodic digital signal.

amplitude spectral components indexed from 0 to $f_s/2$ in terms of $f_s - f$ Hz, where f is in the range from $f_s/2$ to f_s. For convenience, we compute the spectrum over the range from 0 to f_s Hz with nonnegative indices, that is,

$$c_k = \frac{1}{N}\sum_{n=0}^{N-1} x(n)e^{-j2\pi kn/N}, \; k = 0, 1, \ldots, N-1. \tag{4.5}$$

We can apply Eq. (4.4) to find the negative indexed spectral values if they are required.

(c) For the kth harmonic, the frequency is

$$f = kf_0 \text{ Hz}. \tag{4.6}$$

The frequency spacing between the consecutive spectral lines, called the frequency resolution, is f_0 Hz.

EXAMPLE 4.1

The periodic signal

$$x(t) = \sin(2\pi t)$$

is sampled using the sampling rate $f_s = 4$ Hz.

(a) Compute the spectrum c_k using the samples in one period.
(b) Plot the two-sided amplitude spectrum $|c_k|$ over the range from -2 to 2 Hz.

Solution:

(a) From the analog signal, we can determine the fundamental frequency $\omega_0 = 2\pi$ rad/s and $f_0 = \frac{\omega_0}{2\pi} = \frac{2\pi}{2\pi} = 1$ Hz, and the fundamental period $T_0 = 1$ s.
Since using the sampling interval $T = 1/f_s = 0.25$ s, we get the sampled signal as

$$x(n) = x(nT) = \sin(2\pi nT) = \sin(0.5\pi n)$$

and plot the first eight samples as shown in Fig. 4.4.

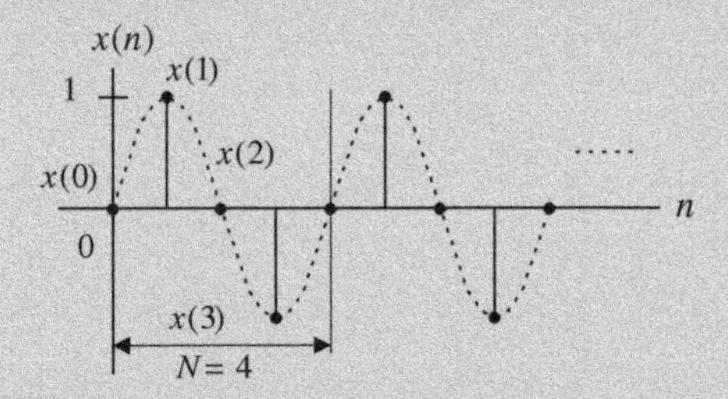

FIG. 4.4

Periodic digital signal.

Choosing the duration of one period, $N = 4$, we have the sample values as follows $x(0) = 0$; $x(1) = 1$; $x(2) = 0$; and $x(3) = -1$.

Using Eq. (4.5),

$$c_0 = \frac{1}{4}\sum_{n=0}^{3} x(n)e^{-j2\pi\times 0n/4} = \frac{1}{4}(x(0)+x(1)+x(2)+x(3))$$
$$= \frac{1}{4}(0+1+0-1) \qquad = 0$$
$$c_1 = \frac{1}{4}\sum_{k=0}^{3} x(n)e^{-j2\pi\times 1n/4} = \frac{1}{4}\left(x(0)+x(1)e^{-j\pi/2}+x(2)e^{-j\pi}+x(3)e^{-j3\pi/2}\right)$$
$$= \frac{1}{4}(x(0)-jx(1)-x(2)+jx(3) = 0-j(1)-0+j(-1)) = -0.5j$$

Similarly, we get

$$c_2 = \frac{1}{4}\sum_{n=0}^{3} x(n)e^{-j2\pi\times 2n/4} = \frac{1}{4}\left(x(0)+x(1)e^{-j2\pi/2}+x(2)e^{-j2\pi}+x(3)e^{-j6\pi/2}\right)$$
$$= \frac{1}{4}(x(0)-x(1)+x(2)-x(3)) = \frac{1}{4}(0-1+0-(-1)) = 0$$

$$c_3 = \frac{1}{4}\sum_{n=0}^{3} x(n)e^{-j2\pi\times 4n/4} = j0.5$$

Using periodicity, it follows that

$$c_{-1} = c_3 = j0.5 \text{ and } c_{-2} = c_2 = 0$$

(b) The amplitude spectrum for the digital signal is sketched in Fig. 4.5.

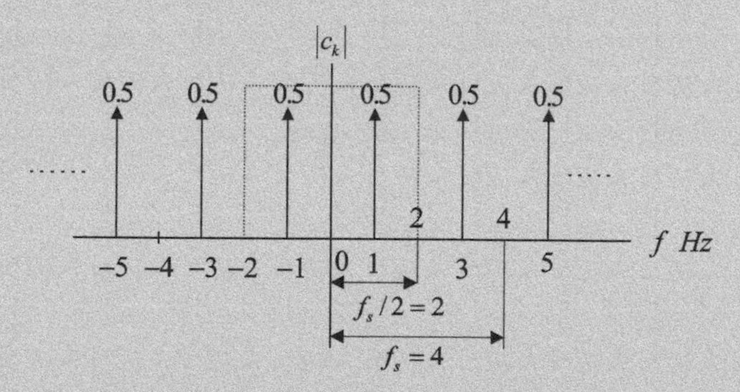

FIG. 4.5

Two-sided spectrum for the periodic digital signal in Example 4.1.

As we know, the spectrum in the range of −2 to 2 Hz presents the information of the sinusoid with a frequency of 1 Hz and a peak value of $2|c_1| = 1$, which is converted from two sided to one sided by doubling the spectral value. Note that we do not double the direct-current (DC) component, that is, c_0.

4.1.2 DISCRETE FOURIER TRANSFORM FORMULAS

Now, let us concentrate on the development of the DFT. Fig. 4.6 shows one way to obtain the DFT formula.

First, we assume that the process acquires data samples from digitizing the interested continuous signal for a duration of T_0 seconds. Next, we assume that a periodic signal $x(n)$ is obtained by copying the acquired N data samples with the duration of T_0 to itself repetitively. Note that we assume continuity between the N data sample frames. This is not true in practice. We will tackle this problem in Section 4.3. Finally, we determine the Fourier series coefficients using one-period N data samples and Eq. (4.5). Then we multiply the Fourier series coefficients by a factor of N to obtain

$$X(k) = Nc_k = \sum_{n=0}^{N-1} x(n)e^{-j2\pi kn/N}, \quad \text{for } k = 0, 1, \ldots, N-1, \tag{4.7}$$

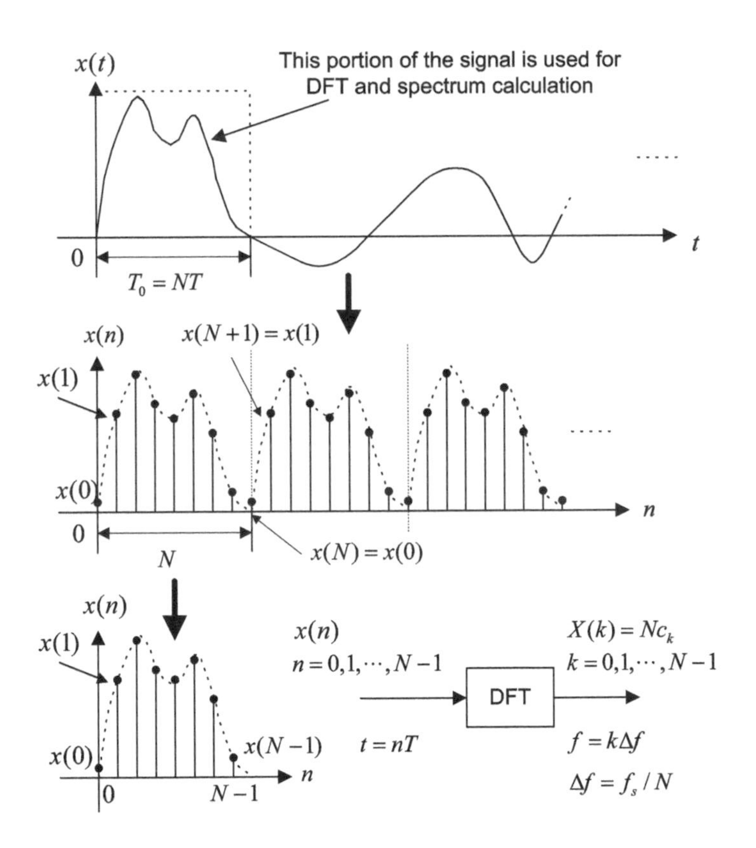

FIG. 4.6

Development of DFT formula.

where $X(k)$ constitutes the DFT coefficients. Note that the factor of N is a constant and does not affect the relative magnitudes of the DFT coefficients $X(k)$. As shown in the last plot, applying DFT with N data samples of $x(n)$ sampled at a sampling rate of f_s (sampling period is $T=1/f_s$) produces N complex DFT coefficients $X(k)$. The index n is the time index representing the sample number of the digital sequence, whereas k is the frequency index indicating each calculated DFT coefficient, and can be further mapped to the corresponding signal frequency in terms of Hz.

Now let us conclude the DFT definition. Given a sequence $x(n), 0 \le n \le N-1$, its DFT is defined as

$$X(k)=\sum_{n=0}^{N-1} x(n)e^{-j2\pi kn/N}=\sum_{n=0}^{N-1} x(n)W_N^{kn},\ \text{for } k=0,1,\cdots,N-1. \tag{4.8}$$

Eq. (4.8) can be expanded as

$$X(k)=x(0)W_N^{k0}+x(1)W_N^{k1}+x(2)W_N^{k2}+\cdots+x(N-1)W_N^{k(N-1)},\ \text{for } k=0,1,\cdots,N-1, \tag{4.9}$$

where the factor W_N (called the twiddle factor in some textbooks) is defined as

$$W_N=e^{-j2\pi/N}=\cos\left(\frac{2\pi}{N}\right)-j\sin\left(\frac{2\pi}{N}\right). \tag{4.10}$$

To determine the inverse of DFT, we can multiply both sides of Eq. (4.8) by $e^{j2\pi km/N}$ and sum from $k=0$ to $k=N-1$, that is

$$\sum_{k=0}^{N-1} X(k)e^{j\frac{2\pi km}{N}}=\sum_{k=0}^{N-1}\sum_{n=0}^{N-1} x(n)e^{-j\frac{2\pi kn}{N}}e^{j\frac{2\pi km}{N}}=\sum_{n=0}^{N-1} x(n)\sum_{k=0}^{N-1} e^{j\frac{2\pi k(m-n)}{N}}. \tag{4.11}$$

Note that

$$\sum_{k=0}^{N-1} e^{j\frac{2\pi k(m-n)}{N}}=\begin{cases} N & m=n \\ 0 & m\neq n \end{cases}. \tag{4.12}$$

With considering the following fact:

$$\sum_{k=0}^{N-1} e^{j\frac{2\pi k(m-n)}{N}}=\sum_{k=0}^{N-1}\left(e^{j\frac{2\pi(m-n)}{N}}\right)^k=\frac{1-e^{j2\pi(m-n)}}{1-e^{j\frac{2\pi(m-n)}{N}}}=\frac{1-1}{1-e^{j\frac{2\pi(m-n)}{N}}}=0.$$

Thus, Eq. (4.11) becomes

$$\sum_{k=0}^{N-1} X(k)e^{j\frac{2\pi km}{N}}=\sum_{n=0}^{N-1} x(n)\sum_{k=0}^{N-1} e^{j\frac{2\pi k(m-n)}{N}}. \tag{4.13}$$

Using Eq. (4.13), we see that

$$\text{for } m=0, \sum_{k=0}^{N-1} X(k)e^{j\frac{2\pi k0}{N}} = \sum_{n=0}^{N-1} x(n) \sum_{k=0}^{N-1} e^{j\frac{2\pi k(0-n)}{N}} = x(0)\sum_{k=0}^{N-1} 1 = Nx(0)$$

$$\text{for } m=1, \sum_{k=0}^{N-1} X(k)e^{j\frac{2\pi k1}{N}} = \sum_{n=0}^{N-1} x(n) \sum_{k=0}^{N-1} e^{j\frac{2\pi k(1-n)}{N}} = x(1)\sum_{k=0}^{N-1} 1 = Nx(1)$$

...

$$\text{for } m=N-1, \sum_{k=0}^{N-1} X(k)e^{j\frac{2\pi k(N-1)}{N}} = \sum_{n=0}^{N-1} x(n) \sum_{k=0}^{N-1} e^{j\frac{2\pi k(N-1-n)}{N}} = x(N-1)\sum_{k=0}^{N-1} 1 = Nx(N-1)$$

As a summary, we conclude

$$x(m) = \frac{1}{N}\sum_{k=0}^{N-1} X(k)e^{j\frac{2\pi km}{N}}. \tag{4.14}$$

The inverse of DFT is given by

$$x(n) = \frac{1}{N}\sum_{k=0}^{N-1} X(k)e^{j2\pi kn/N} = \frac{1}{N}\sum_{k=0}^{N-1} X(k)W_N^{-kn}, \quad \text{for } n=0,1,\cdots,N-1. \tag{4.15}$$

Proof can also be found in Ahmed and Natarajan (1983); Proakis and Manolakis (1996); Oppenheim et al. (1998); and Stearns and Hush (1990).

Similar to Eq. (4.8), the expansion of Eq. (4.15) leads to

$$x(n) = \frac{1}{N}\left(X(0)W_N^{-0n} + X(1)W_N^{-1n} + X(2)W_N^{-2n} + \cdots + X(N-1)W_N^{-(N-1)n}\right), \quad \text{for } n=0,1,\cdots,N-1. \tag{4.16}$$

As shown in Fig. 4.6, in time domain we use the sample number or time index n for indexing the digital sample sequence $x(n)$. However, in frequency domain, we use index k for indexing N calculated DFT coefficients $X(k)$. We also refer k as the frequency bin number in Eqs. (4.8) and (4.9).

We can use MATLAB functions **fft()** and **ifft()** to compute the DFT coefficients and the inverse DFT with the syntax given in Table 4.1.

The following examples serve to illustrate the application of DFT and the inverse of DFT.

Table 4.1 MATLAB FFT Functions

X = fft(x)	**% Calculate DFT coefficients**
x = ifft(X) x = input vector X = DFT coefficient vector	% Inverse of DFT

EXAMPLE 4.2

Given a sequence $x(n)$ for $0 \le n \le 3$, where $x(0)=1$, $x(1)=2$, $x(2)=3$, and $x(3)=4$, evaluate its DFT $X(k)$.

Solution:

Since $N=4$ and $W_4 = e^{-j\frac{\pi}{2}}$, using Eq. (4.8) we have a simplified formula

$$X(k) = \sum_{n=0}^{3} x(n) W_4^{kn} = \sum_{n=0}^{3} x(n) e^{-j\frac{\pi kn}{2}}.$$

Thus, for $k=0$

$$X(0) = \sum_{n=0}^{3} x(n) e^{-j0} = x(0)e^{-j0} + x(1)e^{-j0} + x(2)e^{-j0} + x(3)e^{-j0}$$

$$= x(0) + x(1) + x(2) + x(3)$$

$$= 1 + 2 + 3 + 4 = 10$$

for $k=1$

$$X(1) = \sum_{n=0}^{3} x(n) e^{-j\frac{\pi n}{2}} = x(0)e^{-j0} + x(1)e^{-j\frac{\pi}{2}} + x(2)e^{-j\pi} + x(3)e^{-j\frac{3\pi}{2}}$$

$$= x(0) - jx(1) - x(2) + jx(3)$$

$$= 1 - j2 - 3 + j4 = -2 + j2$$

for $k=2$

$$X(2) = \sum_{n=0}^{3} x(n) e^{-j\pi n} = x(0)e^{-j0} + x(1)e^{-j\pi} + x(2)e^{-j2\pi} + x(3)e^{-j3\pi}$$

$$= x(0) - x(1) + x(2) - x(3)$$

$$= 1 - 2 + 3 - 4 = -2$$

and for $k=3$

$$X(3) = \sum_{n=0}^{3} x(n) e^{-j\frac{3\pi n}{2}} = x(0)e^{-j0} + x(1)e^{-j\frac{3\pi}{2}} + x(2)e^{-j3\pi} + x(3)e^{-j\frac{9\pi}{2}}$$

$$= x(0) + jx(1) - x(2) - jx(3)$$

$$= 1 + j2 - 3 - j4 = -2 - j2$$

Let us verify the result using the MATLAB function **fft()**:

```
>>X=fft([1234])
X=10.0000 -2.0000+2.0000i -2.0000 -2.0000-2.0000i
```

EXAMPLE 4.3

Using the DFT coefficients $X(k)$ for $0 \leq k \leq 3$ computed in Example 4.2, evaluate its inverse DFT to determine the time domain sequence $x(n)$.

Solution:

Since $N=4$ and $W_4^{-1} = e^{j\frac{\pi}{2}}$, using Eq. (4.15) we achieve a simplified formula

$$x(n) = \frac{1}{4}\sum_{k=0}^{3} X(k)W_4^{-nk} = \frac{1}{4}\sum_{k=0}^{3} X(k)e^{j\frac{\pi kn}{2}}.$$

Then for $n=0$

$$x(0) = \frac{1}{4}\sum_{k=0}^{3} X(k)e^{j0} = \frac{1}{4}\left(X(0)e^{j0} + X(1)e^{j0} + X(2)e^{j0} + X(3)e^{j0}\right)$$
$$= \frac{1}{4}(10 + (-2+j2) - 2 + (-2-j2)) = 1$$

for $n=1$

$$x(1) = \frac{1}{4}\sum_{k=0}^{3} X(k)e^{j\frac{k\pi}{2}} = \frac{1}{4}\left(X(0)e^{j0} + X(1)e^{j\frac{\pi}{2}} + X(2)e^{j\pi} + X(3)e^{j\frac{3\pi}{2}}\right)$$
$$= \frac{1}{4}(X(0) + jX(1) - X(2) - jX(3))$$
$$= \frac{1}{4}(10 + j(-2+j2) - (-2) - j(-2-j2)) = 2$$

for $n=2$

$$x(2) = \frac{1}{4}\sum_{k=0}^{3} X(k)e^{jk\pi} = \frac{1}{4}\left(X(0)e^{j0} + X(1)e^{j\pi} + X(2)e^{j2\pi} + X(3)e^{j3\pi}\right)$$
$$= \frac{1}{4}(X(0) - X(1) + X(2) - X(3))$$
$$= \frac{1}{4}(10 - (-2+j2) + (-2) - (-2-j2)) = 3$$

and for $n=3$

$$x(3) = \frac{1}{4}\sum_{k=0}^{3} X(k)e^{j\frac{k\pi 3}{2}} = \frac{1}{4}\left(X(0)e^{j0} + X(1)e^{j\frac{3\pi}{2}} + X(2)e^{j3\pi} + X(3)e^{j\frac{9\pi}{2}}\right)$$
$$= \frac{1}{4}(X(0) - jX(1) - X(2) + jX(3))$$
$$= \frac{1}{4}(10 - j(-2+j2) - (-2) + j(-2-j2)) = 4$$

This example actually verifies the inverse of DFT. Applying the MATLAB function **ifft()** achieves:

```
>> x = ifft([10 -2+2j -2 -2-2j])
x = 1 2 3 4
```

Now we explore the relationship between the frequency bin k and its associated frequency. Omitting the proof, the calculated N DFT coefficients $X(k)$ represent the frequency components ranging from 0 Hz (or rad/s) to f_s Hz (or ω_s rad/s), hence we can map the frequency bin k to its corresponding frequency as follows:

$$\omega = \frac{k\omega_s}{N} \text{ (rad/s)}, \quad (4.17)$$

or in terms of Hz,

$$f = \frac{kf_s}{N} \text{ (Hz)}, \quad (4.18)$$

where $\omega_s = 2\pi f_s$.

We can define the frequency resolution as the frequency step between two consecutive DFT coefficients to measure how fine the frequency domain presentation is and achieve

$$\Delta\omega = \frac{\omega_s}{N} \text{ (rad/s)}, \quad (4.19)$$

or in terms of Hz, it follows that

$$\Delta f = \frac{f_s}{N} \text{ (Hz)}. \quad (4.20)$$

Let us study the following example.

EXAMPLE 4.4

In Example 4.2, given a sequence $x(n)$ for $0 \le n \le 3$, where $x(0)=1, x(1)=2, x(2)=3$, and $x(3)=4$, we have computed four DFT coefficients $X(k)$ for $0 \le k \le 3$ as $X(0)=10$, $X(1)=-2+j2$, $X(2)=-2$, and $X(3)=-2-j2$. If the sampling rate is 10 Hz,

(a) Determine the sampling period, time index, and sampling time instant for a digital sample $x(3)$ in time domain.

(b) Determine the frequency resolution, frequency bin, and mapped frequencies for each of the DFT coefficients $X(1)$ and $X(3)$ in frequency domain.

Solution:

(a) In time domain, we have the sampling period calculated as

$$T = \frac{1}{f_s} = \frac{1}{10} = 0.1 \text{ s}.$$

For data $x(3)$, the time index is $n=3$ and the sampling time instant is determined by

$$t = nT = 3 \times 0.1 = 0.3 \text{ s}.$$

(b) In frequency domain, since the total number of DFT coefficients is four, the frequency resolution is determined by

$$\Delta f = \frac{f_s}{N} = \frac{10}{4} = 2.5 \text{ Hz}.$$

Continued

EXAMPLE 4.4—CONT'D

The frequency bin for $X(1)$ should be $k=1$ and its corresponding frequency is determined by

$$f=\frac{kf_s}{N}=\frac{1\times 10}{4}=2.5\,\text{Hz}.$$

Similarly, for $X(3)$ and $k=3$,

$$f=\frac{kf_s}{N}=\frac{3\times 10}{4}=7.5\,\text{Hz}.$$

Note that from Eq. (4.4), $k=3$ is equivalent to $k-N=3-4=-1$, and $f=7.5$ Hz is also equivalent to the frequency $f=(-1\times 10)/4=-2.5$ Hz, which corresponds to the negative side spectrum. The amplitude spectrum at 7.5 Hz after folding should match the one at $f_s-f=10.0-7.5=2.5$ Hz. We will apply these developed notations in the following section for amplitude and power spectral estimation.

4.2 AMPLITUDE SPECTRUM AND POWER SPECTRUM

One of the DFT applications is transformation of a finite-length digital signal $x(n)$ into the spectrum in frequency domain. Fig. 4.7 demonstrates such an application, where A_k and P_k are the computed amplitude spectrum and the power spectrum, respectively, using the DFT coefficients $X(k)$.

First, we achieve the digital sequence $x(n)$ by sampling the analog signal $x(t)$ and truncating the sampled signal with a data window of a length $T_0=NT$, where T is the sampling period and N is the number of data points. The time for data window is

$$T_0=NT. \tag{4.21}$$

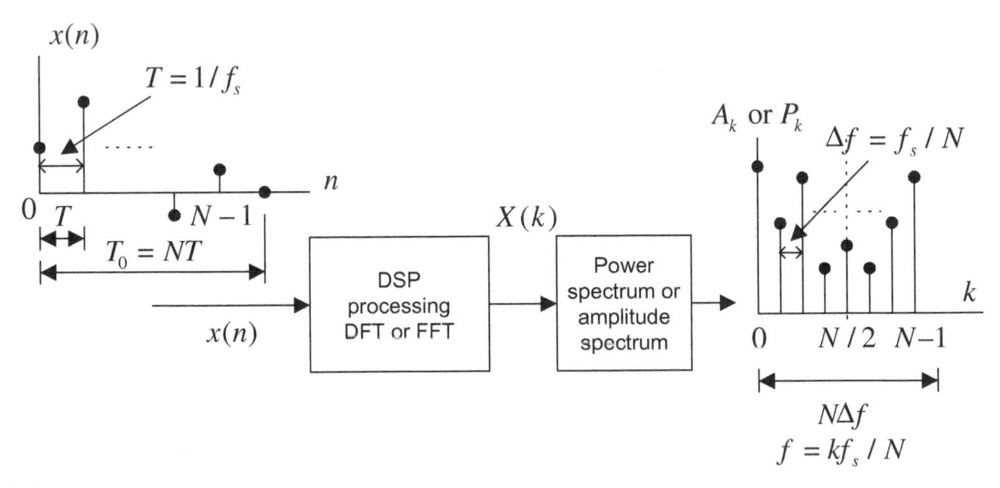

FIG. 4.7

Applications of DFT/FFT.

For the truncated sequence $x(n)$ with a range of $n=0, 1, 2, \cdots, N-1$, we get

$$x(0), x(1), x(2), \cdots, x(N-1). \tag{4.22}$$

Next, we apply the DFT to the obtained sequence, $x(n)$, to get the N DFT coefficients

$$X(k) = \sum_{n=0}^{N-1} x(n) W_N^{nk}, \text{ for } k = 0, 1, 2, \cdots, N-1. \tag{4.23}$$

Since each calculated DFT coefficient is a complex number, it is not convenient to plot it vs. its frequency index. Hence, after evaluating Eq. (4.23), the magnitude and phase of each DFT coefficient (we refer them as the amplitude spectrum and phase spectrum, respectively) can be determined and plotted vs. its frequency index. We define the amplitude spectrum as

$$A_k = \frac{1}{N}|X(k)| = \frac{1}{N}\sqrt{(\text{Real}[X(k)])^2 + (\text{Imag}[X(k)])^2}, \ k = 0, 1, 2, \cdots, N-1. \tag{4.24}$$

We can modify the amplitude spectrum to a one-sided amplitude spectrum by doubling the amplitudes in Eq. (4.24), keeping the original DC term at $k=0$. Thus we have

$$\overline{A}_k = \begin{cases} \frac{1}{N}|X(0)|, & k=0 \\ \frac{2}{N}|X(k)|, & k=1, \cdots, N/2 \end{cases}. \tag{4.25}$$

We can also map the frequency bin k to its corresponding frequency as

$$f = \frac{kf_s}{N}. \tag{4.26}$$

Correspondingly, the phase spectrum is given by

$$\varphi_k = \tan^{-1}\left(\frac{\text{Imag}[X(k)]}{\text{Real}[X(k)]}\right), k = 0, 1, 2, \cdots, N-1. \tag{4.27}$$

Besides the amplitude spectrum, the power spectrum is also used. The DFT power spectrum is defined as

$$P_k = \frac{1}{N^2}|X(k)|^2 = \frac{1}{N^2}\left\{(\text{Real}[X(k)])^2 + (\text{Imag}[X(k)])^2\right\}, \ k = 0, 1, 2, \cdots, N-1. \tag{4.28}$$

Similarly, for a one-sided power spectrum, we get

One-sided power spectrum

$$\overline{P}_k = \begin{cases} \frac{1}{N^2}|X(0)|^2 & k=0 \\ \frac{2}{N^2}|X(k)|^2 & k=1, \cdots, N/2 \end{cases} \tag{4.29}$$

$$\text{and } f = \frac{kf_s}{N}. \tag{4.30}$$

Again, note that the frequency resolution, which denotes the frequency spacing between DFT coefficients in frequency domain, is defined as

$$\Delta f = \frac{f_s}{N} \ (\text{Hz}). \tag{4.31}$$

It follows that better frequency resolution can be achieved using a longer data sequence.

EXAMPLE 4.5

Consider the sequence in Fig. 4.8

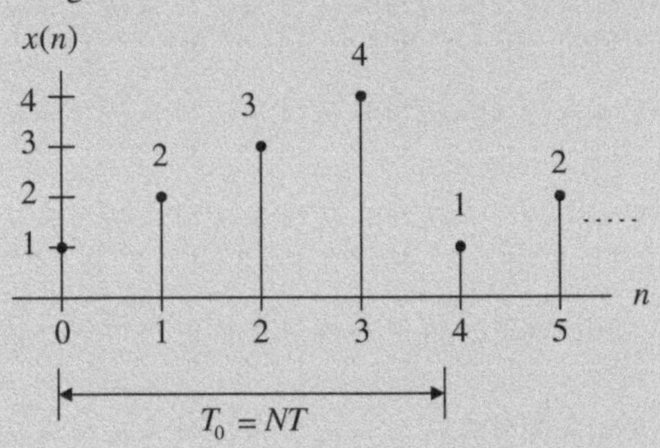

FIG. 4.8

Sampled values in Example 4.5.

Assuming that $f_s = 100$ Hz, compute the amplitude spectrum, phase spectrum, and power spectrum.

Solution:

Since $N = 4$, and using the DFT shown in Example 4.2, we find the DFT coefficients to be

$$X(0) = 10$$
$$X(1) = -2 + j2$$
$$X(2) = -2$$
$$X(3) = -2 - j2.$$

The amplitude spectrum, phase spectrum, and power density spectrum are computed as follows:

$$\text{for } k = 0, f = k \cdot f_s/N = 0 \times 100/4 = 0\,\text{Hz},$$
$$A_0 = \frac{1}{4}|X(0)| = 2.5, \varphi_0 = \tan^{-1}\left(\frac{\text{Imag}[X(0)]}{\text{Real}([X(0)]}\right) = 0^0, P_0 = \frac{1}{4^2}|X(0)|^2 = 6.25$$
$$\text{for } k = 1, f = 1 \times 100/4 = 25\,\text{Hz},$$
$$A_1 = \frac{1}{4}|X(1)| = 0.7071, \varphi_1 = \tan^{-1}\left(\frac{\text{Imag}[X(1)]}{\text{Real}[X(1)]}\right) = 135^0, P_1 = \frac{1}{4^2}|X(1)|^2 = 0.5000$$
$$\text{for } k = 2, f = 2 \times 100/4 = 50\,\text{Hz},$$
$$A_2 = \frac{1}{4}|X(2)| = 0.5, \varphi_2 = \tan^{-1}\left(\frac{\text{Imag}[X(2)]}{\text{Real}[X(2)]}\right) = 180^0, P_2 = \frac{1}{4^2}|X(2)|^2 = 0.2500$$

Similarly,

$$\text{for } k = 3, f = 3 \times 100/4 = 75\,\text{Hz},$$
$$A_3 = \frac{1}{4}|X(3)| = 0.7071, \varphi_3 = \tan^{-1}\left(\frac{\text{Imag}[X(3)]}{\text{Real}[X(3)]}\right) = -135^0, P_3 = \frac{1}{4^2}|X(3)|^2 = 0.5000.$$

Thus, the sketches for the amplitude spectrum, phase spectrum, and power spectrum are given in Fig. 4.9.

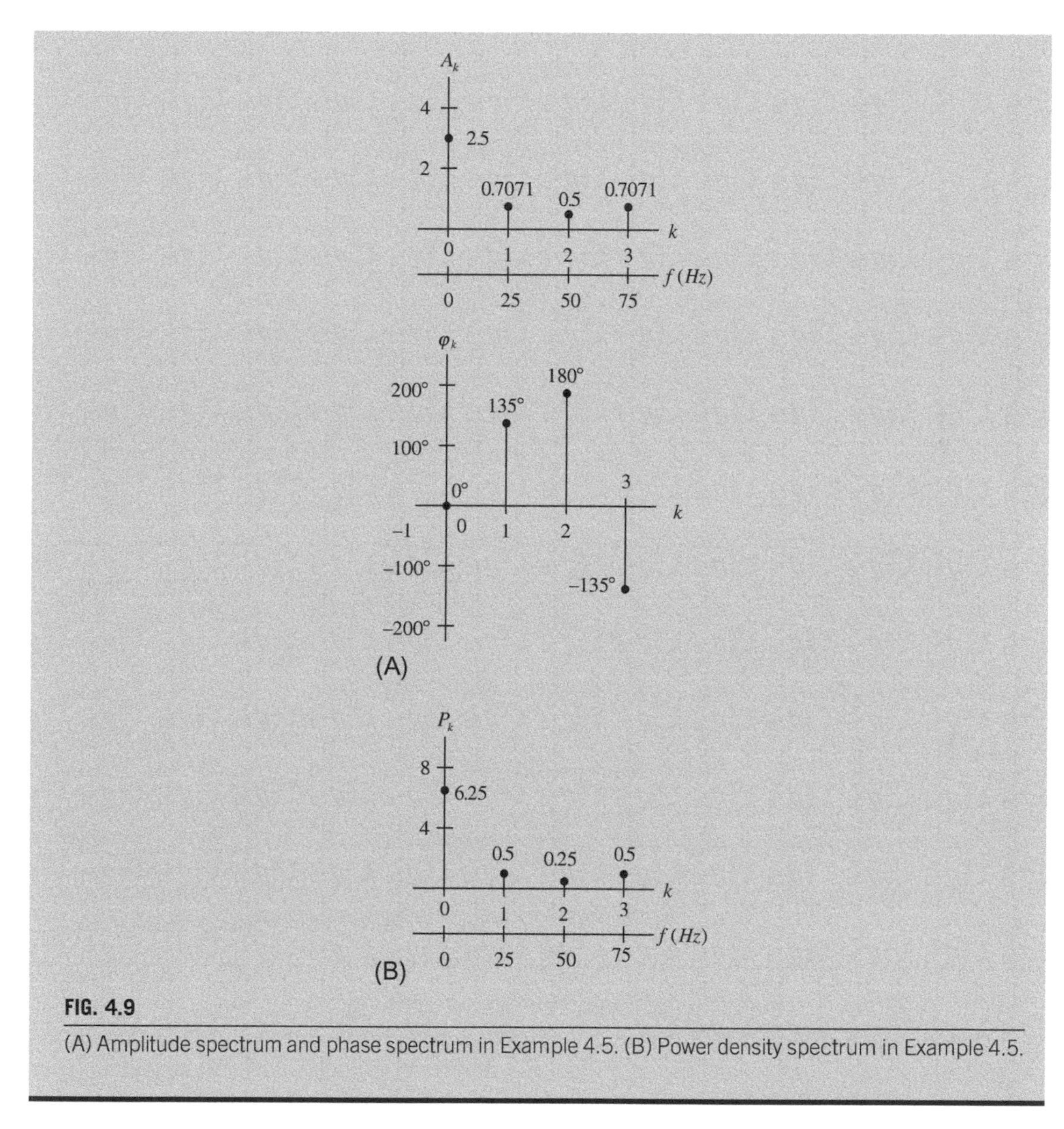

FIG. 4.9

(A) Amplitude spectrum and phase spectrum in Example 4.5. (B) Power density spectrum in Example 4.5.

Note that the folding frequency in this example is 50 Hz and the amplitude and power spectrum values at 75 Hz are each image counterparts (corresponding negative-indexed frequency components), respectively. Thus, values at 0, 25, and 50 Hz are corresponding to the positive-indexed frequency components.

We can easily find the one-sided amplitude spectrum and one-sided power spectrum as

$$\overline{A}_0 = 2.5, \overline{A}_1 = 1.4141, \overline{A}_2 = 1 \text{ and}$$
$$\overline{P}_0 = 6.25, \overline{P}_1 = 2, \overline{P}_2 = 1.$$

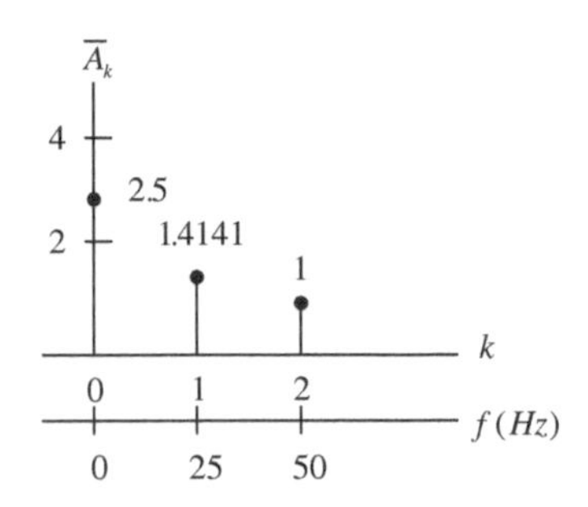

FIG. 4.10

One-sided amplitude spectrum in Example 4.5.

We plot the one-sided amplitude spectrum for comparison as shown in Fig. 4.10.

Note that in the one-sided amplitude spectrum, the negative-indexed frequency components are added back to the corresponding positive-indexed frequency components; thus each amplitude value other than DC term is doubled. It represents the frequency components up to the folding frequency.

EXAMPLE 4.6

Consider a digital sequence sampled at the rate of 10 kHz. If we use a size of 1024 data points and apply the 1024-point DFT to compute the spectrum,

(a) Determine the frequency resolution.

(b) Determine the highest frequency in the spectrum.

Solution:

(a)

$$\Delta f = \frac{f_s}{N} = \frac{10,000}{1024} = 9.776\text{Hz}.$$

(b) The highest frequency is the folding frequency, given by

$$f_{\max} = \frac{N}{2}\Delta f = \frac{f_s}{2}$$
$$= 512 \times 9.776 = 5000\,\text{Hz}.$$

As shown in Fig. 4.7, the DFT coefficients may be computed via a *fast Fourier transform* (FFT) algorithm (see Section 4.5). The FFT is a very efficient algorithm for computing DFT coefficients. In this book, we only focus on the FFT algorithm which requires the time domain sequence $x(n)$ with a length of data points equal to a power of 2; that is, 2^m samples, where m is a positive integer. For example, the number of samples in $x(n)$ can be $N=2, 4, 8, 16$, etc.

In the case of using the FFT algorithm to compute DFT coefficients, where the length of the available data is not equal to a power of 2 (required by the FFT), we can pad the data sequence with zeros to create a new sequence with a larger number of samples, $\overline{N} = 2^m > N$. The modified data sequence for applying FFT, therefore, is

$$\overline{x}(n) = \begin{cases} x(n) & 0 \le n \le N-1 \\ 0 & N \le n \le \overline{N}-1 \end{cases}. \tag{4.32}$$

It is very important to note that the signal spectra obtained via zero padding the data sequence in Eq. (4.32) does not add any new information and does not contain more accurate signal spectral presentation. In this situation, the frequency spacing is reduced due to more DFT points, and the achieved spectrum is a interpolated version with "better display." We illustrate the zero-padding effect via the following example instead of theoretical analysis. A theoretical discussion of zero padding in FFT can be found in Proakis and Manolakis (1996).

Fig. 4.11A shows the 12 data samples from an analog signal containing frequencies of 10 and 25 Hz at a sampling rate of 100 Hz, and the amplitude spectrum obtained by applying the DFT. Fig. 4.11B displays the signal samples with padding of four zeros to the original data to make up a data sequence of 16 samples, along with the amplitude spectrum calculated by FFT. The data sequence padded with 20 zeros and its calculated amplitude spectrum using FFT are shown in Fig. 4.11C. It is evident that increasing the data length via zero padding to compute the signal spectrum does not add basic information and does not change the spectral shape but gives the "interpolated spectrum" with the reduced frequency spacing. We can get a better view of the two spectral peaks described in this case.

The only way to obtain the detailed signal spectrum with a fine frequency resolution is to apply more available data samples, that is, a longer sequence of data. Here, we choose to pad the least number of zeros to satisfy the minimum FFT computational requirement. Let us look at another example.

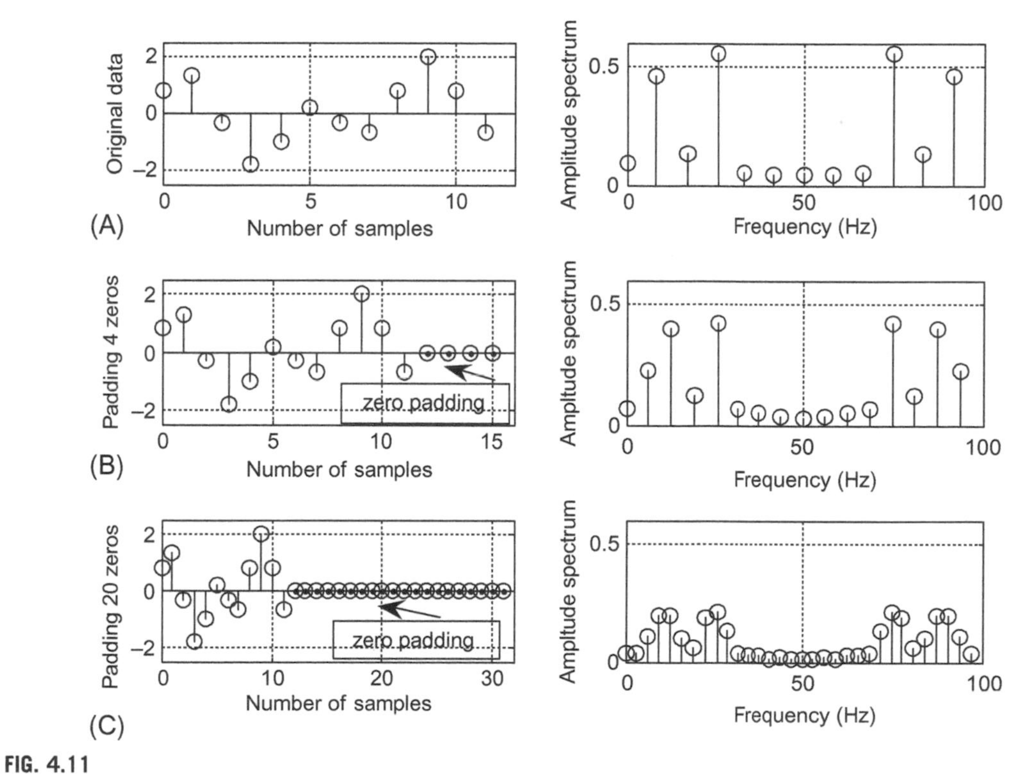

FIG. 4.11

Zero-padding effect by using FFT.

EXAMPLE 4.7

We use the DFT to compute the amplitude spectrum of a sampled data sequence with a sampling rate $f_s = 10\,\text{kHz}$. Given that it requires the frequency resolution to be less than 0.5 Hz, determine the number of data points by using the FFT algorithm, assuming that the data samples are available.

Solution:

$$\Delta f = 0.5\,\text{Hz}$$
$$N = \frac{f_s}{\Delta f} = \frac{10{,}000}{0.5} = 20{,}000$$

Since we use the FFT to compute the spectrum, the number of the data points must be a power of 2, that is,

$$N = 2^{15} = 32{,}768.$$

and the resulting frequency resolution can be recalculated as

$$\Delta f = \frac{f_s}{N} = \frac{10{,}000}{32{,}768} = 0.31\,\text{Hz}.$$

Next, we study a MATLAB example.

EXAMPLE 4.8

Given a sinusoid

$$x(n) = 2 \times \sin\left(2000\pi \frac{n}{8000}\right)$$

obtained by sampling the analog signal

$$x(t) = 2 \times \sin(2000\pi t)$$

with a sampling rate of $f_s = 8000\text{Hz}$,

(a) Use the MATLAB DFT to compute the signal spectrum with the frequency resolution to be equal to or less than 8 Hz.

(b) Use the MATALB FFT and zero padding to compute the signal spectrum, assuming that the data samples are available in (a).

Solution:

(a) The number of data points is found to be

$$N = \frac{f_s}{\Delta f} = \frac{8000}{8} = 1000.$$

There is no zero padding needed if we use the DFT formula. The detailed implementation is given in Program 4.1. The first and second plots in Fig. 4.12 show the two-sided amplitude and power spectra, respectively, using the DFT, where each frequency counterpart at 7000 Hz appears. The third and fourth plots are the one-sided amplitude and power spectra, where the true frequency contents are displayed from 0 Hz to the Nyquist frequency of 4 kHz (folding frequency).

(b) If the FFT is used, the number of data points must be a power of 2. Hence we choose

$$N = 2^{10} = 1024.$$

Assuming that there are only 1000 data samples available in (a), we need to pad 24 zeros to the original 1000 data samples before applying the FFT algorithm, as required.

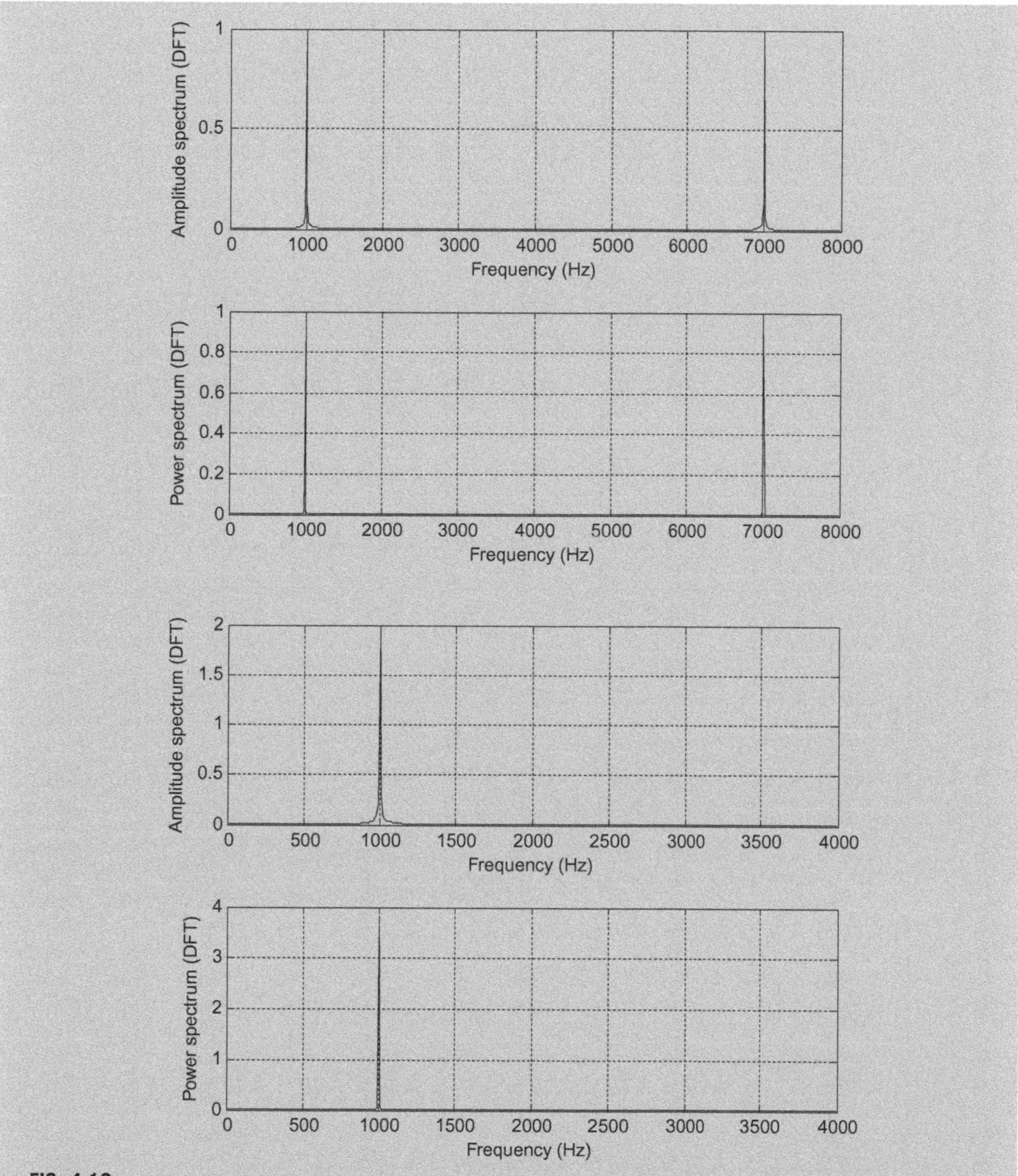

FIG. 4.12

Amplitude spectrum and power spectrum using DFT for Example 4.8.

Thus the calculated frequency resolution is $\Delta f = f_s/N = 8000/1024 = 7.8125$ Hz. Note that this is an interpolated frequency resolution by using zero padding. The zero padding actually interpolates a signal spectrum and carries no additional frequency information. Fig. 4.13 shows the spectral plots using FFT. The detailed implementation is given in Program 4.1.

Continued

EXAMPLE 4.8—CONT'D

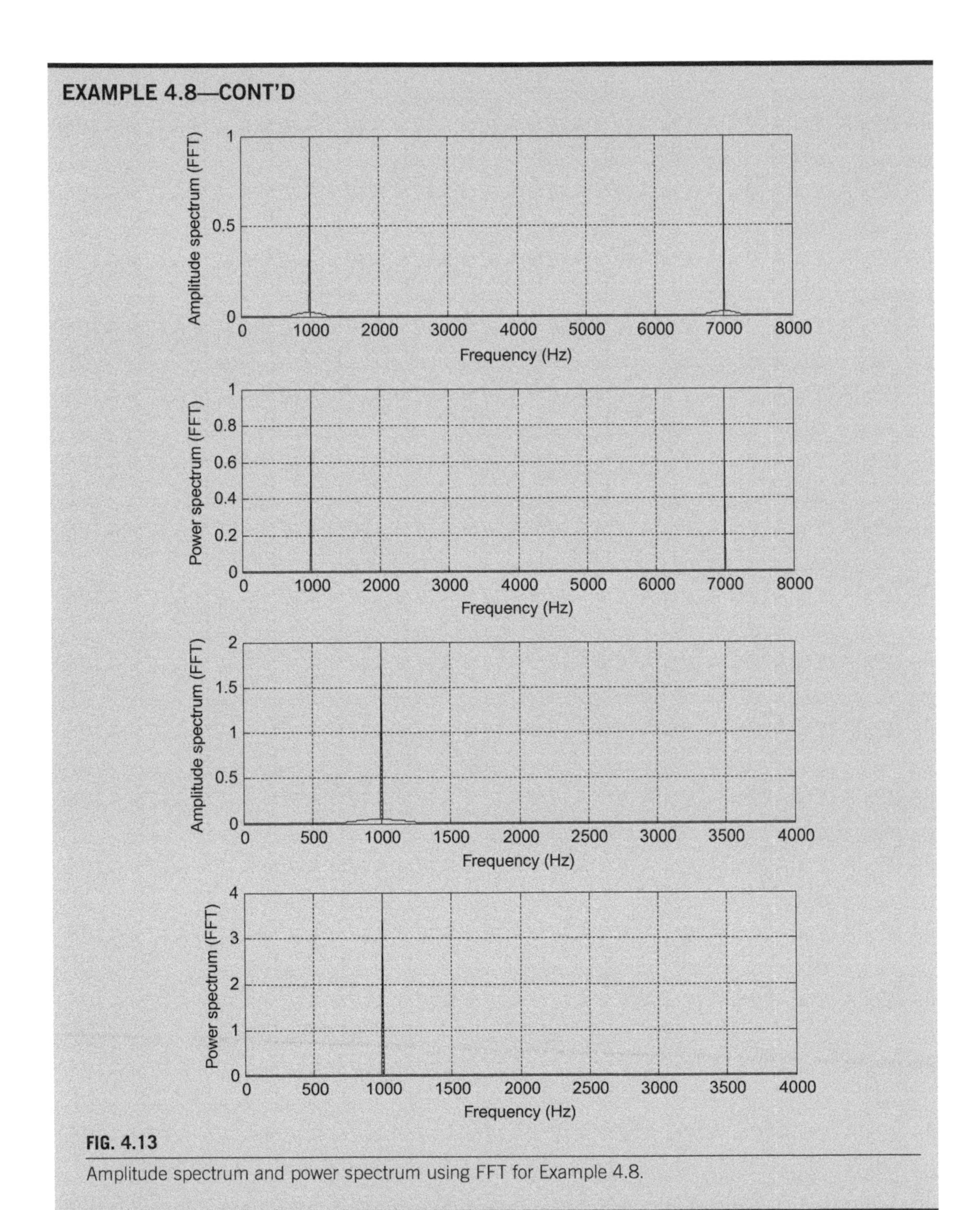

FIG. 4.13

Amplitude spectrum and power spectrum using FFT for Example 4.8.

Program 4.1 MATLAB program for Example 4.8

```
% Example 4.8
close all;clear all
% Generate the sine wave sequence
fs=8000;        %Sampling rate
N=1000;       % Number of data points
x=2*sin(2000*pi*[0:1:N-1]/fs);
% Apply the DFT algorithm
figure(1)
xf=abs(fft(x))/N;        %Compute the amplitude spectrum
P=xf.*xf;           %Compute power spectrum
f=[0:1:N-1]*fs/N;        %Map the frequency bin to frequency (Hz)
subplot(2,1,1); plot(f,xf );grid
xlabel('Frequency (Hz)'); ylabel('Amplitude spectrum (DFT)');
subplot(2,1,2);plot(f,P);grid
xlabel('Frequency (Hz)'); ylabel('Power spectrum (DFT)');
figure(2)
% Convert it to one side spectrum
xf(2:N)=2*xf(2:N);      % Get the single-side spectrum
P=xf.*xf;       % Calculate the power spectrum
f=[0:1:N/2]*fs/N      % Frequencies up to the folding frequency
subplot(2,1,1); plot(f,xf(1:N/2+1));grid
xlabel('Frequency (Hz)'); ylabel('Amplitude spectrum (DFT)');
subplot(2,1,2);plot(f,P(1:N/2+1));grid
xlabel('Frequency (Hz)'); ylabel('Power spectrum (DFT)');
figure (3)
% Zero padding to the length of 1024
x=[x,zeros(1,24)];
N=length(x);
xf=abs(fft(x))/N;      %Compute amplitude spectrum with zero padding
P=xf.*xf;         %Compute power spectrum
f=[0:1:N-1]*fs/N;      %Map frequency bin to frequency (Hz)
subplot(2,1,1); plot(f,xf );grid
xlabel('Frequency (Hz)'); ylabel('Amplitude spectrum (FFT)');
subplot(2,1,2);plot(f,P);grid
xlabel('Frequency (Hz)'); ylabel('Power spectrum (FFT)');
figure(4)
% Convert it to one side spectrum
xf(2:N)=2*xf(2:N);
P=xf.*xf;
f=[0:1:N/2]*fs/N;
subplot(2,1,1); plot(f,xf(1:N/2+1));grid
xlabel('Frequency (Hz)'); ylabel('Amplitude spectrum (FFT)');
subplot(2,1,2);plot(f,P(1:N/2+1));grid
xlabel('Frequency (Hz)'); ylabel('Power spectrum (FFT)');
```

4.3 SPECTRAL ESTIMATION USING WINDOW FUNCTIONS

When we apply DFT to the sampled data in the previous section, we theoretically imply the following assumptions: first, the sampled data are periodic to themselves (repeat themselves), and second, the sampled data is continuous to themselves and band limited to the folding frequency. The second assumption is often violated, thus the discontinuity produces undesired harmonic frequencies. Consider a pure 1-Hz sine wave with 32 samples shown in Fig. 4.14.

As shown in the figure, if we use a window size of $N=16$ samples, which is multiple of the two waveform cycles, the second window repeats with continuity. However, when the window size is chosen to be 18 samples, which is not multiple of the waveform cycles (2.25 cycles), the second window repeats the first window with discontinuity. It is this discontinuity that produces harmonic frequencies that are not present in the original signal. Fig. 4.15 shows the spectral plots for both cases using the DFT/FFT directly.

The first spectral plot contains a single frequency component, as we expected, while the second spectrum has the expected frequency component plus many harmonics, which do not exist in the original signal. We called such an effect *spectral leakage*. The amount of spectral leakage shown in the second plot is due to amplitude discontinuity in time domain. The bigger the discontinuity, the more the leakage. To reduce the effect of spectral leakage, a window function can be used whose amplitude tapers smoothly and gradually toward zero at both ends. Applying the window function $w(n)$ to a data sequence $x(n)$ to obtain the windowed sequence $x_w(n)$ is better illustrated in Fig. 4.16 using Eq. (4.33):

$$x_w(n) = x(n)w(n), \text{ for } n = 0, 1, \cdots, N-1. \tag{4.33}$$

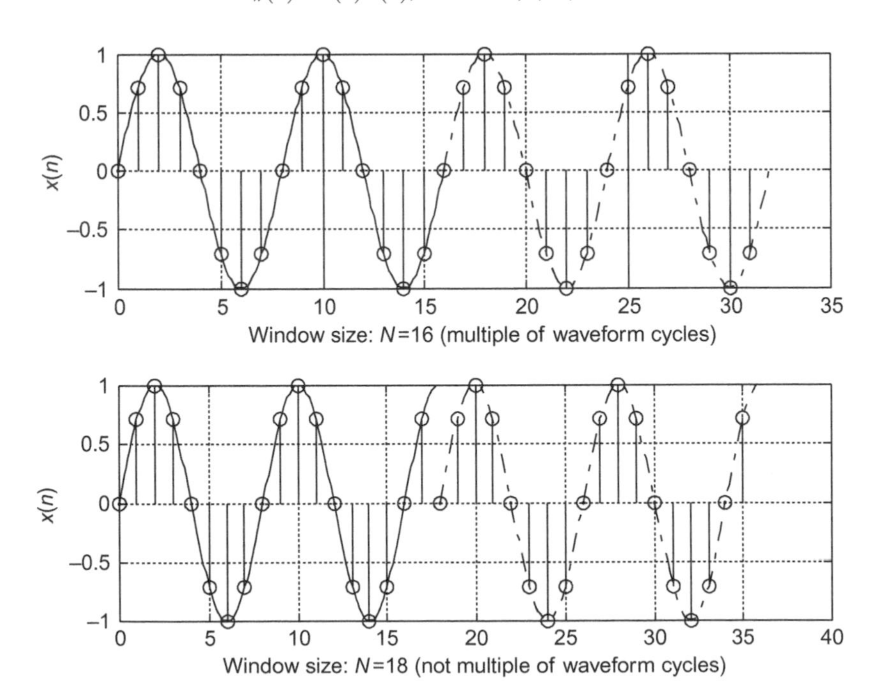

FIG. 4.14

Sampling a 1-Hz sine wave using (top) 16 samples per cycle and (bottom) 18 samples per cycle (the sampling interval: $T=1/8$ s).

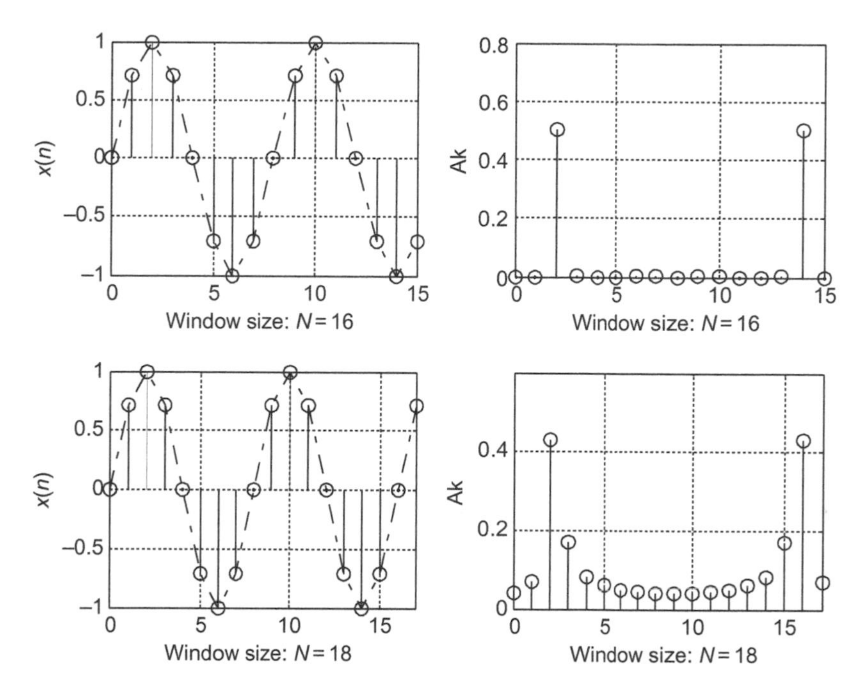

FIG. 4.15

Signal samples and spectra without spectral leakage and with spectral leakage.

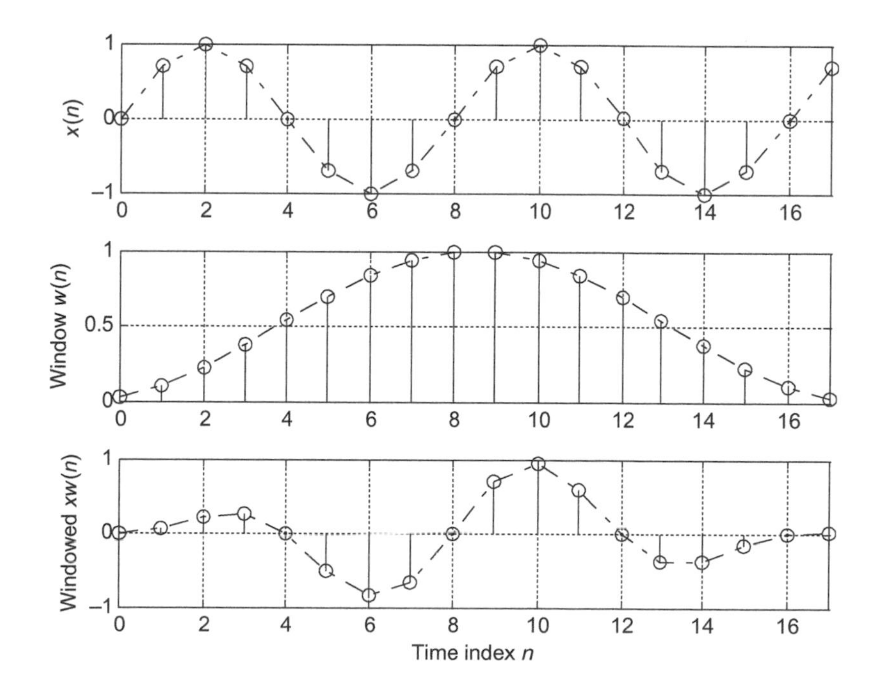

FIG. 4.16

Illustration of the window operation.

The top plot is the data sequence $x(n)$, and the middle plot is the window function $w(n)$. The bottom plot in Fig. 4.16 shows that the windowed sequence $x_w(n)$ is tapped down by a window function to zero at both ends such that the discontinuity is dramatically reduced.

EXAMPLE 4.9

In Fig. 4.16, given

- $x(2)=1$ and $w(2)=0.2265$;
- $x(5)=-0.7071$ and $w(5)=0.7008$,

calculate the windowed sequence data points $x_w(2)$ and $x_w(5)$.

Solution:

Applying the window function operation leads to

$$x_w(2)=x(2)\times w(2)=1\times 0.2265=0.2265 \text{ and}$$
$$x_w(5)=x(5)\times w(5)=-0.7071\times 0.7008=-0.4956,$$

which agree with the values shown in the bottom plot in Fig. 4.16.

Using the windowed function shown in Example 4.9, the spectral plot is reproduced. As a result, the spectral leakage is greatly reduced, as shown in Fig. 4.17.

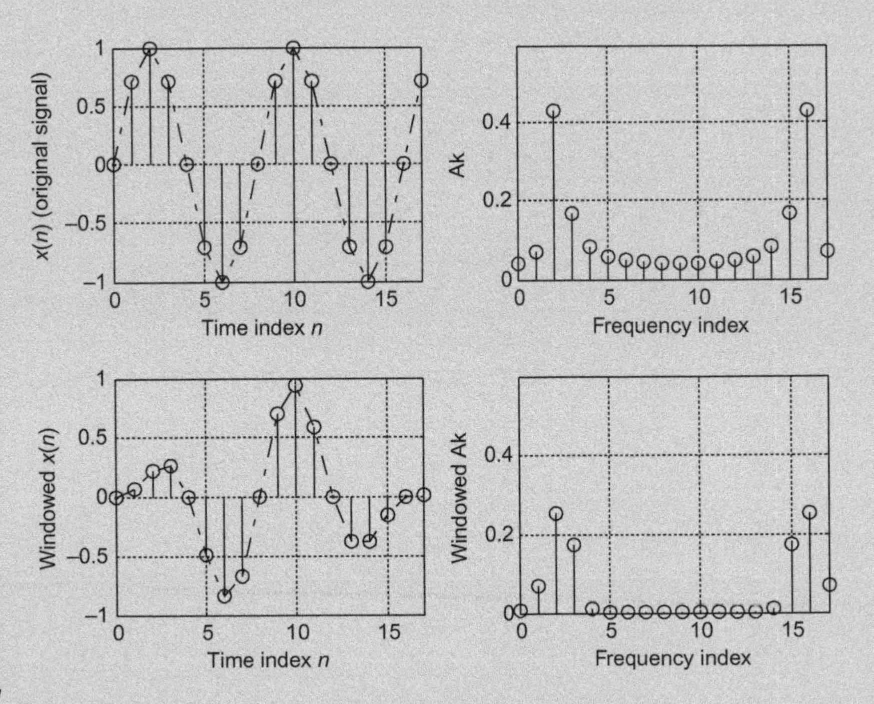

FIG. 4.17

Comparison of spectra calculated without using a window function and using a window function to reduce spectral leakage.

The common window functions are listed as follows:
The rectangular window (no window function):

$$w_R(n) = 1, 0 \leq n \leq N-1. \tag{4.34}$$

The triangular window:

$$w_{tri}(n) = 1 - \frac{|2n-N+1|}{N-1}, 0 \leq n \leq N-1. \tag{4.35}$$

The Hamming window:

$$w_{hm}(n) = 0.54 - 0.46\cos\left(\frac{2\pi n}{N-1}\right), 0 \leq n \leq N-1. \tag{4.36}$$

The Hanning window:

$$w_{hn}(n) = 0.5 - 0.5\cos\left(\frac{2\pi n}{N-1}\right), 0 \leq n \leq N-1. \tag{4.37}$$

Plots for each window function for a size of 20 samples are shown in Fig. 4.18.

The following example details each step for computing the spectral information using the window functions.

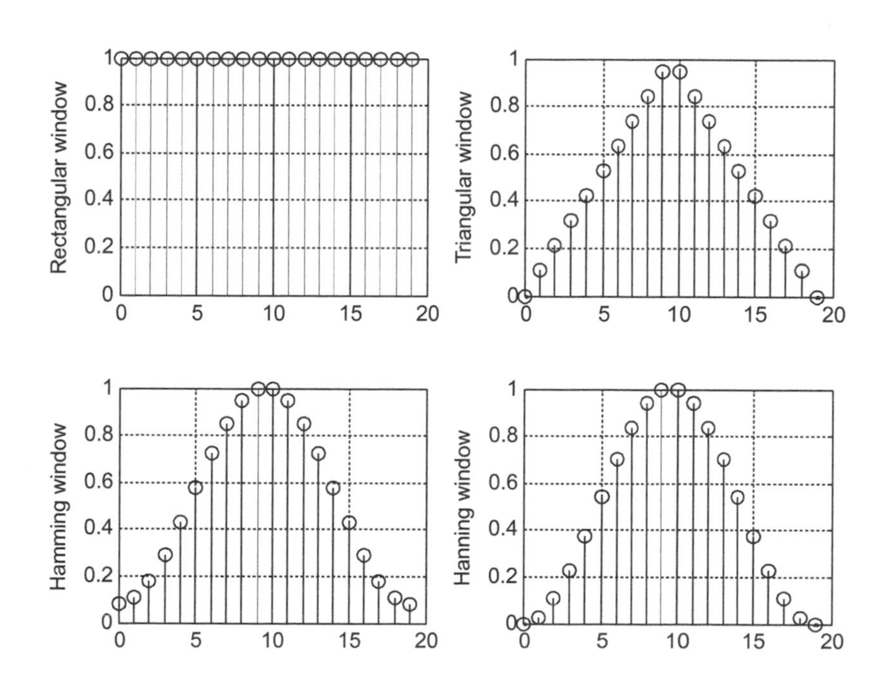

FIG. 4.18

Plots of window sequences.

EXAMPLE 4.10

Considering the sequence $x(0)=1$, $x(1)=2$, $x(2)=3$, $x(3)=4$, and given $f_s=100\,\text{Hz}$, $T=0.01$ s, compute the amplitude spectrum, phase spectrum, and power spectrum

(a) Using the triangle window function.

(b) Using the Hamming window function.

Solution:

(a) Since $N=4$, from the triangular window function, we have

$$w_{tri}(0)=1-\frac{|2\times 0-4+1|}{4-1}=0$$

$$w_{tri}(1)=1-\frac{|2\times 1-4+1|}{4-1}=0.6667.$$

Similarly, $w_{tri}(2)=0.6667, w_{tri}(3)=0$. Next, the windowed sequence is computed as

$$x_w(0)=x(0)\times w_{tri}(0)=1\times 0=0$$

$$x_w(1)=x(1)\times w_{tri}(1)=2\times 0.6667=1.3334$$

$$x_w(2)=x(2)\times w_{tri}(2)=3\times 0.6667=2$$

$$x_w(3)=x(3)\times w_{tri}(3)=4\times 0=0.$$

Apply DFT Eq. (4.8) to $x_w(n)$ for $k=0, 1, 2, 3$, respectively,

$$X(k)=x_w(0)W_4^{k\times 0}+x_w(1)W_4^{k\times 1}+x_w(2)W_4^{k\times 2}+x_w(3)W_4^{k\times 3}.$$

We have the following results:

$$X(0)=3.3334$$

$$X(1)=-2-j1.3334$$

$$X(2)=0.6666$$

$$X(3)=-2+j1.3334$$

$$\Delta f=\frac{1}{NT}=\frac{1}{4\cdot 0.01}=25\,\text{Hz}.$$

Applying Eqs. (4.24), (4.27), and (4.28) leads to

$$A_0=\frac{1}{4}|X(0)|=0.8334,\ \varphi_0=\tan^{-1}\left(\frac{0}{3.3334}\right)=0^0,\ P_0=\frac{1}{4^2}|X(0)|^2=0.6954$$

$$A_1=\frac{1}{4}|X(1)|=0.6009,\ \varphi_1=\tan^{-1}\left(\frac{-1.3334}{-2}\right)=-146.31^0,\ P_1=\frac{1}{4^2}|X(1)|^2=0.3611$$

$$A_2=\frac{1}{4}|X(2)|=0.1667,\ \varphi_2=\tan^{-1}\left(\frac{0}{0.6666}\right)=0^0,\ P_1=\frac{1}{4^2}|X(2)|^2=0.0278$$

Similarly,

$$A_3 = \frac{1}{4}|X(3)| = 0.6009, \varphi_3 = \tan^{-1}\left(\frac{1.3334}{-2}\right) = 146.31^0, P_3 = \frac{1}{4^2}|X(3)|^2 = 0.3611.$$

(b) Since $N=4$, from the Hamming window function, we have

$$w_{hm}(0) = 0.54 - 0.46\cos\left(\frac{2\pi \times 0}{4-1}\right) = 0.08$$

$$w_{hm}(1) = 0.54 - 0.46\cos\left(\frac{2\pi \times 1}{4-1}\right) = 0.77.$$

Similarly, $w_{hm}(2)=0.77$, and $w_{hm}(3)=0.08$. Next, the windowed sequence is computed as

$$x_w(0) = x(0) \times w_{hm}(0) = 1 \times 0.08 = 0.08$$

$$x_w(1) = x(1) \times w_{hm}(1) = 2 \times 0.77 = 1.54$$

$$x_w(2) = x(2) \times w_{hm}(2) = 3 \times 0.77 = 2.31$$

$$x_w(0) = x(3) \times w_{hm}(3) = 4 \times 0.08 = 0.32.$$

Apply DFT Eq. (4.8) to $x_w(n)$ for $k=0$, 1, 2, 3, respectively,

$$X(k) = x_w(0)W_4^{k\times 0} + x_w(1)W_4^{k\times 1} + x_w(2)W_4^{k\times 2} + x_w(3)W_4^{k\times 3}.$$

We yield the following:

$$X(0) = 4.25$$

$$X(1) = -2.23 - j1.22$$

$$X(2) = 0.53$$

$$X(3) = -2.23 + j1.22$$

$$\Delta f = \frac{1}{NT} = \frac{1}{4 \cdot 0.01} = 25\,\text{Hz}.$$

Applying Eqs. (4.24), (4.27), and (4.28), we achieve

$$A_0 = \frac{1}{4}|X(0)| = 1.0625, \varphi_0 = \tan^{-1}\left(\frac{0}{4.25}\right) = 0^0, P_0 = \frac{1}{4^2}|X(0)|^2 = 1.1289$$

$$A_1 = \frac{1}{4}|X(1)| = 0.6355, \varphi_1 = \tan^{-1}\left(\frac{-1.22}{-2.23}\right) = -151.32^0, P_1 = \frac{1}{4^2}|X(1)|^2 = 0.4308$$

$$A_2 = \frac{1}{4}|X(2)| = 0.1325, \varphi_2 = \tan^{-1}\left(\frac{0}{0.53}\right) = 0^0, P_2 = \frac{1}{4^2}|X(2)|^2 = 0.0176.$$

Similarly,

$$A_3 = \frac{1}{4}|X(3)| = 0.6355, \varphi_3 = \tan^{-1}\left(\frac{1.22}{-2.23}\right) = 151.32^0, P_3 = \frac{1}{4^2}|X(3)|^2 = 0.4308.$$

EXAMPLE 4.11

Given the sinusoid

$$x(n) = 2 \times \sin\left(2000\pi \frac{n}{8000}\right)$$

obtained by using a sampling rate of $f_s = 8000$ Hz, use the DFT to compute the spectrum with the following specifications:

(a) Compute the spectrum of a triangular window function with a window size = 50.
(b) Compute the spectrum of a Hamming window function with a window size = 100.
(c) Compute the spectrum of a Hanning window function with a window size = 150 and one-sided spectrum.

The MATLAB program is listed in Program 4.2, and results are plotted in Figs. 4.19–4.21. As compared with the no-windowed (rectangular window) case, all three windows are able to effectively reduce the spectral leakage, as shown in the figures.

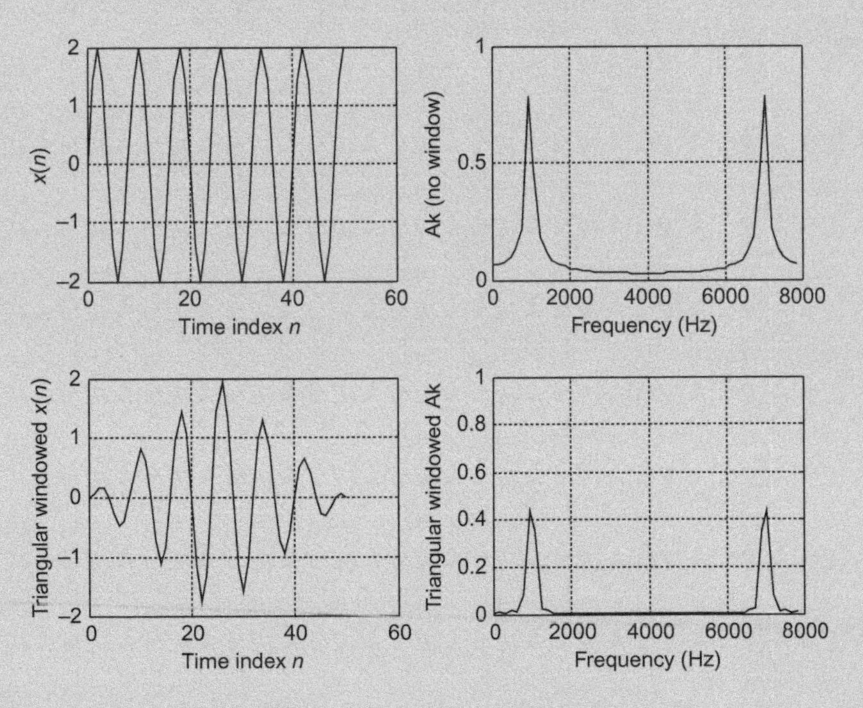

FIG. 4.19

Comparison of a spectrum without using a window function and a spectrum using a triangular window of size of 50 samples in Example 4.11.

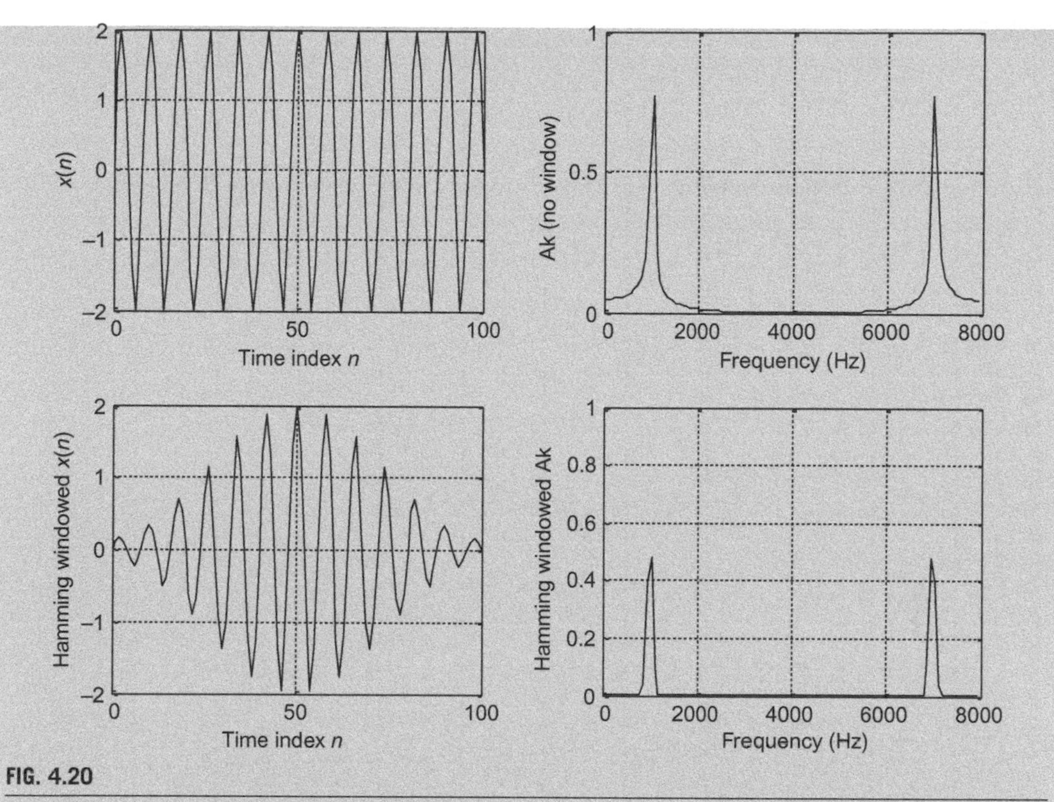

FIG. 4.20

Comparison of a spectrum without using a window function and a spectrum using a Hamming window of size of 100 samples in Example 4.11.

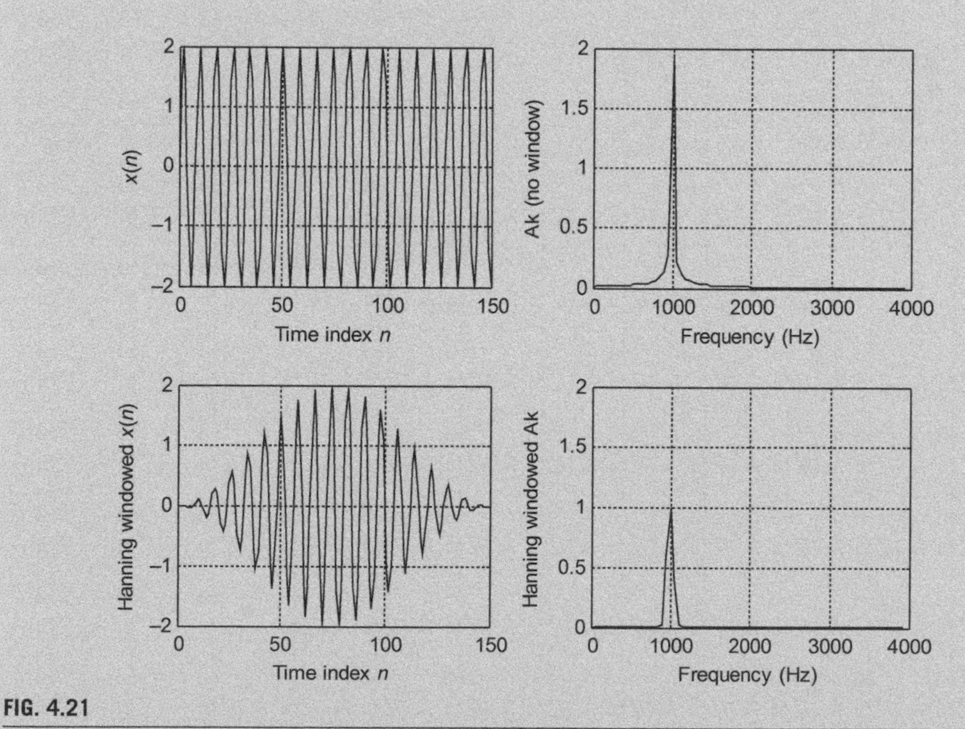

FIG. 4.21

Comparison of a one-sided spectrum without using the window function and a one-sided spectrum using a Hanning window of size of 150 samples in Example 4.11.

Program 4.2 MATLAB program for Example 4.11

```
%Example 4.11
close all;clear all
% Generate the sine wave sequence
fs=8000; T=1/fs;   % Sampling rate and sampling period
x=2*sin(2000*pi*[0:1:50]*T); %Generate 51 2000-Hz samples.
% Apply the FFT algorithm
N=length(x);
index_t=[0:1:N-1];
f=[0:1:N-1]*8000/N;      %Map frequency bin to frequency (Hz)
xf=abs(fft(x))/N;        %Calculate amplitude spectrum
figure(1)
%Using Bartlett window
x_b=x.*bartlett(N)';      %Apply triangular window function
xf_b=abs(fft(x_b))/N;   %Calculate amplitude spectrum
subplot(2,2,1);plot(index_t,x);grid
xlabel('Time index n'); ylabel('x(n)');
subplot(2,2,3); plot(index_t,x_b);grid
xlabel('Time index n'); ylabel('Triangular windowed x(n)');
subplot(2,2,2);plot(f,xf );grid;axis([0 8000 0 1]);
xlabel('Frequency (Hz)'); ylabel('Ak (no window)');
subplot(2,2,4); plot(f,xf_b);grid; axis([0 8000 0 1]);
xlabel('Frequency (Hz)'); ylabel('Triangular windowed Ak');
figure(2)
% Generate the sine wave sequence
x=2*sin(2000*pi*[0:1:100]*T); %Generate 101 2000-Hz samples.
% Apply the FFT algorithm
N=length(x);
index_t=[0:1:N-1];
f=[0:1:N-1]*fs/N;
xf=abs(fft(x))/N;
%Using Hamming window
x_hm=x.*hamming(N)';          %Apply Hamming window function
xf_hm=abs(fft(x_hm))/N;      %Calculate amplitude spectrum
subplot(2,2,1);plot(index_t,x);grid
xlabel('Time index n'); ylabel('x(n)');
subplot(2,2,3); plot(index_t,x_hm);grid
xlabel('Time index n'); ylabel('Hamming windowed x(n)');
subplot(2,2,2);plot(f,xf );grid;axis([0 fs 0 1]);
xlabel('Frequency (Hz)'); ylabel('Ak (no window)');
subplot(2,2,4); plot(f,xf_hm);grid;axis([0 fs 0 1]);
xlabel('Frequency (Hz)'); ylabel('Hamming windowed Ak');
figure(3)
% Generate the sine wave sequence
x=2*sin(2000*pi*[0:1:150]*T); % Generate 151 2-kHz samples
% Apply the FFT algorithm
N=length(x);
index_t=[0:1:N-1];
f=[0:1:N-1]*fs/N;
xf=2*abs(fft(x))/N;xf(1)=xf(1)/2; % Single-sided spectrum
%Using Hanning window
```

```
x_hn=x.*hann(N)';
xf_hn=2*abs(fft(x_hn))/N;xf_hn(1)=xf_hn(1)/2; %Single-sided spectrum
subplot(2,2,1);plot(index_t,x);grid
xlabel('Time index n'); ylabel('x(n)');
subplot(2,2,3); plot(index_t,x_hn);grid
xlabel('Time index n'); ylabel('Hanning windowed x(n)');
subplot(2,2,2);plot(f(1:(N-1)/2),xf(1:(N-1)/2));grid;axis([0 fs/2 0 2]);
xlabel('Frequency (Hz)'); ylabel('Ak (no window)');
subplot(2,2,4); plot(f(1:(N-1)/2),xf_hn(1:(N-1)/2));grid;axis([0 fs/2 0 2]);
xlabel('Frequency (Hz)'); ylabel('Hanning windowed Ak');
```

4.4 APPLICATION TO SIGNAL SPECTRAL ESTIMATION

The following plots show the comparisons of amplitude spectral estimation for a speech data (we.dat) with 2001 samples and a sampling rate of 8000 Hz using the rectangular window (no window) function and the Hamming window function. As demonstrated in Fig. 4.22 (two-sided spectrum) and Fig. 4.23

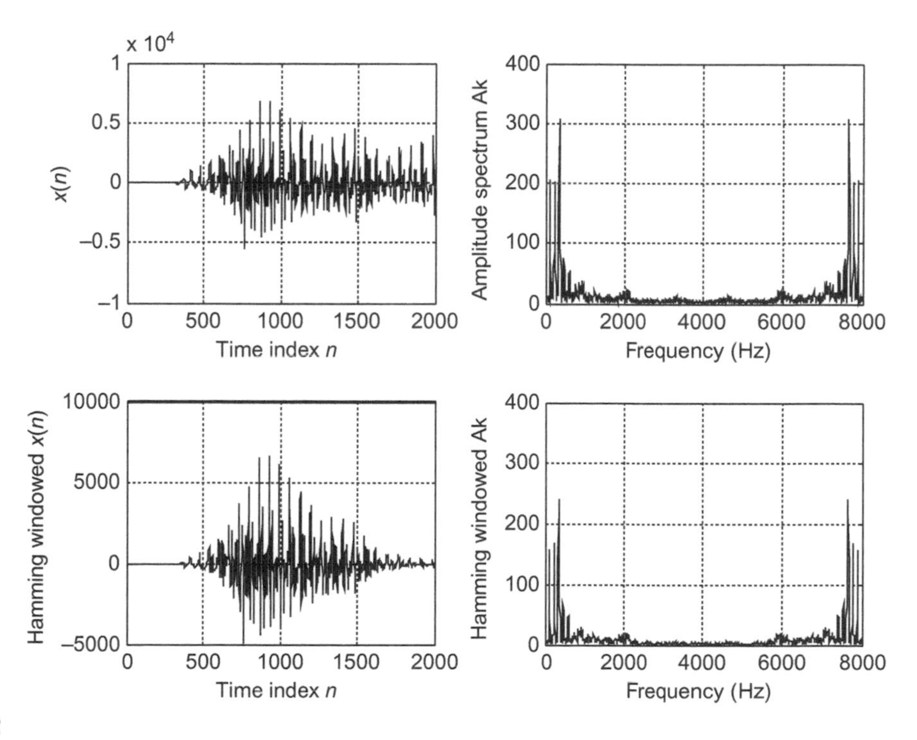

FIG. 4.22

Comparison of a spectrum without using a window function and a spectrum using the Hamming window for speech data.

(one-sided spectrum), there is little difference between the amplitude spectrum using the Hamming window function and the spectrum without using the window function. This is due to the fact that when the data length of the sequence (e.g., 2001 samples) increases, the frequency resolution will be improved and the spectral leakage will become less significant. However, when data length is short, reducing spectral leakage using a window function will become prominent.

Next, we compute the one-sided spectrum for a 32-bit seismic data sampled at 15 Hz (provided by the USGS Albuquerque Seismological Laboratory) with 6700 data samples. The computed spectral plots without using a window function and using the Hamming window are displayed in Fig. 4.24. We can see that most of seismic signal components are below 3 Hz.

We also compute the one-sided spectrum for a standard ECG signal from the MIT-BIH (Massachusetts Institute of Technology-Beth Israel Hospital) Database. The ECG signal contains frequency components ranging from 0.05 to 100 Hz sampled at 500 Hz. As shown in Fig. 4.25, there is a spike located at 60 Hz. This is due to the 60-Hz power line interference when the ECG is acquired via the ADC acquisition process. This 60-Hz interference can be removed by using a digital notch filter which will be studied in Chapter 8.

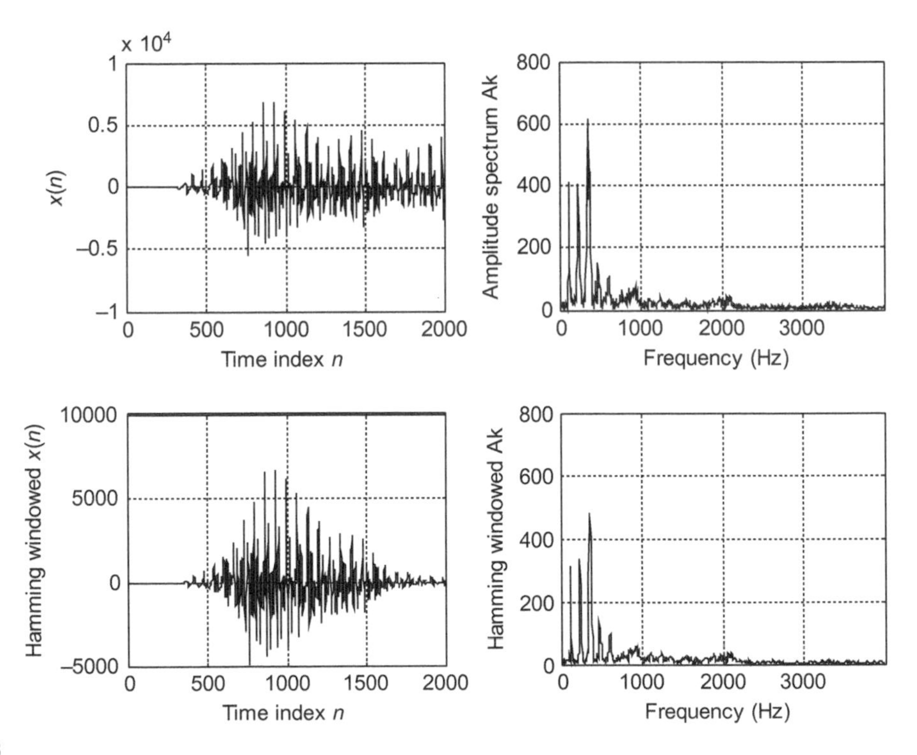

FIG. 4.23

Comparison of a one-sided spectrum without using a window function and a one-sided spectrum using the Hamming window for speech data.

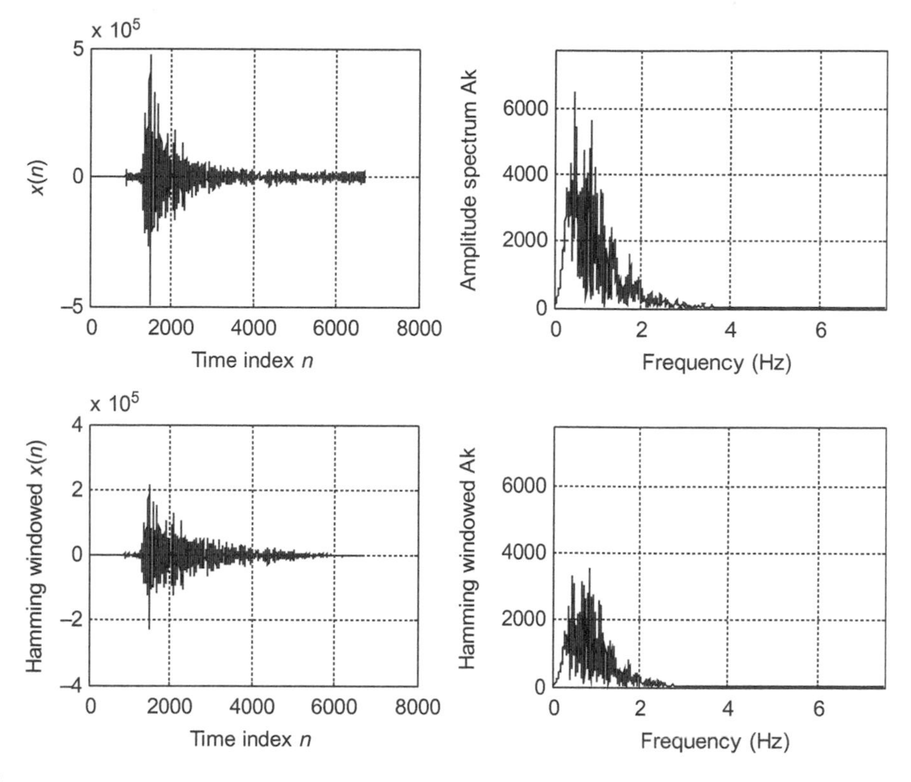

FIG. 4.24

Comparison of a one-sided spectrum without using a window function and a one-sided spectrum using the Hamming window for seismic data.

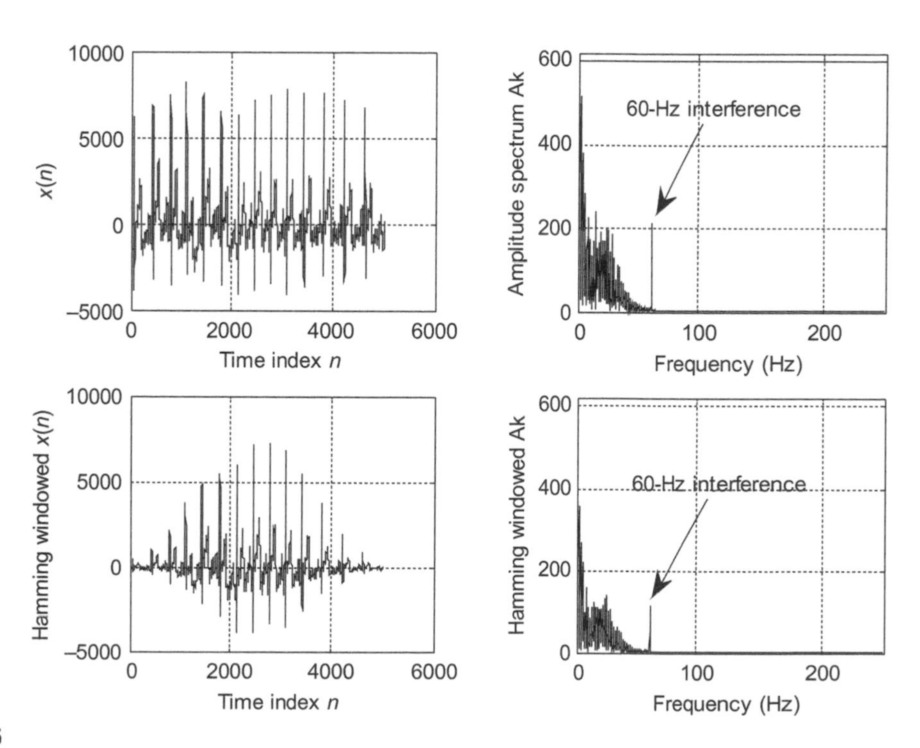

FIG. 4.25

Comparison of a one-sided spectrum without using a window function and a one-sided spectrum using the Hamming window for ECG data.

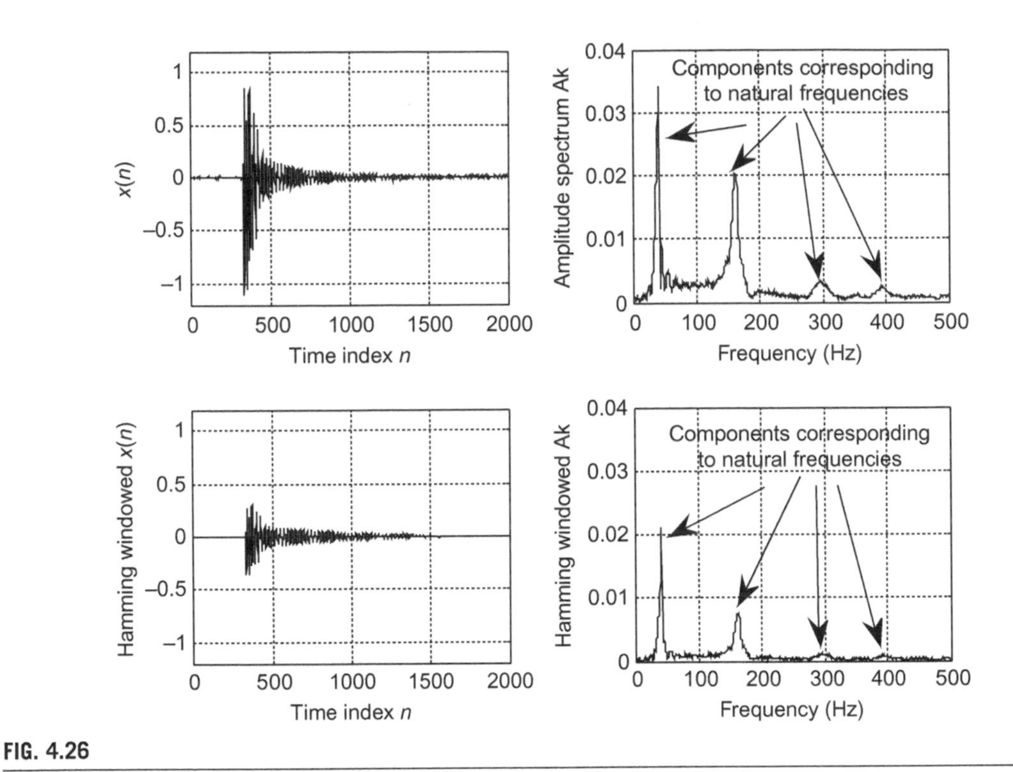

FIG. 4.26

Comparison of a one-sided spectrum without using a window function and a one-sided spectrum using the Hamming window for vibration signal.

Fig. 4.26 shows a vibration signal and its spectrum. The vibration signal is captured using an accelerometer sensor attached to a simple supported beam while an impulse force is applied to the location which is close to the middle of the beam. The sampling rate is 1 kHz. As shown in Fig. 4.26, four dominant modes (natural frequencies corresponding to locations of spectral peaks) can be easily identified from the displayed spectrum.

We now present another practical example for vibration signature analysis of the defected gear tooth described in Section 1.3.5. Fig. 4.27 shows a gearbox containing two straight bevel gears with a transmission ratio of 1.5:1 and the numbers of teeth on the pinion and gear are 18 and 27. The vibration data is collected by an accelerometer installed on the top of the gearbox. The data acquisition system uses a sampling rate of 12.8 kHz. The meshing frequency is determined as $f_m = f_i$(rpm[revolutions per minute]) $\times 18/60 = 300\,\text{Hz}$, where the input shaft frequency is $f_i = 1000\text{rpm} = 16.67\text{Hz}$. Fig. 4.28 shows the baseline vibration signal and its spectrum under the excellent condition, and vibration signal and its spectrum under the different damage severity levels (five levels conducted by SpectraQuest, Inc.). As we can see, the baseline spectrum contains the meshing frequency component of 300 Hz and a sideband frequency component of 283.33 Hz (300 – 16.67). The vibration signatures for the damaged pinions (severity level 1: lightly chipped; severity level 4: heavily chipped; severity level 5: missing tooth) are included in Figs. 4.29–4.31. We can observe that the sidebands

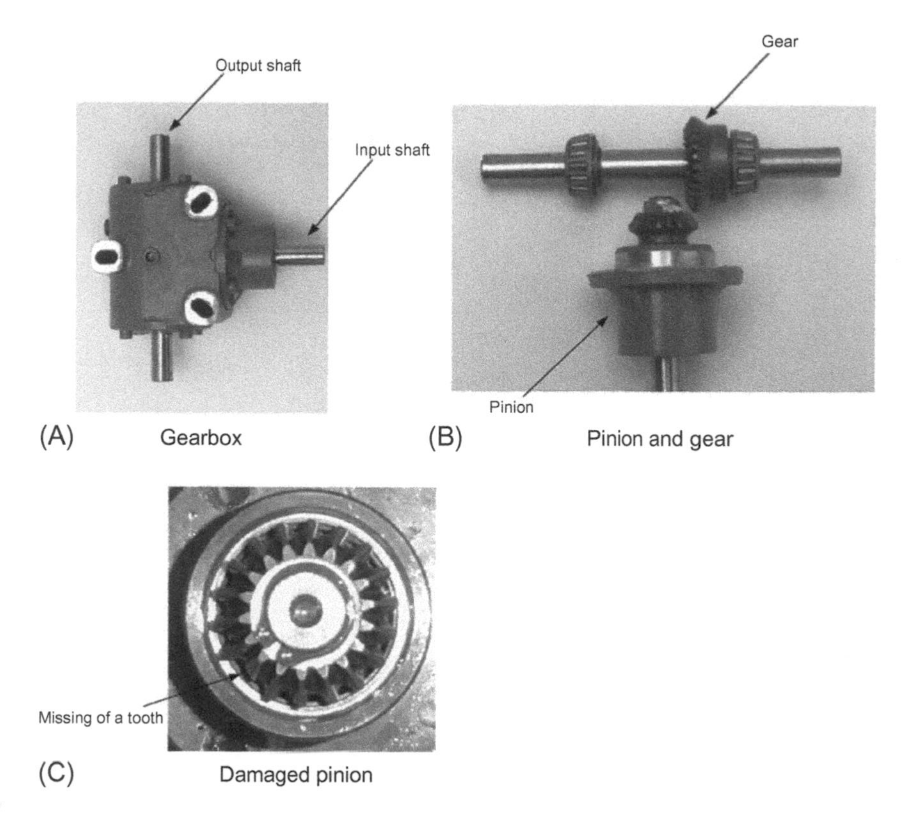

FIG. 4.27

Vibration signature analysis of a gearbox: (A) Gearbox, (B) Pinion and gear, (C) Damaged pinion.

Courtesy of SpectaQuest, Inc.

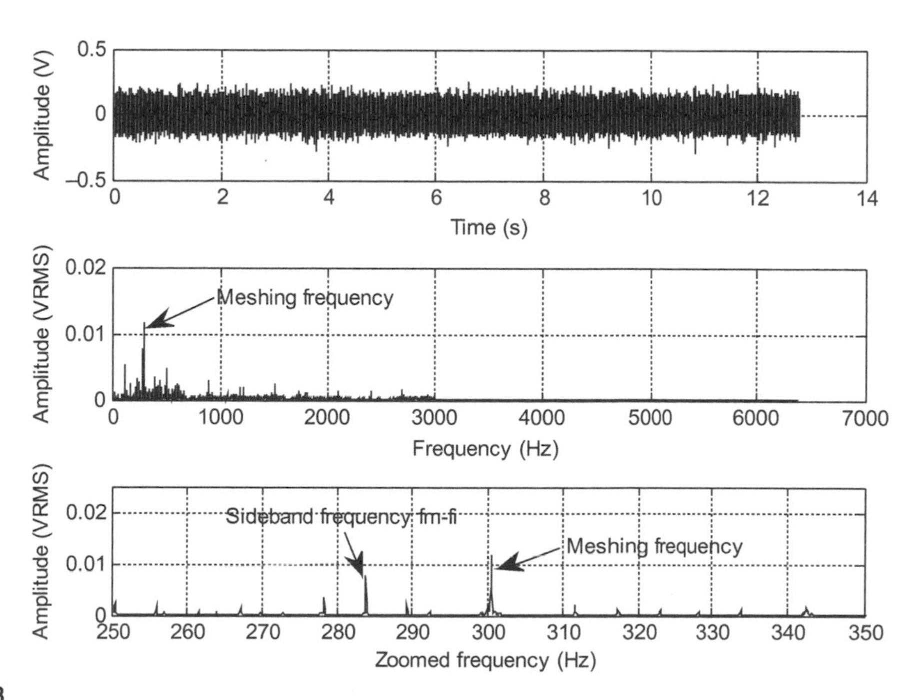

FIG. 4.28

Vibration signal and spectrum from the good condition gearbox.

Data provided by SpectaQuest, Inc.

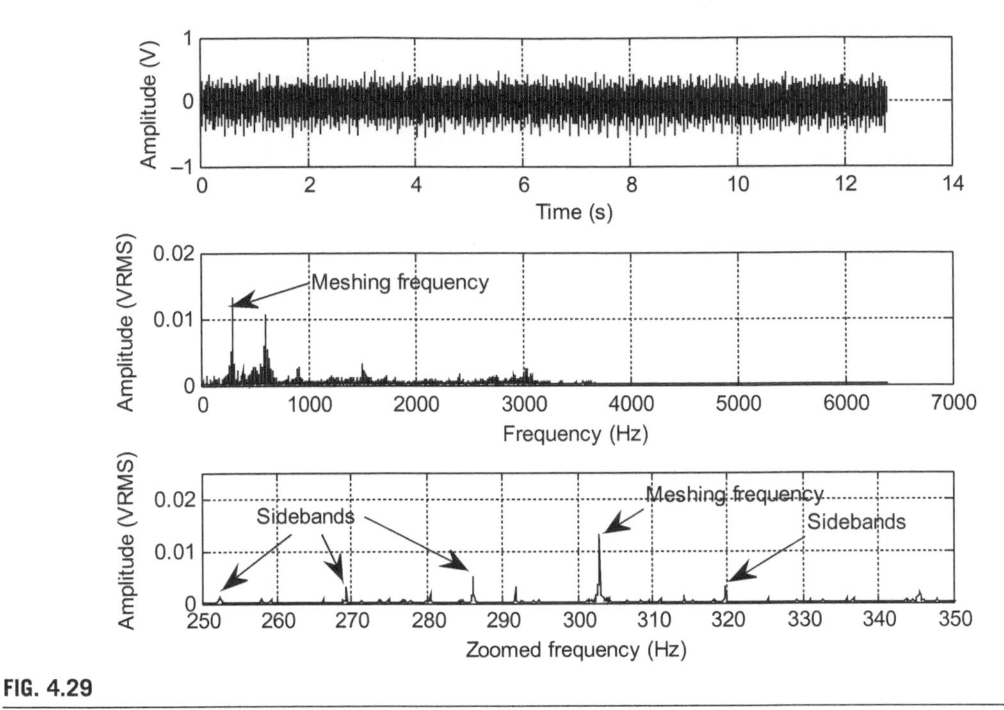

FIG. 4.29

Vibration signal and spectrum for damage severity level 1.

Data provided by SpectaQuest, Inc.

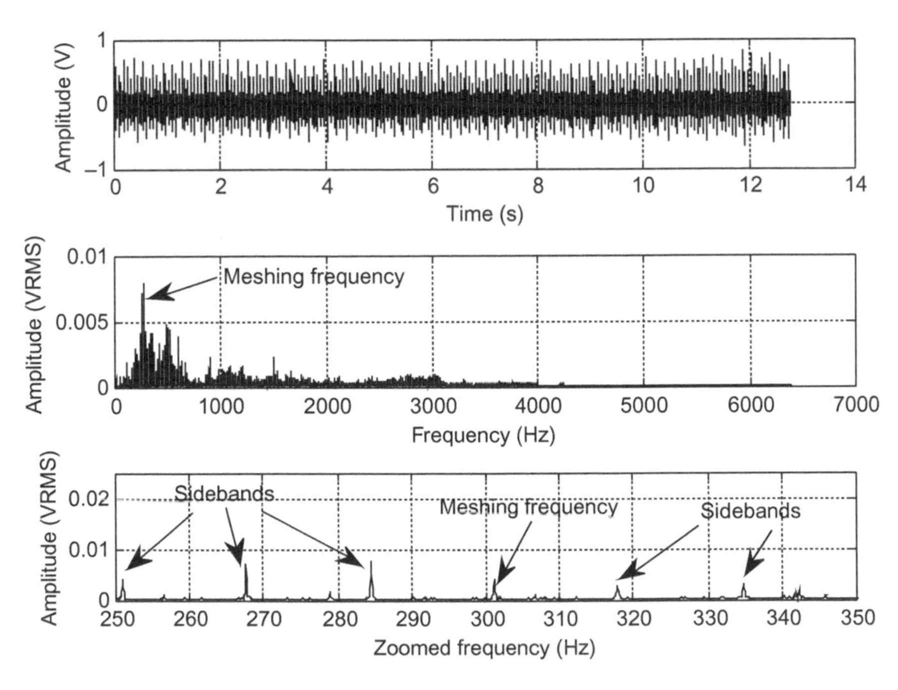

FIG. 4.30

Vibration signal and spectrum for damage severity level 4.

Data provided by SpectaQuest, Inc.

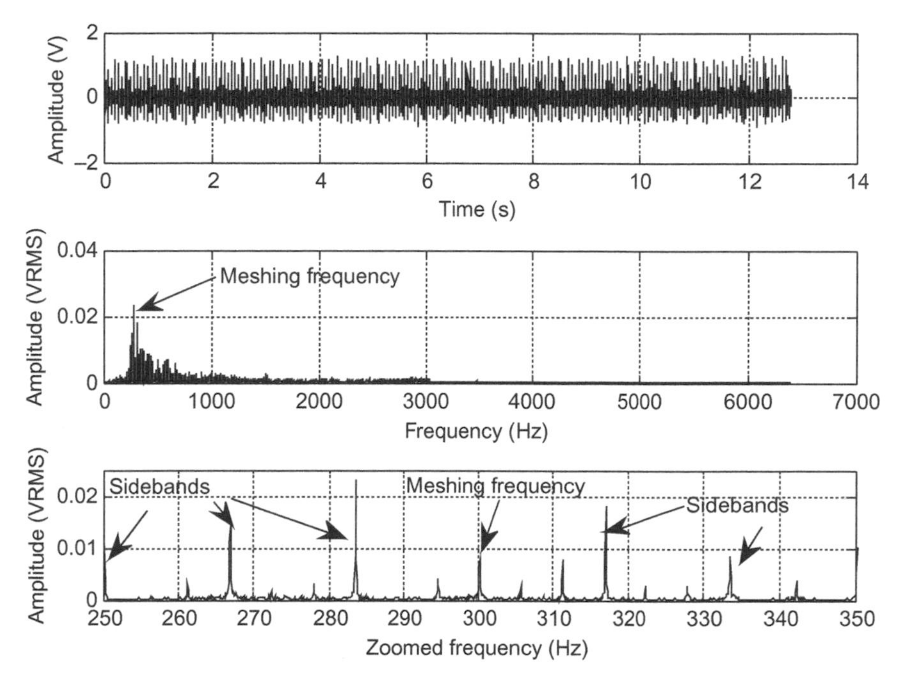

FIG. 4.31

Vibration signal and spectrum for damage severity level 5.

Data provided by SpectaQuest, Inc.

($f_m \pm f_i$, $f_m \pm 2f_i$...) become more dominant when the severity level increases. Hence, the spectral information is very useful for monitoring the health condition of a gearbox.

4.5 FAST FOURIER TRANSFORM

Now we study FFT in detail. The FFT is a very efficient algorithm in computing DFT coefficients and can reduce a very large amount of computational complexity (multiplications). Without loss of generality, we consider the digital sequence $x(n)$ consisting of 2^m samples, where m is a positive integer, that is, the number of samples of the digital sequence $x(n)$ is a power of 2, $N=2, 4, 8, 16$, etc. If $x(n)$ does not contain 2^m samples, then we simply append it with zeros until the number of the appended sequence is a power of 2.

In this section, we focus on two formats. One is called the decimation-in-frequency algorithm, while the other is the decimation-in-time algorithm. They are referred to as the *radix-2* FFT algorithms. Other types of FFT algorithms are the radix-4 and the split radix and their advantages can be explored (see Proakis and Manolakis, 1996).

4.5.1 METHOD OF DECIMATION-IN-FREQUENCY

We begin with the definition of DFT studied in the opening section in this chapter as follows:

$$X(k) = \sum_{n=0}^{N-1} x(n) W_N^{kn} \quad \text{for } k = 0, 1, \cdots, N-1, \tag{4.38}$$

where $W_N = e^{-j\frac{2\pi}{N}}$ is the twiddle factor, and $N = 2, 4, 8, 16, \cdots$, Eq. (4.38) can be expanded as

$$X(k) = x(0) + x(1)W_N^k + \cdots + x(N-1)W_N^{k(N-1)}. \tag{4.39}$$

Again, if we split Eq. (4.39) into

$$\begin{aligned} X(k) &= x(0) + x(1)W_N^k + \cdots + x\left(\frac{N}{2} - 1\right) W_N^{k(N/2-1)} \\ &+ x\left(\frac{N}{2}\right) W^{kN/2} + \cdots + x(N-1) W_N^{k(N-1)}, \end{aligned} \tag{4.40}$$

then we can rewrite as a sum of the following two parts

$$X(k) = \sum_{n=0}^{(N/2)-1} x(n) W_N^{kn} + \sum_{n=N/2}^{N-1} x(n) W_N^{kn}. \tag{4.41}$$

Modifying the second term in Eq. (4.41) yields

$$X(k) = \sum_{n=0}^{(N/2)-1} x(n) W_N^{kn} + W_N^{(N/2)k} \sum_{n=0}^{(N/2)-1} x\left(n + \frac{N}{2}\right) W_N^{kn}. \tag{4.42}$$

Recall $W_N^{N/2} = e^{-j\frac{2\pi(N/2)}{N}} = e^{-j\pi} = -1$; then we have

$$X(k) = \sum_{n=0}^{(N/2)-1} \left(x(n) + (-1)^k x\left(n + \frac{N}{2}\right) \right) W_N^{kn}. \tag{4.43}$$

Now letting $k = 2m$ as an even number achieves

$$X(2m) = \sum_{n=0}^{(N/2)-1} \left(x(n) + x\left(n + \frac{N}{2}\right) \right) W_N^{2mn}, \tag{4.44}$$

while substituting $k = 2m + 1$ as an odd number yields

$$X(2m+1) = \sum_{n=0}^{(N/2)-1} \left(x(n) - x\left(n + \frac{N}{2}\right) \right) W_N^n W_N^{2mn}. \tag{4.45}$$

Using the fact that $W_N^2 = e^{-j\frac{2\pi \times 2}{N}} = e^{-j\frac{2\pi}{(N/2)}} = W_{N/2}$, it follows that

$$X(2m) = \sum_{n=0}^{(N/2)-1} a(n) W_{N/2}^{mn} = \text{DFT}\{a(n) \text{ with } (N/2)\text{points}\}, \tag{4.46}$$

$$X(2m+1) = \sum_{n=0}^{(N/2)-1} b(n) W_N^n W_{N/2}^{mn} = \text{DFT}\{b(n) W_N^n \text{ with } (N/2)\text{points}\}, \tag{4.47}$$

where $a(n)$ and $b(n)$ are introduced and expressed as

$$a(n) = x(n) + x\left(n + \frac{N}{2}\right), \quad \text{for } n = 0, 1 \cdots, \frac{N}{2} - 1, \tag{4.48}$$

$$b(n) = x(n) - x\left(n + \frac{N}{2}\right), \quad \text{for } n = 0, 1, \cdots, \frac{N}{2} - 1. \tag{4.49}$$

Eqs. (4.38), (4.46), and (4.47) can be summarized as

$$\text{DFT}\{x(n)\text{ with } N \text{ points}\} = \begin{cases} \text{DFT}\{a(n)\text{ with}(N/2)\text{points}\} \\ \text{DFT}\{b(n)W_N^n \text{ with}(N/2)\text{points}\} \end{cases} \tag{4.50}$$

The computation process can be illustrated in Fig. 4.32.

As shown in this figure, there are three graphical operations, which are illustrated Fig. 4.33.

If we continue the process described in Fig. 4.32, we obtain the block diagrams shown in Figs. 4.34 and 4.35.

Fig. 4.35 illustrates the FFT computation for the 8-point DFT, where there are 12 complex multiplications. This is a big saving as compared with the eight-point DFT with 64 complex multiplications. For a data length of N, the number of complex multiplications for DFT and FFT, respectively, are determined by

$$\text{Complex multiplications of DFT} = N^2, \text{and}$$

$$\text{Complex multiplications of FFT} = \frac{N}{2}\log_2(N).$$

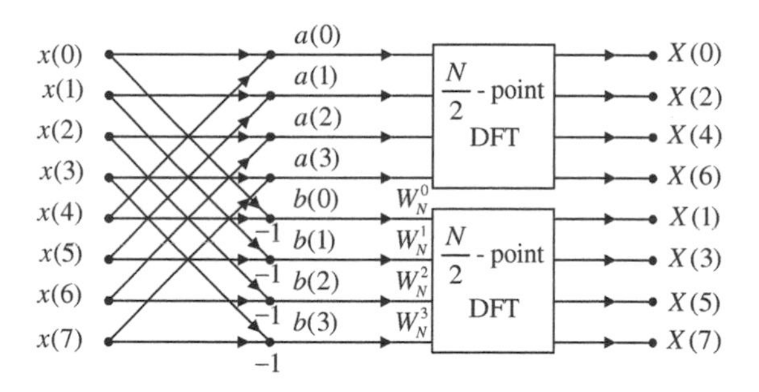

FIG. 4.32

The first iteration of the eight-point FFT.

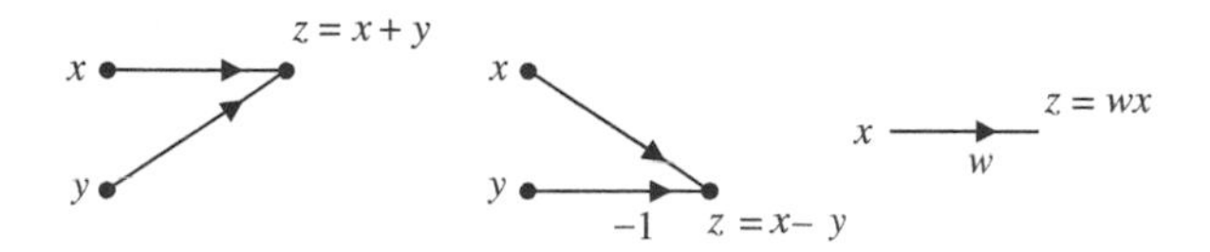

FIG. 4.33

Definitions of the graphical operations.

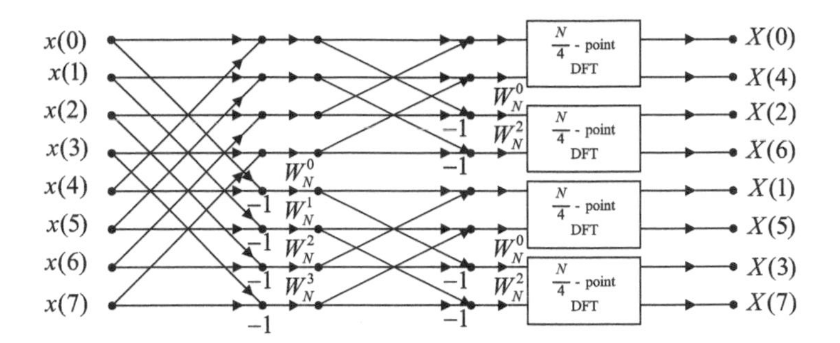

FIG. 4.34

The second iteration of the eight-point FFT.

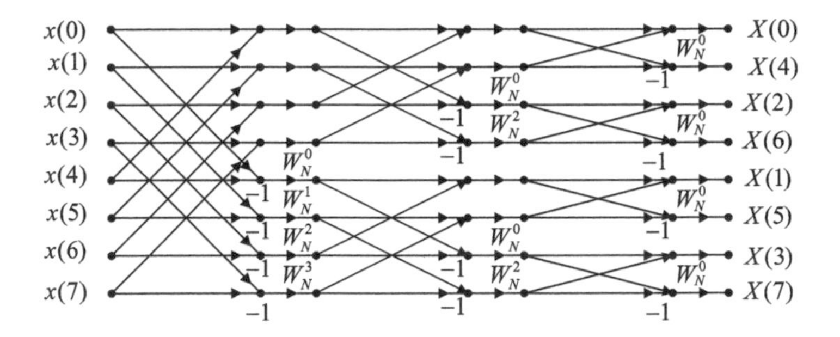

FIG. 4.35

Block diagram for the eight-point FFT (total 12 multiplications).

To see the effectiveness of FFT, let us consider a sequence with 1024 data points. Applying DFT will require $1024 \times 1024 = 1,048,576$ complex multiplications; however, applying FFT will need only $(1024/2)\log_2(1024) = 5120$ complex multiplications. Next, the index (bin number) of the eight-point DFT coefficient $X(k)$ becomes to be 0, 4, 2, 6, 1, 5, 3, and 7, respectively, which are not in the natural order. This can be fixed by index matching. The index matching between the input sequence and output frequency bin number by applying reversal bits is described in Table 4.2.

Table 4.2 Index Mapping for Fast Fourier Transform

Input Data	Index Bits	Reversal Bits	Output Data
$x(0)$	000	000	$X(0)$
$x(1)$	001	100	$X(4)$
$x(2)$	010	010	$X(2)$
$x(3)$	011	110	$X(6)$
$x(4)$	100	001	$X(1)$
$x(5)$	101	101	$X(5)$
$x(6)$	110	011	$X(3)$
$x(7)$	111	111	$X(7)$

Binary	Index	1st split	2nd split	3rd split	Bit reversal
000	0	0	0	0	000
001	1	2	4	4	100
010	2	4	2	2	010
011	3	6	6	6	011
100	4	1	1	1	001
101	5	3	5	5	101
110	6	5	3	3	011
111	7	7	7	7	111

FIG. 4.36

Bit reversal process in FFT.

Fig. 4.36 explains the bit reversal process. First, the input data with indices 0, 1, 2, 3, 4, 5, 6, and 7 are split into two parts. The first half contains even indices—0, 2, 4, 6—while the second half contains odd indices. The first half with indices 0, 2, 4, and 6 at the first iteration continues to split into even indices 0, 4 and odd indices 2, 6 as shown in the second iteration. The second half with indices1, 3, 5, and 7 at the first iteration is split into even indices 1, 5 and odd indices 3, 7 in the second iteration. The splitting process continues till the end at the third iteration. The bit patterns of the output data indices are just the respective reversed bit patterns of the input data indices.

Although Fig. 4.36 illustrates the case of an eight-point FFT, this bit reversal process works as long as N is a power of 2.

MATLAB Program for implementing FFT using decimation-in-frequency method is listed below:

Program 4.3 MATLAB program for FFT with decimation-in-frequency

```
function Xk= fftdinf(x)
% FFT using decimation-in-frequency method
%
XX=fftdinf2(x);
k=bitrev(1:1:length(XX));
Xk=XX(k);
end
function Xk = fftdinf2(x)
% FFT using decimation-in-frequency method
M=ceil(log2(length(x)));
x=[x zeros(1,2^M-length(x))]; % padding zeros to have a length of power of 2
N=length(x);
if (N==1)
Xk=x;
else
a=x(1:N/2)+x(N/2+1:N);
b=x(1:N/2)-x(N/2+1:N);
Xk=[fftdinf2(a) fftdinf2(b.*exp(-2*pi*j*[0:1:N/2-1]/N))];
```

```
end
end
function k = bitrev(x)
% bit reversal in terms of the integer
% x=1:1:2^M
N=length(x);
if N==1
k=x;
else
k=[bitrev(x(1:2:N)) bitrev(x(2:2:N))];
N=N/2;
end
end
```

The inverse FFT is defined as

$$x(n) = \frac{1}{N}\sum_{k=0}^{N-1} X(k)W_N^{-kn} = \frac{1}{N}\sum_{k=0}^{N-1} X(k)\widetilde{W}_N^{kn}, \text{ for } k = 0, 1, \cdots, N-1. \tag{4.51}$$

On comparing Eq. (4.51) with Eq. (4.38), we note the following difference: the twiddle factor W_N is changed to $\widetilde{W}_N = W_N^{-1}$, and the sum is multiplied by a factor of $1/N$. Hence, by modifying the FFT block diagram as shown in Fig. 4.35, we achieve the inverse FFT block diagram shown in Fig. 4.37.

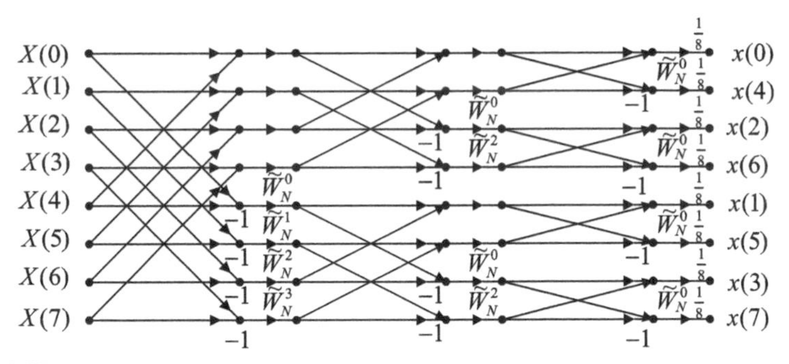

FIG. 4.37

Block diagram for the inverse of eight-point FFT.

EXAMPLE 4.12

Given a sequence $x(n)$ for $0 \le n \le 3$, where $x(0) = 1$, $x(1) = 2$, $x(2) = 3$, and $x(3) = 4$

(a) Evaluate its DFT $X(k)$ using the decimation-in-frequency FFT method.

(b) Determine the number of complex multiplications.

Solution:

(a) Using the FFT block diagram in Fig. 4.35, the result is shown in Fig. 4.38.

(b) From Fig. 4.38, the number of complex multiplications is four, which can also be determined by

$$\frac{N}{2}\log_2(N) = \frac{4}{2}\log_2(4) = 4.$$

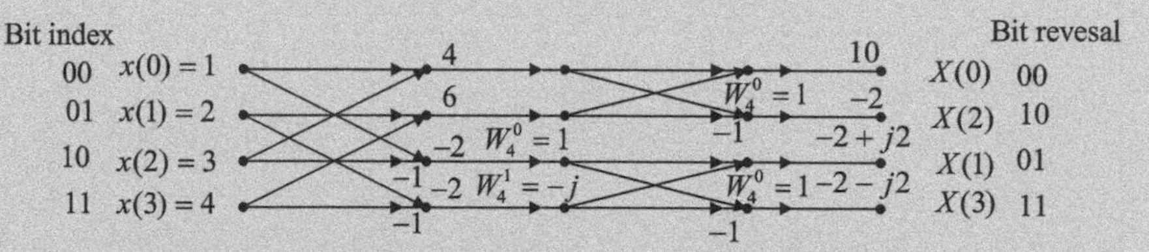

FIG. 4.38

Four-point FFT block diagram in Example 4.12.

EXAMPLE 4.13

Given the DFT sequence $X(k)$ for $0 \le k \le 3$ computed in Example 4.12, evaluate its inverse of DFT $x(n)$ using the decimation-in-frequency FFT method.

Solution:

Using the inverse FFT block diagram in Fig. 4.37, we have the result shown in Fig. 4.39.

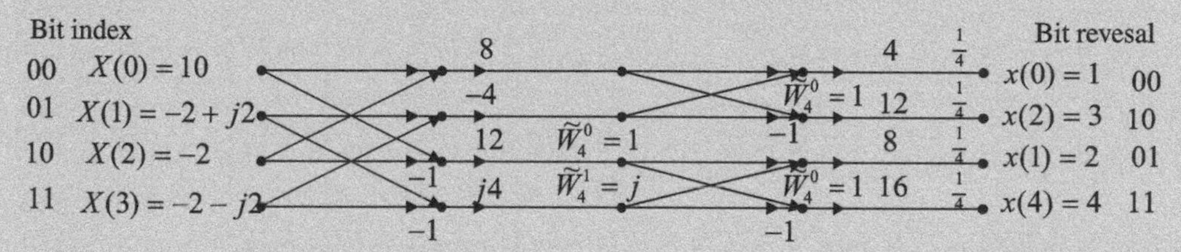

FIG. 4.39

Four-point inverse FFT block diagram in Example 4.13.

4.5.2 METHOD OF DECIMATION-IN-TIME

In this method, we split the input sequence $x(n)$ into the even indexed $x(2m)$ and $x(2m+1)$, each with N data points. Then Eq. (4.38) becomes

$$X(k) = \sum_{m=0}^{(N/2)-1} x(2m)W_N^{2mk} + \sum_{m=0}^{(N/2)-1} x(2m+1)W_N^k W_N^{2mk}, \text{ for } k = 0, 1, \cdots, N-1. \tag{4.52}$$

Using the relation $W_N^2 = W_{N/2}$, it follows that

$$X(k) = \sum_{m=0}^{(N/2)-1} x(2m) W_{N/2}^{mk} + W_N^k \sum_{m=0}^{(N/2)-1} x(2m+1) W_{N/2}^{mk}, \text{ for } k = 0, 1, \cdots, N-1. \tag{4.53}$$

Define new functions as

$$G(k) = \sum_{m=0}^{(N/2)-1} x(2m) W_{N/2}^{mk} = \text{DFT}\{x(2m)\text{with}(N/2)\text{points}\}, \tag{4.54}$$

$$H(k) = \sum_{m=0}^{(N/2)-1} x(2m+1) W_{N/2}^{mk} = \text{DFT}\{x(2m+1)\text{with}(N/2)\text{points}\}. \tag{4.55}$$

Note that

$$G(k) = G\left(k + \frac{N}{2}\right), \quad \text{for } k = 0, 1, \cdots, \frac{N}{2} - 1, \tag{4.56}$$

$$H(k) = H\left(k + \frac{N}{2}\right), \quad \text{for } k = 0, 1, \cdots, \frac{N}{2} - 1. \tag{4.57}$$

Substituting Eqs. (4.56) and (4.57) into Eq. (4.53) yields the first half frequency bins

$$X(k) = G(k) + W_N^k H(k), \quad \text{for } k = 0, 1, \cdots, \frac{N}{2} - 1. \tag{4.58}$$

Considering the following fact and using Eqs. (4.56) and (4.57)

$$W_N^{(N/2+k)} = -W_N^k. \tag{4.59}$$

Then the second half of frequency bins can be computed as follows:

$$X\left(\frac{N}{2} + k\right) = G(k) - W_N^k H(k), \text{ for } k = 0, 1, \cdots, \frac{N}{2} - 1. \tag{4.60}$$

If we perform backward iterations, we can obtain the FFT algorithm. Procedure using Eqs. (4.58) and (4.60) is illustrated in Fig. 4.40, the block diagram for the eight-point FFT algorithm.

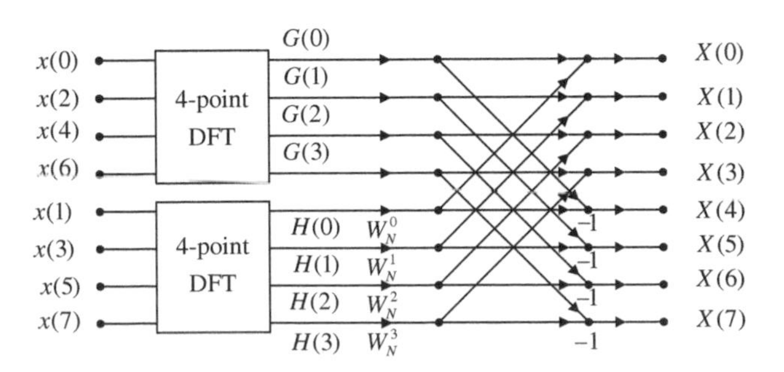

FIG. 4.40

First iteration.

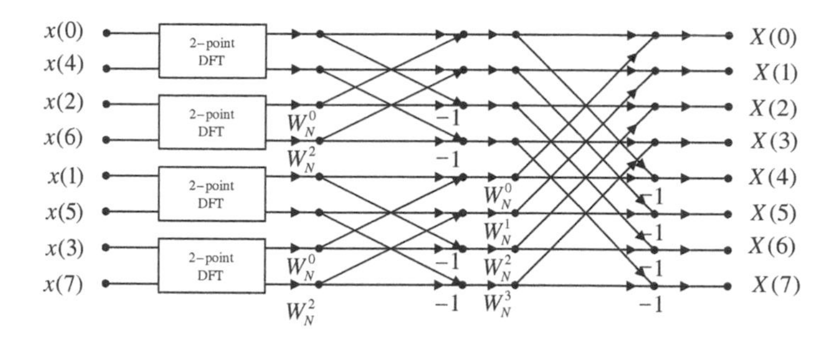

FIG. 4.41

Second iteration.

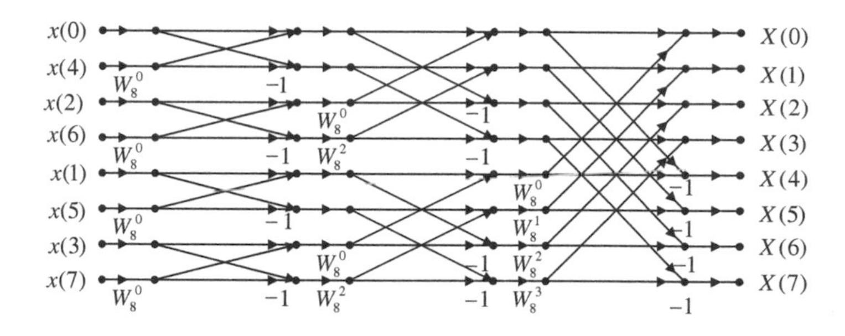

FIG. 4.42

Eight-point FFT algorithm using decimation-in-time (12 complex multiplications).

From a further computation, we obtain Fig. 4.41.

Finally, after three recursions, we end up with the block diagram in Fig. 4.42.

The index for each input sequence element can be achieved by bit reversal of the frequency index in a sequential order. MATLAB program for implementing FFT using decimation-in-time method is listed below:

Program 4.4 MATLAB program for FFT using decimation-in-time method

```
function Xk = fftdint(x)
% FFT using decimation-in-time method, no need for bit reversal algorithm
M=ceil(log2(length(x)));
x=[x zeros(1,2^M-length(x))]; %padding zeros to have a length of power of 2
N=length(x);
if (N==1)
Xk=x;
else
G=fftdint(x(1:2:N));
H=fftdint(x(2:2:N)).*exp(-2*pi*j*[0:1:N/2-1]/N);
Xk=[G+H G-H];
end
end
```

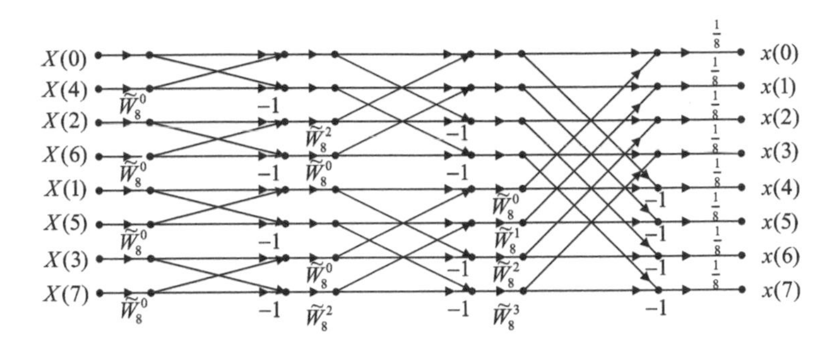

FIG. 4.43

The eight-point IFFT using decimation-in-time.

Similar to the decimation-in-frequency method, after changing W_N to $\widetilde{W}_N$ in Fig. 4.42 and multiplying the output sequence by a factor of $1/N$, we derive the inverse FFT block diagram for the eight-point inverse FFT in Fig. 4.43.

EXAMPLE 4.14

Given a sequence $x(n)$ for $0 \le n \le 3$, where $x(0)=1$, $x(1)=2$, $x(2)=3$, and $x(3)=4$, evaluate its DFT $X(k)$ using the decimation-in-time FFT method.

Solution:

Using the block diagram in Fig. 4.42 leads to the result shown in Fig. 4.44.

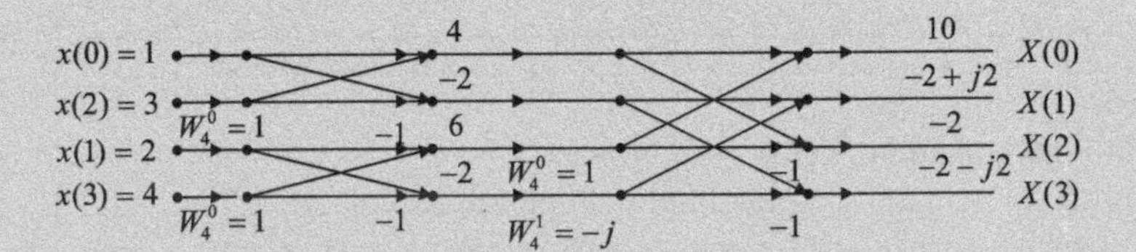

FIG. 4.44

The four-point FFT using decimation-in-time method.

EXAMPLE 4.15

Given the DFT sequence $X(k)$ for $0 \le k \le 3$ computed in Example 4.14, evaluate its inverse of DFT $x(n)$ using the decimation-in-time FFT method.

Solution:

Using the block diagram in Fig. 4.43 yields the result shown in Fig. 4.45

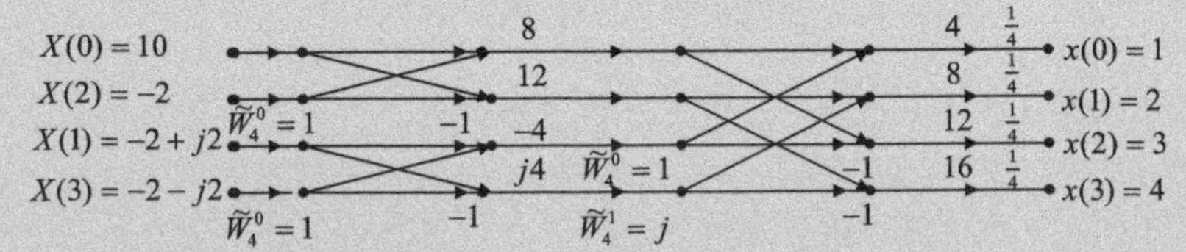

FIG. 4.45

The four-point IFFT using decimation-in-time method.

4.6 SUMMARY

1. The Fourier series coefficients for a periodic digital signal can be used to develop the DFT.
2. The DFT transforms a time sequence to the complex DFT coefficients, while the inverse DFT transforms DFT coefficients back to the time sequence.
3. The *frequency bin number* is the same as the frequency index. *Frequency resolution* is the frequency spacing between two consecutive frequency indices (two consecutive spectrum components).
4. The DFT coefficients for a given digital sequence are applied for computing the amplitude spectrum, power spectrum, or phase spectrum.
5. The spectrum calculated from all the DFT coefficients represents the signal frequency range from 0 Hz to the sampling rate. The spectrum beyond the folding frequency is equivalent to the negative-indexed spectrum from the negative folding frequency to 0 Hz. This two-sided spectrum can be converted into a single-sided spectrum by doubling alternating-current (AC) components from 0 Hz to the folding frequency and retaining the DC component as it is.
6. To reduce the burden of computing DFT coefficients, the FFT algorithm is used, which requires the data length to be a power of 2. Sometime zero padding is used to make up the data length. The zero padding actually does interpolation of the spectrum and does not carry any new information about the signal; even the calculated frequency resolution is smaller due to the zero padded longer length.
7. Applying a window function to the data sequence before DFT reduces the spectral leakage due to abrupt truncation of the data sequence when performing spectral calculation for a short sequence.
8. Two radix-2 FFT algorithms—decimation-in-frequency and decimation-in-time—are developed via the graphical illustrations.

4.7 PROBLEMS

4.1 Given a sequence $x(n)$ for $0 \le n \le 3$, where $x(0) = 1$, $x(1) = 1$, $x(2) = -1$, and $x(3) = 0$, compute its DFT $X(k)$.

4.2 Given a sequence $x(n)$ for $0 \le n \le 3$, where $x(0) = 4$, $x(1) = 3$, $x(2) = 2$, and $x(3) = 1$, evaluate its DFT $X(k)$.

4.3 Given a sequence $x(n)$ for $0 \le n \le 3$, where $x(0) = 0.2$, $x(1) = 0.2$, $x(2) = -0.2$, and $x(3) = 0$, compute its DFT $X(k)$.

4.4 Given a sequence $x(n)$ for $0 \le n \le 3$, where $x(0) = 0.8$, $x(1) = 0.6$, $x(2) = 0.4$, and $x(3) = 0.2$, evaluate its DFT $X(k)$.

4.5 Given the DFT sequence $X(k)$ for $0 \le k \le 3$ obtained in Problem 4.2, evaluate its inverse of DFT $x(n)$.

4.6 Given a sequence $x(n)$, where $x(0) = 4$, $x(1) = 3$, $x(2) = 2$, and $x(3) = 1$ with the last two data zero padded as $x(4) = 0$, and $x(5) = 0$, evaluate its DFT $X(k)$.

4.7 Given the DFT sequence $X(k)$ for $0 \le k \le 3$ obtained in Problem 4.4, evaluate its inverse of DFT $x(n)$.

4.8 Given a sequence $x(n)$, where $x(0)=0.8$, $x(1)=0.6$, $x(2)=0.4$, and $x(3)=0.2$ with the last two data zero padded as $x(4)=0$, and $x(5)=0$, evaluate its DFT $X(k)$.

4.9 Using the DFT sequence $X(k)$ for $0 \leq k \leq 5$ computed in Problem 4.6, evaluate the inverse of DFT for $x(0)$ and $x(4)$.

4.10 Consider a digital sequence sampled at the rate of 20,000 Hz. If we use the 8000-point DFT to compute the spectrum, determine

(a) the frequency resolution.

(b) the folding frequency in the spectrum.

4.11 Using the DFT sequence $X(k)$ for $0 \leq k \leq 5$ computed in Problem 4.8, evaluate the inverse of DFT for $x(0)$ and $x(4)$.

4.12 Consider a digital sequence sampled at the rate of 16,000 Hz. If we use the 4000-point DFT to compute the spectrum, determine

(a) the frequency resolution.

(b) the folding frequency in the spectrum.

4.13 We use the DFT to compute the amplitude spectrum of a sampled data sequence with a sampling rate $f_s=2000$ Hz. It requires the frequency resolution to be less than 0.5 Hz. Determine the number of data points used by the FFT algorithm and actual frequency resolution in Hz, assuming that the data samples are available for selecting the number of data points.

4.14 Given the sequence in Fig. 4.46 and assuming $f_s=100$ Hz, compute the amplitude spectrum, phase spectrum, and power spectrum.

4.15 Compute the following window functions for a size of 8:

(a) Hamming window function.

(b) Hanning window function.

4.16 Given the following data sequence with a length of 6

$$x(0)=0, x(1)=1, x(2)=0, x(3)=-1, x(4)=0, x(5)=1,$$

compute the windowed sequence $x_w(n)$ using

(a) Triangular window function.

(b) Hamming window function.

(c) Hanning window function.

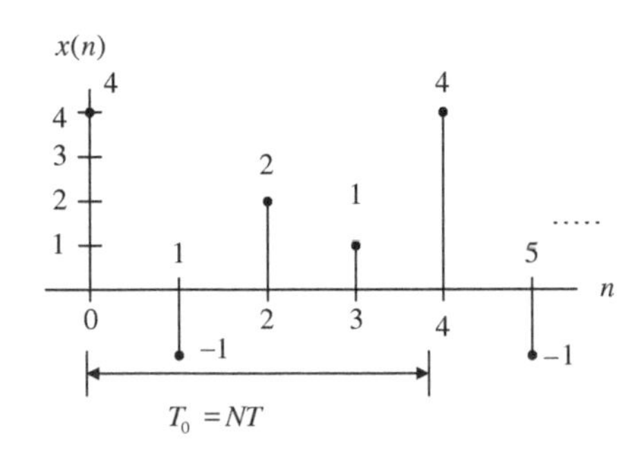

FIG. 4.46

Data sequence in Problem 4.14.

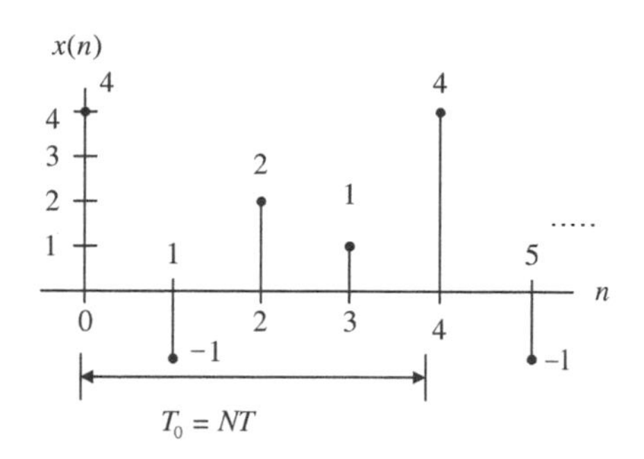

FIG. 4.47

Data sequence in Problem 4.19.

4.17 Compute the following window functions for a size of 10:

(a) Hamming window function.

(b) Hanning window function.

4.18 Given the following data sequence with a length of 6

$$x(0)=0, x(1)=0.2, x(2)=0, x(3)=-0.2, x(4)=0, x(5)=0.2$$

compute the windowed sequence $x_w(n)$ using

(a) Triangular window function.

(b) Hamming window function.

(c) Hanning window function.

4.19 Given a sequence in Fig. 4.47 where $f_s=100\,\text{Hz}$ and $T=0.01$ s, compute the amplitude spectrum, phase spectrum, and power spectrum using

(a) Triangular window.

(b) Hamming window.

(c) Hanning window.

4.20 Given a sinusoid

$$x(n)=2\times\sin\left(2000\times 2\pi\times\frac{n}{8000}\right)$$

obtained by using the sampling rate of $f_s=8000\,\text{Hz}$, we apply the DFT to compute the amplitude spectrum.

(a) Determine the frequency resolution when the data length is 100 samples. Without using the window function, is there any spectral leakage in the computed spectrum? Explain.

(b) Determine the frequency resolution when the data length is 73 samples. Without using the window function, is there any spectral leakage in the computed spectrum? Explain.

4.21 Given a sequence $x(n)$ for $0\le n\le 3$, where $x(0)=4$, $x(1)=3$, $x(2)=2$, and $x(3)=1$, evaluate its DFT $X(k)$ using the decimation-in-frequency FFT method, and determine the number of complex multiplications.

4.22 Given the DFT sequence $X(k)$ for $0\le k\le 3$ obtained in Problem 4.21, evaluate its inverse DFT $x(n)$ using the decimation-in-frequency FFT method.

4.23 Given a sequence $x(n)$ for $0 \le n \le 3$, where $x(0)=0.8$, $x(1)=0.6$, $x(2)=0.4$, and $x(3)=0.2$, evaluate its DFT $X(k)$ using the decimation-in-frequency FFT method, and determine the number of complex multiplications.

4.24 Given the DFT sequence $X(k)$ for $0 \le k \le 3$ obtained in Problem 4.23, evaluate its inverse DFT $x(n)$ using the decimation-in-frequency FFT method.

4.25 Given a sequence $x(n)$ for $0 \le n \le 3$, where $x(0)=4$, $x(1)=3$, $x(2)=2$, and $x(3)=1$, evaluate its DFT $X(k)$ using the decimation-in-time FFT method, and determine the number of complex multiplications.

4.26 Given the DFT sequence $X(k)$ for $0 \le k \le 3$ computed in Problem 4.25, evaluate its inverse DFT $x(n)$ using the decimation-in-time FFT method.

4.27 Given a sequence $x(n)$ for $0 \le n \le 3$, where $x(0)=0.8$, $x(1)=0.4$, $x(2)=-0.4$, and $x(3)=-0.2$, evaluate its DFT $X(k)$ using the decimation-in-time FFT method, and determine the number of complex multiplications.

4.28 Given the DFT sequence $X(k)$ for $0 \le k \le 3$ computed in Problem 4.27, evaluate its inverse DFT $x(n)$ using the decimation-in-time FFT method.

Computer Problems with MATLAB

Use MATLAB to solve Problems 4.29–4.30.

4.29 Given three sinusoids with the following amplitudes and phases:

$$x_1(t) = 5\cos(2\pi(500)t)$$

$$x_2(t) = 5\cos(2\pi(1200)t + 0.25\pi)$$

$$x_3(t) = 5\cos(2\pi(1800)t + 0.5\pi)$$

(a) Create a MATLAB program to sample each sinusoid and generate a sum of three sinusoids, that is, $x(n) = x_1(n) + x_2(n) + x_3(n)$, using a sampling rate of 8000 Hz, and plot the sum $x(n)$ over a range of time that will exhibit approximately 0.1 s.

(b) Use the MATLAB function fft() to compute DFT coefficients, and plot and examine the spectrum of the signal $x(n)$.

4.30 Using the sum of sinusoids in Problem 4.29:

(a) Generate the sum of sinusoids for 240 samples using a sampling rate of 8000 Hz.

(b) Write a MATLAB program to compute and plot the amplitude spectrum of the signal $x(n)$ with the FFT and using each of the following window functions:

(1) Rectangular window (no window).

(2) Triangular window.

(3) Hamming window.

(c) Examine the effect of spectral leakage for each window used in (b).

MATLAB Projects

4.31 Signal spectral analysis:

Given below are four practical signals, compute their one-sided spectra and create their time domain plots and spectral plots, respectively:

(a) Speech signal ("speech.dat"), sampling rate = 8000 Hz.
From the spectral plot, identify the first five formants.

(b) ECG signal ("ecg.dat"), sampling rate = 500 Hz.
From the spectral plot, identify the 60 Hz-interference component.

(c) Seismic data ("seismic.dat"), sampling rate = 15 Hz.
From the spectral plot, determine the dominant frequency component.

(d) Vibration signal of the acceleration response from a simple supported beam ("vbrdata.dat"), sampling rate = 1000 Hz.
From the spectral plot, determine the four dominant frequencies (modes).

4.32 Vibration signature analysis:
The acceleration signals measured from the gearbox can be used for monitoring the condition of the gears inside the gearbox. The early diagnosis of the gear condition can prevent the future catastrophic failure of the system. Given the following measurements and specifications (courtesy of SpectraQuest, Inc.):

(a) The input shaft has a speed of 1000 rpm and meshing frequency is approximately 300 Hz.

(b) Data specifications:

Sampling rate = 12.8 kHz
v0.dat: healthy condition
v1.dat: damage severity level 1 (lightly chipped gear)
v2.dat: damage severity level 2 (moderately chipped gear)
v3.dat: damage severity level 3 (chipped gear)
v4.dat: damage severity level 4 (heavily chipped gear)
v5.dat: damage severity level 5 (missing tooth)
Investigate the spectrum for each measurement and identify sidebands.
For each measurement, determine the ratio of the largest sideband amplitude over the amplitude of meshing frequency and investigate the ratio effect related to the damage severity.

Advanced Problems

4.33 Show that
a. $W_N^N = 1$; b. $W_N^{N/2} = -1$; c. $W_N^{N/4} = -j$; d. $W_N^{-m} = W_N^{N-m}$.

4.34 For a real sequence $x(n)$ defined for $0 \le n < N$, show that

$$X(N-k) = X^*(k).$$

4.35 For $x(n) = \delta(n-m)$ defined for $0 \le n,\ m < N$, show that

$$X(k) = W_N^{km}.$$

4.36 For $x(n) = 1$ defined for $0 \le n < N$, show that

$$X(k) = \begin{cases} N & k = 0 \\ 0 & \text{elsewhere} \end{cases}.$$

4.37 Consider a sequence of $x(n)$ defined for $0 \le n < N$, where $N = $ even

$$x(n) = \begin{cases} 1 & n = \text{even} \\ 0 & n = \text{odd} \end{cases},$$

show that

$$X(k) = \begin{cases} N/2 & k=0, k=N/2 \\ 0 & \text{elsewhere} \end{cases}.$$

4.38 A sequence is shown below:

$$x(n) = \frac{1}{2}\left(1 - \cos\frac{2\pi n}{N}\right) \text{ for } 0 \le n < N,$$

show that the DFT of $x(n)$ is given by

$$X(k) = \begin{cases} N/2 & k=0 \\ -N/4 & k=1, k=N-1 \\ 0 & \text{elsewhere} \end{cases}.$$

4.39 For $x(n) = n$ defined for $0 \le n < N$, show that

$$X(k) = \begin{cases} N(N-1)/2 & k=0 \\ -N/(1-W_N^k) & \text{elsewhere} \end{cases}.$$

4.40 Show that

$$\frac{1}{N}\sum_{n=0}^{N-1} \cos\left[\frac{\pi k(2n+1)}{2N}\right] \cos\left[\frac{\pi m(2n+1)}{2N}\right] = \begin{cases} 1 & k=m=0 \\ 1/2 & k=m\ne 0. \\ 0 & k \ne m \end{cases}$$

4.41 For a real sequence $x(n)$ defined for $0 \le n < N$, its discrete-cosine transform is given by

$$X_{\text{DCT}}(k) = \sum_{n=0}^{N-1} 2x(n)\cos\left[\frac{\pi k(2n+1)}{2N}\right], 0 \le k < N.$$

Show that

$$x(n) = \frac{1}{N}\sum_{k=0}^{N-1} a(k) X_{\text{DCT}}(k)\cos\left[\frac{\pi k(2n+1)}{2N}\right], 0 \le n < N,$$

where $\alpha(k) = \begin{cases} 1/2 & k=0 \\ 1 & 0<k<N \end{cases}$.

4.42 For a real sequence $x(n)$ defined for $0 \le n < N$, its discrete Harley transform is given by

$$X_{\text{DHT}}(k) = \sum_{n=0}^{N-1} 2x(n)\left[\cos\left(\frac{2\pi nk}{N}\right) + \sin\left(\frac{2\pi nk}{N}\right)\right], 0 \le k < N,$$

show that

$$x(n) = \frac{1}{N}\sum_{k=0}^{N-1} X_{\text{DHT}}(k)\left[\cos\left(\frac{2\pi nk}{N}\right) + \sin\left(\frac{2\pi nk}{N}\right)\right], 0 \le n < N.$$

4.43 Modify Program 4.3 to create a MATLAB function [ifftdinf()] for the inverse of FFT using the decimation-in-frequency method.

4.44 Modify Program 4.4 to create a MATLAB function [ifftdint()] for the inverse of FFT using the decimation-in-time method.

CHAPTER

THE Z-TRANSFORM

5

CHAPTER OUTLINE

5.1 DEFINITION

The *z-transform* is a very important tool in describing and analyzing digital systems. It also offers techniques for digital filter design and frequency analysis of digital signals. We begin with the definition of z-transform.

The z-transform of a causal sequence $x(n)$, designated by $X(z)$ or $Z(x(n))$, is defined as

$$\begin{aligned} X(z) &= Z(x(n)) = \sum_{n=0}^{\infty} x(n)z^{-n} \\ &= x(0)z^{-0} + x(1)z^{-1} + x(2)z^{-2} + \cdots \end{aligned} \tag{5.1}$$

where z is the complex variable. Here, the summation taken from $n=0$ to $n=\infty$ is in accordance with the fact that for most situations, the digital signal $x(n)$ is a causal sequence, that is, $x(n)=0$ for $n<0$. Thus, the definition in Eq. (5.1) is referred to as *one-sided z-transform* or *a unilateral transform*. In Eq. (5.1), all the values of z that make the summation to exist form a *region of convergence* in the z-transform domain, while all other values of z outside the region of convergence will cause the summation to diverge. The region of convergence is defined based on the particular sequence $x(n)$ being applied. Note that we deal with the unilateral z-transform first, and hence when performing an inverse z-transform (which we shall study later), we are restricted to the causal sequence. Next, we will briefly study two-sided z-transform for the noncausal sequence. Now let us study the following typical examples.

Digital Signal Processing. https://doi.org/10.1016/B978-0-12-815071-9.00005-1

EXAMPLE 5.1

Given the sequence

$$x(n) = u(n),$$

Find the z-transform of $x(n)$.

Solution:

From the definition of Eq. (5.1), the z-transform is given by

$$X(z) = \sum_{n=0}^{\infty} u(n)z^{-n} = \sum_{n=0}^{\infty} \left(z^{-1}\right)^n = 1 + \left(z^{-1}\right) + \left(z^{-1}\right)^2 + \cdots.$$

This is an infinite geometric series that converges to

$$X(z) = \frac{z}{z-1}$$

with a condition $|z^{-1}| < 1$. Note that for an infinite geometric series, we have $1 + r + r^2 + \cdots = \frac{1}{1-r}$ when $|r| < 1$. The region of convergence for all values of z is given as $|z| > 1$.

EXAMPLE 5.2

Considering the exponential sequence

$$x(n) = a^n u(n),$$

Find the z-transform of the sequence $x(n)$.

Solution:

From the definition of the z-transform in Eq. (5.1), it follows that

$$X(z) = \sum_{n=0}^{\infty} a^n u(n)z^{-n} = \sum_{n=0}^{\infty} \left(az^{-1}\right)^n = 1 + \left(az^{-1}\right) + \left(az^{-1}\right)^2 + \cdots.$$

Since this is a geometric series which will converge for $|az^{-1}| < 1$, it is further expressed as

$$X(z) = \frac{z}{z-a}, \text{ for } |z| > |a|.$$

The z-transforms for common sequences are summarized in Table 5.1.

Example 5.3 illustrates the method of finding the z-transform using Table 5.1.

EXAMPLE 5.3

Find the z-transform for each of the following sequences:

(a) $x(n) = 10u(n)$

(b) $x(n) = 10\sin(0.25\pi n)u(n)$

(c) $x(n) = (0.5)^n u(n)$

(d) $x(n) = (0.5)^n \sin(0.25\pi n)u(n)$

(e) $x(n) = e^{-0.1n}\cos(0.25\pi n)u(n)$

Solution:

(a) From Line 3 in Table 5.1, we get

$$X(z) = Z(10u(n)) = \frac{10z}{z-1}.$$

(b) Line 9 in Table 5.1 leads to

$$X(z) = 10Z(\sin(0.2\pi n)u(n)) = \frac{10\sin(0.25\pi)z}{z^2 - 2z\cos(0.25\pi) + 1} = \frac{7.07z}{z^2 - 1.414z + 1}.$$

(c) From Line 6 in Table 5.1, we yield

$$X(z) = Z((0.5)^n u(n)) = \frac{z}{z-0.5}.$$

(d) From Line 11 in Table 5.1, it follows that

$$X(z) = Z((0.5)^n \sin(0.25\pi n)u(n)) = \frac{0.5 \times \sin(0.25\pi)z}{z^2 - 2 \times 0.5\cos(0.25\pi)z + 0.5^2} = \frac{0.3536z}{z^2 - 0.7071z + 0.25}.$$

(e) From Line 14 in Table 5.1, it follows that

$$X(z) = Z(e^{-0.1n}\cos(0.25\pi n)u(n)) = \frac{z(z - e^{-0.1}\cos(0.25\pi))}{z^2 - 2e^{-0.1}\cos(0.25\pi)z + e^{-0.2}} = \frac{z(z - 0.6397)}{z^2 - 1.2794z + 0.8187}.$$

Table 5.1 Table of z-Transform Pairs

Line No.	$x(n), n \geq 0$	z-Transform $X(z)$	Region of Convergence
1	$x(n)$	$\sum_{n=0}^{\infty} x(n)z^{-n}$	
2	$\delta(n)$	1	$\|z\| > 0$
3	$au(n)$	$\frac{az}{z-1}$	$\|z\| > 1$
4	$nu(n)$	$\frac{z}{(z-1)^2}$	$\|z\| > 1$
5	$n^2u(n)$	$\frac{z(z+1)}{(z-1)^3}$	$\|z\| > 1$
6	$a^n u(n)$	$\frac{z}{z-a}$	$\|z\| > \|a\|$
7	$e^{-na}u(n)$	$\frac{z}{(z-e^{-a})}$	$\|z\| > e^{-a}$
8	$na^n u(n)$	$\frac{az}{(z-a)^2}$	$\|z\| > \|a\|$
9	$\sin(an)u(n)$	$\frac{z\sin(a)}{z^2 - 2z\cos(a) + 1}$	$\|z\| > 1$

Continued

Table 5.1 Table of z-Transform Pairs—cont'd

Line No.	$x(n), n \geq 0$	z-Transform $X(z)$	Region of Convergence
10	$\cos(an)u(n)$	$\frac{z[z-\cos(a)]}{z^2-2z\cos(a)+1}$	$\lvert z\rvert > 1$
11	$a^n \sin(bn)u(n)$	$\frac{[a\sin(b)]z}{z^2-[2a\cos(b)]z+a^2}$	$\lvert z\rvert > \lvert a\rvert$
12	$a^n \cos(bn)u(n)$	$\frac{z[z-a\cos(b)]}{z^2-[2a\cos(b)]z+a^2}$	$\lvert z\rvert > \lvert a\rvert$
13	$e^{-an}\sin(bn)u(n)$	$\frac{[e^{-a}\sin(b)]z}{z^2-[2e^{-a}\cos(b)]z+e^{-2a}}$	$\lvert z\rvert > e^{-a}$
14	$e^{-an}\cos(bn)u(n)$	$\frac{z[z-e^{-a}\cos(b)]}{z^2-[2e^{-a}\cos(b)]z+e^{-2a}}$	$\lvert z\rvert > e^{-a}$
15	$2\lvert A\rvert\lvert P\rvert^n\cos(n\theta+\phi)u(n)$ where P and A are complex constants defined by $P=\lvert P\rvert\angle\theta$, $A=\lvert A\rvert\angle\phi$	$\frac{Az}{z-P}+\frac{A^*z}{z-P^*}$	

5.2 PROPERTIES OF THE Z-TRANSFORM

In this section, we study some important properties of the z-transform. These properties are widely used in deriving the z-transfer functions of difference equations and solving the system output responses of linear digital systems with constant system coefficients, which will be discussed in Chapter 6.

Linearity: The z-transform is a linear transformation, which implies

$$Z(ax_1(n)+bx_2(n)) = aZ(x_1(n)) + bZ(x_2(n)), \tag{5.2}$$

where $x_1(n)$ and $x_2(n)$ denote the sampled sequences, while a and b are the arbitrary constants.

EXAMPLE 5.4

Find the z-transform of the sequence defined by

$$x(n) = u(n) - (0.5)^n u(n).$$

Solution:

Applying the linearity of the z-transform discussed above, we have

$$X(z) = Z(x(n)) = Z(u(n)) - Z(0.5^n(n)).$$

Using Table 5.1 yields

$$Z(u(n)) = \frac{z}{z-1}$$

and

$$Z(0.5^n u(n)) = \frac{z}{z-0.5}$$

Substituting these results in $X(z)$ leads to the final solution

$$X(z) = \frac{z}{z-1} - \frac{z}{z-0.5}.$$

Shift theorem: Given $X(z)$, the z-transform of a sequence $x(n)$, the z-transform of $x(n-m)$, the time-shifted sequence, is given by

$$Z(x(n-m)) = z^{-m}X(z), m \geq 0. \tag{5.3}$$

Note that $m \geq 0$, then $x(n-m)$ is obtained by right shifting $x(n)$ by m samples. Since the shift theorem plays a very important role in developing the transfer function from a difference equation, we verify the shift theorem for the causal sequence. Note that the shift theorem also works for the noncausal sequence.

Verification: Applying the z-transform to the shifted causal signal $x(n-m)$ leads to

$$\begin{aligned} Z(x(n-m)) &= \sum_{n=0}^{\infty} x(n-m)z^{-n} \\ &= x(-m)z^{-0} + \cdots + x(-1)z^{-(m-1)} + x(0)z^{-m} + x(1)z^{-m-1} + \cdots. \end{aligned}$$

Since $x(n)$ is assumed to be a causal sequence, this means that

$$x(-m) = x(-m+1) = \cdots = x(-1) = 0.$$

Then we achieve

$$Z(x(n-m)) = x(0)z^{-m} + x(1)z^{-m-1} + x(2)z^{-m-2} + \cdots. \tag{5.4}$$

Factoring z^{-m} from Eq. (5.4) and applying the definition of z-transform of $X(z)$, we get

$$Z(x(n-m)) = z^{-m}\left(x(0) + x(1)z^{-1} + x(2)z^{-2} + \ldots\right) = z^{-m}X(z).$$

EXAMPLE 5.5

Determine the z-transform of the following sequence:

$$y(n) = (0.5)^{(n-5)} \times u(n-5),$$

where $u(n-5) = 1$ for $n \geq 5$ and $u(n-5) = 0$ for $n < 5$.

Solution:

We first use the shift theorem to have

$$Y(z) = Z\left[(0.5)^{n-5}u(n-5)\right] = z^{-5}Z[(0.5)^n u(n)].$$

Using Table 5.1 leads to

$$Y(z) = z^{-5} \times \frac{z}{z-0.5} = \frac{z^{-4}}{z-0.5}.$$

Convolution: Given two sequences $x_1(n)$ and $x_2(n)$, their convolution can be determined as follows:

$$x(n) = x_1(n) * x_2(n) = \sum_{k=0}^{\infty} x_1(n-k)x_2(k), \tag{5.5}$$

where * designates the linear convolution. In z-transform domain, we have

$$X(z) = X_1(z)X_2(z) \tag{5.6}$$

Here, $X(z) = Z(x(n))$, $X_1(z) = Z(x_1(n))$, and $X_2(z) = Z(x_2(n))$.

EXAMPLE 5.6

Verify Eq. (5.6) using causal sequences $x_1(n)$ and $x_2(n)$.

Solution:

Taking the z-transform of Eq. (5.5) leads to

$$X(z)=\sum_{n=0}^{\infty}x(n)z^{-n}=\sum_{n=0}^{\infty}\sum_{k=0}^{\infty}x_1(n-k)x_2(k)z^{-n}.$$

This expression can be further modified to

$$X(z)=\sum_{n=0}^{\infty}\sum_{k=0}^{\infty}x_2(k)z^{-k}x_1(n-k)z^{-(n-k)}.$$

Now interchanging the order of the previous summation gives

$$X(z)=\sum_{k=0}^{\infty}x_2(k)z^{-k}\sum_{n=0}^{\infty}x_1(n-k)z^{-(n-k)}.$$

Now, let $m=n-k$:

$$X(z)=\sum_{k=0}^{\infty}x_2(k)z^{-k}\sum_{m=-k}^{\infty}x_1(m)z^{-m}=\sum_{k=0}^{\infty}x_2(k)z^{-k}\left(x_1(-k)z^{k}+\cdots x_1(-1)z+\sum_{m=0}^{\infty}x_1(m)z^{-m}\right).$$

Since $x_1(m)$ is a causal sequence, we have

$$X(z)=\sum_{k=0}^{\infty}x_2(k)z^{-k}\sum_{m=0}^{\infty}x_1(m)z^{-m}.$$

By the definition of Eq. (5.1), it follows that

$$X(z)=X_2(z)X_1(z)=X_1(z)X_2(z).$$

EXAMPLE 5.7

Given two sequences,

$$x_1(n)=3\delta(n)+2\delta(n-1)$$
$$x_2(n)=2\delta(n)-\delta(n-1),$$

(a) Find the z-transform of the convolution:

$$X(z)=Z(x_1(n)^*x_2(n)).$$

(b) Determine the convolution sum using the z-transform:

$$x(n)=x_1(n)^*x_2(n)=\sum_{k=0}^{\infty}x_1(k)x_2(n-k).$$

Solution:

(a) Applying z-transform for $x_1(n)$ and $x_2(n)$, respectively, it follows that

$$X_1(z) = 3 + 2z^{-1}$$

$$X_2(z) = 2 - z^{-1}.$$

Using the convolution property, we have

$$X(z) = X_1(z)X_2(z) = (3 + 2z^{-1})(2 - z^{-1}) = 6 + z^{-1} - 2z^{-2}.$$

(b) Applying the inverse z-transform and using the shift theorem and Line 1 of Table 5.1 leads to

$$x(n) = Z^{-1}(6 + z^{-1} - 2z^{-2}) = 6\delta(n) + \delta(n-1) - 2\delta(n-2).$$

Initial value theorem. Given the z-transfer function $X(z)$, then initial value of the causal sequence $x(n)$ can be determined by

$$x(0) = \lim_{z \to \infty} X(z). \tag{5.7}$$

Proof:

According to the definition of z-transform,

$$X(z) = x(0) + x(1)z^{-1} + x(2)z^{-2} + \cdots.$$

Taking a limit for $z \to \infty$ yields the initial theorem.

Final value theorem: Given the z-transfer function $X(z)$, then final value of the causal sequence can be determined by

$$x(\infty) = \lim_{z \to 1} [(z-1)X(z)]. \tag{5.8}$$

Proof:

Let us define

$$\tilde{X}(z) = x(0) + x(1)z^{-1} + x(2)z^{-2} + \cdots + x(N)z^{-N} \tag{5.9}$$

$$\hat{\tilde{X}}(z) = x(0) + x(1)z^{-1} + x(2)z^{-2} + \cdots + x(N)z^{-N} + x(N+1)z^{-(N+1)} \tag{5.10}$$

Multiplying Eq. (5.10) by z leads to

$$z\hat{\tilde{X}}(z) = x(0)z + x(1)z^0 + x(2)z^{-1} + \cdots + x(N)z^{-N+1} + x(N+1)z^{-N}. \tag{5.11}$$

Then, on subtracting Eq. (5.9) from (5.11), we obtain

$$z\hat{\tilde{X}}(z) - \tilde{X}(z) = x(0)(z-1) + x(1)(z-1)/z + \cdots + x(N)(z-1)/z^N + x(N+1)z^{-N}. \tag{5.12}$$

Considering $N \to \infty$, then $\tilde{X}(z) = \hat{\tilde{X}}(z) = X(z)$. Let $z \to 1$. Hence, we yield the final theorem as

$$\lim_{z \to 1} [zX(z) - X(z)] = \lim_{\substack{N \to \infty \\ z \to 1}} \left[x(0)(z-1) + x(1)(z-1)/z + \cdots + x(N)(z-1)/z^N + x(N+1)z^{-N}\right] = x(\infty).$$

EXAMPLE 5.8

Determine the initial and final values for each of the following z-transform functions:

(a) $X(z) = \frac{z}{z - 0.5}$

(b) $X(z) = \frac{z^2}{(z-1)(z-0.5)}$

Solution:

(a) $x(0) = \lim_{z\to\infty} X(z) = \lim_{z\to\infty} \frac{z}{z - 0.5} = 1$

$x(\infty) = \lim_{z\to 1} (z-1)X(z) = \lim_{z\to 1} \frac{(z-1)z}{z - 0.5} = 0$

(b) $x(0) = \lim_{z\to\infty} X(z) = \lim_{z\to\infty} \frac{z^2}{(z-1)(z-0.5)} = 1$

$x(\infty) = \lim_{z\to 1} (z-1)X(z) = \lim_{z\to 1} \frac{(z-1)z^2}{(z-1)(z-0.5)} = 2$

Table 5.2 Properties of z-Transform

Property	Time Domain	z-Transform
Linearity	$ax_1(n) + bx_2(n)$	$aZ(x_1(n)) + bZ(x_2(n))$
Shift theorem	$x(n-m)$	$z^{-m}X(z)$
Linear convolution	$x_1(n)*x_2(n) = \sum_{k=0}^{\infty} x_1(n-k)x_2(k)$	$X_1(z)X_2(z)$
Initial value theorem	$x(0) = \lim_{z\to\infty} X(z)$	$X(z)$
Final value theorem	$x(\infty) = \lim_{z\to 1} [(z-1)X(z)]$	$X(z)$

The properties of the z-transform discussed in this section are listed in Table 5.2.

5.3 INVERSE Z-TRANSFORM

The z-transform of a sequence $x(n)$ and the inverse z-transform of a function $X(z)$ are defined as, respectively

$$X(z) = Z(x(n)) \tag{5.13}$$

$$x(n) = Z^{-1}(X(z)), \tag{5.14}$$

where $Z(\ )$ is the z-transform operator, while $Z^{-1}(\ )$ is the inverse z-transform operator.

The inverse of the z-transform may be obtained by at least three methods:

1. Partial fraction expansion and look-up table (z-transform function must be rational).
2. Power series expansion.
3. Inversion Formula Method.

The first method is widely used, and it is assumed that the reader is well familiar with the partial fraction expansion method in learning Laplace transform. Therefore, we concentrate on the first method in this book. We will also outline the power series expansion and inversion formula methods. The detail of these two methods can also be found in the textbook by Oppenheim and Shafer (1975).

5.3.1 PARTIAL FRACTION EXPANSION AND LOOK-UP TABLE

For simple z-transform functions, we can directly find the inverse z-transform using Table 5.1. The key idea of the partial fraction expansion is that if $X(z)$ is a proper rational function of z, we can expand it to a sum of the first-order factors or higher-order factors using the partial fraction expansion that could be inverted by inspecting the z-transform table. The inverse z-transform using the z-transform table is first illustrated via the following example.

EXAMPLE 5.9

Find the inverse z-transform for each of the following functions:

(a) $X(z) = 2 + \frac{4z}{z-1} - \frac{z}{z-0.5}$

(b) $X(z) = \frac{5z}{(z-1)^2} - \frac{2z}{(z-0.5)^2}$

(c) $X(z) = \frac{10z}{z^2 - z + 1}$

(d) $X(z) = \frac{z^{-4}}{z-1} + z^{-6} + \frac{z^{-3}}{z+0.5}$

Solution:

(a) $x(n) = 2Z^{-1}(1) + 4Z^{-1}\left(\frac{z}{z-1}\right) - Z^{-1}\left(\frac{z}{z-0.5}\right).$

From Table 5.1, we have

$$x(n) = 2\delta(n) + 4u(n) - (0.5)^n u(n).$$

(b) $x(n) = Z^{-1}\left(\frac{5z}{(z-1)^2}\right) - Z^{-1}\left(\frac{2z}{(z-0.5)^2}\right) = 5Z^{-1}\left(\frac{z}{(z-1)^2}\right) - \frac{2}{0.5}Z^{-1}\left(\frac{0.5z}{(z-0.5)^2}\right).$

Then $x(n) = 5nu(n) - 4n(0.5)^n u(n)$.

(c) Since $X(z) = \frac{10z}{z^2 - z + 1} = \left(\frac{10}{\sin(a)}\right)\frac{\sin(a)z}{z^2 - 2z\cos(a) + 1}$,

By coefficient matching, we have

$$-2\cos(a) = -1.$$

Hence, $\cos(a) = 0.5$, and $a = 60°$. Substituting $a = 60°$ into the sine function leads to

$$\sin(a) = \sin(60°) = 0.866.$$

Finally, we have

$$x(n) = \frac{10}{\sin(a)}Z^{-1}\left(\frac{\sin(a)z}{z^2 - 2z\cos(a) + 1}\right) = \frac{10}{0.866}\sin(n \times 60°) = 11.547\sin(n \times 60°).$$

(d) Since $x(n) = Z^{-1}\left(z^{-5}\frac{z}{z-1}\right) + Z^{-1}\left(z^{-6} \times 1\right) + Z^{-1}\left(z^{-4}\frac{z}{z+0.5}\right)$,

Table 5.3 Partial Fraction(s) and Formulas for Constant(s)

Partial fraction with the first-order real pole:
$\frac{R}{z-p}, \; R=(z-p)\frac{X(z)}{z}\Big\vert_{z=p}$
Partial fraction with the first-order complex poles:
$\frac{Az}{(z-P)}+\frac{A^*z}{(z-P^*)}, \; A=(z-P)\frac{X(z)}{z}\Big\vert_{z=P}$
P^* = complex conjugate of P,
A^* = complex conjugate of A
Partial fraction with mth-order real poles:
$\frac{R_m}{(z-p)}+\frac{R_{m-1}}{(z-p)^2}+\cdots+\frac{R_1}{(z-p)^m}, \; R_k=\frac{1}{(k-1)!}\frac{d^{k-1}}{dz^{k-1}}\left((z-p)^m\frac{X(z)}{z}\right)\Big\vert_{z=p}$

Using Table 5.1 and the shift property, we get

$$x(n)=u(n-5)+\delta(n-6)+(-0.5)^{n-4}u(n-4).$$

Now, we are ready to deal with the inverse z-transform using the partial fraction expansion and look-up table. The general procedure is as follows:

(1) Eliminate the negative powers of z for the z-transform function $X(z)$.
(2) Determine the rational function $X(z)/z$ (assuming it is proper), and apply the partial fraction expansion to the determined rational function $X(z)/z$ using the formula in Table 5.3.
(3) Multiply the expanded function $X(z)/z$ by z on both sides of the equation to obtain $X(z)$.
(4) Apply the inverse z-transform using Table 5.1.

The partial fraction format and the formulas for calculating the constants are listed in Table 5.3.

Example 5.10 considers the situation of the z-transform function having first-order poles.

EXAMPLE 5.10

Find the inverse of the following z-transform:

$$X(z)=\frac{1}{(1-z^{-1})(1-0.5z^{-1})}.$$

Solution:

Eliminating the negative power of z by multiplying the numerator and denominator by z^2 yields

$$X(z)=\frac{z^2}{z^2(1-z^{-1})(1-0.5z^{-1})}$$
$$=\frac{z^2}{(z-1)(z-0.5)}.$$

Dividing both sides by z leads to

$$\frac{X(z)}{z}=\frac{z}{(z-1)(z-0.5)}.$$

Again, we write

$$\frac{X(z)}{z} = \frac{A}{(z-1)} + \frac{B}{(z-0.5)}.$$

Then A and B are constants found using the formula in Table 5.3, that is,

$$A = (z-1)\frac{X(z)}{z}\bigg|_{z=1} = \frac{z}{(z-0.5)}\bigg|_{z=1} = 2,$$

$$B = (z-0.5)\frac{X(z)}{z}\bigg|_{z=0.5} = \frac{z}{(z-1)}\bigg|_{z=0.5} = -1.$$

Thus

$$\frac{X(z)}{z} = \frac{2}{(z-1)} + \frac{-1}{(z-0.5)}.$$

Multiplying both sides by z gives

$$X(z) = \frac{2z}{(z-1)} + \frac{-z}{(z-0.5)}.$$

Using Table 5.1 of the z-transform pairs, it follows that

$$x(n) = 2u(n) - (0.5)^n u(n).$$

Tabulating this solution in terms of integer values of n, we obtain the results in Table 5.4.

Table 5.4 Determined Sequence in Example 5.10

n	0	1	2	3	4	…	∞
$x(n)$	1.0	1.5	1.75	1.875	1.9375	…	2.0

The following example considers the case where $X(z)$ has first-order complex poles.

EXAMPLE 5.11

Find $y(n)$ if $Y(z) = \dfrac{z^2(z+1)}{(z-1)(z^2-z+0.5)}$.

Solution:

Dividing $Y(z)$ by z, we have

$$\frac{Y(z)}{z} = \frac{z(z+1)}{(z-1)(z^2-z+0.5)}.$$

Applying the partial fraction expansion leads to

$$\frac{Y(z)}{z} = \frac{B}{z-1} + \frac{A}{(z-0.5-j0.5)} + \frac{A^*}{(z-0.5+j0.5)}.$$

Continued

EXAMPLE 5.11—CONT'D

We first find B:

$$B = (z-1)\frac{Y(z)}{z}\bigg|_{z=1} = \frac{z(z+1)}{(z^2 - z + 0.5)}\bigg|_{z=1} = \frac{1 \times (1+1)}{(1^2 - 1 + 0.5)} = 4.$$

Note that A and A^* are a complex conjugate pair. We determine A as follows:

$$A = (z - 0.5 - j0.5)\frac{Y(z)}{z}\bigg|_{z=0.5+j0.5} = \frac{z(z+1)}{(z-1)(z-0.5+j0.5)}\bigg|_{z=0.5+j0.5}$$

$$= \frac{(0.5+j0.5)(0.5+j0.5+1)}{(0.5+j0.5-1)(0.5+j0.5-0.5+j0.5)} = \frac{(0.5+j0.5)(1.5+j0.5)}{(-0.5+j0.5)j}.$$

Using the polar form, we get

$$A = \frac{(0.707\angle 45°)(1.58114\angle 18.43°)}{(0.707\angle 135°)(1\angle 90°)} = 1.58114\angle -161.57°$$

$$A^* = \overline{A} = 1.58114\angle 161.57°.$$

Assume that a first-order complex pole has form
$P = 0.5 + 0.5j = |P|\angle\theta = 0.707\angle 45°$ and $P^* = |P|\angle -\theta = 0.707\angle -45°$.
We have

$$Y(z) = \frac{4z}{z-1} + \frac{Az}{(z-P)} + \frac{A^* z}{(z-P^*)}$$

Applying the inverse z-transform from Line 15 in Table 5.1 leads to

$$y(n) = 4Z^{-1}\left(\frac{z}{z-1}\right) + Z^{-1}\left(\frac{Az}{(z-P)} + \frac{A^* z}{(z-P^*)}\right).$$

Using the previous formula, the inversion and subsequent simplification yield

$$y(n) = 4u(n) + 2|A|(|P|)^n \cos(n\theta + \phi)u(n)$$
$$= 4u(n) + 3.1623(0.7071)^n \cos(45^\circ n - 161.57^\circ)u(n).$$

The situation dealing with the real repeated poles is presented below.

EXAMPLE 5.12

Find $x(n)$ if $X(z) = \frac{z^2}{(z-1)(z-0.5)^2}$.

Solution:

Dividing both sides of the previous z-transform by z yields

$$\frac{X(z)}{z} = \frac{z}{(z-1)(z-0.5)^2} = \frac{A}{z-1} + \frac{B}{z-0.5} + \frac{C}{(z-0.5)^2},$$

where $A = (z-1)\frac{X(z)}{z}\bigg|_{z=1} = \frac{z}{(z-0.5)^2}\bigg|_{z=1} = 4.$

Using the formulas for mth-order real poles in Table 5.3, where $m=2$ and $p=0.5$, to determine B and C yields

$$B=R_2=\frac{1}{(2-1)!}\frac{d}{dz}\left\{(z-0.5)^2\frac{X(z)}{z}\right\}_{z=0.5}$$

$$=\frac{d}{dz}\left(\frac{z}{z-1}\right)\bigg|_{z=0.5}=\frac{-1}{(z-1)^2}\bigg|_{z=0.5}=-4$$

$$C=R_1=\frac{1}{(1-1)!}\frac{d^0}{dz^0}\left\{(z-0.5)^2\frac{X(z)}{z}\right\}_{z=0.5}$$

$$=\frac{z}{z-1}\bigg|_{z=0.5}=-1.$$

Then

$$X(z)=\frac{4z}{z-1}+\frac{-4z}{z-0.5}+\frac{-1z}{(z-0.5)^2}$$

The inverse z-transform for each term on the right-hand side of the above equation can be achieved by the result listed in Table 5.1, that is,

$$Z^{-1}\left\{\frac{z}{z-1}\right\}=u(n),$$

$$Z^{-1}\left\{\frac{z}{z-0.5}\right\}=(0.5)^n u(n),$$

$$Z^{-1}\left\{\frac{z}{(z-0.5)^2}\right\}=2n(0.5)^n u(n).$$

From these results, it follows that

$$x(n)=4u(n)-4(0.5)^n u(n)-2n(0.5)^n u(n).$$

5.3.2 PARTIAL FRACTION EXPANSION USING MATLAB

The MATLAB function **residue()** can be applied to perform the partial fraction expansion of a z-transform function $X(z)/z$. The syntax is given as

[R, P, K] = residue(B,A).

Here, B and A are the vectors consisting of coefficients for the numerator and denominator polynomials, $B(z)$ and $A(z)$, respectively. Note that $B(z)$ and $A(z)$ are the polynomials with increasing positive powers of z.

$$\frac{B(z)}{A(z)}=\frac{b_0z^M+b_1z^{M-1}+b_2z^{M-2}+\cdots+b_M}{z^N+a_1z^{N-1}+a_2z^{-2}+\cdots+a_N}. \tag{5.15}$$

The function returns the residues in vector R, corresponding poles in vector P, and polynomial coefficients (if any) in vector K. The expansion format is shown as

$$\frac{B(z)}{A(z)} = \frac{r_1}{z-p_1} + \frac{r_2}{z-p_2} + \cdots + k_0 + k_1 z^{-1} + \cdots.$$

For a pole p_j of multiplicity m, the partial fraction includes the following terms:

$$\frac{B(z)}{A(z)} = \cdots + \frac{r_j}{z-p_j} + \frac{r_{j+1}}{(z-p_j)^2} + \cdots + \frac{r_{j+m}}{(z-p_j)^m} + \cdots + k_0 + k_1 z^{-1} + \cdots.$$

EXAMPLE 5.13

Find the partial expansion for each of the following z-transform functions:

(a) $X(z) = \dfrac{1}{(1-z^{-1})(1-0.5z^{-1})}$

(b) $Y(z) = \dfrac{z^2(z+1)}{(z-1)(z^2-z+0.5)}$

(c) $X(z) = \dfrac{z^2}{(z-1)(z-0.5)^2}$

Solution:

(a) From MATLAB, we can show the denominator polynomial as

```
» conv([1 –1], [1 –0.5])
D =
1.0000 –1.5000 0.5000
```

This leads to

$$X(z) = \frac{1}{(1-z^{-1})(1-0.5z^{-1})} = \frac{1}{1-1.5z^{-1}+0.5^{-2}} = \frac{z^2}{z^2-1.5z+0.5}$$

and $\dfrac{X(z)}{z} = \dfrac{z}{z^2-1.5z+0.5}$.

From MATLAB, we have

```
» [R,P,K]=residue([1 0], [1 –1.5 0.5])
R =
2
–1
P =
1.0000
0.5000
K =
[]
»
```

Then the expansion is written as

$$X(z) = \frac{2z}{z-1} - \frac{z}{z-0.5}.$$

(b) From the MATLAB

```
» N=conv([1 0 0], [1 1])
N =
1 1 0 0
» D=conv([1 –1], [1 –1 0.5])
D =
1.0000 –2.0000 1.5000 –0.5000
```

we get

$$Y(z) = \frac{z^2(z+1)}{(z-1)(z^2-z+0.5)} = \frac{z^3+z^2}{z^3-2z^2+1.5z-0.5}$$

and

$$\frac{Y(z)}{z} = \frac{z^2+z}{z^3-2z^2+1.5z-0.5}.$$

Using the MATLAB residue function yields

```
» [R,P,K]=residue([1 1 0], [1 –2 1.5 –0.5])
R =
4.0000
–1.5000 – 0.5000i
–1.5000 + 0.5000i
P =
1.0000
0.5000 + 0.5000i
0.5000 – 0.5000i
K =
[]
»
```

Then the expansion is shown below

$$X(z) = \frac{Bz}{z-p_1} + \frac{Az}{z-p} + \frac{A^*z}{z-p^*},$$

where $B=4$, $p_1=1$,
$A=-1.5-0.5j$, $p=0.5+0.5j$,
$A^*=-1.5+0.5j$, and $p=0.5-0.5j$.
c. Similarly,

```
» D=conv(conv([1 –1], [1 –0.5]),[1 –0.5])
D =
1.0000 –2.0000 1.2500 –0.2500
```

Continued

EXAMPLE 5.13—CONT'D

then $X(z) = \frac{z^2}{(z-1)(z-0.5)^2} = \frac{z^2}{z^3 - 2z^2 + 1.25z - 0.25}$ and we yield

$$\frac{X(z)}{z} = \frac{z}{z^3 - 2z^2 + 1.25z - 0.25}.$$

From MATLAB, we obtain

```
» [R,P,K]=residue([1 0], [1 –2 1.25 –0.25])
R =
4.0000
–4.0000
–1.0000
P =
1.0000
0.5000
0.5000
K =
[]
»
```

Using the previous results leads to

$$X(z) = \frac{4z}{z-1} - \frac{4z}{z-0.5} - \frac{z}{(z-0.5)^2}.$$

5.3.3 POWER SERIES METHOD

By using a long division method, a rational function $X(z)$ can be expressed in the form of power series, that is,

$$X(z) = a_0 + a_1 z^{-1} + a_2 z^{-2} + \cdots. \tag{5.16}$$

Based on the definition of the unilateral z-transform, we also have

$$X(z) = x(0) + x(1)z^{-1} + x(2)z^{-2} + \cdots. \tag{5.17}$$

It can be seen that there is a one-to-one coefficient match between coefficients a_n and the sequence $x(n)$, that is,

$$x(n) = a_n, \text{ for } n \geq 0 \tag{5.18}$$

As an illustration, let us find the inverse of a given rational z-transform function:

$$X(z) = \frac{z}{z^2 - 2z + 1}$$

Using the long division method, we yield

$$
\begin{array}{r}
z^{-1}+2z^{-2}+\cdots \\
z^2-2z+1\overline{)\ z \qquad\qquad\quad} \\
\underline{z-2+z^{-1}} \\
2-z^{-1} \\
\underline{2-4z^{-1}+2z^{-2}} \\
3z^{-1}-2z^{-2} \\
\cdots
\end{array}
$$

This leads to

$$X(z) = 0 + 1z^{-1} + 2z^{-2} + 3z^{-3} + 4z^{-4} + \cdots$$

We see that

$a_0 = 0$, $a_1 = 1$, $a_2 = 2$, $a_3 = 3$, $a_4 = 4$, ...

which can be concluded as

$$x(n) = nu(n).$$

5.3.4 INVERSION FORMULA METHOD

Based on the definition of z-transform for a given sequence:

$$X(z) = \sum_{n=0}^{\infty} x(n)z^{-n}, \text{ for } |z| > R, \tag{5.19}$$

where $|z| > R$ specifies the region of convergence. From the complex variable and function analysis, we have

$$x(n) = \frac{1}{2\pi j} \oint_C z^{n-1} X(z) dz, \text{ for } n \geq 0, \tag{5.20}$$

where C is any simple closed curve enclosing $|z| = R$, and $\oint_C$ designates contour integration along C in the counterclockwise direction.

Proof: Applying the integral $\frac{1}{2\pi j}\oint_C z^{n-1}(.)dz$ with a close path of C on both sides of Eq. (5.20), we obtain

$$\frac{1}{2\pi j} \oint_C z^{n-1} X(z) dz = \sum_{m=0}^{\infty} x(m) \frac{1}{2\pi j} \oint_C z^{n-m} \frac{dz}{z}. \tag{5.21}$$

Let C be a circle of radius $\overline{R}$, where $\overline{R} > R$. Note that

$z = \overline{R}e^{j\theta}$ and $dz = j\overline{R}e^{j\theta}d\theta = jzd\theta$

The integral on the right side of Eq. (5.21) becomes

$$\frac{1}{2\pi j} \oint_C z^{n-m} X(z) \frac{dz}{z} = \frac{\overline{R}^{(n-m)}}{2\pi} \int_0^{2\pi} e^{j(n-m)\theta} d\theta = \begin{cases} 1 & n = m \\ 0 & \text{otherwise} \end{cases}. \tag{5.22}$$

Thus, we have

$$\frac{1}{2\pi j} \oint_C z^{n-1} X(z) dz = x(n).$$

The inversion integral can be evaluated through Cauchy's residue theorem, which is an important subject in the area of complex variables and function analysis. The key result is given by the following residue theorem.

Residue theorem: Given $X(z), |z| > R$, the inverse z-transform can be evaluated by

$$\begin{aligned} x(n) &= \frac{1}{2\pi j}\oint_C z^{n-1}X(z)dz \quad \text{for } n \geq 0 \\ &= \text{Sum of the residues } z^{n-1}X(z) \text{ corresponding to the poles of } z^{n-1}X(z) \\ &\quad \text{that reside inside a simple closed curve } C \text{ enclosing } |z| = R. \end{aligned} \tag{5.23}$$

If the function $z^{n-1}X(z)$ has a simple pole at $z=p$, the residue is evaluated by

$$R_{z=p} = (z-p)z^{n-1}X(z)\big|_{z=p}. \tag{5.24}$$

For a pole with the order of m at $z=p$, the residue can be determined by

$$R_{z=p} = \frac{d^{m-1}}{dz^{m-1}}\left[\frac{(z-p)^m}{(m-1)!}z^{n-1}X(z)\right]\Bigg|_{z=a}, \tag{5.25}$$

where p is the pole value and n is the order of pole.

EXAMPLE 5.14

Given the following z-transfer function, determine $x(n)$ using the inversion formula method.

$$X(z) = \frac{z^2}{(z-1)(z-0.5)^2}$$

Solution:

Since $z^{n-1}X(z) = \dfrac{z^{n+1}}{(z-1)(z-0.5)^2}$
has two poles, the sequence can be written as two residues, that is,

$$x(n) = (R_{z=1} + R_{z=0.5})u(n)$$

where

$$R_{z=1} = (z-1)\frac{z^{n+1}}{(z-1)(z-0.5)^2}\Bigg|_{z=1} = 4$$

$$\begin{aligned} R_{z=0.5} &= \frac{d^{2-1}}{dz^{2-1}}\left[\frac{(z-0.5)^2}{(2-1)!}\frac{z^{n+1}}{(z-1)(z-0.5)^2}\right]\Bigg|_{z=0.5} \\ &= \frac{d}{dz}\left[\frac{z^{n+1}}{(z-1)}\right]\Bigg|_{z=0.5} = \left[\frac{(n+1)z^n(z-1) - z^{n+1}}{(z-1)^2}\right]\Bigg|_{z=0.5} \\ &= -2(n+1)(0.5)^n - 4(0.5)^{n+1} = -2n(0.5)^n - 4(0.5)^n \end{aligned}$$

Thus, we have

$$x(n) = (R_{z=1} + R_{z=0.5})u(n) = 4u(n) - 2n(0.5)^n u(n) - 4(0.5)^n u(n)$$

EXAMPLE 5.15

Given the following z-transfer function, determine $x(n)$ using the inversion formula method.

$$X(z) = \frac{1}{z^2 - z + 0.5}$$

Solution:

$z^{n-1}X(z)$ has three poles for $n=0$ and two poles for $n>0$, that is,

$$z^{n-1}X(z) = \frac{1}{z(z^2 - z + 0.5)} \text{ for } n=0$$

$$z^{n-1}X(z) = \frac{z^{n-1}}{(z^2 - z + 0.5)} \text{ for } n>0$$

We need to apply the inversion formula for two different cases.

For $n=0$

$$x(0) = R_{z=0} + R_{z=0.5+j0.5} + R_{z=0.5-j0.5}$$

$$R_{z=0} = z\frac{1}{z(z^2 - z + 0.5)} = 2$$

$$R_{z=0.5+j0.5} = (z - 0.5 - j0.5)\left.\frac{1}{z(z - 0.5 - j0.5)(z - 0.5 + j0.5)}\right|_{z=0.5+j0.5}$$

$$= \frac{1}{-0.5 + j0.5}$$

$$R_{z=0.5-j0.5} = (z - 0.5 + j0.5)\left.\frac{1}{z(z - 0.5 - j0.5)(z - 0.5 + j0.5)}\right|_{z=0.5-j0.5}$$

$$= \frac{1}{-0.5 - j0.5}$$

$$x(0) = 2 + \frac{1}{-0.5 + j0.5} + \frac{1}{-0.5 - j0.5} = 0$$

For $n>0$

$$x(n) = R_{z=0.5+j0.5} + R_{z=0.5-j0.5}$$

$$R_{z=0.5+j0.5} = (z - 0.5 - j0.5)\left.\frac{z^{n-1}}{(z - 0.5 - j0.5)(z - 0.5 + j0.5)}\right|_{z=0.5+j0.5}$$

$$= \frac{(0.5 + j0.5)^{n-1}}{j} = \frac{(0.5 + j0.5)^n}{-0.5 + j0.5}$$

$$R_{z=0.5-j0.5} = (z - 0.5 + j0.5)\left.\frac{z^{n-1}}{(z - 0.5 - j0.5)(z - 0.5 + j0.5)}\right|_{z=0.5-j0.5}$$

$$= \frac{(0.5 - j0.5)^{n-1}}{-j} = \frac{(0.5 - j0.5)^n}{-0.5 - j0.5}$$

Continued

EXAMPLE 5.15—CONT'D

For $n>0$

$$\begin{aligned} x(n) &= R_{z=0.5+j0.5} + R_{z=0.5-j0.5} \\ &= \frac{(0.707e^{\pi/4})^n}{0.707e^{3\pi/4}} + \frac{(0.707e^{-\pi/4})^n}{0.707e^{-3\pi/4}} \\ &= (0.707)^{n-1}e^{\pi n/4-3\pi/4} + (0.707)^{n-1}e^{-(\pi n/4-3\pi/4)} \\ &= 2(0.707)^{n-1}\cos(\pi n/4 - 3\pi/4) \end{aligned}$$

Finally,

$$x(n) = \begin{cases} 0 & n=0 \\ 2.282(0.707)^n \cos(\pi n/4 - 3\pi/4) & n>0 \end{cases}.$$

5.4 SOLUTION OF DIFFERENCE EQUATIONS USING THE Z-TRANSFORM

To solve a difference equation with initial conditions, we have to deal with time-shifted sequences such as $y(n-1), y(n-2), \ldots, y(n-m)$, and so on. Let us examine the z-transform of these terms. Using the definition of the z-transform, we have

$$\begin{aligned} Z(y(n-1)) &= \sum_{n=0}^{\infty} y(n-1)z^{-n} \\ &= y(-1) + y(0)z^{-1} + y(1)z^{-2} + \cdots \\ &= y(-1) + z^{-1}(y(0) + y(1)z^{-1} + y(2)z^{-2} + \cdots). \end{aligned}$$

It holds that

$$Z(y(n-1)) = y(-1) + z^{-1}Y(z). \tag{5.26}$$

Similarly, we can have

$$\begin{aligned} Z(y(n-2)) &= \sum_{n=0}^{\infty} y(n-2)z^{-n} \\ &= y(-2) + y(-1)z^{-1} + y(0)z^{-2} + y(1)z^{-3} + \cdots \\ &= y(-2) + y(-1)z^{-1} + z^{-2}(y(0) + y(1)z^{-1} + y(2)z^{-2} + \cdots) \end{aligned}$$

$$Z(y(n-2)) = y(-2) + y(-1)z^{-1} + z^{-2}Y(z), \tag{5.27}$$

$$\cdots$$

$$Z(y(n-m)) = y(-m) + y(-m+1)z^{-1} + \cdots + y(-1)z^{-(m-1)} + z^{-m}Y(z), \tag{5.28}$$

where $y(-m), y(-m+1), \ldots, y(-1)$ are the initial conditions. If all initial conditions are considered to be zero, that is,

$$y(-m) = y(-m+1) = \cdots = y(-1) = 0, \tag{5.29}$$

then Eq. (5.28) becomes

$$Z(y(n-m)) = z^{-m}Y(z), \tag{5.30}$$

which is the same as the shift theorem in Eq. (5.3).

The following two examples serve as illustrations of applying the z-transform to find the solutions of the difference equations. The procedure is as follows:

1. Apply z-transform to the difference equation.
2. Substitute the initial conditions.
3. Solve the difference equation in z-transform domain.
4. Find the solution in time domain by applying the inverse z-transform.

EXAMPLE 5.16

A digital signal processing (DSP) system is described by a difference equation

$$y(n)-0.5y(n-1)=5(0.2)^n u(n).$$

Determine the solution when the initial condition is given by $y(-1)=1$.

Solution:

Applying the z-transform on both sides of the difference equation and using Eq. (5.28), we have

$$Y(z)-0.5\left(y(-1)+z^{-1}Y(z)\right)=5Z(0.2^n u(n)).$$

Substituting the initial condition and $Z(0.2^n u(n))=Z(0.2^n u(n))=z/(z-0.2)$, we achieve

$$Y(z)-0.5\left(1+z^{-1}Y(z)\right)=5z/(z-0.2).$$

Simplification leads to

$$Y(z)-0.5z^{-1}Y(z)=0.5+5z/(z-0.2).$$

Factoring out $Y(z)$ and combining the right-hand side of the equation, it follows that

$$Y(z)(1-0.5z^{-1})=(5.5z-0.1)/(z-0.2).$$

Then we obtain

$$Y(z)=\frac{(5.5z-0.1)}{(1-0.5z^{-1})(z-0.2)}=\frac{z(5.5z-0.1)}{(z-0.5)(z-0.2)}.$$

Using the partial fraction expansion method leads to

$$\frac{Y(z)}{z}=\frac{5.5z-0.1}{(z-0.5)(z-0.2)}=\frac{A}{z-0.5}+\frac{B}{z-0.2},$$

where $$A=(z-0.5)\frac{Y(z)}{z}\bigg|_{z=0.5}=\frac{5.5z-0.1}{z-0.2}\bigg|_{z=0.5}=\frac{5.5\times 0.5-0.1}{0.5-0.2}=8.8333,$$

$$B=(z-0.2)\frac{Y(z)}{z}\bigg|_{z=0.2}=\frac{5.5z-0.1}{z-0.5}\bigg|_{z=0.2}=\frac{5.5\times 0.2-0.1}{0.2-0.5}=-3.3333.$$

Thus $$Y(z)=\frac{8.8333z}{(z-0.5)}+\frac{-3.3333z}{(z-0.2)},$$

which gives the solution as

$$y(n)=8.3333(0.5)^n u(n)-3.3333(0.2)^n u(n).$$

EXAMPLE 5.17

A relaxed (zero initial conditions) DSP system is described by a difference equation

$$y(n)+0.1y(n-1)-0.2y(n-2)=x(n)+x(n-1).$$

(a) Determine the impulse response $y(n)$ due to the impulse sequence $x(n)=\delta(n)$.
(b) Determine the system response $y(n)$ due to the unit step function excitation, where $u(n)=1$ for $n\geq 0$.

Solution:

(a) Applying the z-transform on both sides of the difference equation and using Eq. (5.3) or Eq. (5.30), we yield

$$Y(z)+0.1Y(z)z^{-1}-0.2Y(z)z^{-2}=X(z)+X(z)z^{-1}.$$

Factoring out $Y(z)$ on the left-hand side and substituting $X(z)=Z(\delta(n))=1$ on the right-hand side of the above equation achieves

$$Y(z)\left(1+0.1z^{-1}-0.2z^{-2}\right)=1\left(1+z^{-1}\right).$$

Then $Y(z)$ can be expressed as

$$Y(z)=\frac{1+z^{-1}}{1+0.1z^{-1}-0.2z^{-2}}.$$

To obtain the impulse response, which is the inverse z-transform of the transfer function, we multiply the numerator and denominator by z^2.
Thus

$$Y(z)=\frac{z^2+z}{z^2+0.1z-0.2}=\frac{z(z+1)}{(z-0.4)(z+0.5)}.$$

Using the partial fraction expansion method leads to

$$\frac{Y(z)}{z}=\frac{z+1}{(z-0.4)(z+0.5)}=\frac{A}{z-0.4}+\frac{B}{z+0.5},$$

where

$$A=(z-0.4)\frac{Y(z)}{z}\bigg|_{z=0.4}=\frac{z+1}{z+0.5}\bigg|_{z=0.4}=\frac{0.4+1}{0.4+0.5}=1.5556$$

$$B=(z+0.5)\frac{Y(z)}{z}\bigg|_{z=-0.5}=\frac{z+1}{z-0.4}\bigg|_{z=-0.5}=\frac{-0.5+1}{-0.5-0.4}=-0.5556.$$

Thus

$$Y(z)=\frac{1.5556z}{(z-0.4)}+\frac{-0.5556z}{(z+0.5)}$$

which gives the impulse response:

$$y(n)=1.5556(0.4)^n u(n)-0.5556(-0.5)^n u(n).$$

(b) To obtain the response due to a unit step function, the input sequence is set to be

$$x(n)=u(n)$$

and the corresponding z-transform is given by

$$X(z)=\frac{z}{z-1},$$

and note that

$$Y(z)+0.1Y(z)z^{-1}-0.2Y(z)z^{-2}=X(z)+X(z)z^{-1}.$$

Then the z-transform of the output sequence $y(n)$ can be yielded as

$$Y(z)=\left(\frac{z}{z-1}\right)\left(\frac{1+z^{-1}}{1+0.1z^{-1}-0.2z^{-2}}\right)=\frac{z^2(z+1)}{(z-1)(z-0.4)(z+0.5)}.$$

Using the partial fraction expansion method as before gives

$$Y(z)=\frac{2.2222z}{z-1}+\frac{-1.0370z}{z-0.4}+\frac{-0.1852z}{z+0.5},$$

and the system response is found by using Table 5.1:

$$y(n)=2.2222u(n)-1.0370(0.4)^n u(n)-0.1852(-0.5)^n u(n).$$

Since $x(n)=u(n)=\sum_{k=0}^{n}\delta(k)$, we expect $y_b(n)=\sum_{k=0}^{n}y_a(k)$ based on linearity, where $y_a(k)$ is the response from (a) while $y_b(n)$ the response from (b). The verification is shown below: For $n\geq 0$

$$\begin{aligned}y_b(n)&=\sum_{k=0}^{n}\left[1.5556(0.4)^k-0.5556(-0.5)^k\right]=\sum_{k=0}^{n}1.5556(0.4)^k-\sum_{k=0}^{n}0.5556(-0.5)^k\\&=1.5556\frac{1-(0.4)^{n+1}}{1-0.4}-0.5556\frac{1-(-0.5)^{n+1}}{1-(-0.5)}\\&=2.2222-1.0370(0.4)^n-0.1852(-0.5)^n.\end{aligned}$$

5.5 TWO-SIDED Z-TRANSFORM

It is practical to study z-transform for a causal sequence, that is, $x(n)=0$ for $n<0$. However, from a theoretical point of view, we can briefly investigate a two-sized z-transform, which is applied for a noncausal sequence, $x(n)\neq 0$ for $n<0$, that is

$$X(z)=\sum_{n=-\infty}^{\infty}x(n)z^{-n}. \tag{5.31}$$

Note that Eq. (5.31) converges everywhere outside a circle with a radius R_1 if $x(n)=0$ for $n<0$; and on the other hand, Eq. (5.31) converges everywhere inside a circle with a radius R_2 if $x(n)=0$ for $n\geq 0$. Thus, region of convergence for Eq. (5.31) becomes an annular region , that is, $R_1<|z|<R_2$.

The inversion formula according to Cauchy's residue theorem for the two-sized z-transform is listed below

$$x(n)=\frac{1}{2\pi j}\oint_C z^{n-1}X(z)dz \text{ for } |n|<\infty$$
$$=\begin{cases}\text{Sum of the residues } z^{n-1}X(z) \text{ corresponding to the poles of } z^{n-1}X(z) & \\ \text{that reside inside a simple closed curve C enclosing } |z|=R_1 & n\geq 0\\ -\text{Sum of the residues } z^{n-1}X(z) \text{ corresponding to the poles of } z^{n-1}X(z) & \\ \text{that reside outside a simple closed curve C enclosing } |z|=R_2 & n<0\end{cases}. \quad (5.32)$$

Let us examine the following examples.

EXAMPLE 5.18

Given the noncausal sequence defined below

$$x(n)=\begin{cases}(-2)^n & n<0\\ (0.5)^n & n\geq 0\end{cases}$$

Determine $X(z)$.

Solution:

Appling Eq. (5.31), it follows that

$$X(z)=\sum_{n=-\infty}^{-1}\left(-2z^{-1}\right)^n+\sum_{n=0}^{\infty}\left(0.5z^{-1}\right)^2=X_1(z)+X_2(z)$$

where

$$X_1(z)=\sum_{n=-\infty}^{-1}\left(-2z^{-1}\right)^n=\sum_{n=0}^{\infty}\left[(-2)^{-1}z\right]^n-1=\frac{1}{1-(-2)^{-1}z}-1,$$

for $\left|(-2)^{-1}z\right|<1$.

That is,

$$X_1(z)=\frac{1}{1-(-2)^{-1}z}-1=\frac{-0.5z}{1+0.5z} \text{ for } |z|<2$$

and

$$X_2(z)=\sum_{n=0}^{\infty}\left(0.5z^{-1}\right)^2=\frac{1}{1-0.5z^{-1}}=\frac{z}{z-0.5} \text{ for } |z|>0.5.$$

Finally, we achieve

$$X(z)=\frac{-0.5z}{1+0.5z}+\frac{z}{z-0.5}=\frac{1.25z}{(1+0.5z)(z-0.5,)} \text{ for } 0.5<|z|<2.$$

EXAMPLE 5.19

Given

$$X(z)=\frac{-0.5z}{(z-0.5)(z-1)} \text{ for } 0.5<|z|<1,$$

Determine $x(n)$.

Solution:

We first form

$$z^{n-1}X(z) = \frac{-0.5z^n}{(z-0.5)(z-1)}.$$

A closed path C is chosen and plotted in Fig. 5.1 for $n \geq 0$ and $n < 0$ with poles marked correspondingly.

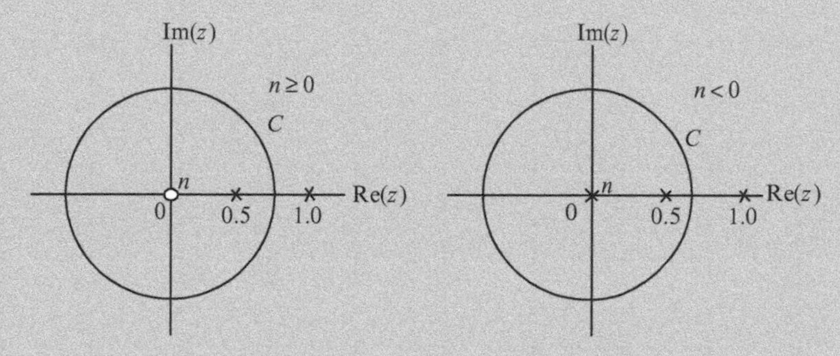

FIG. 5.1

Plots of the poles $z^{n-1}X(z)$ in Example 5.19 for $n \geq 0$ and $n < 0$, respectively.

For $n \geq 0$, C encloses only pole $z = 0.5$. Applying Eq. (5.32), it follows that

$$x(n) = (z-0.5)\frac{-0.5z^n}{(z-0.5)(z-1)}\bigg|_{z=0.5} = (0.5)^n.$$

For $n < 0$, there is only one pole $z = 1$ outside C. The pole at $z = 0.5$ and nth-order pole at $z = 0$ are inside C. Applying Eq. (5.32) leads to

$$x(n) = -(z-1)\frac{-0.5z^n}{(z-0.5)(z-1)}\bigg|_{z=1} = 1.$$

The combined result is given below:

$$x(n) = \begin{cases} 1 & n < 0 \\ (0.5)^n & n \geq 0 \end{cases}.$$

5.6 SUMMARY

1. The one-sided (unilateral) z-transform was defined, which can be used to transform the causal sequence to the z-transform domain.
2. The look-up table of the z-transform determines the z-transform for a simple causal sequence, or the causal sequence from a simple z-transform function.
3. The important properties of the z-transform, such as linearity, shift theorem, convolution, and initial and final value theorems were introduced. The shift theorem can be used to solve a difference equation. The z-transform of a digital convolution of two digital sequences is equal to the product of their z-transforms.

4. The method of finding the inverse of the z-transform, such as the partial fraction expansion, inverses the complicated z-transform function, which can have first-order real poles, multiple-order real poles, and first-order complex poles assuming that the z-transform function is proper. The MATLAB tool for the partial fraction expansion method was introduced. In addition, the power series and inversion formula methods were also described.
5. The z-transform can be applied to solve linear difference equations with nonzero initial conditions and zero initial conditions.
6. Two-sided z-transform was briefly introduced.

5.7 PROBLEMS

5.1 Find the z-transform for each of the following sequences:

(a) $x(n)=4u(n)$
(b) $x(n)=(-0.7)^n u(n)$
(c) $x(n)=4e^{-2n}u(n)$
(d) $x(n)=4(0.8)^n\cos(0.1\pi n)u(n)$
(e) $x(n)=4e^{-3n}\sin(0.1\pi n)u(n)$.

5.2 Using the properties of the z-transform, find the z-transform for each of the following sequences:

(a) $x(n)=u(n)+(0.5)^n u(n)$
(b) $x(n)=e^{-3(n-4)}\cos(0.1\pi(n-4))u(n-4)$, where $u(n-4)=1$ for $n\geq 4$ while $u(n-4)=0$ for $n<4$.

5.3 Find the z-transform for each of the following sequences:

(a) $x(n)=3u(n-4)$
(b) $x(n)=2(-0.5)^n u(n)$
(c) $x(n)=5e^{-2(n-3)}u(n-3)$
(d) $x(n)=6(0.6)^n\cos(0.2\pi n)u(n)$
(e) $x(n)=4e^{-3(n-1)}\sin(0.2\pi(n-1))u(n-1)$.

5.4 Using the properties of the z-transform, find the z-transform for each of the following sequences:

(a) $x(n)=-2u(n)-(0.75)^n u(n)$
(b) $x(n)=e^{-2(n-3)}\sin(0.2\pi(n-3))u(n-3)$, where $u(n-3)=1$ for $n\geq 3$ while $u(n-3)=0$ for $n<3$.

5.5 Given two sequences

$$x_1(n)=5\delta(n)-2\delta(n-2)\ \text{and}\ x_2(n)=3\delta(n-3),$$

(a) Determine the z-transform of convolution of the two sequences using the convolution property of z-transform

$$X(z)=X_1(z)X_2(z);$$

(b) Determine the convolution by the inverse z-transform from the result in (a)

$$x(n)=Z^{-1}(X_1(z)X_2(z)).$$

5.6 Using Table 5.1 and z-transform properties, find the inverse z-transform for each of the following functions:

(a) $X(z)=4-\frac{10z}{z-1}-\frac{z}{z+0.5}$

(b) $X(z)=\frac{-5z}{(z-1)}+\frac{10z}{(z-1)^2}+\frac{2z}{(z-0.8)^2}$

(c) $X(z)=\frac{z}{z^2+1.2z+1}$

(d) $X(z)=\frac{4z^{-4}}{z-1}+\frac{z^{-1}}{(z-1)^2}+z^{-8}+\frac{z^{-5}}{z-0.5}$

5.7 Given two sequences

$$x_1(n)=-2\delta(n)+5\delta(n-2) \text{ and } x_2(n)=4\delta(n-4),$$

(a) Determine the z-transform of convolution of the two sequences using the convolution property of z-transform

$$X(z)=X_1(z)X_2(z);$$

(b) Determine convolution by the inverse z-transform from the result in (a)

$$x(n)=Z^{-1}(X_1(z)X_2(z)).$$

5.8 Using Table 5.1 and z-transform properties, find the inverse z-transform for each of the following functions:

(a) $X(z)=5-\frac{7z}{z+1}-\frac{3z}{z-0.5}$;

(b) $X(z)=\frac{-3z}{(z-0.5)}+\frac{8z}{(z-0.8)}+\frac{2z}{(z-0.8)^2}$

(c) $X(z)=\frac{3z}{z^2+1.414z+1}$

(d) $X(z)=\frac{5z^{-5}}{z-1}-\frac{z^{-2}}{(z-1)^2}+z^{-10}+\frac{z^{-3}}{z-0.75}$

5.9 Using the partial fraction expansion method, find the inverse of the following z-transforms:

(a) $X(z)=\frac{1}{z^2-0.3z-0.04}$

(b) $X(z)=\frac{z}{(z-0.2)(z+0.4)}$

(c) $X(z)=\frac{z}{(z+0.2)(z^2-z+0.5)}$

(d) $X(z)=\frac{z(z+0.5)}{(z-0.1)^2(z-0.6)}$

5.10 A system is described by the difference equation

$$y(n)+0.5y(n-1)=2(0.8)^n u(n).$$

Determine the solution when the initial condition is $y(-1)=2$.

5.11 Using the partial fraction expansion method, find the inverse of the following z-transforms:

(a) $X(z) = \dfrac{1}{z^2 + 0.5z + 0.06}$

(b) $X(z) = \dfrac{z}{(z+0.3)(z-0.5)}$

(c) $X(z) = \dfrac{5z}{(z-0.75)(z^2 - z + 0.5)}$

(d) $X(z) = \dfrac{2z(z-0.4)}{(z-0.2)^2(z+0.8)}$

5.12 A system is described by the difference equation

$$y(n) + 0.2y(n-1) = 4(0.3)^n u(n).$$

Determine the solution when the initial condition is $y(-1) = 1$.

5.13 A system is described by the difference equation

$$y(n) - 0.5y(n-1) + 0.06y(n-2) = (0.4)^{n-1} u(n-1).$$

Determine the solution when the initial conditions are $y(-1) = 1$, and $y(-2) = 2$.

5.14 Given the following difference equation with the input-output relationship of a certain initially relaxed system (all initial conditions are zero),

$$y(n) - 0.7y(n-1) + 0.1y(n-2) = x(n) + x(n-1),$$

(a) Find the impulse response sequence $y(n)$ due to the impulse sequence $\delta(n)$.

(b) Find the output response of the system when the unit step function $u(n)$ is applied.

5.15 A system is described by the difference equation

$$y(n) - 0.6y(n-1) + 0.08y(n-2) = (0.5)^{n-1} u(n-1).$$

Determine the solution when the initial conditions are $y(-1) = 2$, and $y(-2) = 1$.

5.16 Given the following difference equation with the input-output relationship of a certain initially relaxed system (all initial conditions are zero),

$$y(n) - 0.6y(n-1) + 0.25y(n-2) = x(n) + x(n-1),$$

(a) Find the impulse response sequence $y(n)$ due to the impulse sequence $\delta(n)$.

(b) Find the output response of the system when the unit step function $u(n)$ is applied.

5.17 Given the following difference equation with the input-output relationship of a certain initially relaxed DSP system (all initial conditions are zero),

$$y(n) - 0.4y(n-1) + 0.29y(n-2) = x(n) + 0.5x(n-1),$$

(a) Find the impulse response sequence $y(n)$ due to the impulse sequence $\delta(n)$.

(b) Find the output response of the system when the unit step function $u(n)$ is applied.

5.18 Given the following difference equation with the input-output relationship of a certain initially relaxed DSP system (all initial conditions are zero)

$$y(n) - 0.2y(n-1) + 0.17y(n-2) = x(n) + 0.3x(n-1),$$

(a) Find the impulse response sequence $y(n)$ due to the impulse sequence $\delta(n)$.

(b) Find the output response of the system when the unit step function $u(n)$ is applied.

5.19 Use the initial and final value theorems to find $x(0)$ and $x(\infty)$ for Problem 5.11(a), (b), (d).
5.20 Use power series method to find $x(0)$, $x(1)$, $x(2)$, $x(3)$, $x(4)$ for Problem 5.11(a), (b).
5.21 Use the residue formula to find the inverse of the z-transform for Problem 5.11(a), (b), (d).

Advanced Problems:

5.22 If $y(n)=e^{-an}x(n)$, where $a\geq 0$ and $n\geq 0$, show that

$$Y(z)=X(ze^{a}).$$

5.23 If $y(n)=nx(n)$, where $a\geq 0$ and $n\geq 0$, show that

$$Y(z)=-z\frac{dX(z)}{dz}.$$

5.24 Given a difference equation

$$y(n)-y(n-1)+y(n-2)=0,\ n\geq 0$$

with initial conditions $y(-1)=1$ and $y(-2)=0$.

(a) Show that

$$Y(z)=\frac{z(z-1)}{(z-e^{j\pi/3})(z-e^{-j\pi/3})};$$

(b) Use the inversion formula method to show that

$$y(n)=\frac{\sin[(n+1)\pi/3]-\sin(n\pi/3)}{\sin(\pi/3)},\ n\geq 0.$$

5.25 Given a difference equation

$$y(n)=2\cos(\Omega_0)y(n-1)-y(n-2),\ n\geq 0$$

with initial conditions $y(-1)=-\sin(\Omega_0)$ and $y(-2)=-\sin(2\Omega_0)$, show that

$$y(n)=\sin(\Omega_0 n), n\geq 0.$$

5.26 For the following rational z-transfer function

$$X(z)=\frac{b_0+b_1z^{-1}+b_2z^{-2}+b_3z^{-3}+\cdots}{1+a_1z^{-1}+a_2z^{-2}+a_3z^{-3}+\cdots}$$

and the definition of z-transform: $X(z)=x(0)+x(1)z^{-1}+x(2)z^{-2}+x(3)z^{-3}+\cdots$, show that

$$x(0)=b_0 \text{ and } x(n)=b_n-\sum_{k=1}^{n}x(n-k)a_k \text{ for } n>0.$$

5.27 Suppose

$$X(z)=\sum_{n=0}^{\infty}x(n)z^{-n}.$$

Show that

$$\sum_{n=0}^{\infty} x^2(n) = \frac{1}{2\pi j}\oint_C z^{-1}X(z)X\left(z^{-1}\right)dz.$$

5.28 For $z = e^{j\Omega}$, show that

$$\sum_{n=0}^{\infty} x^2(n) = \frac{1}{2\pi}\oint_C \left|X\left(e^{j\Omega}\right)\right|^2 d\Omega.$$

5.29 Given

$$X(z) = \frac{(a-1)}{(1-az^{-1})(z-1)} \text{ and } a < |z| < 1, 0 < a < 1,$$

use the inversion formula to show that

$$x(n) = \begin{cases} 1 & n < 0 \\ a^n & n \geq 0 \end{cases}.$$

5.30 Given

$$X(z) = \frac{(a-b)}{(1-az^{-1})(z-b)} \quad \text{and} \quad a < |z| < b,\ 0 < a < 1 \text{ and } b > 1,$$

use the inversion formula to show that

$$x(n) = \begin{cases} b^n & n < 0 \\ a^n & n \geq 0 \end{cases}.$$

CHAPTER

DIGITAL SIGNAL PROCESSING SYSTEMS, BASIC FILTERING TYPES, AND DIGITAL FILTER REALIZATIONS

6

CHAPTER OUTLINE

6.1 DIFFERENCE EQUATION AND DIGITAL FILTERING

In this chapter, we begin with developing the filtering concept of digital signal processing (DSP) systems. With the knowledge acquired in Chapter 5, z-transform, we will learn how to describe and analyze linear time-invariant systems. We will also become familiar with digital filtering types and their realization structures.

A DSP system (digital filter) is described in Fig. 6.1.

Let $x(n)$ and $y(n)$ be a DSP system's input and output, respectively. We can express the relationship between the input and the output of a DSP system by the following *difference equation*

Digital Signal Processing. https://doi.org/10.1016/B978-0-12-815071-9.00006-3

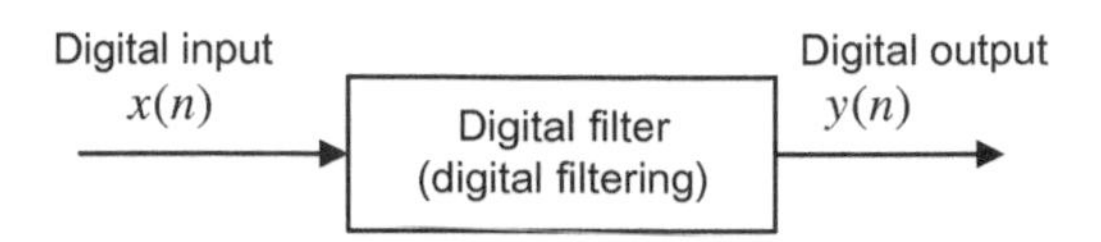

FIG. 6.1

DSP system with input and output.

$$\begin{aligned} y(n) = b_0x(n) + b_1x(n-1) + \cdots + b_Mx(n-M) \\ -a_1y(n-1) - \cdots - a_Ny(n-N) \end{aligned}, \tag{6.1}$$

where b_i, $0 \leq i \leq M$ and a_j, $1 \leq j \leq N$ represent the coefficients of the system and n is the time index. Eq. (6.1) can also be written as

$$y(n) = \sum_{i=0}^{M} b_i x(n-i) - \sum_{j=1}^{N} a_j y(n-j). \tag{6.2}$$

From Eqs. (6.1) and (6.2), we observe that the DSP system output is the weighted summation of the current input value $x(n)$ and its past values: $x(n-1)$, …, $x(n-M)$, and past output sequence: $y(n-1)$, …, $y(n-N)$. The system can be verified as linear, time invariant, and causal. If the initial conditions are specified, we can compute system output (time response) $y(n)$ recursively. This process is referred to as *digital filtering*. We will illustrate filtering operations by Examples 6.1 and 6.2.

EXAMPLE 6.1

Compute the system output

$$y(n) = 0.5y(n-2) + x(n-1)$$

for the first four samples using the following initial conditions:

(a) Initial conditions: $y(-2)=1$, $y(-1)=0$, $x(-1)=-1$, and input $x(n)=(0.5)^n u(n)$.
(b) Zero initial conditions: $y(-2)=0$, $y(-1)=0$, $x(-1)=0$, and input $x(n)=(0.5)^n u(n)$.

Solution:

According to Eq. (6.1), we identify the system coefficients as.
$N=2$, $M=1$, $a_1=0$, $a_2=-0.5$, $b_0=0$, and $b_1=1$.

(a) Setting $n=0$, and using initial conditions, we obtain the input and output as

$$x(0) = (0.5)^0 u(0) = 1$$

$$y(0) = 0.5y(-2) + x(-1) = 0.5 \cdot 1 + (-1) = -0.5.$$

Setting $n=1$, and using the initial condition $y(-1)=0$, we achieve

$$x(1) = (0.5)^1 u(1) = 0.5$$

$$y(1) = 0.5y(-1) + x(0) = 0.5 \cdot 0 + 1 = 1.0.$$

Similarly, using the past output $y(0)=-0.5$, we get

$$x(2) = (0.5)^2 u(2) = 0.25$$

$$y(2) = 0.5y(0) + x(1) = 0.5 \cdot (-0.5) + 0.5 = 0.25$$

and with $y(1)=1.0$, we yield

$$x(3)=(0.5)^3u(3)=0.125$$

$$y(3)=0.5y(1)+x(2)=0.5\cdot 1+0.25=0.75$$

$$\cdots$$

Clearly, $y(n)$ could be recursively computed for $n>3$.

(b) Setting $n=0$, we obtain $x(0)=(0.5)^0u(0)=1$

$$y(0)=0.5y(-2)+x(-1)=0\cdot 1+0=0$$

Setting $n=1$, we achieve

$$x(1)=(0.5)^1u(1)=0.5$$

$$y(1)=0.5y(-1)+x(0)=0\cdot 0+1=1.$$

Similarly, with the past output $y(0)=0$, we determine

$$x(2)=(0.5)^2u(2)=0.25$$

$$y(2)=0.5y(0)+x(1)=0.5\cdot 0+0.5=0.5$$

and with $y(1)=1$, we obtain

$$x(3)=(0.5)^3u(3)=0.125$$

$$y(3)=0.5y(1)+x(2)=0.5\cdot 1+0.25=0.75$$

$$\cdots$$

Clearly, $y(n)$ could be recursively computed for $n>3$.

EXAMPLE 6.2

Given the DSP system

$$y(n)=2x(n)-4x(n-1)-0.5y(n-1)-y(n-2)$$

with initial conditions $y(-2)=1$, $y(-1)=0$, $x(-1)=-1$, and the input $x(n)=(0.8)^n u(n)$, compute the system response $y(n)$ for 20 samples using MATLAB.

Solution:

Program 6.1 lists the MATLAB program for computing the system response $y(n)$. The top plot in Fig. 6.2 shows the input sequence. The middle plot displays the filtered output using the initial conditions, and the bottom plot shows the filtered output for zero initial conditions. As we can see, both system outputs are different at the beginning portion, while they approach the same value later.

Program 6.1 MATLAB program for Example 6.2.

```
% Example 6.2
% Compute y(n)=2x(n)-4x(n 1)-0.5y(n-1)-0.5y(n-2)
% Nonzero initial conditions:
% y(-2)=1, y(-1)=0, x(-1) =-1, and x(n)=(0.8)^n*u(n)
%
```

Continued

EXAMPLE 6.2—CONT'D

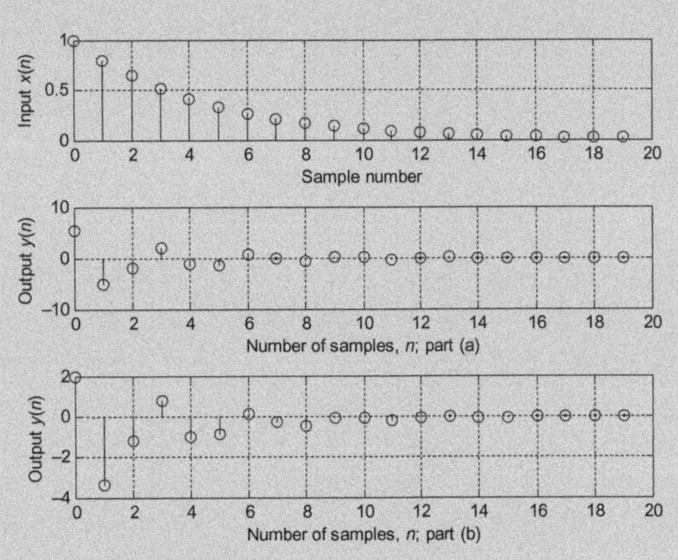

Part (a): response with initial conditions;
Part (b): response with zero initial conditions.

FIG. 6.2

Plots of the input and system outputs $y(n)$ for Example 6.2. Part (a): response with initial conditions; Part (b): response with zero initial conditions.

```
y = zeros(1,20);      %Set up a vector to store y(n)
y = [ 1 0 y];         %Add initial conditions of y(-2) and y(-1)
n=0:1:19;             %Compute time indexes
x=(0.8).^n;           %Compute 20 input samples of x(n)
x = [ 0 -1 x];        %Add initial conditions of x(-2)=0 and x(-1)=1
for n=1:20
  y(n+2)= 2*x(n+2)-4*x(n+1)-0.5*y(n+1)-0.5*y(n); %Compute 20 outputs of y(n)
end
n=0:1:19;
subplot(3,1,1);stem(n,x(3:22));grid;ylabel('Input x(n)');xlabel('Sample number');
Subplot(3,1,2); stem(n,y(3:22)),grid;
xlabel('Number of samples, n; part (a)'); ylabel('Output y(n)');
y(3:22)          %Output y(n)
%Zero- initial conditions:
% y(-2)=0, y(-1)=0, x(-1) =0, and x(n)=(0.8)^n
%
y = zeros(1,20);                %Set up a vector to store y(n)
y = [ 0 0 y];                   %Add zero initial conditions of y(-2) and y(-1)
n=0:1:19;                       %Compute time indexes
x=(0.8).^n;                     %Compute 20 input samples of x(n)
x = [ 0 0 x];                   %Add zero initial conditions of x(-2)=0 and x(-1)=0
```

```
for n=1:20
 y(n+2)= 2*x(n+2)-4*x(n+1)-0.5*y(n+1)-0.5*y(n); %Compute 20 outputs of y(n)
end
n=0:1:19
subplot(3,1,3); stem(n,y(3:22)),grid;
xlabel('Number of samples, n; part (b)'); ylabel('Output y(n)');
y(3:22)                              %Output y(n)
```

MATLAB function **filter()**, developed using a direct-form II realization (which will be discussed in Section 6.6.2), can be used to operate digital filtering, and the syntax is

$$\mathbf{Zi} = \mathbf{filtic}(\mathbf{B}, \mathbf{A}, \mathbf{Yi}, \mathbf{Xi})$$

$$\mathbf{y} = \mathbf{filter}(\mathbf{B}, \mathbf{A}, \mathbf{x}, \mathbf{Zi})$$

where B and A are vectors for the coefficients b_j and a_j whose formats are

$$A = [1 \;\; a_1 \;\; a_2 \;\; \cdots \;\; a_N] \text{ and } B = [b_0 \;\; b_1 \;\; b_2 \;\; \cdots \;\; b_M],$$

and x and y are the input data vector and output data vector, respectively.

Note that the filter function **filtic()** is a MATLAB function used to obtain initial states required by the MATLAB filter function **filter()** (requirement by a direct-form II realization) from initial conditions in the difference equation. Hence, Z_i contains initial states required for operating MATLAB function **filter()**, that is,

$$Z_i = [w(-1) \;\; w(-2) \;\; \cdots \;\;],$$

which can be recovered by another MATLAB function **filtic()**. X_i and Y_i are initial conditions with the length of the greater of M or N, given by.

$$X_i = [x(-1) \;\; x(-2) \;\; \cdots \;\;] \text{ and } Y_i = [y(-1) \;\; y(-2) \;\; \cdots \;\;].$$

Especially for zero initial conditions, the syntax is reduced to.

$$\mathbf{y} = \mathbf{filter}(\mathbf{B}, \mathbf{A}, \mathbf{x}).$$

Let us verify the filter operation results in Example 6.1 using the MATLAB functions. The MATLAB codes and results for Example 6.1(a) with the nonzero initial conditions are listed as

```
» B=[0 1]; A=[1 0 -0.5];
» x=[1 0.5 0.25 0.125];
» Xi=[-1 0];Yi=[0 1];
» Zi=filtic(B,A,Yi,Xi);
» y=filter(B,A,x,Zi)
y =
  -0.5000  1.0000  0.2500  0.7500
»
```

For the case of zero initial conditions in Example 6.1(b), the MATLAB codes and results are as follows:

```
» B=[0 1]; A=[1 0 -0.5];
» x=[1 0.5 0.25 0.125];
```

```
» y=filter(B,A,x)
y =
    0  1.0000  0.5000  0.7500
»
```

As we expected, the filter outputs match the ones in Example 6.1.

6.2 DIFFERENCE EQUATION AND TRANSFER FUNCTION

In this section, given a difference equation, we will derive a z-transfer function which is the ratio of the z-transform of the system output over the z-transform of the system input. To proceed, Eq. (6.1) is rewritten as

$$y(n) = b_0 x(n) + b_1 x(n-1) + \cdots + b_M x(n-M) \\ - a_1 y(n-1) - \cdots - a_N y(n-N).$$

With an assumption that all initial conditions of this system are zero, and with $X(z)$ and $Y(z)$ denoting the z-transforms of $x(n)$ and $y(n)$, respectively, taking the z-transform of Eq. (6.1) yields

$$Y(z) = b_0 X(z) + b_1 X(z) z^{-1} + \cdots + b_M X(z) z^{-M} \\ - a_1 Y(z) z^{-1} - \cdots - a_N Y(z) z^{-N}. \tag{6.3}$$

Rearranging Eq. (6.3), we yield

$$H(z) = \frac{Y(z)}{X(z)} = \frac{b_0 + b_1 z^{-1} + \cdots + b_M z^{-M}}{1 + a_1 z^{-1} + \cdots + a_N z^{-N}} = \frac{B(z)}{A(z)}, \tag{6.4}$$

where $H(z)$ is defined as the transfer function with its numerator and denominator polynomials defined below:

$$B(z) = b_0 + b_1 z^{-1} + \cdots + b_M z^{-M} \tag{6.5}$$

$$A(z) = 1 + a_1 z^{-1} + \cdots + a_N z^{-N}. \tag{6.6}$$

Clearly the z-transfer function is defined as

$$\text{Ratio} = \frac{z - \text{transform of the output}}{z - \text{transform of the input}}.$$

In DSP applications, given the difference equation, we can develop the z-transfer function and represent the digital filter in the z-domain as shown in Fig. 6.3. Then the stability and frequency response can be examined based on the developed transfer function.

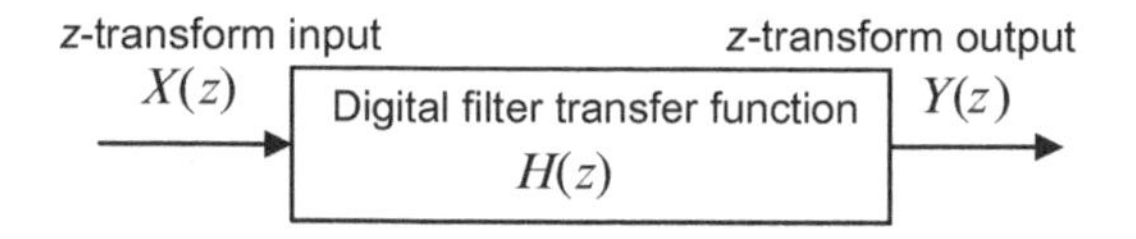

FIG. 6.3

Digital filter transfer function.

EXAMPLE 6.3

A DSP system is described by the following difference equation:

$$y(n) = x(n) - x(n-2) - 1.3y(n-1) - 0.36y(n-2).$$

Find the transfer function $H(z)$, the denominator polynomial $A(z)$, and the numerator polynomial $B(z)$.

Solution:

Taking the z-transform on both sides of the previous difference equation, we achieve

$$Y(z) = X(z) - X(z)z^{-2} - 1.3Y(z)z^{-1} - 0.36Y(z)z^{-2}.$$

Moving the last two terms to the left-hand side of the difference equation and factoring $Y(z)$ on the left-hand side and $X(z)$ on the right-hand side, we obtain

$$Y(z)\left(1 + 1.3z^{-1} + 0.36z^{-2}\right) = \left(1 - z^{-2}\right)X(z).$$

Therefore, the transfer function, which is the ratio of $Y(z)$ over $X(z)$, can be found as

$$H(z) = \frac{Y(z)}{X(z)} = \frac{1 - z^{-2}}{1 + 1.3z^{-1} + 0.36z^{-2}}.$$

From the derived transfer function $H(z)$, we could obtain the denominator polynomial and numerator polynomial as

$$A(z) = 1 + 1.3z^{-1} + 0.36z^{-2}, \text{and}$$

$$B(z) = 1 - z^{-2}.$$

The difference equation and its transfer function, as well as the stability issue of the linear time-invariant system, will be discussed in the following sections.

EXAMPLE 6.4

A digital system is described by the following difference equation:

$$y(n) = x(n) - 0.5x(n-1) + 0.36x(n-2).$$

Find the transfer function $H(z)$, the denominator polynomial $A(z)$, and the numerator polynomial $B(z)$.

Solution:

Taking the z-transform on both sides of the previous difference equation, we achieve

$$Y(z) = X(z) - 0.5X(z)z^{-1} + 0.36X(z)z^{-2}.$$

Therefore, the transfer function, that is the ratio of $Y(z)$ to $X(z)$, can be found as

$$H(z) = \frac{Y(z)}{X(z)} = 1 - 0.5z^{-1} + 0.36z^{-2}.$$

From the derived transfer function $H(z)$, it follows that

$$A(z) = 1$$

$$B(z) = 1 - 0.5z^{-1} + 0.36z^{-2}.$$

In DSP applications, the given transfer function of a digital system can be converted into a difference equation for DSP implementation. The following example illustrates the procedure.

EXAMPLE 6.5

Convert each of the following transfer functions into its difference equations.

(a) $H(z)=\dfrac{z^2-1}{z^2+1.3z+0.36}$

(b) $H(z)=\dfrac{z^2-0.5z+0.36}{z^2}$

Solution:

(a) Dividing the numerator and denominator by z^2 to obtain the transfer function whose numerator and denominator polynomials have the negative power of z, it follows that

$$H(z)=\frac{(z^2-1)/z^2}{(z^2+1.3z+0.36)/z^2}=\frac{1-z^{-2}}{1+1.3z^{-1}+0.36z^{-2}}.$$

We write the transfer function using the ratio of $Y(z)$ to $X(z)$:

$$\frac{Y(z)}{X(z)}=\frac{1-z^{-2}}{1+1.3z^{-1}+0.36z^{-2}}.$$

Then we have

$$Y(z)\left(1+1.3z^{-1}+0.36z^{-2}\right)=X(z)\left(1-z^{-2}\right).$$

By distributing $Y(z)$ and $X(z)$, we yield

$$Y(z)+1.3z^{-1}Y(z)+0.36z^{-2}Y(z)=X(z)-z^{-2}X(z).$$

Applying the inverse z-transform and using the shift property in Eq. (5.3) of Chapter 5, we get

$$y(n)+1.3y(n-1)+0.36y(n-2)=x(n)-x(n-2).$$

Writing the output $y(n)$ in terms of inputs and past outputs leads to

$$y(n)=x(n)-x(n-2)-1.3y(n-1)-0.36y(n-2).$$

(b) Similarly, dividing the numerator and denominator by z^2, we obtain

$$H(z)=\frac{Y(z)}{X(z)}=\frac{(z^2-0.5z+0.36)/z^2}{z^2/z^2}=1-0.5z^{-1}+0.36z^{-2}.$$

Thus, $Y(z)=X(z)(1-0.5z^{-1}+0.36z^{-2})$.

By distributing $X(z)$, we yield

$$Y(z)=X(z)-0.5z^{-1}X(z)+0.36z^{-2}X(z).$$

Applying the inverse z-transform using the shift property in Eq. (5.3), we obtain

$$y(n)=x(n)-0.5x(n-1)+0.36x(n-2).$$

The transfer function $H(z)$ can be factored into the *pole-zero form:*

$$H(z)=\frac{b_0(z-z_1)(z-z_2)\cdots(z-z_M)}{(z-p_1)(z-p_2)\cdots(z-p_N)}, \tag{6.7}$$

where the zeros z_i can be found by solving roots of the numerator polynomial, while the poles p_i can be solved for the roots of the denominator polynomial.

EXAMPLE 6.6

Given the following transfer functions,

$$H(z)=\frac{1-z^{-2}}{1+1.3z^{-1}+0.36z^{-2}},$$

Convert it into the pole-zero form.

Solution:

We first multiply the numerator and denominator polynomials by z^2 to achieve its advanced form in which both the numerator and denominator polynomials have positive powers of z, that is,

$$H(z)=\frac{(1-z^{-2})z^2}{(1+1.3z^{-1}+0.36z^{-2})z^2}=\frac{z^2-1}{z^2+1.3z+0.36}.$$

Letting $z^2-1=0$, we get $z=1$ and $z=-1$. Setting $z^2+1.3z+0.36=0$ leads $z=-0.4$ and $z=-0.9$. We then can write numerator and denominator polynomials in the factored form to obtain the pole-zero form:

$$H(z)=\frac{(z-1)(z+1)}{(z+0.4)(z+0.9)}.$$

6.2.1 IMPULSE RESPONSE, STEP RESPONSE, AND SYSTEM RESPONSE

The impulse response $h(n)$ of the DSP system $H(z)$ can be obtained by solving its difference equation using a unit impulse input $\delta(n)$. With the help of the z-transform and noticing that $X(z)=Z\{\delta(n)\}=1$, we yield

$$h(n)=Z^{-1}\{H(z)X(z)\}=Z^{-1}\{H(z)\}. \quad (6.8)$$

Similarly, for a step input, we can determine step response assuming the zero initial conditions. Letting

$$X(z)=Z[u(n)]=\frac{z}{z-1},$$

the step response can be found as

$$y(n)=Z^{-1}\left\{H(z)\frac{z}{z-1}\right\}. \quad (6.9)$$

Furthermore, the z-transform of the general system response is given by

$$Y(z)=H(z)X(z). \quad (6.10)$$

If we know the transfer function $H(z)$ and z-transform of the input $X(z)$, we are able to determine the system response $y(n)$ by finding the inverse z-transform of the output$Y(z)$:

$$y(n)=Z^{-1}\{Y(z)\}. \quad (6.11)$$

EXAMPLE 6.7

Given a transfer function depicting a DSP system

$$H(z) = \frac{z+1}{z-0.5},$$

Determine

1. The impulse response$h(n)$,
2. Step response$y(n)$, and.
3. The system response $y(n)$ if the input is given as $x(n) = (0.25)^n u(n)$.

Solution:

(a) The transfer function can be rewritten as

$$\frac{H(z)}{z} = \frac{z+1}{z(z-0.5)} = \frac{A}{z} + \frac{B}{z-0.5},$$

where $A = \left.\frac{z+1}{(z-0.5)}\right|_{z=0} = -2$, and $B = \left.\frac{z+1}{z}\right|_{z=0.5} = 3$.

Thus we have

$$\frac{H(z)}{z} = \frac{-2}{z} + \frac{3}{z-0.5} \text{ and}$$

$$H(z) = \left(-\frac{2}{z} + \frac{3}{z-0.5}\right)z = -2 + \frac{3z}{z-0.5}.$$

By taking the inverse z-transform as shown in Eq. (6.8), we yield the impulse response

$$h(n) = -2\delta(n) + 3(0.5)^n u(n).$$

(b) For the step input $x(n) = u(n)$ and its z-transform $X(z) = \frac{z}{z-1}$, we can determine the z-transform of the step response as

$$Y(z) = H(z)X(z) = \frac{z+1}{z-0.5}\frac{z}{z-1}.$$

Applying the partial fraction expansion leads to

$$\frac{Y(z)}{z} = \frac{z+1}{(z-0.5)(z-1)} = \frac{A}{z-0.5} + \frac{B}{z-1},$$

where

$$A = \left.\frac{z+1}{z-1}\right|_{z=0.5} = -3, \text{and } B = \left.\frac{z+1}{z-0.5}\right|_{z=1} = 4.$$

The z-transform step response is therefore

$$Y(z) = \frac{-3z}{z-0.5} + \frac{4z}{z-1}.$$

Applying the inverse z-transform table yields the step response as

$$y(n) = -3(0.5)^n u(n) + 4u(n).$$

(c) To determine the system output response, we first find the z-transform of the input $x(n)$,

$$X(z) = Z\{(0.25)^n u(n)\} = \frac{z}{z-0.25},$$

then $Y(z)$ can be yielded via Eq. (6.10), that is,

$$Y(z) = H(z)X(z) = \frac{z+1}{z-0.5} \cdot \frac{z}{z-0.25} = \frac{z(z+1)}{(z-0.5)(z-0.25)}.$$

Using the partial fraction expansion, we have

$$\frac{Y(z)}{z} = \frac{(z+1)}{(z-0.5)(z-0.25)} = \left(\frac{A}{z-0.5} + \frac{B}{z-0.25}\right)$$

$$Y(z) = \left(\frac{6z}{z-0.5} + \frac{-5z}{z-0.25}\right).$$

Using Eq. (6.11) and Table 5.1 in Chapter 5, we finally yield

$$y(n) = Z^{-1}\{Y(z)\} = 6(0.5)^n u(n) - 5(0.25)^n u(n).$$

The impulse response for (a), step response for (b), and system response for (c) are each plotted in Fig. 6.4.

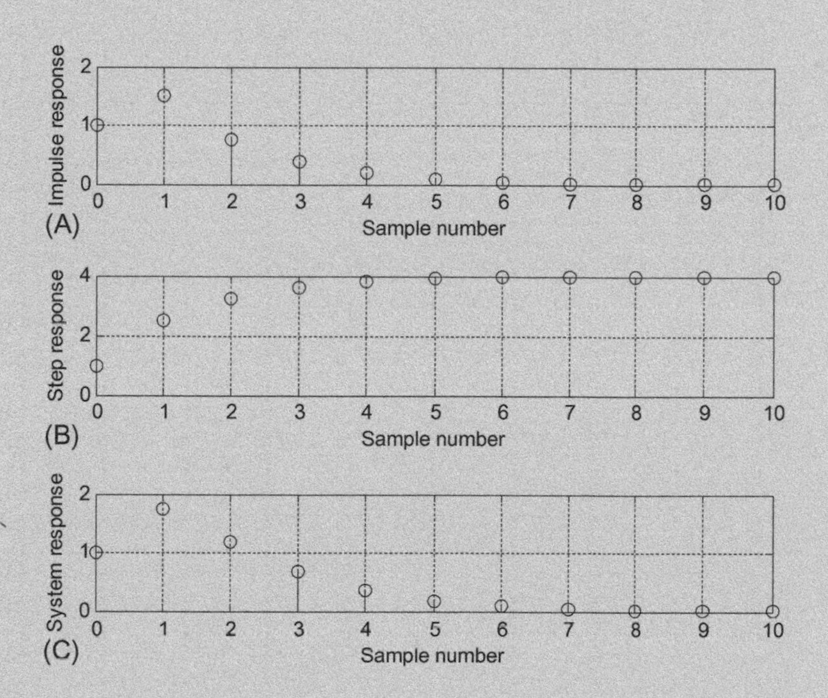

FIG. 6.4

Impulse, step, and system response in Example 6.7.

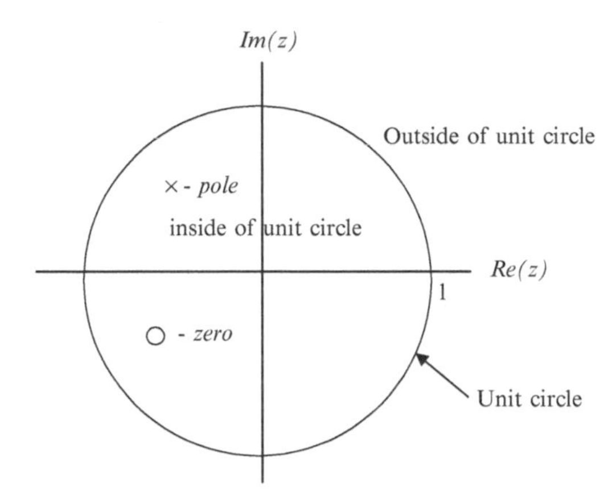

FIG. 6.5

z-Plane and pole-zero plot.

6.3 THE Z-PLANE POLE-ZERO PLOT AND STABILITY

A very useful tool to analyze digital systems is the z-plane pole-zero plot. This graphical technique allows us to investigate characteristics of the digital system shown in Fig. 6.1, including the system stability. In general, a digital transfer function can be written in the pole-zero form as shown in Eq. (6.7), and we can plot the poles and zeros on the z-plane. The z-plane is depicted in Fig. 6.5 and has the following features:

1. The horizontal axis is the real part of the variable z, and the vertical axis represents the imaginary part of the variable z.
2. The z-plane is divided into two parts by a unit circle.
3. Each pole is marked on z-plane using the cross symbol x, while each zero is plotted using the small circle symbol o.

Let us investigate the z-plane pole-zero plot of a digital filter system via the following example.

EXAMPLE 6.8

Given the digital transfer function

$$H(z) = \frac{z^{-1} - 0.5z^{-2}}{1 + 1.2z^{-1} + 0.45z^{-2}},$$

Plot poles and zeros.

Solution:

Converting the transfer function to its advanced form by multiplying both numerator and denominator by z^2, it follows that

$$H(z)=\frac{(z^{-1}-0.5z^{-2})z^2}{(1+1.2z^{-1}+0.45z^{-2})z^2}=\frac{z-0.5}{z^2+1.2z+0.45}.$$

By setting $z^2+1.2z+0.45=0$ and $z-0.5=0$, we obtain two poles

$$p_1=-0.6+j0.3$$

$$p_2=p_1^*=-0.6-j0.3$$

and a zero $z_1=0.5$, which are plotted on the z-plane shown in Fig. 6.6. According to the form of Eq. (6.7), we also yield the pole-zero form as

$$H(z)=\frac{z^{-1}-0.5z^{-2}}{1+1.2z^{-1}+0.45z^{-2}}=\frac{(z-0.5)}{(z+0.6-j0.3)(z+0.6+j0.3)}.$$

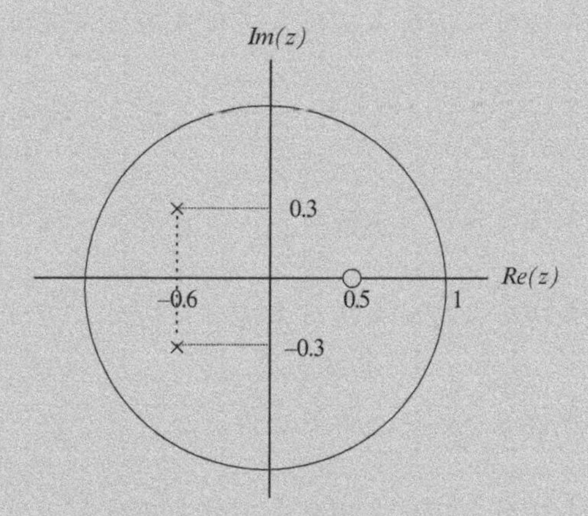

FIG. 6.6

The z-plane pole-zero plot of Example 6.8.

Having zeros and poles plotted on the z-plane, we are able to study the system stability. We first establish the relationship between the s-plane in Laplace domain and the z-plane in z-transform domain, as illustrated in Fig. 6.7.

As shown in Fig. 6.7, the sampled signal, which is not quantized, with a sampling period of T is written as

$$x_s(t)=\sum_{n=0}^{\infty}x(nT)\delta(t-nT)=x(0)\delta(t)+x(T)\delta(t-T)+x(2T)\delta(t-2T)+\cdots. \tag{6.12}$$

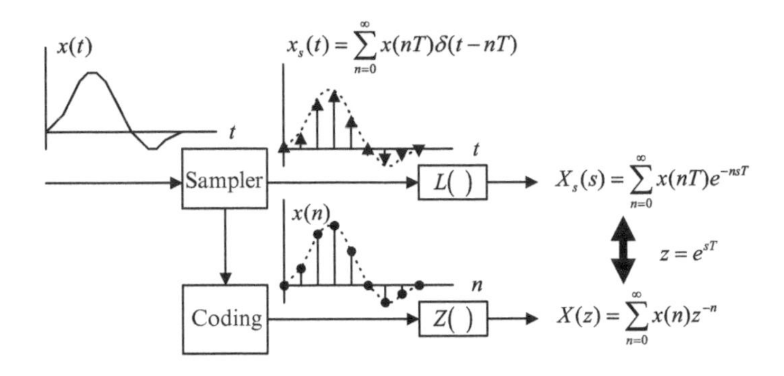

FIG. 6.7

Relationship between Laplace transform and z-transform.

Taking the Laplace transform and using the Laplace shift property (Appendix B, Table B.5) as

$$L(\delta(t-nT))=e^{-nTs} \tag{6.13}$$

leads to

$$X_s(s)=\sum_{n=0}^{\infty}x(nT)e^{-nTs}=x(0)e^{-0\times Ts}+x(T)e^{-Ts}+x(2T)e^{-2Ts}+\cdots. \tag{6.14}$$

Comparing Eq. (6.14) with the definition of a one-sided z-transform of the data sequence $x(n)$ from analog-to-digital conversion (ADC):

$$X(z)=Z(x(n))=\sum_{n=0}^{\infty}x(n)z^{-n}=x(0)z^{-0}+x(1)z^{-1}+x(2)z^{-2}+\cdots. \tag{6.15}$$

Clearly, we see the relationship of the sampled system in Laplace domain and its digital system in z-transform domain by the following mapping:

$$z=e^{sT}. \tag{6.16}$$

Substituting $s=-\alpha\pm j\omega$ into Eq. (6.16), it follows that $z=e^{-\alpha T\pm j\omega T}$. In the polar form, we have

$$z=e^{-\alpha T}\angle\pm\omega T. \tag{6.17}$$

Eqs. (6.16) and (6.17) give the following important conclusions.

If $\alpha>0$, this means $|z|=e^{-\alpha T}<1$. Then the left-hand side half plane (LHHP) of the s-plane is mapped to the inside of the unit circle of the z-plane. When $\alpha=0$, this causes $|z|=e^{-\alpha T}=1$. Thus the $j\omega$ axis of the s-plane is mapped on the unit circle of the z-plane, as shown in Fig. 6.8. Obviously, the right-hand half plane (RHHP) of the s-plane is mapped to the outside of the unit cycle in the z-plane. A stable system

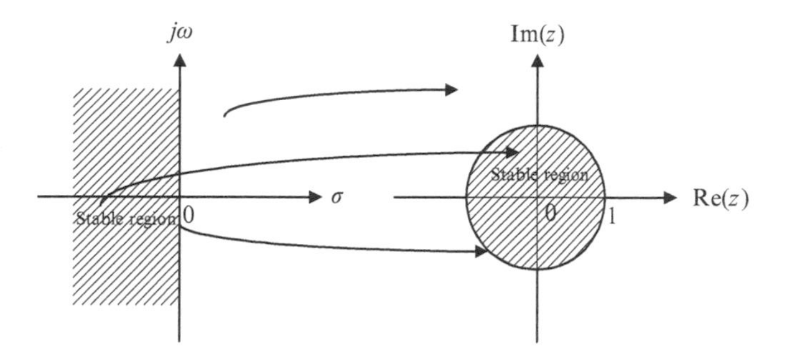

FIG. 6.8

Mapping between *s*-plane and *z*-plane.

means that for a given bounded input, the system output must be bounded. Similar to the analog system, the digital system requires that its all poles plotted on the z-plane must be inside the unit circle. We summarize the rules for determining the stability of a DSP system as follows:

(a) If the outmost pole(s) of the z-transfer function $H(z)$ describing the DSP system is(are) inside the unit circle on the z-plane pole-zero plot, then the system is stable.
(b) If the outmost pole(s) of the z-transfer function $H(z)$ is(are) outside the unit circle on the z-plane pole-zero plot, the system is unstable.
(c) If the outmost pole(s) is(are) first-order pole(s) of the z-transfer function $H(z)$ and on the unit circle on the z-plane pole-zero plot, then the system is marginally stable.
(d) If the outmost pole(s) is(are) multiple-order pole(s) of the z-transfer function $H(z)$ and on the unit circle on the z-plane pole-zero plot, then the system is unstable.
(e) The zeros do not affect the system stability.

Note that the following facts apply to a stable system [bounded-in/bounded-out (BIBO) stability discussed in Chapter 3]:

1. If the input to the system is bounded, then the output of the system will also be bounded, or the impulse response of the system will go to zero in a finite number of steps.
2. An unstable system is one in which the output of the system will grow without bound due to any bounded input, initial condition, or noise, or its impulse response will grow without bound.
3. The impulse response of a marginally stable system stays at a constant level or oscillates between the two finite values.

Examples illustrating these rules are shown in Fig. 6.9.

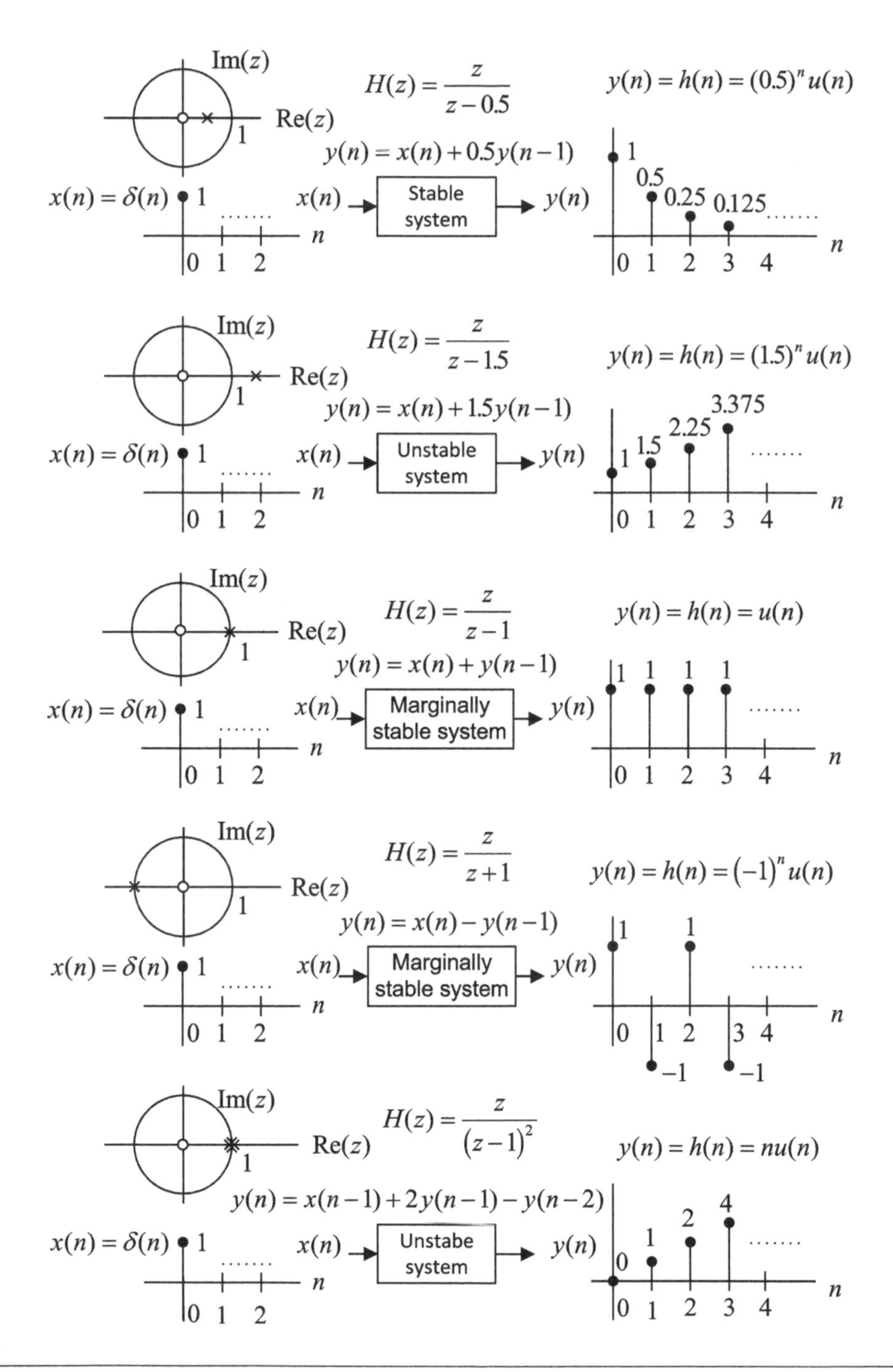

FIG. 6.9

Stability illustrations.

EXAMPLE 6.9

The following transfer functions describe digital systems.

$$H(z)=\frac{z+0.5}{(z-0.5)(z^2+z+0.5)}$$

$$H(z)=\frac{z^2+0.25}{(z-0.5)(z^2+3z+2.5)}$$

$$H(z)=\frac{z+0.5}{(z-0.5)(z^2+1.4141z+1)}$$

$$H(z)=\frac{z^2+z+0.5}{(z-1)^2(z+1)(z-0.6)}$$

For each, sketch the z-plane pole-zero plot and determine the stability status for the digital system.

Solution:

(a) A zero is found to be $z=-0.5$.
Poles: $z=0.5$, $|z|=0.5<1$; $z=-0.5\pm j0.5$,

$$|z|=\sqrt{(-0.5)^2+(\pm 0.5)^2}=0.707<1.$$

The plot of poles and a zero is shown in Fig. 6.10. Since the outmost poles are inside the unit circle, the system is stable.

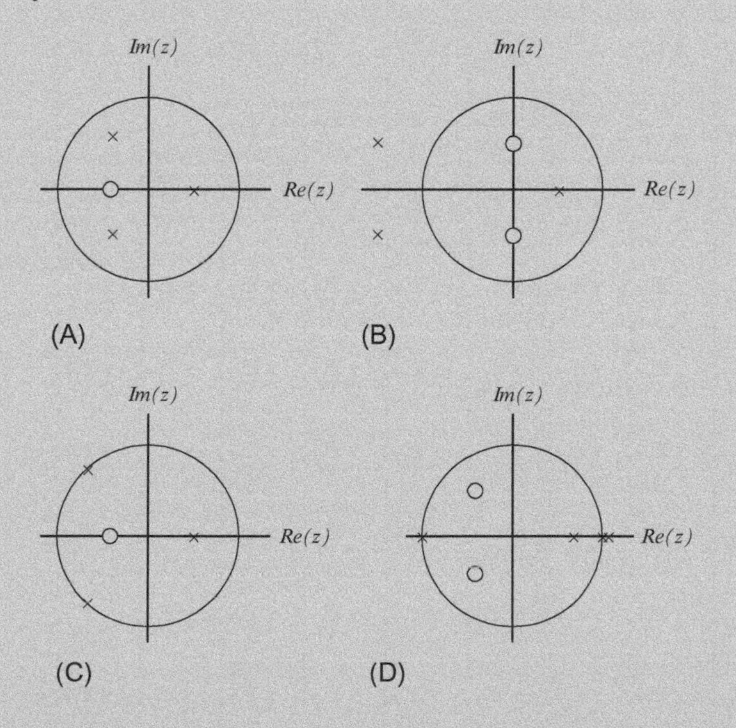

FIG. 6.10

Pole-zero plots for Example 6.9.

Continued

EXAMPLE 6.9—CONT'D

(b) Zeros are $z = \pm j0.5$.
Poles: $z = 0.5$, $|z| = 0.5 < 1$; $z = -1.5 \pm j0.5$,

$$|z| = \sqrt{(1.5)^2 + (\pm 0.5)^2} = 1.5811 > 1.$$

The plot of poles and zeros is shown in Fig. 6.10. Since we have two poles at $z = -1.5 \pm j0.5$, which are outside the unit circle, the system is unstable.

(c) A zero is found to be $z = -0.5$.
Poles: $z = 0.5$, $|z| = 0.5 < 1$; $z = -0.707 \pm j0.707$,

$$|z| = \sqrt{(0.707)^2 + (\pm 0.707)^2} = 1$$

The zero and poles are plotted in Fig. 6.10. Since the outmost poles are first order at $z = -0.707 \pm j0.707$ and are on the unit circle, the system is marginally stable.

(d) Zeros: $z = -0.5 \pm j0.5$.
Poles: $z = 1$, $|z| = 1$; $z = 1$, $|z| = 1$; $z = -1$, $|z| = 1$; $z = 0.6$, $|z| = 0.6 < 1$.
The zeros and poles are plotted in Fig. 6.10. Since the outmost pole is multiple-order (second order) pole at $z = 1$ and is on the unit circle, the system is unstable.

6.4 DIGITAL FILTER FREQUENCY RESPONSE

From the Laplace transfer function, we can achieve the analog filter steady-state frequency response $H(j\omega)$ by substituting $s = j\omega$ into the transfer function $H(s)$, that is,

$$H(s)|_{s=j\omega} = H(j\omega).$$

Then we can study the magnitude frequency response $|H(j\omega)|$ and phase response $\angle H(j\omega)$. Similarly, in a DSP system, using the mapping Eq. (6.16), we substitute $z = e^{sT}|_{s=j\omega} = e^{j\omega T}$ into the z-transfer function $H(z)$ to acquire the digital frequency response, which is converted into the magnitude frequency response $|H(e^{j\omega T})|$ and phase response $\angle|H(e^{j\omega T})|$, that is,

$$H(z)|_{z=e^{j\omega T}} = H\left(e^{j\omega T}\right) = \left|H\left(e^{j\omega T}\right)\right| \angle H\left(e^{j\omega T}\right). \tag{6.18}$$

Let us introduce a normalized digital frequency in radians in digital domain

$$\Omega = \omega T. \tag{6.19}$$

Then the digital frequency response in Eq. (6.18) would become

$$H\left(e^{j\Omega}\right) = H(z)|_{z=e^{j\Omega}} = \left|H\left(e^{j\Omega}\right)\right| \angle H\left(e^{j\Omega}\right). \tag{6.20}$$

The formal derivation for Eqs. (6.20) can be found in Appendix D.

Now we verify the frequency response via the following simple digital filter. Consider a digital filter with a sinusoidal input of the amplitude K (Fig. 6.11):

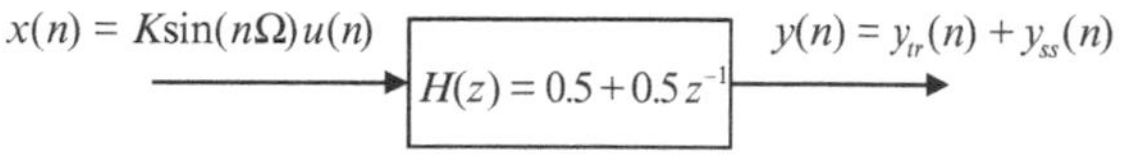

$x(n) = K\sin(n\Omega)u(n)$ → $H(z) = 0.5 + 0.5z^{-1}$ → $y(n) = y_{tr}(n) + y_{ss}(n)$

FIG. 6.11

System transient and steady-state frequency responses.

We can determine the system output $y(n)$, which consists of the transient response $y_{tr}(n)$ and the steady-state response $y_{ss}(n)$. We find the z-transform output as

$$Y(z) = \left(\frac{0.5z + 0.5}{z}\right) \frac{Kz\sin\Omega}{z^2 - 2z\cos\Omega + 1} \tag{6.21}$$

To perform the inverse z-transform to find the system output, we further rewrite Eq. (6.21) as

$$\frac{Y(z)}{z} = \left(\frac{0.5z + 0.5}{z}\right) \frac{K\sin\Omega}{(z - e^{j\Omega})(z - e^{-j\Omega})} = \frac{A}{z} + \frac{B}{z - e^{j\Omega}} + \frac{B^*}{z - e^{-j\Omega}},$$

where A, B, and the complex conjugate B^* are the constants for the partial fractions. Applying the partial fraction expansion leads to

$$A = 0.5K\sin\Omega$$

$$B = \left.\frac{0.5z + 0.5}{z}\right|_{z=e^{j\Omega}} \frac{K}{2j} = \left|H\left(e^{j\Omega}\right)\right| e^{j\angle H\left(e^{j\Omega}\right)} \frac{K}{2j}.$$

Note that the first part of constant B is a complex function, which is obtained by substituting $z = e^{j\Omega}$ into the filter z-transfer function. We can also express the complex function in terms of the polar form:

$$\left.\frac{0.5z + 0.5}{z}\right|_{z=e^{j\Omega}} = 0.5 + 0.5z^{-1}\Big|_{z=e^{j\Omega}} = H(z)\big|_{z=e^{j\Omega}} = H\left(e^{j\Omega}\right) = \left|H\left(e^{j\Omega}\right)\right| e^{j\angle H\left(e^{j\Omega}\right)},$$

where $H(e^{j\Omega}) = 0.5 + 0.5e^{-j\Omega}$, and we call this complex function the steady-state frequency response. Based on the complex conjugate property, we get another residue as

$$B^* = \left|H\left(e^{j\Omega}\right)\right| e^{-j\angle H\left(e^{j\Omega}\right)} \frac{K}{-j2}.$$

The z-transform system output is then given by

$$Y(z) = A + \frac{Bz}{z - e^{j\Omega}} + \frac{B^* z}{z - e^{-j\Omega}}.$$

Taking the inverse z-transform, we achieve the following system transient and steady-state responses:

$$y(n) = \underbrace{0.5K\sin\Omega\delta(n)}_{y_{tr}(n)} + \underbrace{\left|H\left(e^{j\Omega}\right)\right| e^{j\angle H\left(e^{j\Omega}\right)} \frac{K}{j2} e^{jn\Omega} u(n) + \left|H\left(e^{j\Omega}\right)\right| e^{-j\angle H\left(e^{j\Omega}\right)} \frac{K}{-j2} e^{-jn\Omega} u(n)}_{y_{ss}(n)}.$$

Simplifying the response yields the form

$$y(n) = 0.5K\sin\Omega\delta(n) + \left|H\left(e^{j\Omega}\right)\right| K \frac{e^{jn\Omega + j\angle H\left(e^{j\Omega}\right)} u(n) - e^{-jn\Omega - j\angle H\left(e^{j\Omega}\right)} u(n)}{j2}.$$

We can further combine the last term using Euler's formula to express the system response as

$$y(n) = \underbrace{0.5K\sin\Omega\delta(n)}_{y_{tr}(n)\text{ will decay to zero after the first sample}} + \underbrace{\left|H\left(e^{j\Omega}\right)\right|K\sin\left(n\Omega + \angle H\left(e^{j\Omega}\right)\right)u(n)}_{y_{ss}(n)}$$

Finally, the steady-state response is identified as

$$y_{ss}(n) = K\left|H\left(e^{j\Omega}\right)\right|\sin\left(n\Omega + \angle H\left(e^{j\Omega}\right)\right)u(n).$$

For this particular filter, the transient response exists for only the first sample in the system response. By substituting $n=0$ into $y(n)$ and after simplifying algebra, we achieve the response for the first output sample.

$$y(0) = y_{tr}(0) + y_{ss}(0) = 0.5K\sin(\Omega) - 0.5K\sin(\Omega) = 0.$$

Note that the first output sample of the transient response cancels the first output sample of the steady-state response, so the combined first output sample has a value of zero for this particular filter. The system response reaches the steady-state response after the first output sample. At this point, we can conclude:

$$\text{Steady state magnitude frequency response} = \frac{\text{Peak amplitude of steady state response at }\Omega}{\text{Peak amplitude of sinusoidal input at }\Omega}$$

$$= \frac{\left|H\left(e^{j\Omega}\right)\right|K}{K} = \left|H\left(e^{j\Omega}\right)\right|$$

$$\text{Steady state phase frequency response} = \text{Phase difference} = \angle H\left(e^{j\Omega}\right).$$

Fig. 6.12 shows the system response with sinusoidal inputs at $\Omega=0.25\pi$, $\Omega=0.5\pi$, and $\Omega=0.75\pi$, respectively.

Next, we examine the properties of the filter frequency response $H(e^{j\Omega})$. From Euler's identity and trigonometric identity, we know that

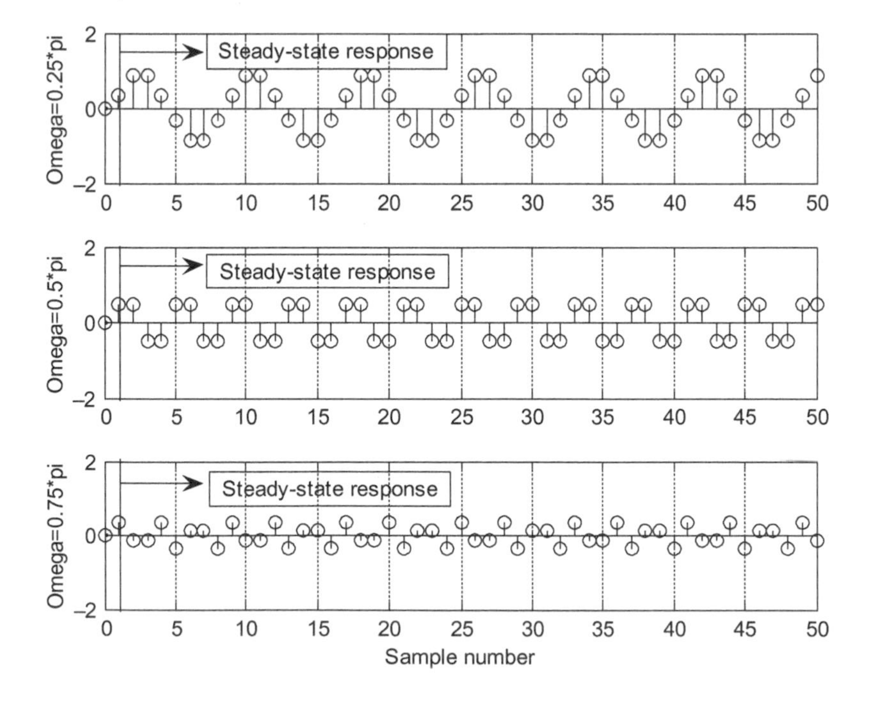

FIG. 6.12

Digital filter responses to different input sinusoids.

$$e^{j(\Omega+k2\pi)} = \cos(\Omega+k2\pi)+j\sin(\Omega+k2\pi)$$
$$= \cos\Omega+j\sin\Omega = e^{j\Omega},$$

where k is an integer taking the values $k=0, \pm 1, \pm 2, \cdots$. Then the frequency response has the following properties (assuming all input sequences are real):

1. Periodicity.
   (a) Frequency response: $H(e^{j\Omega})=H(e^{j(\Omega+k2\pi)})$.
   (b) Magnitude frequency response: $|H(e^{j\Omega})|=|H(e^{j(\Omega+k2\pi)})|$.
   (c) Phase response: $\angle H(e^{j\Omega})=\angle H(e^{j\Omega+k2\pi})$.

The second property is given without proof (see proof in Appendix D).

2. Symmetry.
   (a) Magnitude frequency response: $|H(e^{-j\Omega})|=|H(e^{j\Omega})|$.
   (b) Phase response: $\angle H(e^{-j\Omega})=-\angle H(e^{j\Omega})$.

Since the maximum frequency in a DSP system is the folding frequency, $f_s/2$, where $f_s=1/T$, and T designates the sampling period, the corresponding maximum normalized frequency of the system frequency can be calculated as.

$$\Omega=\omega T=2\pi\frac{f_s}{2}\times T=\pi \text{ radians}. \tag{6.22}$$

The frequency response $H(e^{j\Omega})$ for $|\Omega|>\pi$ consists of the image replicas of $H(e^{j\Omega})$ for $|\Omega|\le\pi$ and will be removed via the reconstruction filter later. Hence, we need to evaluate $H(e^{j\Omega})$ only for the positive normalized frequency range from $\Omega=0$ to $\Omega=\pi$ radians. The frequency, in Hz, can be determined by

$$f=\frac{\Omega}{2\pi}f_s. \tag{6.23}$$

The magnitude frequency response, often expressed in decibels, is defined as

$$\left|H\left(e^{j\Omega}\right)\right|_{dB}=20\times\log_{10}\left(\left|H\left(e^{j\Omega}\right)\right|\right). \tag{6.24}$$

The DSP system stability, magnitude response, and phase response are investigated via the following examples.

EXAMPLE 6.10

Given the following digital system with a sampling rate of 8000 Hz,

$$y(n)=0.5x(n)+0.5x(n-1),$$

Determine the frequency response.

Solution:

Taking the z-transform on both sides of the difference equation leads to

$$Y(z)=0.5X(z)+0.5z^{-1}X(z).$$

Then the transfer function describing the system is easily found as

$$H(z)=\frac{Y(z)}{X(z)}=0.5+0.5z^{-1}.$$

Continued

EXAMPLE 6.10—CONT'D

Substituting $z = e^{j\Omega}$, we have the frequency response as

$$H(e^{j\Omega}) = 0.5 + 0.5e^{-j\Omega}$$
$$= 0.5 + 0.5\cos(\Omega) - j0.5\sin(\Omega).$$

Therefore, the magnitude frequency response and phase response are given by

$$|H(e^{j\Omega})| = \sqrt{(0.5 + 0.5\cos(\Omega))^2 + (0.5\sin(\Omega))^2}$$

and

$$\angle H(e^{j\Omega}) = \tan^{-1}\left(\frac{-0.5\sin(\Omega)}{0.5 + 0.5\cos(\Omega)}\right).$$

Several points for the magnitude response and phase response are calculated and illustrated in Table 6.1.

According to the data, we plot the frequency response and phase response of the DSP system as shown in Fig. 6.13.

Table 6.1 Frequency Response Calculations for Example 6.10

Ω (radians)	$f = \frac{\Omega}{2\pi} f_s$ (Hz)	$\lvert H(e^{j\Omega})\rvert$	$\lvert H(e^{j\Omega})\rvert_{dB}$	$\angle H(e^{j\Omega})$
0	0	1.000	0 dB	0^0
0.25π	1000	0.924	−−0.687 dB	-22.5^0
0.50π	2000	0.707	−−3.012 dB	-45.00^0
0.75π	3000	0.383	−−8.336 dB	-67.50^0
1.00π	4000	0.000	$-\infty$	-90^0

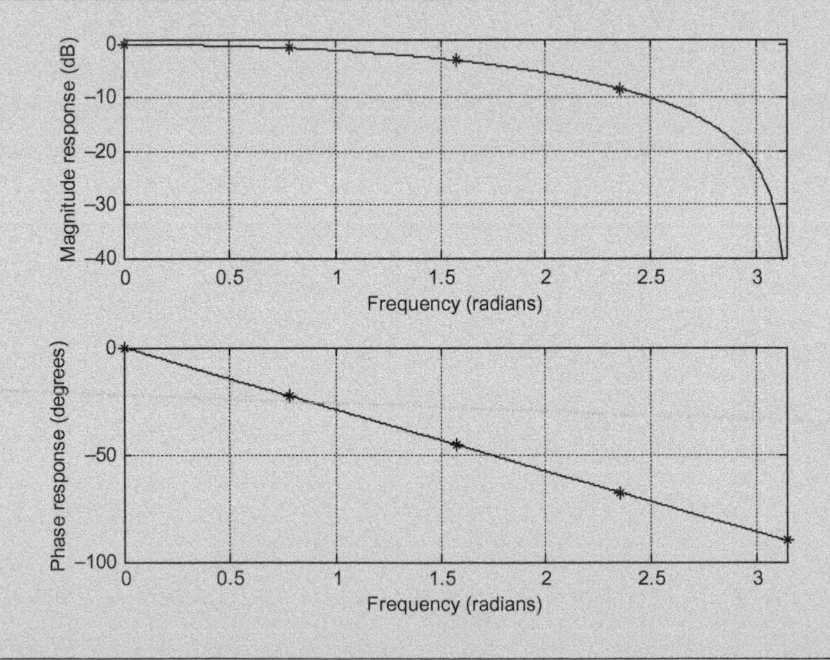

FIG. 6.13

Frequency responses of the digital filter in Example 6.10.

It is observed that when the frequency increases, the magnitude response decreases. The DSP system acts like a digital lowpass filter, and its phase response is linear.

We can also verify the periodicity for $|H(e^{j\Omega})|$ and $\angle H(e^{j\Omega})$ when $\Omega = 0.25\pi + 2\pi$:

$$\left|H\left(e^{j(0.25\pi+2\pi)}\right)\right| = \sqrt{(0.5+0.5\cos(0.25\pi+2\pi))^2 + (0.5\sin(0.25\pi+2\pi))^2}$$
$$= 0.924 = \left|H\left(e^{j0.25\pi}\right)\right|$$

$$\angle H\left(e^{j(0.25\pi+2\pi)}\right) = \tan^{-1}\left(\frac{-0.5\sin(0.25\pi+2\pi)}{0.5+0.5\cos(0.25\pi+2\pi)}\right) = -22.5^0 = \angle H\left(e^{j0.25\pi}\right).$$

For $\Omega = -0.25\pi$, we can verify the symmetry property as

$$\left|H(e^{-j0.25\pi})\right| = \sqrt{(0.5+0.5\cos(-0.25\pi))^2 + (0.5\sin(-0.25\pi))^2}$$
$$= 0.924 = \left|H\left(e^{j0.25\pi}\right)\right|$$

$$\angle H\left(e^{-j0.25\pi}\right) = \tan^{-1}\left(\frac{-0.5\sin(-0.25\pi)}{0.5+0.5\cos(-0.25\pi)}\right) = 22.5^0 = -\angle H\left(e^{j0.25\pi}\right).$$

The properties can be observed in Fig. 6.14, where the frequency range is chosen from $\Omega = -2\pi$ to $\Omega = 4\pi$ radians. As shown in the figure, the magnitude and phase responses are periodic with a period of 2π. For a period between $\Omega = -\pi$ and $\Omega = \pi$, the magnitude responses for the portion $\Omega = -\pi$ to $\Omega = 0$ and the portion $\Omega = 0$ to $\Omega = \pi$ are same, while the phase responses are opposite. Since the magnitude and phase responses calculated for the range from $\Omega = 0$ to $\Omega = \pi$ are sufficient to the frequency response information, this range is only required for generating the frequency response plots.

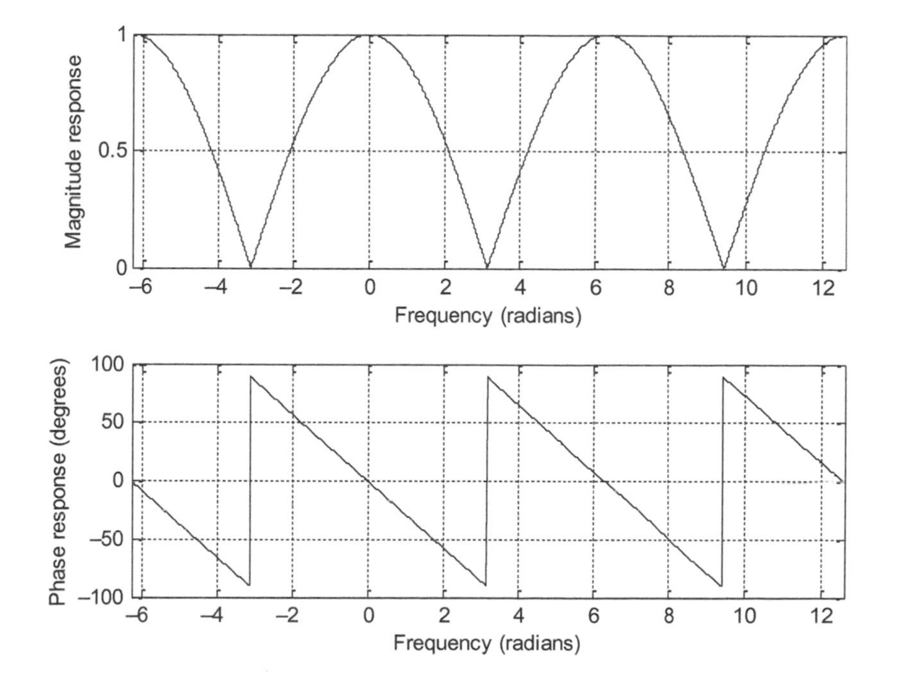

FIG. 6.14

Periodicity of the magnitude response and phase response in Example 6.10.

Again, note that the phase plot shows a sawtooth shape instead of a linear straight line for this particular filter. This is due to the phase wrapping at $\Omega = 2\pi$ radians since $e^{j(\Omega + k2\pi)} = e^{j\Omega}$ is used in the calculation. However, the phase plot shows that the phase is linear in the useful information range from $\Omega = 0$ to $\Omega = \pi$ radians.

EXAMPLE 6.11

Given a digital system with a sampling rate of 8000 Hz

$$y(n) = x(n) - 0.5y(n-1),$$

Determine the frequency response

Solution:

Taking the z-transform on both sides of the difference equation leads to

$$Y(z) = X(z) - 0.5z^{-1}Y(z).$$

Then the transfer function describing the system is easily found as

$$H(z) = \frac{Y(z)}{X(z)} = \frac{1}{1 + 0.5z^{-1}} = \frac{z}{z + 0.5}.$$

Substituting $z = e^{j\Omega}$, we have the frequency response as

$$H(e^{j\Omega}) = \frac{1}{1 + 0.5e^{-j\Omega}}$$

$$= \frac{1}{1 + 0.5\cos(\Omega) - j0.5\sin(\Omega)}.$$

Therefore, the magnitude frequency response and phase response are given by

$$|H(e^{j\Omega})| = \frac{1}{\sqrt{(1 + 0.5\cos(\Omega))^2 + (0.5\sin(\Omega))^2}}$$

and

$$\angle H(e^{j\Omega}) = -\tan^{-1}\left(\frac{-0.5\sin(\Omega)}{1 + 0.5\cos(\Omega)}\right).$$

Several points for the magnitude response and phase response are calculated and illustrated in Table 6.2.

Table 6.2 Frequency Response Calculations in Example 6.11

Ω (radians)	$f = \frac{\Omega}{2\pi}f_s$ (Hz)	$\|H(e^{j\Omega})\|$	$\|H(e^{j\Omega})\|_{dB}$	$\angle H(e^{j\Omega})$
0	0	0.670	−3.479 dB	0^0
0.25π	1000	0.715	−2.914 dB	14.64^0
0.50π	2000	0.894	−0.973 dB	26.57^0
0.75π	3000	1.357	2.652 dB	28.68^0
1.00π	4000	2.000	6.021 dB	0^0

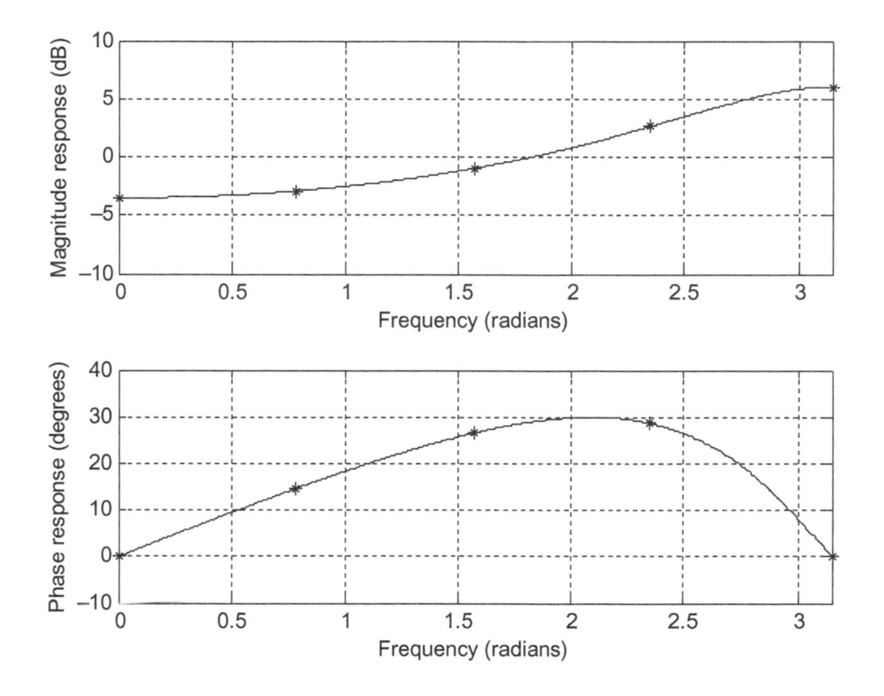

FIG. 6.15

Frequency responses of the digital filter in Example 6.11.

According to the achieved data, the magnitude response and phase response of the DSP system are roughly plotted in Fig. 6.15.

From Table 6.2 and Fig. 6.15, we can see that when the frequency increases, the magnitude response increases. The DSP system actually performs digital highpass filtering.

If all the coefficients a_i for $i=0, 1, \ldots, M$ in Eq. (6.1) are zeros, Eq. (6.2) is reduced to

$$\begin{aligned} y(n) &= \sum_{i=0}^{M} b_i x(n-i) \\ &= b_0 x(n) + b_1 x(n-1) + \cdots + b_K x(n-M). \end{aligned} \tag{6.25}$$

Note that b_i is the ith impulse response coefficient. Also, since M is a finite positive integer, b_i in this particular case is a finite set, $H(z)=B(z)$; note that the denominator $A(z)=1$. Such systems are called *finite impulse response* (FIR) systems. If not all a_i in Eq. (6.1) are zeros, the impulse response $h(i)$ would consist of an infinite number of coefficients. Such systems are called *infinite impulse response* (IIR) systems. The z-transform of the IIR $h(i)$, in general, is given by $H(z)=\frac{B(z)}{A(z)}$, where $A(z)\neq 1$.

6.5 BASIC TYPES OF FILTERING

The basic filter types can be classified into four categories: *lowpass*, *highpass*, *bandpass*, and *bandstop*. Each of them finds a specific application in DSP. One of the objectives in applications may involve the design of digital filters. In general, the filter is designed based on the specifications primarily for the passband, stopband, and transition band of the filter frequency response. The filter passband is

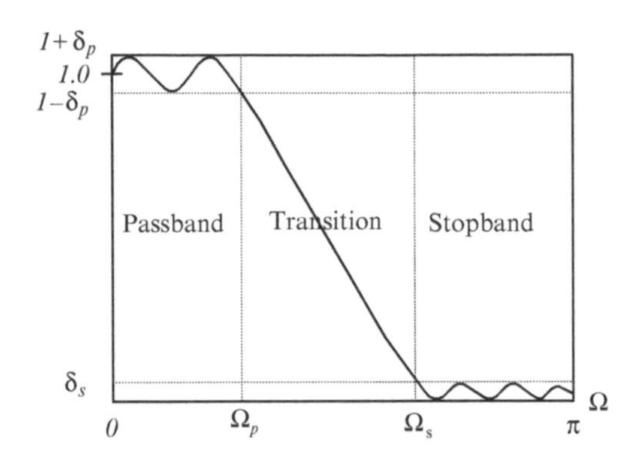

FIG. 6.16

Magnitude response of the normalized lowpass filter.

the frequency range with the amplitude gain of the filter response being approximately unity. The filter stopband is defined as the frequency range over which the filter magnitude response is attenuated to eliminate the input signal whose frequency components are within that range. The transition band denotes the frequency range between the passband and the stopband.

The design specifications of the lowpass filter are illustrated in Fig. 6.16, where the low-frequency components are passed through the filter while the high-frequency components are attenuated. As shown in Fig. 6.16, Ω_p and Ω_s are the passband cutoff frequency and the stopband cutoff frequency, respectively; δ_p is the design parameter to specify the ripple (fluctuation) of the frequency response in the passband, while δ_s specifies the ripple of the frequency response in the stopband.

The highpass filter remains the high-frequency components and rejects low-frequency components. The magnitude frequency response for the highpass filter is demonstrated in Fig. 6.17.

The bandpass filter attenuates both low- and high-frequency components while keeping the middle-frequency components, as shown in Fig. 6.18.

As illustrated in Fig. 6.18, Ω_{pL} and Ω_{sL} are the lower passband cutoff frequency and lower stopband cutoff frequency, respectively. Ω_{pH} and Ω_{sH} are the upper passband cutoff frequency and upper stopband cutoff frequency, respectively. δ_p is the design parameter to specify the ripple of the frequency response in the passband, while δ_s specifies the ripple of the frequency response in the stopband.

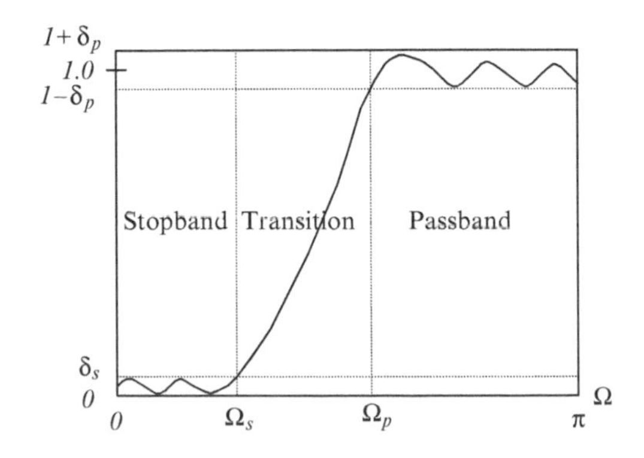

FIG. 6.17

Magnitude response of the normalized highpass filter.

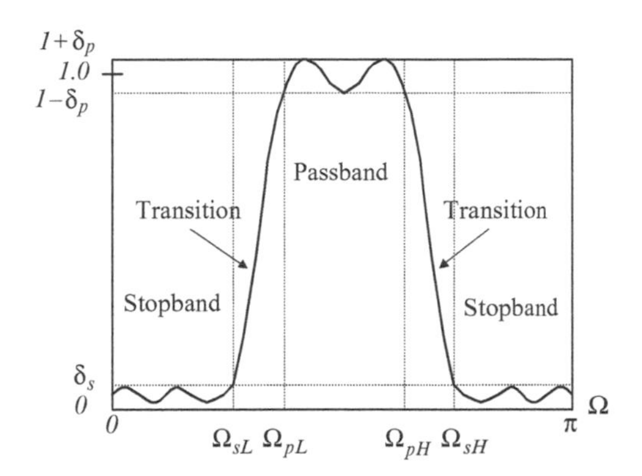

FIG. 6.18

Magnitude response of the normalized bandpass filter.

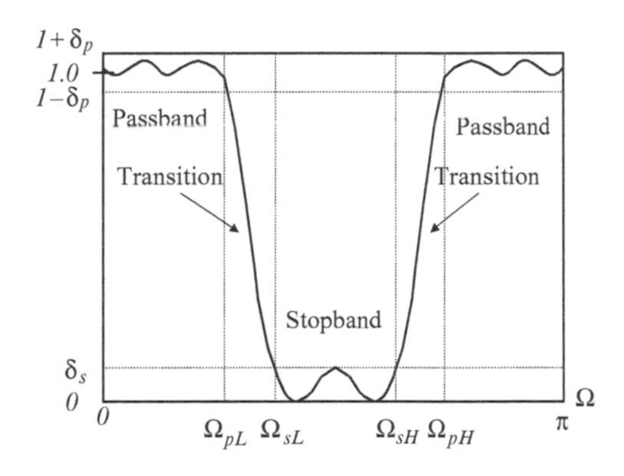

FIG. 6.19

Magnitude of the normalized bandstop filter.

Finally, the bandstop (bandreject or notch) filter shown in Fig. 6.19 rejects the middle-frequency components and accepts both the low- and the high-frequency components.

As a matter of fact, all kinds of digital filters are implemented using FIR or IIR systems. Furthermore, the FIR and IIR system can each be realized by various filter configurations, such as direct-forms, cascade forms, and parallel forms. Such topics will be included in the next section.

Given a transfer function, the MATLAB function **freqz()** can be used to determine the frequency response. The syntax is given by.

$$[\mathbf{h}, \mathbf{w}] = \mathbf{freqz}(\mathbf{B}, \mathbf{A}, \mathbf{N})$$

where all the parameters are defined as

h = an output vector containing frequency response.

w = an output vector containing normalized frequency values distributed in the range from 0 to π radians.

B = an input vector for numerator coefficients.

A = an input vector for denominator coefficients.

N = the number of normalized frequency points used for calculating the frequency response.

Let's consider Example 6.12.

EXAMPLE 6.12

Given each of the following digital transfer functions,

(a) $H(z) = \dfrac{z}{z-0.5}$

(b) $H(z) = 1 - 0.5z^{-1}$

(c) $H(z) = \dfrac{0.5z^2 - 0.32}{z^2 - 0.5z + 0.25}$

(d) $H(z) = \dfrac{1 - 0.9z^{-1} + 0.81z^{-2}}{1 - 0.6z^{-1} + 0.36z^{-2}}$,

1. Plot the poles and zeros on the z-plane.
2. Use MATLAB function **freqz()** to plot the magnitude frequency response and phase response for each transfer function.
3. Identify the corresponding filter type such as lowpass, highpass, bandpass, or bandstop.

Solution:

1. The pole-zero plot for each transfer function is demonstrated in Fig. 6.20. The transfer functions of (a) and (c) need to be converted into the standard form (delay form) required by the MATLAB function **freqz(),** in which both the numerator and denominator polynomials have negative powers of z. Hence, we obtain

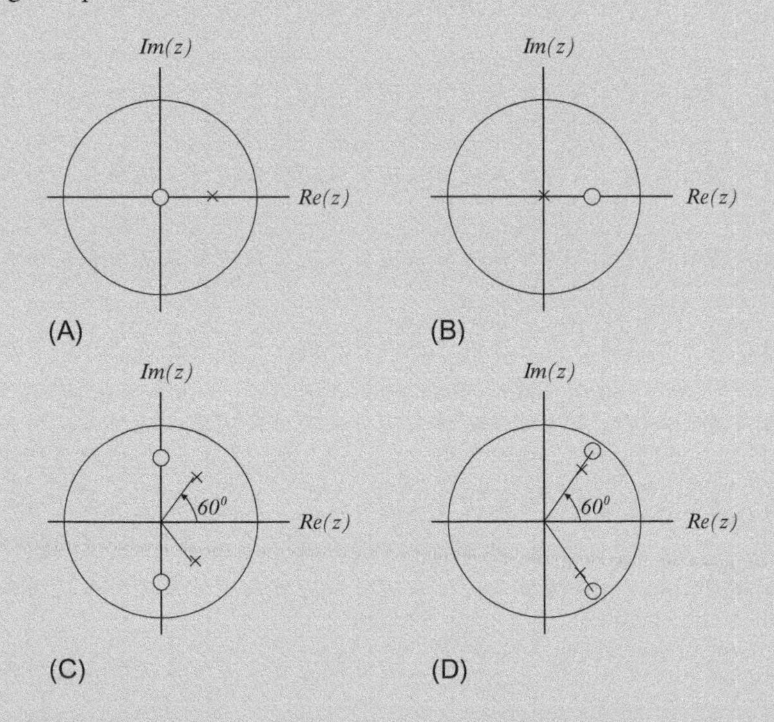

FIG. 6.20

Pole-zero plots of Example 6.12.

$$H(z) = \frac{z}{z-0.5} = \frac{1}{1-0.5z^{-1}}$$

$$H(z) = \frac{0.5z^2-0.32}{z^2-0.5z+0.25} = \frac{0.5-0.32z^{-2}}{1-0.5z^{-1}+0.25z^{-2}},$$

while the transfer functions of (b) and (d) are already in their standard forms (delay forms).

2. The MATLAB program for plotting the magnitude frequency response and the phase response for each case is listed in Program 6.2.
3. From the plots in Figs. 6.21(a–d) of magnitude frequency responses for all cases, we can conclude that case (a) is a lowpass filter, (b) is a highpass filter, (c) is a bandpass filter, and (d) is a bandstop (bandreject) filter.

Program 6.2. MATLAB program for Example 6.12.

```
% Example 6.12
% Plot the frequency response and phase response
% Case a
 figure (1)
 [h w]=freqz([1],[1-0.5],1024);  % Calculate frequency response
 phi=180*unwrap(angle(h))/pi;
 subplot(2,1,1), plot(w,abs(h)),grid;
 xlabel('Frequency (radians)'), ylabel('Magnitude')
 subplot(2,1,2), plot(w,phi),grid;
xlabel('Frequency (radians)'), ylabel('Phase (degrees)')
% Case b
 figure (2)
 [h w]=freqz([1-0.5],[1],1024);  %Calculate frequency response
 phi=180*unwrap(angle(h))/pi;
 subplot(2,1,1), plot(w,abs(h)),grid;
xlabel('Frequency (radians)'), ylabel('Magnitude')
 subplot(2,1,2), plot(w,phi),grid;
xlabel('Frequency (radians)'), ylabel('Phase (degrees)')
% Case c
 figure (3)
 [h w]=freqz([0.5 0-0.32],[1-0.5 0.25],1024); % Calculate frequency response
 phi=180*unwrap(angle(h))/pi;
 subplot(2,1,1), plot(w,abs(h)),grid;
xlabel('Frequency (radians)'), ylabel('Magnitude')
 subplot(2,1,2), plot(w,phi),grid;
xlabel('Frequency (radians)'), ylabel('Phase (degrees)')
% Case d
 figure (4)
 [h w]=freqz([1-0.9 0.81], [1-0.6 0.36],1024); %Calculate frequency response
 phi=180*unwrap(angle(h))/pi;
 subplot(2,1,1), plot(w,abs(h)),grid;
xlabel('Frequency (radians)'), ylabel('Magnitude')
 subplot(2,1,2), plot(w,phi),grid;
xlabel('Frequency (radians)'), ylabel('Phase (degrees)')
%
```

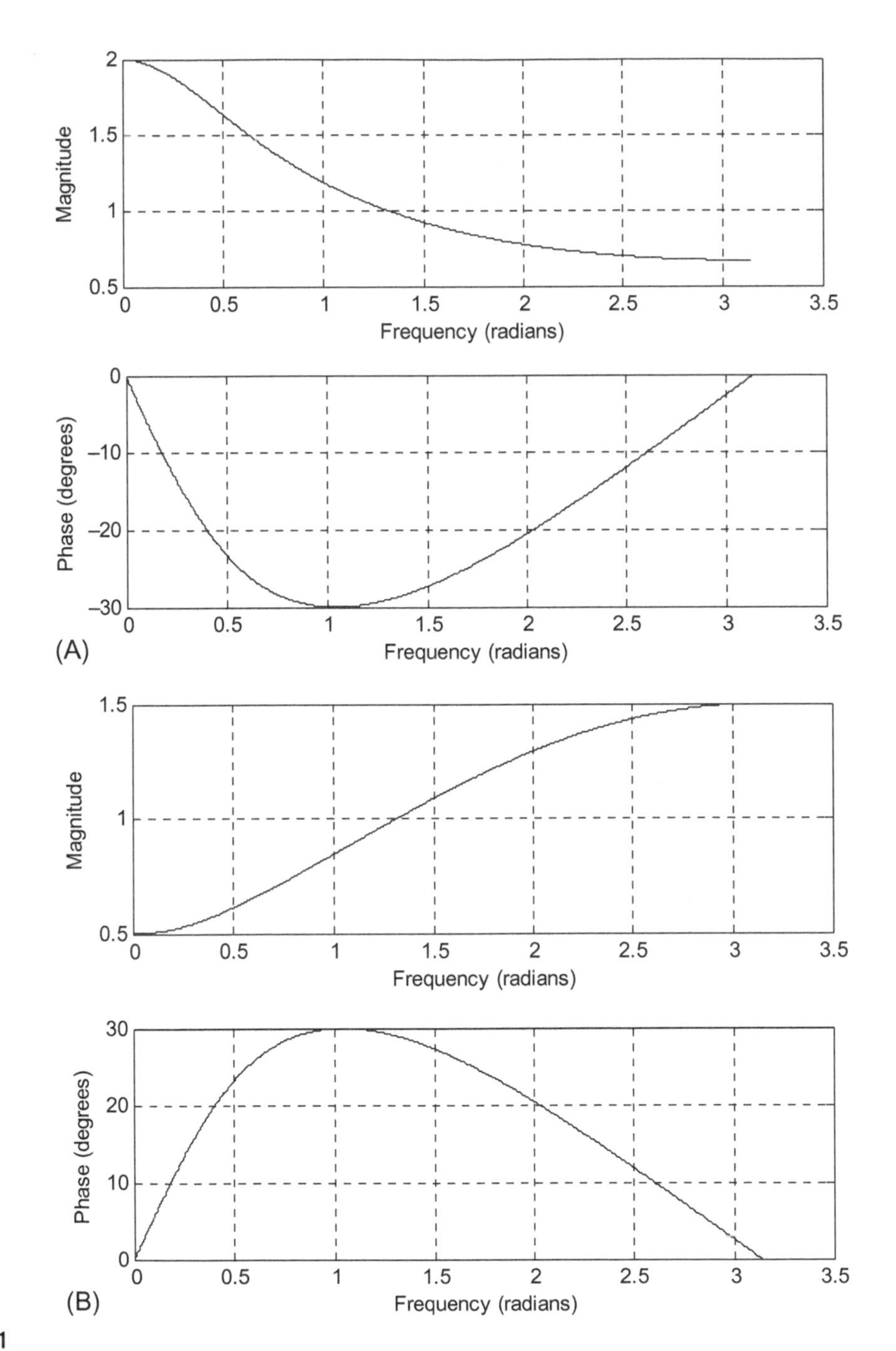

FIG. 6.21

Plots of frequency responses for Example 6.12: (A) for (a), (B) for (b),

(Continued)

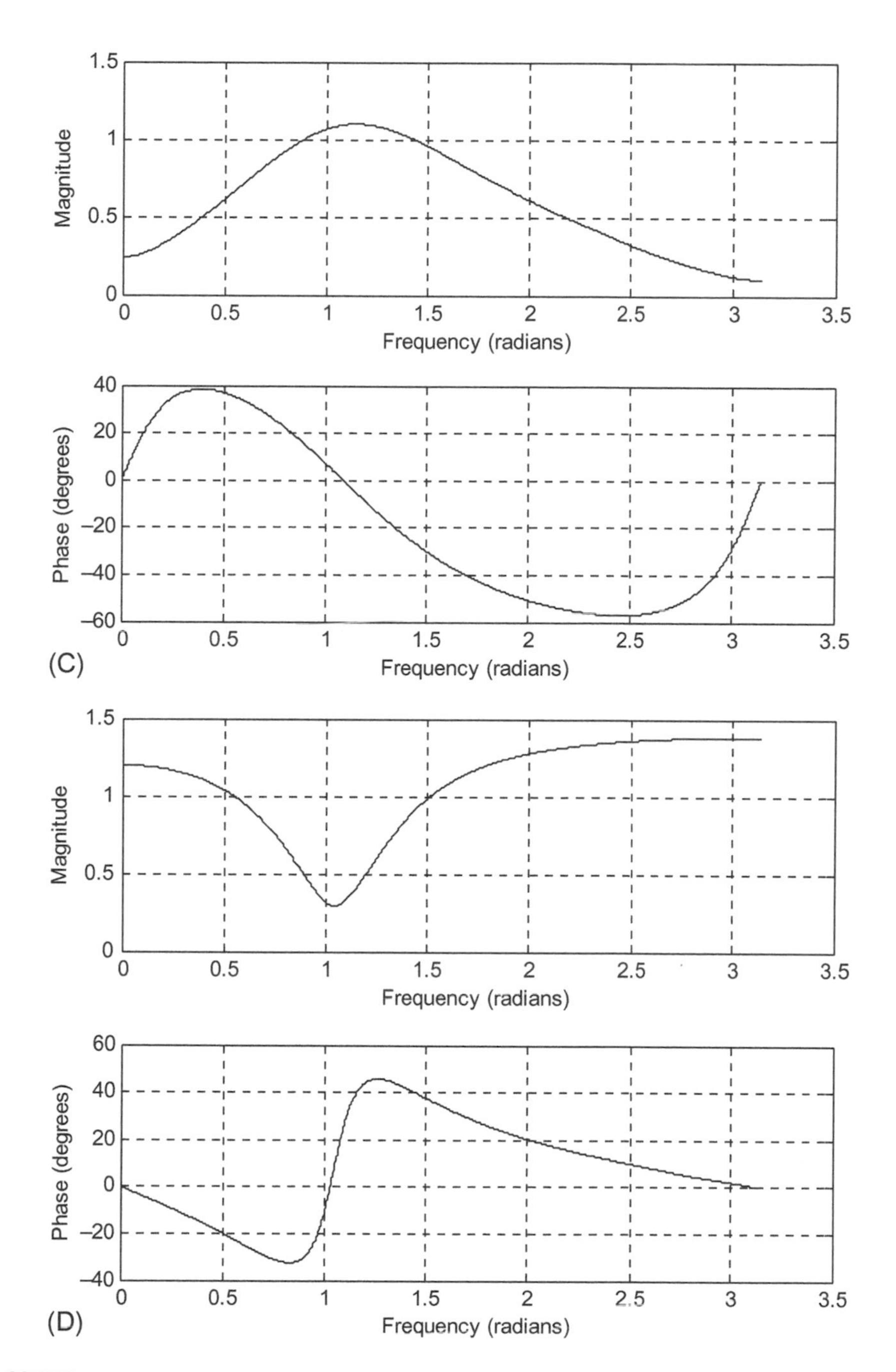

FIG. 6.21, CONT'D

(C) for (c), and (D) for (d).

6.6 REALIZATION OF DIGITAL FILTERS

In this section, basic realization methods for digital filters are discussed. Digital filters described by the transfer function $H(z)$ may be generally realized into the following forms:

- direct-form I
- direct-form II
- cascade
- parallel

The reader can explore various lattice realizations in the textbook by Stearns and Hush [1990]. The lattice filter realization has a good numerical property and easy for the control of filter stability. Lattice filters are widely used in speech processing and a favored structure for implementing adaptive filters.

6.6.1 DIRECT-FORM I REALIZATION

As we know, a digital filter transfer function, $H(z)$, is given by

$$H(z)=\frac{B(z)}{A(z)}=\frac{b_0+b_1z^{-1}+\cdots+b_Mz^{-M}}{1+a_1z^{-1}+\cdots+a_Nz^{-N}}. \tag{6.26}$$

Let $x(n)$ and $y(n)$ be the digital filter input and output, respectively. We can express the relationship in z-transform domain as

$$Y(z)=H(z)X(z), \tag{6.27}$$

where $X(z)$ and $Y(z)$ are the z-transforms of $x(n)$ and $y(n)$, respectively. If we substitute Eq. (6.26) into $H(z)$ in Eq. (6.27), we have

$$Y(z)=\left(\frac{b_0+b_1z^{-1}+\cdots+b_Mz^{-M}}{1+a_1z^{-1}+\cdots+a_Nz^{-N}}\right)X(z). \tag{6.28}$$

Taking the inverse of the z-transform in Eq. (6.28), we yield the relationship between input $x(n)$ and output $y(n)$ in time domain, as follows:

$$\begin{aligned} y(n)&=b_0x(n)+b_1x(n-1)+\cdots+b_Mx(n-M)\\ &\quad-a_1y(n-1)-a_2y(n-2)-\cdots-a_Ny(n-N). \end{aligned} \tag{6.29}$$

This difference equation thus can be implemented by a direct-form I realization shown in Fig. 6.22(a). Fig. 6.22(b) illustrates the realization of the second-order IIR filter ($M=N=2$). Note that the notation used in Figs. 6.22(a) and (b) are defined in Fig. 22(c) and will be applied for discussion of other realizations.

Also, note that any of the a_j and b_i can be zero, thus all the paths are not required to exist for the realization.

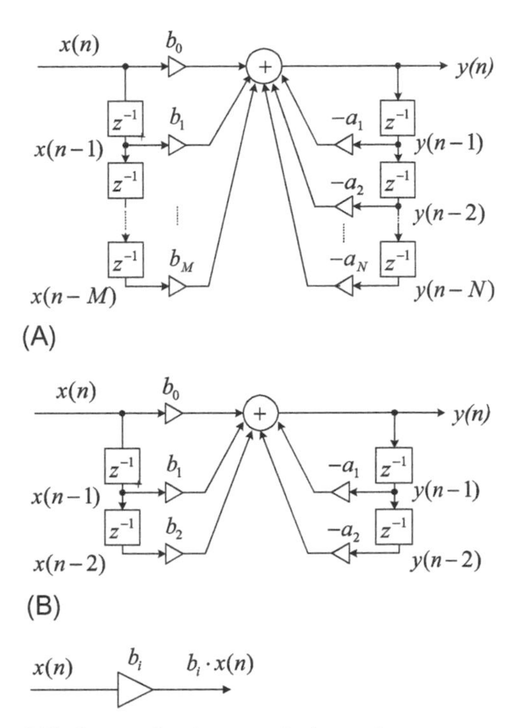

FIG. 6.22

(A) Direct-form I realization, (B) Direct-form I realization with $M=2$, (C) Notation.

6.6.2 DIRECT-FORM II REALIZATION

Considering Eqs. (6.26) and (6.27) with $N=M$, we can express

$$Y(z)=H(z)X(z)=\frac{B(z)}{A(z)}X(z)=B(z)\left(\frac{X(z)}{A(z)}\right)$$
$$=\left(b_0+b_1z^{-1}+\cdots+b_Mz^{-M}\right)\underbrace{\left(\frac{X(z)}{1+a_1z^{-1}+\cdots+a_Mz^{-M}}\right)}_{W(z)}. \tag{6.30}$$

Also, define a new z-transform function as

$$W(z)=\frac{X(z)}{1+a_1z^{-1}+\cdots+a_Mz^{-M}}, \tag{6.31}$$

we have

$$Y(z)=\left(b_0+b_1z^{-1}+\cdots+b_Mz^{-M}\right)W(z). \tag{6.32}$$

The corresponding difference equations for Eqs. (6.31) and (6.32), respectively, become

$$w(n)=x(n)-a_1w(n-1)-a_2w(n-2)-\cdots-a_Mw(n-M) \tag{6.33}$$

and

$$y(n)=b_0w(n)+b_1w(n-1)+\ldots+b_Mw(n-M). \tag{6.34}$$

Realization of Eqs. (6.33) and (6.34) becomes another direct-form II realization, which is demonstrated in Fig. 6.23A. Again, the corresponding realization of the second-order IIR filter is described in Fig. 6.23B. Note that in Fig. 6.23A, the variables $w(n)$, $w(n-1)$, $w(n-2)$, …, $w(n-M)$ are different from the filter inputs $x(n-1)$, $x(n-2)$, …,$x(n-M)$.

On comparing structures between direct-form I realization and direct-form II realization, it can be seen that both realizations require the same number of multiplications while the direct-form II realization requires two accumulators. One of the benefits from the direct-form II structure is the use of the

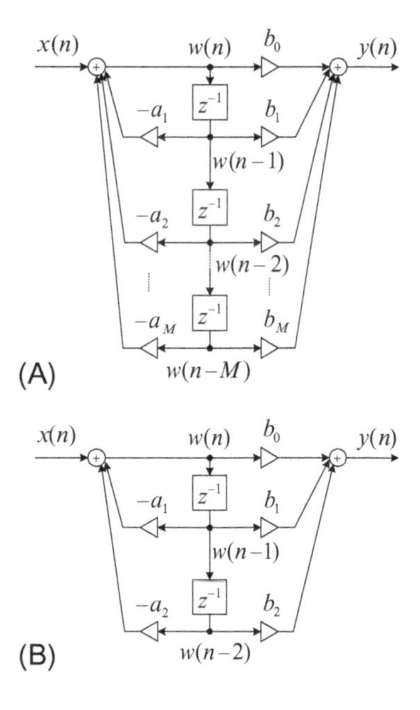

FIG. 6.23

(A) Direct-form II realization, (B) direct-form II realization with $M=2$.

reduced number of delay elements (saving memory). The other benefit relies on the fixed-point realization, where the filtering process involves only integer operations. Scaling of filter coefficients in the fixed-point implementation is required to avoid the overflow in the accumulator. For the direct-form II structure, the numerator and denominator filter coefficients are scaled separately so that the overflow problem for each accumulator can easily be controlled. More details will be illustrated in Chapter 14.

6.6.3 CASCADE (SERIES) REALIZATION

An alternate way to filter realization is to cascade the factorized $H(z)$ in the following form:

$$H(z) = H_1(z) \cdot H_2(z) \cdots H_k(z), \tag{6.35}$$

where $H_k(z)$ is chosen to be the first- or second-order transfer function (section), which is defined by

$$H_k(z) = \frac{b_{k0} + b_{k1}z^{-1}}{1 + a_{k1}z^{-1}} \tag{6.36}$$

or

$$H_k(z) = \frac{b_{k0} + b_{k1}z^{-1} + b_{k2}z^{-2}}{1 + a_{k1}z^{-1} + a_{k2}z^{-2}}, \tag{6.37}$$

respectively. The block diagram of the cascade, or series, realization is depicted in Fig. 6.24.

6.6.4 PARALLEL REALIZATION

Now we convert $H(z)$ into the following form

$$H(z) = H_1(z) + H_2(z) + \cdots + H_k(z), \tag{6.38}$$

where $H_k(z)$ is defined as the first- or second-order transfer function (section) given by

$$H_k(z) = \frac{b_{k0}}{1 + a_{k1}z^{-1}} \tag{6.39}$$

or

$$H_k(z) = \frac{b_{k0} + b_{k1}z^{-1}}{1 + a_{k1}z^{-1} + a_{k2}z^{-2}}, \tag{6.40}$$

respectively. The resulting parallel realization is illustrated in the block diagram in Fig. 6.25.

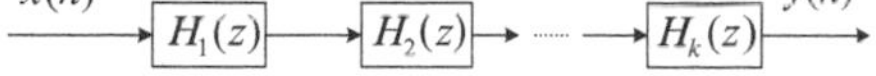

$x(n)$ → $H_1(z)$ → $H_2(z)$ → ...... → $H_k(z)$ → $y(n)$

FIG. 6.24

Cascade realization.

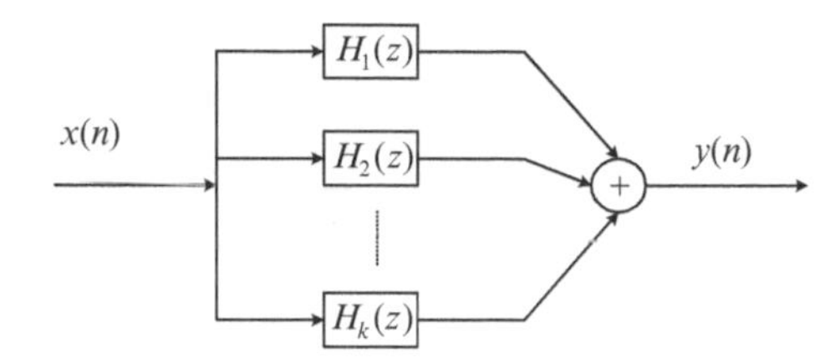

FIG. 6.25

Parallel realization.

EXAMPLE 6.13

Given a second-order transfer function

$$H(z) = \frac{0.5(1 - z^{-2})}{1 + 1.3z^{-1} + 0.36z^{-2}},$$

perform the filter realizations and write difference equations using the following realizations:

1. Direct-form I and direct-form II
2. Cascade form via first-order sections
3. Parallel form via first-order sections

Solution:

1. To perform the filter realizations using the direct-form I and direct-form II, we rewrite the given second-order transfer function as

$$H(z) = \frac{0.5 - 0.5z^{-2}}{1 + 1.3z^{-1} + 0.36z^{-2}}$$

and identify that.

$a_1 = 1.3$, $a_2 = 0.36$, $b_0 = 0.5$, $b_1 = 0$, and $b_2 = -0.5$.

Based on realizations in Fig. 6.22, we sketch the direct-form I realization as shown in Fig. 6.26.

The difference equation for the direct-form I realization is given by

$$y(n) = 0.5x(n) - 0.5x(n-2) - 1.3y(n-1) - 0.36y(n-2).$$

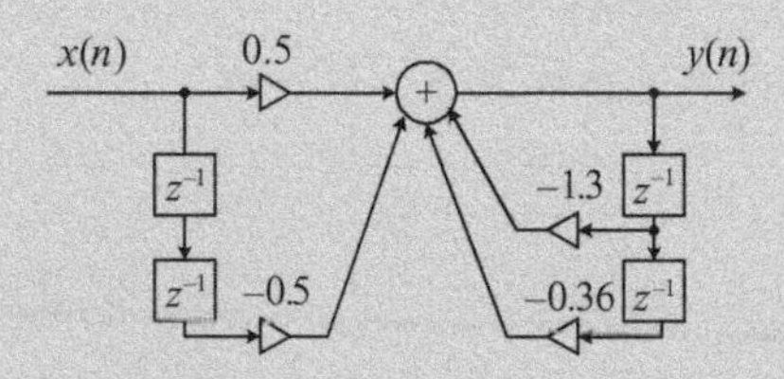

FIG. 6.26

Direct-form I realization for Example 6.13.

Using the direct-form II realization shown in Fig. 6.23, we have the realization depicted in Fig. 6.27.

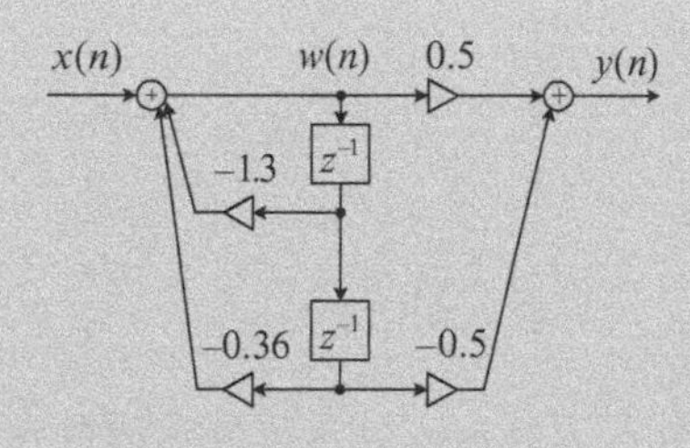

FIG. 6.27

Direct-form II realization for Example 6.13.

The difference equations for the direct-form II realization are expressed as

$$w(n) = x(n) - 1.3w(n-1) - 0.36w(n-2)$$

$$y(n) = 0.5w(n) - 0.5w(n-2)$$

2. To achieve the cascade (series) form realization, we factor $H(z)$ into two first-order sections to yield

$$H(z) = \frac{0.5(1 - z^{-2})}{1 + 1.3z^{-1} + 0.36z^{-2}} = \frac{0.5 - 0.5z^{-1}}{1 + 0.4z^{-1}} \frac{1 + z^{-1}}{1 + 0.9z^{-1}},$$

where $H_1(z)$ and $H_2(z)$ are chosen to be

$$H_1(z) = \frac{0.5 - 0.5z^{-1}}{1 + 0.4z^{-1}}$$

$$H_2(z) = \frac{1 + z^{-1}}{1 + 0.9z^{-1}}.$$

Note that the obtained $H_1(z)$ and $H_2(z)$ are not the unique selections for realization. For example, there is another way of choosing $H_1(z) = \frac{0.5 - 0.5z^{-1}}{1 + 0.9z^{-1}}$ and $H_2(z) = \frac{1 + z^{-1}}{1 + 0.4z^{-1}}$ to yield the same $H(z)$. Using $H_1(z)$ and $H_2(z)$ we have obtained, and with the direct-form II realization, we achieve the cascade form depicted in Fig. 6.28.

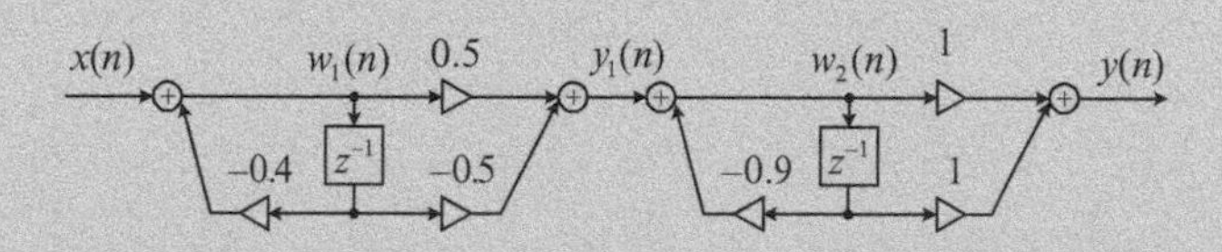

FIG. 6.28

Cascade realization for Example 6.13.

The difference equations for the direct-form II realization have two cascaded sections, expressed as.

Section 1:

$$w_1(n) = x(n) - 0.4w(n-1)$$

$$y_1(n) = 0.5w_1(n) - 0.5w_1(n-1)$$

Continued

EXAMPLE 6.13—CONT'D

Section 2:

$$w_2(n) = y_1(n) - 0.9w_2(n-1)$$

$$y(n) = w_2(n) + w_2(n-1).$$

3. In order to yield the parallel form of realization, we need to make use of the partial fraction expansion, and will first let

$$\frac{H(z)}{z} = \frac{0.5(z^2-1)}{z(z+0.4)(z+0.9)} = \frac{A}{z} + \frac{B}{z+0.4} + \frac{C}{z+0.9},$$

where

$$A = z\left(\frac{0.5(z^2-1)}{z(z+0.4)(z+0.9)}\right)\bigg|_{z=0} = \frac{0.5(z^2-1)}{(z+0.4)(z+0.9)}\bigg|_{z=0} = -1.39$$

$$B = (z+0.4)\left(\frac{0.5(z^2-1)}{z(z+0.4)(z+0.9)}\right)\bigg|_{z=-0.4} = \frac{0.5(z^2-1)}{z(z+0.9)}\bigg|_{z=-0.4} = 2.1$$

$$C = (z+0.9)\left(\frac{0.5(z^2-1)}{z(z+0.4)(z+0.9)}\right)\bigg|_{z=-0.9} = \frac{0.5(z^2-1)}{z(z+0.4)}\bigg|_{z=-0.9} = -0.21.$$

Therefore

$$H(z) = -1.39 + \frac{2.1z}{z+0.4} + \frac{-0.21z}{z+0.9} = -1.39 + \frac{2.1}{1+0.4z^{-1}} + \frac{-0.21}{1+0.9z^{-1}}.$$

Again, using the direct-form II for each section, we obtain the parallel realization shown in Fig. 6.29.

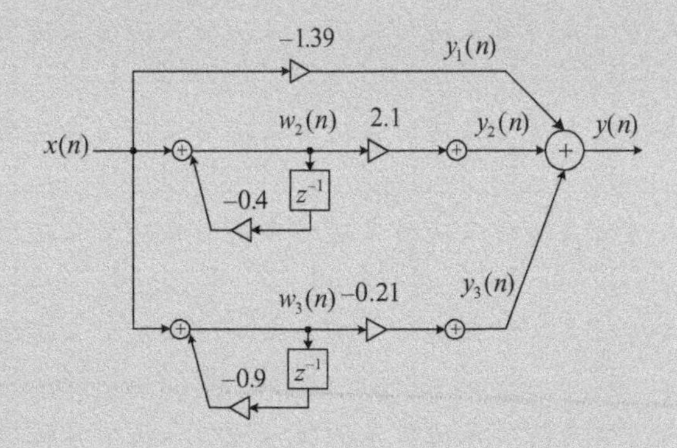

FIG. 6.29

Parallel realization for Example 6.13.

The difference equations for the direct-form II realization have three parallel sections, expressed as

$$y_1(n) = -1.39x(n)$$
$$w_2(n) = x(n) - 0.4w_2(n-1)$$
$$y_2(n) = 2.1w_2(n)$$
$$w_3(n) = x(n) - 0.9w_3(n-1)$$
$$y_3(n) = -0.21w_3(n)$$
$$y(n) = y_1(n) + y_2(n) + y_3(n).$$

In practice, the second-order filter module using the direct-form I or direct-form II is used. The high-order filter can be factored in the cascade form with the first- or second-order sections. In case the first-order filter is required, we can still modify the second-order filter module by setting the corresponding filter coefficients to be zero.

6.7 APPLICATION: SIGNAL ENHANCEMENT AND FILTERING

This section investigates the applications of signal enhancement using a preemphasis filter and speech filtering using a bandpass filter. Enhancement also includes the biomedical signals such as an electrocardiogram (ECG) signal.

6.7.1 PREEMPHASIS OF SPEECH

A speech signal may have frequency components that falloff at high frequencies. In some applications such as speech coding, to avoid overlooking the high frequencies, the high-frequency components are compensated using preemphasis filtering. A simple digital filter used for such compensation is given as

$$y(n) = x(n) - \alpha x(n-1), \tag{6.41}$$

where α is the positive parameter to control the degree of preemphasis filtering and usually is chosen to be <1. The filter described in Eq. (6.41) is essentially a highpass filter. Applying z-transform on both sides of Eq. (6.41) and solving for the transfer function, we have

$$H(z) = 1 - \alpha z^{-1}. \tag{6.42}$$

The magnitude and phase responses adopting the preemphasis parameter $\alpha = 0.9$ and the sampling rate $f_s = 8,000$ Hz are plotted in Fig. 6.30A using MATLAB.

Fig. 6.30B compares the original speech waveform and the preemphasized speech using the filter in Eq. (6.42). Again, we apply the fast Fourier transform (FFT) to estimate the spectrum of the original speech and the spectrum of the preemphasized speech. The plots are displayed in Fig. 6.31.

From Fig. 6.31, we can conclude that the filter does its job to boost the high-frequency components and attenuate the low-frequency components. We can also try this filter with different values of α to examine the degree of preemphasis filtering of the digitally recorded speech. The MATLAB list is given in Program 6.3.

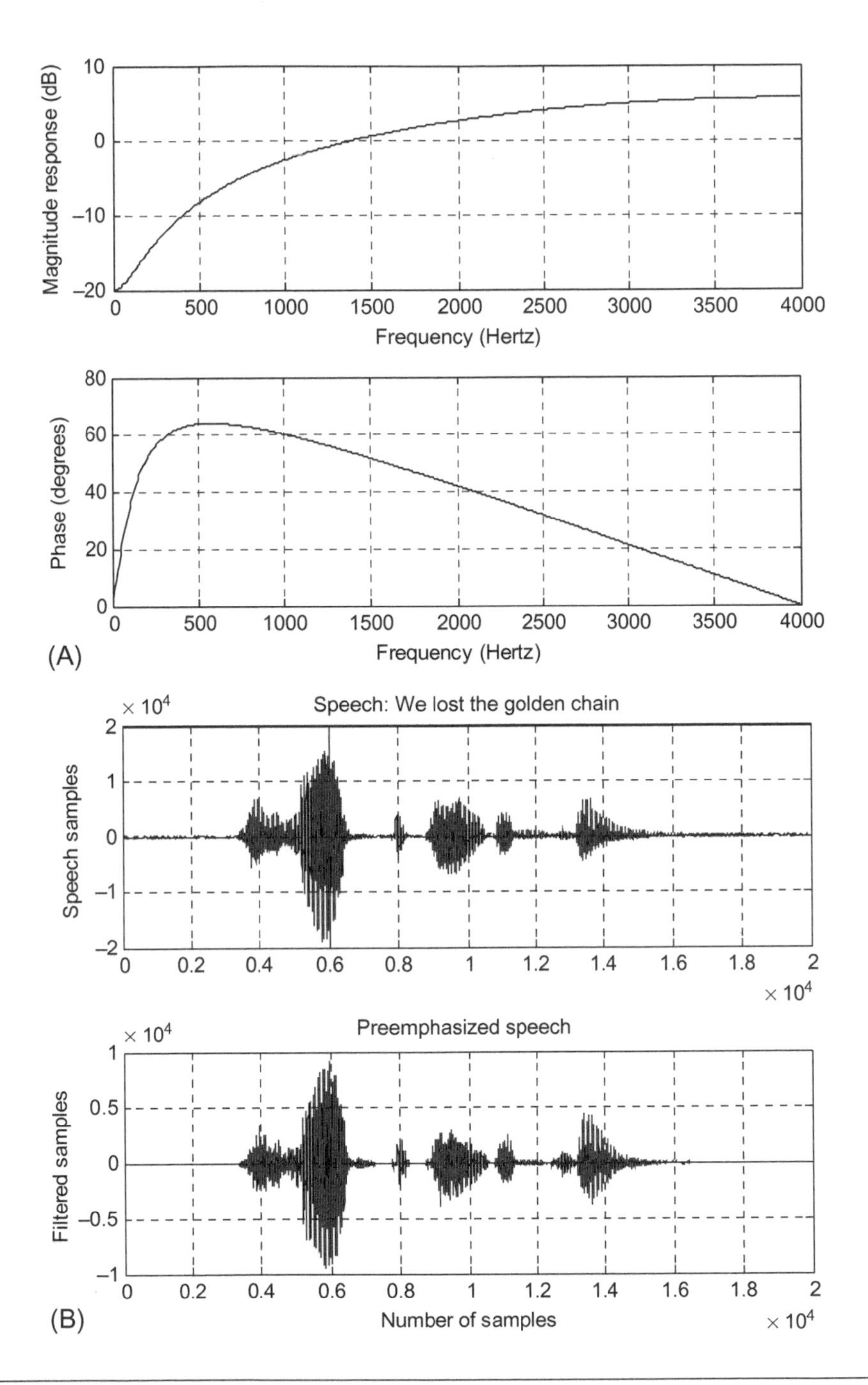

FIG. 6.30

(A) Frequency responses of the preemphasis filter. (B) Original speech and preemphasized speech waveforms.

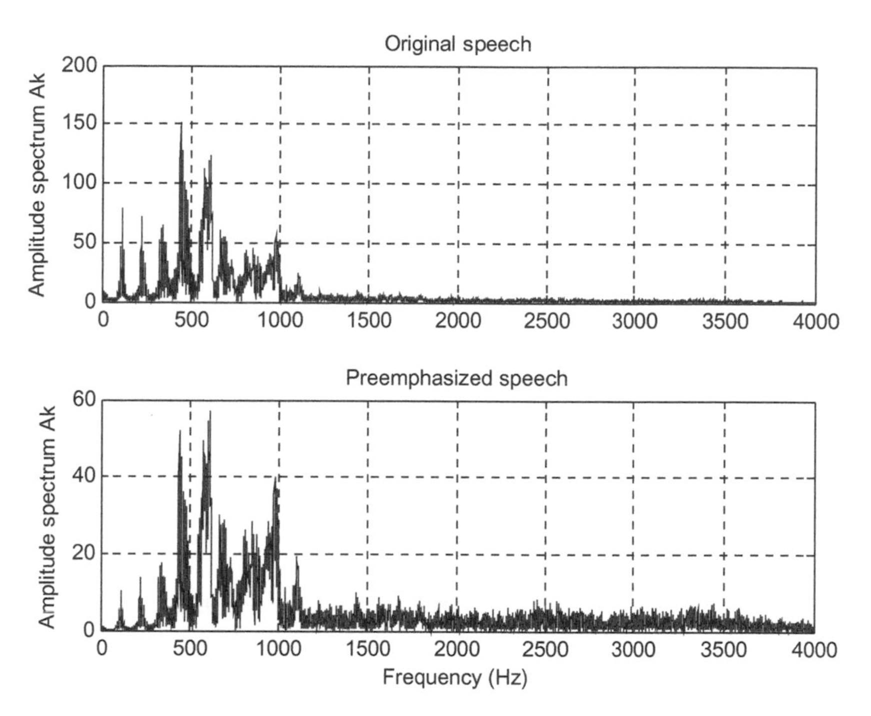

FIG. 6.31

Amplitude spectral plots for the original speech and preemphasized speech.

Program 6.3. MALAB program for preemphasis of speech

```
% Matlab program for Figs. 6.30 and 6.31
close all;clear all
fs=8000;                        % Sampling rate
alpha =0.9;                     % Degree of pre-emphasis
figure(1);
freqz([1 -alpha],1512,fs);       % Calculate and display frequency response
load speech.dat
figure(2);
y=filter([1 -alpha],1,speech);    % Filtering speech
subplot(2,1,1),plot(speech);grid;
ylabel('Speech samples')
title('Speech: We lost the golden chain.')
subplot(2,1,2),plot(y);grid
ylabel('Filtered samples')
xlabel('Number of samples');
title('Preemphasized speech.')
figure(3);
```

```
N=length(speech);                          % Length of speech
Axk=abs(fft(speech.*hamming(N)'))/N;       % Two-sided spectrum of speech
Ayk=abs(fft(y.*hamming(N)'))/N;            % Two-sided spectrum of preemphasized speech
f=[0:N/2]*fs/N;
Axk(2:N)=2*Axk(2:N);                       % Get one-side spectrum of speech
Ayk(2:N)=2*Ayk(2:N);                       % Get one-side spectrum of filtered speech
subplot(2,1,1),plot(f,Axk(1:N/2+1));grid
ylabel('Amplitude spectrum Ak')
title('Original speech');
subplot(2,1,2),plot(f,Ayk(1:N/2+1));grid
ylabel('Amplitude spectrum Ak')
xlabel('Frequency (Hz)');
title('Preemphasized speech');
%
```

6.7.2 BANDPASS FILTERING OF SPEECH

Bandpass filtering plays an important role in DSP applications. It can be used to pass the signals according to the specified frequency passband and reject the frequency other than the passband specification. Then the filtered signal can further be used for the signal feature extraction. Filtering can also be applied to perform applications such as noise reduction, frequency boosting, digital audio equalizing, and digital crossover, among others.

Let us consider the following digital fourth-order bandpass Butterworth filter with a lower cutoff frequency of 1000 Hz, an upper cutoff frequency of 1400 Hz (i.e., the bandwidth is 400 Hz), and a sampling rate of 8000 Hz:

$$H(z)=\frac{0.0201-0.0402z^{-2}+0.0201z^{-4}}{1-2.1192z^{-1}+2.6952z^{-2}-1.6924z^{-3}+0.6414z^{-4}}. \tag{6.43}$$

Converting the z-transfer function into the DSP difference equation yields

$$\begin{aligned}y(n)&=0.0201x(n)-0.0402x(n-2)+0.0201x(n-4)\\&+2.1192y(n-1)-2.6952y(n-2)+1.6924y(n-3)-0.6414y(n-4).\end{aligned} \tag{6.44}$$

The filter frequency responses are computed and plotted in Fig. 6.32A with MATLAB. Fig. 6.32B shows the original speech and filtered speech, while Fig. 6.32C displays the spectral plots for the original speech and filtered speech.

As shown in Fig. 6.32C, the designed bandpass filter significantly reduces low-frequency components, which are <1000 Hz, and the high-frequency components above 1400 Hz, while letting the signals with the frequencies ranging from 1000 to 1400 Hz pass through the filter. Similarly, we can design and implement other types of filters, such as lowpass, highpass, bandpass, and bandreject to filter the signals and examine the performances of their designs. MATLAB implementation detail is given in Program 6.4.

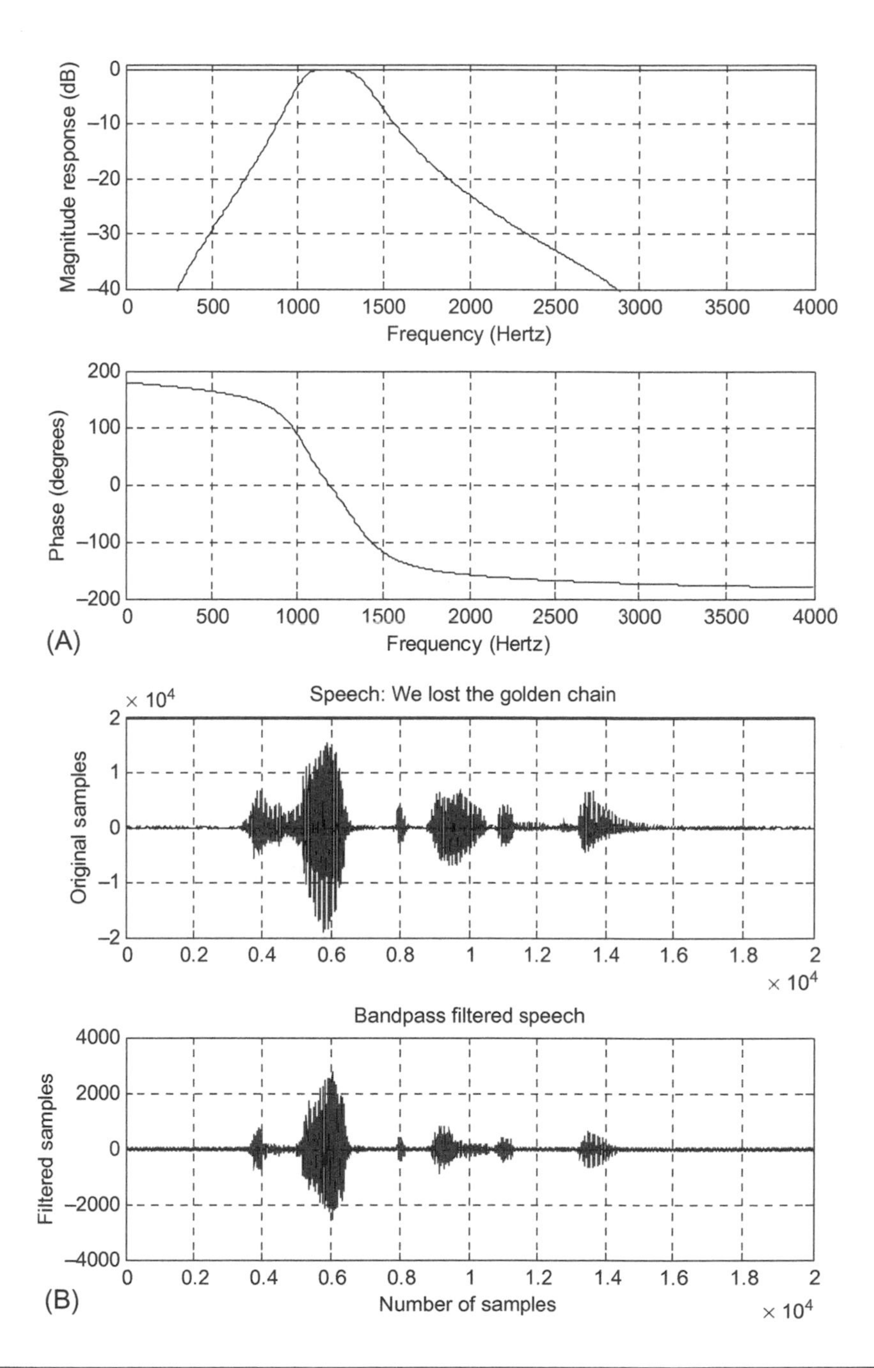

FIG. 6.32

(A) Frequency responses of the designed bandpass filter. (B) Plots of the original speech and filtered speech.

(Continued)

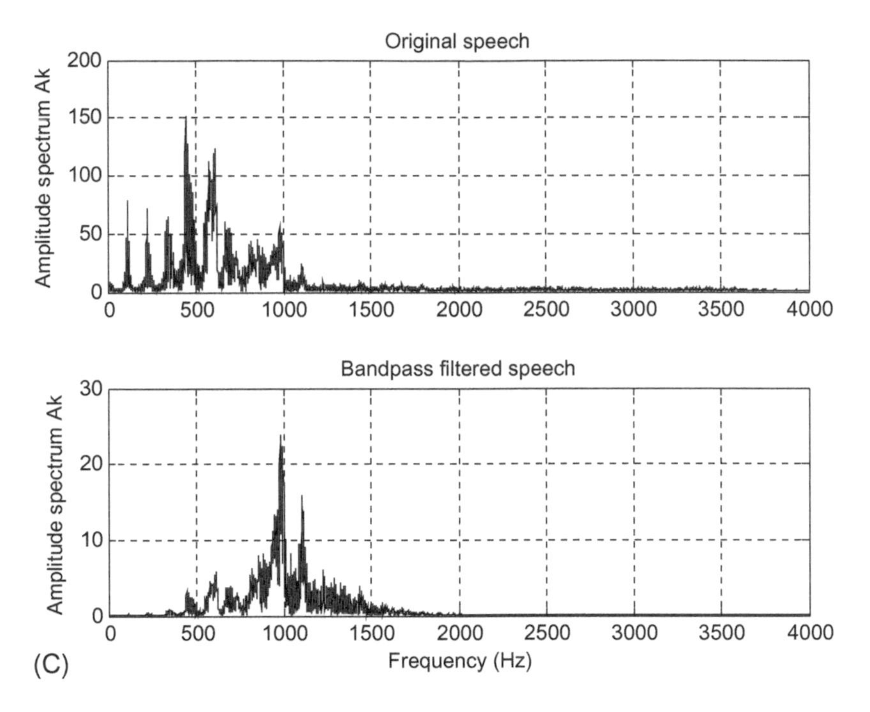

FIG. 6.32, CONT'D

(C) Amplitude spectra of the original speech and bandpass filtered speech.

Program 6.4. MATLAB program for bandpass filtering of speech

```
fs=8000;                    % Sampling rate
freqz([0.0201 0.00-0.0402 0 0.0201],[1-2.1192 2.6952-1.6924 0.6414],512,fs);
axis([0fs/2-40 1]);                  % Frequency response of bandpass filter
figure
load speech.dat
y=filter([0.0201 0.00-0.0402 0.0201],[1-2.1192 2.6952-1.6924 0.6414],speech);
subplot(2,1,1),plot(speech); grid;              % Filtering speech
ylabel('Origibal Samples')
title('Speech: We lost the golden chain.')
subplot(2,1,2),plot(y);grid
xlabel('Number of Samples');ylabel('Filtered Samples')
title('Bandpass filtered speech.')
figure
N=length(speech);
Axk=abs(fft(speech.*hamming(N)'))/N;  % One-sided spectrum of speech
Ayk=abs(fft(y.*hamming(N)'))/N;       % One-sided spectrum of filtered speech
f=[0:N/2]*fs/N;
Axk(2:N)=2*Axk(2:N);Ayk(2:N)=2*Ayk(2:N);  % One-sided spectra
subplot(2,1,1),plot(f,Axk(1:N/2+1));grid
ylabel('Amplitude spectrum Ak')
title('Original speech');
subplot(2,1,2),plot(f,Ayk(1:N/2+1));grid
ylabel('Amplitude spectrum Ak');xlabel('Frequency (Hz)');
title('Bandpass filtered speech');
```

6.7.3 ENHANCEMENT OF ECG SIGNAL USING NOTCH FILTERING

A notch filter is a bandreject filter with a very narrow bandwidth. It can be applied to enhance the ECG signal corrupted during the data acquisition stage, where the signal contains a 60-Hz interference induced from the power line. Let us consider the following digital second-order notch filter with a notch frequency of 60 Hz and the digital system has a sampling frequency of 500 Hz. We obtain a notch filter (details can be found in Chapter 8) as follows:

$$H(z) = \frac{1 - 1.4579z^{-1} + z^{-2}}{1 - 1.3850z^{-1} + 0.9025z^{-2}}. \tag{6.45}$$

The DSP difference equation is expressed as

$$y(n) = x(n) - 1.4579x(n-1) + x(n-2) + 1.3850y(n-1) - 0.9025y(n-2). \tag{6.46}$$

The frequency responses are computed and plotted in Fig. 6.33. Comparison of the raw ECG signal corrupted by the 60-Hz interference with the enhanced ECG signal for both time domain and frequency domain are displayed in Figs. 6.34 and 6.35, respectively. As we can see, the notch filter completely removes the 60-Hz interference.

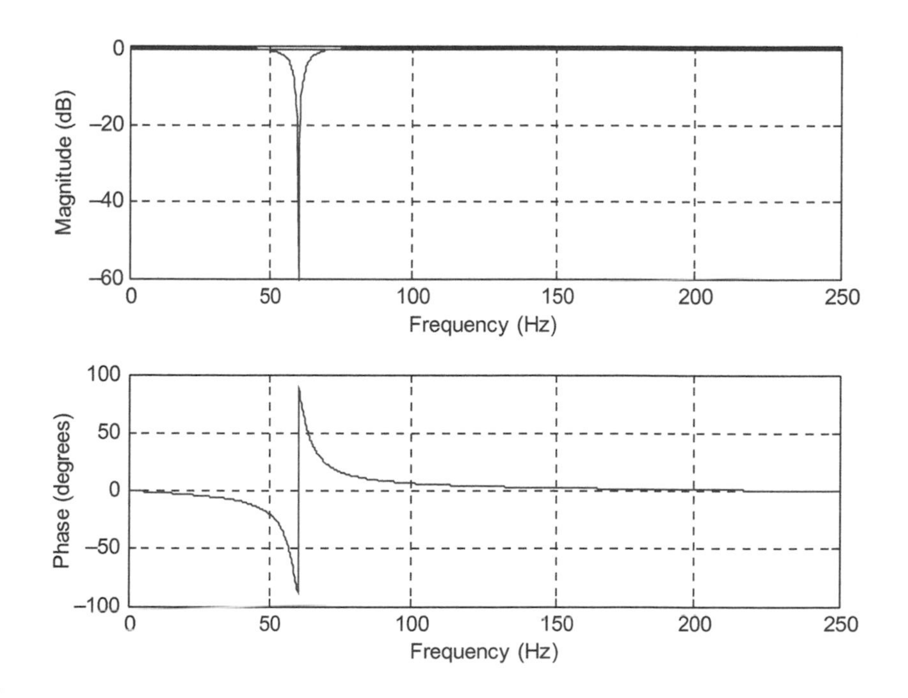

FIG. 6.33

Notch filter frequency responses.

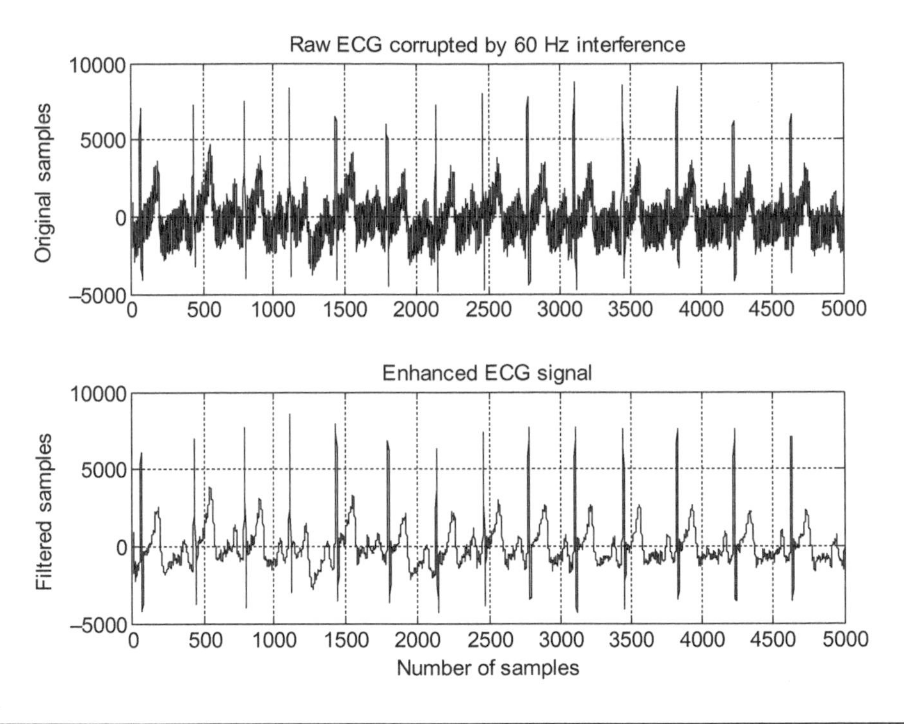

FIG. 6.34

The corrupted ECG signal and the enhanced ECG signal.

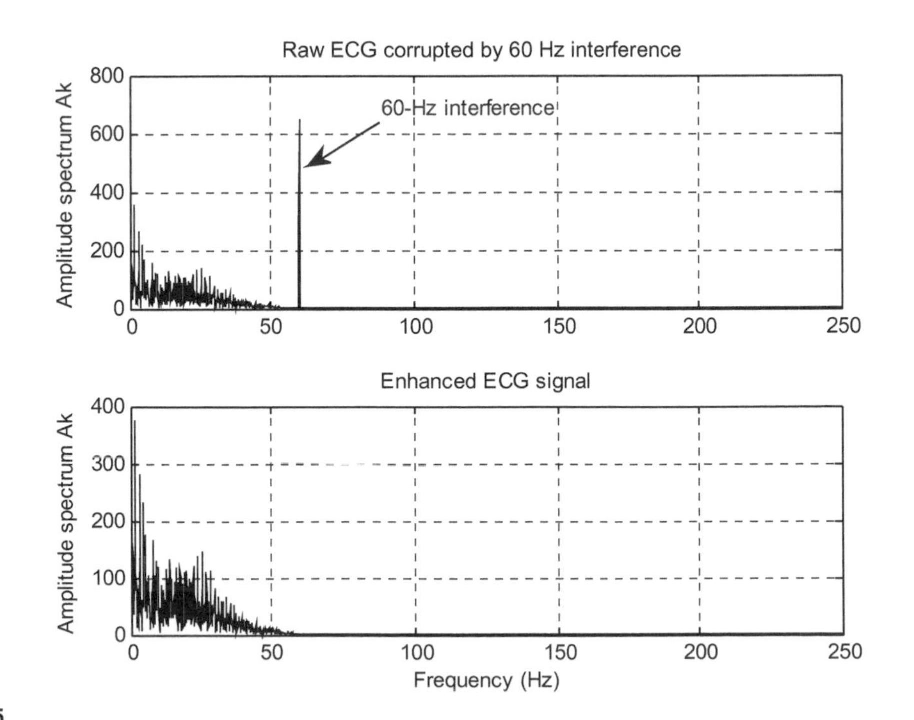

FIG. 6.35

The corrupted ECG signal spectrum and the enhanced ECG signal spectrum.

6.8 SUMMARY

1. The digital filter (DSP system) is represented by a difference equation, which is linear, time invariant.
2. The filter output depends on the filter current input, past input(s), and past output(s) in general. Given the arbitrary inputs and nonzero or zero initial conditions, operating the difference equation can generate the filter output recursively.
3. System responses such as the impulse response and step response can be determined analytically using the z-transform.
4. Transfer function can be obtained by applying z-transform to the difference equation to determine the ratio of the output z-transform over the input z-transform. A digital filter (DSP system) can be represented by its transfer function.
5. System stability can be studied using a very useful tool, a z-plane pole-zero plot.
6. The frequency response of a DSP system was developed and illustrated to investigate magnitude and phase responses. In addition, the FIR and IIR systems were defined.
7. Digital filters and their specifications, such as lowpass, highpass, bandpass, and bandstop, were reviewed.
8. A digital filter can be realized using standard realization methods such as the direct-form I; direct-form II; cascade, or series form; and parallel form.
9. Digital processing of speech using the preemphasis filter and bandpass filter was investigated to study spectral effects of the processed digital speech. The preemphasis filter boosts the high-frequency components, while bandpass filtering keeps the mid-band frequency components and rejects other lower- and upper-band frequency components.

6.9 PROBLEMS

6.1 Given a difference equation

$$y(n) = x(n) - 0.5y(n-1),$$

(a) Calculate the system response $y(n)$ for $n=0, 1, \ldots, 4$ with the input $x(n) = (0.5)^n u(n)$ and initial condition: $y(-1) = 1$.

(b) Calculate the system response $y(n)$ for $n=0, 1, \ldots, 4$ with the input $x(n) = (0.5)^n u(n)$ and zero initial condition: $y(-1) = 0$.

6.2 Given a difference equation

$$y(n) = 0.5x(n-1) + 0.6y(n-1),$$

(a) Calculate the system response $y(n)$ for $n=0, 1, \ldots, 4$ with the input $x(n) = (0.5)^n u(n)$ and initial conditions: $x(-1) = -1$, and $y(-1) = 1$.

(b) Calculate the system response $y(n)$ for $n=0, 1, \ldots, 4$ with the input $x(n) = (0.5)^n u(n)$ and zero initial conditions: $x(-1) = 0$, and $y(-1) = 0$.

6.3 Given a difference equation

$$y(n) = x(n-1) - 0.75y(n-1) - 0.125y(n-2),$$

(a) Calculate the system response $y(n)$ for $n=0, 1, \ldots, 4$ with the input $x(n)=(0.5)^n u(n)$ and initial conditions: $x(-1)=-1$, $y(-2)=2$, and $y(-1)=1$.

(b) Calculate the system response $y(n)$ for $n=0, 1, \ldots, 4$ with the input $x(n)=(0.5)^n u(n)$ and zero initial conditions: $x(-1)=0$, $y(-2)=0$, and $y(-1)=0$.

6.4 Given the following difference equation,

$$y(n)=0.5x(n)+0.5x(n-1),$$

(a) Find $H(z)$.

(b) Determine the impulse response $y(n)$ if the input $x(n)=4\delta(n)$.

(c) Determine the step response $y(n)$ if the input $x(n)=10u(n)$.

6.5 Given the following difference equation,

$$y(n)=x(n)-0.5y(n-1),$$

(a) Find $H(z)$.

(b) Determine the impulse response $y(n)$ if the input$x(n)=\delta(n)$.

(c) Determine the step response $y(n)$ if the input$x(n)=u(n)$.

6.6 A digital system is described by the following difference equation:

$$y(n)=x(n)-0.25x(n-2)-1.1y(n-1)-0.28y(n-2).$$

Find the transfer function $H(z)$, the denominator polynomial $A(z)$, and the numerator polynomial$B(z)$.

6.7 A digital system is described by the following difference equation:

$$y(n)=0.5x(n)+0.5x(n-1)-0.6y(n-2).$$

Find the transfer function $H(z)$, the denominator polynomial $A(z)$, and the numerator polynomial $B(z)$.

6.8 A digital system is described by the following difference equation:

$$y(n)=0.25x(n-2)+0.5y(n-1)-0.2y(n-2).$$

Find the transfer function $H(z)$, the denominator polynomial $A(z)$, and the numerator polynomial $B(z)$.

6.9 A digital system is described by the following difference equation:

$$y(n)=x(n)-0.3x(n-1)+0.28x(n-2).$$

Find the transfer function $H(z)$, the denominator polynomial $A(z)$, and the numerator polynomial $B(z)$.

6.10 Convert each of the following transfer functions into the difference equations:

(a) $H(z)=0.5+0.5z^{-1}$

(b) $H(z)=\dfrac{1}{1-0.3z^{-1}}$

6.11 Convert each of the following transfer functions into the difference equations:

(a) $H(z)=0.1+0.2z^{-1}+0.3z^{-2}$

(b) $H(z)=\dfrac{0.5-0.5z^{-2}}{1-0.3z^{-1}+0.8z^{-2}}$

6.12 Convert each of the following transfer functions into the difference equations:

(a) $H(z)=\frac{z^2-0.25}{z^2+1.1z+0.18}$

(b) $H(z)=\frac{z^2-0.1z+0.3}{z^3}$

6.13 Convert each of the following transfer functions into its pole-zero form:

(a) $H(z)=\frac{z^2+2z+1}{z^2+5z+6}$

(b) $H(z)=\frac{1-0.16z^{-2}}{1+0.7z^{-1}+0.1z^{-2}}$

(c) $H(z)=\frac{z^2+4z+5}{z^3+2z^2+6z}$

6.14 A transfer function depicting a discrete-time system is given by

$$H(z)=\frac{10(z+1)}{(z+0.75)}.$$

(a) Determine the impulse response $h(n)$ and step response.

(b) Determine the system response $y(n)$ if the input is $x(n)=(0.25)^n u(n)$.

6.15 Given each of the following transfer functions that describe the digital systems, sketch the z-plane pole-zero plot and determine the stability for each digital system.

(a) $H(z)=\frac{z-0.5}{(z+0.25)(z^2+z+0.8)}$

(b) $H(z)=\frac{z^2+0.25}{(z-0.5)(z^2+4z+7)}$

(c) $H(z)=\frac{z+0.95}{(z+0.2)(z^2+1.414z+1)}$

(d) $H(z)=\frac{z^2+z+0.25}{(z-1)(z+1)^2(z-0.36)}$

6.16 Given the following digital system with a sampling rate of 8000 Hz,

$$y(n)=0.5x(n)+0.5x(n-2),$$

(a) Determine the frequency response.

(b) Calculate and plot the magnitude and phase frequency responses.

(c) Determine the filter type based on the magnitude frequency response.

6.17 Given the following digital system with a sampling rate of 8000 Hz,

$$y(n)=0.5x(n-1)+0.5x(n-2),$$

(a) Determine the frequency response.

(b) Calculate and plot the magnitude and phase frequency responses.

(c) Determine the filter type based on the magnitude frequency response.

6.18 For the following digital system with a sampling rate of 8000 Hz,

$$y(n) = 0.5x(n) + 0.5y(n-1),$$

(a) Determine the frequency response.
(b) Calculate and plot the magnitude and phase frequency responses.
(c) Determine the filter type based on the magnitude frequency response.

6.19 For the following digital system with a sampling rate of 8000 Hz,

$$y(n) = x(n) - 0.5y(n-2),$$

(a) Determine the frequency response.
(b) Calculate and plot the magnitude and phase frequency responses.
(c) Determine the filter type based on the magnitude frequency response.

6.20 Given the following difference equation:

$$y(n) = x(n) - 2\cdot\cos(\alpha)x(n-1) + x(n-2) + 2\gamma\cdot\cos(\alpha)y(n-1) - \gamma^2 y(n-2),$$

where $\gamma = 0.8$ and $\alpha = 60^0$,
(a) Find the transfer function $H(z)$.
(b) Plot the poles and zeros on the z-plan with the unit circle.
(c) Determine the stability of the system from the pole-zero plot.
(d) Calculate the amplitude (magnitude) response of $H(z)$.
(e) Calculate the phase response of $H(z)$.

6.21 For each of the following difference equations,
(a) $y(n) = 0.5x(n) + 0.5x(n-1)$
(b) $y(n) = 0.5x(n) - 0.5x(n-1)$
(c) $y(n) = 0.5x(n) + 0.5x(n-2)$
(d) $y(n) = 0.5x(n) - 0.5x(n-2)$,
1. Find $H(z)$.
2. Calculate the magnitude response.
3. Specify the filtering type based on the calculated magnitude response.

6.22 An IIR system is expressed as

$$y(n) = 0.5x(n) + 0.2y(n-1), y(-1) = 0.$$

(a) Find $H(z)$.
(b) Find the system response $y(n)$ due to the input $x(n) = (0.5)^n u(n)$.

6.23 Given the following IIR system with zero initial conditions:

$$y(n) = 0.5x(n) - 0.7y(n-1) - 0.1y(n-2),$$

(a) Find $H(z)$.
(b) Find the unit step response.

6.24 Given the first-order IIR system

$$H(z) = \frac{1 + 2z^{-1}}{1 - 0.5z^{-1}},$$

realize $H(z)$ and develop the difference equations using the following forms:

(a) direct-form I

(b) direct-form II

6.25 Given the second-order IIR filter

$$H(z)=\frac{1-0.9z^{-1}-0.1z^{-2}}{1+0.3z^{-1}-0.04z^{-2}},$$

realize $H(z)$ and develop difference equations using the following forms:

(a) direct-form I

(b) direct-form II

(c) cascade (series) form via the first-order sections

(d) parallel form via the first-order sections

6.26 Given the following preemphasis filters:

$$H(z)=1-0.5z^{-1}$$

$$H(z)=1-0.7z^{-1}$$

$$H(z)=1-0.9z^{-1},$$

(a) Write the difference equation for each.

(b) Determine which emphasizes high-frequency components most.

MATLAB Problems

6.27 Given a filter

$$H(z)=\frac{1+2z^{-1}+z^{-2}}{1-0.5z^{-1}+0.25z^{-2}},$$

use MATLAB to plot

(a) its magnitude frequency response;

(b) its phase response.

6.28 Given a difference equation

$$y(n)=x(n-1)-0.75y(n-1)-0.125y(n-2),$$

(a) Use the MATLAB functions **filter()** and **filtic()** to calculate the system response $y(n)$ for $n=0, 1, \dots, 4$ with the input of $x(n)=(0.5)^n u(n)$ and initial conditions: $x(-1)=-1$, $y(-2)=2$, and $y(-1)=1$.

(b) Use the MATLAB function **filter()** to calculate the system response $y(n)$ for $n=0, 1, \dots, 4$ with the input of $x(n)=(0.5)^n u(n)$ and zero initial conditions: $x(-1)=0$, $y(-2)=0$, and $y(-1)=0$.

6.29 Given a filter

$$H(z)=\frac{1-z^{-1}+z^{-2}}{1-0.9z^{-1}+0.81z^{-2}},$$

(a) Plot the magnitude frequency response and phase response using MATLAB.

(b) Specify the type of filtering.

(c) Find the difference equation.

(d) Perform filtering, that is, calculate $y(n)$ for first 1000 samples for each of the following inputs and plot the filter outputs using MATLAB, assuming that all initial conditions are zeros and the sampling rate is 8000 Hz:

$$x(n) = \cos\left(\pi \cdot 10^3 \frac{n}{8{,}000}\right)$$

$$x(n) = \cos\left(\frac{8}{3}\pi \cdot 10^3 \frac{n}{8{,}000}\right)$$

$$x(n) = \cos\left(6\pi \cdot 10^3 \frac{n}{8{,}000}\right);$$

(e) Repeat (d) using the MATLAB function **filter()**.

6.30 Repeat (d) in Problem 6.29 using direct-II form structure.

MATLAB Projects

6.31 Sound effects of preemphasis filtering:
A preemphasis filter is shown in Fig. 6.36 with a selective parameter $0 \le \alpha < 1$, which controls the degree of preemphasis filtering.

Assuming that the system has a sampling rate of 8000 Hz, plot the frequency responses for $\alpha = 0$, $\alpha = 0.4$, $\alpha = 0.8$, $\alpha = 0.95$, and $\alpha = 0.99$, respectively. For each case, apply the preemphasis filter to the given speech ("speech.dat") and discuss the sound effects.

6.32 Echo generation (sound regeneration):
Echo is the repetition of sound due to sound wave reflection from the objects. It can easily be generated using an FIR filter shown in Fig. 6.37:where $|\alpha| < 1$ is an attenuation factor and R is the delay of the echo. However, a single echo generator may not be useful, so a multiple-echo generator using an IIR filter is usually applied, as shown in Fig. 6.38.

As shown in Fig. 6.38, an echo signal is generated by the sum of delayed versions of sound with attenuation and the non-delayed version given by

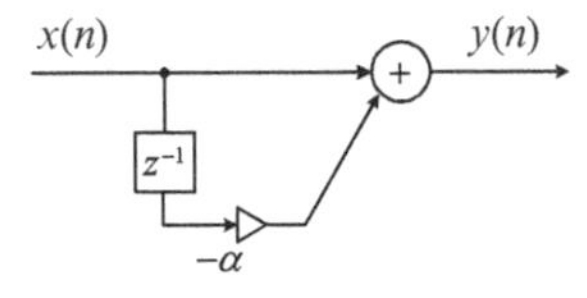

FIG. 6.36

A preemphasis filter.

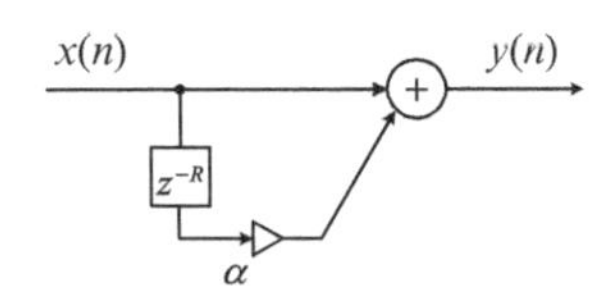

FIG. 6.37

A single echo generator using an FIR filter.

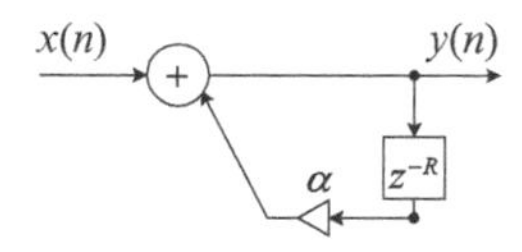

FIG. 6.38

A multiple-echo generator using an IIR filter.

$$y(n) = x(n) + \alpha x(n-R) + \alpha^2 x(n-2R) + \cdots = \sum_{k=0}^{\infty} \alpha^k x(n-kR)$$

where α is the attenuation factor. Applying z-transform, it follows that.
$Y(z) = X(z)\sum_{k=0}^{\infty}(\alpha z^{-R})^k = X(z)\frac{1}{1-\alpha z^{-R}}$ for $|\alpha z^{-R}| < 1$
Thus, we yield the transfer function and difference equation below:

$$H(z) = \frac{1}{1 - \alpha z^{-R}}$$

and $y(n) = x(n) + \alpha y(n-1)$.

(a) Assuming that the system has a sampling rate of 8000 Hz, plot the IIR filter frequency responses for the following cases: $\alpha = 0.5$ and $R = 1$; $\alpha = 0.6$ and $R = 4$; $\alpha = 0.7$ and $R = 10$, and characterize the frequency responses.

(b) After implementing the multiple-echo generator using the following code:

```
y = filter([1], [1 zeros(1, R - 1) - alpha], x)
```

Evaluate the sound effects of processing the speech file ("sppech.dat") for the following cases: $\alpha = 0.5$ and $R = 500$ (62.5 ms); $\alpha = 0.7$ and $R = 1,000$ (125 ms); $\alpha = 0.5$, $R = 2,000$ (250 ms), and $\alpha = 0.5$, $R = 4,000$ (500 ms).

Advanced Problems

6.33 Let $x(n) = \{a_N, a_{N-1}, \cdots, a_0, a_1, \cdots, a_{N-1}, a_N\}$ be a finite-duration sequence, which is real and even. Show that if $z = re^{j\theta}$ is a zero of $X(z)$, then $z = (1/r)e^{-j\theta}$ is also a zero.

6.34 Show that the following two systems are equivalent:

(a) $y(n) = x(n) - 0.5x(n-1)$

(b) $y(n) = 0.2y(n-1) + x(n) - 0.7x(n-1) + 0.1x(n-2)$

6.35 A causal linear time-invariant system has the time-domain input and z-transform output as shown below:

$$x(n) = (0.5)^n u(n)$$

and

$$Y(z) = \frac{z^{-1}}{(1 - 0.5z^{-1})(1 - 0.25z^{-1})}$$

(a) Determine the system transfer function $H(z)$.

(b) Determine the system output $y(n)$.

6.36 Given a linear-phase FIR system as

$$y(n) = b_0 x(n) + b_1 x(n-1) + b_1(n-2)$$

if its frequency response is normalized to $H(e^{j0}) = 1$ and it rejects completely a frequency component at $\Omega_0 = 2\pi/3$, then

(a) determine the transfer function $H(z)$;

(b) compute and sketch the magnitude and phase responses of the filter.

6.37 Let $z = re^{j\Omega}$ be a zero inside of the unit circle, where $0 < r < 1$. Consider a system whose transfer function as

$$H(z) = 1 - re^{j\theta}z^{-1}$$

(a) Show that the magnitude response is

$$\left|H\left(e^{j\Omega}\right)\right| = \sqrt{1 - 2r\cos(\Omega - \theta) + r^2};$$

(b) Show that the phase response is

$$\angle H\left(e^{j\Omega}\right) = \tan^{-1}\left(\frac{r\sin(\Omega - \theta)}{1 - r\cos(\Omega - \theta)}\right);$$

(c) Plot the magnitude response for $r = 0.8$, $\theta = 60^0$, and $\Omega = 0, \pi/4, \pi/2, 3\pi/4, \pi$, respectively.

6.38 Let $z = re^{j\Omega}$ be a zero inside of the unit circle, where $0 < r < 1$. Consider a system whose transfer function as

$$H(z) = \frac{1}{1 - re^{j\theta}z^{-1}}$$

(a) Show that the magnitude response is

$$\left|H\left(e^{j\Omega}\right)\right| = \frac{1}{\sqrt{1 - 2r\cos(\Omega - \theta) + r^2}};$$

(b) Show that the phase response is

$$\angle H\left(e^{j\Omega}\right) = -\tan^{-1}\left(\frac{r\sin(\Omega - \theta)}{1 - r\cos(\Omega - \theta)}\right);$$

(c) Plot the magnitude response for $r = 0.8$, $\theta = 60^0$, and $\Omega = 0, \pi/4, \pi/2, 3\pi/4, \pi$, respectively.

6.39 Determine the 3-dB bandwidth for the following filters:

(a) $H(z) = \dfrac{1 - a}{1 - az^{-1}}$

(b) $H(z) = \dfrac{1 - a}{2}\dfrac{1 + z^{-1}}{1 - az^{-1}}$

where $0 < a < 1$, which is a better lowpass filer.

6.40 Given a system as

$$y(n) = e^{J\Omega_0}y(n-1) + x(n),$$

Show that for $x(n) = \delta(n)$, the response is

$$y(n) = \cos(\Omega_0 n)u(n) + j\sin(\Omega_0 n)u(n).$$

6.41 Convert the first-order highpass filter with the system function

$$H(z) = \frac{1 - z^{-1}}{1 - az^{-1}}$$

into a lowpass filter, where $0 < a < 1$.

(a) Determine the difference equation.

(b) Sketch the pole-zero pattern.

(c) Sketch magnitude and phase responses.

6.42 Determine the magnitude and phase responses of the multipath channel

$$y(n) = x(n) + x(n - M)$$

Find the frequencies when $H(e^{j\Omega}) = 0$.

6.43 Consider a system whose transfer function as

$$H(z) = \frac{(1 - e^{j\theta}z^{-1})}{(1 - re^{j\theta}z^{-1})}$$

(a) Show that the magnitude response is

$$|H(e^{j\Omega})| = \frac{\sqrt{2 - 2\cos(\Omega - \theta)}}{\sqrt{1 - 2r\cos(\Omega - \theta) + r^2}};$$

(b) Show that the phase response is

$$\Omega = \tan^{-1}\left(\frac{\sin(\Omega - \theta)}{1 - \cos(\Omega - \theta)}\right) - \tan^{-1}\left(\frac{r\sin(\Omega - \theta)}{1 - r\cos(\Omega - \theta)}\right);$$

(c) Plot the magnitude response for $r = 0.8$, $\theta = 45^{\circ}$, and $\Omega = 0, \pi/4, \pi/2, 3\pi/4, \pi$, respectively.

6.44 Consider a system whose transfer function as

$$H(z) = \frac{(1 - e^{j\theta}z^{-1})(1 - e^{-j\theta}z^{-1})}{(1 - re^{j\theta}z^{-1})(1 - re^{-j\theta}z^{-1})}$$

(a) Show that

$$H(z) = \frac{1 - 2\cos(\theta)z^{-1} + z^{-2}}{1 - 2r\cos(\theta)z^{-1} + r^2z^{-2}};$$

(b) Show that the magnitude response is

$$|H(e^{j\Omega})| = \frac{\sqrt{2 - 2\cos(\Omega - \theta)}}{\sqrt{1 - 2r\cos(\Omega - \theta) + r^2}} \frac{\sqrt{2 - 2\cos(\Omega + \theta)}}{\sqrt{1 - 2r\cos(\Omega + \theta) + r^2}};$$

(c) Plot the magnitude response for $r = 0.8$, $\theta = 45^{\circ}$, and $\Omega = 0, \pi/4, \pi/2, 3\pi/4, \pi$, respectively.

6.45 Echoes and reverberations can be generated by delaying and scaling the signal $x(n)$ as

$$y(n) = \sum_{k=0}^{\infty} b_k x(n - kD)$$

where $0 < b_k < 1$ and D is the integer delay.

(a) Show that for $b_k = a^k$

$$H(z) = \frac{1}{1 - az^{-D}};$$

(b) Show that if we generate reverberations using

$$y(n) = \left(\frac{1}{a} - a\right)\sum_{k=0}^{\infty} a^k x(n-kD) - \frac{1}{a}x(n)$$

Then

$$H(z) = \frac{z^{-D} - a}{1 - az^{-D}},$$

and the magnitude frequency response is allpass filter, that is,

$$\left|H\left(e^{j\Omega}\right)\right| = 1.$$

CHAPTER

FINITE IMPULSE RESPONSE FILTER DESIGN

7

CHAPTER OUTLINE

7.1 FINITE IMPULSE RESPONSE FILTER FORMAT

In this chapter, we describe techniques of designing *finite impulse response* (FIR) filters. An FIR filter is completely specified by the following input-output relationship:

$$\begin{aligned} y(n) &= \sum_{i=0}^{K} b_i x(n-i) \\ &= b_0 x(n) + b_1 x(n-1) + b_2 x(n-2) + \cdots + b_K x(n-K), \end{aligned} \tag{7.1}$$

Digital Signal Processing. https://doi.org/10.1016/B978-0-12-815071-9.00007-5

where b_i represents FIR filter coefficients and $K+1$ denotes the FIR filter length. Applying the z-transform on both sides of Eq. (7.1) leads to

$$Y(z)=b_0X(z)+b_1z^{-1}X(z)+\cdots+b_Kz^{-K}X(z). \tag{7.2}$$

Factoring out $X(z)$ on the right-hand side of Eq. (7.2) and then dividing by $X(z)$ on both sides, we have the transfer function, which depicts the FIR filter, as

$$H(z)=\frac{Y(z)}{X(z)}=b_0+b_1z^{-1}+\cdots+b_Kz^{-K}. \tag{7.3}$$

The following example serves to illustrate the notations used in Eqs. (7.1) and (7.3) numerically.

EXAMPLE 7.1

Given the following FIR filter:

$$y(n)=0.1x(n)+0.25x(n-1)+0.2x(n-2),$$

Determine the transfer function, filter length, nonzero coefficients, and impulse response.

Solution:

Applying z-transform on both sides of the difference equation yields

$$Y(z)=0.1X(z)+0.25X(z)z^{-1}+0.2X(z)z^{-2}.$$

Then the transfer function is found to be

$$H(z)=\frac{Y(z)}{X(z)}=0.1+0.25z^{-1}+0.2z^{-2}.$$

The filter length is $K+1=3$, and the identified coefficients are
$b_0=0.1$, $b_1=0.25$, and $b_2=0.2$.
Taking the inverse z-transform of the transfer function, we have

$$h(n)=0.1\delta(n)+0.25\delta(n-1)+0.2\delta(n-2).$$

This FIR filter impulse response has only three terms.

The previous example is to help us understand the FIR filter format. We can conclude that

1. The transfer function in Eq. (7.3) has a constant term, all the other terms have negative powers of z, all the poles are at the origin on the z-plane. Hence, the stability of filter is guaranteed. Its impulse response has only a finite number of terms.
2. The FIR filter operations involve only multiplying the filter inputs by their corresponding coefficients and accumulating them; the implementation of this filter type in real time is straightforward.

From the FIR filter format, the design objective is to obtain the FIR filter b_i coefficients such that the magnitude frequency response of the FIR filter $H(z)$ will approximate the desired magnitude frequency response, such as that of a lowpass, highpass, bandpass, or bandstop filter. The following sections will introduce design methods to calculate the FIR filter coefficients.

7.2 FOURIER TRANSFORM DESIGN

We begin with an ideal lowpass filter with a normalized cutoff frequency Ω_c (Chapter 6), whose magnitude frequency response in terms of the normalized digital frequency Ω is plotted in Fig. 7.1 and is characterized by

$$H(e^{j\Omega}) = \begin{cases} 1, & 0 \le |\Omega| \le \Omega_c \\ 0, & \Omega_c \le |\Omega| \le \pi. \end{cases} \tag{7.4}$$

Since the frequency response is periodic with a period of $\Omega = 2\pi$(rad), as we have discussed in Chapter 6, we can extend the frequency response of the ideal filter $H(e^{j\Omega})$, as shown in Fig. 7.2.

The periodic frequency response can be approximated using a complex Fourier series expansion (see this topic in Appendix B) in terms of the normalized digital frequency Ω, that is,

$$H(e^{j\Omega}) = \sum_{n=-\infty}^{\infty} c_n e^{j\omega_0 n\Omega}, \tag{7.5}$$

and the Fourier coefficients are given by

$$\tilde{c}_n = \frac{1}{2\pi}\int_{-\pi}^{\pi} H(e^{j\Omega}) e^{-j\omega_0 n\Omega} d\Omega \text{ for } -\infty < n < \infty. \tag{7.6}$$

Let $n = -n$, Eqs. (7.9) and (7.10) become

$$H(e^{j\Omega}) = \sum_{n=\infty}^{-\infty} \tilde{c}_{-n} e^{-j\omega_0 n\Omega} = \sum_{n=-\infty}^{\infty} \tilde{c}_{-n} e^{-j\omega_0 n\Omega}, \tag{7.7}$$

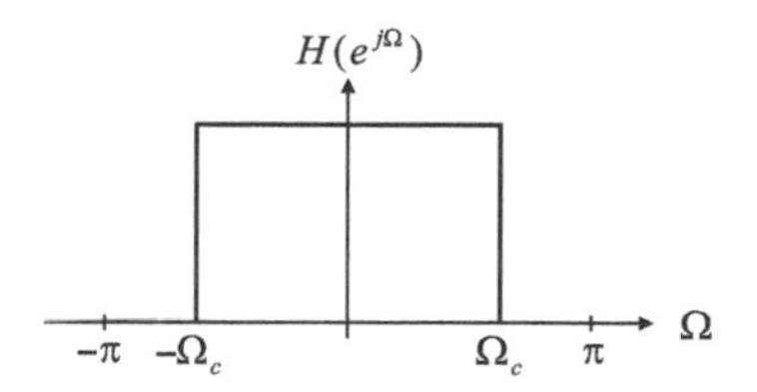

FIG. 7.1

Frequency response of an ideal lowpass filter.

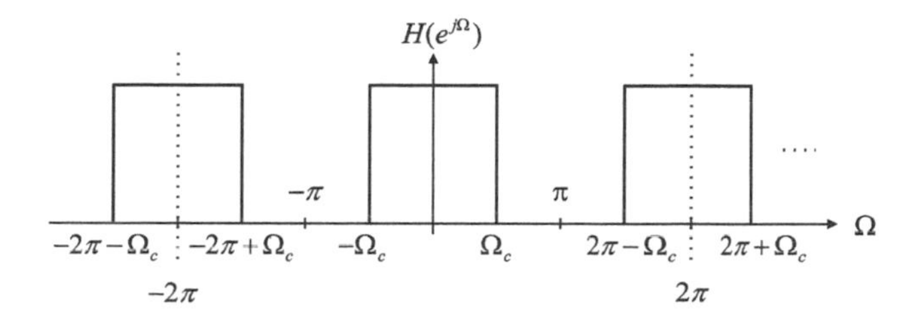

FIG. 7.2

Periodicity of the ideal lowpass frequency response.

$$\widetilde{c}_{-n} = \frac{1}{2\pi}\int_{-\pi}^{\pi} H\left(e^{j\Omega}\right) e^{j\omega_0 n\Omega} d\Omega. \tag{7.8}$$

Defining $c_n = \widetilde{c}_{-n}$, it leads to

$$H\left(e^{j\Omega}\right) = \sum_{n=-\infty}^{\infty} c_n e^{-j\omega_0 n\Omega}, \tag{7.9}$$

$$c_n = \frac{1}{2\pi}\int_{-\pi}^{\pi} H\left(e^{j\Omega}\right) e^{j\omega_0 n\Omega} d\Omega. \tag{7.10}$$

Note that we obtain Eqs. (7.9) and (7.10) simply by treating the Fourier series expansion in time domain with the time variable t replaced by the normalized digital frequency variable Ω.

The fundamental frequency is easily found to be

$$\omega_0 = \frac{2\pi}{\text{Period of waveform}} = \frac{2\pi}{2\pi} = 1. \tag{7.11}$$

Substituting $\omega_0 = 1$ into Eqs. (7.9) and (7.10), and introducing $h(n) = c_n$, called the desired impulse response of the ideal filter, we obtain the filter frequency response as

$$H\left(e^{j\Omega}\right) = \sum_{n=-\infty}^{\infty} h(n) e^{-jn\Omega}, \tag{7.12}$$

and the Fourier transform design equation as

$$h(n) = \frac{1}{2\pi}\int_{-\pi}^{\pi} H\left(e^{j\Omega}\right) e^{jn\Omega} d\Omega \text{ for } -\infty < n < \infty. \tag{7.13}$$

Furthermore, if $H(e^{j\Omega})$ is even, that is, $H(e^{-j\Omega}) = H(e^{j\Omega})$, we see that

$$h(-n) = \frac{1}{2\pi}\int_{-\pi}^{\pi} H\left(e^{j\Omega}\right) e^{-jn\Omega} d\Omega. \tag{7.14}$$

Let $\Omega = -\Omega'$ and $d\Omega = -d\Omega'$. It is easy to verify that $h(n) = h(-n)$, that is,

$$h(-n) = \frac{1}{2\pi}\int_{\pi}^{-\pi} H\left(e^{-j\Omega'}\right) e^{jn\Omega'} (-d\Omega') = \frac{1}{2\pi}\int_{-\pi}^{\pi} H\left(e^{j\Omega'}\right) e^{jn\Omega'} d\Omega' = h(n). \tag{7.15}$$

Now, let us look at the z-transfer function. If we substitute $e^{j\Omega} = z$ and $\omega_0 = 1$ back to Eq. (7.12), we yield a z-transfer function in the following format:

$$\begin{aligned} H(z) &= \sum_{n=-\infty}^{\infty} h(n) z^{-n} \\ &\cdots + h(-2)z^2 + h(-1)z^1 + h(0) + h(1)z^{-1} + h(2)z^{-2} + \cdots \end{aligned} \tag{7.16}$$

This is a noncausal FIR filter. We will deal with this later in this section. Using the Fourier transform design shown in Eq. (7.13), the desired impulse response of the ideal lowpass filter is solved as follows:

$$
\begin{aligned}
\text{For } n=0 \ h(n) &= \frac{1}{2\pi}\int_{-\pi}^{\pi} H\left(e^{j\Omega}\right)e^{j\Omega\times 0}d\Omega \\
&= \frac{1}{2\pi}\int_{-\Omega_c}^{\Omega_c} 1 d\Omega = \frac{\Omega_c}{\pi} \\
\text{For } n\neq 0 \quad h(n) &= \frac{1}{2\pi}\int_{-\pi}^{\pi} H\left(e^{j\Omega}\right)e^{j\Omega n}d\Omega = \frac{1}{2\pi}\int_{-\Omega_c}^{\Omega_c} e^{j\Omega n}d\Omega \\
&= \left.\frac{e^{jn\Omega}}{2\pi jn}\right|_{-\Omega_c}^{\Omega_c} = \frac{1}{\pi n}\frac{e^{jn\Omega_c} - e^{-jn\Omega_c}}{2j} = \frac{\sin\left(\Omega_c n\right)}{\pi n}.
\end{aligned}
\tag{7.17}
$$

The desired impulse response $h(n)$ is plotted vs the sample number n in Fig. 7.3.

Theoretically, $h(n)$ in Eq. (7.13) exists for $-\infty < n < \infty$ and is symmetrical about $n=0$; that is, $h(n)=h(-n)$ since $H(e^{j\Omega})$ shown in Fig. 7.2 is an even function. The amplitude of the impulse response sequence $h(n)$ becomes smaller when n increases in both directions. The FIR filter design must first be completed by truncating the infinite-length sequence $h(n)$ to achieve the $2M+1$ dominant coefficients using the coefficient symmetry, that is,

$$H(z) = h(M)z^{M} + \cdots + h(1)z^{1} + h(0) + h(1)z^{-1} + \cdots + h(M)z^{-M}. \tag{7.18}$$

The obtained filter is a noncausal z-transfer function of the FIR filter, since the filter transfer function contains terms with the positive powers of z, which in turn means that the filter output depends on the future filter inputs. To remedy the noncausal z-transfer function, we delay the truncated impulse response $h(n)$ by M samples to yield the following causal FIR filter:

$$H(z) = b_0 + b_1 z^{-1} + \cdots + b_{2M}(2M)z^{-2M}, \tag{7.19}$$

where the delay operation is given by

$$b_n = h(n-M) \text{ for } n = 0, 1, \cdots, 2M. \tag{7.20}$$

Similarly, we can obtain the design equations for other types of FIR filters, such as highpass, bandpass, and bandstop, using their ideal frequency responses and Eq. (7.13). The derivations are omitted here. Table 7.1 illustrates a summary of all the formulas for FIR filter coefficient calculations.

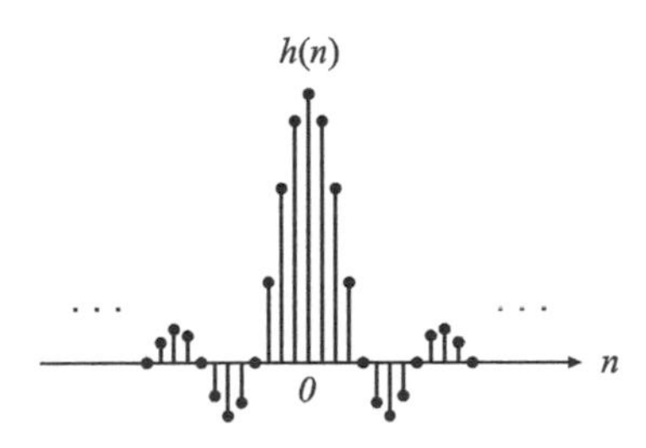

FIG. 7.3

Impulse response of an ideal digital lowpass filter.

Table 7.1 Summary of Truncated Ideal Impulse Responses for Standard FIR Filters

Filter Type	Ideal Impulse Response $h(n)$ (Noncausal FIR Coefficients)
Lowpass:	$h(n)=\begin{cases}\dfrac{\Omega_c}{\pi} & \text{for } n=0\\ \dfrac{\sin(\Omega_c n)}{n\pi} & \text{for } n\neq 0 \quad -M\leq n\leq M\end{cases}$
Highpass:	$h(n)=\begin{cases}\dfrac{\pi-\Omega_c}{\pi} & \text{for } n=0\\ -\dfrac{\sin(\Omega_c n)}{n\pi} & \text{for } n\neq 0 \quad -M\leq n\leq M\end{cases}$
Bandpass:	$h(n)=\begin{cases}\dfrac{\Omega_H-\Omega_L}{\pi} & \text{for } n=0\\ \dfrac{\sin(\Omega_H n)}{n\pi}-\dfrac{\sin(\Omega_L n)}{n\pi} & \text{for } n\neq 0 \quad -M\leq n\leq M\end{cases}$
Bandstop:	$h(n)=\begin{cases}\dfrac{\pi-\Omega_H+\Omega_L}{\pi} & \text{for } n=0\\ -\dfrac{\sin(\Omega_H n)}{n\pi}+\dfrac{\sin(\Omega_L n)}{n\pi} & \text{for } n\neq 0 \quad -M\leq n\leq M\end{cases}$

Causal FIR filter coefficients: shifting $h(n)$ to the right by M samples.
Transfer function:
$H(z)=b_0+b_1z^{-1}+b_2z^{-2}+\cdots b_{2M}z^{-2M}$
where $b_n=h(n-M)$, $n=0, 1, \cdots, 2M$.

The following example illustrates the coefficient calculation for a lowpass FIR filter.

EXAMPLE 7.2

(a) Calculate the filter coefficients for a three-tap (number of coefficients) FIR lowpass filter with a cutoff frequency of 800 Hz and a sampling rate of 8000 Hz using the Fourier transform method.

(b) Determine the transfer function and difference equation of the designed FIR system.

(c) Compute and plot the magnitude frequency response for $\Omega=0$, $\pi/4$, $\pi/2$, $3\pi/4$, and π (rad).

Solution:

(a) Calculating the normalized cutoff frequency leads to

$$\Omega_c=2\pi f_c T_s=\frac{2\pi\times 800}{8000}=0.2\pi\,(\text{rad}).$$

Since $2M+1=3$ in this case, using the equation in Table 7.1 results in

$$h(0)=\frac{\Omega_c}{\pi}\quad \text{for } n=0$$

$$h(n)=\frac{\sin(\Omega_c n)}{n\pi}=\frac{\sin(0.2\pi n)}{n\pi},\quad \text{for } n\neq 0.$$

The computed filter coefficients via the previous expression are listed as

$$h(0)=\frac{0.2\pi}{\pi}=0.2$$

$$h(1) = \frac{\sin[0.2\pi \times 1]}{1 \times \pi} = 0.1871.$$

Using the symmetry leads to

$$h(-1) = h(1) = 0.1871.$$

Thus delaying $h(n)$ by $M = 1$ sample using Eq. (7.20) gives

$$b_0 = h(0-1) = h(-1) = 0.1871,$$
$$b_1 = h(1-1) = h(0) = 0.2,$$
$$\text{and } b_2 = h(2-1) = h(1) = 0.1871.$$

(b) The transfer function is achieved as

$$H(z) = 0.1871 + 0.2z^{-1} + 0.1871z^{-2}.$$

Using the technique described in Chapter 6, we have

$$\frac{Y(z)}{X(z)} = H(z) = 0.1871 + 0.2z^{-1} + 0.1871z^{-2}.$$

Multiplying $X(z)$ leads to

$$Y(z) = 0.1871X(z) + 0.2z^{-1}X(z) + 0.1871z^{-2}X(z).$$

Applying the inverse z-transform on both sides, the difference equation is yielded as

$$y(n) = 0.1871x(n) + 0.2x(n-1) + 0.1871x(n-2).$$

(c) The magnitude frequency response and phase response can be obtained using the technique introduced in Chapter 6. Substituting $z = e^{j\Omega}$ into $H(z)$, it follows that

$$H(e^{j\Omega}) = 0.1871 + 0.2e^{-j\Omega} + 0.1871e^{-j2\Omega}.$$

Factoring term $e^{-j\Omega}$ and using the Euler formula $e^{jx} + e^{-jx} = 2\cos(x)$, we achieve

$$H(e^{j\Omega}) = e^{-j\Omega}(0.1871e^{j\Omega} + 0.2 + 0.1871e^{-j\Omega})$$
$$= e^{-j\Omega}(0.2 + 0.3742\cos(\Omega)).$$

Then the magnitude frequency response and phase response are found to be

$$|H(e^{j\Omega})| = |0.2 + 0.3472\cos\Omega|$$
$$\text{and } \angle H(e^{j\Omega}) = \begin{cases} -\Omega & \text{if } 0.2 + 0.3472\cos\Omega > 0 \\ -\Omega + \pi & \text{if } 0.2 + 0.3472\cos\Omega < 0. \end{cases}$$

Details of the magnitude calculations for several typical normalized frequencies are listed in Table 7.2.

Continued

EXAMPLE 7.2—CONT'D

Table 7.2 Frequency Response Calculation in Example 7.2

Ω (rad)	$f=\Omega f_s/(2\pi)$ (Hz)	$0.2+0.3742\cos\Omega$	$\lvert H(e^{j\Omega})\rvert$	$\lvert H(e^{j\Omega})\rvert_{dB}$ (dB)	$\angle H(e^{j\Omega})$ (deg)
0	0	0.5742	0.5742	−4.82	0
$\pi/4$	1000	0.4646	0.4646	−6.66	−45
$\pi/2$	2000	0.2	0.2	−14.0	−90
$3\pi/4$	3000	−0.0646	0.0646	−23.8	45
π	4000	−0.1742	0.1742	−15.2	0

Due to the symmetry of the coefficients, the obtained FIR filter has a linear phase response as shown in Fig. 7.4. The sawtooth shape is produced by the contribution of the negative sign of the real magnitude term $0.2+0.3742\cos\Omega$ in the three-tap filter frequency response, that is,

$$H(e^{j\Omega}) = e^{-j\Omega}(0.2+0.3742\cos\Omega). \tag{7.21}$$

In general, the FIR filter with symmetric coefficients has a linear phase response (linear function of Ω) as follows:

$$\angle H(e^{j\Omega}) = -M\Omega + \text{possible phase of } 180^\circ. \tag{7.22}$$

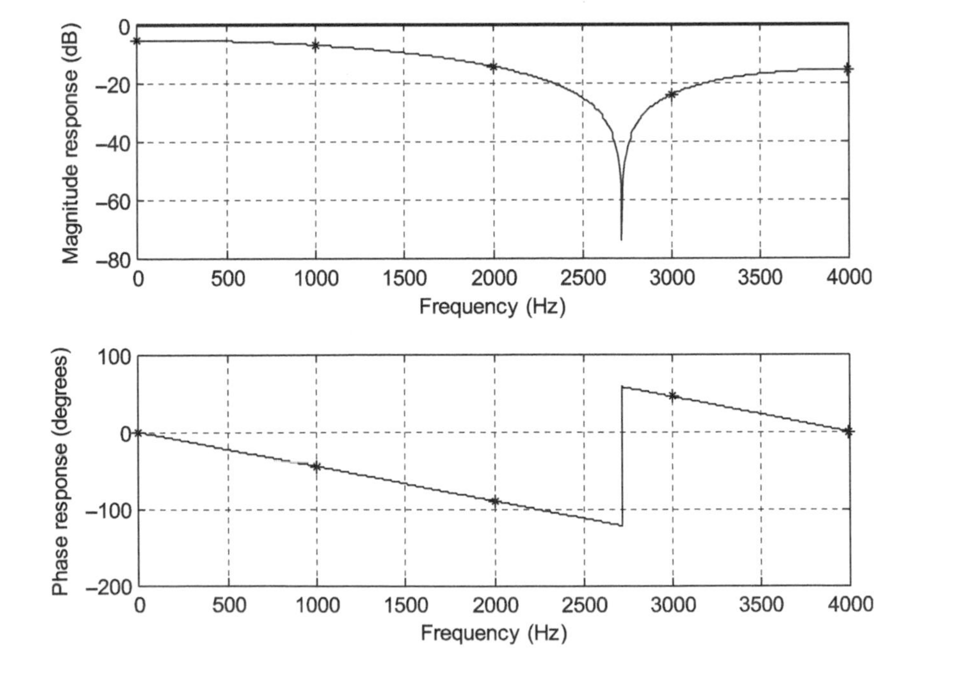

FIG. 7.4

Magnitude and phase frequency responses in Example 7.2.

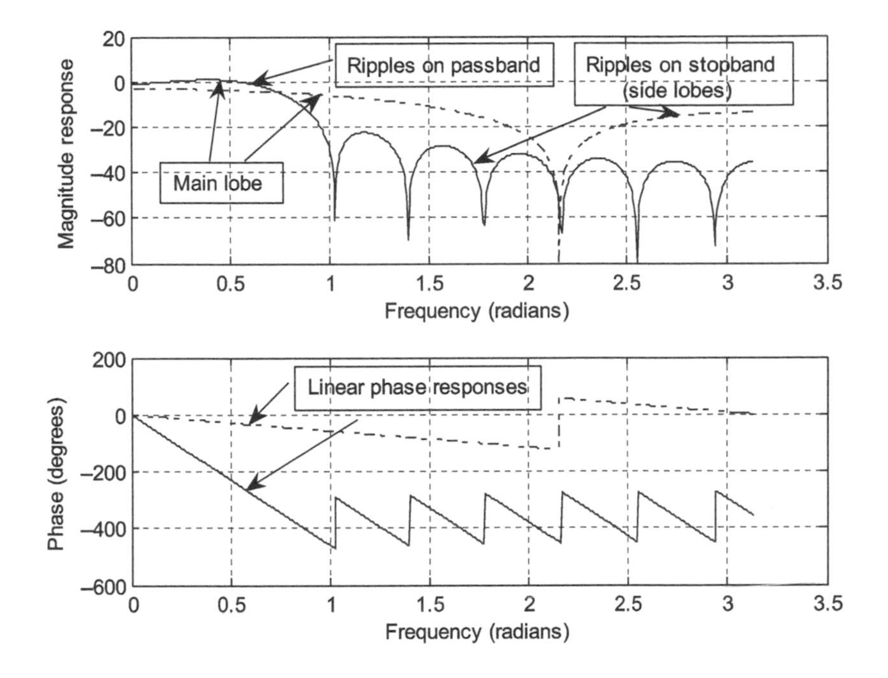

FIG. 7.5

Magnitude and phase frequency responses of the lowpass FIR filters with three coefficients (dash-dotted line) and 17 coefficients (solid line).

Next, we see that the three-tap FIR filter does not give an acceptable magnitude frequency response.

To explore this response further, Fig. 7.5 displays the magnitude and phase responses of three-tap ($M=1$) and 17 tap ($M=8$) FIR lowpass filters with a normalized cutoff frequency $\Omega_c=0.2\pi$ (rad). The calculated coefficients for the 17-tap FIR lowpass filter are listed in Table 7.3.

We can make the following observations at this point:

1. The oscillations (ripples) exhibited in the passband (main lobe), and stopband (side lobes) of the magnitude frequency response constitute the *Gibbs effect*. The Gibbs oscillatory behavior originates from the abrupt truncation of the infinite impulse response (IIR) in Eq. (7.18). To remedy this problem, window functions will be used and will be discussed in the next section.
2. Using a larger number of the filter coefficients will produce the sharp roll-off characteristic of the transition band but may cause increased time delay and increase computational complexity for implementing the designed FIR filter.

Table 7.3 17-Tap FIR Lowpass Filter Coefficients in Example 7.2 ($M=8$)

$b_0=b_{16}=-0.0378$	$b_1=b_{15}=-0.0432$	
$b_2=b_{14}=-0.0312$	$b_3=b_{13}=0.0000$	
$b_4=b_{12}=0.0468$	$b_5=b_{11}=0.1009$	
$b_6=b_{10}=0.1514$	$b_7=b_9=0.1871$	$b_8=0.2000$

3. The phase response is linear in the passband. This is consistent with Eq. (7.22), which means that all frequency components of the filter input within passband are subjected to the same amount of time delay at the filter output. Note that we impose the following linear phase requirement, that is, the FIR coefficients are symmetry about the middle coefficient, and the FIR filter order is an odd number. If the design methods cannot produce the symmetric coefficients or generate antisymmetric coefficients (Proakis and Manolakis, 2007), the resultant FIR filter does not have the linear phase property. [Linear phase even-order FIR filters and FIR filters using the antisymmetry of coefficients are discussed in Proakis and Manolakis (2007).]

To further probe the linear phase property, we consider a sinusoidal sequence $x(n)=A\sin(n\Omega)$ as the FIR filter input, with the output neglecting the transient response expected to be

$$y(n)=A|H|\sin(n\Omega+\varphi),$$

where $\varphi=-M\Omega$. Substituting $\varphi=-M\Omega$ into $y(n)$ leads to

$$y(n)=A|H|\sin[\Omega(n-M)].$$

This clearly indicates that within the passband, all frequency components passing through the FIR filter will have the same constant delay at the output, which equals M samples. Hence, phase distortion is avoided.

Fig. 7.6 verifies the linear phase property using an FIR filter with 17 taps. Two sinusoids of the normalized digital frequencies 0.05π and 0.15π rad, respectively, are used as inputs. These two input

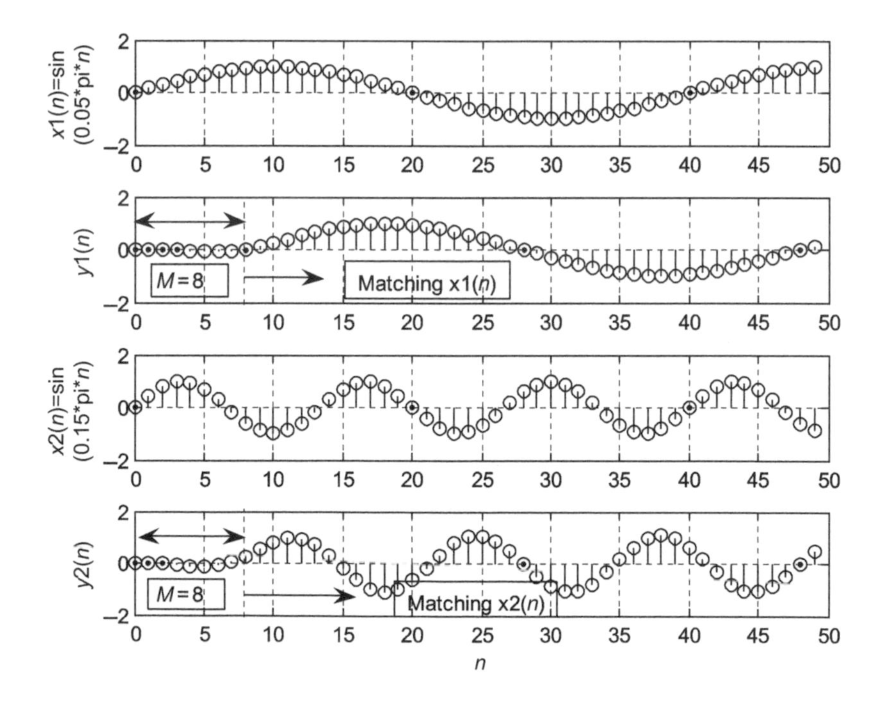

FIG. 7.6

Illustration of FIR filter linear phase property (constant delay of eight samples).

signals are within passband, so their magnitudes are not changed. As shown in Fig. 7.6, beginning at the ninth sample the output matches the input, which is delayed by eight samples for each case.

What would happen if the filter phase were nonlinear? This can be illustrated using the following combined sinusoids as the filter input:

$$x(n) = x_1(n) + x_2(n) = \sin(0.05\pi n)u(n) - \frac{1}{3}\sin(0.15\pi n)u(n).$$

The original $x(n)$ is shown in the top plot of Fig. 7.7. If the linear phase response of a filter is considered, such as $\varphi = -M\Omega_0$, where $M=8$ in our illustration, we have the filtered output as

$$y_1(n) = \sin[0.05\pi(n-8)] - \frac{1}{3}\sin[0.15\pi(n-8)].$$

The linear phase effect is shown in the middle plot of Fig. 7.7. We see that $y_1(n)$ is the eight sample-delayed version of $x(n)$. However, considering a unit gain filter with a phase delay of 90° for all the frequency components, we have the filtered output as

$$y_2(n) = \sin(0.05\pi n - \pi/2) - \frac{1}{3}\sin(0.15\pi n - \pi/2),$$

where the first term has a phase shift of 10 samples (see $\sin[0.05\pi(n-10)]$), while the second term has a phase shift of 10/3 samples (see $\frac{1}{3}\sin\left[0.15\pi\left(n-\frac{10}{3}\right)\right]$). Certainly, we do not have the linear phase feature. The signal $y_2(n)$ plotted in Fig. 7.7 shows that the waveform shape is different from that

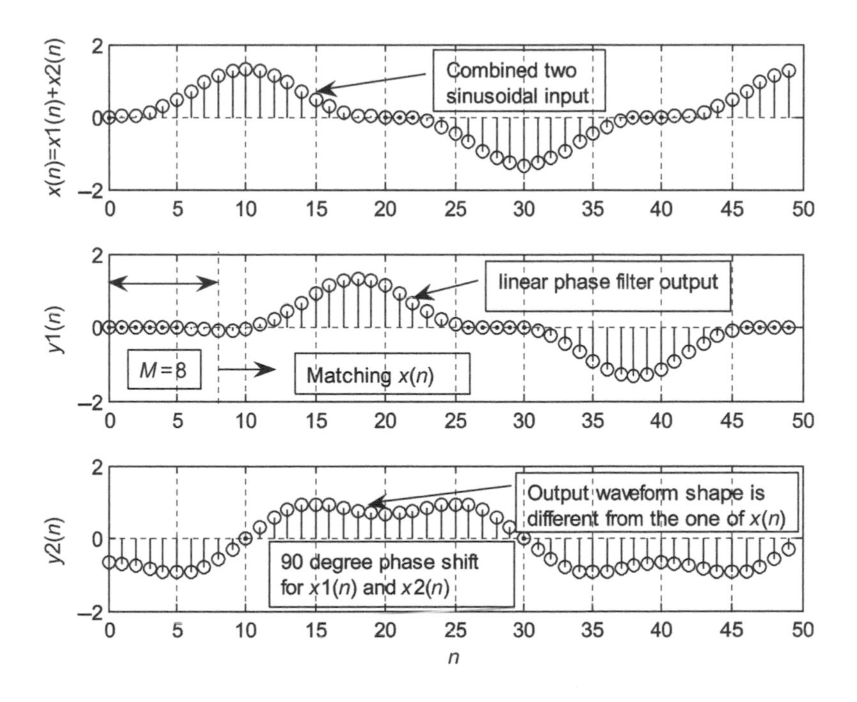

FIG. 7.7

Comparison of linear and nonlinear phase responses.

of the original signal $x(n)$, hence has significant phase distortion. This phase distortion is audible for audio applications and can be avoided by using an FIR filter, which has the linear phase feature.

We now have finished discussing the coefficient calculation for the FIR lowpass filter, which has a good linear phase property. To explain the calculation of filter coefficients for the other types of filters and examine the Gibbs effect, we look at another simple example.

EXAMPLE 7.3

(a) Calculate the filter coefficients for a five-tap FIR bandpass filter with a lower cutoff frequency of 2000 Hz and an upper cutoff frequency of 2400 Hz with a sampling rate of 8000 Hz.

(b) Determine the transfer function and plot the frequency responses with MATLAB.

Solution:

(a) Calculating the normalized cutoff frequencies leads to

$$\Omega_L = \frac{2\pi f_L}{f_s} = 2\pi \times \frac{2000}{8000} = 0.5\pi \text{ (rad)}$$

$$\Omega_H = \frac{2\pi f_H}{f_s} = 2\pi \times \frac{2400}{8000} = 0.6\pi \text{ (rad)}.$$

Since $2M+1=5$ in this case, using the equation in Table 7.1 yields

$$h(n) = \begin{cases} \dfrac{\Omega_H - \Omega_L}{\pi} & n=0 \\ \dfrac{\sin(\Omega_H n)}{n\pi} - \dfrac{\sin(\Omega_L n)}{n\pi} & n \neq 0 \quad -2 \leq n \leq 2 \end{cases}.$$

Calculations for noncausal FIR coefficients are listed as

$$h(0) = \frac{\Omega_H - \Omega_L}{\pi} = \frac{0.6\pi - 0.5\pi}{\pi} = 0.1.$$

The other computed filter coefficients via Eq. (7.22) are

$$h(1) = \frac{\sin[0.6\pi \times 1]}{1 \times \pi} - \frac{\sin[0.5\pi \times 1]}{1 \times \pi} = -0.01558$$

$$h(2) = \frac{\sin[0.6\pi \times 2]}{2 \times \pi} - \frac{\sin[0.5\pi \times 2]}{2 \times \pi} = -0.09355.$$

Using the symmetry leads to

$$h(-1) = h(1) = -0.01558$$
$$h(-2) = h(2) = -0.09355.$$

Thus, delaying $h(n)$ by $M=2$ samples gives

$$b_0 = b_4 = -0.09355,$$
$$b_1 = b_3 = -0.01558, \quad \text{and } b_2 = 0.1.$$

(b) The transfer function is achieved as

$$H(z) = -0.09355 - 0.01558z^{-1} + 0.1z^{-2} - 0.01558z^{-3} - 0.09355z^{-4}.$$

To complete Example 7.3, the magnitude frequency response plotted in terms of $|H(e^{j\Omega})|_{dB} = 20\log_{10}|H(e^{j\Omega})|$ using the MATLAB Program 7.1 is displayed in Fig. 7.8.

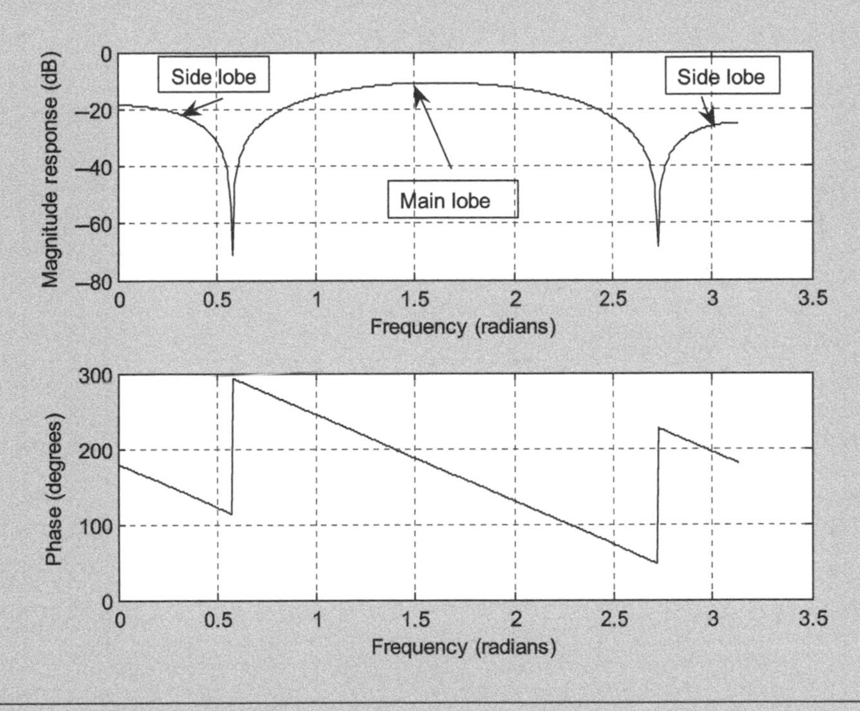

FIG. 7.8

Frequency responses for Example 7.3.

Program 7.1. MATLAB program for Example 7.3.

```
% Example 7.3
% MATLAB program to plot frequency response
%
[hz,w]=freqz([-0.09355 -0.01558 0.1 -0.01558 -0.09355], [1], 512);
phi=180*unwrap(angle(hz))/pi;
subplot(2,1,1), plot(w,20*log10(abs(hz))),grid;
xlabel('Frequency (radians)');
ylabel('Magnitude Response (dB)')
subplot(2,1,2), plot(w, phi); grid;
xlabel('Frequency (radians)');
ylabel('Phase (degrees)');
```

To summarize Example 7.3, the magnitude frequency response demonstrates the Gibbs oscillatory behavior existing in the passband and stopband. The peak of the main lobe in the passband is dropped from 0 dB to approximately −10 dB, while for the stopband, the lower side lobe in the magnitude response plot swings approximately between −18 and −70 dB, and the upper side lobe swings between −25 and −68 dB. As we have pointed out, this is due to the abrupt truncation of the infinite impulse sequence $h(n)$. The oscillations can be reduced by increasing the number of coefficient and using a window function, which will be studied next.

7.3 WINDOW METHOD

In this section, the *window method* (Fourier transform design with window functions) is developed to remedy the undesirable Gibbs oscillations in the passband and stopband of the designed FIR filter. Recall that the Gibbs oscillations originate from the abrupt truncation of the infinite-length coefficient sequence. Then it is natural to seek a window function, which is symmetrical and can gradually weight the designed FIR coefficients down to zeros at both ends for the range of $-M \leq n \leq M$. Applying the window sequence to the filter coefficients gives

$$h_w(n) = h(n) \cdot w(n),$$

where $w(n)$ designates the window function. Common window functions used in the FIR filter design are as follows:

1. Rectangular window:

$$w_{\text{rec}}(n) = 1, \ -M \leq n \leq M. \tag{7.23}$$

2. Triangular(Bartlett) window:

$$w_{\text{tri}}(n) = 1 - \frac{|n|}{M}, \ -M \leq n \leq M. \tag{7.24}$$

3. Hanning window:

$$w_{\text{han}}(n) = 0.5 + 0.5\cos\left(\frac{n\pi}{M}\right), \ -M \leq n \leq M. \tag{7.25}$$

4. Hamming window:

$$w_{\text{ham}}(n) = 0.54 + 0.46\cos\left(\frac{n\pi}{M}\right), \ -M \leq n \leq M. \tag{7.26}$$

5. Blackman window:

$$w_{\text{black}}(n) = 0.42 + 0.5\cos\left(\frac{n\pi}{M}\right) + 0.08\cos\left(\frac{2n\pi}{M}\right), \ -M \leq n \leq M. \tag{7.27}$$

In addition, there is another popular window function, called the Kaiser window [its detailed information can be found in Oppenheim et al. (1998)]. As we expected, the rectangular window function has a constant value of 1 within the window, hence does only truncation. As a comparison, shapes of the other window functions from Eqs. (7.23) to (7.27) are plotted in Fig. 7.9 for the case of $2M+1=81$.

We apply the Hamming window function in Example 7.4.

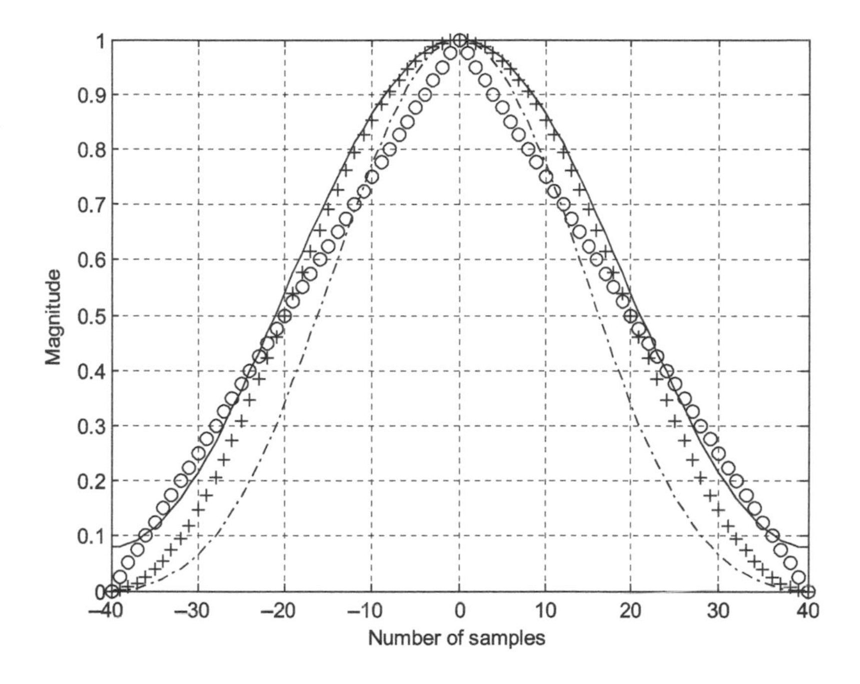

FIG. 7.9

Shapes of window functions for the case of $2M+1=81$. "o" line, triangular window; "+" line, Hanning window; Solid line, Hamming window; Dashed line, Blackman window.

EXAMPLE 7.4

Given the calculated filter coefficients

$$h(0)=0.25, h(-1)=h(1)=0.22508, h(-2)=h(2)=0.15915, h(-3)=h(3)=0.07503,$$

a. Apply the Hamming window function to obtain windowed coefficients $h_w(n)$.
b. Plot the impulse response $h(n)$ and windowed impulse response $h_w(n)$.

Solution:

(a) Since $M=3$, applying Eq. (7.18) leads to the window sequence

$$w_{\text{ham}}(-3)=0.54+0.46\cos\left(\frac{-3\times\pi}{3}\right)=0.08$$

$$w_{\text{ham}}(-2)=0.54+0.46\cos\left(\frac{-2\times\pi}{3}\right)=0.31$$

$$w_{\text{ham}}(-1)=0.54+0.46\cos\left(\frac{-1\times\pi}{3}\right)=0.77$$

$$w_{\text{ham}}(0)=0.54+0.46\cos\left(\frac{0\times\pi}{3}\right)=1,$$

Continued

EXAMPLE 7.4—CONT'D

$$w_{\mathrm{ham}}(1) = 0.54 + 0.46\cos\left(\frac{1 \times \pi}{3}\right) = 0.77$$

$$w_{\mathrm{ham}}(2) = 0.54 + 0.46\cos\left(\frac{2 \times \pi}{3}\right) = 0.31$$

$$w_{\mathrm{ham}}(3) = 0.54 + 0.46\cos\left(\frac{3 \times \pi}{3}\right) = 0.08.$$

Applying the Hamming window function and its symmetric property to the filter coefficients, we get

$$h_w(0) = h(0) \cdot w_{\mathrm{ham}}(0) = 0.25 \times 1 = 0.25$$
$$h_w(1) = h(1) \cdot w_{\mathrm{ham}}(1) = 0.22508 \times 0.77 = 0.17331 = h_w(-1)$$
$$h_w(2) = h(2) \cdot w_{\mathrm{ham}}(2) = 0.15915 \times 0.31 = 0.04934 = h_w(-2)$$
$$h_w(3) = h(3) \cdot w_{\mathrm{ham}}(3) = 0.07503 \times 0.08 = 0.00600 = h_w(-3).$$

(b) Noncausal impulse responses $h(n)$ and $h_w(n)$ are plotted in Fig. 7.10.

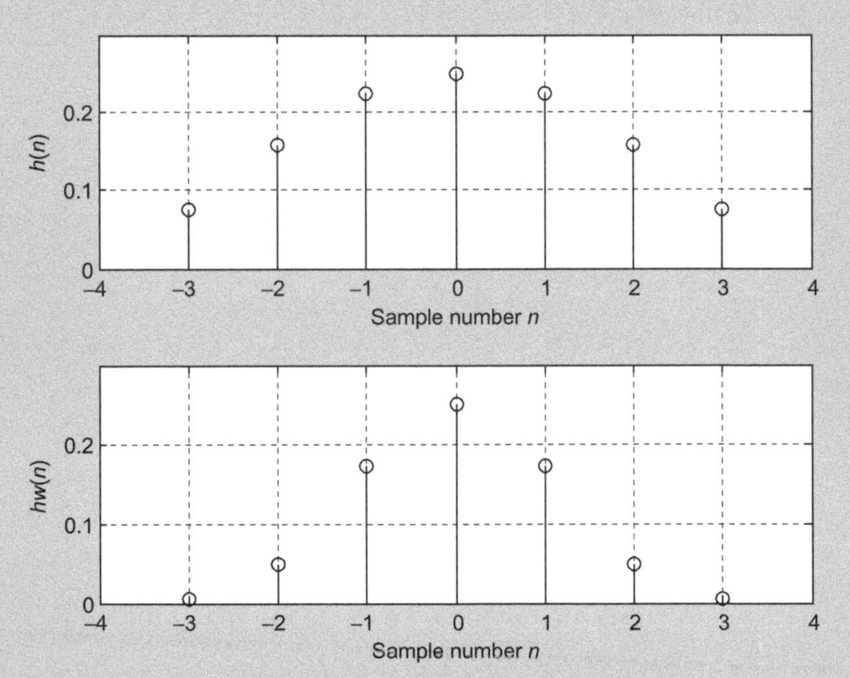

FIG. 7.10

Plots of FIR noncausal coefficients and windowed FIR coefficients in Example 7.4.

We observe that the Hamming window does its job to weight the FIR filter coefficients to zero gradually at both ends. Hence, we can expect a reduced Gibbs effect in the magnitude frequency response.

Now the lowpass FIR filter design via the window design method can be achieved. The design procedure includes three steps. The first step is to obtain the truncated impulse response $h(n)$, where $-M \le n \le M$; then we multiply the obtained sequence $h(n)$ by the selected window data sequence to yield the windowed noncausal FIR filter coefficients $h_w(n)$; and the final step is to delay the windowed noncausal sequence $h_w(n)$ by M samples to achieve the causal FIR filter coefficients, $b_n = h_w(n - M)$. The design procedure of the FIR filter via windowing is summarized as follows:

1. Obtain the FIR filter coefficients $h(n)$ via the Fourier transform method (Table 7.1).
2. Multiply the generated FIR filter coefficients by the selected window sequence

$$h_w(n) = h(n)w(n), \quad n = -M, \ldots 0, 1, \ldots, M, \tag{7.28}$$

where $w(n)$ is chosen to be one of the window functions listed in Eqs. (7.23)–(7.27).

3. Delay the windowed impulse sequence $h_w(n)$ by M samples to get the windowed FIR filter coefficients:

$$b_n = h_w(n - M), \quad \text{for } n = 0, 1, \cdots, 2M. \tag{7.29}$$

Let us study the following design examples.

EXAMPLE 7.5

(a) Design a three-tap FIR lowpass filter with a cutoff frequency of 800 Hz and a sampling rate of 8000 Hz using the Hamming window function.

(b) Determine the transfer function and difference equation of the designed FIR system.

(c) Compute and plot the magnitude frequency response for $\Omega = 0$, $\pi/4$, $\pi/2$, $3\pi/4$, and π (rad).

Solution:

(a) The normalized cutoff frequency is calculated as

$$\Omega_c = 2\pi f_c T_s = 2\pi \times \frac{800}{8000} = 0.2\pi \text{ (rad)}.$$

Since $2M + 1 = 3$ in this case, FIR coefficients obtained by using the equation in Table 7.1 are listed as

$$h(0) = 0.2 \text{ and } h(-1) = h(1) = 0.1871$$

(see Example 7.2). Applying Hamming window function defined in Eq. (7.26), we have

$$w_{\text{ham}}(0) = 0.54 + 0.46\cos\left(\frac{0\pi}{1}\right) = 1$$

$$w_{\text{ham}}(1) = 0.54 + 0.46\cos\left(\frac{1 \times \pi}{1}\right) = 0.08.$$

Continued

EXAMPLE 7.5—CONT'D

Using the symmetry of the window function gives

$$w_{\mathrm{ham}}(-1) = w_{\mathrm{ham}}(1) = 0.08.$$

The windowed impulse response is calculated as

$$\begin{aligned} h_w(0) &= h(0)w_{\mathrm{ham}}(0) = 0.2 \times 1 = 0.2 \\ h_w(1) &= h(1)w_{\mathrm{ham}}(1) = 0.1871 \times 0.08 = 0.01497 \\ h_w(-1) &= h(-1)w_{\mathrm{ham}}(-1) = 0.1871 \times 0.08 = 0.01497. \end{aligned}$$

Thus delaying $h_w(n)$ by $M=1$ sample gives

$$b_0 = b_2 = 0.01496 \text{ and } b_1 = 0.2.$$

(b) The transfer function is achieved as

$$H(z) = 0.01497 + 0.2z^{-1} + 0.01497z^{-2}.$$

Using the technique described in Chapter 6, we have

$$\frac{Y(z)}{X(z)} = H(z) = 0.01497 + 0.2z^{-1} + 0.01497z^{-2}.$$

Multiplying $X(z)$ leads to

$$Y(z) = 0.01497X(z) + 0.2z^{-1}X(z) + 0.01497z^{-2}X(z).$$

Applying the inverse z-transform on both sides, the difference equation is yielded as

$$y(n) = 0.01497x(n) + 0.2x(n-1) + 0.01497x(n-2).$$

(c) The magnitude frequency response and phase response can be obtained using the technique introduced in Chapter 6. Substituting $z = e^{j\Omega}$ into $H(z)$, it follows that

$$\begin{aligned} H(e^{j\Omega}) &= 0.01497 + 0.2e^{-j\Omega} + 0.01497e^{-j2\Omega} \\ &= e^{-j\Omega}(0.01497e^{j\Omega} + 0.2 + 0.01497e^{-j\Omega}). \end{aligned}$$

Using Euler formula leads to

$$H(e^{j\Omega}) = e^{-j\Omega}(0.2 + 0.02994\cos\Omega).$$

Then the magnitude frequency response and phase response are found to be

$$|H(e^{j\Omega})| = |0.2 + 0.2994\cos\Omega|.$$

$$\text{and } \angle H(e^{j\Omega}) = \begin{cases} -\Omega & \text{if } 0.2 + 0.02994\cos\Omega > 0 \\ -\Omega + \pi & \text{if } 0.2 + 0.02994\cos\Omega < 0. \end{cases}$$

The calculation details of the magnitude response for several normalized values are listed in Table 7.4. Fig. 7.11 shows the plots of the frequency responses.

Table 7.4 Frequency Response Calculation in Example 7.5

Ω (rad)	$f=\Omega f_s/(2\pi)$ (Hz)	$0.2+0.02994\cos\Omega$	$\lvert H(e^{j\Omega})\rvert$	$\lvert H(e^{j\Omega})\rvert_{dB}$ (dB)	$\angle H(e^{j\Omega})$ (deg)
0	0	0.2299	0.2299	−12.77	0
$\pi/4$	1000	0.1564	0.2212	−13.11	−45
$\pi/2$	2000	0.2000	0.2000	−13.98	−90
$3\pi/4$	3000	0.1788	0.1788	−14.95	−135
π	4000	0.1701	0.1701	−15.39	−180

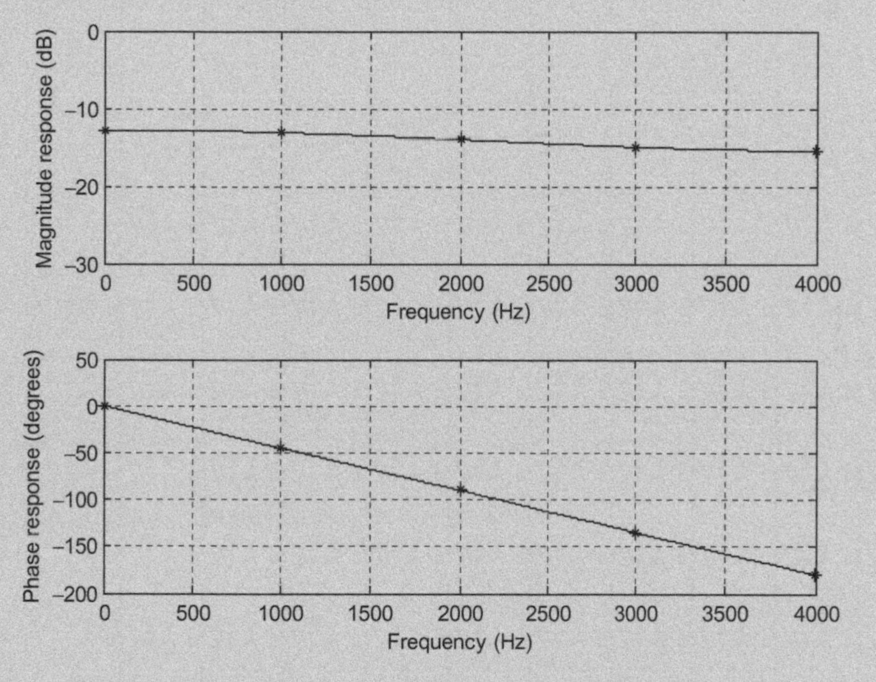

FIG. 7.11

The frequency responses in Example 7.5.

EXAMPLE 7.6

(a) Design a five-tap FIR bandreject filter with a lower cutoff frequency of 2000 Hz, an upper cutoff frequency of 2400 Hz, and a sampling rate of 8000 Hz using the Hamming window method.
(b) Determine the transfer function.

Solution:

(a) Calculating the normalized cutoff frequencies leads to

$$\Omega_L = 2\pi f_L T = 2\pi \times \frac{2000}{8000} = 0.5\pi \text{ (rad)}$$
$$\Omega_H = 2\pi f_H T = 2\pi \times \frac{2400}{8000} = 0.6\pi \text{ (rad)}.$$

Since $2M+1=5$ in this case, using the equation in Table 7.1 yields

$$h(n) = \begin{cases} \dfrac{\pi - \Omega_H + \Omega_L}{\pi} & n=0 \\ -\dfrac{\sin(\Omega_H n)}{n\pi} + \dfrac{\sin(\Omega_L n)}{n\pi} & n \neq 0 \quad -2 \leq n \leq 2. \end{cases}$$

when $n=0$, we have

$$h(0) = \frac{\pi - \Omega_H + \Omega_L}{\pi} = \frac{\pi - 0.6\pi + 0.5\pi}{\pi} = 0.9.$$

The other computed filter coefficients via the previous expression are listed below

$$h(1) = \frac{\sin[0.5\pi \times 1]}{1 \times \pi} - \frac{\sin[0.6\pi \times 1]}{1 \times \pi} = 0.01558$$
$$h(2) = \frac{\sin[0.5\pi \times 2]}{2 \times \pi} - \frac{\sin[0.6\pi \times 2]}{2 \times \pi} = 0.09355.$$

Using the symmetry leads to

$$h(-1) = h(1) = 0.01558$$
$$h(-2) = h(2) = 0.09355.$$

Applying Hamming window function in Eq. (7.25), we have

$$w_{\text{ham}}(0) = 0.54 + 0.46\cos\left(\frac{0 \times \pi}{2}\right) = 1.0$$
$$w_{\text{ham}}(1) = 0.54 + 0.46\cos\left(\frac{1 \times \pi}{2}\right) = 0.54$$
$$w_{\text{ham}}(2) = 0.54 + 0.46\cos\left(\frac{2 \times \pi}{2}\right) = 0.08.$$

Using the symmetry of the window function gives

$$w_{\text{ham}}(-1) = w_{\text{ham}}(1) = 0.54$$
$$w_{\text{ham}}(-2) = w_{\text{ham}}(2) = 0.08.$$

The windowed impulse response is calculated as

$$h_w(0) = h(0)w_{ham}(0) = 0.9 \times 1 = 0.9$$
$$h_w(1) = h(1)w_{ham}(1) = 0.01558 \times 0.54 = 0.00841$$
$$h_w(2) = h(2)w_{ham}(2) = 0.09355 \times 0.08 = 0.00748$$
$$h_w(-1) = h(-1)w_{ham}(-1) = 0.00841$$
$$h_w(-2) = h(-2)w_{ham}(-2) = 0.00748.$$

Thus, delaying $h_w(n)$ by $M=2$ samples gives

$$b_0 = b_4 = 0.00748, \quad b_1 = b_3 = 0.00841, \quad \text{and} \quad b_2 = 0.9.$$

(b) The transfer function is achieved as

$$H(z) = 0.00748 + 0.00841z^{-1} + 0.9z^{-2} + 0.00841z^{-3} + 0.00748z^{-4}.$$

The following design examples are demonstrated using MATLAB programs. The MATLAB function **firwd(N, Ftype, WnL, WnH, Wtype)** is listed in the "MATLAB Programs" section at the end of this chapter. Table 7.5 lists comments to show how the function is used.

Table 7.5 Illustration of the MATLAB Function for FIR Filter Design Using the Window Methods

```
Function B=firwd(N,Ftype,WnL,WnH,Wtype)
% B = firwd(N,Ftype,WnL,WnH,Wtype)
% FIR filter design using the window function method.
% Input parameters:
% N: the number of the FIR filter taps.
% Note: It must be odd number.
% Ftype: the filter type
%       1. Lowpass filter;
%       2. Highpass filter;
%       3. Bandpass filter;
%       4. Bandreject filter.
% WnL: lower cutoff frequency in radians. Set WnL=0 for the highpass filter.
% WnH: upper cutoff frequency in radians. Set WnH=0 for the lowpass filter.
% Wtype: window function type
%       1. Rectangular window;
%       2. Triangular window;
%       3. Hanning window;
%       4. Hamming window;
%       5. Blackman window;
```

EXAMPLE 7.7

(a) Design a lowpass FIR filter with 25 taps using the MATLAB function listed in the "MATLAB Programs" section at the end of this chapter. The cutoff frequency of the filter is 2000 Hz, assuming a sampling frequency of 8000 Hz. The rectangular window and Hamming window functions are used for each design.
(b) Plot the frequency responses along with those obtained using the rectangular window and Hamming window for comparison.
(c) List FIR filter coefficients for each window design method.

Solution:

(a) With a given sampling rate of 8000 Hz, the normalized cutoff frequency can be found as

$$\Omega_c = \frac{2000 \times 2\pi}{8000} = 0.5\pi \text{ (rad)}.$$

Now we are ready to design FIR filters via the MATLAB program. The function, firwd(N, Ftype,WnL,WnH,Wtype), listed in the "MATLAB Programs" section at the end of this chapter, has five input parameters, which are described as follows:

- "N" is the number of specified filter coefficients (the number of filter taps).
- "Ftype" denotes the filter type, that is, input "1" for the lowpass filter design, input "2" for the highpass filter design, input "3" for the bandpass filter design, and input "4" for the bandreject filter design.
- "WnL" and "WnH" are the lower and upper cutoff frequency inputs, respectively. Note that WnH=0 when specifying WnL for the lowpass filter design, while WnL=0 when specifying WnH for the highpass filter design.
- "Wtype" specifies the window data sequence to be used in the design, that is, input "1" for the rectangular window, input "2" for the triangular window, input "3" for the Hanning window, input "4" for the Hamming window, and input "5" for the Blackman window.

(b) The following application program (Program 7.2) is used to generate FIR filter coefficients using the rectangular window. Its frequency responses will be plotted together with that obtained using the Hamming window for comparison, as shown in Program 7.3.

As a comparison, the frequency responses achieved from the rectangular window and the Hamming window are plotted in Fig. 7.12, where the dash-dotted line indicates the frequency response via the rectangular window, while the solid line indicates the frequency response via the Hamming window.

(c) The FIR filter coefficients for both methods are listed in Table 7.6.

For comparison with other window functions, Fig. 7.13 shows the magnitude frequency responses using the Hanning, Hamming, and Blackman windows, with 25 taps and a cutoff frequency of 2000 Hz. The Blackman window offers the lowest side lobe, but with an increased width of the main lobe. The Hamming window and Hanning window have a similar narrow width of the main lobe, but the Hamming window accommodates a lower side lobe than the Hanning window. Next, we will study how to choose a window in practice.

Program 7.2. MATLAB program for Example 7.7.

```
% Example 7.7
% MATLAB program to generate FIR coefficients
% using the rectangular window.
%
N=25; Ftype=1; WnL=0.5*pi; WnH=0; Wtype=1;
B=firwd(N,Ftype,WnL,WnH,Wtype);
```

Results of the FIR filer design using the Hamming window are illustrated in Program 7.3.

Program 7.3. MATLAB program for Example 7.7.

```
% Fig. 7.12
% MATLAB program to create Fig. 7.12
%
N=25; Ftype=1; WnL=0.5*pi; WnH=0; Wtype=1;fs=8000;
%design using the rectangular window;
Brec=firwd(N,Ftype,WnL,WnH,Wtype);
N=25; Ftype=1; WnL=0.5*pi; WnH=0; Wtype=4;
%design using the Hamming window;
Bham=firwd(N,Ftype,WnL,WnH,Wtype);
[hrec,f]=freqz(Brec,1,512,fs);
[hham,f]=freqz(Bham,1,512,fs);
prec=180*unwrap(angle(hrec))/pi;
pham=180*unwrap(angle(hham))/pi;
subplot(2,1,1);
plot(f,20*log10(abs(hrec)),'-.',f,20*log10(abs(hham)));grid
axis([0 4000 -100 10]);
xlabel('Frequency (Hz)'); ylabel('Magnitude Response (dB)');
subplot(2,1,2);
plot(f,prec,'-.',f,pham);grid
xlabel('Frequency (Hz)'); ylabel('Phase (degrees)');
```

Applying the window to remedy Gibbs effect will change the characteristics of the magnitude frequency response of the FIR filter, where the width of the main lobe becomes wider while more attenuation of side lobes are achieved.

Next, we illustrate the design for customer specifications in practice. Given the required stopband attenuation and passband ripple specifications shown in Fig. 7.14, where the lowpass filter specifications are given for illustrative purpose, the appropriate window can be selected based on performances of the window functions listed in Table 7.7. For example, the Hamming window offers the passband ripple of 0.0194 dB and stopband attenuation of 53 dB. With the selected Hamming window and the calculated normalized transition band defined in Table 7.7.

$$\Delta f = \frac{|f_{\text{stop}} - f_{\text{pass}}|}{f_s}, \tag{7.30}$$

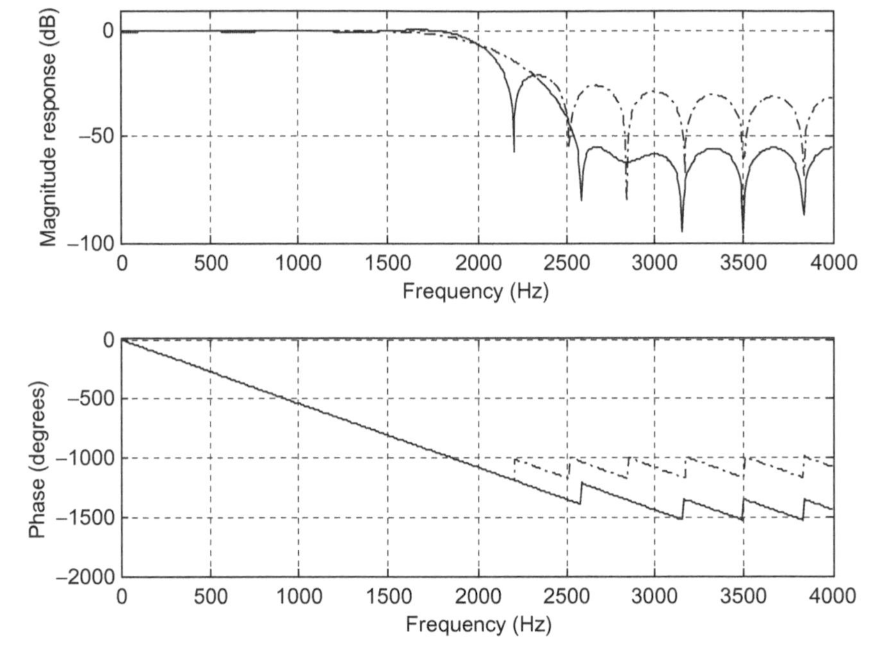

FIG. 7.12

Frequency responses using the rectangular and Hamming windows.

Table 7.6 FIR Filter Coefficients in Example 7.7 (Rectangular and Hamming Windows)

B: FIR Filter Coefficients (Rectangular Window)	Bham: FIR Filter Coefficients (Hamming Window)
$b_0 = b_{24} = 0.000000$	$b_0 = b_{24} = 0.000000$
$b_1 = b_{23} = -0.028937$	$b_1 = b_{23} = -0.002769$
$b_2 = b_{22} = 0.000000$	$b_2 = b_{22} = 0.000000$
$b_3 = b_{21} = 0.035368$	$b_3 = b_{21} = 0.007595$
$b_4 = b_{20} = 0.000000$	$b_4 = b_{20} = 0.000000$
$b_5 = b_{19} = -0.045473$	$b_5 = b_{19} = -0.019142$
$b_6 = b_{18} = 0.000000$	$b_6 = b_{18} = 0.000000$
$b_7 = b_{17} = 0.063662$	$b_7 = b_{17} = 0.041957$
$b_8 = b_{16} = 0.000000$	$b_8 = b_{16} = 0.000000$
$b_9 = b_{15} = -0.106103$	$b_9 = b_{15} = -0.091808$
$b_{10} = b_{14} = 0.000000$	$b_{10} = b_{14} = 0.000000$
$b_{11} = b_{13} = 0.318310$	$b_{11} = b_{13} = 0.313321$
$b_{12} = 0.500000$	$b_{12} = 0.500000$

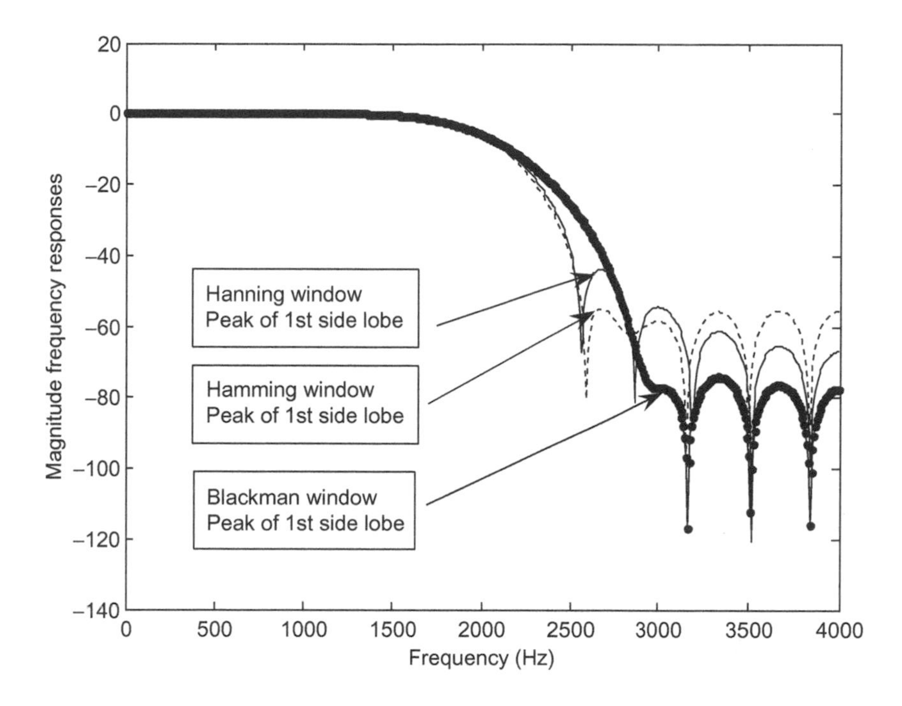

FIG. 7.13

Comparisons of magnitude frequency responses for the Hanning, Hamming, and Blackman windows.

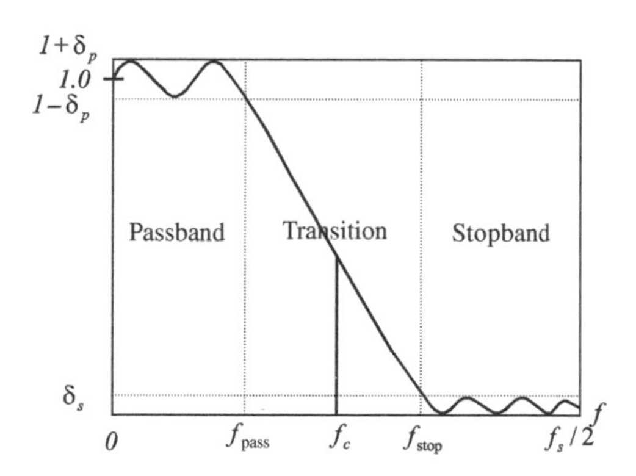

FIG. 7.14

Lowpass filter frequency domain specifications.

where f_{pass} and f_{stop} are the passband frequency edge and stop frequency edge. The filter length using the Hamming window can be determined by

$$N = \frac{3.3}{\Delta f}. \tag{7.31}$$

Note that the passband ripple is defined as

$$\delta_p dB = 20 \times \log_{10}(1 + \delta_p), \tag{7.32}$$

Table 7.7 FIR Filter Length Estimation Using Window Functions (Normalized Transition Width $\Delta f = \frac{|f_{stop} - f_{pass}|}{f_s}$)

Window Type	Window Function $w(n)$, $-M \le n \le M$	Window Length (N)	Passband Ripple (dB)	Stopband Attenuation (dB)
Rectangular	1	$N = 0.9/\Delta f$	0.7416	21
Hanning	$0.5 + 0.5\cos\left(\frac{\pi n}{M}\right)$	$N = 3.1/\Delta f$	0.0546	44
Hamming	$0.54 + 0.46\cos\left(\frac{\pi n}{M}\right)$	$N = 3.3/\Delta f$	0.0194	53
Blackman	$0.42 + 0.5\cos\left(\frac{n\pi}{M}\right) + 0.08\cos\left(\frac{2n\pi}{M}\right)$	$N = 5.5/\Delta f$	0.0017	74

while the stopband attenuation is defined as

$$\delta_s dB = -20 \times \log_{10}(\delta_s). \tag{7.33}$$

The cutoff frequency used for design will be chosen at the middle of the transition band, as illustrated for the lowpass filter case shown in Fig. 7.14.

As a rule of thumb, the cutoff frequency used for design is determined by

$$f_c = \frac{f_{pass} + f_{stop}}{2}. \tag{7.34}$$

Note that Eq. (7.31) and formulas for other window lengths in Table 7.7 are empirically derived based on the normalized spectral transition width of each window function. The spectrum of each window function appears to be shaped like the lowpass filter magnitude frequency response with ripples in the passband and side lobes in the stopband. The passband frequency edge of the spectrum is the frequency where the magnitude just begins to drop below the passband ripple and where the stop frequency edge is at the peak of the first side lobe in spectrum. With the passband ripple and stopband attenuation specified for a particular window, the normalized transition width of the window is in inverse proportion to the window length N multiplied by a constant. For example, the normalized spectral transition Δf for the Hamming window is $3.3/N$. Hence, matching the FIR filter transition width with the transition width of the window spectrum gives the filter length estimation listed in Table 7.7.

The following examples illustrate the determination of each filter length and cutoff frequency/frequencies for the design of lowpass, highpass, bandpass, and bandstop filters. Application of each designed filter to the processing of speech data is included, along with an illustration of filtering effects in both time domain and frequency domain.

EXAMPLE 7.8

A lowpass FIR filter has the following specifications:

Passband	0–1850 Hz
Stopband	2150–4000 Hz
Stopband attenuation	20 dB
Passband ripple	1 dB
Sampling rate	8000 Hz

Determine the FIR filter length and the cutoff frequency to be used in the design equation.

Solution:

The normalized transition band as defined in Eq. (7.30) and Table 7.7 is given by

$$\Delta f = \frac{|2150 - 1850|}{8000} = 0.0375.$$

Again, based on Table 7.7, selecting the rectangular window will result in a passband ripple of 0.74 dB and stopband attenuation of 21 dB. Thus, this window selection would satisfy the design requirement for the passband ripple of 1 dB and stopband attenuation of 20 dB (Although all the other windows satisfy the requirement as well but this one results in a small number of coefficients). Next, we determine the length of the filter as

$$N = \frac{0.9}{\Delta f} = \frac{0.9}{0.0375} = 24.$$

We choose the odd number $N=25$ [requirement in Eq. (7.18)]. The cutoff frequency is determined by $(1850+2150)/2=2000$ Hz. Such a filter has been designed in Example 7.7, its filter coefficients is listed in Table 7.6, and its frequency responses can be found in Fig. 7.12 (dashed lines).

Now we look at the time domain and frequency domain results from filtering a speech signal by using the lowpass filter we have just designed. Fig. 7.15A shows the original speech and lowpass filtered speech. The spectral comparison is given in Fig. 7.15B, where, as we can see, the frequency components beyond 2 kHz are filtered. The lowpass filtered speech would sound muffled.

We will continue to illustrate the determination of the filter length and cutoff frequency for other types of filters via the following examples.

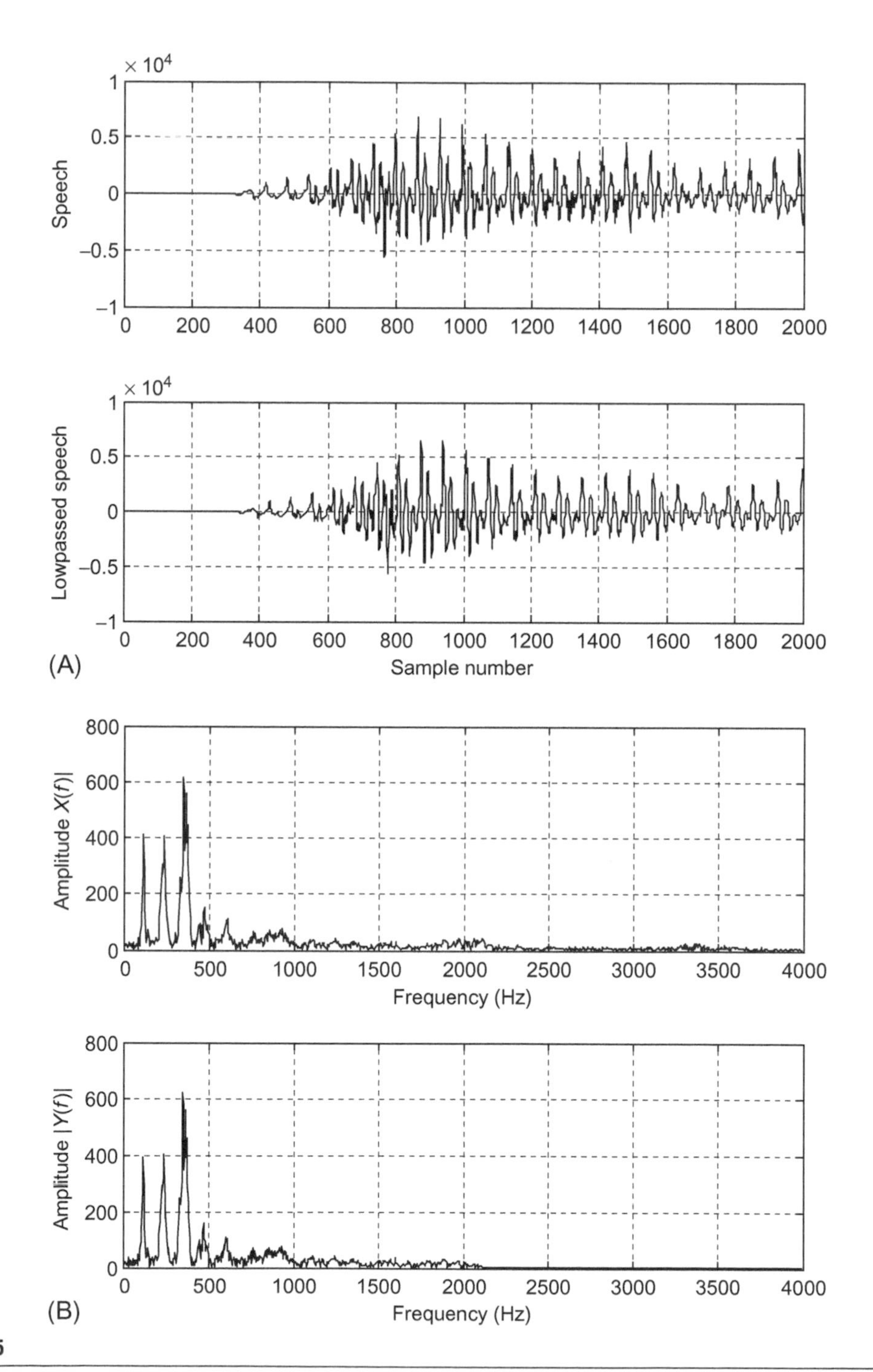

FIG. 7.15

(A) Original speech and processed speech using the lowpass filter. (B) Spectral plots of the original speech and processed speech by the lowpass filter.

EXAMPLE 7.9

Design a highpass FIR filter with the following specifications:

Stopband	0–1500 Hz
Passband	2500–4000 Hz
Stopband attenuation	40 dB
Passband ripple	0.1 dB
Sampling rate	8000 Hz

Solution:

Based on the specification, the Hanning window will do the job since it has passband ripple of 0.0546 dB and stopband attenuation of 44 dB.

Then

$$\Delta f = \frac{|1500 - 2500|}{8000} = 0.125$$

$$N = \frac{3.1}{\Delta f} = 24.2 \text{ Choose } N = 25.$$

Hence, we choose 25 filter coefficients using the Hanning window method. The cutoff frequency is (1500+2500)/2=2000 Hz. The normalized cutoff frequency can be easily found as

$$\Omega_c = \frac{2000 \times 2\pi}{8000} = 0.5\pi \text{ (rad)}.$$

and note that $2M+1=25$. The application program and design results are listed in Program 7.4 and Table 7.8.

Table 7.8 FIR Filter Coefficients in Example 7.9 (Hanning Window)

Bhan: FIR Filter Coefficients (Hanning Window)	
$b_0 = b_{24} = 0.000000$	$b_1 = b_{23} = 0.000493$
$b_2 = b_{22} = 0.000000$	$b_3 = b_{21} = -0.005179$
$b_4 = b_{20} = 0.000000$	$b_5 = b_{19} = 0.016852$
$b_6 = b_{18} = 0.000000$	$b_7 = b_{17} = -0.040069$
$b_8 = b_{16} = 0.0000000$	$b_9 = b_{15} = 0.090565$
$b_{10} = b_{14} = 0.000000$	$b_{11} = b_{13} = -0.312887$
$b_{12} = 0.500000$	

Program 7.4. MATLAB program for Example 7.9.

```
% Fig. 7.16 (Example 7.9)
% MATLAB program to create Fig. 7.16
%
N=25; Ftype=2; WnL=0; WnH=0.5*pi; Wtype=3;fs=8000;
Bhan=firwd(N,Ftype,WnL,WnH,Wtype);
freqz(Bhan,1,512,fs);
axis([0 fs/2 -120 10]);
```

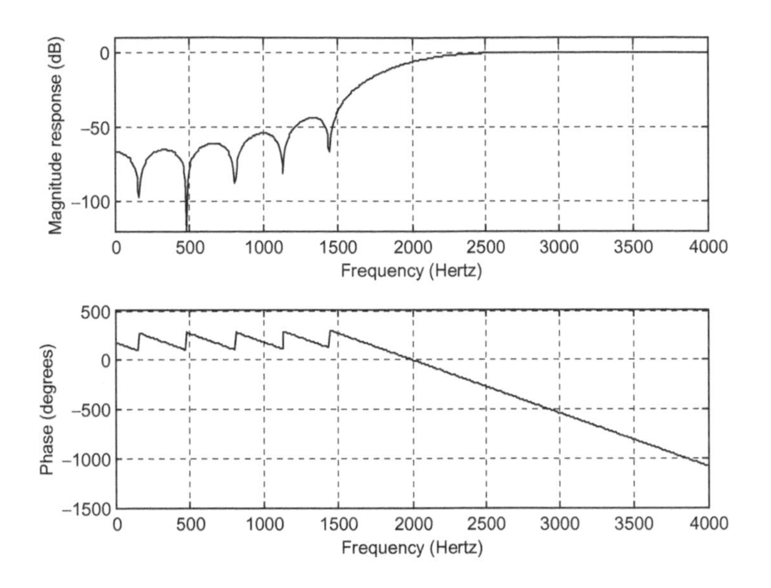

FIG. 7.16

Frequency responses of the designed highpass filter using the Hanning window.

The corresponding frequency responses of the designed highpass FIR filter are displayed in Fig. 7.16.

Comparisons are depicted in Fig. 7.17A, where the original speech and processed speech using the highpass filter are plotted, respectively. The high-frequency components of speech generally contain small amount of energy. Fig. 7.17B displays the spectral plots, where clearly the frequency components less than 1.5 kHz are filtered. The processed speech would sound crisp.

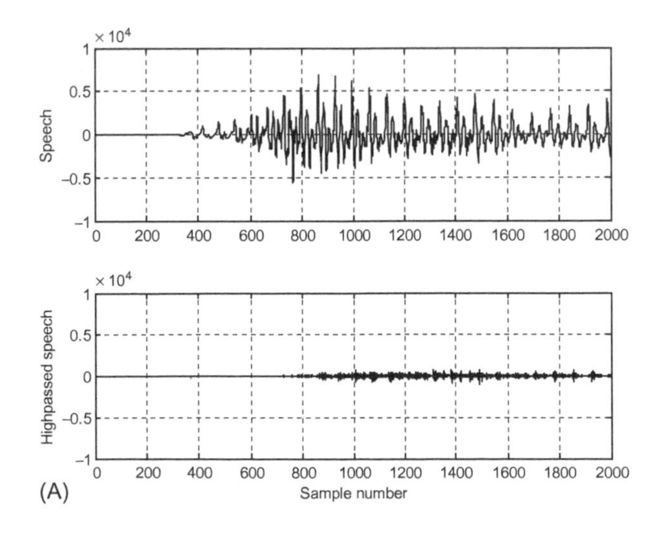

FIG. 7.17

(A) Original speech and processed speech using the highpass filter.

(Continued)

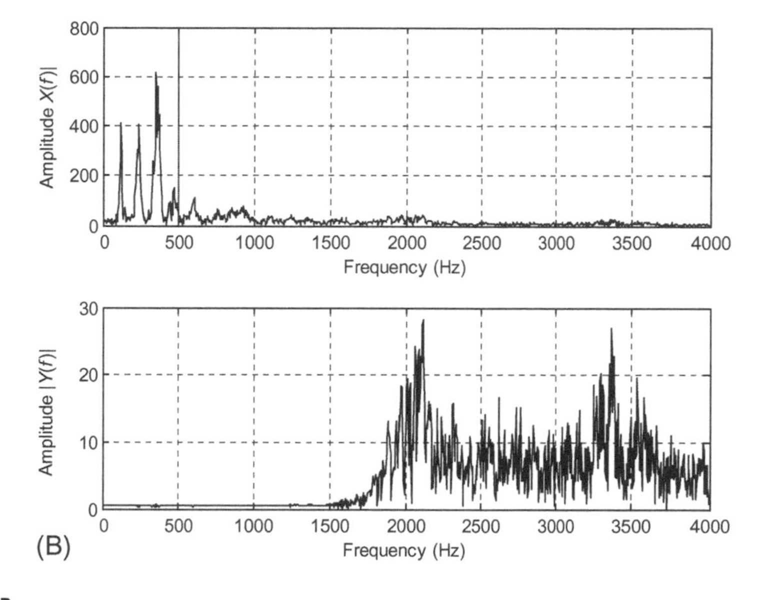

FIG. 7.17, CONT'D

(B) Spectral comparison of the original speech and processed speech using the highpass filter.

EXAMPLE 7.10

Design a bandpass FIR filter with the following specifications:

Lower stopband	0–500 Hz
Passband	1600–2300 Hz
Upper stopband	3500–4000 Hz
Stopband attenuation	50 dB
Passband ripple	0.05 dB
Sampling rate	8000 Hz

Solution:

$$\Delta f_1 = \frac{|1600-500|}{8000} = 0.1375 \text{ and } \Delta f_2 = \frac{|3500-2300|}{8000} = 0.15$$

$$N_1 = \frac{3.3}{0.1375} = 24 \text{ and } N_2 = \frac{3.3}{0.15} = 22$$

Choosing $N=25$ filter coefficients using the Hamming window method:

$$f_1 = \frac{1600+500}{2} = 1050\,\text{Hz} \text{ and } f_2 = \frac{3500+2300}{2} = 2900\,\text{Hz}.$$

The normalized lower and upper cutoff frequencies are calculated as

$$\Omega_L = \frac{1050 \times 2\pi}{8000} = 0.2625\pi\,(\text{rad}) \text{ and}$$

Continued

EXAMPLE 7.10—CONT'D

$$\Omega_H = \frac{2900 \times 2\pi}{8000} = 0.725\pi \text{ (rad)},$$

and $N = 2M + 1 = 25$. Using the MATLAB program, design results are achieved as shown in Program 7.5.

Program 7.5. MATLAB program for Example 7.10

```
% Fig. 7.18 (Example 7.10)
% MATLAB program to create Fig. 7.18
%
N=25; Ftype=3; WnL=0.2625*pi; WnH=0.725*pi; Wtype=4;fs=8000;
Bham=firwd(N,Ftype,WnL,WnH,Wtype);
freqz(Bham,1,512,fs);
axis([0 fs/2 -130 10]);
```

Fig. 7.18 depicts the frequency responses of the designed bandpass FIR filter. Table 7.9 lists the designed FIR filter coefficients.

For comparison, the original speech and bandpass filtered speech are plotted in Fig. 7.19A, where the bandpass frequency components contains a small portion of speech energy. Fig. 7.19B shows a comparison indicating that the low-frequency and high-frequency components are removed by the bandpass filter.

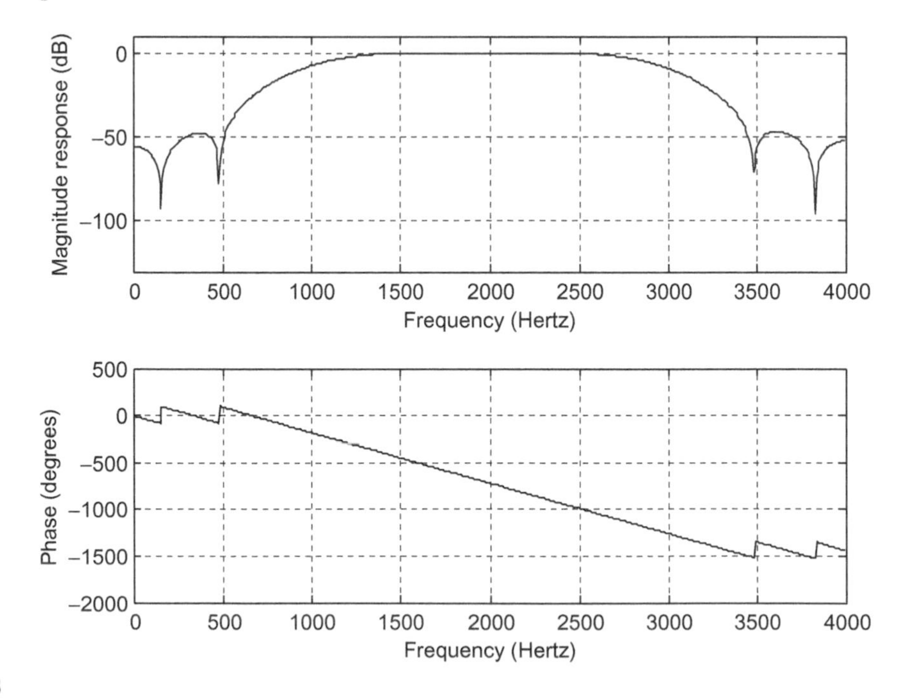

FIG. 7.18

Frequency responses of the designed bandpass filter using the Hamming window.

Table 7.9 FIR Filter Coefficients in Example 7.10 (Hamming Window)

Bham: FIR Filter Coefficients (Hamming Window)	
$b_0=b_{24}=0.002680$	$b_1=b_{23}=-0.001175$
$b_2=b_{22}=-0.007353$	$b_3=b_{21}=0.000674$
$b_4=b_{20}=-0.011063$	$b_5=b_{19}=0.004884$
$b_6=b_{18}=0.053382$	$b_7=b_{17}=-0.003877$
$b_8=b_{16}=0.028520$	$b_9=b_{15}=-0.008868$
$b_{10}=b_{14}=-0.296394$	$b_{11}=b_{13}=0.008172$
$b_{12}=0.462500$	

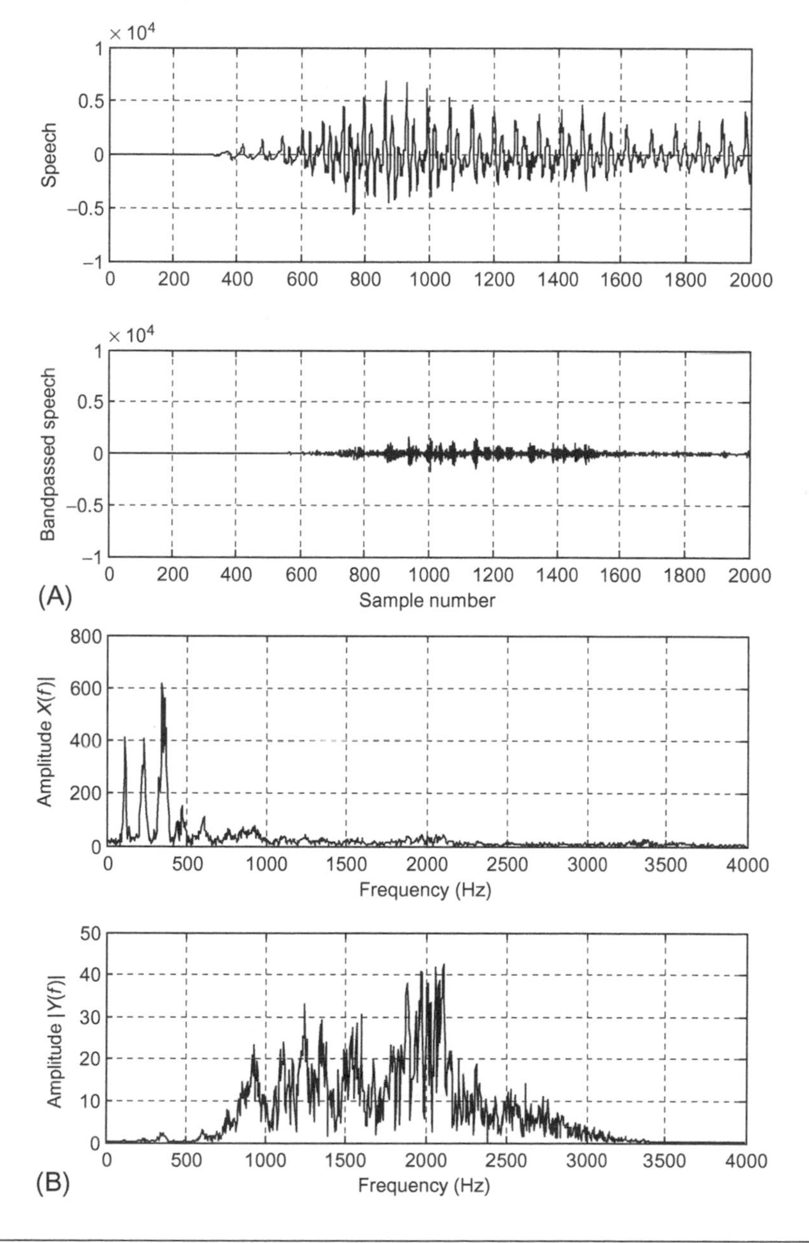

FIG. 7.19

(A) Original speech and processed speech using the bandpass filter. (B) Spectral comparison of the original speech and processed speech using the bandpass filter.

EXAMPLE 7.11

Design a bandstop FIR filter with the following specifications:

Lower cutoff frequency	1250 Hz
Lower transition width	1500 Hz
Upper cutoff frequency	2850 Hz
Upper transition width	1300 Hz
Stopband attenuation	60 dB
Passband ripple	0.02 dB
Sampling rate	8000 Hz

Solution:

We can directly compute the normalized transition width:

$$\Delta f_1 = \frac{1500}{8000} = 0.1875, \text{ and } \Delta f_2 = \frac{1300}{8000} = 0.1625.$$

The filter lengths are determined, respectively, using the Blackman windows as

$$N_1 = \frac{5.5}{0.1875} = 29.33, \text{ and } N_2 = \frac{5.5}{0.1625} = 33.8.$$

We choose an odd number $N=35$. The normalized lower and upper cutoff frequencies are calculated as

$$\Omega_L = \frac{2\pi \times 1250}{8000} = 0.3125\pi \text{ (rad) and}$$

$$\Omega_H = \frac{2\pi \times 2850}{8000} = 0.7125\pi \text{ (rad)},$$

and $N=2M+1=35$. Using the MATLAB program, the design results are demonstrated in Program 7.6.

Program 7.6. MATLAB program for Example 7.11.

```
% Fig. 7.20 (Example 7.11)
% MATLAB program to create Fig. 7.20
%
N=35; Ftype=4; WnL=0.3125*pi; WnH=0.7125*pi; Wtype=5;fs=8000;
Bblack=firwd(N,Ftype,WnL,WnH,Wtype);
freqz(Bblack,1,512,fs);
axis([0 fs/2 -120 10]);
```

Fig. 7.20 shows the plot of the frequency responses of the designed bandstop filter. The designed filter coefficients are listed in Table 7.10.

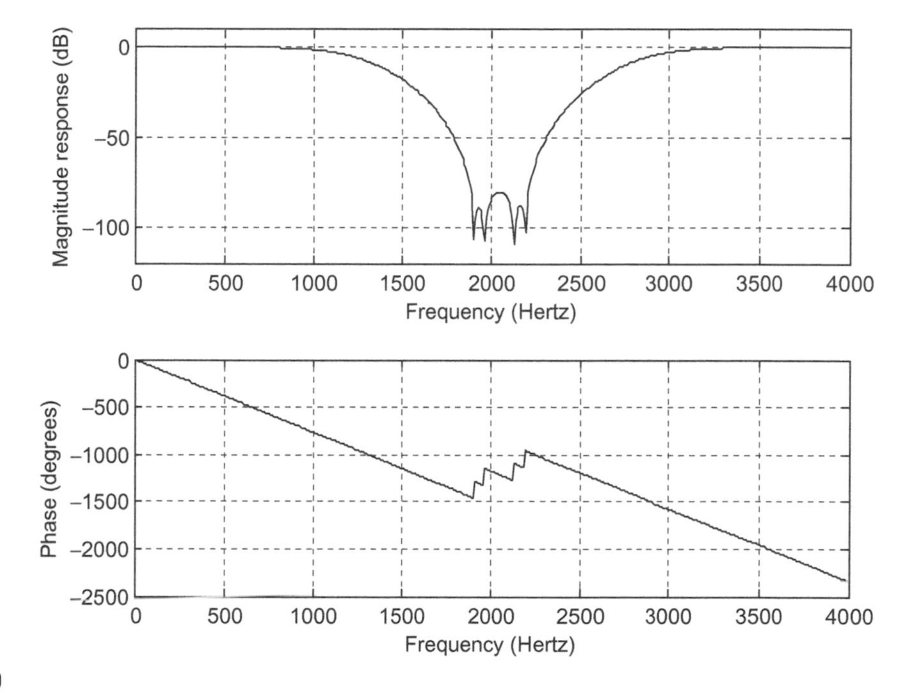

FIG. 7.20

Frequency responses of the designed bandstop filter using the Blackman window.

Table 7.10 FIR Filter Coefficients in Example 7.11 (Blackman Window)

Black: FIR Filter Coefficients (Blackman Window)	
$b_0=b_{34}=0.000000$	$b_1=b_{33}=0.000059$
$b_2=b_{32}=0.000000$	$b_3=b_{31}=0.000696$
$b_4=b_{30}=0.001317$	$b_5=b_{29}=-0.004351$
$b_6=b_{28}=-0.002121$	$b_7=b_{27}=0.000000$
$b_8=b_{26}=-0.004249$	$b_9=b_{25}=0.027891$
$b_{10}=b_{24}=0.011476$	$b_{11}=b_{23}=-0.036062$
$b_{12}=b_{22}=0.000000$	$b_{13}=b_{21}=-0.073630$
$b_{14}=b_{20}=-0.020893$	$b_{15}=b_{19}=0.285306$
$b_{16}=b_{18}=0.014486$	$b_{17}=0.600000$

Comparisons of filtering effects are illustrated in Figs. 7.21A and B. In Fig. 7.21A, the original speech and processed speech by the bandstop filter are plotted. The processed speech contains most of the energy of the original speech because most energy of the speech signal exists in the low-frequency band. Fig. 7.21B verifies the filtering frequency effects. The frequency components ranging from 2000 to 2200 Hz have been greatly attenuated.

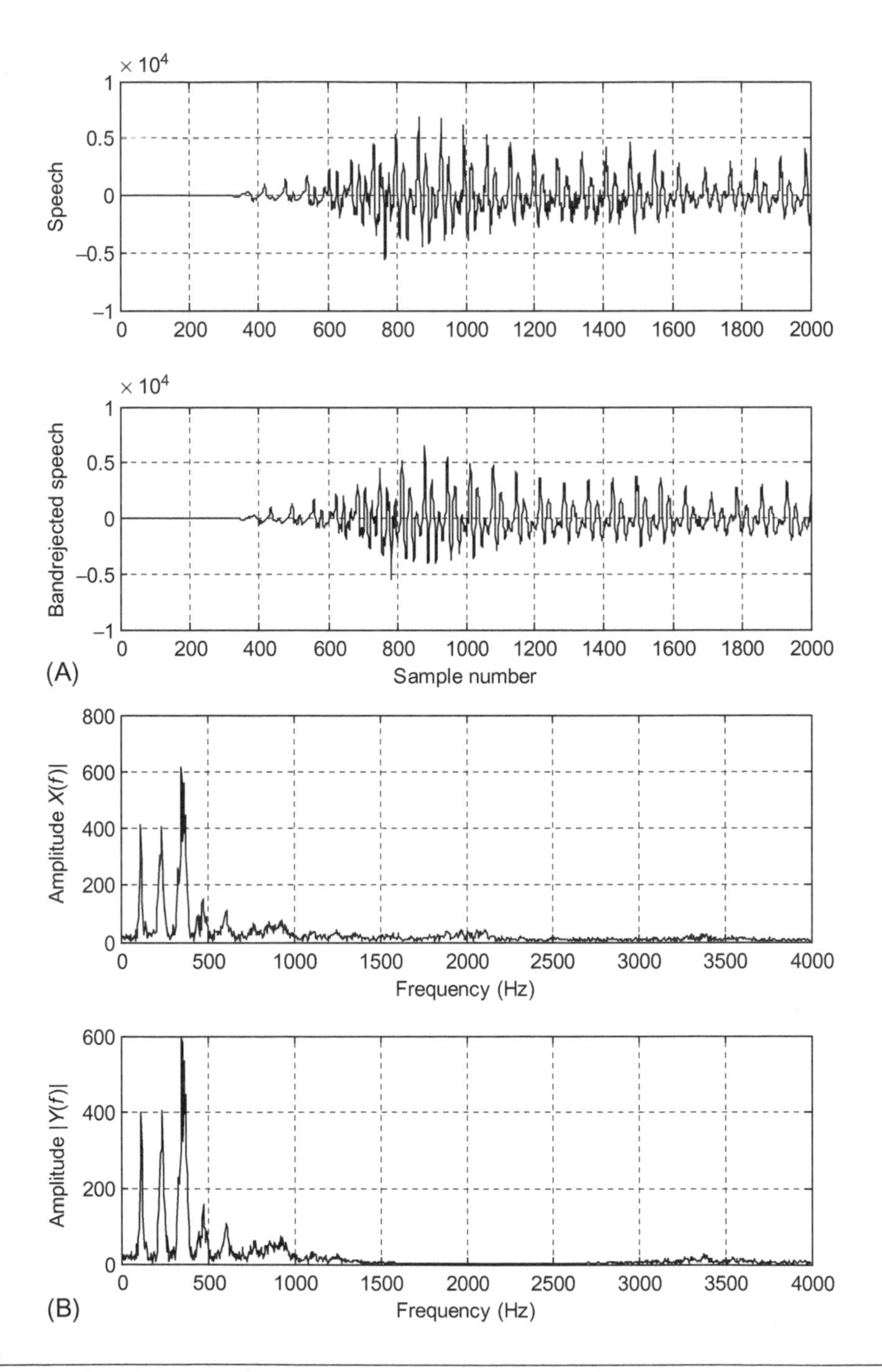

FIG. 7.21

(A) Original speech and processed speech using the bandstop filter. (B) Spectral comparison of the original speech and processed speech using the bandstop filter.

7.4 APPLICATIONS: NOISE REDUCTION AND TWO-BAND DIGITAL CROSSOVER

In this section, we will investigate the noise reduction and digital crossover design using the FIR filters.

7.4.1 NOISE REDUCTION

One of the key digital signal processing (DSP) applications is noise reduction. In this application, a digital FIR filter removes noise in a signal that is contaminated by noise existing in a broad frequency range. For example, such noise often appears during the data acquisition process. In real-world applications, the desired signal usually occupies a certain frequency range. We can design a digital filter to remove frequency components other than the desired frequency range.

In a data acquisition system, we record a 500-Hz sine wave at a sampling rate of 8000 samples per second. The signal is corrupted by broadband noise $v(n)$:

$$x(n) = 1.4141 \times \sin\left(\frac{2\pi \times 500n}{8000}\right) + v(n).$$

The 500-Hz signal with noise and its spectrum are plotted in Fig. 7.22, from which it is obvious that the digital sine wave contains noise. The spectrum is also displayed to give better understanding of the

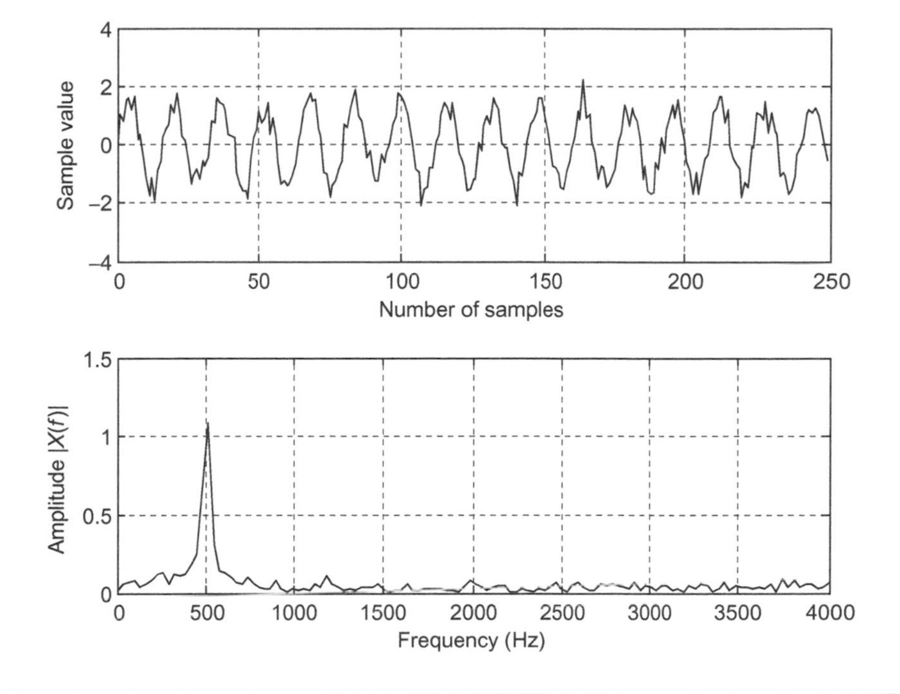

FIG. 7.22

Signal with noise and its spectrum.

noise frequency level. We can see that noise is broadband, existing from 0 Hz to the folding frequency of 4000 Hz. Assuming that the desired signal has a frequency range of only 0–800 Hz, we can filter noise from 800 Hz and beyond. A lowpass filter would complete such a task. Then we develop the filter specifications:

Passband frequency range: 0–800 Hz with the passband ripple less than 0.02 dB.

Stopband frequency range: 1–4 kHz with 50 dB attenuation.

As we will see, lowpass filtering will remove the noise ranging from 1000 to 4000 Hz, and hence the signal quality is improved.

Based on the specifications, we design an FIR filter with the Hamming window, a cutoff frequency of 900 Hz, and an estimated filter length of 133 taps using Table 7.7. The enhanced signal is depicted in Fig. 7.23, where the clean signal can be observed. The amplitude spectrum for the enhanced signal is also plotted. As shown in the spectral plot, the noise level is almost neglected between 1 and 4 kHz. Note that since we use the higher-order FIR filter, the signal experiences a linear phase delay of 66 samples, as is expected. We also see the transient response effects in this example. However, the transient response effects will be ended totally after first 132 samples due to the length of the FIR filter. Also shown in Fig. 7.23, in the frequency range between 400 and 700 Hz, there are shoulders in the spectral display. This is due to the fact that the noise in the passband cannot be removed. MATLAB implementation is given in Program 7.7.

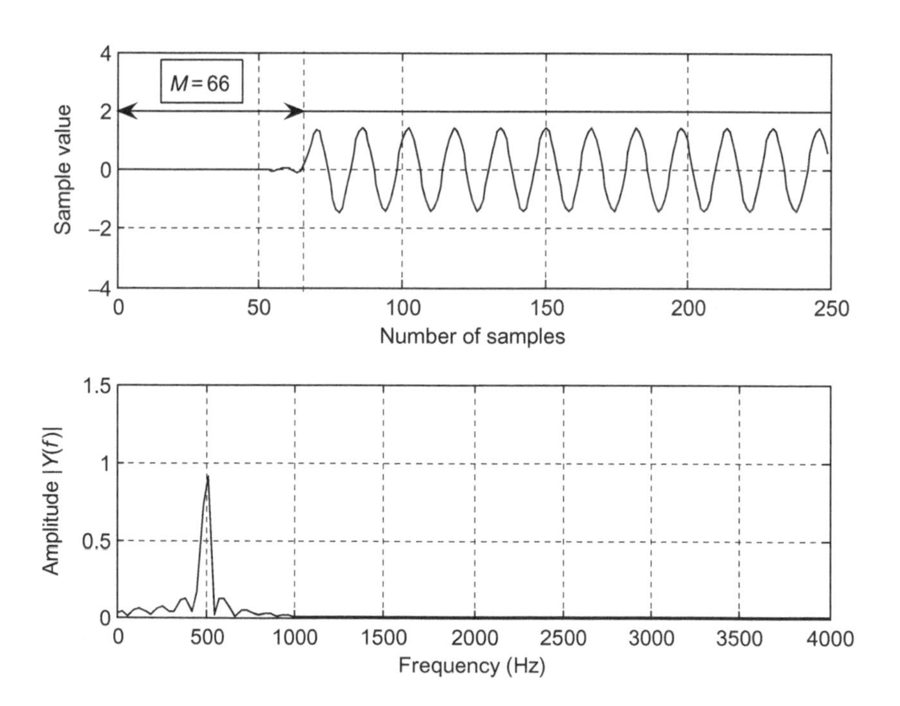

FIG. 7.23

The noise-removed clean signal and spectrum.

Program 7.7. MATLAB program for the application of noise filtering.

```
close all; clear all
fs=8000;                                 % Sampling rate
T=1/fs;                                  % Sampling period
v=sqrt(0.1)*randn(1,250);                % Generate Gaussian random noise
n=0:1:249;                               % Indexes
x=sqrt(2)*sin(2*pi*500*n*T)+v;           % Generate 500-Hz sinusoid plus noise
subplot(2,1,1);plot(n,x);
xlabel('Number of samples');ylabel('Sample value');grid;
N=length(x);
f=[0:N/2]*fs/N;
Axk=2*abs(fft(x))/N;Axk(1)=Axk(1)/2;     % calculate single side spectrum
for x(n)
subplot(2,1,2); plot(f,Axk(1:N/2+1));
xlabel('Frequency (Hz)'); ylabel('Amplitude |X(f )|');grid;
figure
Wnc=2*pi*900/fs;                 % determine the normalized digital cutoff frequency
B=firwd(133,1,Wnc,0,4);          % design FIR filter
y=filter(B,1,x);                 % perform digital filtering
Ayk=2*abs(fft(y))/N;Ayk(1)=Ayk(1)/2; % single-side spectrum of the filtered data
subplot(2,1,1); plot(n,y);
xlabel('Number of samples');ylabel('Sample value');grid;
subplot(2,1,2);plot(f,Ayk(1:N/2+1)); axis([0 fs/2 0 1.5]);
xlabel('Frequency (Hz)'); ylabel('Amplitude |Y(f )|');grid;
```

7.4.2 SPEECH NOISE REDUCTION

In a speech recording system, we digitally record speech in a noisy environment at a sampling rate of 8000 Hz. Assuming that the recorded speech contains information within 1800 Hz, we can design a lowpass filter to remove the noise between 1800 Hz and the Nyquist limit (the folding frequency of 4000 Hz). Therefore, we have the filter specifications listed below:

Filter type: lowpass FIR filter
Passband frequency range: 0–1800 Hz
Passband ripple: 0.02 dB
Stopband frequency range: 2000–4000 Hz
Stopband attenuation: 50 dB

According to these specifications, we can determine the following parameters for filter design:

Window type = Hamming window
Number of filter tap = 133
Lowpass cutoff frequency = 1900 Hz.

Fig. 7.24A shows the plots of the recorded noisy speech and its spectrum. As we can see in the noisy spectrum, the noise level is high and broadband. After applying the designed lowpass filter, we plot the filtered speech and its spectrum shown in Fig. 7.24B, where the clean speech is clearly identified, while the spectrum shows that the noise components above 2 kHz have been completely removed.

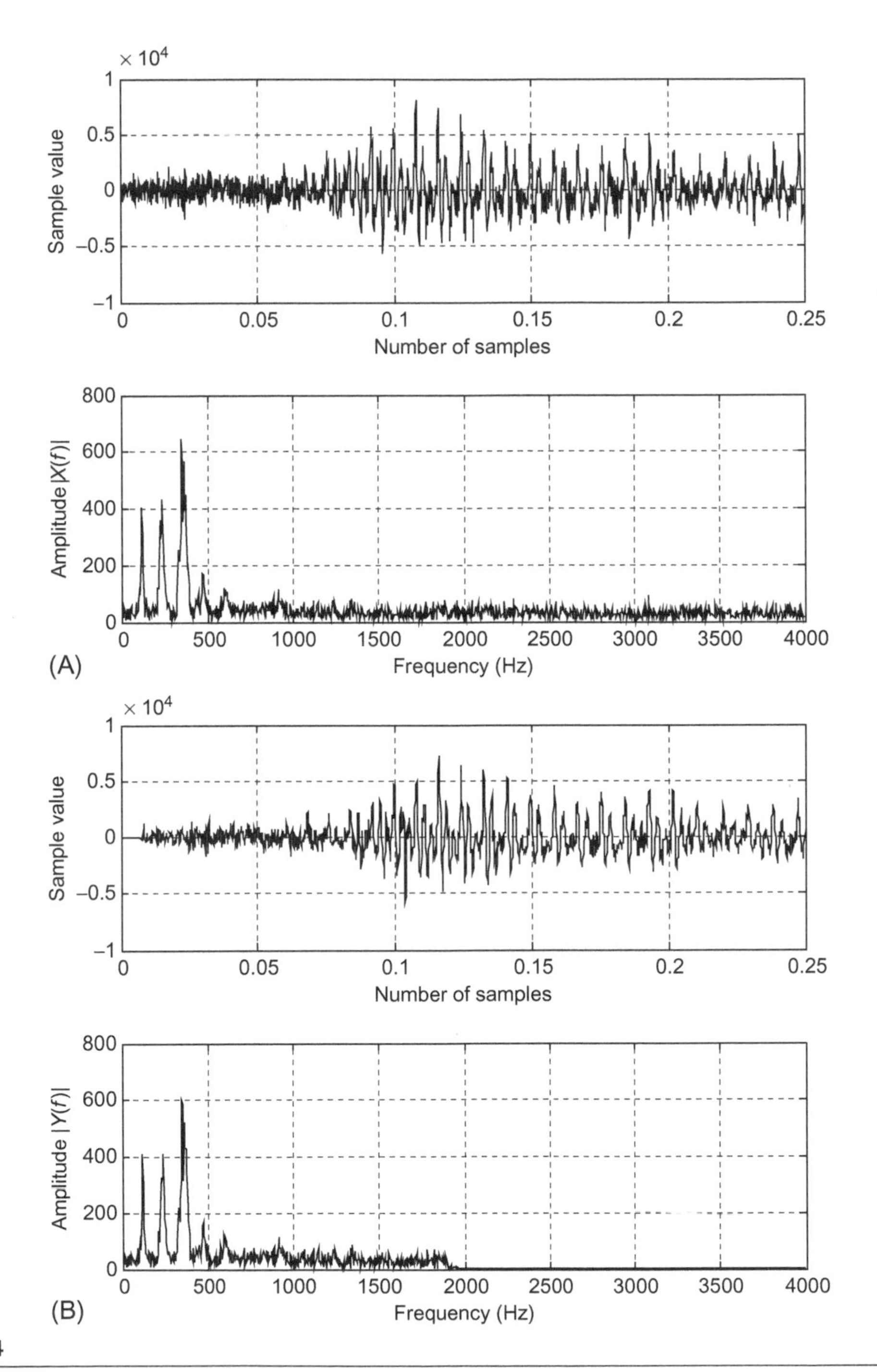

FIG. 7.24

(A) Noisy speech and its spectrum. (B) Enhanced speech and its spectrum.

7.4.3 NOISE REDUCTION IN VIBRATION SIGNAL

In a data acquisition system for vibration analysis, a vibration signal is captured using the accelerometer sensor in the noisy environment. The sampling rate is 1000 Hz. The captured signal is significantly corrupted by a broadband noise. In vibration analysis, the first dominant frequency component in the range from 35 to 50 Hz is required to be retrieved. We list the filter specifications below:

Filter type = bandpass FIR filter
Passband frequency range: 35–50 Hz
Passband ripple: 0.02 dB
Stopband frequency ranges: 0–15 and 70–500 Hz
Stopband attenuation: 50 dB.

According to these specifications, we can determine the following parameters for filter design:

Window type = Hamming window
Number of filter tap = 167
Low cutoff frequency = 25 Hz
High cutoff frequency = 60 Hz.

Fig. 7.25 displays the plots of the recorded noisy vibration signal and its spectrum. Fig. 7.26 shows the retrieved vibration signal with noise reduction by a bandpass filter.

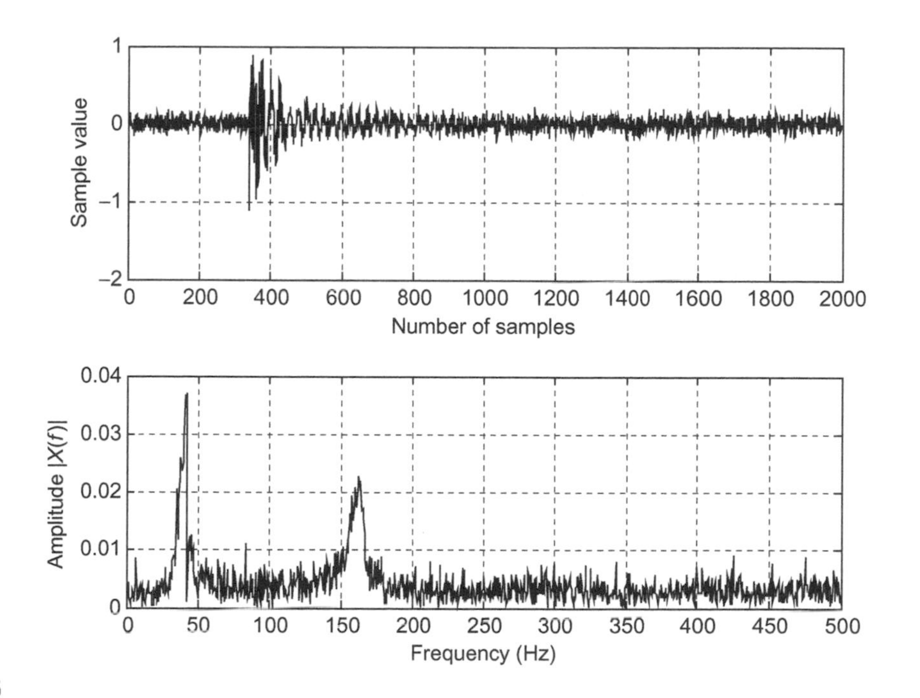

FIG. 7.25

Noisy vibration signal and its spectrum.

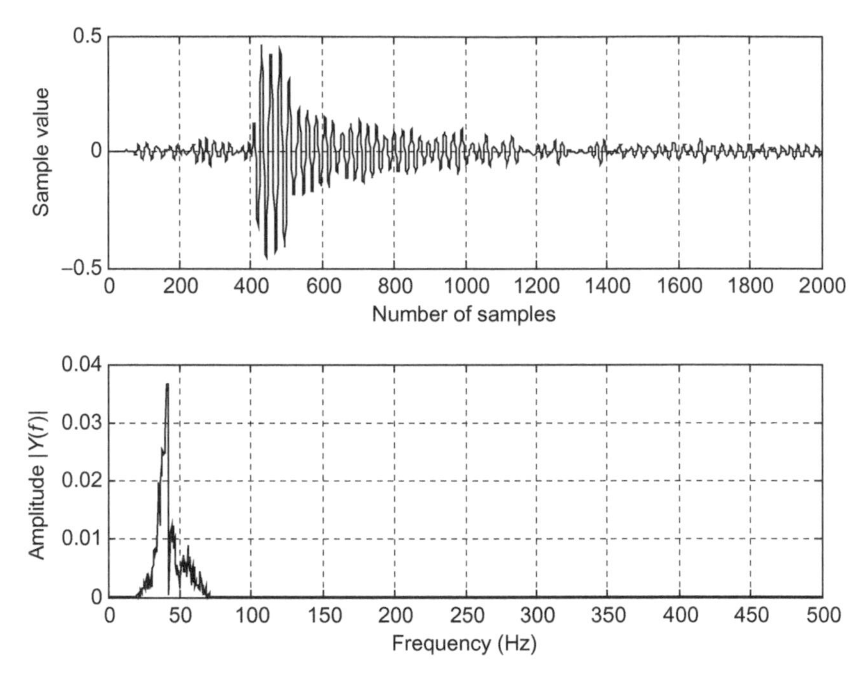

FIG. 7.26

Retrieved vibration signal and its spectrum.

7.4.4 TWO-BAND DIGITAL CROSSOVER

In audio systems, there is often a situation where the application requires the entire audible range of frequencies, but this is beyond the capability of any single speaker driver. So, we combine several drivers, such as the speaker cone and horns, each covering different frequency range, to reproduce the full audio frequency range.

A typical two-band digital crossover can be designed as shown in Fig. 7.27. There are two speaker drivers. The woofer responds to low frequencies, and the tweeter responds to high frequencies.

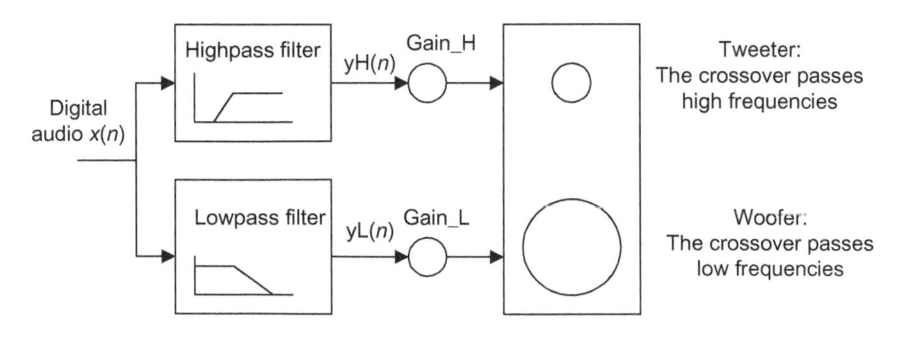

FIG. 7.27

Two-band digital crossover.

The incoming digital audio signal is split into two bands by using a lowpass filter and a highpass filter in parallel. We then amplify the separated audio signals and send them to their respective corresponding speaker drivers. Hence, the objective is to design the lowpass filter and the highpass filter so that their combined frequency response is flat, while keeping transition as sharp as possible to prevent audio signal distortion in the transition frequency range. Although traditional crossover systems are designed using active circuits (analog systems) or passive circuits, the digital crossover system provides a cost-effective solution with programmable ability, flexibility, and high quality.

A crossover system has the following specifications:

Sampling rate = 44,100 Hz

Crossover frequency = 1000 Hz (cutoff frequency)

Transition band = 600 Hz to1400 Hz

Lowpass filter = passband frequency range from 0 to 600 Hz with a ripple of 0.02 dB and stopband edge at 1400 Hz with the attenuation of 50 dB.

Highpass filter = passband frequency range from 1.4 to 44.1 kHz with ripple of 0.02 dB and stopband edge at 600 Hz with the attenuation of 50 dB.

In the design of this crossover system, one possibility is to use an FIR filter, since it provides a linear phase for the audio system. However, an IIR filter (which will be discussed in Chapter 8) can be an alternative. Based on the transition band of 800 Hz and the passband ripple and stopband attenuation requirements, the Hamming window is chosen for both lowpass and highpass filters, we can determine the number of filter taps as 183, each with a cutoff frequency of 1000 Hz.

The frequency responses for the designed lowpass filter and highpass filter are shown in Fig. 7.28A, and for the lowpass filter, highpass filter, and combined responses appear in Fig. 7.28B. As we can see, the crossover frequency for both filters is at 1000 Hz, and the combined frequency response is perfectly flat. The impulse responses (filter coefficients) for lowpass and highpass filters are plotted in Fig. 7.28C.

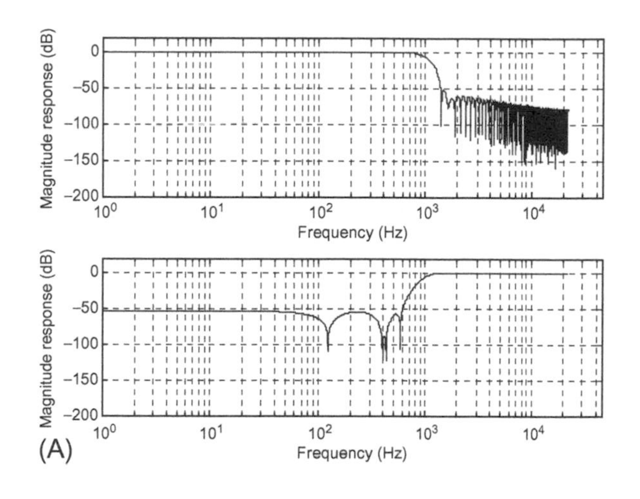

FIG. 7.28

(A) Magnitude frequency responses for the lowpass filter and highpass filter.

(Continued)

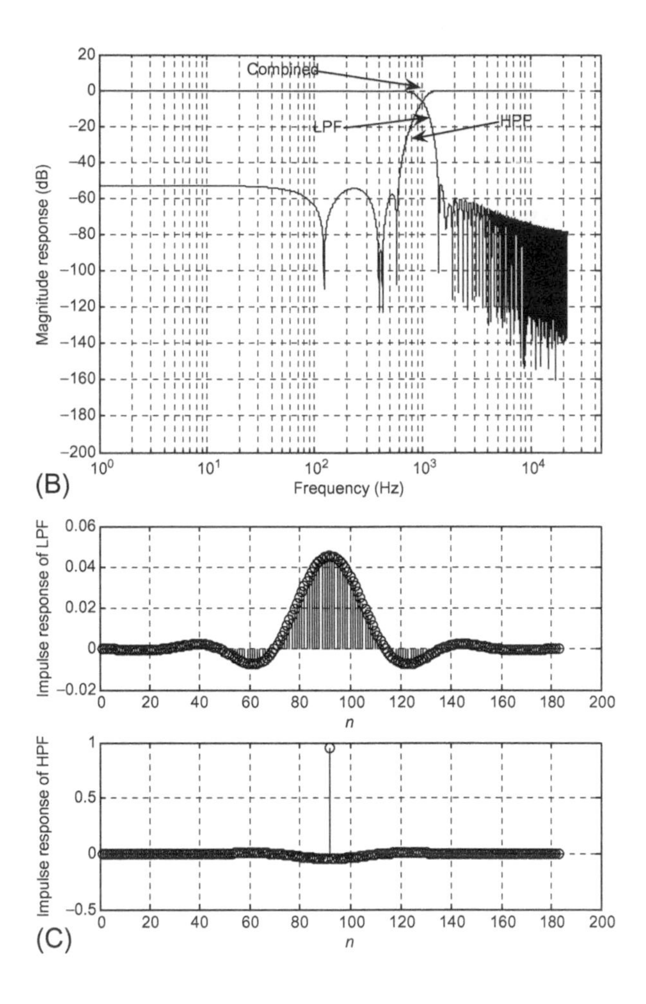

FIG. 7.28, CONT'D

(B) Magnitude frequency responses for both the lowpass filter and highpass filter, and the combined magnitude frequency response for the digital audio crossover system. (C) Impulse responses of both the FIR lowpass filter and the FIR highpass filter for the digital audio crossover system.

7.5 FREQUENCY SAMPLING DESIGN METHOD

In addition to methods of Fourier transform design and Fourier transform with windowing discussed in the previous section, *frequency sampling* is another alternative. The key feature of frequency sampling is that the filter coefficients can be calculated based on the specified magnitudes of the desired filter frequency response uniformly sampled in the frequency domain. Hence, it has design flexibility.

To begin with development, we let $h(n)$, for $n=0, 1, \cdots, N-1$, be the causal impulse response (FIR filter coefficients) that approximates the FIR filter, and we let $H(k)$, for $k=0, 1, \cdots, N-1$, represent the corresponding discrete Fourier transform (DFT) coefficients. We obtain $H(k)$ by sampling the desired

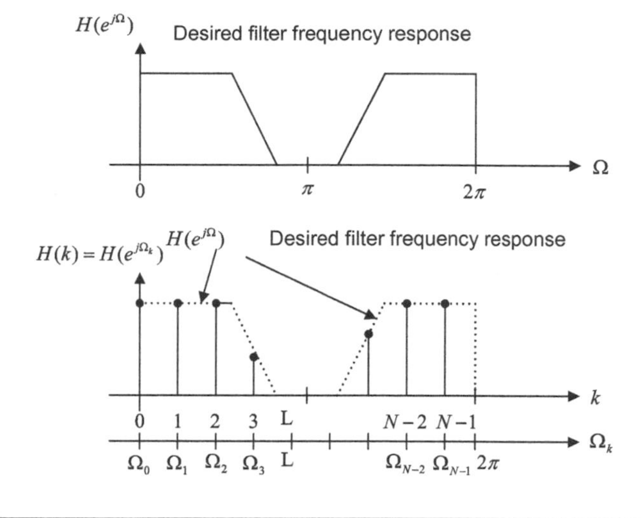

FIG. 7.29

Desired filter frequency response and sampled frequency response.

frequency filter response $H(k)=H(e^{j\Omega_k})$ at equally spaced instants in frequency domain, as shown in Fig. 7.29.

Consider the FIR filter transfer function:

$$H(z)=h(0)+h(1)z^{-1}+\cdots+h(N-1)z^{-(N-1)}. \tag{7.35}$$

Its frequency response is then given by

$$H\left(e^{j\Omega}\right)=h(0)+h(1)e^{-j1\Omega}+\cdots+h(N-1)e^{-j(N-1)\Omega}. \tag{7.36}$$

Sampling $H(e^{j\Omega})$ at equally spaced *instants* $\Omega_k=2\pi k/N$ for $k=0, 1, \cdots, N-1$, as shown in Fig. 7.29, we have

$$H\left(e^{j\Omega_k}\right)=h(0)+h(1)e^{-j1\times 2\pi k/N}+\cdots+h(N-1)e^{-j(N-1)\times 2\pi k/N}. \tag{7.37}$$

Using the definition of DFT, it is easy to verify the expansion of Eq. (7.37) as follows:

for $k=0$:

$$H\left(e^{j\Omega_0}\right)=h(0)+h(1)e^{-j1\times 2\pi 0/N}+\cdots+h(N-1)e^{-j(N-1)\times 2\pi 0/N}=\sum_{n=0}^{N-1}h(n)W_N^{0\times n}=H(0),$$

for $k=1$:

$$H\left(e^{j\Omega_1}\right)=h(0)+h(1)e^{-j1\times 2\pi 1/N}+\cdots+h(N-1)e^{-j(N-1)\times 2\pi 1/N}=\sum_{n=0}^{N-1}h(n)W_N^{1\times n}=H(1),$$

$$\cdots$$

for $k=N-1$:

$$H\left(e^{j\Omega_{N-1}}\right)=h(0)+h(1)e^{-j1\times 2\pi (N-1)/N}+\cdots+h(N-1)e^{-j(N-1)\times 2\pi (N-1)/N}=\sum_{n=0}^{N-1}h(n)W_N^{k\times (N-1)}=H(N-1).$$

It is observed that

$$H\left(e^{j\Omega_k}\right) = H(k) = \mathrm{DFT}\{h(n)\}. \tag{7.38}$$

Thus, the general equation to obtain $h(n)$ is

$$h(n) = \mathrm{IDFT}\left\{H\left(e^{j\Omega_k}\right)\right\}. \tag{7.39}$$

Note that the obtained formula (see Eq. 7.39) does not impose the linear phase, thus it can be used for nonlinear phase FIR filter design. Second, the sequence $h(n)$ is guaranteed to be real valued.

To simplify the design algorithm in Eq. (7.39), we begin with

$$H(k) = \sum_{n=0}^{N-1} h(n) W_N^{nk}.$$

Note that

$$H(N-k) = \sum_{n=0}^{N-1} h(n) W_N^{n(N-k)} = \sum_{n=0}^{N-1} h(n) W_N^{-nk} = \left(\sum_{n=0}^{N-1} h(n) W_N^{nk}\right)^* = \overline{H}(k).$$

Then

$$\begin{aligned} h(n) &= \frac{1}{N}\sum_{k=0}^{N-1} H(k) W_N^{-kn} \\ &= \frac{1}{N}\left(H(0) + \sum_{k=1}^{M} H(k) W_N^{-kn} + \sum_{k=M+1}^{2M} H(k) W_N^{-kn}\right). \end{aligned}$$

For the second summation, let $j = N - k = 2M + 1 - k$. We yield

$$\begin{aligned} h(n) &= \frac{1}{N}\left(H(0) + \sum_{k=1}^{M} H(k) W_N^{-kn} + \sum_{j=M}^{1} H(N-j) W_N^{-(N-j)n}\right) \\ &= \frac{1}{N}\left(H(0) + \sum_{k=1}^{M} H(k) W_N^{-kn} + \sum_{k=1}^{M} \overline{H}(k) W_N^{-(N-k)n}\right). \end{aligned}$$

Furthermore,

$$h(n) = \frac{1}{N}\left(H(0) + \sum_{k=1}^{M} H(k) W_N^{-kn} + \left(\sum_{k=1}^{M} H(k) W_N^{-kn}\right)^*\right). \tag{7.40}$$

Combining two summations in Eq. (7.40) yields the design equation as

$$h(n) = \frac{1}{N}\left(H(0) + 2\,\mathrm{Re}\left\{\sum_{k=1}^{M} H(k) W_N^{-kn}\right\}\right). \tag{7.41}$$

Now, let us consider the linear phase FIR filter design in Eq. (7.41) with $N = 2M + 1$ and symmetric coefficients

$$H\left(e^{j\Omega}\right) = h(0) + h(1)e^{-j1\Omega} + \cdots + h(M)e^{-jM\Omega} + h(M-1)e^{-j(M+1)\Omega} + \cdots + h(0)e^{-j2M\Omega}. \tag{7.42a}$$

We can express Eq. (7.42a) as

$$H\left(e^{j\Omega}\right) = e^{-jM\Omega}\left[h(M) + \sum_{n=1}^{M-1} 2h(n)\cos\left[2\pi(M-n)\Omega\right]\right] \tag{7.42b}$$

Since the term in the bracket in Eq. (7.42b) is real valued (magnitude frequency response), we can specify the frequency response as

$$H\left(e^{j\Omega_k}\right) = H_k e^{-jM\frac{2\pi k}{N}} = H(k) \tag{7.43}$$

Clearly, $H(e^{j\Omega_k}) = H_k e^{-jk\frac{2\pi}{N}M} = H(k)$ for $k=0, 1, 2, \cdots, M$ holds linear phase.

Now, on substituting the magnitude and its corresponding linear phase in Eq. (7.43), Eq. (7.41) leads to

$$\begin{aligned} h(n) &= \frac{1}{N}\left(H_0 + 2\mathrm{Re}\left\{\sum_{k=1}^{M} H_k e^{-j\frac{2\pi}{N}kM} W_N^{-kn}\right\}\right) \\ &= \frac{1}{N}\left(H_0 + 2\mathrm{Re}\left\{\sum_{k=1}^{M} H_k e^{-j\frac{2\pi}{N}kM} e^{j\frac{2\pi}{N}kn}\right\}\right) \\ &= \frac{1}{N}\left(H_0 + 2\mathrm{Re}\left\{\sum_{k=1}^{M} H_k e^{j\frac{2\pi}{N}k(n-M)}\right\}\right). \end{aligned} \tag{7.44}$$

Finally, we obtain the design formula for FIR filter with linear phase (only magnitudes specification are required) as

$$h(n) = \frac{1}{N}\left(H_0 + 2\sum_{k=1}^{M} H_k \cos\left(\frac{2\pi k(n-M)}{2M+1}\right)\right) \text{ for } 0 \le n \le M. \tag{7.45}$$

The rigor derivation can be found in Appendix E. The design procedure is therefore simply summarized as follows:

1. Given the filter length of $2M+1$, specify the magnitude frequency response for the normalized frequency range from 0 to π:

$$H_k \text{ at } \Omega_k = \frac{2\pi k}{(2M+1)} \quad \text{for} \quad k = 0, 1, \cdots, M. \tag{7.46}$$

2. Calculate FIR filter coefficients:

$$b_n = h(n) = \frac{1}{2M+1}\left\{H_0 + 2\sum_{k=1}^{M} H_k \cos\left(\frac{2\pi k(n-M)}{2M+1}\right)\right\} \tag{7.47}$$

for $n=0, 1, \cdots, M$.

3. Using the symmetry (linear phase requirement) to determine the rest of coefficients:

$$h(n) = h(2M-n) \quad \text{for} \quad n = M+1, \cdots, 2M. \tag{7.48}$$

Example 7.12 illustrates the design procedure.

EXAMPLE 7.12

Design a linear phase lowpass FIR filter with seven taps and a cutoff frequency of $\Omega_c = 0.3\pi$ (rad) using the frequency sampling method.

Solution:

Since $N = 2M + 1 = 7$ and $M = 3$, the sampled frequencies are given by

$$\Omega_k = \frac{2\pi}{7}k \text{ (rad)}, \quad k = 0, 1, 2, 3.$$

Next we specify the magnitude values H_k at the specified frequencies as follows:

$$\text{for } \Omega_0 = 0 \text{ (rad)}, H_0 = 1.0$$

$$\text{for } \Omega_1 = \frac{2}{7}\pi \text{ (rad)}, H_1 = 1.0$$

$$\text{for } \Omega_2 = \frac{4}{7}\pi \text{ (rad)}, H_2 = 0.0$$

$$\text{for } \Omega_3 = \frac{6}{7}\pi \text{ (rad)}, H_3 = 0.0.$$

Fig. 7.30 shows the specifications.

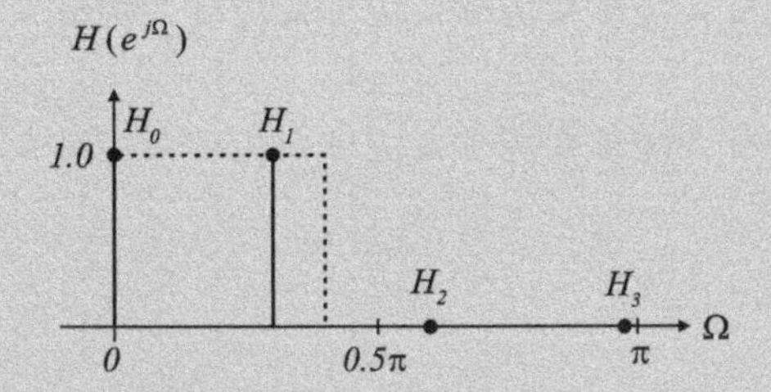

FIG. 7.30

Sampled values of the frequency response in Example 7.12.

Using Eq. (7.47), we achieve

$$h(n) = \frac{1}{7}\left\{1 + 2\sum_{k=1}^{3} H_k \cos\left[\frac{2\pi k(n-3)}{7}\right]\right\} \quad , n = 0, 1, 2, 3.$$

$$= \frac{1}{7}\left\{1 + 2\cos\left[\frac{2\pi(n-3)}{7}\right]\right\}$$

Thus, computing the FIR filter coefficients yields

$$h(0)=\frac{1}{7}\left\{1+2\cos\left(\frac{-6\pi}{7}\right)\right\}=-0.11456$$
$$h(1)=\frac{1}{7}\left\{1+2\cos\left(\frac{-4\pi}{7}\right)\right\}=0.07928$$
$$h(2)=\frac{1}{7}\left\{1+2\cos\left(\frac{-2\pi}{7}\right)\right\}=0.32100$$
$$h(3)=\frac{1}{7}\left\{1+2\cos\left(\frac{-0\times\pi}{7}\right)\right\}=0.42857.$$

By symmetry, we obtain the rest of the coefficients as follows:

$$h(4)=h(2)=0.32100,$$
$$h(5)=h(1)=0.07928,$$
$$h(6)=h(0)=-0.11456.$$

The following two examples are devoted to illustrate the FIR filter design using the frequency sampling method. A MATLAB program, **firfs(N,Hk),** is provided in the "MATLAB Programs" section at the end of this chapter (see its usage in Table 7.11) to implement the design in Eq. (7.47) with the input parameters of $N=2M+1$ (number of taps) and a vector Hk containing the specified magnitude values H_k, $k=0, 1, \ldots, M$. Finally, the MATLAB function will return the calculated FIR filter coefficients.

Table 7.11 Illustrative Usage for MATLAB Function firfs(N,Hk)

```
Function B=firfs(N,Hk)
% B=firfs(N,Hk)
% Fir filter design using the frequency sampling method.
% Input parameters:
% N: the number of filter coefficients.
% note: N must be odd number.
% Hk: sampled frequency response for k=0,1,2,...,M=(N-1)/2.
% Output:
% B: FIR filter coefficients.
```

EXAMPLE 7.13

(a) Design a linear phase lowpass FIR filter with 25 coefficients using the frequency sampling method. Let the cutoff frequency be 2000 Hz and assume a sampling frequency of 8000 Hz.
(b) Plot the magnitude and phase frequency responses.
(c) List FIR filter coefficients.

Solution:

(a) The normalized cutoff frequency for the lowpass filter is $\Omega_c = \omega T = 2\pi \times 2000/8000 = 0.5\pi$(rad), $N = 2M + 1 = 25$, and the specified values of the sampled frequency response are chosen to be

$$H_k = [1111111000000].$$

The MATLAB Program 7.8 produces the design results.

Program 7.8. MATLAB program for Example 7.13.

```
% Fig. 7.31 (Example 7.13)
% MATLAB program to create Fig. 7.31
fs=8000;                                % sampling frequency
H1=[1 1 1 1 1 1 1 0 0 0 0 0 0];         % magnitude specifications
B1=firfs(25,H1);                        % design filter
[h1,f]=freqz(B1,1,512,fs);              % calculate magnitude frequency response
H2=[1 1 1 1 1 1 1 0.5 0 0 0 0 0];       % magnitude specifications
B2=firfs(25,H2);                        % Design filter
[h2,f]=freqz(B2,1,512,fs);              % calculate magnitude frequency response
p1=180*unwrap(angle(h1))/pi;
p2=180*unwrap(angle(h2))/pi
subplot(2,1,1); plot(f,20*log10(abs(h1)),'-.',f,20*log10(abs(h2)));grid
axis([0 fs/2 -80 10]);
xlabel('Frequency (Hz)'); ylabel('Magnitude Response (dB)');
subplot(2,1,2); plot(f,p1,'-.',f,p2);grid
xlabel('Frequency (Hz)'); ylabel('Phase (degrees)');
```

(b) The magnitude frequency response plotted using the dash-dotted line is displayed in Fig. 7.31, where it is observed that oscillations (shown as the dash-dotted line) occur in the passband and stopband of the designed FIR filter. This is due to the abrupt change of the specification in transition band (between the passband and the stopband). To reduce this ripple effect, the modified specification with a smooth transition band, H_k, $k = 0, 1, \cdots, 13$, is used

$$H_k = [11111110.500000].$$

The improved magnitude frequency response is shown in Fig. 7.31 via the solid line.

(c) The calculated FIR coefficients for both filters are listed in Table 7.12.

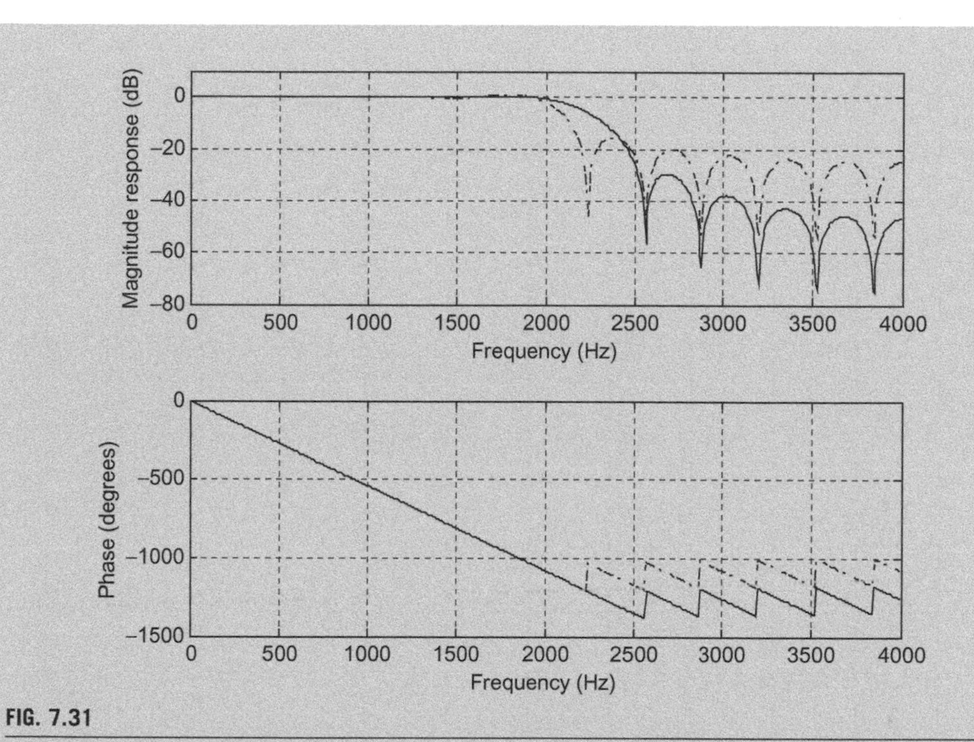

FIG. 7.31

Frequency responses using the frequency sampling method in Example 7.13.

Table 7.12 FIR Filter Coefficients in Example 7.13 (Frequency Sampling Method)

B1: FIR Filter Coefficients	B2: FIR Filter Coefficients
$b_0=b_{24}=0.027436$	$b_0=b_{24}=0.001939$
$b_1=b_{23}=-0.031376$	$b_1=b_{23}=0.003676$
$b_2=b_{22}=-0.024721$	$b_2=b_{22}=-0.012361$
$b_3=b_{21}=0.037326$	$b_3=b_{21}=-0.002359$
$b_4=b_{20}=0.022823$	$b_4=b_{20}=0.025335$
$b_5=b_{19}=-0.046973$	$b_5=b_{19}=-0.008229$
$b_6=b_{18}=-0.021511$	$b_6=b_{18}=-0.038542$
$b_7=b_{17}=0.064721$	$b_7=b_{17}=0.032361$
$b_8=b_{16}=0.020649$	$b_8=b_{16}=0.049808$
$b_9=b_{15}=-0.106734$	$b_9=b_{15}=-0.085301$
$b_{10}=b_{14}=-0.020159$	$b_{10}=b_{14}=-0.057350$
$b_{11}=b_{13}=0.318519$	$b_{11}=b_{13}=0.311024$
$b_{12}=0.520000$	$b_{12}=0.560000$

EXAMPLE 7.14

(a) Design a linear phase bandpass FIR filter with 25 coefficients using the frequency sampling method. Let the lower and upper cutoff frequencies be 1000 and 3000 Hz, respectively, and assume a sampling frequency of 8000 Hz.
(b) List the FIR filter coefficients.
(c) Plot the frequency responses.

Solution:

(a) First we calculate the normalized lower and upper cutoff frequencies for the bandpass filter; that is, $\Omega_L = 2\pi \times 1000/8000 = 0.25\pi$ (rad) and $\Omega_H = 2\pi \times 3000/8000 = 0.75\pi$ (rad), respectively. The sampled values of the bandpass frequency response are specified by the following vector:

$$H_k = [0\,0\,0\,0\,1\,1\,1\,1\,1\,0\,0\,0\,0].$$

As a comparison, the second specification of H_k with smooth transition bands is used; that is,

$$H_k = [0\,0\,0\,0.5\,1\,1\,1\,1\,1\,0.5\,0\,0\,0].$$

(b) The MATLAB list is shown in Program 7.9. The generated FIR coefficients are listed in Table 7.13.

Table 7.13 FIR Filter Coefficients in Example 7.14 (Frequency Sampling Method)

B1: FIR Filter Coefficients	B2: FIR Filter Coefficients
$b_0 = b_{24} = 0.055573$	$b_0 = b_{24} = 0.001351$
$b_1 = b_{23} = -0.030514$	$b_1 = b_{23} = -0.008802$
$b_2 = b_{22} = 0.000000$	$b_2 = b_{22} = -0.020000$
$b_3 = b_{21} = -0.027846$	$b_3 = b_{21} = 0.009718$
$b_4 = b_{20} = -0.078966$	$b_4 = b_{20} = -0.011064$
$b_5 = b_{19} = 0.042044$	$b_5 = b_{19} = 0.023792$
$b_6 = b_{18} = 0.063868$	$b_6 = b_{18} = 0.077806$
$b_7 = b_{17} = 0.000000$	$b_7 = b_{17} = -0.020000$
$b_8 = b_{16} = 0.094541$	$b_8 = b_{16} = 0.017665$
$b_9 = b_{15} = -0.038728$	$b_9 = b_{15} = -0.029173$
$b_{10} = b_{14} = -0.303529$	$b_{10} = b_{14} = -0.308513$
$b_{11} = b_{13} = 0.023558$	$b_{11} = b_{13} = 0.027220$
$b_{12} = 0.400000$	$b_{12} = 0.480000$

Program 7.9 MATLAB program for Example 7.14.

```
% Fig. 7.32 (Example 7.14)
% MATLAB program to create Fig. 7.32
%
fs=8000;
H1=[0 0 0 0 1 1 1 1 1 0 0 0 0];              % magnitude specifications
B1=firfs(25,H1);                             % design filter
[h1,w]=freqz(B1,1,512);                      % calculate magnitude frequency response
H2=[0 0 0 0.5 1 1 1 1 1 0.5 0 0 0];          % magnitude specification
B2=firfs(25,H2);                             % design filter
[h2,w]=freqz(B2,1,512);                      % calculate magnitude frequency response
p1=180*unwrap(angle(h1)')/pi;
p2=180*unwrap(angle(h2)')/pi
subplot(2,1,1); plot(f,20*log10(abs(h1)),'-.',f,20*log10(abs(h2)));grid
axis([0 fs/2 -100 10]);
xlabel('Frequency (Hz)'); ylabel('Magnitude Response (dB)');
subplot(2,1,2); plot(f,p1,'-.',f,p2);grid
xlabel('Frequency (Hz)'); ylabel('Phase (degrees)');
```

(c) Similar to the preceding example, Fig. 7.32 shows the frequency responses. Focusing on the magnitude frequency responses depicted in Fig. 7.32, the dash-dotted line indicates the magnitude frequency response obtained without specifying the smooth transition band, while the solid line indicates the magnitude frequency response achieved with the specification of the smooth transition band, hence resulting in the reduced ripple effect.

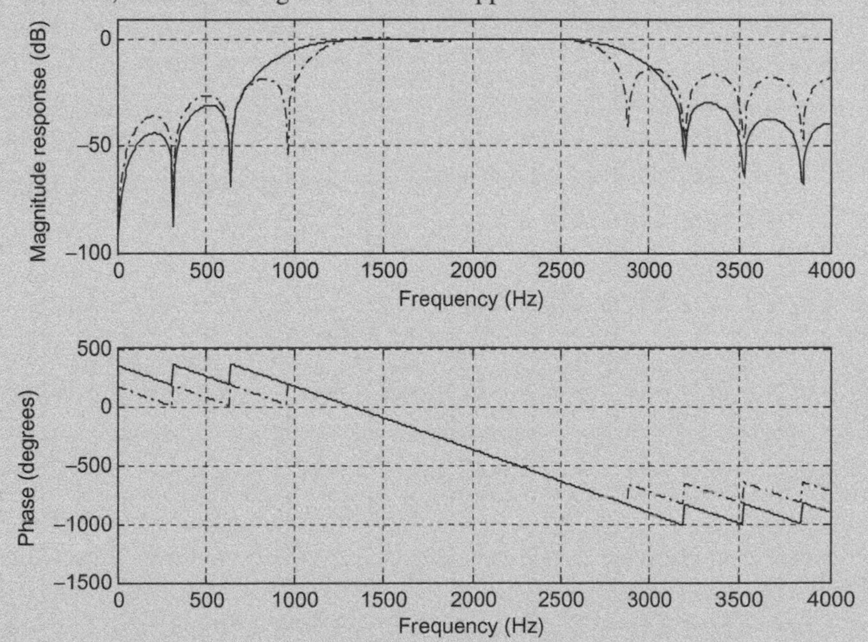

FIG. 7.32

Frequency responses using the frequency sampling method in Example 7.14.

Observations can be made by examining Examples 7.13 and 7.14. First, the oscillations (Gibbs behavior) in the passband and stopband can be reduced at the expense of increasing the width of the main lobe. Second, we can modify the specification of the magnitude frequency response with a smooth transition band to reduce the oscillations and hence to improve the performance of the FIR filter. Third, the magnitude values H_k, $k=0, 1, \ldots, M$, in general can be arbitrarily specified. This indicates that the frequency sampling method is more flexible and can be used to design the FIR filter with an arbitrary specification of the magnitude frequency response.

7.6 OPTIMAL DESIGN METHOD

This section introduces the Parks-McClellan algorithm, which is a popular optimal design method used in industry due to its efficiency and flexibility. The FIR filter design using the Parks-McClellan algorithm is developed based on the idea of minimizing the maximum approximation error in a Chebyshev polynomial approximation to the desired filter magnitude frequency response. The details of this design development are beyond the scope of this text and can be found in Ambardar (1999) and Porat (1997). We will outline the design criteria and notation and then focus on the design procedure.

Given an ideal magnitude response $H_d(e^{j\omega T})$, the approximation error $E(\omega)$ is defined as

$$E(\omega) = W(\omega)\left[H\left(e^{j\omega T}\right) - H_d\left(e^{j\omega T}\right)\right], \tag{7.49}$$

where $H(e^{j\omega T})$ is the frequency response of the linear phase FIR filter to be designed, and $W(\omega)$ is the weight function for emphasizing certain frequency band over others during the optimization process. The process is to minimize the error shown in Eq. (7.50):

$$\min\left(\max|E(\omega)|\right) \tag{7.50}$$

over the set of FIR coefficients. With the help of the Remez exchange algorithm, which is also beyond the scope of this book and can be found in Textbooks (Ambardar, 1999; Porat, 1997), we can obtain the best FIR filter whose magnitude response has an equiripple approximation to the ideal magnitude response. The achieved filters are optimal in the sense that the algorithms minimize the maximum error between the desired frequency response and actual frequency response. These are often called *minimax filters*.

Next, we establish notations that will be used in the design procedure. Fig. 7.33 shows the characteristics of the FIR filter designed by the Parks-McClellan and Remez exchange algorithms. As illustrated in the top graph of Fig. 7.33, the passband frequency response and stopband frequency response have equiripples. δ_p is used to specify the magnitude ripple in the passband, while δ_s specifies the stopband magnitude attenuation. In terms of dB value specification, we have $\delta_p dB = 20 \times \log_{10}(1+\delta_p)$ and $\delta_s dB = 20 \times \log_{10}\delta_s$.

The middle graph in Fig. 7.33 describes the error between the ideal frequency response and the actual frequency response. In general, the error magnitudes in the passband and stopband are different. This makes optimization unbalanced, since the optimization process involves an entire band. When the error magnitude in a band dominates the other(s), the optimization process may deemphasize the contribution due to a small magnitude error. To make the error magnitudes balanced, a weight function can

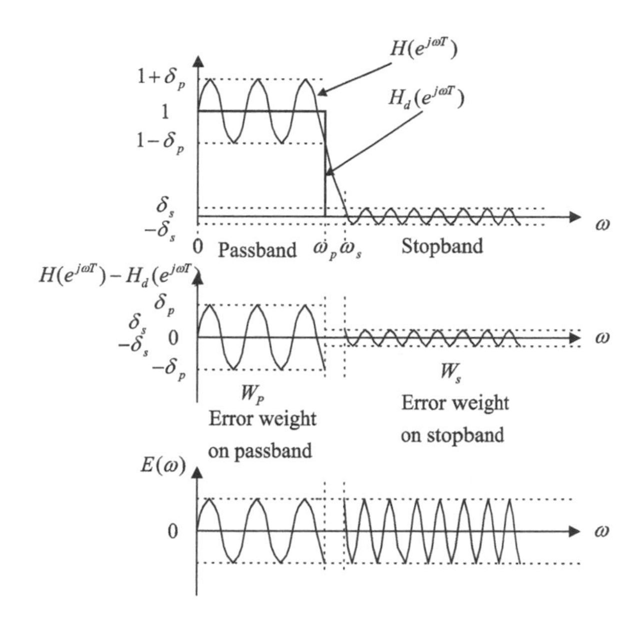

FIG. 7.33

(Top) Magnitude frequency response of an ideal lowpass filter and a typical lowpass filter designed using Parks-McClellan algorithm. (Middle) Error between the ideal and practical responses. (Bottom) Weighted error between the ideal and practical responses.

be introduced. The idea is to weight the band with a bigger magnitude error with a small weight factor and to weight the band with a smaller magnitude error with a big weight factor. We use a weight factor W_p for weighting the passband error and W_s for weighting the stopband error. The bottom graph in Fig. 7.33 shows the weighted error, and clearly, the error magnitudes on both bands are at the same level. Selection of the weighting factors is further illustrated in the following design procedure.

Optimal FIR Filter Design Procedure for Parks-McClellan Algorithm

1. Specify the band edge frequencies such as passband and stopband frequencies, passband ripple, stopband attenuation, filter order, and sampling frequency of the DSP system.
2. Normalize band edge frequencies to the Nyquist limit (folding frequency $=f_s/2$) and specify the ideal magnitudes.
3. Calculate absolute values of the passband ripple and stopband attenuation if they are given in terms of dB values:

$$\delta_p = 10^{\left(\frac{\delta_p dB}{20}\right)} - 1 \tag{7.51}$$

$$\delta_s = 10^{\left(\frac{\delta_s dB}{20}\right)}. \tag{7.52}$$

Then calculate the ratio and put it into fraction form:

$$\frac{\delta_p}{\delta_s} = \text{fraction form} = \frac{\text{numerator}}{\text{denominator}} = \frac{W_s}{W_p}. \tag{7.53}$$

Next, set the error weight factors for passband and stopband, respectively:

$$
\begin{aligned}
W_s &= \text{numerator} \\
W_p &= \text{denominator}
\end{aligned} \tag{7.54}
$$

4. Apply the Remez algorithm to calculate filter coefficients.
5. If the specifications are not met, then increase the filter order and repeat steps 1 to 4.

The following two examples are given to illustrate the design procedure.

EXAMPLE 7.15

Design a lowpass filter with the following specifications:

DSP system sampling rate	800 Hz
Passband	0–80 Hz
Stopband	1000–400 Hz
Passband ripple	1 dB
Stopband attenuation	40 dB
Filter order	53

Solution:

From the specifications, we have two bands: a low passband and a stopband. We perform normalization and specify ideal magnitudes as follows:

$$\text{Folding frequency}: \frac{f_s}{2} = \frac{8000}{2} = 4000\,\text{Hz}$$

$$\text{For } 0\,\text{Hz}: \quad \frac{0}{4000} = 0 \qquad \text{magnitude}: 1$$

$$\text{For } 800\,\text{Hz}: \quad \frac{800}{4000} = 0.2 \qquad \text{magnitude}: 1$$

$$\text{For } 1000\,\text{Hz}: \quad \frac{1000}{4000} = 0.25 \qquad \text{magnitude}: 0$$

$$\text{For } 4000\,\text{Hz}: \quad \frac{4000}{4000} = 1 \qquad \text{magnitude}: 0$$

Next, let us determine the weights:

$$\delta_p = 10^{\left(\frac{1}{20}\right)} - 1 = 0.1220$$

$$\delta_s = 10^{\left(\frac{-40}{20}\right)} = 0.01.$$

Then, applying Eq. (7.53) gives

$$\frac{\delta_p}{\delta_s} = 12.2 \approx \frac{12}{1} = \frac{W_s}{W_p}.$$

Hence, we have

$$W_s = 12 \text{ and } W_p = 1.$$

Applying **firpm()** routine provided by MATLAB, we list MATLAB codes in Program 7.10. The filter coefficients are listed in Table 7.14.

Table 7.14 FIR Filter Coefficients in Example 7.15

B: FIR Filter Coefficients (Optimal Design Method)	
$b_0 = b_{53} = -0.006075$	$b_1 = b_{52} = -0.00197$
$b_2 = b_{51} = 0.001277$	$b_3 = b_{50} = 0.006937$
$b_4 = b_{49} = 0.013488$	$b_5 = b_{48} = 0.018457$
$b_6 = b_{47} = 0.019347$	$b_7 = b_{46} = 0.014812$
$b_8 = b_{45} = 0.005568$	$b_9 = b_{44} = -0.005438$
$b_{10} = b_{43} = -0.013893$	$b_{11} = b_{42} = -0.015887$
$b_{12} = b_{41} = -0.009723$	$b_{13} = b_{40} = 0.002789$
$b_{14} = b_{39} = 0.016564$	$b_{15} = b_{38} = 0.024947$
$b_{16} = b_{37} = 0.022523$	$b_{17} = b_{36} = 0.007886$
$b_{18} = b_{35} = -0.014825$	$b_{19} = b_{34} = -0.036522$
$b_{20} = b_{33} = -0.045964$	$b_{21} = b_{32} = -0.033866$
$b_{22} = b_{31} = 0.003120$	$b_{23} = b_{30} = 0.060244$
$b_{24} = b_{29} = 0.125252$	$b_{25} = b_{28} = 0.181826$
$b_{26} = b_{27} = 0.214670$	

Program 7.10. MATLAB program for Example 7.15.

```
% Fig. 7.34 (Example 7.15)
% MATLAB program to create Fig. 7.34
%
fs=8000;
f=[ 0 0.2 0.25 1]; % edge frequencies
m=[ 1 1 0 0] ; % ideal magnitudes
w=[ 1 12 ]; % error weight factors
b=firpm(53,f,m,w); % (53+1)Parks-McClellen algorithm and Remez exchange
format long
freqz(b,1,512,fs) % plot the frequency response
axis([0 fs/2 -80 10]);
```

Fig. 7.34 shows the frequency responses.

Clearly, the stopband attenuation is satisfied. We plot the details for the filter passband in Fig. 7.35.

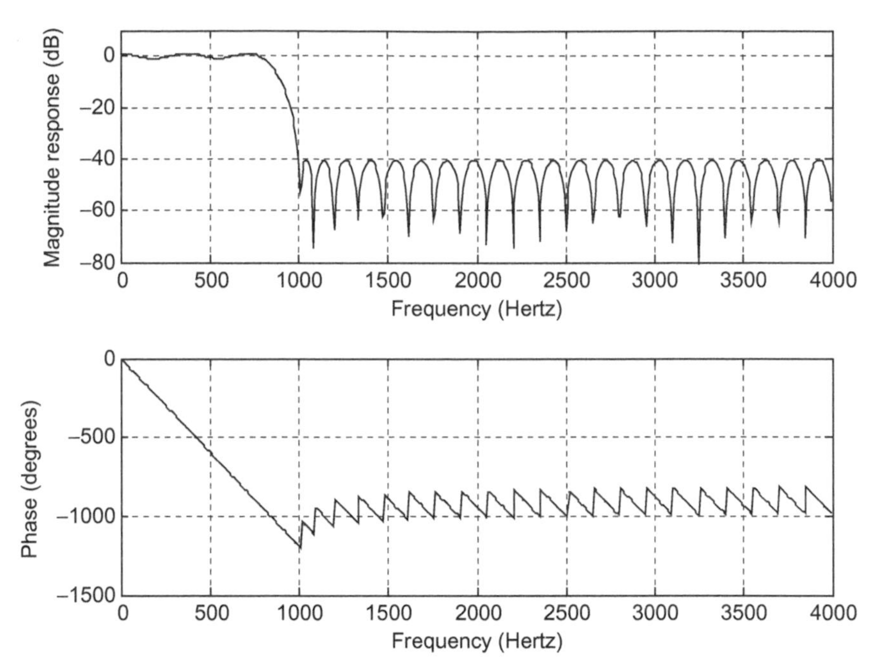

FIG. 7.34

Frequency and phase responses for Example 7.15.

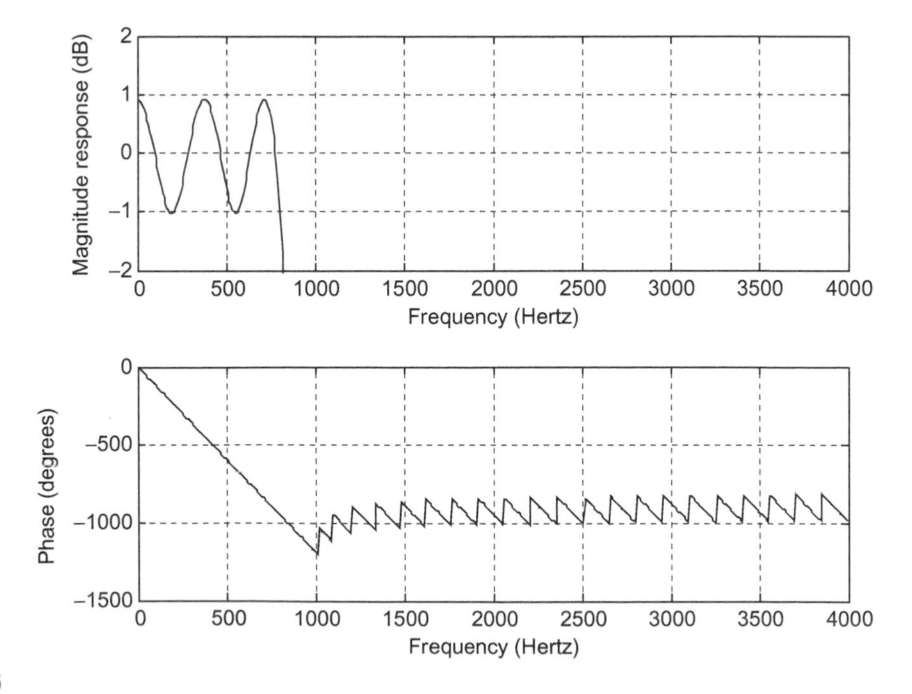

FIG. 7.35

Frequency response details for passband in Example 7.15.

As shown in Fig. 7.35, the ripples in the passband are between -1 and 1 dB. Hence, all the specifications are met. Note that if the specification is not satisfied, we will increase the order until the stopband attenuation and passband ripple are met.

The next example illustrates the bandpass filter design.

EXAMPLE 7.16

Design a bandpass filter with the following specifications:

DSP system sampling rate	800 Hz
Passband	1000–160 Hz
Stopband	0–60 Hz and 2000–400 Hz
Passband ripple	1 dB
Stopband attenuation	30 dB
Filter order	25

Solution:

From the specifications, we have three bands: a passband, a lower stopband, and a upper stopband. We perform normalization and specify ideal magnitudes as follows:

$$\text{Folding frequency}: \frac{f_s}{2} = \frac{8000}{2} = 4000\,\text{Hz}$$

$$\text{For } 0\,\text{Hz}: \quad \frac{0}{4000} = 0 \quad \text{magnitude}: 0$$

$$\text{For } 600\,\text{Hz}: \quad \frac{600}{4000} = 0.15 \quad \text{magnitude}: 0$$

$$\text{For } 1000\,\text{Hz}: \quad \frac{1000}{4000} = 0.25 \quad \text{magnitude}: 1$$

$$\text{For } 1600\,\text{Hz}: \quad \frac{1600}{4000} = 0.4 \quad \text{magnitude}: 1$$

$$\text{For } 2000\,\text{Hz}: \quad \frac{2000}{4000} = 0.5 \quad \text{magnitude}: 0$$

$$\text{For } 4000\,\text{Hz}: \quad \frac{4000}{4000} = 1 \quad \text{magnitude}: 0$$

Next, let us determine the weights:

$$\delta_p = 10^{\left(\frac{1}{20}\right)} - 1 = 0.1220$$

$$\delta_s = 10^{\left(\frac{-30}{20}\right)} = 0.0316.$$

Continued

EXAMPLE 7.16—CONT'D

Then applying Eq. (7.53), we get

$$\frac{\delta_p}{\delta_s} = 3.86 \approx \frac{39}{10} = \frac{W_s}{W_p}.$$

Hence, we have

$$W_s = 39 \text{ and } W_p = 10.$$

Applying the **firpm()** routine provided by MATLAB and check performance, we have Program 7.11. Table 7.15 lists the filter coefficients.

Table 7.15 FIR Filter Coefficients in Example 7.16

B: FIR Filter Coefficients (Optimal Design Method)	
$b_0 = b_{25} = -0.022715$	$b_1 = b_{24} = -0.012753$
$b_2 = b_{23} = 0.005310$	$b_3 = b_{22} = 0.009627$
$b_4 = b_{21} = -0.004246$	$b_5 = b_{20} = 0.006211$
$b_6 = b_{19} = 0.057515$	$b_7 = b_{18} = 0.076593$
$b_8 = b_{17} = -0.015655$	$b_9 = b_{16} = -0.156828$
$b_{10} = b_{15} = -0.170369$	$b_{11} = b_{14} = 0.009447$
$b_{12} = b_{13} = 0.211453$	

Program 7.11. MATLAB program for Example 7.16

```
% Fig. 7.36 (Example 7.16)
% MATLAB program to create Fig. 7.36
%
fs=8000;
f=[ 0 0.15 0.25 0.4 0.5 1];          % edge frequencies
m=[ 0 0 1 1 0 0];                    % ideal magnitudes
w=[ 39 10 39 ];                      % error weight factors
format long
b=firpm(25,f,m,w)   % (25+1) taps Parks-McClellen algorithm and Remez exchange
freqz(b,1,512,fs);  % plot the frequency response
axis([0 fs/2 -80 10])
```

The frequency responses are depicted in Fig. 7.36.

Clearly, the stopband attenuation is satisfied. We also check the details for the passband as shown in Fig. 7.37.

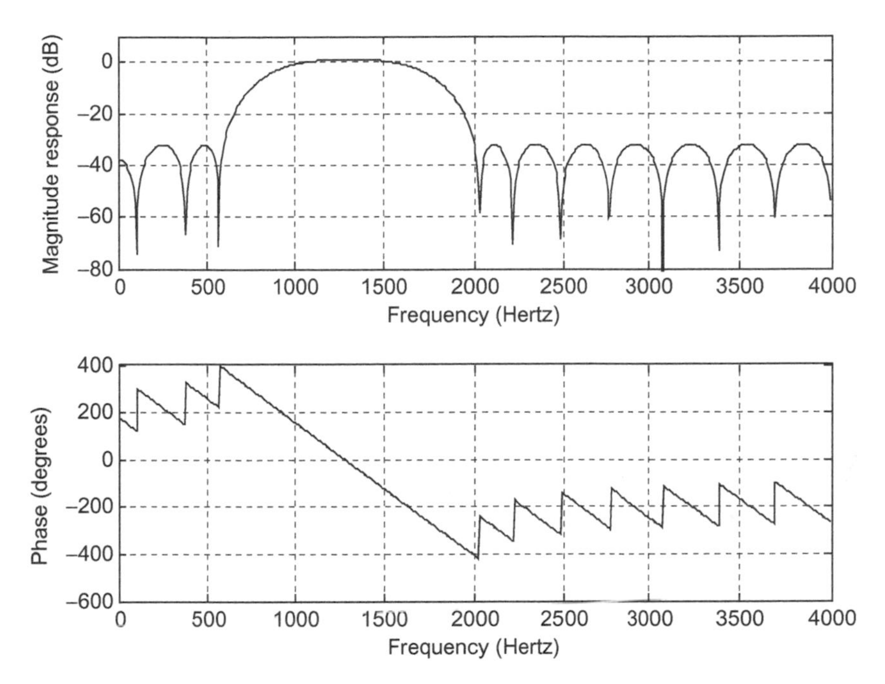

FIG. 7.36

Frequency and phase responses for Example 7.16.

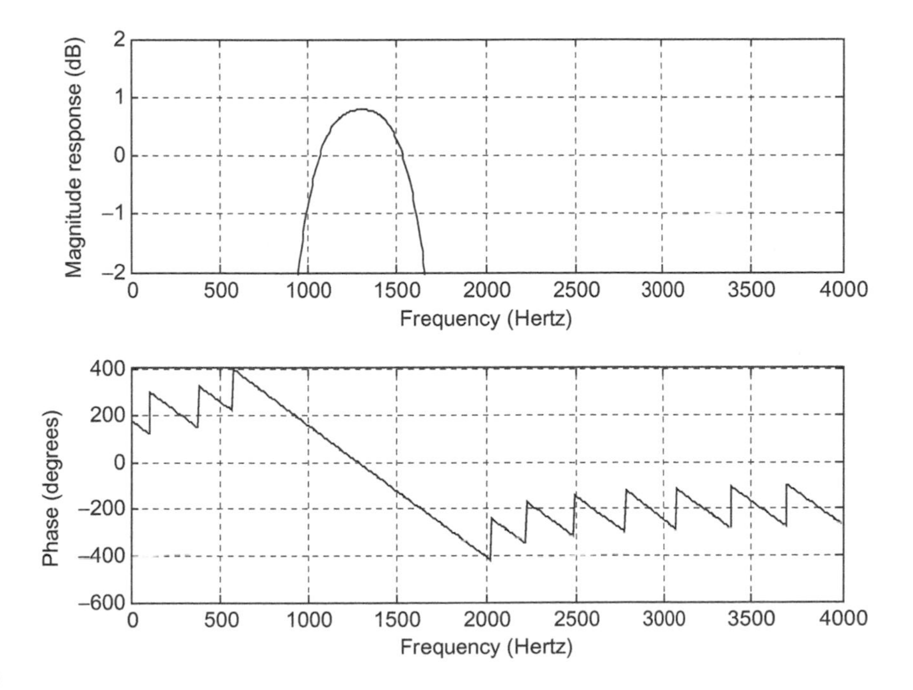

FIG. 7.37

Frequency response details for passband in Example 7.16.

As shown in Fig. 7.37, the ripples in the passband between 1000 and 1600 Hz are between –1 and 1 dB. Hence, all specifications are satisfied.

EXAMPLE 7.17

Now we show how the Remez exchange algorithm in Eq. (7.49) is processed using a linear phase three-tap FIR filter as

$$H(z) = b_0 + b_1 z^{-1} + b_0 z^{-2}.$$

The ideal frequency response specifications are shown in Fig. 7.38A, where the filter gain increases linearly from the gain of 0.5 at $\Omega = 0$ (rad) to the gain of 1 at $\Omega = \pi/4$ (rad). The band between $\Omega = \pi/4$ (rad) and $\Omega = \pi/2$ (rad) is a transition band. Finally, the filter gain decreases linearly from the gain 0.75 at $\Omega = \pi/2$ (rad) to the gain of 0 at $\Omega = \pi$ (rad).

For simplicity, we use all the weight factors as 1, that is, $W(\Omega) = 1$. Eq. (7.49) is simplified to be

$$E(\Omega) = H\left(e^{j\Omega}\right) - H_d\left(e^{j\Omega}\right).$$

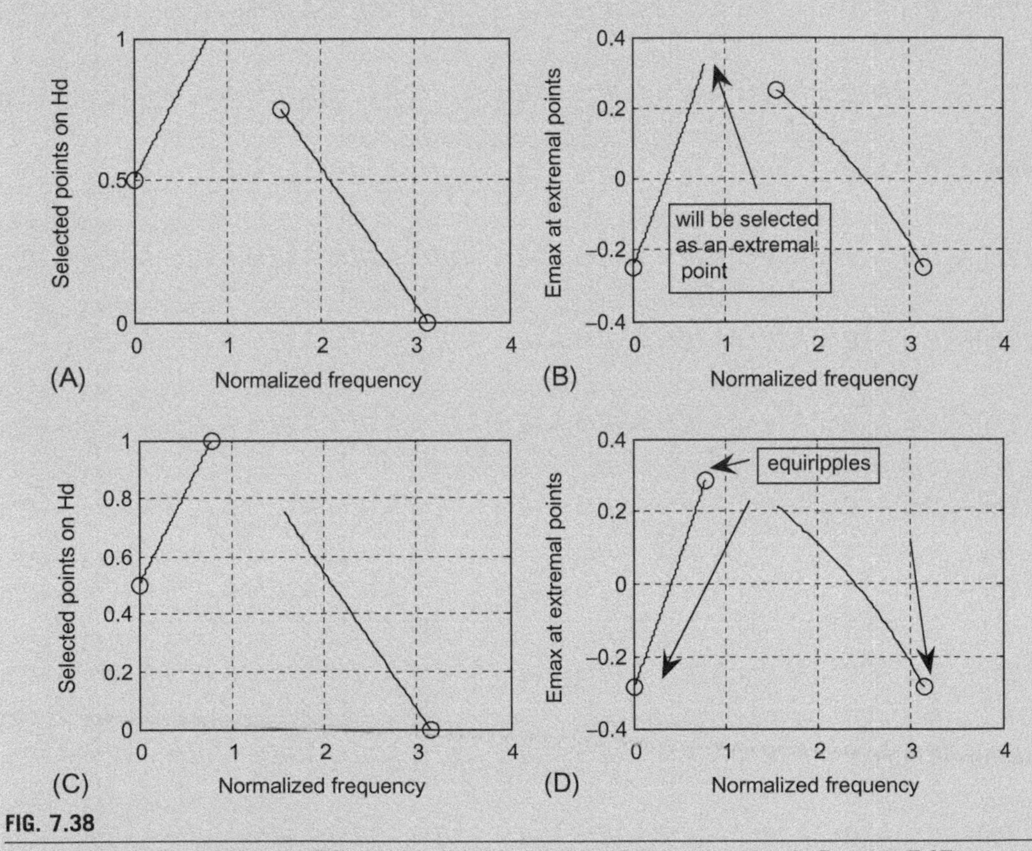

FIG. 7.38

Determination of the three-tap FIR filter coefficients using the Remez algorithm in Example 7.17.

Substituting $z = e^{j\Omega}$ into the transfer function $H(z)$ gives

$$H(e^{j\Omega}) = b_0 + b_1 e^{-j\Omega} + b_0 e^{-j2\Omega}.$$

After simplification using Euler's identity $e^{j\Omega} + e^{-j\Omega} = 2\cos\Omega$, the filter frequency response is given by

$$H(e^{j\Omega}) = e^{j\Omega}(b_1 + 2b_0 \cos\Omega).$$

Disregarding the linear phase shift term $e^{j\Omega}$ for the time being, we have a Chebyshev real magnitude function (there are few other types as well) as

$$H(e^{j\Omega}) = b_1 + 2b_0 \cos\Omega.$$

The *alternation theorem* (Ambardar, 1999; Porat, 1997) must be used. The alternation theorem states that given Chebyshev polynomial $H(e^{j\Omega})$ to approximate the ideal magnitude response $H_d(e^{j\Omega})$, we can find at least $M+2$ (where $M=1$ for our case) frequencies $\Omega_0, \Omega_1, \ldots, \Omega_{M+1}$, called the extremal frequencies, so that signs of the error at the extremal frequencies alternate and the absolute error value at each extremal point reaches the maximum absolute error, that is,

$$E(\Omega_k) = -E(\Omega_{k+1}) \text{ for } \Omega_0, \Omega_1, \ldots \Omega_{M+1}$$

and

$$|E(\Omega_k)| = E_{\max}.$$

But the alternation theorem does not tell us how to do the algorithm. The Remez exchange algorithm actually is employed to solve this problem. The equations and steps (Ambardar, 1999; Porat, 1997) are briefly summarized for our illustrative example:

1. Given the order of $N = 2M+1$, choose initial extremal frequencies: $\Omega_0, \Omega_1, \ldots, \Omega_{M+1}$ (can be uniformly distributed first).
2. Solve the following equation to satisfy the alternate theorem:

$$-(-1)^k E = W(\Omega_k)\left(H_d\left(e^{j\Omega_k}\right) - H\left(e^{j\Omega_k}\right)\right) \text{ for } \Omega_0, \Omega_1, \ldots, \Omega_{M+1}.$$

Note that since $H(e^{j\Omega}) = b_1 + 2b_0 \cos\Omega$, for this example, the solution will include solving for three unknowns: b_0, b_1, and $E_{\max}$.

3. Determine the extremal points including band edges (can be more than $M+2$ points), and retain $M+2$ extremal points with the largest error values $E_{\max}$.
4. Output the coefficients, if the extremal frequencies are not changed, otherwise, go to step 2 using the new set of extremal frequencies.

Now let us apply the Remez exchange algorithm.

First iteration:

1. We use the uniformly distributed extremal points: $\Omega_0 = 0$, $\Omega_1 = \pi/2$, $\Omega_2 = \pi$ whose ideal magnitudes are marked by the symbol "o" as shown in Fig. 7.38A.
2. Alternate theorem requires: $-(-1)^k E = H_d(e^{j\Omega}) - (b_1 + 2b_0 \cos\Omega)$.

Continued

EXAMPLE 7.17—CONT'D

Applying extremal points yields the following three simultaneous equations with three unknowns: b_0, b_1, and E:

$$\begin{cases} -E = & 0.5 - b_1 - 2b_0 \\ E = & 0.75 - b_1 \\ -E = & 0 - b_1 + 2b_0 \end{cases}.$$

We solve these three equations to get

$$b_0 = 0.125, \quad b_1 = 0.5, \quad E = 0.25, H(e^{j\Omega}) = 0.5 + 0.25\cos\Omega.$$

3. We then determine the extremal points, including at the band edge, with their error values from Fig. 7.38B using the following error function:

$$E(\Omega) = H_d(e^{j\Omega}) - 0.5 - 0.25\cos\Omega.$$

These extremal points are marked by symbol "o" and their error values are listed in Table 7.16.

4. Since the band edge $\Omega = \pi/4$ has an error lager than others, it must be chosen as the extremal frequency. After deleting the extremal point at $\Omega = \pi/2$, a new set of extremal points are found according the largest error values as

$$\Omega_0 = 0$$
$$\Omega_1 = \pi/4$$
$$\Omega_2 = \pi.$$

The ideal magnitudes at these three extremal points are given in Fig. 7.38C, that is, 0.5, 1, and 0. Now let us examine the second iteration.

Second iteration:

Applying the alternation theorem at the new set of extremal points, we have

$$\begin{cases} -E = 0.5 - b_1 - 2b_0 \\ E = 1 - b_1 - 1.4142b_0 \\ -E = 0 - b_1 + 2b_0. \end{cases}$$

Solving these three simultaneous equations leads to

$$b_0 = 0.125, \quad b_1 = 0.537, \quad E = 0.287, \quad \text{and} \quad H(e^{j\Omega}) = 0.537 + 0.25\cos\Omega.$$

Table 7.16 Extremal Points and Band Edges with Their Error Values for the First Iteration

Ω	0	$\pi/4$	$\pi/2$	π
E_{max}	−0.25	0.323	0.25	0.25

The determined extremal points and band edge with their error values are listed in Table 7.17 and shown in Fig. 7.38D, where the determined extremal points are marked by the symbol "o." Since at the extremal points in Table 7.17, their maximum absolute error values are the same, that is, 0.287; and these extremal points are found to be $\Omega_0=0$, $\Omega_1=\pi/4$, and $\Omega_2=\pi$, and are unchanged in comparison with the ones in Table 7.16. Then we stop the iteration and output the filter transfer function as

$$H(z)=0.125+0.537z^{-1}+0.125z^{-2}.$$

As shown in Fig. 7.37D, we achieve the equiripples of error at the extremal points: $\Omega_0=0, \Omega_1=\pi/4, \Omega_2=\pi$; their signs are alternating, and the maximum absolute error of 0.287 is obtained at each point. It takes two iterations to determine the coefficients for this simplified example.

Table 7.17 Error Values at Extremal Frequencies and Band edge

Ω	0	$\pi/4$	$\pi/2$	π
E_{max}	−0.287	0.287	0.213	−0.287

As we have mentioned, the Parks-McClellen algorithm is one of the popular filter design methods in industry due to its flexibility and performance. However, there are two disadvantages. The filter length has to be estimated by the empirical method. Once the frequency edges, magnitudes, and weighting factors are specified, applying the Remez exchange algorithm cannot control over the actual ripple obtained from the design. We may often need to try a longer length of filter or different weight factors to remedy the situations where the ripple is unacceptable.

7.7 DESIGN OF FIR DIFFERENTIATOR AND HILBERT TRANSFORMER

In many signal processing applications such as radar and sonar signal processing as well as vibration signal analysis, digital differentiators are often applied to estimate velocity and acceleration from position measurements. Taking time derivative of a signal $x(t)=e^{j\omega t}$ with the frequency ω results in $y(t)=j\omega e^{j\omega t}=j\omega x(t)$. This indicates that an ideal differentiator has a frequency response which is linearly proportional to its frequency. The ideal digital differentiator has the frequency response defined below:

$$H\left(e^{j\Omega}\right)=j\Omega \text{ for } -\pi<\Omega<\pi. \tag{7.55}$$

Using the Fourier transform design method, we can obtain the filter coefficients below:

$$\begin{aligned} h(n) &= \frac{1}{2\pi}\int_{-\pi}^{\pi} H\left(e^{j\Omega}\right)e^{jn\Omega}d\Omega \\ &= \frac{1}{2\pi}\int_{-\pi}^{\pi} j\Omega e^{jn\Omega}d\Omega \text{ for } -\infty<n<\infty \text{ and } n\neq 0 \\ &= \frac{\cos(\pi n)}{n}. \end{aligned} \tag{7.56}$$

It is easy to verify that $h(0)=0$ and $h(-n)=-h(n)$, which means the ideal digital differentiator has antisymmetric unit impulse response. Converting noncausal differentiator to causal one with $(2M+1)$ coefficients yields the following relation:

$$b_n = \begin{cases} -h(M-n) & n=0,1,2,\cdots,M-1 \\ 0 & n=M \\ h(n-M) & n=M+1,\cdots,2M \end{cases} \tag{7.57}$$

Fig. 7.39 depicts the frequency response plots using 101 coefficients $(2M+1=101)$. The differentiator has a linear phase and operates at a sampling rate of 8000 Hz. Similarly, the higher-order differentiator (Tan and Wang, 2011) can be developed.

Hilbert transformers are frequently used in communication and signal processing systems. An ideal Hilbert transformer is an allpass filter with a 90° phase shift on the input signal. The frequency response of the ideal digital Hilbert transformer is specified below:

$$H(e^{j\Omega}) = \begin{cases} -j & 0 \le \Omega < \pi \\ j & -\pi \le \Omega < 0 \end{cases}. \tag{7.58}$$

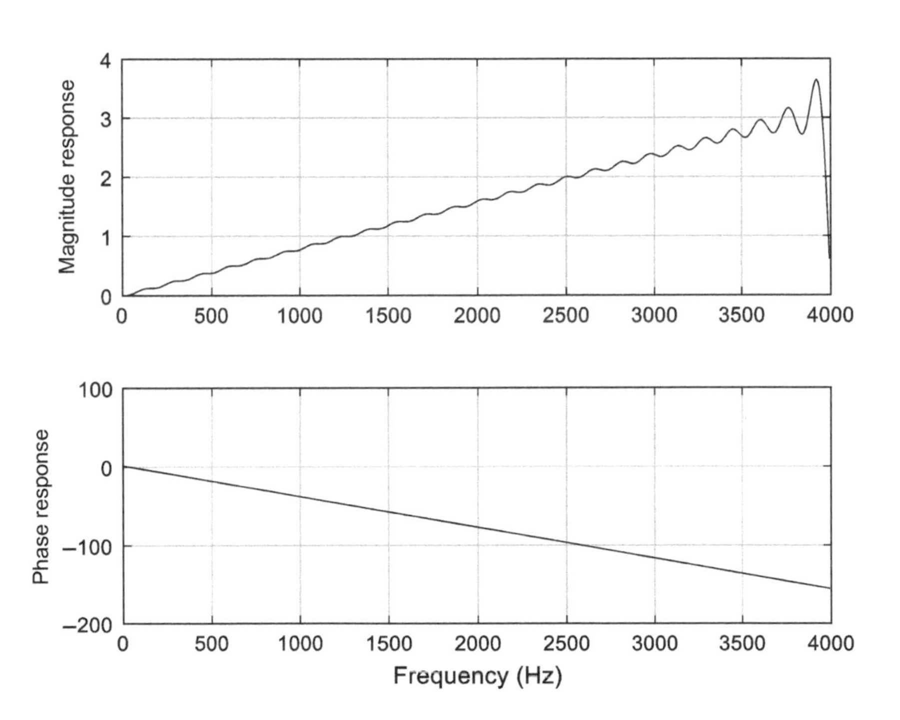

FIG. 7.39

Frequency response plots of the digital differentiator.

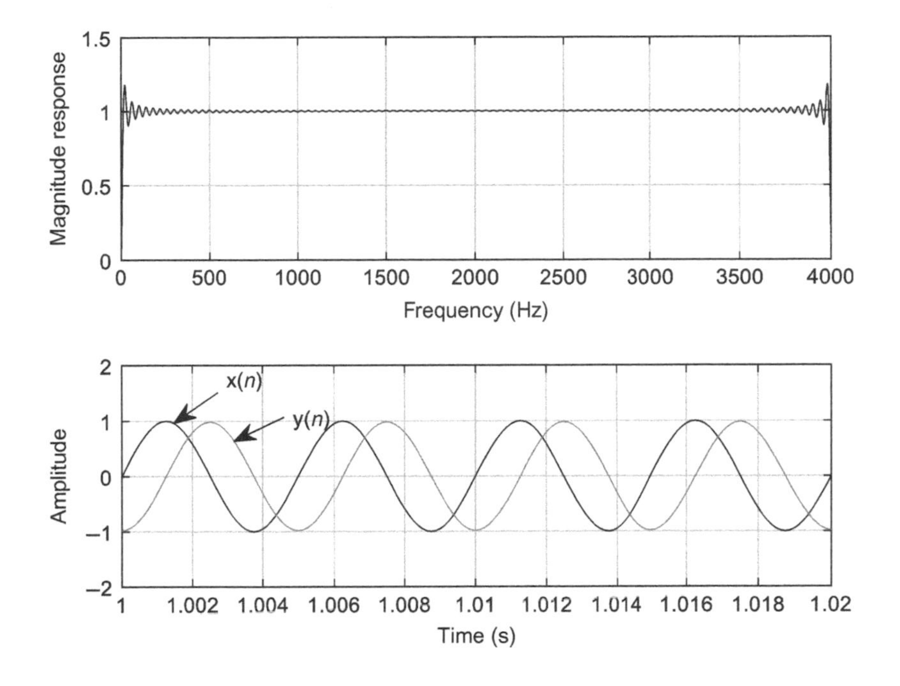

FIG. 7.40

Magnitude frequency response of the Hilbert transformer; and displays of input and output signals.

Applying the Fourier transform design method, it follows that

$$\begin{aligned} h(n) &= \frac{1}{2\pi}\int_{-\pi}^{0} je^{jn\Omega}d\Omega - \frac{1}{2\pi}\int_{0}^{\pi} je^{jn\Omega}d\Omega \text{ for} -\infty < n < \infty \\ &= \frac{2\sin^2(\pi n/2)}{\pi n}. \end{aligned} \tag{7.59}$$

Similarly, it is easy to show the antisymmetric impulse response, that is, $h(-n) = -h(n)$. The causal coefficients can be obtained via Eq. (7.57). Fig. 7.40 shows the magnitude frequency response using 401 coefficients. The Hilbert transformer processes a 200-Hz sinusoid at a sampling rate of 8000 Hz. From Fig. 7.40, it is seen that the processed output has a 90° phase shift on the input signal.

7.8 REALIZATION STRUCTURES OF FINITE IMPULSE RESPONSE FILTERS

Using the direct-I form (discussed in Chapter 6), we will get a special realization form, called the *transversal form*. Using the linear phase property will produce a linear phase realization structure.

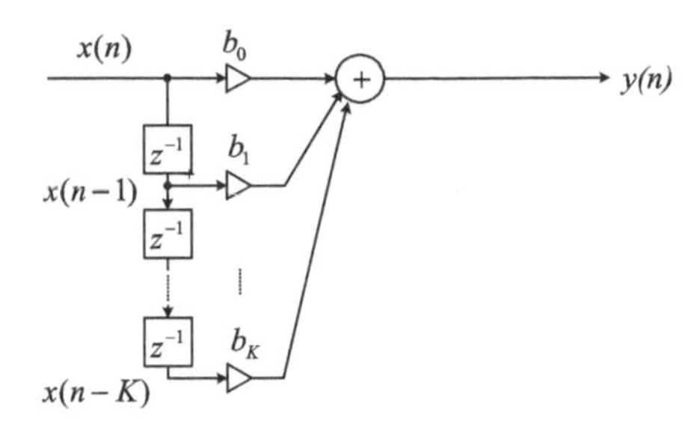

FIG. 7.41

FIR filter realization (transversal form).

7.8.1 TRANSVERSAL FORM

Given the transfer function of an FIR filter in Eq. (7.60),

$$H(z) = b_0 + b_1 z^{-1} + \cdots + b_K z^{-K}, \tag{7.60}$$

we obtain the difference equation as

$$y(n) = b_0 x(n) + b_1 x(n-1) + b_2 x(n-2) + \cdots + b_K x(n-K).$$

Realization of such a transfer function is the transversal form, displayed in Fig. 7.41.

EXAMPLE 7.18

Given an FIR filter transfer function

$$H(z) = 1 + 1.2z^{-1} + 0.36z^{-2},$$

perform the FIR filter realization.

Solution:

From the transfer function, we can identify that $b_0 = 1$, $b_1 = 1.2$, and $b_2 = 0.36$. Using Fig. 7.39, we find the FIR realization as follows (Fig. 7.42):

We determine the DSP equation for implementation as

$$y(n) = x(n) + 1.2x(n-1) + 0.36x(n-2).$$

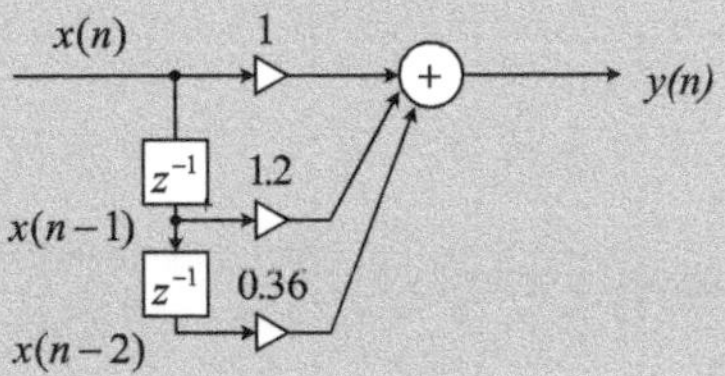

FIG. 7.42

FIR filter realization for Example 7.18.

Program 7.12 shows the MATLAB implementation.

Program 7.12. MATLAB program for Example 7.18.

```
%Sample MATLAB code
sample =1:1:10;                         % Input test array
x=[ 0 0 0];                             % Input buffer [x(n) x(n-1) ...]
y=[0];                                  % Output buffer [y(n) y(n-1) ... ]
b=[1.0 1.2 0.36];                       % FIR filter coefficients [b0 b1 ...]
KK=length(b);
for n=1:1:length(sample)                 % Loop processing
  for k=KK:-1:2                          % Shift input by one sample
    x(k)=x(k-1);
  end
  x(1)=sample(n); % Get new sample
  y(1)=0;  % Perform FIR filtering
  for k=1:1:KK
    y(1)=y(1)+b(k)*x(k);
  end
  out(n)=y(1); %send filtered sample to the output array
end
  out
```

7.8.2 LINEAR PHASE FORM

We illustrate the linear phase structure using the following simple example.

Considering the transfer function with five taps obtained from the design as follows:

$$H(z) = b_0 + b_1 z^{-1} + b_2 z^{-2} + b_1 z^{-3} + b_0 z^{-4}, \tag{7.61}$$

we can see that the coefficients are symmetrical and the difference Equation is

$$y(n) = b_0 x(n) + b_1 x(n-1) + b_2 x(n-2) + b_1 x(n-3) + b_0 x(n-4).$$

This DSP equation can further be combined to be

$$y(n) = b_0(x(n) + x(n-4)) + b_1(x(n-1) + x(n-3)) + b_2 x(n-2).$$

Then we obtain the realization structure in a linear phase form as follows (Fig. 7.43):

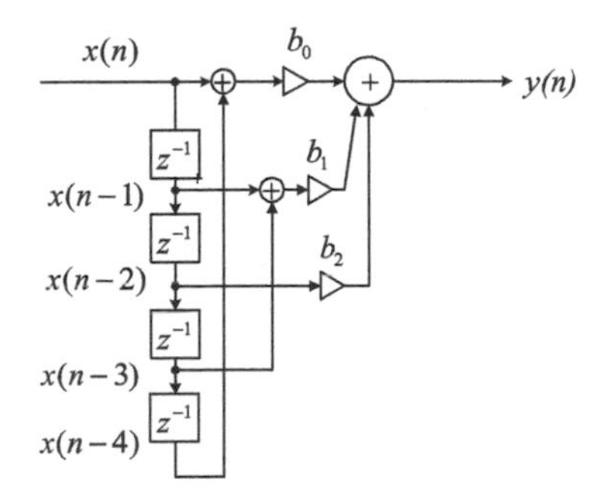

FIG. 7.43

Linear phase FIR filter realization.

7.9 COEFFICIENT ACCURACY EFFECTS ON FINITE IMPULSE RESPONSE FILTERS

In practical applications, the filter coefficients achieved through high-level software such as MATLAB must be quantized using finite word length. This may have two effects. First, the locations of zeros are changed; second, due to the location change of zeros, the filter frequency response will change correspondingly. In practice, there are two types of digital signal (DS) processors: *fixed-point processors* and *floating-point processors*. The fixed-point DS processor uses integer arithmetic, and the floating-point processor employs floating-point arithmetic. Such effects of filter coefficient quantization will be covered in Chapter 14.

In this section, we will study effects of FIR filter coefficient quantization in general, since during practical filter realization, obtaining filter coefficients with infinite precision is impossible. Filter coefficients are usually truncated or rounded off for an application. Assume that the FIR filter transfer function with infinite precision is given by

$$H(z) = \sum_{n=0}^{K} b_n z^{-n} = b_0 + b_1 z^{-1} + \cdots + b_{2M} z^{-K}, \tag{7.62}$$

where each filter coefficient b_n has infinite precision. Now let the quantized FIR filter transfer function be

$$H^q(z) = \sum_{n=0}^{K} b_n^q z^{-n} = b_0^q + b_1^q z^{-1} + \cdots + b_K^q z^{-K}, \tag{7.63}$$

where each filter coefficient b_n^q is quantized (round off) using the specified number of bits. Then the error of the magnitude frequency response can be bounded as

$$\begin{aligned} \left|H(e^{j\Omega}) - H^q(e^{j\Omega})\right| &= \sum_{n=0}^{K} \left|(b_n - b_n^q)e^{j\Omega}\right| \\ &< \sum_{n=0}^{K} \left|b_n - b_n^q\right| < (K+1)\cdot 2^{-B}, \end{aligned} \tag{7.64}$$

where B is the number of bits used to encode each magnitude of the filter coefficient. Look at Example 7.19.

EXAMPLE 7.19

In Example 7.7, a lowpass FIR filter with twenty-five taps using a Hamming window is designed, and FIR filter coefficients are listed for comparison in Table 7.18. One sign bit is used, and 7 bits are used for fractional parts, since all FIR filter coefficients are less than 1. We would multiply each filter coefficient by a scale factor of 2^7 and round off each scaled magnitude to an integer whose magnitude could be encoded using 7 bits. When the coefficient integer is scaled back, the coefficient with finite precision (quantized filter coefficient) using 8 bits, including the sign bit, will be achieved.

Table 7.18 FIR Filter Coefficients and Their Quantized Filter Coefficients in Example 7.19 (Hamming Window)

Bham: FIR Filter Coefficients	BhamQ: FIR Filter Coefficients
$b_0=b_{24}=0.00000000000000$	$b_0=b_{24}=0.0000000$
$b_1=b_{23}=-0.00276854711076$	$b_1=b_{23}=-0.0000000$
$b_2=b_{22}=0.00000000000000$	$b_2=b_{22}=0.0000000$
$b_3=b_{21}=0.00759455135346$	$b_3=b_{21}=0.0078125$
$b_4=b_{20}=0.00000000000000$	$b_4=b_{20}=0.0000000$
$b_5=b_{19}=-0.01914148493949$	$b_5=b_{19}=-0.0156250$
$b_6=b_{18}=0.00000000000000$	$b_6=b_{18}=0.0000000$
$b_7=b_{17}=0.04195685650042$	$b_7=b_{17}=0.0390625$
$b_8=b_{16}=0.00000000000000$	$b_8=b_{16}=0.0000000$
$b_9=b_{15}=-0.09180790496577$	$b_9=b_{15}=-0.0859375$
$b_{10}=b_{14}=0.00000000000000$	$b_{10}=b_{14}=0.0000000$
$b_{11}=b_{13}=0.31332065886015$	$b_{11}=b_{13}=0.3125000$
$b_{12}=0.50000000000000$	$b_{12}=0.5000000$

To understand quantization, we take look at one of the infinite precision coefficients Bham(3) = 0.00759455135346, for illustration. The quantization using 7 magnitude bits is shown as

$$0.00759455135346 \times 2^7 = 0.9721 = 1 \text{ (round up to the nearest integer)}.$$

Then the quantized filter coefficient is obtained as

$$\text{Bham}Q(3) = \frac{1}{2^7} = 0.0078125.$$

Since the poles for both FIR filters always reside at origin, we need to examine only their zeros. The z-plane zero plots for both FIR filters are shown in Fig. 7.44A, where the circles are zeros from the FIR filter with infinite precision, while the crosses are zeros from the FIR filter with the quantized coefficients.

Most importantly, Fig. 7.44B shows the difference of the frequency responses for both filters obtained using Program 7.13. In the figure, the solid line represents the frequency response with infinite filter coefficient precision, and the dot-dashed line indicates the frequency response with finite filter coefficients. It is observed that the stopband performance is degraded due to the filter coefficient quantization. The degradation in the passband is not severe.

Continued

EXAMPLE 7.19—CONT'D

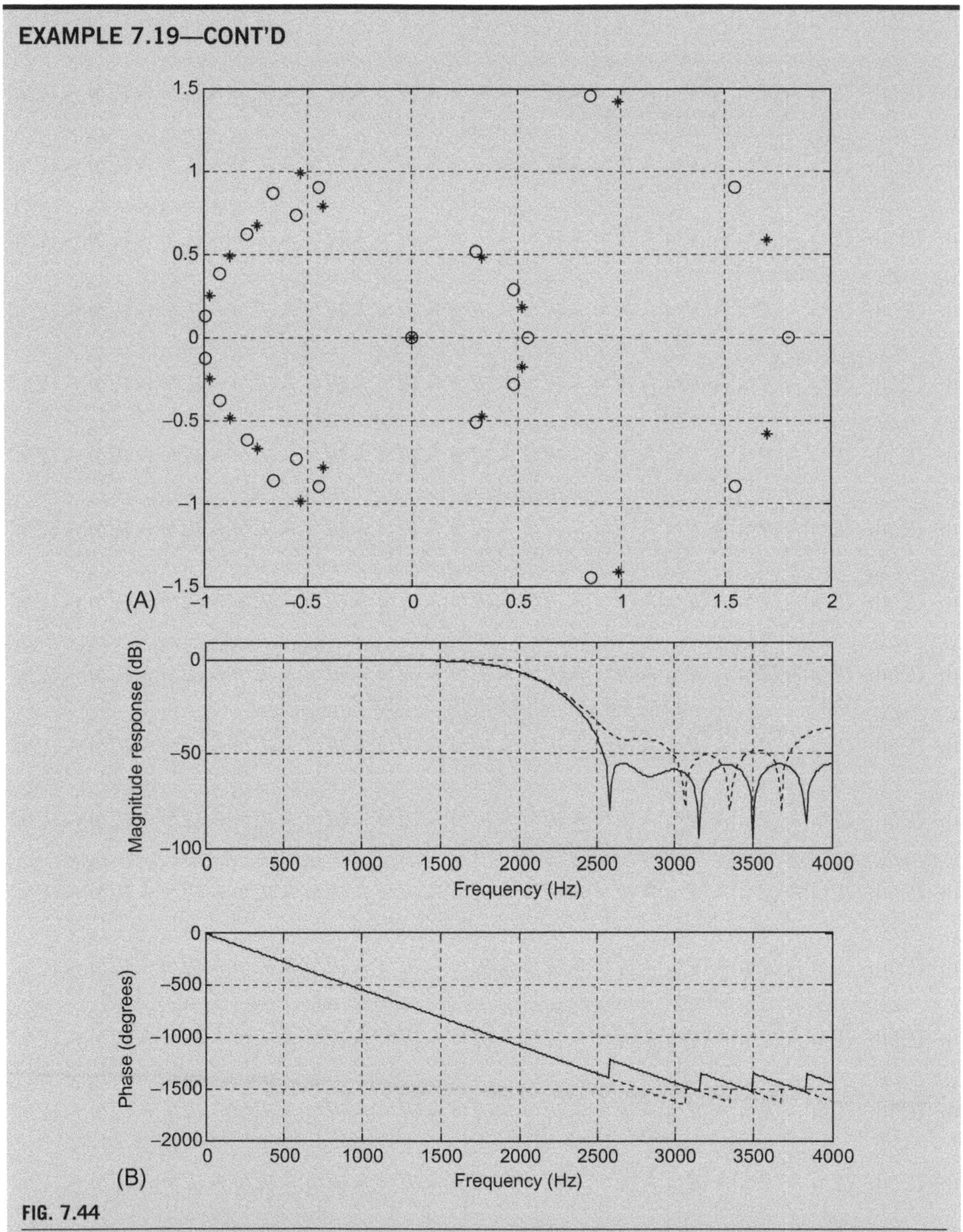

FIG. 7.44

(A) The z-plane zero plots for both FIR filters. The circles are zeros for infinite precision; the crosses are zeros for round-off coefficients. (B) Frequency responses. The solid line indicates the FIR filter with infinite precision; the dashed line indicates the FIR filter with the round-off coefficients.

Program 7.13. MATLAB program for Example 7.19.

```
fs=8000;
[hham,f]=freqz(Bham,1,512,fs);
[hhamQ,f]=freqz(BhamQ,1,512,fs);
p=180*unwrap(angle(hham))/pi;
pQ=180*unwrap(angle(hhamQ))/pi
subplot(2,1,1);
plot(f,20*log10(abs(hham)),f,20*log10(abs(hhamQ)),':');grid
axis([0 4000 -100 10]);
xlabel('Frequency (Hz)'); ylabel('Magnitude Response (dB)');
subplot(2,1,2); plot(f,p,f,pQ,':');grid
xlabel('Frequency (Hz)'); ylabel('Phase (degrees)');
```

Using Eq. (7.64), the error of the magnitude frequency response due to quantization is bounded by

$$\left|H\left(e^{j\Omega}\right) - H^q\left(e^{j\Omega}\right)\right| < 25/256 = 0.0977.$$

This can easily be verified at the stopband of the magnitude frequency response for the worst condition as follows:

$$\left|H\left(e^{j\Omega}\right) - H^q\left(e^{j\Omega}\right)\right| = \left|10^{-100/20} - 10^{-30/20}\right| = 0.032 < 0.0977.$$

In practical situations, the similar procedure can be used to analyze the effects of filter coefficient quantization to make sure that the designed filter meets the requirements.

7.10 SUMMARY OF FIR DESIGN PROCEDURES AND SELECTION OF THE FIR FILTER DESIGN METHODS IN PRACTICE

In this section, we first summarize the design procedures of the window design, frequency sampling design, and optimal design methods, and then discuss the selection of the particular filter for typical applications.

The window method (Fourier transform design using windows):

1. Given the filter frequency specifications, determine the filter order (odd number used in this book) and the cutoff frequency/frequencies using Table 7.7 and Eq. (7.30).
2. Compute the impulse sequence $h(n)$ via the Fourier transform method using the appropriate equations (in Table 7.1).
3. Multiply the generated FIR filter coefficients $h(n)$ in (2) by the selected window sequence using Eq. (7.28) to obtain the windowed impulse sequence $h_w(n)$.
4. Delay the windowed impulse sequence $h_w(n)$ by M samples to get the causal windowed FIR filter coefficients $b_n = h_w(n-M)$ using Eq. (7.29).
5. Output the transfer function and plot the frequency responses.
6. If the frequency specifications are satisfied, output the difference equation. If the frequency specifications are not satisfied, increase the filter order and repeat beginning with step 2.

The frequency sampling method:

1. Given the filter frequency specifications, choose the filter order (odd number used in the book), and specify the equally spaced magnitudes of the frequency response for the normalized frequency range from 0 to π using Eq. (7.46).
2. Calculate FIR filter coefficients using Eq. (7.47).
3. Use the symmetry, in Eq. (7.48), and linear phase requirement to determine the rest of coefficients.
4. Output the transfer function and plot the frequency responses.
5. If the frequency specifications are satisfied, output the difference equation. If the frequency specifications are not satisfied, increase the filter order and repeat beginning with step 2.

The optimal design method (Parks-McClellan Algorithm):

1. Given the band edge frequencies, choose the filter order, normalize each band edge frequency to the Nyquist limit (folding frequency $=f_s/2$), and specify the ideal magnitudes.
2. Calculate absolute values of the passband ripple and stopband attenuation, if they are given in terms of dB values using Eqs. (7.51) and (7.52).
3. Determine the error weight factors for passband and stopband, respectively, using Eqs. (7.53) and (7.54).
4. Apply the Remez algorithm to calculate the filter coefficients.
5. Output the transfer function and check the frequency responses.
6. If the frequency specifications are satisfied, output the difference equation. If the frequency specifications are not satisfied, increase the filter order and repeat beginning with step 4.

Table 7.19 illustrates the comparisons for the window, frequency sampling, and optimal methods. The table can be used as a selection guide for each design method in this book.

Table 7.19 Comparisons of Three Design Methods

Design Method	Window	Frequency Sampling	Optimal Design
Filter type	1. Lowpass, highpass, bandpass, bandstop 2. Formulas are not valid for arbitrary frequency selectivity.	1. Any type filter 2. The formula is valid for arbitrary frequency selectivity	1. Any type filter 2. Valid for arbitrary frequency selectivity.
Linear phase	Yes	Yes	Yes
Ripple and stopband specifications	Used for determining the filter order and cutoff frequency/cies	Needed to be checked after each design trial	Used in the algorithm; need to be checked after each deign trial
Algorithm complexity for coefficients	Moderate 1. Impulse sequence calculation 2. Window function weighting	Simple: Single equation	Complicated: 1. Parks-McClellan Algorithm 2. Rmez exchange algorithm
Minimal design tool	Calculator	Calculator	Software

Example 7.20 describes the possible selection of the design method by a DSP engineer to solve a real-world problem.

EXAMPLE 7.20 Determine the appropriate FIR filter design method for each of the following DSP applications.

1. A DSP engineer implements a digital two-band crossover system as described in this section. The FIR filters are selected to satisfy the following specifications:

Sampling rate = 44,100 Hz
Crossover frequency = 1000 Hz (cutoff frequency)
Transition band: 600–1400 Hz
Lowpass filter: passband frequency range from 0 to 600 Hz with a ripple of 0.02 dB and stopband edge at 1400 Hz with the attenuation of 50 dB.
Highpass filter: passband frequency range from 1.4 to 44.1 kHz with a ripple of 0.02 dB and stopband edge at 1600 Hz with the attenuation of 50 dB.
The engineer does not have the software routine for the Remez algorithm.

2. An audio engineer tries to equalize the speech signal sampled at 8000 Hz using a linear phase FIR filter based on the magnitude specifications in Fig. 7.45. The engineer does not have the software routine for the Remez algorithm.

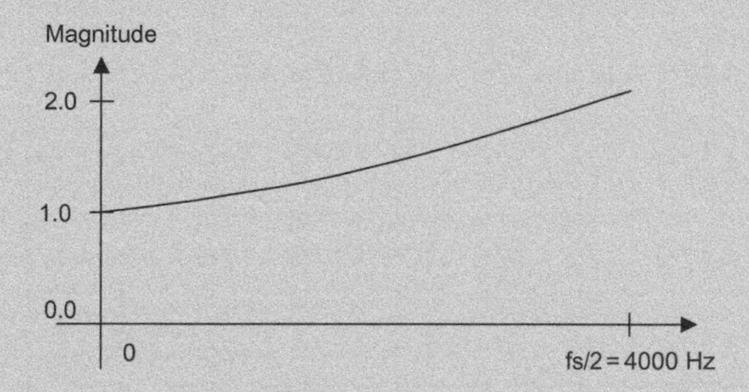

FIG. 7.45

Magnitude frequency response in Example 7.20 (2).

Solution:

1. The window design method is the first choice, since the window design formula is in terms of the cutoff frequency (crossover frequency), the filter order is based on the transient band, and filter types are standard lowpass and highpass. The ripple and stopband specifications can be satisfied by selecting the Hamming window. The optimal design method will also do the job with a challenge to satisfy the combined unity gains at the crossover frequency of 1000 Hz if the Remez algorithm is available.
2. Since the magnitude frequency response is not a standard filter type of lowpass, highpass, bandpass, or bandreject, and the Remez algorithm is not available, the first choice should be the frequency sampling method.

7.11 SUMMARY

1. The Fourier transform method is used to compute noncausal FIR filter coefficients, including those of lowpass, highpass, bandpass, and bandstop filters.
2. Converting noncausal FIR filter coefficients to causal FIR filter coefficients only introduces linear phase, which is a good property for audio application. The linear phase filter output has the same amount of delay for all the input signals whose frequency components are within passband.
3. The causal FIR filter using the Fourier transform method generates ripple oscillations (Gibbs effect) in the passband and stopband in its filter magnitude frequency response due to abrupt truncation of the FIR filter coefficient sequence.
4. To reduce the oscillation effect, the window method is introduced to tap down the coefficient values toward both ends. A substantial improvement in the magnitude frequency response is achieved.
5. Real-life DSP applications such as noise reduction system and two-band digital audio crossover system were investigated.
6. Frequency sampling design is feasible for an FIR filter with an arbitrary magnitude response specification.
7. An optimal design method, the Parks-McClellan algorithm using the Remez exchange algorithm, offers flexibility for filter specifications. The Remez exchange algorithm was explained using a simplified example.
8. Realization structures of FIR filters have special forms, such as the transversal form and the linear phase form.
9. The effect of quantizing FIR filter coefficients for implementation changes zero locations of the FIR filter. More effects on the stopband in the magnitude and phase responses are observed.
10. Guidelines for selecting an appropriate design method in practice were summarized with consideration of the filter type, linear phase, ripple and stopband specifications, algorithm complexity, and design tools.

7.12 MATLAB PROGRAMS

Program 7.14 enables one to design FIR filters via the window method using the window functions such as the rectangular window, triangular window, Hanning window, Hamming window, and Blackman window. Filter types of the design include lowpass, highpass, bandpass, and bandstop.

Program 7.14. MATLAB function for FIR filter design using the window method

```
function B=firwd(N,Ftype,WnL,WnH,Wtype)
% B = firwd(N,Ftype,WnL,WnH,Wtype)
% FIR filter design using the window function method.
% Input parameters:
% N: the number of the FIR filter taps.
%   Note: It must be odd number.
% Ftype: the filter type
%     1. Lowpass filter;
```

```
%     2. Highpass filter;
%     3. Bandpass filter;
%     4. Bandreject filter.
% WnL: lower cutoff frequency in radians. Set WnL=0 for the highpass filter.
% WnH: upper cutoff frequency in radians. Set WnL=0 for the lowpass filter.
% Wtype: window function type
%     1. Rectangular window;
%     2. Triangular window;
%     3. Hanning window;
%     4. Hamming window;
%     5. Blackman window;
% Output:
% B: FIR filter coefficients.
      M=(N-1)/2;
      hH=sin(WnH*[-M:1:-1])./([-M:1:-1]*pi);
      hH(M+1)=WnH/pi;
      hH(M+2:1:N)=hH(M:-1:1);
      hL=sin(WnL*[-M:1:-1])./([-M:1:-1]*pi);
      hL(M+1)=WnL/pi;
      hL(M+2:1:N)=hL(M:-1:1);
      if Ftype == 1
      h(1:N)=hL(1:N);
     end
     if Ftype == 2
      h(1:N)=-hH(1:N);
      h(M+1)=1+h(M+1);
     end
     if Ftype ==3
      h(1:N)=hH(1:N)-hL(1:N);
     end
     if Ftype == 4
      h(1:N)=hL(1:N)-hH(1:N);
      h(M+1)=1+h(M+1);
   end
% window functions;
     if Wtype ==1
      w(1:N)=ones(1,N);
     end
     if Wtype ==2
      w=1-abs([-M:1:M])/M;
     end
   if Wtype ==3
      w= 0.5+0.5*cos([-M:1:M]*pi/M);
     end
     if Wtype ==4
      w=0.54+0.46*cos([-M:1:M]*pi/M);
     end
     if Wtype ==5
      w=0.42+0.5*cos([-M:1:M]*pi/M)+0.08*cos(2*[-M:1:M]*pi/M);
   end
     B=h .* w
```

Program 7.15 enables one to design FIR filters using the frequency sampling method. Note that values of the frequency response, which correspond to the equally spaced DFT frequency components, must be specified for design. Besides the lowpass, highpass, bandpass, and bandstop filter designs, the method can be used to design FIR filters with an arbitrarily specified magnitude frequency response.

Program 7.15. MATLAB function for FIR filter design using the frequency sampling method.

```
function B=firfs(N,Hk)
% B=firfs(N,Hk)
% Fir filter design using the frequency sampling method.
% Input parameters:
% N: the number of filter coefficients.
%   note: N must be odd number.
% Hk: sampled frequency response for k=0,1,2,...,M=(N-1)/2.
% Output:
% B: FIR filter coefficients.
        M=(N-1)/2;
        for n=1:1:N
        B(n)=(1/N)*(Hk(1)+...
        2*sum(Hk(2:1:M+1)...
        .*cos(2*pi*([1:1:M])*(n-1-M)/N)));
  End
```

7.13 PROBLEMS

7.1 Design a three-tap FIR lowpass filter with a cutoff frequency of 1500 Hz and a sampling rate of 8000 Hz using

(a) Rectangular window function

(b) Hamming window function.

Determine the transfer function and difference equation of the designed FIR system, and compute and plot the magnitude frequency response for $\Omega=0$, $\pi/4$, $\pi/2$, $3\pi/4$, and π (rad).

7.2 Design a three-tap FIR highpass filter with a cutoff frequency of 1600 Hz and a sampling rate of 8000 Hz using

(a) Rectangular window function

(b) Hamming window function.

Determine the transfer function and difference equation of the designed FIR system, and compute and plot the magnitude frequency response for $\Omega=0$, $\pi/4$, $\pi/2$, $3\pi/4$, and π (rad).

7.3 Design a five-tap FIR lowpass filter with a cutoff frequency of 100 Hz and a sampling rate of 1000 Hz using

(a) Rectangular window function

(b) Hamming window function.

Determine the transfer function and difference equation of the designed FIR system, and compute and plot the magnitude frequency response for $\Omega=0$, $\pi/4$, $\pi/2$, $3\pi/4$, and π (rad).

7.4 Design a five-tap FIR highpass filter with a cutoff frequency of 250 Hz and a sampling rate of 1000 Hz using

(a) Rectangular window function

(b) Hamming window function.

Determine the transfer function and difference equation of the designed FIR system, and compute and plot the magnitude frequency response for $\Omega=0$, $\pi/4$, $\pi/2$, $3\pi/4$, and π (rad).

7.5 Design a five-tap FIR bandpass filter with a lower cutoff frequency of 1600 Hz, an upper cutoff frequency of 1800 Hz, and a sampling rate of 8000 Hz using

(a) Rectangular window function

(b) Hamming window function.

Determine the transfer function and difference equation of the designed FIR system, and compute and plot the magnitude frequency response for $\Omega=0$, $\pi/4$, $\pi/2$, $3\pi/4$, and π (rad).

7.6 Design a five-tap FIR bandreject filter with a lower cutoff frequency of 1600 Hz, an upper cutoff frequency of 1800 Hz, and a sampling rate of 8000 Hz using

(a) Rectangular window function

(b) Hamming window function.

Determine the transfer function and difference equation of the designed FIR system, and compute and plot the magnitude frequency response for $\Omega=0$, $\pi/4$, $\pi/2$, $3\pi/4$, and π (rad).

7.7 Given an FIR lowpass filter design with the following specifications:

Passband = 0 –800 Hz

Stopband = 1200–4000 Hz

Passband ripple = 0.1 dB

Stopband attenuation = 40 dB

Sampling rate = 8000 Hz

Determine the following:

(a) Window method

(b) Length of the FIR filter

(c) Cutoff frequency for the design equation.

7.8 Given an FIR highpass filter design with the following specifications:

Stopband = 0–1500 Hz

Passband = 2000–4000 Hz

Passband ripple = 0.02 dB

Stopband attenuation = 60 dB

Sampling rate = 8000 Hz

Determine the following:

(a) Window method

(b) Length of the FIR filter

(c) Cutoff frequency for the design equation.

7.9 Given an FIR bandpass filter design with the following specifications:

Lower cutoff frequency = 1500 Hz

Lower transition width = 600 Hz

Upper cutoff frequency = 2300 Hz

Upper transition width = 600 Hz

Passband ripple = 0.1 dB

Stopband attenuation = 50 dB
Sampling rate: 8000 Hz
Determine the following:
(a) Window method
(b) Length of the FIR filter
(c) Cutoff frequencies for the design equation.

7.10 Given an FIR bandstop filter design with the following specifications:
Lower passband = 0–1200 Hz
Stopband = 1600–2000 Hz
Upper passband = 2400–4000 Hz
Passband ripple = 0.05 dB
Stopband attenuation = 60 dB
Sampling rate = 8000 Hz
Determine the following:
(a) Window method
(b) Length of the FIR filter
(c) Cutoff frequencies for the design equation.

7.11 Given an FIR system

$$H(z) = 0.25 - 0.5z^{-1} + 0.25z^{-2},$$

realize $H(z)$ using each of the following specified methods:
(a) Transversal form, and write the difference equation for implementation.
(b) Linear phase form, and write the difference equation for implementation.

7.12 Given an FIR filter transfer function

$$H(z) = 0.2 + 0.5z^{-1} - 0.3z^{-2} + 0.5z^{-3} + 0.2z^{-4},$$

perform the linear phase FIR filter realization, and write the difference equation for implementation.

7.13 Determine the transfer function for a three-tap FIR lowpass filter with a cutoff frequency of 150 Hz and a sampling rate of 1000 Hz using the frequency sampling method.

7.14 Determine the transfer function for a three-tap FIR highpass filter with a cutoff frequency of 250 Hz and a sampling rate of 1000 Hz using the frequency sampling method.

7.15 Determine the transfer function for a five-tap FIR lowpass filter with a cutoff frequency of 2000 Hz and a sampling rate of 8000 Hz using the frequency sampling method.

7.16 Determine the transfer function for a five-tap FIR highpass filter with a cutoff frequency of 3000 Hz and a sampling rate of 8000 Hz using the frequency sampling method.

7.17 Given the following specifications:
- a seven-tap FIR bandpass filter
- a lower cutoff frequency of 1500 Hz and an upper cutoff frequency of 3000 Hz
- a sampling rate of 8000 Hz
- the frequency sampling design method

Determine the transfer function.

7.18 Given the following specifications:
- a seven-tap FIR bandreject filter
- a lower cutoff frequency of 1500 Hz and an upper cutoff frequency of 3000 Hz

- a sampling rate of 8000 Hz
- the frequency sampling design method

Determine the transfer function.

7.19 A lowpass FIR filter to be designed has the following specifications:

Design method: Packs-McClellan algorithm
Sampling rate = 1000 Hz
Passband = 0–200 Hz
Stopband = 300–500 Hz
Passband ripple = 1 dB
Stopband attenuation = 40 dB
Determine the error weights W_p and W_s for passband and stopband in the Packs-McClellan algorithm.

7.20 A bandpass FIR filter to be designed has the following specifications:

Design method: Packs-McClellan algorithm
Sampling rate = 1000 Hz
Passband = 200–250 Hz
Lower stopband = 0–150 Hz
Upper stopband = 300–500 Hz
Passband ripple = 1 dB
Stopband attenuation = 30 dB
Determine the error weights W_p and W_s for passband and stopband in the Packs-McClellan algorithm.

7.21 A highpass FIR filter to be designed has the following specifications:

Design method: Packs-McClellan algorithm
Sampling rate = 1000 Hz
Passband = 350–500 Hz
Stopband = 0–250 Hz
Passband ripple = 1 dB
Stopband attenuation = 60 dB
Determine the error weights W_p and W_s for passband and stopband in the Packs-McClellan algorithm.

7.22 A bandstop FIR filter to be designed has the following specifications:

Design method: Packs-McClellan algorithm
Sampling rate = 1000 Hz
Stopband = 250–350 Hz
Lower passband = 0–200 Hz
Upper passband = 400–500 Hz
Passband ripple = 1 dB
Stopband attenuation = 25 dB
Determine the error weights W_p and W_s for passband and stopband in the Packs-McClellan algorithm.

7.23 In a speech recording system with a sampling rate of 10,000 Hz, the speech is corrupted by broadband random noise. To remove the random noise while preserving speech information, the following specifications are given:

Speech frequency range = 0–3000 Hz
Stopband range = 4000–5000 Hz
Passband ripple = 0.1 dB
Stopband attenuation = 45 dB
FIR filter with Hamming window
Determine the FIR filter length (number of taps) and the cutoff frequency; use MATLAB to design the filter; and plot frequency response.

7.24 Given a speech equalizer shown in Fig. 7.46 to compensate midrange frequency loss of hearing:
Sampling rate = 8000 Hz
Bandpass FIR filter with Hamming window
Frequency range to be emphasized = 1500–2000 Hz
Lower stopband = 0–1000 Hz
Upper stopband = 2500–4000 Hz
Passband ripple = 0.1 dB
Stopband attenuation = 45 dB
Determine the filter length and the lower and upper cutoff frequencies.

7.25 A digital crossover can be designed as shown in Fig. 7.47.
Given the following audio specifications:
Sampling rate = 44,100 Hz
Crossover frequency = 2000 Hz
Transition band range = 1600 Hz
Passband ripple = 0.1 dB

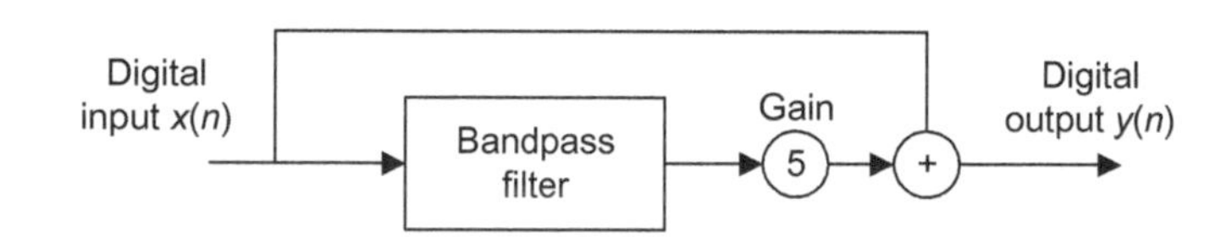

FIG. 7.46

Speech equalizer in Problem 7.24.

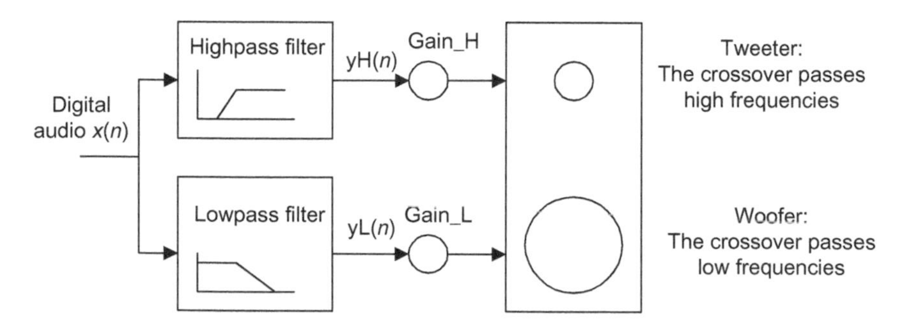

FIG. 7.47

Two-band crossover in Problem 7.25.

Stopband attenuation = 50 dB
Filter type = FIR
Determine the following for each filter:

(a) window function
(b) filter length
(c) cutoff frequency.

Use MATLAB to design both filters and plot frequency responses for both filters.

Computer Problems with MATLAB

Use the MATLAB programs provided in Section 7.11 to design the following FIR filters.

7.26 Design a 41-tap lowpass FIR filter whose cutoff frequency is 1600 Hz using the following window functions. Assume that the sampling frequency is 8000 Hz.

(a) Rectangular window function
(b) Triangular window function
(c) Hanning window function
(d) Hamming window function
(e) Blackman window function.

List the FIR filter coefficients and plot the frequency responses for each case.

7.27 Design a lowpass FIR filter whose cutoff frequency is 1000 Hz using the Hamming window function for the following specified filter length. Assume that the sampling frequency is 8000 Hz.

(a) 21 filter coefficients
(b) 31 filter coefficients
(c) 41 filter coefficients.

List FIR filter coefficients for each design and compare the magnitude frequency responses.

7.28 Design a 31-tap highpass FIR filter whose cutoff frequency is 2500 Hz using the following window functions. Assume that the sampling frequency is 8000 Hz.

(a) Hanning window function
(b) Hamming window function
(c) Blackman window function.

List the FIR filter coefficients and plot the frequency responses for each design.

7.29 Design a 41-tap bandpass FIR filter with the lower and upper cutoff frequencies being 2500 and 3000 Hz, respectively, using the following window functions. Assume a sampling frequency of 8000 Hz.

(a) Hanning window function
(b) Blackman window function.

List the FIR filter coefficients and plot the frequency responses for each design.

7.30 Design 41-tap bandreject FIR filter with the cutoff frequencies of 2500 and 3000 Hz, respectively, using the Hamming window function. Assume a sampling frequency of 8000 Hz. List the FIR filter coefficients and plot the frequency responses.

7.31 Use the frequency sampling method to design a linear phase lowpass FIR filter with 17 coefficients. Let the cutoff frequency be 2000 Hz and assume a sampling frequency of 8000 Hz. List FIR filter coefficients and plot the frequency responses.

7.32 Use the frequency sampling method to design a linear phase bandpass FIR filter with 21 coefficients. Let the lower and upper cutoff frequencies be 2000 and 2500 Hz, respectively, and assume a sampling frequency of 8000 Hz. List the FIR filter coefficients and plot the frequency responses.

7.33 Given an input data sequence:

$$x(n) = 1.2 \times \sin\left(2\pi \times \frac{1000n}{8000}\right) - 1.5 \times \cos\left(2\pi \times \frac{2800n}{8000}\right),$$

assuming a sampling frequency of 8000 Hz, use the designed FIR filter with Hamming window in Problem 7.26 to filter 400 data points of $x(n)$, and plot the 400 samples of the input and output data.

7.34 Design a lowpass FIR filter with the following specifications:

Design method: Packs-McClellan algorithm
Sampling rate = 8000 Hz
Passband = 0–1200 Hz
Stopband = 1500–4000 Hz
Passband ripple = 1 dB
Stopband attenuation = 40 dB
List the filter coefficients and plot the frequency responses.

7.35 Design a bandpass FIR filter with the following specifications:

Design method: Packs-McClellan algorithm
Sampling rate = 8000 Hz
Passband = 1200–1600 Hz
Lower stopband = 0–800 Hz
Upper stopband = 2000–4000 Hz
Passband ripple = 1 dB
Stopband attenuation = 40 dB
List the filter coefficients and plot the frequency responses.

MATLAB Projects

7.36 Speech enhancement:

A digitally recorded speech in the noisy environment can be enhanced using a lowpass filter if the recorded speech with a sampling rate of 8000 Hz contains information within 1600 Hz. Design a lowpass filter to remove the high-frequency noise above 1600 Hz with following filter specifications: passband frequency range: 0–1600 Hz; passband ripple: 0.02 dB; stopband frequency range: 1800–4000 Hz; and stopband attenuation: 50 dB.

Use the designed lowpass filter to filter the noisy speech and adopt the following code to simulate the noisy speech:

```
load speech.dat
t=[0:length(speech)-1]*T;
th=mean(speech.*speech)/4; %Noise power =(1/4) speech power
noise=sqrt(th)*randn([1,length(speech)]); %Generate Gaussian noise
nspeech=speech+noise; % Generate noisy speech
```

In this project, plot the speech samples and spectra for both noisy speech and the enhanced speech and use MATLAB sound() function to evaluate the sound qualities. For example, to hear the noisy speech:

sound(nspeech/max(abs(nspeech)),8000);

7.37 Digital crossover system:

Design a two-band digital crossover system with the following specifications:

Sampling rate = 44,100 Hz;
Crossover frequency = 1200 Hz (cutoff frequency);
Transition band = 800–1600 Hz;
Lowpass filter: passband frequency range from 0 to 800 Hz with a ripple of 0.02 dB and stopband edge at 1600 Hz with the attenuation of 50 dB;
Highpass filter: passband frequency range from 1.6 to 22.05 kHz with a ripple of 0.02 dB and stopband edge at 800 Hz with the attenuation of 50 dB.
In this project, plot the magnitude frequency responses for both lowpass and highpass filters. Use the following MATLAB code to read stereo audio data ("No9seg.wav").

```
[x fs Nbits]=audioread('No9seg.wav');
```

Process the given stereo audio segment. Listen and experience the processed audio in the following sequences:
Channel 1: original, lowband, and highband
Channel 2: original, lowband, and highband
Stereo (both channels): original, lowband, and highband.

Advanced Problems

7.38 The frequency response of a half-band digital differentiator is given below:

$$H(e^{j\Omega}) = j\Omega \text{ for } |\Omega| < \pi/2.$$

Design the FIR differentiator with $(2M+1)$ coefficients using the Fourier transform method.

7.39 The desired frequency response is given as

$$H(e^{j\Omega}) = \cos(\Omega/2) \text{ for } |\Omega| < \pi.$$

Design the FIR filter with $(2M+1)$ coefficients using the Fourier transform method.

7.40 The frequency response of an ideal digital differentiator is given below:

$$H(e^{j\Omega}) = j\Omega \text{ for } |\Omega| < \pi.$$

Derive formula for $h(n)$ with $(2M+1)$ coefficients using the frequency sampling method.

7.41 Derive formula for $h(n)$ for the FIR filter with $(2M+1)$ coefficients in Problem 7.39 using the frequency sampling method.

7.42 The second-order derivative transfer function is given by

$$H(e^{j\Omega}) = (j\Omega)^2 \text{ for } |\Omega| < \pi.$$

Design the FIR second-order differentiator with $(2M+1)$ coefficients using the Fourier transform method.

7.43 Derive the formula for $h(n)$ for the FIR differentiator filter with $(2M+1)$ coefficients in Problem 7.42 using the frequency sampling method.

7.44 The frequency response of the Hilbert transformer is given by

$$H(e^{j\Omega}) = \begin{cases} -j & 0 \le \Omega < \pi/2 \\ j & -\pi/2 \le \Omega < 0 \end{cases}.$$

Derive the formula for $h(n)$ with $(2M+1)$ coefficients using the Fourier transform method.

CHAPTER

INFINITE IMPULSE RESPONSE FILTER DESIGN

8

CHAPTER OUTLINE

Digital Signal Processing. https://doi.org/10.1016/B978-0-12-815071-9.00008-7

8.1 INFINITE IMPULSE RESPONSE FILTER FORMAT

In this chapter, we will study several methods for infinite impulse response (IIR) filter design. An IIR filter is described using the difference equation, as discussed in Chapter 6:

$$\begin{aligned} y(n) = & b_0 x(n) + b_1 x(n-1) + \cdots + b_M x(n-M) \\ & - a_1 y(n-1) - \cdots - a_N y(n-N). \end{aligned}$$

Chapter 6 also gives the IIR filter transfer function as

$$H(z) = \frac{Y(z)}{X(z)} = \frac{b_0 + b_1 z^{-1} + \cdots + b_M z^{-M}}{1 + a_1 z^{-1} + \cdots + a_N z^{-N}},$$

where b_i and a_i are the $(M+1)$ numerator and N denominator coefficients, respectively. $Y(z)$ and $X(z)$ are the z-transform functions of the input $x(n)$ and output $y(n)$. To become familiar with the form of the IIR filter, let us look at the following example.

EXAMPLE 8.1

Given the following IIR filter:

$$y(n) = 0.2x(n) + 0.4x(n-1) + 0.5y(n-1),$$

determine the transfer function, nonzero coefficients, and impulse response.

Solution:

Applying the z-transform and solving for a ratio of the z-transform output over input, we have

$$H(z) = \frac{Y(z)}{X(z)} = \frac{0.2 + 0.4z^{-1}}{1 - 0.5z^{-1}}.$$

We also identify the nonzero numerator coefficients and denominator coefficient as

$$b_0 = 0.2, b_1 = 0.4, \text{and } a_1 = -0.5.$$

To solve the impulse response, we rewrite the transfer function as

$$H(z) = \frac{0.2}{1 - 0.5z^{-1}} + \frac{0.4z^{-1}}{1 - 0.5z^{-1}}.$$

Using the inverse z-transform and shift theorem, we obtain the impulse response as

$$h(n) = 0.2(0.5)^n u(n) + 0.4(0.5)^{n-1} u(n-1).$$

The obtained impulse response has an infinite number of terms, where the first several terms are calculated as

$$h(0) = 0.2, h(1) = 0.5, h(2) = 0.25, \cdots$$

At this point, we can make the following remarks:

1. The IIR filter output $y(n)$ depends not only on the current input $x(n)$ and past inputs $x(n-1), \ldots$, but also on the past output(s) $y(n-1) \ldots$, (recursive terms). Its transfer function is a ratio of the numerator polynomial over the denominator polynomial, and its impulse response has an infinite number of terms.
2. Since the transfer function has the denominator polynomial, the pole(s) of a designed IIR filter must be inside the unit circle on the z-plane to ensure its stability.
3. Comparing with the finite impulse response (FIR) filter (see Chapter 7), the IIR filter offers a much smaller filter size. Hence, the filter operation requires a fewer number of computations, but the linear phase is not easily obtained. The IIR filter is preferred when a small filter size is called for but the application does not require a linear phase.

The objective of IIR filter design is to determine the filter numerator and denominator coefficients to satisfy filter specifications such as passband gain and stopband attenuation, as well as cutoff frequency/frequencies for the lowpass, highpass, bandpass, and bandstop filters.

We will first focus on bilinear transformation (BLT) design method. Then we will introduce other design methods such as the impulse invariant design and the pole-zero placement design.

8.2 BILINEAR TRANSFORMATION DESIGN METHOD

Fig. 8.1 illustrates a flow chart of the BLT design used in this book. The design procedure includes the following steps: (1) transforming digital filter specifications into analog filter specifications, (2) performing analog filter design, and (3) applying BLT (which will be introduced in the next section) and verifying its frequency response.

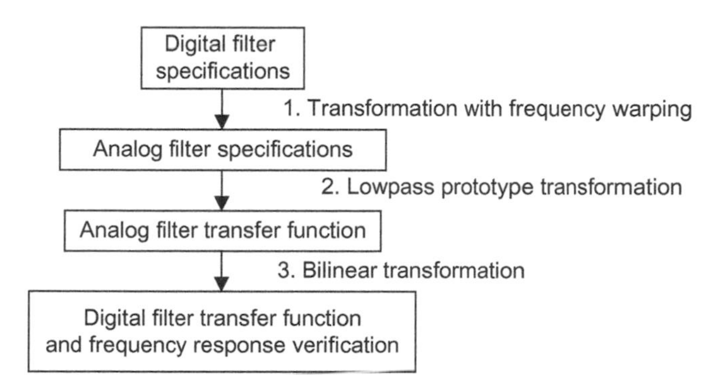

FIG. 8.1

General procedure for IIR filter design using BLT.

8.2.1 ANALOG FILTERS USING LOWPASS PROTOTYPE TRANSFORMATION

Before we begin to develop the BLT design, let us review analog filter design using *lowpass prototype transformation*. This method converts the analog lowpass filter with a cutoff frequency of 1 rad/s, called the lowpass prototype, into practical analog lowpass, highpass, bandpass, and bandstop filters with their frequency specifications.

Letting $H_P(s)$ be a transfer function of the lowpass prototype, the transformation of the lowpass prototype into a lowpass filter is shown in Fig. 8.2.

As shown in Fig. 8.2, $H_{LP}(s)$ designates the analog lowpass filter with a cutoff frequency ω_c rad/s. The lowpass prototype to lowpass filter transformation substitutes s in the lowpass prototype function $H_P(s)$ with s/ω_c, where v is the normalized frequency of the lowpass prototype and ω_c is the cutoff frequency of the lowpass filter to be designed. Let us consider the following first-order lowpass prototype:

$$H_P(s) = \frac{1}{s+1}. \tag{8.1}$$

Its frequency response is obtained by substituting $s = jv$ into Eq. (8.1), that is,

$$H_P(jv) = \frac{1}{jv+1}$$

with the magnitude gain given in Eq. (8.2):

$$|H_P(jv)| = \frac{1}{\sqrt{1+v^2}}. \tag{8.2}$$

We compute the gains at $v=0$, $v=1$, $v=100$, and $v=10,000$ to obtain 1, $1/\sqrt{2}$, 0.0995, and 0.01, respectively. The cutoff frequency gain at $v=1$ equals $1/\sqrt{2}$, which is equivalent to -3 dB, and the direct-current (DC) gain is 1. The gain approaches zero when the frequency tends to $v=+\infty$. This verifies that the lowpass prototype is a normalized lowpass filter with a normalized cutoff frequency of 1. Applying the prototype transformation $s = s/\omega_c$ in Fig. 8.2, we get an analog lowpass filter with a cutoff frequency of ω_c as

$$H(s) = \frac{1}{s/\omega_c + 1} = \frac{\omega_c}{s+\omega_c}. \tag{8.3}$$

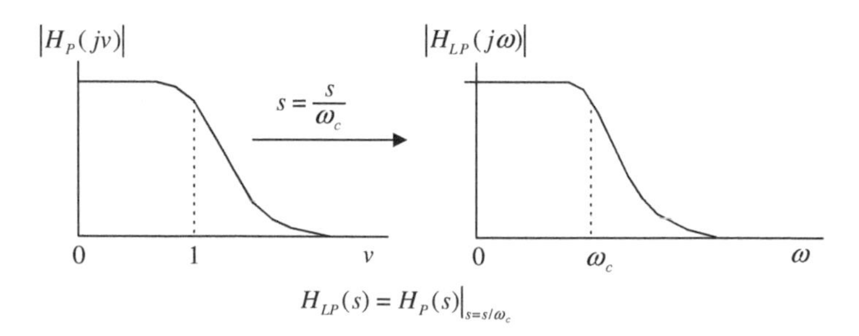

FIG. 8.2

Analog lowpass prototype transformation into a lowpass filter.

We can obtain the analog frequency response by substituting $s = j\omega$ into Eq. (8.3), that is,

$$H(j\omega) = \frac{1}{j\omega/\omega_c + 1}.$$

The magnitude response is determined by

$$|H(j\omega)| = \frac{1}{\sqrt{1 + \left(\frac{\omega}{\omega_c}\right)^2}}. \tag{8.4}$$

Similarly, we verify the gains at $\omega = 0$, $\omega = \omega_c$, $\omega = 100\omega_c$, and $\omega = 10,000\omega_c$ to be 1, $1/\sqrt{2}$, 0.0995, and 0.01, respectively. The filter gain at the cutoff frequency ω_c equals $1/\sqrt{2}$, and the DC gain is 1. The gain approaches zero when $\omega = +\infty$. We note that filter gains do not change but the filter frequency is scaled up by a factor of ω_c. This verifies that the prototype transformation converts the lowpass prototype to the analog lowpass filter with the specified cutoff frequency of ω_c without an effect on the filter gain.

This first-order prototype function is used here for an illustrative purpose. We will obtain general functions for Butterworth and Chebyshev lowpass prototypes in a later section.

The highpass, bandpass, and bandstop filters using the specified lowpass prototype transformation can be easily verified. We review them in Figs. 8.3–8.5, respectively.

The transformation from the lowpass prototype to the highpass filter $H_{HP}(s)$ with a cutoff frequency ω_c rad/s is shown in Fig. 8.3, where $s = \omega_c/s$ in the lowpass prototype transformation.

The transformation of the lowpass prototype function to a bandpass filter with a center frequency ω_0, a lower cutoff frequency ω_l, and an upper cutoff frequency ω_h in the passband is depicted in Fig. 8.4, where $s = (s^2 + \omega_0^2)/(sW)$ is substituted into the lowpass prototype.

As shown in Fig. 8.4, ω_0 is the geometric center frequency, which is defined as $\omega_0 = \sqrt{\omega_l \omega_h}$ while the passband bandwidth is given by $W = \omega_h - \omega_l$. Similarly, the transformation from the lowpass prototype to a bandstop (bandreject) filter is illustrated in Fig. 8.5 with $s = sW/(s^2 + \omega_0^2)$ substituted into the lowpass prototype.

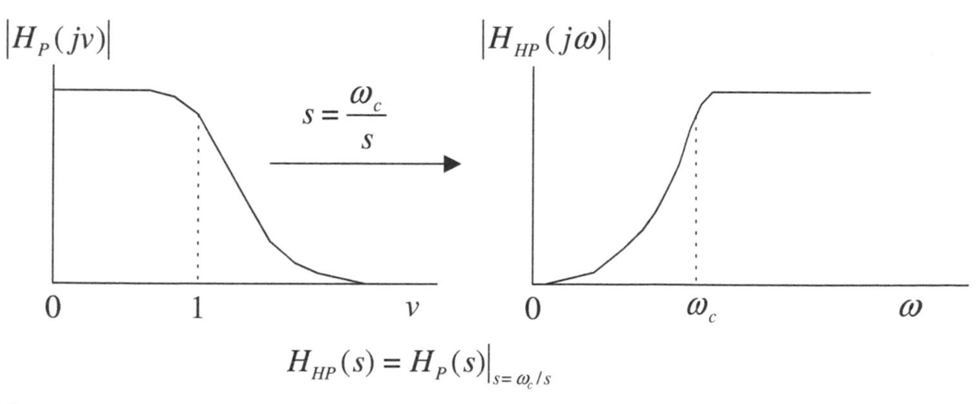

FIG. 8.3

Analog lowpass prototype transformation to the highpass filter.

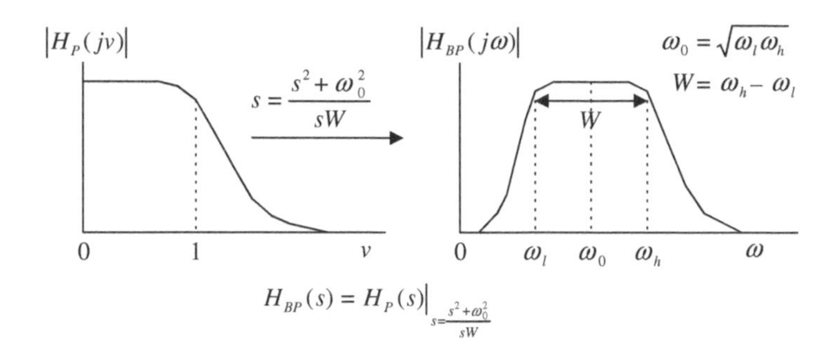

FIG. 8.4

Analog lowpass prototype transformation to the bandpass filter.

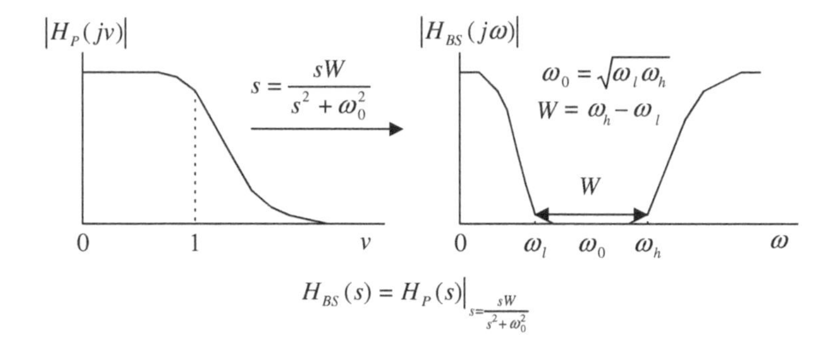

FIG. 8.5

Analog lowpass prototype transformation to a bandstop filter.

Finally, the lowpass prototype transformations are summarized in Table 8.1.

MATLAB function **freqs ()** can be used to plot analog filter frequency responses for verification with the following syntax:

H = freqs(B,A,W)

B = the vector containing the numerator coefficients

A = the vector containing the denominator coefficients

W = the vector containing the specified analog frequency points (rad/s)

H = the vector containing the frequency response.

The following example verifies the lowpass prototype transformation.

Table 8.1 Analog Lowpass Prototype Transformations

Filter Type	Prototype Transformation
Lowpass	$\frac{s}{\omega_c}$, ω_c is the cutoff frequency
Highpass	$\frac{\omega_c}{s}$, ω_c is the cutoff frequency
Bandpass	$\frac{s^2+\omega_0^2}{sW}$, $\omega_0 = \sqrt{\omega_l \omega_h}$, $W = \omega_h - \omega_l$
Bandstop	$\frac{sW}{s^2+\omega_0^2}$, $\omega_0 = \sqrt{\omega_l \omega_h}$, $W = \omega_h - \omega_l$

EXAMPLE 8.2

Given a lowpass prototype

$$H_P(s) = \frac{1}{s+1},$$

Determine each of the following analog filters and plot their magnitude responses from 0 to 200 rad/s.

1. The highpass filter with a cutoff frequency of 40 rad/s.
2. The bandpass filter with a center frequency of 10 rad/ s and bandwidth of 20 rad/s.

Solution:

(1) Applying the lowpass prototype transformation by substituting $s = 40/s$ into the lowpass prototype, we have an analog highpass filter as

$$H_{HP}(s) = \frac{1}{\frac{40}{s}+1} = \frac{s}{s+40}.$$

(2) Similarly, substituting the lowpass to bandpass transformation $s = (s^2+100)/(20s)$ into the lowpass prototype leads to

$$H_{BP}(s) = \frac{1}{\frac{s^2+100}{20s}+1} = \frac{20s}{s^2+20s+100}.$$

The program for plotting the magnitude responses for highpass filter and bandpass filter is shown in Program 8.1, and Fig. 8.6 displays the magnitude responses for the highpass filter and bandpass filter, respectively.

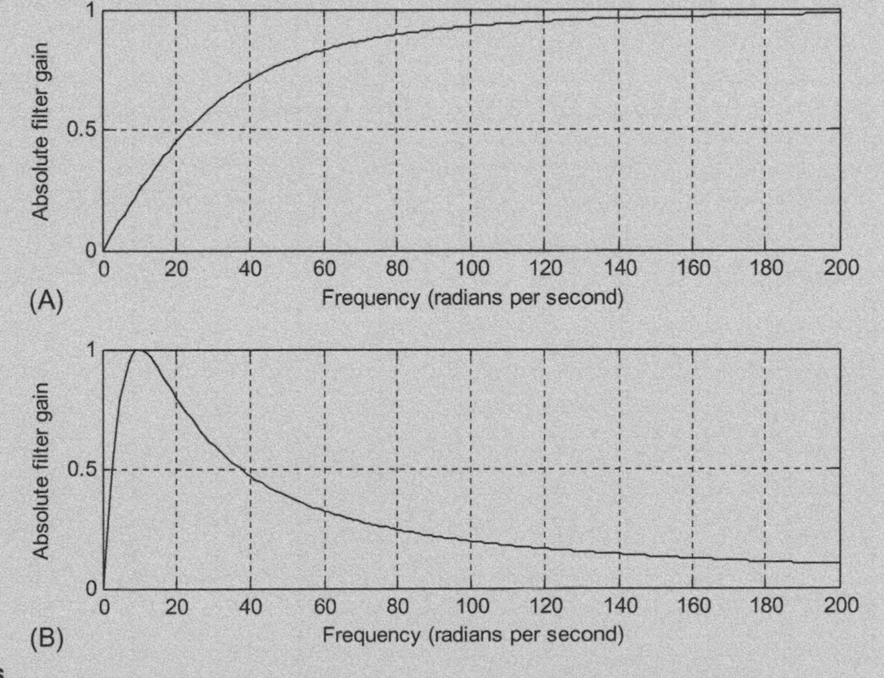

FIG. 8.6

Magnitude responses for the analog highpass filter and bandpass filter in Example 8.2.

Program 8.1. MATLAB program in Example 8.2.

```
W=0:1:200; %Aanalog frequency points for computing the filter gains
Ha=freqs([1 0][1 4 0],W);                    %Frequency response for the highpass filter
Hb=freqs([20 0][1 20 100],W);                %Frequency response for the bandpass filter
subplot(2,1,1);plot(W, abs(Ha),'k');grid % The filter gain plot for highpass filter
xlabel('(a) Frequency (radians per second)')
ylabel('Absolute filter gain');
subplot(2,1,2);plot(W,abs(Hb),'k');grid  % The filter gain plot for bandpass filter
xlabel('(b) Frequency (radians per second)')
ylabel('Absolute filter gain');
```

Fig. 8.6 confirms the transformation of the lowpass prototype into a highpass filter and a bandpass filter, respectively. To obtain the transfer function of an analog filter, we always begin with lowpass prototype and apply the corresponding lowpass prototype transformation. To transfer from a lowpass prototype to a bandpass or bandstop filter, the resultant order of the analog filter is twice the lowpass prototype order.

8.2.2 BILINEAR TRANSFORMATION AND FREQUENCY WARPING

In this section, we develop the BLT, which converts an analog filter to a digital filter. We begin by finding the area under a curve using the integration of calculus and the numerical recursive method in order to determine the BLT. The area under the curve is a common problem in early calculus course. As shown in Fig. 8.7, the area under the curve can be determined using the following integration:

$$y(t)=\int_0^t x(t)dt, \tag{8.5}$$

where $y(t)$ (area under the curve) and $x(t)$ (curve function) are the output and input of the analog integrator, respectively, and t is the upper limit of the integration.

Applying Laplace transform in Eq. (8.5), we have

$$Y(s)=\frac{X(s)}{s} \tag{8.6}$$

and find the Laplace transfer function as

$$G(s)=\frac{Y(s)}{X(s)}=\frac{1}{s}. \tag{8.7}$$

Now we examine the numerical integration method shown in Fig. 8.7 to approximate the integration of Eq. (8.5) using the following difference equation:

$$y(n)=y(n-1)+\frac{x(n)+x(n-1)}{2}T, \tag{8.8}$$

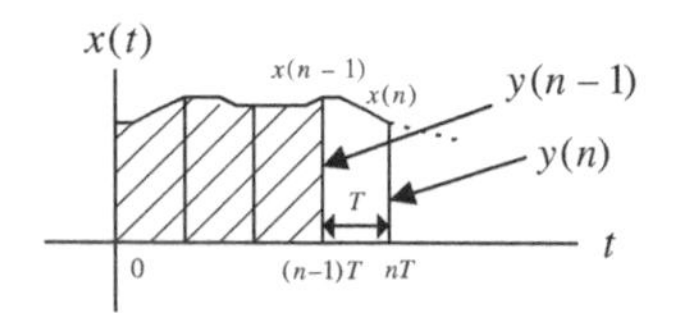

FIG. 8.7

Digital integration method to calculate the area under the curve.

where T denotes the sampling period. $y(n)=y(nT)$ is the output sample that is the whole area under the curve, while $y(n-1)=y(nT-T)$ is the previous output sample from the integrator indicating the previously computed area under the curve (the shaded area in Fig. 8.7). Note that $x(n)=x(nT)$ and $x(n-1)=x(nT-T)$, sample amplitudes from the curve, are the current input sample and the previous input sample in Eq. (8.8). Applying the z-transform on both sides of Eq. (8.8) leads to

$$Y(z)=z^{-1}Y(z)+\frac{T}{2}\left(X(z)+z^{-1}X(z)\right).$$

Solving for the ratio of $Y(z)/X(z)$, we achieve the z-transfer function as

$$H(z)=\frac{Y(z)}{X(z)}=\frac{T}{2}\frac{1+z^{-1}}{1-z^{-1}}. \tag{8.9}$$

Next, on comparing Eq. (8.9) with Eq. (8.7), it follows that

$$\frac{1}{s}=\frac{T}{2}\frac{1+z^{-1}}{1-z^{-1}}=\frac{T}{2}\frac{z+1}{z-1}. \tag{8.10}$$

Solving for s in Eq. (8.10) gives the BLT

$$s=\frac{2}{T}\frac{z-1}{z+1}. \tag{8.11}$$

The BLT method is a mapping or transformation of points on the s-plane to the z-plane. Eq. (8.11) can be alternatively written as

$$z=\frac{1+sT/2}{1-sT/2}. \tag{8.12}$$

The general mapping properties are summarized as follows:

1. The left half of the s-plane is mapped onto the inside area of the unit circle of the z-plane.
2. The right half of the s-plane is mapped onto the outside area of the unit circle of the z-plane.
3. The positive $j\omega$ axis portion in the s-plane is mapped onto the positive half circle (the dashed line arrow in Fig. 8.8) on the unit circle, while the negative $j\omega$ axis is mapped onto the negative half circle (the dotted line arrow in Fig. 8.8) on the unit circle.

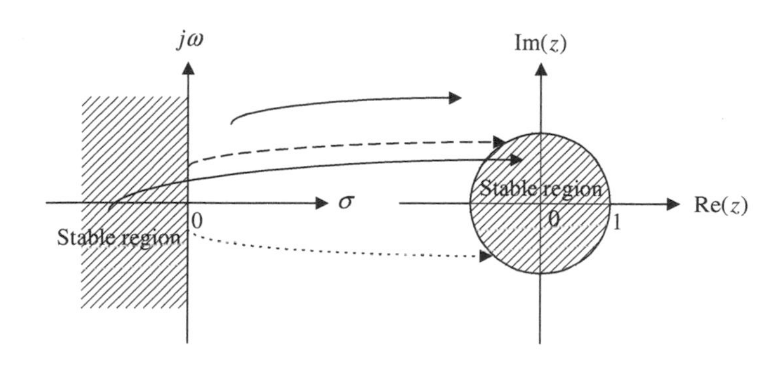

FIG. 8.8

Mapping between the s-plane and the z-plane by the BLT.

To verify these features, let us look at the following illustrative example:

EXAMPLE 8.3

Assuming that $T=2$ s in Eq. (8.12), and given the following points:
(1) $s=-1+j$, on the left half of s-plane
(2) $s=1-j$, on the right half of s-plane
(3) $s=j$, on the positive $j\omega$ on the s-pane
(4) $s=-j$, on the negative $j\omega$ on the s-plane

Convert each of the points in the s-plane to the z-plane, and verify the mapping properties (1) to (3).

Solution:

Substituting $T=2$ into Eq. (8.12) leads to

$$z=\frac{1+s}{1-s}.$$

We can carry out mapping for each point as follows:

(1) $z=\frac{1+(-1+j)}{1-(-1+j)}=\frac{j}{2-j}=\frac{1\angle 90^\circ}{\sqrt{5}\angle -26.57^\circ}=0.4472\angle 116.57^\circ$,

since $|z|=0.4472<1$, which is inside the unit circle on the z-plane.

(2) $z=\frac{1+(1-j)}{1-(1-j)}=\frac{2-j}{j}=\frac{\sqrt{5}\angle -26.57^\circ}{1\angle 90^\circ}=2.2361\angle -116.57^\circ$,

since $|z|=2.2361>1$, which is outside the unit circle on the z-plane.

(3) $z=\frac{1+j}{1-j}=\frac{\sqrt{2}\angle 45^\circ}{\sqrt{2}\angle -45^\circ}=1\angle 90^\circ$, since $|z|=1$ and $\theta=90^\circ$, which is on the positive half circle on the unit circle in the z-plane.

(4) $z=\frac{1-j}{1-(-j)}=\frac{1-j}{1+j}=\frac{\sqrt{2}\angle -45^\circ}{\sqrt{2}\angle 45^\circ}=1\angle -90^\circ$,

since $|z|=1$ and $\theta=-90^\circ$, which is on the negative half circle on the unit circle in the z-plane.

As shown in Example (8.3), the BLT offers conversion of an analog transfer function to a digital transfer function. Example (8.4) shows how to perform the BLT.

EXAMPLE 8.4

Given an analog filter whose transfer function is

$$H(s)=\frac{10}{s+10},$$

convert it to the digital filter transfer function and difference equation, respectively, when a sampling period is given as $T=0.01$ s.

Solution:

Applying the BLT, we have

$$H(z)=H(s)\Big|_{s=\frac{2}{T}\frac{z-1}{z+1}}=\frac{10}{s+10}\Big|_{s=\frac{2}{T}\frac{z-1}{z+1}}.$$

Substituting $T=0.01$, it follows that

$$H(z)=\frac{10}{\frac{200(z-1)}{z+1}+10}=\frac{0.05}{\frac{z-1}{z+1}+0.05}=\frac{0.05(z+1)}{z-1+0.05(z+1)}=\frac{0.05z+0.05}{1.05z-0.95}.$$

Finally, we get

$$H(z)=\frac{(0.05z+0.05)/(1.05z)}{(1.05z-0.95)/(1.05z)}=\frac{0.0476+0.0476z^{-1}}{1-0.9048z^{-1}}.$$

Applying the technique in Chapter 6, we yield the difference equation as

$$y(n)=0.0476x(n)+0.0476x(n-1)+0.9048y(n-1).$$

Next, we examine frequency mapping between the s-plane and the z-plane. As illustrated in Fig. 8.9, the analog frequency ω_a is marked on the $j\omega$ axis on the s-plane, whereas ω_d is the digital frequency labeled on the unit circle in the z-plane.

We substitute $s=jw_a$ and $z=e^{j\omega T}$ into the BLT in Eq. (8.11) to get

$$j\omega_a=\frac{2}{T}\frac{e^{j\omega_d T}-1}{e^{j\omega_d T}+1}. \tag{8.13}$$

Simplifying Eq. (8.13) leads to

$$\omega_a=\frac{2}{T}\tan\left(\frac{\omega_d T}{2}\right). \tag{8.14}$$

Eq. (8.14) explores the relation between the analog frequency on $j\omega$ axis and the corresponding digital frequency ω_d on the unit circle. We can also write its inverse as

$$\omega_d=\frac{2}{T}\tan^{-1}\left(\frac{\omega_a T}{2}\right). \tag{8.15}$$

The range of the digital frequency ω_d is from 0 rad/s to the folding frequency $\omega_s/2$ rad/ s, where ω_s is the sampling frequency in terms of radians per second. We make a plot of Eq. (8.14) in Fig. 8.10.

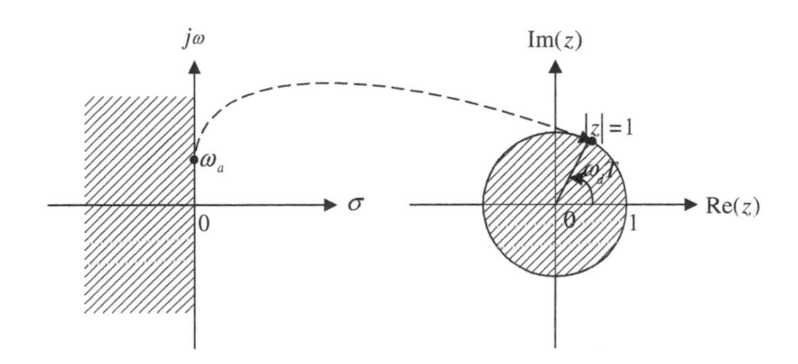

FIG. 8.9

Frequency mapping from the analog domain to the digital domain.

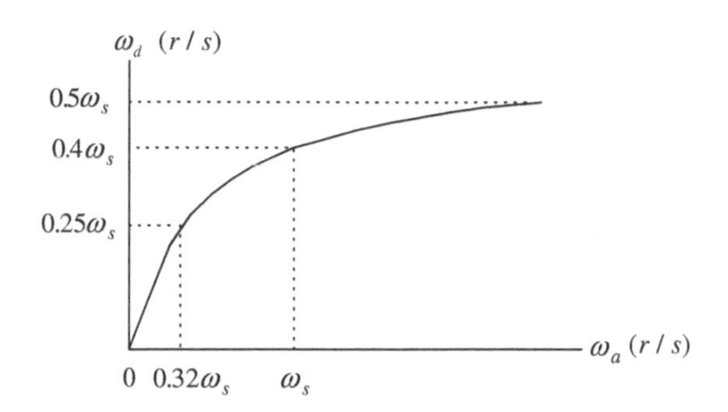

FIG. 8.10

Frequency warping from BLT.

From Fig. 8.10, when the digital frequency range $0 \le \omega_d \le 0.25\omega_s$ is mapped to the analog frequency range $0 \le \omega_a \le 0.32\omega_s$, the transformation appears to be linear; however, the digital frequency range $0.25\omega_s \le \omega_d \le 0.5\omega_s$ is mapped to the analog frequency range for $\omega_a > 0.32\omega_s$, the transformation is nonlinear. The analog frequency range for $\omega_a > 0.32\omega_s$ is compressed into the digital frequency range $0.25\omega_s \le \omega_d \le 0.5\omega_s$. This nonlinear frequency mapping effect is called *frequency warping*. We must incorporate the frequency warping into the IIR filter design.

The following example will illustrate the frequency warping effect in the BLT.

EXAMPLE 8.5

Assume the following analog frequencies:

$\omega_a = 10$ rad/s

$\omega_a = \omega_s/4 = 50\pi = 157$ rad/s

$\omega_a = \omega_s/2 = 100\pi = 314$ rad/s.

Find their digital frequencies using the BLT with a sampling period of 0.01 s, given the analog filter in Example 8.4 and the developed digital filter.

Solution:

From Eq. (8.15), we can calculate digital frequency ω_d as follows:

When $\omega_a = 10$ (rad/s) and $T = 0.01$ (s)

$$\omega_d = \frac{2}{T}\tan^{-1}\left(\frac{\omega_a T}{2}\right) = \frac{2}{0.01}\tan^{-1}\left(\frac{10 \times 0.01}{2}\right) = 9.99\ (\text{rad/s}),$$

which is close to the analog frequency of 10 rad/s. When $\omega_a = 157$ (rad/s) and $T = 0.01$ (s),

$$\omega_d = \frac{2}{0.01}\tan^{-1}\left(\frac{157 \times 0.01}{2}\right) = 133.11\ (\text{rad/s}),$$

which has an error as compared with the desired value 157. When $\omega_a = 314$ (rad/s) and $T = 0.01$ (s),

$$\omega_d = \frac{2}{0.01}\tan^{-1}\left(\frac{314 \times 0.01}{2}\right) = 252.5\ (\text{rad/s}),$$

which gives a bigger error compared with the digital folding frequency of 314 rad/s.

Fig. 8.11 shows how to correct the frequency warping error. First, given the digital frequency specification, we prewarp the digital frequency specification to the analog frequency specification by Eq. (8.14).

Second, we obtain the analog lowpass filter $H(s)$ using the prewarped analog frequency ω_a and the lowpass prototype. For the lowpass analog filter, we have

$$H(s) = H_P(s)|_{s=\frac{s}{\omega_a}} = H_P\left(\frac{s}{\omega_a}\right). \tag{8.16}$$

Finally, substituting the BLT Eq. (8.11) into Eq. (8.16) yields the digital filter as

$$H(z) = H(s)|_{s=\frac{2}{T}\frac{z-1}{z+1}}. \tag{8.17}$$

Similarly, this approach can be extended to other types of filter designs.

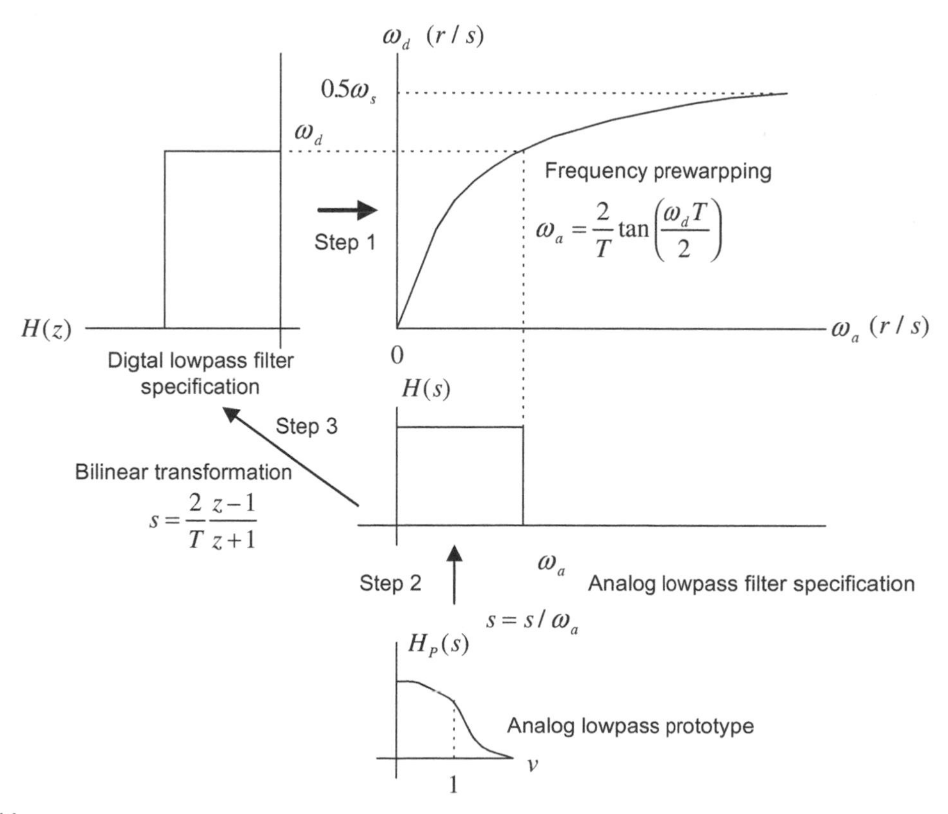

FIG. 8.11

Graphical representation of IIR filter design using the BLT.

8.2.3 BILINEAR TRANSFORMATION DESIGN PROCEDURE

Now we can summarize the BLT design procedure.

1. Given the digital filter frequency specifications, prewarp the digital frequency specifications to the analog frequency specifications.

   For the lowpass filter and the highpass filter:

$$\omega_a = \frac{2}{T}\tan\left(\frac{\omega_d T}{2}\right). \tag{8.18}$$

   For the bandpass filter and the bandstop filter :

$$\omega_{al} = \frac{2}{T}\tan\left(\frac{\omega_l T}{2}\right), \omega_{ah} = \frac{2}{T}\tan\left(\frac{\omega_h T}{2}\right), \tag{8.19}$$

   and $\omega_0 = \sqrt{\omega_{al}\omega_{ah}}$, $W = \omega_{ah} - \omega_{al}$.

2. Perform the prototype transformation using the lowpass prototype $H_P(s)$.

$$\text{From lowpass to lowpass}: H(s) = H_P(s)|_{s=\frac{s}{\omega_a}} \tag{8.20}$$

$$\text{From lowpass to highpass}: H(s) = H_P(s)|_{s=\frac{\omega_a}{s}} \tag{8.21}$$

$$\text{From lowpass to bandpass}: H(s) = H_P(s)|_{s=\frac{s^2+\omega_0^2}{sW}} \tag{8.22}$$

$$\text{From lowpass to bandstop}: H(s) = H_P(s)|_{s=\frac{sW}{s^2+\omega_0^2}} \tag{8.23}$$

3. Substitute the BLT to obtain the digital filter

$$H(z) = H(s)|_{s=\frac{2}{T}\frac{z-1}{z+1}}. \tag{8.24}$$

Table 8.2 lists MATLAB functions for the BLT design.

Table 8.2 MATLAB Functions for the Bilinear Transformation Design

Lowpass to lowpass: $H(s) = H_P(s)|_{s=\frac{s}{\omega_a}}$

>>[B,A] = lp2lp(Bp,Ap,wa)

Lowpass to highpass: $H(s) = H_P(s)|_{s=\frac{\omega_a}{s}}$

>>[B,A] = lp2hp(Bp,Ap,wa)

Lowpass to bandpass: $H(s) = H_P(s)|_{s=\frac{s^2+\omega_0^2}{sW}}$

Table 8.2 MATLAB Functions for the Bilinear Transformation Design—cont'd

>>[B,A] = lp2bp(Bp,Ap,w0,W)
Lowpass to bandstop: $H(s) = H_P(s)\|_{s=\frac{sW}{s^2+\omega_0^2}}$
>>[B,A] = lp2bs(Bp,Ap,w0,W)
Bilinear transformation to achieve the digital filter:
>>[b,a] = bilinear(B,A,fs)
Plot of the magnitude and phase frequency responses of the digital filter:
>>**freqz(b,a,512,fs)**
Definitions of design parameters:
Bp = vector containing the numerator coefficients of the lowpass prototype. Ap = vector containing the denominator coefficients of the lowpass prototype. wa = cutoff frequency for the lowpass or highpass analog filter (rad/s). w0 = center frequency for the bandpass or bandstop analog filter (rad/s). W = bandwidth for the bandpass or banstop analog filter (rad/s). B = vector containing the numerator coefficients of the analog filter. A = vector containing the denominator coefficients of the analog filter. b = vector containing the numerator coefficients of the digital filter. a = vector containing the denominator coefficients of the digital filter. fs = sampling rate (samples/s).

We illustrate the lowpass filter design procedure in Example 8.6. Other types of filters, such as high-pass, bandpass, and bandstop will be illustrated in the next section.

EXAMPLE 8.6

The normalized lowpass filter with a cutoff frequency of 1 rad/s is given as

$$H_P(s) = \frac{1}{s+1}.$$

(a) Use the given $H_P(s)$ and the BLT to design a corresponding digital IIR lowpass filter with a cutoff frequency of 15 Hz and a sampling rate of 90 Hz.
(b) Use MATLAB to plot the magnitude response and phase response of $H(z)$.

Solution:

(a) First, we obtain the digital frequency as
$\omega_d = 2\pi f = 2\pi(15) = 30\pi$ (rad/s), and $T = \frac{1}{f_s} = \frac{1}{90}$ (s).
We then follow the design procedure:

1. First calculate the prewarped analog frequency as

$$\omega_a = \frac{2}{T}\tan\left(\frac{\omega_d T}{2}\right) = \frac{2}{1/90}\tan\left(\frac{30\pi/90}{2}\right),$$

that is, $\omega_a = 180 \times \tan(\pi/6) = 180 \times \tan(30^\circ) = 103.92$ (rad/s).

Continued

EXAMPLE 8.6—CONT'D

2. Then perform the prototype transformation (lowpass to lowpass) as follows:

$$H(s) = H_P(s)\Big|_{s=\frac{s}{\omega_a}} = \frac{1}{\frac{s}{\omega_a}+1} = \frac{\omega_a}{s+\omega_a},$$

which yields an analog filter:

$$H(s) = \frac{103.92}{s+103.92}.$$

3. Applying the BLT yields

$$H(z) = \frac{103.92}{s+103.92}\Big|_{s=\frac{2}{T}\frac{z-1}{z+1}}.$$

We simplify the algebra by dividing both the numerator and the denominator by 180:

$$H(z) = \frac{103.92}{180\times\frac{z-1}{z+1}+103.92} = \frac{103.92/180}{\frac{z-1}{z+1}+103.92/180} = \frac{0.5773}{\frac{z-1}{z+1}+0.5773}.$$

then we multiply both the numerator and the denominator by $(z+1)$ to obtain

$$H(z) = \frac{0.5773(z+1)}{\left(\frac{z-1}{z+1}+0.5773\right)(z+1)} = \frac{0.5773z+0.5773}{(z-1)+0.5773(z+1)} = \frac{0.5773z+0.5773}{1.5773z-0.4227}.$$

Finally, we divide both the numerator and the denominator by $1.5773z$ to get the transfer function in the standard format:

$$H(z) = \frac{(0.5773z+0.5773)/(1.5773z)}{(1.5773z-0.4227)/(1.5773z)} = \frac{0.3660+0.3660z^{-1}}{1-0.2679z^{-1}}.$$

(b) The corresponding MATLAB design is listed in Program 8.2. Fig. 8.12 shows the magnitude and phase frequency responses.

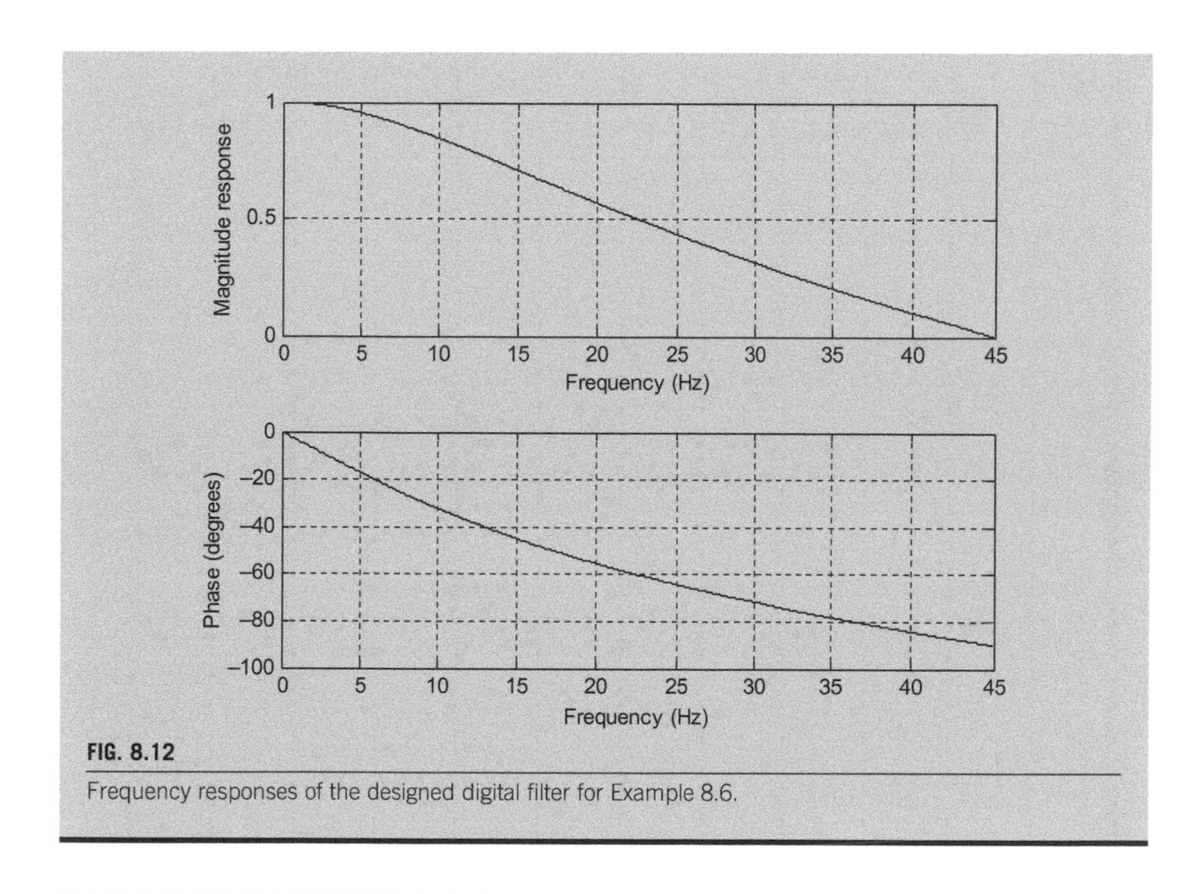

FIG. 8.12

Frequency responses of the designed digital filter for Example 8.6.

Program 8.2. MATLAB program for Example 8.6.

```
%Example 8.6.
% Plot the magnitude and phase responses
 fs=90;   % Sampling rate (Hz)
 [B, A]=lp2lp([1][1 1],103.92);
 [b,a]=bilinear(B,A,fs)
 % b= [0.3660 0.3660] numerator coefficients of the digital filter from MATLAB
 % a= [1 -0.2679]  denominator coefficients of the digital filter from MATLAB
[hz, f]=freqz([0.3660 0.3660][1 -0.2679],512,fs); %the frequency response
phi=180*unwrap(angle(hz))/pi;
subplot(2,1,1), plot(f, abs(hz)),grid;
axis([0 fs/2 0 1]);
xlabel('Frequency (Hz)'); ylabel('Magnitude Response')
subplot(2,1,2), plot(f, phi); grid;
axis([0 fs/2 -100 0]);
xlabel('Frequency (Hz)'); ylabel('Phase (degrees)')
```

8.3 DIGITAL BUTTERWORTH AND CHEBYSHEV FILTER DESIGNS

In this section, we design various types of digital Butterworth and Chebyshev filters using the BLT design method discussed in the previous section.

8.3.1 LOWPASS PROTOTYPE FUNCTION AND ITS ORDER

As described in Section 8.2 (The BLT Design Procedure), BLT design requires obtaining the analog filter with prewarped frequency specifications. These analog filter design requirements include the ripple specification at the passband frequency edge, the attenuation specification at the stopband frequency edge, and the type of lowpass prototype (which we shall discuss) and its order.

Table 8.3 lists the Butterworth prototype functions with 3-dB passband ripple specification. Tables 8.4 and 8.5 contain the Chebyshev prototype functions (type I) with 1- and 0.5-dB passband ripple specifications, respectively. Other lowpass prototypes with different ripple specifications and order can be computed using the methods described in Appendix C.

Table 8.3 3-dB Butterworth Lowpass Prototype Transfer Functions ($\varepsilon = 1$)

n	$H_P(s)$
1	$\frac{1}{s+1}$
2	$\frac{1}{s^2+1.4142s+1}$
3	$\frac{1}{s^3+2s^2+2s+1}$
4	$\frac{1}{s^4+2.6131s^3+3.4142s^2+2.6131s+1}$
5	$\frac{1}{s^5+3.2361s^4+5.2361s^3+5.2361s^2+3.2361s+1}$
6	$\frac{1}{s^6+3.8637s^5+7.4641s^4+9.1416s^3+7.4641s^2+3.8637s+1}$

Table 8.4 Chebyshev Lowpass Prototype Transfer Functions with 0.5 dB Ripple ($\varepsilon = 0.3493$)

n	$H_P(s)$
1	$\frac{2.8628}{s+2.8628}$
2	$\frac{1.4314}{s^2+1.4256s+1.5162}$
3	$\frac{0.7157}{s^3+1.2529s^2+1.5349s+0.7157}$
4	$\frac{0.3579}{s^4+1.1974s^3+1.7169s^2+1.0255s+0.3791}$
5	$\frac{0.1789}{s^5+1.1725s^4+1.9374s^3+1.3096s^2+0.7525s+0.1789}$
6	$\frac{0.0895}{s^6+1.1592s^5+2.1718s^4+1.5898s^3+1.1719s^2+0.4324s+0.0948}$

Table 8.5 Chebyshev Lowpass Prototype Transfer Functions with 1 dB Ripple ($\varepsilon = 0.5088$)

n	$H_P(s)$
1	$\frac{1.9652}{s+1.9652}$
2	$\frac{0.9826}{s^2+1.0977s+1.1025}$
3	$\frac{0.4913}{s^3+0.9883s^2+1.2384s+0.4913}$
4	$\frac{0.2456}{s^4+0.9528s^3+1.4539s^2+0.7426s+0.2756}$
5	$\frac{0.1228}{s^5+0.9368s^4+1.6888s^3+0.9744s^2+0.5805s+0.1228}$
6	$\frac{0.0614}{s^6+0.9283s^5+1.9308s^4+1.20121s^3+0.9393s^2+0.3071s+0.0689}$

In this section, we will focus on Chebyshev type I filter. The Chebyshev type II filter design can be found in Proakis and Manolakis (2007) and Porat (1997).

The magnitude response function of the Butterworth lowpass prototype with order n is shown in Fig. 8.13, where the magnitude response $|H_P(v)|$ vs. the normalized frequency v is given by Eq. (8.25):

$$|H_P(v)| = \frac{1}{\sqrt{1+\varepsilon^2 v^{2n}}}. \tag{8.25}$$

With the given passband ripple A_P dB at the normalized passband frequency edge $v_p = 1$, and the stopband attenuation A_s dB at the normalized stopband frequency edge v_s, the following two equations must be satisfied to determine the prototype filter order:

$$A_P\text{dB} = -20 \times \log_{10}\left(\frac{1}{\sqrt{1+\varepsilon^2}}\right), \tag{8.26}$$

$$A_s\text{dB} = -20 \times \log_{10}\left(\frac{1}{\sqrt{1+\varepsilon^2 v_s^{2n}}}\right). \tag{8.27}$$

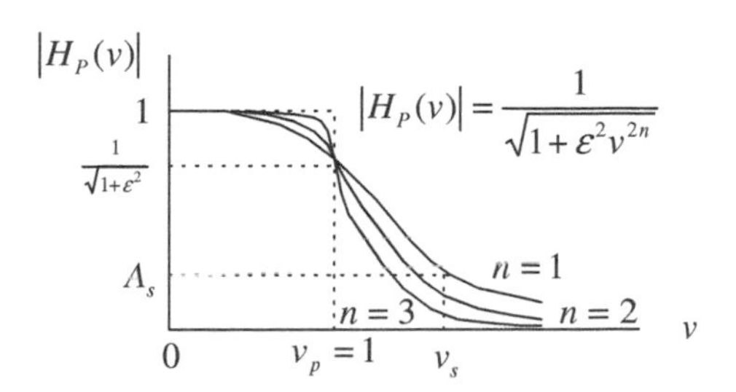

FIG. 8.13

Normalized Butterworth magnitude response function.

Solving Eqs. (8.26) and (8.27), we determine the lowpass prototype order as

$$\varepsilon^2 = 10^{0.1A_p} - 1, \tag{8.28}$$

$$n \geq \frac{\log_{10}\left(\frac{10^{0.1A_s} - 1}{\varepsilon^2}\right)}{[2 \times \log_{10}(v_s)]}, \tag{8.29}$$

where ε is the absolute ripple specification.

The magnitude response function of Chebyshev lowpass prototype with order n is shown in Fig. 8.14, where the magnitude response $|H_P(v)|$ vs. the normalized frequency v is given by

$$|H_P(v)| = \frac{1}{\sqrt{1 + \varepsilon^2 C_n^2(v)}}, \tag{8.30}$$

where

$$C_n(v_s) = \cosh\left[n \cosh^{-1}(v_s)\right] \tag{8.31}$$

$$\cosh^{-1}(v_s) = \ln\left(v_s + \sqrt{v_s^2 - 1}\right). \tag{8.32}$$

As shown in Fig. 8.14, the magnitude response for the Chebyshev lowpass prototype with an odd-numbered order begins with a filter DC gain of 1. In the case of a Chebyshev lowpass prototype an even-numbered order, the magnitude starts at a filter DC gain of $1/\sqrt{1+\varepsilon^2}$. For both cases, the filter gain at the normalized cutoff frequency $v_p = 1$ is $1/\sqrt{1+\varepsilon^2}$.

Similarly, Eqs. (8.33) and (8.34) must be satisfied:

$$A_P\text{dB} = -20 \times \log_{10}\left(\frac{1}{\sqrt{1+\varepsilon^2}}\right), \tag{8.33}$$

$$A_s\text{dB} = -20 \times \log_{10}\left(\frac{1}{\sqrt{1+\varepsilon^2 C_n^2(v_s)}}\right). \tag{8.34}$$

The lowpass prototype order can be solved in Eqs. (8.35a) and (8.35b):

$$\varepsilon^2 = 10^{0.1A_p} - 1 \tag{8.35a}$$

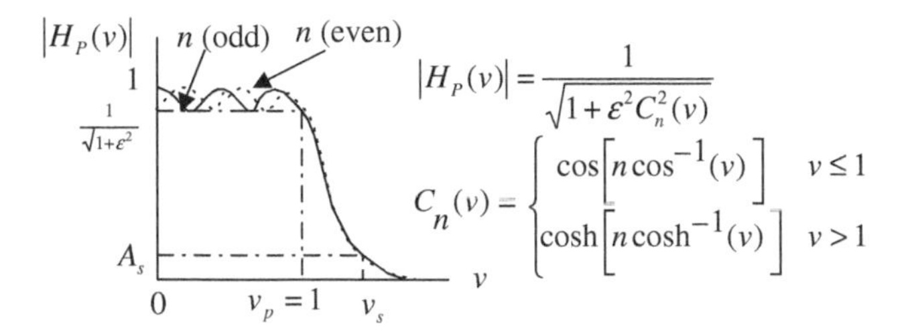

FIG. 8.14

Normalized Chebyshev magnitude response function.

Table 8.6 Conversion from Analog Filter Specifications to Lowpass Prototype Specifications

Analog Filter Specifications	Lowpass Prototype Specifications
Lowpass: ω_{ap}, ω_{as}	$v_p = 1$, $v_s = \omega_{as}/\omega_{ap}$
Highpass: ω_{ap}, ω_{as}	$v_p = 1$, $v_s = \omega_{ap}/\omega_{as}$
Bandpass: ω_{apl}, ω_{aph}, ω_{asl}, ω_{ash}, $\omega_0 = \sqrt{\omega_{apl}\omega_{aph}}$, $\omega_0 = \sqrt{\omega_{asl}\omega_{ash}}$	$v_p = 1$, $v_s = \dfrac{\omega_{ash} - \omega_{asl}}{\omega_{aph} - \omega_{apl}}$
Bandstop: ω_{apl}, ω_{aph}, ω_{asl}, ω_{ash}, $\omega_0 = \sqrt{\omega_{apl}\omega_{aph}}$, $\omega_0 = \sqrt{\omega_{asl}\omega_{ash}}$	$v_p = 1$, $v_s = \dfrac{\omega_{aph} - \omega_{apl}}{\omega_{ash} - \omega_{asl}}$

ω_{ap}, *passband frequency edge;* ω_{as}, *stopband frequency edge;*
ω_{apl}, *lower cutoff frequency in passband;* ω_{aph}, *upper cutoff frequency in passband;*
ω_{asl}, *lower cutoff frequency in stopband;* ω_{ash}, *upper cutoff frequency in stopband;*
ω_0, *geometric center frequency.*

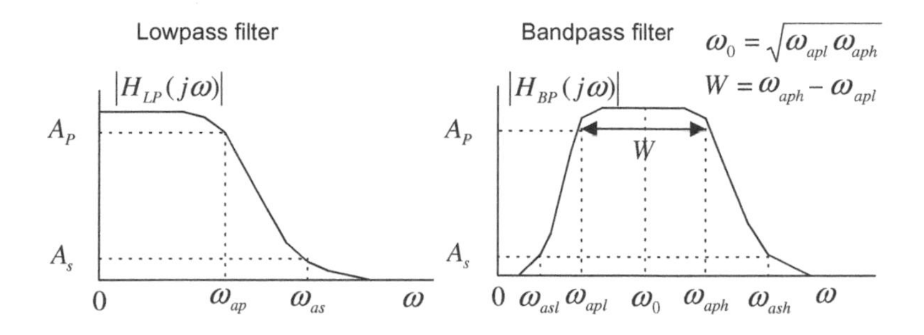

FIG. 8.15

Specifications for analog lowpass and bandpass filters.

$$n \geq \frac{\cosh^{-1}\left[\left(\dfrac{10^{0.1A_s} - 1}{\varepsilon^2}\right)^{0.5}\right]}{\cosh^{-1}(v_s)}, \tag{8.35b}$$

where $\cosh^{-1}(x) = \ln\left(x + \sqrt{x^2 - 1}\right)$, and ε is the absolute ripple parameter.

The normalized stopband frequency v_s can be determined from the frequency specifications of an analog filter illustrated in Table 8.6. Then the order of the lowpass prototype can be determined by Eqs. (8.29) for the Butterworth function and by Eq. (8.35b) for the Chebyshev function. Fig. 8.15 gives frequency edge notations for analog lowpass and bandpass filters. The notations for analog highpass and bandstop filters can be defined correspondingly.

8.3.2 LOWPASS AND HIGHPASS FILTER DESIGN EXAMPLES

The following examples illustrate various designs for the Butterworth and Chebyshev lowpass and highpass filters.

EXAMPLE 8.7

(a) Design a digital lowpass Butterworth filter with the following specifications:
  (1) 3-dB attenuation at the passband frequency of 1.5 kHz.
  (2) 10-dB stopband attenuation at the frequency of 3 kHz.
  (3) Sampling frequency at 8000 Hz.

(b) Use MATLAB to plot the magnitude and phase responses.

Solution:

(a) First, we obtain the digital frequencies in radians per second:

$$\omega_{dp} = 2\pi f = 2\pi(1500) = 3000\pi \text{ (rad/s)}$$

$$\omega_{ds} = 2\pi f = 2\pi(3000) = 6000\pi \text{ (rad/s)}$$

$$T = \frac{1}{f_s} = \frac{1}{8000} \text{ (s)}$$

Following the steps of the design procedure,

(1) We apply the warping equation as

$$\omega_{ap} = \frac{2}{T}\tan\left(\frac{\omega_d T}{2}\right) = 16{,}000 \times \tan\left(\frac{3000\pi/8000}{2}\right) = 1.0691 \times 10^4 \text{ (rad/s)}.$$

$$\omega_{as} = \frac{2}{T}\tan\left(\frac{\omega_d T}{2}\right) = 16{,}000 \times \tan\left(\frac{6000\pi/8000}{2}\right) = 3.8627 \times 10^4 \text{ (rad/s)}.$$

We then find the lowpass prototype specifications using Table 8.6 as follows:

$$v_s = \frac{\omega_{as}}{\omega_{ap}} = \frac{3.8627 \times 10^4}{1.0691 \times 10^4} = 3.6130 \text{ and } A_s = 10 \text{ dB}.$$

The filter order is computed as

$$\varepsilon^2 = 10^{0.1 \times 3} - 1 = 1$$

$$n = \frac{\log_{10}\left(10^{0.1 \times 10} - 1\right)}{2 \times \log_{10}(3.6130)} = 0.8553.$$

(2) Rounding n up, we choose $n = 1$ for the lowpass prototype. From Table 8.3, we have

$$H_P(s) = \frac{1}{s+1}.$$

Applying the prototype transformation (lowpass to lowpass) yields the analog filter

$$H(s) = H_P(s)\Big|_{\frac{s}{\omega_{ap}}} = \frac{1}{\frac{s}{\omega_{ap}} + 1} = \frac{\omega_{ap}}{s + \omega_{ap}} = \frac{1.0691 \times 10^4}{s + 1.0691 \times 10^4}.$$

(3) Finally, using the BLT, we have

$$H(z) = \frac{1.0691 \times 10^4}{s + 1.0691 \times 10^4}\Bigg|_{s = 16000(z-1)/(z+1)}.$$

Substituting the BLT leads to

$$H(z) = \frac{1.0691 \times 10^4}{\left(16,000\frac{z-1}{z+1}\right) + 1.0691 \times 10^4}.$$

To simplify the algebra, we divide both numerator and denominator by 16,000 to get

$$H(z) = \frac{0.6682}{\left(\frac{z-1}{z+1}\right) + 0.6682}.$$

Then multiplying both numerator and denominator by $(z+1)$ leads to

$$H(z) = \frac{0.6682(z+1)}{(z-1) + 0.6682(z+1)} = \frac{0.6682z + 0.6682}{1.6682z - 0.3318}.$$

Dividing both numerator and denominator by $(1.6682 \times z)$ leads to

$$H(z) = \frac{0.4006 + 0.4006z^{-1}}{1 - 0.1989z^{-1}}.$$

(b) Steps 2 and 3 can be carried out using the MATLAB Program 8.3, as shown in the first three lines of the MATLAB codes. Fig. 8.16 describes the filter frequency responses.

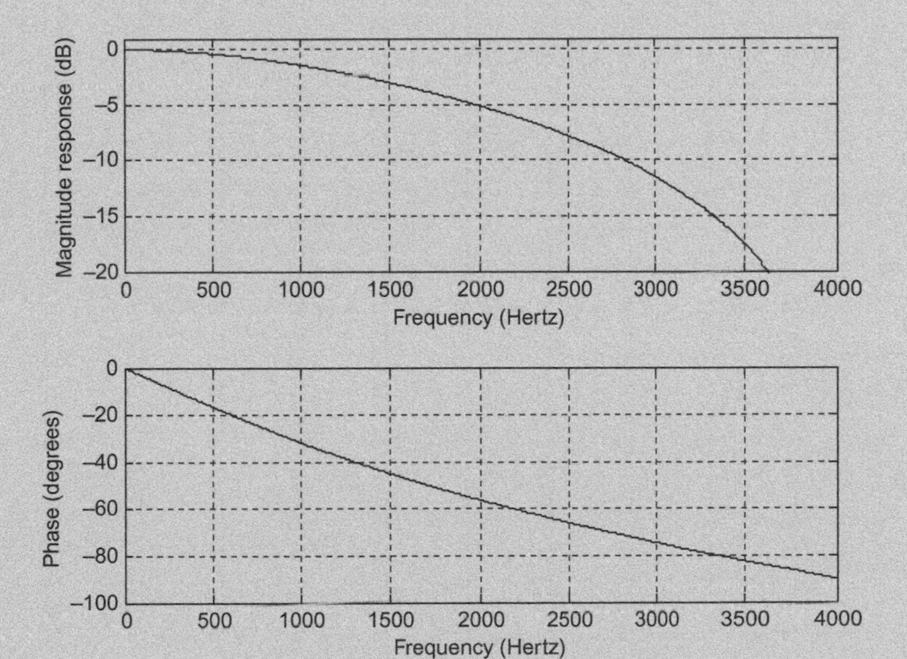

FIG. 8.16

Frequency responses of the designed digital filter for Example 8.7.

Program 8.3. MATLAB program for Example 8.7.

```
%Example 8.7
% Design of the digital lowpass Butterworth filter
format long
fs=8000;                    % Sampling rate
[B A]=lp2lp([1][1 1], 1.0691*10^4) % Complete step 2
[b a]=bilinear(B,A,fs) % Complete step 3
% Plot the magnitude and phase responses |H(z)|
% b=[0.4005 0.4005]; numerator coefficients from MATLAB
% a=[1 -0.1989]; denominator coefficients from MATLAB
freqz(b,a,512,fs);
axis([0 fs/2 -20 1])
```

EXAMPLE 8.8

(a) Design a first-order highpass digital Chebyshev filter with a cutoff frequency of 3 kHz and 1 dB ripple on passband using a sampling frequency of 8000 Hz.
(b) Use MATLAB to plot the magnitude and phase responses.

Solution:

(a) First, we obtain the digital frequency in radians per second:

$$\omega_d = 2\pi f = 2\pi(3000) = 6000\pi \text{ (rad/s), and } T = \frac{1}{f_s} = \frac{1}{8000} \text{ (s)}.$$

Following the steps of the design procedure, we have

1. $\omega_a = \frac{2}{T}\tan\left(\frac{\omega_d T}{2}\right) = 16{,}000 \times \tan\left(\frac{6000\pi/8000}{2}\right) = 3.8627 \times 10^4 \text{ (rad/s)}.$
2. Since the filter order is given as 1, we select the first-order lowpass prototype from Table 8.5 as

$$H_P(s) = \frac{1.9652}{s + 1.9652}.$$

Applying the prototype transformation (lowpass to highpass), we obtain

$$H(s) = H_P(s)|_{\frac{\omega_a}{s}} = \frac{1.9652}{\frac{\omega_a}{s} + 1.9652} = \frac{1.9652s}{1.9652s + 3.8627 \times 10^4}.$$

Dividing both numerator and denominator by 1.9652 gives

$$H(s) = \frac{s}{s + 1.9656 \times 10^4}.$$

3. Using the BLT, we have

$$H(z) = \left.\frac{s}{s + 1.9656 \times 10^4}\right|_{s=16{,}000(z-1)/(z+1)}.$$

Algebra work is demonstrated as follows:

$$H(z) = \frac{16{,}000\frac{z-1}{z+1}}{16{,}000\frac{z-1}{z+1} + 1.9656 \times 10^4}.$$

Simplifying the transfer function yields

$$H(z) = \frac{0.4487 - 0.4487z^{-1}}{1 + 0.1025z^{-1}}.$$

(b) Steps 2 and 3 and frequency response plots shown in Fig. 8.17 can be carried out using the MATLAB Program 8.4.

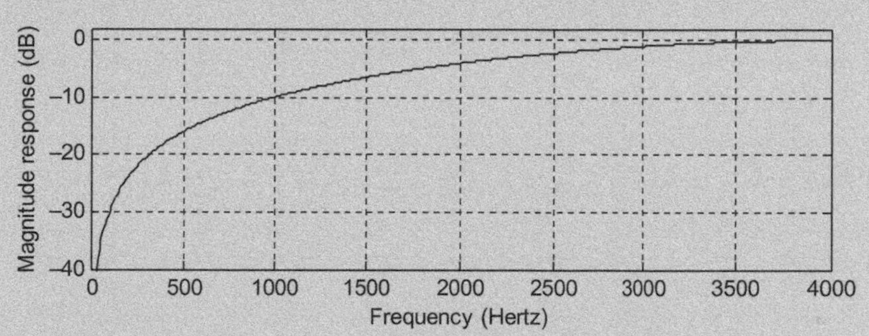

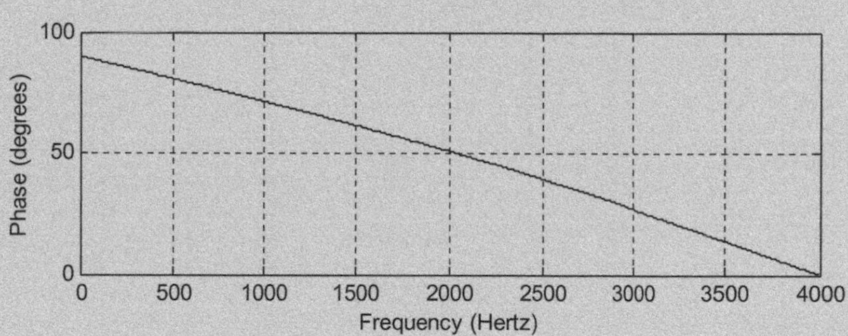

FIG. 8.17

Frequency responses of the designed digital filter for Example 8.8.

Program 8.4. MATLAB program for Example 8.8

```
%Example 8.8
% Design of the digital highpass Butterworth filter
format long
fs=8000; % Sampling rate
[B A]=lp2hp([1.9652][1 1.9652], 3.8627*10^4) % Complete step 2
```

```
[b a]=bilinear(B,A,fs) % Complete step 3
% Plot the magnitude and phase responses |H(z)|
% b=[0.4487 -0.4487 ]; numerator coefficients from MATLAB
% a=[1 0.1025]; denominator coefficients from MATLAB
freqz(b,a,512,fs);
axis([0 fs/2 -40 2])
```

EXAMPLE 8.9

(a) Design a second-order lowpass digital Butterworth filter with a cutoff frequency of 3.4kHz at a sampling frequency of 8000Hz.
(b) Use MATLAB to plot the magnitude and phase responses.

Solution:

(a) First, we obtain the digital frequency in radians per second:

$$\omega_d = 2\pi f = 2\pi(3400) = 6800\pi \text{ (rad/s), and } T = \frac{1}{f_s} = \frac{1}{8000} \text{ (s)}.$$

Following the steps of the design procedure, we compute the prewarped analog frequencies as
1.

$$\omega_a = \frac{2}{T}\tan\left(\frac{\omega_d T}{2}\right) = 16{,}000 \times \tan\left(\frac{6800\pi/8000}{2}\right) = 6.6645 \times 10^4 \text{ (rad/s)}.$$

2. Since the order of 2 is given in the specification, we directly pick the second-order lowpass prototype from Table 8.3:

$$H_P(s) = \frac{1}{s^2 + 1.4142s + 1}.$$

After applying the prototype transformation (lowpass to lowpass), we have

$$H(s) = H_P(s)|_{\frac{s}{\omega_a}} = \frac{4.4416 \times 10^9}{s^2 + 9.4249 \times 10^4 s + 4.4416 \times 10^9}.$$

3. Carrying out the BLT yields

$$H(z) = \left.\frac{4.4416 \times 10^9}{s^2 + 9.4249 \times 10^4 s + 4.4416 \times 10^9}\right|_{s=16000(z-1)/(z+1)}.$$

Let us work on algebra:

$$H(z) = \frac{4.4416 \times 10^9}{\left(16{,}000\frac{z-1}{z+1}\right)^2 + 9.4249 \times 10^4\left(16{,}000\frac{z-1}{z+1}\right) + 4.4416 \times 10^9}.$$

To simplify, we divide both numerator and denominator by $(16,000)^2$ to get

$$H(z) = \frac{17.35}{\left(\frac{z-1}{z+1}\right)^2 + 5.8906\left(\frac{z-1}{z+1}\right) + 17.35}.$$

Then multiplying both numerator and denominator by $(z+1)^2$ leads to

$$H(z) = \frac{17.35(z+1)^2}{(z-1)^2 + 5.8906(z-1)(z+1) + 17.35(z+1)^2}.$$

Using identities, we have

$$H(z) = \frac{17.35(z^2+2z+1)}{(z^2-2z+1) + 5.8906(z^2-1) + 17.35(z^2+2z+1)} = \frac{17.35z^2 + 34.7z + 17.35}{24.2406z^2 + 32.7z + 12.4594}.$$

Dividing both numerator and denominator by $(24.2406z^2)$ leads to

$$H(z) = \frac{0.7157 + 1.4314z^{-1} + 0.7151z^{-2}}{1 + 1.3490z^{-1} + 0.5140z^{-2}}.$$

(b) Steps 2 and 3 require a certain amount of algebra work and can be verified using MATLAB Program 8.5, as shown in the first three lines of the code. Fig. 8.18 plots the filter magnitude and phase frequency responses.

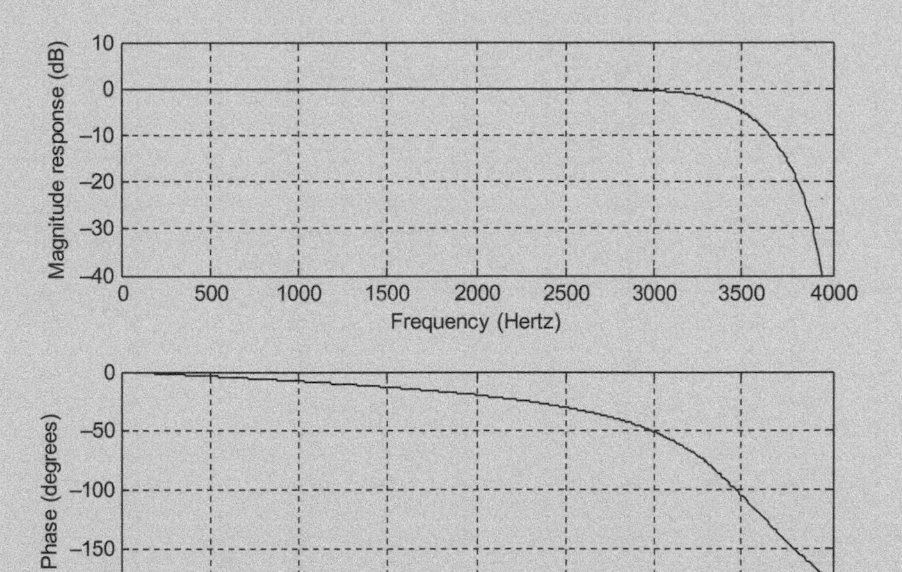

FIG. 8.18

Frequency responses of the designed digital filter for Example 8.9.

Program 8.5. MATLAB program for Example 8.9.

```
%Example 8.9
% Design of the digital lowpass Butterworth filter
format long
fs=8000;    % Sampling rate
[B A]=lp2lp([1][1 1.4142 1], 6.6645*10^4) % Complete step 2
[b a]=bilinear(B,A,fs) % Complete step 3
% Plot the magnitude and phase responses |H(z)|
% b=[0.7157 1.4315 0.7157]; numerator coefficients from MATLAB
%a=[1 1.3490 0.5140]; denominator coefficients from MATLAB
freqz(b,a,512,fs);
axis([0 fs/2 -40 10])
```

EXAMPLE 8.10

(a) Design a highpass digital Chebyshev filter with the following specifications:
1. 0.5-dB ripple on passband at frequency of 3000 Hz.
2. 25-dB attenuation at frequency of 1000 Hz.
3. Sampling frequency at 8000 Hz.

(b) Use MATLAB to plot the magnitude and phase responses.

Solution:

(a) From the specifications, the digital frequencies are

$$\omega_{dp} = 2\pi f = 2\pi(3000) = 6000\pi \text{ (rad/s)}$$

$$\omega_{ds} = 2\pi f = 2\pi(1000) = 2000\pi \text{ (rad/s)}$$

$$\text{and } T = \frac{1}{f_s} = \frac{1}{8000} \text{ (s)}.$$

Using the design procedure, it follows that

$$\omega_{ap} = \frac{2}{T}\tan\left(\frac{\omega_{dp}T}{2}\right) = 16{,}000 \times \tan\left(\frac{6000\pi/8000}{2}\right) = 3.8627 \times 10^4 \text{ (rad/s)}$$

$$\omega_{as} = 16{,}000 \times \tan\left(\frac{\omega_{ds}T}{2}\right) = 16{,}000 \times \tan\left(\frac{2000\pi/8000}{2}\right) = 6.6274 \times 10^3 \text{ (rad/s)}.$$

We find the lowpass prototype specification as follows:

$$v_s = \frac{\omega_{ps}}{\omega_{sp}} = \frac{3.8627 \times 10^4}{6.6274 \times 10^3} = 5.8284 \text{ and } A_s = 25 \text{ dB},$$

then the filter order is computed as

$$\varepsilon^2 = 10^{0.1 \times 0.5} - 1 = 0.1220$$

$$\frac{(10^{0.1 \times 25} - 1)}{0.1220} = 2583.8341$$

$$n = \frac{\cosh^{-1}\left[(2583.8341)^{0.5}\right]}{\cosh^{-1}(5.8284)} = \frac{\ln\left(50.8314 + \sqrt{50.8314^2 - 1}\right)}{\ln\left(5.8284 + \sqrt{5.8284^2 - 1}\right)} = 1.8875.$$

We select $n=2$ for the lowpass prototype function. Following the steps of the design procedure, it follows that

1. $\omega_p = 3.8627 \times 10^4$ (rad/s).
2. Performing the prototype transformation (lowpass to lowpass) using the prototype filter illustrated in Table 8.4, we have

$$H_P(s) = \frac{1.4314}{s^2 + 1.4256s + 1.5162} \text{ and}$$

$$H(s) = H_P(s)|_{\frac{s}{\omega_a}} = \frac{1.4314}{\left(\frac{\omega_p}{s}\right)^2 + 1.4256\left(\frac{\omega_p}{s}\right) + 1.5162} = \frac{0.9441s^2}{s^2 + 3.6319 \times 10^4 s + 9.8407 \times 10^8}.$$

3. Hence, applying the BLT, we convert the analog filter to the digital filter as

$$H(z) = \left.\frac{0.9441s^2}{s^2 + 3.6319 \times 10^4 s + 9.8407 \times 10^8}\right|_{s=16{,}000(z-1)/(z+1)}.$$

After algebraic simplification, it follows that

$$H(z) = \frac{0.1327 - 0.2654z^{-1} + 0.1327z^{-2}}{1 + 0.7996z^{-1} + 0.3618z^{-2}}.$$

(b) MATLAB Program 8.6 is listed for this example, and frequency responses are given in Fig. 8.19.

Program 8.6. MATLAB program for Example 8.10.

```
%Example 8.10
% Design of the digital highpass Chebyshev filter
format long
 fs=8000;  % Sampling rate
 % BLT design
[B A]=lp2hp([1.4314][1 1.4256 1.5162], 3.8627*10^4) % Complete step 2
[b a]=bilinear(B,A,fs) % Complete step 3
% Plot the magnitude and phase responses |H(z)|
% b=[0.1327 -0.2654 0.1327]; numerator coefficients from MATLAB
% a=[1 0.7996 0.3618]; denominator coefficients from MATLAB
freqz(b,a,512,fs);
axis([0 fs/2 -40 10])
```

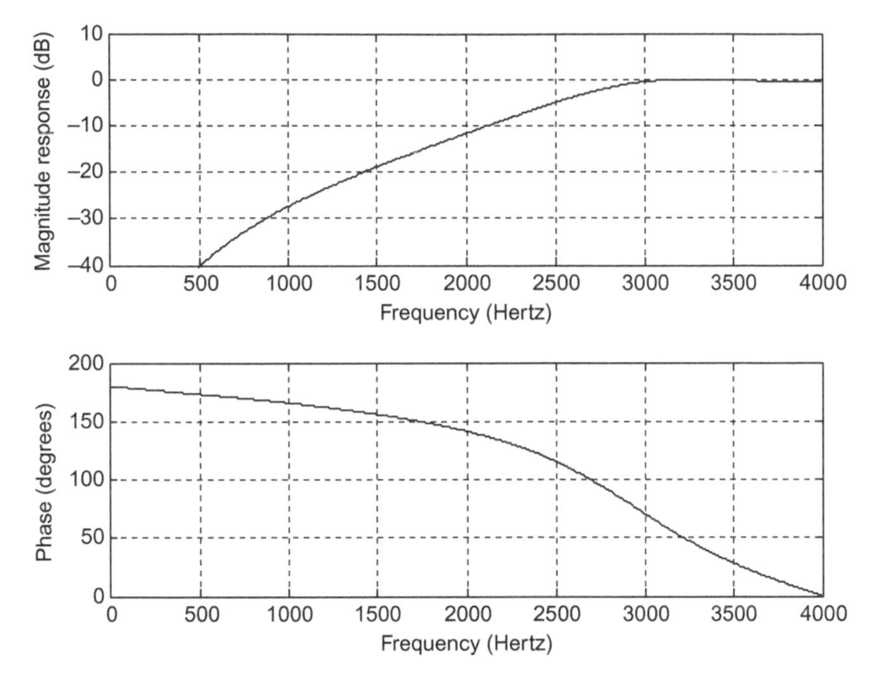

FIG. 8.19

Frequency responses of the designed digital filter for Example 8.10.

8.3.3 BANDPASS AND BANDSTOP FILTER DESIGN EXAMPLES

EXAMPLE 8.11

(a) Design a second-order digital bandpass Butterworth filter with the following specifications:
- an upper cutoff frequency of 2.6kHz,
- a lower cutoff frequency of 2.4kHz, and
- a sampling frequency of 8000Hz.

(b) Use MATLAB to plot the magnitude and phase responses.

Solution:

(a) Let us find the digital frequencies in radians per second:
- $\omega_h = 2\pi f_h = 2\pi(2600) = 5200\pi$ (rad/s)
- $\omega_l = 2\pi f_l = 2\pi(2400) = 4800\pi$ (rad/s), and $T = \frac{1}{f_s} = \frac{1}{8000}$ (s) s.

Following the steps of the design procedure, we have the following:

1. $\omega_{ah} = \frac{2}{T}\tan\left(\frac{\omega_h T}{2}\right) = 16{,}000 \times \tan\left(\frac{5200\pi/8000}{2}\right) = 2.6110 \times 10^4$ (rad/s)

$$\omega_{al} = 16{,}000 \times \tan\left(\frac{\omega_l T}{2}\right) = 16{,}000 \times \tan(0.3\pi) = 2.2022 \times 10^4 \text{ (rad/s)}$$

$$W = \omega_{ah} - \omega_{al} = 26{,}110 - 22{,}022 = 4088 \text{ (rad/s)}$$

$$\omega_0^2 = \omega_{ah} \times \omega_{al} = 5.7499 \times 10^8.$$

2. We perform the prototype transformation (lowpass to bandpass) to obtain $H(s)$. From Table 8.3, we pick the lowpass prototype with order of 1 to produce a bandpass filter with order of 2 as

$$H_P(s) = \frac{1}{s+1},$$

and applying the lowpass to bandpass transformation, it follows that

$$H(s) = H_P(s)\Big|_{\frac{s^2+\omega_0^2}{sW}} = \frac{Ws}{s^2 + Ws + \omega_0^2} = \frac{4088s}{s^2 + 4088s + 5.7499 \times 10^8}.$$

3. Hence we apply the BLT to yield

$$H(z) = \frac{4088s}{s^2 + 4088s + 5.7499 \times 10^8}\Bigg|_{s=16,000(z-1)/(z+1)}.$$

Via algebra work, we obtain the digital filter as

$$H(z) = \frac{0.0730 - 0.0730z^{-2}}{1 + 0.7117z^{-1} + 0.8541z^{-2}}.$$

(b) MATLAB Program 8.7 is given for this example, and the corresponding frequency response plots are illustrated in Fig. 8.20.

Program 8.7. MATLAB program for Example 8.11

```
%Example 8.11
% Design of the digital bandpass Butterworth filter
format long
fs=8000;
[B A]=lp2bp([1][1 1],sqrt(5.7499*10^8),4088) % Complete step 2
[b a]=bilinear(B,A,fs) % Complete step 3
% Plot the magnitude and phase responses |H(z)|
% b=[0.0730 0 -0.0730]; numerator coefficients from MATLAB
% a=[1 0.7117 0.8541]; denominator coefficients form MATLAB
freqz(b, a,512,fs);
axis([0 fs/2 -40 10])
```

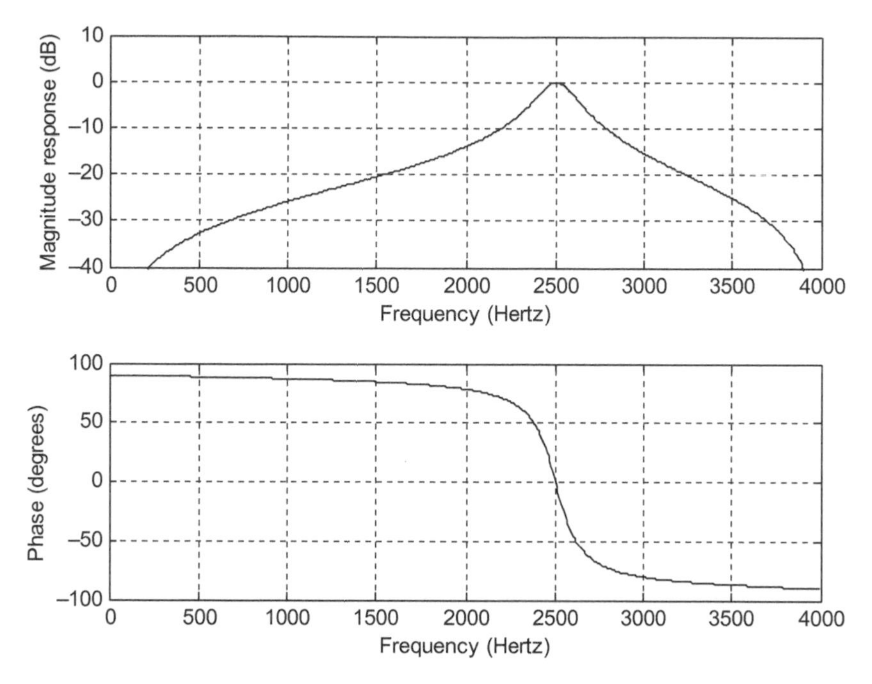

FIG. 8.20

Frequency responses of the designed digital filter for Example 8.11.

EXAMPLE 8.12

Now let us examine the bandstop Butterworth filter design.

(a) Design a digital bandstop Butterworth filter with the following specifications:
- Center frequency of 2.5 kHz
- Passband width of 200 Hz and ripple of 3 dB
- Stopband width of 50 Hz and attenuation of 10 dB
- Sampling frequency of 8000 Hz

(b) Use MATLAB to plot the magnitude and phase responses.

Solution:

(a) The digital frequencies of the digital filter are as follows:

$$\omega_h = 2\pi f_h = 2\pi(2600) = 5200\pi\,(\text{rad/s}),$$

$$\omega_l = 2\pi f_l = 2\pi(2400) = 4800\pi\,(\text{rad/s}),$$

$$\omega_{d0} = 2\pi f_0 = 2\pi(2500) = 5000\pi\,(\text{rad/s}), \text{and}\, T = 1/f_s = 1/8000\,(\text{s}).$$

Applying the steps of the IIR filter design approach, it follows that

1.

$$\omega_{ah} = \frac{2}{T}\tan\left(\frac{\omega_h T}{2}\right) = 16{,}000 \times \tan\left(\frac{5200\pi/8000}{2}\right) = 2.6110 \times 10^4\,(\text{rad/s})$$

$$\omega_{al} = 16,000 \times \tan\left(\frac{\omega_l T}{2}\right) = 16,000 \times \tan(0.3\pi) = 2.2022 \times 10^4 \text{ (rad/s)}$$

$$\omega_0 = 16,000 \times \tan\left(\frac{\omega_{d0} T}{2}\right) = 16,000 \times \tan(0.3125\pi) = 2.3946 \times 10^4 \text{ (rad/s)}$$

$$\omega_{sh} = \frac{2}{T}\tan\left(\frac{2525 \times 2\pi/8000}{2}\right) = 16,000 \times \tan(56.8125^\circ) = 2.4462 \times 10^4 \text{ (rad/s)}$$

$$\omega_{sl} = 16,000 \times \tan\left(\frac{2475 \times 2\pi/8000}{2}\right) = 16,000 \times \tan(55.6875^\circ) = 2.3444 \times 10^4 \text{ (rad/s)}$$

To adjust the unit passband gain at the center frequency of 2500 Hz, we perform the following:
Fixing $\omega_{al} = 2.2022 \times 10^4$, we compute $\omega_{ah} = \frac{\omega_0^2}{\omega_{al}} = \frac{(2.3946 \times 10^4)^2}{2.2022 \times 10^4} = 2.6037 \times 10^4$
and the passband bandwidth: $W = \omega_{ah} - \omega_{al} = 4015$
Fixing $\omega_{sl} = 2.3444 \times 10^4$, $\omega_{sh} = \omega_0^2/\omega_{sl} = \frac{(2.3946 \times 10^4)^2}{2.3444 \times 10^4} = 2.4459 \times 10^4$
and the stopband bandwidth: $W_s = \omega_{sh} - \omega_{sl} = 1015$
Again, fixing $\omega_{ah} = 2.6110 \times 10^4$, we get $\omega_{al} = \frac{\omega_0^2}{\omega_{ah}} = \frac{(2.3946 \times 10^4)^2}{2.6110 \times 10^4} = 2.1961 \times 10^4$
and the passband bandwidth: $W = \omega_{ah} - \omega_{al} = 4149$
Fixing $\omega_{sh} = 2.4462 \times 10^4$, $\omega_{sl} = \frac{\omega_0^2}{\omega_{sh}} = \frac{(2.3946 \times 10^4)^2}{2.4462 \times 10^4} = 2.3441 \times 10^4$
and the stopband bandwidth: $W_s = \omega_{sh} - \omega_{sl} = 1021$
For an aggressive bandstop design, we choose $\omega_{al} = 2.6110 \times 10^4$, $\omega_{ah} = 2.1961 \times 10^4$, $\omega_{sl} = 2.3441 \times 10^4$, $\omega_{sh} = 2.4462 \times 10^4$, and $\omega_0 = 2.3946 \times 10^4$ to satisfy a larger bandwidth.
Thus we develop the prototype specification

$$v_s = \frac{26,110 - 21,916}{24,462 - 23,441} = 4.0177$$

$$n = \left(\frac{\log_{10}\left(10^{0.1 \times 10} - 1\right)}{2 \times \log_{10}(4.0177)}\right) = 0.7899, \text{ choose } n = 1.$$

$W = \omega_{ah} - \omega_{al} = 26,110 - 21,961 = 4149 \text{ (rad/s)}$, $\omega_0^2 = 5.7341 \times 10^8$.

2. Then, carrying out the prototype transformation (lowpass to bandstop) using the first-order lowpass prototype filter given by

$$H_P(s) = \frac{1}{s+1},$$

it follows that

$$H(s) = H_P(s)\Big|_{\frac{sW}{s^2+\omega_0^2}} = \frac{(s^2 + \omega_0^2)}{s^2 + Ws + \omega_0^2}.$$

Substituting the values of ω_0^2 and W yields

$$H(s) = \frac{s^2 + 5.7341 \times 10^8}{s^2 + 4149s + 5.7341 \times 10^8}.$$

Continued

EXAMPLE 8.12—CONT'D

3. Hence, applying the BLT leads to

$$H(z) = \left. \frac{s^2 + 5.7341 \times 10^8}{s^2 + 4149s + 5.73411 \times 10^8} \right|_{s=16,000(z-1)/(z+1)}.$$

After BLT, we get

$$H(z) = \frac{0.9259 + 0.7078z^{-1} + 0.9249z^{-2}}{1 + 0.7078z^{-1} + 0.8518z^{-2}}.$$

(b) MATLAB Program 8.8 includes the design steps. Fig. 8.21 shows the filter frequency responses.

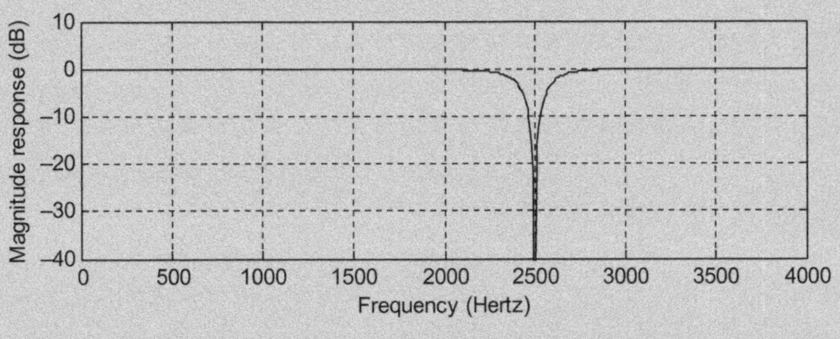

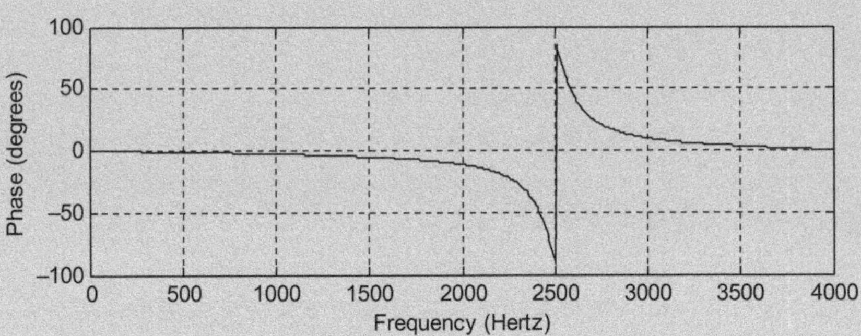

FIG. 8.21

Frequency responses of the designed digital filter for Example 8.12.

Program 8.8. MATLAB program for Example 8.12.

```
%Example 8.12
% Design of the digital bandstop Butterworth filter
format long
```

```
fs=8000; % Sampling rate
[B A]=lp2bs([1][1 1],sqrt(5.7341*10^8),4149) % Complete step 2
[b a]=bilinear(B,A,fs) % Complete step 3
% Plot the magnitude and phase responses |H(z)|
% b=[0.9259 0.7078 0.9259]; numerator coefficients from MATLAB
% a=[1 0.7078 0.8518]; denominator coefficients from MATLAB
freqz(b,a,512,fs);
axis([0 fs/2 -40 10])
```

EXAMPLE 8.13

(a) Design a digital bandpass Chebyshev filter with the following specifications:
- Center frequency of 2.5 kHz
- Passband bandwidth of 200 Hz, 0.5 dB ripple on passband
- Lower stop frequency of 1.5 kHz, upper stop frequency of 3.5 kHz
- Stopband attenuation of 10 dB
- Sampling frequency of 8000 Hz

(b) Use MATLAB to plot the magnitude and phase responses.

Solution:

(a) The digital frequencies are given as

$$\omega_{dph} = 2\pi f_{dph} = 2\pi(2600) = 5200\pi \text{ (rad/s)},$$

$$\omega_{dpl} = 2\pi f_{dpl} = 2\pi(2400) = 4800\pi \text{ (rad/s)},$$

$$\omega_{d0} = 2\pi f_0 = 2\pi(2500) = 5000\pi \text{ (rad/s)}, \text{and } T = 1/f_s = 1/8000 \text{ (s)}.$$

Applying the frequency prewarping equation, it follows that

$$\omega_{aph} = \frac{2}{T}\tan\left(\frac{\omega_{dph}T}{2}\right) = 16{,}000 \times \tan\left(\frac{5200\pi/8000}{2}\right) = 2.6110 \times 10^4 \text{ (rad/s)}$$

$$\omega_{apl} = 16{,}000 \times \tan\left(\frac{\omega_{dpl}T}{2}\right) = 16{,}000 \times \tan(0.3\pi) = 2.2022 \times 10^4 \text{ (rad/s)}$$

$$\omega_0 = 16{,}000 \times \tan\left(\frac{\omega_{d0}T}{2}\right) = 16{,}000 \times \tan(0.3125\pi) = 2.3946 \times 10^4 \text{ (rad/s)}$$

$$\omega_{ash} = 16{,}000 \times \tan\left(\frac{3500 \times 2\pi/8000}{2}\right) = 16{,}000 \times \tan(78.75^\circ) = 8.0437 \times 10^4 \text{ (rad/s)}$$

$$\omega_{asl} = 16{,}000 \times \tan\left(\frac{1500 \times 2\pi/8000}{2}\right) = 1.0691 \times 10^4 \text{ (rad/s)}$$

Now, adjusting the unit gain for the center frequency of 2500 Hz leads to the following:

Fixing $\omega_{apl} = 2.2022 \times 10^4$, we have $\omega_{aph} = \frac{\omega_0^2}{\omega_{apl}} = \frac{(2.3946 \times 10^4)^2}{2.2022 \times 10^4} = 2.6038 \times 10^4$

and the passband bandwidth: $W = \omega_{aph} - \omega_{apl} = 4016$

Continued

EXAMPLE 8.13—CONT'D

Fixing $\omega_{asl}=1.0691\times10^4$, $\omega_{ash}=\frac{\omega_0^2}{\omega_{asl}}=\frac{(2.3946\times10^4)^2}{2.10691\times10^4}=5.3635\times10^4$

and the stopband bandwidth: $W_s=\omega_{ash}-\omega_{asl}=42,944$

Again, fixing $\omega_{aph}=2.6110\times10^4$, we have $\omega_{apl}=\frac{\omega_0^2}{\omega_{aph}}=\frac{(2.3946\times10^4)^2}{2.6110\times10^4}=2.1961\times10^4$

and the passband bandwidth: $W=\omega_{aph}-\omega_{apl}=4149$

Fixing $\omega_{ash}=8.0437\times10^4$, $\omega_{asl}=\frac{\omega_0^2}{\omega_{ash}}=\frac{(2.3946\times10^4)^2}{8.0437\times10^4}=0.7137\times10^4$

and the stopband bandwidth: $W_s=\omega_{ash}-\omega_{asl}=73,300$

For an aggressive bandpass design, we select $\omega_{apl}=2.2022\times10^4$, $\omega_{aph}=2.6038\times10^4$, $\omega_{asl}=1.0691\times10^4$, $\omega_{ash}=5.3635\times10^4$ for a smaller bandwidth for passband.
Thus, we obtain the prototype specifications:

$$v_s=\frac{53,635-10,691}{26,038-22,022}=10.6932$$

$$\varepsilon^2=10^{0.1\times0.5}-1=0.1220$$

$$\frac{(10^{0.1\times10}-1)}{0.1220}=73.7705$$

$$n=\frac{\cosh^{-1}\left[(73.7705)^{0.5}\right]}{\cosh^{-1}(10.6932)}=\frac{\ln\left(8.5890+\sqrt{8.5890^2-1}\right)}{\ln\left(10.6932+\sqrt{10.6932^2-1}\right)}=0.9280$$

rounding up n leads to $n=1$.
Applying the design steps leads to

1. $\omega_{aph}=2.6038\times10^4$(rad/s), $\omega_{apl}=2.2022\times10^4$(rad/s)

$$W=4016\,(\text{rad/s}), \omega_0^2=5.7341\times10^8$$

2. Performing the prototype transformation (lowpass to bandpass), we obtain

$$H_P(s)=\frac{2.8628}{s+2.8628}$$

and

$$H(s)=H_P(s)\Big|_{s=\frac{s^2+\omega_0^2}{sW}}=\frac{2.8628Ws}{s^2+2.8628Ws+\omega_0^2}=\frac{1.1497\times10^4 s}{s^2+1.1497\times10^4 s+5.7341\times10^8}.$$

3. Applying the BLT, the analog filter is converted to a digital filter as follows:

$$H(z)=\frac{1.1497\times10^4 s}{s^2+1.1497\times10^4 s+5.7341\times10^8}\Bigg|_{s=16,000(z-1)/(z+1)},$$

which is simplified and arranged to be

$$H(z)=\frac{0.1815-0.1815z^{-2}}{1+0.6264z^{-1}+0.6396z^{-2}}.$$

(a) Program 8.9 lists the MATLAB details. Fig. 8.22 displays the frequency responses.

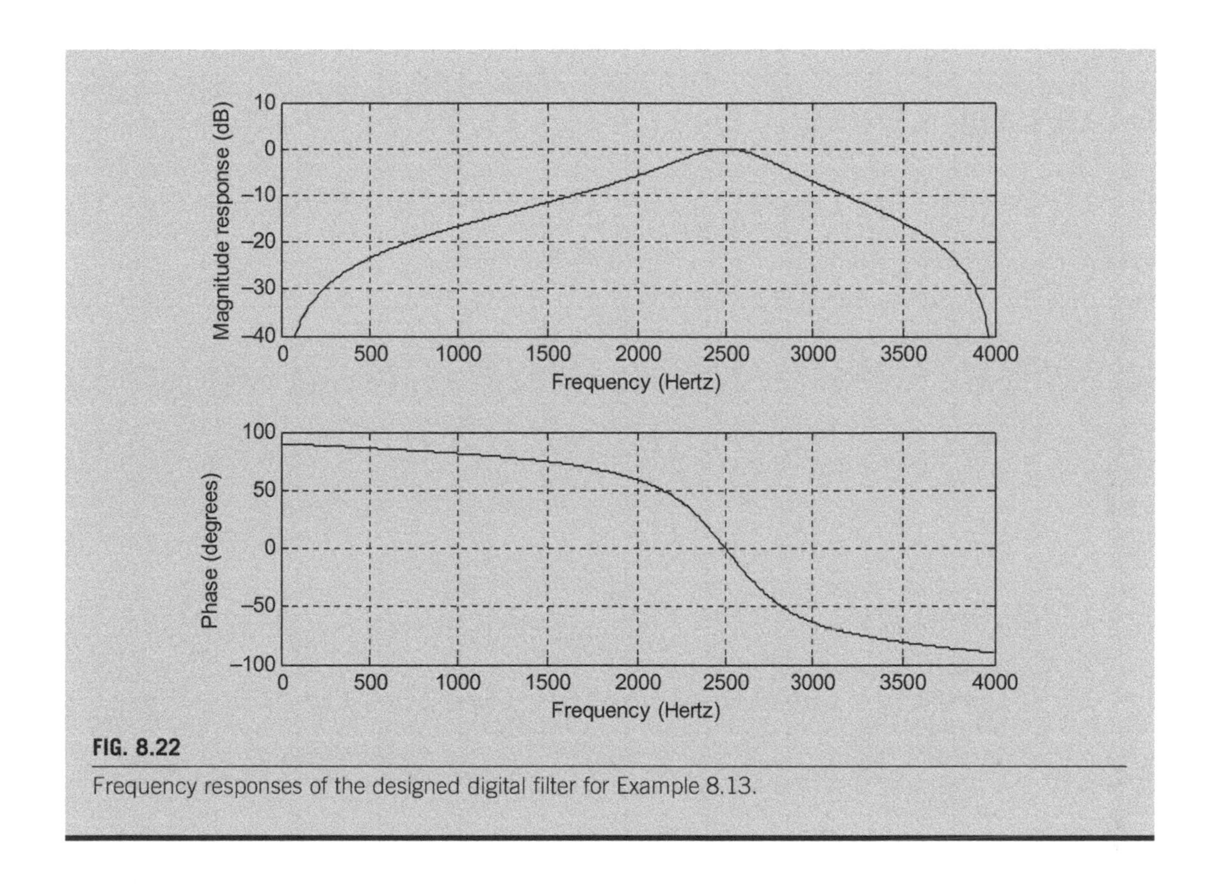

FIG. 8.22

Frequency responses of the designed digital filter for Example 8.13.

Program 8.9. MATLAB program for Example 8.13.

```
%Example 8.13
% Design of the digital bandpass Chebyshev filter
format long
fs=8000;
[B A]=lp2bp([2.8628][1 2.8628],sqrt(5.7341*10^8),4016) % Complete step 2
[b a]=bilinear(B,A,fs) % Complete step 3
% Plot the magnitude and phase responses |H(z)|
% b=[0.1815 0.0 -0.1815]; numerator coefficients from MATLAB
% a=[1 0.6264 0.6396]; denominator coefficients from MATLAB
freqz(b,a,512,fs);
axis([0 fs/2 -40 10])
```

8.4 HIGHER-ORDER INFINITE IMPULSE RESPONSE FILTER DESIGN USING THE CASCADE METHOD

For higher-order IIR filter design, use of a cascade transfer function is preferred. The factored forms for the lowpass prototype transfer functions for Butterworth and Chebyshev filters are provided in Tables 8.7–8.9. A Butterworth filter design example will be provided and the similar procedure can be adopted for Chebyshev filters.

Table 8.7 3dB Butterworth Prototype Functions in the Cascade Form

n	$H_P(s)$
3	$\frac{1}{(s+1)(s^2+s+1)}$
4	$\frac{1}{(s^2+0.7654s+1)(s^2+1.8478s+1)}$
5	$\frac{1}{(s+1)(s^2+0.6180s+1)(s^2+1.6180s+1)}$
6	$\frac{1}{(s^2+0.5176s+1)(s^2+1.4142s+1)(s^2+1.9319s+1)}$

Table 8.8 Chebyshev Prototype Functions in the Cascade Form with 0.5dB Ripple ($\varepsilon = 0.3493$)

n	$H_P(s)$ 0.5dB Ripple ($\varepsilon = 0.3493$)
3	$\frac{0.7157}{(s+0.6265)(s^2+0.6265s+1.1425)}$
4	$\frac{0.3579}{(s^2+0.3507s+1.0635)(s^2+0.8467s+0.3564)}$
5	$\frac{0.1789}{(s+0.3623)(s^2+0.2239s+1.0358)(s^2+0.5862s+0.4768)}$
6	$\frac{0.0895}{(s^2+0.1553s+1.0230)(s^2+0.4243s+0.5900)(s^2+0.5796s+0.1570)}$

Table 8.9 Chebyshev Prototype Functions in the Cascade Form with 1dB Ripple ($\varepsilon = 0.5088$)

n	$H_P(s)$ 1dB Ripple ($\varepsilon = 0.5088$)
3	$\frac{0.4913}{(s+0.4942)(s^2+0.4942s+0.9942)}$
4	$\frac{0.2456}{(s^2+0.2791s+0.9865)(s^2+0.6737s+0.2794)}$
5	$\frac{0.1228}{(s+0.2895)(s^2+0.1789s+0.9883)(s^2+0.4684s+0.4293)}$
6	$\frac{0.0614}{(s^2+0.1244s+0.9907)(s^2+0.3398s+0.5577)(s^2+0.4641s+0.1247)}$

EXAMPLE 8.14

(a) Design a fourth-order digital lowpass Butterworth filter with a cutoff frequency of 2.5 kHz at a sampling frequency of 8000 Hz.
(b) Use MATLAB to plot the magnitude and phase responses.

Solution:

(a) First, we obtain the digital frequency in radians per second:

$$\omega_d = 2\pi f = 2\pi(2500) = 5000\pi \text{ (rad/s), and } T = \frac{1}{f_s} = \frac{1}{8000} \text{ (s)}.$$

Following the design steps, we compute the specifications for the analog filter.

1.

$$\omega_a = \frac{2}{T}\tan\left(\frac{\omega_d T}{2}\right) = 16{,}000 \times \tan\left(\frac{5000\pi/8000}{2}\right) = 2.3946 \times 10^4 \text{ (rad/s)}$$

2. From Table 8.7, we have the fourth-order factored prototype transfer function as

$$H_P(s) = \frac{1}{(s^2 + 0.7654s + 1)(s^2 + 1.8478s + 1)}.$$

Applying the prototype transformation, we obtain

$$H(s) = H_P(s)\big|_{s = \frac{s}{\omega_a}} = \frac{\omega_a^2 \times \omega_a^2}{(s^2 + 0.7654\omega_a s + \omega_a^2)(s^2 + 1.8478\omega_a s + \omega_a^2)}.$$

Substituting $\omega_a = 2.3946 \times 10^4$ (rad/s) yields

$$H(s) = \frac{(5.7340 \times 10^8) \times (5.7340 \times 10^8)}{(s^2 + 1.8328s + 5.7340 \times 10^8)(s^2 + 4.4247 \times 10^4 s + 5.7340 \times 10^8)}.$$

3. Hence, after applying BLT, we have

$$H(z) = \left.\frac{(5.7340 \times 10^8) \times (5.7340 \times 10^8)}{(s^2 + 1.8328s + 5.7340 \times 10^8)(s^2 + 4.4247 \times 10^4 s + 5.7340 \times 10^8)}\right|_{s = 16{,}000(z-1)/(z+1)}.$$

Simplifying the algebra, we have the digital filter as

$$H(z) = \frac{0.5108 + 1.0215z^{-1} + 0.5108z^{-2}}{1 + 0.5654z^{-1} + 0.4776z^{-2}} \times \frac{0.3730 + 0.7460z^{-1} + 0.3730z^{-2}}{1 + 0.4129z^{-1} + 0.0790z^{-2}}.$$

(b) A MATLAB program is better to carry out the algebra and is listed in Program 8.10. Fig. 8.23 shows the filter magnitude and phase frequency responses.

Continued

EXAMPLE 8.14—CONT'D

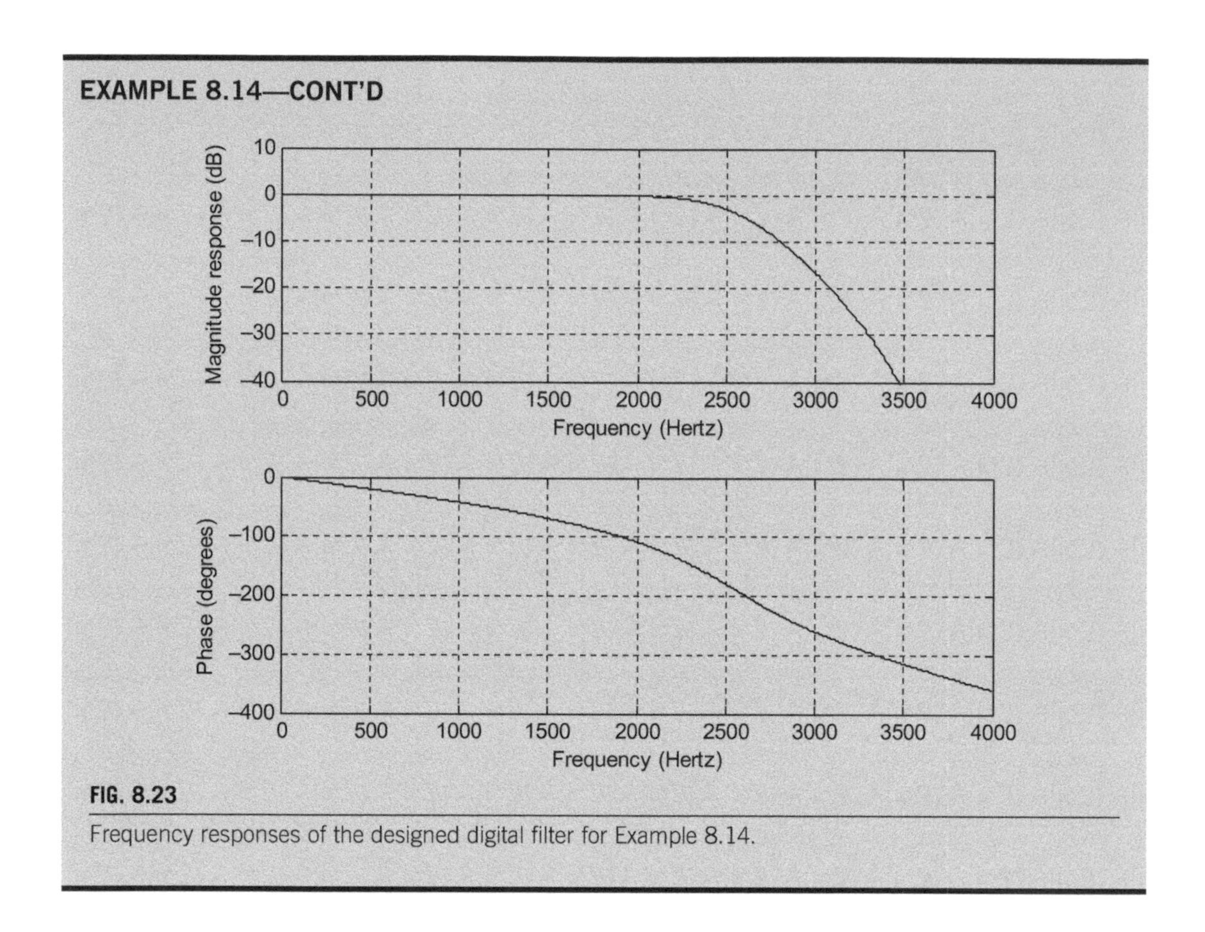

FIG. 8.23

Frequency responses of the designed digital filter for Example 8.14.

Program 8.10. MATLAB program for Example 8.14.

```
%Example 8.14
% Design of the fourth-order digital lowpass Butterworth filter
% in the cascade form
format long
fs=8000;    % Sampling rate
[B1 A1]=lp2lp([1][1 0.7654 1], 2.3946*10^4) % Complete step 2
[b1 a1]=bilinear(B1,A1,fs) % complete step 3
[B2 A2]=lp2lp([1][1 1.8478 1], 2.3946*10^4) % Complete step 2
[b2 a2]=bilinear(B2,A2,fs) % complete step 3
% Plot the magnitude and phase responses |H(z)|
% b1=[0.5108 1.0215 0.5108]; a1=[1 0.5654 0.4776]; coefficients from MATLAB
% b2=[0.3730 0.7460 0.3730]; a2=[1 0.4129 0.0790]; coefficients from MATLAB
freqz(conv(b1,b2),conv(a1,a2),512,fs); % Combined filter responses
axis([0 fs/2 -40 10]);
```

The higher-order bandpass, highpass, and bandstop filters using the cascade form can be designed similarly.

8.5 APPLICATION: DIGITAL AUDIO EQUALIZER

In this section, the design of a digital audio equalizer is introduced. For an audio application such as the CD player, the digital audio equalizer is used to make the sound as one desires by changing filter gains for different audio frequency bands. Other applications include adjusting the sound source to take room acoustics into account, removing undesired noise, and boosting the desired signal in the specified passband. The simulation is based on the consumer digital audio processor—such as CD player—handling the 16-bit digital samples with a sampling rate of 44.1 kHz and an audio signal bandwidth at 22.05 kHz. A block diagram of the digital audio equalizer is depicted in Fig. 8.24.

A seven-band audio equalizer is adopted for discussion. The center frequencies are listed in Table 8.10. The 3-dB bandwidth for each bandpass filter is chosen to be 50% of the center frequency. As shown in Fig. 8.24, g_0 through g_6 are the digital gains for each bandpass filter output and can be adjusted to make sound effects, while $y_0(n)$ through $y_6(n)$ are the digital amplified bandpass filter outputs. Finally, the equalized signal is the sum of the amplified bandpass filter outputs and itself. By changing the digital gains of the equalizer, many sound effects can be produced.

To complete the design and simulation, second-order IIR bandpass Butterworth filters are chosen for the audio equalizer, the coefficients are achieved using the BLT method, and are provided in Table 8.11.

The magnitude response for each filter bank is plotted in Fig. 8.25 for design verification. As shown in Fig. 8.25, after careful examination, the magnitude response of each filter band meets the design specification. We will perform simulation next.

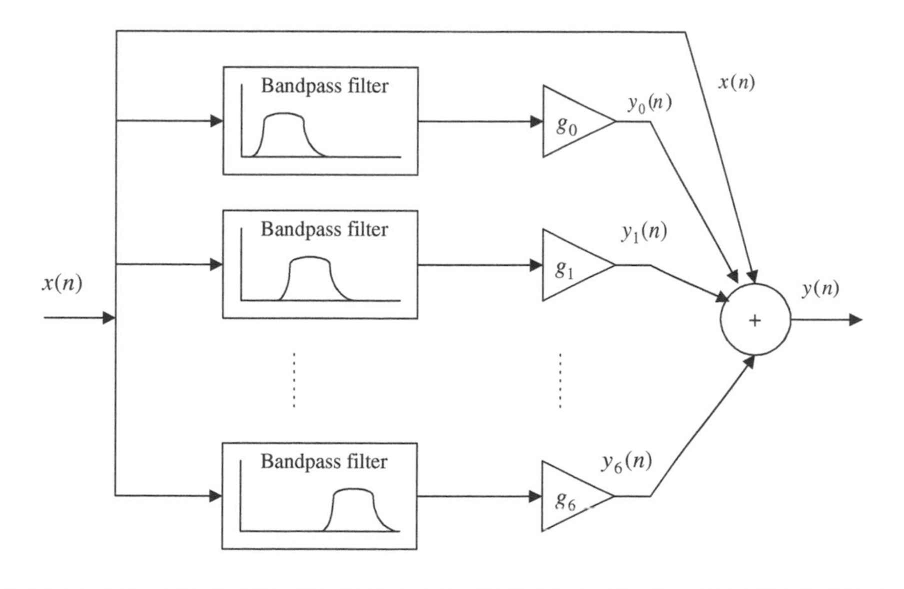

FIG. 8.24

Simplified block diagram of the audio equalizer.

Table 8.10 Specifications for an Audio Equalizer to be Designed

Center frequency (Hz)	100	200	400	1000	2500	6000	15,000
Bandwidth (Hz)	50	100	200	500	1250	3000	7500

Table 8.11 Designed Filter Banks

Filter Banks	Coefficients for the Numerator	Coefficients for the Denominator
Bandpass filter 0	0.0031954934, 0, −0.0031954934	1, −1.9934066716, 0.9936090132
Bandpass filter 1	0.0063708102, 0, −0.0063708102	1, −1.9864516324, 0.9872583796
Bandpass filter 2	0.0126623878, 0, −0.0126623878	1, −1.9714693192, 0.9746752244
Bandpass filter 3	0.0310900413, 0, −0.0310900413	1, −1.9181849043, 0.9378199174
Bandpass filter 4	0.0746111954, 0, −0.0746111954	1, −1.7346085867, 0.8507776092
Bandpass filter 5	0.1663862883, 0, −0.1663862884	1, −1.0942477187, 0.6672274233
Bandpass filter 6	0.3354404899, 0, −0.3354404899	1, 0.7131366534, 0.3291190202

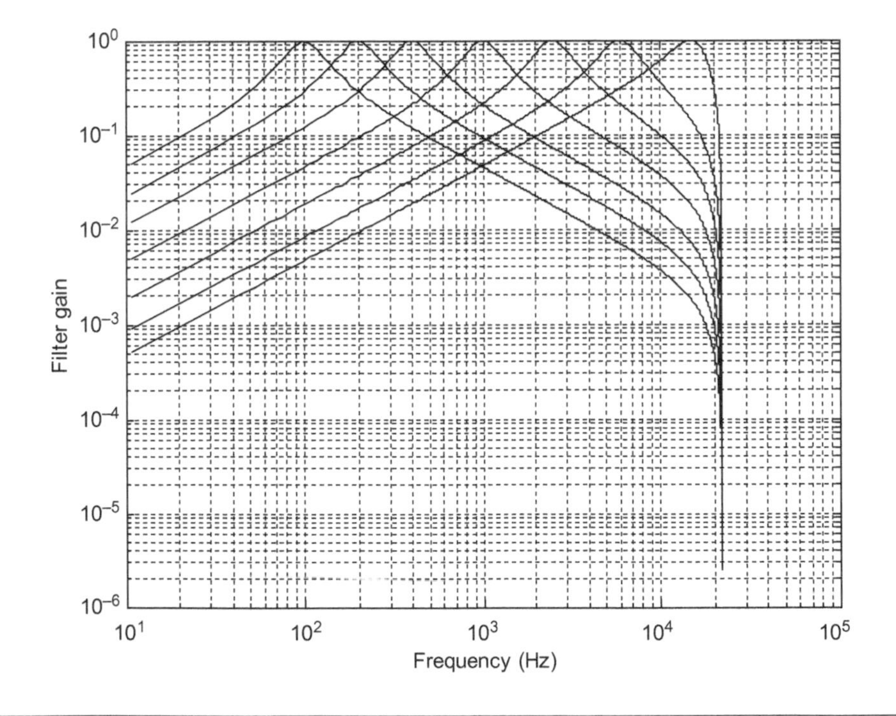

FIG. 8.25

Magnitude frequency responses for the audio equalizer.

Simulation in the MATLAB environment is based on the following setting. The audio test signal having frequency components of 100, 200, 400, 1000, 2500, 6000, and 15,000 Hz is generated from Eq. (8.36):

$$\begin{aligned} x(n) = &\sin(200\pi n/44,100) + \sin(400\pi n/44,100 + \pi/14) \\ &+ \sin(800\pi n/44,100 + \pi/7) + \sin(2,000\pi n/44,100 + 3\pi/14) \\ &+ \sin(5,000\pi n/44,100 + 2\pi/7) + \sin(12,000\pi n/44,100 + 5\pi/14) \\ &+ \sin(30,000\pi n/44,100 + 3\pi/7). \end{aligned} \tag{8.36}$$

The gains set for the filter banks are as follows:

$$g_0 = 10,\ g_1 = 10,\ g_2 = 0,\ g_3 = 0,\ g_4 = 0,\ g_5 = 10,\ g_6 = 10.$$

After simulation, we note that the frequency components at 100, 200, 6000, and 15,000 Hz will be boosted by $20 \times \log_{10} 10 = 20\,(\text{dB})$. The top plot in Fig. 8.26 shows the spectrum for the audio test signal, while the bottom plot depicts the spectrum for the equalized audio test signal. As shown in the plots, before audio digital equalization, the spectral peaks at all bands are at the same level; after audio digital equalization, the frequency components at bank 0, bank 1, bank 5, and bank 6 are amplified. Therefore, as we expected, the operation of the digital equalizer boosts the low-frequency and high-frequency components. The MATLAB list for the simulation is shown in Program 8.11.

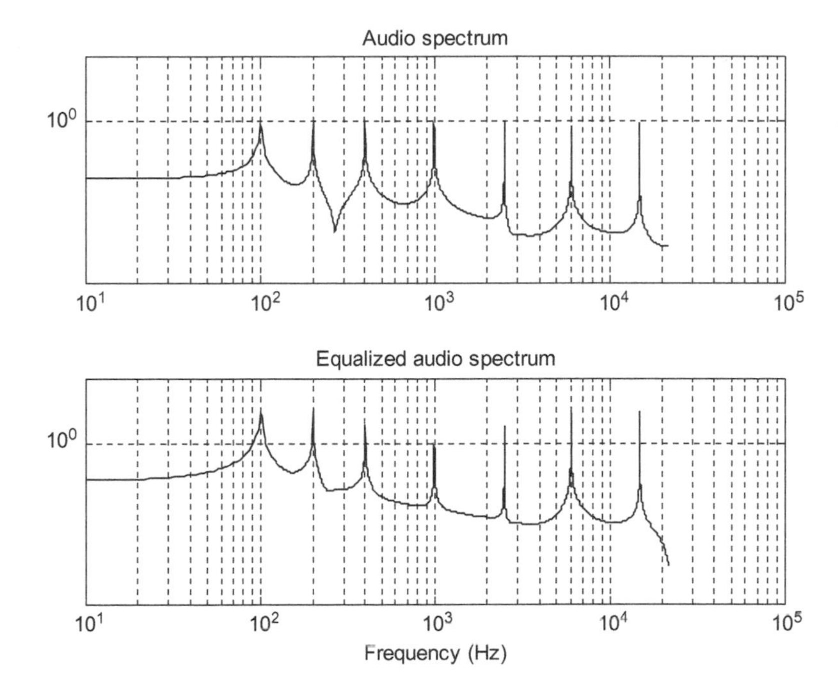

FIG. 8.26

Audio spectrum and equalized audio spectrum.

Program 8.11. MATLAB program for the digital audio equalizer.

```
close all; clear all
% Filter Coefficients (Butterworth type designed using the BLT)
B0=[0.0031954934 0 -0.0031954934]; A0=[1.0000000000 -1.9934066716 0.9936090132];
B1=[0.0063708102 0 -0.0063708102]; A1=[1.0000000000 -1.9864516324 0.9872583796];
B2=[0.0126623878 0 -0.0126623878]; A2=[1.0000000000 -1.9714693192 0.9746752244];
B3=[0.0310900413 0 -0.0310900413]; A3=[ 1.0000000000 -1.9181849043 0.9378199174];
B4=[ 0.0746111954 0.000000000 -0.0746111954];
A4=[1.0000000000 -1.7346085867 0.8507776092];
B5=[0.1663862883 0.0000000000 -0.1663862884];
A5=[1.0000000000 -1.0942477187 0.6672274233];
B6=[0.3354404899 0.0000000000 -0.3354404899];
A6=[1.0000000000 0.7131366534 0.3291190202];
[h0,f]=freqz(B0,A0,2048,44100);
[h1,f]=freqz(B1,A1,2048,44100);
[h2,f]=freqz(B2,A2,2048,44100);
[h3,f]=freqz(B3,A3,2048,44100);
[h4,f]=freqz(B4,A4,2048,44100);
[h5,f]=freqz(B5,A5,2048,44100);
[h6,f]=freqz(B6,A6,2048,44100);
loglog(f,abs(h0),f,abs(h1), f,abs(h2), ...
f,abs(h3),f,abs(h4),f,abs(h5),f,abs(h6));
xlabel('Frequency (Hz)');
ylabel('Filter Gain');grid
axis([10 10^5 10^(-6) 1]);
figure(2)
g0=10;g1=10;g2=0;g3=0;g4=0;g5=10;g6=10;
p0=0;p1=pi/14;p2=2*p1;p3=3*p1;p4=4*p1;p5=5*p1;p6=6*p1;
n=0:1:20480;                    % Indices of samples
fs=44100;                       % Sampling rate
x=sin(2*pi*100*n/fs)+sin(2*pi*200*n/fs+p1)+...
 sin(2*pi*400*n/fs+p2)+sin(2*pi*1000*n/fs+p3)+...
 sin(2*pi*2500*n/fs+p4)+sin(2*pi*6000*n/fs+p5)+...
 sin(2*pi*15000*n/fs+p6); % Generate test audio signals
y0=filter(B0,A0,x);             % Bandpass filter 0
y1=filter(B1,A1,x);             % Bandpass filter 1
y2=filter(B2,A2,x);             % Bandpass filter 2
y3=filter(B3,A3,x);             % Bandpass filter 3
y4=filter(B4,A4,x);             % Bandpass filter 4
y5=filter(B5,A5,x);             % Bandpass filter 5
y6=filter(B6,A6,x);             % Bandpass filter 6
y=g0.*y0+g1.*y1+g2.*y2+g3.*y3+g4.*y4+g5.*y5+g6.*y6+x; % Equalizer output
N=length(x);
Axk=2*abs(fft(x))/N;Axk(1)=Axk(1)/2; % One-sided amplitude spectrum of the input
f=[0:N/2]*fs/N;
subplot(2,1,1);loglog(f,Axk(1:N/2+1));
title('Audio spectrum');
axis([10 100000 0.00001 100]);grid;
Ayk=2*abs(fft(y))/N; Ayk(1)=Ayk(1)/2; % One-sided amplitude spectrum of the output
subplot(2,1,2);loglog(f,Ayk(1:N/2+1));
xlabel('Frequency (Hz)');
title('Equalized audio spectrum');
axis([10 100000 0.00001 100]);grid;
```

8.6 IMPULSE INVARIANT DESIGN METHOD

We illustrate the concept of the impulse invariant design shown in Fig. 8.27. Given the transfer function of a designed analog filter, an analog impulse response can be easily found by the inverse Laplace transform of the transfer function. To replace the analog filter by equivalent digital filter, we apply an approximation in time domain in which the digital filter impulse response must be equivalent to the analog impulse response. Therefore, we can sample the analog impulse response to get the digital impulse response, and take the z-transform of the sampled analog impulse response to obtain the transfer function of the digital filter.

The analog impulse response can be achieved by taking the inverse Laplace transform of the analog filter $H(s)$, that is,

$$h(t) = L^{-1}(H(s)). \tag{8.37}$$

Now, if we sample the analog impulse response with a sampling interval of T and use T as a scale factor, it follows that

$$T \times h(n) = T \times h(t)|_{t=nT}, n \geq 0. \tag{8.38}$$

Taking the z-transform on both sides of Eq. (8.38) yields the digital filter as

$$H(z) = Z[T \times h(n)]. \tag{8.39}$$

The effect of the scale factor T in Eq. (8.38) can be explained as follows. We approximate the area under the curve specified by the analog impulse function $h(t)$ using a digital sum given by

$$\text{Area} = \int_0^{\infty} h(t)dt \approx T \times h(0) + T \times h(1) + T \times h(2) + \cdots. \tag{8.40}$$

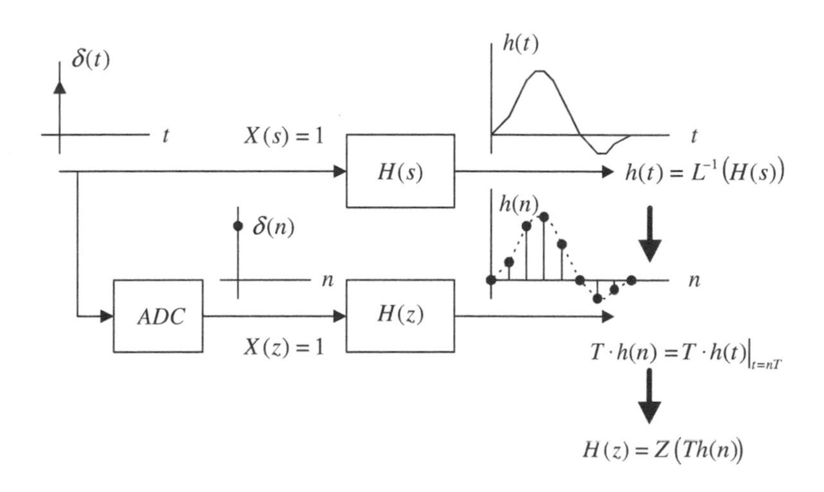

FIG. 8.27

Impulse invariant design method.

Note that the area under the curve indicates the DC gain of the analog filter while the digital sum in Eq. (8.40) is the DC gain of the digital filter.

The rectangular approximation is used, since each sample amplitude is multiplied by the sampling interval T. Due to the interval size for approximation in practice, we cannot guarantee that the digital sum has exactly the same value as the one from the integration unless the sampling interval T in Eq. (8.40) approaches zero. This means that the higher the sampling rate, that is, the smaller the sampling interval, the more accurately the digital filter gain matches the analog filter gain. Hence, in practice, we need to further apply gain scaling for adjustment if it is a requirement. We look at the following examples.

EXAMPLE 8.15

Consider the following Laplace transfer function:

$$H(s) = \frac{2}{s+2}.$$

(a) Determine $H(z)$ using the impulse invariant method if the sampling rate $f_s = 10$ (Hz).

(b) Use MATLAB to plot

1. The magnitude response $|H(f)|$ and phase response $\phi(f)$ with respect to $H(s)$ for the frequency range from 0 to $f_s/2$ (Hz).
2. The magnitude response $|H(e^{j\Omega})| = |H(e^{j2\pi fT})|$ and phase response $\phi(f)$ with respect to $H(z)$ for the frequency range from 0 to $f_s/2$ (Hz).

Solution:

(a) Taking the inverse Laplace transform the analog transfer function, the impulse response therefore is found to be

$$h(t) = L^{-1}\left[\frac{2}{s+2}\right] = 2e^{-2t}u(t).$$

Sampling the impulse response $h(t)$ with $T = 1/f_s = 0.1$ (s), we have

$$Th(n) = T2e^{-2nT}u(n) = 0.2e^{-0.2n}u(n).$$

Using the z-transform table in Chapter 5, we have

$$Z[e^{-an}u(n)] = \frac{z}{z - e^{-a}}.$$

and noting that $e^{-a} = e^{-0.2} = 0.8187$, the digital filter transfer function $H(z)$ is finally given by

$$H(z) = \frac{0.2z}{z - 0.8187} = \frac{0.2}{1 - 0.8187z^{-1}}.$$

(b) The MATLAB list is given in Program 8.12. The first and third plots in Fig. 8.28 show comparisons of the magnitude and phase frequency responses. The shape of the magnitude response (first plot) closely matches that of the analog filter, while the phase response (third plot) differs from the analog phase response in this example.

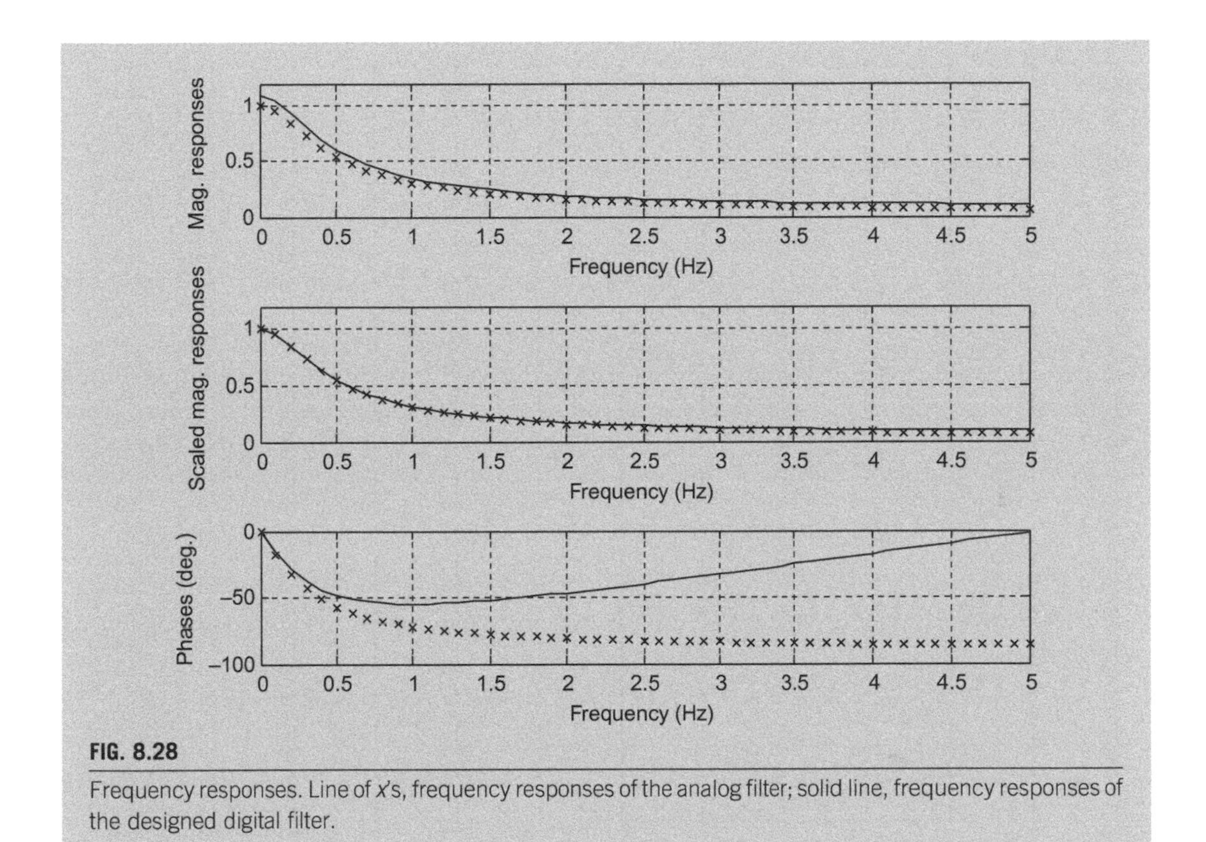

FIG. 8.28

Frequency responses. Line of *x*'s, frequency responses of the analog filter; solid line, frequency responses of the designed digital filter.

Program 8.12. MATLAB program for Example 8.15.

```
%Example 8.15.
% Plot the magnitude responses |H(s)| and |H(z)|
% For the Laplace transfer function H(s)
f=0:0.1:5;T=0.1              % Frequency range and sampling interval
w=2*pi*f;                    % Frequency range in rad/sec
hs=freqs([2], [1 2],w);      % Analog frequency response
phis=180*angle(hs)/pi;
% For the z-transfer function H(z)
hz=freqz([0.2][1 -0.8187],length(w));  % Digital frequency response
hz_scale=freqz([0.1813][1 -0.8187],length(w)); % Scaled digital mag. response
 phiz=180*angle(hz)/pi;
 %Plot magnitude and phase responses.
subplot(3,1,1), plot(f,abs(hs),'kx',f, abs(hz),'k-'),grid; axis([0 5 0 1.2]);
xlabel('Frequency (Hz)'); ylabel('Mag. Responses')
subplot(3,1,2), plot(f,abs(hs),'kx',f, abs(hz_scale),'k-'),grid; axis([0 5 0 1.2]);
xlabel('Frequency (Hz)'); ylabel('Scaled Mag. Responses')
subplot(3,1,3), plot(f,phis,'kx',f, phiz,'k-'); grid;
xlabel('Frequency (Hz)'); ylabel('Phases (deg.)');
```

The filter DC gain is given by

$$H(e^{j\Omega})|_{\Omega=0} = H(1) = 1.1031.$$

we can further scale the filter to have a unit gain of

$$H(z) = \frac{1}{1.1031} \frac{0.2}{1 - 0.8187z^{-1}} = \frac{0.1813}{1 - 0.8187z^{-1}}.$$

The scaled magnitude frequency response is shown in the middle plot along with that of analog filter in Fig. 8.28, where the magnitudes are matched very well below 1.8 Hz.

Example 8.15 demonstrates the design procedure using the impulse invariant design. The filter performance depends on the sampling interval (Lynn and Fuerst, 1999). As shown in Fig. 8.27, the analog impulse response $h(t)$ is not a band-limited signal whose frequency components are generally larger than the Nyquist limit (folding frequency); hence, sampling $h(t)$ could cause aliasing. Fig. 8.29A shows the analog impulse response $Th(t)$ in Example 8.15 and its sampled version $Th(nT)$, where the sampling

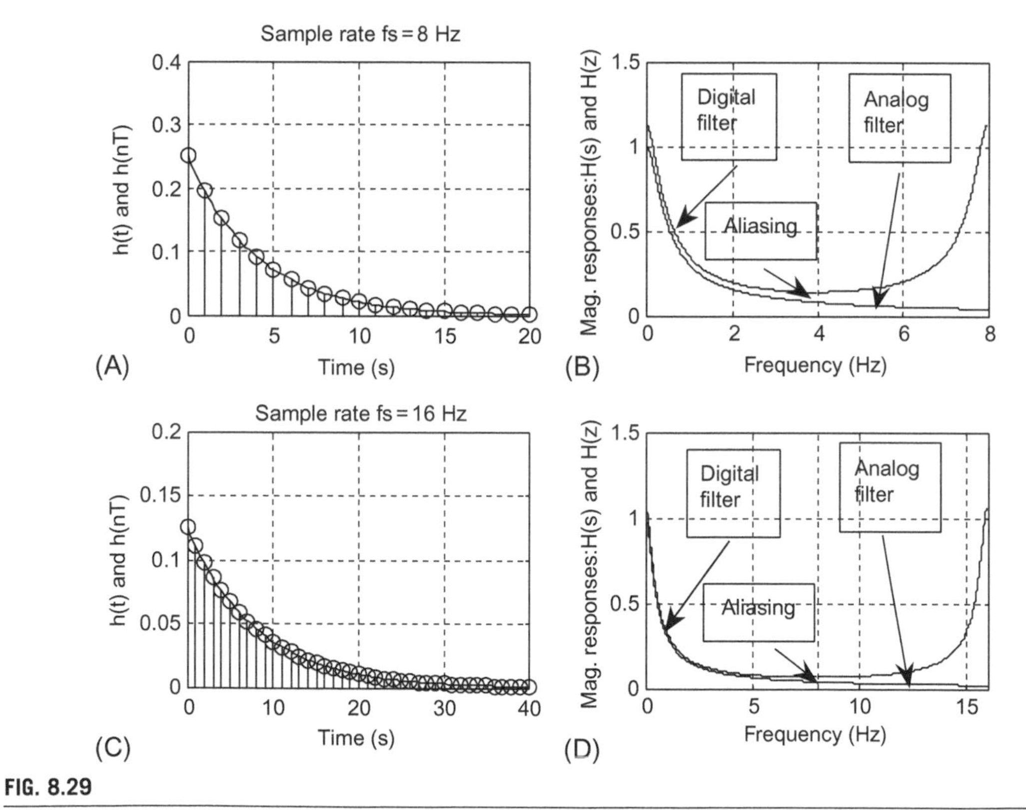

FIG. 8.29

Sampling interval effect in the impulse invariant IIR filter design. (A) Analog impulse response and its sampled version at a sampling rate of 8 Hz. (B) Magnitude frequency responses of the analog filter and the digital filter at a sampling rate of 8 Hz. (C) Analog impulse response and its sampled version at a sample rate of 16 Hz. (D) Magnitude frequency responses of the analog filter and the digital filter at a sampling rate of 16 Hz.

interval is 0.125 s. The magnitude frequency responses of the analog and digital filters are plotted in Fig. 8.29B. The aliasing occurs, since the impulse response contains the frequency components beyond the Nyquist limit, that is, 4 Hz, in this case. Furthermore, using lower sampling rate of 8 Hz causes less accuracy in the magnitude frequency response of the digital filter, so more aliasing develops.

Fig. 8.29C shows the analog impulse response and its sampled version using a higher sampling rate of 16 Hz. Fig. 8.29D displays the more accurate magnitude frequency response of the digital filter. Hence, we can obtain a reduced aliasing level. Note that the aliasing cannot be avoided, due to sampling of the analog impulse response. The only way to reduce the aliasing is to use a higher sampling frequency or design a filter with a very low cutoff frequency to reduce the aliasing to a minimum level.

Investigation of the sampling interval effect leads us to the following conclusions. Note that the analog impulse response for a highpass filter or a bandstop filter contains frequency components at the maximum level at the Nyquist limit (folding frequency), even assuming that the sampling rate is much higher than the cutoff frequency of a highpass filter or the upper cutoff frequency of a bandstop filter. Hence, sampling the analog impulse response always produces the maximum aliasing level. Without using an additional anti-aliasing filter or the advanced method (Nelatury, 2007) employing the frequency prewarping technique, the impulse invariant method alone are not suggested for designing the highpass filter or bandstop filter.

Instead, in practice, we should apply the BLT design method. The impulse invariant design method is only appropriate for designing a lowpass filter or bandpass filter with a sampling rate much larger than the lower cutoff frequency of the lowpass filter or the upper cutoff frequency of the bandpass filter.

Next, let us focus on the second-order filter design via Example 8.16.

EXAMPLE 8.16

Consider the following Laplace transfer function:

$$H(s)=\frac{s}{s^2+2s+5}.$$

(a) Determine $H(z)$ using the impulse invariant method if the sampling rate $f_s=10$ Hz.
(b) Use MATLAB to plot:
1. The magnitude response $|H(f)|$ and phase response $\phi(f)$ with respect to $H(s)$ for the frequency range from 0 to $f_s/2$ (Hz).
2. The magnitude response $|H(e^{j\Omega})|=|H(e^{j2\pi fT})|$ and phase response $\phi(f)$ with respect to $H(z)$ for the frequency range from 0 to $f_s/2$ (Hz).

Solution:

(a) Since $H(s)$ has complex poles located at $s=-1\pm 2j$, we can write it in a quadratic form as

$$H(s)=\frac{s}{s^2+2s+5}=\frac{s}{(s+1)^2+2^2}.$$

We can further write the transfer function as

$$H(s)=\frac{(s+1)-1}{(s+1)^2+2^2}=\frac{(s+1)}{(s+1)^2+2^2}-0.5\times\frac{2}{(s+1)^2+2^2}.$$

Continued

EXAMPLE 8.16—CONT'D

From the Laplace transform table (Appendix B), the analog impulse response can easily be found as

$$h(t) = e^{-t}\cos(2t)u(t) - 0.5e^{-t}\sin(2t)u(t).$$

Sampling the impulse response $h(t)$ using a sampling interval $T=0.1$ and using the scale factor of $T=0.1$, we have

$$Th(n) = Th(t)|_{t=nT} = 0.1e^{-0.1n}\cos(0.2n)u(n) - 0.05e^{-0.1n}\sin(0.2n)u(n).$$

Applying the z-transform (Chapter 5) leads to

$$\begin{aligned} H(z) &= Z[0.1e^{-0.1n}\cos(0.2n)u(n) - 0.05e^{-0.1n}\sin(0.2n)u(n)] \\ &= \frac{0.1z(z-e^{-0.1}\cos(0.2))}{z^2 - 2e^{-0.1}\cos(0.2)z + e^{-0.2}} - \frac{0.05e^{-0.1}\sin(0.2)z}{z^2 - 2e^{-0.1}\cos(0.2)z + e^{-0.2}}. \end{aligned}$$

After algebraic simplification, we obtain the second-order digital filter as

$$H(z) = \frac{0.1 - 0.09767z^{-1}}{1 - 1.7735z^{-1} + 0.8187z^{-2}}.$$

(b) The magnitude and phase frequency responses are shown in Fig. 8.30 and MATLAB Program 8.13 is given. The passband gain of the digital filter is higher than that of the analog filter, but their shapes are same.

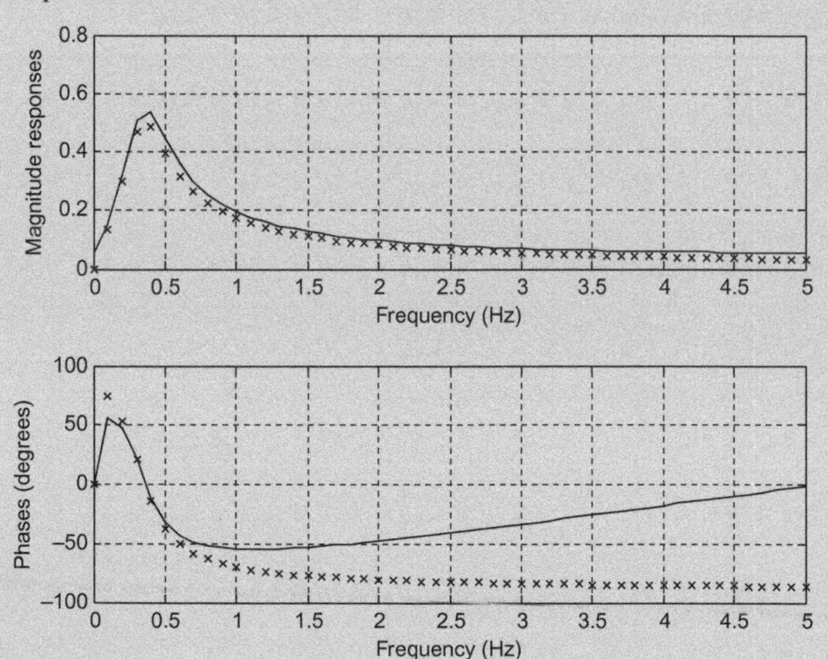

FIG. 8.30

Frequency responses. Line of x's, frequency responses of the analog filter; solid line, frequency responses of the designed digital filter.

Program 8.13. MATLAB program for Example 8.16.

```
%Example 8.16
% Plot the magnitude responses |H(s)| and |H(z)|
% For the Laplace transfer function H(s)
f=0:0.1:5;T=0.1;  % Initialize analog frequency range in Hz and sampling interval
w=2*pi*f;         % Convert the frequency range to radians/second
hs=freqs([1 0], [1 2 5],w);     % Calculate analog filter frequency responses
phis=180*angle(hs)/pi;
% For the z-transfer function H(z)
% Calculate digital filter frequency responses
hz=freqz([0.1 -0.09766][1 -1.7735 0.8187],length(w));
 phiz=180*angle(hz)/pi;
 % Plot magnitude and phase responses
subplot(2,1,1), plot(f,abs(hs),'x',f, abs(hz),'-'),grid;
xlabel('Frequency (Hz)'); ylabel('Magnitude Responses')
subplot(2,1,2), plot(f,phis,'x',f, phiz,'-'); grid;
xlabel('Frequency (Hz)'); ylabel('Phases (degrees)')
```

8.7 POLE-ZERO PLACEMENT METHOD FOR SIMPLE INFINITE IMPULSE RESPONSE FILTERS

This section introduces a pole-zero placement method for a simple IIR filter design. Let us first examine the effects of the pole-zero placement on the magnitude response in the z-plane shown in Fig. 8.31.

In the z-plane, when we place a pair of complex conjugate zeros at a given point on the unit circle with an angle θ (usually we do), we will have a numerator factor of $(z-e^{j\theta})(z-e^{-j\theta})$ in the transfer function. Its magnitude contribution to the frequency response at $z=e^{j\Omega}$ is $(e^{j\Omega}-e^{j\theta})(e^{j\Omega}-e^{-j\theta})$. When $\Omega=\theta$,

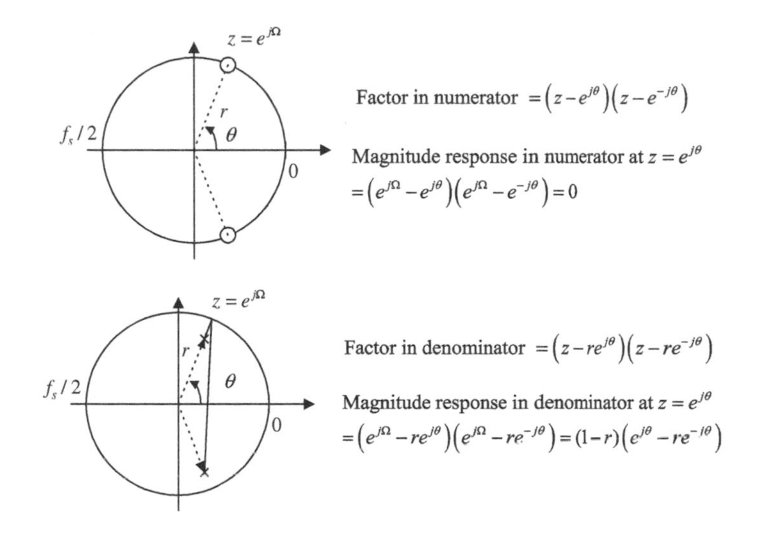

FIG. 8.31

Effects of the pole-zero placement on the magnitude response.

the magnitude will reach zero, since the first factor $(e^{j\theta}-e^{j\theta})=0$ contributes zero magnitude. When a pair of complex conjugate poles are placed at a given point within the unit circle, we have a denominator factor of $(z-re^{j\theta})(z-re^{-j\theta})$, where r is the radius chosen to be less than and close to 1 to place the poles inside the unit circle. The magnitude contribution to the frequency response at $\Omega=\theta$ will rise to a large magnitude, since the first factor $(e^{j\theta}-re^{j\theta})=(1-r)e^{j\theta}$ gives a small magnitude of $1-r$, which is the length between the pole and the unit circle at the angle $\Omega=\theta$. Note that the magnitude of $e^{j\theta}$ is 1.

Therefore, we can reduce the magnitude response using zero placement, while we increase the magnitude response using pole placement. Placing a combination of poles and zeros will result in different frequency responses, such as lowpass, highpass, bandpass, and bandstop. The method is intuitive and approximate. However, it is easy to compute filter coefficients for simple IIR filters. Here, we describe the design procedures for second-order bandpass and bandstop filters, as well first-order lowpass and highpass filters. Practically, the pole-zero placement method has a good performance when the bandpass and bandstop filters have very narrow bandwidth requirements and the lowpass and highpass filters have either very low cutoff frequency close to DC or very high cutoff frequency close to the folding frequency (Nyquist limit).

8.7.1 SECOND-ORDER BANDPASS FILTER DESIGN

Typical pairs of poles and zeros for a bandpass filter are placed in Fig. 8.32. Poles are complex conjugate, with the magnitude r controlling the bandwidth and the angle θ controlling the center frequency. The zeros are placed at $z=1$ corresponding to DC, and at $z=-1$ corresponding to the folding frequency.

The poles will raise the magnitude response at the center frequency while the zeros will cause zero gains at DC (zero frequency) and at the folding frequency.

From Fig. 8.32A, we see that the angles for pole locations are $\pm\theta$. The angle can be determined by

$$\theta=\left(\frac{f_0}{f_s}\right)\times 360^{\circ} \quad (8.41)$$

Fig. 8.32B describes how the radius of complex conjugate poles is determined.

As shown in Fig. 8.32B, the magnitude at the center frequency can be approximated by

$$\left|H\left(e^{j\theta}\right)\right|=\frac{ab}{(1-r)r_1}$$

where a and b are the distances from $z=e^{j\theta}$ to the locations of zeros while $(1-r)$ and r_1 are the distances from $z=e^{j\theta}$ to the locations of two complex conjugate poles, respectively. r is the desired radius to be determined. Now, we can approximate the magnitude at the 3-dB cutoff frequency as

$$\left|H\left(e^{j\theta_{3dB}}\right)\right|=\frac{a'b'}{r_2r_3}$$

where a' and b' are the distances from $z=e^{j\theta_{3dB}}$ to the locations of zeros; and r_2 and r_3 are the distances from $z=e^{j\theta_{3dB}}$ to the locations of two complex conjugate poles, respectively. For r is close to 1, assuming that $a\approx a'$, $b\approx b'$, $r_1\approx r_2$, and $\left|H\left(e^{j\theta_{3dB}}\right)\right|/\left|H\left(e^{j\theta}\right)\right|=1/\sqrt{2}$ due to 3-dB magnitude attenuation, we obtain the relation $r_3=\sqrt{2}(1-r)$, which indicates 45° between r_3 and $1-r$. The bandwidth is then approximated to $2(1-r)$. Normalizing BW_{3dB} in Hz to radians, it follows that

$$2\pi\times BW_{3dB}T=2(1-r).$$

Solving this equation, we obtain the desired radius for complex conjugate poles, that is,

$$r=1-\pi\times\frac{BW_{3dB}}{f_s} \quad (8.42)$$

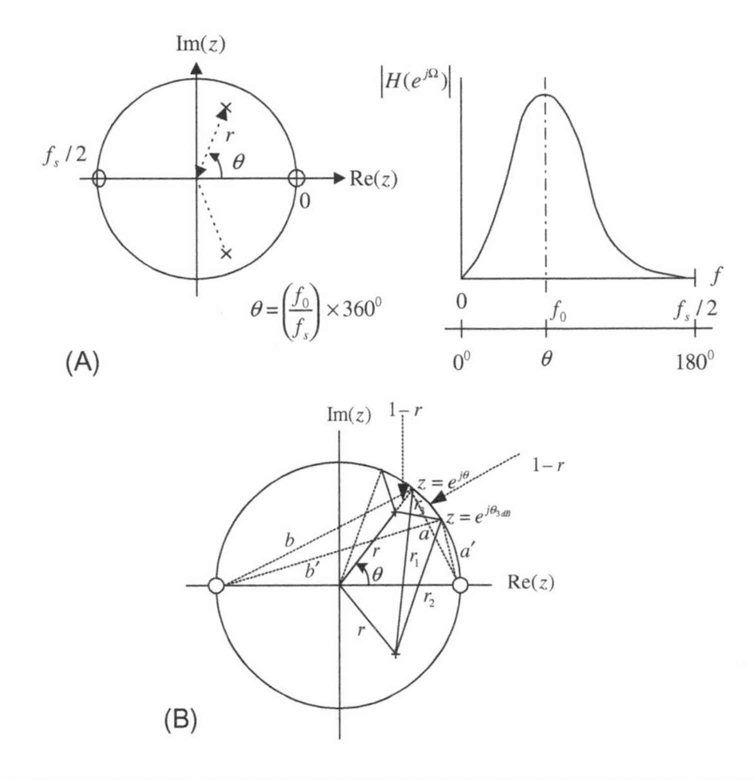

FIG. 8.32

(A) Pole-zero placement for a second-order narrow bandpass filter. (B) Radius of the complex conjugate poles for a second-order narrow bandpass filter.

The design equations for a bandpass filter using pole-zero placement are summarized as

$$\theta = \left(\frac{f_0}{f_s}\right) \times 360^{\circ} \tag{8.43}$$

$$r \approx 1 - (BW_{3dB}/f_s) \times \pi, \text{ good for } 0.9 \leq r < 1 \tag{8.44}$$

$$H(z) = \frac{K(z-1)(z+1)}{(z - re^{j\theta})(z - re^{-j\theta})} = \frac{K(z^2 - 1)}{(z^2 - 2rz\cos\theta + r^2)}, \tag{8.45}$$

where K is a scale factor to adjust the bandpass filter to have a unit passband gain given by

$$K = \frac{(1-r)\sqrt{1 - 2r\cos 2\theta + r^2}}{2|\sin\theta|}. \tag{8.46}$$

EXAMPLE 8.17

A second-order bandpass filter is required to satisfy the following specifications:

- Sampling rate = 8000 Hz
- 3-dB bandwidth: $BW = 200$ Hz
- Narrow passband centered at $f_0 = 1000$ Hz

zero gain at 0 and 4000 Hz

Continued

EXAMPLE 8.17—CONT'D

Find the transfer function using the pole-zero placement method.

Solution:

First, we calculate the required magnitude of the poles.

$$r = 1 - \frac{200}{8000} \times \pi = 0.9215,$$

which is a good approximation. Use the center frequency to obtain the angle of the pole location:

$$\theta = \left(\frac{1000}{8000}\right) \times 360 = 45^\circ.$$

Compute the unit-gain scale factor as

$$K = \frac{(1 - 0.9215)\sqrt{1 - 2 \times 0.9215 \times \cos 2 \times 45^\circ + 0.9215^2}}{2|\sin 45^\circ|} = 0.0755$$

Finally, the transfer function is given by

$$H(z) = \frac{0.0755(z^2 - 1)}{(z^2 - 2 \times 0.9215 z \cos 45^\circ + 0.9215^2)} = \frac{0.0755 - 0.0755 z^{-2}}{1 - 1.3031 z^{-1} + 0.8491 z^{-2}}.$$

8.7.2 SECOND-ORDER BANDSTOP (NOTCH) FILTER DESIGN

For this type of filter, the pole placement is the same as the bandpass filter (Fig. 8.33). The zeros are placed on the unit circle with the same angles with respect to poles. This will improve passband performance. The magnitude and the angle of the complex conjugate poles determine the 3-dB bandwidth and center frequency, respectively.

Through a similar derivation, the design formulas for bandstop filters can be obtained via the following equations:

$$\theta = \left(\frac{f_0}{f_s}\right) \times 360^\circ \tag{8.47}$$

$$r \approx 1 - (BW_{3dB}/f_s) \times \pi, \text{ good for } 0.9 \le r < 1 \tag{8.48}$$

$$H(z) = \frac{K(z - e^{j\theta})(z + e^{-j\theta})}{(z - re^{j\theta})(z - re^{-j\theta})} = \frac{K(z^2 - 2z\cos\theta + 1)}{(z^2 - 2rz\cos\theta + r^2)}. \tag{8.49}$$

The scale factor to adjust the bandstop filter to have a unit passband gain is given by

$$K = \frac{(1 - 2r\cos\theta + r^2)}{(2 - 2\cos\theta)}. \tag{8.50}$$

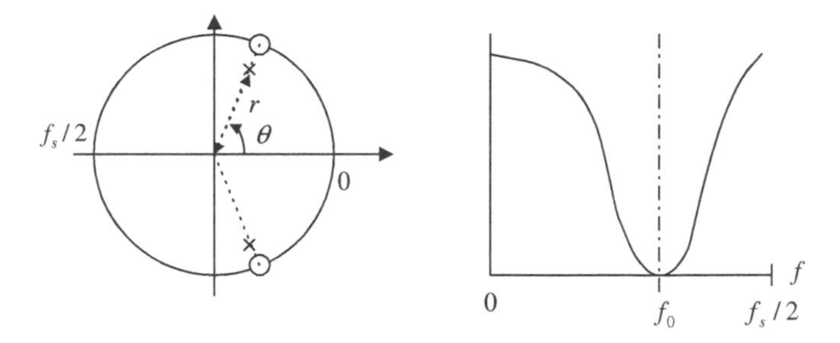

FIG. 8.33

Pole-zero placement for a second-order notch filter.

EXAMPLE 8.18

A second-order notch filter is required to satisfy the following specifications:

- sampling rate = 8000 Hz
- 3-dB bandwidth: $BW = 100\,\text{Hz}$
- narrow passband centered at $f_0 = 1500\,\text{Hz}$

Find the transfer function using the pole-zero placement approach.

Solution:

We first calculate the required magnitude of the poles

$$r \approx 1 - \left(\frac{100}{8000}\right) \times \pi = 0.9607,$$

which is a good approximation. We use the center frequency to obtain the angle of the pole location:

$$\theta = \left(\frac{1500}{8000}\right) \times 360^\circ = 67.5^\circ.$$

The unit-gain scale factor is calculated as

$$K = \frac{(1 - 2 \times 0.9607 \cos 67.5^\circ + 0.9607^2)}{(2 - 2\cos 67.5^\circ)} = 0.9620.$$

Finally, we obtain the transfer function:

$$H(z) = \frac{0.9620(z^2 - 2z\cos 67.5^\circ + 1)}{(z^2 - 2 \times 0.9607 z \cos 67.5^\circ + 0.9607^2)} = \frac{0.9620 - 0.7363z^{-1} + 0.9620z^{-2}}{1 - 0.7353z^{-1} + 0.9229}.$$

8.7.3 FIRST-ORDER LOWPASS FILTER DESIGN

The first-order pole-zero placement can be operated in two cases. The first situation is when the cutoff frequency is less than $f_s/4$. Then the pole-zero placement is as shown in Fig. 8.34A.

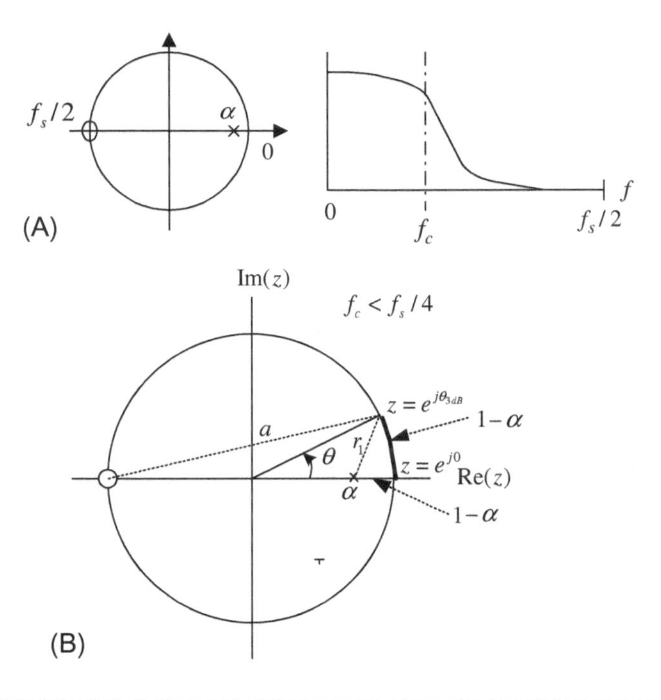

FIG. 8.34

(A) Pole-zero placement for the first-order lowpass filter with $f_c < f_s/4$. (B) Determination of pole location for the first-order lowpass filter with $f_c < f_s/4$.

As shown in Fig. 8.34A, the pole $z=\alpha$ is placed in real axis. The zero is placed at $z=-1$ to ensure zero gain at the folding frequency (Nyquist limit). Fig. 8.34B illustrates how to determine α.

As depicted in Fig. 8.34B, the DC magnitude response is given by

$$\left|H\left(e^{j0}\right)\right| = \frac{\text{distance from } e^{j0} \text{ to } z=-1}{\text{distance from } e^{j0} \text{ to } z=\alpha} = \frac{2}{1-\alpha}$$

Magnitude at the 3-dB cutoff frequency can be approximated by

$$\left|H\left(e^{j\theta_{3dB}}\right)\right| = \frac{\text{distance from } e^{j\theta_{3dB}} \text{ to } z=-1}{\text{distance from } e^{j\theta_{3dB}} \text{ to } z=\alpha} = \frac{a}{r_1}$$

For r is close to 1, we assume $a \approx 2$. Then $\left|H\left(e^{j\theta_{3dB}}\right)\right| / \left|H\left(e^{j\theta}\right)\right| = 1/\sqrt{2}$, that is, $r_1 \approx \sqrt{2}(1-\alpha)$, which indicates 45° between r_1 and Re(z) axis. We can approximate the bandwidth can be $(1-\alpha)$. Normalizing the bandwidth f_c in Hz to radians, we have

$$2\pi \times f_c T = 1 - \alpha$$

Therefore, the pole location can be solved as

$$\alpha = 1 - 2\pi \times \frac{f_c}{f_s}$$

When the cutoff frequency is above $f_s/4$, the pole-zero placement is adopted as shown in Fig. 8.35. The similar derivation can be performed to achieve the pole location.

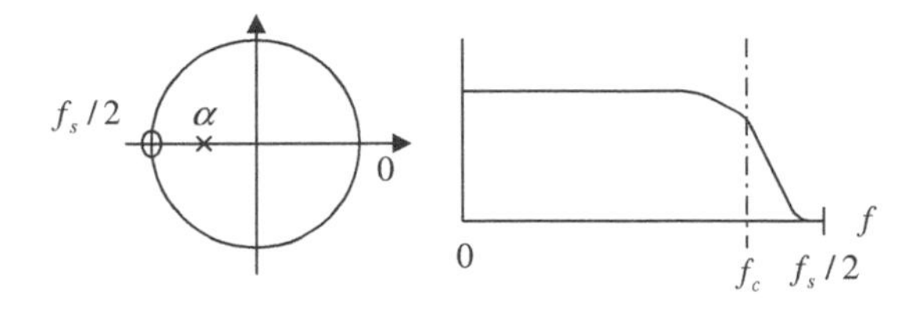

FIG. 8.35

Pole-zero placement for the first-order lowpass filter with $f_c > f_s/4$.

Design formulas for lowpass filters using the pole-zero placement are given in the following equations:

$$\text{When } f_c < f_s/4, \alpha \approx 1 - 2 \times (f_c/f_s) \times \pi, \text{ good for } 0.9 \le \alpha < 1 \tag{8.51}$$

$$\text{When } f_c > f_s/4, \alpha \approx -(1 - \pi + 2 \times (f_c/f_s) \times \pi), \text{ good for } -1 < \alpha \le -0.9 \tag{8.52}$$

The transfer function is

$$H(z) = \frac{K(z+1)}{(z-\alpha)}, \tag{8.53}$$

and the unit passband gain scale factor is given by

$$K = \frac{(1-\alpha)}{2}. \tag{8.54}$$

EXAMPLE 8.19

A first-order lowpass filter is required to satisfy the following specifications:

- sampling rate = 8000 Hz
- 3-dB cutoff frequency: $f_c = 100$ Hz
- zero gain at 4000 Hz

Find the transfer function using the pole-zero placement method.

Solution:

Since the cutoff frequency of 100 Hz is much less than $f_s/4 = 2000$ Hz, we determine the pole as

$$\alpha \approx 1 - 2 \times \left(\frac{100}{8000}\right) \times \pi = 0.9215,$$

which is above 0.9. Hence, we have a good approximation. The unit-gain scale factor is calculated by

$$K = \frac{(1-0.9215)}{2} = 0.03925.$$

Last, we can develop the transfer function as

$$H(z) = \frac{0.03925(z+1)}{(z-0.9215)} = \frac{0.03925 + 0.03925z^{-1}}{1 - 0.9215z^{-1}}.$$

Continued

EXAMPLE 8.19—CONT'D

Note that we can also determine the unit-gain factor K by substituting $z=e^{j0}=1$ to the transfer function $H(z)=\frac{(z+1)}{(z-\alpha)}$, then find a DC gain. Set the scale factor to be a reciprocal of the DC gain. This can be easily done, that is,

$$\text{DCgain}=\left.\frac{z+1}{z-0.9215}\right|_{z=1}=\frac{1+1}{1-0.9215}=25.4777.$$

Hence, $K=1/25.4777=0.03925$.

8.7.4 FIRST-ORDER HIGHPASS FILTER DESIGN

Similar to the lowpass filter design, the pole-zero placements for the first-order highpass filters in two cases are shown in Fig. 8.36A and B.

Formulas for designing highpass filters using the pole-zero placement are listed in the following equations:

$$\text{When } f_c < f_s/4, \alpha \approx 1 - 2\times(f_c/f_s)\times\pi, \text{ good for } 0.9 \le \alpha < 1. \tag{8.55}$$

$$\text{When } f_c > f_s/4, \alpha \approx -(1 - \pi + 2\times(f_c/f_s)\times\pi), \text{ good for } -1 < \alpha \le -0.9. \tag{8.56}$$

$$H(z)=\frac{K(z-1)}{(z-\alpha)} \tag{8.57}$$

$$K=\frac{(1+\alpha)}{2}. \tag{8.58}$$

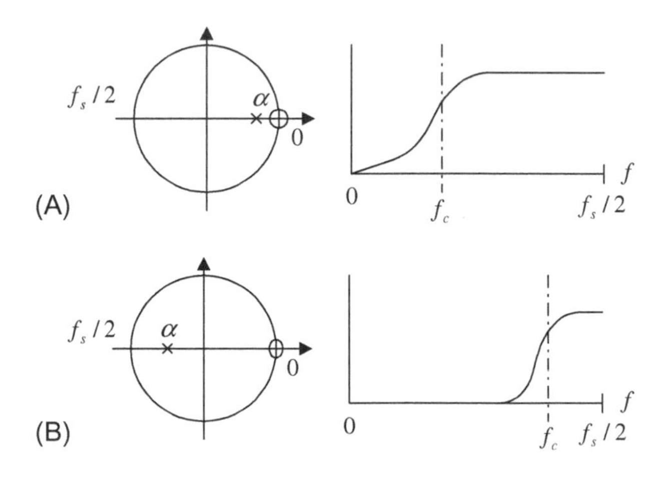

FIG. 8.36

(A) Pole-zero placement for the first-order highpass filter with $f_c < f_s/4$. (B) Pole-zero placement for the first-order highpass filter with $f_c > f_s/4$.

EXAMPLE 8.20

A first-order highpass filter is required to satisfy the following specifications:

- sampling rate = 8000 Hz
- 3-dB cutoff frequency: $f_c = 3800$ Hz
- zero gain at 0 Hz

Find the transfer function using the pole-zero placement method.

Solution:

Since the cutoff frequency of 3800 Hz is much larger than $f_s/4 = 2000$ Hz, we determine the pole as

$$\alpha \approx -\left(1 - \pi + 2 \times \left(\frac{3800}{8000}\right) \times \pi\right) = -0.8429,$$

The unit-gain scale factor and transfer function are obtained as

$$K = \frac{(1 - 0.8429)}{2} = 0.07854$$

$$H(z) = \frac{0.07854(z-1)}{(z+0.8429)} = \frac{0.07854 - 0.07854z^{-1}}{1 + 0.8429z^{-1}}.$$

Note that we can also determine the unit-gain scale factor K by substituting $z = e^{j180^0} = -1$ to the transfer function $H(z) = \frac{(z-1)}{(z-\alpha)}$, finding a passband gain at the Nyquist limit $f_s/2 = 4000$ Hz. We then set the scale factor to be a reciprocal of the passband gain, that is,

$$\text{Passband gain} = \left.\frac{z-1}{z+0.8429}\right|_{z=1} = \frac{-1-1}{-1+0.8429} = 12.7307.$$

Hence, $K = 1/12.7307 = 0.07854$.

8.8 REALIZATION STRUCTURES OF INFINITE IMPULSE RESPONSE FILTERS

In this section, we will realize the designed IIR filter using direct-form I as well as direct-form II. We will then realize a higher-order IIR filter using a cascade form.

8.8.1 REALIZATION OF INFINITE IMPULSE RESPONSE FILTERS IN DIRECT-FORM I AND DIRECT-FORM II

EXAMPLE 8.21

Realize the first-order digital highpass Butterworth filter

$$H(z) = \frac{0.1936 - 0.1936z^{-1}}{1 + 0.6128z^{-1}}$$

using a direct-form I.

Continued

EXAMPLE 8.21—CONT'D

Solution:

From the transfer function, we can identify

$$b_0 = 0.1936, b_1 = -0.1936, \text{and } a_1 = 0.6128.$$

Applying the direct-form I developed in Chapter 6 results in the diagram in Fig. 8.37. The digital signal processing (DSP) equation for implementation is then given by

$$y(n) = -0.6128y(n-1) + 0.1936x(n) - 0.1936x(n-1).$$

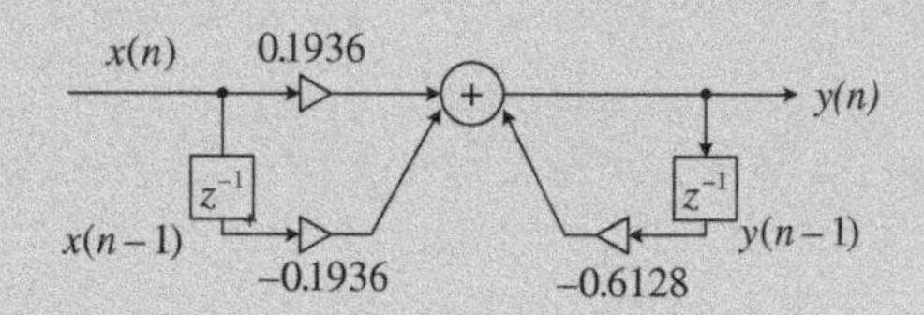

FIG. 8.37

Realization of IIR filter in Example 8.21 in direct-form I.

Program 8.14 lists the MATLAB implementation.

Program 8.14. *m*-File for Example 8.21.

```
%Sample MATLAB code
sample = 2:2:20; %Input test array
x=[ 0 0 ]; % Input buffer [x(n) x(n-1) ...]
y=[ 0 0]; % Output buffer [y(n) y(n-1) ... ]
b=[0.1936 -0.1936]; % Numerator coefficients [b0 b1 ... ]
a=[1 0.6128]; % Denominator coefficients [1 a0 a1 ...]
for n=1:1:length(sample) % Processing loop
  for k=2:-1:2
     x(k)=x(k-1);     % Shift input by one sample
     y(k)=y(k-1);     % Shift output by one sample
  end
  x(1)=sample(n);  % Get new sample
  y(1)=0;       % Digital filtering
  for k=1:1:2
     y(1)=y(1)+x(k)*b(k);
  end
  for k=2:2
     y(1)=y(1)-a(k)*y(k);
  end
  out(n)=y(1); %Output the filtered sample to output array
end
  out
```

EXAMPLE 8.22

Realize the following digital filter using a direct-form II.

$$H(z) = \frac{0.7157 + 1.4314z^{-1} + 0.7157z^{-2}}{1 + 1.3490z^{-1} + 0.5140z^{-2}}.$$

Solution:

First, we can identify

$$b_0 = 0.7157, b_1 = 1.4314, b_2 = 0.7157$$

$$\text{and } a_1 = 1.3490, a_2 = 0.5140.$$

Applying the direct-form II developed in Chapter 6 leads to Fig. 8.38.
There are two difference equations required for implementation:

$$w(n) = x(n) - 1.3490w(n-1) - 0.5140w(n-2)$$

$$y(n) = 0.7157w(n) + 1.4314w(n-1) + 0.7157w(n-2).$$

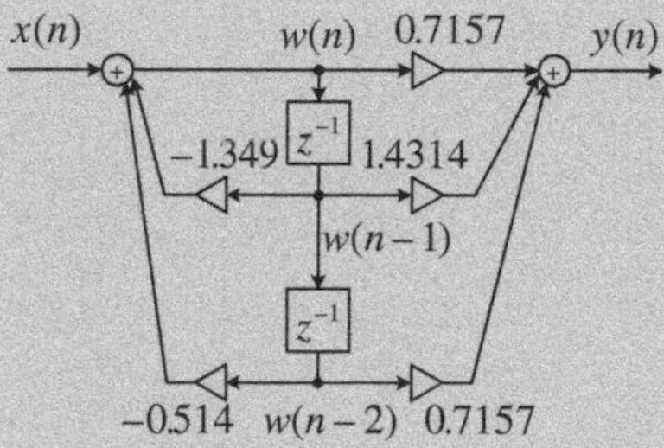

FIG. 8.38

Realization of IIR filter in Example 8.22 in direct-form II.

The MATLAB implementation is listed in Program 8.15.

Program 8.15. *m*-File for Example 8.22.

```
%Sample MATLAB code
sample =2:2:20; % Input test array
x=[0];                          %Input buffer [x(n) ]
y=[0];                          %Output buffer [y(n)]
w=[0 0 0]; % Buffer for w(n) [w(n) w(n-1) ...]
b=[0.7157 1.4314 0.7157];  % Numerator coefficients [b0 b1 ...]
a=[1 1.3490 0.5140]; % Denominator coefficients [1 a1 a2 ...]
for n=1:1:length(sample)   % Processing loop
  for k=3:-1:2
    w(k)=w(k-1);              %Shift w(n) by one sample
  end
  x(1)=sample(n);   % Get new sample
  w(1)=x(1);        % Perform IIR filtering
```

```
    for k=2:1:3
       w(1)=w(1)-a(k)*w(k);
    end
    y(1)=0;              % Perform FIR filtering
    for k=1:1:3
       y(1)=y(1)+b(k)*w(k);
    end
    out(n)=y(1);         % Send the filtered sample to output array
  end
    out
```

8.8.2 REALIZATION OF HIGHER-ORDER INFINITE IMPULSE RESPONSE FILTERS VIA THE CASCADE FORM

EXAMPLE 8.23

Given a fourth-order filter transfer function designed as

$$H(z)=\frac{0.5108z^2+1.0215z+0.5108}{z^2+0.5654z+0.4776}\times\frac{0.3730z^2+0.7460z+0.3730}{z^2+0.4129z+0.0790},$$

realize the digital filter using the cascade (series) form via second-order sections.

Solution:

Since the filter is designed using the cascade form, we have two sections of the second-order filters, whose transfer functions are

$$H_1(z)=\frac{0.5108z^2+1.0215z+0.5108}{z^2+0.5654z+0.4776}=\frac{0.5180+1.0215z^{-1}+0.5108z^{-2}}{1+0.5654z^{-1}+0.4776z^{-2}}$$

and

$$H_2(z)=\frac{0.3730z^2+0.7460z+0.3730}{z^2+0.4129z+0.0790}=\frac{0.3730+0.7460z^{-1}+0.3730z^{-2}}{1+0.4129z^{-1}+0.0790z^{-2}}.$$

Each filter section is developed using the direct-form I, shown in Fig. 8.39.

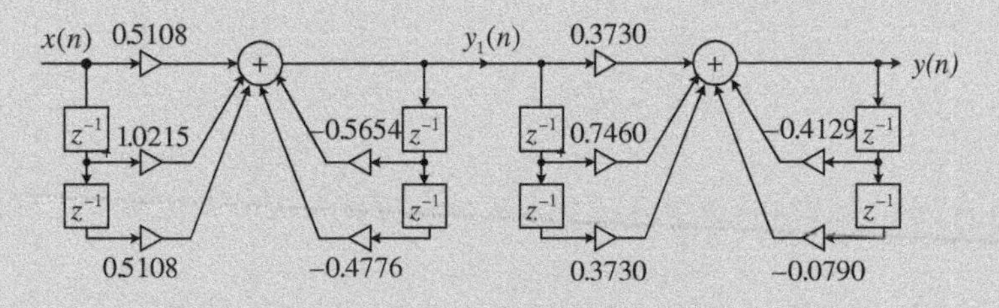

FIG. 8.39

Cascade realization of IIR filter in Example 8.23 in direct-form I.

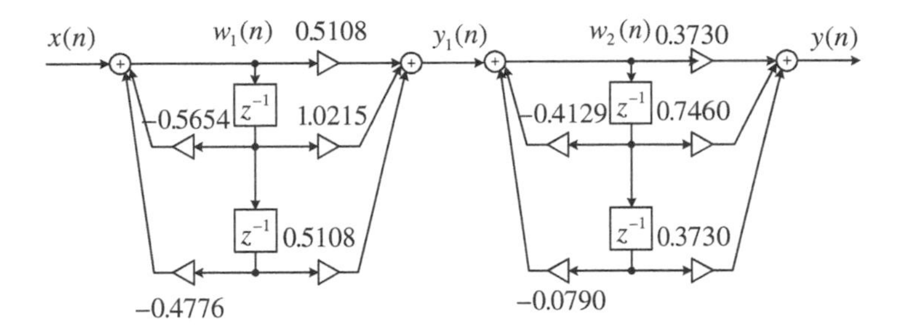

FIG. 8.40

Cascade realization of IIR filter in Example 8.23 in direct-form II.

There are two sets of DSP equations for the implementation of the first and second sections, respectively.

First section:

$$y_1(n) = -0.5654y_1(n-1) - 0.4776y_1(n-2)$$
$$+0.5108x(n) + 1.0215x(n-1) + 0.5108x(n-2)$$

Second section:

$$y(n) = -0.4129y(n-1) - 0.0790y(n-2)$$
$$+0.3730y_1(n) + 0.7460y_1(n-1) + 0.3730y_1(n-2).$$

Again, after we use the direct-form II for realizing each second-order filter, the realization shown in Fig. 8.40 is developed.

The difference equations for the implementation of the first section are:

$$w_1(n) = x(n) - 0.5654w_1(n-1) - 0.4776w_1(n-2)$$

$$y_1(n) = 0.5108w_1(n) + 1.0215w_1(n-1) + 0.5108w_1(n-2).$$

The difference equations for the implementation of the second section are:

$$w_2(n) = y_1(n) - 0.4129w_2(n-1) - 0.0790w_2(n-2)$$

$$y(n) = 0.3730w_2(n) + 0.7460w_2(n-1) + 0.3730w_2(n-2).$$

Note that for both direct-form I and direct-form II, the output from the first filter section becomes the input for the second filter section.

8.9 APPLICATION: 60-HZ HUM ELIMINATOR AND HEART RATE DETECTION USING ELECTROCARDIOGRAPHY

Hum noise created by poor power suppliers, transformers, or electromagnetic interference sourced by a main power supply is characterized by a frequency of 60-Hz and its harmonics. If this noise interferes with a desired audio or biomedical signal [e.g., in electrocardiography (ECG)], the desired signal could be corrupted. It is sufficient to eliminate the 60-Hz hum frequency with its second and third harmonics in most practical applications. We can complete this by cascading with notch filters with notch frequencies of 60, 120, and 180 Hz, respectively. Fig. 8.41 depicts the functional block diagram.

Now let us apply the 60-Hz hum eliminator to an ECG recording system. The ECG is a small electrical signal captured from an ECG sensor. The ECG signal is produced by the activity of the human heart, thus can be used for heart rate detection, fetal monitoring, and diagnostic purposes. The single pulse of the ECG is depicted in Fig. 8.42, which shows that an ECG signal is characterized by five peaks and valleys, labeled P, Q, R, S, and T. The highest positive wave is the R wave. Shortly before and after the R wave are negative waves called Q wave and S wave. The P wave comes before the Q wave, while the T wave comes after the S wave. The Q, R, and S waves together are called the QRS complex.

The properties of the QRS complex, with its rate of occurrence and times, highs, and widths, provide information to cardiologists concerning various pathological conditions of the heart. The reciprocal of the time period between R wave peaks (in milliseconds) multiplied by 60,000 gives instantaneous heart rate in beats per minute. On a modern ECG monitor, the acquired ECG signal is displayed for the diagnostic purpose.

However, a major source of frequent interference is the electric-power system. Such interference appears on the recorded ECG data due to electrical-field coupling between the power lines and the electrocardiograph or patient, which is the cause of the electrical field surrounding mains power lines. Another cause is magnetic induction in the power line, whereby current in the power line generates a magnetic field around the line. Sometimes, the harmonics of 60-Hz hum exist due to the nonlinear sensor and signal amplifier effects. If such interference is severe, the recorded ECG data becomes useless.

In this application, we focus on ECG enhancement for heart rate detection. To significantly reduce the 60-Hz interference, we apply signal enhancement to the ECG recording system, as shown in Fig. 8.43.

The 60-Hz eliminator removes the 60-Hz interference and has the capability to reduce its second harmonic of 120 Hz and third harmonic of 180 Hz.

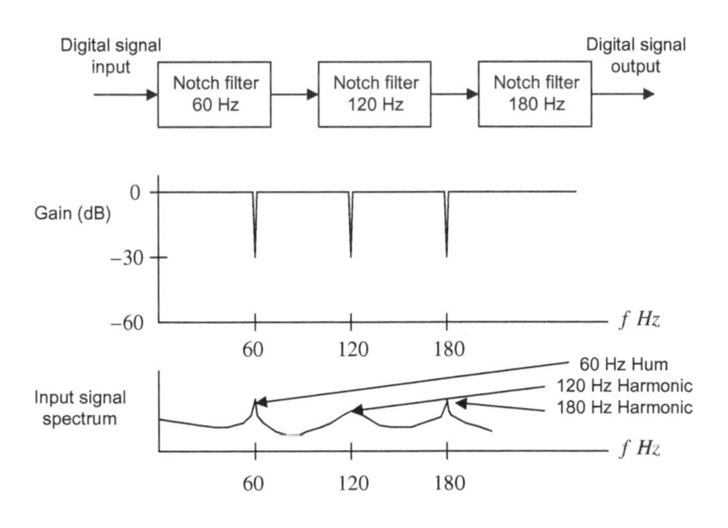

FIG. 8.41

(Top) 60-Hz Hum eliminator; (middle) the filter frequency response of the eliminator; (bottom) the input signal spectrum corrupted by the 60-Hz hum and its second and third harmonics.

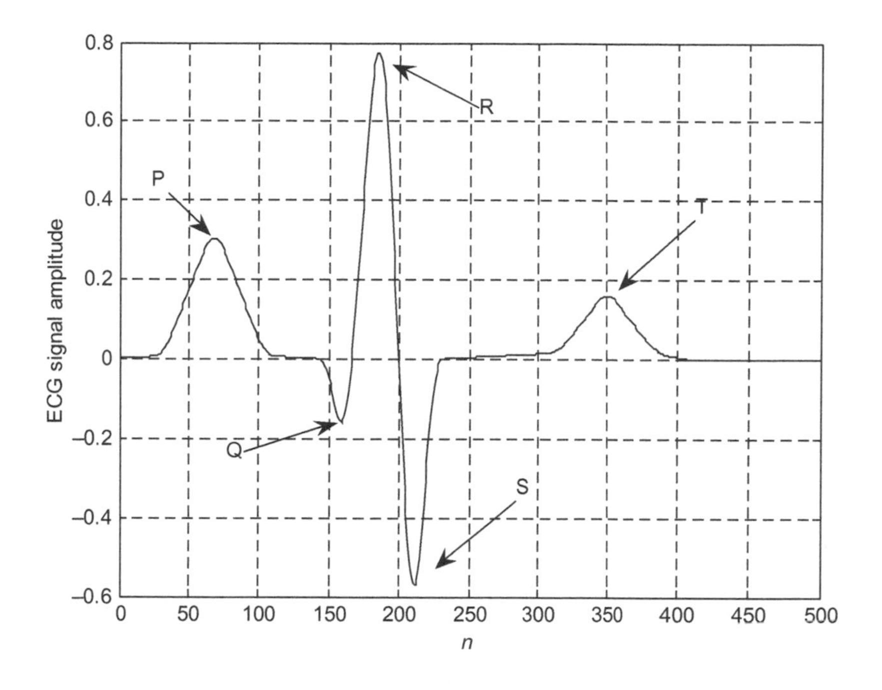

FIG. 8.42

Characteristics of the ECG pulse.

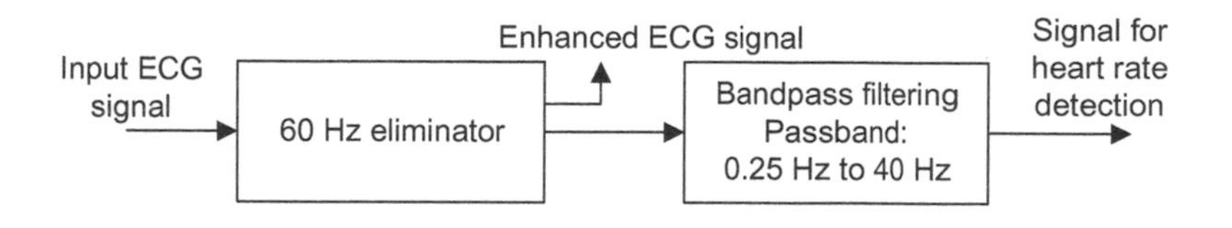

FIG. 8.43

ECG signal enhancement system.

The next objective is to detect the heart rate using the enhanced ECG signal. We need to remove DC drift and filter muscle noise, which may occur at approximately 40 Hz or more. If we consider the lowest heart rate as 30 beats per minute, the corresponding frequency is 30/60 = 0.5 Hz. Choosing the lower cutoff frequency of 0.25 Hz should be reasonable.

Thus, a bandpass filter with a passband from 0.25 to 40 Hz [range from 0.67 to 40 Hz, discussed in Webster (2009)], either FIR or IIR type, can be designed to reduce such effects. The resultant ECG signal is valid only for the detection of heart rate. Note that the ECG signal after bandpass filtering with a passband from 0.25 to 40 Hz is no longer valid for general ECG applications, since the original ECG signal occupies the frequency range from 0.01 to 250 Hz (diagnostic-quality ECG), as discussed in Carr and Brown (2001) and Webster (2009). The enhanced ECG signal from the 60-Hz hum eliminator can serve for general ECG signal analysis (which is beyond the scope of this book). We summarize the design specifications for the heart rate detection application as follows:

System outputs: Enhanced ECG signal with the 60-Hz elimination
Processed ECG signal for heart rate detection

60-Hz eliminator:	
Harmonics to be removed:	60 Hz (fundamental)
	120 Hz (second harmonic)
	180 Hz (third harmonic)
3-dB bandwidth for each filter:	4 Hz
Sampling rate:	600 Hz
Notch filter type:	Second-order IIR
Design method:	Pole-zero placement
Bandpass filter:	
Passband frequency range:	0.25–40 Hz
Passband ripple:	0.5 dB
Filter type:	Chebyshev fourth order
Design method:	BLT method
DSP sampling rate:	600 Hz

Let us carry out the 60-Hz eliminator design and determine the transfer function and difference equation for each notch filter and bandpass filter. For the first section with the notch frequency of 60 Hz, applying Eqs. (8.47)–(8.50) leads to

$$r = 1 - \left(\frac{4}{600}\right) \times \pi = 0.9791$$

$$\theta = \left(\frac{60}{600}\right) \times 360^{\circ} = 36^{\circ}.$$

We calculate $2\cos(36^{\circ}) = 1.6180$, $2r\cos(36^{\circ}) = 1.5842$, and

$$K = \frac{(1 - 2r\cos\theta + r^2)}{(2 - 2\cos\theta)} = 0.9803.$$

Hence it follows that

$$H_1(z) = \frac{0.9803 - 1.5862z^{-1} + 0.9803z^{-2}}{1 - 1.5842z^{-1} + 0.9586z^{-2}}$$

$$y_1(n) = 0.9803x(n) - 1.5862x(n-1) + 0.9802x(n-2) + 1.5842y_1(n-1) - 0.9586y_1(n-2).$$

Similarly, we yield the transfer functions and difference equations for the second section and the third section as

Second section:

$$H_2(z) = \frac{0.9794 - 0.6053z^{-1} + 0.9794z^{-2}}{1 - 0.6051z^{-1} + 0.9586z^{-2}}$$

$$y_2(n) = 0.9794y_1(n) - 0.6053y_1(n-1) + 0.9794y_1(n-2) + 0.6051y_2(n-1) - 0.9586y_2(n-2)$$

Third section:

$$H_3(z) = \frac{0.9793 + 0.6052z^{-1} + 0.9793z^{-2}}{1 + 0.6051z^{-1} + 0.9586z^{-2}}$$

$$y_3(n) = 0.9793y_2(n) + 0.6052y_2(n-1) + 0.9793y_2(n-2) - 0.6051y_3(n-1) - 0.9586y_3(n-2).$$

The cascaded frequency responses are plotted in Fig. 8.44. As we can see, the rejection for each notch frequency is below 50 dB.

The second-stage design using the BLT gives the bandpass filter transfer function and difference equation

$$H_4(z) = \frac{0.0464 - 0.0927z^{-2} + 0.0464z^{-4}}{1 - 3.3523z^{-1} + 4.2557z^{-2} - 2.4540z^{-3} + 0.5506z^{-4}}$$

$$y_4(n) = 0.0464y_3(n) - 0.0927y_3(n-2) + 0.0464y_3(n-4)$$
$$+3.3523y_4(n-1) - 4.25571y_4(n-2) + 2.4540y_4(n-3) - 0.5506y_4(n-4).$$

Fig. 8.45 depicts the processed results at each stage. In Fig. 8.45, plot A shows the initial corrupted ECG data, which includes 60-Hz interference and its 120 and 180 Hz harmonics, along with muscle noise. Plot B shows that the interferences of 60 Hz and its harmonics of 120 and 180 Hz have been removed. Finally, plot C displays the result after the bandpass filter. As we have expected, the muscle noise has been removed; and the enhanced ECG signal has been observed.

MATLAB simulation is listed in Program 8.16.

With the processed ECG signal, a simple zero-cross algorithm can be designed to detect the heart rate. Based on plot C in Fig. 8.45, we use a threshold value of 0.5 and continuously compare each of two

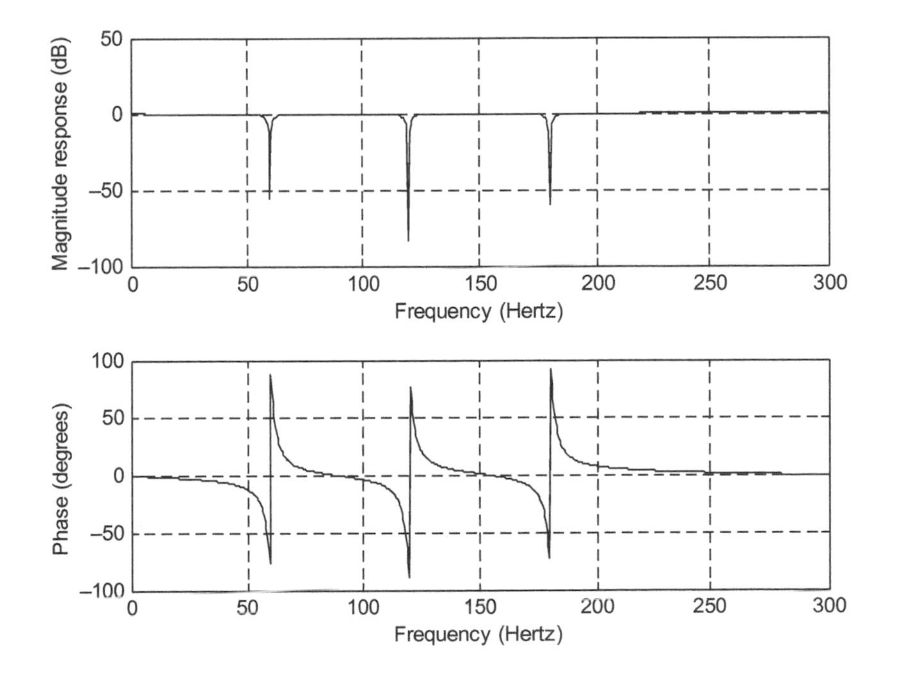

FIG. 8.44

Frequency responses of three cascaded notch filters.

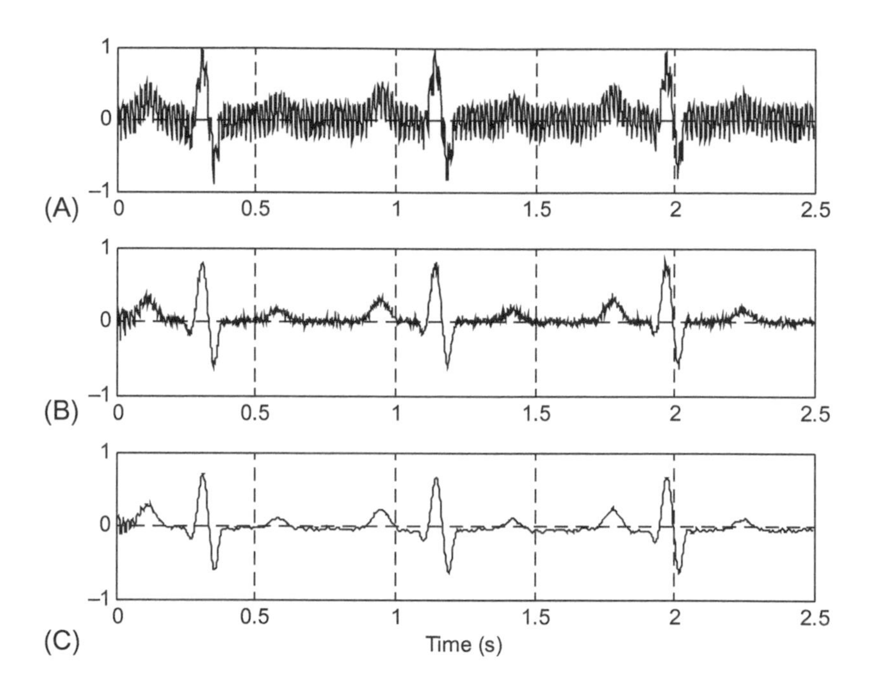

FIG. 8.45

Results of ECG signal processing: (A) initial corrupted ECG data; (B) ECG data enhanced by removing 60 Hz interference; (C) ECG data with DC blocking and noise removal for heart rate detection.

consecutive samples with the threshold. If both results are opposite, then a zero cross is detected. Each zero-crossing measure is given by

$$\text{Zero crossing} = \frac{|cur\_sign - pre\_sign|}{2},$$

where cur_sign and pre_sign are determined based on the current input $x(n)$, the past input $x(n-1)$, and the threshold value, given as

$$\begin{aligned} &\text{if } x(n) \geq threshold\ cur\_sign = 1 \text{ else } cur\_sign = -1 \\ &\text{if } x(n-1) \geq threshold\ pre\_sign = 1 \text{ else } pre\_sign = -1. \end{aligned}$$

Fig. 8.46 summarizes the algorithm.

After detecting the total number of zero crossings, the number of the peaks will be half the number of the zero crossings. The heart rate in terms of pulses per minute can be determined by

$$\text{Heart rate} = \frac{60}{\left(\frac{\text{Number of enhanced } ECG \text{ data}}{f_s}\right)} \times \left(\frac{\text{zero crossing number}}{2}\right).$$

In our simulation, we have detected six zero-crossing points using 1500 captured data at a sampling rate of 600 samples per second. Hence,

$$\text{Heart rate} = \frac{60}{\left(\frac{1500}{600}\right)} \times \left(\frac{6}{2}\right) = 72 \text{ pulses per minute.}$$

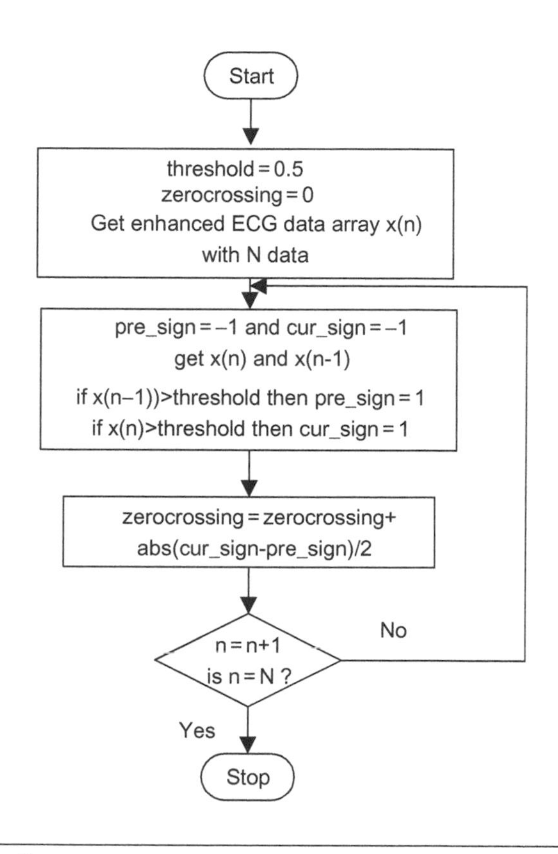

FIG. 8.46

A simple zero-crossing algorithm.

The MATLAB implementation of the zero-crossing detection can be found in the last part in Program 8.16.

Program 8.16. MATLAB program for heart rate detection using an ECG signal.

```
load ecgbn.dat; % Load noisy ECG recording
b1=[0.9803 -1.5862 0.9803];    %Notch filter with a notch frequency of 60 Hz
a1=[1 -1.5842 0.9586];
b2=[0.9794 -0.6053 0.9794];    % Notch filter with a notch frequency 120 Hz
a2=[1 -0.6051 0.9586];
b3=[0.9793 0.6052 0.9793];     % Notch filter with a notch frequency of 180 Hz
a3=[1 0.6051 0.9586];
y1=filter(b1,a1,ecgbn);  % First section filtering
y2=filter(b2,a2,y1);     % Second section filtering
y3=filter(b3,a3,y2);     % Third section filtering
%Bandpass filter
fs=600;                % Sampling rate
T=1/600;               % Sampling interval
```

```
% BLT design
wd1=2*pi*0.25;
wd2=2*pi*40;
wa1=(2/T)*tan(wd1*T/2);
wa2=(2/T)*tan(wd2*T/2);
[B,A]=lp2bp([1.4314], [1 1.4652 1.5162],sqrt(wa1*wa2),wa2-wa1);
[b,a]=bilinear(B,A,fs);
b =[ 0.046361 0 -0.092722 0 0.046361]
a =[1 -3.352292 4.255671 -2.453965 0.550587]
y4=filter(b,a,y3);                  %Bandpass filtering
t=0:T:1499*T;                       % Recover time
subplot(3,1,1);plot(t,ecgbn);grid;ylabel('(a)');
subplot(3,1,2);plot(t,y3);grid;ylabel('(b)');
subplot(3,1,3);plot(t,y4);grid;ylabel('(c)');
xlabel('Time (sec.)');
%Zero cross algorithm
zcross=0.0;threshold=0.5
for n=2:length(y4)
  pre_sign=-1;cur_sign=-1;
 if y4(n-1)>threshold
    pre_sign=1;
 end
 if y4(n)>threshold
    cur_sign=1;
 end
zcross=zcross+abs(cur_sign-pre_sign)/2;
end
zcross                                  % Output the number of zero crossings
rate=60*zcross/(2*length(y4)/600)       % Output the heart rate
```

8.10 COEFFICIENT ACCURACY EFFECTS ON INFINITE IMPULSE RESPONSE FILTERS

In practical applications, the IIR filter coefficients with infinite precision may be quantized due to the finite word length. Quantization of infinite precision filter coefficients changes the locations of the zeros and poles of the designed filter transfer function, hence changes the filter frequency responses. Since analysis of filter coefficient quantization for the IIR filter is very complicated and beyond the scope of this textbook, we pick only a couple of simple cases for discussion. Filter coefficient quantization for specific processors such as the fixed-point DSP processor and floating-point processor will be included in Chapter 14. To illustrate this effect, we look at the following first-order IIR filter transfer function having filter coefficients with infinite precision,

$$H(z)=\frac{b_0+b_1z^{-1}}{1+a_1z^{-1}}. \tag{8.59}$$

After filter coefficient quantization, we have the quantized digital IIR filter transfer function

$$H^q(z)=\frac{b_0^q+b_1^qz^{-1}}{1+a_1^qz^{-1}}. \tag{8.60}$$

Solving for pole and zero, we achieve

$$z_1 = -\frac{b_1^q}{b_0^q} \tag{8.61}$$

$$p_1 = -a_1^q. \tag{8.62}$$

Now considering a second-order IIR filter transfer function as

$$H(z) = \frac{b_0 + b_1 z^{-1} + b_2 z^{-2}}{1 + a_1 z^{-1} + a_2 z^{-2}}, \tag{8.63}$$

and its quantized IIR filter transfer function

$$H^q(z) = \frac{b_0^q + b_1^q z^{-1} + b_2^q z^{-2}}{1 + a_1^q z^{-1} + a_2^q z^{-2}}, \tag{8.64}$$

solving for poles and zeros finds:

$$z_{1,2} = -0.5 \times \frac{b_1^q}{b_0^q} \pm j\left(\frac{b_2^q}{b_0^q} - 0.25 \times \left(\frac{b_1^q}{b_0^q}\right)^2\right)^{\frac{1}{2}} \tag{8.65}$$

$$p_{1,2} = -0.5 \times a_1^q \pm j\left(a_2^q - 0.25 \times \left(a_1^q\right)^2\right)^{\frac{1}{2}}. \tag{8.66}$$

With the developed Eqs. (8.61) and (8.62) for the first-order IIR filter, and Eqs. (8.65) and (8.66) for the second-order IIR filter, we can study the effects of the location changes of the poles and zeros, and the frequency responses due to filter coefficient quantization.

EXAMPLE 8.24

Given the following first-order IIR filter

$$H(z) = \frac{1.2341 + 0.2126z^{-1}}{1 - 0.5126z^{-1}},$$

and assuming that we use 1 sign bit and 6 bits for encoding the magnitude of the filter coefficients, find the quantized transfer function and pole-zero locations.

Solution:

Let us find the pole and zero for infinite precision filter coefficients:

Solving $1.2341z + 0.2126 = 0$ leads a zero location $z_1 = -0.17227$.

Solving $z - 0.5126 = 0$ gives a pole location $p_1 = 0.5126$.

Now let us quantize the filter coefficients. Quantizing 1.2341 can be illustrated as

$1.2341 \times 2^5 = 39.4912 = 39$(rounded to integer).

Since the maximum magnitude of the filter coefficients is 1.2341, which is between 1 and 2, we scale all coefficient magnitudes by a factor of 2^5 and round off each value to an integer whose magnitude is encoded using 6 bits. As shown in the quantization, 6 bits are required to encode the integer 39. When the coefficient integer is scaled back by the same scale factor, the corresponding quantized coefficient with finite precision (7 bits, including the sign bit) is found to be

Continued

EXAMPLE 8.24—CONT'D

$$b_0^q = \frac{39}{2^5} = 1.21875.$$

Following the same procedure, we can obtain

$$b_1^q = 0.1875$$

and

$$a_1^q = -0.5.$$

Thus we achieve the quantized transfer function

$$H^q(z) = \frac{1.21875 + 0.1875z^{-1}}{1 - 0.5z^{-1}}.$$

Solving for pole and zero leads to

$$p_1 = 0.5$$

and

$$z_1 = -0.1538.$$

It is clear that the pole and zero locations change after the filter coefficients are quantized. This effect can change the frequency response of the designed filter as well. In Example 8.25, we study the quantization of the filter coefficients for the second-order IIR filter and examine the pole/zero location changes and magnitude/phase frequency responses.

EXAMPLE 8.25

A second-order digital lowpass Chebyshev filter with a cutoff frequency of 3.4 kHz and 0.5-dB ripple on passband at a sampling frequency of 8000 Hz is designed. Assume that we use 1 sign bit and 7 bits for encoding the magnitude of each filter coefficient. The z-transfer function is given by

$$H(z) = \frac{0.7434 + 1.4865z^{-1} + 0.7434z^{-2}}{1 + 1.5149z^{-1} + 0.6346z^{-2}}.$$

(a) Find the quantized transfer function and pole and zero locations.
(b) Plot the magnitude and phase responses, respectively.

Solution:

(a) Since the maximum magnitude of the filter coefficients is between 1 and 2, the scale factor for quantization is chosen to be 2^6, so that the coefficient integer can be encoded using 7 bits.

After performing filter coefficient encoding, we have

$$H^q(z) = \frac{0.7500 + 1.484375z^{-1} + 0.7500z^{-2}}{1 + 1.515625z^{-1} + 0.640625z^{-2}}.$$

For comparison, the uncoded zeros and encoded zeros of the transfer function $H(z)$ are as follows:
Uncoded zeros: –1, –1;
Coded zeros: –0.9896 + 0.1440i, –0.9896 –0.1440i.
Similarly, the uncoded poles and coded poles of the transfer function $H^q(z)$ are as follows:
Uncoded poles: –0.7574 + 0.2467i, –0.7574 –0.2467i;
Coded poles: –0.7578 + 0.2569i, –0.7578 –0.2569i.

(b) The comparisons for the magnitude responses and phase responses are listed in Program 8.17 and plotted in Fig. 8.47.

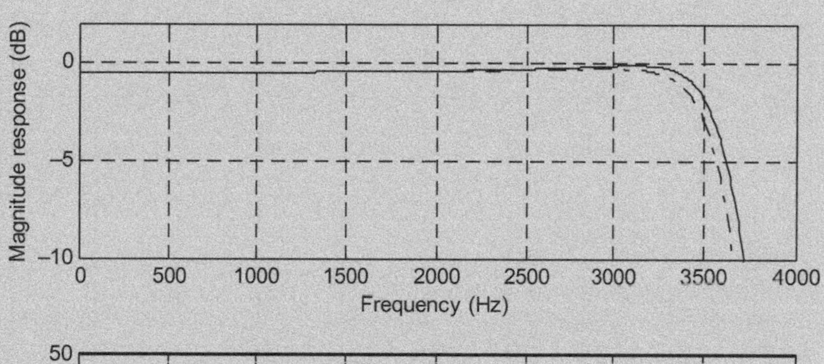

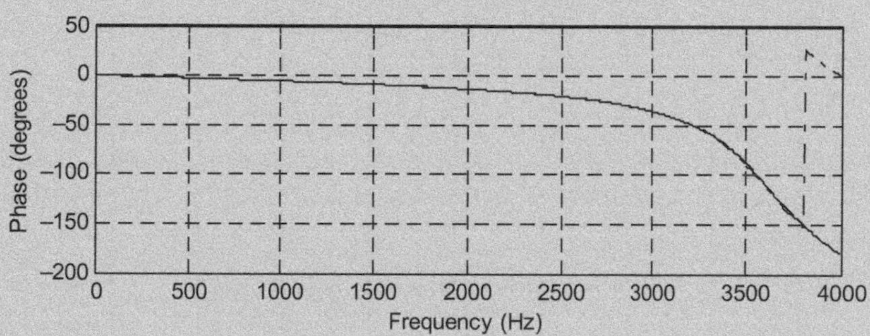

FIG. 8.47

Frequency responses (dash-dotted line, quantized coefficients; solid line, unquantized coefficients).

Program 8.17. MATLAB *m*-file for Example 8.25.

```
%Example 8.25
% Plot the magnitude and phase responses
 fs=8000;                          % Sampling rate
 B=[0.7434 1.4868 0.7434];
 A=[1 1.5149 0.6346];
 [hz,f]=freqz(B,A,512,fs);  % Calculate responses without coefficient quantization
 phi=180*unwrap(angle(hz))/pi;
 Bq=[0.750 1.4834375 0.75000];
 Aq=[1 1.515625 0.640625];
```

```
[hzq,f]=freqz(Bq,Aq,512,fs); % Calculate responses with coefficient quantization
phiq=180*unwrap(angle(hzq))/pi;
subplot(2,1,1), plot(f,20*log10(abs(hz)),f,20*log10(abs(hzq)),'-.');grid;
axis([0 4000 -10 2])
xlabel('Frequency (Hz)');
ylabel('Magnitude Response (dB)');
subplot(2,1,2), plot(f, phi, f, phiq,'-.'); grid;
xlabel('Frequency (Hz)');
ylabel('Phase (degrees)');
```

From Fig. 8.47, we observe that the quantization of IIR filter coefficients has more effect on magnitude response and less effect on the phase response in the passband. In practice, one needs to verify this effect to make sure that the magnitude frequency response meets the filter specifications.

8.11 APPLICATION: GENERATION AND DETECTION OF DTMF TONES USING THE GOERTZEL ALGORITHM

In this section, we study an application of the digital filters for the generation and detection of dual-tone multifrequency (DTMF) signals used for telephone touch keypads. In our daily life, DTMF touch tones produced by telephone keypads on handsets are applied to dial telephone numbers routed to telephone companies, where the DTMF tones are digitized and processed and the detected dialed telephone digits are used for the telephone switching system to ring the party to be called. A telephone touch keypad is shown in Fig. 8.48, where each key is represented by two tones with their specified frequencies. For example, if the key "7" is pressed, the DTMF signal containing the designated frequencies of 852 and 1209 Hz is generated, which is sent to the central office at the telephone company for processing. At the

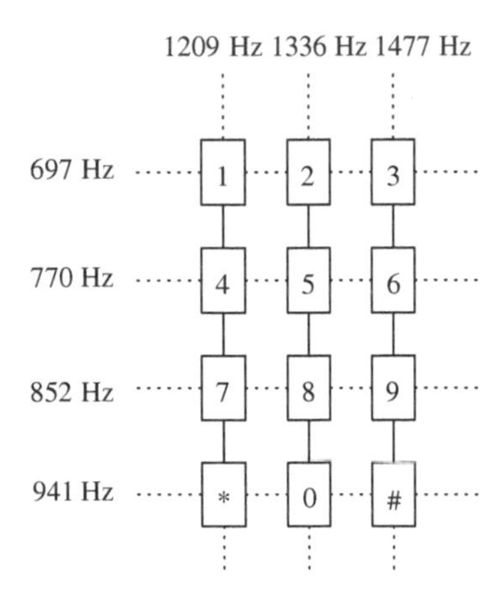

FIG. 8.48

DTMF tone specifications.

central office, the received DTMF tones are detected through the digital filters and some logic operations are used to decode the dialed signal consisting of 852 and 1209 Hz as key "7." The frequencies defined for each key are in Fig. 8.48.

8.11.1 SINGLE-TONE GENERATOR

Now, let us look at a digital tone generator whose transfer function is obtained from the z-transform function of a sinusoidal sequence $\sin(n\Omega_0)$ as

$$H(z) = \frac{z\sin\Omega_0}{z^2 - 2z\cos\Omega_0 + 1} = \frac{z^{-1}\sin\Omega_0}{1 - 2z^{-1}\cos\Omega_0 + z^{-2}}, \tag{8.67}$$

where Ω_0 is the normalized digital frequency. Given the sampling rate of the DSP system and the frequency of the tone to be generated, we have the relationship

$$\Omega_0 = \frac{2\pi f_0}{f_s}. \tag{8.68}$$

Applying the inverse z-transform to the transfer function leads to the difference equation

$$y(n) = \sin\Omega_0 x(n-1) + 2\cos\Omega_0 y(n-1) - y(n-2), \tag{8.69}$$

since

$$Z^{-1}(H(z)) = Z^{-1}\left(\frac{z\sin\Omega_0}{z^2 - 2z\cos\Omega_0 + 1}\right) = \sin(\Omega_0 n) = \sin\left(\frac{2\pi f_0 n}{f_s}\right),$$

which is the impulse response. Hence, to generate a pure tone with the amplitude of A, an impulse function $x(n) = A\delta(n)$ must be used as the input to the digital filter, as illustrated in Fig. 8.49.

Now, we illustrate implementation. Assuming that the sampling rate of the DSP system is 8000 Hz, we need to generate a digital tone of 1 kHz. Then we compute

$$\Omega_0 = 2\pi \times \frac{1000}{8000} = \frac{\pi}{4}, \ \sin\Omega_0 = 0.707107, \text{ and } 2\cos\Omega_0 = 1.414214.$$

The required filter transfer function is determined as

$$H(z) = \frac{0.707107z^{-1}}{1 - 1.414214z^{-1} + z^{-2}}.$$

The MATLAB simulation using the input $x(n) = \delta(n)$ is displayed in Fig. 8.50, where the top plot is the generated tone of 1 kHz, and the bottom plot shows its spectrum. The corresponding MATLAB list is given in Program 8.18.

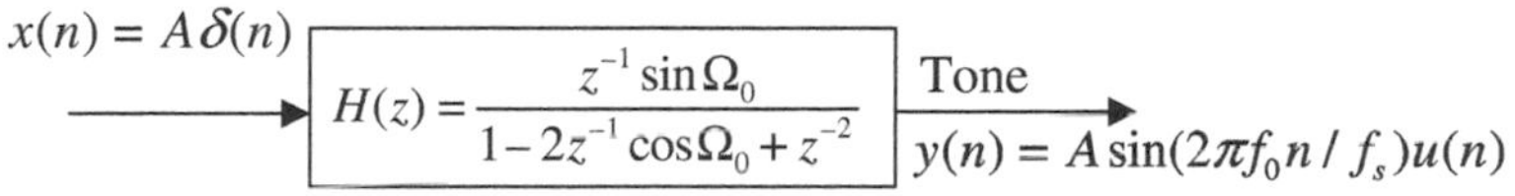

FIG. 8.49

Single-tone generator.

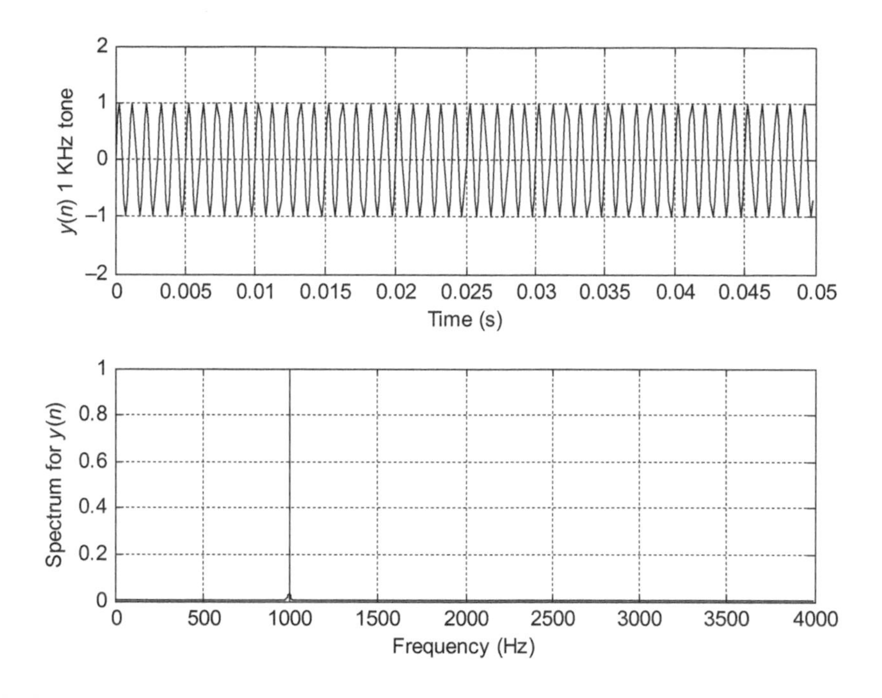

FIG. 8.50

Plots of a generated single tone of 1000 Hz and its spectrum.

Note that if we replace the filter $H(z)$ with the z-transform of other sequences such as a cosine function and use the impulse sequence as the filter input, the filter will generate the corresponding digital wave such as the digital cosine wave.

Program 8.18. MATLAB program for generating a sinusoid.

```
fs=8000;                                     % Sampling rate
t=0:1/fs:1;                                  % Time vector for 1 s
x=zeros(1,length(t));                        % Initialize input to be zero
x(1)=1;                                      % Set up impulse function
y=filter([0 0.707107][1 -1.414214 1],x);     % Perform filtering
subplot(2,1,1);plot(t(1:400),y(1:400));grid
ylabel('y(n) 1 kHz tone'); xlabel('time (second)')
Ak=2*abs(fft(y))/length(y);Ak(1)=Ak(1)/2; % One-sides amplitude spectrum
f=[0:1:(length(y)-1)/2]*fs/length(y);        % Indices to frequencies (Hz) for plot
subplot(2,1,2);plot(f,Ak(1:(length(y)+1)/2));grid
ylabel('Spectrum for y(n)'); xlabel('frequency (Hz)')
```

8.11.2 DUAL-TONE MULTIFREQUENCY TONE GENERATOR

Now that the principle of a single-tone generator is illustrated, we can extent it to develop the DTMF tone generator using two digital filters in parallel. The DTMF tone generator for key "7" is depicted in Fig. 8.51.

Here we generate the DTMF tone for key "7" for a duration of 1 s, assuming the sampling rate of 8000 Hz. The generated tone and its spectrum are plotted in Fig. 8.52 for verification, while the MATLAB implementation is given in Program 8.19.

Program 8.19. MATLAB program for DTMF tone generation.

```
close all; clear all
fs=8000;                    % Sampling rate
t=0:1/fs:1;                 % 1-s time vector
x=zeros(1,length(t));       % Initialize input to be zero
x(1)=1;                     % Set-up impulse function
% Generate 852-Hz tone
y852=filter([0 sin(2*pi*852/fs)][1 -2*cos(2*pi*852/fs) 1],x);
% Generate 1209-Hz tone
y1209=filter([0 sin(2*pi*1209/fs) ][1 -2*cos(2*pi*1209/fs) 1],x); % Filtering
y7=y852+y1209;              % Generate DTMF tone
subplot(2,1,1);plot(t(1:400),y7(1:400));grid
ylabel('y(n) DTMF: number 7');
xlabel('time (second)')
Ak=2*abs(fft(y7))/length(y7);Ak(1)=Ak(1)/2; % One-sided amplitude spectrum
f=[0:1:(length(y7)-1)/2]*fs/length(y7); % Map indices to frequencies (Hz) for plot
subplot(2,1,2);plot(f,Ak(1:(length(y7)+1)/2));grid
ylabel('Spectrum for y7(n)');
xlabel('frequency (Hz)');
```

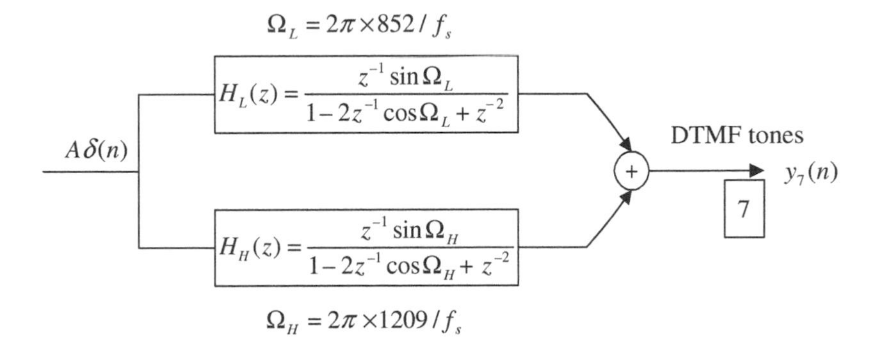

FIG. 8.51

Digital DTMF tone generator for the keypad digit "7."

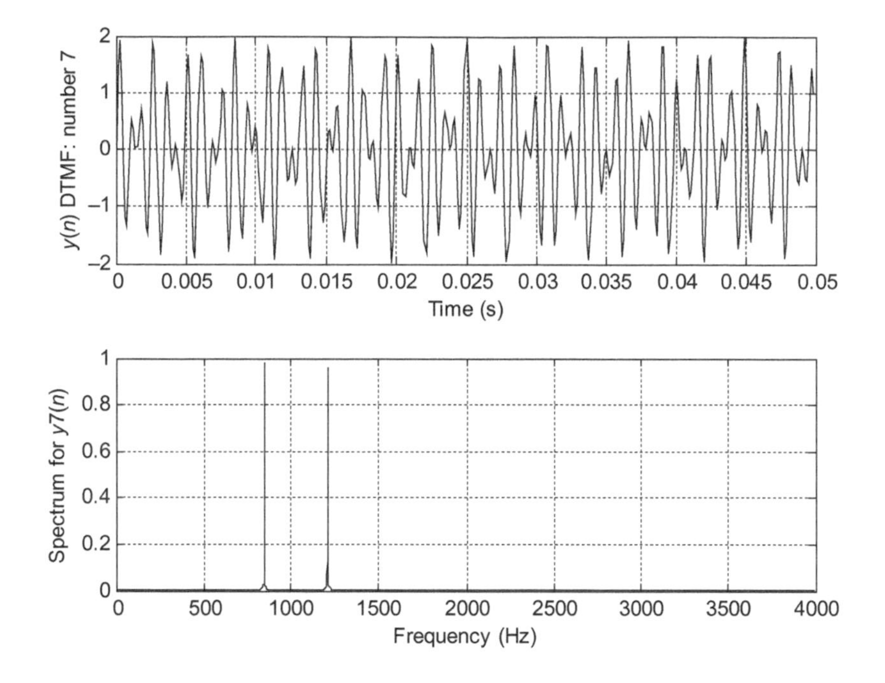

FIG. 8.52

Plots of the generated DTMF tone of "7" and its spectrum.

8.11.3 GOERTZEL ALGORITHM

In practice, the DTMF tone detector is designed using the Goertzel algorithm. This is a special and powerful algorithm used for computing discrete Fourier transform (DFT) coefficients and signal spectra using a digital filtering method. The modified Goertzel algorithm can be used for computing signal spectra without involving complex algebra like the DFT algorithm.

To devise the Goertzel algorithm, we begin with applying DFT to a sequence $x(n)$ as

$$X(k)=\sum_{n=0}^{N-1}x(n)W_N^{kn}=\sum_{n=0}^{N-1}x(n)e^{-j\frac{2\pi kn}{N}}=e^{j\frac{2\pi kN}{N}}\sum_{n=0}^{N-1}x(n)e^{-j\frac{2\pi kn}{N}} \tag{8.70}$$

Note that $W_N^k=e^{-2\pi k/N}$ and adding $e^{j2\pi kN/N}=1$ does not affect Eq. (8.70). We rewrite Eq. (8.70) in the convolution format. It follows that

$$X(k)=\sum_{n=0}^{N-1}x(n)e^{j\frac{2\pi k}{N}(N-n)} \tag{8.71}$$

In Eq. (8.71), if we define $h_k(n)=e^{j2\pi kn/N}$ and note that $h_k(n)$ is periodic of N, we can express $X(k)$ as

$$X(k)=\sum_{n=0}^{N-1}x(n)h_k(N-n)=x(0)h_k(N)+x(1)h_k(N-1)+\cdots+x(N-1)h_k(1). \tag{8.72}$$

The results of Eq. (8.72) will be used later. If we treat $h_k(n)$ as the FIR filter coefficients, its output can be obtained by convolution sum as shown below:

$$y_k(m) = \sum_{n=0}^{m} x(n)h_k(m-n), \tag{8.73}$$

where $x(n)$ is a causal sequence. The convolution is depicted below:

After filtering $N+1$ times, we obtain

$$y_k(0) = \sum_{n=0}^{0} x(n)h_k(0-n) = x(0)h_k(0),$$

$$y_k(1) = \sum_{n=0}^{1} x(n)h_k(1-n) = x(0)h_k(1) + x(1)h_k(0),$$

$$y_k(2) = \sum_{n=0}^{2} x(n)h_k(2-n) = x(0)h_k(2) + x(1)h_k(1) + x(2)h_k(0),$$

$$\cdots\cdot$$

$$\begin{aligned} y_k(N) &= \sum_{n=0}^{N} x(n)h_k(N-n) \\ &= x(0)h_k(N) + x(1)h_k(N-1) + \cdots + x(N-1)h_k(1) + x(N)h_k(0). \end{aligned} \tag{8.74}$$

Now, if we set $x(N)=0$, then

$$y_k(N) = x(0)h_k(N) + x(1)h_k(N-1) + \cdots + x(N-1)h_k(1) = X(k). \tag{8.75}$$

In general, the convolution form is expressed as

$$y_k(n) = x(n)*h_k(n). \tag{8.76}$$

But we can find a recursive solution by considering $h_k(n) = e^{j2\pi kn/N}$ and $n \to \infty$. We see that

$$\begin{aligned} H_k(z) &= h_k(0) + h_k(1)z^{-1} + h_k(2)z^{-2} + \cdots \\ &= \left(e^{j\frac{2\pi k}{N}}\right)^0 + \left(e^{j\frac{2\pi k}{N}}\right)^1 z^{-1} + \left(e^{j\frac{2\pi k}{N}}\right)^2 z^{-2} + \cdots. \end{aligned} \tag{8.77}$$

For $\left|e^{j\frac{2\pi k}{N}} z^{-1}\right| < 1$, we achieve the filter transfer function as

$$H_k(z) = \frac{1}{1 - e^{j\frac{2\pi k}{N}} z^{-1}} = \frac{1}{1 - W_N^{-k} z^{-1}}. \tag{8.78}$$

Through simple manipulations, we yield the Goertzel filter as follows:

$$H_k(z) = \frac{1}{1 - W_N^{-k} z^{-1}} \frac{1 - W_N^k z^{-1}}{1 - W_N^k z^{-1}} = \frac{1 - W_N^k z^{-1}}{1 - 2\cos\left(\frac{2\pi k}{N}\right) z^{-1} + z^{-2}}. \tag{8.79}$$

Note that the Goertzel filter $H_k(z)$ is a marginally stable filter and has a resonance response at $\Omega = 2\pi k/N$.

Comparing Eqs. (8.72), (8.74), and (8.75), we can compute DFT coefficient $X(k)$ at the specified frequency bin k with the given input data $x(n)$ for $n=0, 1, \ldots, N-1$, and the last element set to be $x(N)=0$ using Goertzel filter $H_k(z)$. We can process the data sequence $N+1$ times to achieve the filter output as $y_k(n)$ for $n=0, 1, \ldots, N$. The DFT coefficient $X(k)$ is the last datum from the Goertzel filter, that is,

$$X(k) = y_k(N). \tag{8.80}$$

The implementation of the Goertzel filter is presented by direct-form II realization in Fig. 8.53.

According to the direct-form II realization, we can write the Goertzel algorithm as

$$x(N) = 0 \tag{8.81}$$

$$\text{for } n = 0, 1, \cdots, N$$

$$v_k(n) = 2\cos\left(\frac{2\pi k}{N}\right) v_k(n-1) - v_k(n-2) + x(n). \tag{8.82}$$

$$y_k(n) = v_k(n) - W_N^k v_k(n-1) \tag{8.83}$$

with initial conditions: $v_k(-2)=0$, $v_k(-1)=0$.

Then the DFT coefficient $X(k)$ is given as

$$X(k) = y_k(N) \tag{8.84}$$

The squared magnitude of $x(k)$ is computed as

$$|X(k)|^2 = v_k^2(N) + v_k^2(N-1) - 2\cos\left(\frac{2\pi k}{N}\right) v_k(N) v_k(N-1). \tag{8.85}$$

We show the derivation of Eq. (8.85) as follows. Note that Eq. (8.83) involves complex algebra, since the equation contains only one complex number, a factor

$$W_N^k = e^{-j\frac{2\pi k}{N}} = \cos\left(\frac{2\pi k}{N}\right) - j\sin\left(\frac{2\pi k}{N}\right)$$

discussed in Chapter 4. If our objective is to compute the spectrum value, we can substitute $n=N$ into Eq. (8.83) to obtain $X(k)$ and multiply $X(k)$ by its conjugate $X^*(k)$ to achieve the squared magnitude of the DFT coefficient. It follows (Ifeachor and Jervis, 2002) that

$$|X(k)|^2 = X(k) X^*(k).$$

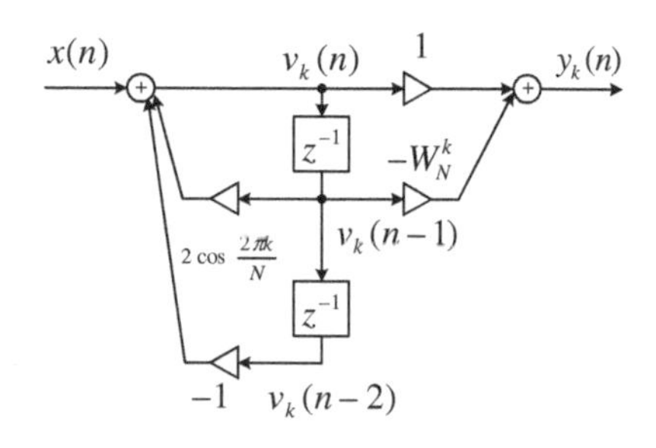

FIG. 8.53

Second-order Goertzel IIR filter.

Since $X(k)=y_k(N)-W_N^k v_k(N-1)$,

$$X^*(k)=y_k(N)-W_N^{-k}v_k(N-1)$$

then

$$\begin{aligned}|X(k)|^2&=\left(y_k(N)-W_N^k y_k(N-1)\right)\left(y_k(N)-W_N^{-k}y_k(N-1)\right)\\&=y_k^2(N)+y_k^2(N-1)-\left(W_N^k+W_N^{-k}\right)y_k(N)y_k(N-1).\end{aligned} \tag{8.86}$$

Using Euler's identity yields

$$W_N^k+W_N^{-k}=e^{-j\frac{2\pi}{N}k}+e^{j\frac{2\pi}{N}k}=2\cos\left(\frac{2\pi k}{N}\right). \tag{8.87}$$

Substituting Eq. (8.87) into Eq. (8.86) leads to Eq. (8.85).

We can see that the DSP equation for $v_k(k)$ and computation of the squared magnitude of the DFT coefficient $|X(k)|^2$ do not involve any complex algebra. Hence, we will use this advantage for later development. To illustrate the algorithm, let us consider Example 8.26.

EXAMPLE 8.26

Given the digital data sequence of length 4 as $x(0)=1$, $x(1)=2$, $x(2)=3$, and $x(3)=4$, use the Goertzel algorithm to compute DFT coefficient $X(1)$ and the corresponding spectral amplitude at the frequency bin $k=1$.

Solution:

We have $k=1$, $N=4$, $x(0)=1$, $x(1)=2$, $x(2)=3$, and $x(3)=4$. Note that

$$2\cos\left(\frac{2\pi}{4}\right)=0 \text{ and } W_4^1=e^{-j\frac{2\pi\times 1}{4}}=\cos\left(\frac{\pi}{2}\right)-j\sin\left(\frac{\pi}{2}\right)=-j.$$

We first write the simplified difference equations:

$$x(4)=0$$

$$\text{for } n=0,1,\cdots,4$$

$$v_1(n)=-v_1(n-2)+x(n)$$

$$y_1(n)=v_1(n)+jv_1(n-1)$$

then

$$X(1)=y_1(4)$$

$$|X(1)|^2=v_1^2(4)+v_1^2(3).$$

The digital filter process is demonstrated in the following:

$$v_1(0)=-v_1(-2)+x(0)=0+1=1$$

$$y_1(0)=v_1(0)+jv_1(-1)=1+j\times 0=1$$

$$v_1(1)=-v_1(-1)+x(1)=0+2=2$$

$$y_1(1)=v_1(1)+jv_1(0)=2+j\times 1=2+j$$

Continued

EXAMPLE 8.26—CONT'D

$$v_1(2) = -v_1(0) + x(2) = -1 + 3 = 2$$

$$y_1(2) = v_1(2) + jv_1(1) = 2 + j \times 2 = 2 + j2$$

$$v_1(3) = -v_1(1) + x(3) = -2 + 4 = 2$$

$$y_1(3) = v_1(3) + jv_1(2) = 2 + j \times 2 = 2 + j2$$

$$v_1(4) = -v_1(2) + x(4) = -2 + 0 = -2$$

$$y_1(4) = v_1(4) + jv_1(3) = -2 + j \times 2 = -2 + j2.$$

Then the DFT coefficient and its squared magnitude are determined as

$$X(1) = y_1(4) = -2 + j2$$

$$|X(1)|^2 = v_1^2(4) + v_1^2(3) = (-2)^2 + (2)^2 = 8.$$

Thus, the two-side amplitude spectrum is computed as

$$A_1 = \frac{1}{4}\sqrt{\left(|X(1)|^2\right)} = 0.7071$$

and the corresponding single-sided amplitude spectrum is $A_1 = 2 \times 0.707 = 1.4141$.

From this simple illustrative example, we see that the Goertzel algorithm has the following advantages:

1. We can apply the algorithm for computing the DFT coefficient $X(k)$ at a specified frequency bin k; unlike the fast Fourier transform (FFT) algorithm, all the DFT coefficients are computed once it is applied.
2. If we want to compute the spectrum at frequency bin k, that is, $|X(k)|$, Eq. (8.82) shows that we need to process $v_k(n)$ for $N+1$ times and then compute $|X(k)|^2$. The operations avoid complex algebra.

If we use the modified Goertzel filter in Fig. 8.54, then the corresponding transfer function is given by

$$G_k(z) = \frac{V_k(z)}{X(z)} = \frac{1}{1 - 2\cos\left(\frac{2\pi k}{N}\right)z^{-1} + z^{-2}}. \tag{8.88}$$

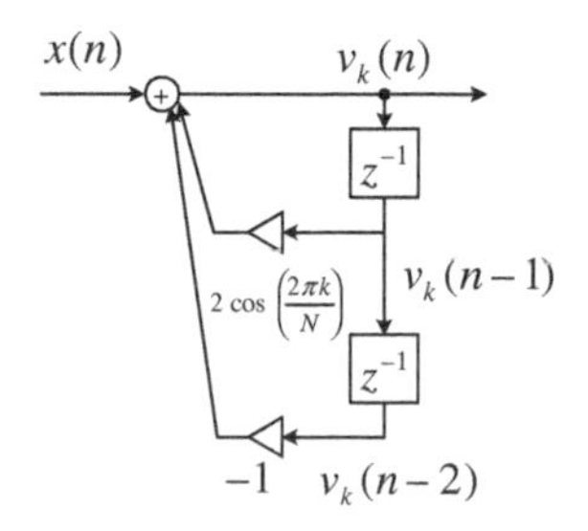

FIG. 8.54

Modified second-order Goertzel IIR filter.

The modified Goertzel algorithm becomes

$$x(N) = 0$$

$$\text{for } n = 0, 1, \cdots, N,$$

$$v_k(n) = 2\cos\left(\frac{2\pi k}{N}\right) v_k(n-1) - v_k(n-2) + x(n)$$

with initial conditions: $v_k(-2) = 0$, and $v_k(-1) = 0$.

Then the squared magnitude of the DFT coefficient is given by

$$|X(k)|^2 = v_k^2(N) + v_k^2(N-1) - 2\cos\left(\frac{2\pi k}{N}\right) v_k(N) v_k(N-1).$$

EXAMPLE 8.27

Given the digital data sequence of length 4 as $x(0) = 1$, $x(1) = 2$, $x(2) = 3$, and $x(3) = 4$, use the Goertzel algorithm to compute the spectral amplitude at the frequency bin $k = 0$.

Solution:

$$k = 0, N = 4, x(0) = 1, x(1) = 2, x(2) = 3, \text{and } x(3) = 4.$$

Using the modified Goertzel algorithm and noting that $2 \cdot \cos\left(\frac{2\pi}{4} \times 0\right) = 2$, we get the simplified difference equations as

$$x(4) = 0$$

$$\text{for } n = 0, 1, \cdots, 4$$

$$v_0(n) = 2v_0(n-1) - v_0(n-2) + x(n)$$

$$\text{then } |X(0)|^2 = v_0^2(4) + v_0^2(3) - 2v_0(4)v_0(3).$$

The digital filtering is performed as

$$v_0(0) = 2v_0(-1) - v_0(-2) + x(0) = 0 + 0 + 1 = 1$$

$$v_0(1) = 2v_0(0) - v_0(-1) + x(1) = 2 \times 1 + 0 + 2 = 4$$

$$v_0(2) = 2v_0(1) - v_0(0) + x(2) = 2 \times 4 - 1 + 3 = 10$$

$$v_0(3) = 2v_0(2) - v_0(1) + x(3) = 2 \times 10 - 4 + 4 = 20$$

$$v_0(4) = 2v_0(3) - v_0(2) + x(4) = 2 \times 20 - 10 + 0 = 30.$$

Then the squared magnitude is determined by

$$|X(0)|^2 = v_0^2(4) + v_0^2(3) - 2v_0(4)v_0(3) = (30)^2 + (20)^2 - 2 \times 30 \times 20 = 100.$$

Thus, the amplitude spectrum is computed as

$$A_0 = \frac{1}{4}\sqrt{\left(|X(0)|^2\right)} = 2.5.$$

We can write a MATLAB function for the Geortzel algorithm shown in Program 8.20:

Program 8.20. MATLAB function for Geortzel Algorithm.

```
function [ Xk, Ak] = galg(x,k)
% Geortzel Algorithm
% [ Xk, Ak] = galg(x,k)
% x=input vetcor; k=frequency index
% Xk= kth DFT coeficient; Ak=magnitude of the kth DFT coefficient
N=length(x); x=[x 0];
vk=zeros(1,N+3);
for n=1:N+1
    vk(n+2)=2*cos(2*pi*k/N)*vk(n+1)-vk(n)+x(n);
end
Xk=vk(N+3)-exp(-2*pi*j*k/N)*vk(N+2);
Ak=vk(N+3)*vk(N+3)+vk(N+2)*vk(N+2)-2*cos(2*pi*k/N)*vk(N+3)*vk(N+2);
Ak=sqrt(Ak)/N;
End
```

EXAMPLE 8.28

Use MATLAB function (Program 8.20) to verify the results in Examples 8.26 and 8.27.

Solution:

(a) For Example 8.26, we obtain

```
>> x=[1 2 3 4]
x = 1 2 3 4
>> [X1, A1]=galg(x,1)
X1 = -2.0000 + 2.0000i
A1 =0.7071
```

(b) For Example 8.27, we obtain

```
>> x=[1 2 3 4]
x = 1 2 3 4
>> [X0, A0]=galg(x,1)
X0 = 10
A0 = 2.5000
```

8.11.4 DUAL-TONE MULTIFREQUENCY TONE DETECTION USING THE MODIFIED GOERTZEL ALGORITHM

Based on the specified frequencies of each DTMF tone shown in Fig. 8.48 and the modified Goertzel algorithm, we can develop the following design principles for DTMF tone detection:

1. When a digitized DTMF tone $x(n)$ is received, it has two nonzero frequency components from the following seven numbers: 697, 770, 852, 941, 1209, 1336, and 1477 Hz.
2. We can apply the modified Goertzel algorithm to compute seven spectral values, which correspond to the seven frequencies in (1). The single-sided amplitude spectrum is computed as

$$A_k = \frac{2}{N}\sqrt{|X(k)|^2}. \tag{8.89}$$

Table 8.12 DTMF Frequencies and Their Frequency Bins

DTMF Frequency (Hz)	Frequency bin: $k = \frac{f}{f_s} \times N$
697	18
770	20
852	22
941	24
1209	31
1336	34
1477	38

Since the modified Goertzel algorithm is used, there is no complex algebra involved.

3. Ideally, there are two nonzero spectral components. We will use these two nonzero spectral components to determine which key is pressed.
4. The frequency bin number (frequency index) can be determined based on the sampling rate f_s, and the data size of N via the following relation:

$$k = \frac{f}{f_s} \times N \text{ (round off to an integer)}. \tag{8.90}$$

Given the key frequency specification in Table 8.12, we can determine the frequency bin k for each DTMF frequency with $f_s = 8000$ Hz and $N = 205$. The results are summarized in Table 8.12. The DTMF detector block diagram is shown in Fig. 8.55.

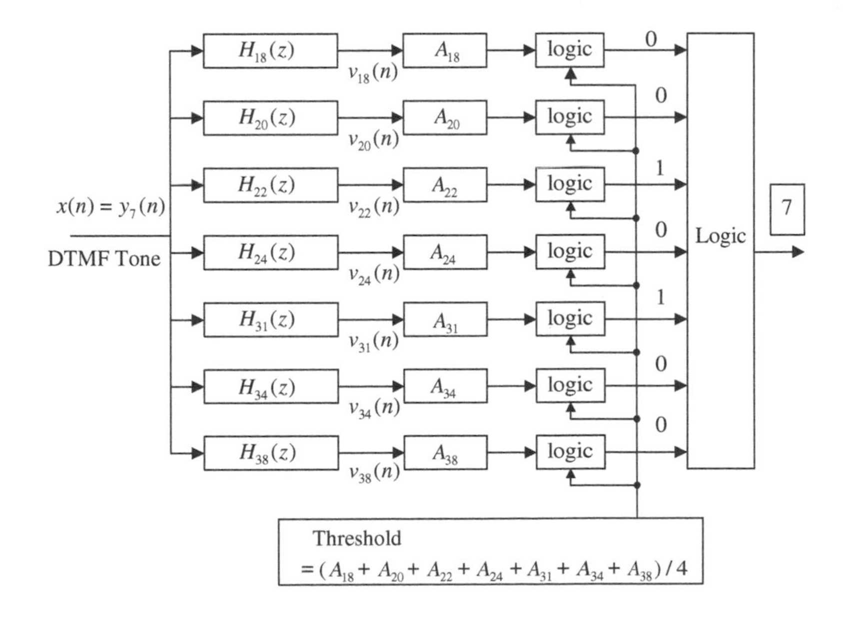

FIG. 8.55

DTMF detector using the Goertzel algorithm.

5. The threshold value can be the sum of all seven spectral values divided by a factor of 4. Note that there are only two nonzero spectral values and the others are zeros; hence, the threshold value should ideally be half of the individual nonzero spectral value. If the detected spectral value is larger than the threshold value, then the logic operation outputs logic 1; otherwise, it outputs logic 0. Finally, the logic operation at the last stage is to decode the key information based on the 7-bit binary pattern.

EXAMPLE 8.29

Given a DSP system with $f_s = 8000\,\text{Hz}$ and data size $N = 205$, seven Goertzel IIR filters are implemented for DTMF tone detection.

Determine the following for the frequencies corresponding to key 7:

1. Frequency bin numbers
2. The Goertzel filter transfer functions and DSP equations
3. Equations for calculating amplitude spectral values

Solution:

For key 7, we have $f_L = 852\,\text{Hz}$ and $f_H = 1209\,\text{Hz}$.

1. Using Eq. (8.90), we get

$$k_L = \frac{852}{8{,}000} \times 205 \approx 22, \text{and } k_H = \frac{1{,}209}{8{,}000} \times 205 \approx 31.$$

2. Since $2\cos\left(\frac{2\pi \times 22}{205}\right) = 1.5623$, and $2\cos\left(\frac{2\pi \times 31}{205}\right) = 1.1631$, it follows that

$$H_{22}(z) = \frac{1}{1 - 1.5623z^{-1} + z^{-2}}$$

and

$$H_{31}(z) = \frac{1}{1 - 1.1631z^{-1} + z^{-2}}.$$

The DSP equations are therefore given by

$$v_{22}(n) = 1.5623v_{22}(n-1) - v_{22}(n-2) + x(n) \text{ with } x(205) = 0, \text{ for } n = 0, 1, \cdots, 205$$
$$v_{31}(n) = 1.1631v_{31}(n-1) - v_{31}(n-2) + x(n) \text{ with } x(205) = 0, \text{ for } n = 0, 1, \cdots, 205.$$

3. The amplitude spectral values are determined by

$$|X(22)|^2 = (v_{22}(205))^2 + (v_{22}(204))^2 - 1.5623(v_{22}(205)) \times (v_{22}(204))$$

$$A_{22} = \frac{2\sqrt{|X(22)|^2}}{205}$$

and

$$|X(31)|^2 = (v_{31}(205))^2 + (v_{31}(204))^2 - 1.1631(v_{31}(205)) \times (v_{31}(204))$$

$$A_{31} = \frac{2\sqrt{|X(31)|^2}}{205}.$$

The MATLAB simulation for decoding the key 7 is shown in Program 8.21. Fig. 8.56A shows the frequency responses of the second-order Goertzel bandpass filters. The input is generated as shown in Fig. 8.52. After filtering, the calculated spectral values and threshold value for decoding key 7 are displayed in Fig. 8.56B, where only two spectral values corresponding to the frequencies of 770 and 1209 Hz are above the threshold, and are encoded as logic 1. According to the key information in Fig. 8.55, the final logic operation decodes the key as 7.

The principle can easily be extended to transmit the ASCII (American Standard Code for Information Interchange) code or other types of code using the parallel Goertzel filter bank.

If the calculated spectral value is larger than the threshold value, then the logic operation outputs logic 1; otherwise, it outputs logic 0. Finally, the logic operation at the last stage decodes the key information based on the 7-bit binary pattern.

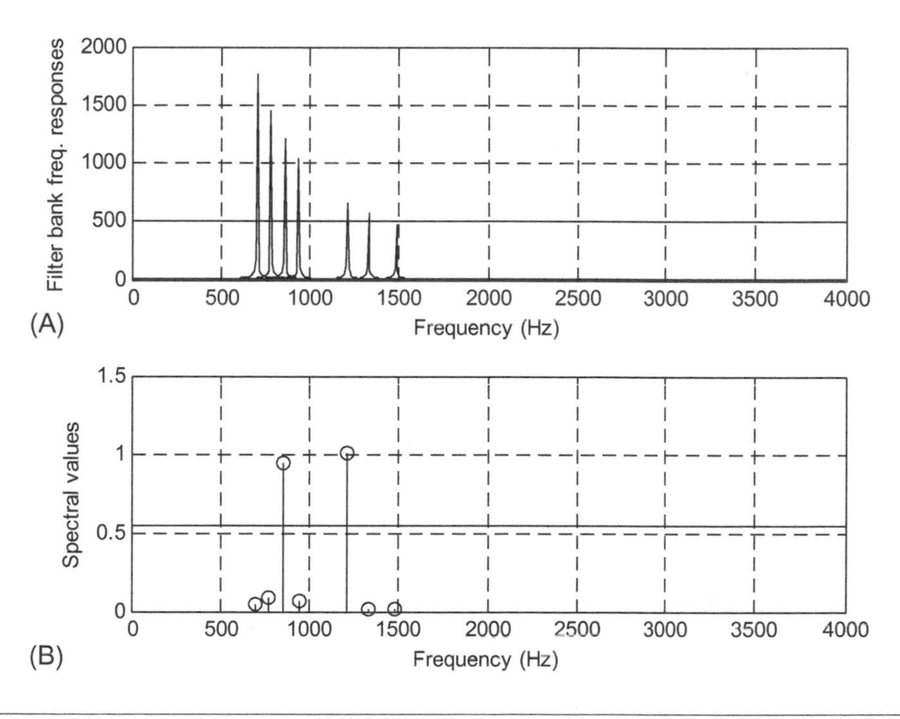

FIG. 8.56

(A) Goertzel filter bank frequency responses; (B) Display of spectral values and threshold for key 7.

Program 8.21. DTMF detection using the Goertzel algorithm.

```
close all;clear all;
% DTMF tone generator
N=205;
fs=8000; t=[0:1:N-1]/fs;                % Sampling rate and time vector
x=zeros(1,length(t));x(1)=1;            % Generate the impulse function
%Generation of tones
y697=filter([0 sin(2*pi*697/fs)][1 -2*cos(2*pi*697/fs) 1],x);
y770=filter([0 sin(2*pi*770/fs)][1 -2*cos(2*pi*770/fs) 1],x);
y852=filter([0 sin(2*pi*852/fs)][1 -2*cos(2*pi*852/fs) 1],x);
y941=filter([0 sin(2*pi*941/fs)][1 -2*cos(2*pi*941/fs) 1],x);
y1209=filter([0 sin(2*pi*1209/fs) ][1 -2*cos(2*pi*1209/fs) 1],x);
y1336=filter([0 sin(2*pi*1336/fs)][1 -2*cos(2*pi*1336/fs) 1],x);
y1477=filter([0 sin(2*pi*1477/fs)][1 -2*cos(2*pi*1477/fs) 1],x);
key=input('input of the following keys: 1,2,3,4,5,6,7,8,9,*,0,# =>','s');
yDTMF=[];
if key=='1' yDTMF=y697+y1209; end
if key=='2' yDTMF=y697+y1336; end
if key=='3' yDTMF=y697+y1477; end
if key=='4' yDTMF=y770+y1209; end
if key=='5' yDTMF=y770+y1336; end
if key=='6' yDTMF=y770+y1477; end
if key=='7' yDTMF=y852+y1209; end
if key=='8' yDTMF=y852+y1336; end
if key=='9' yDTMF=y852+y1477; end
if key=='*' yDTMF=y941+y1209; end
if key=='0' yDTMF=y941+y1336; end
if key=='#' yDTMF=y941+y1477; end
if size(yDTMF)==0 disp('Invalid input key'); return; end
yDTMF=[yDTMF 0];          % DTMF signal appended with a zero
% DTMF detector (use Goertzel algorithm)
a697=[1 -2*cos(2*pi*18/N) 1];
a770=[1 -2*cos(2*pi*20/N) 1];
a852=[1 -2*cos(2*pi*22/N) 1];
a941=[1 -2*cos(2*pi*24/N) 1];
a1209=[1 -2*cos(2*pi*31/N) 1];
a1336=[1 -2*cos(2*pi*34/N) 1];
a1477=[1 -2*cos(2*pi*38/N) 1];
% Filter bank frequency responses
[w1, f]=freqz(1,a697,512,fs);
[w2, f]=freqz(1,a770,512,fs);
[w3, f]=freqz(1,a852,512,fs);
[w4, f]=freqz(1,a941,512,fs);
[w5, f]=freqz(1,a1209,512,fs);
[w6, f]=freqz(1,a1336,512,fs);
[w7, f]=freqz(1,a1477,512,fs);
subplot(2,1,1);plot(f,abs(w1),f,abs(w2),f,abs(w3), ...
f,abs(w4),f,abs(w5),f,abs(w6),f,abs(w7));grid
xlabel('Frequency (Hz)'); ylabel('(a) Filter bank freq. responses');
% Filter bank bandpass filtering
```

```
y697=filter(1,a697,yDTMF);
y770=filter(1,a770,yDTMF);
y852=filter(1,a852,yDTMF);
y941=filter(1,a941,yDTMF);
y1209=filter(1,a1209,yDTMF);
y1336=filter(1,a1336,yDTMF);
y1477=filter(1,a1477,yDTMF);
% Determine the absolute magnitude of DFT coefficents
m(1)=sqrt(y697(206)^2+y697(205)^2- ...
  2*cos(2*pi*18/205)*y697(206)*y697(205));
m(2)=sqrt(y770(206)^2+y770(205)^2- ...
  2*cos(2*pi*20/205)*y770(206)*y770(205));
m(3)=sqrt(y852(206)^2+y852(205)^2- ...
  2*cos(2*pi*22/205)*y852(206)*y852(205));
m(4)=sqrt(y941(206)^2+y941(205)^2- ...
  2*cos(2*pi*24/205)*y941(206)*y941(205));
m(5)=sqrt(y1209(206)^2+y1209(205)^2- ...
  2*cos(2*pi*31/205)*y1209(206)*y1209(205));
m(6)=sqrt(y1336(206)^2+y1336(205)^2- ...
  2*cos(2*pi*34/205)*y1336(206)*y1336(205));
m(7)=sqrt(y1477(206)^2+y1477(205)^2- ...
  2*cos(2*pi*38/205)*y1477(206)*y1477(205));
% Convert the magnitude of DFT coefficients to the single-side spectrum
m=2*m/205;
% Determine the threshold
th=sum(m)/4;
% Plot the DTMF spectrum with the threshold
f=[ 697 770 852 941 1209 1336 1477];
f1=[0 fs/2];
th=[ th th];
subplot(2,1,2);stem(f,m);grid;hold; plot(f1,th);
xlabel('Frequency (Hz)'); ylabel(' (b) Spectral values');
m=round(m); % Round to the binary pattern
if m== [ 1 0 0 0 1 0 0] disp('Detected Key 1'); end
if m== [ 1 0 0 0 0 1 0] disp('Detected Key 2'); end
if m== [ 1 0 0 0 0 0 1] disp('Detected Key 3'); end
if m== [ 0 1 0 0 1 0 0] disp('Detected Key 4'); end
if m== [ 0 1 0 0 0 1 0] disp('Detected Key 5'); end
if m== [ 0 1 0 0 0 0 1] disp('Detected Key 6'); end
if m== [ 0 0 1 0 1 0 0] disp('Detected Key 7'); end
if m== [ 0 0 1 0 0 1 0] disp('Detected Key 8'); end
if m== [ 0 0 1 0 0 0 1] disp('Detected Key 9'); end
if m== [ 0 0 0 1 1 0 0] disp('Detected Key *'); end
if m== [ 0 0 0 1 0 1 0] disp('Detected Key 0'); end
if m== [ 0 0 0 1 0 0 1] disp('Detected Key #'); end
```

8.12 SUMMARY OF INFINITE IMPULSE RESPONSE (IIR) DESIGN PROCEDURES AND SELECTION OF THE IIR FILTER DESIGN METHODS IN PRACTICE

In this section, we first summarize the design procedures of the BLT design, impulse invariant design, and pole-zero placement design methods, and then discuss the selection of the particular filter for typical applications.

The BLT design method:

1. Given the digital filter frequency specifications, prewarp each of the digital frequency edge to the analog frequency edge using Eqs. (8.18) and (8.19).
2. Determine the prototype filter order using Eq. (8.29) for the Butterworth filter or Eq. (8.35b) for the Chebyshev filter, and perform lowpass prototype transformation using the lowpass prototype in Table 8.3 (Butterworth function) or Tables 8.4 and 8.5 (Chebyshev function) using Eqs. (8.20)–(8.23).
3. Apply the BLT to the analog filter using Eq. (8.24) and output the transfer function.
4. Verify the frequency responses, and output the difference equation.

The impulse invariant design method:

1. Given the lowpass or bandpass filter frequency specifications, perform analog filter design. For the highpass or bandstop filter design, quit this method and use the BLT.
   (a) Determine the prototype filter order using Eq. (8.29) for the Butterworth filter or Eq. (8.35b) for the Chebyshev filter.
   (b) Perform lowpass prototype transformation using the lowpass prototype in Table 8.3 (Butterworth function) or Tables 8.4 and 8.5 (Chebyshev functions) using Eqs. (8.20)–(8.23).
   (c) Skip step (1) if the analog filter transfer function is given to begin with.
2. Determine the impulse response by applying the partial fraction expansion technique to the analog transfer function and inverse Laplace transform using Eq. (8.37).
3. Sample the analog impulse response using Eq. (8.38) and apply the z-transform to the digital impulse function to obtain the digital filter transfer function.
4. Verify the frequency response, and output the difference equation. If the frequency specifications are not net, quit the design method and use the BLT.

The pole-zero placement method:

1. Given the filter cutoff frequency specifications, determine the pole-zero locations using the corresponding equations:
   (a) Second-order bandpass filter: Eqs. (8.43) and (8.44).
   (b) Second-order notch filter: Eqs. (8.47) and (8.48).
   (c) First-order lowpass filter: Eqs. (8.51) or (8.52).
   (d) First-order highpass filter: Eqs. (8.55) or (8.56).
2. Apply the corresponding equation and scale factor to obtain the digital filter transfer function:
   (a) Second-order bandpass filter: Eqs. (8.45) and (8.46).
   (b) Second-order notch filter: Eqs. (8.49) and (8.50).
   (c) First-order lowpass filter: Eqs. (8.53) and (8.54).
   (d) First-order highpass filter: Eqs. (8.57) and (8.58).

3. Verify the frequency response, and output the difference equation. If the frequency specifications are not net, quit the design method and use BLT.

Table 8.13 compares the design parameters of the three design methods.

BLT = bilinear transformation; LPF = lowpass filter; BPF = bandpass filter; HPF = highpass filter.

Performance comparisons using the three methods are displayed in Fig. 8.57, where the bandpass filter is designed using the following specifications:

Passband ripple = −3 dB
Center frequency = 400 Hz
Bandwidth = 200 Hz
Sampling rate = 2000 Hz
Butterworth IIR filter = second-order

As expected, the BLT method satisfies the design requirement, and the pole-zero placement method has little performance degradation because $r = 1-(f_0/f_s)\pi = 0.6858 < 0.9$, and this effect will also cause the center frequency to be shifted. For the bandpass filter designed using the impulse invariant method, the gain at the center frequency is scaled to 1 for a frequency response shape comparison. The performance of the impulse invariant method is satisfied in passband. However, it has significant performance degradation in stopband when compared with the other two methods. This is due to aliasing when sampling the analog impulse response in time domain. The advanced method (Nelatury, 2007), which incorporates the frequency prewarping technique, can be applied to correct the aliasing effect.

Table 8.13 Comparisons of Three IIR Design Methods

	Design Method		
	Bilinear Transformation	Impulse Invariant	Pole-zero Placement
Filter type	Lowpass, highpass, bandpass, bandstop	Appropriate for lowpass and bandpass	Second-order for bandpass and bandstop; first-order for lowpass and highpass
Linear phase	No	No	No
Ripple and stopband specifications	Used for determining the filter order	Used for determining the filter order	Not required; 3 dB on passband offered.
Special requirement	None	Very high sampling relative to the cutoff frequency (LPF) or to upper cutoff frequency (BPF)	narrow band for BPF or Notch filter; lower cutoff frequency or higher cutoff frequency for LPF or HPF.
Algorithm complexity	High: Frequency prewarping, analog filter design, BLT	Moderate: Analog filter design determining digital impulse response. Apply z-transform	Simple: Simple design equations
Minimum design tool	Calculator, algebra	Calculator, algebra	Calculator

BLT, bilinear transformation; *LPF*, *lowpass filter*; *BPF*, *band pass filter*; *HPF*, *highpass filter*.

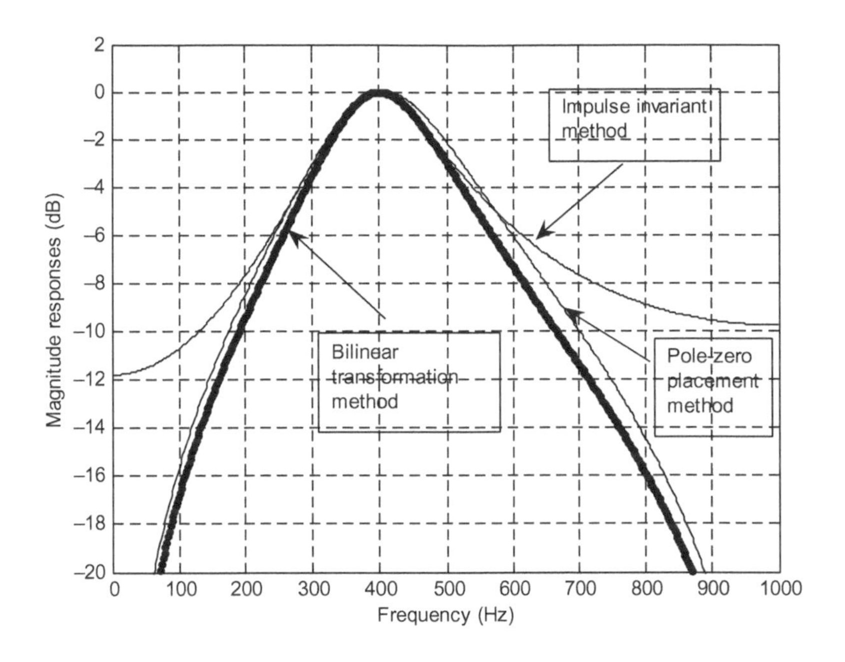

FIG. 8.57

Performance comparisons for the BLT, pole-zero placement, and impulse invariant methods.

Improvement in using the pole-zero placement and impulse invariant methods can be achieved by using a very high sampling rate. Example 8.30 describes the possible selection of the design method by a DSP engineer to solve the real-world problem.

EXAMPLE 8.30

Determine an appropriate IIR filter design method for each of the following DSP applications. As described in the previous section, we apply a notch filter to remove the 60-Hz interference and cascade a bandpass filter to remove noise in an ECG signal for heart rate detection. The following specifications are required:

Notch filter:
Harmonics to be removed = 60 Hz
3-dB bandwidth for the notch filter = 4 Hz
Bandpass filter:
Passband frequency range = 0.25–40 Hz
Passband ripple = 0.5 dB
Sampling rate = 600 Hz.

The pole-zero placement method is the best choice, since the notch filter to be designed has a very narrow 3-dB bandwidth of 4 Hz. This simple design gives a quick solution. Since the bandpass filter requires a passband ripple of 0.5 dB from 0.25 to 40 Hz, the BLT can also be an appropriate choice. Even though the impulse invariant method could work for this case, since the sampling rate of 600 Hz is much larger than 40 Hz, aliasing cannot be prevented completely. Hence, the BLT is a preferred design method for the bandpass filter.

8.13 SUMMARY

1. The BLT method is able to transform the transfer function of an analog filter to the transfer function of the corresponding digital filter in general.
2. The BLT maps the left half of an s-plane to the inside unit circle of the z-plane. Stability of mapping is guaranteed.
3. The BLT causes analog frequency warping. The analog frequency range from 0 Hz to infinite is warped to a digital frequency range from 0 Hz to the folding frequency.
4. Given the digital frequency specifications, analog filter frequency specifications must be developed using the frequency warping equation before designing the corresponding analog filter and applying the BLT.
5. An analog filter transfer function can be obtained by lowpass prototype, which can be selected from the Butterworth and Chebyshev functions.
6. The higher-order IIR filter can be designed using a cascade form.
7. The impulse invariant design method maps the analog impulse response to the digital equivalent impulse response. The method works for lowpass and bandpass filter design with a very high sampling rate. It is not appropriate for the highpass and bandstop filter design.
8. The pole-zero placement method can be applied for a simple IIR filter design such as the second-order bandpass and bandstop filters with narrow band specifications, first-order lowpass and highpass filters with the cutoff frequency close to either DC or the folding frequency.
9. Quantizing IIR filter coefficients explore the fact that the quantization of the filter coefficients has more effect on the magnitude frequency response than on the phase frequency response. It may cause the quantized IIR filter to be unstable.
10. A simple audio equalizer uses bandpass IIR filter banks to create sound effects.
11. The 60-Hz interference eliminator is designed to enhance biomedical ECG signals for heart rate detection. It can also be adapted for audio humming noise elimination.
12. A single tone or a DTMF tone can be generated using the IIR filter with the impulse sequence as the filter input.
13. The Goertzel algorithm is derived and applied for DTMF tone detection. This is an important application in the telecommunication industry.
14. The procedures for the BLT, impulse invariant, and pole-zero placement design methods were summarized, and their design feasibilities were compared, including the filter type, linear phase, ripple and stopband specifications, special requirements, algorithm complexity, and design tool(s).

8.14 PROBLEMS

8.1 Given an analog filter with the transfer function

$$H(s) = \frac{1000}{s + 1000},$$

convert it to the digital filter transfer function and difference equation using the BLT if the DSP system has a sampling period of $T = 0.001$ s.

8.2 The lowpass filter with a cutoff frequency of 1 rad/s is given as

$$H_P(s) = \frac{1}{s+1}.$$

(a) Use $H_P(s)$ and the BLT to obtain a corresponding IIR digital lowpass filter with a cutoff frequency of 30 Hz, assuming a sampling rate of 200 Hz.
(b) Use MATLAB to plot the magnitude and phase frequency responses of $H(z)$.

8.3 The normalized lowpass filter with a cutoff frequency of 1 rad/s is given as

$$H_P(s) = \frac{1}{s+1}.$$

(a) Use $H_P(s)$ and the BLT to obtain a corresponding IIR digital highpass filter with a cutoff frequency of 30 Hz, assuming a sampling rate of 200 Hz.
(b) Use MATLAB to plot the magnitude and phase frequency responses of $H(z)$.

8.4 Consider the normalized lowpass filter with a cutoff frequency of 1 rad/s:

$$H_P(s) = \frac{1}{s+1}.$$

(a) Use $H_P(s)$ and the BLT to design a corresponding IIR digital notch (bandstop) filter with a lower cutoff frequency of 20 Hz, an upper cutoff frequency of 40 Hz, and a sampling rate of 120 Hz.
(b) Use MATLAB to plot the magnitude and phase frequency responses of $H(z)$.

8.5 Consider the following normalized lowpass filter with a cutoff frequency of 1 rad/s:

$$H_P(s) = \frac{1}{s+1}.$$

(a) Use $H_P(s)$ and the BLT to design a corresponding IIR digital bandpass filter with a lower cutoff frequency of 15 Hz, an upper cutoff frequency of 25 Hz, and a sampling rate of 120 Hz.
(b) Use MATLAB to plot the magnitude and phase frequency responses of $H(z)$.

8.6 Design a first-order digital lowpass Butterworth filter with a cutoff frequency of 1.5 kHz and a passband ripple of 3 dB at a sampling frequency of 8000 Hz.
(a) Determine the transfer function and difference equation.
(b) Use MATLAB to plot the magnitude and phase frequency responses.

8.7 Design a second-order digital lowpass Butterworth filter with a cutoff frequency of 1.5 kHz and a passband ripple of 3 dB at a sampling frequency of 8000 Hz.
(a) Determine the transfer function and difference equation.
(b) Use MATLAB to plot the magnitude and phase frequency responses.

8.8. Design a third-order digital highpass Butterworth filter with a cutoff frequency of 2 kHz and a passband ripple of 3 dB at a sampling frequency of 8000 Hz.
(a) Determine the transfer function and difference equation.
(b) Use MATLAB to plot the magnitude and phase frequency responses.

8.9 Design a second-order digital bandpass Butterworth filter with a lower cutoff frequency of 1.9 kHz, an upper cutoff frequency 2.1 kHz, and a passband ripple of 3 dB at a sampling frequency of 8000 Hz.
(a) Determine the transfer function and difference equation.
(b) Use MATLAB to plot the magnitude and phase frequency responses.

8.10 Design a second-order digital bandstop Butterworth filter with a center frequency of 1.8 kHz, a bandwidth of 200 Hz, and a passband ripple of 3 dB at a sampling frequency of 8000 Hz.

(a) Determine the transfer function and difference equation.

(b) Use MATLAB to plot the magnitude and phase frequency responses.

8.11 Design a first-order digital lowpass Chebyshev filter with a cutoff frequency of 1.5 kHz and 1 dB ripple on passband at a sampling frequency of 8000 Hz.

(a) Determine the transfer function and difference equation.

(b) Use MATLAB to plot the magnitude and phase frequency responses.

8.12 Design a second-order digital lowpass Chebyshev filter with a cutoff frequency of 1.5 kHz and 0.5 dB ripple on passband at a sampling frequency of 8000 Hz. Use MATLAB to plot the magnitude and phase frequency responses.

8.13 Design a third-order digital highpass Chebyshev filter with a cutoff frequency of 2 kHz and 1 dB ripple on the passband at a sampling frequency of 8000 Hz.

(a) Determine the transfer function and difference equation.

(b) Use MATLAB to plot the magnitude and phase frequency responses.

8.14 Design a second-order digital bandpass Chebyshev filter with the following specifications:
Center frequency of 1.5 kHz
Bandwidth of 200 Hz
0.5-dB ripple on passband
Sampling frequency of 8000 Hz.

(a) Determine the transfer function and difference equation.

(b) Use MATLAB to plot the magnitude and phase frequency responses.

8.15 Design a second-order bandstop digital Chebyshev filter with the following specifications:
Center frequency of 2.5 kHz
Bandwidth of 200 Hz
1-dB ripple on stopband
Sampling frequency of 8000 Hz.

(a) Determine the transfer function and difference equation.

(b) Use MATLAB to plot the magnitude and phase frequency responses.

8.16 Design a fourth-order lowpass digital Butterworth filter with a cutoff frequency of 2 kHz, and the passband ripple of 3 dB at a sampling frequency at 8000 Hz.

1. Determine transfer function and difference equation.

2. Use MATLAB to plot the magnitude and phase frequency responses.

8.17 Design a fourth-order digital lowpass Chebyshev filter with a cutoff frequency of 1.5 kHz and a 0.5-dB ripple at a sampling frequency of 8000 Hz.

(a) Determine the transfer function and difference equation.

(b) Use MATLAB to plot the magnitude and phase frequency responses.

8.18 Design a fourth-order digital bandpass Chebyshev filter with a center frequency of 1.5 kHz, a bandwidth of 200 Hz, and a 0.5-dB ripple at a sampling frequency of 8000 Hz.

(a) Determine the transfer function and difference equation.

(b) Use MATLAB to plot the magnitude and phase frequency responses.

8.19 Consider the following Laplace transfer function:

$$H(s) = \frac{10}{s+10}.$$

(a) Determine $H(z)$ and the difference equation using the impulse invariant method if the sampling rate $f_s = 10\,\text{Hz}$.
(b) Use MATLAB to plot the magnitude frequency response $|H(f)|$ and the phase frequency response $\phi(f)$ with respect to $H(s)$ for the frequency range from 0 to $f_s/2\,\text{Hz}$.
(c) Use MATLAB to plot the magnitude frequency response $|H(e^{j\Omega})| = |H(e^{j2\pi fT})|$ and the phase frequency response $\phi(f)$ with respect to $H(z)$ for the frequency range from 0 to $f_s/2\,\text{Hz}$.

8.20 Consider the following Laplace transfer function:

$$H(s) = \frac{1}{s^2 + 3s + 2}.$$

(a) Determine $H(z)$ and the difference equation using the impulse invariant method if the sampling rate $f_s = 10\,\text{Hz}$.
(b) Use MATLAB to plot the magnitude frequency response $|H(f)|$ and the phase frequency response $\phi(f)$ with respect to $H(s)$ for the frequency range from 0 to $f_s/2\,\text{Hz}$.
(c) Use MATLAB to plot the magnitude frequency response $|H(e^{j\Omega})| = |H(e^{j2\pi fT})|$ and the phase frequency response $\phi(f)$ with respect to $H(z)$ for the frequency range from 0 to $f_s/2\,\text{Hz}$.

8.21 Consider the following Laplace transfer function:

$$H(s) = \frac{s}{s^2 + 4s + 5}.$$

(a) Determine $H(z)$ and the difference equation using the impulse invariant method if the sampling rate $f_s = 10\,\text{Hz}$.
(b) Use MATLAB to plot the magnitude frequency response $|H(f)|$ and the phase frequency response $\phi(f)$ with respect to $H(s)$ for the frequency range from 0 to $f_s/2\,\text{Hz}$;
(c) Use MATLAB to plot the magnitude frequency response $|H(e^{j\Omega})| = |H(e^{j2\pi fT})|$ and the phase frequency response $\phi(f)$ with respect to $H(z)$ for the frequency range from 0 to $f_s/2\,\text{Hz}$.

8.22 A second-order bandpass filter is required to satisfy the following specifications:
Sampling rate $= 8000\,\text{Hz}$
3-dB bandwidth: $BW = 100\,\text{Hz}$ Hz
Narrow passband centered at $f_0 = 2000\,\text{Hz}$
Zero gain at 0 and 4000 Hz.
Find the transfer function and difference equation by the pole-zero placement method.

8.23 A second-order notch filter is required to satisfy the following specifications:
Sampling rate $= 8000\,\text{Hz}$
3-dB bandwidth: $BW = 200\,\text{Hz}$
Narrow passband centered at $f_0 = 1000\,\text{Hz}$.
Find the transfer function and difference equation by the pole-zero placement method.

8.24 A first-order lowpass filter is required to satisfy the following specifications:
Sampling rate $= 8000\,\text{Hz}$
3-dB cutoff frequency: $f_c = 200\,\text{Hz}$
Zero gain at 4000 Hz.
Find the transfer function and difference equation using the pole-zero placement method.

8.25 A first-order lowpass filter is required to satisfy the following specifications:
Sampling rate $= 8000\,\text{Hz}$
3-dB cutoff frequency: $f_c = 3800\,\text{Hz}$

Zero gain at 4000 Hz.

Find the transfer function and difference equation by the pole-zero placement method.

8.26 A first-order highpass filter is required to satisfy the following specifications:

Sampling rate = 8000 Hz

3-dB cutoff frequency: $f_c = 3850$ Hz

Zero gain at 0 Hz.

Find the transfer function and difference equation by the pole-zero placement method.

8.27 A first-order highpass filter is required to satisfy the following specifications:

Sampling rate = 8000 Hz

3-dB cutoff frequency: $f_c = 100$ Hz

Zero gain at 0 Hz.

Find the transfer function and difference equation by the pole-zero placement method.

8.28 Given a filter transfer function

$$H(z) = \frac{0.3430z^2 + 0.6859z + 0.3430}{z^2 + 0.7075z + 0.7313},$$

(a) Realize the digital filter using direct-form I and direct-form II.

(b) Determine the difference equations for each implementation.

8.29 Given a fourth-order filter transfer function

$$H(z) = \frac{0.3430z^2 + 0.6859z + 0.3430}{z^2 + 0.7075z + 0.7313} \times \frac{0.4371z^2 + 0.8742z + 0.4371}{z^2 - 0.1316z + 0.1733},$$

(a) Realize the digital filter using the cascade (series) form via second-order sections using the direct-form II.

(b) Determine the difference equations for implementation.

8.30 Given a DSP system with a sampling rate of 1000 Hz, develop a 200-Hz single tone generator using the digital IIR filter by completing the following steps:

(a) Determine the digital IIR filter transfer function.

(b) Determine the DSP equation (difference equation).

8.31 Given a DSP system with a sampling rate of 8000 Hz, develop a 250-Hz single tone generator using the digital IIR filter by completing the following steps:

(a) Determine the digital IIR filter transfer function.

(b) Determine the DSP equation (difference equation).

8.32 Given a DSP system with a sampling rate of 8000 Hz, develop a DTMF tone generator for key 9 using the digital IIR filters by completing the following steps:

(a) Determine the digital IIR filter transfer functions.

(b) Determine the DSP equations (difference equation).

8.33 Given a DSP system with a sampling rate 8000 Hz, develop a DTMF tone generator for key 3 using the digital IIR filters by completing the following steps:

(a) Determine the digital IIR filter transfer functions.

(b) Determine the DSP equations (difference equation).

8.34 Given $x(0) = 1$, $x(1) = 2$, $x(2) = 0$, $x(3) = -1$, using the Goertzel algorithm to compute the following DFT coefficients and their amplitude spectrum:

(a) $X(0)$
(b) $|X(0)|^2$
(c) A_0 (single side)
(d) $X(1)$
(e) $|X(1)|^2$
(f) A_1 (single side)

8.35 Repeat Problem 8.34 for $X(1)$ and $X(3)$.

8.36 Given the digital data sequence of length 4 as $x(0)=4$, $x(1)=3$, $x(2)=2$, and $x(3)=1$, use the modified Goertzel algorithm to compute the spectral amplitude at the frequency bin $k=0$ and $k=2$.

8.37 Repeat Problem 8.36 for $X(1)$ and $X(3)$.

Use **MATLAB** to solve Problems 8.38–8.50.

8.38 A speech sampled at 8000 Hz is corrupted by a sine wave of 360 Hz. Design a notch filter to remove the noise with the following specifications:
Chebyshev notch filter
Center frequency: 360 Hz
Bandwidth: 60 Hz
Passband ripple: 0.5 dB
Stopband attenuation: 5 dB at 355 and 365 Hz, respectively
Determine the transfer function and difference equation.

8.39 In Problem 8.38, if the speech is corrupted by a sine wave of 360 Hz and its third harmonics, cascading two notch filters can be applied to remove noise signals. The possible specifications are given as follows:
Chebyshev notch filter 1
Center frequency: 360 Hz
Bandwidth: 60 Hz
Passband ripple: 0.5 dB
Stopband attenuation: 5 dB at 355 and 365 Hz, respectively
Chebyshev notch filter 2
Center frequency: 1080 Hz
Bandwidth: 60 Hz
Passband and ripple: 0.5 dB
Stopband attenuation: 5 dB at 1075 and 1085 Hz, respectively
Determine the transfer function and difference equation for each filter (Fig. 8.58).

8.40 In a speech recording system with a sampling frequency of 10,000 Hz, the speech is corrupted by random noise. To remove the random noise while preserving speech information, the following specifications are given:

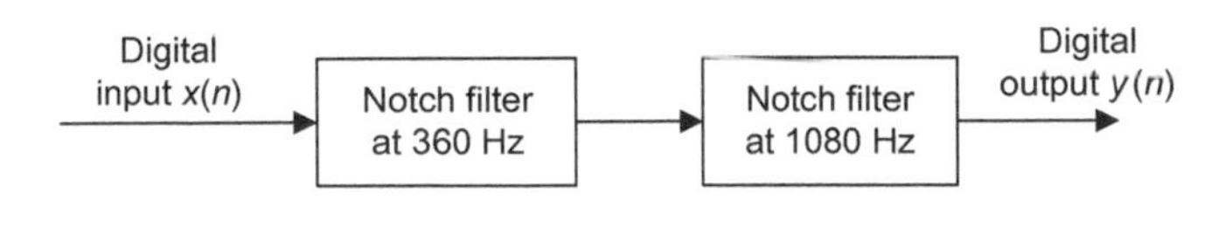

FIG. 8.58

Cascaded notch filter in Problem 8.39.

Speech frequency range: 0–3000 Hz
Stopband range: 4000–5000 Hz
Passband ripple: 3 dB
Stopband attenuation: 25 dB
Butterworth IIR filter
Determine the filter order and transfer function.

8.41 In Problem 8.40, if we use a Chebyshev IIR filter with the following specifications:
Speech frequency range: 0–3000 Hz
Stopband range: 4000–5000 Hz
Passband ripple: 1 dB
Stopband attenuation: 35 dB
Chebyshev IIR filter
Determine the filter order and transfer function.

8.42 Given a speech equalizer to compensate midrange frequency loss of hearing (Fig. 8.59) and the following specifications:
Sampling rate: 8000 Hz
Second-order bandpass IIR filter
Frequency range to be emphasized: 1500–2000 Hz
Passband ripple: 3 dB
Pole-zero placement design method
Determine the transfer function.

8.43 In Problem 8.42, if we use an IIR filter with following specifications:
Sampling rate: 8000 Hz
Butterworth IIR filter
Frequency range to be emphasized: 1500–2000 Hz
Lower stopband: 0–1000 Hz
Upper stopband: 2500–4000 Hz
Passband ripple: 3 dB
Stopband attenuation: 20 dB
Determine the filter order and filter transfer function.

8.44 A digital crossover can be designed as shown in Fig. 8.60.
Given audio specifications as
Sampling rate: 44,100 Hz
Crossover frequency: 1000 Hz
Highpass filter: third-order Butterworth type at a cutoff frequency of 1000 Hz
Lowpass filter: third-order Butterworth type at a cutoff frequency of 1000 Hz

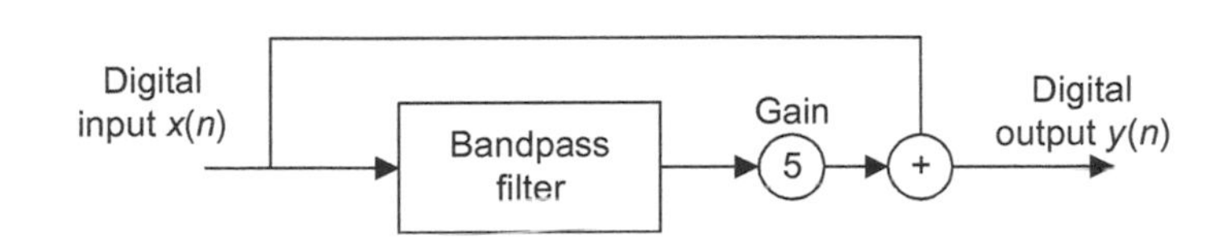

FIG. 8.59

Speech equalizer in Problem 8.42.

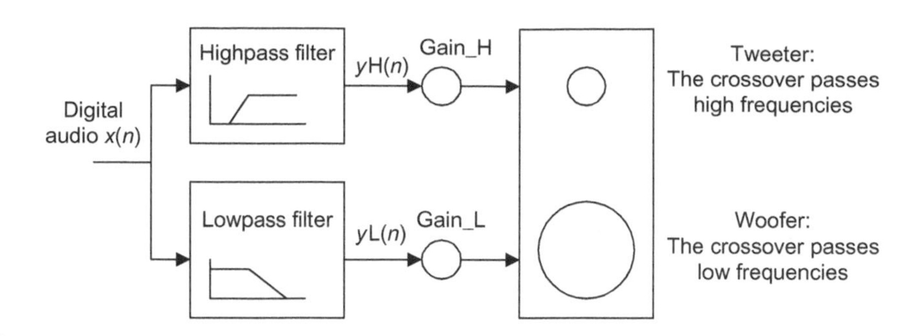

FIG. 8.60

Two-band digital crossover system in Problem 8.44.

Use the MATLAB BLT design method to determine:

(a) the transfer functions and difference equations for the highpass and lowpass filters;
(b) frequency responses for the highpass filter and the lowpass filter;
(c) combined frequency response for both filters.

8.45 Given a DSP system with a sampling rate of 8000 Hz, develop an 800 Hz single-tone generator using a digital IIR filter by completing the following steps:

(a) Determine the digital IIR filter transfer function.
(b) Determine the DSP equation (difference equation).
(c) Write a MATLAB program using the MATLAB function filter() to generate and plot the 800-Hz tone for a duration of 0.01 s.

8.46 Given a DSP system with a sampling rate set up to be 8000 Hz, develop a DTMF tone generator for key "5" using digital IIR filters by completing the following steps:

(a) Determine the digital IIR filter transfer functions.
(b) Determine the DSP equations (difference equation).
(c) Write a MATLAB program using the MATLAB function filter() to generate and plot the DTMF tone for key 5 for 205 samples.

8.47 Given $x(0)=1$, $x(1)=1$, $x(2)=0$, $x(3)=-1$, using the Goertzel algorithm to compute the following DFT coefficients and their amplitude spectra:

(a) $X(0)$
(b) $|X(0)|^2$
(c) A_0 (single sided)
(d) $X(1)$
(e) $|X(1)|^2$
(f) A_1 (single sided)

8.48 Repeat Problem 8.47 for single-side spectra: A_2 and A_3.

8.49 Given a DSP system with a sampling rate set up to be 8000 Hz and data size of 205 ($N=205$), seven Goertzel IIR filters are implemented for DTMF tone detection. For the frequencies corresponding to key 5, determine:

(a) the modified Goertzel filter transfer functions
(b) the filter DSP equations for $v_k(n)$

(c) the DSP equations for the squared magnitudes

$$|X(k)|^2 = |y_k(205)|^2$$

(d) using the data generated in Problem 8.46 (c), Write a program using the MATLAB function filter() and Goertzel algorithm to detect the spectral values of the DTMF tone for key 5.

8.50 Given an input data sequence:

$$x(n) = 1.2 \cdot \sin(2\pi(1000)n/10,000)) - 1.5 \cdot \cos(2\pi(4000)n/10,000)$$

assuming a sampling frequency of 10 kHz, implement the designed IIR filter in Problem 8.41 to filter 500 data points of $x(n)$ with the following specified method, and plot the 500 samples of the input and output data.

(a) Direct-form I implementation

(b) Direct-form II implementation.

MATLAB Projects

8.51 The 60-Hz hum eliminator with harmonics and heart rate detection

Given the recorded ECG data (ecgbn.dat) which is corrupted by 60-Hz interference with its harmonics and the sampling rate is 600 Hz, plot its spectrum and determine the harmonics. With the harmonic frequency information, design a notch filter to enhance the ECG signal. Then use the designed notch filter to process the given ECG signal and apply the zero-cross algorithm to determine the heart rate.

8.52 Digital speech and audio equalizer

Design a seven-band audio equalizer using fourth-order bandpass filters with a sampling rate of 44.1 kHz. The center frequencies are listed in Table 8.14.

In this project, use the designed equalizer to process a stereo audio ("No9seg.wav").

Plot the magnitude response for each filter bank.

Listen and evaluate the processed audio with the following gain settings:

(a) each filter bank gain = 0 (no equalization)

(b) lowpass filtered

(c) bandpass filtered

(d) highpass filtered

8.53 DTMF tone generation and detection

Implement the DTMF tone generation and detection according to Section 8.11 with the following specifications:

(a) Input keys: 1,2,3,4,5,6,7,8,9,*,0,#, A, B, C, D (key frequencies are given in Fig. 8.61).

(b) Sampling frequency is 8000 Hz.

(c) Program will respond each input key with its DTMF tone and display the detected key.

Table 8.14 Specification for Center Frequencies and Bandwidths

Center Frequency (Hz)	160	320	640	1280	2560	5120	10,240
Bandwidth (Hz)	80	160	320	640	1280	2560	5120

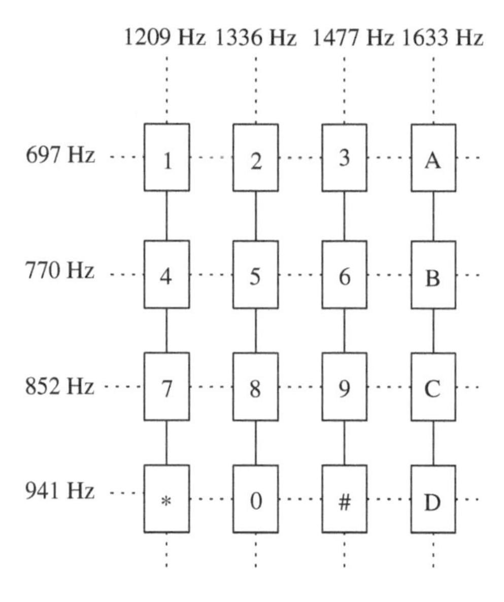

FIG. 8.61

DTMF key frequencies.

Advanced Problems

8.54 For the second-order IIR notch filter design using the pole-placement method show that the pole placement for 3-dB bandwidth is

$$\theta = \frac{f_0}{f_s} \times 360^\circ$$
$$r = 1 - \left(\frac{BW_{3dB}}{f_s}\right)\pi \text{ for } 0.9 < r < 1.$$

8.55 For the first-order IIR lowpass filter design with $f_s/4 < f_c < f_s/2$ using the pole-placement method show that the pole placement for 3-dB bandwidth is

$$\alpha = -\left(1 - \pi + \frac{2\pi f_c}{f_s}\right) \text{ for } -1 < \alpha < -0.9.$$

8.56 For the first-order IIR highpass filter design with $0 < f_c < f_s/4$ using the pole-placement method show that the pole placement for 3-dB bandwidth is

$$\alpha = 1 - 2\pi f_c / f_s \text{ for } 0.9 < \alpha < 1.$$

8.57 For the first-order IIR highpass filter design with $f_s/4 < f_c < f_s/2$ using the pole-placement method show that the pole placement for 3-dB bandwidth is

$$\alpha = -\left(1 - \pi + \frac{2\pi f_c}{f_s}\right) \text{ for } -1 < \alpha < -0.9.$$

8.58 Filter design by Padé approximation

Given the desired impulse response $h_d(n)$, $n \geq 0$ for an IIR filter, the IIR filter to be designed can be written as

$$H(z) = \frac{\sum_{k=0}^{M} b_k z^{-k}}{1 + \sum_{k=1}^{N} a_k z^{-k}} = \sum_{k=0}^{\infty} h(k) z^{-k}.$$

Let $x(n) = \delta(n)$, the impulse response $y(n) = h(n)$ can be expressed as

$$h(n) = -a_1 h(n-1) - a_2 h(n-2) - \cdots - a_N h(n-N)$$
$$+ b_0 \delta(n) + b_1 \delta(n-1) + \cdots + b_M \delta(n-M)$$

(a) Show that for $0 \leq n \leq M$

$$h(n) = -a_1 h(n-1) - a_2 h(n-2) - \cdots - a_N h(n-N) + b_n$$

(b) Show that for $n > M$

$$h(n) = -a_1 h(n-1) - a_2 h(n-2) - \cdots - a_N h(n-N)$$

(c) Let $h(n)$ match $h_d(n)$ for $0 \leq n \leq N+M$, that is, $h(n) = h_d(n)$, for $0 \leq n \leq N+M$. Show that

$$\begin{bmatrix} -h_d(M) & -h_d(M-1) & \cdots & -h_d(M+1-N) \\ -h_d(M+1) & -h_d(M) & \cdots & -h_d(M+2-N) \\ \cdots & \cdots & \ddots & \cdots \\ -h_d(M+N-1) & -h_d(M+N-2) & \cdots & -h_d(M) \end{bmatrix} \begin{bmatrix} a_1 \\ a_2 \\ \vdots \\ a_N \end{bmatrix} = \begin{bmatrix} h_d(M+1) \\ h_d(M+2) \\ \vdots \\ h_d(M+N) \end{bmatrix}$$

$$\begin{bmatrix} b_0 \\ b_1 \\ \vdots \\ b_M \end{bmatrix} = \begin{bmatrix} h_d(0) & 0 & \cdots & 0 \\ h_d(1) & h_d(0) & \cdots & 0 \\ \cdots & \cdots & \ddots & \cdots \\ h_d(M) & h_d(M-1) & \cdots & h_d(0) \end{bmatrix} \begin{bmatrix} 1 \\ a_1 \\ \vdots \\ a_M \end{bmatrix}.$$

8.59 Assume the desired unit impulse response is

$$h_d(n) = \frac{1}{n+1} u(n).$$

Design an IIR filter using the Pade approximation method with the following form:

$$H(z) = \frac{\sum_{k=0}^{5} b_k z^{-k}}{1 + \sum_{k=1}^{5} a_k z^{-k}}.$$

8.60 Shanks methods for least-mean squares filter design

Given the desired impulse response $h_d(n)$, $n \geq 0$ for an IIR filter, the IIR filter to be designed can be written as

$$H(z) = \frac{\sum_{k=0}^{M} b_k z^{-k}}{1 + \sum_{k=1}^{M} a_k z^{-k}} = \sum_{k=0}^{\infty} h(k) z^{-k}.$$

Find b_k and a_k such that the sum of squared errors between $h_d(n)$ and $h(n)$ is minimized.

(a) From Padé approximation, we can set

$$\tilde{h}_d(n) = -\sum_{k=1}^{N} a_k h_d(n-k).$$

Show that the equations determine a_k by minimizing the following squared errors:

$$E_1 = \sum_{n=M+1}^{\infty} \left[h_d(n) - \tilde{h}_d(n)\right]^2,$$

That is, for $m = 1, 2, \cdots, N$

$$\sum_{k=1}^{N} a_k \sum_{n=M+1}^{\infty} h_d(n-k) h_d(n-m) = -\sum_{n=M+1}^{\infty} h_d(n) h_d(n-m).$$

(b) To determine b_k, we first split $H(z)$ into $H_1(z)$ and $H_2(z)$, that is, $H(z) = H_1(z)H_2(z)$ where

$$H_1(z) \frac{1}{1 + \sum_{k=1}^{N} \tilde{a}_k z^{-k}} \text{ and } H_2(z) = \sum_{k=0}^{M} b_k z^{-k}.$$

Let $v(n)$ be the impulse response of $H_1(z)$, that is,

$$v(n) = -\sum_{k=1}^{N} \tilde{a}_k v(n-k) + \delta(n).$$

Then

$$\hat{h}_d(n) = \sum_{k=0}^{M} b_k v(n-k).$$

After minimizing the sum of squared error

$$E_2 = \sum_{n=0}^{\infty} \left[h_d(n) - \hat{h}_d(n)\right]^2,$$

Show that for $m = 0, 1, \cdots, M$

$$\sum_{k=0}^{M} b_k \sum_{n=0}^{\infty} v(n-k)v(n-m) = \sum_{n=0}^{\infty} h_d(n)v(n-m).$$

8.61 Assume that the desired unit impulse response is

$$h_d(n) = 2\left(\frac{1}{2}\right)^n u(n).$$

Design an IIR filter using the Shanks method with the following form:

$$H(z) = \frac{b_0 + b_1 z^{-1}}{1 + a_1 z^{-1}}.$$

CHAPTER

ADAPTIVE FILTERS AND APPLICATIONS

9

CHAPTER OUTLINE

9.1 INTRODUCTION TO LEAST MEAN SQUARE ADAPTIVE FINITE IMPULSE RESPONSE FILTERS

An *adaptive filter* is a digital filter that has self-adjusting characteristics. It is capable of adjusting its filter coefficients automatically to adapt the input signal via an adaptive algorithm. Adaptive filters play an important role in modern digital signal processing (DSP) products in areas such as telephone echo cancellation, noise cancellation, equalization of communications channels, biomedical signal enhancement, active noise control (ANC), and adaptive control systems. Adaptive filters work generally for the adaptation of signal-changing environments, spectral overlap between noise and signal, and unknown or time-varying noise. For example, when the interference noise is strong and its spectrum overlaps that of the desired signal, removing the interference using a traditional filter such as a notch filter with the fixed filter coefficients will fail to preserve the desired signal spectrum, as shown in Fig. 9.1.

Digital Signal Processing. https://doi.org/10.1016/B978-0-12-815071-9.00009-9

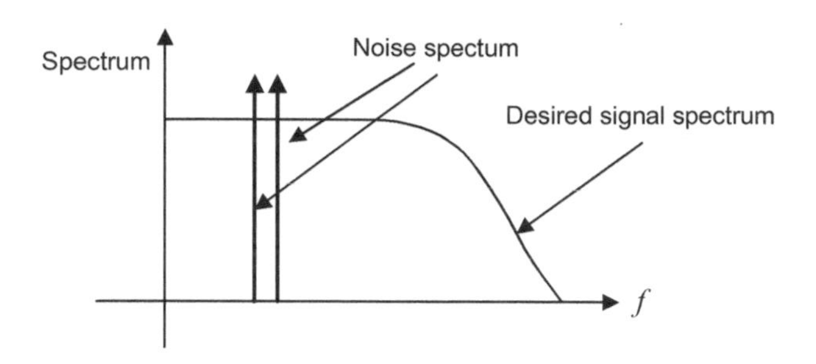

FIG. 9.1

Spectrum illustration for using adaptive filters.

However, an adaptive filter will do the job. Note that adaptive filtering, with its applications, has existed more than two decades in the research community and is still active. This chapter introduces some fundamentals of the subject, adaptive finite impulse response (FIR) filters with a simple and popular least mean square (LMS) algorithm and recursive least squares (RLS) algorithm. Further exploration into adaptive infinite impulse response (IIR) filters, adaptive lattice filters, their associated algorithms and applications, and so on, can be found in comprehensive texts by Haykin (2014), Stearns and Hush (2011), and Widrow and Stearns (1985).

To understand the concept of adaptive filtering, we will first look at an illustrative example of the simplest noise canceller to see how it works before diving into detail. The block diagram for such a noise canceller is shown in Fig. 9.2.

As shown in Fig. 9.2, first, the DSP system consists of two analog-to-digital conversion (ADC) channels. The first microphone with ADC is used to capture the desired speech $s(n)$. However, due to a noisy environment, the signal is contaminated and the ADC channel produces a signal with the noise; that is, $d(n)=s(n)+n(n)$. The second microphone is placed where only noise is picked up and the second ADC channel captures noise $x(n)$, which is fed to the adaptive filter.

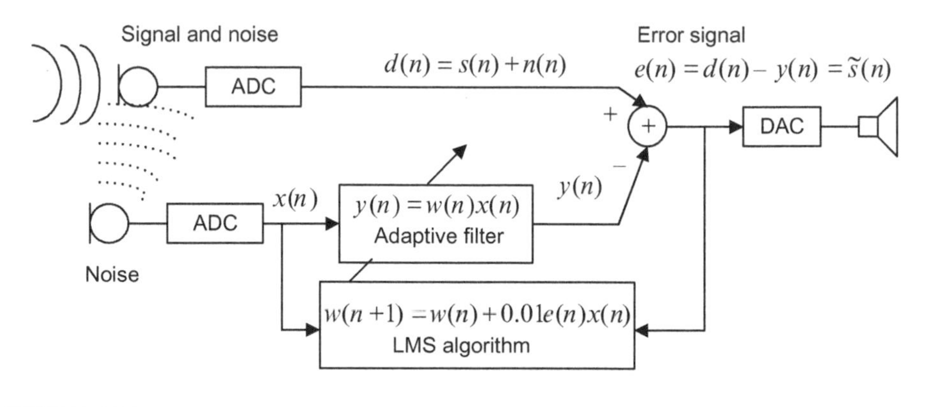

FIG. 9.2

Simplest noise canceller using a one-tap adaptive filter.

Note that the corrupting noise $n(n)$ in the first channel is uncorrelated to the desired signal $s(n)$, so that separation between them is possible. The noise signal $x(n)$ from the second channel is correlated to the corrupting noise $n(n)$ in the first channel, since both come from the same noise source. Similarly, the noise signal $x(n)$ is not correlated to the desired speech signal $s(n)$.

We assume that the corrupting noise in the first channel is a linear filtered version of the second-channel noise, since it has a different physical path from the second-channel noise, and the noise source is time varying, so that we can estimate the corrupting noise $n(n)$ using an adaptive filter. The adaptive filter contains a digital filter with adjustable coefficient(s) and the LMS algorithm to modify the value(s) of coefficient(s) for filtering each sample. The adaptive filter then produces an estimate of noise $y(n)$, which will be subtracted from the corrupted signal $d(n)=s(n)+n(n)$. When the noise estimate $y(n)$ equals or approximates the noise $n(n)$ in the corrupted signal, that is, $y(n)\approx n(n)$, the error signal $e(n)=s(n)+n(n)-y(n)\approx \tilde{s}(n)$ will approximate the clean speech signal $s(n)$. Hence, the noise is cancelled.

In our illustrative numerical example, the adaptive filter is set to be one-tap FIR filter to simplify numerical algebra. The filter adjustable coefficient $w(n)$ is adjusted based on the LMS algorithm (discussed later in detail) in the following:

$$w(n+1)=w(n)+0.01\times e(n)\times x(n),$$

where $w(n)$ is the coefficient used currently, while $w(n+1)$ is the coefficient obtained from the LMS algorithm and will be used for the next coming input sample. The value of 0.01 controls the speed of the coefficient change. To illustrate the concept of the adaptive filter in Fig. 9.2, the LMS algorithm has the initial coefficient set to be $w(0)=0.3$ and leads to

$$\begin{aligned}
&y(n)=w(n)x(n)\\
&e(n)=d(n)-y(n)\\
&w(n+1)=w(n)+0.01e(n)x(n).
\end{aligned}$$

The corrupted signal is generated by adding noise to a sine wave. The corrupted signal and noise reference are shown in Fig. 9.3, and their first 16 values are listed in Table 9.1.

Let us perform adaptive filtering for several samples using the values for the corrupted signal and reference noise in Table 9.1. We see that

$$\begin{aligned}
n=0,y(0)&=w(0)x(0)=0.3\times(-0.5893)=-0.1768\\
e(0)&=d(0)-y(0)=-0.2947-(-0.1768)=-0.1179=\tilde{s}(0)\\
w(1)&=w(0)+0.01e(0)x(0)=0.3+0.01\times(-0.1179)\times(-0.5893)=0.3007\\
n=1,y(1)&=w(1)x(1)=0.3007\times 0.5893=0.1772\\
e(1)&=d(1)-y(1)=1.0017-0.1772=0.8245=\tilde{s}(1)\\
w(2)&=w(1)+0.01e(1)x(1)=0.3007+0.01\times 0.8245\times 0.5893=0.3056\\
n=2,y(2)&=w(2)x(2)=0.3056\times 3.1654=0.9673\\
e(2)&=d(2)-y(2)=2.5827-0.9673=1.6155=\tilde{s}(2)\\
w(3)&=w(2)+0.01e(2)x(2)=0.3056+0.01\times 1.6155\times 3.1654=0.3567\\
n&=3,\cdots
\end{aligned}$$

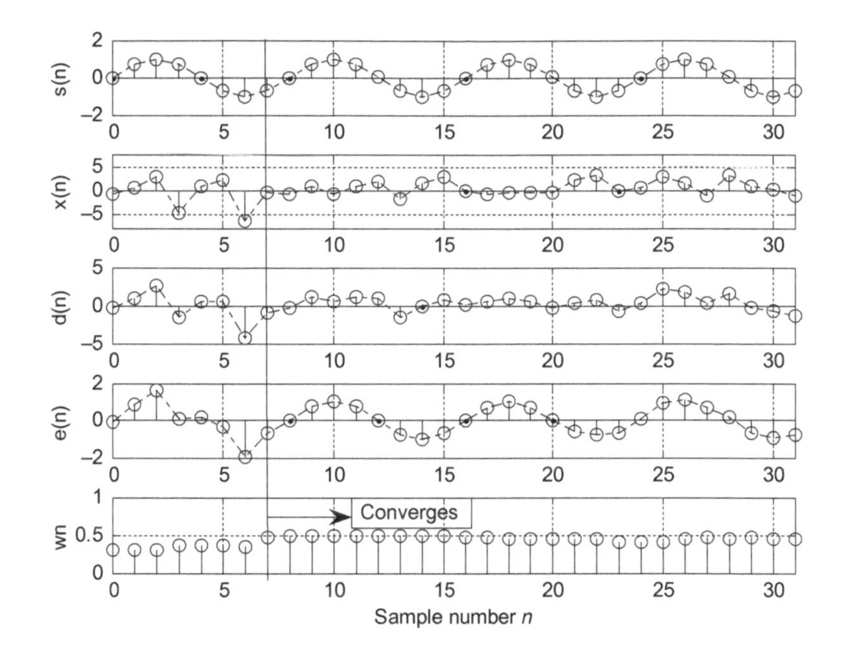

FIG. 9.3

Original signal $s(n)$, reference noise $x(n)$, corrupted signal $d(n)$, enhanced signal $e(n)$, and adaptive coefficient $w(n)$ in the noise cancellation.

Table 9.1 Adaptive Filtering Results for the Simplest Noise Canceller Example

n	$d(n)$	$x(n)$	$\tilde{s}(n)=e(n)$	Original $s(n)$	$w(n+1)$
0	−0.2947	−0.5893	−0.1179	0	0.3
1	1.0017	0.5893	0.8245	0.7071	0.3007
2	2.5827	3.1654	1.6155	1	0.3056
3	−1.6019	−4.6179	0.0453	0.7071	0.3567
4	0.5622	1.1244	0.1635	0	0.3546
5	0.4456	2.3054	−0.3761	−0.7071	0.3564
6	−4.2674	−6.5348	−1.9948	−1.0000	0.3478
7	−0.8418	−0.2694	−0.7130	−0.7071	0.4781
8	−0.3862	−0.7724	−0.0154	−0.0000	0.48
9	1.2274	1.0406	0.7278	0.7071	0.4802
10	0.6021	−0.7958	0.9902	1	0.4877
11	1.1647	0.9152	0.7255	0.7071	0.4799
12	0.963	1.926	0.026	0	0.4865
13	−1.5065	−1.5988	−0.7279	−0.7071	0.487
14	−0.1329	1.7342	−0.9976	−1.0000	0.4986
15	0.8146	3.0434	−0.6503	−0.7071	0.4813

For comparison, results of the first 16 processed output samples, original samples, and filter coefficient values are also included in Table 9.1. Fig. 9.3 also shows the original signal samples, reference noise samples, corrupted signal samples, enhanced signal samples, and filter coefficient values for each incoming sample, respectively.

As shown in Fig. 9.3, after seven adaptations, the adaptive filter learns noise characteristics and cancels the noise in the corrupted signal. The adaptive coefficient is close to the optimal value of 0.5. The processed output is close to the original signal. The first 16 processed values for corrupted signal, reference noise, clean signal, original signal, and adaptive filter coefficient used at each step are listed in Table 9.1.

Clearly, the enhanced signal samples look much like the sinusoid input samples. Now our simplest one-tap adaptive filter works for this particular case. In general, an FIR filter with multiple taps is used and has the following format:

$$y(n)=\sum_{k=0}^{N-1} w_k(n)x(n-k)=w_0(n)x(n)+w_1(n)x(n-1)+\cdots+w_{N-1}(n)x(n-N+1). \tag{9.1}$$

The LMS algorithm for the adaptive FIR filter will be developed next.

9.2 BASIC WIENER FILTER THEORY AND ADAPTIVE ALGORITHMS

In this section, we will first study Wiener filter theory and linear prediction. Then we will develop standard adaptive algorithms such as steepest decent algorithm, LMS algorithm, and RLS algorithm.

9.2.1 WIENER FILTER THEORY AND LINEAR PREDICTION

9.2.1.1 Basic Wiener Filter Theory

Many adaptive algorithms can be viewed as approximations of the discrete Wiener filter shown in Fig. 9.4, where the Wiener filter output $y(n)$ is a sum of its N weighted inputs, that is,

$$y(n)=w_0x(n)+w_1x(n-1)+\cdots+w_{N-1}x(n-N+1).$$

The Wiener filter adjusts its weight(s) to produce filter output $y(n)$, which would be as close as possible to the noise $n(n)$ contained in the corrupted signal $d(n)$. Hence, at the subtracted output, the noise is cancelled and the output $e(n)$ contains clean signal.

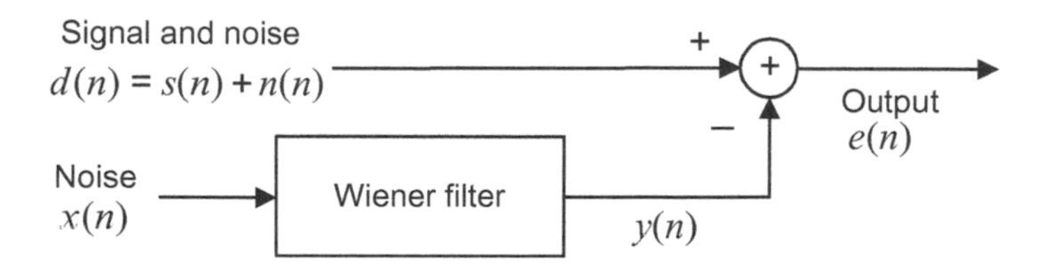

FIG. 9.4

Wiener filter for noise cancellation.

Consider a single-weight case of $y(n) = wx(n)$, and note that the error signal $e(n)$ is given by

$$e(n) = d(n) - wx(n). \tag{9.2}$$

Now let us solve the best weight w^*. Taking the square of the output error leads to

$$e^2(n) = (d(n) - wx(n))^2 = d^2(n) - 2d(n)wx(n) + w^2x^2(n). \tag{9.3}$$

Taking the statistical expectation of Eq. (9.3), we have

$$E(e^2(n)) = E(d^2(n)) - 2wE(d(n)x(n)) + w^2E(x^2(n)). \tag{9.4}$$

Using the notations in statistics, we define

$$\begin{aligned}
&J = E(e^2(n)) = MSE = \text{mean squared error}\\
&\sigma^2 = E(d^2(n)) = \text{power of corrupted signal}\\
&P = E(d(n)x(n)) = \text{cross-correlation between } d(n) \text{ and } x(n)\\
&R = E(x^2(n)) = \text{auto-correlation}
\end{aligned}$$

We can view the statistical expectation as an average of the N signal terms, each being a product of two individual samples:

$$E(e^2(n)) = \frac{e^2(0) + e^2(1) + \cdots + e^2(N-1)}{N}$$

or

$$E(d(n)x(n)) = \frac{d(0)x(0) + d(1)x(1) + \cdots + d(N-1)x(N-1)}{N}.$$

For a sufficiently large sample number of N, we can write Eq. (9.4) as

$$J = \sigma^2 - 2wP + w^2R. \tag{9.5}$$

Since σ^2, P, and R are constants, J is a quadratic function of w which may be plotted in Fig. 9.5A.

The best weight (optimal) w^* is at the location where the minimum MSE $J_{\min}$ is achieved. To obtain w^*, taking the derivative of J and setting it to zero leads to

$$\frac{dJ}{dw} = -2P + 2wR = 0. \tag{9.6}$$

Solving Eq. (9.6), we get the best weight solution as

$$w^* = R^{-1}P. \tag{9.7}$$

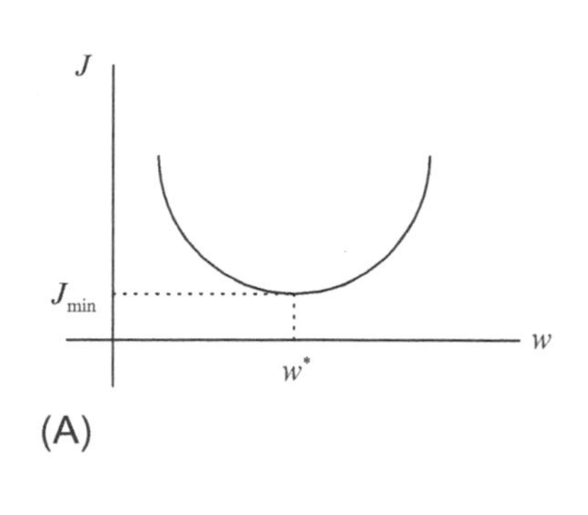

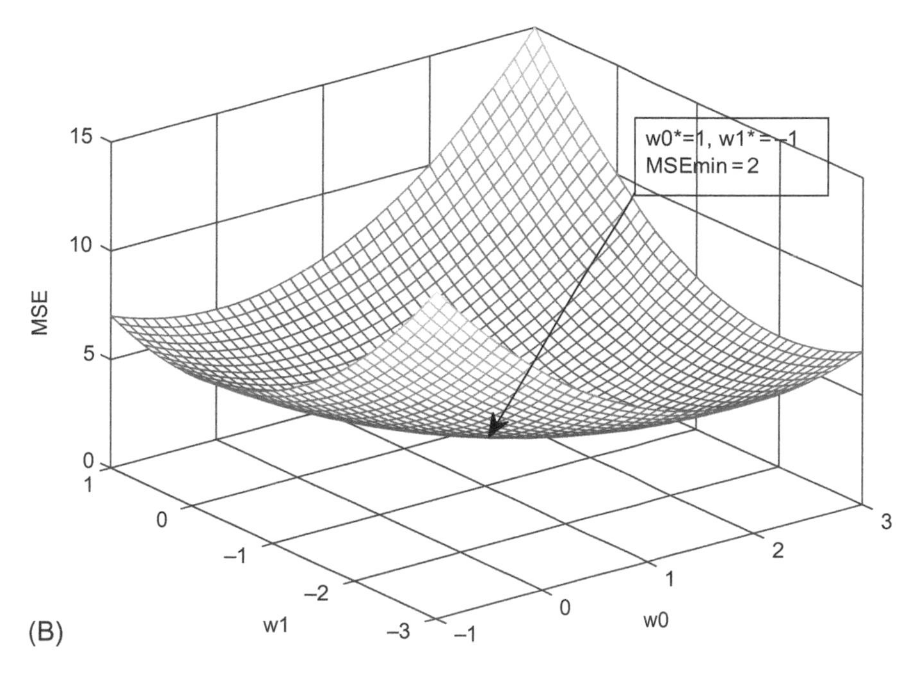

FIG. 9.5

(A) Mean square error quadratic function. (B) Plot of the MSE function vs. two weights.

EXAMPLE 9.1

Given a quadratic MSE function for the Wiener filter:

$$J = 40 - 20w + 10w^2,$$

Find the optimal solution for w^* to achieve the minimum MSE $J_{\min}$ and determine $J_{\min}$.

Solution:

Taking the derivative of the MSE function and setting it to zero, we have

$$\frac{dJ}{dw} = -20 + 10 \times 2w = 0.$$

Continued

EXAMPLE 9.1—CONT'D

Solving the equation leads to

$$w^* = 1.$$

Finally, substituting $w^* = 1$ into the MSE function, we get the minimum $J_{\min}$ as

$$J_{\min} = J|_{w=w^*} = 40 - 20w + 10w^2\Big|_{w=1} = 40 - 20\times 1 + 10\times 1^2 = 30.$$

In general, the Wiener filter has N coefficients. The filter output from Eq. (9.1) in a matrix form can be written as

$$y(n) = \sum_{k=0}^{N-1} w_k x(n-k) = W^T X(n), \tag{9.8}$$

where $X(n) = [x(n)x(n-1)\cdots x(n-N+1)]^T$ and $W = [w_0 w_1 \cdots w_{N-1}]^T$ are the input and coefficient vectors, respectively. Taking this modification into the MSE function, it follows that

$$\begin{aligned} J &= E\{e^2(n)\} = E\left\{\left(d(n) - W^T X(n)\right)^2\right\} \\ &= E\{d^2(n)\} - 2W^T E\{X(n)d(n)\} + W^T E\{X^T(n)X(n)\}W. \end{aligned} \tag{9.9}$$

The MSE function of Eq. (9.9) can be rewritten as

$$J = \sigma^2 - 2W^T P + W^T R W, \tag{9.10}$$

where $R = E\{X^T(n)X(n)\}$ and $P = E\{X(n)d(n)\}$ are the autocorrelation matrix and cross-correlation vector, respectively. Setting the derivative of the MSE function over W to zero, we obtain

$$\frac{\partial J}{\partial W} = -2P + RW = 0. \tag{9.11}$$

Finally, the optimal solution is obtained by

$$W^* = R^{-1}P, \tag{9.12}$$

and the minimum MSE can be derived by substituting Eq. (9.12) into Eq. (9.10):

$$\begin{aligned} J_{\min} &= \sigma^2 - 2W^{*T}P + W^{*T}RW^* \\ &= \sigma^2 - 2W^{*T}P + W^{*T}RR^{-1}P \\ &= \sigma^2 - W^{*T}P. \end{aligned} \tag{9.13}$$

Next, we examine the MSE function assuming the following statistical data:

$$\begin{aligned} &\sigma^2 = E[d^2(n)] = 6,\ \ E[x^2(n)] = E[x^2(n-1)] = 1,\ \ E[x(n)x(n-1)] = 1/2, \\ &E[d(n)x(n)] = 1/2,\ \text{and}\ E[d(n)x(n)] = 1/2, \end{aligned}$$

for the two-tap adaptive filter $y(n) = w_0 x(n) + w_1 x(n-1)$. Notice that

$$R = \begin{bmatrix} 1 & 0 \\ 0 & 1 \end{bmatrix} \text{and}\, P = \begin{bmatrix} 1/2 \\ -1/2 \end{bmatrix}.$$

Applying Eq. (9.12), we yield

$$\begin{bmatrix} w_0^* \\ w_1^* \end{bmatrix} = R^{-1}P = \begin{bmatrix} 1 & 1/2 \\ 1/2 & 1 \end{bmatrix}^{-1} \begin{bmatrix} 1/2 \\ -1/2 \end{bmatrix} = \begin{bmatrix} 1 \\ -1 \end{bmatrix}.$$

Using Eq. (9.10), it follows that

$$\begin{aligned} J &= \sigma^2 - 2W^T P + W^T R W \\ &= 4 - 2[w_0 \quad w_1] \begin{bmatrix} 1/2 \\ -1/2 \end{bmatrix} + [w_0 \quad w_1] \begin{bmatrix} 1 & 1/2 \\ 1/2 & 1 \end{bmatrix} \begin{bmatrix} w_0 \\ w_1 \end{bmatrix} = 4 + w_0^2 + w_0 w_1 + w_1^2 - w_0 + w_1. \end{aligned}$$

Eq. (9.13) gives

$$J_{\min} = \sigma^2 - W^{*T} P = 4 - [2 \quad -2] \begin{bmatrix} 1/2 \\ -1/2 \end{bmatrix} = 2.$$

Fig. 9.5B shows the MSE function vs. the weights, where the optimal weights and the minimum MSE are $w_0^* = 1$, $w_1^* = -1$, and $J_{\min} = 2$. The plot also indicates that the MSE function is a quadratic function and that there exists only one minimum of the MSE surface.

Notice that a few points need to be clarified for Eq. (9.12):

1. Optimal coefficient (s) can be different for every block of data, since the corrupted signal and reference signal are unknown. The autocorrelation and cross-correlation may vary.
2. If a larger number of coefficients (weights) are used, the inverse matrix of R^{-1} may require a larger number of computations and may become ill conditioned. This will make real-time implementation impossible.
3. The optimal solution is based on the statistics, assuming that the size of the data block, N, is sufficiently long. This will cause a long processing delay that will hinder real-time implementation.

9.2.1.2 Forward Linear Prediction

Linear prediction deals with the problem of predicting a future value of a stationary random process using the observed past values. Let us consider a one-step forward linear predictor in which the prediction of the value $x(n)$ by using a weighted linear combination of the past values, that is, $x(n-1)$, $x(n-2)$, …, $x(n-P)$. We will only focus on the treatment of the forward linear predictor, and others such as backward linear prediction, and forward and backward prediction can be found in Haykin (2014) and Proakis and Manolakis (2007). The output of the one-step forward linear predictor can be written as

$$\hat{x}(n) = -\sum_{k=1}^{P} a_k x(n-k), \tag{9.14}$$

where $-a_k$ are the prediction coefficients and P is the order of the forward linear predictor. Fig. 9.6 shows the block diagram of the one-step forward linear prediction.

As shown in Fig. 9.6, the difference between the value $x(n)$ and the predicted value $\hat{x}(n)$ is referred to as the forward prediction error, which is defined as

$$\begin{aligned} e_f(n) &= x(n) - \hat{x}(n) \\ &= x(n) + \sum_{k=1}^{P} a_k x(n-k). \end{aligned} \tag{9.15}$$

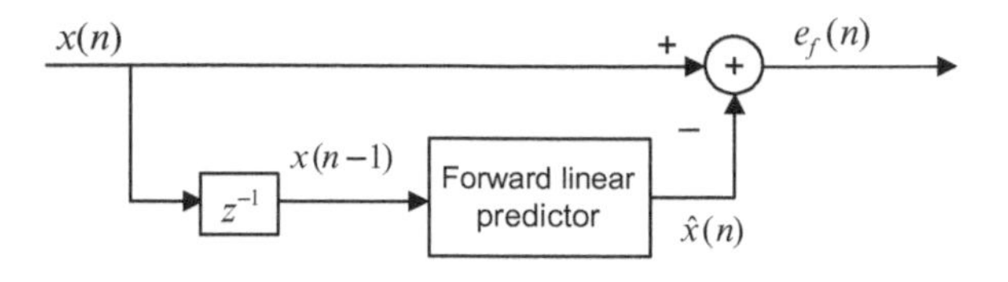

FIG. 9.6

One-step forward linear prediction.

We can minimize $E\{e_f^2(n)\}$ to find the prediction coefficients. Taking the derivative of $E\{e_f^2(n)\}$ to a_p, it follows that

$$\begin{aligned} E\left\{\frac{\partial e_f^2(n)}{\partial a_p}\right\} &= E\left\{e_f(k)\frac{\partial e_f(n)}{\partial a_p}\right\} \\ &= E\{e_f(k)x(n-p)\} \text{ for } p=1,2,\cdots,P \\ &= E\left\{\left[x(n)+\sum_{k=1}^{P} a_k x(n-k)\right]x(n-p)\right\}. \end{aligned} \tag{9.16}$$

Distribute the expectation in Eq. (9.16), we obtain

$$E\left\{\frac{\partial e_f^2(n)}{\partial a_p}\right\} = E\{x(n)x(n-p)\} + \sum_{k=1}^{P} a_k E\{x(n-k)x(n-p)\} \text{ for } p=1,2,\cdots,P. \tag{9.17}$$

Setting Eq. (9.17) to zero, we obtain

$$\begin{bmatrix} r_{xx}(0) & r_{xx}(-1) & \cdots & r_{xx}(-P+1) \\ r_{xx}(1) & r_{xx}(0) & \cdots & r_{xx}(-P+2) \\ \vdots & \vdots & \ddots & \vdots \\ r_{xx}(P-1) & r_{xx}(P-2) & \cdots & r_{xx}(0) \end{bmatrix} \begin{bmatrix} a_1 \\ a_2 \\ \vdots \\ a_P \end{bmatrix} = -\begin{bmatrix} r_{xx}(1) \\ r_{xx}(2) \\ \vdots \\ r_{xx}(P) \end{bmatrix}. \tag{9.18}$$

where $r_{xx}(p)=E\{x(n)x(n-p)\}$.

For the stationary process of $x(n)$, we use $r_{xx}(p)=r_{xx}(-p)$. Eq. (9.18) becomes

$$\begin{bmatrix} r_{xx}(0) & r_{xx}(1) & \cdots & r_{xx}(P-1) \\ r_{xx}(1) & r_{xx}(0) & \cdots & r_{xx}(P-2) \\ \vdots & \vdots & \ddots & \vdots \\ r_{xx}(P-1) & r_{xx}(P-2) & \cdots & r_{xx}(0) \end{bmatrix} \begin{bmatrix} a_1 \\ a_2 \\ \vdots \\ a_P \end{bmatrix} = -\begin{bmatrix} r_{xx}(1) \\ r_{xx}(2) \\ \vdots \\ r_{xx}(P) \end{bmatrix}. \tag{9.19}$$

We refer the following matrix:

$$R = \begin{bmatrix} r_{xx}(0) & r_{xx}(1) & \cdots & r_{xx}(P-1) \\ r_{xx}(1) & r_{xx}(0) & \cdots & r_{xx}(P-2) \\ \vdots & \vdots & \ddots & \vdots \\ r_{xx}(P-1) & r_{xx}(P-2) & \cdots & r_{xx}(0) \end{bmatrix}, \tag{9.20}$$

as the autocorrelation matrix. The solution for the prediction coefficients is given by

$$\begin{bmatrix} a_1 \\ a_2 \\ \vdots \\ a_P \end{bmatrix} = -R^{-1} \begin{bmatrix} r_{xx}(1) \\ r_{xx}(2) \\ \vdots \\ r_{xx}(P) \end{bmatrix}. \tag{9.21}$$

We can also extend the one-step forward linear prediction to a Δ-step forward linear prediction by considering the following predictor output:

$$\hat{x}(n) = -\sum_{k=1}^{P} a_k x(n - \Delta - k + 1), \tag{9.22}$$

where $\Delta \geq 1$. Following the similar derivation, we achieve

$$\begin{bmatrix} r_{xx}(0) & r_{xx}(-1) & \cdots & r_{xx}(-P+1) \\ r_{xx}(1) & r_{xx}(0) & \cdots & r_{xx}(-P+2) \\ \vdots & \vdots & \ddots & \vdots \\ r_{xx}(P-1) & r_{xx}(P-2) & \cdots & r_{xx}(0) \end{bmatrix} \begin{bmatrix} a_1 \\ a_2 \\ \vdots \\ a_P \end{bmatrix} = -\begin{bmatrix} r_{xx}(\Delta) \\ r_{xx}(\Delta+1) \\ \vdots \\ r_{xx}(\Delta+P-1) \end{bmatrix} \tag{9.23}$$

$$\begin{bmatrix} a_1 \\ a_2 \\ \vdots \\ a_P \end{bmatrix} = -R^{-1} \begin{bmatrix} r_{xx}(\Delta) \\ r_{xx}(\Delta+1) \\ \vdots \\ r_{xx}(\Delta+P-1) \end{bmatrix}. \tag{9.24}$$

EXAMPLE 9.2

A second-order forward linear predictor is used to predict the sinusoid $x(n) = \sqrt{2}\sin(\pi n/6)$.

(a) For the one-step forward linear predictor, find the predictor coefficients and show that the sinusoid can be fully predicted by showing $E\{e_f^2(n)\} = 0$

(b) Compute $x(n)$ and $\hat{x}(n)$ for 10 samples for verification.

Solution:

(a) The second-order predictor is given by

$$\hat{x}(n) = -a_1 x(n-1) - a_2 x(n-2).$$

Since

$$r_{xx}(0) = E\left\{\left[\sqrt{2}\sin(\pi n/6)\right]^2\right\} = 1$$
$$r_{xx}(1) = E\left\{\left[\sqrt{2}\sin(n\pi/6)\right]\left[\sqrt{2}\sin((n-1)\pi/6)\right]\right\} = \cos(\pi/6) = \sqrt{3}/2$$
$$r_{xx}(2) = E\left\{\sqrt{2}\sin(n\pi/6)\left[\sqrt{2}\sin((n-2)\pi/6)\right]\right\} = \cos(2\pi/6) = 1/2.$$

Therefore,

Continued

EXAMPLE 9.2—CONT'D

$$R = \begin{bmatrix} 1 & \sqrt{3}/2 \\ \sqrt{3}/2 & 1 \end{bmatrix}$$

$$\begin{bmatrix} a_1 \\ a_2 \end{bmatrix} = -\begin{bmatrix} 1 & \sqrt{3}/2 \\ \sqrt{3}/2 & 1 \end{bmatrix}^{-1} \begin{bmatrix} \sqrt{3}/2 \\ 1/2 \end{bmatrix}$$

$$= -\begin{bmatrix} 4 & -2\sqrt{3} \\ -2\sqrt{3} & 4 \end{bmatrix} \begin{bmatrix} \sqrt{3}/2 \\ 1/2 \end{bmatrix} = \begin{bmatrix} -\sqrt{3} \\ 1 \end{bmatrix}.$$

Thus, the predictor is found to be

$$\hat{x}(n) = \sqrt{3}x(n-1) - x(n-2).$$

To show the performance of the predictor, we can apply the Wiener filter results by considering

$$\sigma^2 = r_{xx}(0) = 1,\ R = \begin{bmatrix} 1 & \sqrt{3}/2 \\ \sqrt{3}/2 & 1 \end{bmatrix},\ \text{and}\ P = \begin{bmatrix} \sqrt{3}/2 \\ 1/2 \end{bmatrix}$$

$$W^* = -\begin{bmatrix} a_1 \\ a_2 \end{bmatrix} = \begin{bmatrix} \sqrt{3} \\ -1 \end{bmatrix}.$$

Finally, the MSE of prediction results in

$$J_{\min} = \sigma^2 - W^{*T}P = 1 - \begin{bmatrix} \sqrt{3} & -1 \end{bmatrix} \begin{bmatrix} \sqrt{3}/2 \\ 1/2 \end{bmatrix} = 0.$$

(b) Using MATLAB, we have

```
>>x=sin(pi*(0:9)/6)
x =
0 0.5000 0.8660 1.0000 0.8660 0.5000 0.0000 -0.5000 -0.8.660 -1.0000
xhat=filter([0 sqrt(3) -1],1,x)
xhat = 0 0 0.8660 1.0000 0.8660 0.5000 -0.0000 -0.5000 -0.8660 -1.0000
```

It is observed that a sinusoid can fully be predicted by a second-order one-step forward predictor after P samples ($P=2$).

In general, for predicting a sinusoid $x(n) = A\sin(n\Omega)$, a forward predictor can be used as

$$x(n) = -a_1x(n-n_1) - a_2x(n-n_2) \text{ for } 0 < n_1 < n_2. \tag{9.25}$$

We can show that the predictor coefficients are

$$\begin{bmatrix} a_1 \\ a_2 \end{bmatrix} = -\begin{bmatrix} \dfrac{\sin(n_2\Omega)}{\sin[(n_2-n_1)\Omega]} \\ -\dfrac{\sin(n_1\Omega)}{\sin[(n_2-n_1)\Omega]} \end{bmatrix} \text{ for } (n_2-n_1)\Omega \neq k\pi \text{ for } k = 0, \pm 1, \pm 2, \ldots \tag{9.26}$$

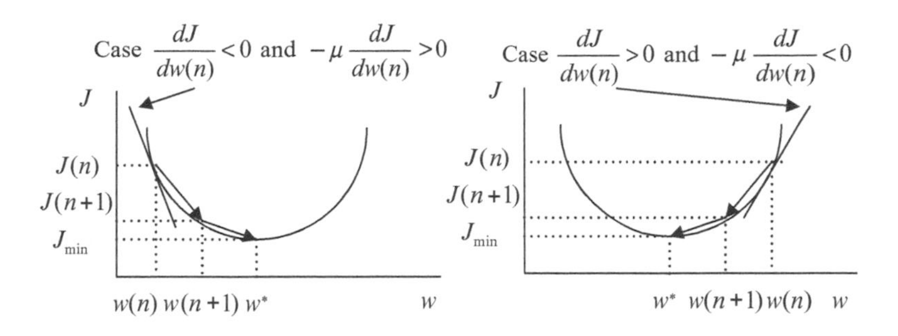

FIG. 9.7

Illustration of the steepest descent algorithm.

9.2.2 STEEPEST DESCENT ALGORITHM

As we pointed out, solving the Wiener solution Eq. (9.12) requires a lot of computations, including matrix inversion for a general multi-tap FIR filter. The well-known textbook authored by Widrow and Stearns (1985) described a powerful LMS algorithm by using the steepest descent algorithm to minimize the MSE sample by sample to locate the filter coefficient(s). We first study the steepest descent algorithm using a single weight as illustrated in Eq. (9.27):

$$w(n+1) = w(n) - \mu \frac{dJ}{dw(n)}, \tag{9.27}$$

where μ = constant controlling the speed of convergence.

The illustration of the steepest decent algorithm for solving the optimal coefficient(s) is described in Fig. 9.7.

As shown in the first plot of Fig. 9.7, if $\frac{dJ}{dw(n)} < 0$, notice that $-\mu\frac{dJ}{dw(n)} > 0$. The new coefficient $w(n+1)$ will be increased to approach the optimal value w^* by Eq. (9.27). On the other hand, if $\frac{dJ}{dw(n)} > 0$, as shown in the second plot of Fig. 9.7, we see that $-\mu\frac{dJ}{dw(n)} < 0$. The new coefficient $w(n+1)$ will be decreased to approach the optimal value w^*. When $\frac{dJ}{dw(n)} = 0$, the best coefficient $w(n+1)$ is reached.

EXAMPLE 9.3

Given a quadratic MSE function for the Wiener filter:

$$J = 40 - 20w + 10w^2,$$

Use the steepest decent method with an initial guess as $w(0) = 0$ and $\mu = 0.04$ to find the optimal solution for w^* and determine J_{min} by iterating three times.

Solution:

Taking a derivative of the MSE function, we have

$$\frac{dJ}{dw(n)} = -20 + 10 \times 2w(n).$$

Continued

EXAMPLE 9.3—CONT'D

When $n=0$, we calculate

$$\mu\frac{dJ}{dw(0)}=0.04\times(-20+10\times 2w(0))\Big|_{w(0)=0}=-0.8.$$

Applying the steepest decent algorithm, it follows that

$$w(1)=w(0)-\mu\frac{dJ}{dw(0)}=0-(-0.8)=0.8.$$

Similarly, for $n=1$, we get

$$\mu\frac{dJ}{dw(1)}=0.04\times(-20+10\times 2w(1))\Big|_{w(1)=0.8}=-0.16$$
$$w(2)=w(1)-\mu\frac{dJ}{dw(1)}=0.8-(-0.16)=0.96,$$

and for $n=2$, it follows that

$$\mu\frac{dJ}{dw(2)}=0.04\times(-20+10\times 2w(2))\Big|_{w(2)=0.96}=-0.032$$
$$w(3)=w(2)-\mu\frac{dJ}{dw(2)}=0.96-(-0.032)=0.992.$$

Finally, substituting $w^*\approx w(3)=0.992$ into the MSE function, we get the minimum $J_{\min}$ as

$$J_{\min}\approx 40-20w+10w^2\big|_{w=0.992}=40-20\times 0.992+10\times 0.992^2=30.0006.$$

As we can see, after three iterations, the filter coefficient and minimum MSE values are very close to the theoretical values obtained in Example 9.1.

For general case, that is, $y(n)=W(n)^TX(n)$, we have

$$\begin{bmatrix} w_0(n+1) \\ w_1(n+1) \\ \vdots \\ w_{N-1}(n+1) \end{bmatrix} = \begin{bmatrix} w_0(n) \\ w_1(n) \\ \vdots \\ w_{N-1}(n) \end{bmatrix} - \mu \begin{bmatrix} \partial J/\partial w_0(n) \\ \partial J/\partial w_1(n) \\ \vdots \\ \partial J/\partial w_{N-1}(n) \end{bmatrix}. \tag{9.28}$$

In a vector form, we yield

$$W(n+1)=W(n)-\mu\nabla J_{W(n)}, \tag{9.29}$$

where $\nabla J_{W(n)}$ is the gradient vector.

EXAMPLE 9.4

Given a quadratic MSE function for the Wiener filter:

$$J = 4 + w_0^2 + w_0 w_1 + w_1^2 - w_0 + w_1,$$

(a) Set up the steepest decent algorithm.

(b) For initial weights as $w_0(0) = 0$ and $w_1(0) = -3$, and $\mu = 0.2$, use MATLAB to find the optimal solution by iteration 100 times and plot the trajectory of $w_0(n)$ and $w_1(n)$ on the MSE contour surface.

Solution:

(a) Taking partial derivatives of the MSE function, we obtain the gradients as

$$\frac{\partial J}{\partial w_0(n)} = 2w_0(n) + w_1(n) - 1$$

and

$$\frac{\partial J}{\partial w_1(n)} = 2w_1(n) + w_0(n) + 1.$$

Then we obtain the coefficients updating equations from Eq. (9.28) as follows:

$$w_0(n+1) = w_0(n) - \mu[2w_0(n) + w_1(n) - 1]$$
$$w_1(n+1) = w_1(n) - \mu[2w_1(n) + w_0(n) + 1]$$

(b) The trajectory of $w_0(n)$ and $w_1(n)$ are plotted in Fig. 9.8.

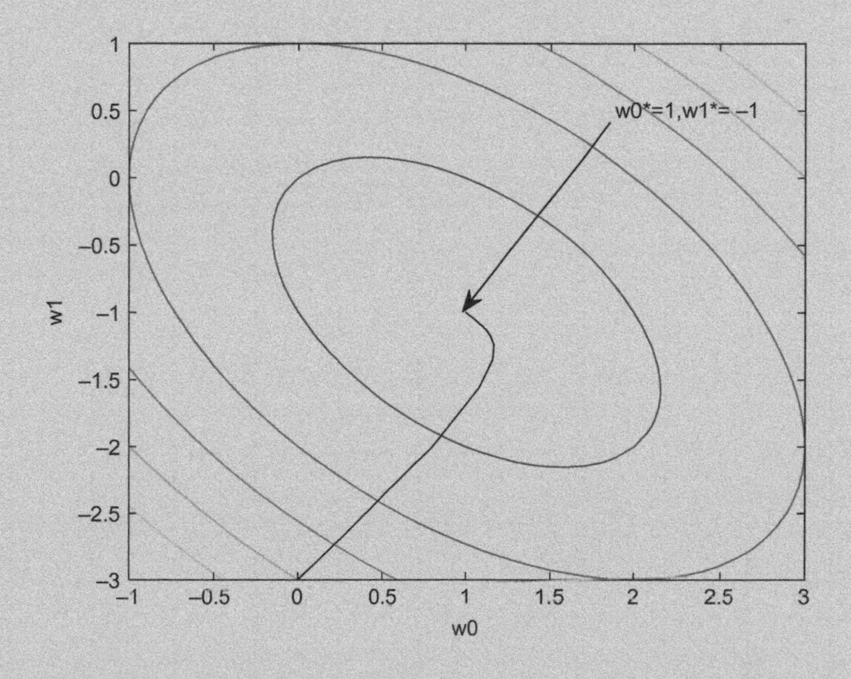

FIG. 9.8

Trajectory of coefficients in Example 9.4.

9.2.3 LEAST MEAN SQUARE ALGORITHM

Application of the steepest descent algorithm still needs an estimation of the derivative of the MSE function that could include statistical calculation of a block of data. To change the algorithm to do sample-based processing, an LMS algorithm must be used. To develop the LMS algorithm in terms of sample-based processing, we take the statistical expectation out of J and then take the derivative to obtain an approximate of $\frac{dJ}{dw(n)}$, that is,

$$J=e^2(n)=(d(n)-w(n)x(n))^2 \tag{9.30}$$

$$\frac{dJ}{dw(n)}=2(d(n)-w(n)x(n))\frac{d(d(n)-w(n)x(n))}{dw(n)}=-2e(n)x(n). \tag{9.31}$$

Substituting $\frac{dJ}{dw(n)}$ into the steepest descent algorithm in Eq. (9.27), we achieve the LMS algorithm for updating a single-weight case as

$$w(n+1)=w(n)+2\mu e(n)x(n), \tag{9.32}$$

where μ is the convergence parameter controlling the speed of the convergence and must satisfy $0<\mu<1/\lambda_{max}$. λ_{max} is the maximum eigenvalue of the autocorrelation matrix $R=E\{X^T(n)X(n)\}$. In our case, $R=E\{x^2(n)\}$ and $\lambda_{max}=1/R$. In general, with an adaptive FIR filter of length N, we extend the single-tap LMS algorithm without going through derivation, as shown in the following equations:

$$y(n)=w_0(n)x(n)+w_1(n)x(n-1)+\cdots+w_{N-1}(n)x(n-N+1) \tag{9.33}$$

$$\text{for } k=0,\cdots,N-1$$

$$w_k(n+1)=w_k(n)+2\mu e(n)x(n-k). \tag{9.34a}$$

$$\text{vector form}: W(n+1)=W(n)+2\mu e(n)X(n) \tag{9.34b}$$

The convergence factor is empirically chosen to be

$$0<\mu<\frac{1}{NP_x}, \tag{9.35}$$

where P_x is the input signal power. In practice, if the ADC has 16-bit data, the maximum signal amplitude should be $A=2^{15}$. Then the maximum input power is less than

$$P_x<\left(2^{15}\right)^2=2^{30}.$$

Hence, we may make a selection of the convergence parameter as

$$\mu=\frac{1}{N\times 2^{30}}\approx\frac{9.3\times 10^{-10}}{N}. \tag{9.36}$$

We further neglect time index for $w_k(n)$ and use the notation $w_k=w_k(n)$, since only the current updated coefficients are needed for next sample adaptation. We conclude the implementation of the LMS algorithm by the following steps:

1. Initialize w_0, w_1, … w_{N-1} to arbitrary values.
2. Read $d(n)$, $x(n)$, and perform digital filtering:
   $y(n)=w_0x(n)+w_1x(n-1)+\cdots+w_{N-1}x(n-N+1)$.

3. Compute the output error:
$e(n) = d(n) - y(n)$.
4. Update each filter coefficient using the LMS algorithm:
for $k = 0, \cdots, N-1$
$w_k = w_k + 2\mu e(n)x(n-k)$.
vector from: $W = W + 2\mu e(n)X(n)$.

9.2.4 RECURSIVE LEAST SQUARES ALGORITHM

The RLS algorithms aim to minimize the sum of the squares of the difference between the desired signal and the filter output signal using the new samples of the incoming signal. The RLS algorithms compute filter coefficients in a recursive form at each iteration. The RLS algorithms are well known for their fast convergence even when the eigenvalue spread of the input signal correlation matrix is large. These algorithms offer excellent performance with the cost of larger computational complexity and problem of stability in comparison with the LMS algorithm. In next section, we will study the standard RLS algorithm.

Given a linear adaptive filter $y(n) = W^T(n)X(n)$ with its coefficient and input vectors defined below:

$$W(n) = [w_0(n)w_1(n)\cdots w_{N-1}(n)]^T \tag{9.37}$$

$$X(i) = [x(i)x(i-1)\cdots x(i-N+1)]^T. \tag{9.38}$$

We begin with the exponential weighted squared-error function with a forgetting factor of λ given by

$$\zeta(n) = \sum_{i=1}^{n} \lambda^{n-i} e^2(i), \tag{9.39}$$

where $0 < \lambda < 1$, and the error at past time index i is expressed as

$$e(i) = d(i) - W^T(n)X(i) \text{ for } 1 \le i \le n. \tag{9.40}$$

Taking the derivative of the exponential weighted squared-error function to $W(n)$ leads to

$$\frac{\partial \zeta(n)}{\partial W(n)} = 2\sum_{i=1}^{n} \lambda^{n-i}\{(d(i) - W^T(n)X(i))X(i)\}. \tag{9.41}$$

Setting Eq. (9.41) to zero yield the following:

$$\sum_{i=1}^{n} \lambda^{n-i} X(i)X^T(i)W(n) = \sum_{i=1}^{n} \lambda^{n-i} d(i)X(i). \tag{9.42}$$

Now, the term on the right-hand side of Eq. (9.42) is the cross-correlation vector, which can be expanded to be

$$\begin{aligned} P(n) &= \sum_{i=1}^{n} \lambda^{n-i} d(i)X(i) \\ &= d(n)X(n) + \lambda \sum_{i=1}^{n-1} \lambda^{n-1-i} d(i)X(i). \end{aligned} \tag{9.43}$$

We modify Eq. (9.43) to a recursion form, that is,

$$P(n) = \lambda P(n-1) + d(n)X(n). \tag{9.44}$$

According to Eq. (9.42), we define the autocorrelation matrix below:

$$R(n) = \sum_{i=1}^{n} \lambda^{n-i} X(i) X^T(i). \tag{9.45}$$

Expanding Eq. (9.45), we obtain the recursion relation as follows:

$$\begin{aligned} R(n) &= \sum_{i=1}^{n} \lambda^{n-i} X(i) X^T(i) \\ &= \lambda \sum_{i=1}^{n-1} \lambda^{n-1-i} X(i) X^T(i) + X(n) X^T(n) \\ &= \lambda R(n-1) + X(n) X^T(n). \end{aligned} \tag{9.46}$$

In order to determine the filter coefficients $W(n)$, we need to compute the inverse of the autocorrelation matrix, $R^{-1}(n)$. This can be completed by using the matrix inversion lemma (Haykin, 2014) listed below:

$$A^{-1} = B - BC\left(D + C^T BC\right)^{-1} C^T B. \tag{9.47}$$

To use the matrix inversion lemma, we set up the following:
$A = R(n)$, $B^{-1} = \lambda R(n-1)$, $C = X(n)$, $D = 1$, and $Q(n) = R^{-1}(n)$, we can conclude

$$Q(n) = \lambda^{-1} Q(n-1) - \lambda^{-1} Q(n-1) X(n) \left[1 + X^T(n) \lambda^{-1} Q(n-1) X(n)\right]^{-1} X^T(n) \lambda^{-1} Q(n-1). \tag{9.48}$$

To simplify Eq. (9.48), define the gain vector as

$$k(n) = \frac{\lambda^{-1} Q(n-1) X(n)}{1 + \lambda^{-1} X^T(n) Q(n-1) X(n)}. \tag{9.49}$$

Substituting Eq. (9.49) into Eq. (9.48), we have

$$Q(n) = \lambda^{-1} Q(n-1) - \lambda^{-1} k(n) X^T(n) Q(n-1). \tag{9.50}$$

From Eq. (9.49), we explore a new modified relation for $k(n)$, that is,

$$k(n) = \lambda^{-1} Q(n-1) X(n) - \lambda^{-1} k(n) X^T(n) Q(n-1) X(n). \tag{9.51}$$

Substituting Eq. (9.50) into Eq. (9.51) yields

$$k(n) = Q(n) X(n). \tag{9.52}$$

Based on the Wiener solution for the filter coefficient vector and using Eq. (9.44), it follows that

$$\begin{aligned} W(n) &= Q(n) P(n) \\ &= \lambda Q(n) P(n-1) + Q(n) d(n) X(n). \end{aligned} \tag{9.53}$$

Applying Eqs. (9.50) and (9.52) to Eq. (9.53), we obtain

$$\begin{aligned} W(n) &= \lambda Q(n)P(n-1) + k(n)d(n) \\ &= \left(Q(n-1) - k(n)X^T(n)Q(n-1)\right)P(n-1) + k(n)d(n). \end{aligned} \tag{9.54}$$

Distributing $P(n-1)$ in Eq. (9.54) and noting that $W(n-1)=Q(n-1)P(n-1)$, it follows that

$$W(n) = W(n-1) - k(n)X^T(n)W(n-1) + d(n)k(n). \tag{9.55}$$

Factoring out the gain vector $k(n)$ in Eq. (9.55), it leads to

$$W(n) = W(n-1) + k(n)\left(d(n) - X^T(n)W(n-1)\right). \tag{9.56}$$

Define the innovation $\alpha(n)$ as

$$\alpha(n) = d(n) - X^T(n)W(n-1). \tag{9.57}$$

Substituting Eq. (9.57) into Eq. (9.56), we yield the following weight update equation:

$$W(n) = W(n-1) + k(n)\alpha(n). \tag{9.58}$$

Finally, we can summarize the RLS algorithm:

Initialize $Q(-1)=\delta I$, $I =$ the identity matrix and $\delta=$ the inverse of the input signal power

$$\begin{aligned} \alpha(n) &= d(n) - X^T(n)W(n-1) \\ k(n) &= \frac{\lambda^{-1}Q(n-1)X(n)}{1+\lambda^{-1}X^T(n)Q(n-1)X(n)} \\ W(n) &= W(n-1) + k(n)\alpha(n) \\ Q(n) &= \lambda^{-1}Q(n-1) - \lambda^{-1}k(n)X^T(n)Q(n-1) \\ y(n) &= W^T(n)X(n) \\ e(n) &= d(n) - y(n). \end{aligned}$$

Similar to the LMS algorithm, by omitting time index for $w_k(n)$, that is, $w_k=w_k(n)$, we conclude the implementation of the RLS algorithm by the following steps:

1. Initialize $W = \left[w_0 \ w_1 \ \cdots \ w_{N-1}\right]^T$ to arbitrary values
   Initialize $Q=\delta I$, $I =$ the identity matrix and $\delta=$the inverse of the input signal power
2. Compute the innovation and gain vector:

$$\begin{aligned} \alpha &= d(n) - X^T(n)W \\ k &= \frac{QX(n)}{\lambda + X^T(n)QX(n)} \end{aligned}$$

3. Update the filter coefficients:

$$W = W + k\alpha$$

4. Update Q matrix, produce the filter output, and measure the error:

$$\begin{aligned} Q &= \left[Q - kX^T(n)Q\right]/\lambda \\ y(n) &= W^T X(n) \\ e(n) &= d(n) - y(n) \end{aligned}$$

We will apply the adaptive filter to solve real-world problems in the following section.

9.3 APPLICATIONS: NOISE CANCELLATION, SYSTEM MODELING, AND LINE ENHANCEMENT

We now examine several applications of the adaptive filters, such as noise cancellation, system modeling, and line enhancement via application examples. First, we begin with the noise cancellation problem to illustrate operations of the adaptive FIR filter.

9.3.1 NOISE CANCELLATION

The concept of noise cancellation was introduced in the previous section. Fig. 9.9A shows the main idea.

The DSP system consists of two ADC channels. The first microphone with ADC captures the noisy speech, $d(n)=s(n)+n(n)$, which contains the clean speech $s(n)$ and noise $n(n)$ due to a noisy environment, while the second microphone with ADC resides where it picks up only the correlated noise and feeds the noise reference $x(n)$ to the adaptive filter. The adaptive filter uses the adaptive algorithm to adjust its coefficients to produce the best estimate of noise $y(n)\approx n(n)$, which will be subtracted from the corrupted signal $d(n)=s(n)+n(n)$. The output of the error signal $e(n)=s(n)+n(n)-y(n)\approx \tilde{s}(n)$ is expected to be the best estimate of the clean speech signal. Through digital-to-analog conversion (DAC), the cleaned digital speech becomes analog voltage, which drives the speaker.

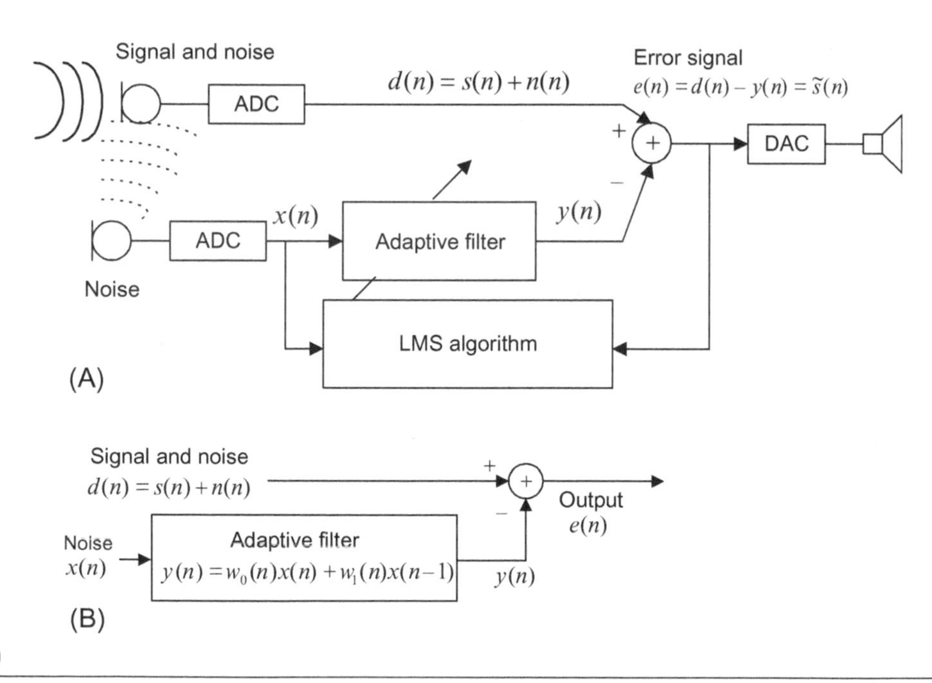

FIG. 9.9

(A) Simplest noise canceller using a one-tap adaptive filter. (B) Noise cancellation in Example 9.5.

We first study the noise cancellation problem using a simple two-tap adaptive filter via Example 9.5 and assumed data. The purpose of doing so is to become familiar with the setup and operations of the adaptive filter algorithms. The simulation for real adaptive noise cancellation follows.

EXAMPLE 9.5

Given the DSP system for the noise cancellation application using an adaptive filter with two coefficients shown in Fig. 9.9B,

(a) Set up the LMS algorithm for the adaptive filter.
(b) Perform adaptive filtering to obtain outputs $e(n)$ for $n=0, 1, 2$ given the following inputs and outputs:

$$x(0)=1,\ x(1)=1,\ x(2)=-1,\ d(0)=2,\ d(1)=1,\ d(2)=-2$$

and initial weights: $w_0=w_1=0$,
convergence factor is set to be $\mu=0.1$.
(c) Set up the RLS algorithm.
(d) Repeat (a) using $\lambda=0.96$.

Solution:

(a) The adaptive LMS algorithm is set up as
Initialization: $w_0=0,\ w_1=0$
Digital filtering:

$$y(n)=w_0x(n)+w_1x(n-1)$$

Computing the output error = output:

$$e(n)=d(n)-y(n).$$

Updating each weight for the next coming sample:

$$w_k=w_k+2\mu e(n)x(n-k),\ \text{for}\,k=0,1$$

or

$$w_0=w_0+2\mu e(n)x(n)$$
$$w_1=w_1+2\mu e(n)x(n-1).$$

(b) We can see the adaptive filtering operations as follows:
For $n=0$
Digital filtering:

$$y(0)=w_0x(0)+w_1x(-1)=0\times 1+0\times 0=0$$

Computing the output:

$$e(0)=d(0)-y(0)=2-0=2$$

Updating coefficients:

$$w_0=w_0+2\times 0.1\times e(0)x(0)=0+2\times 0.1\times 2\times 1=0.4$$
$$w_1=w_1+2\times 0.1\times e(0)x(-1)=0+2\times 0.1\times 2\times 0=0.0$$

Continued

EXAMPLE 9.5—CONT'D

For $n=1$

Digital filtering:

$$y(1)=w_0x(1)+w_1x(0)=0.4\times1+0\times1=0.4$$

Computing the output:

$$e(1)=d(1)-y(1)=1-0.4=0.6$$

Updating coefficients:

$$\begin{aligned} w_0&=w_0+2\times0.1\times e(1)x(1)=0.4+2\times0.1\times0.6\times1=0.52 \\ w_1&=w_1+2\times0.1\times e(1)x(0)=0+2\times0.1\times0.6\times1=0.12 \end{aligned}$$

For $n=2$

Digital filtering:

$$y(2)=w_0x(2)+w_1x(1)=0.52\times(-1)+0.12\times1=-0.4$$

Computing the output:

$$e(2)=d(2)-y(2)=-2-(-0.4)=-1.6$$

Updating coefficients:

$$\begin{aligned} w_0&=w_0+2\times0.1\times e(2)x(2)=0.52+2\times0.1\times(-1.6)\times(-1)=0.84 \\ w_1&=w_1+2\times0.1\times e(2)x(1)=0.12+2\times0.1\times(-1.6)\times1=-0.2 \end{aligned}$$

Hence, the adaptive filter outputs for the first three samples are listed as $e(0)=2$, $e(1)=0.6$, $e(2)=-1.6$.

(c) The RLS algorithm is set up as

$$\delta\approx1/[(x^2(0)+x^2(1)+x^2(2))/3]=1;\ \lambda=0.96$$

$$Q=\begin{bmatrix}1&0\\0&1\end{bmatrix},\ \begin{bmatrix}w_0\\w_1\end{bmatrix}=\begin{bmatrix}0\\0\end{bmatrix}$$

$$\alpha=d(n)-[x(n)x(n-1)]\begin{bmatrix}w_0\\w_1\end{bmatrix}$$

$$k=\frac{Q\begin{bmatrix}x(n)\\x(n-1)\end{bmatrix}}{\lambda+[x(n)\ \ x(n-1)]Q\begin{bmatrix}x(n)\\x(n-1)\end{bmatrix}}$$

$$\begin{bmatrix}w_0\\w_1\end{bmatrix}=\begin{bmatrix}w_0\\w_1\end{bmatrix}+k\alpha$$

$$Q=(Q-k[x(n)\ \ x(n-1)]Q)/\lambda$$

$$y(n)=[w_0\ \ w_1]\begin{bmatrix}x(n)\\x(n-1)\end{bmatrix}$$

$$e(n)=d(n)-y(n).$$

(d) For $n=0$

$$\alpha = d(0) - [x(0)\ \ x(-1)]\begin{bmatrix} w_0 \\ w_1 \end{bmatrix} = 2 - [1\ \ 0]\begin{bmatrix} 0 \\ 0 \end{bmatrix} = 2$$

$$k = \frac{Q\begin{bmatrix} x(0) \\ x(-1) \end{bmatrix}}{\lambda + [x(0)\ \ x(-1)]Q\begin{bmatrix} x(0) \\ x(-1) \end{bmatrix}} = \frac{\begin{bmatrix} 1 & 0 \\ 0 & 1 \end{bmatrix}\begin{bmatrix} 1 \\ 0 \end{bmatrix}}{0.96 + [1\ \ 0]\begin{bmatrix} 1 & 0 \\ 0 & 1 \end{bmatrix}\begin{bmatrix} 1 \\ 0 \end{bmatrix}} = \begin{bmatrix} 0.5102 \\ 0 \end{bmatrix}$$

$$\begin{bmatrix} w_0 \\ w_1 \end{bmatrix} = \begin{bmatrix} w_0 \\ w_1 \end{bmatrix} + k\alpha = \begin{bmatrix} 0 \\ 0 \end{bmatrix} + \begin{bmatrix} 0.5102 \\ 0 \end{bmatrix} \times 2 = \begin{bmatrix} 1.0204 \\ 0 \end{bmatrix}$$

$$Q = \{Q - k[x(0)\ \ x(-1)]Q\}/\lambda$$

$$= \left\{\begin{bmatrix} 1 & 0 \\ 0 & 1 \end{bmatrix} - \begin{bmatrix} 0.5102 \\ 0 \end{bmatrix}[1\ \ 0]\begin{bmatrix} 1 & 0 \\ 0 & 1 \end{bmatrix}\right\}\Big/ 0.96 = \begin{bmatrix} 0.5102 & 0 \\ 0 & 1.0417 \end{bmatrix}$$

$$y(0) = [w_0\ \ w_1]\begin{bmatrix} x(0) \\ x(-1) \end{bmatrix} = [1.0204\ \ 0]\begin{bmatrix} 1 \\ 0 \end{bmatrix} = 1.0204$$

$$e(0) = d(0) - y(0) = 2 - 1.0204 = 0.9796.$$

For $n=1$

$$\alpha = d(1) - [x(1)x(0)]\begin{bmatrix} w_0 \\ w_1 \end{bmatrix} = 1 - [1\ \ 1]\begin{bmatrix} 1.0204 \\ 0 \end{bmatrix} = -0.0204$$

$$k = \frac{Q\begin{bmatrix} x(1) \\ x(0) \end{bmatrix}}{\lambda + [x(1)\ \ x(0)]Q\begin{bmatrix} x(1) \\ x(0) \end{bmatrix}} = \frac{\begin{bmatrix} 0.5102 & 0 \\ 0 & 1.0417 \end{bmatrix}\begin{bmatrix} 1 \\ 1 \end{bmatrix}}{0.96 + [1\ \ 1]\begin{bmatrix} 0.5102 & 0 \\ 0 & 1.0417 \end{bmatrix}\begin{bmatrix} 1 \\ 1 \end{bmatrix}} = \begin{bmatrix} 0.2031 \\ 0.4147 \end{bmatrix}$$

$$\begin{bmatrix} w_0 \\ w_1 \end{bmatrix} = \begin{bmatrix} w_0 \\ w_1 \end{bmatrix} + k\alpha = \begin{bmatrix} 1.0204 \\ 0 \end{bmatrix} + \begin{bmatrix} 0.2031 \\ 0.4147 \end{bmatrix} \times (-0.0204) = \begin{bmatrix} 1.0163 \\ -0.0085 \end{bmatrix}$$

$$Q = \{Q - k[x(1)\ \ x(0)]Q\}/\lambda$$

$$= \left\{\begin{bmatrix} 0.5102 & 0 \\ 0 & 1.0417 \end{bmatrix} - \begin{bmatrix} 0.2031 \\ 0.4147 \end{bmatrix}[1\ \ 1]\begin{bmatrix} 0.5102 & 0 \\ 0 & 1.0417 \end{bmatrix}\right\}\Big/ 0.96 = \begin{bmatrix} 0.4235 & -0.2204 \\ -0.2204 & 0.6351 \end{bmatrix}$$

$$y(1) = [w_0\ \ w_1]\begin{bmatrix} x(1) \\ x(0) \end{bmatrix} = [1.0163\ \ -0.0085]\begin{bmatrix} 1 \\ 1 \end{bmatrix} = 1.0078$$

$$e(1) = d(1) - y(1) = 1 - 1.0078 = \ \ 0.0078.$$

Continued

EXAMPLE 9.5—CONT'D

For $n=2$

$$\alpha = d(2) - [x(2)x(1)]\begin{bmatrix} w_0 \\ w_1 \end{bmatrix} = 1 - [-1 \ \ 1]\begin{bmatrix} 1.0163 \\ -0.0085 \end{bmatrix} = -0.9753$$

$$k = \frac{Q\begin{bmatrix} x(2) \\ x(1) \end{bmatrix}}{\lambda + [x(2) \ \ x(1)]Q\begin{bmatrix} x(2) \\ x(1) \end{bmatrix}} = \frac{\begin{bmatrix} 0.4235 & -0.2204 \\ -0.2204 & 0.6351 \end{bmatrix}\begin{bmatrix} -1 \\ 1 \end{bmatrix}}{0.96 + [-1 \ \ 1]\begin{bmatrix} 0.4235 & -0.2204 \\ -0.2204 & 0.6351 \end{bmatrix}\begin{bmatrix} -1 \\ 1 \end{bmatrix}} = \begin{bmatrix} -0.2618 \\ 0.3478 \end{bmatrix}$$

$$\begin{bmatrix} w_0 \\ w_1 \end{bmatrix} = \begin{bmatrix} w_0 \\ w_1 \end{bmatrix} + k\alpha = \begin{bmatrix} 1.0163 \\ -0.0085 \end{bmatrix} + \begin{bmatrix} -0.2618 \\ 0.3478 \end{bmatrix} \times (-0.9753) = \begin{bmatrix} 1.2716 \\ -0.3477 \end{bmatrix}$$

$$Q = \{Q - k[x(2) \ \ x(1)]Q\}/\lambda$$

$$= \left\{\begin{bmatrix} 0.4235 & -0.2204 \\ -0.2204 & 0.6351 \end{bmatrix} - \begin{bmatrix} -0.2618 \\ 0.3478 \end{bmatrix}[-1 \ \ 1]\begin{bmatrix} 0.4235 & -0.2204 \\ -0.2204 & 0.6351 \end{bmatrix}\right\} \Big/ 0.96$$

$$= \begin{bmatrix} 1.1556 & -0.6013 \\ -0.6013 & 1.7382 \end{bmatrix}$$

$$y(2) = [w_0 \ \ w_1]\begin{bmatrix} x(2) \\ x(1) \end{bmatrix} = [1.2716 \ \ -0.3477]\begin{bmatrix} -1 \\ 1 \end{bmatrix} = -1.6193$$

$$e(2) = d(2) - y(2) = -2 - (-1.6139) = -0.3807.$$

Next, a simulation example is given to illustrate this idea and its results. The noise cancellation system is assumed to have the following specifications:

Sample rate = 8000 Hz

- Original speech data: wen.dat
- Speech corrupted by Gaussian noise with a power of 1 delayed by five samples from the noise reference.
- Noise reference containing Gaussian noise with a power of 1.
- Adaptive FIR filter used to remove the noise.
- Number of FIR filter tap = 21.
- Convergence factor for the LMS algorithm is chosen to be 0.01 ($<1/21$).

The speech waveforms and spectral plots for the original, corrupted, and reference noise and for the cleaned speech are plotted in Fig. 9.10A and B. From the figures, it is observed that the enhanced speech waveform and spectrum are very close to the original ones. The LMS algorithm converges after approximately 400 iterations. The method is a very effective approach for noise canceling. MATLAB implementation is detailed in Program 9.1A.

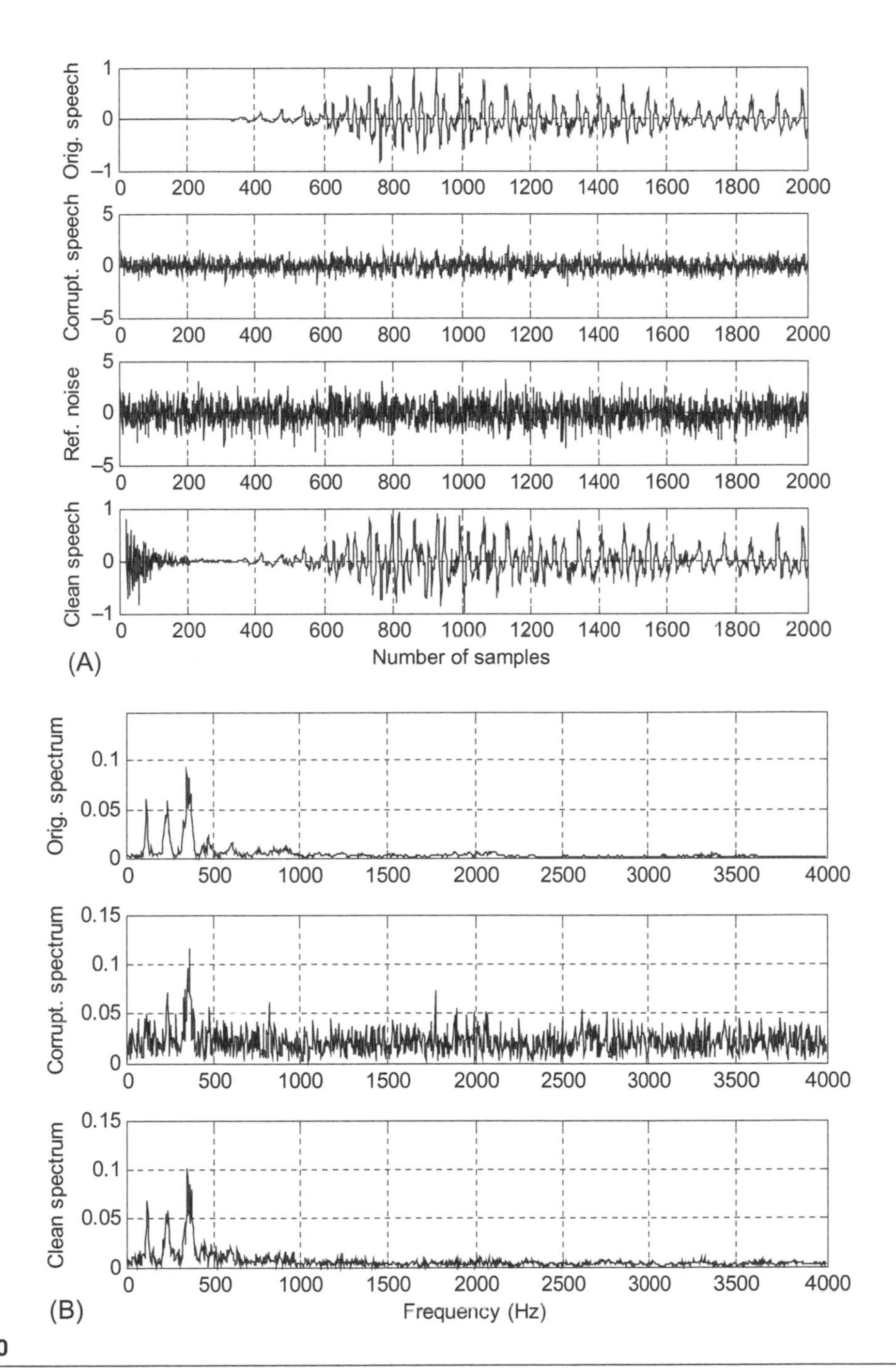

FIG. 9.10

(A) Waveforms for original speech, corrupted speech, reference noise, and clean speech. (B) Spectra for original speech, corrupted speech, and clean speech.

Continued

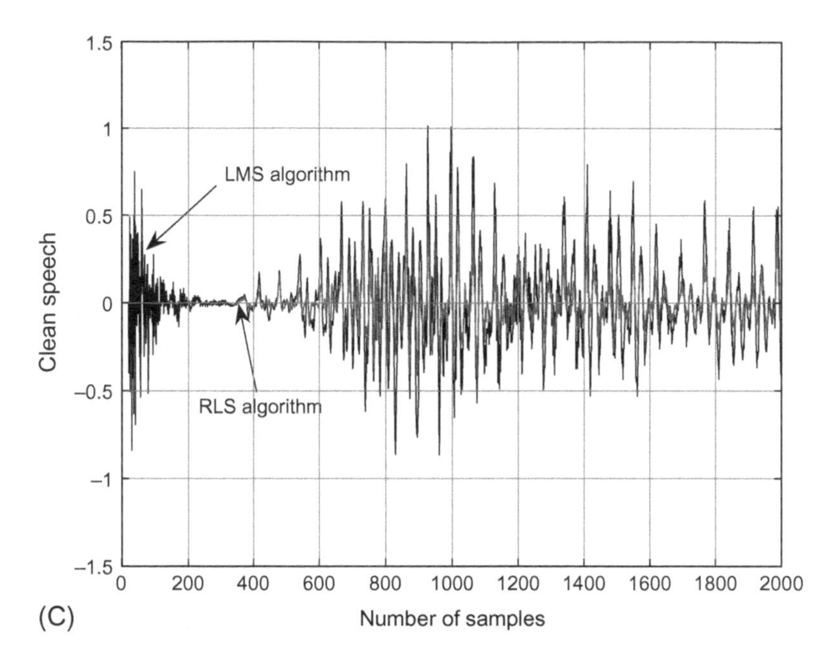

FIG. 9.10, CONT'D

(C) Comparison of the enhanced speech signals by the LMS (converges after 400 iterations) and RLS (converges immediately) algorithms.

Program 9.1A. MATLAB program for adaptive noise cancellation using the LMS algorithm.

```
close all; clear all
load wen.dat                        % Given by the instructor
fs=8000;                            % Sampling rate
t=0:1:length(wen)-1;                % Create index array
t=t/fs;                             % Convert indices to time instant
x=randn(1,length(wen));             % Generate random noise
n=filter([ 0 0 0 0 0 0.5 ],1,x);    % Generate the corruption noise
d=wen+n;                            % Generate signal plus noise
mu=0.01;                            % Initialize step size
w=zeros(1,21);                      % Initialize adaptive filter coefficients
y=zeros(1,length(t));               % Initialize the adaptive filter output array
e=y;                                % Initialize the output array
% Adaptive filtering using LMS algorithm
for m=22:1:length(t)-1
    sum=0;
    for k=1:1:21
    sum=sum+w(k)*x(m-k);
    end
    y(m)=sum;
    e(m)=d(m)-y(m);
    for k=1:1:21
    w(k)=w(k)+2*mu*e(m)*x(m-k);
    end
```

```
end
% Calculate the single-sided amplitude spectrum for the original signal
WEN=2*abs(fft(wen))/length(wen);WEN(1)=WEN(1)/2;
% Calculate the single-sided amplitude spectrum for the corrupted signal
D=2*abs(fft(d))/length(d);D(1)=D(1)/2;
f=[0:1:length(wen)/2]*8000/length(wen);
% Calculate the single-sided amplitude spectrum for the noise-cancelled signal
E=2*abs(fft(e))/length(e);E(1)=E(1)/2;
% Plot signals and spectrums
subplot(4,1,1), plot(wen);grid; ylabel('Orig. speech');
subplot(4,1,2),plot(d);grid; ylabel('Corrupt. speech')
subplot(4,1,3),plot(x);grid;ylabel('Ref. noise');
subplot(4,1,4),plot(e);grid; ylabel('Clean speech');
xlabel('Number of samples');
figure
subplot(3,1,1),plot(f,WEN(1:length(f)));grid
ylabel('Orig. spectrum')
subplot(3,1,2),plot(f,D(1:length(f)));grid; ylabel('Corrupt. spectrum')
subplot(3,1,3),plot(f,E(1:length(f)));grid
ylabel('Clean spectrum'); xlabel('Frequency (Hz)');
```

Program 9.1B shows a program segment for the RLS algorithm with $\lambda = 0.96$. The comparison for the enhanced speech signals via both the LMS and RLS algorithms is displayed in Fig. 9.10C. It can be seen that the RLS algorithm converges much faster (converges immediately) and has better performance but the algorithm has the large computational complexity.

Program 9.1B. MATLAB program for adaptive noise cancellation using the RLS algorithm.

```
w=zeros(1,21);        % Initialize adaptive filter coefficients
Q=eye(21)*mean(sum(x.*x)); % Initialize the inverse of auto-correlation matrix
% Adaptive filtering using the RLS algorithm
lamda=0.96;
% Perform adaptive filtering using the RLS algorithm
for n=21:length(t)
  xx=x(n:-1:n-20)'; % Obtain input vector
  alpha=d(n)-w*xx;
  k=Q*xx/(lamda+xx'*Q*xx);
  w=w+k'*alpha;
  Q=(Q-k*xx'*Q)/lamda;
  y(n)=w*xx;
  e(n)=d(n)-y(n);
end
```

Other interference cancellations include that of 60-Hz interference cancellation in electrocardiography (ECG) (Chapter 8) and Echo cancellation in long-distance telephone circuits, which will be described in a later section.

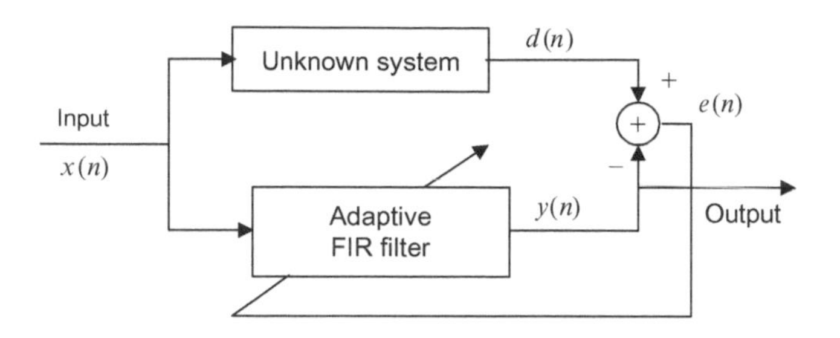

FIG. 9.11

Adaptive filter for system modeling.

9.3.2 SYSTEM MODELING

Another application of the adaptive filter is system modeling. The adaptive filter can keep tracking the behavior of an unknown system by using the unknown system input and output, as depicted in Fig. 9.11.

As shown in the figure, after the adaptive filter converges, the adaptive filter output $y(n)$ will be as close as possible to the unknown system's output. Since both the unknown system and the adaptive filter respond the same input, the transfer function of the adaptive filter will approximate that of the unknown system.

EXAMPLE 9.6

Given system modeling described in Fig. 9.11, use a single-weight adaptive filter $y(n)=wx(n)$ to perform the system-modeling task.

(a) Set up the LMS algorithm to implement the adaptive filter assuming the initial $w=0$ and $\mu=0.3$.

(b) Perform adaptive filtering to obtain $y(0)$, $y(1)$, $y(2)$, and $y(3)$, given

$$d(0)=1,\ d(1)=2,\ d(2)=-2,\ d(3)=2,$$
$$x(0)=0.5,\ x(1)=1,\ x(2)=-1,\ x(3)=1.$$

(c) Set up the RLS algorithm $\lambda=0.96$ and $\delta=1$.

(d) Repeat (b) using the RLS algorithm.

Solution:

(a) Adaptive filtering equations are set up as

$$w=0 \text{ and } 2\mu=2\times 0.3=0.6$$
$$y(n)=wx(n)$$
$$e(n)=d(n)-y(n)$$
$$w=w+0.6\times e(n)x(n)$$

(b) Adaptive filtering:

$$
\begin{aligned}
n=0,\ & y(0)=wx(0)=0\times 0.5=0 \\
& e(0)=d(0)-y(0)=1-0=1 \\
& w=w+0.6\times e(0)x(0)=0+0.6\times 1\times 0.5=0.3 \\
n=1,\ & y(1)=wx(1)=0.3\times 1=0.3 \\
& e(1)=d(1)-y(1)=2-0.3=1.7 \\
& w=w+0.6\times e(1)x(1)=0.3+0.6\times 1.7\times 1=1.32 \\
n=2,\ & y(2)=wx(2)=1.32\times(-1)=-1.32 \\
& e(2)=d(2)-y(2)=-2-(-1.32)=-0.68 \\
& w=w+0.6\times e(2)x(2)=1.32+0.6\times(-0.68)\times(-1)=1.728 \\
n=3,\ & y(3)=wx(3)=1.728\times 1=1.728 \\
& e(3)=d(3)-y(3)=2-1.728=0.272 \\
& w=1.728+0.6\times e(3)x(3)=1.728+0.6\times 0.272\times 1=1.8912.
\end{aligned}
$$

For this particular case, the system is actually a digital amplifier with a gain of 2.

(c) The RLS algorithm is set up as

$$
\begin{aligned}
& \delta=1;\lambda=0.96 \\
& Q=\delta\times 1=1, w=0 \\
& \alpha=d(n)-x(n)w \\
& k=\frac{Qx(n)}{\lambda+x(n)Qx(n)} \\
& w=w+k\alpha \\
& Q=[Q-kx(n)Q]/\lambda \\
& y(n)=x(n)w \\
& e(n)=d(n)-y(n)
\end{aligned}
$$

(d) For $n=0$

$$
\begin{aligned}
& \alpha=d(0)-x(0)w=1-0.5\times 0=1 \\
& k=\frac{Qx(0)}{\lambda+x(0)Qx(0)}=1\times 0.5/(0.96+0.5\times 1\times 0.5)=0.4132 \\
& w=w+k\alpha=0+0.4132\times 1=0.4132 \\
& Q=[Q-kx(0)Q]/\lambda=(1-0.4132\times 0.5\times 1)/0.96=0.8265 \\
& y(0)=wx(0)=0.4132\times 0.5=0.2066 \\
& e(0)=d(0)-y(0)=0.7934.
\end{aligned}
$$

Continued

EXAMPLE 9.6—CONT'D

For $n=1$

$$\alpha = d(1) - x(1)w = 1.5868$$
$$k = \frac{Qx(1)}{\lambda + x(1)Qx(1)} = \frac{0.8265 \times 1}{0.96 + 1 \times 0.8265 \times 1} = 0.4626$$
$$w = w + k\alpha = 0.4123 + 0.4626 \times 1.5868 = 1.1473$$
$$Q = [Q - kx(1)Q]/\lambda = [0.8265 - 0.4626 \times 1 \times 0.8265]/0.96 = 0.4626$$
$$y(1) = wx(1) = 1.1473 \times 1 = 1.1473$$
$$e(1) = d(1) - y(1) = 2 - 1.1473 = 0.8527.$$

For $n=2$

$$\alpha = d(2) - wx(2) = -2 - 1.1473 \times (-1) = -0.8527$$
$$k = \frac{Qx(2)}{\lambda + x(2)Qx(2)} = \frac{0.4626 \times (-1)}{0.96 + (-1) \times 0.4626 \times (-1)} = -0.3252$$
$$w = w + k\alpha = 1.1473 + (-0.3252) \times (-0.8527) = 2.1708$$
$$Q = [Q - kx(2)Q]/\lambda = [0.2626 - (-0.3252) \times (-1) \times 0.4626]/0.96 = 0.3252$$
$$y(2) = wx(2) = 2.1708 \times (-1) = -2.1708$$
$$e(2) = d(2) - y(2) = -2 - (-2.1708) = 0.1708.$$

For $n=3$

$$\alpha = d(3) - x(3)w = 2 - 1 \times 2.1708 = -0.1708$$
$$k = \frac{Qx(3)}{\lambda + x(3)Qx(3)} = \frac{0.3252 \times 1}{0.96 + 1 \times 0.3252 \times 1} = 0.2530$$
$$w = w + k\alpha = 2.1708 + 0.2530 \times (-0.1708) = 2.1276$$
$$Q = [Q - kx(3)Q]/\lambda = [0.3252 - 0.2530 \times 1 \times 0.3252]/0.96 = 0.2530$$
$$y(3) = wx(3) = 2.1276 \times 1 = 2.1276$$
$$e(3) = d(3) - y(3) = 2 - 2.1276 = -0.1276.$$

Next, we assume the unknown system is a fourth-order bandpass IIR filter whose 3-dB lower and upper cutoff frequencies are 1400 and 1600 Hz operating at 8000 Hz. We use an input consisting of tones of 500, 1500, and 2500 Hz. The unknown system's frequency responses are shown in Fig. 9.12.

The input waveform $x(n)$ with three tones is shown in the first plot of Fig. 9.13A. We can predict that the output of the unknown system will contain a 1500-Hz tone only, since the other two tones are rejected by the unknown system. Now, let us look at adaptive filter results. We use an FIR adaptive filter with the number of taps being 21, and a convergence factor set to be 0.01. In time domain, the output waveforms of the unknown system $d(n)$ and adaptive filter output $y(n)$ are almost identical after 70 samples when the LMS algorithm converges. The error signal $e(n)$ is also plotted to show the adaptive filter keeps tracking the unknown system's output with no difference after the first 50 samples.

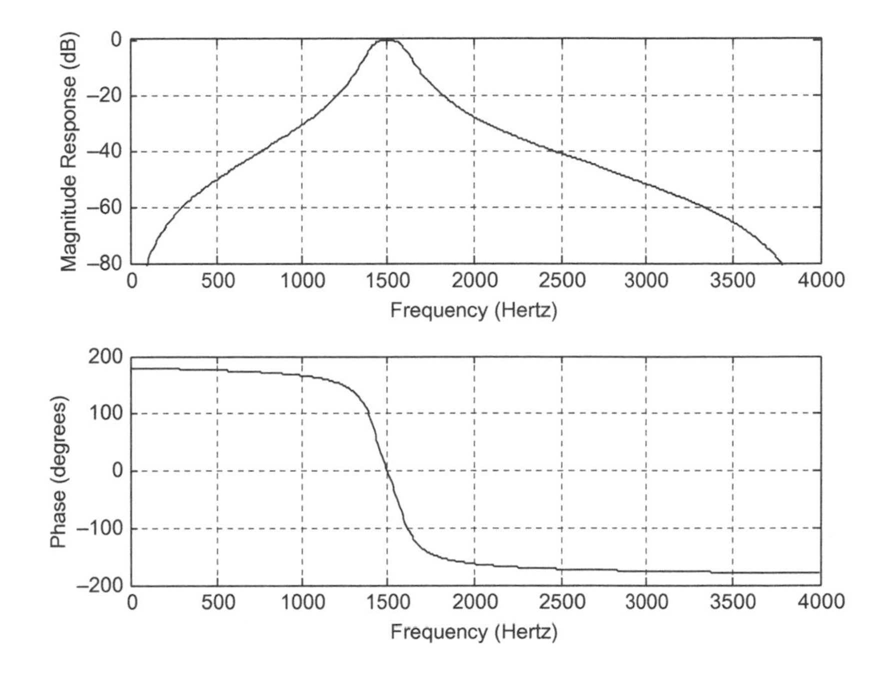

FIG. 9.12

Unknown system's frequency responses.

Fig. 9.13B depicts the frequency domain comparisons. The first plot displays the frequency components of the input signal, which clearly shows 500, 1500, and 2500 Hz. The second plot shows the unknown system's output spectrum, which contains only 1500 Hz tone, while the third plot displays the spectrum of the adaptive filter output. As we can see, in frequency domain, the adaptive filter tracks the characteristics of the unknown system. The MATLAB implementation is given in Program 9.2.

Program 9.2. MATLAB program for adaptive system identification.

```
close all; clear all
%Create the unknown system
fs=8000; T=1/fs;                    % Sampling rate and sampling period
% Bandpass filter design
%for the assumed unknown system using the bilinear transformation
%(BLT) method (see Chapter 8)
wd1=1400*2*pi; wd2=1600*2*pi;
wa1=(2/T)*tan(wd1*T/2); wa2=(2/T)*tan(wd2*T/2);
BW-wa2 wa1;
w0=sqrt(wa2*wa1);
[B,A]=lp2bp([1]
[b,a]=bilinear(B,A,fs);
freqz(b,a,512,fs); axis([0 fs/2 -80 1]); % Frequency response plots
```

```
figure
t=0:T:0.1;           % Generate the time vector
x=cos(2*pi*500*t)+sin(2*pi*1500*t)+cos(2*pi*2500*t+pi/4);
d=filter(b,a,x);  % Produce unknown system output
mu=0.01;             % Convergence factor
w=zeros(1,21); y=zeros(1,length(t)); % Initialize the coefficients and output
e=y;                                 % Initialize the error vector
% Perform adaptive filtering using LMS algorithm
for m=22:1:length(t)-1
    sum=0;
    for k=1:1:21
    sum=sum+w(k)*x(m-k);
    end
    y(m)=sum;
    e(m)=d(m)-y(m);
    for k=1:1:21
    w(k)=w(k)+2*mu*e(m)*x(m-k);
 end
end
% Calculate the single-sided amplitude spectrum for the input
X=2*abs(fft(x))/length(x);X(1)=X(1)/2;
% Calculate the single-sided amplitude spectrum for the unknown system output
D=2*abs(fft(d))/length(d);D(1)=D(1)/2;
% Calculate the single-sided amplitude spectrum for the adaptive filter output
Y=2*abs(fft(y))/length(y);Y(1)=Y(1)/2;
% Map the frequency index to its frequency in Hz
f=[0:1:length(x)/2]*fs/length(x);
% Plot signals and spectra
subplot(4,1,1), plot(x);grid; axis([0 length(x) -3 3]);
ylabel('System input');
subplot(4,1,2), plot(d);grid; axis([0 length(x) -1.5 1.5]);
ylabel('System output');
subplot(4,1,3),plot(y);grid; axis([0 length(y) -1.5 1.5]);
ylabel('ADF output')
subplot(4,1,4),plot(e);grid; axis([0 length(e) -1.5 1.5]);
ylabel('Error'); xlabel('Number of samples')
figure
subplot(3,1,1),plot(f,X(1:length(f )));grid; ylabel('Syst. input spect.')
subplot(3,1,2),plot(f,D(1:length(f )));grid; ylabel('Syst. output spect.')
subplot(3,1,3),plot(f,Y(1:length(f )));grid
ylabel('ADF output spect.'); xlabel('Frequency (Hz)');
```

Fig. 9.14 displays the error comparison of the system modeling by the LMS and RLS algorithms for first 160 samples. As shown in Fig. 9.14, the LMS algorithm takes 40 samples for the absolute value of its system error to reach less than 0.2. The RLS algorithms offer much better performance.

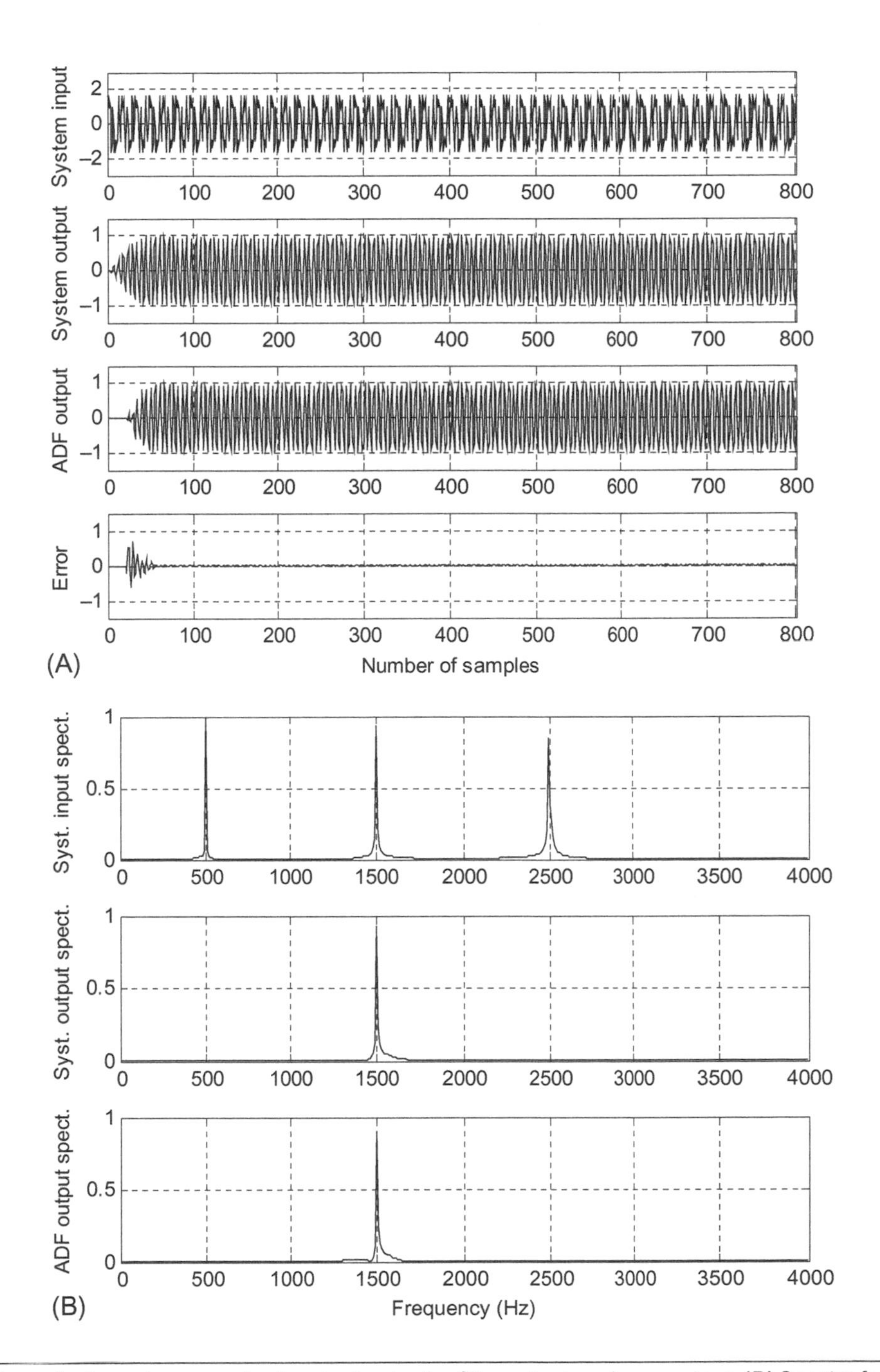

FIG. 9.13

(A) Waveforms for the unknown system's output, adaptive filter output, and error output. (B) Spectra for the input signal, unknown system output, and the adaptive filter output.

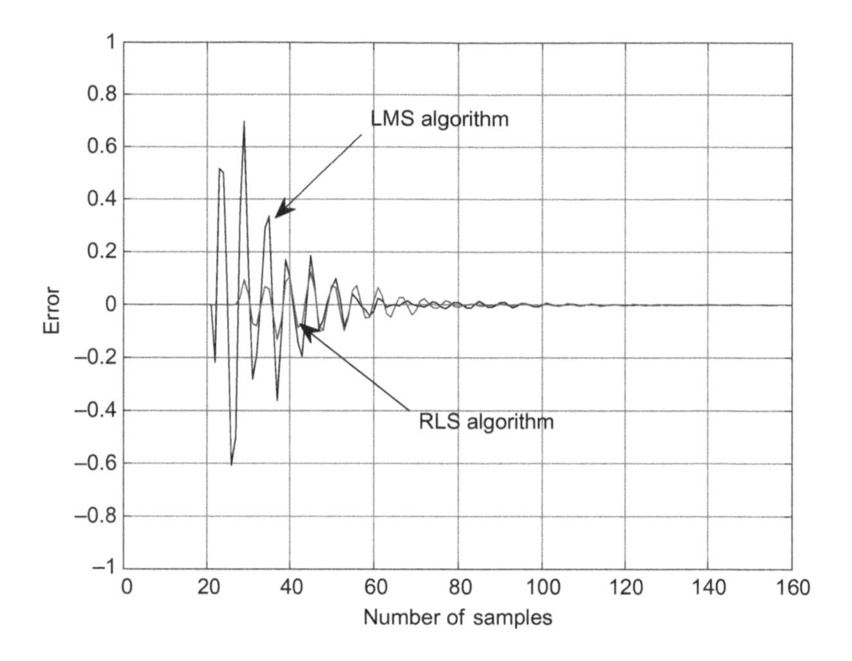

FIG. 9.14

Modeling error comparison of the system modeling by the LMS and RLS algorithms.

9.3.3 LINE ENHANCEMENT USING LINEAR PREDICTION

We study adaptive filtering via another application example: line enhancement. If a signal frequency content is very narrow compared with the bandwidth and changes with time, then the signal can efficiently be enhanced by the adaptive filter, which is line enhancement. Fig. 9.15 shows line enhancement using the adaptive filter where the LMS algorithm is used. As illustrated in the figure, the signal $d(n)$ is the corrupted sine wave signal by the white Gaussian noise $n(n)$. The enhanced line consists of

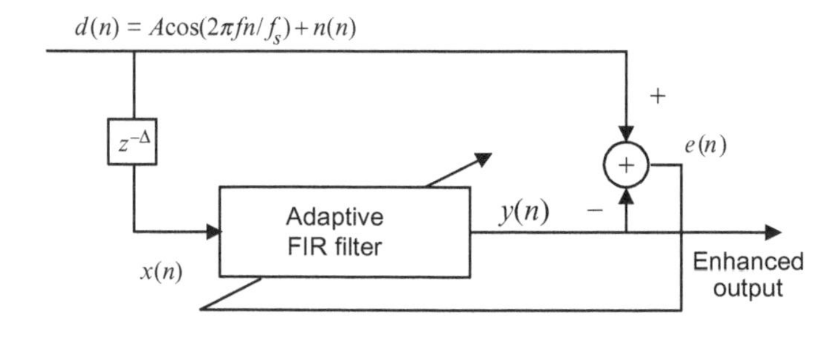

FIG. 9.15

Line enhancement using an adaptive filter.

a delay element to delay the corrupted signal by Δ samples to produce an input to the adaptive filter. The adaptive filter is actually a linear predictor of the desired narrow band signal. A two-tap FIR adaptive filter can predict one sinusoid. The value of Δ is usually determined by experiments or experience in practice to achieve the best-enhanced signal.

Our simulation example has the following specifications:

- Sampling rate = 8000 Hz
- Corrupted signal = 500 Hz tone with the unit amplitude added with white Gaussian noise
- Adaptive filter = FIR type, 21 taps
- Convergence factor = 0.001
- Delay value Δ = 7
- LMS algorithm is applied

Fig. 9.16 shows time domain results. The first plot is the noisy signal, while the second plot clearly demonstrates the enhanced signal. Fig. 9.17 describes the frequency domain point of view. The spectrum of the noisy signal is shown in the top plot, where we can see the white noise is populated over the entire bandwidth. The bottom plot is the enhanced signal spectrum. Since the method is adaptive, it is especially effective when the enhanced signal frequency changes with time. Program 9.3 lists the MATLAB program for this simulation.

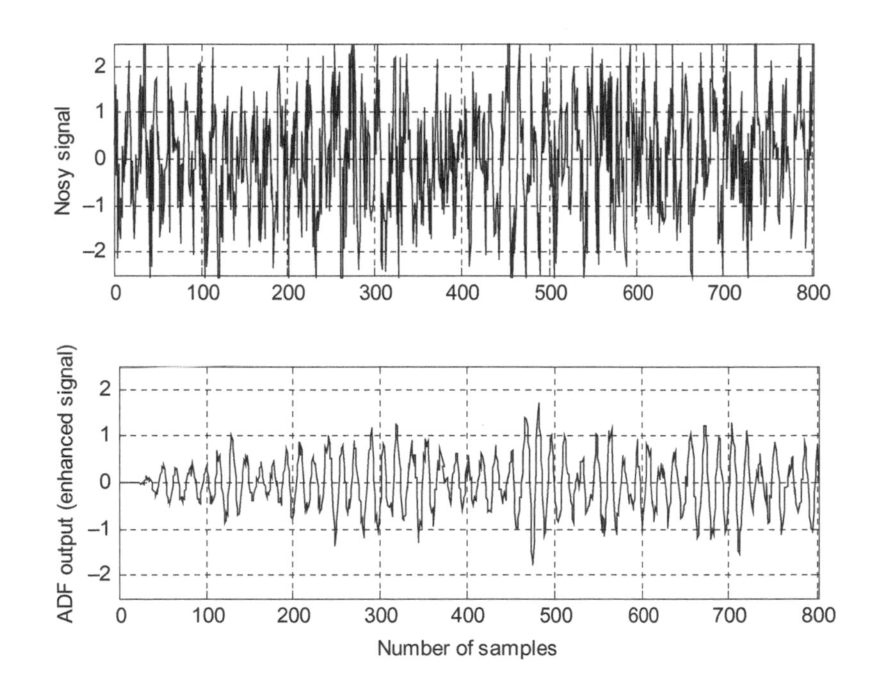

FIG. 9.16

Noisy signal and enhanced signal.

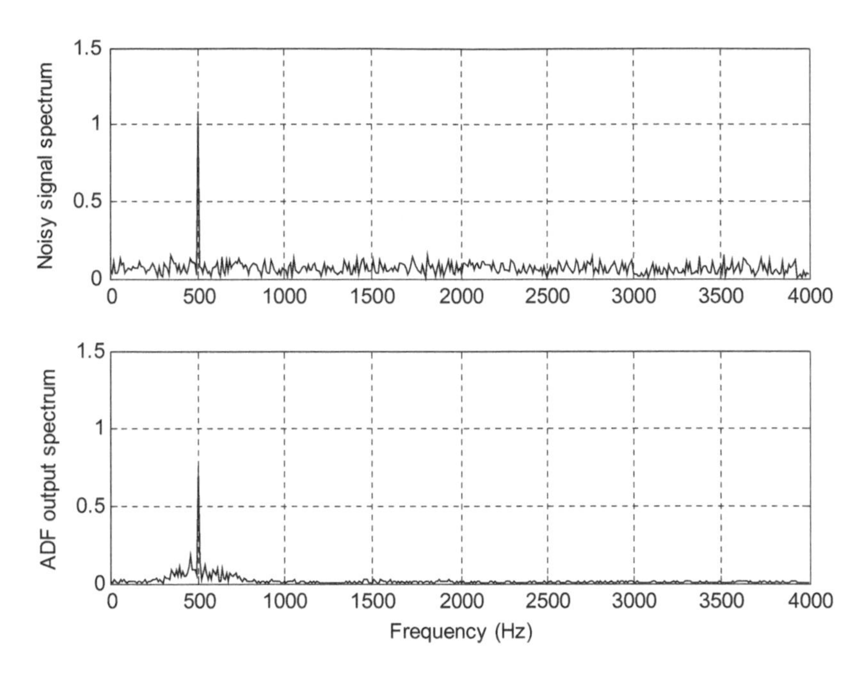

FIG. 9.17

Spectrum plots for the noisy signal and enhanced signal.

Program 9.3. MATLAB program for adaptive line enhancement.

```
close all; clear all
fs=8000; T=1/fs;                   % Sampling rate and sampling period
t=0:T:0.1;                         % 1 second time instants
n=randn(1,length(t));              % Generate Gaussian random noise
d=cos(2*pi*500*t)+n;               % Generate 500-Hz tone plus noise
x=filter([ 0 0 0 0 0 0 0 1 ],1,d); %Delay filter
mu=0.001;                          % Initialize the step size for LMS algorithms
w=zeros(1,21);                     % Initialize the adaptive filter coefficients
y=zeros(1,length(t));              % Initialize the adaptive filter output
e=y;                               % Initialize the error vector
%Perform adaptive filtering using the LMS algorithm
for m=22:1:length(t)-1
    sum=0;
    for i=1:1:21
    sum=sum+w(i)*x(m-i);
    end
    y(m)=sum;
    e(m)=d(m)-y(m);
    for k=1:1:21
    w(k)=w(k)+2*mu*e(m)*x(m-k);
    end
end
```

```
% Calculate the single-sided amplitude spectrum for corrupted signal
D=2*abs(fft(d))/length(d);D(1)=D(1)/2;
% Calculate the single-sided amplitude spectrum for enhanced signal
Y=2*abs(fft(y))/length(y);Y(1)=Y(1)/2;
% Map the frequency index to its frequency in Hz
f=[0:1:length(x)/2]*8000/length(x);
% Plot the signals and spectra
subplot(2,1,1), plot(d);grid; axis([0 length(x) -2.5 2.5]); ylabel('Noisy signal');
subplot(2,1,2),plot(y);grid; axis([0 length(y) -2.5 2.5]);
ylabel('ADF output (enhanced signal)'); xlabel('Number of samples')
figure
subplot(2,1,1),plot(f,D(1:length(f )));grid; axis([0 fs/2 0 1.5]);
ylabel('Noisy signal spectrum')
subplot(2,1,2),plot(f,Y(1:length(f )));grid; axis([0 fs/2 0 1.5]);
ylabel('ADF output spectrum'); xlabel('Frequency (Hz)');
```

9.4 OTHER APPLICATION EXAMPLES

This section continues to explore other adaptive filter applications briefly, without showing computer simulations. The topics include periodic interference cancellation, ECG interference cancellation, and echo cancellation in long-distance telephone circuits. Detailed information can also be explored in Haykin (2014), Ifeachor and Jervis (2002), Stearns and Hush (2011), and Widrow and Stearns (1985).

9.4.1 CANCELING PERIODIC INTERFERENCES USING LINEAR PREDICTION

An audio signal may be corrupted by periodic interference and no noise reference available. Such examples include the playback of speech or music with the interference of tape hum, or tunable rumble, or vehicle engine, or power line interference. We can use the modified line enhancement structure as shown in Fig. 9.18.

The adaptive filter uses the delayed version of the corrupted signal $x(n)$ to predict the periodic interference. The number of delayed samples is selected through experiments that determine the

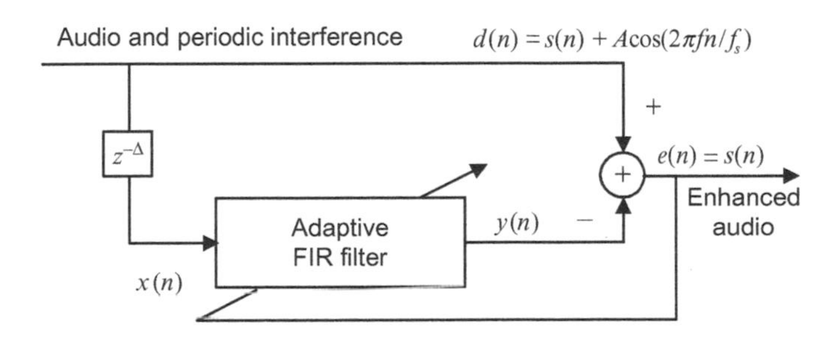

FIG. 9.18

Canceling periodic interference using the adaptive filter.

performance of the adaptive filter. Note that a two-tap FIR adaptive filter can predict one sinusoid, as noted earlier. After convergence, the adaptive filter will predict the interference as

$$y(n) = \sum_{k=0}^{N-1} w_k(n)x(n-k) \approx A\cos(2\pi fn/f_s). \tag{9.59}$$

Therefore, the error signal contains only the desired audio signal

$$e(n) \approx s(n). \tag{9.60}$$

9.4.2 ELECTROCARDIOGRAPHY INTERFERENCE CANCELLATION

As discussed in Chapters 1 and 8, in recording of electrocardiograms (ECG), there often exists unwanted 60-Hz interference, along with its harmonics, in the recorded data. This interference comes from the power line, including effects from magnetic induction, displacement currents in leads or in the body of the patient, and equipment interconnections and imperfections.

Fig. 9.19 illustrates the application of adaptive noise canceling in ECG. The primary input is taken from the ECG preamplifier, while a 60-Hz reference input is taken from a wall outlet with proper attenuation. After proper signal conditioning, the digital interference $x(n)$ is acquired by the digital signal (DS) processor. The digital adaptive filter uses this reference input signal to produce an estimate, which approximates the 60-Hz interference $n(n)$ sensed from the ECG amplifier:

$$y(n) \approx n(n). \tag{9.61}$$

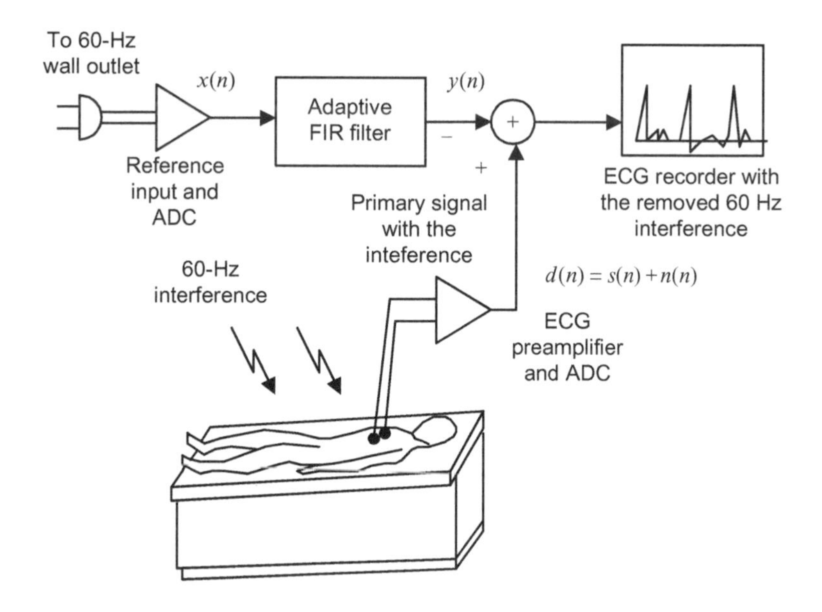

FIG. 9.19

Illustration of canceling 60-Hz interference in ECG.

Here, an adaptive FIR filter with N taps and the LMS algorithm can be used for this application:

$$y(n) = w_0(n)x(n) + w_1(n)x(n-1) + \cdots + w_{N-1}(n)x(n-N+1). \tag{9.62}$$

Then after convergence of the adaptive filter, the estimated interference is subtracted from the primary signal of the ECG preamplifier to produce the output signal $e(n)$, in which the 60-Hz interference is cancelled:

$$e(n) = d(n) - y(n) = s(n) + n(n) - x(n) \approx s(n). \tag{9.63}$$

With enhanced ECG recording, doctors in clinics can give more accurate diagnoses for patients.

Cancelling maternal ECG in fetal monitoring is another important application. The block diagram is shown in Fig. 9.20A. Fetal ECG plays important roles in monitoring the condition of the baby before or during birth. However, the ECG acquired from the mother's abdomen is contaminated by the noise such as muscle activity and fetal motion, as well as mother's own ECG. In order to reduce the effect of the mother's ECG, four (or more) chest leads (electrodes) are used to acquire the reference inputs: $x_0(n)$, $x_1(n)$, $x_2(n)$, and $x_3(n)$, assuming these channels only contain mother's ECG [see Fig. 9.20B]. One lead (electrode) placed on the mother's abdomen is used to capture the fetal information $d(n)$,

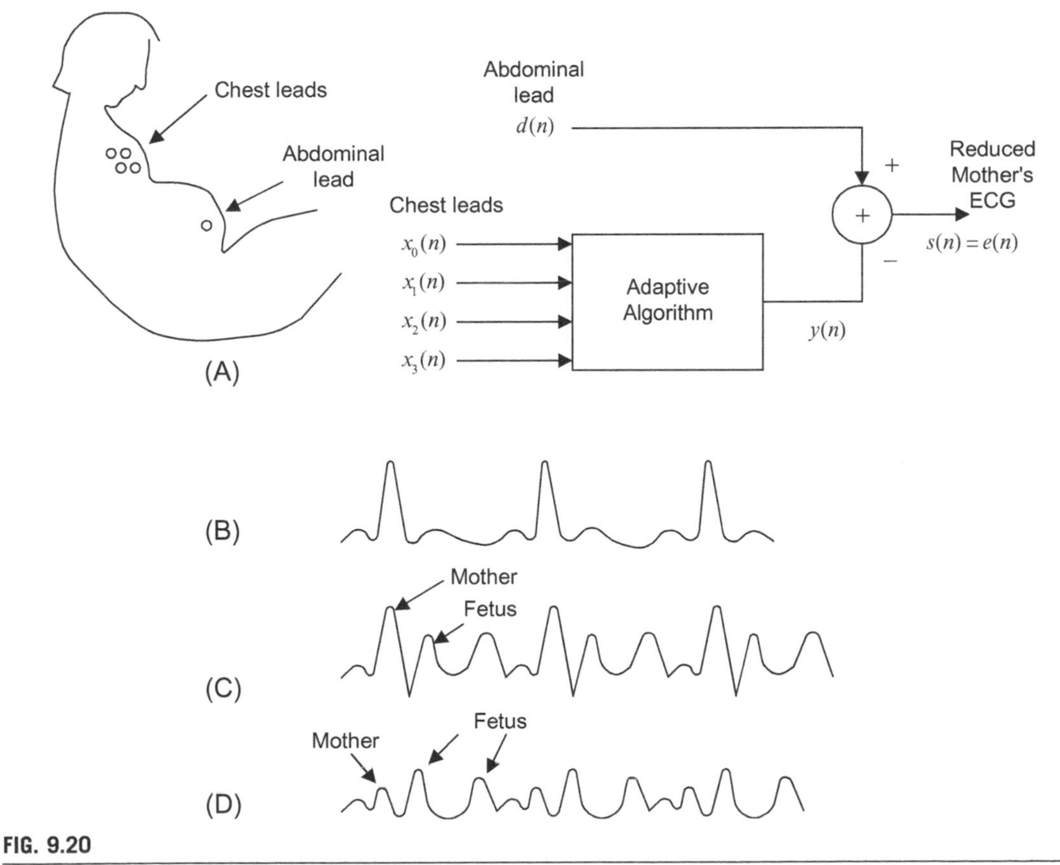

FIG. 9.20

Fetal monitoring with cancelling maternal ECG.

which may be corrupted by the mother's ECG as shown in Fig. 9.20C. An adaptive filter uses its references to predict the mother ECG, which will be subtracted from the corrupted fetus signal. Then the fetal ECG with the reduced mother's ECG is expected, as depicted in Fig. 9.20D. One possible LMS algorithm is listed below:

For $k=0, 1, 2, 3$

$$y_k(n)=w_{k,0}(n)x_k(n)+w_{k,1}(n)x_k(n-1)+\cdots+w_{k,N-1}(n)x_k(n-N+1)$$
$$y(n)=y_0(n)+y_1(n)+y_2(n)+y_3(n)$$
$$s(n)=e(n)=d(n)-y(n)$$

For $k=0, 1, 2, 3$

$$w_{k,i}(n+1)=w_{k,i}(n)+2\mu e(n)x_k(n-i), \text{ for } i=0,1,\cdots N-1.$$

9.4.3 ECHO CANCELLATION IN LONG-DISTANCE TELEPHONE CIRCUITS

Long-distance telephone transmission often suffers from impedance mismatches. This occurs primarily at the hybrid circuit interface. Balancing electric networks within the hybrid can never perfectly match the hybrid to the subscriber loop due to temperature variations, degradation of transmission lines, and so on. As a result, a small portion of the received signal is leaked for transmission. For example, in Fig. 9.21A, if speaker B talks, the speech indicated as $x_B(n)$ will pass the transmission line to reach user A, and a portion of $x_B(n)$ at site A is leaked and transmitted back to the user B, forcing caller B to hear his or her own voice. This is known as an echo for speaker B. A similar echo illustration can be conducted for speaker A. When the telephone call is made over a long distance (more than 1000 miles, such as geostationary satellites), the echo can be delayed by as much as 540 ms. The echo impairment can be annoying to the customer and increases with the distance.

To circumvent the problem of echo in long-distance communications, an adaptive filter is applied at each end of the communication system, as shown in Fig. 9.21B. Let us examine the adaptive filter installed at the speaker A site. The coming signal is $x_B(n)$ from the speaker B, while the outgoing signal contains the speech from the speaker A and a portion of leakage from the hybrid circuit $d_A(n)=x_A(n)+\overline{x}_B(n)$. If the leakage $\overline{x}_B(n)$ returns back to speaker B, it becomes an annoying echo. To prevent the echo, the adaptive filter at the speaker A site uses the incoming signal from speaker B as an input and makes its output approximate to the leaked speaker B signal by adjusting its filter coefficients; that is,

$$y_A(n)=\sum_{k=0}^{N-1} w_k(n)x_B(n-k)\approx\overline{x}_B(n). \tag{9.64}$$

As shown in Fig. 9.21B, the estimated echo $y_A(n)\approx\overline{x}_B(n)$ is subtracted from the outgoing signal, thus producing the signal that contains only speech A; that is, $e_A(n)\approx x_A(n)$. As a result, the echo of the speaker B is removed. We can illustrate similar operations for the adaptive filter used at the speaker B site. In practice, an adaptive FIR filter with several hundred coefficients or more is commonly used to effectively cancel the echo. If nonlinearities are concerned in the echo path, a corresponding nonlinear adaptive canceller can be used to improve the performance of the echo cancellation.

Other forms of adaptive filters and other applications are beyond the scope of this book. The reader is referred to the references for further development.

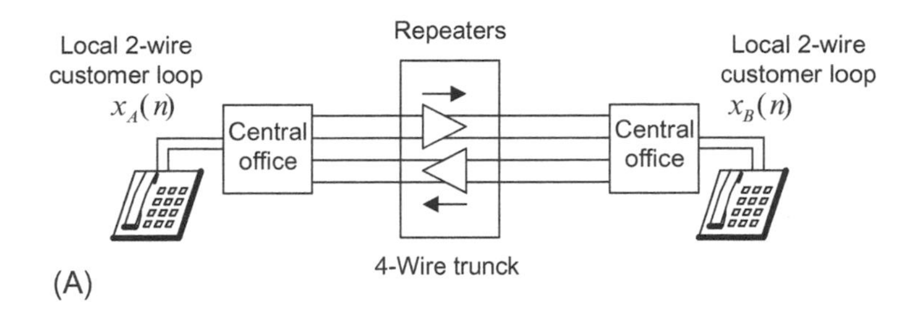

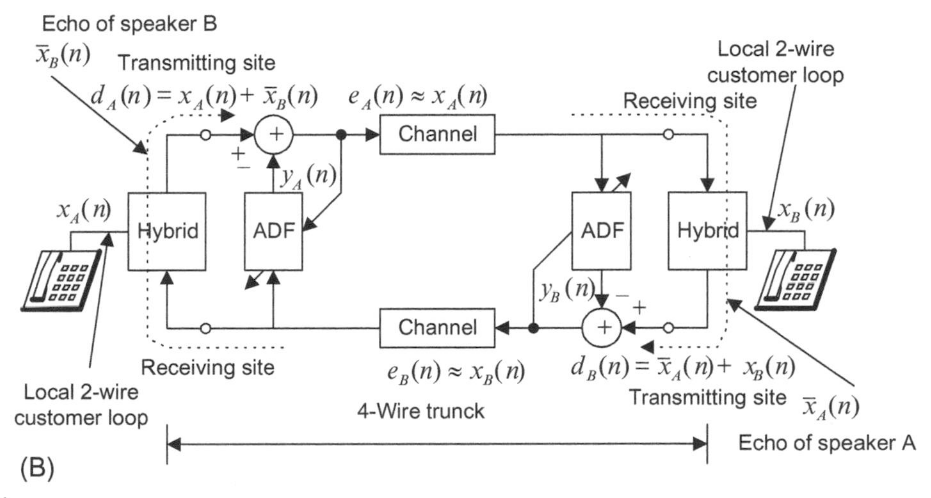

FIG. 9.21

(A) Simplified long-distance circuit. (B) Adaptive echo cancellers.

9.5 SUMMARY

1. Adaptive filters can be applied to signal-changing environments, spectral overlap between noise and signal, and unknown or time-varying noise.
2. Wiener filter theory provides optimal weight solution based on statistics. It involves collection of a large block of data, calculation of an autocorrelation matrix and a cross-correlation matrix, and inversion of a large size of the autocorrelation matrix.
3. The steepest decent algorithm can find the optimal weight solution using an iterative method, so a large matrix inversion is not needed. But it still requires calculating an autocorrelation matrix and cross-correlation matrix.
4. The LMS is the sample-based algorithm, which does not need collection of data or computation of statistics and does not involve matrix inversion.
5. The convergence factor for the LMS algorithm is bounded by the reciprocal of the product of the number of filter coefficients and input signal power.

6. The RLS adaptive FLR filter is introduced. The RLS algorithm offers fast convergence rate but requires a large computational load.
7. The adaptive FIR filter can effectively be applied for noise cancellation, system modeling, and line enhancement.
8. Further exploration includes other applications such as cancellation of periodic interference, biomedical ECG signal enhancement, and adaptive telephone echo cancellation.

9.6 PROBLEMS

9.1 Given a quadratic MSE function for the Wiener filter:

$$J = 50 - 40w + 10w^2,$$

find the optimal solution for w^* to achieve the minimum MSE $J_{\min}$ and determine $J_{\min}$.

9.2 Given a quadratic MSE function for the Wiener filter:

$$J = 15 + 20w + 10w^2,$$

find the optimal solution for w^* to achieve the minimum MSE $J_{\min}$ and determine $J_{\min}$.

9.3 Given a quadratic MSE function for the Wiener filter:

$$J = 100 + 20w + 2w^2,$$

find the optimal solution for w^* to achieve the minimum MSE $J_{\min}$ and determine $J_{\min}$.

9.4 Given a quadratic MSE function for the Wiener filter:

$$J = 10 - 30w + 15w^2,$$

find the optimal solution for w^* to achieve the minimum MSE $J_{\min}$ and determine $J_{\min}$.

9.5 Given a quadratic MSE function for the Wiener filter:

$$J = 50 - 40w + 10w^2,$$

use the steepest decent method with an initial guess as $w(0) = 0$ and the convergence factor $\mu = 0.04$ to find the optimal solution for w^* and determine $J_{\min}$ by iterating three times.

9.6 Given a quadratic MSE function for the Wiener filter:

$$J = 15 + 20w + 10w^2,$$

use the steepest decent method with an initial guess as $w(0) = 0$ and the convergence factor $\mu = 0.04$ to find the optimal solution for w^* and determine $J_{\min}$ by iterating three times.

9.7 Given a quadratic MSE function for the Wiener filter:

$$J = 100 + 20w + 2w^2,$$

use the steepest decent method with an initial guess as $w(0) = -4$ and the convergence factor $\mu = 0.2$ to find the optimal solution for w^* and determine $J_{\min}$ by iterating three times.

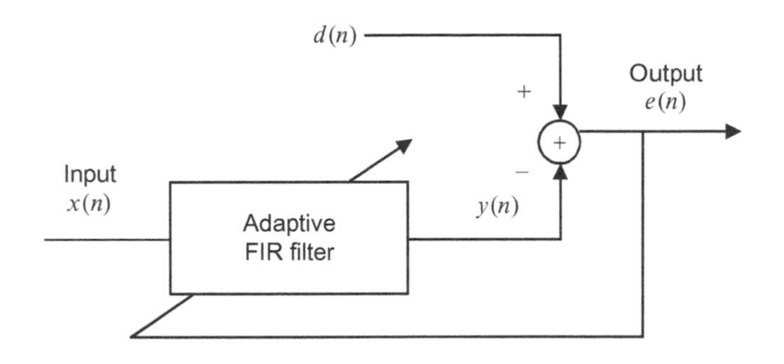

FIG. 9.22

Noise cancellation in Problem 9.9.

9.8 Given a quadratic MSE function for the Wiener filter:

$$J = 10 - 30w + 15w^2,$$

use the steepest decent method with an initial guess as $w(0) = 2$ and the convergence factor $\mu = 0.02$ to find the optimal solution for w^* and determine $J_{\min}$ by iterating three times.

9.9 Given the following adaptive filter used for noise cancellation application (Fig. 9.22), in which $d(0) = 3$, $d(1) = -2$, $d(2) = 1$, $x(0) = 3$, $x(1) = -1$, $x(2) = 2$, and an adaptive filter with two taps: $y(n) = w_0 x(n) + w_1 x(n-1)$ with initial values $w_0 = 0$, $w_1 = 1$, and $\mu = 0.1$

(a) Determine the LMS algorithm equations

$y(n) =$

$e(n) =$

$w_0 =$

$w_1 =$

(b) Perform adaptive filtering for each $n = 0, 1, 2$.

(c) Determine equations using the RLS algorithm.

(d) Repeat (b) using the RLS algorithm with $\delta = 1/2$ and $\lambda = 0.96$.

9.10 Given a DSP system with a sampling rate set up to 8000 samples per second, implement adaptive filter with five taps for system modeling.
As shown in Fig. 9.23, assume that the unknown system transfer function is

$$H(z) = \frac{0.25 + 0.25z^{-1}}{1 - 0.5z^{-1}}.$$

(a) Determine the DSP equations using the LMS algorithm

$y(n) =$

$e(n) =$

$w_k =$

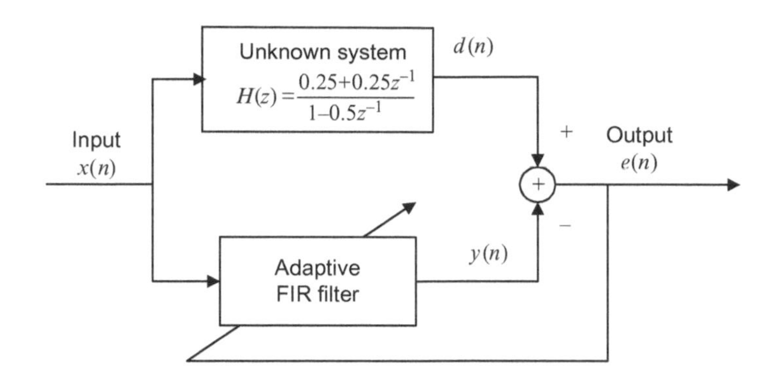

FIG. 9.23

System modeling in Problem 9.9.

for $i = 0, 1, 2, 3, 4$; that is, write the equations for all adaptive coefficients:

$w_0 =$

$w_1 =$

$w_2 =$

$w_3 =$

$w_4 =$

(b) Determine equations using the RLS algorithm.

9.11 Given an adaptive filter used for noise cancellation application in Problem 9.9, in which $d(0) = 3$, $d(1) = -2$, $d(2) = 1$, $x(0) = 3$, $x(1) = -1$, $x(2) = 2$, and an adaptive filter with three taps:

$y(n) = w_0x(n) + w_1x(n-1) + w_2x(n-2)$ with initial values $w_0 = 0$, $w_1 = 0$, $w_2 = 0$ and $\mu = 0.2$

(a) Determine the LMS algorithm equations

$y(n) =$

$e(n) =$

$w_0 =$

$w_1 =$

$w_2 =$

(b) Perform adaptive filtering for each of $n = 0, 1, 2$.

(c) Determine equations using the RLS algorithm.

(d) Repeat (b) using the RLS algorithm with $\delta = 1/2$ and $\lambda = 0.96$.

9.12 Given a DSP system with a sampling rate set up to 8000 samples per second in Problem 9.10, implement adaptive filter with five taps for system modeling, assuming that the unknown system transfer function is

$$H(z) = 0.2 + 0.3z^{-1} + 0.2z^{-2},$$

(a) Determine the DSP equations using the LMS algorithm

$y(n) =$
$e(n) =$
$w_k =$

for $i = 0, 1, 2, 3, 4$; that is, write the equations for all adaptive coefficients:

$w_0 =$
$w_1 =$
$w_2 =$
$w_3 =$
$w_4 =$

(b) Determine equations using the RLS algorithm.

9.13 Given a DSP system set up for noise cancellation application with a sampling rate set up to 8000 Hz, as shown in Fig. 9.24, the desired signal of a 1000-Hz tone is generated internally via a tone generator; and the generated tone is corrupted by the noise captured from a microphone. An FIR adaptive filter with 25 taps is applied to reduce the noise in the corrupted tone.

(a) Determine the DSP equation for the channel noise $n(n)$.
(b) Determine the DSP equation for signal tone $yy(n)$.
(c) Determine the DSP equation for the corrupted tone $d(n)$.
(d) Set up the LMS algorithm for the adaptive FIR filter.
(e) Set up equations using the RLS algorithm.

9.14 Given a DSP system for noise cancellation application with two (2) taps in Fig. 9.25:

(a) Set up the LMS algorithm for the adaptive filter.
(b) Given the following inputs and outputs:

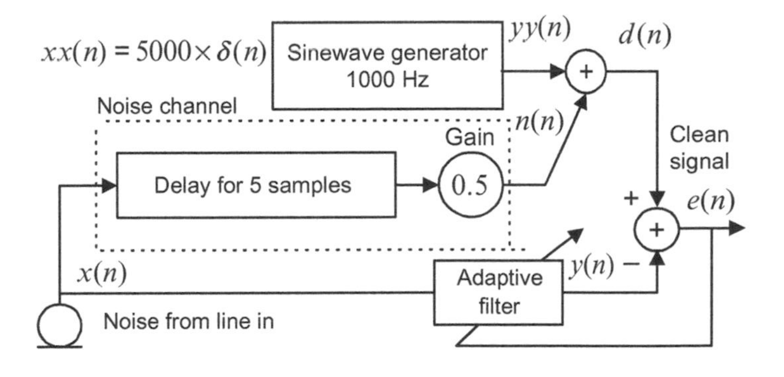

FIG. 9.24

Noise cancellation in Problem 9.13.

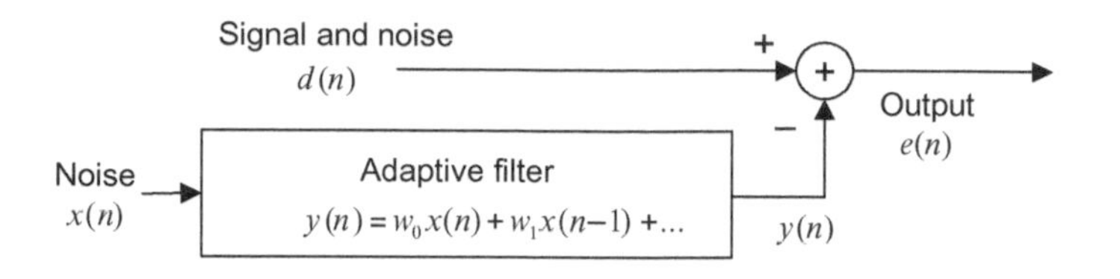

FIG. 9.25

Noise cancellation in Problem 9.14.

$x(0)=1$, $x(1)=1$, $x(2)=-1$, $x(3)=2$, $d(0)=0$, $d(1)=2$, $d(2)=-1$, and $d(3)=1$ and initial weights: $w_0=w_1=0$, convergence factor is set to be $\mu=0.1$, perform adaptive filtering to obtain outputs $e(n)$ for $n=0, 1, 2$.

9.15 Given a DSP system for noise cancellation application with three (3) taps in Fig. 9.25:

(a) Set up the LMS algorithm for the adaptive filter.

(b) Given the following inputs and outputs:

$x(0)=1$, $x(1)=1$, $x(2)=-1$, $x(3)=2$, $d(0)=0$, $d(1)=2$, $d(2)=-1$, and $d(3)=1$ and initial weights: $w_0=w_1=w_2=0$, convergence factor is set to be $\mu=0.1$, perform adaptive filtering to obtain outputs $e(n)$ for $n=0, 1, 2$.

(c) Determine equations using the RLS algorithm.

(d) Repeat (b) using the RLS algorithm with $\delta=1$ and $\lambda=0.96$.

9.16 For a line enhancement application using the FIR adaptive filter depicted in Fig. 9.26,

(a) Set up the LMS algorithm for the adaptive filter using two filter coefficients and delay $\Delta=2$.

(b) Given the following inputs and outputs:

$d(0)=-1$, $d(1)=1$, $d(2)=-1$, $d(3)=1$, $d(4)=-1$, $d(5)=1$and $d(6)=-1$ and initial weights: $w_0=w_1=0$, convergence factor is set to be $\mu=0.1$, perform adaptive filtering to obtain outputs $y(n)$ for $n=0, 1, 2, 3, 4$.

(c) Determine equations using the RLS algorithm.

(d) Repeat (b) using the RLS algorithm with $\delta=1$ and $\lambda=0.96$.

9.17 Repeat Problem 9.16 using three-tap FIR filter and $\Delta=3$.

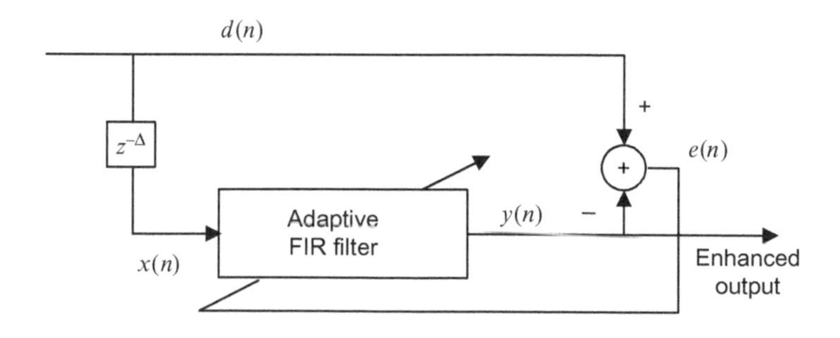

FIG. 9.26

Line enhancement in Problem 9.16.

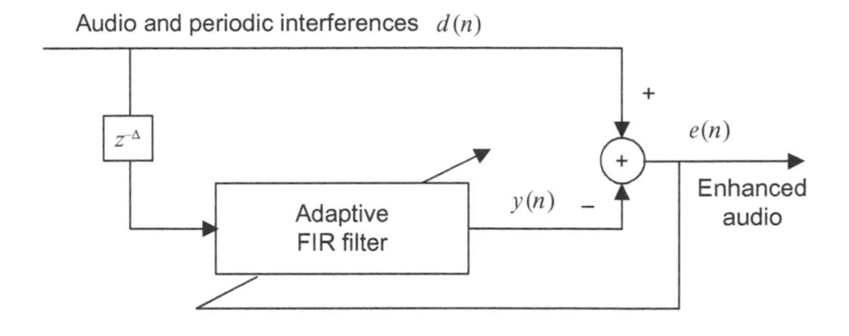

FIG. 9.27

Interference cancellation in Problem 9.18.

9.18 An audio playback application is described in Fig. 9.27.

Due to the interference environment, the audio is corrupted by 15 different periodic interferences. The DSP engineer uses an FIR adaptive filter to remove such interferences as shown in Fig. 9.27.

(a) What is the minimum number of filter coefficients?

(b) Set up the LMS algorithm for the adaptive filter using the number of taps obtained in (a).

(c) Set up equations using the RLS algorithm.

9.19 Repeat Problem 9.18 for the corrupted audio which contains five different periodic interferences.

9.20 In a noisy ECG acquisition environment, the DSP engineer uses an adaptive FIR filter with 20 coefficients to remove 60-Hz interferences. The system is set up as shown in Fig. 9.19, where the corrupted ECG and enhanced ECG are represented as $d(n)$ and $e(n)$, respectively; $x(n)$ is the captured reference signal from the 60-Hz interference; and $y(n)$ is the adaptive filter output. Determine all difference equations to implement the adaptive filter.

9.21 Given an application of the echo cancellation as shown in Fig. 9.21B, determine all the difference equations to implement the adaptive filter with four adaptive coefficients at the speaker A site.

9.22 Given an application of the echo cancellation as shown in Fig. 9.21B,

(a) Explain the concepts and benefits using the echo canceller.

(b) Explain the operations of the adaptive filter at the speaker B site.

(c) Determine all difference equations to implement the adaptive filter at the speaker A site.

Computer Problems with MATLAB:

Use MATLAB to solve Problems 9.23–9.26.

9.23 Write a MATLAB program for minimizing the two-weight mean squared error (MSE) function

$$J = 100 + 100w_1^2 + 4w_2^2 - 100w_1 - 8w_2 + 10w_1w_2$$

by applying the steepest descent algorithm for 500 iterations.

The derivatives are derived as

$$\frac{dJ}{dw_1} = 200w_1 - 100 + 10w_2 \text{ and } \frac{dJ}{dw_2} = 8w_2 - 8 + 10w_1$$

and the initial weights are assumed as $w_1(0)=0$, $w_2(0)=0$, $\mu=0.001$.

Plot $w_1(k)$, $w_2(k)$, and $J(k)$ vs. the number of iterations, respectively, and summarize your results.

9.24 In Problem 9.10, the unknown system is assumed as a fourth-order Butterworth bandpass filter with a lower cutoff frequency of 700 Hz and an upper cutoff frequency of 900 Hz. Design a bandpass filter by the bilinear transformation method for simulating the unknown system with a sampling rate of 8000 Hz.

(a) Generate the input signal for 0.1 s using a sum of three sinusoids having 100, 800, and 1500 Hz with a sampling rate of 8000 Hz.

(b) Use the generated input as the unknown system input to produce the system output.

The adaptive FIR filter is then applied to model the designed bandpass filter. The following parameters are assumed:

Adaptive FIR filter
Number of taps: 15 coefficients
Algorithm: LMS algorithm
Convergence factor: 0.01

(c) Implement the adaptive FIR filter, plot the system input, system output, adaptive filter output, and the error signal, respectively.

(d) Plot the input spectrum, system output spectrum, and adaptive filter output spectrum, respectively.

(e) Repeat (a)–(d) using the RLS algorithm with $\delta=1$ and $\lambda=0.96$.

9.25 Use the following MATLAB code to generate the reference noise and the signal of 300 Hz corrupted by the noise with a sampling rate of 8000 Hz.

```
fs=8000; T=1/fs;                        % Sampling rate and sampling period
t=0:T:1;                                % Create time instants
x=randn(1,length(t));                   % Generate reference noise
n=filter([ 0 0 0 0 0 0 0 0 0 0.8 ],1,x); % Generate the corruption noise
d=sin(2*pi*300*t)+n;                    % Generate the corrupted signal
```

(a) Implement an adaptive FIR filter to remove the noise. The adaptive filter specifications are as follows:

Sample rate = 8000 Hz
Signal corrupted by Gaussian noise delayed by nine samples from the reference noise
Reference noise: Gaussian noise with a power of 1
Number of FIR filter tap: 16
Convergence factor for the LMS algorithm: 0.01

(b) Plot the corrupted signal, reference noise, and enhanced signal, respectively.

(c) Compare the spectral plots between the corrupted signal and the enhanced signal.

(d) Repeat (a)–(c) using the RLS algorithm with $\delta=1$ and $\lambda=0.96$.

9.26 A line enhancement system (Fig. 9.28) has following specifications:

Sampling rate = 1000 Hz
Corrupted signal: 100 Hz tone with the unit amplitude added with the unit power white Gaussian noise
Adaptive filter: FIR type, 16 taps
Convergence factor = 0.001
Delay value Δ = to be decided according to the experiment

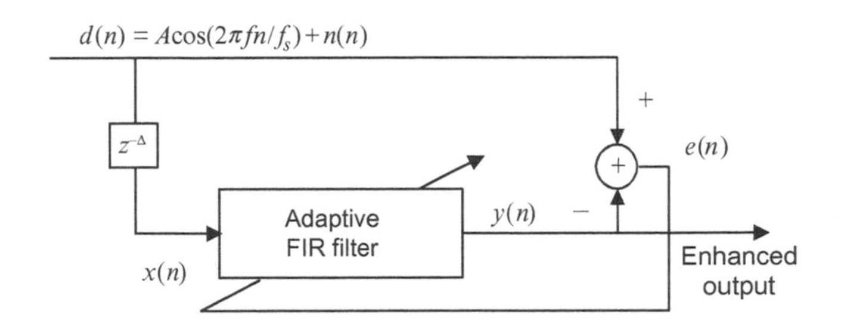

FIG. 9.28

A line enhancement system in Problem 9.26.

LMS algorithm is applied

(a) Write a MATLAB program to perform the line enhancement for 1-s corrupted signal. Run the developed program with a trail of delay value Δ and plot the noisy signal, the enhanced signals and their spectra.

(b) Run your simulation to find the best delay value Δ, which achieves the largest noise reduction.

(c) Repeat (a)–(b) using the RLS algorithm with $\delta=1$ and $\lambda=0.96$.

MATLAB Projects

9.27 Active noise control:

The ANC system is based on the principle of superposition of the primary noise source and secondary source with its acoustic output being of the same amplitude but the opposite phase of the primary noise source, as shown in Fig. 9.29. The primary noise is captured using the reference microphone, which is located close to the noise source. The ANC system uses the sensed reference signal $x(n)$ to generate a canceling signal $y(n)$, which drives the secondary speaker to destructively attenuate the primary noise. An error microphone is used to detect the residue noise $e(n)$, which is fed back to the ANC system to monitor the system performance. The residue noise $e(n)$ together with the reference signal $x(n)$ are used by the linear adaptive controller whose coefficients are adjusted via an adaptive algorithm to minimize the measured error signal $e(n)$, or the residue acoustic noise. $P(z)$ designates the physical primary path between the reference sensor and the error sensor, and $S(z)$ the physical secondary path between the ANC adaptive filter output and the error sensor. To control the noise at the cancelling point, the instantaneous power $e(n)$ must be minimized. Note that

$$E(z)=D(z)-Y(z)S(z)=D(z)-[W(z)X(z)]S(z),$$

where $W(z)$ denotes the adaptive control filter. Exchange of the filter order (since the filters are linear filters) gives

$$E(z)=D(z)-W(z)[S(z)X(z)]=D(z)-W(z)U(z).$$

Assuming that $\overline{S}(z)$ is the secondary path estimate and noticing that $u(n)$ and $y(n)$ are the filtered reference signal and adaptive filter output, applying the LMS algorithm gives the filtered-x LMS algorithm:

$$w_k=w_k+2\mu e(n)u(n-k).$$

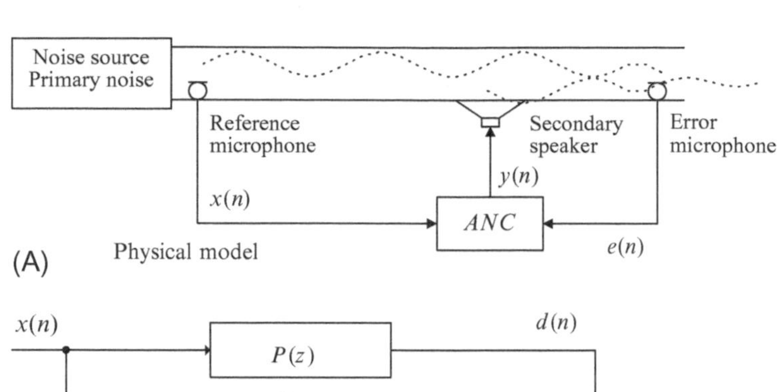

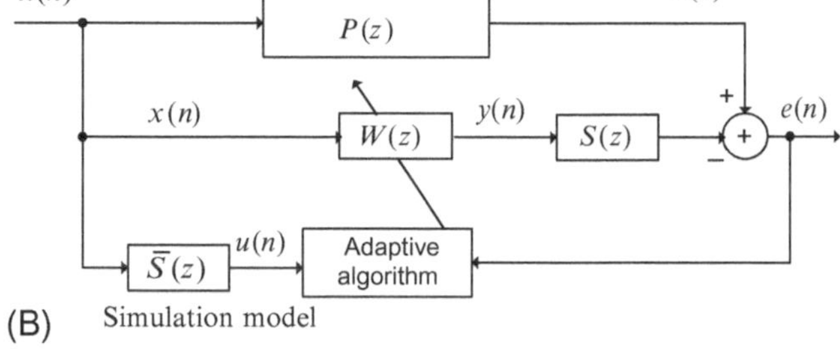

FIG. 9.29

An active noise control system. (A) Physical model, (B) Simulation model.

Note that $e(n)$ is measured from the error microphone.

The completed filtered-x LMS algorithm is summarized below:

---

(1) Initialize w_0, w_1, … w_{N-1} to arbitrary values

(2) Read $x(n)$ and perform digital filtering

$y(n) = w_0 x(n) + w_1 x(n-1) + \cdots + w_{N-1} x(n-N+1)$

(3) Compute filtered inputs

$u(n) = \bar{s}_0 x(n) + \bar{s}_1 x(n-1) + \cdots + \bar{s}_{M-1} x(n-M+1)$

(4) Read $e(n)$and update each filter coefficient

for $k=0, \ldots, N-1$, $w_k = w_k + 2\mu e(n) u(n-k)$

Assuming the following:

Sampling rate = 8000 Hz and simulation duration = 10 s

Primary noise: $x(n)$=500-Hz sine wave

Primary path: $P(z) = 0.2 + 0.25z^{-1} + 0.2z^{-2}$

Secondary path: $S(z) = 0.25 + 0.2z^{-1}$

Secondary path estimate: $\bar{S}(z) = S(z) = 0.2 + 0.2z^{-1}$ [can have slight error as compared to $S(z)$]

Residue error signal:

$$e(n) = d(n) - \text{filtering } y(n) \text{ using coefficients of the secondary path } s(n) = d(n) - y(n) * s(n),$$

where the symbol "*" denotes the filter convolution.

Implement the ANC system and plot the residue sensor signal to verify the effectiveness.

The primary noise at cancelling point $d(n)$, and filtered reference signal $u(n)$ can be generated in MATLAB as follows:

```
d = filter([0.2 0.25 0.2],1,x); % Simulate physical media
u=filter([0.2 0.2],1,x);
```

The residue error signal $e(n)$ should be generated sample by sample and embedded into the adaptive algorithm, that is,

```
e(n)=d(n)-(s(1)*y(n)+s(2)*y(n-1)); % Simulate the residue error
```

whereas s(1)=0.25 and s(2)=0.2.

Details of active control systems can be found in the textbook (Kuo and Morgan, 1996).

9.28 Frequency tracking:

An adaptive filter can be applied for real-time frequency tracking (estimation). In this application, a special second notch IIR filter structure, as shown in Fig. 9.30, is preferred for simplicity. The notch filter transfer function

$$H(z) = \frac{1 - 2\cos(\theta)z^{-1} + z^{-2}}{1 - 2r\cos(\theta)z^{-1} + r^2 z^{-2}}$$

has only one adaptive parameter θ. It has two zeros on the unit circle resulting in an infinite-depth notch. The parameter r controls the notch bandwidth. It requires $0 << r < 1$ for achieving a narrowband notch. When r is close to 1, the 3-dB notch filter bandwidth can be approximated as $BW \approx 2(1-r)$ (see Chapter 8). The input sinusoid whose frequency f needs to be estimated and tracked is given below:

$$x(n) = A\cos(2\pi fn/f_s + \alpha)$$

where A and α are the amplitude and phase angle. The filter output is expressed as

$$y(n) = x(n) - 2\cos[\theta(n)]x(n-1) + x(n-2) + 2r\cos[\theta(n)]y(n-1) - r^2 y(n-2).$$

The objective is to minimize the filter instantaneous output power $y^2(n)$. Once the output power is minimized, the filter parameter $\theta = 2\pi f/f_s$ will converge to its corresponding frequency f(Hz). The LMS algorithm to minimize the instantaneous output power $y^2(n)$ is given as

$$\theta(n+1) = \theta(n) - 2\mu y(n)\beta(n),$$

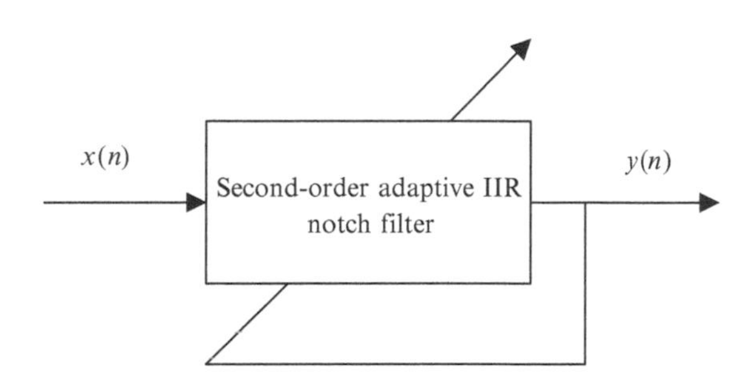

FIG. 9.30

A frequency tracking system.

where the gradient function $\beta(n)=\partial y(n)/\partial\theta(n)$ can be derived as follows:

$$\beta(n)=2\sin[\theta(n)]x(n-1)-2r\sin[\theta(n)]y(n-1)+2r\cos[\theta(n)]\beta(n-1)-r^2\beta(n-2)$$

and μ is the convergence factor which controls the speed of algorithm convergence.

In this project, plot and verify the notch frequency response by setting $f_s=8000\,\text{Hz}$, $f=1000\,\text{Hz}$, and $r=0.95$. Then generate the sinusoid with duration of 10s, frequency of 1000Hz, and amplitude of 1. Implement the adaptive algorithm using an initial guess $\theta(0)=2\pi\times 2000/f_s=0.5\pi$ and plot the tracked frequency $f(n)=\theta(n)f_s/2\pi$ for tracking verification.

Notice that this particular notch filter only works for a single frequency tracking, since the mean squared error function $E[y^2(n)]$ has a one global minimum (one best solution when the LMS algorithm converges). Details of adaptive notch filter can be found in the reference (Tan and Jiang, 2012). Notice that the general IIR adaptive filter suffers from local minima, that is, the LMS algorithm converges to local minimum and the nonoptimal solution results in.

9.29 Channel equalization:

Channel equalization or inverse modeling is depicted in Fig. 9.31. An adaptive filter (equalizer) is utilized to compensate for the linear distortion caused by the channel. For noise-free case: $n(n)=0$,
we wish

$$H(z)W(z)=z^{-L}$$

where $H(z)$ and $W(z)$ are the unknown channel transfer function and the adaptive filter. L is the time delay, which is best chosen to be half of time span of the equalizer. In this project, only the FIR adaptive filter is considered. The channel $H(z)$ usually causes the symbols $s(n)$ dispersed after they arrive at the receiver. $x(n)$, the received symbols, may interfere with each other. This effect will cause the wrong information after detection. Equalizer uses $x(n)$ to recover $s(n-L)$. This type equalization usually involves two steps: training step and decision step. In the training step, a previously chosen training signal (pseudo-noise sequence long enough to allow the equalizer to compensate for channel distortion) is transmitted through the channel $H(z)$. The received symbols $x(n)$ are used for reference signal for the adaptive filter while the properly delayed symbols $s(n-L)$ are employed for the designed signal. After the adaptive filter converges, the optimal

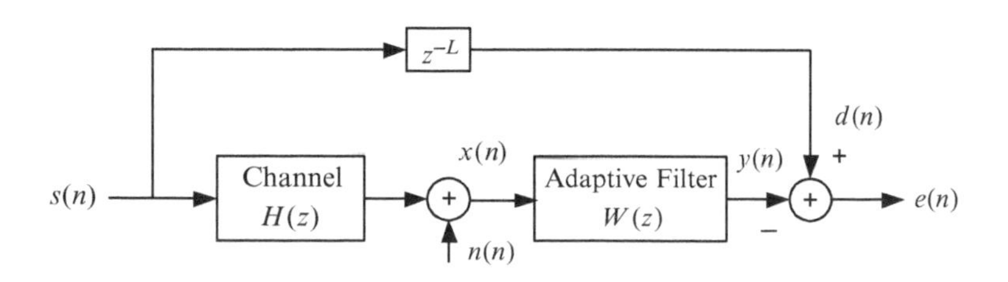

FIG. 9.31

Channel equalization system.

coefficients $W(z)$ are found. Step 2 is the decision stage. The equalizer operates using real communication symbols.
In this project, assuming

$$H(z)=\frac{1}{z-0.8},\ n(n)=0,\ \text{and adaptive filter length}=32,$$

Stage 1: Generate 1000 symbols (+1 and −1): s=2*round(rand(1,1000))-1
(a) Choose the best delay L to design the equalizer using the LMS algorithm.
(b) Choose the best delay L to design the equalizer using the RLS algorithm.
Stage 2: Generate another 1000 symbols (+1 and −1): s=2*round(rand(1,1000))-1
Use the designed equalizer (a) or (b) to obtain the equalized symbols

$$\hat{s}(n)=\begin{cases}1 & y(n)\geq 0\\ -1 & y(n)<0\end{cases}.$$

Delay the test symbols: $s_d(n)=s(n-L)$,

Verify the error rate: $0.5\sum_{n=L+1}^{1000}|s_d(n)-\hat{s}(n)|/(1000-L)$.

Advanced Problems

9.30 Determine the following
(a) $E\{\sin n\Omega\}$
(b) $E\{e^{jn\Omega}\}$
(c) $E\{(A\sin n\Omega)^2\}$
(d) $E\{(A\cos n\Omega)^2\}$
(e) $E\{(A\cos n\Omega)(A\sin n\Omega)\}$
(f) $E\{[A\cos(n-n_0)\Omega]^2\}$
(g) $E\{[A\sin(n-n_0)\Omega][A\cos(n)\Omega]\}$
where n is the time index and n_0 is the fixed number of samples, and $\Omega\neq 0$.

9.31 The following Wiener filter is used to predict the sinusoid:

$$d(n)=A\sin(n\Omega),$$

where the Wiener filter predictor with delays of n_1, n_2, and n_3 is given as

$$y(n)=w_1d(n-n_1)+w_2d(n-n_2)+w_3d(n-n_3).$$

Find the Wiener filter coefficients: w_1, w_2, and w_3.

9.32 Given the MSE function as

$$J=16+2w_0^2+w_1^2-2w_0+2w_1.$$

Determine the optimal weights and $J_{\min}$.

9.33 Given the MSE function as

$$J=16+2w_0^2+w_1^2-2w_0+2w_1-2w_0w_1.$$

Determine the optimal weights and $J_{\min}$.

9.34 Given the MSE function as

$$J=16+2w_0^2+w_1^2-2w_0+2w_1.$$

(a) Set up the steepest decent algorithm.

(b) For initial weights as $w_0(0)=0$ and $w_1(0)=0$, and $\mu=0.1$, use MATLAB to find the optimal solution by iterating 100 times and plot the trajectory of $w_0(n)$ and $w_1(n)$ on the MSE contour surface.

9.35 Given the MSE function as

$$J=16+2w_0^2+w_1^2-2w_0+2w_1-2w_0w_1.$$

(a) Set up the steepest decent algorithm.

(b) For initial weights as $w_0(0)=1$ and $w_1(0)=0$, and $\mu=0.1$, use MATLAB to find the optimal solution by iterating 100 times and plot the trajectory of $w_0(n)$ and $w_1(n)$ on the MSE contour surface.

9.36 For predicting a sinusoid $x(n)=A\sin(n\Omega)$ using a forward predictor

$$x(n)=-a_1x(n-n_1)-a_2x(n-n_2) \text{ for } 0<n_1<n_2.$$

Show that the predictor coefficients are as follows:

$$\begin{bmatrix} a_1 \\ a_2 \end{bmatrix} = -\begin{bmatrix} \dfrac{\sin(n_2\Omega)}{\sin[(n_2-n_1)\Omega)]} \\ -\dfrac{\sin(n_1\Omega)}{\sin[(n_2-n_1)\Omega)]} \end{bmatrix} \text{ for } (n_2-n_1)\Omega \neq k\pi \text{ for } k=0, \pm 1, \pm 2, \cdots$$

9.37 A fourth-order forward linear predictor is used to predict the sinusoids: $x(n)=\sqrt{2}\sin(\pi n/8)+\cos(\pi n/16)$.

For the one-step forward linear predictor, find the predictor coefficients and show the sinusoid can be fully predicted by showing $E\{e_f^2(n)\}=0$.

9.38 Given the matrix inversion lemma: $A^{-1}=B-BC(D+C^TBC)^{-1}C^TB$

Verify the matrix inversion lemma using the following:

$$A=\begin{bmatrix} 2 & -1/2 \\ 0 & 1/2 \end{bmatrix},\quad B=\begin{bmatrix} 1 & 1 \\ 0 & 2 \end{bmatrix},\quad C=\begin{bmatrix} 1 \\ 0 \end{bmatrix},\quad D=1.$$

9.39 For predicting $x(n)=A\sin^2(n\Omega)$ using a forward predictor with a constant

$$\hat{x}(n)=-a_0-a_1x(n-n_1)-a_2x(n-n_2) \text{ for } 0<n_1<n_2.$$

Determine the predictor coefficients.

CHAPTER

WAVEFORM QUANTIZATION AND COMPRESSION 10

CHAPTER OUTLINE

10.1 LINEAR MIDTREAD QUANTIZATION

As we discussed in Chapter 2, in a digital signal processing (DSP) system, the first step is to sample and quantize the continuous signal. Quantization is the process of rounding off the sampled signal voltage to the predetermined levels that will be encoded by analog-to-digital conversion (ADC). We have described the quantization process in Chapter 2, in which we studied unipolar and bipolar linear quantizers in detail. In this section, we focus on a linear midtread quantizer, which is used in the digital communications (Roddy and Coolen, 1997; Tomasi, 2004), and its use to quantize the speech waveform. The linear midtread quantizer is similar to the bipolar linear quantizer discussed in Chapter 2 except that the midtread quantizer offers the same decoded magnitude range for both positive and negative voltages.

Let us look at a midtread quantizer. The characteristics and binary codes for a 3-bit midtread quantizer are depicted in Fig. 10.1, where the code is in a sign magnitude format. Positive voltage is coded

Digital Signal Processing. https://doi.org/10.1016/B978-0-12-815071-9.00010-5

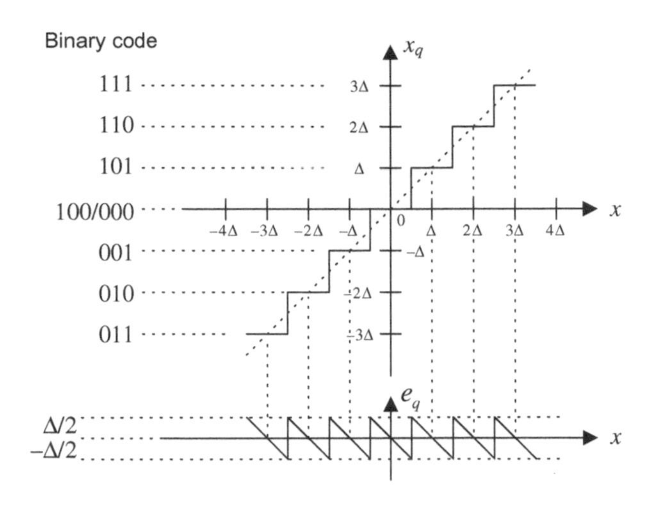

FIG. 10.1

Characteristics of a 3-bit midtread quantizer.

using a sign bit of logic 1, while negative voltage is coded by a sign bit of logic 0; the next two bits are the magnitude bits. The key feature of the linear midtread quantizer is noted as follows: when $0 \le x < \Delta/2$, the binary code of 100 is produced; when $-\Delta/2 \le x < 0$, the binary code of 000 is generated, where Δ is the quantization step size. However, the quantized values for both codes of 100 and 000 are the same and equal to $x_q = 0$. We can also see details in Table 10.1. For the 3-bit midtread quantizer, we expect seven quantized values instead of 8; that is; there are $2^n - 1$ quantization levels for the n-bit midtread quantizer. Note that the quantization signal range is $(2^n - 1)\Delta$ and the magnitudes of the quantized values are symmetric, as shown in Table 10.1. We apply the midtread quantizer particularly for speech waveform coding.

The following example serves to illustrate coding principles of the 3-bit midtread quantizer.

Table 10.1 Quantization Table for the 3-Bit Midtread Quantizer

Binary Code	Quantization Level x_q (V)	Input Signal Subrange (V)
0 1 1	-3Δ	$-3.5\Delta \le x < -2.5\Delta$
0 1 0	-2Δ	$-2.5\Delta \le x < -1.5\Delta$
0 0 1	$-\Delta$	$-1.5\Delta \le x < -0.5\Delta$
0 0 0	0	$-0.5\Delta \le x < 0$
1 0 0	0	$0 \le x < 0.5\Delta$
1 0 1	Δ	$0.5\Delta \le x < 1.5\Delta$
1 1 0	2Δ	$1.5\Delta \le x < 2.5\Delta$
1 1 1	3Δ	$2.5\Delta \le x < 3.5\Delta$

Note that: step size $=\Delta=(x_{max}-x_{min})/(2^3-1)$; x_{max} *= maximum voltage; and* $x_{min} = -x_{max}$; *and coding format: (a) sign bit: 1 = plus; 0 = minus; (b) 2 magnitude bits.*

EXAMPLE 10.1

For the 3-bit midtread quantizer described in Fig. 10.1 and the analog signal with a range from −5 to 5 V,

(a) Determine the quantization step size, and
(b) Determine the binary codes, recovered voltages, and quantization errors when the input is −3.6 and 0.5 V, respectively.

Solution:

(a) The quantization step size is calculated as

$$\Delta = \frac{5-(-5)}{2^3-1} = 1.43\text{ V}.$$

(b) For $x = -3.6$ V, we have $x = \frac{-3.6}{1.43} = -2.52\Delta$. From quantization characteristics, it follows that the binary code = 011 and the recovered voltage is $x_q = -3\Delta = -4.29$ V. Thus the quantization error is computed as

$$e_q = x_q - x = -4.28 - (-3.6) = -0.69\text{ V}.$$

For $x = 0.5 = \frac{0.5}{1.43}\Delta = 0.35\Delta$, we get binary code = 100. Based on Fig. 10.1, the recovered voltage and quantization error are found to be $x_q = 0$ and

$$e_q = 0 - 0.5 = -0.5\text{ V}.$$

As discussed in Chapter 2, the linear midtread quantizer introduces quantization noise, as shown in Fig. 10.1; and the signal-to-noise power ratio (SNR) is given by

$$\text{SNR dB} = 10.79 + 20 \times \log_{10}\left(\frac{x_{rms}}{\Delta}\right), \tag{10.1}$$

where x_{rms} designates the root-mean-squared value of the speech data to be quantized. The practical equation for estimating the SNR for the speech data sequence $x(n)$ of N data points is written as

$$\text{SNR} = \left(\frac{\sum_{n=0}^{N-1} x^2(n)}{\sum_{n=0}^{N-1} \left(x_q(n) - x(n)\right)^2} \right) \tag{10.2}$$

$$\text{SNR dB} = 10 \cdot \log_{10}(\text{SNR}). \tag{10.3}$$

Note that $x(n)$ and $x_q(n)$ are the speech data to be quantized and the quantized speech data, respectively. Eq. (10.2) gives the absolute SNR, and Eq. (10.3) produces the SNR in terms of decibel (dB). Quantization error is the difference between the quantized speech data (or quantized voltage level) and speech data (or analog voltage), that is, $(x_q(n) - x(n))$. Also note that from Eq. (10.1), increasing 1 bit to the linear quantizer would improve SNR by approximately 6 dB. Let us examine performance of the 5-bit linear midtread quantizer.

In the following simulation, we use a 5-bit midtread quantizer to quantize the speech data. After quantization, the original speech, quantized speech, and quantized error are plotted in Fig. 10.2. Since the program calculates $x_{rms}/x_{\max} = 0.203$, we yield x_{rms} as $x_{rms} = 0.203 \times x_{\max} = 0.0203 \times 5 = 1.015$ and

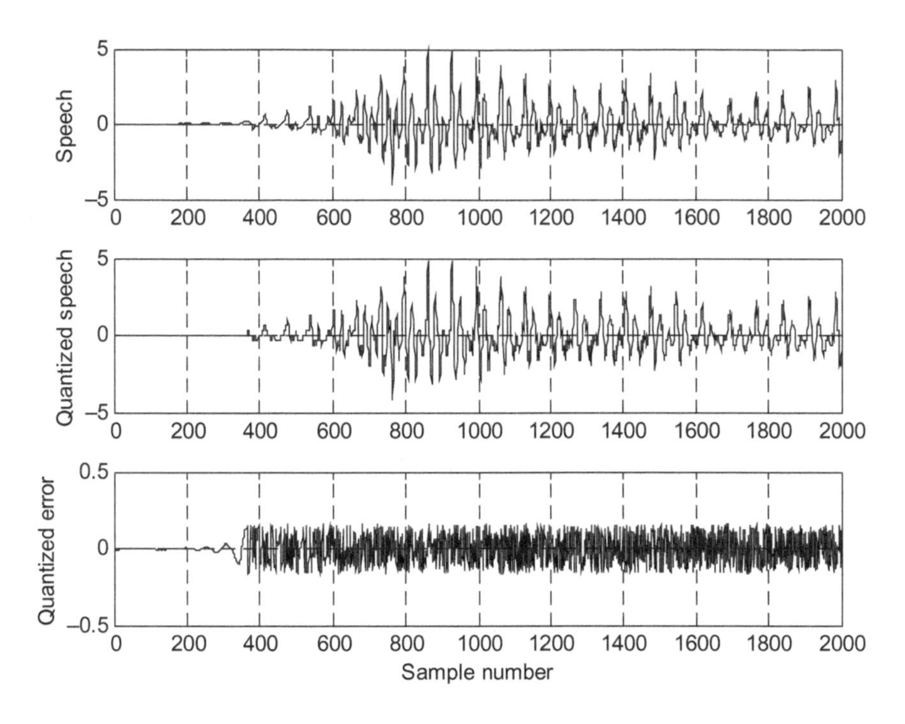

FIG. 10.2

Plots of original speech, quantized speech, and quantization error.

$\Delta = 10/(2^5 - 1) = 0.3226$. Applying Eq. (10.1) gives SNR = 21.02 dB. The SNR using Eqs. (10.2) and (10.3) is approximately 21.6 dB.

The first plot in Fig. 10.2 is the original speech, and the second plot shows the quantized speech. Quantization error is displayed in the third plot, where the error amplitude interval is uniformly distributed between −0.1613 and 0.1613, indicating the bounds of the quantized error ($\Delta/2$). The details of the MATLAB implementation is given in Programs 10.1–10.3, 10.7 in Section 10.6.

To improve the SNR, the number of bits must be increased. However, increasing the number of encoding bits will cost an expansive ADC device, larger storage media for storing the speech data, and a larger bandwidth for transmitting the digital data. To gain a more efficient quantization approach, we study the μ-law companding in the next section.

10.2 μ-LAW COMPANDING

In this section, we study the analog μ-law companding, which takes an analog input signal, and digital μ-law companding, which deals with linear pulse-code modulation (PCM) codes.

10.2.1 ANALOG μ-LAW COMPANDING

To reduce the number of bits to encode each speech datum, μ-law companding, called log-PCM coding, is applied. μ-Law companding (Roddy and Coolen, 1997; Tomasi, 2004) was first used in the United States and Japan in the telephone industry (G.711 standard). μ-Law companding is a compression process. It explores the principle that the higher amplitudes of analog signals are compressed before ADC while expanded after digital-to-analog conversion (DAC). As studied in the linear quantizer, the quantization error is uniformly distributed. This means that the maximum quantization error stays the same no matter how big or small the speech samples are. μ-Law companding can be employed to make the quantization error smaller when the sample amplitude is smaller and to make the quantization error bigger when the sample amplitude is bigger, using the same number of bits per sample. It is described in Fig. 10.3.

As shown in Fig. 10.3, x is the original speech sample, which is the input to the compressor, while y is the output from the μ-law compressor; then the output y is uniformly quantized. Assuming that the quantized sample y_q is encoded and sent to the μ-law expander, the expander will perform the reverse process to obtain the quantized speech sample x_q. The compression and decompression processes cause the maximum quantization error $|x_q - x|_{max}$ to be small for the smaller sample amplitudes and large for the larger sample amplitudes.

The equation for the μ-law compressor is given by

$$y = \text{sign}(x) \frac{\ln\left(1 + \mu \frac{|x|}{|x|_{max}}\right)}{\ln(1+\mu)}, \tag{10.4}$$

where $|x|_{max}$ is the maximum amplitude of the inputs, while μ is a positive parameter to control the degree of the compression. $\mu = 0$ corresponds to no compression, while $\mu = 255$ is adopted in the industry. The compression curve with $\mu = 255$ is plotted in Fig. 10.4. Note that the sign function *sign*(x) shown Eq. (10.4) is defined as

$$\text{sign}(x) = \begin{cases} 1 & x \geq 0 \\ -1 & x < 0 \end{cases}. \tag{10.5}$$

Solving Eq. (10.4) by substituting the quantized value, that is, $y - y_q$ we achieve the expander equation as

$$x_q = |x|_{max} \, \text{sign}(y_q) \frac{(1+\mu)^{|y_q|} - 1}{\mu}. \tag{10.6}$$

For the case $\mu = 255$, the expander curve is plotted in Fig. 10.5.

Let us look at Example 10.2 for the μ-law compression.

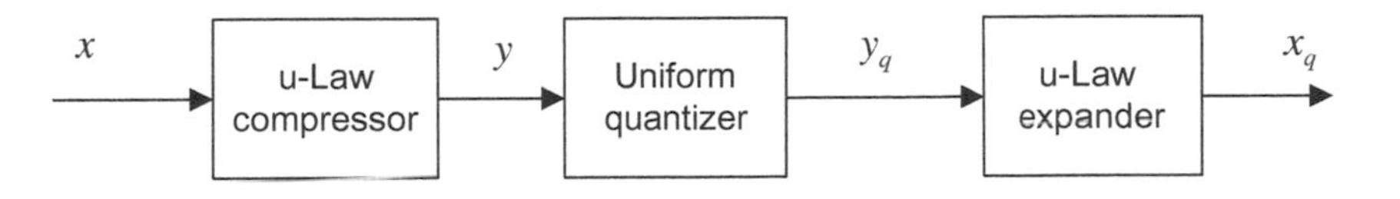

FIG. 10.3

Block diagram for μ-law compressor and μ-law expander.

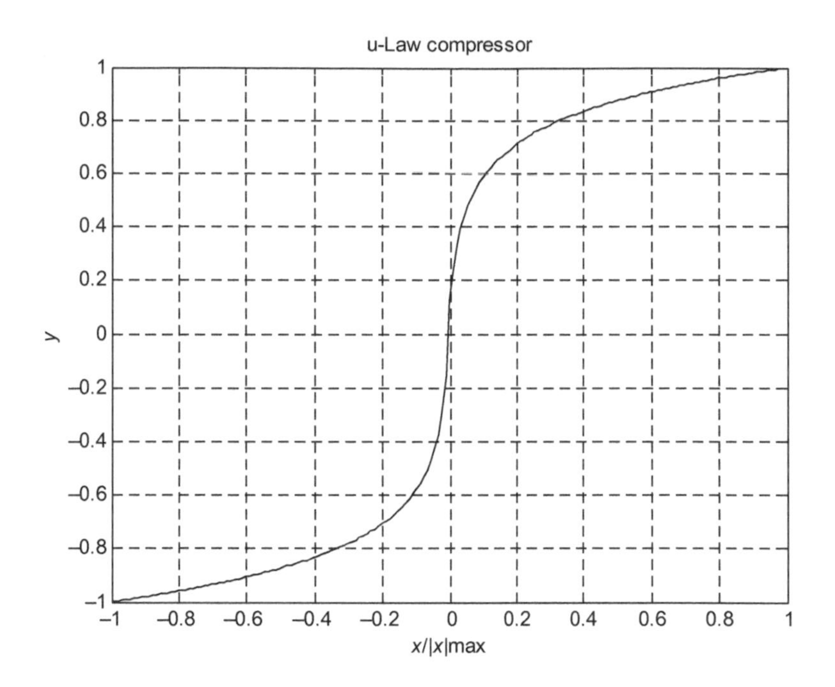

FIG. 10.4

Characteristics for the μ-law compander.

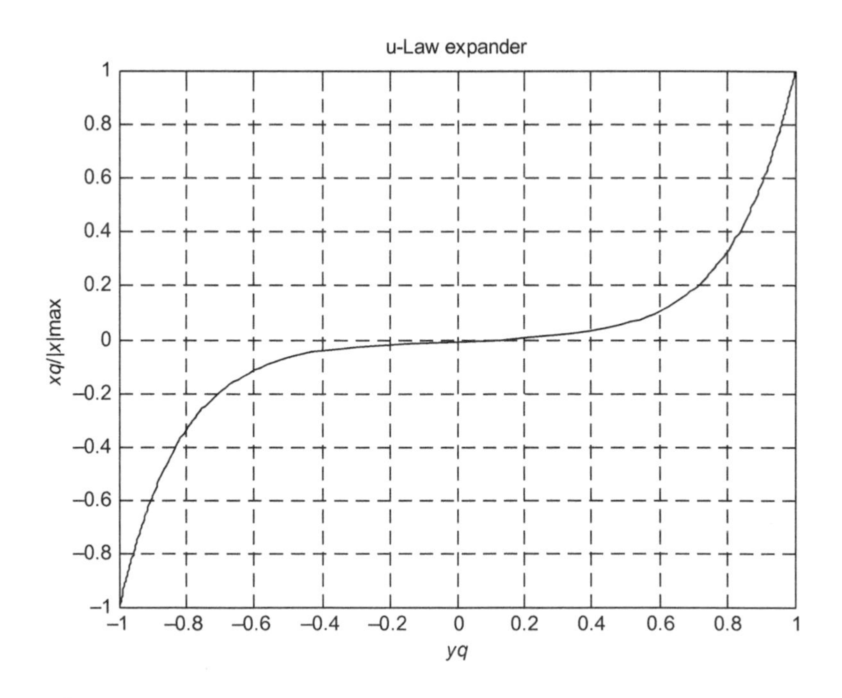

FIG. 10.5

Characteristics for the μ-law expander.

EXAMPLE 10.2

For the μ-law compression and expansion process shown in Fig. 10.3, with $\mu=255$, the 3-bit midtread quantizer described in Fig. 10.1, and an analog signal ranging from -5 to 5V, determine the binary codes, recovered voltages, and quantization errors when the input is

(a) -3.6V

(b) 0.5 V.

Solution:

(a) For the μ-law compression and $x=-3.6$ V, we can determine the quantization input as

$$y=\text{sign}(-3.6)\frac{\ln\left(1+255\frac{|-3.6|}{|5|_{\max}}\right)}{\ln(1+255)}=-0.94.$$

As shown in Fig. 10.4, the range of y is 2, thus the quantization step size is calculated as

$$\Delta=\frac{2}{2^3-1}=0.286 \text{ and } y=\frac{-0.94}{0.286}=-3.28\Delta.$$

From quantization characteristics, it follows that the binary code$=$011 and the recovered signal is $y_q=-3\Delta=-0.858$.
Applying the μ-law expander leads to

$$x_q=|5|_{\max}\text{sign}(-0.858)\frac{(1+255)^{|-0.858|}-1}{255}=-2.264.$$

Thus the quantization error is computed as

$$e_q=x_q-x=-2.264-(-3.6)=1.336 \text{ volts}.$$

(b) Similarly, for $x=0.5$, we get

$$y=\text{sign}(0.5)\frac{\ln\left(1+255\frac{|0.5|}{|5|_{\max}}\right)}{\ln(1+255)}=0.591$$

In terms of the quantization step, we get

$$y=\frac{0.519}{0.286}\Delta=2.1\Delta \text{ and binary code}=110.$$

Based on Fig. 10.1, the recovered signal is.
$y_q=2\Delta=0.572$
and the expander gives

$$x_q=|5|_{\max}\text{sign}(0.572)\frac{(1+255)^{|0.572|}-1}{255}=0.448 \text{ volts}.$$

Finally, the quantization error is given by

$$e_q=0.448-0.5=-0.052 \text{ volts}.$$

As we can see, with 3 bits per sample, the strong signal is encoded with bigger quantization error, while the weak signal is quantized with less quantization error.

In the following simulation, we apply a 5-bit μ-law compander with $\mu = 255$ in order to quantize and encode the speech data used in the last section. Fig. 10.6 is a block diagram of compression and decompression.

Fig. 10.7 shows the original speech data, the quantized speech data using μ-law compression, and the quantization error for comparisons. The quantized speech wave is very close to the original speech wave. From the plots in Fig. 10.7, we can observe that the amplitude of the quantization error changes according to the amplitude of the speech being quantized. The bigger quantization error is introduced when the amplitude of speech data is larger; on the other hand, the smaller quantization error is produced when the amplitude of speech data is smaller.

Compared with the quantized speech using the linear quantizer shown in Fig. 10.2, the decompressed signal using the μ-law compander looks and sounds much better, since the quantized signal

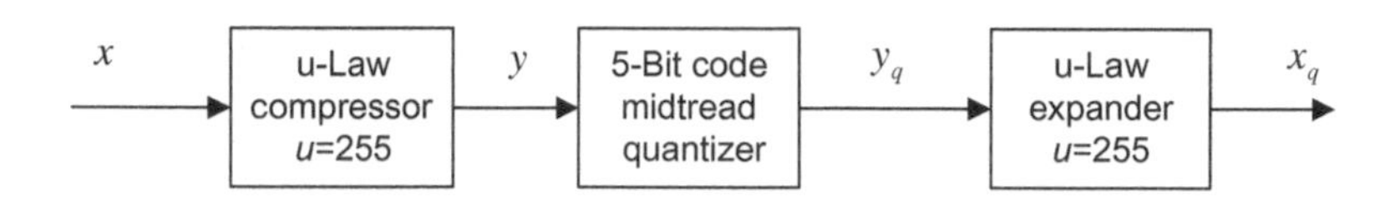

FIG. 10.6

The 5-bit midtread uniform quantizer with $\mu = 255$ used for simulation.

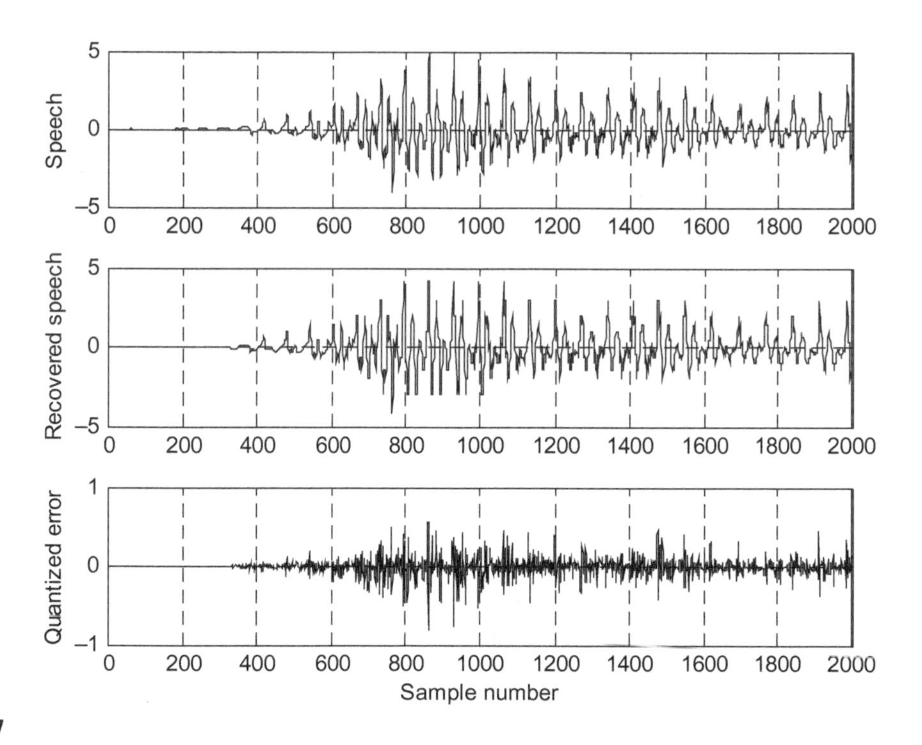

FIG. 10.7

Plots of the original speech, quantized speech, and quantization error with the μ-law compressor and expander.

can better keep tracking the original large amplitude signal and original small amplitude signal as well. MATLAB implementation is shown in Programs 10.4-7 in Section 10.6.

10.2.2 DIGITAL μ-LAW COMPANDING

In many multimedia applications, the analog signal is first sampled and then it is digitized into a linear PCM code with a larger number of bits per sample. The digital μ-law companding further compresses the linear PCM code using the compressed PCM code with a smaller number of bits per sample without losing sound quality. The block diagram of a digital μ-law compressor and expander is shown in Fig. 10.8.

The typical digital μ-law companding system compresses a 12-bit linear PCM code to an 8-bit compressed code. This companding characteristic is depicted in Fig. 10.9, where it closely resembles an analog compression curve with $\mu = 255$ by approximating the curve using a set of eight straight-line segments. The slope of each successive segment is exactly one-half of the previous segment. Fig. 10.9 shows the 12-bit to 8-bit digital companding curve for the positive portion only. There are 16 segments, accounting for both positive and negative portions.

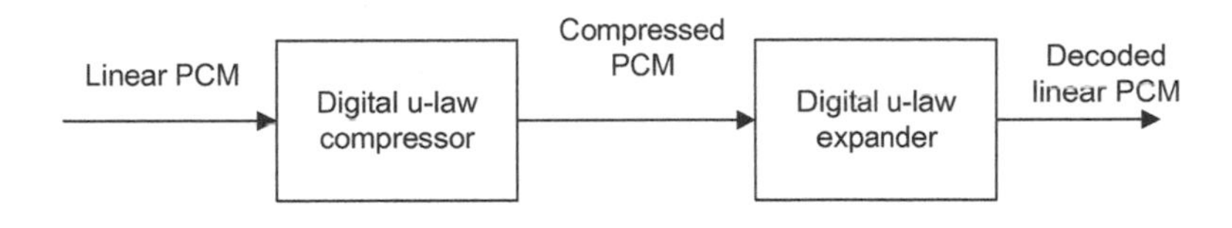

FIG. 10.8

The block diagram for the μ-law compressor and expander.

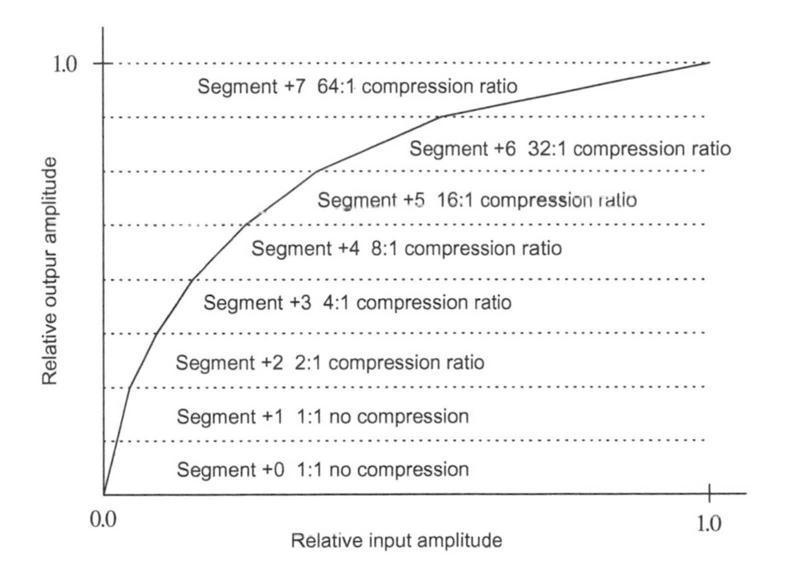

FIG. 10.9

μ-255 compression characteristics (for positive portion only).

Table 10.2 The Format of 8-Bit Compressed PCM Code

Sign Bit	3-Bit	4-Bit
1=+ 0=−	Segment identifier 000 to 111	Quantization interval A B C D 0000 to 1111

However, like the midtread quantizer discussed in the first section, only 13-segments are used, since segments +0, −0, +1, and −1 form a straight line with a constant slope and are considered as one segment. As shown in Fig. 10.9, when the relative input is very small, such as in segment 0 or segment 1, there is no compression, while when the relative input is larger such that it is in segment 3 or segment 4, the compression occurs with the compression ratios (CRs) of 2:1 and 4:1, respectively. The format of the 12-bit linear PCM code is in the sign-magnitude form with the most significant bit (MSB) as the sign bit (1 = positive value and 0 = negative value) plus 11 magnitude bits. The compressed 8-bit code has a format shown in Table 10.2, where it consists of a sign bit, a 3-bit segment identifier, and a 4-bit quantization interval within the specified segment. Encoding and decoding procedures are very simple, as illustrated in Tables 10.3 and 10.4, respectively.

As shown in those two tables, the prefix "S" is used to indicate the sign bit, which could be either 1 or 0; A, B, C, and D, are transmitted bits; and the bit position with an "X" is the truncated bit during the compression and hence would be lost during decompression. For the 8-bit compressed PCM code in Table 10.3, the 3 bits between "S" and "ABCD" indicate the segment number that is obtained by subtracting the number of consecutive zeros (less than or equal to 7) after the "S" bit in the original 12-bit PCM code from 7. Similarly, to recover the 12-bit linear code in Table 10.4, the number of consecutive zeros after the "S" bit can be determined by subtracting the segment number in the 8-bit compressed code from 7. We will illustrate the encoding and decoding processes in Examples 10.3 and 10.4.

EXAMPLE 10.3

In a digital companding system, encode each of the following 12-bit linear PCM codes into to an 8-bit compressed PCM code.

1. 1 0 0 0 0 0 0 0 0 1 0 1
2. 0 0 0 0 1 1 1 0 1 0 1 0

Solution:

1. Based on Table 10.3, we identify the 12-Bit PCM Code as S = 1, A = 0, B = 1, C = 0, and D = 1, which is in segment 0. From the fourth column in Table 10.3, we get the 8-bit compressed code as 1 0 0 0 0 1 0 1.
2. For the second 12-bit PCM code, we note that S = 0, A = 1, B = 1, C = 0, D = 1, and XXX = 010, and the code belongs to segment 4. Thus, from the fourth column in Table 10.3, we have 0 1 0 0 1 1 0 1.

Table 10.3 $\mu - 255$ **Encoding Table**

Segment	12-Bit Linear Code	12-Bit Amplitude Range in Decimal	8-bit Compressed Code
0	S0000000ABCD	0 to 15	S000ABCD
1	S0000001ABCD	16 to 31	S001ABCD
2	S000001ABCDX	32 to 63	S010ABCD
3	S00001ABCDXX	64 to 127	S011ABCD
4	S0001ABCDXXX	128 to 255	S100ABCD
5	S001ABCDXXXX	256 to 511	S101ABCD
6	S01ABCDXXXXX	512 to 1023	S110ABCD
7	S1ABCDXXXXXX	1023 to 2047	S111ABCD

Table 10.4 $\mu - 255$ **Decoding Table**

8-Bit Compressed Code	8-Bit Amplitude Range in Decimal	Segment	12-Bit Linear Code
S000ABCD	0–15	0	S0000000ABCD
S001ABCD	16–31	1	S0000001ABCD
S010ABCD	32–47	2	S000001ABCD1
S011ABCD	48–63	3	S00001ABCD10
S100ABCD	64–79	4	S0001ABCD100
S101ABCD	80–95	5	S001ABCD1000
S110ABCD	96–111	6	S01ABCD10000
S111ABCD	112–127	7	S1ABCD100000

EXAMPLE 10.4

In a digital companding system, decode each of the following 8-bit compressed PCM codes into a 12-bit linear PCM code.

1. 1 0 0 0 0 1 0 1
2. 0 1 0 0 1 1 0 1

Solution:

1. Using the decoding Table 10.4, we note that S=1, A=0, B=1, C=0, and D=1, and the code is in segment 0. Decoding leads to: 1 0 0 0 0 0 0 0 0 1 0 1, which is identical to the 12-bit PCM code in (1) in Example 10.3. We expect this result, since there is no compression for segment 0 and segment 1.
2. Applying Table 10.4, it follows that S=0, A=1, B=1, C=0, and D=1, and the code resides in segment 4. Decoding achieves: 0 0 0 0 1 1 1 0 1 1 0 0.
   As expected, this code is the approximation of the code in (2) in Example 10.3. Since segment 4 has compression, the last 3 bits in the original 12-bit linear code, that is, XXX – 010 = 2 in decimal, are discarded during transmission or storage. When we recover these 3 bits, the best guess should be the middle value: XXX = 100 = 4 in decimal for the 3-bit coding range from 0 to 7.

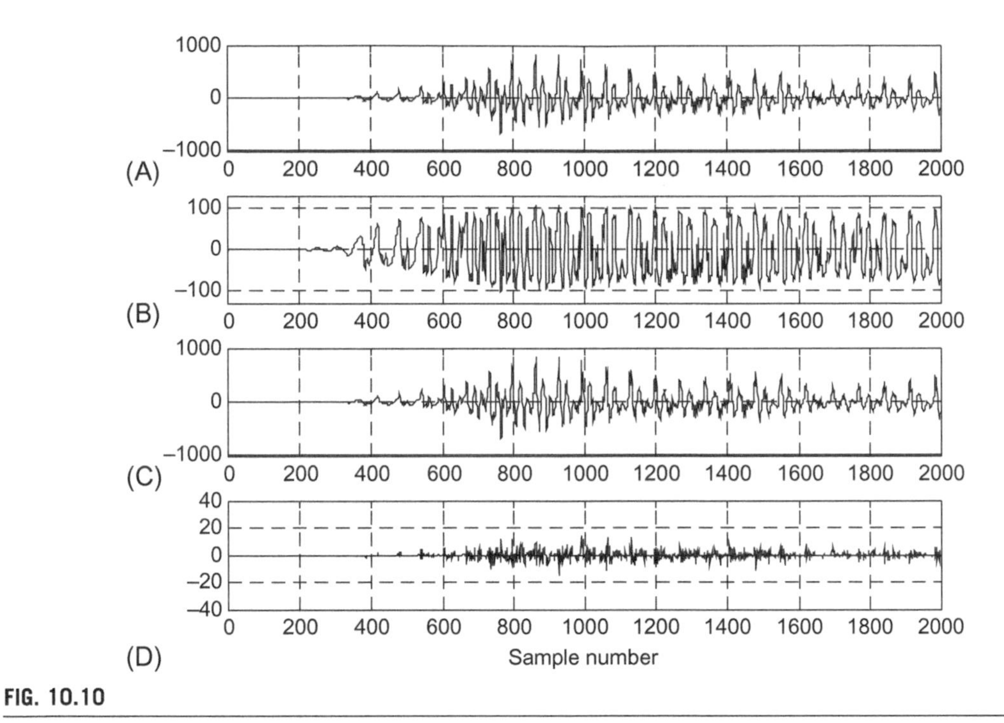

FIG. 10.10

The μ-255 compressor and expander: (A) 12-bit speech data; (B) 8-bit compressed data; (C) 12-bit decoded speech; (D) quantization error.

Now we apply the μ-255 compander to compress the 12-bit speech data as shown in Fig. 10.10A. The 8-bit compressed code is plotted in Fig. 10.10B. Plots C and D in the figure show the 12-bit speech after decoding and quantization error, respectively. We can see that the quantization error follows the amplitude of speech data relatively. The decoded speech sounds no difference when compared with the original speech. Programs 10.8–10.10 in Section 10.6 show the detail of the MATLAB implementation.

10.3 EXAMPLES OF DIFFERENTIAL PULSE CODE MODULATION (DPCM), DELTA MODULATION, AND ADAPTIVE DPCM G.721

Data compression can be further achieved using *differential pulse code modulation* (DPCM). The general idea is to use past recovered values as the basis to predict the current input data and then encode the difference between the current input and the predicted input. Since the difference has a significantly reduced signal dynamic range, it can be encoded with fewer bits per sample. Therefore, we obtain data compression. First, we study the principles of the DPCM concept that will help us understand adaptive DPCM (ADPCM) in the next subsection.

10.3.1 EXAMPLES OF DPCM AND DELTA MODULATION

Fig. 10.11 shows a schematic diagram for the DPCM encoder and decoder. We denote the original signal $x(n)$; the predicted signal $\tilde{x}(n)$; the quantized, or recovered signal $\hat{x}(n)$; the difference signal to be quantized $d(n)$; and the quantized difference signal $d_q(n)$. The quantizer can be chosen as a uniform quantizer, a midtread quantizer (e.g., see Table 10.5), or others available. The encoding block produces binary bit stream in the DPCM encoding. The predictor uses the past predicted signal and quantized difference signal to predict the current input value $x(n)$ as close as possible. The digital filter or adaptive filter can serve as the predictor. On the other hand, the decoder recovers the quantized difference signal, which can be added to the predictor output signal to produce the quantized and recovered signal, as shown in Fig. 10.11B.

In Example 10.5, we examine a simple DPCM coding system via the process of encoding and decoding numerical actual data.

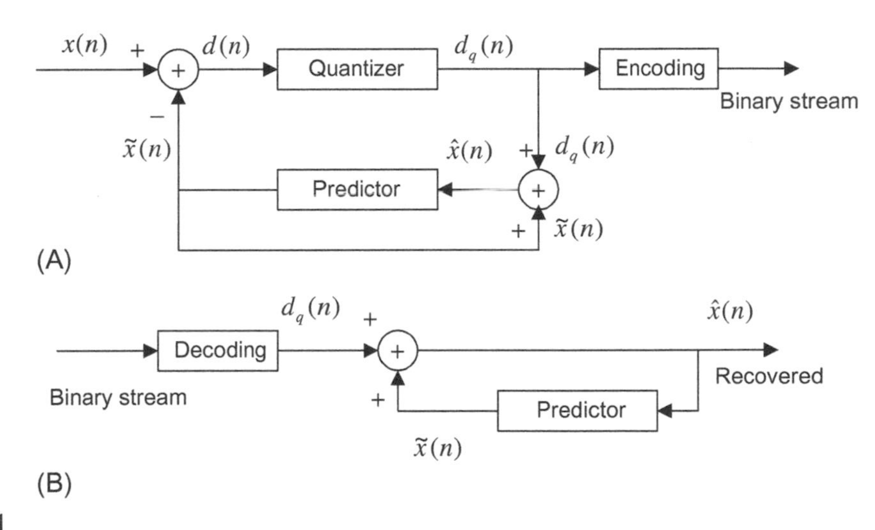

FIG. 10.11

DPCM block diagram: (A) encoder; and (B) decoder.

Table 10.5 Quantization Table for the 3-Bit Quantizer in Example 10.5

Binary Code	Quantization Value $d_q(n)$	Subrange in $d(n)$
0 1 1	−11	$-15 \leq d(n) < -7$
0 1 0	−5	$-7 \leq d(n) < -3$
0 0 1	−2	$-3 \leq d(n) < -1$
0 0 0	0	$-1 \leq d(n) < 0$
1 0 0	0	$0 \leq d(n) \leq 1$
1 0 1	2	$1 < d(n) \leq 3$
1 1 0	5	$3 < d(n) \leq 7$
1 1 1	11	$7 < d(n) \leq 15$

EXAMPLE 10.5

A DPCM system has the following specifications:

Encoder scheme: $\tilde{x}(n) = \hat{x}(n-1)$; predictor
$d(n) = x(n) - \tilde{x}(n)$
$d_q(n) = Q[d(n)] =$ quantizer in Table 10.5
$\hat{x}(n) = \tilde{x}(n) + d_q(n)$

Decoding scheme: $\tilde{x}(n) = \hat{x}(n-1)$; predictor
$d_q(n) =$ quantizer in Table 10.5.
$\hat{x}(n) = \tilde{x}(n) + d_q(n)$

The 5-bit input data: $x(0) = 6$, $x(1) = 8$, $x(2) = 13$.

(a) Perform DPCM encoding to produce the binary code for each input datum
(b) Perform DCPM decoding to recover the data using the binary code in (1).

Solution:

(a) Let us perform encoding according to the encoding scheme.
For $n = 0$, we have.
$\tilde{x}(0) = \hat{x}(-1) = 0$
$d(0) = x(0) - \tilde{x}(0) = 6 - 0 = 6$
$d_q(0) = Q[d(0)] = 5$
$\hat{x}(0) = \tilde{x}(0) + d_q(0) = 0 + 5 = 5$
Binary code $= 110$.
For $n = 1$, it follows that.
$\tilde{x}(1) = \hat{x}(0) = 5$
$d(1) = x(1) - \tilde{x}(1) = 8 - 5 = 3$
$d_q(1) = Q[d(1)] = 2$
$\hat{x}(1) = \tilde{x}(1) + d_q(1) = 5 + 2 = 7$
Binary code $= 101$.
For $n = 2$, results are.
$\tilde{x}(2) = \hat{x}(1) = 7$
$d(2) = x(2) - \tilde{x}(2) = 13 - 7 = 6$
$d_q(2) = Q[d(2)] = 5$
$\hat{x}(2) = \tilde{x}(2) + d_q(2) = 7 + 5 = 12$
Binary code $= 110$.

(b) We conduct the decoding scheme as follows.
For $n = 0$, we get:
Binary code $= 110$
$d_q(0) = 5$; from Table 10.5
$\tilde{x}(0) = \hat{x}(-1) = 0$
$\hat{x}(0) = \tilde{x}(0) + d_q(0) = 0 + 5 = 5$ (recovered).
For $n = 1$, decoding shows:
Binary code $= 101$
$d_q(1) = 2$; from Table 10.5
$\tilde{x}(1) = \hat{x}(0) = 5$
$\hat{x}(1) = \tilde{x}(1) + d_q(1) = 5 + 2 = 7$ (recovered).

For $n=2$, we have:

Binary code = 110

$d_q(2)=5$; from Table 10.5

$\tilde{x}(2)=\hat{x}(1)=7$

$\hat{x}(2)=\tilde{x}(2)+d_q(2)=7+5=12$ (recovered).

From this example, we could verify that the 5-bit code is compressed to the 3-bit code. However, we can see that each recovered data has a quantization error. Hence, the DPCM is a lossy data compression scheme.

DPCM for which a single bit is used in the quantization table becomes *delta modulation* (DM). The quantization table contains two quantized values, A and $-A$, where A is the quantization step size. DM quantizes the difference of the current input sample and the previous input sample using a 1-bit code word. To conclude the idea, we list the equations for encoding and decoding as follows:

Encoder scheme: $\tilde{x}(n)=\hat{x}(n-1)$; predictor

$$d(n)=x(n)-\tilde{x}(n)$$

$$d_q(n)=\begin{cases} +A, \text{ for } & d_q(n)\geq 0, \text{ output bit}: 1 \\ -A, \text{ for } & d_q(n)<0, \text{ output bit}: 0 \end{cases}$$

$$\hat{x}(n)=\tilde{x}(n)+d_q(n)$$

Decoding scheme: $\tilde{x}(n)=\hat{x}(n-1)$; predictor

$$d_q(n)=\begin{cases} +A, \text{ input bit}: 1 \\ -A, \text{ input bit}: 0 \end{cases}$$

$$\hat{x}(n)=\tilde{x}(n)+d_q(n)$$

Note that the predictor has one sample delay.

EXAMPLE 10.6

For a DM system with 5-bit input data

$$x(0)=6,\ x(1)=8,\ x(2)=13$$

and the quantized constant as $A = 7$,

(a) Perform the DM encoding to produce the binary code for each input datum

(b) Perform the DM decoding to recover the data using the binary code in (a).

Solution:

(a) Applying encoding according, we have

For $n=0$,

$\tilde{x}(0)=\hat{x}(-1)=0$, $d(0)=x(0)-\tilde{x}(0)=6-0=6$.

$d_q(0)=7$, $\hat{x}(0)=\tilde{x}(0)+d_q(0)=0+7=7$.

Binary code = 1.

For $n=1$,

$\tilde{x}(1)=\hat{x}(0)=7$, $d(1)=x(1)-\tilde{x}(1)=8-7=1$.

$d_q(1)=7$, $\hat{x}(1)=\tilde{x}(1)+d_q(1)=7+7=14$.

Binary code = 1.

Continued

EXAMPLE 10.6—CONT'D

For $n=2$,

$\tilde{x}(2)=\hat{x}(1)=14$, $d(2)=x(2)-\tilde{x}(2)=13-14=-1$.
$d_q(2)=-7$, $\hat{x}(2)=\tilde{x}(2)+d_q(2)=14-7=7$.
Binary code $=0$.

(b) Applying the decoding scheme leads to

For $n=0$,

Binary code $=1$.
$d_q(0)=7$, $\tilde{x}(0)=\hat{x}(-1)=0$.
$\hat{x}(0)=\tilde{x}(0)+d_q(0)=0+7=7$ (recovered).

For $n=1$,

Binary code 1.
$d_q(1)=7$, $\tilde{x}(1)=\hat{x}(0)=7$.
$\hat{x}(1)=\tilde{x}(1)+d_q(1)=7+7=14$ (recovered).

For $n=2$,

Binary code 0.
$d_q(2)=-7$, $\tilde{x}(2)=\hat{x}(1)=14$.
$\hat{x}(2)=\tilde{x}(2)+d_q(2)=14-7=7$ (recovered).

We can see that the DM coding causes a larger quantization error for each recovered sample. In practice, this can be solved using a very high sampling rate (much larger than the Nyquist rate), and making the quantization step size A adaptive. The quantization step size increases by a factor when the slope magnitude of the input sample curve becomes bigger, that is, the condition in which the encoder produces continuous logic 1's or generates continuous logic 0's in the coded bit stream. Similarly, the quantization step decreases by a factor when the encoder generates logic 1 and logic 0 alternatively. Hence, the resultant DM is called *adaptive* DM. In practice, the DM chip replaces the predictor, feedback path, and adder (see Fig. 10.11) with an integrator for both the encoder and the decoder. Detailed information can be found in Li et al. (2014), Roddy and Coolen (1997), and Tomasi (2004).

10.3.2 ADAPTIVE DIFFERENTIAL PULSE CODE MODULATION G.721

In this subsection, an efficient compression technique for speech waveform is described, that is, adaptive DPCM (ADPCM), per recommendation G.721 of the CCITT (the Comité Consultatif International Téléphonique et Télégraphique). General discussion can be found in Li et al. (2014), Roddy and Coolen (1997), and Tomasi (2004). The simplified block diagrams of the ADPCM encoder and decoder are shown in Fig. 10.12A and B.

As shown in Fig. 10.12A for the ADPCM encoder, first a difference signal $d(n)$ is obtained, by subtracting an estimate of the input signal $\tilde{x}(n)$ from the input signal $x(n)$. An adaptive 16-level quantizer is used to assign four binary digits $I(n)$ to the value of the difference signal for transmission to the decoder. At the same time, an inverse quantizer produces a quantized difference signal $d_q(n)$ from the same four binary digits $I(n)$. The adaptive quantizer and inverse quantizer operate based on the quantization table and the scale factor obtained from the quantizer adaptation to keep tracking the energy change of the difference signal to be quantized. The input signal estimate from the adaptive predictor is then added to

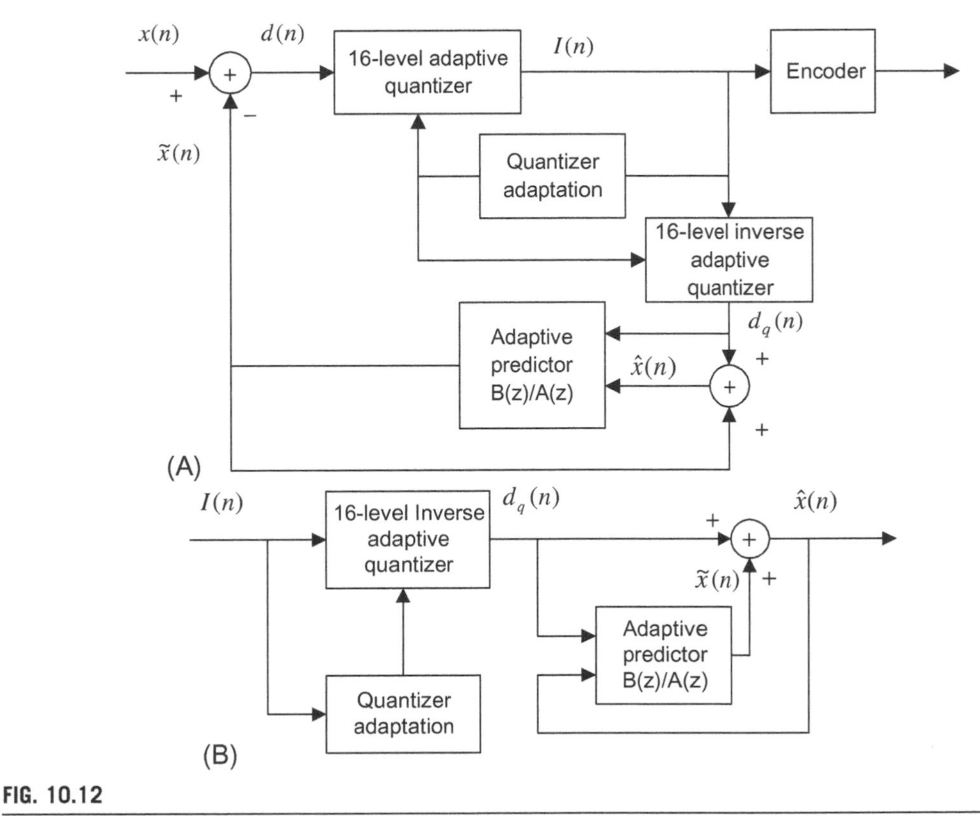

FIG. 10.12

(A) The ADPCM encoder, (B) ADPCM decoder.

this quantized difference signal to produce the reconstructed version of the input $\hat{x}(n)$. Both the reconstructed signal and the quantized difference signal are operated on by an adaptive predictor, which generates the estimate of the input signal, thereby completing the feedback loop.

The decoder shown in Fig. 10.12B includes a structure identical to the feedback part of the encoder as depicted in Fig. 10.12A. It first converts the received 4-bit data $I(n)$ to the quantized difference signal $d_q(n)$ using the adaptive quantizer. Then, at the second stage, the adaptive predictor uses the recovered quantized difference signal $d_q(n)$ and recovered current output $\tilde{x}(n)$ to generate the next output. Note that the adaptive predictors of both the encoder and the decoder change correspondingly based on the signal to be quantized. The details of the adaptive predictor will be discussed.

Now, let us examine the ADPCM encoder principles. As shown in Fig. 10.12A, the difference signal is computed as

$$d(n) = x(n) - \tilde{x}(n). \tag{10.7}$$

A 16-level nonuniform adaptive quantizer is used to quantize the difference signal $d(n)$. Before quantization, $d(n)$ is converted to a base-2 logarithmic representation and scaled by $y(n)$, which is computed by the scale-factor algorithm. Four binary codes $I(n)$ are used to specify the quantized signal level representing $d_q(n)$, and the quantized difference $d_q(n)$ is also fed to the inverse adaptive quantizer. Table 10.6 shows the quantizer normalized input and output characteristics.

Table 10.6 Quantizer Normalized Input and Output Characteristics

Normalized Quantizer Input Range: $\log_2\|d(n)\| - y(n)$	Magnitude $\|I(n)\|$	Normalized Quantizer Output: $\log_2\|d_q(n)\| - y(n)$
[3.12, +∞)	7	3.32
[2.72, 3.12)	6	2.91
[2.34, 2.72)	5	2.52
[1.91, 2.34)	4	2.13
[1.38, 1.91)	3	1.66
[0.62, 1.38)	2	1.05
[−0.98, 0.62)	1	0.031
(−∞, −0.98)	0	−∞

The scaling factor for the quantizer and the inverse quantizer $y(n)$ are computed according to the 4-bit quantizer output $I(n)$ and the adaptation speed control parameter $a_l(n)$, the fast (unlocked) scale factor $y_u(n)$, the slow (locked)-scale factor $y_l(n)$, and the discrete function $W(I)$, defined in Table 10.7:

$$y_u(n) = \left(1 - 2^{-5}\right)y(n) + 2^{-5}W(I(n)), \tag{10.8}$$

where $1.06 \le y_u(n) \le 10.00$.

The slow scale factor $y_l(n)$ is derived from the fast scale factor $y_u(n)$ using a lowpass filter as following:

$$y_l(n) = \left(1 - 2^{-6}\right)y_l(n-1) + 2^{-6}y_u(n). \tag{10.9}$$

The fast and slow scale factors are then combined to compute the scale factor

$$y(n) = a_l(n)y_u(n-1) + (1 - a_l(n))y_l(n-1). \tag{10.10}$$

Next the controlling parameter $0 \le a_l(n) \le 1$ tends toward unity for speech signals and toward zero for voice band data signals and tones. It is updated based on the following parameters: $d_{ms}(n)$, which is the relatively short term average of $F(I(n))$; $d_{ml}(n)$, which is the relatively long-term average of $F(I(n))$; and the variable $a_p(n)$, where $F(I(n))$ is defined as in Table 10.8.

Hence, we have

$$d_{ms}(n) = \left(1 - 2^{-5}\right)d_{ms}(n-1) + 2^{-5}F(I(n)) \tag{10.11}$$

and

$$d_{ml}(n) = \left(1 - 2^{-7}\right)d_{ml}(n-1) + 2^{-7}F(I(n)), \tag{10.12}$$

Table 10.7 Discrete Function $W(I)$

$\|I(n)\|$	7	6	5	4	3	2	1	0
$W(I)$	70.13	22.19	12.38	7.00	4.0	2.56	1.13	−0.75

Table 10.8 Discrete Function $F(I(n))$

$\|I(n)\|$	7	6	5	4	3	2	1	0
$F(I(n))$	7	3	1	1	1	0	0	0

while the variable $a_p(n)$ is given by

$$a_p(n)=\begin{cases}(1-2^{-4})a_p(n-1)+2^{-3} & \text{if } |d_{ms}(n)-d_{ml}(n)|\geq 2^{-3}d_{ml}(n)\\ (1-2^{-4})a_p(n-1)+2^{-3} & \text{if } y(n)<3\\ (1-2^{-4})a_p(n)+2^{-3} & \text{if } t_d(n)=1\\ 1 & \text{if } t_r(n)=1\\ (1-2^{-4})a_p(n) & \text{otherwise}\end{cases}. \tag{10.13}$$

$a_p(n)$ approaches 2 when the difference between $d_{ms}(n)$ and $d_{ml}(n)$ is large and approaches 0 when the difference is small. Also $a_p(n)$ approaches 2 for an idle channel (indicated by $y(n)<3$) or partial band signals (indicated by $t_d(n)=1$). Finally, $a_p(n)$ is set to 1 when the partial band signal transition is detected ($t_r(n)=1$).

$a_l(n)$ used in Eq. (10.10) is defined as

$$a_l(n)=\begin{cases}1 & a_p(n-1)>1\\ a_p(n-1) & a_p(n-1)\leq 1\end{cases}. \tag{10.14}$$

The partial band signal $t_d(n)$ and the partial band signal transition $t_r(n)$ that appear in Eq. (10.13) will be discussed later.

The predictor is to compute the signal estimate $\tilde{x}(n)$ from the quantized difference signal $d_q(n)$. The predictor z-transfer function, which is effectively suitable for a variety of input signals, is given by

$$\frac{B(z)}{A(z)}=\frac{b_0+b_1z^{-1}+b_2z^{-2}+b_3z^{-3}+b_4z^{-4}+b_5z^{-5}}{1-a_1z^{-1}-a_2z^{-2}}. \tag{10.15}$$

It consists of a fifth-order portion that models the zeros and a second-order portion that models poles of the input signals. The input signal estimate is expressed in terms of the processed signal $\hat{x}(n)$ and the signal $x_z(n)$ processed by the finite impulse response (FIR) filter as follows:

$$\tilde{x}(n)=a_1(n)\hat{x}(n-1)+a_2(n)\hat{x}(n-2)+x_z(n), \tag{10.16}$$

where

$$\hat{x}(n-i)=\tilde{x}(n-i)+d_q(n-i) \tag{10.17}$$

$$x_z(n)=\sum_{i=0}^{5}b_i(n)d_q(n-i). \tag{10.18}$$

Both sets of predictor coefficients are updated using a simplified gradient algorithm:

$$a_1(n)=(1-2^{-8})a_1(n-1)+3\cdot 2^{-8}\,\mathrm{sign}(p(n))\,\mathrm{sign}(p(n-1)) \tag{10.19}$$

$$a_2(n)=(1-2^{-7})a_2(n-1)+2^{-7}\{\mathrm{sign}(p(n)\,\mathrm{sign}(p(n-2)-f(a_1(n-1))\,\mathrm{sign}(p(n)\,\mathrm{sign}(p(n-1))\}, \tag{10.20}$$

where $p(n)=d_q(n)+x_z(n)$ and

$$f(a_1(n))=\begin{cases}4a_1(n) & |a_1(n)|\le 2^{-1}\\ 2\,\mathrm{sign}(a_1(n)) & |a_1(n)|>2^{-1}\end{cases}. \tag{10.21}$$

Note that the function *sign*(x) is defined in Eq. (10.5), while the function signn(x)$=1$ when $x>0$; signn(x)$=0$ when $x=0$; and signn(x)$=-1$ when $x<0$ with stability constrains as

$$|a_2(n)|\le 0.75 \text{ and } |a_1(n)|\le 1-2^{-4}-a_2(n) \tag{10.22}$$

$$a_1(n)=a_2(n)=0 \;\text{ if }\; t_r(n)=1. \tag{10.23}$$

Also, the equations for updating the coefficients for the zero-order portion are given by

$$b_i(n)=\left(1-2^{-8}\right)b_i(n-1)+2^{-7}\,\mathrm{sign}\left(d_q(n)\right)\mathrm{sign}\left(d_q(n-i)\right) \tag{10.24}$$

for $i=0, 1, 2, \ldots, 5$ with the following constrains:

$$b_0(n)=b_1(n)=b_2(n)=b_3(n)=b_4(n)=b_5(n)=0 \;\text{ if }\; t_r(n)=1. \tag{10.25}$$

$$t_d(n)=\begin{cases}1 & a_2(n)<-0.71875\\ 0 & \text{otherwise}\end{cases} \tag{10.26}$$

$$t_r(n)=\begin{cases}1 & a_2(n)<-0.71875 \text{ and } \left|d_q(n)\right|>24\cdot 2^{y_l}\\ 0 & \text{otherwise}\end{cases}. \tag{10.27}$$

$t_d(n)$ is the indicator of detecting a partial band signal (tone). If a tone is detected ($t_d(n)=1$); Eq. (10.13) is invoked to drive the quantizer into the fast mode of adaptation. $t_r(n)$ is the indicator for a transition from a partial band signal. If it is detected ($t_r(n)=1$), setting the predictor coefficients to be zero as shown in Eqs. (10.23) and (10.25) will force the quantizer into the fast mode of adaptation.

Simulation Example

To illustrate the performance, we apply the ADPCM encoder to the speech data used in Section 10.1 and then operate the ADPCM decoder to recover the speech signal. As described, the ADPCM uses 4 bits to encode each speech sample. The MATLAB implementations for the encoder and decoder are listed in Programs 10.11–10.13 in Section 10.6. Fig. 10.13 plots the original speech samples, decoded speech samples, and the quantization errors. From the figure, we see that the decoded speech data are very close to the original speech data; the quantization error is very small as compared with the speech sample, and its amplitude follows the change in amplitude of the speech data. In practice, we cannot tell the difference between the original speech and the decoded speech by listening to them. However, the ADPCM encodes each speech sample using 4 bit per sample, while the original data are presented using 16 bits, thus the CR of 4:1 is achieved.

In practical applications, data compression can reduce the storage media and bit rate for the efficient digital transmission. To measure the performance of data compression, we use

- the data CR, which is the ratio of original data file size to the compressed data file size, or ratio of the original code size in bits to the compressed code size in bits for the fixed length coding, and
- the bit rate, which is in terms of bits per second (bps) and can be calculated by:

$$\text{bit rate}=m\times f_s(\text{bps}), \tag{10.28}$$

where $m=$ number of bits per sample (bits) and $f_s=$ sampling rate (samples per second).

Now, let us look at an application example.

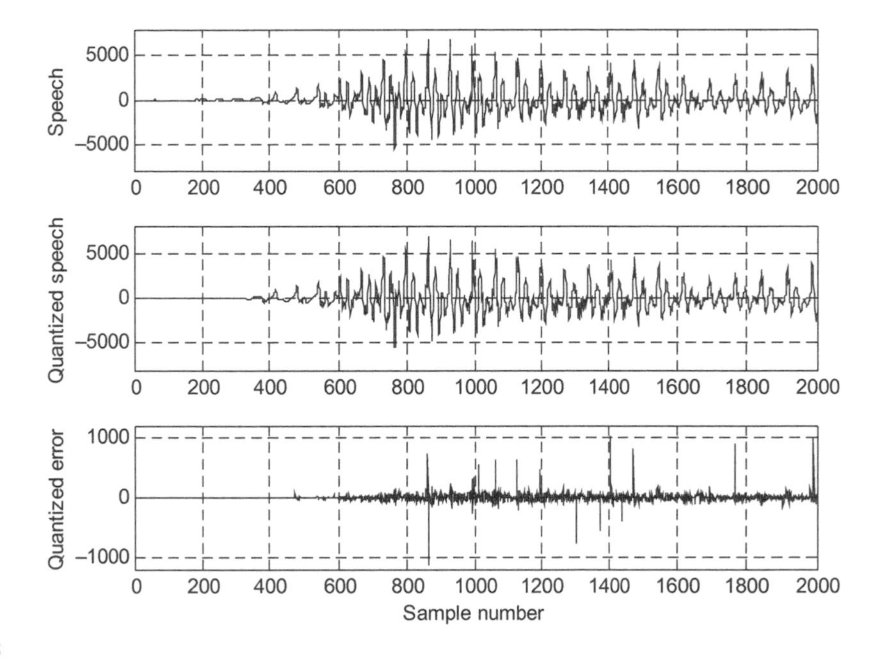

FIG. 10.13

Original speech, quantized speech, and quantization error using the ADPCM.

EXAMPLE 10.7

Speech is sampled at 8 kHz and each sample is encoded by 12 bits per sample. Using the following encoding methods:

1. Non compression
2. Standard μ-law compression
3. Standard ADPCM encoding (4 bits per sample),
   (a) Determine the CR and the bit rate for each of the encoders and decoders.
   (b) Determine the number of channels that the phone company can carry if a telephone system can transport the digital voice channel over the digital link having a capacity of 1.536 Mbps.

Solution:

a.1 For noncompression:
CR = 1:1

$$\text{Bit rate} = 12\frac{\text{bits}}{\text{sample}} \times 8000\frac{\text{sample}}{\text{second}} = 96\,(\text{kbps}).$$

b.1 Number of channels $= \dfrac{1.536\,\text{MBPS}}{96\ \ \text{KBPS}} = 16.$

a.2 For the standard μ-law compression, each sample is encoded using 8 bits per sample. Hence, we have

Continued

EXAMPLE 10.7—CONT'D

$$CR = \frac{12\,\text{bits/sample}}{8\ \text{bits/sample}} = 1.5:1.$$

$$\text{Bit rate} = 8\frac{\text{bits}}{\text{sample}} \times 8000\frac{\text{sample}}{\text{second}} = 64\ (\text{kbps}).$$

b.2 Number of channels $= \dfrac{1.536\,\text{MBPS}}{64\ \text{KBPS}} = 24.$

a.3 For the standard ADPCM with 4 bits per sample, it follows that

$$CR = \frac{12\,\text{bits/sample}}{4\ \text{bits/sample}} = 3:1$$

$$\text{Bit rate} = 4\frac{\text{bits}}{\text{sample}} \times 8000\frac{\text{sample}}{\text{second}} = 32\ (\text{kbps}).$$

b.3 Number of channels $= \dfrac{1.536\,\text{MBPS}}{32\ \text{KBPS}} = 48.$

10.4 DISCRETE COSINE TRANSFORM, MODIFIED DISCRETE COSINE TRANSFORM, AND TRANSFORM CODING IN MPEG AUDIO

This section introduces *discrete cosine transform* (DCT) and explains how to apply it in transform coding. This section also shows how to remove the block effects in transform coding using a modified DCT (MDCT). Finally, we examine how MDCT coding is used in the MPEG (Motion Picture Experts Group) audio format, which is used as a part of MPEG audio, such as MP3 (MPEG-1 layer 3).

10.4.1 DISCRETE COSINE TRANSFORM

Given N data samples, we define the one-dimensional (1D) DCT pair given:

Forward transform:

$$X_{DCT}(k) = \sqrt{\frac{2}{N}} C(k) \sum_{n=0}^{N-1} x(n) \cos\left[\frac{(2n+1)k\pi}{2N}\right], \quad k = 0, 1, \ldots, N-1. \tag{10.29}$$

Inverse transform:

$$x(n) = \sqrt{\frac{2}{N}} \sum_{k=0}^{N-1} C(k) X_{DCT}(k) \cos\left[\frac{(2n+1)k\pi}{2N}\right], \quad n = 0, 1, \ldots, N-1 \tag{10.30}$$

$$C(k) = \begin{cases} \dfrac{\sqrt{2}}{2} & k = 0 \\ 1 & \text{otherwise,} \end{cases} \tag{10.31}$$

where $x(n)$ is the input data sample and $X_{DCT}(k)$ is the DCT coefficient. The DCT transforms the time domain signal to frequency domain coefficients. However, unlike the discrete Fourier transform (DFT), there are no complex number operations for both forward and inverse transforms. Both forward and inverse transforms use the same scale factor:

$$\sqrt{\frac{2}{N}}C(k).$$

In terms of transform coding, the DCT decomposes a block of data into the direct-current (DC) coefficient corresponding to the average of the data samples and the alternating-current (AC) coefficients, each corresponding to the frequency component (fluctuation). The terms "DC" and "AC" come from basic electrical engineering. In transform coding, we can quantize the DCT coefficients and encode them into binary information. The inverse DCT can transform the DCT coefficients back to the input data. Let us proceed to Examples 10.8 and 10.9.

EXAMPLE 10.8

Assuming that the following input data each can be encoded by 5 bits, including a sign bit:

$$x(0)=10,\ x(1)=8,\ x(2)=10, \text{ and } x(3)=12,$$

(a) Determine the DCT coefficients.
(b) Use the MATLAB function **dct()** to verify all the DCT coefficients.

Solution:

(a) Using Eq. (10.29) leads to

$$X_{DCT}(k)=\sqrt{\frac{1}{2}}C(k)\left[x(0)\cos\left(\frac{\pi k}{8}\right)+x(1)\cos\left(\frac{3\pi k}{8}\right)+x(2)\cos\left(\frac{5\pi k}{8}\right)+x(3)\cos\left(\frac{7\pi k}{8}\right)\right].$$

When $k=0$, we see that the DC component is calculated as

$$X_{DCT}(0)=\sqrt{\frac{1}{2}}C(0)\left[x(0)\cos\left(\frac{\pi\times 0}{8}\right)+x(1)\cos\left(\frac{3\pi\times 0}{8}\right)+x(2)\cos\left(\frac{5\pi\times 0}{8}\right)+x(3)\cos\left(\frac{7\pi\times 0}{8}\right)\right]$$
$$=\sqrt{\frac{1}{2}}\times\frac{\sqrt{2}}{2}[x(0)+x(1)+x(2)+x(3)]=\frac{1}{2}(10+8+10+12)=20.$$

We clearly see that the first DCT coefficient is a scaled average value.

For $k=1$,

$$X_{DCT}(1)=\sqrt{\frac{1}{2}}C(1)\left[x(0)\cos\left(\frac{\pi\times 1}{8}\right)+x(1)\cos\left(\frac{3\pi\times 1}{8}\right)+x(2)\cos\left(\frac{5\pi\times 1}{8}\right)+x(3)\cos\left(\frac{7\pi\times 1}{8}\right)\right]$$
$$=\sqrt{\frac{1}{2}}\times 1\times\left[10\times\cos\left(\frac{\pi\times 1}{8}\right)+8\times\cos\left(\frac{3\pi\times 1}{8}\right)+10\times\cos\left(\frac{5\pi\times 1}{8}\right)+12\times\cos\left(\frac{7\pi\times 1}{8}\right)\right]=-1.8478.$$

Similarly, we have the rest as

$$X_{DCT}(2)=2 \text{ and } X_{DCT}(3)=0.7654.$$

(b) Using the MATLAB 1D-DCT function **dct()**, we can verify that
>>dct([10 8 10 12])
ans=20.0000 – 1.8478 2.0000 0.7654.

EXAMPLE 10.9

Assuming the following DCT coefficients:

$$X_{DCT}(0) = 20, X_{DCT}(1) = -1.8478, X_{DCT}(2) = 2, \text{ and } X_{DCT}(3) = 0.7654,$$

(a) Determine $x(0)$.
(b) Use the MATLAB function **idct()** to verify all the recovered data samples.

Solution:

(a) Applying Eqs. (10.30) and (10.31), we have

$$x(0) = \sqrt{\frac{1}{2}} \times \left\{ C(0)X_{DCT}(0)\cos\left(\frac{\pi}{8}\right) + C(1)X_{DCT}(1)\cos\left(\frac{3\pi}{8}\right) + C(2)X_{DCT}(2)\cos\left(\frac{5\pi}{8}\right) + C(3)X_{DCT}(3)\cos\left(\frac{7\pi}{8}\right) \right\}$$

$$= \sqrt{\frac{1}{2}} \times \left\{ \frac{\sqrt{2}}{2} \times 20 \times \cos\left(\frac{\pi}{8}\right) + 1 \times (-1.8474) \times \cos\left(\frac{3\pi}{8}\right) + 1 \times 2 \times \cos\left(\frac{5\pi}{8}\right) + 1 \times 0.7654 \times \cos\left(\frac{7\pi}{8}\right) \right\} = 10.$$

(b) With the MATLAB 1D inverse DCT function **idct()**, we obtain

```
>>idct([20 -1.8478 2 0.7654])
ans = 10.0000 8.0000 10.0000 12.0000.
```

We verify that the input data samples are as the ones in Example 10.8.

In Example 10.9, we obtained an exact recovery of the input data from the DCT coefficients, since infinite precision of each DCT coefficient is preserved. However, in transform coding, each DCT coefficient is quantized using the number of bits per sample assigned by a bit allocation scheme. Usually the DC coefficient requires a larger number of bits to encode, since it carries more energy of the signal, while each AC coefficient requires a smaller number of bits to encode. Hence, the quantized DCT coefficients approximate the DCT coefficients in infinite precision, and the recovered input data with the quantized DCT coefficients will certainly have quantization errors.

EXAMPLE 10.10

Assuming the following DCT coefficients in infinite precision:

$$X_{DCT}(0) = 20, X_{DCT}(1) = -1.8478, X_{DCT}(2) = 2, \text{ and } X_{DCT}(3) = 0.7654,$$

we had exact recovered data as: 10, 8, 10, 12; this was verified in Example 10.9. If a bit allocation scheme quantizes the DCT coefficients using a scale factor of 4 in the following form:

$$X_{DCT}(0) = 4 \times 5 = 20, X_{DCT}(1) = 4 \times (-0) = 0, X_{DCT}(2) = 4 \times 1 = 4, \text{ and } X_{DCT}(3) = 4 \times 0 = 0.$$

We can code the scale factor of 4 by 3 bits (magnitude bits only), the scaled DC coefficient of 5 with 4 bits (including a sign bit), and the scaled AC coefficients of 0, 1, 0 using 2 bits each. 13 bits in total are required.

Use the MATLAB **idct()** to recover the input data samples.

Solution:
Using the MATLAB function **idct()** and the quantized DCT coefficients, it follows that.
```
>> idct([20 0 4 0])
ans = 12 8 8 12.
```
As we see, the original sample requires 5 bits (4 magnitude bits and 1 sign bit) to encode each of 10, 8, 10, and 12 for a total of 20 bits. Hence, 7 bit are saved for coding this data block using the DCT. We expect many more bits to be saved in practice, in which a longer frame of the correlated data samples is used. However, quantization errors are introduced.

For comprehensive coverage of the topics on DCT, see Li et al. (2014), Nelson (1992), Sayood (2012), and Stearns and Hush (2011).

10.4.2 MODIFIED DISCRETE COSINE TRANSFORM

In the previous section, we have seen how a 1D-DCT is adopted for coding a block of data. When we apply the 1D-DCT to audio coding, we first divide the audio samples into blocks and then transform each block of data with DCT. The DCT coefficients for each block are quantized according to the bit allocation scheme. However, when we decode DCT blocks back, we encounter edge artifacts at boundaries of the recovered DCT blocks, since the DCT coding is block based. This effect of edge artifacts produces periodic noise and is annoying in the decoded audio. To solve for such a problem, the windowed MDCT has been developed (described in Pan, 1995; Princen and Bradley, 1986). The principles are illustrated in Fig. 10.14. As we shall see, the windowed MDCT is used in MPEG-1 MP3 audio coding.

We describe and discuss only main steps for coding data blocks using the windowed MDCT (W-MDCT) based on Fig. 10.14.

Encoding stage:

1. Divide data samples into blocks that each have N (must be an even number) samples, and further divide each block into two subblocks, each with $N/2$ samples for the data overlap purpose.
2. Apply the window function for the overlapped blocks. As shown in Fig. 10.14, if one block contains the subblocks A and B, the next one would consist of subblocks B and C. The subblock B is the overlapped block. This procedure continues. A window function $h(n)$ is applied to each N sample block to reduce the possible edge effects. Next, the W-MDCT is applied. The W-MDCT is given by

$$X_{MDCT}(k) = 2\sum_{n=0}^{N-1} x(n)h(n)\cos\left[\frac{2\pi}{N}(n+0.5+N/4)(k+0.5)\right] \text{ for } k=0,1,\cdots,N/2-1. \tag{10.32}$$

Note that we need to compute and encode only half of the MDCT coefficients (since the other half can be reconstructed based on the first half of the MDCT coefficients).

3. Quantize and encode the MDCT coefficients.

Decoding stage:

1. Receive the $N/2$ MDCT coefficients, and use Eq. (10.33) to recover the second half of the coefficients:

$$X_{MDCT}(k) = (-1)^{\frac{N}{2}+1} X_{MDCT}(N-1-k), \text{ for } k=N/2, N/2+1, \cdots, N-1. \tag{10.33}$$

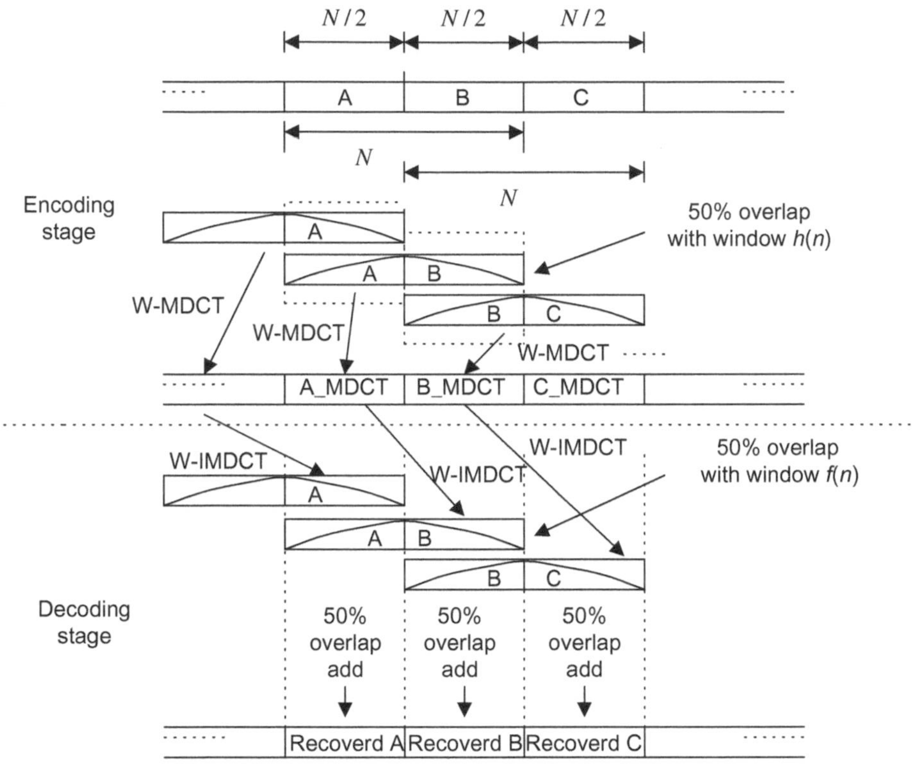

FIG. 10.14

Modified discrete cosine transform (MDCT).

2. Apply the windowed inverse MDCT (W-IMDCT) to each N MDCT coefficient block using Eq. (10.34) and then apply a decoding window function $f(n)$ to reduce the artifacts at the block edges:

$$x(n) = \frac{1}{N} f(n) \sum_{k=0}^{N-1} X_{MDCT}(k) \cos\left[\frac{2\pi}{N}(n + 0.5 + N/4)(k + 0.5)\right] \text{ for } n = 0, 1, \cdots, N-1. \tag{10.34}$$

Note that the recovered sequence contains the overlap portion. As shown in Fig. 10.14, if a decoded block has the decoded subblocks A and B, the next one would have subblocks B and C, where the subblock B is an overlapped block. The procedure continues.

3. Reconstruct the subblock B using the overlap and add operation, as shown in Fig. 10.14, where two subblocks labeled B are overlapped and added to generate the recovered subblock B. Note that the first subblock B comes from the recovered N samples containing A and B, while the second subblock B belongs to the next recovered N samples containing B and C.

In order to obtain the perfect reconstruction, that is, the full cancellation of all aliasing introduced by the MDCT, the following two conditions must be met for selecting the window functions, in which one is used for encoding while the other is used for decoding (Princen and Bradley, 1986):

$$f\left(n+\frac{N}{2}\right)h\left(n+\frac{N}{2}\right)+f(n)h(n)=1 \tag{10.35}$$

$$f\left(n+\frac{N}{2}\right)h(N-n-1)-f(n)h\left(\frac{N}{2}-n-1\right)=0. \tag{10.36}$$

Here, we choose the following simple function for the W-MDCT given by

$$f(n)=h(n)=\sin\left(\frac{\pi}{N}(n+0.5)\right). \tag{10.37}$$

Eq. (10.37) must satisfy the conditions described in Eqs. (10.35) and (10.36). This will be left for an exercise in the problem section at the end of this chapter. The MATLAB functions **wmdcth()** and **wimdctf()** relate to this topic are listed in the MATLAB program section (Section 10.6). Now, let us examine the W-MDCT in Example 10.11.

EXAMPLE 10.11

Given the data sequence: 1, 2, −3, 4, 5, −6, 4, 5 ...,

(a) Determine the W-MDCT coefficients for the first three blocks using a block size of 4.
(b) Determine the first two overlapped subblocks, and compare the results with the original data sequence using the W-MDCT coefficients in (a).

Solution:

(a) We divided the first two data blocks using the overlapping of two samples:
First block data: 1 2 −3 4
Second block data: −3 4 5 −6
Third block data: 5 −6 4 5
We apply the W-MDCT to get.

```
>> wmdct([1 2 −3 4])
ans = 1.1716 3.6569
>> wmdct([−3 4 5 −6])
ans = −8.0000 7.1716
>> wmdct([5 −6 4 5])
ans =−4.6569 −18.0710.
```

(b) We show the results from W-IWDCT as:

```
>> x1=wimdct([1.1716 3.6569])
x1 =−0.5607 1.3536 −1.1465 −0.4749
>> x2=wimdct([−8.0000 7.1716])
x2 = −1.8536 4.4749 2.1464 0.8891
>> x3=wimdct([−4.6569 −18.0711])
x3 =2.8536 −6.8891 5.1820 2.1465
```

Applying the overlap and add, we have.

```
>> [x1 0 0 0 0]+ [0 0 x2 0 0 ]+ [ 0 0 0 0 x3]
ans = −0.5607 1.3536 −3.0000 4.0000 5.0000 −6.0000 5.1820 2.1465
```

The recovered first two subblocks have values −3, 4, 5, −6 which are consistent with the input data.

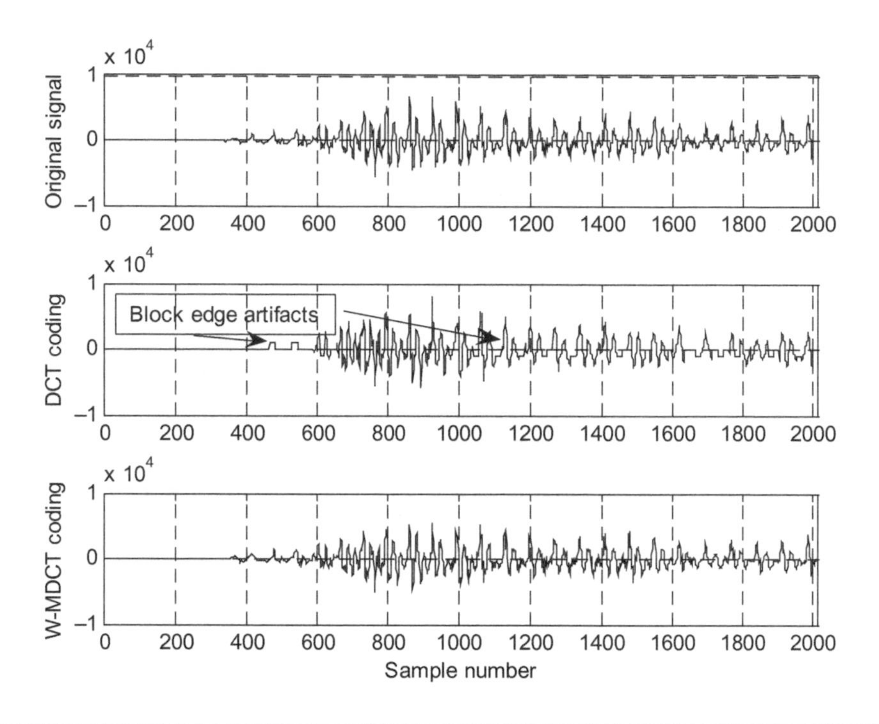

FIG. 10.15

Waveform coding using DCT and W-MDCT.

Fig. 10.15 shows coding of speech data we.dat using the DCT transform and W-MDCT transform. To be able to see the block edge artifacts, the following parameters are used for both DCT and W-MDCT transform coding:

Speech data: 16 bits per sample, 8000 samples per second

Block size: 16 samples

Scale factor: 2-bit nonlinear quantizer

Coefficients: 3-bit linear quantizer

Note that we assume a lossless scheme will further compress the quantized scale factors and coefficients. This stage does not affect the simulation results.

We use a 2-bit nonlinear quantizer with four levels to select the scale factor so that the block artifacts can be clearly displayed in Fig. 10.15. We also apply a 3-bit linear quantizer to the scaled coefficients for both DCT and W-MDCT coding. As shown in Fig. 10.15, the W-MDCT demonstrates significant improvement in smoothing out the block edge artifacts. The MATLAB simulation list is given in Programs 10.14–10.16 (Section 10.6), where Program 10.16 is the main program.

10.4.3 TRANSFORM CODING IN MPEG AUDIO

With the DCT and MDCT concepts developed, we now explore the MPEG audio data format, where the DCT plays a key role. MPEG was established in 1988 to develop a standard for delivery of digital video and audio. Since MPEG audio compression contains so many topics, we focus here

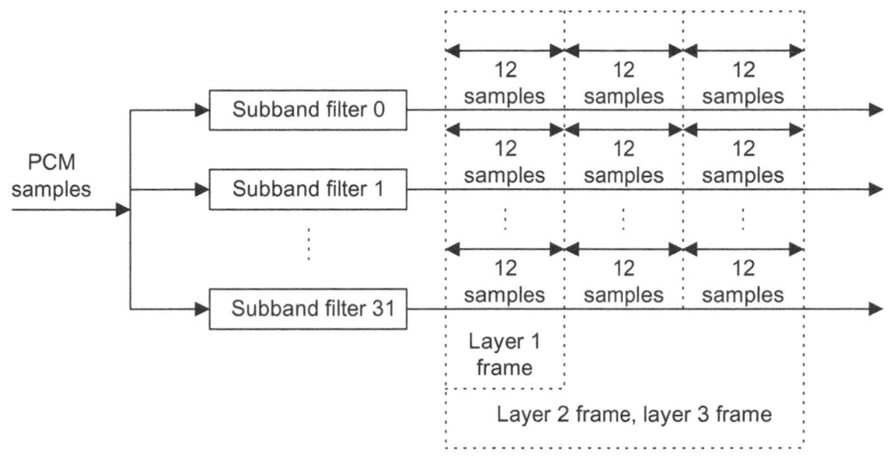

FIG. 10.16

MPEG audio frame size.

on examining its data format briefly, using the basic concepts developed in this book. Readers can further explore this subject by reading Pan's (1995) tutorial on MPEG audio compression, as well as Li et al. (2014).

Fig. 10.16 shows the MPEG audio frame. First, the input PCM samples—with a possible selection of sampling rates of 32, 44.1, and 48 kHz—are divided into 32 frequency subbands. All the subbands have equal bandwidths. The sum of their bandwidths covers up to the folding frequency, that is, $f_s/2$, which is the Nyquist limit in the DSP system. The subband filters are designed to minimized aliasing in frequency domain. Each subband filter outputs one sample for every 32 input PCM samples continuously, and forms a data segment for every 12 output samples. The purpose of the filter banks is to separate the data into different frequency bands so that the psychoacoustic model of the human auditory system (Yost, 2000) can be applied to activate the bit allocation scheme for a particular subband. The data frames are formed before quantization.

There are three types of data frames, as shown in Fig. 10.16. Layer 1 contains 32 data segments, each coming from one subband with 12 samples, so the total frame has 384 data samples. As we see, layers 2 and 3 have the same size data frame, consisting of 96 data segments, where each filter outputs 3 data segments of 12 samples. Hence, layer 2 or layer 3 each have 1152 data samples.

Next, let us examine briefly the content of each data frame, as shown in Fig. 10.17. Layer 1 contains 384 audio samples from 32 subbands, each having 12 samples. It begins with a header followed by a cyclic redundancy check (CRC) code. The numbers within parentheses indicate the possible number of bits to encode each field. The bit allocation informs the decoder of the number of bits used for each encoded sample in the specific band. Bit allocation can also be set to zero number of bits for a particular subband if analysis of the psychoacoustic model finds that the data in the band can be discarded without affecting the audio quality. In this way, the encoder can achieve more data compression. Each scale factor is encoded with 6 bits. The decoder will multiply the scale factor by the decoded quantizer output to get the quantized subband value. Use of the scale factor allows for utilization of the full range of the quantizer. The field "ancillary data" is reserved for "extra" information.

Layer 2 encoder takes 1152 samples per frame, with each subband channel having 3 data segments of 12 samples. These 3 data segments may have a bit allocation and up to three scale factors. Using one

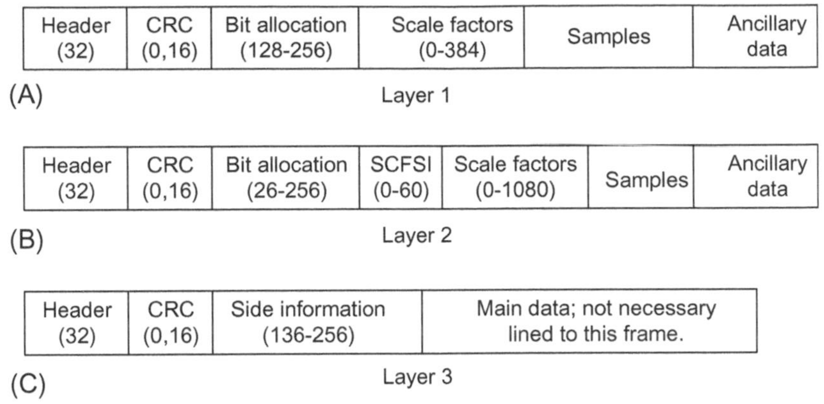

Header (32)	CRC (0,16)	Bit allocation (128-256)	Scale factors (0-384)	Samples	Ancillary data

(A) Layer 1

Header (32)	CRC (0,16)	Bit allocation (26-256)	SCFSI (0-60)	Scale factors (0-1080)	Samples	Ancillary data

(B) Layer 2

Header (32)	CRC (0,16)	Side information (136-256)	Main data; not necessary lined to this frame.

(C) Layer 3

FIG. 10.17

MPEG audio frame formats.

scale factor for 3 data segments would be called when values of the scale factors per subband are sufficiently close and the encoder applies temporal noise masking (a type noise masking by human auditory system) to hide any distortion. In Fig. 10.17, the field "SCFSI" (scale-factor selection information) contains the information to inform the decoder. A different scale factor is used for each suabband channel when avoidance of audible distortion is required. The bit allocation can also provide a possible single compact code word to represent three consecutive quantized values.

The layer 3 frame contains side information and main data that come from Huffman encoding (lossless coding having an exact recovery) of the W-MDCT coefficients to gain improvement over the layers 1 and 2.

Fig. 10.18 shows the MPEG-1 layers 1 and 2 encoder, layer 3 encoder. For the MPEG-1 layers 1 and 2, the encoder examines the audio input samples using a 1024-point fast Fourier transform (FFT). The psychoacoustic model is analyzed based on the FFT coefficients. This includes the possible frequency masking (hiding noise in frequency domain) and the noise temporal masking (hiding noise in time domain). The result of the analysis of the psychoacoustic model instructs the bit allocation scheme.

The major difference in layer 3, called MP3 (the most popular format in the multimedia industry), is that it adopts the MDCT. First, the encoder can gain further data compression by transforming the data segments using DCT from each subband channel and then quantize the DCT coefficients, which, again, are losslessly compressed using Huffman encoding. As shown in Examples 10.8–10.11, since the DCT uses block-based processing, it produces block edge effects, where the beginning samples and ending samples show discontinuity and causes audible periodic noise. This periodic edge noise can be alleviated, as discussed in the previous section, by using the W-MDCT, in which there are 50% overlap between successive transform windows.

There are two sizes of windows. One has 36 samples and another has 12 samples used in MPEG-1 layer 3 (MP3) audio. The larger block length offers the better frequency resolution for low-frequency tonelike signals, hence it is used for the lowest two subbands. For the rest of the subbands, the shorter block is used, since it allows better time resolution for noise-like transient signals. Other improvements of MP3 over layers 1 and 2 include use of the scale-factor band, where the W-MDCT coefficients are regrouped from the original 32 uniformly divided subbands into 25 actual critical bands based on the human auditory system. Then the corresponding scale factors are assigned, and a nonlinear quantizer is used.

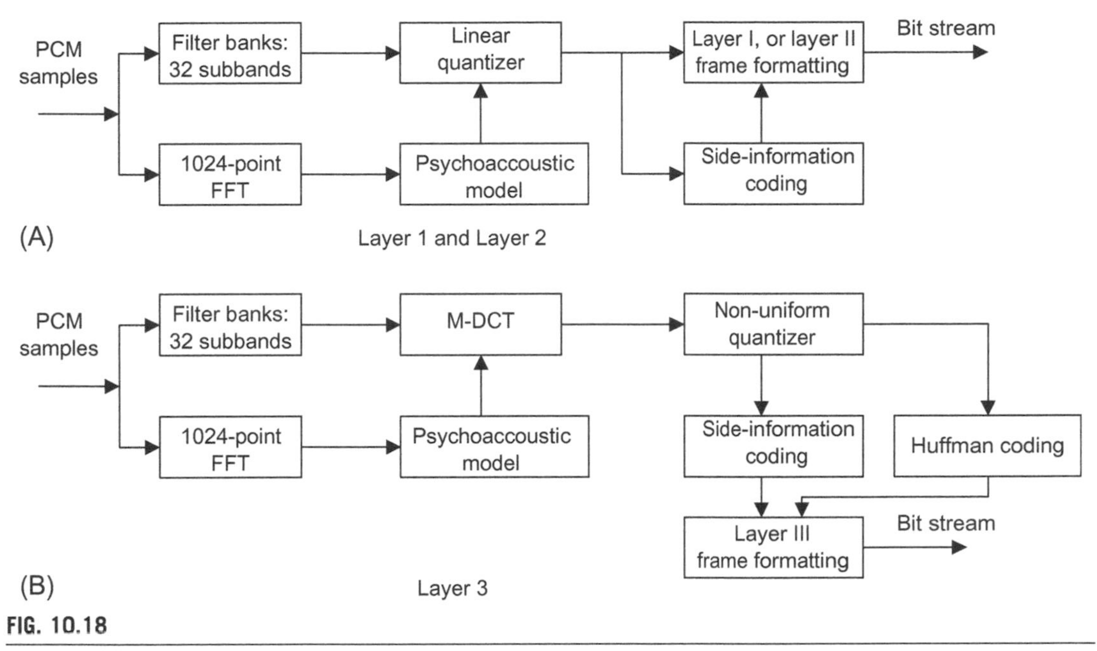

FIG. 10.18

Encoder block diagrams for layers 1 and 2 and for layer 3.

Finally, Huffman coding is applied to the quantizer outputs to obtain more compression. Particularly in CD-quality audio, MP3 (MPEG-1 layer 3) can achieve CRs varying from 12:1 to 8:1, corresponding to the bit rates from 128 to 192 kbps. Besides the use of DCT in MP3, MPEG-2 audio coding methods such as AC-2, AC-3, ATRAC, and PAC/MPAC also use W-MDCT coding. Readers can further explore these subjects in Brandenburg (1997) and Li et al. (2014).

10.5 SUMMARY

1. The linear midtread quantizer used in the PCM coding has an odd number of quantization levels, that is, $2^n - 1$. It accommodates the same decoded magnitude range for quantizing the positive and negative voltages.
2. Analog or digital μ-law compression improves coding efficiency. 8-bit μ-law compression of speech is equivalent to 12-bit linear PCM coding, with no difference in the sound quality. These methods are widely used in the telecommunications industry and multimedia system applications.
3. DPCM encodes the difference between the input sample and predicted sample using a predictor to achieve the coding efficiency.
4. DM coding is essentially a 1-bit DPCM.
5. ADPCM is similar to DPCM except that the predictor transfer function has six zeros and two poles and is an adaptive filter. ADPCM is superior to 8-bit μ-law compression, since it provides the same sound quality with only 4 bits per code.
6. Data compression performance is measured in terms of the data CR and the bit rate.

7. The DCT decomposes a block of data to the DC coefficient (average) and AC coefficients (fluctuation) so that different numbers of bits are assigned to encode DC coefficients and AC coefficients to achieve data compression.
8. W-MDCT alleviates the block effects introduced by the DCT.
9. The MPEG-1 audio formats such as MP3 (MPEG-1, layer 3) include W-MDCT, filter banks, a psychoacoustic model, bit allocation, a nonlinear quantizer, and Huffman lossless coding.

10.6 MATLAB PROGRAMS

Program 10.1 MATLAB program for the linear midtread quantizer.

```
clear all;close all
disp('load speech: We');
load we.dat;                      % Provided by your instructor
sig=we;
lg=length(sig);                   % Length of the speech data
t=[0:1:lg-1];                     % Time index
sig=5*sig/max(abs(sig));          % Normalize signal to the range between -5 to 5
Emax=max(abs(sig));
Erms=sqrt(sum(sig. * sig) / length(sig))
k=Erms/Emax
disp('20*log10(k)≥');
k=20*log10(k)
bits=input('input number of bits≥');
lg=length(sig);
% Encoding
for x=1:lg
  [indx(x) qy]=mtrdenc(bits, 5, sig(x));
end
disp('Finished and transmitted');
% Decoding
for x=1:lg
  qsig(x)=mtrddec(bits, 5, indx(x));
end
disp('decoding finished');
  qerr=sig-qsig;               % Calculate quantization errors
subplot(3,1,1);plot(t, sig);grid
ylabel('Speech');axis([0 length(we) -5 5]);
subplot(3,1,2);plot(t, qsig);grid
ylabel('Quantized speech');axis([0 length(we) -5 5]);
subplot(3,1,3);plot(qerr);grid
axis([0 length(we) -0.5 0.5]);
ylabel('Qunatized error');xlabel('Sample number');
disp('signal to noise ratio due to quantization noise')
snr(sig,qsig);                % Calculate signal to noise ratio due to quantization
```

Program 10.2. MATLAB function for midread quantizer encoding.

```
function [indx, pq]=mtrdenc(NoBits,Xmax,value)
% function pq=mtrdenc(NoBits, Xmax, value)
% This routine is created for simulation of midtread uniform quatizer.
%
% NoBits: number of bits used in quantization.
% Xmax: overload value.
% value: input to be quantized.
% pq: output of quantized value
% indx: integer index
%
% Note: the midtread method is used in this quantizer.
%
  if NoBits == 0
   pq=0;
   indx=0;
  else
   delta=2*abs(Xmax)/(2^NoBits-1);
   Xrmax=delta*(2^NoBits/2-1);
   if abs(value)>= Xrmax
     tmp=Xrmax;
   else
     tmp=abs(value);
   end
   indx=round(tmp/delta);
   pq=indx*delta;
   if value <0
     pq=-pq;
     indx=-indx;
   end
end
```

Program 10.3. MATLAB function for midtread quantizer decoding.

```
function pq=mtrddec(NoBits,Xmax,indx)
% function pq=mtrddec(NoBits, Xmax, value)
% This routine is dequantizer.
%
% NoBits: number of bits used in quantization.
% Xmax: overload value.
% pq: output of quantized value
% indx: integer index
%
% Note: the midtread method is used in this quantizer.
%
  delta=2*abs(Xmax)/(2^NoBits-1);
  pq=indx*delta;
```

Program 10.4. MATLAB program for μ-law encoding and decoding.

```
close all; clear all
disp('load speech file');
load we.dat;                    % Provided by your instructor
lg=length(we);                  % Length of the speech data
we=5*we/max(abs(we));           % Normalize the speech data
we_nor=we/max(abs(we));         % Normalization
t=[0:1:lg-1];                   % Time index
disp('mulaw companding')
mu=input('input mu≥');
for x=1:lg
  ymu(x)=mulaw(we_nor(x),1,mu);
end
disp('finished mu-law companding');
disp('start to quantization')
bits=input('input bits≥');
% Midtread quantization and encoding
for x=1:lg
  [indx(x) qy]=mtrdenc(bits, 1, ymu(x));
end
disp('finished and transmitted');
%
% Midtread decoding
for x=1:lg
  qymu(x)=mtrddec(bits, 1, indx(x));
end
disp('expander');
for x=1:lg
  dymu(x)=muexpand(qymu(x),1,mu)*5;
end
disp('finished')
qerr=dymu-we; % Quantization error
subplot(3,1,1);plot(we);grid
ylabel('Speech');axis([0 length(we) -5 5]);
subplot(3,1,2);plot(dymu);grid
ylabel('recovered speech');axis([0 length(we) -5 5]);
subplot(3,1,3);plot(qerr);grid
ylabel('Quantized error');xlabel('Sample number');
axis([0 length(we) -1 1]);
snr(we,dymu);               % Calculate signal to noise ratio due to quantization
```

Program 10.5. MATLAB function for μ-law companding.

```
Function qvalue=mulaw(vin, vmax, mu)
% This function performs mu-law companding
% Usage:
% function qvalue=mulaw(vin, vmax, mu)
% vin=input value
```

```
% vmax=maximum input amplitude
% mu=parameter for controlling the degree of compression which must be the same as
mu-law expander
% qvalue=output value from the mu-law compander
%
        vin=vin/vmax;       % Normalization
% mu-law companding
        qvalue=vmax*sign(vin)*log(1+mu*abs(vin))/log(1+mu);
```

Program 10.6. MATLAB program for μ-law expanding.

```
function rvalue=muexpand(y,vmax, mu)
% This function performs mu-law expanding
% Usage:
% function rvalue=muexpand(y,vmax, mu)
% y=input signal
% vmax=maximum input amplitude
% mu=parameter for controlling the degree of compression, which must be the same
% as the mu-law compander
% rvalue=output value from the mu-law expander
%
        y=y/vmax;               % Normalization
% mu-law expanding
        rvalue=sign(y)*(vmax/mu)*((1+mu)âbs(y) -1);
```

Program 10.7. MATLAB function for calculation of signal to quantization noise ratio (SNR).

```
function snr=calcsnr(speech, qspeech)
% function snr=calcsnr(speech, qspeech)
% This routine is created for calculation of SNR
%
% speech: original speech waveform.
% qspeech: quantized speech.
% snr: output SNR in dB.
%
% Note: midrise method is used in this quantizer.
%
 qerr=speech-qspeech;
 snr=10*log10(sum(speech.*speech)/sum(qerr.*qerr))
```

Program 10.8. Main program for the digital μ-law encoding and decoding.

```
load we12b.dat
for i=1:1:length(we12b)
    code8b(i)=dmuenc(12, we12b(i)); % Encoding
    qwe12b(i)=dmudec(code8b(i));    % Decoding
end
subplot(4,1,1),plot(we12b);grid
ylabel('a');axis([0 length(we12b) -1024 1024]);
subplot(4,1,2),plot(code8b);grid
ylabel('b');axis([0 length(we12b) -128,128]);
subplot(4,1,3),plot(qwe12b);grid
ylabel('c');axis([0 length(we12b) -1024 1024]);
subplot(4,1,4),plot(qwe12b-we12b);grid
ylabel('d');xlabel('Sample number');axis([0 length(we12b) -40 40]);
```

Program 10.9. The digital μ-law compressor.

```
function [cmp_code]=dmuenc(NoBits, value)
% This routine is created for simulation of 12-bit mu law compression.
% function [cmp_code]=dmuenc(NoBits, value)
% NoBits=number of bits for the data
% value=input value
% cmp_code=output code
%
 scale=NoBits-12;
 value=value*2^(-scale); % Scale to 12 bit
 if (abs(value)>=0) & (abs(value)<16)
     cmp_code=value;
 end
 if (abs(value)>=16) & (abs(value)<32)
     cmp_code=sgn(value)*(16+fix(abs(value)-16));
 end
 if (abs(value)>=32) & (abs(value)<64)
     cmp_code=sgn(value)*(32+fix((abs(value) -32)/2));
 end
 if (abs(value)>=64) & (abs(value)<128)
     cmp_code=sgn(value)*(48+fix((abs(value) -64)/4));
 end
 if (abs(value)>=128) & (abs(value)<256)
     cmp_code=sgn(value)*(64+fix((abs(value) -128)/8));
 end
 if (abs(value)>=256) & (abs(value)<512)
     cmp_code=sgn(value)*(80+fix((abs(value) -256)/16));
end
if (abs(value)>=512) & (abs(value)<1024)
```

```
    cmp_code=sgn(value)*(96+fix((abs(value) -512)/32));
end
if (abs(value)>=1024) & (abs(value)<2048)
    cmp_code=sgn(value)*(112+fix((abs(value) -1024)/64));
end
```

Program 10.10. The digital μ-law expander.

```
function [value]=dmudec(cmp_code)
% This routine is created for simulation of 12-bit mu law decoding.
% Usage:
% unction [value]=dmudec(cmp_code)
% cmp_code=input mu-law encoded code
% value=recovered output value
%
 if (abs(cmp_code)>=0) & (abs(cmp_code)<16)
     value=cmp_code;
 end
 if (abs(cmp_code)>=16) & (abs(cmp_code)<32)
     value=sgn(cmp_code)*(16+(abs(cmp_code)-16));
 end
 if (abs(cmp_code)>=32) & (abs(cmp_code)<48)
     value=sgn(cmp_code)*(32+(abs(cmp_code)-32)*2+1);
 end
 if (abs(cmp_code)>=48) & (abs(cmp_code)<64)
     value=sgn(cmp_code)*(64+(abs(cmp_code)-48)*4+2);
 end
 if (abs(cmp_code)>=64) & (abs(cmp_code)<80)
     value=sgn(cmp_code)*(128+(abs(cmp_code)-64)*8+4);
 end
 if (abs(cmp_code)>=80) & (abs(cmp_code)<96)
     value=sgn(cmp_code)*(256+(abs(cmp_code)-80)*16+8);
 end
 if (abs(cmp_code)>=96) & (abs(cmp_code)<112)
     value=sgn(cmp_code)*(512+(abs(cmp_code)-96)*32+16);
 end
 if (abs(cmp_code)>=112) & (abs(cmp_code)<128)
     value=sgn(cmp_code)*(1024+(abs(cmp_code)-112)*64+32);
 end
```

Program 10.11. Main program for ADPCM coding.

```
% This program is written for off-line simulation
% file: adpcm.m
clear all; close all
load we.dat                 % Provided by the instructor
```

```
speech=we;
desig=speech;
lg=length(desig);         % Length of speech data
enc=adpcmenc(desig);      % ADPCM encoding
%ADPCM finished
dec=adpcmdec(enc);        % ADPCM decoding
snrvalue=snr(desig,dec) % Calculate signal to noise ratio due to quantization
subplot(3,1,1);plot(desig);grid;
ylabel('Speech');axis([0 length(we) -8000 8000]);
subplot(3,1,2);plot(dec);grid;
ylabel('Quantized speech');axis([0 length(we) -8000 8000]);
subplot(3,1,3);plot(desig-dec);grid
ylabel('Quantized error');xlabel('Sample number');
axis([0 length(we) -1200 1200]);
```

Program 10.12 MATLAB function for ADPCM encoding.

```
function iiout = adpcmenc(input)
% This function performs ADPCM encoding
% function iiout = adpcmenc(input)
% Usage:
% input = input value
% iiout = output index
%
% Quantization tables
fitable = [0 0 0 1 1 1 1 3 7];
witable = [-0.75 1.13 2.56 4.00 7.00 12.38 22.19 70.13 ];
qtable = [ -0.98 0.62 1.38 1.91 2.34 2.72 3.12 ];
invqtable = [0.031 1.05 1.66 2.13 2.52 2.91 3.32 ];
lgth = length(input);
sr = zeros(1,2); pk = zeros(1,2);
a = zeros(1,2); b = zeros(1,6);
dq = zeros(1,6); ii= zeros(1,lgth);
y=0; ap = 0; al = 0; yu=0; yl = 0; dms = 0; dml = 0; tr = 0; td = 0;
for k = 1:lgth
sl = input(k);
%
% predict zeros
%
sez = b(1)*dq(1);
for i=2:6
sez = sez + b(i)*dq(i);
end
se = a(1)*sr(1)+a(2)*sr(2)+ sez;
d = sl - se;
%
% Perform quantization
%
dqq = log10(abs(d))/log10(2.0)-y;
```

```
  ik=0;
  for i=1:7
  if dqq > qtable(i)
        ik = i;
  end
end
if d < 0
 ik = -ik;
end
ii(k) = ik;
yu = (31.0/32.0)*y + witable(abs(ik)+1)/32.0;
if yu > 10.0
  yu = 10.0;
end
if yu < 1.06
  yu = 1.06;
end
yl = (63.0/64.0)*yl+yu/64.0;
%
%Inverse quantization
%
if ik == 0
  dqq = 2^(-y);
else
  dqq = 2^(invqtable(abs(ik))+y);
end
if ik < 0
  dqq = -dqq;
end
srr = se + dqq;
dqsez = srr+sez-se;
%
% Update state
%
pk1 = dqsez;
%
% Obtain adaptive predictor coefficients
%
  if tr == 1
  a = zeros(1,2); b = zeros(1,6);
  tr = 0;
  td = 0; %Set for the time being
  else
% Update predictor poles
% Update a2 first
  a2p = (127.0/128.0)*a(2);
  if abs(a(1)) <= 0.5
        fa1 = 4.0*a(1);
else
        fa1 = 2.0*sgn(a(1));
end
  a2p=a2p+(sign(pk1)*sgn(pk(1))-fa1*sign(pk1)*sgn(pk(2)))/128.0;
```

```
  if abs(a2p) > 0.75
      a2p = 0.75*sgn(a2p);
end
      a(2) = a2p;
%
% Update a1
  a1p = (255.0/256.0)*a(1);
       a1p = a1p + 3.0*sign(pk1)*sgn(pk(2))/256.0;
  if abs(a1p) > 15.0/16.0-a2p
       a1p = 15.0/16.0 -a2p;
end
   a(1) = a1p;
%
% Update b coefficients
%
for i=1:6
  b(i) = (255.0/256.0)*b(i)+sign(dqq)*sgn(dq(i))/128.0;
end
if a2p < -0.7185
  td = 1;
else
   td = 0;
end
if a2p < -0.7185 & abs(dq(6)) > 24.0*2^(yl)
  tr = 1;
else
   tr = 0;
end
for i=6:-1:2
  dq(i) = dq(i-1);
end
dq(1) = dqq; pk(2) = pk(1); pk(1) = pk1; sr(2) = sr(1); sr(1) = srr;
%
% Adaptive speed control
%
dms = (31.0/32.0)*dms; dms = dms + fitable(abs(ik)+1)/32.0;
dml = (127.0/128.0)*dml; dml = dml + fitable(abs(ik)+1)/128.0;
if ap > 1.0
  al = 1.0;
else
  al = ap;
end
ap = (15.0/16.0)*ap;
if abs(dms-dml) >= dml/8.0
   ap = ap + 1/8.0;
end
if y < 3
   ap = ap +1/8.0;
end
if td == 1
   ap = ap + 1/8.0;
end
```

```
if tr == 1
   ap = 1.0;
end
y = al*yu + (1.0-al)*yl;
end
end
iiout = ii;
```

Program 10.13. MATLAB function for ADPCM decoding.

```
function iiout = adpcmdec(ii)
% This function performs ADPCM decoding
% function iiout = adpcmdec(ii)
% Usage:
% ii = input ADPCM index
% iiout = decoded output value
%
% Quantization tables:
fitable = [0 0 0 1 1 1 1 3 7];
witable = [-0.75 1.13 2.56 4.00 7.00 12.38 22.19 70.13 ];
qtable = [ -0.98 0.62 1.38 1.91 2.34 2.72 3.12 ];
invqtable = [0.031 1.05 1.66 2.13 2.52 2.91 3.32 ];

lgth = length(ii);
sr = zeros(1,2); pk = zeros(1,2);
a = zeros(1,2); b = zeros(1,6);
dq = zeros(1,6); out= zeros(1,lgth);
y=0; ap = 0; al = 0; yu=0; yl = 0; dms = 0; dml = 0; tr = 0; td = 0;
for k = 1:lgth
%
sez = b(1)*dq(1);
for i=2:6
sez = sez + b(i)*dq(i);
end
se = a(1)*sr(1)+a(2)*sr(2)+ sez;
%
%Inverse quantization
%
ik = ii(k);
yu = (31.0/32.0)*y + witable(abs(ik)+1)/32.0;
if yu > 10.0
   yu = 10.0;
end
if yu < 1.06
   yu = 1.06;
end
yl = (63.0/64.0)*yl+yu/64.0;
if ik == 0
   dqq = 2^(-y);
```

```
else
   dqq = 2^(invqtable(abs(ik))+y);
end
if ik < 0
   dqq = -dqq;
end
srr = se + dqq;
dqsez = srr+sez-se;
out(k) =srr;
%
% Update state
%
pk1 = dqsez;
%
% Obtain adaptive predictor coefficients
%
  if tr == 1
          a = zeros(1,2);
          b = zeros(1,6);
          tr = 0;
          td = 0; %Set for the time being
          else
% Update predictor poles
% Update a2 first;
   a2p = (127.0/128.0)*a(2);
   if abs(a(1)) <= 0.5
       fa1 = 4.0*a(1);
   else
       fa1 = 2.0*sgn(a(1));
   end
   a2p=a2p+(sign(pk1)*sgn(pk(1))-fa1*sign(pk1)*sgn(pk(2)))/128.0;
   if abs(a2p) > 0.75
      a2p = 0.75*sgn(a2p);
   end
      a(2) = a2p;
%
% Update a1
  a1p = (255.0/256.0)*a(1);
      a1p = a1p + 3.0*sign(pk1)*sgn(pk(2))/256.0;
  if abs(a1p) > 15.0/16.0-a2p
        a1p = 15.0/16.0-a2p;
  end
a(1) = a1p;
%
% Update b coefficients
%
for i=1: 6
  b(i) = (255.0/256.0)*b(i)+sign(dqq)*sgn(dq(i))/128.0;
end
if a2p < -0.7185
  td = 1;
else
```

```
  td = 0;
end
if a2p < -0.7185 & abs(dq(6)) > 24.0*2^(yl)
  tr = 1;
else
  tr = 0;
end
for i=6:-1:2
  dq(i) = dq(i-1);
end
dq(1) = dqq; pk(2) = pk(1); pk(1) = pk1; sr(2) = sr(1); sr(1) = srr;
%
% Adaptive speed control
%
dms = (31.0/32.0)*dms;
dms = dms + fitable(abs(ik)+1)/32.0;
dml = (127.0/128.0)*dml;
dml = dml + fitable(abs(ik)+1)/128.0;
if ap > 1.0
  al = 1.0;
else
  al = ap;
end
ap = (15.0/16.0)*ap;
if abs(dms-dml) >= dml/8.0
   ap = ap + 1/8.0;
end
if y < 3
   ap = ap +1/8.0;
end
if td == 1
  ap = ap + 1/8.0;
end
if tr == 1
  ap = 1.0;
end
y = al*yu + (1.0-al)*yl;
end
end
iiout = out;
```

Program 10.14. W-MDCT function.

```
function [ tdac_coef ] = wmdct(ipsig)
%
% This function transforms the signal vector using the W-MDCT
% Usage:
% ipsig: inpput signal block of N samples (N=even number)
% tdac_coe: W-MDCT coefficients (N/2 coefficients)
```

```
%
N = length(ipsig);
NN =N;
for i=1:NN
  h(i) = sin((pi/NN)*(i-1+0.5));
end
for k=1:N/2
  tdac_coef(k) = 0.0;
  for n=1:N
    tdac_coef(k) = tdac_coef(k) + ...
    h(n)*ipsig(n)*cos((2*pi/N)*(k-1+0.5)*(n-1+0.5+N/4));
    end
end
tdac_coef=2*tdac_coef;
```

Program 10.15. Inverse W-IMDCT function.

```
function [ opsig ] = wimdct(tdac_coef )
%
% This function transform the W-MDCT coefficients back to the signal
% Usage:
% tdac_coeff: N/2 W-MDCT coefficients
% opsig: output signal block with N samples
%
N = length(tdac_coef );
tmp_coef = ((-1)^(N+1))*tdac_coef(N:-1:1);
tdac_coef = [ tdac_coef tmp_coef];
N = length(tdac_coef );
 NN =N;
for i=1:NN
  f(i) = sin((pi/NN)*(i-1+0.5));
end
for n=1:N
  opsig(n) = 0.0;
  for k=1:N
   opsig(n) = opsig(n) + ...
   tdac_coef(k)*cos((2*pi/N)*(k-1+0.5)*(n-1+0.5+N/4));
  end
  opsig(n) = opsig(n)*f(n)/N;
end
```

Program 10.16. Waveform coding using DCT and W-MDCT.

```
% Waveform coding using DCT and MDCT for a block size of 16 samples
% Main program
close all; clear all
```

```
load we.dat               % Provided by the instructor
% Create a simple 3 bit scale fcator
scalef4bits=[1 2 4 8 16 32 64 128 256 512 1024 2048 4096 8192 16384 32768];
scalef3bits=[256 512 1024 2048 4096 8192 16384 32768];
scalef2bits=[4096 8192 16384 32768];
scalef1bit=[16384 32768];
scalef=scalef1bit;
nbits =3;
% Ensure the block size to be 16 samples.
x=[we zeros(1,16-mod(length(we),16))];
Nblock=length(x)/16;
DCT_code=[]; scale_code=[];
% DCT transform coding
% Encoder
 for i=1:Nblock
  xblock_DCT=dct(x((i-1)*16+1:i*16));
  diff=abs(scalef-(max(abs(xblock_DCT))));
  iscale(i)=min(find(diff<=min(diff ))); %find a scale factor
  xblock_DCT=xblock_DCT/scalef(iscale(i)); % scale the input vetor
   for j=1:16
    [DCT_coeff(j) pp]=biquant(nbits,-1,1,xblock_DCT(j));
   end
   DCT_code=[DCT_code DCT_coeff ];
 end
%Decoder
Nblock=length(DCT_code)/16;
xx=[];
for i=1:Nblock
     DCT_coefR=DCT_code((i-1)*16+1:i*16);
   for j=1:16
       xrblock_DCT(j)=biqtdec(nbits,-1,1,DCT_coefR(j));
   end
      xrblock=idct(xrblock_DCT.*scalef(iscale(i)));
   xx=[xx xrblock];
end
% Transform coding using MDCT
xm=[zeros(1,8) we zeros(1,8-mod(length(we),8)), zeros(1,8)];
Nsubblock=length(x)/8;
MDCT_code=[];
% Encoder
 for i=1:Nsubblock
  xsubblock_DCT=wmdct(xm((i-1)*8+1:(i+1)*8));
  diff=abs(scalef-max(abs(xsubblock_DCT)));
  iscale(i)=min(find(diff<=min(diff ))); %find a scale factor
  xsubblock_DCT=xsubblock_DCT/scalef(iscale(i)); % scale the input vetor
  for j=1:8
   [MDCT_coeff(j) pp]=biquant(nbits,-1,1,xsubblock_DCT(j));
   end
  MDCT_code=[MDCT_code MDCT_coeff];
end
%Decoder
% Recover thr first subblock
```

```
Nsubblock=length(MDCT_code)/8;
xxm=[];
MDCT_coeffR=MDCT_code(1:8);
for j=1:8
      xmrblock_DCT(j)=biqtdec(nbits,-1,1,MDCT_coeffR(j));
end
xmrblock=wimdct(xmrblock_DCT*scalef(iscale(1)));
xxr_pre=xmrblock(9:16) % recovered first block for overlap and add
for i=2:Nsubblock
   MDCT_coeffR=MDCT_code((i-1)*8+1:i*8);
   for j=1:8
      xmrblock_DCT(j)=biqtdec(nbits,-1,1,MDCT_coeffR(j));
   end
      xmrblock=wimdct(xmrblock_DCT*scalef(iscale(i)));
   xxr_cur=xxr_pre+xmrblock(1:8); % overlap and add
   xxm=[xxm xxr_cur];
   xxr_pre=xmrblock(9:16);          % set for the next overlap
end

subplot(3,1,1);plot(x,'k');grid; axis([0 length(x) -10000 10000])
ylabel('Original signal');
subplot(3,1,2);plot(xx,'k');grid;axis([0 length(xx) -10000 10000]);
ylabel('DCT coding')
subplot(3,1,3);plot(xxm,'k');grid;axis([0 length(xxm) -10000 10000]);
ylabel('W-MDCT coding');
xlabel('Sample number');
```

Program 10.17. Sign function.

```
function sgn = sgn(sgninp)
%
% Sign function
% if signp >=0 then sign=1
% else sign =-1
%
 if sgninp >= 0
    opt = 1;
 else
    opt = -1;
 end
 sgn = opt;
```

10.7 PROBLEMS

10.1 For the 3-bit midtread quantizer described in Fig. 10.1, and the analog signal range from −2.5 to 2.5 V, determine
(a) the quantization step size
(b) the binary codes, recovered voltages, and quantization errors when each input is 1.6 and −0.2 V.

10.2 For the 3-bit midtread quantizer described in Fig. 10.1, and the analog signal range from −4 to 4 V, determine
(a) the quantization step size
(b) the binary codes, recovered voltages, and quantization errors when each input is −2.6 and 0.1 V.

10.3 For the 3-bit midtread quantizer described in Fig. 10.1, and the analog signal range from −5 to 5 V, determine
(a) the quantization step size
(b) the binary codes, recovered voltages, and quantization errors when each input is −2.6 and 3.5 V.

10.4 For the 3-bit midtread quantizer described in Fig. 10.1, and the analog signal range from −10 to 10 V, determine
(a) the quantization step size
(b) the binary codes, recovered voltages, and quantization errors when each input is −5, 0, and 7.2 V.

10.5 For the μ-law compression and expanding process shown in Fig. 10.3 with $u=255$ and the 3-bit midtread quantizer described in Fig. 10.1, with the analog signal range from −2.5 to 2.5 V, determine the binary codes, recovered voltages, and quantization errors when each input is 1.6 and −0.2 V.

10.6 For the μ-law compression and expanding process shown in Fig. 10.3 with $u=255$ and the 3-bit midtread quantizer described in Fig. 10.1, with the analog signal range from −4 to 4 V, determine the binary codes, recovered voltages, and quantization errors when each input is −2.6 and 0.1 V.

10.7 For the μ-law compression and expanding process shown in Fig. 10.3 with $u=255$ and the 3-bit midtread quantizer described in Fig. 10.1, with the analog signal range from −5 to 5 V, determine the binary codes, recovered voltages, and quantization errors when each input is −2.6 and 3.5 V.

10.8 For the μ-law compression and expanding process shown in Fig. 10.3 with $u=255$ and the 3-bit midtread quantizer described in Fig. 10.1, with the analog signal range from −10 to 10 V, determine the binary codes, recovered voltages, and quantization errors when each input is −5, 0, and 7.2 V.

10.9 In a digital companding system, encode each of the following 12-linear PCM codes into the 8-bit compressed PCM code.
(a) 0 0 0 0 0 0 0 1 0 1 0 1
(b) 1 0 1 0 1 1 1 0 1 0 1 0

10.10 In a digital companding system, decode each of the following 8-bit compressed PCM codes into the 12-bit linear PCM code.

(a) 0 0 0 0 0 1 1 1

(b) 1 1 1 0 1 0 0 1

10.11 In a digital companding system, encode each of the following 12-linear PCM codes into the 8-bit compressed PCM code.

(a) 0 0 1 0 1 0 1 0 1 0 1 0

(b) 1 0 0 0 0 0 0 0 1 1 0 1

10.12 In a digital companding system, decode each of the following 8-bit compressed PCM codes into the 12-bit linear PCM code.

(a) 0 0 1 0 1 1 0 1

(b) 1 0 0 0 0 1 0 1

10.13 For a 3-bit DPCM encoding system with the following specifications (Fig. 10.19):

Encoder scheme: $\tilde{x}(n) = \hat{x}(n-1)$ (predictor)

$d(n) = x(n) - \tilde{x}(n)$

$d_q(n) = Q[d(n)] =$ Quantizer in Table 10.9

$\hat{x}(n) = \tilde{x}(n) + d_q(n)$

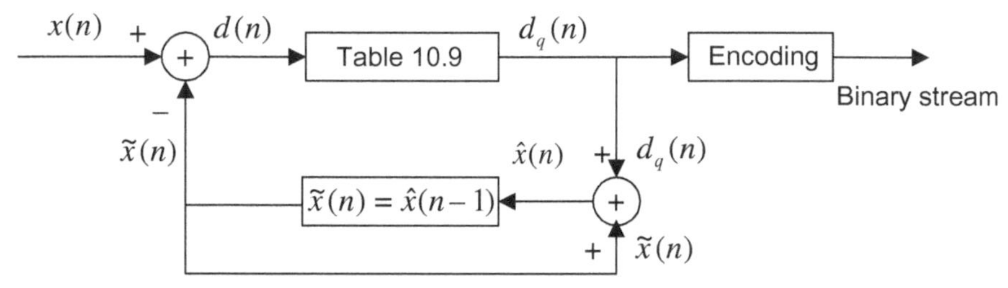

FIG. 10.19

DPCM encoding in Problem 10.13.

Table 10.9 Quantization Table for the 3-Bit Quantizer in Problem 10.13

Binary Code	Quantization Value $d_q(n)$	Subrange in $d(n)$
0 1 1	−11	$-15 \le d(n) < -7$
0 1 0	−5	$-7 \le d(n) < -3$
0 0 1	−2	$-3 \le d(n) < -1$
0 0 0	0	$-1 \le d(n) < 0$
1 0 0	0	$0 \le d(n) \le 1$
1 0 1	2	$1 < d(n) \le 3$
1 1 0	5	$3 < d(n) \le 7$
1 1 1	11	$7 < d(n) \le 15$

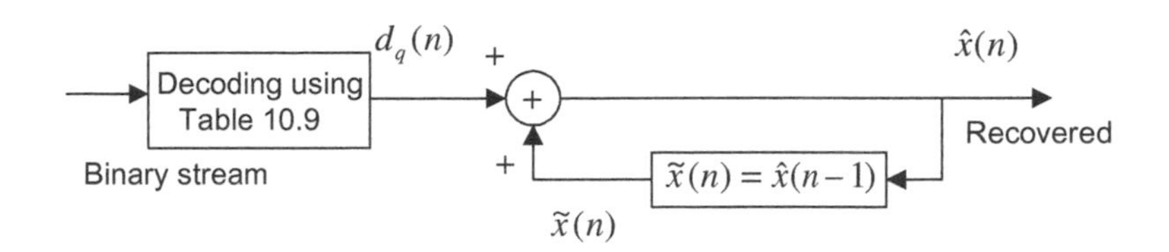

FIG. 10.20

DPCM decoding in Problem 10.14.

The 5-bit input data $x(0) = -6$, $x(1) = -8$, and $x(2) = -13$, perform DPCM encoding to produce the binary code for each input data.

10.14 For a 3-bit DPCM decoding system as shown in Fig. 10.20 with the following specifications:

Decoding scheme: $\tilde{x}(n) = \hat{x}(n-1)$ (predictor)
$d_q(n) =$ Quantizer in Table 10.9.
$\hat{x}(n) = \tilde{x}(n) + d_q(n)$

Received 3 binary codes: 110, 100, 101, perform DCPM decoding to recover each digital value using its binary code.

10.15 For a 3-bit DPCM encoding system shown in Problem 10.13 and the given data: the 5-bit input data: $x(0) = 6$, $x(1) = 8$, and $x(2) = 13$, perform DPCM encoding to produce the binary code for each input data.

10.16 For a 3-bit DPCM decoding system shown in Problem 10.14 and the received data: 010, 000, 001, perform DCPM decoding to recover each digital value using its binary code.

10.17 Assume that a speech waveform is sampled at 8 kHz and each sample is encoded by 16 bits, determine the CR for each of the encoding methods.
(a) Noncompression;
(b) Standard μ-law compression (8 bits per sample);
(c) Standard ADPCM encoding (4 bits per sample).

10.18 Suppose that a speech waveform is sampled at 8 kHz and each sample is encoded by 16 bits, determine the bit rate for each of the encoding methods.
(a) Noncompression;
(b) Standard μ-law companding (8 bits per sample);
(c) Standard ADPCM encoding (4 bits per sample).

10.19 Assume that an audio waveform is sampled at 44.1 kHz and each sample is encoded by 16 bits, determine the CR for each of the encoding methods.
(a) noncompression;
(b) standard μ-law compression (8 bits per sample);
(c) standard ADPCM encoding (4 bits per sample).

10.20 Suppose that an audio waveform is sampled at 44.1 kHz and each sample is encoded by 12 bits, determine the bit rate for each of the encoding methods.
(a) noncompression;
(b) standard μ-law companding (8 bits per sample);
(c) standard ADPCM encoding (4 bits per sample).

10.21 Speech is sampled at 8 kHz and each sample is encoded by 16 bits. The telephone system can transport the digital voice channel over a digital link having a capacity of 1.536 MBPS. Determine the number of channels that the phone company can carry for each of the following encoding methods:

(a) noncompression;

(b) standard 8-bit μ-law companding (8 bits per sample);

(c) standard ADPCM encoding (4 bits per sample).

10.22 Given the following input data:

$$x(0)=25,\ x(1)=30,\ x(2)=28,\ \text{and}\ x(3)=25,$$

determine the DCT coefficients.

10.23 Given the following input data:

$$x(0)=25,\ \text{and}\ x(1)=30,$$

determine the DCT coefficients.

10.24 Given the following input data:

$$x(0)=25, x(1)=30,\ x(2)=28,\ x(3)=25,$$

$$x(4)=10,\ x(5)=0,\ x(6)=0,\ \text{and}\ x(7)=0,$$

determine the DCT coefficient $X_{DCT}(0)$, $X_{DCT}(2)$, $X_{DCT}(4)$, and $X_{DCT}(6)$.

10.25 Given the following input data:

$$x(0)=25,\ x(1)=30,\ x(2)=28,\ x(3)=25,$$

$$x(4)=10,\ x(5)=0,\ x(6)=0,\ \text{and}\ x(7)=0,$$

determine the DCT coefficient $X_{DCT}(1)$, $X_{DCT}(3)$, $X_{DCT}(5)$, and $X_{DCT}(7)$.

10.26 Assuming the following DCT coefficients with infinite precision:

$$X_{DCT}(0)=14,\ X_{DCT}(1)=6,\ X_{DCT}(2)=-6,\ \text{and}\ X_{DCT}(3)=8,$$

(a) determine the input data using the MATLAB function **idct();**

(b) recover the input data samples using the MATLAB function **idct()** if a bit allocation scheme quantizes the DCT coefficients as follows: 2 magnitude bits plus 1 sign bit (3 bits) for the DC coefficient, 1 magnitude bit plus 1 sign bit (2 bits) for each AC coefficient and a scale factor of 8, that is,

$$X_{DCT}(0)=8\times 2=16,\ X_{DCT}(1)=8\times 1=8,\ X_{DCT}(2)=8\times(-1)=-8,\ \text{and}\ X_{DCT}(3)=8\times 1=8;$$

(c) compute the quantized error in part b in this problem.

10.27 Assuming the following DCT coefficients with infinite precision:

$$X_{DCT}(0)=11, X_{DCT}(1)=5, X_{DCT}(2)=7, \text{and}\ X_{DCT}(3)=-3,$$

(a) determine the input data using the MATLAB function **idct()**.
(b) recover the input data samples using the MATLAB function **idct()** if a bit allocation scheme quantizes the DCT coefficients as follows: 2 magnitude bits plus 1 sign bit (3 bits) for the DC coefficient, 1 magnitude bit plus 1 sign bit (2 bits) for each AC coefficient and a scale factor of 8, that is,

$$X_{DCT}(0) = 8 \times 1 = 8, X_{DCT}(1) = 8 \times 1 = 8, X_{DCT}(2) = 8 \times 1 = -8, \text{and } X_{DCT}(3) = 8 \times 0 = 0;$$

(c) compute the quantized error in part b in this problem.

10.28 **(a)** Verify the window function

$$f(n) = h(n) = \sin\left(\frac{\pi}{N}(n + 0.5)\right)$$

used in MDCT is satisfied with Eqs. (10.35) and (10.36).
(b) Verify W-MDCT coefficients

$$X_{MDCT}(k) = (-1)^{\frac{N}{2}+1} X_{MDCT}(N-1-k) \text{ for } k = N/2, N/2+1, \cdots, N-1.$$

10.29 Given a data sequence: 1, 2, 3, 4, 5, 4, 3, 2, ...,
(a) Determine the W-MDCT coefficients for the first three blocks using a block size of 4;
(b) Determine the first two overlapped subblocks and compare the results with the original data sequence using the W-MDCT coefficients in part (a).

10.30 Given a data sequence: 1, 2, 3, 4, 5, 4, 3, 2, 1, 2, 3, 4, 5, ...,
(a) Determine the W-MDCT coefficients for the first three blocks using a block size of 6;
(b) Determine the first two overlapped subblocks and compare the results with the original data sequence using the W-MDCT coefficients in part (a).

Computer Problems with MATLAB

Use the MATLAB programs in the program section for Problems 10.31–10.33.

10.31 Given the data file "speech.dat" with 16 bits per sample and a sampling rate of 8 kHz,
(a) Use the PCM coding (midtread quantizer) to perform compression and decompression and apply the MATLAB function **sound()** to evaluate the sound quality in terms of "excellent," "good," "intelligent," and "unacceptable" for the following bit rates
1. 4 bits/sample (32 kbits per second)
2. 6 bits/sample (48 kbits per second)
3. 8 bits/sample (64 kbits per second)

(b) Use the μ-law PCM coding to perform compression and decompression and apply the MATLAB function **sound()** to evaluate the sound quality.
1. 4 bits/sample (32 kbits per second)
2. 6 bits/sample (48 kbits per second)
3. 8 bits/sample (64 kbits per second)

10.32 Given the data file "speech.dat" with 16 bits per sample, a sampling rate of 8 kHz, and ADCPM coding, perform compression, and decompression and apply the MATLAB function **sound()** to evaluate the sound quality.

10.33 Given the data file "speech.dat" with 16 bits per sample, a sampling rate of 8 kHz, and DCT and M-DCT coding in the program section, perform compression, and decompression using the following specified parameters in Program 10.14 to compare the sound quality.

(a) nbits = 3, scalef = scalef2bits

(b) nbits = 3, scalef = scalef3bits

(c) nbits = 4, scalef = scalef2bits

(d) nbits = 4, scalef = scalef3bits

Advanced Problems

10.34 Given the companding function $y = \text{sign}(x)\dfrac{\ln\left(1+\mu\dfrac{|x|}{|x|_{max}}\right)}{\ln(1+\mu)}$, show that

(a) $y = x/|x|_{max}$, when $\mu \to 0$,

(b) $y = \text{sign}(x)$, when $\mu \to \infty$.

10.35 Given the companding function as defined in Eq. (10.4), prove Eq. (10.6).

10.36 For ADPCM encoding, the predictor has the following IIR transfer function:

$$\frac{B(z)}{A(z)} = \frac{b_0 + b_1 z^{-1} + b_2 z^{-2} + b_3 z^{-3} + b_4 z^{-4} + b_5 z^{-5}}{1 - a_1 z^{-1} - a_2 z^{-2}}.$$

Show that the stability of the predictor is guaranteed if the following constraint is satisfied:

$$|a_2| < 0.75 \text{ and } |a_1| < 1 - a_2.$$

That is, the poles of the predictor are inside the unit circle in the z-plane.

10.37 Given the DCT transform as defined in Eq. (10.29), prove the inverse DCT transform, that is, Eq. (10.30).

10.38 Given an N-point sequence of $x(n)$, define a $2N$-point even symmetric sequence of $s(n)$ as

$$s(n) = \begin{cases} x(n) & 0 \le n \le N-1 \\ x(2N-n-1) & N < n \le 2N-1 \end{cases}.$$

Show that DFT coefficients of $s(n)$ are given by

$$S(k) = W_{2N}^{-k/2} 2\sum_{n=0}^{N-1} x(n)\cos\left[\frac{\pi(2n+1)k}{2N}\right] = W_{2N}^{-k/2} V(k) \text{ for } 0 \le k \le 2N-1,$$

where $W_{2N} = e^{-j2\pi/2N}$.

10.39 In Problem 10.38, show that

(a) $S(N)=0$.

(b) $s(n)=\frac{1}{N}Re\left\{\frac{S(0)}{2}+\sum_{k=1}^{N-1}S(k)W_{2N}^{-kn}\right\}$ for $0\leq k\leq 2N-1$.

10.40 Using results from Problems 10.38 and 10.39, show that

$$x(n)=\frac{1}{N}\left\{\frac{V(0)}{2}+\sum_{k=1}^{N-1}V(k)\cos\left[\frac{\pi(2n+1)k}{2N}\right]\right\} \text{for } 0\leq k\leq N-1.$$

CHAPTER

MULTIRATE DIGITAL SIGNAL PROCESSING, OVERSAMPLING OF ANALOG-TO-DIGITAL CONVERSION, AND UNDERSAMPLING OF BANDPASS SIGNALS

11

CHAPTER OUTLINE

11.1 MULTIRATE DIGITAL SIGNAL PROCESSING BASICS

In many areas of digital signal processing (DSP) applications—such as communications, speech, and audio processing—raising or lowering of a sampling rate is required. The principle that deals with changing the sampling rate belongs essentially to *multirate signal processing* (Ifeachor and Jervis, 2002; Proakis and Manolakis, 2007; Porat, 1997; Sorensen and Chen, 1997). As an introduction, we focus on the sampling rate conversion; that is, sampling rate reduction or increase.

Digital Signal Processing. https://doi.org/10.1016/B978-0-12-815071-9.00011-7

11.1.1 SAMPLING RATE REDUCTION BY AN INTEGER FACTOR

The process of reducing the sampling rate by an integer factor is referred to as *downsampling* of a data sequence. We also refer downsampling as "decimation" (not taking one of ten). The term "decimation" used for the downsampling process has been accepted and used in many textbooks and fields. To downsample a data sequence $x(n)$ by an integer factor of M, we use the following notation:

$$y(m) = x(mM), \tag{11.1}$$

where $y(m)$ is the downsampled sequence, obtained by taking a sample from the data sequence $x(n)$ for every M samples (discarding $M-1$ samples for every M samples). As an example, if the original sequence with a sampling period $T=0.1$ s (sampling rate $=10$ samples per second) is given by

$$x(n): 8\ 7\ 4\ 8\ 9\ 6\ 4\ 2\ -2\ -5\ -7\ -7\ -6\ -4\ldots$$

and we downsample the data sequence by a factor of 3, we obtain the downsampled sequence as

$$y(m): 8\ 8\ 4\ -5\ -6\ldots,$$

with the resultant sampling period $T=3\times 0.1=0.3$ s (the sampling rate now is 3.33 samples per second). Although the example is straightforward, there is a requirement to avoid aliasing noise. We illustrate this next.

From the Nyquist sampling theorem, it is known that aliasing can occur in the downsampled signal due to reduced sampling rate. After downsampling by a factor of M, the new sampling period becomes MT, and therefore the new sampling frequency is

$$f_{sM} = \frac{1}{MT} = \frac{f_s}{M}, \tag{11.2}$$

where f_s is the original sampling rate.

Hence, the folding frequency after downsampling becomes

$$f_{sM}/2 = \frac{f_s}{2M}. \tag{11.3}$$

This tells us that after downsampling by a factor of M, the new folding frequency will be decreased by M times. If the signal to be downsampled has frequency components larger than the new folding frequency, $f>f_s/(2M)$, aliasing noise will be introduced into the downsampled data.

To overcome this problem, it is required that the original signal $x(n)$ be processed by a lowpass filter $H(z)$ before downsampling, which should have a stop frequency edge at $f_s/(2M)$ (Hz). The corresponding normalized stop frequency edge is then converted to be

$$\Omega_{stop} = 2\pi \frac{f_s}{2M} T = \frac{\pi}{M} \text{ radians.} \tag{11.4}$$

In this way, before downsampling, we can guarantee the maximum frequency of the filtered signal satisfies

$$f_{\max} < \frac{f_s}{2M}, \tag{11.5}$$

such that no aliasing noise is introduced after downsampling. A general block diagram of decimation is given in Fig. 11.1, where the filtered output in terms of the z-transform can be written as

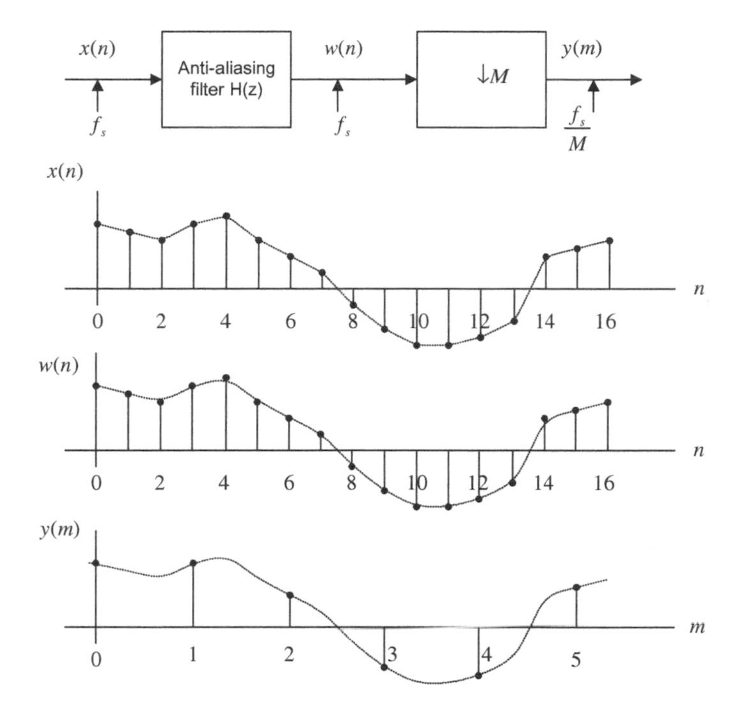

FIG. 11.1

Block diagram of the downsampling process with $M=3$.

$$W(z) = H(z)X(z), \tag{11.6}$$

where $X(z)$ is the z-transform of the sequence to be decimated, $x(n)$, and $H(z)$ is the lowpass filter transfer function. After anti-aliasing filtering, the downsampled signal $y(m)$ takes its value from the filter output as:

$$y(m) = w(mM). \tag{11.7}$$

The process of reducing the sampling rate by a factor of 3 is shown in Fig. 11.1. The corresponding spectral plots for $x(n)$, $w(n)$, and $y(m)$ in general are shown in Fig. 11.2.

To verify this principle, let us consider a signal $x(n)$ generated by the following:

$$x(n) = 5\sin\left(\frac{2\pi \times 1000n}{8000}\right) + \cos\left(\frac{2\pi \times 2500}{8000}\right) = 5\sin\left(\frac{n\pi}{4}\right) + \cos\left(\frac{5n\pi}{8}\right), \tag{11.8}$$

with a sampling rate of $f_s = 8000$ Hz, the spectrum of $x(n)$ is plotted in the first graph in Fig. 11.3A, where we observe that the signal has components at frequencies of 1000 and 2500 Hz. Now we downsample $x(n)$ by a factor of 2, that is, $M=2$. According to Eq. (11.3), we know that the new folding frequency is $4000/2 = 2000$ Hz. Hence, without using the anti-aliasing lowpass filter, the spectrum

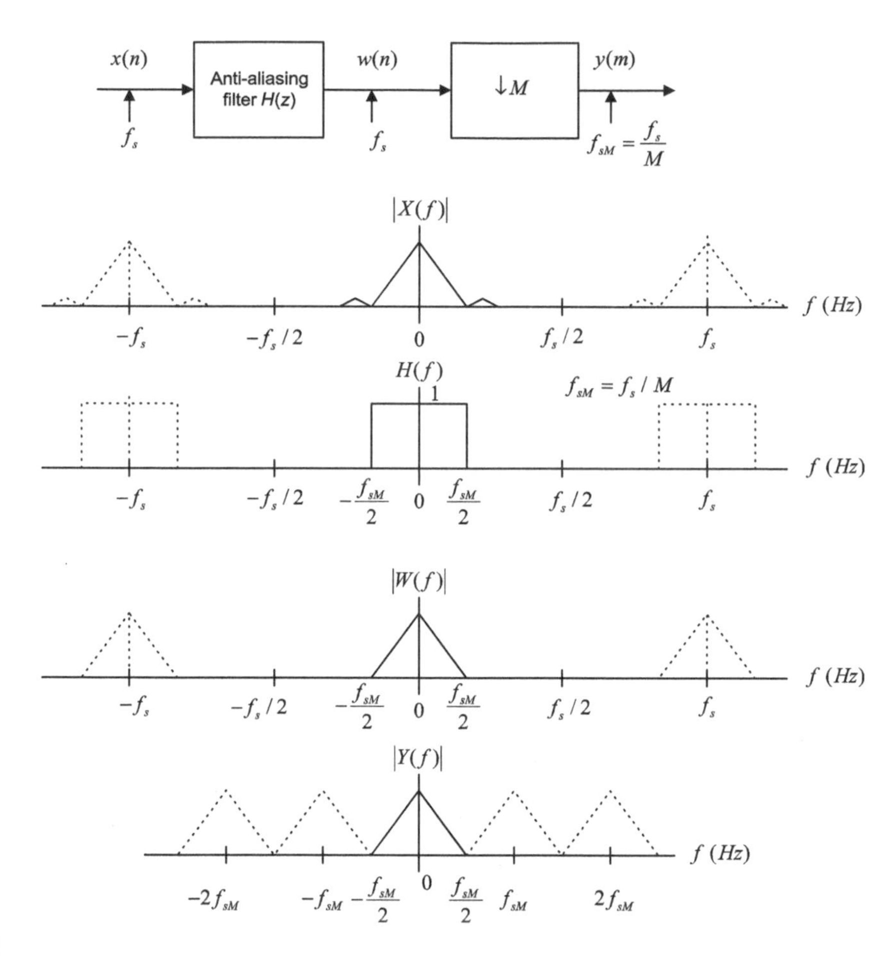

FIG. 11.2

Spectrum after downsampling.

would contain the aliasing frequency of 4–2.5 kHz = 1.5 kHz introduced by 2.5 kHz, plotted in the second graph in Fig. 11.3A.

Now we apply a finite impulse response (FIR) lowpass filter designed with a filter length of $N = 27$ and a cutoff frequency of 1.5 kHz to remove the 2.5 kHz signal before downsampling to avoid aliasing. How to obtain such specifications will be discussed in the later example. The normalized cutoff frequency used for design is given by

$$\Omega_c = 2\pi \times 1500 \times (1/8000) = 0.375\pi.$$

Thus, the aliasing noise is avoided. The spectral plots are given in Fig. 11.3B, where the first plot shows the spectrum of $w(n)$ after anti-aliasing filtering, while the second plot describes the spectrum of $y(m)$ after downsampling. Clearly, we prevent aliasing noise in the downsampled data by sacrificing the original 2.5-kHz signal. Program 11.1 gives the detail of MATLAB implementation.

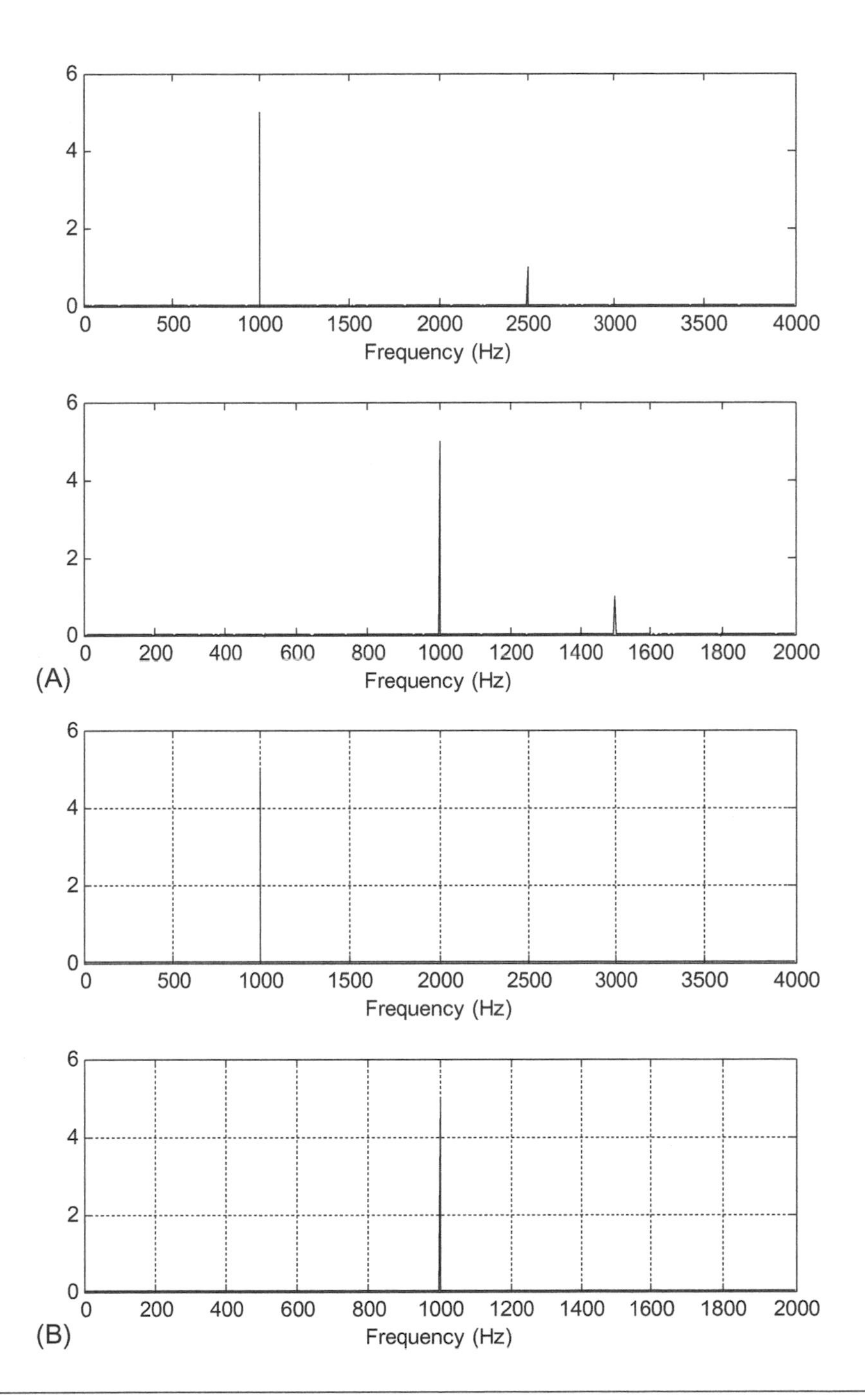

FIG. 11.3

(A) Spectrum before downsampling and spectrum after downsampling without using the anti-aliasing filter.
(B) Spectrum before downsampling and spectrum after downsampling using the anti-aliasing filter.

Program 11.1. MATLAB program for signal decimation.

```
close all; clear all;
% Downsampling filter (see Chapter 7 for FIR filter design)
B=[0.00074961181416 0.00247663033476 0.00146938649416 -0.00440446121505 ...
 -0.00910635730662 0.00000000000000 0.02035676831506 0.02233710562885...
 -0.01712963672810 -0.06376620649567 -0.03590670035210 0.10660980550088...
 0.29014909103794 0.37500000000000 0.29014909103794 0.10660980550088...
 -0.03590670035210 -0.06376620649567 -0.01712963672810 0.02233710562885...
 0.02035676831506 0.00000000000000 -0.00910635730662 -0.00440446121505...
 0.00146938649416 0.00247663033476 0.00074961181416];
% Generate 2048 samples
fs=8000;                    % Sampling rate
N=2048;                     % Number of samples
M=2;                        % Downsample factor
n=0:1:N-1;
x=5*sin(n*pi/4)+cos(5*n*pi/8);
% Compute single-side amplitude spectrum
% AC component will be doubled, and DC component will be kept as the same value
X=2*abs(fft(x,N))/N;X(1)=X(1)/2;
% Map the frequency index up to the folding frequency in Hz
f=[0:1:N/2-1]*fs/N;
%Downsampling
y=x(1:M:N);
NM=length(y);               % Length of the down sampled data
% Compute the single-sided amplitude spectrum for the downsampled signal
Y=2*abs(fft(y,NM))/length(y);Y(1)=Y(1)/2;
% Map the frequency index to the frequency in Hz
fsM=[0:1:NM/2-1]*(fs/M)/NM;
subplot(2,1,1);plot(f,X(1:1:N/2));grid; xlabel('Frequency (Hz)');
subplot(2,1,2);plot(fsM,Y(1:1:NM/2));grid; xlabel('Frequency (Hz)');
figure
w=filter(B,1,x);            % Anti-aliasing filtering
% Compute the single-sided amplitude spectrum for the filtered signal
W=2*abs(fft(w,N))/N;W(1)=W(1)/2;
% Downsampling
y=w(1:M:N);
NM=length(y);
% Compute the single-sided amplitude spectrum for the downsampled signal
Y=2*abs(fft(y,NM))/NM;Y(1)=Y(1)/2;
% Plot spectra
subplot(2,1,1);plot(f,W(1:1:N/2));grid; xlabel('Frequency (Hz)');
subplot(2,1,2);plot(fsM,Y(1:1:NM/2));grid; xlabel('Frequency (Hz)');
```

Now we focus on how to design an anti-aliasing FIR filter, or decimation filter. We discuss this topic via the following example.

EXAMPLE 11.1

Given a DSP downsampling system with the following specifications:
Sampling rate = 6000 Hz
Input audio frequency range = 0–800 Hz
Passband ripple = 0.02 dB
Stopband attenuation = 50 dB
Downsample factor $M=3$,
Determine the FIR filter length, cutoff frequency, and window type if the window method is used.

Solution:

Specifications are reorganized as:
Anti-aliasing filter operating at the sampling rate = 6000 Hz
Passband frequency range = 0–800 Hz
Stopband frequency range = 1–3 kHz
Passband ripple = 0.02 dB
Stopband attenuation = 50 dB
Filter type = FIR.
The block diagram and specifications are depicted in Fig. 11.4.
The Hamming window is selected, since it provides 0.019 dB ripple and 53 dB stopband attenuation. The normalized transition band is given by

$$\Delta f = \frac{f_{stop} - f_{pass}}{f_s} = \frac{1000 - 800}{6000} = 0.033.$$

The length of the filter and the cutoff frequency can be determined by

$$N = \frac{3.3}{\Delta f} = \frac{3.3}{0.033} = 100.$$

We choose an odd number; that is, $N=101$, and

$$f_c = \frac{f_{pass} + f_{stop}}{2} = \frac{800 + 1000}{2} = 900\,\text{Hz}.$$

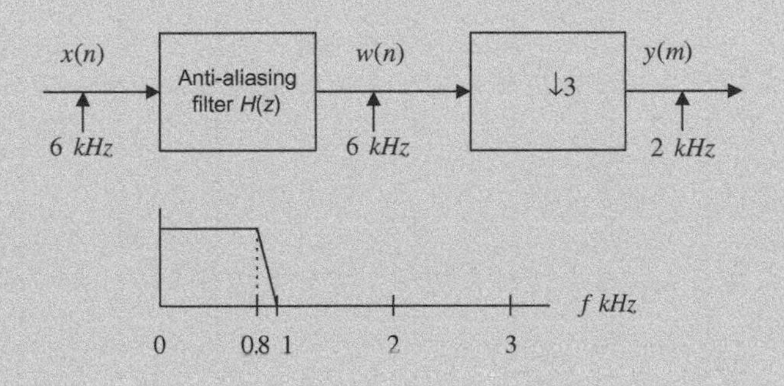

FIG. 11.4

Filter specifications for Example 11.1.

11.1.2 SAMPLING RATE INCREASE BY AN INTEGER FACTOR

Increasing a sampling rate is a process of upsampling by an integer factor of L. This process is described as follows:

$$y(m) = \begin{cases} x\left(\frac{m}{L}\right) & m = nL \\ 0 & \text{otherwise} \end{cases}, \tag{11.9}$$

where $n = 0, 1, 2, \ldots, x(n)$ is the sequence to be upsampled by a factor of L, and $y(m)$ is the upsampled sequence. As an example, suppose that the data sequence is given as follows:

$$x(n): 8\ 8\ 4\ -5\ -6\ldots$$

After upsampling the data sequence $x(n)$ by a factor of 3 (adding L-1 zeros for each sample), we have the upsampled data sequence $w(m)$ as:

$$w(m): 8\ 0\ 0\ 8\ 0\ 0\ 4\ 0\ 0\ -5\ 0\ 0\ -6\ 0\ 0\ldots$$

The next step is to smooth the upsampled data sequence via an interpolation filter. The process is illustrated in Fig. 11.5A.

Similar to the downsampling case, assuming that the data sequence has the current sampling period of T, the Nyquist frequency is given by $f_{\max} = f_s/2$. After upsampling by a factor of L, the new sampling period becomes T/L, thus the new sampling frequency is changed to be

$$f_{sL} = Lf_s. \tag{11.10}$$

This indicates that after upsampling, the spectral replicas originally centered at $\pm f_s$, $\pm 2f_s$, … are included in the frequency range from 0 Hz to the new Nyquist limit $Lf_s/2$ Hz, as shown in Fig. 11.5B. To remove those included spectral replicas, an interpolation filter with a stop frequency edge of $f_s/2$ in Hz must be attached, and the normalized stop frequency edge is given by

$$\Omega_{stop} = 2\pi\left(\frac{f_s}{2}\right) \times \left(\frac{T}{L}\right) = \frac{\pi}{L} \text{ radians.} \tag{11.11}$$

After filtering via the interpolation filter, we will achieve the desired spectrum for $y(n)$, as shown in Fig. 11.5B. Note that since the interpolation is to remove the high-frequency images that are aliased by the upsampling operation, it is essentially an anti-aliasing lowpass filter.

To verify the upsampling principle, we generate the signal $x(n)$ with 1 and 2.5 kHz as follows:

$$x(n) = 5\sin\left(\frac{2\pi \times 1000n}{8000}\right) + \cos\left(\frac{2\pi \times 2500n}{8000}\right),$$

with a sampling rate of $f_s = 8000$ Hz. The spectrum of $x(n)$ is plotted in Fig. 11.6. Now we upsample $x(n)$ by a factor of 3, that is, $L = 3$. We know that the sampling rate is increased to be $3 \times 8000 = 24{,}000$ Hz. Hence, without using the interpolation filter, the spectrum would contain the image frequencies originally centered at the multiple frequencies of 8 kHz. The top plot in Fig. 11.6 shows the spectrum for the sequence after upsampling and before applying the interpolation filter.

Now we apply an FIR lowpass filter designed with a length of 53, a cutoff frequency of 3250 Hz, and a new sampling rate of 24,000 Hz as the interpolation filter, whose normalized frequency should be

$$\Omega_c = 2\pi \times 3{,}250 \times \left(\frac{1}{24{,}000}\right) = 0.2708\pi.$$

The bottom plot in Fig. 11.6 shows the spectrum for $y(m)$ after applying the interpolation filter, where only the original signals with frequencies of 1 and 2.5 kHz are presented. Program 11.2 shows the implementation detail in MTALAB.

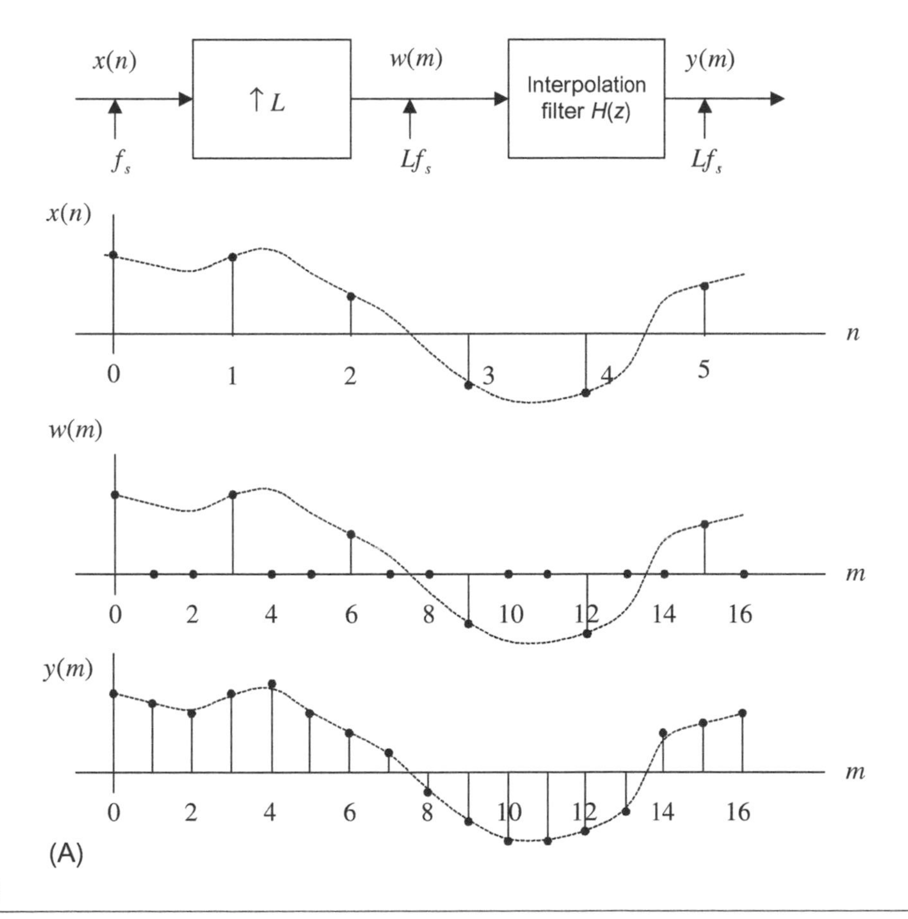

FIG. 11.5

(A) Block diagram for the upsampling process with $L=3$.

Continued

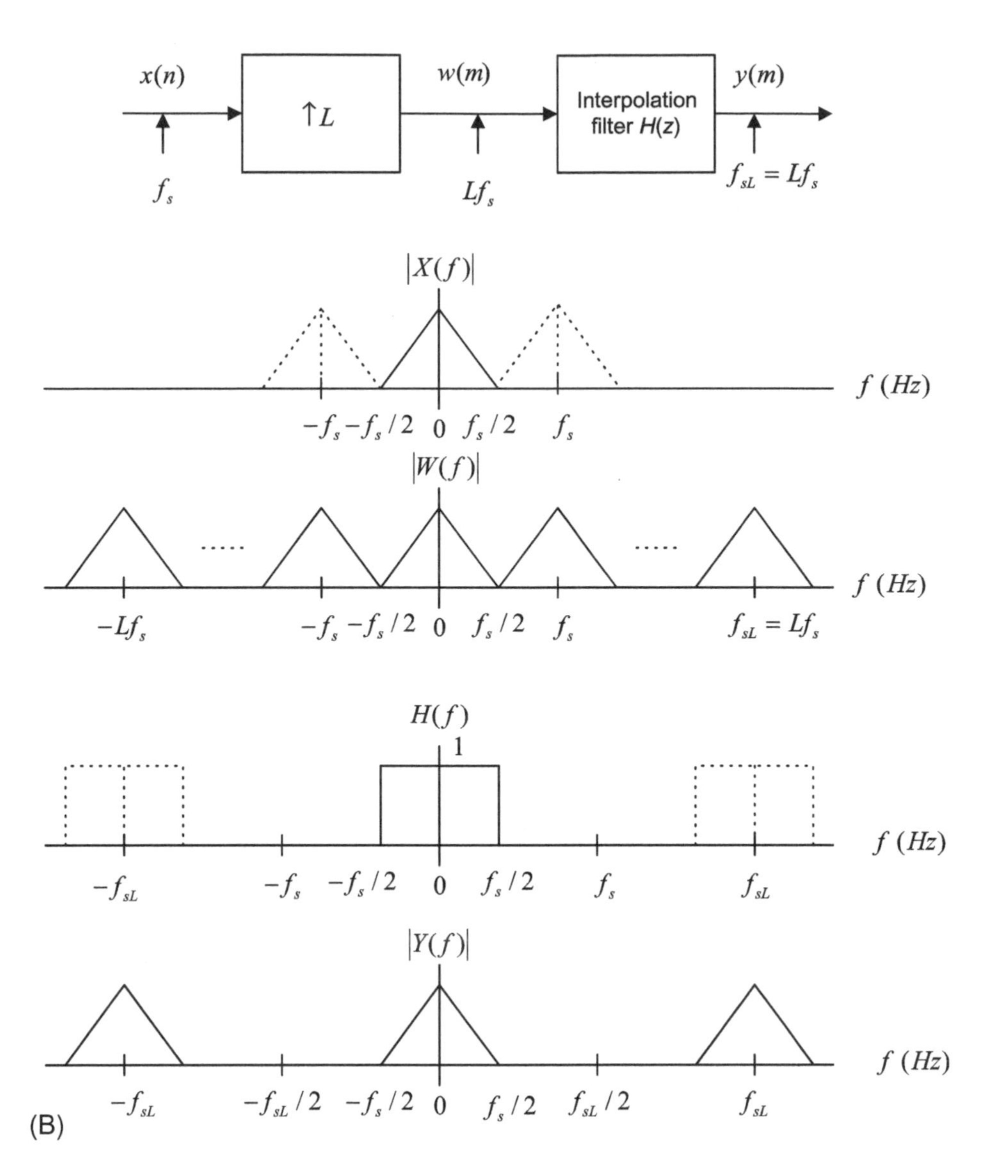

FIG. 11.5, CONT'D

(B) Spectra before and after upsampling.

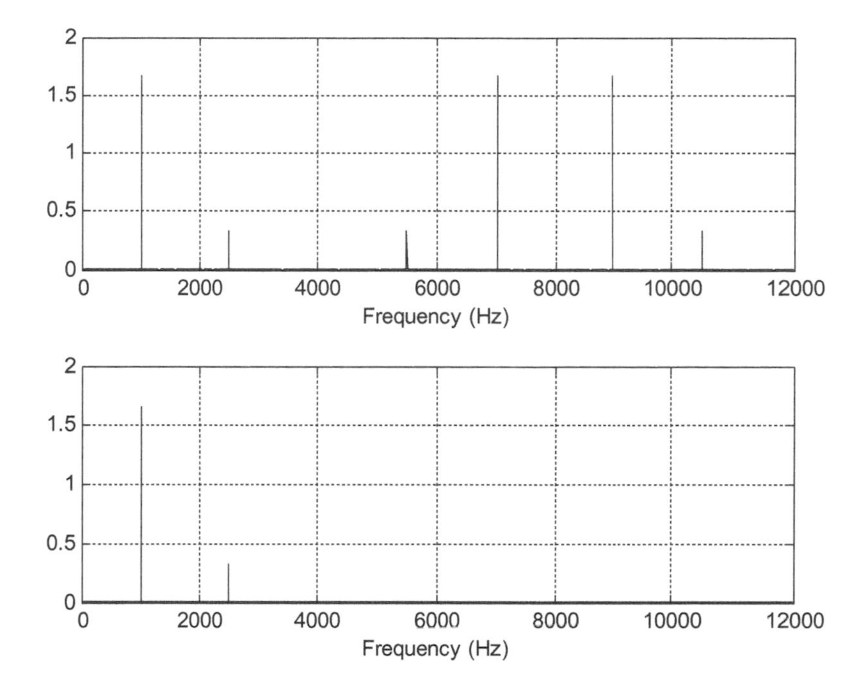

FIG. 11.6

(Top) The spectrum after upsampling and before applying the interpolation filter; (bottom) spectrum after applying the interpolation filter.

Program 11.2. MATLAB program for signal interpolation.

```
close all; clear all
%Upsampling filter (see Chapter 7 for FIR filter design)
B=[-0.00012783931504 0.00069976044649 0.00123831516738 0.00100277549136...
 -0.00025059018468 -0.00203448515158 -0.00300830295487 -0.00174101657599...
  0.00188598835011  0.00578414933758  0.00649330625041  0.00177982369523...
 -0.00670672686935 -0.01319379342716 -0.01116855281442  0.00123034314117...
  0.01775600060894  0.02614700427364  0.01594155162392 -0.01235169936557...
 -0.04334322148505 -0.05244745563466 -0.01951094855292  0.05718573279009...
  0.15568416401644  0.23851539047347  0.27083333333333  0.23851539047347...
  0.15568416401644  0.05718573279009 -0.01951094855292 -0.05244745563466...
 -0.04334322148505 -0.01235169936557  0.01594155162392  0.02614700427364...
  0.01775600060894  0.00123034314117 -0.01116855281442 -0.01319379342716...
 -0.00670672686935  0.00177982369523  0.00649330625041  0.00578414933758...
  0.00188598835011 -0.00174101657599 -0.00300830295487 -0.00203448515158...
 -0.00025059018468  0.00100277549136  0.00123831516738  0.00069976044649...
 -0.00012783931504];
% Generate the 2048 samples with fs=8000Hz
fs=8000;                    % Sampling rate
N=2048;                     % Number of samples
L=3;                        % Upsampling factor
n=0:1:N-1;
x=5*sin(n*pi/4)+cos(5*n*pi/8);
```

```
% Upsampling by a factor of L
w=zeros(1,L*N);
for n=0:1:N-1
w(L*n+1)=x(n+1);
end
NL=length(w);                    % Length of the upsampled data
W=2*abs(fft(w,NL))/NL;W(1)=W(1)/2; %Compute one-sided amplitude spectrum
f=[0:1:NL/2-1]*fs*L/NL; % Map the frequency index to the frequency (Hz)
%Interpolation
y=filter(B,1,w); % Apply interpolation filter
Y=2*abs(fft(y,NL))/NL;Y(1)=Y(1)/2; %Compute the one-sided amplitude spectrum
fsL=[0:1:NL/2-1]*fs*L/NL; % Map the frequency index to the frequency (Hz)
subplot(2,1,1);plot(f,W(1:1:NL/2));grid; xlabel('Frequency (Hz)');
subplot(2,1,2);plot(fsL,Y(1:1:NL/2));grid; xlabel('Frequency (Hz)');
```

Now let us study how to design the interpolation filter via Example 11.2.

EXAMPLE 11.2

Given a DSP upsampling system with the following specifications:
Sampling rate = 6000 Hz
Input audio frequency range = 0–800 Hz
Passband ripple = 0.02 dB
Stopband attenuation = 50 dB
Upsample factor $L=3$,

Determine the FIR filter length, cutoff frequency, and window type if the window design method is used.

Solution:

The specifications are reorganized as follows:
Interpolation filter operating at the sampling rate = 18,000 Hz
Passband frequency range = 0–800 Hz
Stopband frequency range = 3–9 kHz
Passband ripple = 0.02 dB
Stopband attenuation = 50 dB
Filter type: FIR filter
The block diagram and filter frequency specifications are given in Fig. 11.7.
We choose the Hamming window for this application. The normalized transition band is

$$\Delta f = \frac{f_{stop} - f_{pass}}{f_{sL}} = \frac{3000 - 800}{18,000} = 0.1222.$$

The length of the filter and the cutoff frequency can be determined by

$$N = \frac{3.3}{\Delta f} = \frac{3.3}{0.1222} = 27,$$

and the cutoff frequency is given by

$$f_c = \frac{f_{pass} + f_{stop}}{2} = \frac{3000 + 800}{2} = 1900\,\text{Hz}.$$

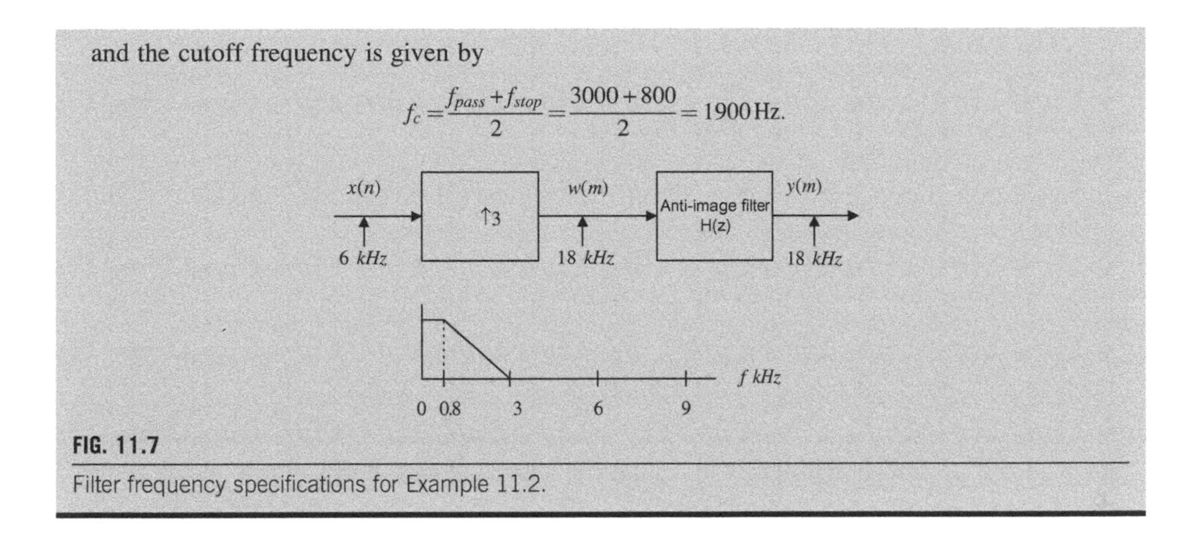

FIG. 11.7

Filter frequency specifications for Example 11.2.

11.1.3 CHANGING SAMPLING RATE BY A NON-INTEGER FACTOR *L/M*

With an understanding of the downsampling and upsampling processes, we now study the sampling rate conversion by a non-integer L/M. This can be viewed as two sampling conversion processes. In step 1, we perform the upsampling process by a factor of integer L following application of an interpolation filter $H_1(z)$; in step 2, we continue filtering the output from the interpolation filter via an anti-aliasing filter $H_2(z)$, and finally execute downsampling. The entire process is illustrated in Fig. 11.8.

Since the interpolation and anti-aliasing filters are in a cascaded form and operate at the same rate, we can select one of them. We choose the one with the lower stop frequency edge and choose the most demanding requirements for passband gain and stopband attenuation for the filter design. A lot of computational savings can be achieved by using one lowpass filter. We illustrate the procedure via the following simulation. Let us generate the signal $x(n)$ by:

$$x(n) = 5\sin\left(\frac{2\pi \times 1000n}{8000}\right) + \cos\left(\frac{2\pi \times 2500n}{8000}\right),$$

with a sampling rate of $f_s = 8000$ Hz and frequencies of 1 and 2.5 kHz.

Now we resample $x(n)$ to 3000 Hz by a non-integer factor of 0.375, that is,

$$\left(\frac{L}{M}\right) = 0.375 = \frac{3}{8}.$$

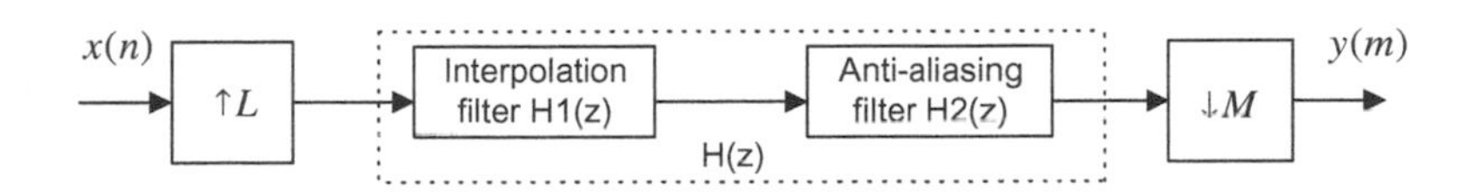

FIG. 11.8

Block diagram for the sampling rate conversion.

Upsampling is at a factor of $L=3$ and the upsampled sequence is filtered by an FIR lowpass filter designed with the filter length $N=53$ and a cutoff frequency of 3250Hz at the sampling rate of $3 \times 8000 = 24{,}000$ Hz. The spectrum for the upsampled sequence and the spectrum after application of the interpolation filter are plotted in Fig. 11.9A.

The sequence from step 1 can be filtered via another FIR lowpass filter designed with the filter length $N=159$ and a cutoff frequency of 1250Hz, following by downsampling by a factor of $M=8$. The spectrum after the anti-aliasing filter and the spectrum for the final output $y(m)$ are plotted in Fig. 11.9B. Note that the anti-aliasing filter removes the frequency component of 2.5kHz to avoid aliasing. This is because after downsampling, the Nyquist limit is 1.5kHz. As we discussed previously, we can select one filter for implementation. We choose the FIR lowpass filter with $N=159$ and a cutoff frequency of 1250Hz because its bandwidth is smaller than that of the interpolation filter. The MATLAB implementation is listed in Program 11.3.

Program 11.3. MATLAB program for changing sampling rate with a non-integer factor.

```
close all; clear all;clc;
% Down sampling filter
Bdown=firwd(159,1,2*pi*1250/24000,0,4);
% Generate 2048 samples with fs=8000Hz
fs=8000;                                  % Original sampling rate
N=2048;                                   % The number of samples
L=3;                                      % Upsampling factor
M=8;                                      % Downsampling factor
n=0:1:N-1;                                % Generate the time index
x=5*sin(n*pi/4)+cos(5*n*pi/8);           % Generate the test signal
% Up sampling by a factor of L
w1=zeros(1,L*N);
for n=0:1:N-1
 w1(L*n+1)=x(n+1);
end
NL=length(w1);                            % Length of up sampled data
W1=2*abs(fft(w1,NL))/NL;W1(1)=W1(1)/2;   % Compute the one-sided
%amplitude spectrum
f=[0:1:NL/2-1]*fs*L/NL; % Map frequency index to its frequency in Hz
subplot(3,1,1);plot(f,W1(1:1:NL/2));grid
xlabel('Frequency (Hz)');
w2=filter(Bdown,1,w1); % Perform the combined anti-aliasing filter
W2=2*abs(fft(w2,NL))/NL;W2(1)=W2(1)/2; % Compute the one-sided
%amplitude spectrum
y2=w2(1:M:NL);
NM=length(y2);                            % Length of the downsampled data
Y2=2*abs(fft(y2,NM))/NM;Y2(1)=Y2(1)/2;% Compute the one-sided
%amplitude spectrum
% Map frequency index to its frequency in Hz before downsampling
fbar=[0:1:NL/2-1]*24,000/NL;
% Map frequency index to its frequency in Hz
fsM=[0:1:NM/2-1]*(fs*L/M)/NM;
subplot(3,1,2);plot(f,W2(1:1:NL/2));grid; xlabel('Frequency (Hz)');
subplot(3,1,3);plot(fsM,Y2(1:1:NM/2));grid; xlabel('Frequency (Hz)');
```

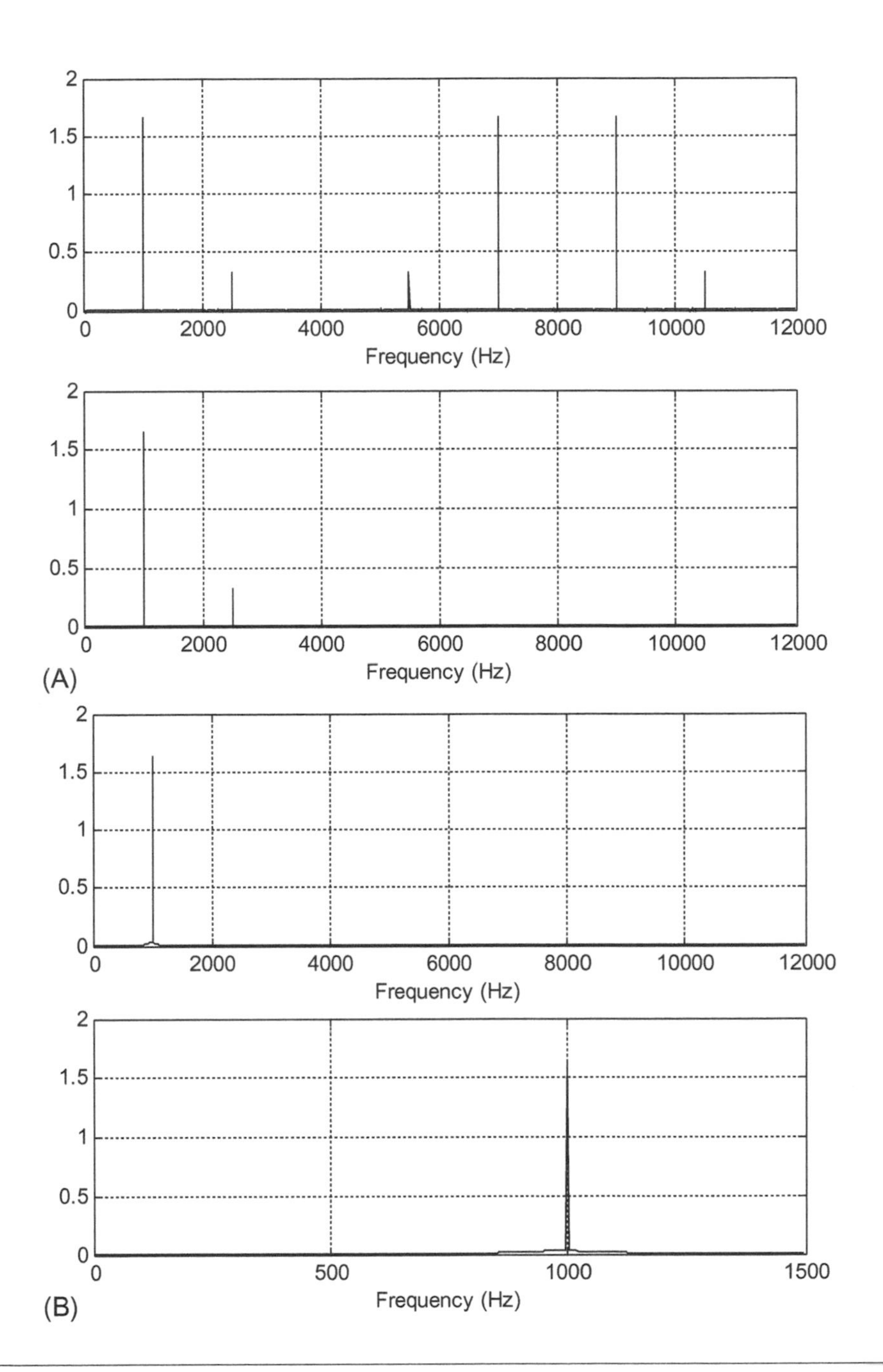

FIG. 11.9

(A) (Top) Spectrum after upsampling and (bottom) spectrum after interpolation filtering. (B) (Top) Spectrum after anti-aliasing filtering and (bottom) spectrum after downsampling.

Therefore, three steps are required to accomplish the process:

1. Upsampling by a factor of $L=3$
2. Filtering the upsampled sequence by an FIR lowpass filter designed with the filter length $N=159$ and a cutoff frequency of 1250 Hz at the sampling rate of $3 \times 8000 = 24{,}000$ Hz
3. Downsampling by a factor of $M=8$.

EXAMPLE 11.3

Given a sampling conversion DSP system (Fig. 11.10A) with the following specifications:
Audio input $x(n)$ is sampled at the rate of 6000 Hz;
Audio output $y(m)$ is operated at the rate of 9000 Hz.
Determine the filter length and cutoff frequency for the combined anti-aliasing filter $H(z)$, and window types, respectively, if the window design method is used.

Solution:

(a) The filter frequency specifications and corresponding block diagram are developed in Fig. 11.10B.

Specifications for the interpolation filter $H_1(z)$:
Passband frequency range = 0–2500 Hz.
Passband ripples for $H_1(z) = 0.04$ dB
Stopband frequency range = 3000–9000 Hz
Stopband attenuation = 42 dB
Specifications for the anti-aliasing filter $H_2(z)$:
Passband frequency range = 0–2500 Hz.
Passband ripples for $H_2(z) = 0.02$ dB
Stopband frequency range = 4500 Hz–9000 Hz
Stopband attenuation = 46 dB
Combined specifications $H(z)$:
Passband frequency range = 0–2500 Hz.
Passband ripples for $H(z) = 0.02$ dB
Stopband frequency range = 3000 Hz–9000 Hz
Stopband attenuation = 46 dB.
We use an FIR filter with a Hamming window. Since

$$\Delta f = \frac{f_{stop} - f_{pass}}{f_{sL}} = \frac{3{,}000 - 2{,}500}{18{,}000} = 0.0278,$$

the length of the filter and the cutoff frequency can be determined by

$$N = \frac{3.3}{\Delta f} = \frac{3.3}{0.0278} = 118.8.$$

we choose $N=119$, and

$$f_c = \frac{f_{pass} + f_{stop}}{2} = \frac{3000 + 2500}{2} = 2750\,\text{Hz}.$$

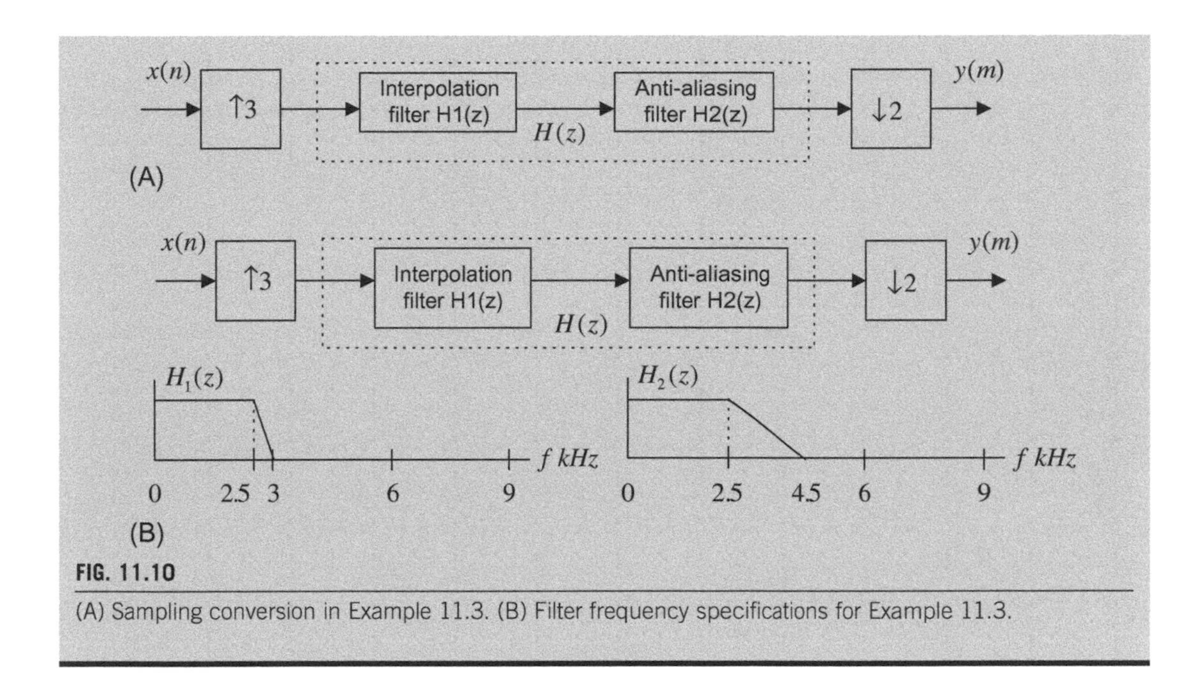

FIG. 11.10

(A) Sampling conversion in Example 11.3. (B) Filter frequency specifications for Example 11.3.

11.1.4 APPLICATION: CD AUDIO PLAYER

In this application example, we discuss principles of the upsampling process and interpolation-filter processes used in the CD audio system to help the reconstruction filter design.

Each raw digital sample recorded on the CD audio system contains 16 bits and is sampled at the rate of 44.1 kHz. Fig. 11.11 describes a portion of one channel of the CD player in terms of a simplified block diagram.

Let us consider the situation without upsampling and application of a digital interpolation filter. We know that the audio signal has a bandwidth of 22.05 kHz, that is, the Nyquist frequency, and digital-to-analog conversion (DAC) produces the sample-and-hold signals that contain the desired audio band and images thereof. To achieve the audio band signal, we need to apply a *reconstruction filter* (also called a smooth filter or anti-image filter) to remove all image frequencies beyond the Nyquist frequency of 22.05 kHz. Due to the requirement of the sharp transition band, a higher-order analog filter design becomes a requirement.

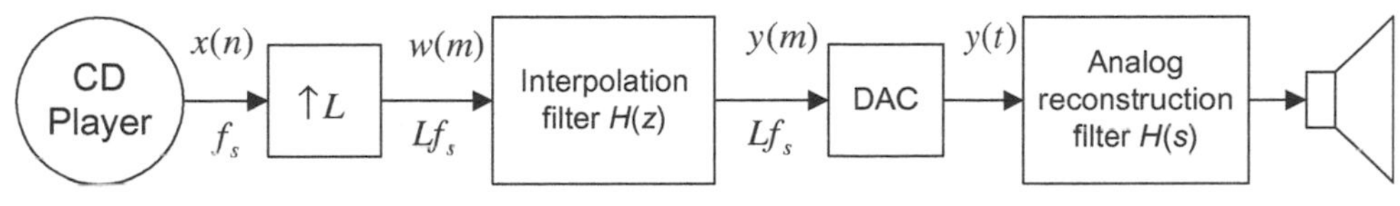

FIG. 11.11

Sample rate conversion in the CD audio player system.

The design of the higher-order analog filter is complex and expensive to implement. In order to relieve such design constraints, we can add the upsampling process before DAC, followed by application of the digital interpolation filter (assume $L=4$). Certainly, the interpolation filter design must satisfy the specifications studied in the previous section on increasing the sampling rate by an integer factor. Again, after digital interpolation, the audio band is kept the same, while the sampling frequency is increased by fourfold ($L=4$), that is, $44.1 \times 4 = 176.4$ kHz.

Since the audio band of 22.05 kHz is now relatively low compared with the new folding frequency (176.4/2 = 88.2 kHz), the use of a simple first-order or second-order analog anti-image filter may be sufficient. Let us look the following simulation.

A test audio signal with a frequency of 16 kHz and a sampling rate of 44.1 kHz are generated using the formula

$$x(n) = \sin\left(\frac{2\pi \times 16{,}000n}{44{,}100}\right).$$

If we use an upsampling factor of 4, then the bandwidth would increase to 88.2 kHz. Based on the audio frequency of 16 kHz, the original Nyquist frequency of 22.05 kHz, and the new sampling rate of 176.4 kHz, we can determine the filter length as

$$\Delta f = \frac{22.05 - 16}{176.4} = 0.0343.$$

Using the Hamming window for FIR filter design leads to

$$N = \frac{3.3}{\Delta f} = 96.2.$$

We choose $N=97$. The cutoff frequency therefore is

$$f_c = \frac{16 + 22.05}{2} = 19.025\,\text{kHz}.$$

The spectrum of the interpolated audio test signal is shown in Fig. 11.12, where the top plot illustrates that after the upsampling, the audio test signal has the frequency of 16 kHz, along with the image frequencies coming from 44.1–16 = 28.1 kHz, 44.1 + 16 = 60.1 kHz, 88.2–16 = 72.2 kHz, and so on. The bottom graph describes the spectrum after the interpolation filter. From lowpass FIR filtering, an interpolated audio signal with a frequency of 16 kHz is observed.

Let us examine the corresponding process in time domain, as shown in Fig. 11.13. The upper left plot shows the original samples. The upper right plot describes the upsampled signals. The lower left plot shows the signals after the upsampling process and digital interpolation filter. Finally, the lower right plot shows the sample-and-hold signals after DAC. Clearly, we can easily design a reconstruction filter to smooth the sample-and-hold signals and obtain the original audio test signal. The advantage of reducing hardware is illustrated. The MATLAB implementation can be seen in Program 11.4.

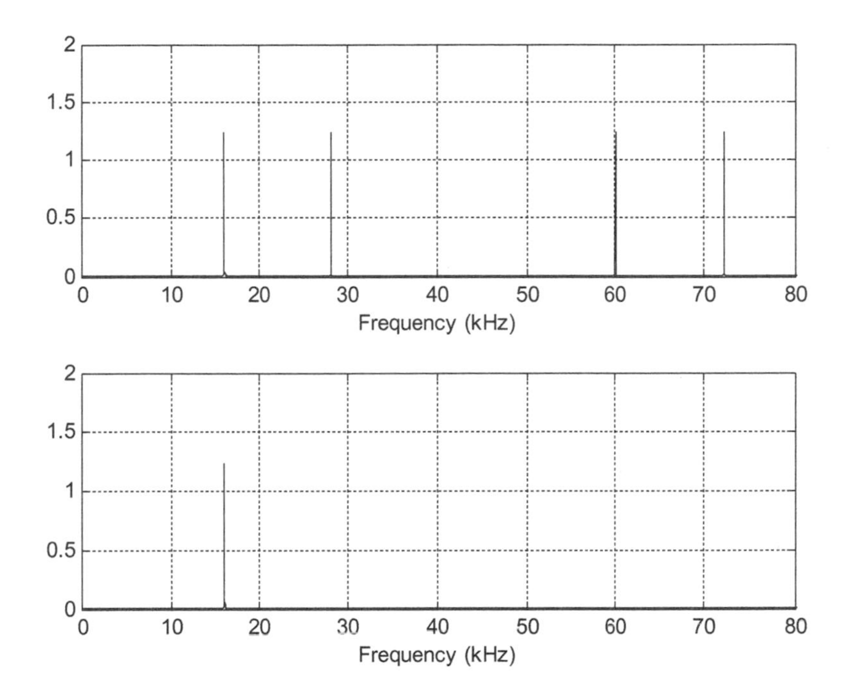

FIG. 11.12

(Top) The spectrum after upsampling and (bottom) the spectrum after applying the interpolation filter.

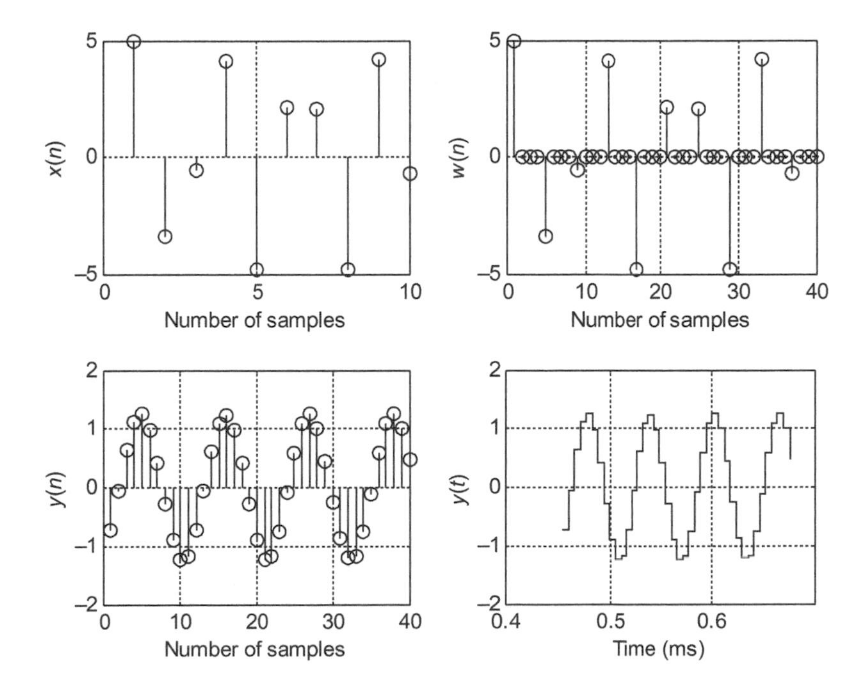

FIG. 11.13

Plots of signals at each stage according to the block diagram in Fig. 11.11.

Program 11.4. MATLAB program for CD player example.

```
close all; clear all; clc
% Generate the 2048 samples with fs=44,100Hz
fs=44,100;                              % Original sampling rate
T=1/fs;                                 % Sampling period
N=2048;                                 % Number of samples
L=4;
fsL=fs*L;                               % Upsampling rate
%Upsampling filter (see Chapter 7 for FIR filter design)
Bup=firwd(97,1,2*19025*pi/fsL,0,4);
n=0:1:N-1;                              % Generate the time indices
x=5*sin(2*pi*16000*n*T);                % Generate the test signal
% Upsampling by a factor of L
w=zeros(1,L*N);
for n=0:1:N-1
w(L*n+1)=x(n+1);
end
NL=length(w);                           % Number of the upsampled data
W=2*abs(fft(w,NL))/NL;W(1)=W(1)/2; % Compute the one-sided
%amplitude spectrum
f=[0:1:NL/2-1]*fs*L/NL; % Map the frequency index to its frequency in Hz
f=f/1000; % Convert to kHz
%Interpolation
y=filter(Bup,1,w);      % Perform the interpolation filter
Y=2*abs(fft(y,NL))/NL;Y(1)=Y(1)/2; % Compute the one-sided
%amplitude spectrum
subplot(2,1,1);plot(f,W(1:1:NL/2));grid;
xlabel('Frequency (kHz)'); axis([0 f(length(f)) 0 2]);
subplot(2,1,2);plot(f,Y(1:1:NL/2));grid;
xlabel('Frequency (kHz)');axis([0 f(length(f)) 0 2]);
figure
subplot(2,2,1);stem(x(21:30));grid
xlabel('Number of Samples');ylabel('x(n)');
subplot(2,2,2);stem(w(81:120));grid
xlabel('Number of Samples'); ylabel('w(n)');
subplot(2,2,3);stem(y(81:120));grid
xlabel('Number of Samples'); ylabel('y(n)')
subplot(2,2,4);stairs([80:1:119]*1000*T,y(81:120));grid
xlabel('Time (ms)'); ylabel('y(t)')
```

11.1.5 MULTISTAGE DECIMATION

The multistage approach for the downsampling rate conversion can be used to dramatically reduce the anti-aliasing filter length. Fig. 11.14 describes a two-stage decimator.

As shown in Fig. 11.14, a total decimation factor is $M=M_1\times M_2$. Here, even though we develop a procedure for a two-stage case, a similar principle can be applied to general multistage cases.

Using the two-stage decimation in Fig. 11.15, the final Nyquist limit is $\frac{f_s}{2M}$ after final downsampling. So our useful information bandwidth should stop at the frequency edge of $\frac{f_s}{2M}$. Next, we need to

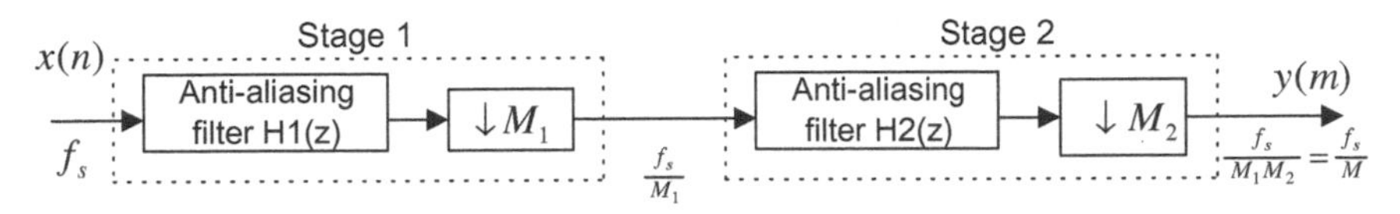

FIG. 11.14

Multistage decimation.

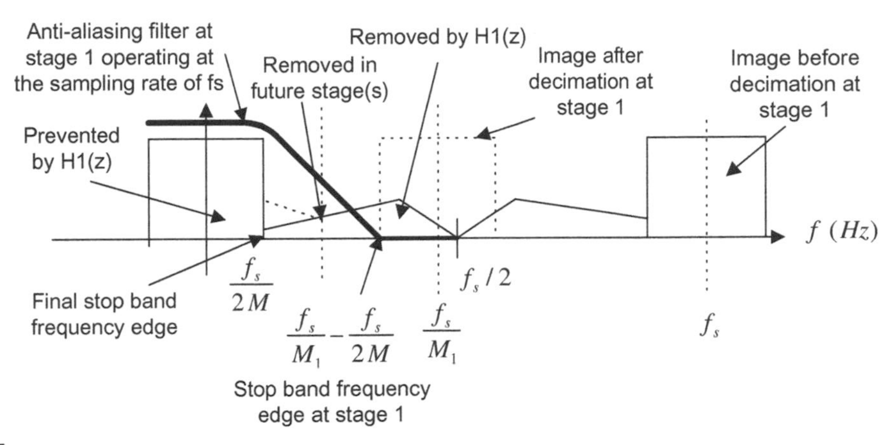

FIG. 11.15

Stopband frequency edge for the anti-aliasing filter at stage 1 for the two-stage decimation.

determine the stop frequency edge for the anti-aliasing lowpass filter at stage 1 before the first decimation process begins. This stop frequency edge is actually the lower frequency edge of the first image replica centered at the sampling frequency of $\frac{f_s}{M_1}$ after the stage 1 decimation. This lower frequency edge of the first image replica is then determined by

$$\frac{f_s}{M_1} - \frac{f_s}{2M}.$$

After downsampling, we expect that some frequency components from $\frac{f_s}{2M_1}$ to $\frac{f_s}{M_1} - \frac{f_s}{2M}$ to be folded over to the frequency band between $\frac{f_s}{2M}$ and $\frac{f_s}{2M_1}$. However, these aliased frequency components do not affect the final useful band between 0 Hz to $\frac{f_s}{2M}$ and will be removed by the anti-aliasing filter(s) in the future stage(s). As illustrated in Fig. 11.15, any frequency components beyond the edge $\frac{f_s}{M_1} - \frac{f_s}{2M}$ can fold over into the final useful information band to create aliasing distortion. Therefore, we can use this frequency as the lower stop frequency edge of the anti-aliasing filter to prevent the aliasing distortion at the final stage. The upper stopband edge (Nyquist limit) for the anti-image filter at stage 1 is clearly $\frac{f_s}{2}$, since the filter operates at f_s samples per second. So the stopband frequency range at stage 1 is therefore from $\frac{f_s}{M_1} - \frac{f_s}{2M}$ to $\frac{f_s}{2}$. The aliasing distortion, introduced into the frequency band from $\frac{f_s}{2M}$ to $\frac{f_s}{2M_1}$, will be filtered out after future decimation stage(s).

Similarly, for stage 2, the lower frequency edge of the first image developed after stage 2 downsampling is

$$\frac{f_s}{M_1 M_2} - \frac{f_s}{2M} = \frac{f_s}{2M}.$$

As is evident in our two-stage scheme, the stopband frequency range for the second anti-aliasing filter at stage 2 should be from $\frac{f_s}{2M}$ to $\frac{f_s}{2M_1}$.

We summarize specifications for the two-stage decimation as follows:

Filter requirement for stage 1:

- Passband frequency range $= 0$ to f_p
- Stopband frequency range $= \frac{f_s}{M_1} - \frac{f_s}{2M}$ to $\frac{f_s}{2}$
- Passband ripple $= \delta_p/2$, where δ_p is the combined absolute ripple on passband
- Stopband attenuation $= \delta_s$

Filter requirement for stage 2:

- Passband frequency range $= 0$ to f_p
- Stopband frequency range $= \frac{f_s}{M_1 \times M_2} - \frac{f_s}{2M}$ to $\frac{f_s}{2M_1}$
- Passband ripple $= \delta_p/2$, where δ_p is the combined absolute ripple on passband
- Stopband attenuation $= \delta_s$.

Example 11.4 illustrates the two-stage decimator design.

EXAMPLE 11.4

Determine the anti-aliasing FIR filter lengths and cutoff frequencies for the two-stage decimator with the following specifications and block diagram (Fig. 11.16A):

Original sampling rate: $f_s = 240$ kHz

Audio frequency range: 0–3400 Hz

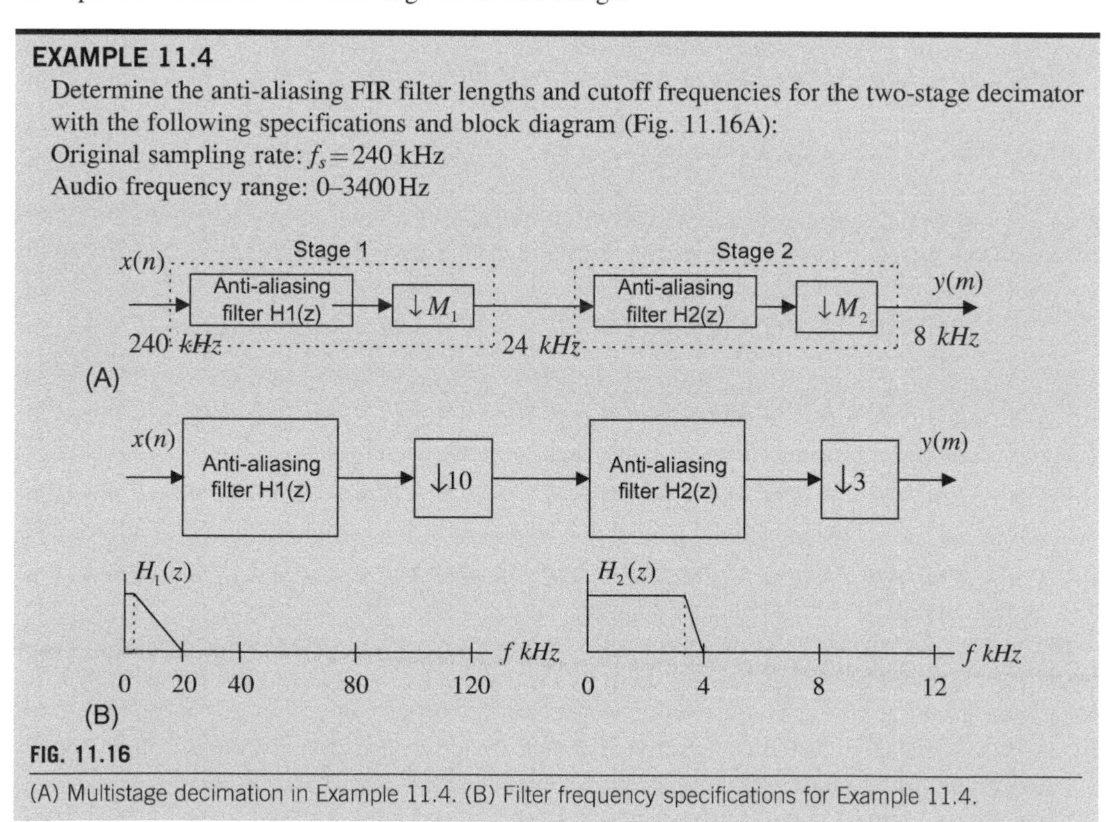

FIG. 11.16

(A) Multistage decimation in Example 11.4. (B) Filter frequency specifications for Example 11.4.

Passband ripple: $\delta_p = 0.05$ (absolute)
Stopband attenuation: $\delta_s = 0.005$ (absolute)
FIR filter design using the window method
New sampling rage: $f_{sM} = 8$ kHz

Solution:

$$M = \frac{240\text{kHz}}{8\text{kHz}} = 30 = 10 \times 3$$

We choose $M_1 = 10$ and $M_2 = 3$; there could be other choices. Fig. 11.16B shows the block diagram and filter frequency specifications.
Filter specification for $H_1(z)$:
Passband frequency range: 0–3400 Hz.
Passband ripples: 0.05/2 = 0.025 ($\delta_p\ dB = 20\log_{10}(1+\delta_p) = 0.212\,\text{dB}$)
Stopband frequency range: 20,000–120,000 Hz
Stopband attenuation: 0.005, $\delta_s\ dB = -20 \times \log_{10}(\delta_s) = 46$ dB
Filter type: FIR, Hamming window.
Note that the lower stopband edge can be determined as

$$f_{stop} = \frac{f_s}{M_1} - \frac{f_s}{2 \times M} = \frac{240{,}000}{10} - \frac{240{,}000}{2 \times 30} = 20{,}000Hz$$

$$\Delta f = \frac{f_{stop} - f_{pass}}{f_s} = \frac{20{,}000 - 3400}{240{,}000} = 0.06917.$$

The length of the filter and the cutoff frequency can be determined by

$$N = \frac{3.3}{\Delta f} = 47.7.$$

We choose $N = 49$, and

$$f_c = \frac{f_{pass} + f_{stop}}{2} = \frac{20{,}000 + 3400}{2} = 11{,}700\,\text{Hz}.$$

Filter specification for $H_2(z)$:
Passband frequency range: 0–3400 Hz.
Passband ripples: 0.05/2 = 0.025 (0.212 dB)
Stopband frequency range: 4000–12,000 Hz
Stopband attenuation: 0.005, $\delta_s\ dB = 46$ dB
Filter type: FIR, Hamming window
Note that

$$\Delta f = \frac{f_{stop} - f_{pass}}{f_{sM1}} = \frac{4000 - 3400}{24{,}000} = 0.025.$$

The length of the filter and the cutoff frequency can be determined by

Continued

EXAMPLE 11.4—CONT'D

$$N = \frac{3.3}{\Delta f} = 132.$$

We choose $N=133$, and

$$f_c = \frac{f_{pass} + f_{stop}}{2} = \frac{4000 + 3400}{2} = 3700\,\text{Hz}.$$

Reader can verify for the case by using one-stage with a decimation factor of $M=30$. Using the Hamming window for the FIR filter, the resulting number of taps is 1321, and the cutoff frequency is 3700 Hz. Thus, such a filter requires a huge number of computations and causes a large delay during implementation compared with the two-stage case.

The multistage scheme is very helpful for the sampling rate conversion between audio systems. For example, to convert the CD audio at the sampling rate of 44.1 kHz to the MP3 or digital audio type system (professional audio rate), in which the sampling rate of 48 kHz is used, the conversion factor $L/M=48/44.1=160/147$ is required. Using the single-stage scheme may cause impractical FIR filter sizes for interpolation and downsampling. However, since $L/M=160/147=(4/3)(8/7)(5/7)$, we may design an efficient 3-stage system, in which stages 1, 2, and 3 use the conversion factors as $L/M=8/7$, $L/M=5/7$, and $L/M=4/3$, respectively.

11.2 POLYPHASE FILTER STRUCTURE AND IMPLEMENTATION

Due to the nature of the decimation and interpolation processes, polyphase filter structures can be developed to efficiently implement the decimation and interpolation filters (using fewer number of multiplications and additions). As we will explain, these filters are all-pass filters with different phase shifts (Proakis and Manolakis, 2007), thus we call them *polyphase filters*.

Here, we skip their derivations and illustrate implementations of decimation and interpolation using simple examples. Consider the interpolation process shown in Fig. 11.17, where $L=2$.

We assume that the FIR interpolation filter has four taps, shown as

$$H(z) = h(0) + h(1)z^{-1} + h(2)z^{-2} + h(3)z^{-3}$$

and the filter output is

$$y(m) = h(0)w(m) + h(1)w(m-1) + h(2)w(m-2) + h(3)w(m-3).$$

For the purpose of comparison, the direct interpolation process shown in Fig. 11.17 is summarized in Table 11.1, where $w(m)$ is the upsampled signal and $y(m)$ the interpolated output. Processing each input

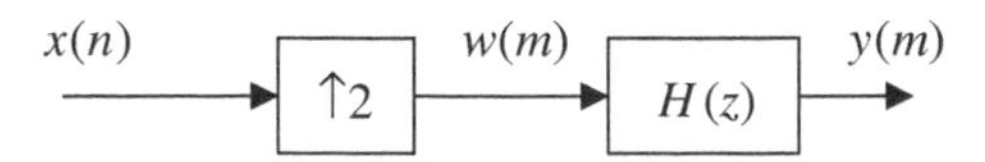

FIG. 11.17

Upsampling by a factor of 2 and a four-tap interpolation filter.

Table 11.1 Results of the Direct Interpolation Process in Fig. 11.17 (8 Multiplications and 6 Additions for Processing Each Input Sample $x(n)$)

n	$x(n)$	m	$w(m)$	$y(m)$
$n=0$	$x(0)$	$m=0$	$w(0)=x(0)$	$y(0)=h(0)x(0)$
		$m=1$	$w(1)=0$	$y(1)=h(1)x(0)$
$n=1$	$x(1)$	$m=2$	$w(2)=x(1)$	$y(2)=h(0)x(1)+h(2)x(0)$
		$m=3$	$w(3)=0$	$y(3)=h(1)x(1)+h(3)x(0)$
$n=2$	$x(2)$	$m=4$	$w(4)=x(2)$	$y(4)=h(0)x(2)+h(2)x(1)$
		$m=5$	$w(5)=0$	$y(5)=h(1)x(2)+h(3)x(1)$
…	…	…	…	…

sample $x(n)$ requires applying the difference equation twice to obtain $y(0)$ and $y(1)$. Hence, for this example, we need eight multiplications and six additions.

The output results in Table 11.1 can be easily obtained using the polyphase filters shown in Fig. 11.18.

In general, there are L polyphase filters. Having the designed interpolation filter $H(z)$ of N taps, we can determine each bank of filter coefficients as follows:

$$\rho_k(n) = h(k+nL) \text{ for } k=0,1,\ldots,L-1 \text{ and } n=0,1,\ldots,\frac{N}{L}-1. \tag{11.12}$$

For our example, $L=2$ and $N=4$, we have $L-1=1$, and $N/L-1=1$, respectively. Hence, there are two filter banks, $\rho_0(z)$ and $\rho_1(z)$, each having a length of 2, as illustrated in Fig. 11.18. When $k=0$ and $n=1$, the upper limit of time index required for $h(k+nL)$ is $k+nL=0+1\times 2=2$. When $k=1$ and $n=1$, the upper limit of the time index for $h(k+nL)$ is 3. Hence, the first filter $\rho_0(z)$ has the coefficients $h(0)$ and $h(2)$. Similarly, the second filter $\rho_1(z)$ has coefficients $h(1)$ and $h(3)$. In fact, the filter coefficients of $\rho_0(z)$ are a decimated version of $h(n)$ starting at $k=0$, while the filter coefficients of $\rho_1(z)$ are a decimated version of $h(n)$ starting at $k=1$, and so on.

As shown in Fig. 11.18, we can reduce the computational complexity from eight multiplications and six additions down to four multiplications and three additions for processing each input sample $x(n)$. Generally, the computation can be reduced by a factor of L as compared with the direct process.

The commutative model for the polyphase interpolation filter is given in Fig. 11.19.

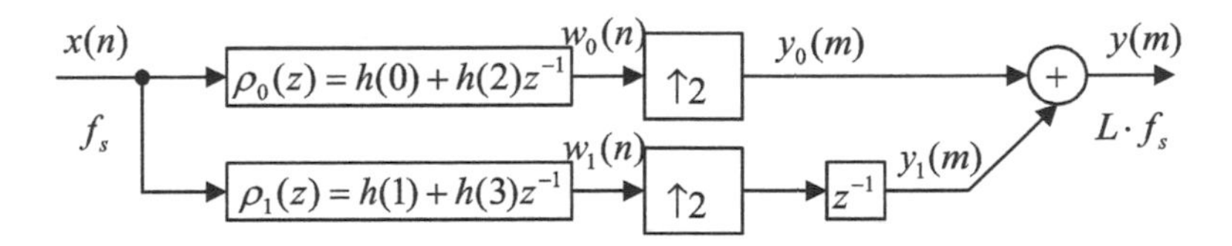

FIG. 11.18

Polyphase filter implementation for the interpolation in Fig. 11.17 (4 multiplications and 3 additions for processing each input sample $x(n)$).

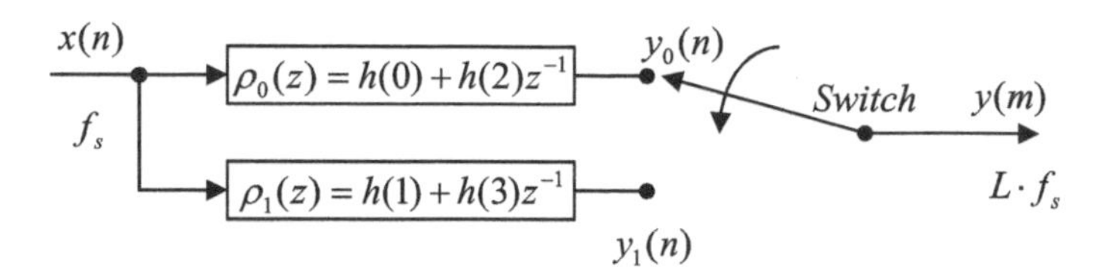

FIG. 11.19

Commutative model for the polyphase interpolation filter.

EXAMPLE 11.5

Verify $y(1)$ in Table 11.1 using the polyphase filter implementation in Figs. 11.18 and 11.19, respectively.

Solution:

Applying the polyphase interpolation filter as shown in Fig. 11.18 leads to

$$w_0(n) = h(0)x(n) + h(2)x(n-1)$$
$$w_1(n) = h(1)x(n) + h(3)x(n-1);$$

when $n = 0$,

$$w_0(0) = h(0)x(0)$$
$$w_1(0) = h(1)x(0).$$

After interpolation, we have

$$y_0(m) : w_0(0)\ 0\ \cdots$$

and

$$y_1(m) : 0\ \ w_1(0)\ \ 0\ \ \cdots$$

Note: there is a unit delay for the second filter bank. Hence

$$m = 0, y_0(0) = h(0)x(0), y_1(0) = 0$$
$$m = 1, y_0(1) = 0, y_1(1) = h(1)x(0).$$

Combining two channels, we finally get

$$m = 0, y(0) = y_0(0) + y_1(0) = h(0)x(0),$$
$$m = 1, y(1) = y_0(1) + y_1(1) = h(1)x(0).$$

Therefore, $y(1)$ matches that in the direct interpolation process given in Table 11.1.
Applying the polyphase interpolation filter using the commutative model in Fig. 11.19, we have

$$y_0(n) = h(0)x(n) + h(2)x(n-1)$$
$$y_1(n) = h(1)x(n) + h(3)x(n-1);$$

when $n=0$,

$$m=0, y(0)=y_0(0)=h(0)x(0)+h(2)x(-1)=h(0)x(0)$$

$$m=1, y(1)=y_1(0)=h(1)x(0)+h(3)x(-1)=h(1)x(0).$$

Clearly, $y(1)=h(1)x(0)$ matches $y(1)$ the result in Table 11.1.

Next, we explain the properties of polyphase filters (i.e., they have all-pass gain and possible different phases). Each polyphase filter $\rho_k(n)$ operating at the original sampling rate f_s (assuming 8 kHz) is a downsampled version of the interpolation filter $h(n)$ operating at the upsampling rate Lf_s (32 kHz assuming an interpolation factor of $L=4$). Considering that the designed interpolation FIR filter coefficients $h(n)$ are the impulse response sequence with a flat frequency spectrum up to a bandwidth of $f_s/2$ (assume a bandwidth of 4 kHz with a perfect flat frequency magnitude response, theoretically) at a sampling rate of Lf_s (32 kHz), we then downsample $h(n)$ to obtain polyphase filters by a factor of $L=4$ and operate them at a sampling rate of f_s (8 kHz).

The Nyquist frequency after downsampling should be $(Lf_s/2)/L=f_s/2$ (4 kHz); at the same time, each downsampled sequence $\rho_k(n)$ operating at f_s (8 kHz) has a flat spectrum up to $f_s/2$ (4 kHz) due to the $f_s/2$ (4 kHz) bandlimited sequence of $h(n)$ at the sampling rate of Lf_s (32 kHz). Hence, all of the polyphase filters are all-pass filters. Since each polyphase $\rho_k(n)$ filter has different coefficients, each may have a different phase. Therefore, these polyphase filters are the all-pass filters having possible different phases, theoretically.

Next, consider the following decimation process in Fig. 11.20.

Assuming a three-tap decimation filter, we have

$$H(z)=h(0)+h(1)z^{-1}+h(2)z^{-2}$$

$$w(n)=h(0)x(n)+h(1)x(n-1)+h(2)x(n-2).$$

The direct decimation process is shown in Table 11.2 for the purpose of comparison. Obtaining each output $y(m)$ requires processing filter difference equations twice, resulting in six multiplications and four additions for this particular example.

The efficient way to implement a polyphase filter is given in Fig. 11.21.

Similarly, there are M polyphase filters. With the designed decimation filter $H(z)$ of N taps, we can obtain filter bank coefficients by

$$\rho_k(n)=h(k+nM) \text{ for } k=0,1,\cdots,M-1 \text{ and } n=0,1,\cdots,\frac{N}{M}-1. \tag{11.13}$$

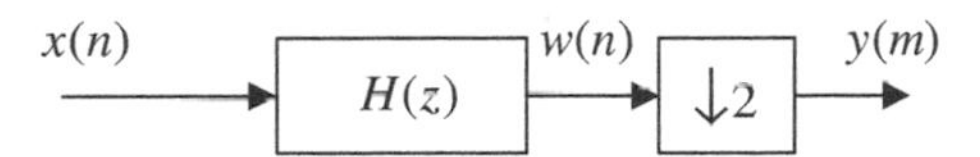

FIG. 11.20

Decimation by a factor of 2 and a three-tap anti-aliasing filter.

Table 11.2 Results of Direct Decimation Process in Fig. 11.20 (6 Multiplications and 4 Additions for Obtaining Each Output $y(m)$).

n	$x(n)$	$w(n)$	m	$y(m)$
$n=0$	$x(0)$	$w(0)=h(0)x(0)$	$m=0$	$y(0)=h(0)x(0)$
$n=1$	$x(1)$	$w(1)=h(0)x(1)+h(1)x(0)$ discard		
$n=2$	$x(2)$	$w(2)=h(0)x(2)+h(1)x(1)+h(2)x(0)$	$m=1$	$y(1)=h(0)x(2)+h(1)x(1)+h(2)x(0)$
$n=3$	$x(3)$	$w(3)=h(0)x(3)+h(1)x(2)+h(2)x(1)$ discard		
$n=4$	$x(5)$	$w(4)=h(0)x(4)+h(1)x(3)+h(2)x(2)$	$m=2$	$y(2)=h(0)x(4)+h(1)x(3)+h(2)x(2)$
$n=5$	$x(6)$	$w(5)=h(0)x(5)+h(1)x(4)+h(2)x(3)$ discard		
…	…	…	…	…

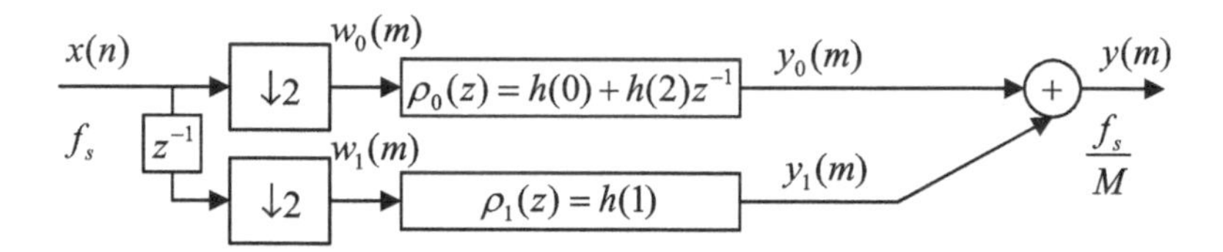

FIG. 11.21

Polyphase filter implementation for the decimation in Fig. 11.20 (3 multiplications and 1 addition for obtaining each output $y(m)$).

For our example, we see that $M-1=1$ and $N/M-1=1$(roundedup). Thus, we have two filter banks. Since $k=0$ and $n=1$, $k+nM=0+1\times 2=2$. The time index upper limit required for $h(k+nM)$ is 2 for the first filter bank $\rho_0(z)$. Hence $\rho_0(z)$ has filter coefficients $h(0)$ and $h(2)$. However, when $k=1$ and $n=1$, $k+nM=1+1\times 2=3$, the time index upper limit required for $h(k+nM)$ is 3 for the second filter bank, and the corresponding filter coefficients are required to be $h(1)$ and $h(3)$. Since our direct interpolation filter $h(n)$ does not contain the coefficient $h(3)$, we set $h(3)=0$ to get the second filter bank with one tap only, as shown in Fig. 11.21. Also as shown in that figure, achieving each $y(m)$ needs three multiplications and one addition. In general, the number of multiplications is reduced by a factor of M.

The commutative model for the polyphase decimator is shown in Fig. 11.22.

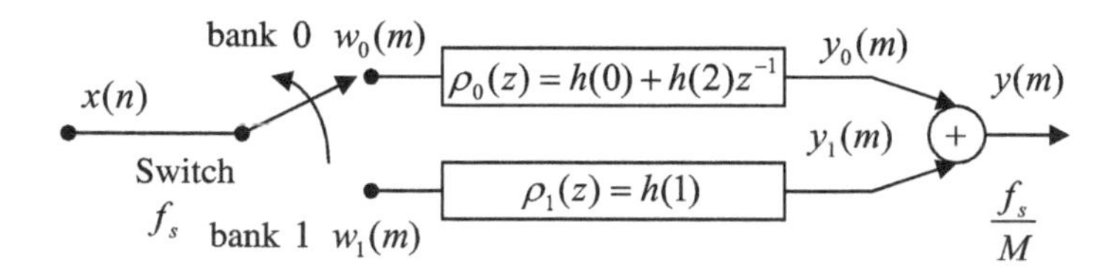

FIG. 11.22

Commutative model for the polyphase decimation filter.

EXAMPLE 11.6

Verify $y(1)$ in Table 11.2 using the polyphase decimation filter implementation in Fig. 11.21.

Solution:

Using Fig. 11.21, we write the difference equations as

$$y_0(m) = h(0)w_0(m) + h(2)w_0(m-1)$$

$$y_1(m) = h(1)w_1(m).$$

Assume $n=0$, $n=1$, $n=2$, and $n=3$, we have the inputs as $x(0)$, $x(1)$, $x(2)$, and $x(3)$, and.
$w_0(m)$: $x(0)$ $x(2)$ $\cdots$
Delaying $x(n)$ by one sample and decimate it by a factor or 2 leads to.
$w_1(m)$: 0 $x(1)$ $x(3)$ $\cdots$.
Hence, applying the filter banks yields the following:
$m=0$, we have inputs for each filter as

$$w_0(0) = x(0) \text{ and } w_1(0) = 0$$

then

$$y_0(0) = h(0)w_0(0) + h(2)w_0(-1) = h(0)x(0)$$

$$y_1(0) = h(1)w_1(0) = h(1) \times 0 = 0.$$

Combining two channels, we obtain

$$y(1) = y_0(1) + y_1(1) = h(0)x(0) + 0 = h(0)x(0)$$

$m=1$, we get inputs for each filter as

$$w_0(1) = x(2) \text{ and } w_1(1) = x(1),$$

then

$$y_0(1) = h(0)w_0(1) + h(2)w_0(0) = h(0)x(2) + h(2)x(0)$$

$$y_1(1) = h(1)w_1(1) = h(1)x(1).$$

Combining two channels leads to

$$y(1) = y_0(1) + y_1(1) = h(0)x(2) + h(2)x(0) + h(1)x(1).$$

We note that $y(1)$ is the same as that shown in Table 11.2. Similar analysis can be done for the commutative model shown in Fig. 11.22.

Program 11.5 demonstrates the polyphase implementation of decimation. The program is modified based on Program 11.1.

Program 11.5. Decimation using polyphase implementation.

```
close all; clear all;
% Downsampling filter (see Chapter 7 for FIR filter design)
B =[0.00074961181416  0.00247663033476  0.00146938649416  -0.00440446121505 ...
-0.00910635730662  0.00000000000000  0.02035676831506  0.02233710562885...
-0.01712963672810  -0.06376620649567  -0.03590670035210  0.10660980550088...
0.29014909103794  0.37500000000000  0.29014909103794  0.10660980550088...
-0.03590670035210  -0.06376620649567  -0.01712963672810  0.02233710562885...
0.02035676831506  0.00000000000000  -0.00910635730662  -0.00440446121505...
0.00146938649416 0.00247663033476 0.00074961181416];
% Generate 2048 samples
fs=8000;                                                    % Sampling rate
N=2048;                                                     % Number of samples
M=2;                                                        % Down sample factor
n=0:1:N-1;
x=5*sin(n*pi/4)+cos(5*n*pi/8);
% Compute the single-sided amplitude spectrum
% AC component will be doubled, and DC component will be kept the
% same value
X=2*abs(fft(x,N))/N;X(1)=X(1)/2;
% Map the frequency index up to the folding frequency in Hz
f=[0:1:N/2-1]*fs/N;
% Decimation
w0=x(1:M:N); p0=B(1:2:length(B)); % Downsampling
w1=filter([0 1],1,x); % Delay one sample
w1=w1(1:M:N); p1=B(2:M:length(B)) % Downsampling
y=filter(p0,1,w0)+filter(p1,1,w1);
NM=length(y);                                  % Length of the downsampled data
% Compute the single-sided amplitude spectrum for the downsampled
% signal
Y=2*abs(fft(y,NM))/NM;Y(1)=Y(1)/2;
% Map the frequency index to the frequency in Hz
fsM=[0:1:NM/2-1]*(fs/M)/NM;
% Plot spectra
subplot(2,1,1);plot(f,X(1:1:N/2));grid; xlabel('Frequency (Hz)');
subplot(2,1,2);plot(fsM,Y(1:1:NM/2));grid; xlabel('Frequency (Hz)');
```

Program 11.6 demonstrates polyphase implementation of interpolation using the information in Program 11.2.

Program 11.6. Interpolation using polyphase implementation.

```
close all; clear all
%Upsampling filter (see Chapter 7 for FIR filter design)
B =[-0.00012783931504  0.00069976044649  0.00123831516738  0.00100277549136...
   -0.00025059018468  -0.00203448515158 -0.00300830295487 -0.00174101657599...
   0.00188598835011    0.00578414933758  0.00649330625041  0.00177982369523...
   -0.00670672686935  -0.01319379342716 -0.01116855281442  0.00123034314117...
   0.01775600060894    0.02614700427364  0.01594155162392 -0.01235169936557...
   -0.04334322148505  -0.05244745563466 -0.01951094855292  0.05718573279009...
```

```
0.15568416401644   0.23851539047347   0.27083333333333   0.23851539047347...
0.15568416401644   0.05718573279009  -0.01951094855292  -0.05244745563466...
-0.04334322148505 -0.01235169936557   0.01594155162392   0.02614700427364...
0.01775600060894   0.00123034314117  -0.01116855281442  -0.01319379342716...
-0.00670672686935  0.00177982369523   0.00649330625041   0.00578414933758...
0.00188598835011  -0.00174101657599  -0.00300830295487  -0.00203448515158...
-0.00025059018468  0.00100277549136   0.00123831516738   0.00069976044649...
-0.00012783931504];
% Generate 2048 samples with fs=8000Hz
fs=8000;                                                    % Sampling rate
N=2048;                                                     % Number of samples
L = 3;                                                      % Upsampling factor
n=0:1:N-1;
x=5*sin(n*pi/4)+cos(5*n*pi/8);
p0=B(1:L:length(B)); p1=B(2:L:length(B)); p2=B(3:L:length(B));
% Interpolation
w0=filter(p0,1,x);
w1=filter(p1,1,x);
w2=filter(p2,1,x);
y0=zeros(1,L*N);y0(1:L:length(y0))=w0;
y1=zeros(1,L*N);y1(1:L:length(y1))=w1;
y1=filter([0 1],1,y1);
y2=zeros(1,L*N);y2(1:L:length(y2))=w2;
y2=filter([001],1,y2);
y=y0+y1+y2; % Interpolated signal
NL = length(y);                                       % Length of the upsampled data
X=2*abs(fft(x,N))/N;X(1)=X(1)/2; %Compute the one-sided amplitude
% spectrum
f=[0:1:N/2-1]*fs/N; % Map the frequency index to the frequency (Hz)
Y=2*abs(fft(y,NL))/NL;Y(1)=Y(1)/2; %Compute the one-sided amplitude %spectrum
fsL=[0:1:NL/2-1]*fs*L/NL; % Map the frequency index to the frequency %(Hz)
subplot(2,1,1);plot(f,X(1:1:N/2));grid; xlabel('Frequency (Hz)');
subplot(2,1,2);plot(fsL,Y(1:1:NL/2));grid; xlabel('Frequency (Hz)');
```

Note that wavelet transform and subband coding are also in the area of multirate signal processing. We discuss these subjects in Chapter 12.

11.3 OVERSAMPLING OF ANALOG-TO-DIGITAL CONVERSION

Oversampling of the analog signal has become more popular in DSP industry to improve resolution of analog-to-digital conversion (ADC). Oversampling uses a sampling rate, which is much higher than the Nyquist rate. We can define an oversampling ratio as

$$\frac{f_s}{2f_{\max}} >> 1. \tag{11.14}$$

The benefits from an oversampling ADC include:

1. helping to design a simple analog anti-aliasing filter before ADC, and
2. reducing the ADC noise floor with possible noise shaping so that a low resolution ADC can be used.

11.3.1 OVERSAMPLING AND ADC RESOLUTION

To begin with developing the relation between oversampling and ADC resolution, we first summarize the regular ADC and some useful definitions discussed in Chapter 2:

$$\text{Quantization noise power} = \sigma_q^2 = \frac{\Delta^2}{12} \tag{11.15}$$

$$\text{Quantization step} = \Delta = \frac{A}{2^n} \tag{11.16}$$

A= full range of the analog signal to be digitized
n= number of bits per sample (ADC resolution).

Substituting Eq. (11.16) into Eq. (11.15), we have:

$$\text{Quantization noise power} = \sigma_q^2 = \frac{A^2}{12} \times 2^{-2n} \tag{11.17}$$

The power spectral density of the quantization noise with an assumption of uniform probability distribution is shown in Fig. 11.23. Note that this assumption is true for quantizing a uniformly distributed signal in a full range with a sufficiently long duration. It is not generally true in practice, see research papers authored by Lipshiz et al. (1992) and Maher (1992). However, using the assumption will guide us for some useful results for oversampling systems.

The quantization noise power is the area obtained from integrating the power spectral density function in the range from $-f_s/2$ to $f_s/2$. Now let us examine the oversampling ADC, where the sampling rate is much bigger than that of the regular ADC; that is $f_s >> 2f_{\max}$. The scheme is shown in Fig. 11.24.

As we can see, oversampling can reduce the level of noise power spectral density. After the decimation process with the decimation filter, only a portion of quantization noise power in the range from $-f_{\max}$ and $f_{\max}$ is kept in the DSP system. We call this an *in-band frequency range*.

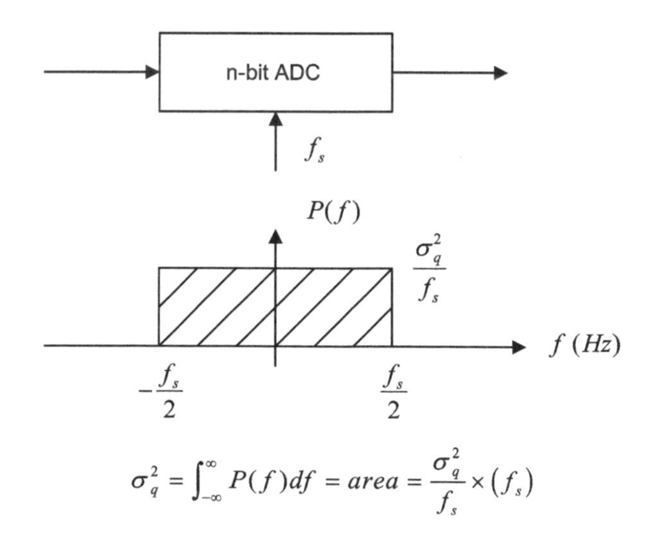

FIG. 11.23

Regular ADC system.

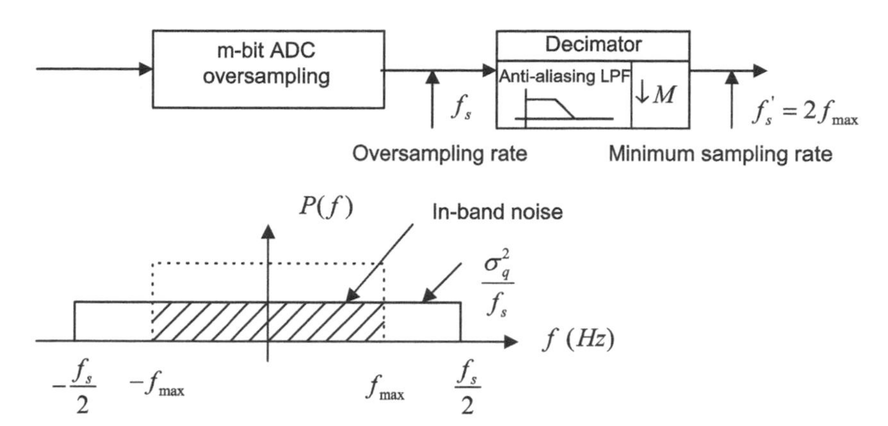

FIG. 11.24

Oversampling ADC system.

In Fig. 11.24, the shaded area, which is the quantization noise power, is given by

$$\text{Quantization Noise Power} = \int_{-\infty}^{\infty} P(f)df = \frac{2f_{max}}{f_s} \times \sigma_q^2 = \frac{2f_{max}}{f_s} \times \frac{A^2}{12} \times 2^{-2m}. \tag{11.18}$$

Assuming that the regular ADC shown in Fig. 11.23 and the oversampling ADC shown in Fig. 11.24 are equivalent, we set their quantization noise powers to be the same to obtain

$$\frac{A^2}{12} \times 2^{-2n} = \frac{2f_{max}}{f_s} \times \frac{A^2}{12} \times 2^{-2m}. \tag{11.19}$$

Eq. (11.19) leads to two useful equations for applications:

$$n = m + 0.5 \times \log_2\left(\frac{f_s}{2f_{max}}\right) \text{ and} \tag{11.20}$$

$$f_s = 2f_{max} \times 2^{2(n-m)}, \tag{11.21}$$

where

f_s = sampling rate in the oversampling DSP system,
f_{max} = maximum frequency of the analog signal,
m = number of bits per sample in the oversampling DSP system,
n = number of bits per sample in the regular DSP system using the minimum sampling rate.

From Eq. (11.20) and given the number of bits (m) used in the oversampling scheme, we can determine the number of bits per sample equivalent to the regular ADC. On the other hand, given the number of bits in the oversampling ADC, we can determine the required oversampling rate so that the oversampling ADC is equivalent to the regular ADC with the larger number of bits per sample (n). Let us look at the following examples.

EXAMPLE 11.7

Given an oversampling audio DSP system with maximum audio input frequency of 20 kHz and ADC resolution of 14 bits, determine the oversampling rate to improve the ADC resolution to 16-bit resolution.

Solution:

Based on the specifications, we have

$$f_{max} = 20\,\text{kHz}, m = 14\,\text{bits and}\, n = 16\,\text{bits}.$$

Using Eq. (11.21) leads to

$$f_s = 2f_{max} \times 2^{2(n-m)} = 2 \times 20 \times 2^{2(16-14)} = 640\,\text{kHz}.$$

Since $fs/(2f_{max}) = 2^4$, we see that each doubling of the minimum sampling rate ($2f_{max} = 40$ kHz) will increase the resolution by a half bit.

EXAMPLE 11.8

Given an oversampling audio DSP system with the following specifications:
Maximum audio input frequency = 4 kHz
ADC resolution = 8 bits.
Sampling rate = 80 MHz,
Determine the equivalent ADC resolution.

Solution:

Since $f_{max} = 4$ kHz, $f_s = 80$ kHz, and $m = 8$ bits, applying Eq. (11.20) yields

$$n = m + 0.5 \times \log_2\left(\frac{f_s}{2f_{max}}\right) = 8 + 0.5 \times \log_2\left(\frac{80{,}000\text{kHz}}{2 \times 4\text{kHz}}\right) \approx 15\,\text{bits}.$$

The MATLAB program shown in Program 11.7 validates the oversampling technique. We consider the following signal

$$x(t) = 1.5\sin(2\pi \times 150t) + 0.9\sin(2\pi \times 175t + \pi/6) + 0.6\sin(2\pi \times 200t + \pi/4) \tag{11.22}$$

with a regular sampling rate of 1 kHz. The oversampling rate of 4 kHz is used and each sample is quantized using a 3-bit code. The anti-aliasing lowpass filter is designed with a cutoff frequency of $\Omega = 2\pi f_{max} T = 2\pi \times 500/4000 = 0.25\pi$ rad. Fig. 11.25 shows the frequency responses of the designed filter while Fig. 11.26 compares the signals in the time and frequency domains, respectively, where $x(t)$ denotes the continuous version, $x_q(n)$ is the quantized version using a regular sampling rate of 1 kHz, and $y_q(n)$ is the enhanced version using the oversampling system with $L = 4$. The detailed amplitude comparisons are given in Fig. 11.27. The measured SNRs are 14.3 dB using the regular sampling system and 21.0 dB using the oversampling system. Since $L = 4$, the achieved signal is expected to have a 4-bit quality ($0.5 \times \log_2 4 = 1$ bit improvement). From simulation, we achieve approximately 6-dB SNR improvement. The improvement will stop when L increases due to the fact that when the

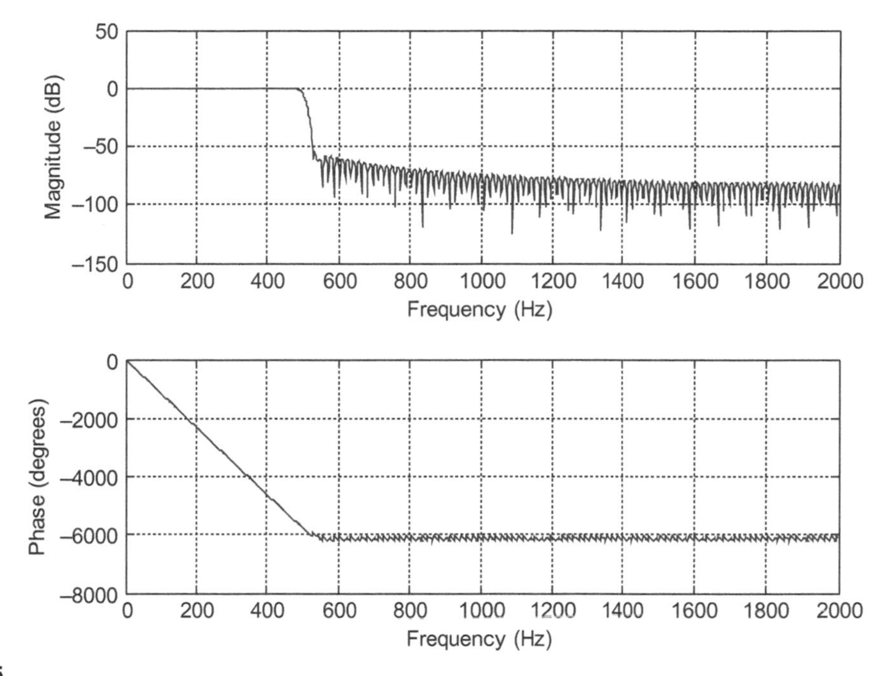

FIG. 11.25

Frequency responses of the designed filter.

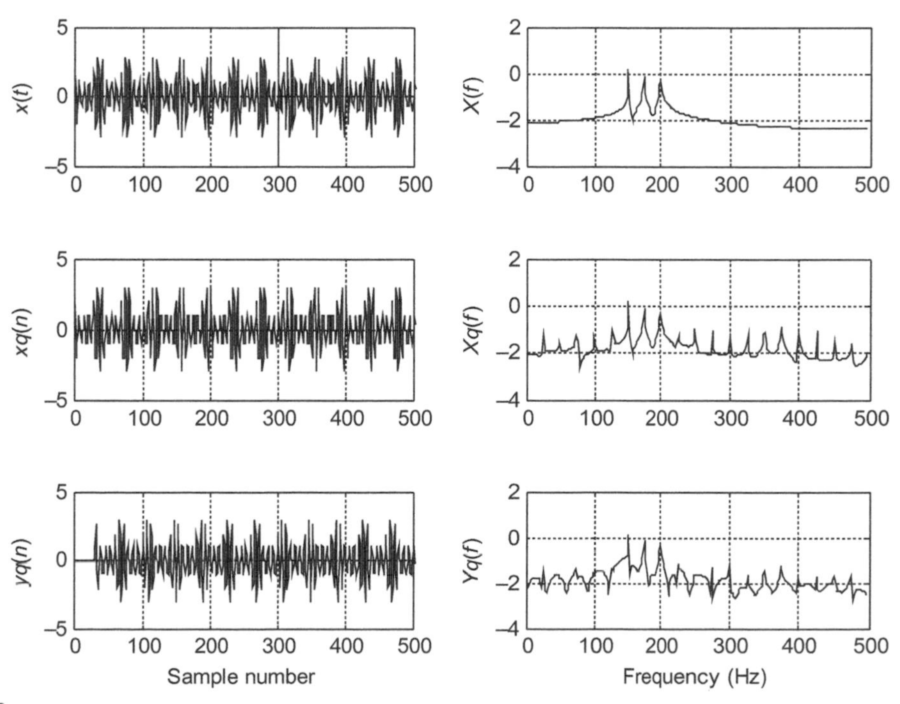

FIG. 11.26

Signal comparisons in both time and frequency domains.

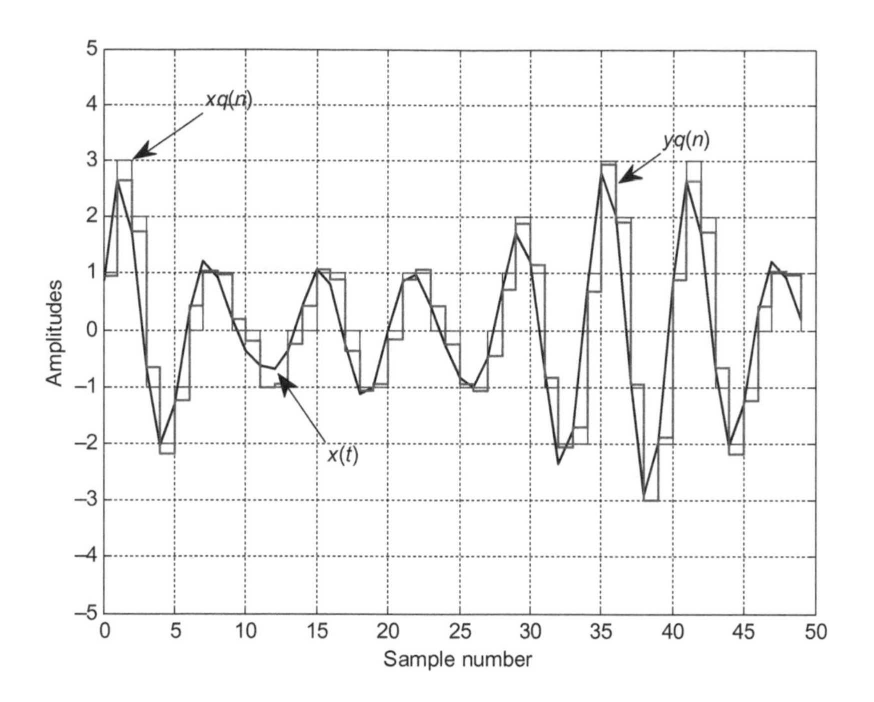

FIG. 11.27

Comparisons of continuous, regular sampled, and oversampled signal amplitudes.

sampling increases the quantization error may have correlation with the sinusoidal signal. The degradation performance can be cured using the dithering technique (Tan and Wang, 2011), which is beyond our scope.

Program 11.7. Oversampling implementation.

```
clear all; close all,clc
ntotal=512;
n=0:ntotal; % Number of samples
L=4; % Oversampling factor
nL=0:ntotal*L; % Number of samples for oversampling
numb=3; % Number of bits
A=2^(numb-1)-1; %Peak value
f1=150;C1=0.5*A;f2=175;C2=A*0.3;f3=200;C3=A*0.2; %Frequencies and amplitudes
fmax=500;fs=1000;T=1/fs; % Maximum frequency, sampling rate, sampling period
fsL=L*fs;TL=1/fsL;%Oversampling rate, and oversampling period
% Sampling at fs=1000Hz
x=C1*sin(2*pi*f1*n*T)+C2*sin(2*pi*f2*T*n+pi/6)+C3*sin(2*pi*f3*T*n+pi/4);
xq=round(x); %Quantized signal at the minimum sampling rate
NN=length(n);
f=[0:ntotal-1]*fs/NN;
M=32*L;nd=M/L; %Number of delay in samples due to anti-aliasing filtering
B=firwd(2*M+1,1,2*pi*fmax/fsL,0,4); % Anti-aliasing filter design (5% transition
```

```
bandwidth)
Figure(1);
freqz(B,1,1000,fsL)
% Oversampling
xx=C1*sin(2*pi*f1*nL*TL)+C2*sin(2*pi*f2*nL*TL+pi/6)+C3*sin(2*pi*f3*nL*TL+
pi/4);
xxq=round(xx); % Quantized signal
% Down sampling
y=filter(B,1,xxq);%Anti-aliasing filtering
yd.=y(1:L:length(y));% Down sample
Figure (2)
subplot(3,2,1);plot(n,x,'k');grid;axis([0500-5 5]);ylabel('x(t)')
Ak=2*abs(fft(x))/NN; Ak(1)=Ak(1)/2;
subplot(3,2,2);plot(f(1:NN/2),log10(Ak(1:NN/2)),'k');grid;ylabel('X(f)'); axis
([0500-4 2])
subplot(3,2,3);plot(n,xq,'k');grid;axis([0500-5 5]);ylabel('xq(n)');
Ak=2*abs(fft(xq))/NN; Ak(1)=Ak(1)/2;
subplot(3,2,4);plot(f(1:NN/2),log10(Ak(1:NN/2)),'k');grid;ylabel('Xq(f)'); axis
([0500-4 2])
subplot(3,2,5);plot(n,yd.,'k');grid;axis([0500-5 5]);ylabel('yq(n)');
xlabel('Sample number');
Ak=2*abs(fft(yd))/NN; Ak(1)=Ak(1)/2;
subplot(3,2,6);plot(f(1:NN/2),log10(Ak(1:NN/2)),'k');grid;ylabel('Yq(f)'); axis
([0500-4 2])
xlabel('Frequency (Hz)');
Figure (3)
plot(n(1:50),x(1:50),'k','LineWidth',2); hold % Plot of first 50 samples
stairs(n(1:50),xq(1:50),'b');
stairs(n(1:50),yd.(1+nd:50+nd),'r','LineWidth',2);grid
axis([0 50-5 5]);xlabel('Sample number');ylabel('Amplitudes')
snr(x,xq);
snr(x(1:ntotal-nd),yd.(1+nd:ntotal));
```

11.3.2 SIGMA-DELTA MODULATION ADC

To further improve ADC resolution, *sigma-delta modulation* (SDM) ADC is used. The principles of the first-order SDM are described in Fig. 11.28.

First, the analog signal is sampled to obtain the discrete-time signal $x(n)$. This discrete-time signal is subtracted by the analog output from the m-bit DAC, converting the m bit oversampled digital signal $y(n)$. Then the difference is sent to the discrete-time analog integrator, which is implemented by the switched-capacitor technique, for example. The output from the discrete-time analog integrator is converted using an m-bit ADC to produce the oversampled digital signal. Finally, the decimation filter removes outband quantization noise. Further decimation process can change the oversampling rate back to the desired sampling rate for the output digital signal $w(m)$.

To examine the SDM, we need to develop a DSP model for the discrete-time analog filter described in Fig. 11.29.

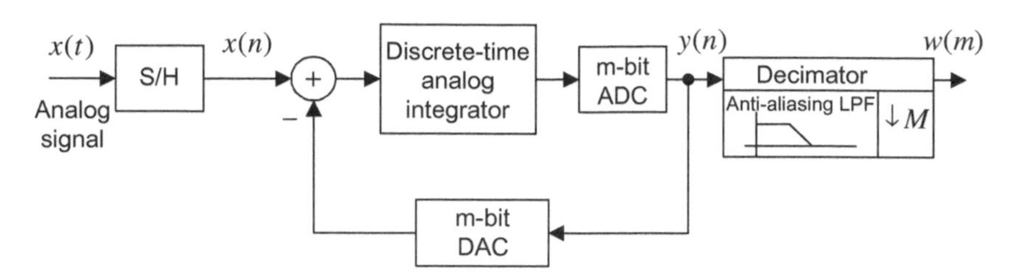

FIG. 11.28

Block diagram of SDM ADC.

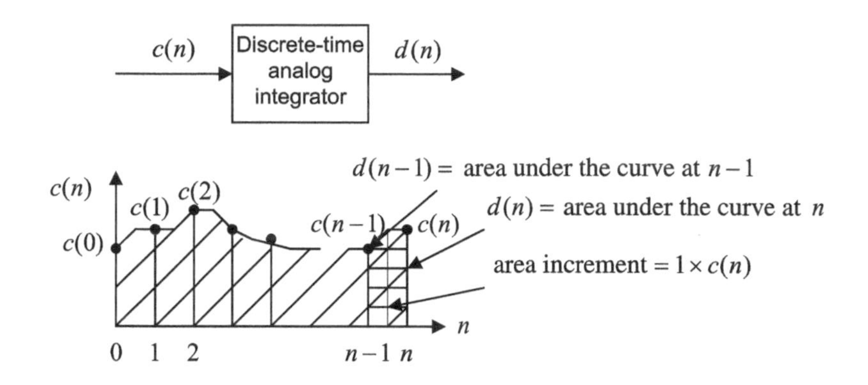

FIG. 11.29

Illustration of discrete-time analog integrator.

As shown in Fig. 11.29, the input signal $c(n)$ designates the amplitude at time instant n, while the output $d(n)$ is the area under the curve at time instant n, which can be expressed as a sum of the area under the curve at time instant $n-1$ and area increment:

$$d(n) = d(n-1) + \text{area incremetal}.$$

Using the extrapolation method, we have

$$d(n) = d(n-1) + 1 \times c(n). \tag{11.23}$$

Applying the z-transform to Eq. (11.23) leads to a transfer function of the discrete-time analog filter as

$$H(z) = \frac{D(z)}{C(z)} = \frac{1}{1 - z^{-1}}. \tag{11.24}$$

Again, considering that the m-bit quantization requires one sample delay, we get the DSP model for the first-order SDM depicted in Fig. 11.30, where $y(n)$ is the oversampling data encoded by m bits each, and $e(n)$ represents quantization error.

The SDM DSP model represents a feedback control system. Appling the z-transform leads to

$$Y(z) = \frac{1}{1 - z^{-1}}\left(X(z) - z^{-1}Y(z)\right) + E(z). \tag{11.25}$$

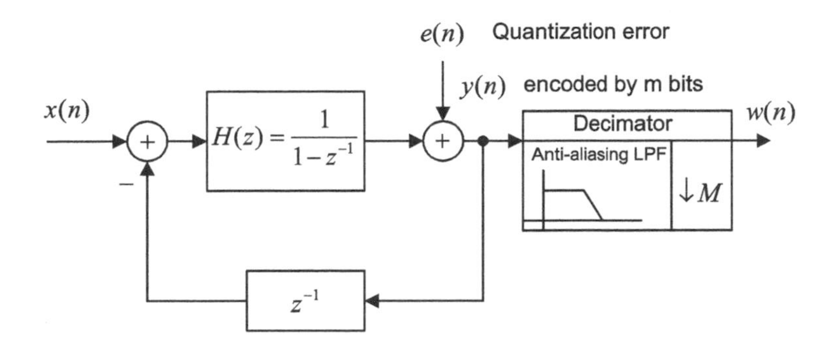

FIG. 11.30

DSP model for the first-order SDM ADC.

After simple algebra, we have

$$Y(z) = \underbrace{X(z)}_{\substack{\text{Original}\\\text{digitalsignal}\\\text{transform}}} + \underbrace{\left(1 - z^{-1}\right)}_{\substack{\text{Highpass}\\\text{filter}}} \times \underbrace{E(z)}_{\substack{\text{Quantization}\\\text{error}\\\text{transform}}} \tag{11.26}$$

In Eq. (11.26), the indicated highpass filter pushes quantization noise to the high-frequency range, where later the quantization noise can be removed by the decimation filter. Thus we call this highpass filter $(1 - z^{-1})$ as the *noise shaping filter,* illustrated in Fig. 11.31.

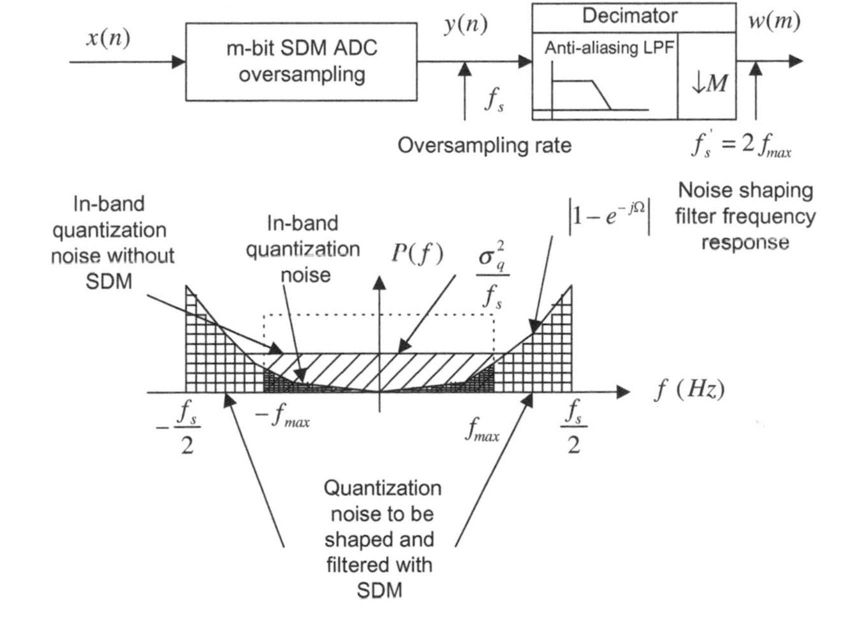

FIG. 11.31

Noise shaping of quantization noise for SDM ADC.

Shaped-in-band noise power after use of decimation filter can be estimated by the solid area under the curve. We have

$$\text{Shaped-in-band noise power} = \int_{-\Omega_{max}}^{\Omega_{max}} \frac{\sigma_q^2}{2\pi} \left|1 - e^{-j\Omega}\right|^2 d\Omega \tag{11.27}$$

Using the Maclaurin series expansion and neglecting the higher-order terms due to the small value of Ω_{max}, we yield

$$1 - e^{-j\Omega} = 1 - \left(1 + \frac{(-j\Omega)}{1!} + \frac{(-j\Omega)^2}{2!} + \cdots\right) \approx j\Omega.$$

Applying this approximation to Eq. (11.27) leads to

$$\text{Shaped-in-band noise power} \approx \int_{-\Omega_{max}}^{\Omega_{max}} \frac{\sigma_q^2}{2\pi} |j\Omega|^2 d\Omega = \frac{\sigma_q^2}{3\pi} \Omega_{max}^3. \tag{11.28}$$

After simple algebra, we have

$$\text{Shaped-in-band noise power} \approx \frac{\pi^2 \sigma_q^2}{3} \left(\frac{2f_{max}}{f_s}\right)^3 = \frac{\pi^2}{3} \times \frac{A^2 2^{-2m}}{12} \left(\frac{2f_{max}}{f_s}\right)^3. \tag{11.29}$$

If we let the shaped-in-band noise power equal the quantization noise power from the regular ADC using a minimum sampling rate, we have

$$\frac{\pi^2}{3} \times \frac{A^2 2^{-2m}}{12} \left(\frac{2f_{max}}{f_s}\right)^3 = \frac{A^2}{12} \times 2^{-2n}. \tag{11.30}$$

We modify Eq. (11.30) into the following useful formats for applications:

$$n = m + 1.5 \times \log_2\left(\frac{f_s}{2f_{max}}\right) - 0.86 \tag{11.31}$$

$$\left(\frac{f_s}{2f_{max}}\right)^3 = \frac{\pi^2}{3} \times 2^{2(n-m)}. \tag{11.32}$$

EXAMPLE 11.9

Given the following DSP system specifications:
Over sampling rate system
First-order SDM with 2-bit ADC
Sampling rate = 4 MHz
Maximum audio input frequency = 4 kHz,
Determine the equivalent ADC resolution.

Solution:
Since $m=2$ bits, and

$$\frac{f_s}{2f_{\max}}=\frac{4000kHz}{2\times 4kHz}=500.$$

we calculate

$$n=m+1.5\times\log_2\left(\frac{f_s}{2f_{\max}}\right)-0.86=2+1.5\times\log_2(500)-0.86\approx 15\ \text{bits}.$$

We can also extend the first-order SDM DSP model to the second-order SDM DSP model by cascading one section of the first-order discrete-time analog filter as depicted in Fig. 11.32.

Similarly to the first-order SDM DSP model, applying the z-transform leads the following relationship:

$$Y(z)=\underbrace{X(z)}_{\substack{Original\\ digitalsignal\\ transform}}+\underbrace{\left(1-z^{-1}\right)^2}_{\substack{Highpass\\ noiseshaping\\ filter}}\times\underbrace{E(z)}_{\substack{Quantization\\ error\\ transform}}. \tag{11.33}$$

Note that the noise shape filter becomes to a second-order highpass filter; hence, the more quantization noise is pushed to the high-frequency range, the better ADC resolution is expected to be. In a similar analysis to the first-order SDM, we get the following useful formulas:

$$n=m+2.5\times\log_2\left(\frac{f_s}{2f_{\max}}\right)-2.14 \tag{11.34}$$

$$\left(\frac{f_s}{2f_{\max}}\right)^5=\frac{\pi^4}{5}\times 2^{2(n-m)}. \tag{11.35}$$

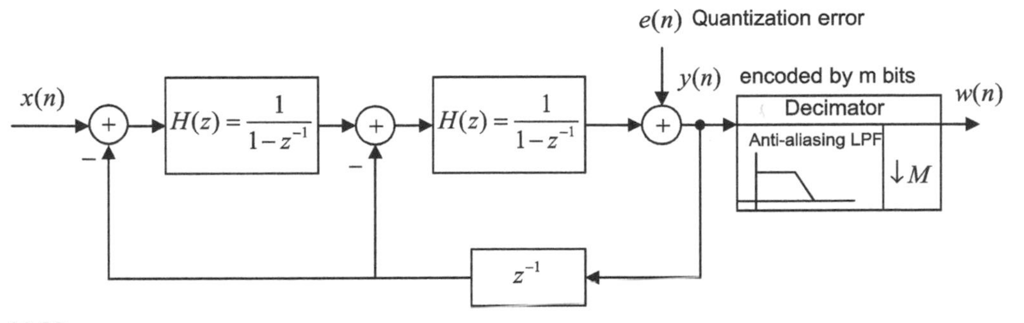

FIG. 11.32

DSP model for the second-order SDM ADC.

In general, the Kth-order SDM DSP model and ADC resolution formulas are given as

$$Y(z) = \underbrace{X(z)}_{\substack{\text{Original}\\ \text{digital signal}\\ \text{transform}}} + \underbrace{\left(1-z^{-1}\right)^{K}}_{\substack{\text{Highpass}\\ \text{noise shaping}\\ \text{filter}}} \times \underbrace{E(z)}_{\substack{\text{Quantization}\\ \text{error}\\ \text{transform}}} \tag{11.36}$$

$$n = m + 0.5 \times (2K+1) \times \log_2\left(\frac{f_s}{2f_{\max}}\right) - 0.5 \times \log_2\left(\frac{\pi^{2K}}{2K+1}\right) \tag{11.37}$$

$$\left(\frac{f_s}{2f_{\max}}\right)^{2K+1} = \frac{\pi^{2K}}{2K+1} \times 2^{2(n-m)}. \tag{11.38}$$

EXAMPLE 11.10

Given the oversampling rate DSP system with the following specifications:
second-order SDM = 1 bit ADC
sampling rate = 1 MHz
maximum audio input frequency = 4 kHz,
Determine the effective ADC resolution.

Solution:

$$n = 1 + 2.5 \times \log_2\left(\frac{1{,}000kHz}{2 \times 4kHz}\right) - 2.14 \approx 16\,\text{bits}.$$

We implement the first-order SDM system using the same continuous signal in Eq. (11.22). The continuous signal is originally sampled at 1 kHz and each sample is encoded using 3 bits. The SDM system uses an oversampling rate of 8 kHz ($L = 8$) and each sample is quantized using a 3-bit code. The anti-aliasing lowpass filter is designed with a cutoff frequency of $\Omega = 2\pi f_{\max} T = 2\pi \times 500/8,000 = \pi/8$ rad. Fig. 11.33 shows the frequency responses of the designed filter while Fig. 11.34 compares the time and frequency domain signals, where $x(t)$ designates the continuous version, $x_q(n)$ denotes the quantized version using a regular sampling rate ($L = 1$) while $y_q(n)$ is the enhanced version using $L = 8$. The detailed amplitude comparisons are given in Fig. 11.35. The measured SNRs are 14.3 dB in the regular sampling system while 33.83 dB in the oversampling SDM system. We can observe a significant SNR improvement with 19.5 dB. The detailed implementation using MATLAB is given in Program 11.8.

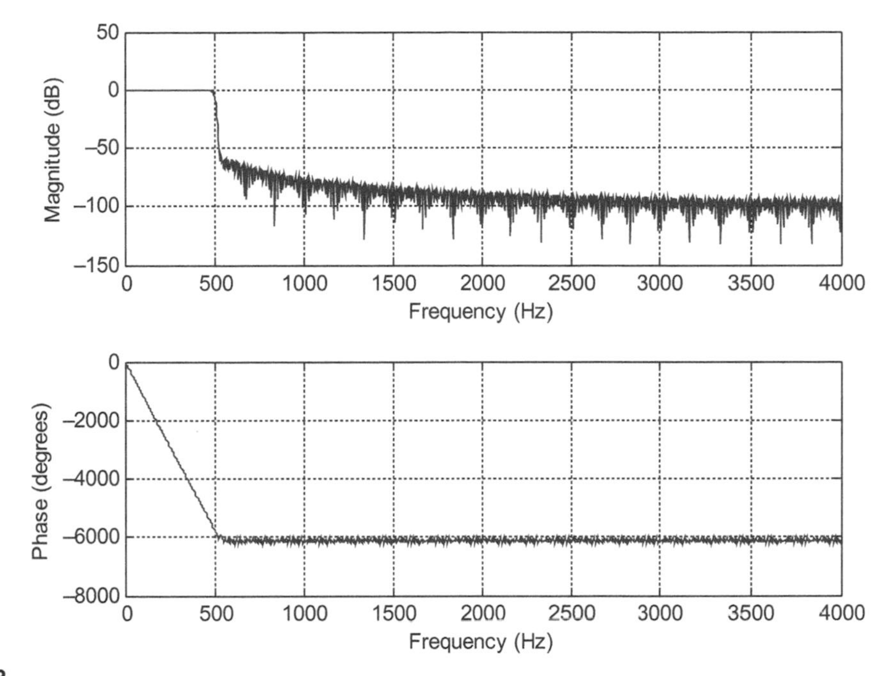

FIG. 11.33

Frequency responses of the designed filter.

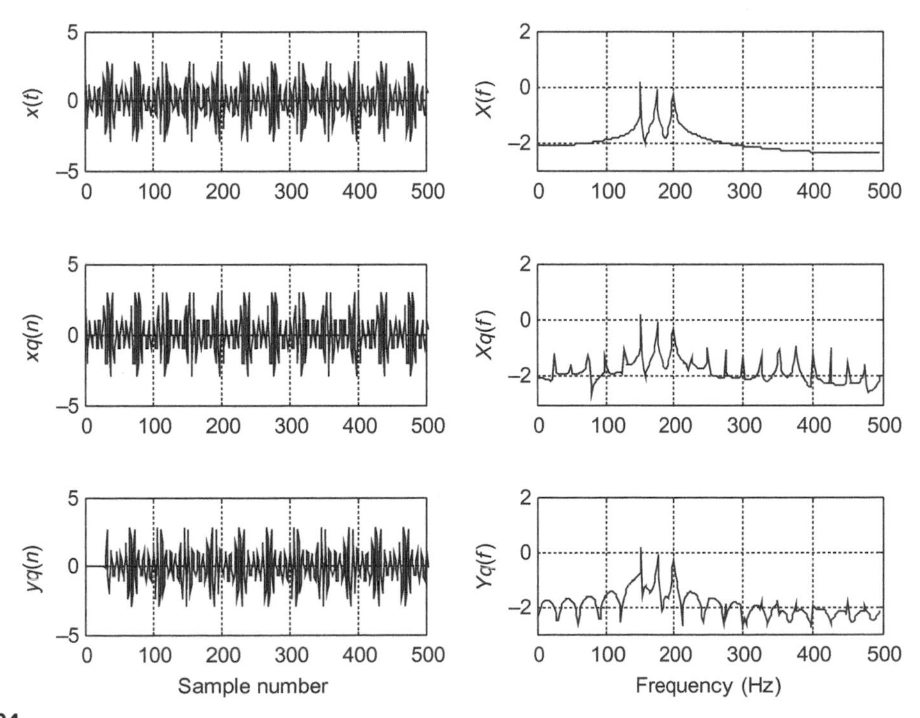

FIG. 11.34

Signal comparisons in both time and frequency domains.

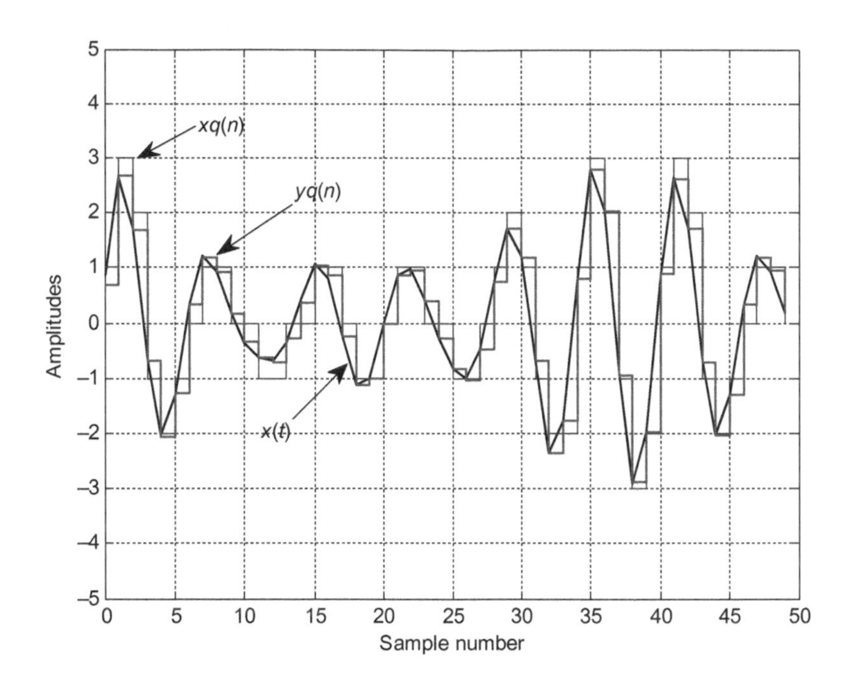

FIG. 11.35

Comparisons of continuous, regular sampled, and oversampled signal amplitudes.

Program 11.8. First-order SDM oversampling implementation.

```
clear all; close all;clc
ntotal=512; % Number of samples
n=0:ntotal;
L=8; %Oversampling factor
nL=0:ntotal*L;numb=3;A=2^(numb-1)-1; %Peak value
f1=150;C1=0.5*A;f2=175;C2=A*0.3;f3=200;C3=A*0.2;%Frequencies and amplitudes
fmax=500;fs=1000; T=1/fs; % Sampling rate and sampling period
fsL=L*fs;TL=1/fsL; % Oversampling rate and oversampling period
% Sampling at fs=1000Hz
x=C1*sin(2*pi*f1*n*T)+C2*sin(2*pi*f2*T*n+pi/6)+C3*sin(2*pi*f3*T*n+pi/4);
xq=round(x); %Quantization
NN=length(n);
M=32*L;nd=M/L; %Delay in terms of samples for anti-aliasing filtering
B=firwd(2*M+1,1,2*pi*fmax/fsL,0,4);% Design of an anti-aliasing filter
Figure(1)
freqz(B,1,1000,fsL);
% Oversampling
xx=C1*sin(2*pi*f1*nL*TL)+C2*sin(2*pi*f2*nL*TL+pi/6)+C3*sin(2*pi*f3*nL*TL+pi/4);
% The first-order SDM processing
  yq=zeros(1,ntotal*L+1+1); %Initializing the buffer
  y=yq;
```

```
for i=1:ntotal*L
  y(i+1)=(xx(i+1)-yq(i))+y(i);
  yq(i+1)=round(y(i+1));
end
xxq=yq(1:ntotal*L+1); %Signal Quantization
% Downsampling
y=filter(B,1,xxq);
yd.=y(1:L:length(y));
f=[0:ntotal-1]*fs/NN;
Figure (2)
subplot(3,2,1);plot(n,x,'k');grid;axis([0500-5 5]);ylabel('x(t)');
Ak=2*abs(fft(x))/NN; Ak(1)=Ak(1)/2;
subplot(3,2,2);plot(f(1:NN/2),log10(Ak(1:NN/2)),'k');grid;
axis([0500-3 2]);ylabel('X(f)');
subplot(3,2,3);plot(n,xq,'k');grid;axis([0500-5 5]);ylabel('xq(n)');
Ak=2*abs(fft(xq))/NN; Ak(1)=Ak(1)/2;
subplot(3,2,4);plot(f(1:NN/2),log10(Ak(1:NN/2)),'k');grid
axis([0500-3 2]);ylabel('Xq(f)');
subplot(3,2,5);plot(n,yd.,'k');grid;axis([0500-5 5]);ylabel('yq(n)');
xlabel('Sample number');
Ak=2*abs(fft(yd))/NN; Ak(1)=Ak(1)/2;
subplot(3,2,6);plot(f(1:NN/2),log10(Ak(1:NN/2)),'k');grid
axis([0500-3 2]);ylabel('Yq(f)');xlabel('Frequency (Hz)');
Figure (3)
plot(n(1:50),x(1:50),'k','LineWidth',2); hold
stairs(n(1:50),xq(1:50),'b');
stairs(n(1:50),yd.(1+nd:50+nd),'r','LineWidth',2);
axis([0 50-5 5]);grid;xlabel('Sample number');ylabel('Amplitudes');
snr(x,xq);
snr(x(1:ntotal-nd),yd.(1+nd:ntotal));
```

Next, we review the application of the oversampling ADC used in the industry. Fig. 11.36 illustrates a function diagram for the MAX1402 low-power, multichannel oversampling sigma-delta ADC used in industry. It applies a sigma-delta modulator with a digital decimation filter to achieve 16-bit accuracy. The device offers three fully differential input channels, which can be independently programmed. It can also be configured as five pseudo-differential input channels. It comprises two chopper buffer amplifiers and a programmable gain amplifier, a DAC unit with predicted input subtracted from the analog input to acquire the differential signal, and a second-order switched-capacitor sigma-delta modulator.

The chip produces a 1-bit data stream, which will be filtered by the integrated digital filter to complete ADC. The digital filter's user-selectable decimation factor offers flexibility as conversion resolution can be reduced in exchange for a higher data rate or vice versa. The integrated digital lowpass filter is the first-order or third-order Sinc infinite impulse response filter. Such a filter offers notches corresponding to its output data rate and its frequency harmonics, so it can effectively reduce the developed image noises in the frequency domain. (The Sinc filter is beyond the scope of our discussion.) The MAX1402 can provide 16-bit accuracy at 480 samples per second and 12-bit accuracy at 4800 samples per second. The chip finds a wide application in sensors and instrumentation. Its detailed features can be found in the MAX1402 data sheet (Maxim Integrated Products, 2018).

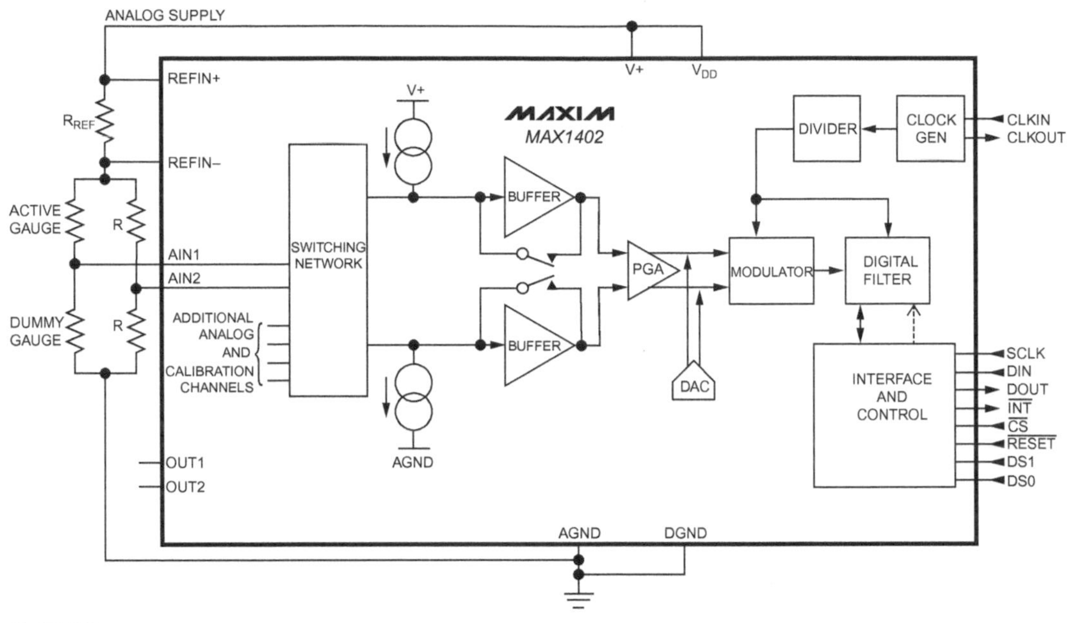

FIG. 11.36

Functional Diagram for the Sigma-delta ADC.

11.4 APPLICATION EXAMPLE: CD PLAYER

Fig. 11.37 illustrates a CD playback system, also described earlier. A laser optically scans the tracks on a CD to produce a digital signal. The digital signal is then demodulated, and parity bits are used to detect bit errors due to manufacturing defects, dust, and so on, and to correct them. The demodulated signal is again oversampled by a factor of 4 and hence the sampling rate is increased to 176.4 kHz for each channel. Each digital sample then passes through a 14-bit DAC, which produces the sample-and-hold voltage signals that pass the anti-image lowpass filter. The output from each analog filter is fed to its corresponding loudspeaker. Oversampling relaxes the design requirements of the analog anti-image lowpass filter, which is used to smooth out the voltage steps.

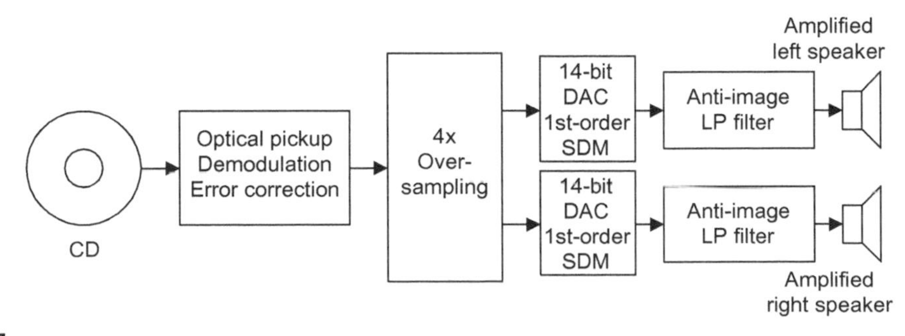

FIG. 11.37

Simplified decoder of a CD recording system.

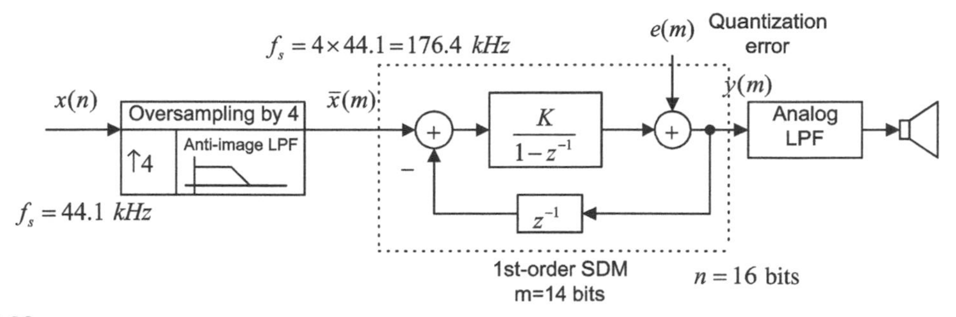

FIG. 11.38

Illustration of oversampling and SDM ADC used in the decoder of a CD recording system.

The earliest system used a third-order Bessel filter with a 3-dB gain attenuation at 30 kHz. Note that the first-order SDM is added to the 14-bit DAC unit to further improve the 14-bit DAC to 16-bit DAC.

Let us examine the single-channel DSP portion as shown in Fig. 11.38.

The spectral plots for the oversampled and interpolated signal $\bar{x}(n)$, the 14-bit SDM output $y(n)$, and the final analog output audio signal are given in Fig. 11.39. As we can see in plot (a) in the figure, the quantization noise is uniformly distributed, and only in-band quantization noise (0–22.05 kHz) is

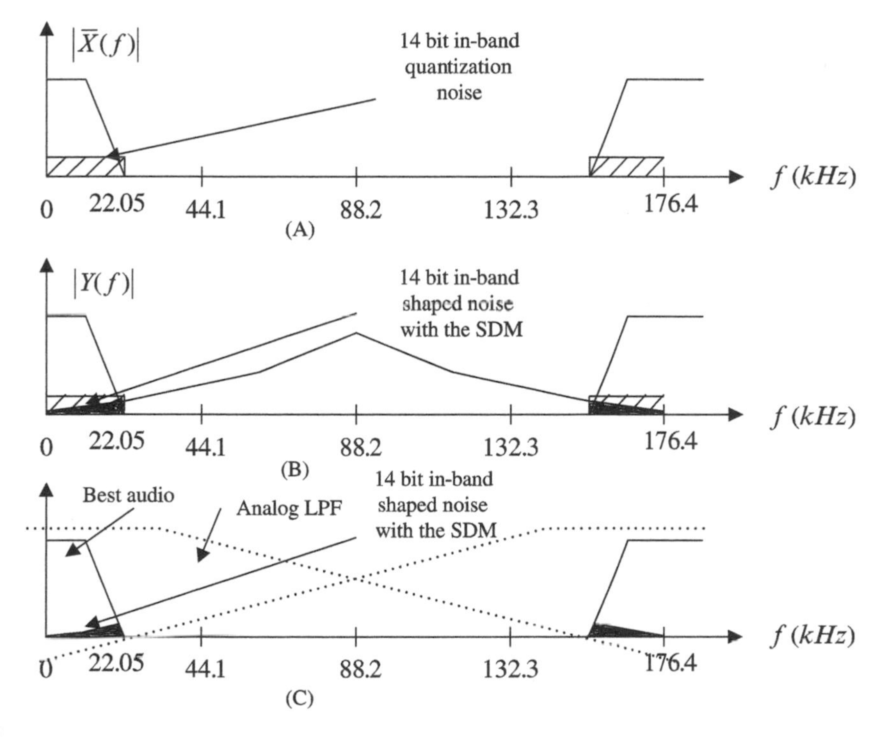

FIG. 11.39

Spectral illustrations for the oversampling and SDM ADC used in the decoder of a CD recording system.

expected. Again, 14 bits for each sample are kept after oversampling. Without using the first-order SDM, we expect the effective ADC resolution due to oversampling to be

$$n = 14 + 0.5 \times \log_2\left(\frac{176.4}{44.1}\right) = 15\,\text{bits},$$

which is fewer than 16 bits. To improve quality further, the first-order SDM is used. The in-band quantization noise is then shaped. The first-order SDM pushes quantization noise to the high-frequency range, as illustrated in plot (b) in Fig. 11.39. The effective ADC resolution now becomes

$$n = 14 + 1.5 \times \log_2\left(\frac{176.4}{44.1}\right) - 0.86 \approx 16\,\text{bits}.$$

Hence, 16-bit ADC audio quality is preserved. On the other hand, from plot (c) in Fig. 11.39, the audio occupies a frequency range up to 22.05 kHz, while the DSP Nyquist limit is 88.2 kHz, so the low-order analog anti-image filter can satisfy the design requirement.

11.5 UNDERSAMPLING OF BANDPASS SIGNALS

As we discussed in Chapter 2, the sampling theorem requires that the sampling rate be twice as large as the highest frequency of the analog signal to be sampled. The sampling theorem ensures the complete reconstruction of the analog signal without aliasing distortion. In some applications, such as the modulated signals in communications systems, the signal exists in only a small portion of the bandwidth. Fig. 11.40 shows an amplitude modulated (AM) signal in both time domain and frequency domain. Assuming that the message signal has a bandwidth of 4 kHz and a carrier frequency of 96 kHz, the upper frequency edge the AM signal is therefore 100 kHz ($f_c + B$). Then the traditional sampling process

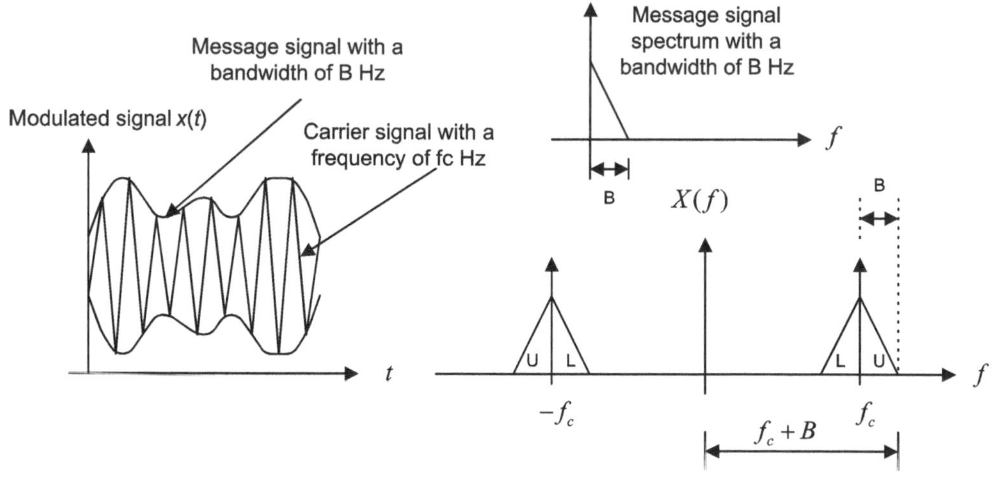

FIG. 11.40

Message signal, modulated signal, and their spectra.

requires that the sampling rate be larger than 200 kHz $2(f_c+B)$, resulting in at a high processing cost. Note that sampling the baseband signal of 4 kHz only requires a sampling rate of 8 kHz ($2B$).

If a certain condition is satisfied at the undersampling stage, we are able to make use of the aliasing signal to recover the message signal, since the aliasing signal contains the folded original message information (which we used to consider as distortion). The reader is referred to the undersampling technique discussed in Ifeachor and Jervis (2002) and Porat (1997). Let the message to be recovered have a bandwidth of B, the theoretical minimum sampling rate be $f_s=2B$, and the carrier frequency of the modulated signal be f_c. We discuss the following cases.

Case 1.

If f_c = even integer $\times B$ and $f_c = 2B$, the sampled spectrum with all the replicas will be as shown in Fig. 11.41A.

As an illustrative example in time domain for Case 1, suppose we have a bandpass signal with a carrier frequency of 20 Hz; that is,

$$x(t) = \cos(2\pi \times 20t)m(t), \tag{11.39}$$

where $m(t)$ is the message signal with a bandwidth of 2 Hz. Using a sampling rate of 4 Hz by substituting $t=nT$, where $T=1/f_s$ into Eq. (11.39), we get the sampled signal as

$$x(nT) = \cos(2\pi \times 20t)m(t)|_{t=nT} = \cos(2\pi \times 20n/4)m(nT). \tag{11.40}$$

Since $10n\pi = 5n(2\pi)$ is the multiple of 2π,

$$\cos(2\pi \times 20n/4) = \cos(10\pi n) = 1, \tag{11.41}$$

we obtain the undersampled signal as

$$x(nT) = \cos(2\pi \times 20n/4)m(nT) = m(nT), \tag{11.42}$$

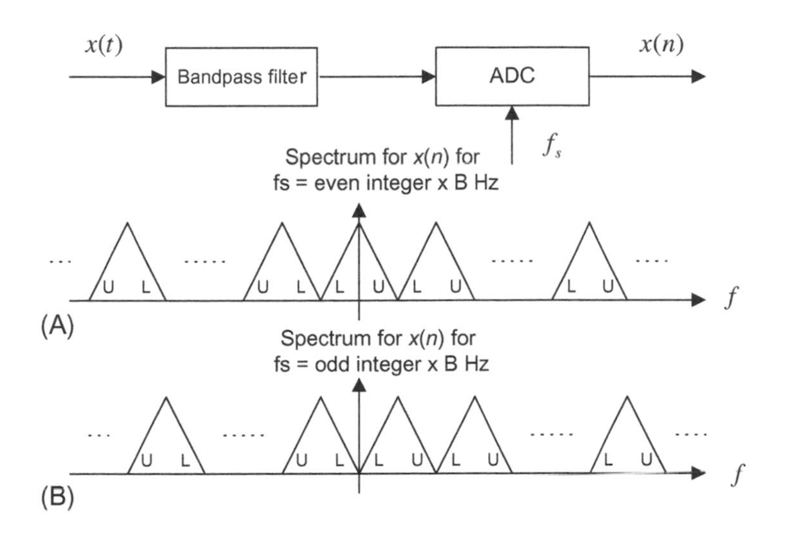

FIG. 11.41

Spectrum of the undersampled signal.

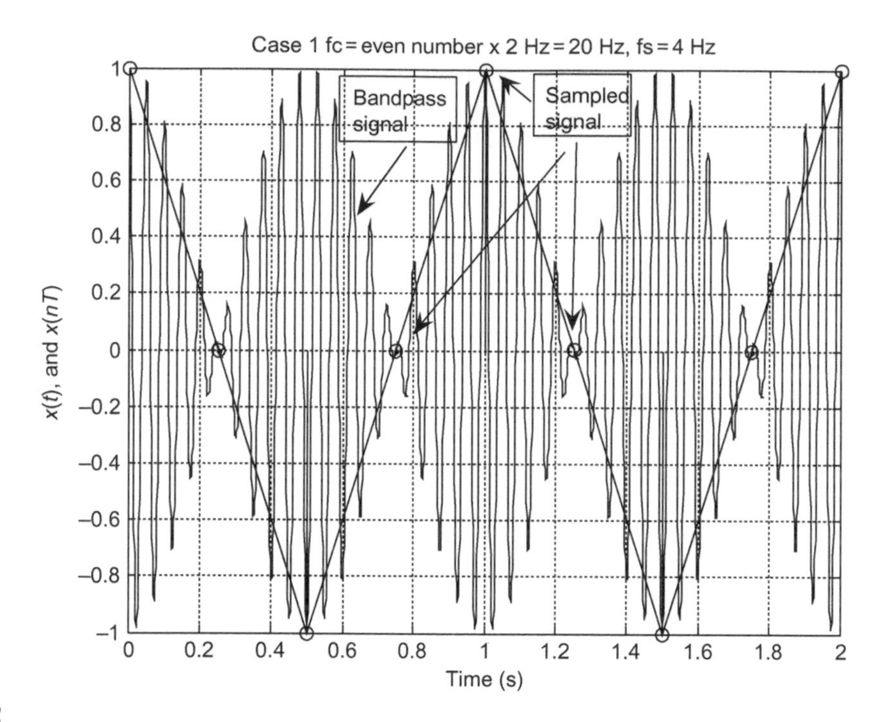

FIG. 11.42

Plots of the bandpass signal and sampled signal for Case 1.

which is a perfect digital message signal. Fig. 11.42 shows the bandpass signal and its sampled signals when the message signal is 1 Hz, given as

$$m(t) = \cos(2\pi t). \tag{11.43}$$

Case 2.

If f_c = odd integer × B and $f_c = 2B$, the sampled spectrum with all the replicas will be as shown in Fig. 11.41B, where the spectral portions L and U are reversed. Hence, the frequency reversal will occur. Then a further digital modulation in which the signal is multiplied by the digital oscillator with a frequency of B Hz can be used to adjust the spectrum to be the same as that in Case 1.

As another illustrative example for Case 2, let us sample the following the bandpass signal with a carrier frequency of 22 Hz, given by

$$x(t) = \cos(2\pi \times 22t)m(t). \tag{11.44}$$

Applying undersampling using a sampling rate of 4 Hz, it follows that

$$x(nT) = \cos(2\pi \times 22n/4)m(nT) = \cos(11n\pi)m(nT). \tag{11.45}$$

Since $11n\pi$ can be either an odd or an even integer multiple of π, we have

$$\cos(11\pi n) = \begin{cases} -1 & n = \text{odd} \\ 1 & n = \text{even.} \end{cases} \tag{11.46}$$

We see that Eq. (11.46) causes the message samples to change sign alternatively with a carrier frequency of 22 Hz, which is the odd integer multiple of the message bandwidth of 2 Hz. This in fact will reverse the baseband message spectrum. To correct the spectrum reversal, we multiply an oscillator with a frequency of $B = 2$ Hz by the bandpass signal, that is

$$x(t)\cos(2\pi \times 2t) = \cos(2\pi \times 22t)m(t)\cos(2\pi \times 2t). \tag{11.47}$$

Then the undersampled signal is then given by

$$\begin{aligned} x(nT)\cos(2\pi \times 2n/4) &= \cos(2\pi \times 22n/4)m(nT)\cos(2\pi \times 2n/4) \\ &= \cos(11n\pi)m(nT)\cos(n\pi) \end{aligned} \tag{11.48}$$

Since

$$\cos(11\pi n)\cos(\pi n) = 1, \tag{11.49}$$

it follows that

$$x(nT)\cos(2\pi \times 2n/4) = \cos(\pi \times 11n)m(nT)\cos(\pi \times n) = m(nT), \tag{11.50}$$

which is the recovered message signal. Fig. 11.43 shows the sampled bandpass signals with the reversed message spectrum and the corrected message spectrum, respectively, for a message signal having a frequency of 0.5 Hz, that is,

$$m(t) = \cos(2\pi \times 0.5t). \tag{11.51}$$

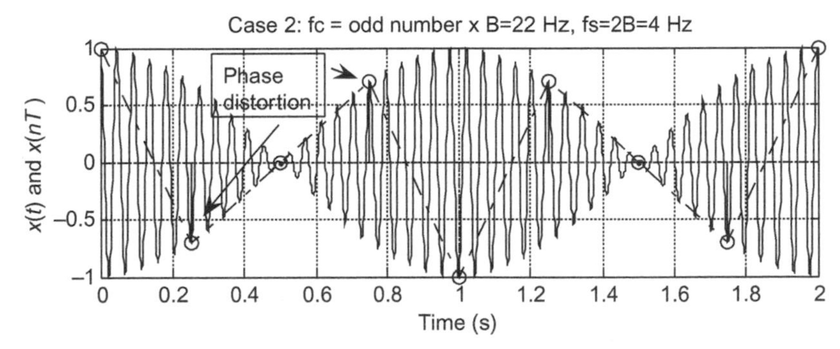

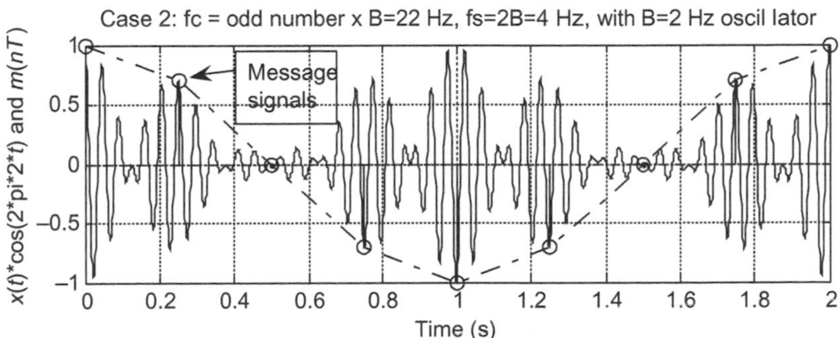

FIG. 11.43

Plots of the bandpass signals and sampled signals for Case 2.

Case 3.

If $f_c =$ non interger $\times B$, we can extend the bandwidth B to $\overline{B}$ such that

$$f_c = \text{integer} \times \overline{B} \text{ and } f_s = 2\overline{B}. \tag{11.52}$$

Then we can apply Case 1 or Case 2. An illustration of Case 3 is included in the following example.

EXAMPLE 11.11

Given a bandpass signal with the spectrum and the carry frequency f_c shown in Fig. 11.44A–C, respectively, and assuming the baseband bandwidth $B = 4$ kHz, select the sampling rate and sketch the sampled spectrum ranging from 0 Hz to the carrier frequency for each of the following carrier frequencies:

1. $f_c = 16$ kHz
2. $f_c = 12$ kHz
3. $f_c = 18$ kHz

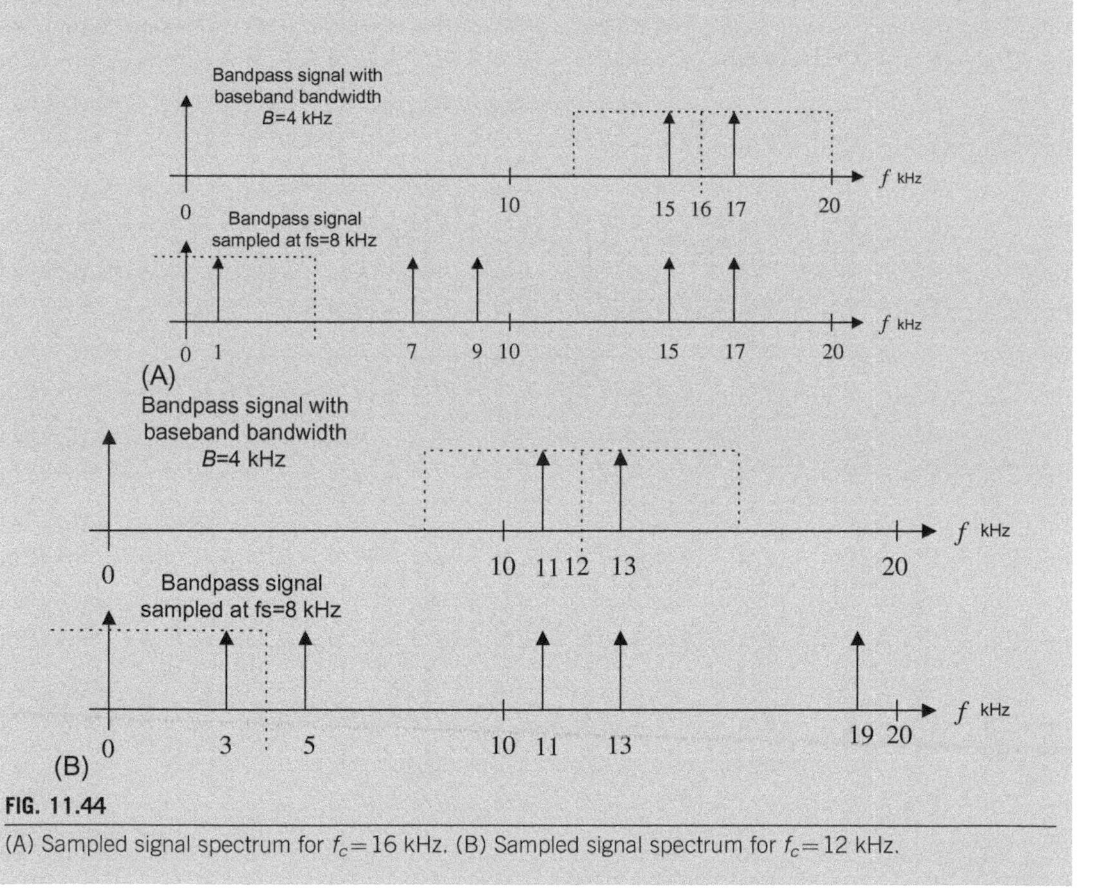

FIG. 11.44

(A) Sampled signal spectrum for $f_c = 16$ kHz. (B) Sampled signal spectrum for $f_c = 12$ kHz.

Continued

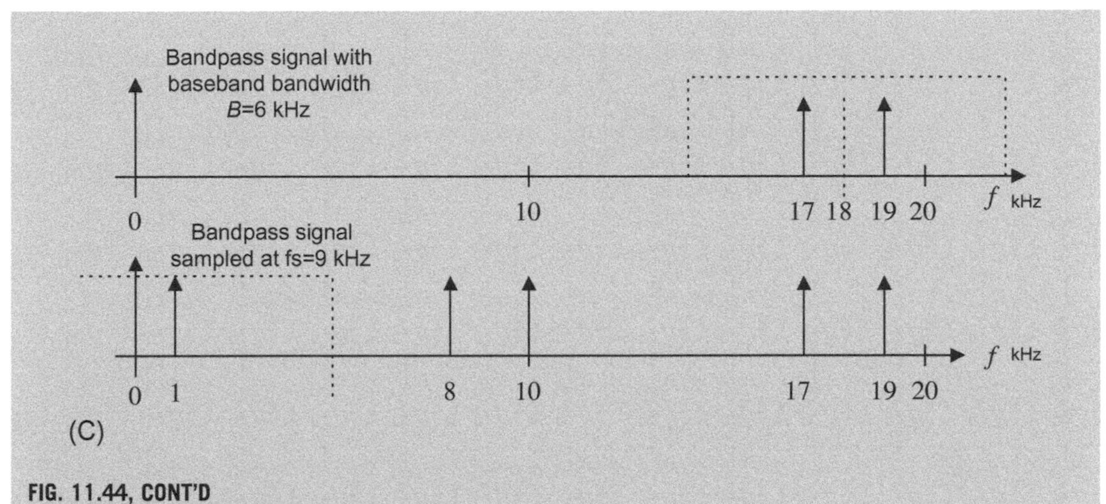

FIG. 11.44, CONT'D

(C) Sampled signal spectrum for $f_c = 18$ kHz.

Solution:

1. Since $f_c/B = 4$ is an even number, which is Case 1, we select $f_s = 8$ kHz and sketch the sampled spectrum shown in Fig. 11.44A.
2. Since $f_c/B = 3$ is an odd number, we select $f_s = 8$ kHz and sketch the sampled spectrum shown in Fig. 11.44B.
3. Now, $f_c/B = 4.5$ which is a non-integer. We extend the band width $\overline{B} = 4.5$ kHz, so $f_c/\overline{B} = 4$ and $f_s = 2\overline{B} = 9$ kHz. Then the sketched spectrum is shown in Fig. 11.44C.

Simulation Example

An AM with a 1-kHz message signal is given as:

$$x(t) = [1 + 0.8 \times \sin(2\pi \times 1000t)]\cos(2\pi \times f_c t). \tag{11.53}$$

Assuming a message bandwidth of 4 kHz, determine the sampling rate, use MATLAB to sample the AM signal, and sketch the sampled spectrum up to the sampling frequency for each the following carrier frequencies:

1. $f_c = 96$ kHz
2. $f_c = 100$ kHz
3. $f_c = 99$ kHz

1. For this case, $f_c/B = 24$ is an even number. We select $f_s = 8$ kHz. Fig. 11.45A describes the simulation, where the upper left plot is the AM signal, the upper right plot is the spectrum of the AM signal, the lower left plot is the undersampled signal, and the lower right plot is the spectrum of the undersampled signal displayed from 0 to 8 kHz.

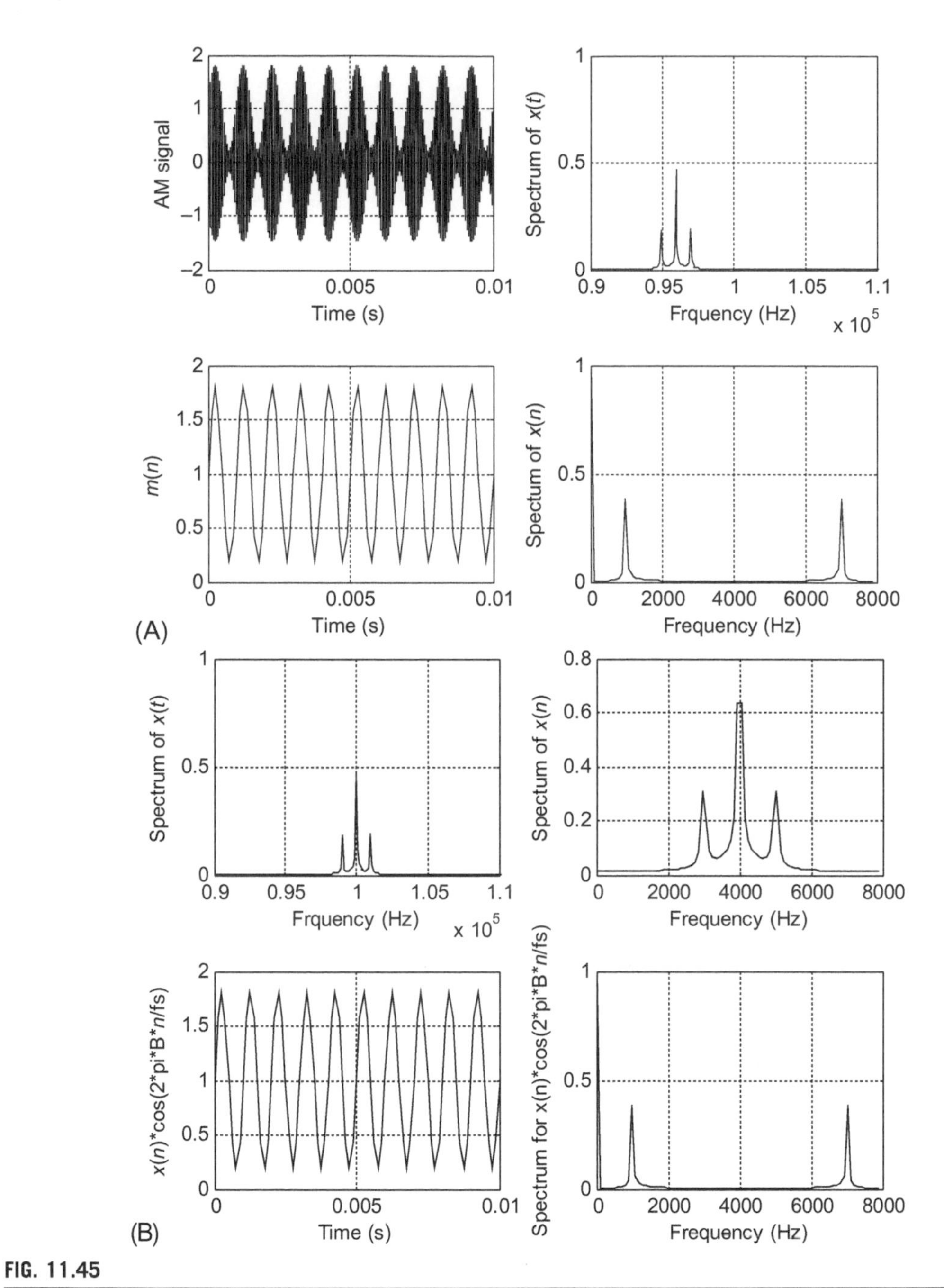

FIG. 11.45

(A) Sampled AM signal and spectrum for $f_c = 96$ kHz. (B) Sampled AM signal and spectrum for $f_c = 100$ kHz.

(Continued)

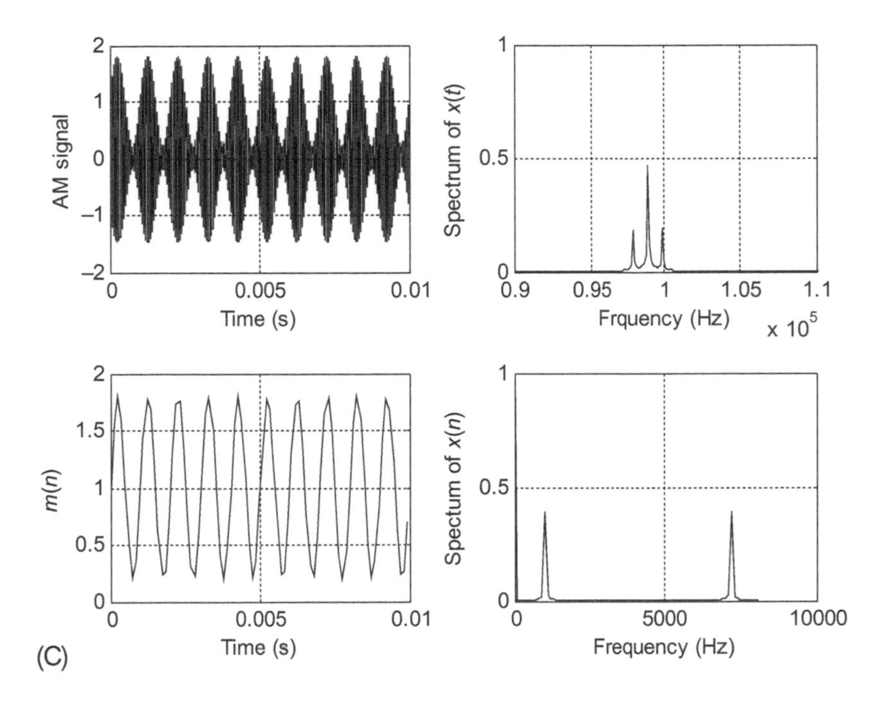

FIG. 11.45, CONT'D

(C) Sampled AM signal and spectrum for $f_c = 99$ kHz.

2. $f_c/B = 25$ is an odd number, we choose $f_s = 8$ kHz, and a further process is needed. We can multiply the undersampled signal by a digital oscillator with a frequency of $B = 4$ kHz to achieve the 1-kHz baseband signal. The plots of the AM signal spectrum, undersampled signal spectrum, and the oscillator mixed signal and its spectrum are shown in Fig. 11.45B.
3. For $f_c = 99$ kHz, $f_c/B = 24.75$. We extend the bandwidth to $\overline{B} = 4.125$ so that $f_c/\overline{B} = 24$. Hence, the undersampling rate is used as $f_s = 8.25$ kHz. Fig. 11.45C shows the plots of the AM signal, the AM signal spectrum, the undersampled signal based on the extended baseband width, and sampled signal spectrum ranging from 0 to 8.25 kHz, respectively.

This example verifies principles of undersampling of bandpass signals.

11.6 SUMMARY

1. Downsampling (decimation) by an integer factor of M means taking one sample from the data sequence $x(n)$ for every M samples and discard the last $M - 1$ samples.
2. Upsampling (Interpolation) by an integer factor of L means inserting $L - 1$ zeros for every sample in the data sequence $x(n)$.
3. Downsampling requires a decimation (anti-aliasing) filter to avoid frequency aliasing before downsampling.

4. Upsampling requires an interpolation (anti-image) filter to remove the images after interpolation.
5. Changing the sampling rate by a non-integer factor of L/M requires two stages: an interpolation stage and a downsampling stage.
6. Two-stage decimation can dramatically reduce the anti-aliasing filter length.
7. Polyphase implementations of the decimation filter and interpolation filter can reduce complexity of the filter operations, that is, fewer multiplications and additions.
8. Using oversampling can improve the regular ADC resolution. SDM ADC can achieve even higher ADC resolution, using noise-shaping effect for further reduction of quantization noise.
9. The audio CD player uses multirate signal processing and oversampling.
10. Undersampling can be used to sample the bandpass signal and find its application in communications.

11.7 PROBLEMS

11.1 For a single-stage decimator with the following specifications:
Original sampling rate = 1 kHz
Decimation factor $M=2$
Frequency of interest = 0–100 Hz
Passband ripple = 0.015 dB
Stopband attenuation = 40 dB,
(a) Draw the block diagram for the decimator;
(b) Determine the window type, filter length, and cutoff frequency if the window method is used for the anti-aliasing FIR filter design.

11.2 For a single-stage interpolator with the following specifications:
Original sampling rate = 1 kHz
Interpolation factor $L=2$
Frequency of interest = 0–150 Hz
Passband ripple = 0.02 dB
Stopband attenuation = 45 dB,
(a) Draw the block diagram for the interpolator;
(b) Determine the window type, filter length, and cutoff frequency if the window method is used for the anti-image FIR filter design.

11.3 For a single-stage decimator with the following specifications:
Original sampling rate = 8 kHz.
Decimation factor $M=4$
Frequency of interest = 0–800 Hz
Passband ripple = 0.02 dB
Stopband attenuation = 46 dB,
(a) Draw the block diagram for the decimator;
(b) Determine the window type, filter length, and cutoff frequency if the window method is used for the anti-aliasing FIR filter design.

11.4 For a single-stage interpolator with the following specifications:
Original sampling rate = 8 kHz.
Interpolation factor $L = 3$
Frequency of interest = 0–3400 Hz
Passband ripple = 0.02 dB
Stopband attenuation = 46 dB,
(a) Draw the block diagram for the interpolator;
(b) Determine the window type, filter length, and cutoff frequency if the window method is used for the anti-image FIR filter design.

11.5 For the sampling conversion from 4 to 3 kHz with the following specifications:
Original sampling rate = 4 kHz.
Interpolation factor $L = 3$
Decimation factor $M = 2$
Frequency of interest = 0–400 Hz
Passband ripple = 0.02 dB
Stopband attenuation = 46 dB,
(a) Draw the block diagram for the interpolator;
(b) Determine the window type, filter length, and cutoff frequency if the window method is used for the combined FIR filter $H(z)$.

11.6 For the design of a two-stage decimator with the following specifications:
Original sampling rate = 32 kHz
Frequency of interest = 0–250 Hz
Passband ripple = 0.05 (absolute)
Stopband attenuation = 0.005 (absolute)
Final sampling rate = 1000 Hz,
(a) Draw the decimation block diagram;
(b) Specify the sampling rate for each stage;
(c) Determine the window type, filter length, and cutoff frequency for the first stage if the window method is used for anti-aliasing FIR filter design ($H_1(z)$);
(d) Determine the window type, filter length, and cutoff frequency for the second stage if the window method is used for the anti-aliasing FIR filter design ($H_2(z)$).

11.7 For the sampling conversion from 6 to 8 kHz with the following specifications:
Original sampling rate = 6 kHz
Interpolation factor $L = 4$
Decimation factor $M = 3$
Frequency of interest = 0–2400 Hz
Passband ripple = 0.02 dB
Stopband attenuation = 46 dB,
(a) Draw the block diagram for the processor;
(b) Determine the window type, filter length, and cutoff frequency if the window method is used for the combined FIR filter $H(z)$.

11.8 For the design of a two-stage decimator with the following specifications:
Original sampling rate = 320 kHz
Frequency of interest = 0–3400 Hz

Passband ripple = 0.05 (absolute)
Stopband attenuation = 0.005 (absolute)
Final sampling rate = 8000 Hz

(a) Draw the decimation block diagram;
(b) Specify the sampling rate for each stage;
(c) determine the window type, filter length, and cutoff frequency for the first stage if the window method is used for anti-aliasing FIR filter design ($H_1(z)$);
(d) determine the window type, filter length, and cutoff frequency for the second stage if the window method is used for anti-aliasing FIR filter design ($H_2(z)$).

11.9 **(a)** Given an interpolator filter as

$$H(z) = 0.25 + 0.4z^{-1} + 0.5z^{-2},$$

draw the block diagram for interpolation polyphase filter implementation for the case of $L = 2$.

(b) Given a decimation filter as

$$H(z) = 0.25 + 0.4z^{-1} + 0.5z^{-2} + 0.6z^{-3},$$

draw the block diagram for the decimation polyphase filter implementation for the case of $M = 2$.

11.10 Using the commutative models for the polyphase interpolation and decimation filters,
(a) draw the block diagram for the interpolation polyphase filter implementation for the case of $L = 2$, and $H(z) = 0.25 + 0.4z^{-1} + 0.5z^{-2}$;
(b) draw the block diagram for the decimation polyphase filter implementation for the case of $M = 2$, and $H(z) = 0.25 + 0.4z^{-1} + 0.5z^{-2} + 0.6z^{-3}$.

11.11 **(a)** Given an interpolator filter as

$$H(z) = 0.25 + 0.4z^{-1} + 0.5z^{-2} + 0.6z^{-3} + 0.7z^{-4} + 0.6z^{-5},$$

draw the block diagram for the interpolation polyphase filter implementation for the case of $L = 4$.

(b) Given a decimation filter as

$$H(z) = 0.25 + 0.4z^{-1} + 0.5z^{-2} + 0.6z^{-3} + 0.5z^{-3} + 0.4z^{-4},$$

draw the block diagram for decimation polyphase filter implementation for the case of $M = 4$.

11.12 Using the commutative models for the polyphase interpolation and decimation filters,
(a) Draw the block diagram for interpolation polyphase filter implementation for the case of $L = 4$, and $H(z) = 0.25 + 0.4z^{-1} + 0.5z^{-2} + 0.6z^{-3} + 0.7z^{-4} + 0.6z^{-5}$;
(b) Draw the block diagram for decimation polyphase filter implementation for the case of $M = 4$, and $H(z) = 0.25 + 0.4z^{-1} + 0.5z^{-2} + 0.6z^{-3} + 0.5z^{-3} + 0.4z^{-4}$.

11.13 Given a speech system with the following specifications:
Speech input frequency range: 0–4 kHz.
ADC resolution = 16 bits.
Current sampling rate = 8 kHz,
(a) Determine the oversampling rate if a 12-bit ADC chip is used to replace the speech system;
(b) Draw the block diagram.

11.14 Given a speech system with the following specifications:
Speech input frequency range: 0–4 kHz.
ADC resolution = 6 bits.
Oversampling rate = 4 MHz,
(a) Draw the block diagram;
(b) Determine the actual effective ADC resolution (number of bits per sample).

11.15 Given an audio system with the following specifications:
Audio input frequency range: 0–15 kHz.
ADC resolution = 16 bits.
Current sampling rate = 30 kHz,
(a) Determine the oversampling rate if a 12-bit ADC chip is used to replace the audio system;
(b) Draw the block diagram.

11.16 Given an audio system with the following specifications:
Audio input frequency range: 0–15 kHz.
ADC resolution = 6 bits.
Oversampling rate = 45 MHz,
(a) Draw the block diagram;
(b) Determine the actual effective ADC resolution (number of bits per sample).

11.17 Given the following specifications of an oversampling DSP system:
Audio input frequency range: 0–4 kHz
First-order SDM with a sampling rate of 128 kHz
ADC resolution in SDM = 1 bit,
(a) draw the block diagram using the DSP model;
(b) determine the equivalent (effective) ADC resolution.

11.18 Given the following specifications of an oversampling DSP system:
Audio input frequency range: 0–20 kHz
Second-order SDM with a sampling rate of 160 kHz
ADC resolution in SDM = 10 bits,
(a) Draw the block diagram using the DSP model;
(b) Determine the equivalent (effective) ADC resolution.

11.19 Given the following specifications of an oversampling DSP system:
Signal input frequency range: 0–500 Hz
First-order SDM with a sampling rate of 128 kHz
ADC resolution in SDM = 1 bit,
(a) Draw the block diagram using the DSP model;
(b) Determine the equivalent (effective) ADC resolution.

11.20 Given the following specifications of an oversampling DSP system:
Signal input frequency range: 0–500 Hz
Second-order SDM with a sampling rate of 16 kHz
ADC resolution in SDM = 8 bits,
(a) Draw the block diagram using the DSP model;
(b) Determine the equivalent (effective) ADC resolution.

11.21 Given a bandpass signal with its spectrum shown in Fig. 11.46, and assuming the bandwidth $B = 5$ kHz, select the sampling rate, and sketch the sampled spectrum ranging from 0 Hz to the carrier frequency for each of the following carrier frequencies:

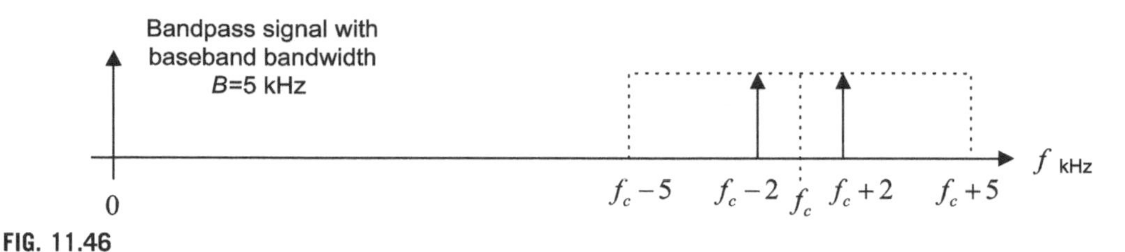

FIG. 11.46

Spectrum of the bandpass signal in Problem 11.21.

(a) $f_c=30$ kHz
(b) $f_c=25$ kHz
(c) $f_c=33$ kHz

11.22 Given a bandpass signal with a spectrum shown in Fig. 11.46, and assuming $f_s=10$ kHz, select the sampling rate and sketch the sampled spectrum ranging from 0 Hz to the carrier frequency for each of the following carrier frequency f_c
(a) $f_c=15$ kHz
(b) $f_c=20$ kHz

11.23 Given a bandpass signal with a spectrum shown in Fig. 11.46, and assuming $B=5$ kHz, select the sampling rate and sketch the sampled spectrum ranging from 0 Hz to the carrier frequency for each of the following carrier frequency f_c
(a) $f_c=35$ kHz,
(b) $f_c=40$ kHz,
(c) $f_c=22$ kHz.

11.8 MATLAB PROBLEMS

Use MATLAB to solve Problems 11.24–11.30.

11.24 Generate a sinusoid with a 1000 Hz for 0.05 s using a sampling rate of 8 kHz,
(a) Design a decimator to change the sampling rate to 4 kHz with specifications below:
Signal frequency range: 0–1800 Hz.
Hamming window required for FIR filter design
(b) Write a MATLAB program to implement the downsampling scheme, and plot the original signal and the downsampled signal versus the sample number, respectively.

11.25 Generate a sinusoid with a 1000 Hz for 0.05 s using a sampling rate of 8 kHz,
(a) Design an interpolator to change the sampling rate to 16 kHz with following specifications:
Signal frequency range: 0–3600 Hz
Hamming window required for FIR filter design
(b) Write a MATLAB program to implement the upsampling scheme, and plot the original signal and the upsampled signal versus the sample number, respectively.

FIG. 11.47

Decimators in Problem 11.31.

11.26 Generate a sinusoid with a frequency of 500 Hz for 0.1 s using a sampling rate of 8 kHz,

(a) Design an interpolation and decimation processing algorithm to change the sampling rate to 22 kHz
Signal frequency range: 0–3400 Hz.
Hamming window required for FIR filter design

(b) Write a MATLAB program to implement the scheme, and plot the original signal and the sampled signal at the rate of 22 kHz versus the sample number, respectively.

11.27 Repeat Problem 11.24 using the polyphase form for the decimator.

11.28 Repeat Problem 11.25 using the polyphase form for the interpolator.

11.29 **(a)** Use MATLAB to create a 1-s sinusoidal signal using the sampling rate of $f_s = 1000$ Hz

$$x(t) = 1.8\cos(2\pi \times 100t) + 1.0\sin(2\pi \times 150t + \pi/4),$$

where each sample $x(t)$ can be round off using 3-bit signed integer (directly round off the calculated $x(t)$) and evaluate the SQNR.

(b) Use MATLAB to design an oversampling system including the anti-aliasing filter with a selectable integer factor L using the same equation for the input $x(t)$.

(c) Recover the signal using the quantized 3-bit signal and measure the SQNRs for the following integer factors: $L=2$, $L=4$, $L=8$, $L=16$, and $L=32$. From the results, explain which one offers better quality for the recovered signals.

11.30 **(a)** Use MATLAB to create a 1-s sinusoidal signal using the sampling rate $f_s = 1000$ Hz

$$x(t) = 1.8\cos(2\pi \times 100t) + 1.0\sin(2\pi \times 150t + \pi/4),$$

where each sample $x(t)$ can be round off using 3-bit signed integer (directly round off the calculated $x(t)$) and evaluate the SQNR.

(b) Use MATLAB to implement the first-order SDM system including anti-aliasing filter with an oversampling factor of 16. Measure the SQNR.

(c) Use MATLAB to implement the second-order SDM system including anti-aliasing filter with an oversampling factor of 16. Measure the SQNR. Compare the SQNR with the one obtained in (b).

11.31 Show that the two decimators are equivalent (Vaidyanathan, 1990, 1993) in Fig. 11.47.

11.32 Show that the following two interpolators are equivalent (Vaidyanathan, 1990, 1993) in Fig. 11.48.

FIG. 11.48

Interpolators in Problem 11.32.

MATLAB PROJECT

Problem 11.33 Audio-Rate Conversion System

Given a 16-bit stereo audio file ("No9seg.wav") with a sampling rate of 44.1 kHz, design a multi-stage conversion system and implement the designed system to convert the audio file from 44.1 to 48 kHz. Listen and compare the quality of the original audio with the converted audio.

CHAPTER

SUBBAND AND WAVELET-BASED CODING 12

CHAPTER OUTLINE

12.1 SUBBAND CODING BASICS

In many applications such as speech and audio analysis, synthesis, and compression, digital filter banks are often used. The filter bank system consists of two stages. The first stage, called the analysis stage, is in the form of filter bank decomposition, in which the signal is filtered into subbands along with a sampling rate decimation; the second stage interpolates the decimated subband signals to reconstruct the original signal. For the purpose of data compression, spectral information from each subband channel can be used to quantize the subband signal efficiently to achieve efficient coding.

Fig. 12.1 illustrates the basic framework for a four-channel filter bank analyzer and synthesizer. At the analysis stage, the input signal $x(n)$ at the original sampling rate f_s is divided via an analysis filter bank into four channels, $x_0(m)$, $x_1(m)$, $x_2(m)$, and $x_3(m)$, each at the decimated sampling rate f_s/M, where $M=4$. For the synthesizer, these four decimated signals are interpolated via a synthesis filter bank. The outputs from all four channels $[\bar{x}_0(n), \bar{x}_1(n), \bar{x}_2(n)$, and $\bar{x}_3(n)]$ of the synthesis filter bank are then combined to reconstruct the original signal $\bar{x}(n)$ at the original sampling rate f_s. Each channel essentially generates a bandpass signal. The decimated signal spectrum for channel 0 can be achieved via a standard downsampling process, while the decimated spectra of other channels can be obtained using the principle of undersampling of bandpass signals with the integer band discussed in Section 11.5, where the inherent frequency aliasing or image properties of decimation and interpolation are involved.

Digital Signal Processing. https://doi.org/10.1016/B978-0-12-815071-9.00012-9

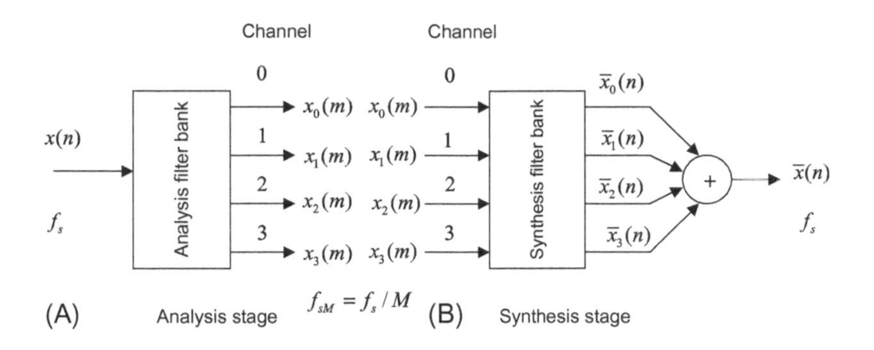

FIG. 12.1

Filter bank framework with an analyzer and a synthesizer.

The theoretical development will follow next. With a proper design of analysis and synthesis filter banks, we are able to achieve perfect reconstruction of the original signal.

Let us examine the spectral details of each band (subband). Fig. 12.2 depicts the spectral information of the analysis and synthesis stages, as shown in Fig. 12.2A and B. $H_0(z)$ and $G_0(z)$ are the analysis and synthesis filters of channel 0, respectively. At the analyzer (C–E), $x(n)$ is bandlimited by a lowpass filter $H_0(z)$ to get $w_0(n)$ and decimated by $M=4$ to obtain $x_0(m)$. At the synthesizer (F–H), $x_0(m)$ is upsampled by a factor of 4 to obtain $\overline{w}_0(n)$ and then goes through an anti-aliasing (synthesis) filter $G_0(z)$ to achieve the lowpass signal $\bar{x}_0(n)$.

Fig. 12.3 depicts the analysis and synthesis stages for channel 1 (see Fig. 12.3A and B). $H_1(z)$ and $G_1(z)$ are the bandpass analysis and synthesis filters, respectively. Similarly, at the analyzer (C–E), $x(n)$

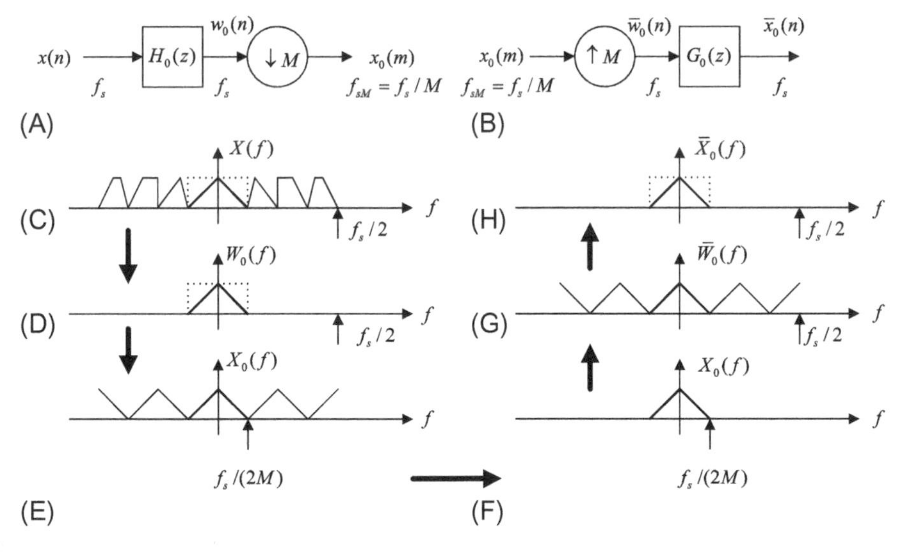

FIG. 12.2

Analysis and synthesis stages for channel 0.

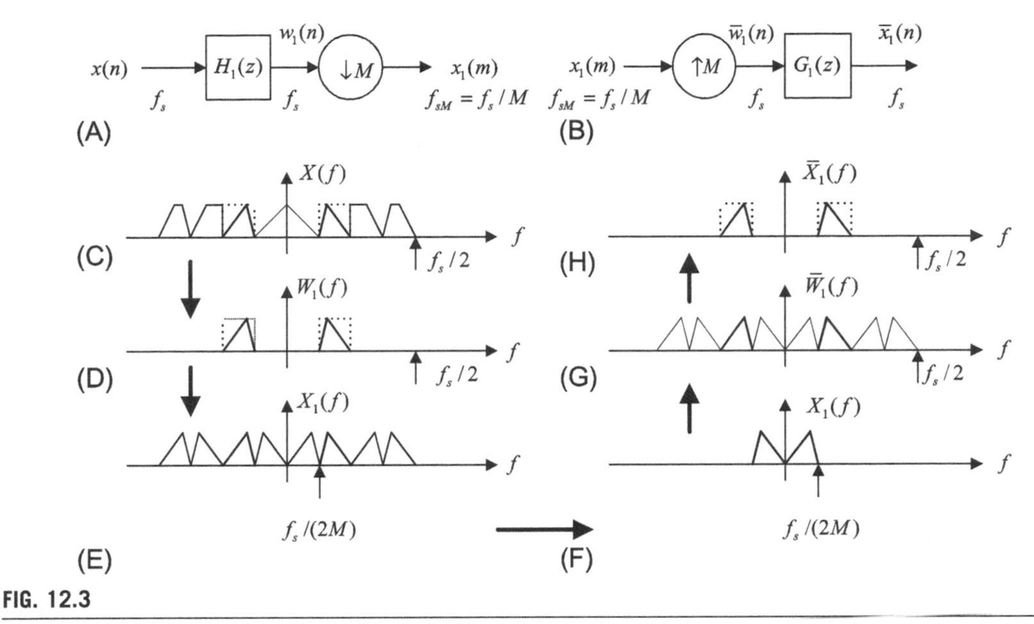

FIG. 12.3

Analysis and synthesis stages for channel 1.

is filtered by a bandpass filter $H_1(z)$ to get $w_1(n)$ and decimated by $M=4$ to obtain $x_1(m)$. Since the low-frequency edge of $W_1(z)$ is $f_c/B=1=$odd number, where $f_c=f_s/(2M)=B, f_c$ corresponds to the carrier frequency and B is the baseband bandwidth as described in Section 11.5, the reversed spectrum in the baseband occurs as shown in Fig. 12.3E. However, this is not a problem, since at the synthesizer as shown in Fig. 12.3 F and G, the spectral reversal occurs again so that $\overline{W}_1(z)$ will have the same spectral components as $W_1(z)$ at the analyzer. After $\overline{w}_1(n)$ goes through the anti-aliasing (synthesis) filter $G_1(z)$, we achieve the reconstructed bandpass signal $\overline{x}_1(n)$.

Fig. 12.4 describes the analysis and synthesis stages for channel 2. At the analyzer (C–E), $x(n)$ is filtered by a bandpass filter $H_2(z)$ to get $w_2(n)$ and decimated by $M=4$ to obtain $x_2(m)$. Similarly, considering the low-frequency edge of $W_2(z)$ as $f_c=2(f_s/(2M))=2B$, $f_c/B=2=$even, we obtain the non-reversed spectrum in the baseband as shown in Fig. 12.4F. At the synthesizer shown in Fig. 12.4G, the spectrum $\overline{W}_2(z)$ has the same spectral components as $W_2(z)$ at the analyzer. After $\overline{w}_2(n)$ is filtered by the synthesis bandpass filter, $G_2(z)$, we get the reconstructed bandpass signal $\overline{x}_2(n)$.

The process in channel 3 is similar to that in channel 1 with the spectral reversal effect and is illustrated in Fig. 12.5.

Now let us examine the theory. Without quantization of subband channels, perfect reconstruction of the filter banks (see Fig. 12.1) depends on the analysis and syntheses filter effects. To develop the perfect reconstruction required for the analysis and synthesis filters, consider a signal in a single channel flowing up to the synthesis filter in general as depicted in Fig. 12.6.

As shown in Fig. 12.6, $w(n)$ is the output signal from the analysis filter at the original sampling rate, that is,

$$W(z)=H(z)X(z). \tag{12.1}$$

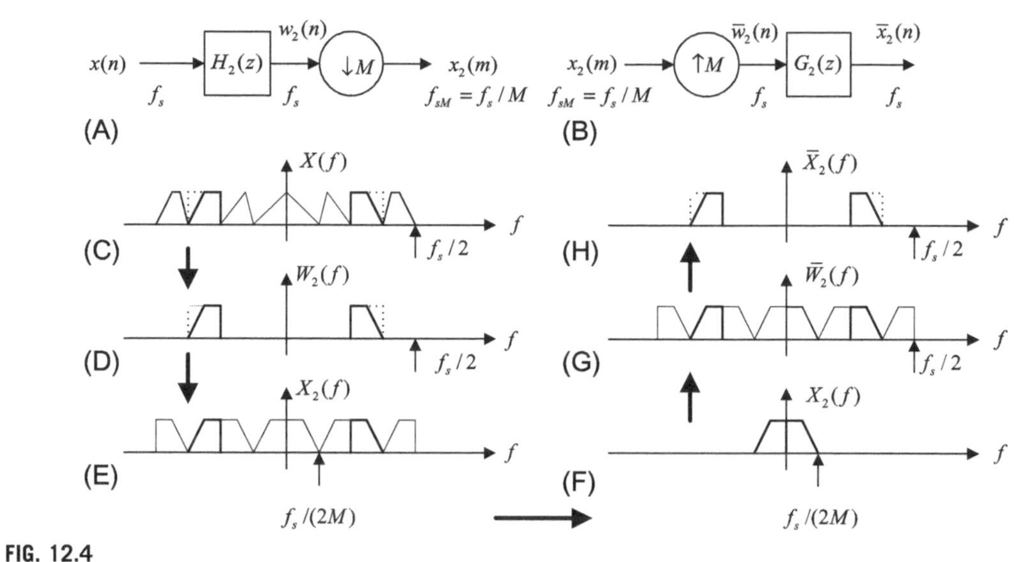

FIG. 12.4

Analysis and synthesis stages for channel 2.

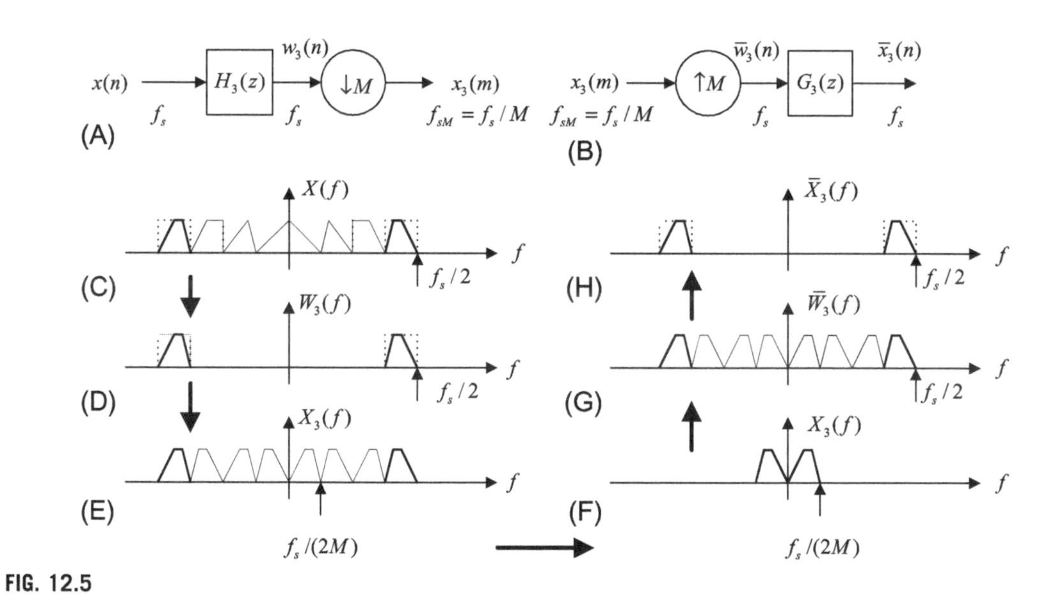

FIG. 12.5

Analysis and synthesis stages for channel 3.

$x_d(m)$ is the downsampled version of $w(n)$ while $\overline{w}(n)$ is the interpolated version of $w(n)$ prior to the synthesis filter and can be expressed as

$$\overline{w}(n) = \begin{cases} w(n) & n = 0, M, 2M, \ldots \\ 0 & \text{otherwise} \end{cases}. \tag{12.2}$$

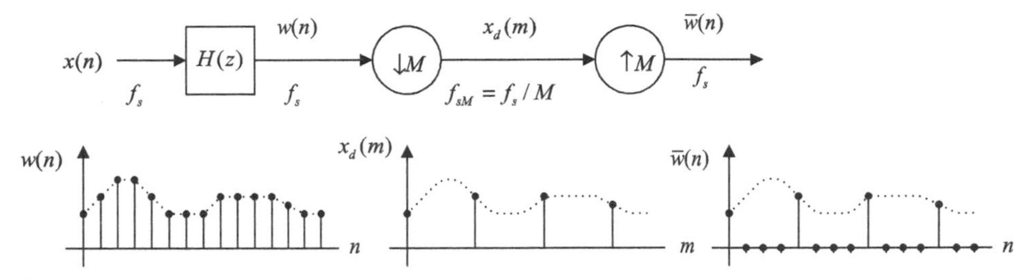

FIG. 12.6

Signal flow in one channel.

Using a delta function $\delta(n)$, that is, $\delta(n)=1$ for $n=0$ and $\delta(n)=0$ for $n\neq 0$, we can write $\overline{w}(n)$ as

$$\overline{w}(n)=\left[\sum_{k=0}^{\infty}\delta(n-kM)\right]w(n)=i(n)w(n). \tag{12.3}$$

where $i(n)$ is defined as

$$i(n)=\sum_{k-0}^{\infty}\delta(n-kM)=\delta(n)+\delta(n-M)+\delta(n-2M)+\cdots.$$

Clearly, $i(n)$ is a periodic function (impulse train with a period of M samples) as shown in Fig. 12.7.

We can determine the discrete Fourier transform of the impulse train with a period of M samples as

$$I(k)=\sum_{n=0}^{M-1}i(n)e^{-j\frac{2\pi kn}{M}}=\sum_{n=0}^{M-1}\delta(n)e^{-j\frac{2\pi kn}{M}}=1. \tag{12.4}$$

Hence, using the inverse of discrete Fourier transform, $i(n)$ can be expressed as

$$i(n)=\frac{1}{M}\sum_{k=0}^{M-1}I(k)e^{j\frac{2\pi kn}{M}}=\frac{1}{M}\sum_{k=0}^{M-1}e^{j\frac{2\pi kn}{M}}. \tag{12.5}$$

Substituting Eq. (12.5) into Eq. (12.3) leads to

$$\overline{w}(n)=\frac{1}{M}\sum_{k=0}^{M-1}w(n)e^{j\frac{2\pi kn}{M}}. \tag{12.6}$$

Applying the z-transform to Eq. (12.6), we achieve the fundamental relationship between $W(z)$ and $\overline{W}(z)$:

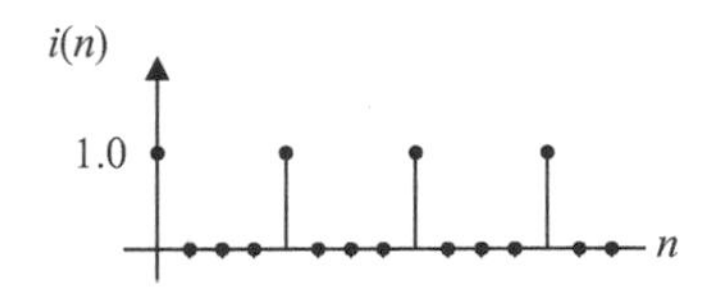

FIG. 12.7

Impulse train.

$$\overline{W}(z)=\frac{1}{M}\sum_{k=0}^{M-1}\sum_{n=0}^{\infty}w(n)e^{j\frac{2\pi kn}{M}}z^{-n}=\frac{1}{M}\sum_{k=0}^{M-1}\sum_{n=0}^{\infty}w(n)\left(e^{-j\frac{2\pi k}{M}}z\right)^{-n}$$
$$=\frac{1}{M}\sum_{k=0}^{M-1}W\left(e^{-j\frac{2\pi k}{M}}z\right) \quad (12.7)$$
$$=\frac{1}{M}\left[W\left(e^{-j\frac{2\pi\times 0}{M}}z\right)+W\left(e^{-j\frac{2\pi\times 1}{M}}z\right)+\cdots+W\left(e^{-j\frac{2\pi\times(M-1)}{M}}z\right)\right].$$

Eq. (12.7) indicates that the signal spectrum $\overline{W}(z)$ before the synthesis filter is an average of the various modulated spectrum $W(z)$. Note that both $\overline{W}(z)$ and $W(z)$ are at the original sampling rate f_s. We will use this result for further development in the following section.

12.2 SUBBAND DECOMPOSITION AND TWO-CHANNEL PERFECT RECONSTRUCTION-QUADRATURE MIRROR FILTER BANK

To explore Eq. (12.7), let us begin with a two-band case as illustrated in Fig. 12.8.

Substituting $M=2$ in Eq. (12.7), it follows that

$$\overline{W}(z)=\frac{1}{2}\sum_{k=0}^{1}W\left(e^{-j\frac{2\pi k}{2}}z\right)=\frac{1}{2}[W(z)+W(-z)]. \quad (12.8)$$

Applying for each band in Fig. 12.8 by substituting Eq. (12.1) into Eq. (12.8), we have

$$Y_0(z)=\frac{1}{2}G_0(z)(H_0(z)X(z)+H_0(-z)X(-z)). \quad (12.9)$$

$$Y_1(z)=\frac{1}{2}G_1(z)(H_1(z)X(z)+H_1(-z)X(-z)). \quad (12.10)$$

Since the synthesized signal $\overline{X}(z)$ is a sum of $Y_0(z)$ and $Y_1(z)$, it can be expressed as

$$\overline{X}(z)=\frac{1}{2}(G_0(z)H_0(z)+G_1(z)H_1(z))X(z)$$
$$+\frac{1}{2}(G_0(z)H_0(-z)+G_1(z)H_1(-z))X(-z) \quad (12.11)$$
$$=A(z)X(z)+S(z)X(-z).$$

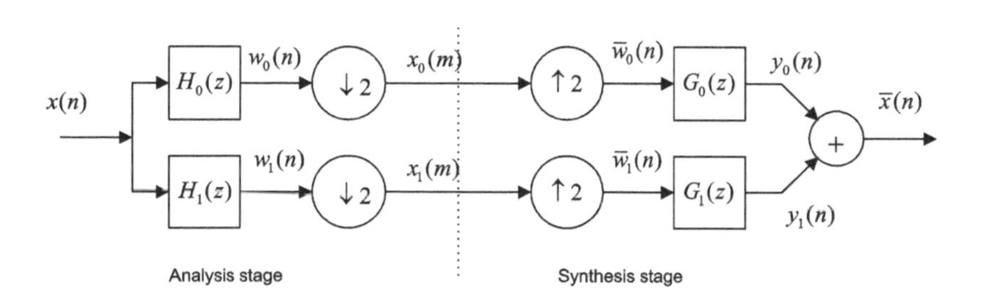

FIG. 12.8

Two-band filter bank system.

For perfect reconstruction, the recovered signal $\overline{x}(n)$ should be a scaled and delayed version of the original signal $x(n)$, that is, $\overline{x}(n)=cx(n-n_0)$. Hence, to achieve a perfect reconstruction, it is required that

$$S(z)=\frac{1}{2}(G_0(z)H_0(-z)+G_1(z)H_1(-z))=0 \tag{12.12}$$

$$A(z)=\frac{1}{2}(G_0(z)H_0(z)+G_1(z)H_1(z))=cz^{-n_0}, \tag{12.13}$$

where c is the constant while n_0 is the delay introduced by the analysis and synthesis filters.

Forcing $S(z)=0$ leads to the following relationship:

$$\frac{G_0(z)}{G_1(z)}=-\frac{H_1(-z)}{H_0(-z)}. \tag{12.14}$$

It follows that

$$G_0(z)=-H_1(-z) \tag{12.15}$$

$$G_1(z)=H_0(-z). \tag{12.16}$$

Substituting $G_0(z)$ and $G_1(z)$ into Eq. (12.13) gives

$$A(z)=\frac{1}{2}(H_0(-z)H_1(z)-H_0(z)H_1(-z)). \tag{12.17}$$

Assume $H_0(z)$ and $H_1(z)$ be N-tap FIR filters, where N is even, and let

$$H_1(z)=z^{-(N-1)}H_0(-z^{-1}) \tag{12.18}$$

and note that

$$H_1(-z)=-z^{-(N-1)}H_0(z^{-1}). \tag{12.19}$$

Substituting Eqs. (12.18) and (12.19) in Eq. (12.17), we can simplify Eq. (12.17) as

$$A(z)=\frac{1}{2}z^{-(N-1)}(H_0(z)H_0(z^{-1})+H_0(-z)H_0(-z^{-1})). \tag{12.20}$$

Finally, for perfect reconstruction, it requires that

$$H_0(z)H_0(z^{-1})+H_0(-z)H_0(-z^{-1})=R(z)+R(-z)=\text{constant}, \tag{12.21}$$

where

$$R(z)=H_0(z)H_0(z^{-1})=a_{N-1}z^{N-1}+a_{N-2}z^{N-2}+\cdots+a_0z^0+\cdots+a_{N-1}z^{-(N-1)} \tag{12.22}$$

$$R(-z)=H_0(-z)H_0(-z^{-1})=-a_{N-1}z^{N-1}+a_{N-2}z^{N-2}+\cdots+a_0z^0+\cdots-a_{N-1}z^{-(N-1)}. \tag{12.23}$$

It is important to note that the sum of $R(z)+R(-z)$ only consists of even order of powers of z, since the terms with odd powers of z cancel each other. Using algebraic simplification, we conclude that the coefficients of $R(z)=H(z)H(z^{-1})$ are essentially samples of the autocorrelation function given by

$$\rho(n)=\sum_{k=0}^{N-1}h_0(k)h_0(k+n)=\rho(-n)=h_0(n)\odot h_0(n), \tag{12.24}$$

where $\odot$ denotes the correlation operation. Hence, we require $\rho(n)=0$ for $n=$even and $n\neq 0$, that is,

$$\rho(2n)=\sum_{k=0}^{N-1} h_0(k)h_0(k+2n)=0. \tag{12.25}$$

For normalization, when $n=0$, we require

$$\sum_{k=0}^{N-1} |h_0(k)|^2=0.5. \tag{12.26}$$

We then obtain the filter design constraint as

$$\rho(2n)=\sum_{k=0}^{N-1} h_0(k)h_0(k+2n)=\delta(n). \tag{12.27}$$

For a two-band filter bank, $h_0(k)$ and $h_1(k)$ are designed as lowpass and highpass filters, respectively, which are essentially the quadrature mirror filters (QMF). Their expected frequency responses must satisfy Eq. (12.28) and are shown in Fig. 12.9.

$$\left|H_0\left(e^{j\Omega}\right)\right|^2+\left|H_1\left(e^{j\Omega}\right)\right|^2=1. \tag{12.28}$$

Eq. (12.28) implies that

$$R(z)+R(-z)=1. \tag{12.29}$$

To verify Eq. (12.29), we use

$$\left|H\left(e^{j\Omega}\right)\right|^2=H\left(e^{j\Omega}\right)H\left(e^{-j\Omega}\right)=H(z)H\left(z^{-1}\right)\big|_{z=e^{j\Omega}}.$$

Eq. (12.28) becomes

$$H_0(z)H_0\left(z^{-1}\right)+H_1(z)H_1\left(z^{-1}\right)\big|_{z=e^{j\Omega}}=1,$$

which is equivalent to

$$H_0(z)H_0\left(z^{-1}\right)+H_1(z)H_1\left(z^{-1}\right)=1.$$

$$\left|H_0(e^{j\Omega})\right|^2+\left|H_1(e^{j\Omega})\right|^2=1.$$

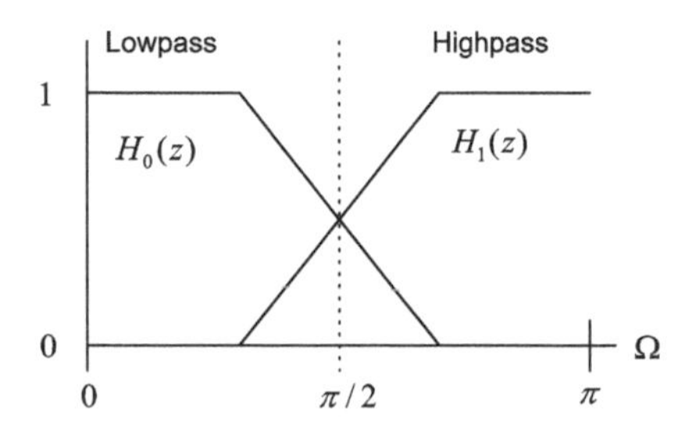

FIG. 12.9

Frequency responses for quadrature mirror filters.

From Eq. (12.18), we can verify that

$$H_1(z)H_1(z^{-1}) = H_0(-z)H_0(-z^{-1}).$$

Finally, we see that

$$H_0(z)H_0(z^{-1}) + H_1(z)H_1(z^{-1}) = H_0(z)H_0(z^{-1}) + H_0(-z)H_0(-z^{-1}) = R(z) + R(-z) = 1.$$

Once the lowpass analysis filter $H_0(z)$ is designed, the highpass filter can be obtained using the developed relationship in Eq. (12.18). The key equations are summarized as follows:

Filter design constraint equations for the lowpass filter $H_0(z)$:

$$R(z) = H_0(z)H_0(z^{-1})$$
$$R(z) + R(-z) = 1$$
$$\rho(2n) = 0.5\delta(n).$$

Equations for the other filters:

$$H_1(z) = z^{-(N-1)}H_0(-z^{-1})$$
$$G_0(z) = -H_1(-z)$$
$$G_1(z) = H_0(-z).$$

Design of the analysis and synthesis filters to satisfy the above conditions is very challenging. Smith and Barnwell (1984) were first to show that perfect reconstruction in a two-band filter bank is possible when the linear phase of the FIR filter requirement is relaxed. The Smith-Barnwell filters are called the conjugate quadrature filters (PR-CQF). 8- and 16-tap PR-CQF coefficients are listed in Table 12.1. As illustrated in Table 12.1, the filter coefficients are not symmetric; hence, the obtained analysis filter

Table 12.1 Smith-Barnwell PR-CQF Filters

8 Taps	16 Taps
0.0348975582178515	0.02193598203004352
−0.01098301946252854	0.001578616497663704
−0.06286453934951963	−0.06025449102875281
0.223907720892568	−0.0118906596205391
0.556856993531445	0.137537915636625
0.357976304997285	0.05745450056390939
−0.02390027056113145	−0.321670296165893
−0.07594096379188282	−0.528720271545339
	−0.295779674500919
	0.0002043110845170894
	0.0290669978946796
	−0.03533486088708146
	−0.006821045322743358
	0.02606678468264118
	0.001033363491944126
	−0.01435930957477529

does not have a linear phase. The detailed design of Smith-Barnwell filters can be found in their research paper (Smith and Barnwell, 1984) and the design of other types of analysis and synthesis filters can be found in Akansu and Haddad (2001).

Now let us verify the filter constraint in the following example.

EXAMPLE 12.1

Use the 8-tap PR-CQF coefficients (Table 12.1) and MATLAB to verify the following conditions:

$$\rho(2n) = \sum_{k=0}^{N-1} h_0(k)h_0(k+2n) = 0.5\delta(n)$$

$$R(z) + R(-z) = 1,$$

and plot the magnitude frequency responses of the analysis and synthesis filters.

Solution:

Since $\rho(n) = \sum_{k=0}^{N-1} h_0(k)h_0(k+n)$, we perform the following:

$$\text{For } n=0,\ \rho(0) = \sum_{k=0}^{8-1} h_0(k)h_0(k) = h_0^2(0) + h_0^2(1) + \cdots + h_0^2(7) = 0.5$$

$$\text{For } n=1,\ \rho(1) = \sum_{k=0}^{8-1} h_0(k)h_0(k+1) = h_0(0)h_0(1) + h_0(1)h_0(2) + \cdots + h_0(6)h_0(7) = 0.3035$$

$$\text{For } n=-1,\ \rho(-1) = \sum_{k=0}^{8-1} h_0(k)h_0(k-1) = h_0(1)h_0(0) + h_0(2)h_0(1) + \cdots + h_0(7)h_0(6) = 0.3035$$

$$\text{For } n=2,\ \rho(2) = \sum_{k=0}^{8-1} h_0(k)h_0(k+2) = h_0(0)h_0(2) + h_0(1)h_0(3) + \cdots + h_0(5)h_0(7) = 0.0$$

$$\text{For } n=-2,\ \rho(-2) = \sum_{k=0}^{8-1} h_0(k)h_0(k-2) = h_0(2)h_0(0) + h_0(3)h_0(1) + \cdots + h_0(7)h_0(5) = 0.0.$$

....

We can easily verify that $\rho(n)=0$ for $n \neq 0$ and $n=$even number.

Next, we use the MATLAB built-in function **xcorr()** to compute the autocorrelation coefficients. The results are listed as follows:

```
>>h0=[0.0348975582178515 -0.01098301946252854 -0.06286453934951963 ...
  0.223907720892568 0.556856993531445 0.357976304997285 ...
  -0.02390027056113145 -0.07594096379188282];
>>p=xcorr(h0,h0)
p = -0.0027 -0.0000 0.0175 0.0000 -0.0684 -0.0000 0.3035 0.5000
  0.3035 -0.0000  -0.0684 0.0000 0.0175 -0.0000 -0.0027
```

We can observe that there are 15 coefficients. The middle one is $\rho(0)=0.5$ and. $\rho(\pm 2)=\rho(\pm 4)=\rho(\pm 6)=0$ as well as $\rho(\pm 1)=0.3035$, $\rho(\pm 3)=-0.0684$, $\rho(\pm 5)=0.0175$, and $\rho(\pm 7)=-0.0027$.

Next, we write.

$$R(z) = -0.0027z^{7} - 0.0000z^{6} + 0.0175z^{5} + 0.0000z^{4} - 0.0684z^{3} - 0.0000z^{2} + 0.3035z^{1} + 0.5000z^{0} + 0.3035z^{-1} - 0.0000z^{-2} - 0.0684z^{-3} + 0.0000z^{-4} + 0.0175z^{-5} - 0.0000z^{-6} - 0.0027z^{-7}.$$

Substituting $z = -z$ in $R(z)$, it yields:

$$R(-z) = 0.0027z^{7} - 0.0000z^{6} - 0.0175z^{5} + 0.0000z^{4} + 0.0684z^{3} - 0.0000z^{2} - 0.3035z^{1} + 0.5000z^{0} - 0.3035z^{-1} - 0.0000z^{-2} + 0.0684z^{-3} + 0.0000z^{-4} - 0.0175z^{-5} - 0.0000z^{-6} + 0.0027z^{-7}.$$

Clearly, by adding the expressions $R(z)$ and $R(-z)$, we can verify that

$$R(z) + R(-z) = 1.$$

Using MATLAB, the PR-CQF frequency responses are plotted and shown in Fig. 12.10.

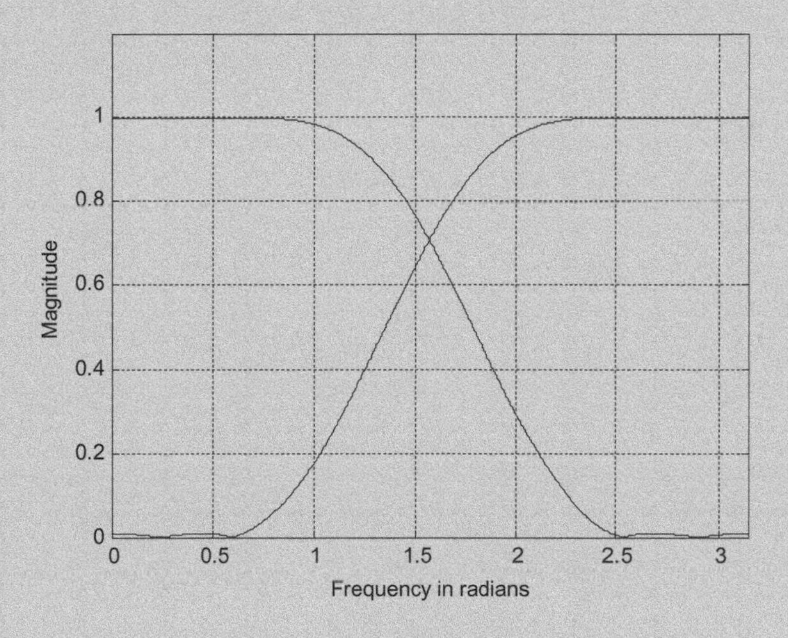

FIG. 12.10

Magnitude frequency responses of the analysis and synthesis filters in Example 12.1.

Fig. 12.11 shows the perfect reconstruction for performing the two-band system shown in Fig. 12.8 using two-band CQF filters. The MATLAB program is listed in Program 12.1, in which the quantization is deactivated. Since the obtained signal-to-noise ratio (SNR) = 135.5803 dB, a perfect reconstruction is achieved. Note that both $x_0(m)$ and $x_1(m)$ have half of the data samples, where $x_0(m)$ contains low-frequency components with more signal energy while $x_1(m)$ possesses high-frequency components with less signal energy.

MATLAB Program 12.1. Two-band subband system implementation.

```
% This program is for implementing the analysis and synthesis using two subbands.
close all; clear all;clc
%Smith-Barnwell PR-CQF 8-taps
h0=[0.0348975582178515-0.01098301946252854 -0.06286453934951963 ...
0.223907720892568 0.556856993531445 0.357976304997285 ...
-0.02390027056113145 -0.07594096379188282];
% Read data file 'orig.dat' with a sampling rate of 8 kHz
load orig.dat; % Load speech data
M=2; % Downsample factor
N=length(h0); PNones=ones(1,N); PNones(2:2:N)=-1;
h1=h0.*PNones; h1=h1(N:-1:1);
g0=-h1.*PNones; g1=h0.*PNones;
disp('check R(z)+R(-z)≥');
xcorr(h0,h0)
sum(h0.*h0)
w=0:pi/1000:pi;
fh0=freqz(h0,1,w); fh1=freqz(h1,1,w);
plot(w,abs(fh0),'k',w,abs(fh1),'k');grid; axis([0 pi 0 1.2]);
xlabel('Frequency in radians');ylabel('Magnitude)')
figure
speech=orig;
%Analysis
sb_low=filter(h0,1,speech); sb_high=filter(h1,1,speech);
% Downsampling
sb_low=sb_low(1:M:length(sb_low)); sb_high=sb_high(1:M:length(sb_high));
% Quantization
sb_low=round((sb_low/2^15)*2^9)*2^(15-9); %Quantization with 10 bits
sb_high=round((sb_high/2^15)*2^5)*2^(15-5); % Quantization with 6 bits
% Syntheiss
low_sp=zeros(1,M*length(sb_low)); % Upsampling
low_sp(1:M:length(low_sp))=sb_low;
high_sp=zeros(1,M*length(sb_high)); high_sp(1:M:length(high_sp))=sb_high;
low_sp=filter(g0,1,low_sp); high_sp=filter(g1,1,high_sp);
rec_sig=2*(low_sp+high_sp);
% Signal alignment for SNR caculations
speech=[zeros(1,N-1) speech]; %Align the signal
subplot(4,1,1);plot(speech);grid,ylabel('x(n)');axis([0 20,000-20,000 20,000]);
subplot(4,1,2);plot(sb_low);grid,ylabel('0(m)'); axis([0 10,000-20,000 20,000]);
subplot(4,1,3);plot(sb_high);grid, ylabel('1(m)'); axis([0 10,000-2000 2000]);
subplot(4,1,4);plot(rec_sig);grid, ylabel('xbar(n)'),xlabel('Sample Number');
axis([0 20,000-20,000 20,000]);
NN=min(length(speech),length(rec_sig));
err=rec_sig(1:NN)-speech(1:NN);
SNR=sum(speech.*speech)/sum(err.*err);
disp('PR reconstruction SNR dB≥');
SNR=10*log10(SNR)
```

This two-band composition method can easily be extended to a multiband filter bank using a binary tree structure. Fig. 12.12 describes a four-band implementation. As shown in Fig. 12.12, the filter banks divide an input signal into two equal subbands, resulting in the low (L) and high (H) bands using PR-QMF. This two-band PR-QMF again splits L and H into half bands to produce quarter bands:

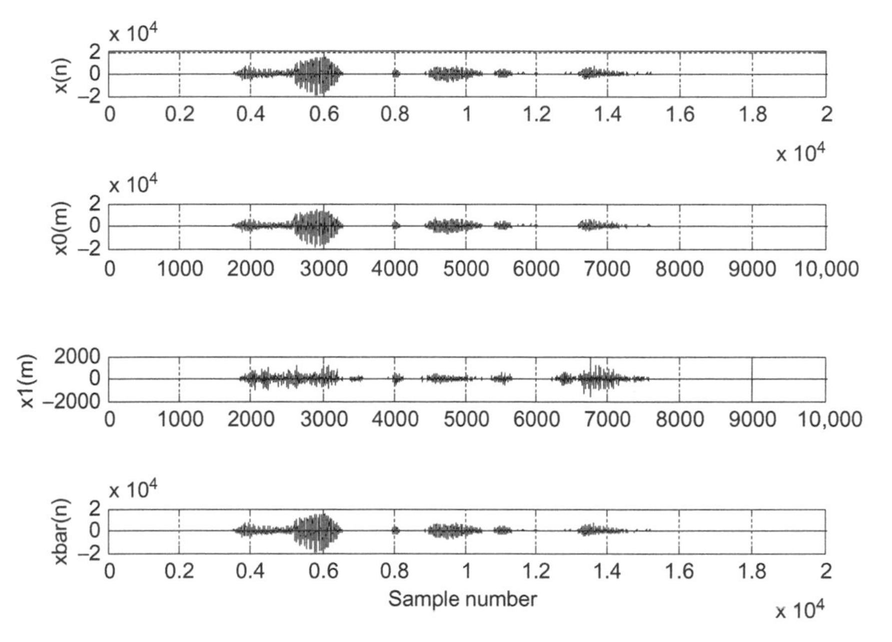

FIG. 12.11

Two-band analysis and synthesis.

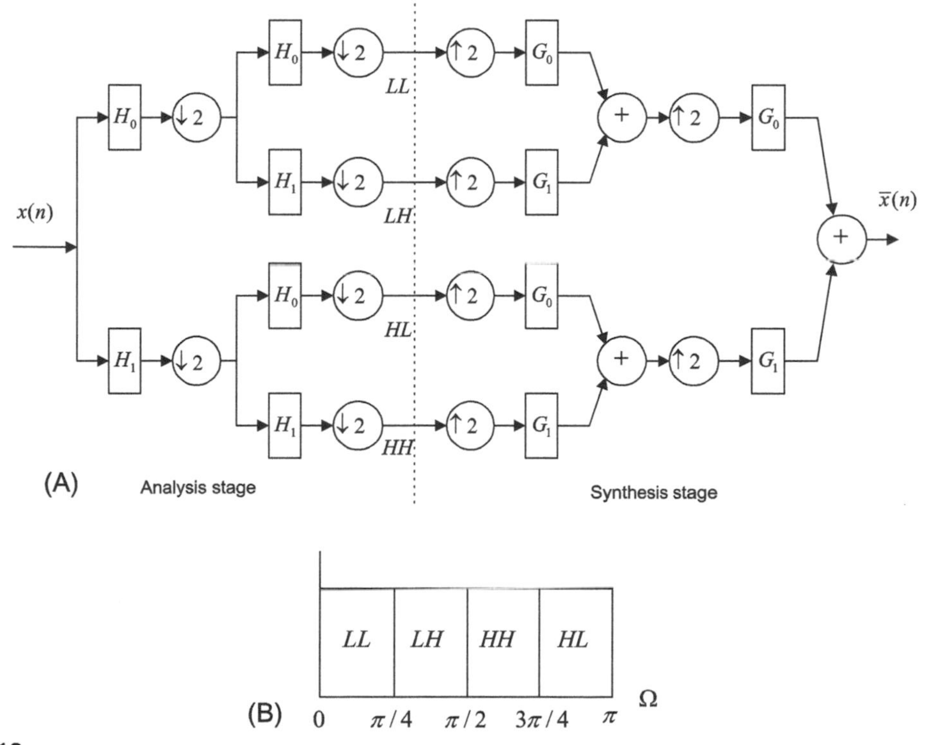

FIG. 12.12

Four-band implementation based on binary tree structure.

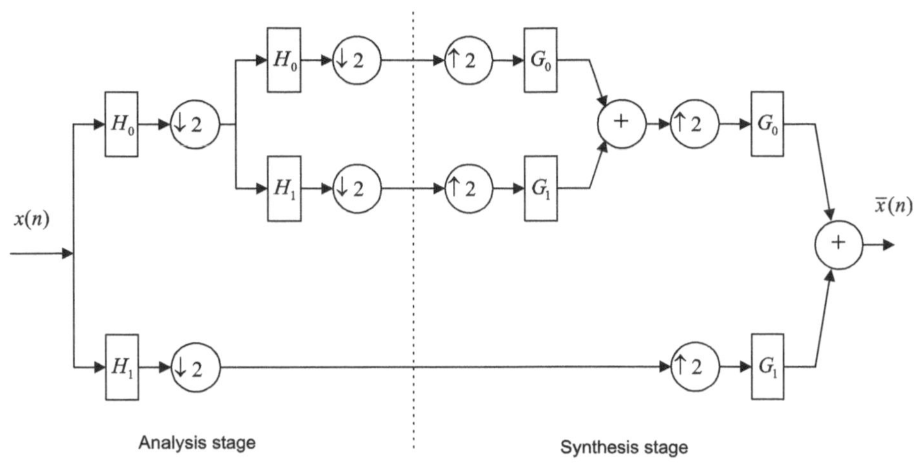

FIG. 12.13

Four-band implementation based on the dyadic tree structure.

LL, LH, HL, and HH. The four-band spectrum is labeled in Fig. 12.12B. Note that the HH band is actually centered in $[\pi/2, 3\pi/4]$ instead of $[3\pi/4, \pi]$.

In applications of signal coding, a dyadic subband tree structure is often used, as shown in Fig. 12.13, where the PR-QMF bank splits only the lower half of the spectrum into two equal bands at any level. Through continuation of splitting, we can achieve a coarser-and-coarser version of the original signal.

12.3 SUBBAND CODING OF SIGNALS

Subband analysis and synthesis can be successfully applied to signal coding, shown in Fig. 12.14 as an example for a two-band case. The analytical signals from each channel are filtered by the analysis filter, downsampled by a factor of 2, and quantized using quantizers Q_0 and Q_1 each with its assigned number of bits. The quantized codes are multiplexed for transmission or storage. At the synthesis stage, the

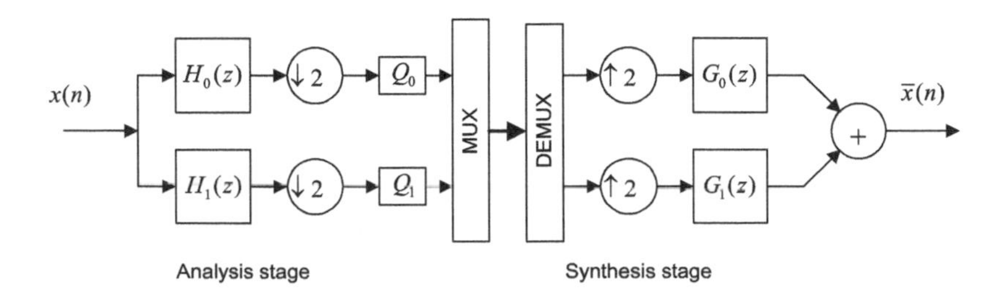

FIG. 12.14

Two-band filter bank system used for signal compression.

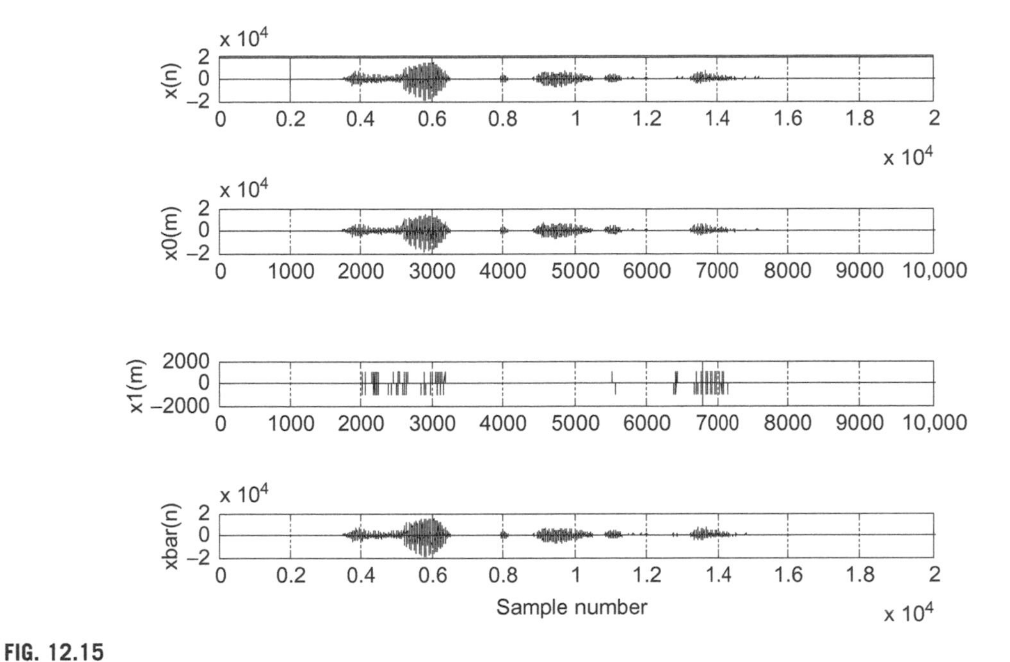

FIG. 12.15

Two-subband compression for speech data.

received or recovered quantized signals are demultiplexed, upsampled by a factor of 2, and processed by the synthesis filters. Then the output signals from all the channels are added to reconstruct the original signal. Since the signal from each analytical channel is quantized, the resultant scheme is a lossy compression one. The coding quality can be measured using the SNR.

Fig. 12.15 shows the speech coding results using the subband coding system (two-band) with the following specifications:

Sampling rate is 8 ksps (kilosamples per second)

Sample size = 16 bits/sample

Original data rate = 8 kHz × 16 bits = 128 kbps (kilobits per second)

We assign 10 bits for Q_0 (since the low-band signal contains more energy) and 6 bits for Q_1. We obtain a new data rate as (10+6) bits × 8 ksps/2 = 64 kbps. The implementation is shown in MATLAB Program 12.1 with the activated quantizers. Note that $x_0(m)$ and $x_1(m)$ shown in Fig. 12.15 are the quantized versions using Q_0 and Q_1. The measured SNR is 24.51 dB.

Fig. 12.16 shows the results using a four-band system. We designate both Q_0 and Q_1 as 11 bits, Q_3 as 10 bits and Q_2 as 0 bits (discarded). Note that the HL band contains the highest frequency components with the lowest signal energy level (see Fig. 12.12B). Hence, we discard HL band information to increase the coding efficiency. Therefore, we obtain the data rate as (11+11+10+0) bits × 8 ksps/4 = 64 kbps. The measured SNR is 27.06 dB. Four-band system offers possibility of signal quality improvement over the two-band system. Plots for the original speech, reconstructed speech, and quantized signal version for each subband are displayed in Fig. 12.17.

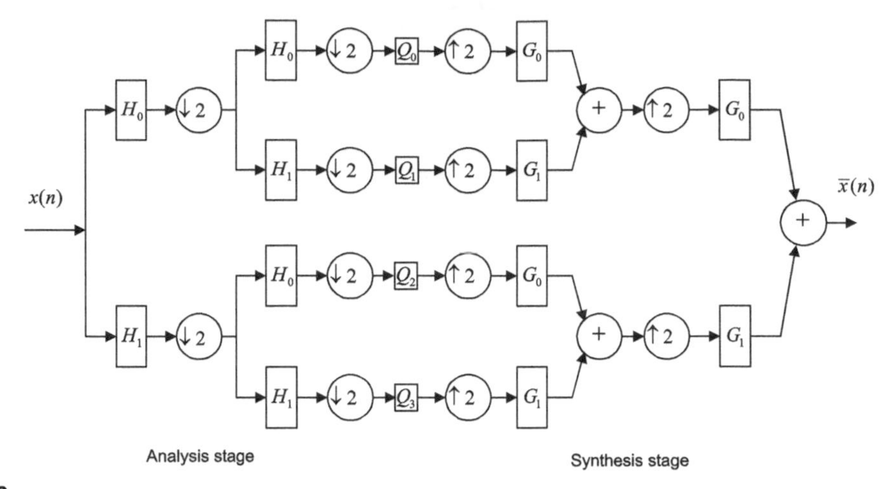

FIG. 12.16

Four-subband compression for speech data.

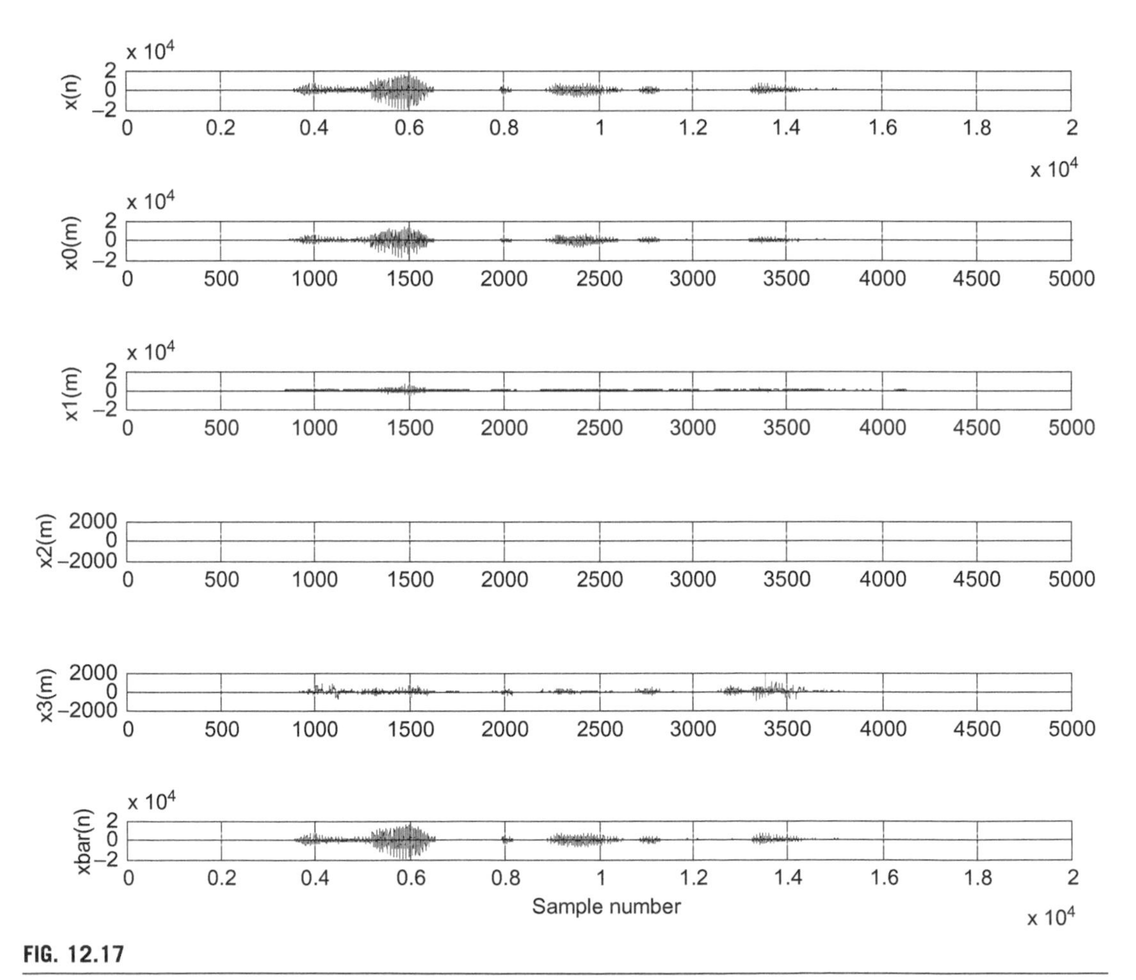

FIG. 12.17

Four-subband compression for 16-bit speech data and SNR=27.5dB.

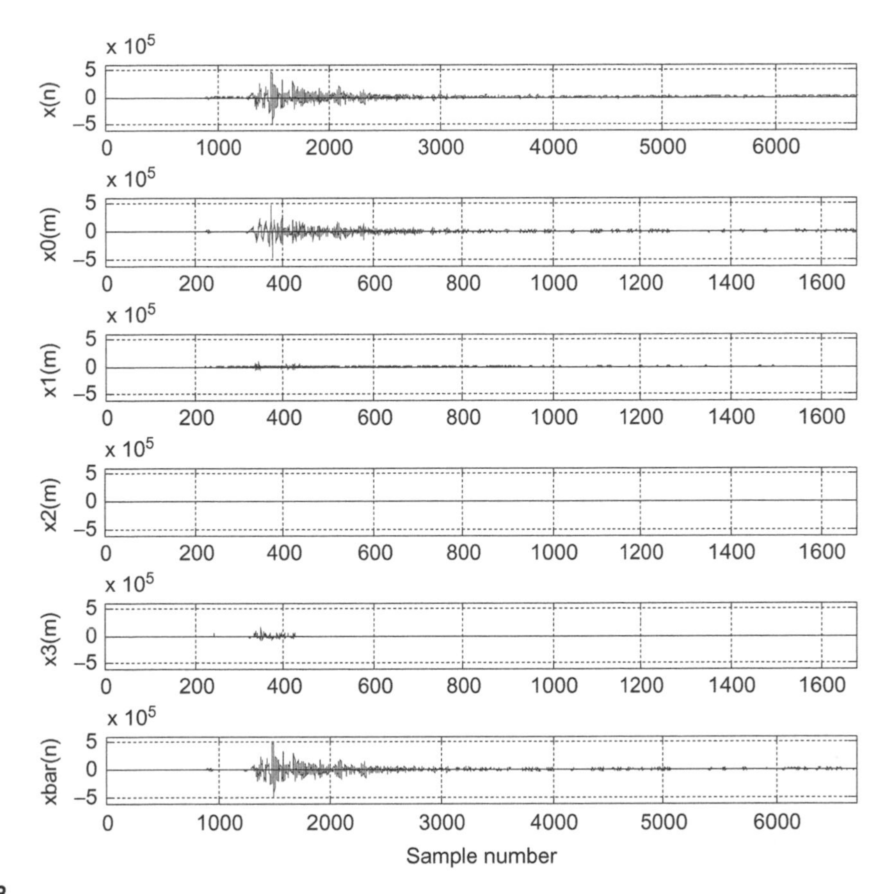

FIG. 12.18

Four-subband compression for 32-bit seismic data and SNR = 36 dB.

Fig. 12.18 shows the results using a four-band system (Fig. 12.16) for encoding seismic data. The seismic signal (provided by the USGS Albuquerque Seismological Laboratory) has a sampling rate of 15 Hz with 6700 data samples, and each sample is encoded in 32 bits. For a four-band system, the bit allocations for all bands are listed below: Q_0 (LL) = 22 bits, Q_1 (LH) = 22 bits, Q_3 (HL) = 0 (discarded), and Q_4 (HH) = 20 bits. We achieve a compression ratio of 2:1 with SNR = 36.00 dB.

12.4 WAVELET BASICS AND FAMILIES OF WAVELETS

Wavelet transform has become a powerful tool for signal processing. It offers time-frequency analysis to decompose the signal in terms of a family of wavelets or a set of basic functions, which have a fixed shape but can be shifted and dilated in time. The wavelet transform can present a signal with a good time resolution or a good frequency resolution. There are two types of wavelet transforms: the continuous wavelet transform (CWT) and the discrete wavelet transform (DWT). Specifically, the DWT provides an efficient tool for signal coding. It operates on discrete samples of the signal and has a relation

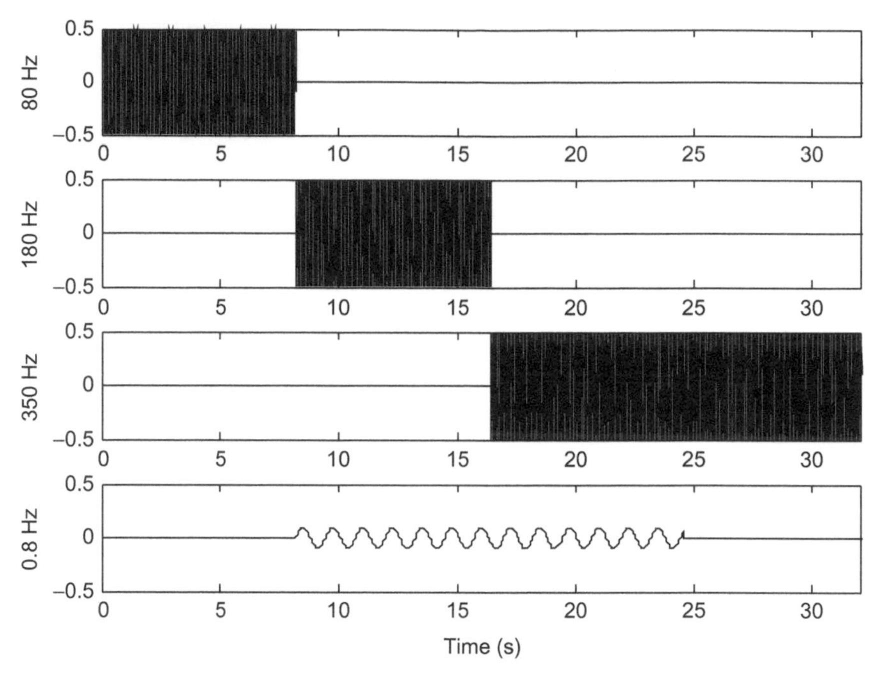

FIG. 12.19

Individual signal components.

with the dyadic subband coding as described in Section 12.2. The DWT resembles other discrete transform, such as the DFT or the DCT. In this section, without going too in-depth with mathematics, we review basics of CWT, which will lay out the foundation. Next, we emphasize the DWT for applications of signal coding.

Let us examine a signal sampled at 1 kHz with 1024×32 (32678) samples given by

$$\begin{aligned} x(t) = {} & 0.5\cos(2\pi \times 80t)[u(t) - u(t-8)] + \sin(2\pi \times 180t)[u(t-8) - u(t-16)] \\ & + \sin(2\pi \times 250t)[u(t-16) - u(t-32)] + 0.1\sin(2\pi \times 0.8t)[u(t-8) - u(t-24)]. \end{aligned} \tag{12.30}$$

The signal contains four sinusoids: 80 Hz for $0 \le t < 8$ s, 180 Hz for $8 \le t < 16$ s, 350 Hz for $16 \le t \le 32$ s, and finally 0.8 Hz for $8 \le t \le 24$ s. All the signals are plotted separately in Fig. 12.19 while Fig. 12.20 shows the combined signal and its DFT spectrum.

Based on the traditional spectral analysis shown in Fig. 12.20, we can identify the frequency components of 80, 180, and 350 Hz. However, the 0.8-Hz component and transient behaviors such as the start and stop time instants of the sinusoids (discontinuity) cannot be observed from the spectrum. Fig. 12.21 depicts the wavelet transform of the same signal. The horizontal axis is time in seconds while the vertical axis is index j, which is inversely proportional to the scale factor ($a = 2^{-j}$). As will be discussed, the larger the scale factor (the smaller index j), the smaller the frequency value. The amplitudes of the wavelet transform are displayed according to the intensity. The brighter the intensity, the larger the amplitude. The areas with brighter intensities indicate the strongest resonances between the signal and wavelets of various frequency scales and time shifts. In Fig. 12.21, four different frequency

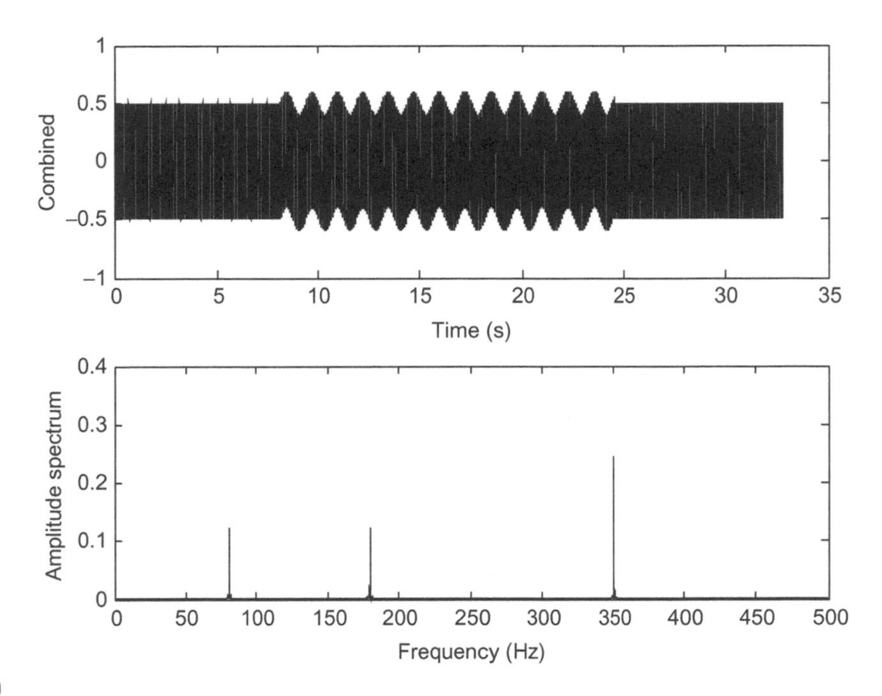

FIG. 12.20

Combined signal and its spectrum.

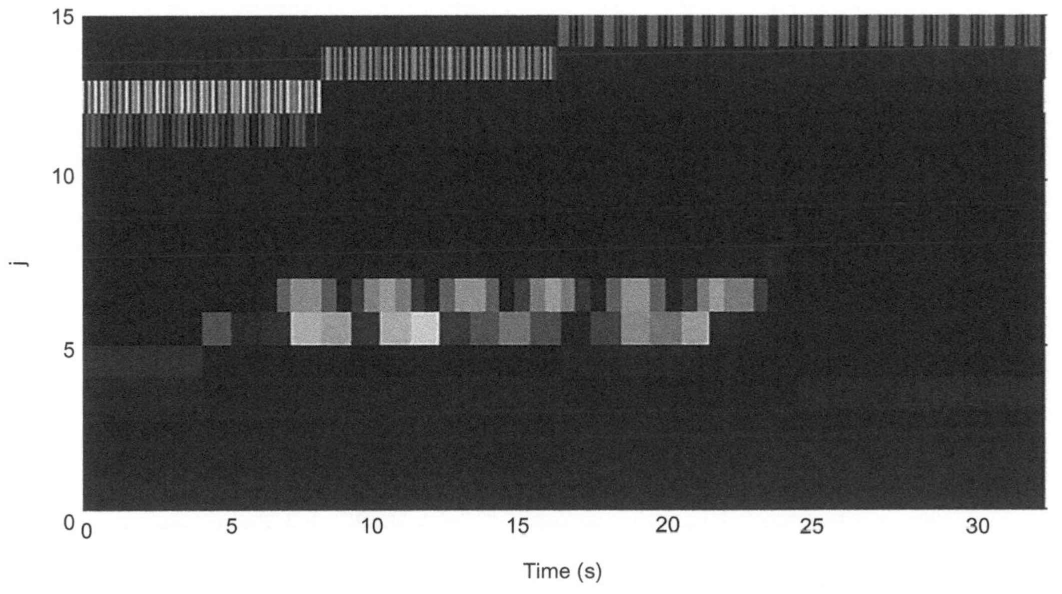

FIG. 12.21

Wavelet transform amplitudes.

components and the discontinuities of the sinusoids are displayed as well. We can further observe the fact that the finer the frequency resolution, the coarser the time resolution. For example, we can clearly identify the start and stop times for 80-, 180-, and 350-Hz frequency components, but frequency resolution is coarse, since index j has larger frequency spacing. However, for the 0.8-Hz sinusoid, we have fine frequency resolution (small frequency spacing so we can see the 0.8-Hz sinusoid) and coarse time resolution as evidence in which the start and stop times are blurred.

The CWT is defined as

$$W(a,b) = \int_{-\infty}^{\infty} f(t)\psi_{ab}(t)dt, \tag{12.31}$$

where $W(a,b)$ is the wavelet transform and $\psi_{ab}(t)$ is called the mother wavelet, which is defined as:

$$\psi_{ab}(t) = \frac{1}{\sqrt{a}}\psi\left(\frac{t-b}{a}\right). \tag{12.32}$$

The wavelet function consists of two important parameters: scaling a and translation b. A scaled version of a function $\psi(t)$ with a scale factor of a is defined as $\psi(t/a)$. Consider a basic function $\psi(t)=\sin(\omega t)$ when $a=1$. When $a>1$, $\psi(t)=\sin(\omega t/a)$ is the scaled function with a frequency less than ω rad/s. When $a<1$, $\psi(t)=\sin(\omega t/a)$ has a frequency larger than ω. Fig. 12.22 shows the scaled wavelet functions.

A translated version of a function $\psi(t)$ with a shifted time constant b is defined as $\psi(t-b)$.

Fig. 12.23 shows several translated versions of the wavelet. A scaled and translated function $\psi(t)$ is given by $\psi((t-b)/a)$. This means that $\psi((t-b)/a)$ changes frequency and time shift. Several combined scaling and translated wavelets are displayed in Fig. 12.24.

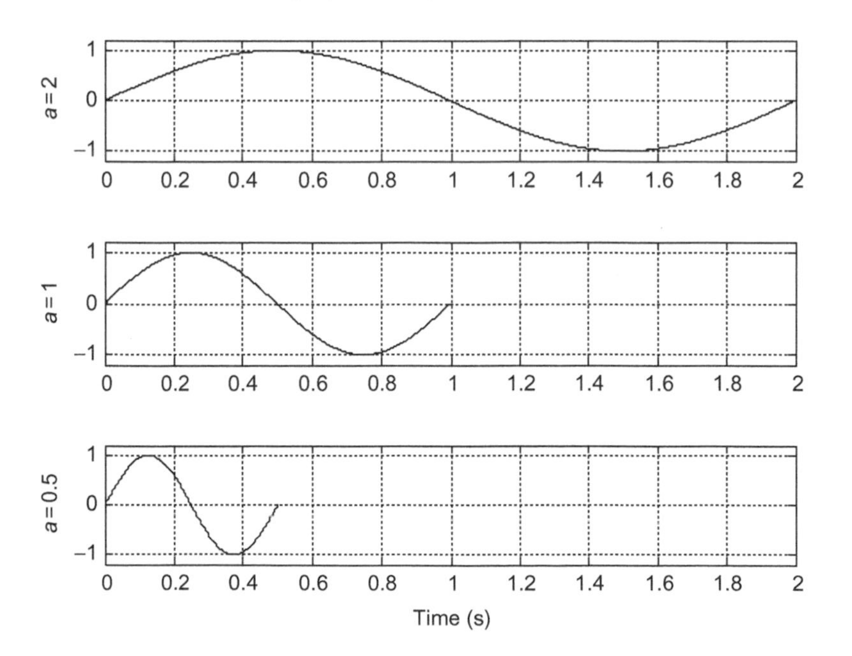

FIG. 12.22

Scaled wavelet functions.

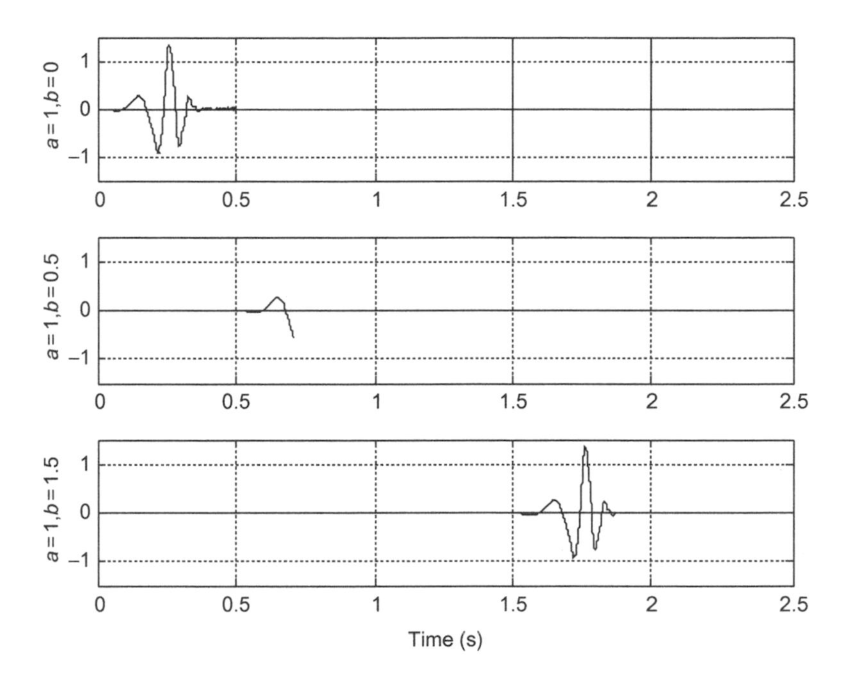

FIG. 12.23

Translated wavelet functions.

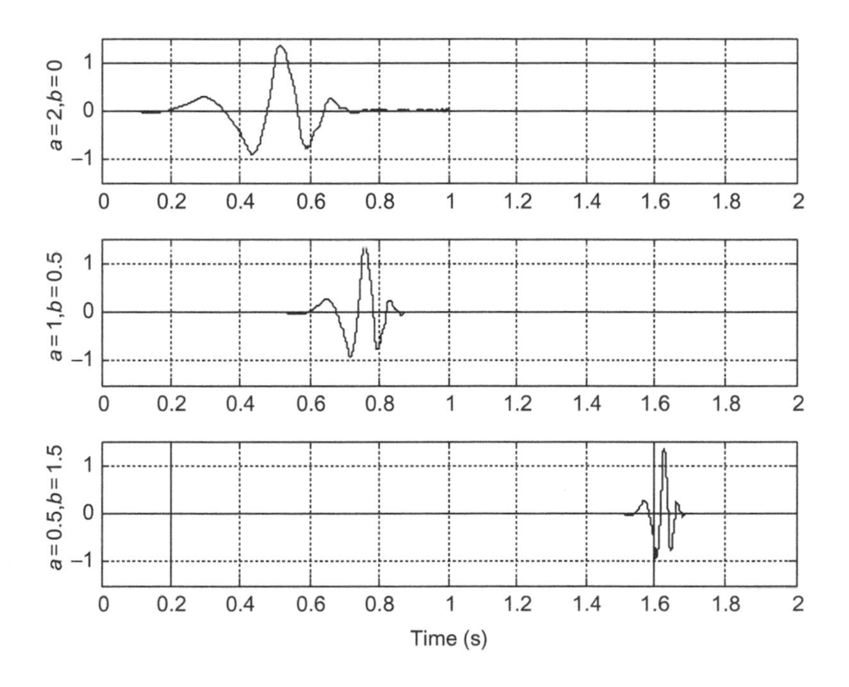

FIG. 12.24

Scaled and translated wavelet functions.

Besides these two properties, a wavelet function must satisfy admissibility and regularity conditions (vanishing moment up to certain order). Admissibility requires that the wavelet (mother wavelet) have a bandpass-limited spectrum and a zero average in the time domain, which means that wavelets must be oscillatory. Regularity requires that wavelets have some smoothness and concentration in both time and frequency domains. This topic is beyond the scope of this book and the details can be found in Akansu and Haddad (2001). There exists a pair of wavelet functions: the father wavelet (also called the scaling function) and mother wavelet. Fig. 12.25 shows a simplest pair of wavelets: the Haar father wavelet and mother wavelet.

To devise an efficient wavelet transform algorithm, we let the scale factor be a power of two, that is,

$$a = 2^{-j}. \tag{12.33}$$

Note that the larger the index j, the smaller the scale factor $a = 2^{-j}$. The time shift becomes

$$b = k2^{-j} = ka. \tag{12.34}$$

Substituting Eqs. (12.33), (12.34) into the base function gives

$$\psi\left(\frac{t-b}{a}\right) = \psi\left(\frac{t-ka}{a}\right) = \psi\left(a^{-1}t - k\right) = \psi\left(2^{j}t - k\right). \tag{12.35}$$

We can define a mother wavelet at scale j and translation k as

$$\psi_{jk}(t) = 2^{j/2}\psi\left(2^{j}t - k\right). \tag{12.36}$$

Similarly, a father wavelet (scaling function) at scale j and translation k is defined as

$$\phi_{jk}(t) = 2^{j/2}\phi\left(2^{j}t - k\right). \tag{12.37}$$

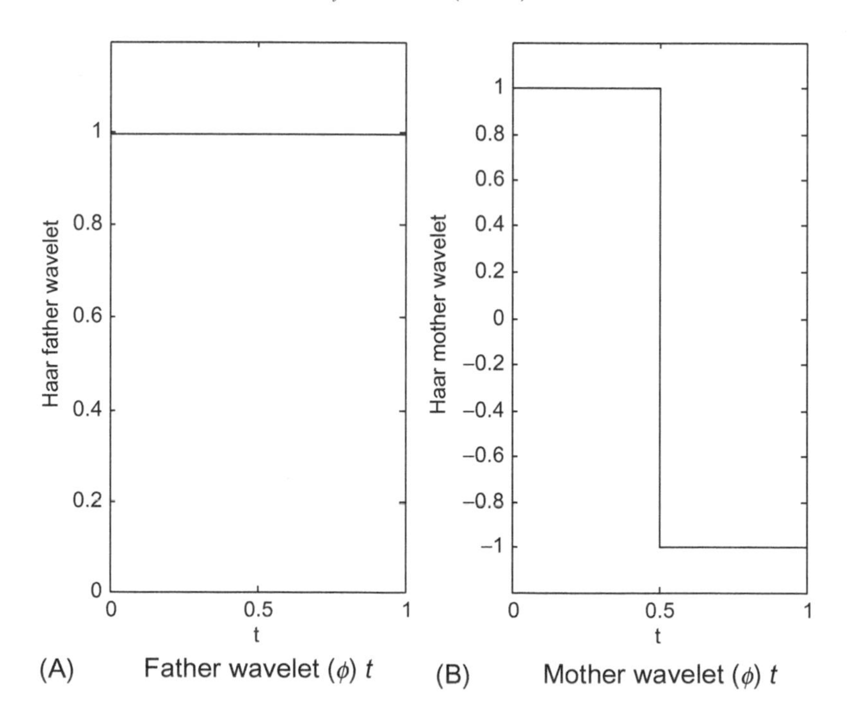

FIG. 12.25

Haar father and mother wavelets. (A) Father wavelet $\phi(t)$. (B) Mother wavelet $\psi(t)$.

EXAMPLE 12.2

Sketch the Haar father wavelet families for four different scales: $j=0, 1, 2, 3$ for a period of 1 s.

Solution:

Based on Eq. (12.37), we can determine the wavelet at each required scale as follows:
For $j=0$, $\phi_{0k}(t)=\phi(t-k)$, only $k=0$ is required to cover a second-s duration
For $j=1$, $\phi_{1k}(t)=\sqrt{2}\phi(2t-k)$, $k=0$ and $k=1$ are required
For $j=2$, $\phi_{2k}(t)=2\phi(4t-k)$, we need $k=0, 1, 2, 3$
For $j=3$, $\phi_{3k}(t)=2\sqrt{2}\phi(8t-k)$, we need $k=0, 1, 2, \cdots, 7$.
Using Fig. 12.25A, we obtain the plots as shown in Fig. 12.26.

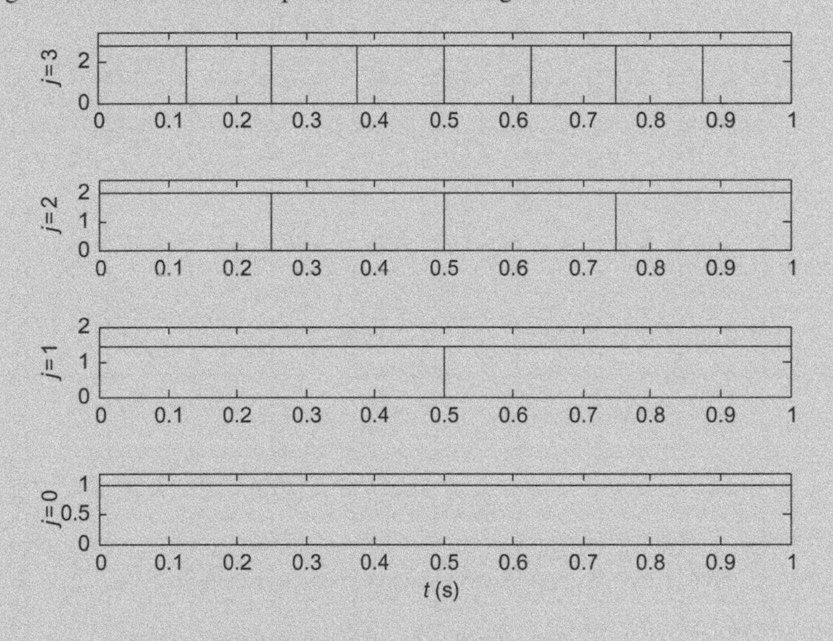

FIG. 12.26

Haar father wavelets at different scales and translations.

EXAMPLE 12.3

Sketch the Haar mother wavelet families for four different scales: $j=0, 1, 2, 3$ for a period of 1 s.

Solution:

Based on Eq. (12.36), we have.
For $j=0$, $\psi_{0k}=\psi(t-k)$, $k=0$ and $k=1$ are required
For $j=1$, $\psi_{1k}=\sqrt{2}\psi(2t-k)$, $k=0$ and $k=1$ are required
For $j=2$, $\psi_{2k}=2\psi(4t-k)$, we need $k=0, 1, 2, 3$
For $j=3$, $\psi_{3k}(t)=2\sqrt{2}\psi(8t-k)$, we need $k=0, 1, 2, \cdots, 7$.
Using Fig. 12.24B, we obtain the plots shown in Fig. 12.27.

Continued

EXAMPLE 12.3—CONT'D

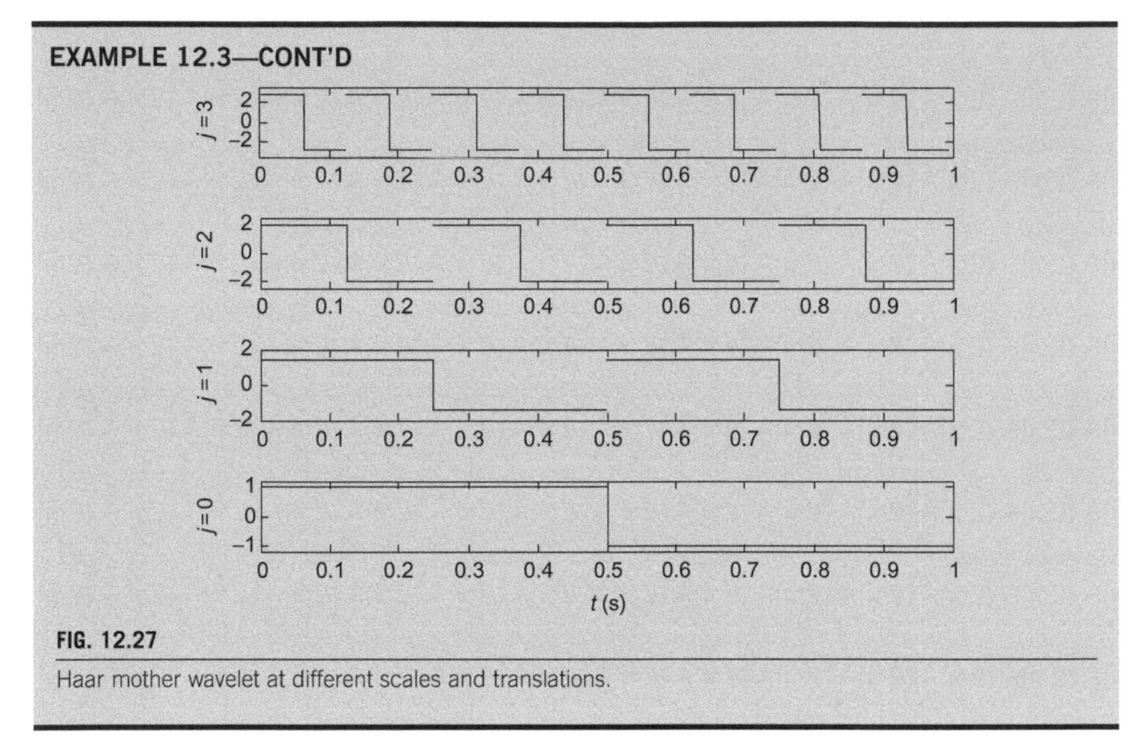

FIG. 12.27

Haar mother wavelet at different scales and translations.

A signal can be expanded by father wavelets (scaling function and its translations) at level j. More accuracy can be achieved by using larger j. The expanded function is approximated by

$$f(t) \approx f_j(t) = \sum_{k=-\infty}^{\infty} c_j(k) 2^{j/2} \phi(2^j t - k) = \sum_{k=-\infty}^{\infty} c_j(k) \phi_{jk}(t), \tag{12.38}$$

where the wavelet coefficients $c_j(k)$ can be determined by an inner product:

$$c_j(k) = f(t)\phi_{jk}(t) = \int f(t) 2^{j/2} \phi(2^j t - k) dt. \tag{12.39}$$

EXAMPLE 12.4

Approximate the following function using the Haar scaling function at level $j=1$.

$$f(t) = \begin{cases} 2 & 0 \le t < 0.5 \\ 1 & 0.5 \le t \le 1 \end{cases}.$$

Solution:

Substituting $j=1$ in Eq. (12.38), it leads to

$$f(t) \approx f_1(t) = \sum_{k=-\infty}^{\infty} c_1(k) \phi_{1k}(t) = \sum_{k=-\infty}^{\infty} c_1(k) 2^{1/2} \phi(2t - k).$$

We only need $k=0$ and $k=1$ to cover the range: $0 \le t \le 1$, that is,

$$f(t) = c_1(0)2^{1/2}\phi(2t) + c_1(1)2^{1/2}\phi(2t-1).$$

Note that

$$\phi(2t) = \begin{cases} 1 & \text{for } 0 \le t \le 0.5 \\ 0 & \text{elsewhere} \end{cases} \text{ and } \phi(2t-1) = \begin{cases} 1 & \text{for } 0.5 \le t \le 1 \\ 0 & \text{elsewhere} \end{cases}.$$

Applying Eq. (12.39) yields

$$c_1(0) = \int_0^{1/2} f(t)2^{1/2}\phi(2t)dt = \int_0^{1/2} 2 \times 2^{1/2} \times 1dt = 2^{1/2}.$$

Similarly,

$$c_1(1) = \int_{1/2}^{1} f(t)2^{1/2}\phi(2t-1)dt = \int_{1/2}^{1} 1 \times 2^{1/2} \times 1dt = 0.5 \times 2^{1/2}.$$

Then substituting the coefficients $c_1(0)$ and $c_1(1)$ leads to

$$f_1(t) = 2^{1/2} \times 2^{1/2}\phi(2t) + 0.5 \times 2^{1/2}2^{1/2}\phi(2t-1) = 2\phi(2t) + \phi(2t-1) = f(t).$$

Eq. (12.39) can also be approximated numerically:

$$c_j(k) \approx \sum_{m=0}^{M-1} f(t_m)\phi_{jk}(t_m)\Delta t = \sum_{m=0}^{M-1} f(t_m)2^{j/2}\phi(2^j t_m - k)\Delta t,$$

where $t_m = m\Delta t$ is the time instant, Δt denotes the time step, and M is the number of intervals.

In this example, if we chose $\Delta t = 0.2$, then $M=5$ and $t_m = m\Delta t$. The numerical calculations are listed:

$$\begin{aligned} c_1(0) \approx \sum_{m=0}^{4} f(t_m)2^{1/2}\phi(2t_m)\Delta t &= 2^{1/2} \times [f(0) \times \phi(0) + f(0.2) \times \phi(0.4) \\ &+ f(0.4) \times \phi(0.8) + f(0.6) \times \phi(1.2) + f(0.8) \times \phi(1.6)]\Delta t \\ &= 2^{1/2}(2 \times 1 + 2 \times 1 + 2 \times 1 + 1 \times 0 + 1 \times 0) \times 0.2 = 1.2 \times 2^{1/2} \end{aligned}$$

$$\begin{aligned} c_1(1) \approx \sum_{m=0}^{4} f(t_m)2^{1/2}\phi(2t_m - 1)\Delta t &= 2^{1/2} \times [f(0) \times \phi(-1) + f(0.2) \times \phi(-0.6) \\ &+ f(0.4) \times \phi(-0.2) + f(0.6) \times \phi(0.2) + f(0.8) \times \phi(0.6)]\Delta t \\ &= 2^{1/2} \times (2 \times 0 + 2 \times 0 + 2 \times 0 + 1 \times 1 + 1 \times 1 \times 0.2 = 0.4 \times 2^{1/2}. \end{aligned}$$

Finally, we have

$$f_1(t) = 1.2 \times 2^{1/2} \times 2^{1/2}\phi(2t) + 0.4 \times 2^{1/2}2^{1/2}\phi(2t-1) = 2.4\phi(2t) + 0.8\phi(2t-1) \approx f(t).$$

It is clear that there is a numerical error. The error can be reduced when a smaller time interval Δt is adopted.

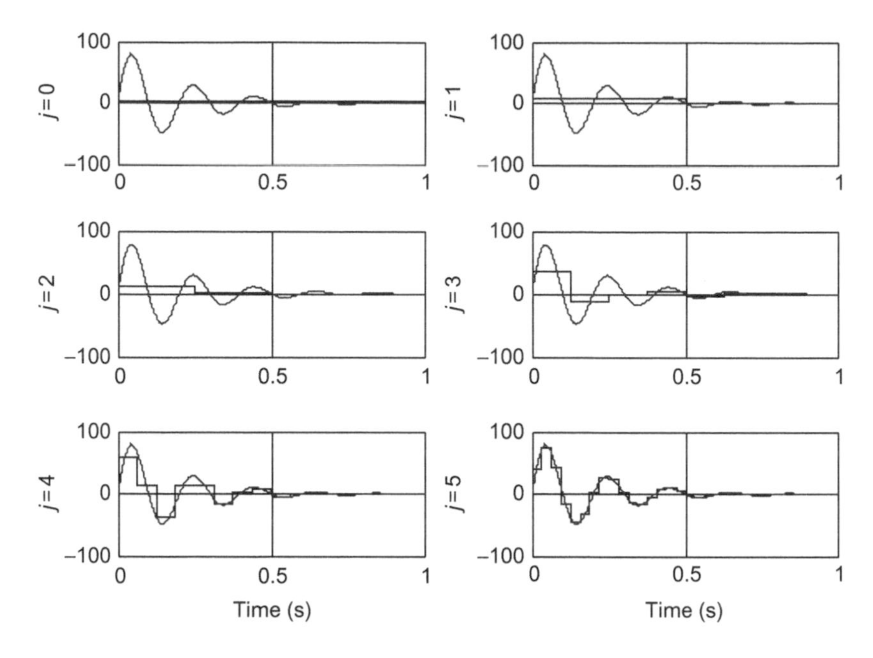

FIG. 12.28

Signal expanded by Haar father wavelets.

Fig. 12.28 demonstrates the approximation of a sinusoidal delaying function using the scaling functions (Haar father wavelets) at different scales, that is, $j=0$, 1, 2, 4, 5.

Now, let us examine the function approximation at resolution $j=1$:

$$\begin{aligned} f_1(t) &\approx \sum_{k=-\infty}^{\infty} c_1(k)\phi_{1k}(2t) = \sum_{k=-\infty}^{\infty} c_1(k)\sqrt{2}\phi(2t-k) \\ &= c_1(0)\phi_{10}(2t) + c_1(1)\phi_{11}(2t) = c_1(0)\sqrt{2}\phi(2t) + c_1(1)\sqrt{2}\phi(2t-1). \end{aligned}$$

We also look at another possibility at a coarser scale with both the scaling functions (father wavelets) and mother wavelets, that is, $j=0$:

$$\begin{aligned} f_1(t) &\approx \sum_{k=-\infty}^{\infty} c_0(k)\phi_{0k}(t) + \sum_{k=-\infty}^{\infty} d_0(k)\psi_{0k}(t) \\ &= c_0(0)\phi_{00}(t) + d_0(0)\psi_{00}(t) = c_0(0)\phi(t) + d_0(0)\psi(t). \end{aligned}$$

Furthermore, we see that

$$\begin{aligned} f_1(t) &\approx c_0(0)\phi(t) + d_0(0)\psi(t) \\ &= c_0(0)(\phi(2t) + \phi(2t-1)) + d_0(0)(\phi(2t) - \phi(2t-1)) \\ &= \frac{1}{\sqrt{2}}(c_0(0) + d_0(0))\phi_{00}(2t) + \frac{1}{\sqrt{2}}(c_0(0) - d_0(0))\phi_{01}(2t-1) \\ &= c_1(0)\phi_{10}(2t) + c_1(1)\phi_{10}(2t). \end{aligned}$$

We observe that

$$c_1(0) = \frac{1}{\sqrt{2}}(c_0(0) + d_0(0))$$
$$c_1(1) = \frac{1}{\sqrt{2}}(c_0(0) - d_0(0)).$$

This means that

$$S_1 = S_0 \cup W_0.$$

where S_1 contains functions in terms of basis scaling functions at $\phi_{1k}(t)$, and the function can also be expanded using the scaling functions $\phi_{0k}(t)$ and wavelet functions $\psi_{0k}(t)$ at a coarser level $j-1$. In general, the following statement is true:

$$\begin{aligned} S_j &= S_{j-1} \cup W_{j-1} = \left[S_{j-2} \cup W_{j-2}\right] \cup W_{j-1} \\ &= \left\{\left[S_{j-3} \cup W_{j-3}\right] \cup W_{j-2}\right\} \cup W_{j-1} \\ \cdots &= S_0 \cup W_0 \cup W_1 \cup \cdots \cup W_{j-1}. \end{aligned}$$

Hence, the approximation of $f_j(t)$ can be expressed as

$$\begin{aligned} f(t) \approx f_j(t) &= \sum_{k=-\infty}^{\infty} c_j(k)\phi_{jk}(t) \\ &= \sum_{k=-\infty}^{\infty} c_{j-1}(k)\phi_{(j-1)k}(t) + \sum_{k=-\infty}^{\infty} d_{j-1}(k)\psi_{(j-1)k}(t) \\ &= \sum_{k=-\infty}^{\infty} c_{(j-1)}(k)2^{(j-1)/2}\phi\left(2^{(j-1)}t - k\right) + \sum_{k=-\infty}^{\infty} d_{(j-1)}(k)2^{(j-1)/2}\psi\left(2^{(j-1)}t - k\right). \end{aligned}$$

Repeating the expansion of the first sum leads to

$$\begin{aligned} f(t) \approx f_J(t) &= \sum_{k=-\infty}^{\infty} c_0(k)\phi_{0k}(t) + \sum_{j=0}^{J-1}\sum_{k=-\infty}^{\infty} d_j(k)\psi_{jk}(t) \\ &= \sum_{k=-\infty}^{\infty} c_0(k)\phi(t-k) + \sum_{j=0}^{J-1}\sum_{k=-\infty}^{\infty} d_j(k)2^{j/2}\psi(2^j t - k), \end{aligned} \tag{12.40}$$

where the mother wavelet coefficients $d_j(k)$ can also be determined by the inner product:

$$d_j(k) = \langle f(t)\psi_{jk}(t)\rangle = \int f(t)2^{j/2}\psi(2^j t - k)dt. \tag{12.41}$$

Fig. 12.29 demonstrates the function approximation (Fig. 12.28) with the base scaling function at resolution $j=0$, and mother wavelets at scales $j=0, 1, 2, 3, 4$. The combined approximation ($J=5$) using Eq. (12.40) is shown in Fig. 12.30.

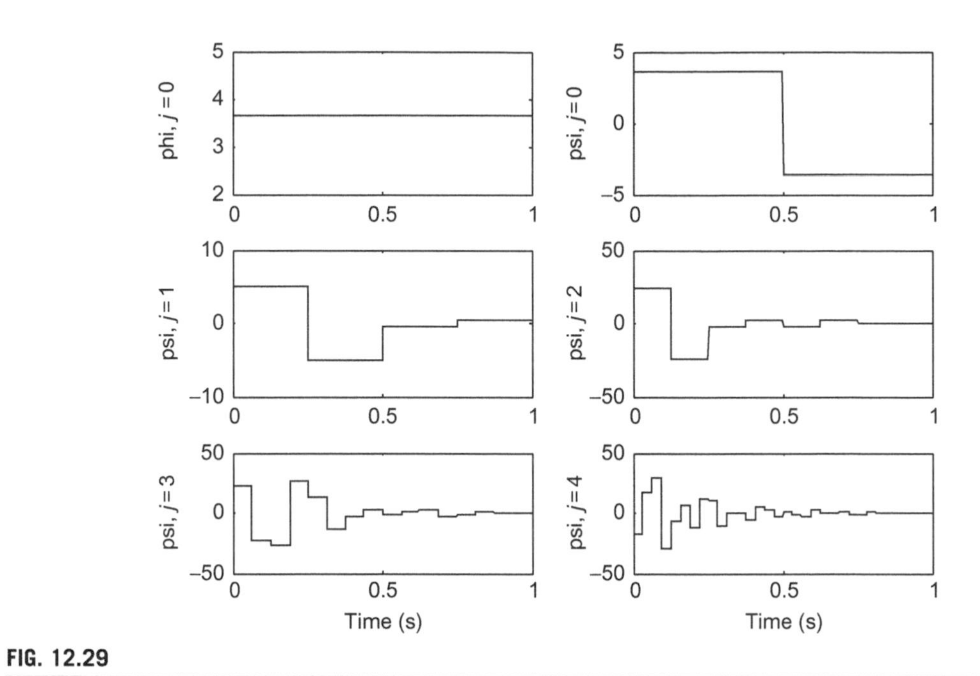

FIG. 12.29

Approximations using Haar scaling functions and mother wavelets.

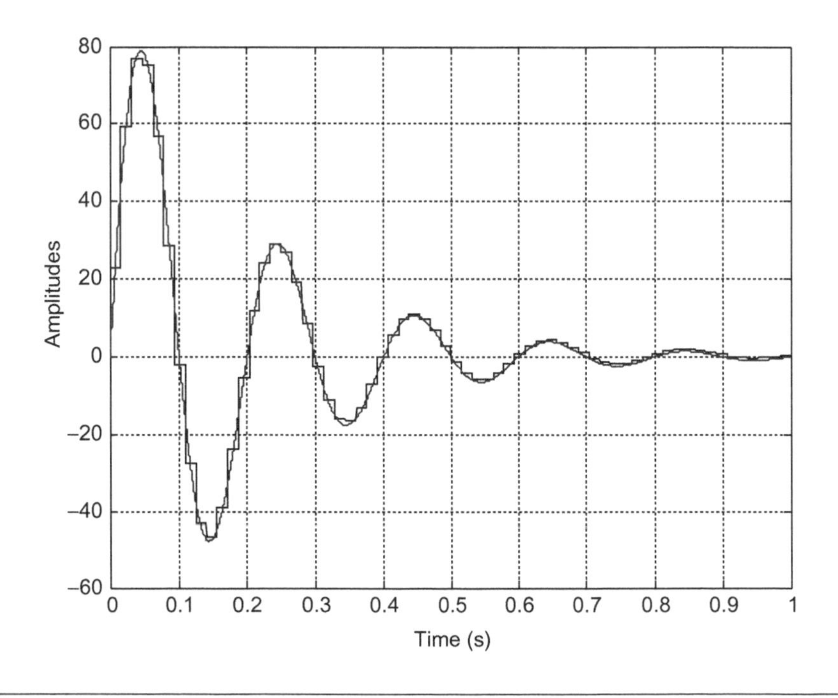

FIG. 12.30

Signal coded using the wavelets at resolution $J=5$.

12.5 MULTIRESOLUTION EQUATIONS

There are two very important equations for multiresolution analysis. Each scaling function can be constructed by a linear combination of translations with the doubled frequency of a base scaling function $\phi(2t)$, that is,

$$\phi(t) = \sum_{k=-\infty}^{\infty} \sqrt{2} h_0(k) \phi(2t-k), \tag{12.42}$$

where $h_0(k)$ is a set of the scaling function coefficients (wavelet filter coefficients). The mother wavelet function can also be built by a sum of translations with the double frequency of the base scaling function $\phi(2t)$, that is,

$$\psi(t) = \sum_{k=-\infty}^{\infty} \sqrt{2} h_1(k) \phi(2t-k), \tag{12.43}$$

where $h_1(k)$ is another set of wavelet filter coefficients. Let us verify these two relationships via Example 12.5.

EXAMPLE 12.5

Determine $h_0(k)$ for the Haar father wavelet.

Solution:

From Eq. (12.42), we can express

$$\phi(t) = \sqrt{2} h_0(0) \phi(2t) + \sqrt{2} h_0(1) \phi(2t-1).$$

Then we deduce that

$$h_0(0) = h_0(1) = 1/\sqrt{2}.$$

Fig. 12.31 shows that the Haar father wavelet is a sum of two scaling functions at scale $j=1$.

EXAMPLE 12.6

Determine $h_1(k)$ for the Haar mother wavelet.

Solution:

From Eq. (12.43), we can write

$$\psi(t) = \sqrt{2} h_1(0) \phi(2t) + \sqrt{2} h_1(1) \phi(2t-1).$$

Hence, we deduce that

$$h_1(0) = 1/\sqrt{2} \text{ and } h_1(1) = -1/\sqrt{2}.$$

Fig. 12.32 shows that the Haar mother wavelet is a difference of two scaling functions at scale $j=1$.

Note that the relation between $H_0(z)$ and $H_1(z)$ exists and is given by

$$h_1(k) = (-1)^k h_0(N-1-k). \tag{12.44}$$

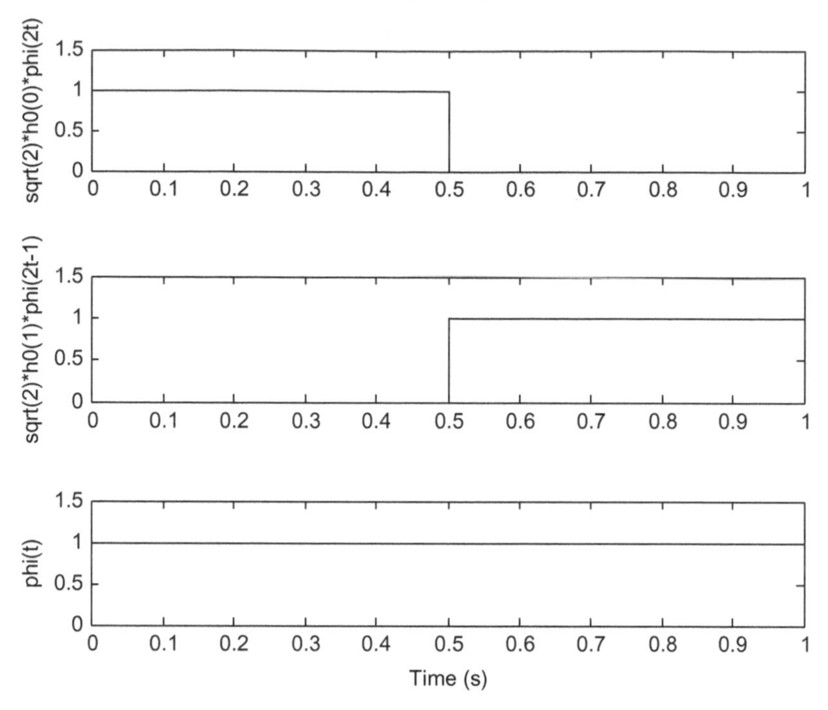

FIG. 12.31

Haar wavelets in Example 12.5.

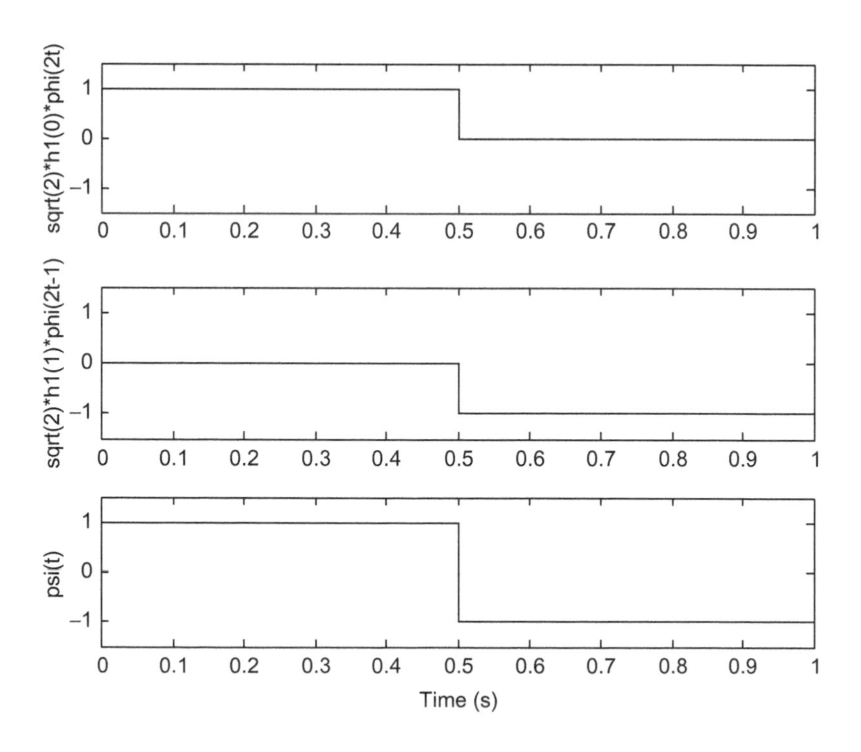

FIG. 12.32

Haar wavelets in Example 12.6.

We can verify Eq. (12.44) for the Haar wavelet:

$$h_1(k) = (-1)^k h_0(1-k).$$

Then

$$h_1(0) = (-1)^0 h_0(1-0) = h_0(1) = 1/\sqrt{2}$$
$$h_1(1) = (-1)^1 h_0(1-1) = -h_0(0) = -1/\sqrt{2}.$$

This means that once we obtain the coefficients of $h_0(k)$, the coefficients $h_1(k)$ can be determined via Eq. (12.44). We do not aim to obtain wavelet filter coefficients here. The topic is beyond the scope of this book and the details are given in Akansu and Haddad (2001). Instead, some typical filter coefficients for Haar and Doubechies are presented in Table 12.2.

We can apply the Daubechies-4 filter coefficients to examine multiresolution Eqs. (12.42), (12.43). From Table 12.2, we have

$$h_0(0) = 0.4830, h_0(1) = 0.8365, h_0(2) = 0.2241, \text{ and } h_0(3) = -0.1294.$$

We then expand Eq. (12.42) as

$$\phi(t) = \sqrt{2}h_0(0)\phi(2t) + \sqrt{2}h_0(1)\phi(2t-1) + \sqrt{2}h_0(2)\phi(2t-2) + \sqrt{2}h_0(3)\phi(2t-3).$$

Fig. 12.33 shows each component at resolution $j=1$ and the constructed scaling function $\phi(t)$. The original scaling function $\phi(t)$ is also included as shown in the last plot for comparison.

With the given coefficients $h_0(k)$ and applying Eq. (12.44), we can obtain the wavelet coefficients $h_1(k)$ as

$$h_1(0) = -0.1294, h_1(1) = -0.2241, h_1(2) = 0.8365, \text{ and } h_1(3) = -0.4830.$$

Expanding Eq. (12.43) leads to

$$\psi(t) = \sqrt{2}h_1(0)\phi(2t) + \sqrt{2}h_1(1)\phi(2t-1) + \sqrt{2}h_1(2)\phi(2t-2) + \sqrt{2}h_1(3)\phi(2t-3).$$

Similarly, Fig. 12.34 displays each component at resolution $j=1$ and the constructed mother wavelet function $\psi(t)$. The last plot displays the original mother wavelet function $\psi(t)$ for comparison.

Table 12.2 Typical Wavelet Filter Coefficients $h_0(k)$

Haar	Daubechies 4	Daubechies 6	Daubechies 8
0.707106781186548	0.482962913144534	0.332670552950083	0.230377813308896
0.707106781186548	0.836516303737808	0.806891509311093	0.714846570552915
	0.224143868042013	0.459877502118492	0.630880767929859
	−0.129409522551260	−0.135011020010255	−0.027983769416859
		−0.085441273882027	−0.187034811719093
		0.035226291885710	0.030841381835561
			0.032883011666885
			−0.010597401785069

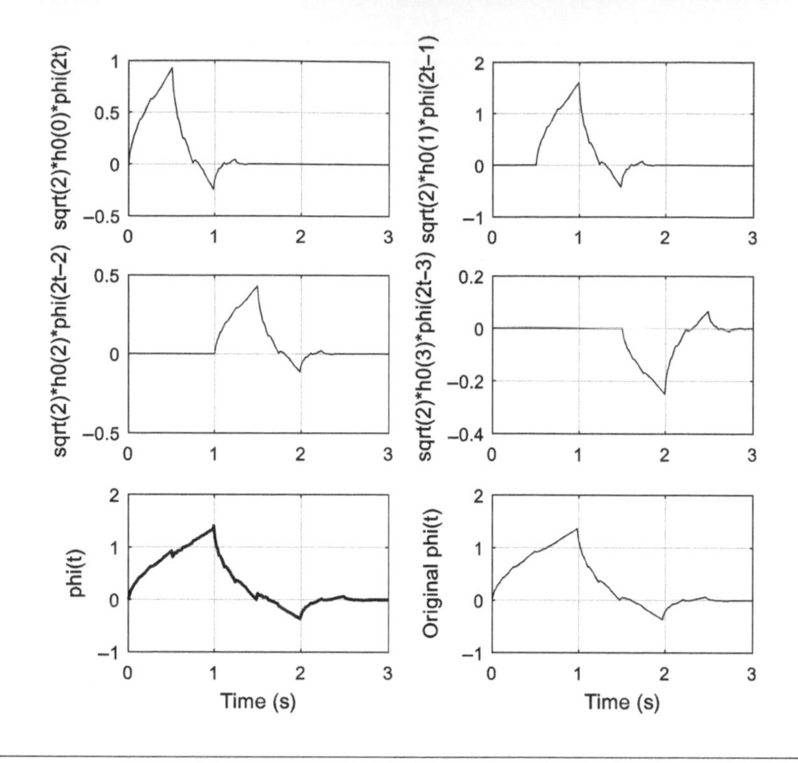

FIG. 12.33

Constructed four-tap Daubechies father wavelet.

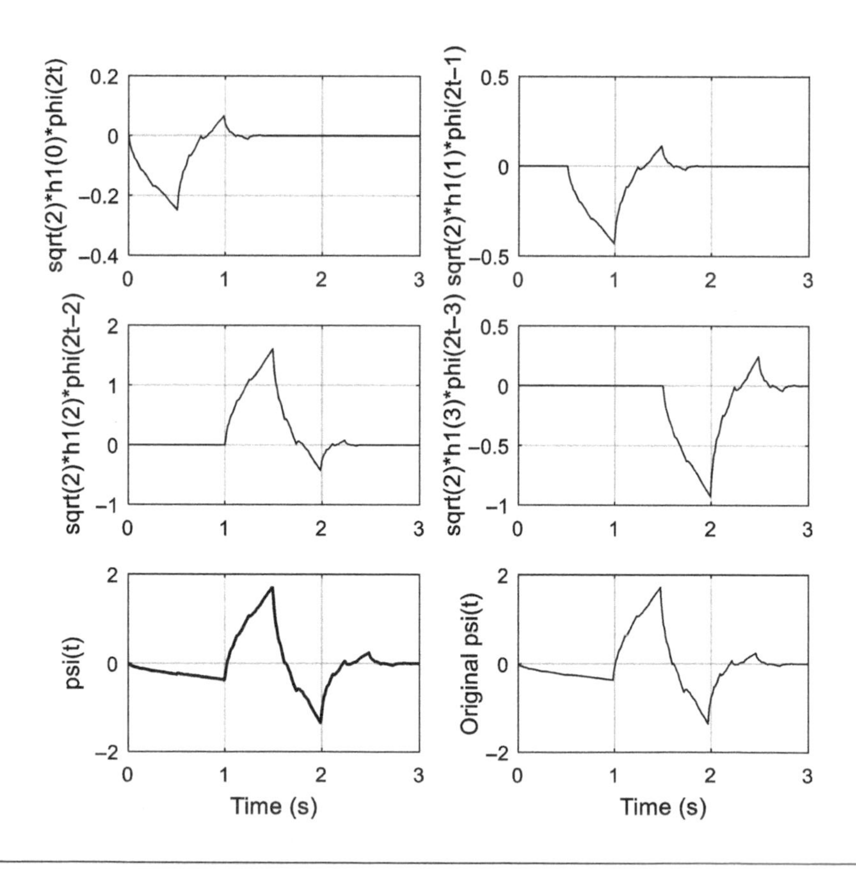

FIG. 12.34

Constructed four-tap Daubechies mother wavelet.

12.6 DISCRETE WAVELET TRANSFORM

Now let us examine the discrete wavelet transform (DWT). We begin with coding a signal using a wavelet expansion as shown in Eq. (12.45):

$$f(t) \approx f_{j+1}(t) = \sum_{k=-\infty}^{\infty} c_j(k) 2^{j/2} \phi(2^j t - k) + \sum_{k=-\infty}^{\infty} d_j(k) 2^{j/2} \psi(2^j t - k). \tag{12.45}$$

By applying and continuing to apply Eq. (12.45), $f(t)$ can be coded at any level we wish. Furthermore, by recursively applying Eq. (12.45) until $j=0$, we can obtain the signal expansion using all the mother wavelets as well as a one scaling function at scale $j=0$, that is,

$$f(t) \approx f_J(t) = \sum_{k=-\infty}^{\infty} c_0(k) \phi(t-k) + \sum_{j=0}^{J-1} \sum_{k=-\infty}^{\infty} d_j(k) 2^{j/2} \psi(2^j t - k). \tag{12.46}$$

All $c_j(k)$ and all $d_j(k)$ are called the wavelet coefficients. They are essentially weights for the scaling function(s) and wavelet functions (mother wavelets). The DWT computes these wavelet coefficients. On the other hand, given the wavelet coefficients, we are able to reconstruct the original signal by applying the inverse discrete wavelet transform (IDWT).

Based on the wavelet theory without proof (see Appendix F), we can perform the DWT using the analysis equations as follows:

$$c_j(k) = \sum_{m=-\infty}^{\infty} c_{j+1}(m) h_0(m - 2k) \tag{12.47}$$

$$d_j(k) = \sum_{m=-\infty}^{\infty} c_{j+1}(m) h_1(m - 2k), \tag{12.48}$$

where $h_0(k)$ are the lowpass wavelet filter coefficients listed in Table 12.2, while $h_1(k)$, the highpass filter coefficients, can be determined by

$$h_1(k) = (-1)^k h_0(N - 1 - k). \tag{12.49}$$

These lowpass and highpass filters are called the quadrature mirror filters (QMF). As an example, the frequency responses of the four-tap Daubechies wavelet filters are plotted in Fig. 12.35.

Next, we need to determine the filter inputs $c_{j+1}(k)$ in Eqs. (12.47), (12.48). In practice, since j is a large number, the function $\phi(2^j t - k)$ appears to be close to an impulse-like function, that is, $\phi(2^j t - k) \approx 2^{-j}\delta(t - k2^{-j})$. For example, the Haar scaling function can be expressed as $\phi(t) = u(t) - u(t-1)$, where $u(t)$ is the step function. We can easily get $\phi(2^5 t - k) = u(2^5 t - k) - u(2^5 t - 1 - k) = u(t - k2^{-5}) - u(t - (k+1)2^{-5})$ for $j=5$, which is a narrow pulse with a unit height and a width 2^{-5} located at $t = k2^{-5}$. The area of the pulse is therefore equal to 2^{-5}. When j approaches to a bigger positive integer, $\phi(2^j t - k) \approx 2^{-j}\delta(t - k2^{-j})$. Therefore, $f(t)$ approximated by the scaling function at level j is rewritten as

$$\begin{aligned} f(t) \approx f_j(t) &= \sum_{k=-\infty}^{\infty} c_j(k) 2^{j/2} \phi(2^j t - k) \\ &= \cdots + c_j(0) 2^{j/2} \phi(2^j t) + c_j(1) 2^{j/2} \phi(2^j t - 1) + c_j(2) 2^{j/2} \phi(2^j t - 2) + \cdots \\ &\approx \cdots + c_j(0) 2^{-j/2} \delta(t) + c_j(1) 2^{-j/2} \delta(t - 1 \times 2^{-j}) + c_j(2) 2^{-j/2} \delta(t - 2 \times 2^{-j}) + \cdots. \end{aligned} \tag{12.50}$$

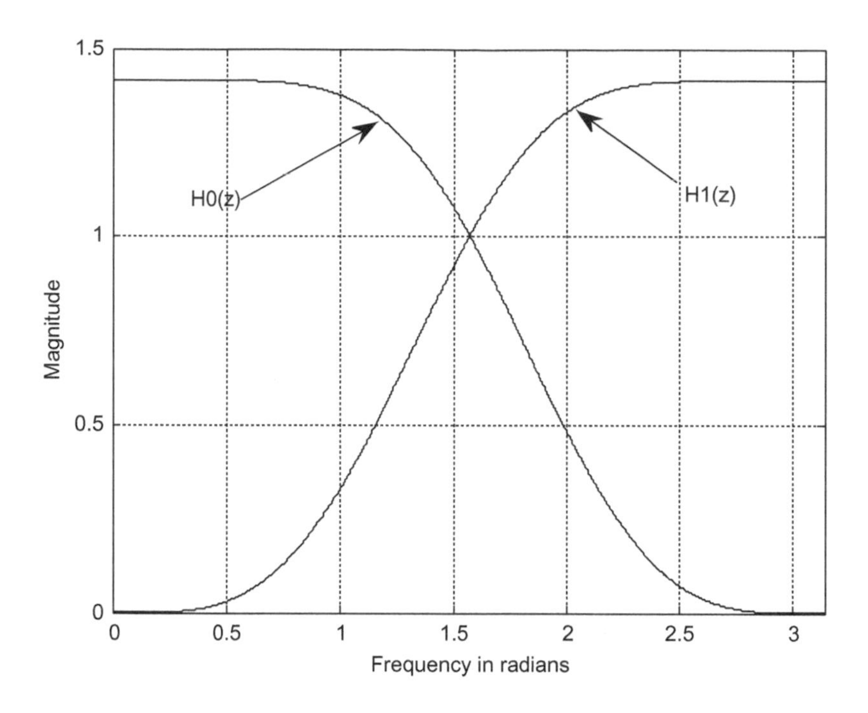

FIG. 12.35

Frequency responses for 4-tap Daubechies filters.

On the other hand, if we sample $f(t)$ using the same sample interval $T_s = 2^{-j}$ (time resolution), the discrete-time function can be expressed as

$$f(n) = f(nT_s) = \cdots + f(0T_s)T_s\delta(t - T_s) + f(T_s)T_s\delta(t - T_s) + f(2T_s)T_s\delta(n - 2T_s) + \cdots. \tag{12.51}$$

Hence, comparing Eq. (12.50) with the discrete-time version in Eq. (12.51), it follows that

$$c_j(k)2^{-j/2} = f(k)T_s. \tag{12.52}$$

Substituting $T_s = 2^{-j}$ in Eq. (12.52) leads to

$$c_j(k) = 2^{-j/2} f(k). \tag{12.53}$$

With the obtained sequence $c_j(k)$ using sample values $f(k)$, we can perform the DWT using Eqs. (12.47), (12.48). Furthermore, Eqs. (12.47) and (12.48) can be implemented using a dyadic tree structure similar to the subband coding case. Fig. 12.36 depicts the case for $j = 2$.

Note that the reversed sequences $h_0(-k)$ and $h_1(-k)$ are used in the analysis stage. Similarly, the IDWT (synthesis equation) can be developed (see Appendix F) and expressed as

$$c_{j+1}(k) = \sum_{m=-\infty}^{\infty} c_j(m)h_0(k - 2m) + \sum_{m=-\infty}^{\infty} d_j(m)h_1(k - 2m). \tag{12.54}$$

Finally, the signal amplitude can be rescaled by

$$f(k) = 2^{j/2} c_j(k). \tag{12.55}$$

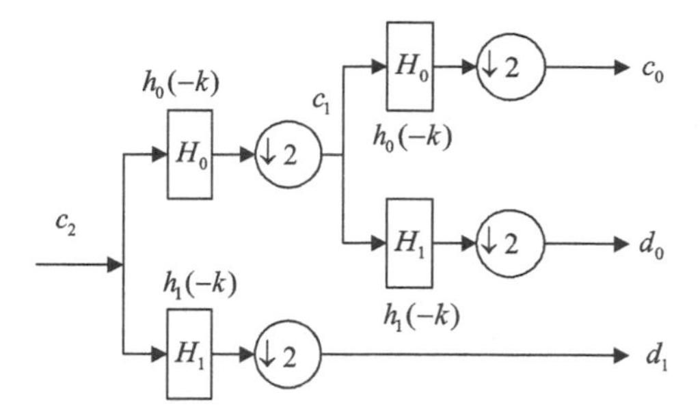

FIG. 12.36

Analysis using the dyadic subband coding structure.

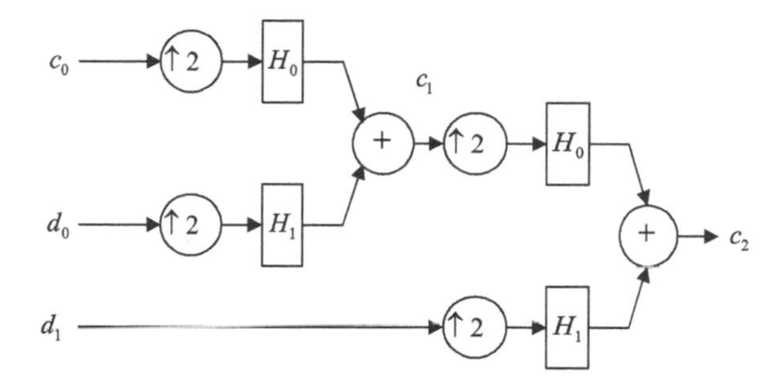

FIG. 12.37

Synthesis using the dyadic subband coding structure.

An implementation for $j=2$ using the dyadic subband coding structure is illustrated in Fig. 12.37.

Now, let us study the DWT and IDWT in the following examples.

EXAMPLE 12.7

Given the sample values as [4 2 −1 0], use the Haar wavelets to determine the wavelet coefficients.

Solution:

Form the filter inputs: $c_2(k)=2^{-2/2}\times[4\ \ 2\ \ -1\ \ 0]=\left[2\ \ 1\ \ -\frac{1}{2}\ \ 0\right]$.

The acquired Haar wavelet filter coefficients are listed as

$$h_0(k)=\left[\frac{1}{\sqrt{2}}\ \ \frac{1}{\sqrt{2}}\right] \text{ and } h_1(k)=\left[\frac{1}{\sqrt{2}}\ \ -\frac{1}{\sqrt{2}}\right].$$

The function is expanded by the scaling functions as

$$\begin{aligned} f(t)\approx f_2(t)&=\sum_{k=-\infty}^{\infty} c_j(k)2^{j/2}\phi(2^jt-k)\\ &=4\times\phi(4t)+2\times\phi(4t-1)-1\times\phi(4t-2)+0\times\phi(4t-3). \end{aligned}$$

Continued

EXAMPLE 12.7—CONT'D

We will verify this expression later. Applying the wavelet analysis equations, we have

$$c_1(k) = \sum_{m=-\infty}^{\infty} c_2(m)h_0(m-2k)$$

$$d_1(k) = \sum_{m=-\infty}^{\infty} c_2(m)h_1(m-2k).$$

Specifically,

$$c_1(0) = \sum_{m=-\infty}^{\infty} c_2(m)h_0(m) = c_2(0)h_0(0) + c_2(1)h_0(1) = 2\times\frac{1}{\sqrt{2}} + 1\times\frac{1}{\sqrt{2}} = \frac{3\sqrt{2}}{2}$$

$$c_1(1) = \sum_{m=-\infty}^{\infty} c_2(m)h_0(m-2) = c_2(2)h_0(0) + c_2(3)h_0(1) = \left(-\frac{1}{2}\right)\times\frac{1}{\sqrt{2}} + 0\times\frac{1}{\sqrt{2}} = -\frac{1}{2\sqrt{2}}$$

$$d_1(0) = \sum_{m=-\infty}^{\infty} c_2(m)h_1(m) = c_2(0)h_1(0) + c_2(1)h_1(1) = 2\times\frac{1}{\sqrt{2}} + 1\times\left(-\frac{1}{\sqrt{2}}\right) = \frac{1}{\sqrt{2}}$$

$$d_1(1) = \sum_{m=-\infty}^{\infty} c_2(m)h_1(m-2) = c_2(2)h_1(0) + c_2(3)h_1(1) = \left(-\frac{1}{2}\right)\times\frac{1}{\sqrt{2}} + 0\times\left(-\frac{1}{\sqrt{2}}\right) = -\frac{1}{2\sqrt{2}}.$$

Using the subband coding method in Fig. 12.36 yields.

```
>> x0=rconv([1 1]/sqrt(2),[2 1 -0.5 0])
x0 =2.1213 0.3536 -0.3536 1.4142
>> c1=x0(1:2:4)
c1 = 2.1213 -0.3536
>> x1=rconv([1 -1]/sqrt(2),[2 1 -0.5 0])
x1 = 0.7071 1.0607 -0.3536 -1.4142
>> d1=x1(1:2:4)
d1 =0.7071 -0.3536
```

where MATLAB function **rconv()** for filter operations with the reversed filter coefficients is listed in Section 12.8. Repeating for the next level, we have

$$c_0(k) = \sum_{m=-\infty}^{\infty} c_1(m)h_0(m-2k)$$

$$d_0(k) = \sum_{m=-\infty}^{\infty} c_1(m)h_1(m-2k).$$

Thus,

$$c_0(0) = \sum_{m=-\infty}^{\infty} c_1(m)h_0(m) = c_1(0)h_0(0) + c_1(1)h_0(1) = \frac{3\sqrt{2}}{2}\times\frac{1}{\sqrt{2}} + \left(-\frac{1}{2\sqrt{2}}\right)\times\frac{1}{\sqrt{2}} = \frac{5}{4}$$

$$d_0(0)=\sum_{m=-\infty}^{\infty}c_1(m)h_1(m)=c_1(0)h_1(0)+c_1(1)h_1(1)=\frac{3\sqrt{2}}{2}\times\frac{1}{\sqrt{2}}+\left(-\frac{1}{2\sqrt{2}}\right)\times\left(-\frac{1}{\sqrt{2}}\right)=\frac{7}{4}.$$

MATLAB verifications are shown below:

```
>> xx0=rconv([1 1]/sqrt(2),c1)
xx0 =1.2500 1.2500
>> c0=xx0(1:2:2)
c0 =1.2500
>> xx1=rconv([1 -1]/sqrt(2),c1)
xx1 =1.7500 -1.7500
>> d0=xx1(1:2:2)
d0 =1.7500
```

Finally, we pack the wavelet coefficients $w_2(k)$ at $j=2$ together as

$$w_2(k)=[c_0(0)d_0(0)d_1(0)d_1(1)]=\left[\frac{5}{4}\frac{7}{4}\frac{1}{\sqrt{2}}-\frac{1}{2\sqrt{2}}\right].$$

Then the function can be expanded using one scaling function and three mother wavelet functions.

$$f(t)\approx f_2(t)=\sum_{k=-\infty}^{\infty}c_0(k)\phi(t-k)+\sum_{j=0}^{1}\sum_{k=-\infty}^{\infty}d_j(k)2^{j/2}\psi(2^jt-k)$$
$$=\frac{5}{4}\phi(t)+\frac{7}{4}\psi(t)+\psi(2t)-\frac{1}{2}\psi(2t-1).$$

Fig. 12.38 shows the plots for each function and the combined function to verify that $f(t)$ does have amplitudes of 4, 2, -1, and 0.
We can use the MATLAB function **dwt()** provided in Section 12.8 to compute the DWT coefficients.

```
function w = dwt(h0,c,kLevel)
% h0 = wavelet filter coefficients (lowpass filter)
% c = input vector
% kLevel = level
% w= wavelet coefficients
```

The results are verified as follows:

```
>> w=dwt([1/sqrt(2) 1/sqrt(2)][4 2 -1 0]/2,2)'
w = 1.2500 1.7500 0.7071 -0.3536
```

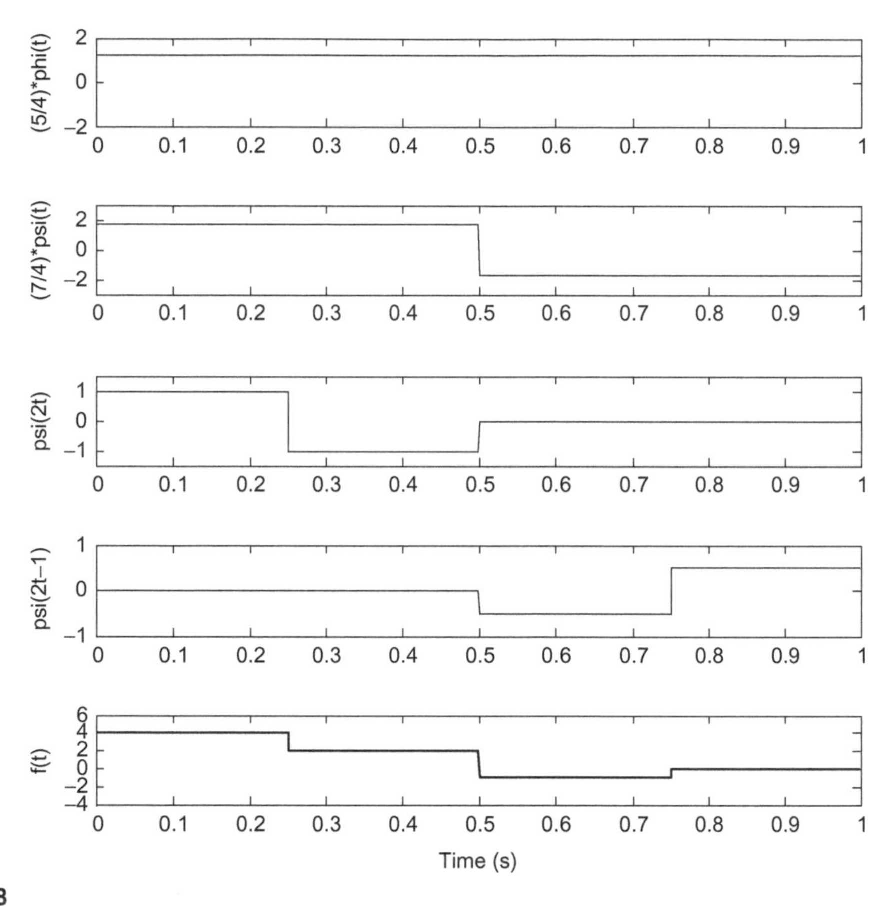

FIG. 12.38

Signal reconstructed using the Haar wavelets in Example 12.7.

From Example 12.7, we can create a time-frequency plot of the DWT amplitudes in two dimensions as shown in Fig. 12.39. Assuming that the sampling frequency is f_s, we have the smallest frequency resolution as $f_s/N=f_s/4$, where $N=4$. When j ($j=0$) is small, we achieve a small frequency resolution $\Delta f=f_s/4$ and each wavelet presents four samples. In this case, we have a good frequency resolution but a poor time resolution. Similarly, when j ($j=1$) is a large value, the frequency resolution becomes $\Delta f=2f_s/4$ and each wavelet presents two samples (more details in time domain). Hence, we achieve a good time resolution but a poor frequency resolution. Note that the DWT cannot achieve good resolutions in both frequency and time at the same time. The time-frequency plot of the DWT amplitudes in terms of their intensity is shown in Fig. 12.39B, and the time and frequency plane for the DWT is shown in Fig. 12.40.

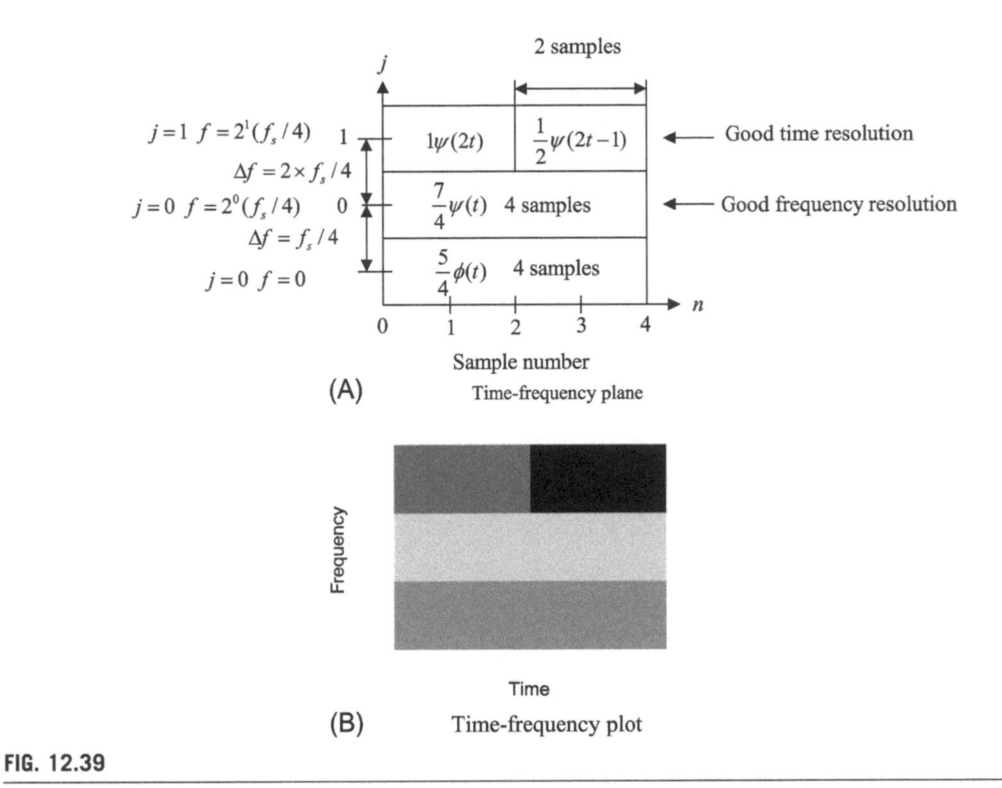

FIG. 12.39

Time-frequency plot of the DWT amplitudes. (A) Time-frequency plane, (B) time-frequency plot.

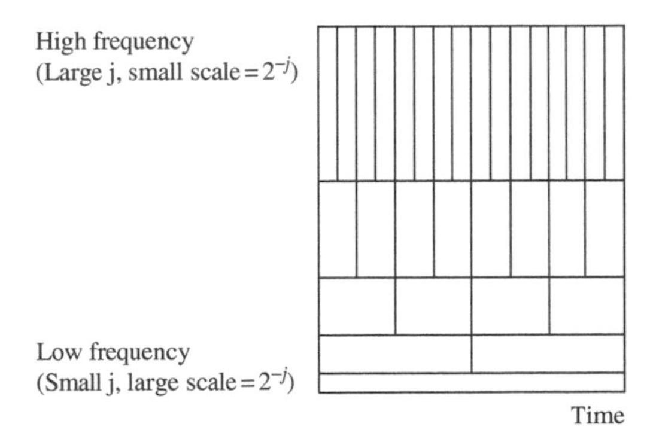

FIG. 12.40

Time-frequency plane.

EXAMPLE 12.8

Given the wavelet coefficients obtained using the Haar wavelet filters

$$[c_0(0)d_0(0)d_1(0)d_1(1)] = \left[\frac{5}{4}\,\frac{7}{4}\,\frac{1}{\sqrt{2}}\,-\frac{1}{2\sqrt{2}}\right],$$

perform the IDWT.

Solution:

From Eq. (12.54), we get

$$c_1(k) = \sum_{m=-\infty}^{\infty} c_0(m)h_0(k-2m) + \sum_{m=-\infty}^{\infty} d_0(m)h_1(k-2m).$$

Then we recover coefficients $c_1(k)$ as

$$\begin{aligned}
c_1(0) &= \sum_{m=-\infty}^{\infty} c_0(m)h_0(-2m) + \sum_{m=-\infty}^{\infty} d_0(m)h_1(-2m) \\
&= c_0(0)h_0(0) + d_0(0)h_1(0) = \frac{5}{4}\times\frac{1}{\sqrt{2}} + \frac{7}{4}\times\frac{1}{\sqrt{2}} = \frac{3\sqrt{2}}{2} \\
c_1(1) &= \sum_{m=-\infty}^{\infty} c_0(m)h_0(1-2m) + \sum_{m=-\infty}^{\infty} d_0(m)h_1(1-2m) \\
&= c_0(0)h_0(1) + d_0(0)h_1(1) = \frac{5}{4}\times\frac{1}{\sqrt{2}} + \frac{7}{4}\times\left(-\frac{1}{\sqrt{2}}\right) = -\frac{1}{2\sqrt{2}}.
\end{aligned}$$

MATLAB verification is given as.

```
>> c1=fconv([1 1]/sqrt(2),[5/4 0])+fconv([1 -1]/sqrt(2),[7/4 0])
  c1 = 2.1213 -0.3536
```

where the MATLAB function **fconv()** for filter operations with the forward filter coefficients is listed in Section 12.8. Again, from Eq. (12.54), we obtain

$$c_2(k) = \sum_{m=-\infty}^{\infty} c_1(m)h_0(k-2m) + \sum_{m=-\infty}^{\infty} d_1(m)h_1(k-2m).$$

Substituting the achieved wavelet coefficients $c_2(k)$, we yield

$$\begin{aligned}
c_2(0) &= \sum_{m=-\infty}^{\infty} c_1(m)h_0(-2m) + \sum_{m=-\infty}^{\infty} d_1(m)h_1(-2m) \\
&= c_1(0)h_0(0) + d_1(0)h_1(0) = \frac{3\sqrt{2}}{2}\times\left(\frac{1}{\sqrt{2}}\right) + \frac{1}{\sqrt{2}}\times\frac{1}{\sqrt{2}} = 2 \\
c_2(1) &= \sum_{m=-\infty}^{\infty} c_1(m)h_0(1-2m) + \sum_{m=-\infty}^{\infty} d_1(m)h_1(1-2m) \\
&= c_1(0)h_0(1) + d_1(0)h_1(1) = \frac{3\sqrt{2}}{2}\times\frac{1}{\sqrt{2}} + \frac{1}{\sqrt{2}}\left(-\frac{1}{\sqrt{2}}\right) = 1
\end{aligned}$$

$$c_2(2)=\sum_{m=-\infty}^{\infty}c_1(m)h_0(2-2m)+\sum_{m=-\infty}^{\infty}d_1(m)h_1(2-2m)$$

$$=c_1(1)h_0(0)+d_1(1)h_1(0)=\left(-\frac{1}{2\sqrt{2}}\right)\times\frac{1}{\sqrt{2}}+\left(-\frac{1}{2\sqrt{2}}\right)\times\frac{1}{\sqrt{2}}=-\frac{1}{2}$$

$$c_2(3)=\sum_{m=-\infty}^{\infty}c_1(m)h_0(3-2m)+\sum_{m=-\infty}^{\infty}d_1(m)h_1(3-2m)$$

$$=c_1(1)h_0(1)+d_1(1)h_1(1)=\left(-\frac{1}{2\sqrt{2}}\right)\times\frac{1}{\sqrt{2}}+\left(-\frac{1}{2\sqrt{2}}\right)\times\left(-\frac{1}{\sqrt{2}}\right)=0.$$

We can verify the results using the MATLAB program as follows:

```
>> c2=fconv([1 1]/sqrt(2),[3*sqrt(2)/2 0 -1/(2*sqrt(2)) 0])+fconv([1 -1]/sqrt(2),[1/
sqrt(2) 0 -1/(2*sqrt(2)) 0])
c2 = 2.0000 1.0000 -0.5000  0
```

Scaling the wavelet coefficients, we finally recover the original sample values as

$$f(k)=2^{2/2}[2\ 1\ -0.5\ 0]=[4\ 2\ -1\ 0].$$

Similarly, we can use the MATLAB function **idwt()** provided in Section 12.8 to perform the IDWT.

idwt.m

```
function c = idwt(h0,w,kLevel)
% h0 = wavelet filter coefficients (lowpass filter)
% w= wavelet coefficients
% kLevel = level
% c = input vector
```

Appling the MATLAB function **idwt()** leads to.

```
f=2*idwt([1/sqrt(2) 1/sqrt(2)][5/4 7/4 1/sqrt(2) -1/(2*sqrt(2))],2)'
f = 4.0000 2.0000 -1.0000 0.0000
```

Since $2^{j/2}$ scales signal amplitudes down in the analysis stage while scales them up back in the synthesis stage, we can omit $2^{j/2}$ by using $c(k)=f(k)$ directly in practice.

EXAMPLE 12.9

Given the sample values [4 2 – 1 0], use the provided MATLAB DWT (dwt.m) and IDWT (idwt.m) and specified wavelet filter to perform the DWT and IWDT without using the scale factor $2^{j/2}$.

(a) Haar wavelet filter
(b) four-tap Daubechies wavelet filter.

Solution:

(a) From Table 12.2, the Haar wavelet filter coefficients are

$$h_0 = \left(\frac{1}{\sqrt{2}} \frac{1}{\sqrt{2}}\right).$$

Applying the MATLAB function **dwt()** and **idwt()**, we have.

```
>> w=dwt([1/sqrt(2) 1/sqrt(2)][4 2 -1 0],2)'
w = 2.5000 3.5000 1.4142 -0.7071
>> f=idwt([1/sqrt(2) 1/sqrt(2)],w,2)'
w = 4.0000 2.0000 -1.0000  0
```

(b) From Table 12.2, the 4-tap Duabechies wavelet filter coefficients are.

```
h0=[0.482962913144534 0.836516303737808 0.224143868042013 -0.129409522551260];
```

MATLAB program verification is demonstrated below:

```
>> w=dwt([0.482962913144534 0.836516303737808 0.224143868042013 -
0.129409522551260][4 2 -1 0],2)'
w = 2.5000 2.2811 -1.8024 2.5095
>> f=idwt([0.482962913144534 0.836516303737808 0.224143868042013 -
0.129409522551260],w,2)'
f = 4.0000 2.0000 -1.0000  0
```

12.7 WAVELET TRANSFORM CODING OF SIGNALS

We can apply the DWT and IWDT for data compression and decompression. The compression and decompression involves two stages, that is, the analysis stage and the synthesis stage. At analysis stage, the wavelet coefficients are quantized based on their significance. Usually, we assign more bits to the coefficient in a coarser scale, since the corresponding subband has larger signal energy and low-frequency components. We assign a small number of bits to the coefficient, which resides in a finer scale, since the corresponding subband has lower signal energy and high-frequency components. The quantized coefficients can be efficiently transmitted. The DWT coefficients are laid out in a format described in Fig. 12.41. The coarse coefficients are placed toward the left side. For example, in Example 12.7, we organized the DWT coefficient vector as

$$w_2(k) = [c_0(0)d_0(0)d_1(0)d_1(1)] = \left[\frac{5}{4} \ \frac{7}{4} \ \frac{1}{\sqrt{2}} \ -\frac{1}{2\sqrt{2}}\right].$$

$$w_2(k) = [c_0(0)\ d_0(0)\ d_1(0)\ d_1(1)] = \left[\frac{5}{4}\ \frac{7}{4}\ \frac{1}{\sqrt{2}}\ -\frac{1}{2\sqrt{2}}\right]$$

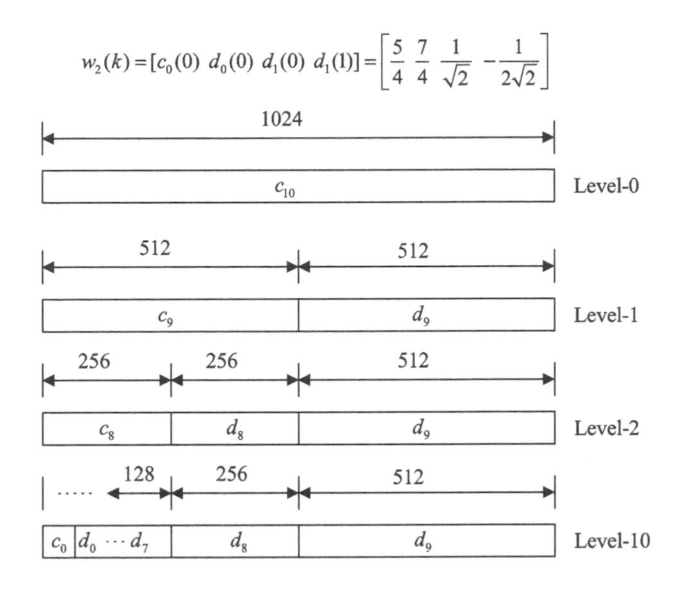

FIG. 12.41

DWT coefficient layout.

Let us look at the following simulation examples.

EXAMPLE 12.10

Given a 40-Hz sinusoidal signal plus random noise sampled at 8000 Hz with 1024 samples,

$$x(n) = 100\cos(2\pi \times 40nT) + 10 \times randn,$$

where $T = 1/8000$ s and *randn* is a random noise generator with a unit power and Gaussian distribution. Use a 16-bit code for each wavelet coefficient and write a MATLAB program to perform data compressions for each of the following ratios: 2:1, 4:1, 8:1, and 16:1. Plot the reconstructed waveforms.

Solution:

We use the 8-tap Daubechies filter as listed in Table 12.2. We achieve the data compression by dropping the high subband coefficients for each level consecutively and coding each wavelet coefficient in the lower subband using 16 bits. For example, we achieve the 2:1 compression ratio by omitting 512 high-frequency coefficients at the first level, 4:1 by omitting 512 high-frequency coefficients at first level and 256 high-frequency coefficients at the second level, and so on. The recovered signals are plotted in Fig. 12.42. SNR =21 dB is achieved for the 2:1 compression ratio. As we can see, when more and more higher frequency coefficients are dropped, the reconstructed signal contains less and less details. The recovered signal with the compression of 16:1 presents the least details but shows the smoothest signal. On the other hand, omitting the high-frequency wavelet coefficients can be very useful for a signal denoising application, in which the high-frequency noise contaminating the clean signal is removed. A complete MATLAB program is given in Program 12.2.

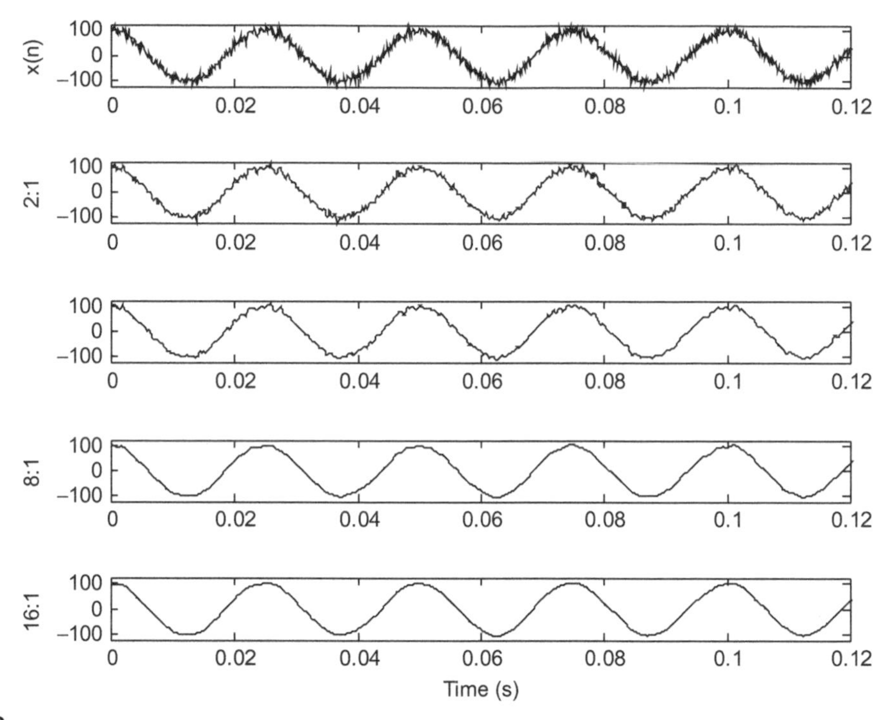

FIG. 12.42

Reconstructed signal at various compression ratios.

Program 12.2. Wavelet data compression.

```
close all; clear all;clc
t=0:1:1023;t=t/8000;
x=100*cos(40*2*pi*t)+10*randn(1,1024);
h0=[0.230377813308896  0.714846570552915  0.630880767929859 ...
  -0.027983769416859  -0.187034811719092  0.030841381835561 ....
  0.032883011666885  -0.010597401785069];
N=1024; nofseg=1
rec_sig=[]; rec_sig2t1=[]; rec_sig4t1=[]; rec_sig8t1=[]; rec_sig16t1=[];
for i=1:nofseg
  sp.=x((i-1)*1024+1:i*1024);
  w=dwt(h0,sp.,10);
% Quantization
  wmax=round(max(abs(w)));
  wcode=round(2^15*w/wmax); % 16-bit code for storage
  w=wcode*wmax/2^15; % Recovered wavelet coefficients
  w(513:1024)=zeros(1512); % 2:1 compression ratio
  sig_rec2t1=idwt(h0,w,10);
  rec_sig2t1=[rec_sig2t1 sig_rec2t1'];
  w(257:1024)=0; % 4:1 compression ratio
  sig_rec4t1=idwt(h0,w,10);
  rec_sig4t1=[rec_sig4t1 sig_rec4t1'];
  w(129:1024)=0; % 8:1 compression ratio
```

```
    sig_rec8t1=idwt(h0,w,10);
    rec_sig8t1=[rec_sig8t1 sig_rec8t1'];
    w(65:1024)=0; % 16:1 compression ratio
    sig_rec16t1=idwt(h0,w,10);
    rec_sig16t1=[rec_sig16t1 sig_rec16t1'];
end
subplot(5,1,1),plot(t,x,'k'); axis([0 0.12-120,120]);ylabel('x(n)');
subplot(5,1,2),plot(t,rec_sig2t1,'k'); axis([0 0.12-120,120]);ylabel('2:1');
subplot(5,1,3),plot(t,rec_sig4t1,'k'); axis([0 0.12-120,120]);ylabel(4:1);
subplot(5,1,4),plot(t,rec_sig8t1,'k'); axis([0 0.12-120,120]);ylabel('8:1');
subplot(5,1,5),plot(t,rec_sig16t1,'k'); axis([0 0.12-120,120]);ylabel('16:1');
xlabel('Time (Sec.)')
NN=min(length(x),length(rec_sig2t1)); axis([0 0.12-120,120]);
err=rec_sig2t1(1:NN)-x(1:NN);
SNR=sum(x.*x)/sum(err.*err);
disp('PR reconstruction SNR dB≥');
SNR=10*log10(SNR)
```

Fig. 12.43 shows the wavelet compression for 16-bit speech data sampled at 8 kHz. The original speech data is divided into speech segments, each with 1024 samples. After applying the DWT to each segment, the coefficients, which correspond to high frequency components indexed from 513 to 1024, are discarded in order to achieve the coding efficiency. The reconstructed speech data has a compression ratio 2:1 with SNR = 22 dB. The MATLAB program is given in Program 12.3.

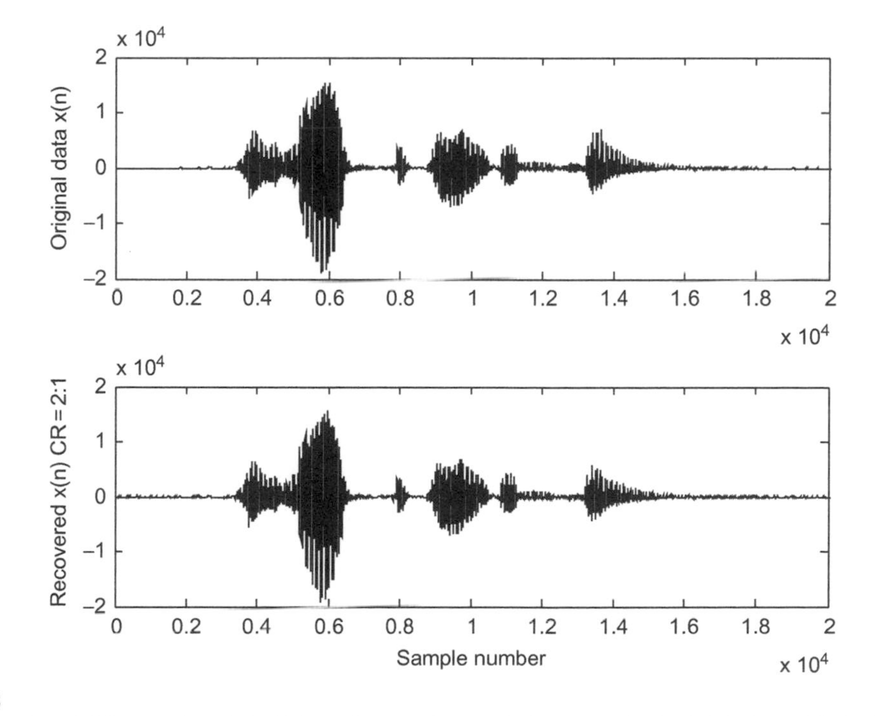

FIG. 12.43

Reconstructed speech signal with compression ratio of 2 and SNR = 22 dB.

Program 12.3. Wavelet data compression for speech segments.

```
close all; clear all;clc
load orig.dat; %Load speech data
h0=[0.230377813308896 0.714846570552915 0.630880767929859 ...
  -0.027983769416859 -0.187034811719092 0.030841381835561 ....
  0.032883011666885 -0.010597401785069];
N=length(orig);
nofseg=ceil(N/1024);
speech=zeros(1,nofseg*1024);
speech(1:N)=orig(1:N);% Making the speech length to be multiple of 1024 samples
rec_sig=[];
for i=1:nofseg
  sp.=speech((i-1)*1024+1:i*1024);
  w=dwt(h0,sp.,10);
% Quantization
  w=(round(2^15*w/2^15))*2^(15-15);
  w(513:1024)=zeros(1512); % Omitting the high frequency coefficients
  sp_rec=idwt(h0,w,10);
  rec_sig=[rec_sig sp_rec'];
end
subplot(2,1,1),plot([0:length(speech)-1],speech,'k');axis([0 20,000-20,000
20,000]);
ylabel('Original data x(n)');
subplot(2,1,2),plot([0:length(rec_sig)-1],rec_sig,'k');axis([0 20,000-20,000
20,000]);
xlabel('Sample number');ylabel('Recovered x(n) CR=2:1');
NN=min(length(speech),length(rec_sig));
err=rec_sig(1:NN)-speech(1:NN);
SNR=sum(speech.*speech)/sum(err.*err);
disp('PR reconstruction SNR dB≥');
SNR=10*log10(SNR)
```

Fig. 12.44 displays the wavelet compression for 16-bit ECG data using Program 12.3. The reconstructed ECG data has a compression ratio of 2:1 with SNR = 33.8 dB.

Fig. 12.45 illustrates an application of signal denoising using the DWT with coefficient threshold. During the analysis stage, the obtained DWT coefficient (quantization is not necessary) is set to zero if its value is less than the predefined threshold as depicted in Fig. 12.45. This simple technique is called the hard threshold. Usually, the small wavelet coefficients are related to the high-frequency components in signals. Therefore, setting high-frequency components to zero is the same as lowpass filtering.

An example is shown in Fig. 12.46. The first plot depicts a 40-Hz noisy sinusoidal signal (sine wave plus noise with SNR = 18 dB) and the clean signal with a sampling rate of 8000 Hz. The second plot shows that after zero threshold operations, 67% of coefficients are set to zero and the recovered signal

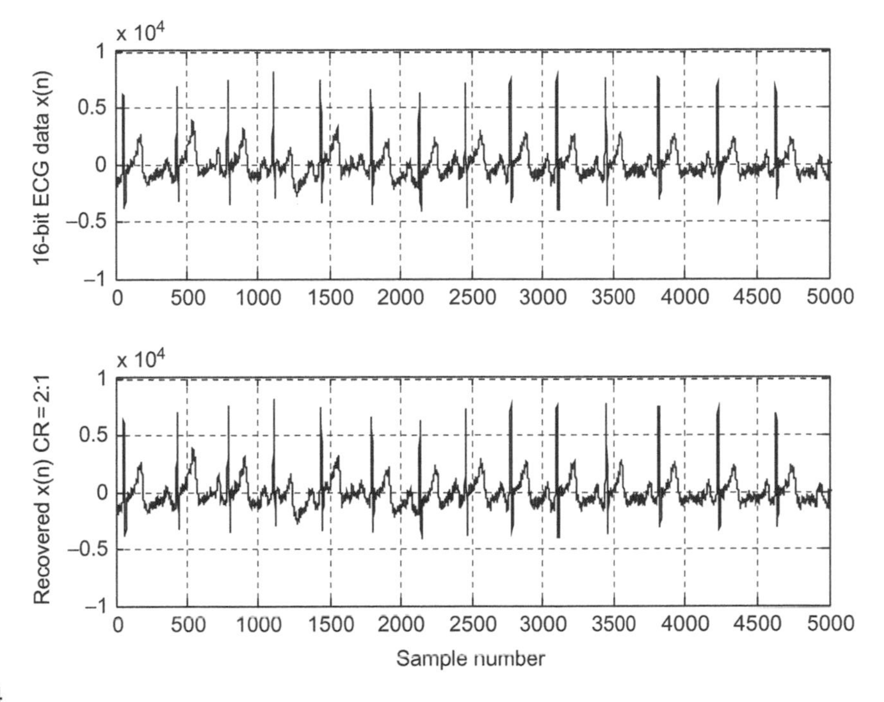

FIG. 12.44

Reconstructed ECG signal with the compression ratio of 2 and SNR = 33.8 dB.

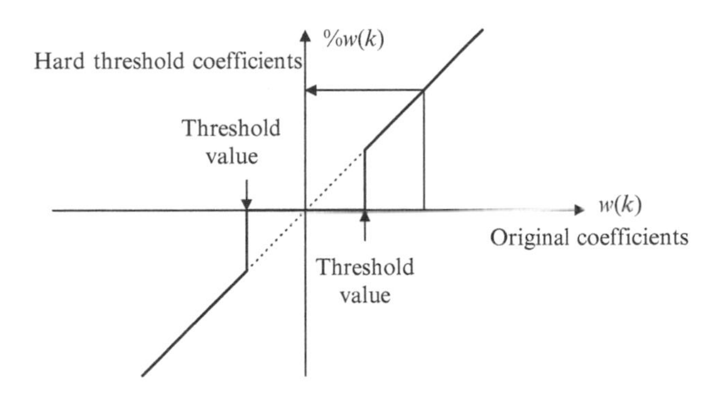

FIG. 12.45

Hard threshold for the DWT coefficients.

has a SNR = 19 dB. Similarly, the third and fourth plots illustrate that 93% and 97% of coefficients are set to zero after threshold operations and the recovered signals have the SNR = 23 and 28 dB, respectively. As an evidence that the signal is smoothed, that is, the high frequency noise is attenuated, the wavelet denoise technique is equivalent to lowpass filtering.

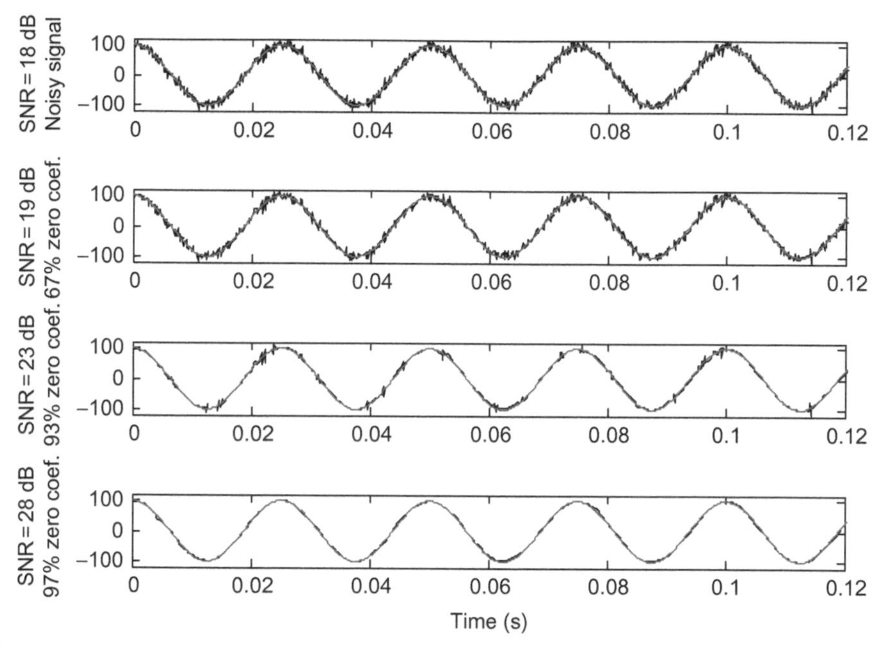

FIG. 12.46

Signal denoising using wavelet transform coding.

12.8 MATLAB PROGRAMS

In this section, four key MATLAB programs are included. **rconv()** and **fconv()** perform circular convolutions with the reversed filter coefficients and the forward filter coefficients, respectively. **dwt()** and **idwt()** are the programs to compute the DWT coefficients and IDWT coefficients. The resolution level can be specified.

Program 12.4. Circular convolution with the reversed filter coefficients (rconv.m).

```
function [y]=rconv(h,c)
% Circular convolution using the reversed filter coefficients h(-k)
% h=filter coefficients
% c=input vector
% y=output vector
N=length(c); M=length(h);
xx=zeros(1,M+N-1);
xx(1:N)=c;
xx(N+1:N+M-1)=c(1:M-1); % Use periodized input
for n=1:N;
  y(n)=0;
  for m=1:M
    y(n)=y(n)+h(m)*xx(n+m-1);
  end
end
```

Program 12.5. Circular convolution with the forward filter coefficients (fconv.m).

```
function [y]=fconv(h,c)
% Circular convolution using the forward filter coefficients h(k)
% h=filter coefficients
% c=input vector
% y=output vector
N=length(c); M=length(h);
x(1:N+M-1)=zeros(1,N+M-1);
  for j=1:N
  x(j:M+(j-1))=x(j:M+(j-1))+c(j)*h;
  end
  for i=N+M-1:-1:N+1
  x(i-N)=x(i-N)+x(i); % Circular convolution
  end
  y=x(1:N);
```

Program 12.6. DWT coefficients (dwt.m).

```
function w=dwt(h0,c,kLevel)
% w=dwt(h,c,k)
% Computes wavelet transform coefficients for a vector c using the
% orthonormal wavelets defined by the coefficients h
% h=wavelet coefficients
% c=input vector
% kLelvel=level
% w=wavelet coefficients
n=length(c); m=length(h0);
h1=h0(m:-1:1); h1(2:2:m)=-h1(2:2:m);
h0=h0(:)'; h1=h1(:)';
c=c(:); w=c;
x=zeros(n+m-2,1);
% Perform decomposition through k levels
% at each step, x=periodized version of x-coefficients
for j=1:kLevel
  x(1:n)=w(1:n);
  for i=1:m-2
  x(n+i)=x(i);
  end
  for i=1:n/2
  w(i)=h0 * x(1+2*(i-1):m+2*(i-1));
  w(n/2+i)=h1* x(1+2*(i-1):m+2*(i-1));
  end
  n=n/2;
end
```

Program 12.7. IDWT coefficients (idwt.m).

```
function c=idwt(h0,w,kLevel)
% c=idwt(h0,w,kLevel)
% Computes the inverse fast wavelet transform from data W using the
% orthonormal wavelets defined by the coefficients.
% h0=wavelet filter coefficients
% w=wavelet coefficients
% kLevel=level
% c=IDWT coefficients
n=length(w); m=length(h0);
h1=h0(m:-1:1); h1(2:2:m)=-h1(2:2:m);
h0=h0(:); h1=h1(:);
w=w(:); c=w;
x=zeros(n+m-2,1);
% Perform the reconstruction through k levels
% x=periodized version of x-coefficients
n=n/2^kLevel;
for i=1:kLevel
  x(1:2*n+m-2)=zeros(2*n+m-2,1);
  for j=1:n
  x(1+2*(j-1):m+2*(j-1))=x(1+2*(j-1):m+2*(j-1))+c(j)*h0+w(n+j)*h1;
  end
  for i=2*n+m-2:-1:2*n+1
  x(i-2*n)=x(i-2*n)+x(i);
  end
  c(1:2*n)=x(1:2*n);
  n=2 * n;
end
```

12.9 SUMMARY

1. A signal can be decomposed using a filter bank system. The filter bank contains two stages: the analysis stage and the synthesis stage. The analysis stage applies analysis filters to decompose the signal into multichannels. The signal from each channel is downsampled and coded. At the synthesis stage, the recovered signal from each channel is upsampled and processed using its synthesis filter. Then the outputs from all the synthesis filters are combined to produce the recovered signal.
2. Perfect reconstruction conditions for two-band case are derived to design the analysis and synthesis filters. The conditions consist of a half-band filter requirement and normalization. Once the lowpass analysis filter coefficients are obtained, the coefficients for other filters can be achieved from the derived relationships.
3. In a binary tree structure, a filter bank divides an input signal into two equal subbands, resulting in the low and high bands. Each band again splits into low and high bands to produce quarter bands. The process continues in this form.
4. The dyadic structure implementation of the filter bank first splits the input signal into low and high bands and then continues into split the low band only each time.

5. By quantizing each subband channel using the assigned number bits based on the signal significance (more bits are assigned to code the sample for a channel with large signal energy while less bits are assigned to code the samples with small signal energy), the subband-coding method demonstrates efficiency for data compression.
6. The wavelet transform can identify the frequencies in a signal and the times when the signal elements occur and end.
7. The wavelet transform provides either good frequency resolution or good time resolution, but not both.
8. The wavelet has two important properties: scaling and translation. The scaling process is related to changes of wavelet frequency (oscillation) while the translation is related to the time localization.
9. A family of wavelets contains a father wavelet and a mother wavelet, and their scaling and translation versions. The father wavelet and its scaling and translation are called the scaling functions while the mother wavelet and its scaling and translation are called the wavelet function.
   Each scaling function and wavelet function can be presented using the scaling functions in the next finer scale.
10. A signal can be approximated from a sum of weighted scaling functions and wavelet functions. The weights are essentially the DWT coefficients. A signal can also be coded at any desired level using the smaller-scale wavelets.
11. Implementation of DWT and IDWT consists of the analysis and synthesis stages, which are similar to the subband-coding scheme. The implementation uses the dyadic structure but with each analysis filter coefficients in a reversed format.
12. The DWT and IDWT are very effective for data compression or signal denoising by eliminating smaller DWT coefficients, which correspond to higher frequency components.

12.10 PROBLEMS

12.1 Given each of the following downsampling systems (Fig. 12.47A and B) and input spectrum $W(f)$, sketch the downsampled spectrum $X(f)$.

12.2 Given each of the following upsampling systems (Fig. 12.48A and B) and input spectrum $X(f)$, sketch the upsampled spectrum $\overline{W}(f)$. Note that the sampling rate for input $x(m)$ $f_{sM}=f_s/4$ and the output-sampling rate $\overline{w}(n)$ isf_s.

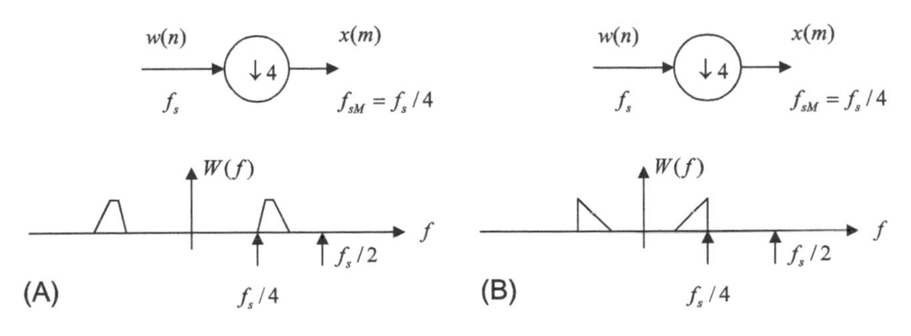

FIG. 12.47

Downsampling systems in Problem 12.1.

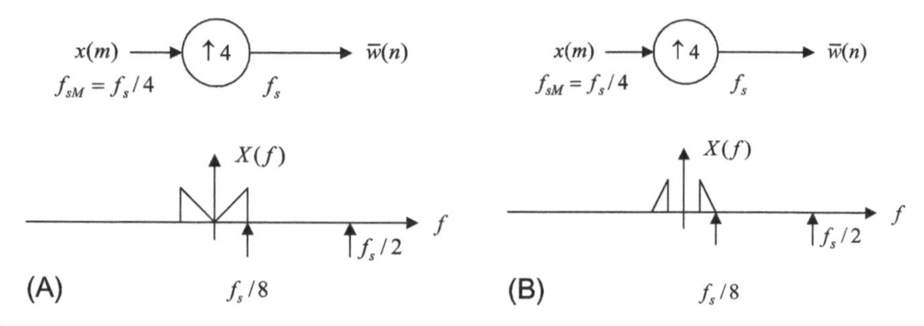

FIG. 12.48

Upsampling systems in Problem 12.2.

12.3 Given each of the following down- and upsampling systems (Fig. 12.49A and B) and the input spectrum $W(f)$, sketch output spectrum $\overline{Y}(f)$ and express $\overline{Y}(f)$ in terms of $W(f)$.

12.4 Given each of the following down- and upsampling systems (Fig. 12.50A and B) and the input spectrum $W(f)$, sketch output spectrum $\overline{Y}(f)$ and express $\overline{Y}(f)$ in terms of $W(f)$.

12.5 Given $H_0(z) = \frac{1}{\sqrt{2}} + \frac{1}{\sqrt{2}}z^{-1}$, determine $H_1(z)$, $G_0(z)$, and $G_1(z)$.

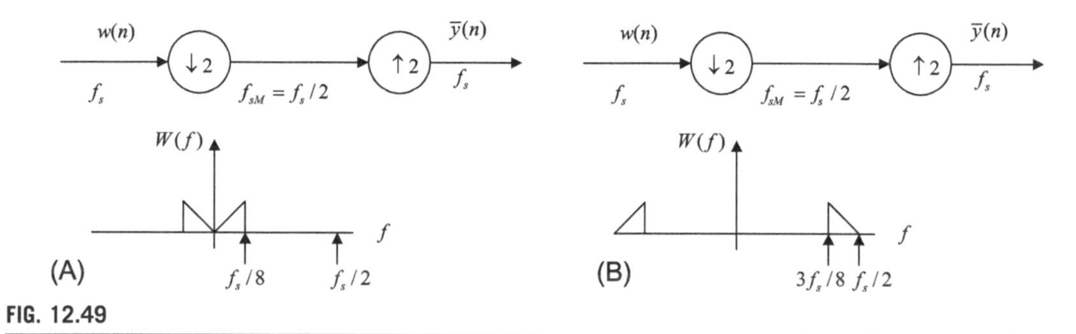

FIG. 12.49

Down- and upsampling systems in Problem 12.3.

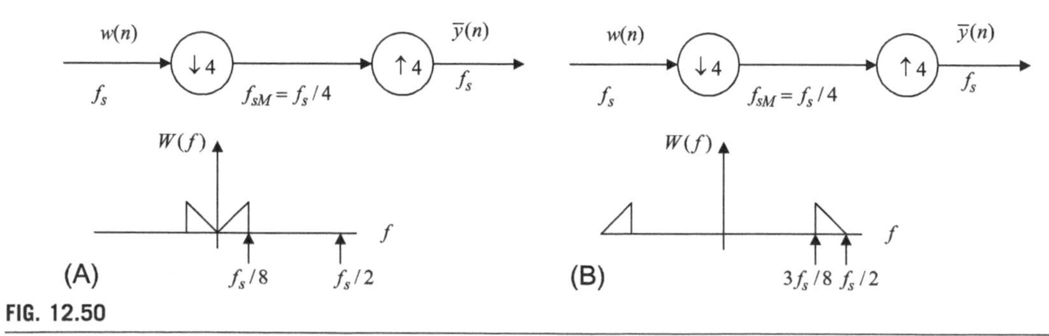

FIG. 12.50

Down- and upsampling systems in Problem 12.4.

12.6 Given $H_0(z) = \frac{1}{\sqrt{2}} + \frac{1}{\sqrt{2}}z^{-1}$, verify the following conditions:

$$\rho(2n) = \sum_{k=0}^{N-1} h_0(k)h_0(k+2n) = \delta(n)$$
$$R(z) + R(-z) = 2,$$

and plot the magnitude frequency responses of analysis and synthesis filters.

12.7 Given

$$H_0(z) = 0.483 + 0.837z^{-1} + 0.224z^{-2} - 0.129z^{-3},$$

determine $H_1(z)$, $G_0(z)$, and $G_1(z)$.

12.8 Given

$$H_0(z) = 0.483 + 0.837z^{-1} + 0.224z^{-2} - 0.129z^{-3},$$

verify the following conditions:

$$\rho(2n) = \sum_{k=0}^{N-1} h_0(k)h_0(k+2n) = \delta(n)$$
$$R(z) + R(-z) = 2.$$

12.9 Draw a four-band dyadic tree structure of a subband system including the analyzer and the synthesizer.

12.10 Draw an eight-band dyadic tree structure of a subband system including the analyzer and the synthesizer.

12.11 Given a function in Fig. 12.51

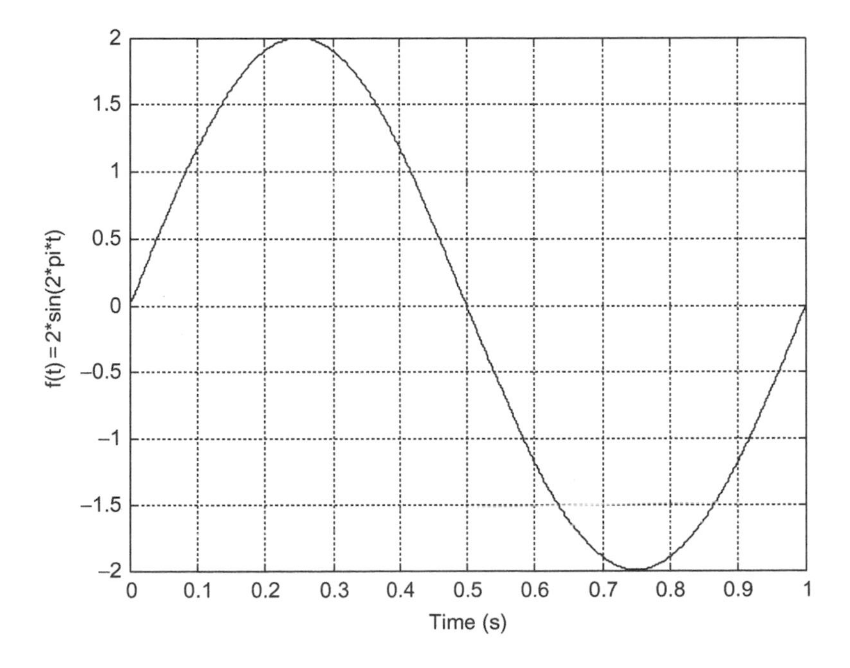

FIG. 12.51

A sine function in Problem 12.11.

Sketch
(a) $f(4t)$
(b) $f(t-2)$
(c) $f(2t-3)$
(d) $f(t/2)$
(e) $f(t/4-0.5)$.

12.12 Given a father wavelet (base scaling function) in base scale plotted in Fig. 12.52A, determine a and b for each of the wavelets plotted in Fig. 12.52B and C.

12.13 Given the signal as depicted in Fig. 12.53,

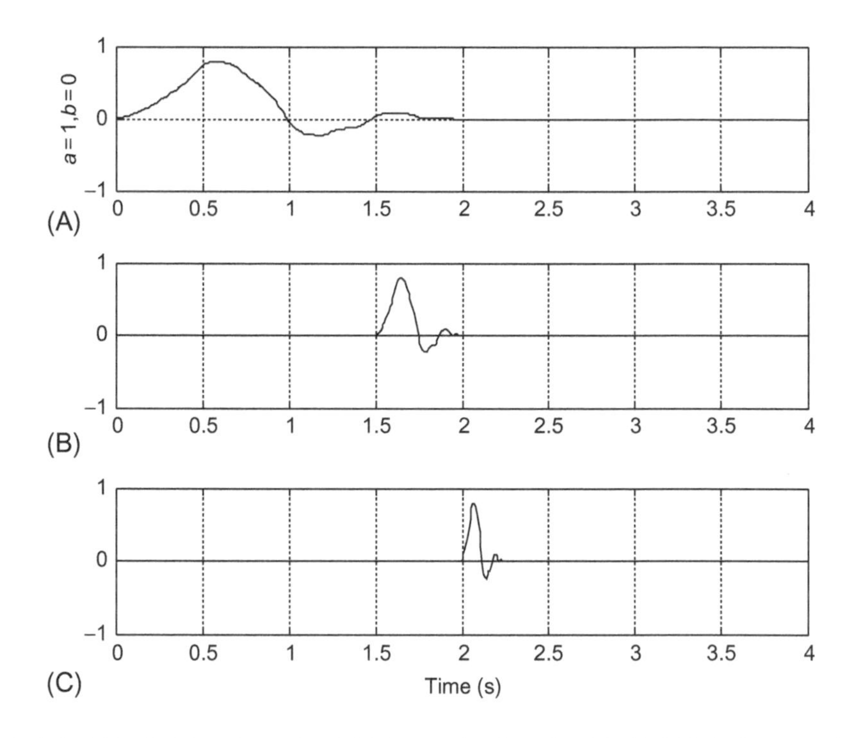

FIG. 12.52
Wavelets in Problem 12.12.

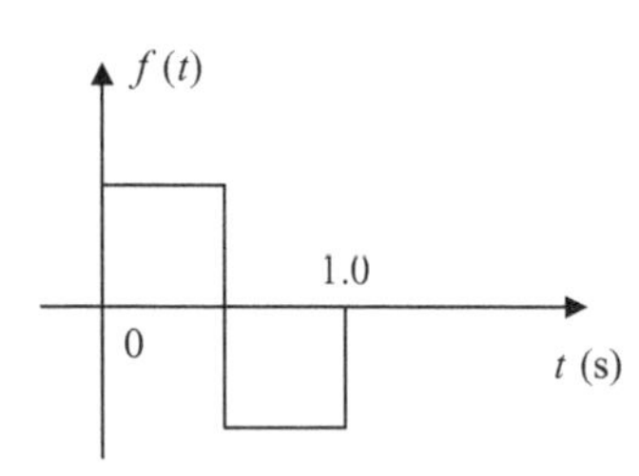

FIG. 12.53
The function in Problem 12.13.

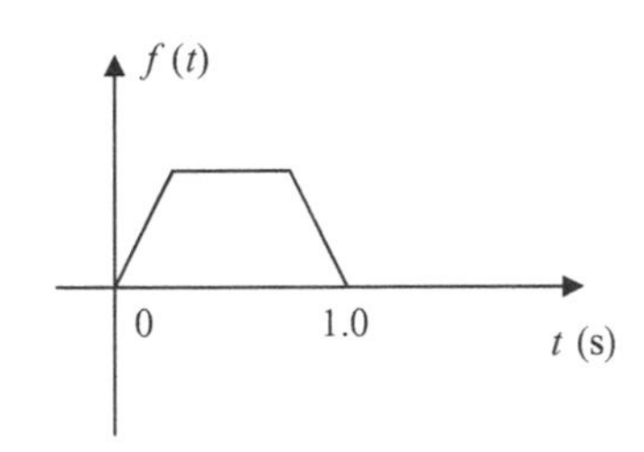

FIG. 12.54

A trapezoidal function in Problem 12.14.

Sketch

(a) $f(4t)$
(b) $f(t-2)$
(c) $f(2t-3)$
(d) $f(t/2)$
(e) $f(t/4-0.5)$.

12.14 Given the signal as shown in Fig. 12.54

Sketch

(a) $f(4t)$
(b) $f(t-2)$
(c) $f(2t-3)$
(d) $f(t/2)$
(e) $f(t/4-1)$.

12.15 Sketch the Haar father wavelet families for three different scales: $j=0, 1, 2$ for a period of 2 s.

12.16 Sketch the Haar mother wavelet families for three different scales: $j=0, 1, 2$ for a period of 2 s.

12.17 Use the Haar wavelet family to expand the signal depicted in Fig. 12.55.

(a) Use only scaling functions $\phi(2t-k)$.
(b) Use scaling functions and wavelets: $\phi(t)$ and $\psi(t-k)$.

12.18 Use the Haar wavelet family to expand the signal depicted in Fig. 12.56.

(a) Use only scaling functions $\phi(4t-k)$.
(b) Use scaling functions and wavelets: $\phi(2t-k)$ and $\psi(2t-k)$.
(c) Using the scaling function and wavelets: $\phi(t)$, $\psi(2t-k)$, and $\psi(t-k)$.

12.19 Use the Haar wavelet family to expand the signal

$$x(t) = \sin(2\pi t) \text{for} 0 \le t \le 1.$$

(a) Use only scaling functions $\phi(2t-k)$.
(b) Use scaling functions and wavelets: $\phi(t)$ and $\psi(t-k)$.

12.20 Use the Haar wavelet family to expand the signal

$$x(t) = e^{-5t} \text{for} 0 \le t \le 1.$$

(a) Use only scaling functions $\phi(2t-k)$.
(b) Use scaling functions and wavelets: $\phi(t)$ and $\psi(t-k)$.

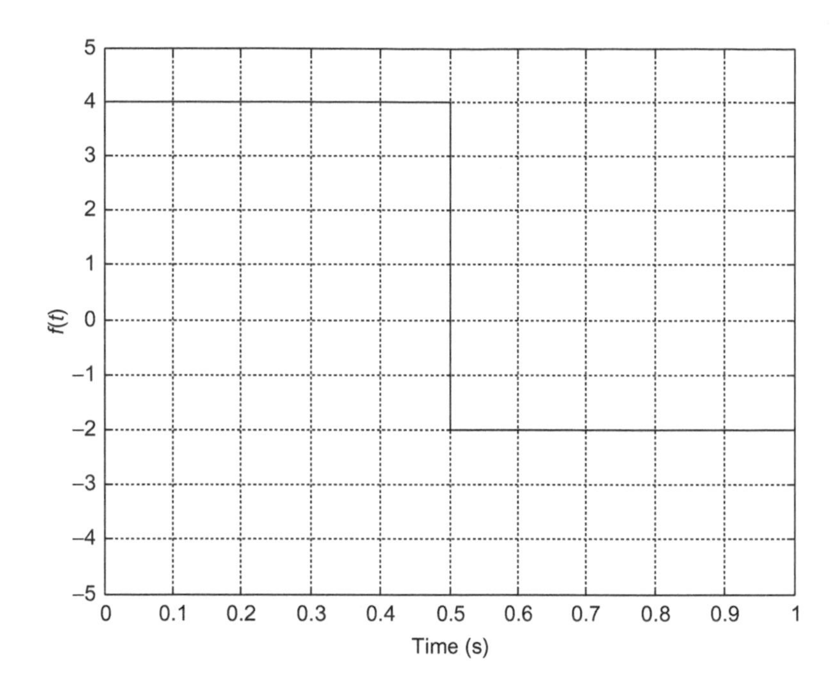

FIG. 12.55

A gate function in Problem 12.17.

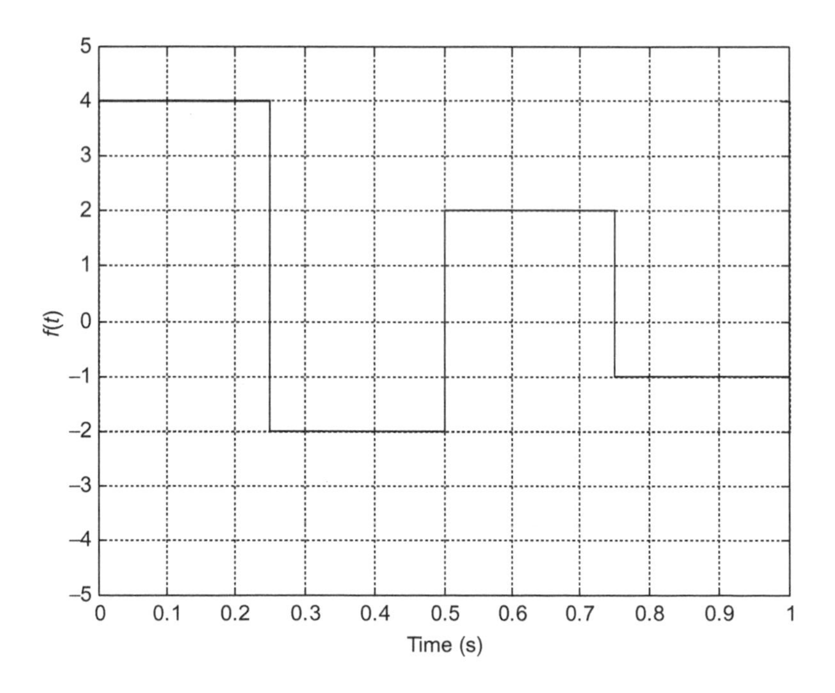

FIG. 12.56

A piecewise function in Problem 12.18.

12.21 Verify the following equations using the Haar wavelet families:

(a) $\phi(2t) = \sum_{k=-\infty}^{\infty} \sqrt{2} h_0(k)\phi(4t-k)$

(b) $\psi(2t) = \sum_{k=-\infty}^{\infty} \sqrt{2} h_1(k)\phi(4t-k)$.

12.22 Given the four-tap Daubechies wavelet coefficients

$$h_0(k) = [0.483\ 0.837\ 0.224\ -0.129].$$

Determine $h_1(k)$ and plot magnitude frequency responses for both $h_0(k)$ and $h_1(k)$.

12.23 Given the sample values [8 −2 4 1], use the Haar wavelet to determine the level-2 wavelet coefficients.

12.24 Given the sample values [8 −2 4 3 0 −1 −2 0], use the Haar wavelet to determine level-3 wavelet coefficients.

12.25 Given the level-2 wavelet coefficients [4 2 −1 2], use the Haar wavelet to determine the sampled signal vector $f(k)$.

12.26 Given the level-3 wavelet coefficients [4 2 −1 2 0 0 0 0], use the Haar wavelet to determine the sampled signal vector $f(k)$.

12.27 Given the level-1 wavelet coefficients [4 2 −1 2], use the Haar wavelet to determine the sampled signal vector $f(k)$.

12.28 The four-level DWT coefficients are packed below:

W = [100 20 16 –5 –3 4 2 –6 4 6 1 2 –3 0 2 –1].

List the wavelet coefficients to achieve each of the following compression ratios:

(a) 2:1

(b) 4:1

(c) 8:1

(d) 16:1.

MATLAB Problems

Use MATLAB to solve Problems 12.29–12.31.

12.29 Use the 16-tap PR-CQF coefficients and MATLAB to verify the following conditions:

$$\rho(2n) = \sum_{k=0}^{N-1} h_0(k)h_0(k+2n) = \delta(n);$$

$$R(z) + R(-z) = 2.$$

Plot the frequency responses for $h_0(k)$ and $h_1(k)$.

12.30 Use MATLAB functions provided in Section 12.8 [**dwt(), idwt()**] to verify Problems 12.23–12.27.

12.31 Given a 20-Hz sinusoidal signal plus random noise sampled at 8000 Hz with 1024 samples,

$$x(n) = 100\cos(2\pi \times 20nT) + 50 \times randn,$$

where T = 1/8000 s and *randn* is a random noise generator with a unit power and Gaussian distribution.

(a) Use a 16-bit code for each wavelet coefficient and write a MATLAB program to perform data compressions with the following ratios: 2:1, 4:1, 8:1, 16:1, and 32:1;

(b) Measure the SNR in dB for each case.

(c) Plot the reconstructed waveform for each case.

MATLAB Projects

12.32 Data compression using subband coding:

Given 16-bit speech data ("speech.dat") and using the four-band subband-coding method, write a MATLAB program to compress speech signal with the following specifications:

(a) 16 bits for each of subband coefficients, coding LL, LH, HL, HH subbands, and measure the SNR in dB;
(b) 16 bits for each of subband coefficients, coding LL band, discarding LH, HL, and HH subbands;
(c) 16 bits for each of subband coefficients, coding LL, LH band, discarding HL, and HH subbands;
(d) 16 bits for each of subband coefficients, coding LL, LH, HL band, discarding HH subband;
(e) Measure SNR in dB for (a), (b), (c), and (d);
(f) Determine the achieved compression ratios for (a), (b), (c), (d);
(g) Repeat (a) to (f) for seismic data ("seismic.dat") in which each sample is encoded using 32 bits instead of 16 bits.

12.33 Wavelet-based data compression:

Given 16-bit speech data ("speech.dat") and using the wavelet-coding method with 16-tap Daubechies wavelet filters, write a MATLAB program to compress the speech signal with the following specification:

(a) 16 bits for each of wavelet coefficients, compression 2:1;
(b) 16 bits for each of wavelet coefficients, compression 4:1;
(c) 16 bits for each of wavelet coefficients, compression 8:1;
(d) 16 bits for each of wavelet coefficients, compression 16:1;
(e) 16 bits for each of wavelet coefficients, compression 32:1;
(f) Measure SNR in dB for (a), (b), (c), (d), and (f);
(g) Repeat (a) to (f) for seismic data ("seismic.dat") in which each sample is encoded using 32 bits instead of 16 bits.

CHAPTER

IMAGE PROCESSING BASICS 13

CHAPTER OUTLINE

Digital Signal Processing. https://doi.org/10.1016/B978-0-12-815071-9.00013-0

13.1 IMAGE PROCESSING NOTATION AND DATA FORMATS

A digital image is picture information in a digital form. The image could be filtered to remove noise to obtain enhancement. It can also be transformed to extract features for pattern recognition. The image can be compressed for storage and retrieval as well as transmitted via a computer network or a communication system.

The digital image consists of pixels. The position of each pixel is specified in terms of an index for the number of columns and another for the number of rows. Fig. 13.1 shows that a pixel $p(2,8)$ has a level of 86 and is located in the second row, eighth column. We express it in notation as

$$p(2,8) = 86. \tag{13.1}$$

The number of pixels in the presentation of a digital image is referred to *spatial resolution*, which relates to the image quality. The higher the spatial resolution, the better quality the image has. The resolution can be fairly high, for instance, as high as 1600×1200 (1,920,000 pixels = 1.92 megapixels), or as low as 320×200 (64,000 pixels = 64 kpixels). In notation, the number to the left of the multiplication symbol represents the width, and that to the right of the symbol represents the height in terms of pixels. Image quality also depends on the numbers of bits used in encoding each pixel level, which is discussed in the next section.

13.1.1 8-BIT GRAY LEVEL IMAGES

If a pixel is encoded on a gray scale from 0 to 255 (256 values), where 0 = black and 255 = white, the numbers in between represent levels of gray forming a *grayscale image*. For a 640×480 8-bit image, 307.2 kbytes are required for storage. Fig. 13.2 shows a grayscale image format. As shown in Fig. 13.2, the pixel indicated in the box has an 8-bit value of 25.

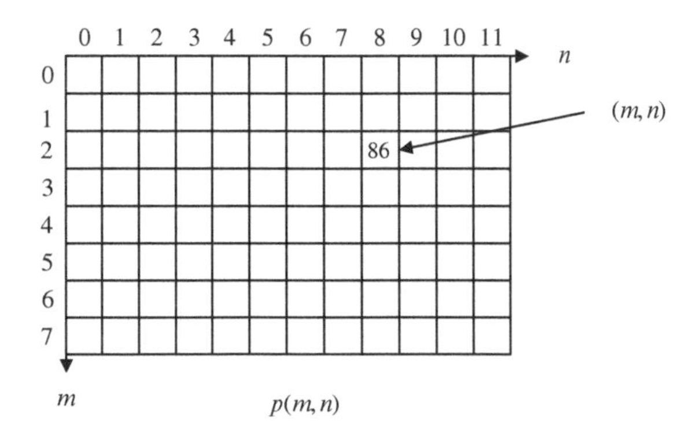

FIG. 13.1

Image pixel notation.

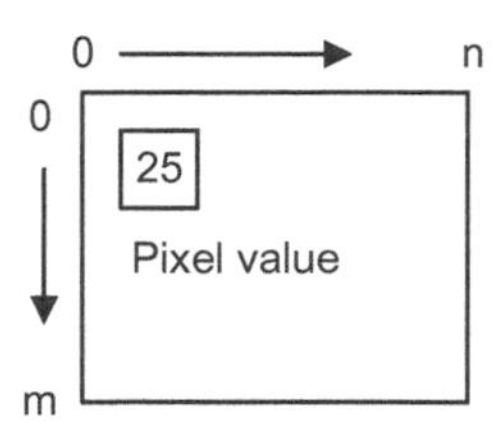

FIG. 13.2

Grayscale image format.

FIG. 13.3

Grayscale image (8-bit 320 × 240).

The image of a cruise ship with a spatial resolution of 320 × 240 in an 8-bit grayscale level is shown in Fig. 13.3. The tagged image file format (TIFF) is the common encoding file format, which is described in the textbook by Li et al. (2014).

13.1.2 24-BIT COLOR IMAGES

In a 24-bit color image representation, each pixel is recoded with the red, green, and blue (RGB) components (RGB channels). With each component value encoded in 8 bits, resulting in 24 bits in total, we achieve a full color RGB image. With such an image, we can have $2^{24} = 16.777216 \times 10^6$ different colors. A 640 × 480 24-bit color image requires 921.6 kbytes for storage. Fig 13.4 shows the format for the 24-bit color image where the indicated pixel has 8-bit RGB components.

Fig. 13.5 shows a 24-bit color image of the Grand Canyon, along with its grayscale displays for the 8-bit RGB component images. The common encoding file format is TIFF.

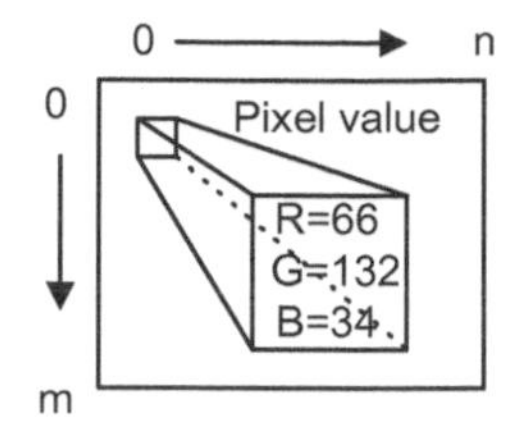

FIG. 13.4

The 24-bit color image format.

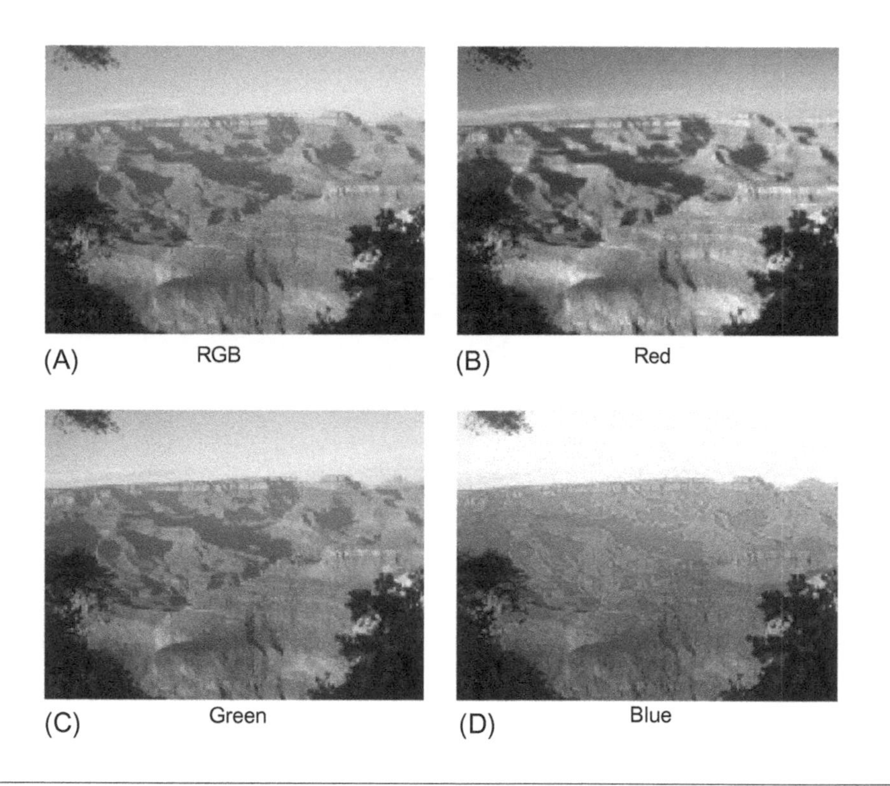

FIG. 13.5

The 24-bit color image and its respective RGB components.

13.1.3 8-BIT COLOR IMAGES

The 8-bit color image is also a popular image format. Its pixel value is a color index that points to a color look-up table containing RGB components. We call this a *color indexed image*, and its format is shown in Fig. 13.6. As an example in the figure, the color indexed image has a pixel index value of 5, which is the index for the entry of the color table, called the *color map*. At the index of location 5, there are three color components with RGB values of 66, 132, and 34, respectively. Each color component is encoded in 8 bits. There are only 256 different colors in the image. A 640×480 8-bit color image

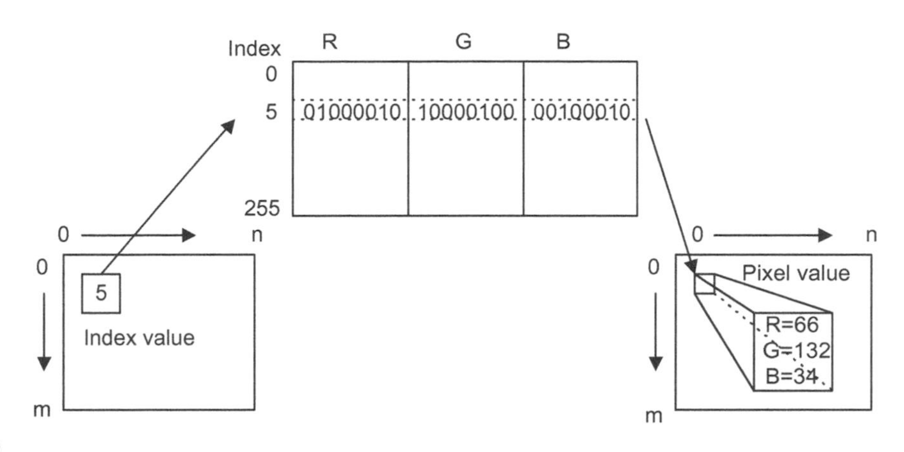

FIG. 13.6

The 8-bit color indexed image format.

FIG. 13.7

The 8-bit color indexed image (For interpretation of the references to colour in this figure legend, the reader is referred to the web version of this article.)

requires memory of 307.2 kbytes for data storage and $3 \times 256 = 768$ bytes for color map storage. The 8-bit color image for the cruise ship shown in Fig. 13.3 is displayed in Fig. 13.7. The common encoding file format is the graphics interchange format (GIF), which described in the textbook by Li et al. (2014).

13.1.4 INTENSITY IMAGES

As we noted in the first section, the grayscale image uses a pixel value ranging from 0 to 255 to present luminance, or the light intensity. A pixel value of 0 designates black, and a value 255 encodes for white.

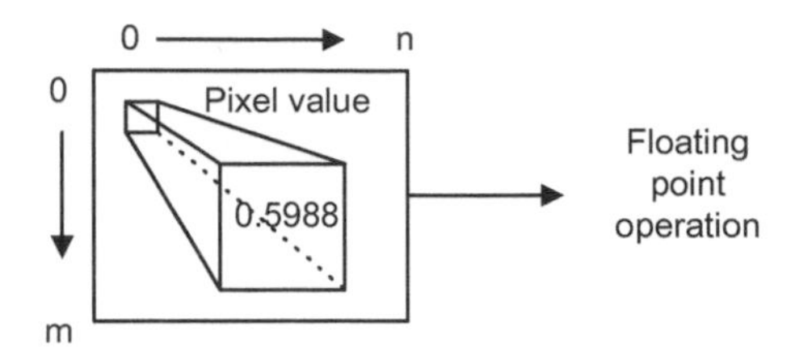

FIG. 13.8

The grayscale intensity image format.

In some processing environment such as MATLAB, floating-point operations are used. The grayscale intensity image has the intensity value that is normalized to the range from 0 to 1.0, where 0 represents black and 1 represents white. We often change the pixel value to the normalized range to get the grayscale intensity image before processing it, and then scale it back to the standard 8-bit range after processing for display. With the intensity image in the floating-point format, the digital filter implementation can be easily applied. Fig. 13.8 shows the format of the grayscale intensity image, where the indicated pixel shows the intensity value of 0.5988.

13.1.5 RGB COMPONENTS AND GRAYSCALE CONVERSION

In some applications, we need to convert a color image to a grayscale image so that storage space can be saved. As an example, fingerprint images are stored in the grayscale format in a database system. In color image compression as another example, the transformation converts the RGB color space to YIQ color space (Li et al., 2014; Rabbani and Jones, 1991), where Y is the luminance (Y) channel representing light intensity while the I (in-space) and Q (quadrature) chrominance channels represent color details.

The luminance $Y(m,n)$ carries grayscale information with most of the signal energy (as much as 93%), and the chrominance channels $I(m,n)$ and $Q(m,n)$ carry color information with much less energy (as little as 7%). The transformation in terms of the standard matrix notion is given by

$$\begin{bmatrix} Y(m,n) \\ I(m,n) \\ Q(m,n) \end{bmatrix} = \begin{bmatrix} 0.299 & 0.587 & 0.114 \\ 0.596 & -0.274 & -0.322 \\ 0.212 & -0.523 & 0.311 \end{bmatrix} \begin{bmatrix} R(m,n) \\ G(m,n) \\ B(m,n) \end{bmatrix}. \tag{13.2}$$

As an example of data compression, after transformation, we can encode $Y(m,n)$ with a higher resolution using a larger number of bits, since it contains most of the signal energy, while we encode chrominance channels $I(m,n)$ and $Q(m,n)$ with less resolution using a smaller number of bits. Inverse transformation can be solved as

$$\begin{bmatrix} R(m,n) \\ G(m,n) \\ B(m,n) \end{bmatrix} = \begin{bmatrix} 1.000 & 0.956 & 0.621 \\ 1.000 & -0.272 & -0.647 \\ 1.000 & -1.106 & 1.703 \end{bmatrix} \begin{bmatrix} Y(m,n) \\ I(m,n) \\ Q(m,n) \end{bmatrix}. \tag{13.3}$$

To obtain the grayscale image, we simply convert each RGB pixel to the YIQ pixel; and then keep its luminance channel and discard IQ channel chrominance. The conversion formula is hence given by

$$Y(m,n) = 0.299 \times R(m,n) + 0.587 \times G(m,n) + 0.114 \times B(m,n). \tag{13.4}$$

Note that $Y(m,n)$, $I(m,n)$, and $Q(m,n)$ can be matrices that represents the luminance image two color component images, respectively. Similarly, $R(m,n)$, $G(m,n)$, and $B(m,n)$ can be matrices for the RGB component images.

EXAMPLE 13.1

Given a pixel in an RGB image as follows:

$$R=200,\ G=10,\ B=100,$$

convert the pixel values to the YIQ values.

Solution:

Applying Eq. (13.2), it follows that

$$\begin{bmatrix} Y \\ I \\ Q \end{bmatrix} = \begin{bmatrix} 0.299 & 0.587 & 0.114 \\ 0.596 & -0.274 & -0.322 \\ 0.212 & -0.523 & 0.311 \end{bmatrix} \begin{bmatrix} 200 \\ 10 \\ 100 \end{bmatrix}.$$

Carrying out the matrix operations leads

$$\begin{bmatrix} Y \\ I \\ Q \end{bmatrix} = \begin{bmatrix} 0.299\times 200 & 0.587\times 10 & 0.114\times 100 \\ 0.596\times 200 & -0.274\times 10 & -0.322\times 100 \\ 0.212\times 200 & -0.523\times 10 & 0.311\times 100 \end{bmatrix} = \begin{bmatrix} 77.07 \\ 84.26 \\ 68.27 \end{bmatrix}.$$

Round the values to integer, we have

$$\begin{bmatrix} Y \\ I \\ Q \end{bmatrix} = round \begin{bmatrix} 77.07 \\ 84.26 \\ 68.27 \end{bmatrix} = \begin{bmatrix} 77 \\ 84 \\ 68 \end{bmatrix}.$$

Now let us study the following example to convert the YIQ values back to the RGB values.

EXAMPLE 9.2

Given a pixel of an image in the YIQ color format as follows:

$$Y=77,\ I=84,\ Q=68,$$

convert the pixel values back to the RGB values.

Solution:

Applying Eq. (13.3) yields

$$\begin{bmatrix} R \\ G \\ B \end{bmatrix} = \begin{bmatrix} 1.000 & 0.956 & 0.621 \\ 1.000 & -0.272 & -0.647 \\ 1.000 & -1.106 & 1.703 \end{bmatrix} \begin{bmatrix} 77 \\ 84 \\ 68 \end{bmatrix} = \begin{bmatrix} 199.53 \\ 10.16 \\ 99.90 \end{bmatrix}.$$

After rounding, it follows that

$$\begin{bmatrix} R \\ G \\ B \end{bmatrix} = round \begin{bmatrix} 199.53 \\ 10.16 \\ 99.9 \end{bmatrix} = \begin{bmatrix} 200 \\ 10 \\ 100 \end{bmatrix}.$$

EXAMPLE 13.3

Given the following 2 × 2 RGB image,

$$R=\begin{bmatrix}100 & 50\\200 & 150\end{bmatrix} G=\begin{bmatrix}10 & 25\\20 & 50\end{bmatrix} B=\begin{bmatrix}10 & 5\\20 & 15\end{bmatrix},$$

convert the RGB color image into a grayscale image.

Solution:

Since only Y components are kept in the grayscale image, we apply Eq. (13.4) to each pixel in the 2×2 image and round the results to integers as follows:

$$Y=0.299\times\begin{bmatrix}100 & 50\\200 & 150\end{bmatrix}+0.587\times\begin{bmatrix}10 & 25\\20 & 50\end{bmatrix}+0.114\times\begin{bmatrix}10 & 5\\20 & 15\end{bmatrix}=\begin{bmatrix}37 & 30\\74 & 76\end{bmatrix}.$$

Fig. 13.9 shows that the grayscale image converted from the 24-bit color image in Fig. 13.5 using the RGB-to-YIQ transformation, where only the luminance information is retained.

13.1.6 MATLAB FUNCTIONS FOR FORMAT CONVERSION

The following list summarizes MATLAB functions for image format conversion:

imread = read image data file with the specified format

X = 8-bit grayscale image, 8-bit indexed image, or 24-bit RGB color image.

map = color map table for the indexed image (256 entries)

imshow(X,map) = 8-bit image display

imshow(X) = 24-bit RGB color image display if image X is in a 24-bit RGB color format; grayscale image display if image X is in an 8-bit grayscale format.

FIG. 13.9

Grayscale image converted from the 24-bit color image in Fig. 13.5 using RGB- to-YIQ transformation.

ind2gray = 8-bit indexed color image to 8-bit grayscale image conversion
ind2rgb = 8-bit indexed color image to 24-bit RGB color image conversion.
rgb2ind = 24-bit RGB color image to 8-bit indexed color image conversion
rgb2gray = 24-bit RGB color image to 8-bit grayscale image conversion
im2double = 8-bit image to intensity image conversion
mat2gray = image data to intensity image conversion
im2uint8 = intensity image to 8-bit grayscale image conversion
Fig. 13.10 outlines the applications of image format conversions.

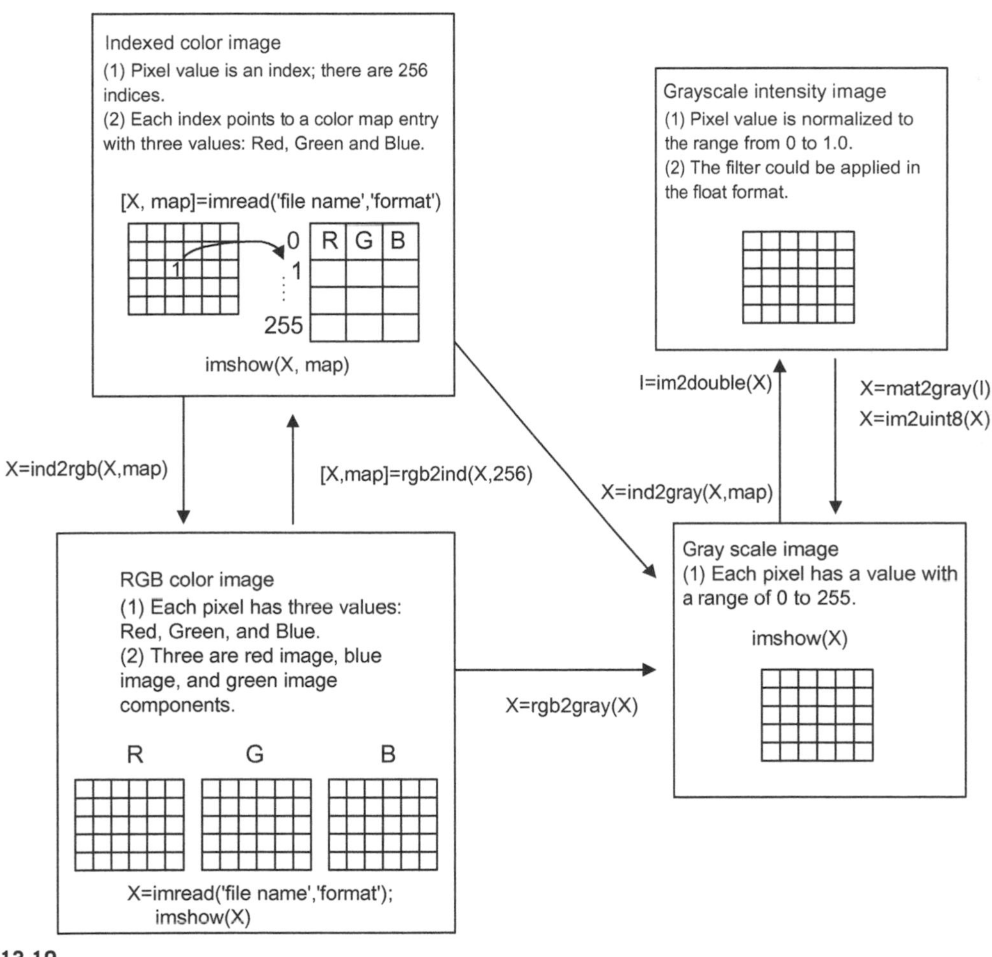
Image format and conversion
Indexed color image
(1) Pixel value is an index; there are 256 indices.
(2) Each index points to a color map entry with three values: Red, Green and Blue.
[X, map]=imread('file name','format')
0
R G B
1
1
255
imshow(X, map)
Grayscale intensity image
(1) Pixel value is normalized to the range from 0 to 1.0.
(2) The filter could be applied in the float format.
I=im2double(X)
X=mat2gray(I)
X=im2uint8(X)
X=ind2rgb(X,map)
[X,map]=rgb2ind(X,256)
X=ind2gray(X,map)
Gray scale image
(1) Each pixel has a value with a range of 0 to 255.
imshow(X)
RGB color image
(1) Each pixel has three values: Red, Green, and Blue.
(2) Three are red image, blue image, and green image components.
X=rgb2gray(X)
R
G
B
X=imread('file name','format');
imshow(X)

FIG. 13.10

Image format conversions.

13.2 IMAGE HISTOGRAM AND EQUALIZATION

An image histogram is a graph to show how many pixels are at each scale level or at each index for the indexed color image. The histogram contains information needed for image equalization, where the image pixels are stretched to give a reasonable contrast.

13.2.1 GRAYSCALE HISTOGRAM AND EQUALIZATION

We can obtain a histogram by plotting pixel value distribution over the full grayscale range.

EXAMPLE 13.4

Produce a histogram given the following image (a matrix filled with integers) with the grayscale value ranging from 0 to 7, that is, with each pixel encoded into 3 bits.

$$\begin{bmatrix} 0 & 1 & 2 & 2 & 6 \\ 2 & 1 & 1 & 2 & 1 \\ 1 & 3 & 4 & 3 & 3 \\ 0 & 2 & 5 & 1 & 1 \end{bmatrix}$$

Solution:

Since the image is encoded using 3 bits for each pixel, we have the pixel value ranging from 0 to 7. The count for each grayscale is listed in Table 13.1

Based on the grayscale distribution counts, the histogram is created as shown in Fig. 13.11.

As we can see, the image has pixels whose levels are more concentrated in the dark scale in this example.

Table 13.1 Pixel Counts Distribution

Pixel $p(m,n)$ Level	Number of Pixels
0	2
1	7
2	5
3	3
4	1
5	1
6	1
7	0

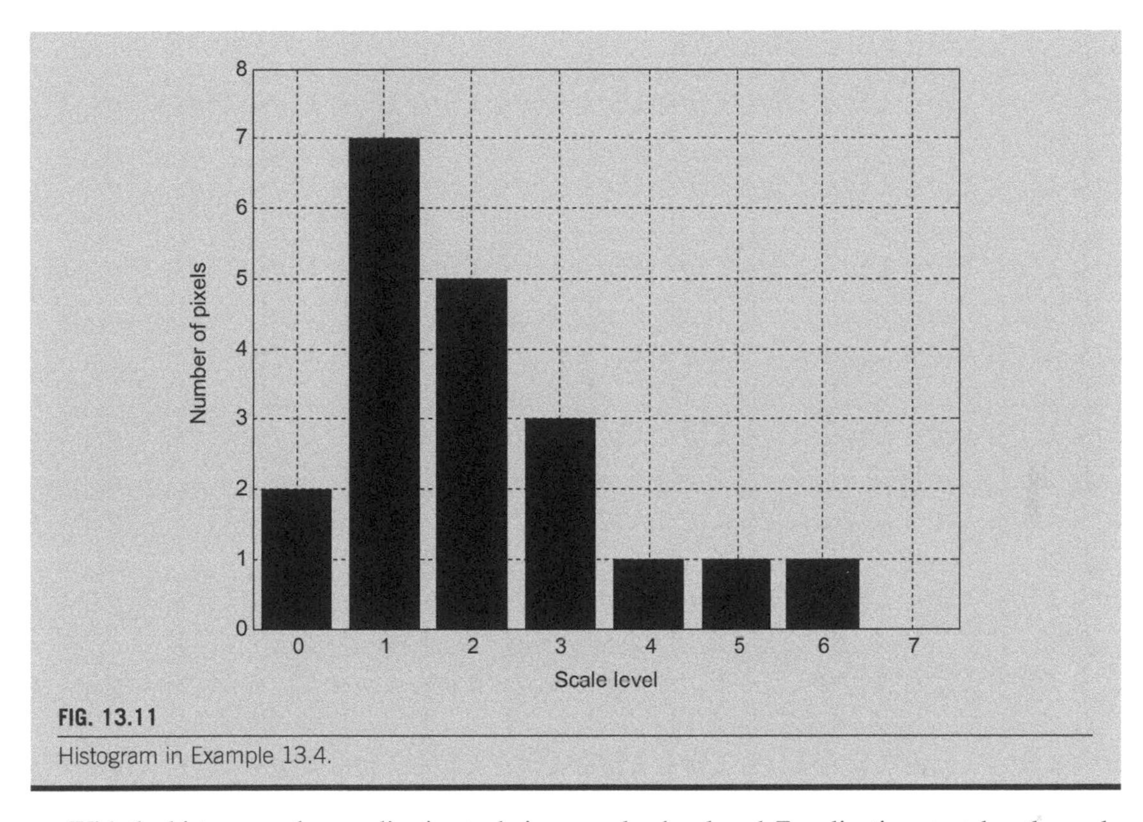

FIG. 13.11

Histogram in Example 13.4.

With the histogram, the equalization technique can be developed. Equalization stretches the scale range of the pixel levels to the full range to give an improved contrast for the given image. To do so, the equalized new pixel value is redefined as

$$p_{eq}(m,n) = \frac{\text{Number of pixels with scale level} \le p(mn)}{\text{Total number of pixels}} \times (\text{maximum scale level}) \tag{13.5}$$

The new pixel value is reassigned using the value obtained by multiplying maximum scale level by the scaled ratio of the accumulative counts up to the current image pixel value over the total number of the pixels. Clearly, since the accumulate counts can range from 0 up to the total number of pixels, the equalized pixel value can vary from 0 to the maximum scale level. It is due to the accumulation procedure that the pixel values are spread over the whole range from 0 to the maximum scale level (255). Let us look at a simplified equalization example.

EXAMPLE 13.5

Given the following image (matrix filled with integers) with a grayscale value ranging from 0 to 7, that is, with each pixel is encoded in 3 bits,

$$\begin{bmatrix} 0 & 1 & 2 & 2 & 6 \\ 2 & 1 & 1 & 2 & 1 \\ 1 & 3 & 4 & 3 & 3 \\ 0 & 2 & 5 & 1 & 1 \end{bmatrix}$$

Continued

EXAMPLE 13.5—CONT'D

Perform equalization using the histogram in Example 13.4, and plot the histogram for the equalized image.

Solution:

Using the histogram result in Table 13.1, we can compute an accumulative count for each gray-scale level as shown in Table 13.2. The equalized pixel level using Eq. (13.5) is given in the last column.

To see how the old pixel level $p(m,n)=4$ is equalized to the new pixel level $p_{eq}(m,n)=6$, we apply Eq. (13.5):

$$p_{eq}(m,n)=round\left(\frac{18}{20}\times 7\right)=6.$$

The equalized image using Table 13.2 is finally obtained by replacing each old pixel value in the old image with its corresponding equalized new pixel value and given by

$$\begin{bmatrix} 1 & 3 & 5 & 5 & 7 \\ 5 & 3 & 3 & 5 & 3 \\ 3 & 6 & 6 & 6 & 6 \\ 1 & 5 & 7 & 3 & 3 \end{bmatrix}.$$

To see how the histogram is changed, we compute the pixel level counts according to the equalized image. The result is given in Table 13.3, and Fig. 13.12 shows the new histogram for the equalized image.

As we can see, the pixel levels in the equalized image are stretched to the larger scale levels. This technique works for underexposed images.

Table 13.2 Image Equalization in Example 13.5

Pixel $p(m,n)$ Level	Number of Pixels	Number of Pixels $\le p(m,n)$	Equalized Pixel Level
0	2	2	1
1	7	9	3
2	5	14	5
3	3	17	6
4	1	18	6
5	1	19	7
6	1	20	7
7	0	20	7

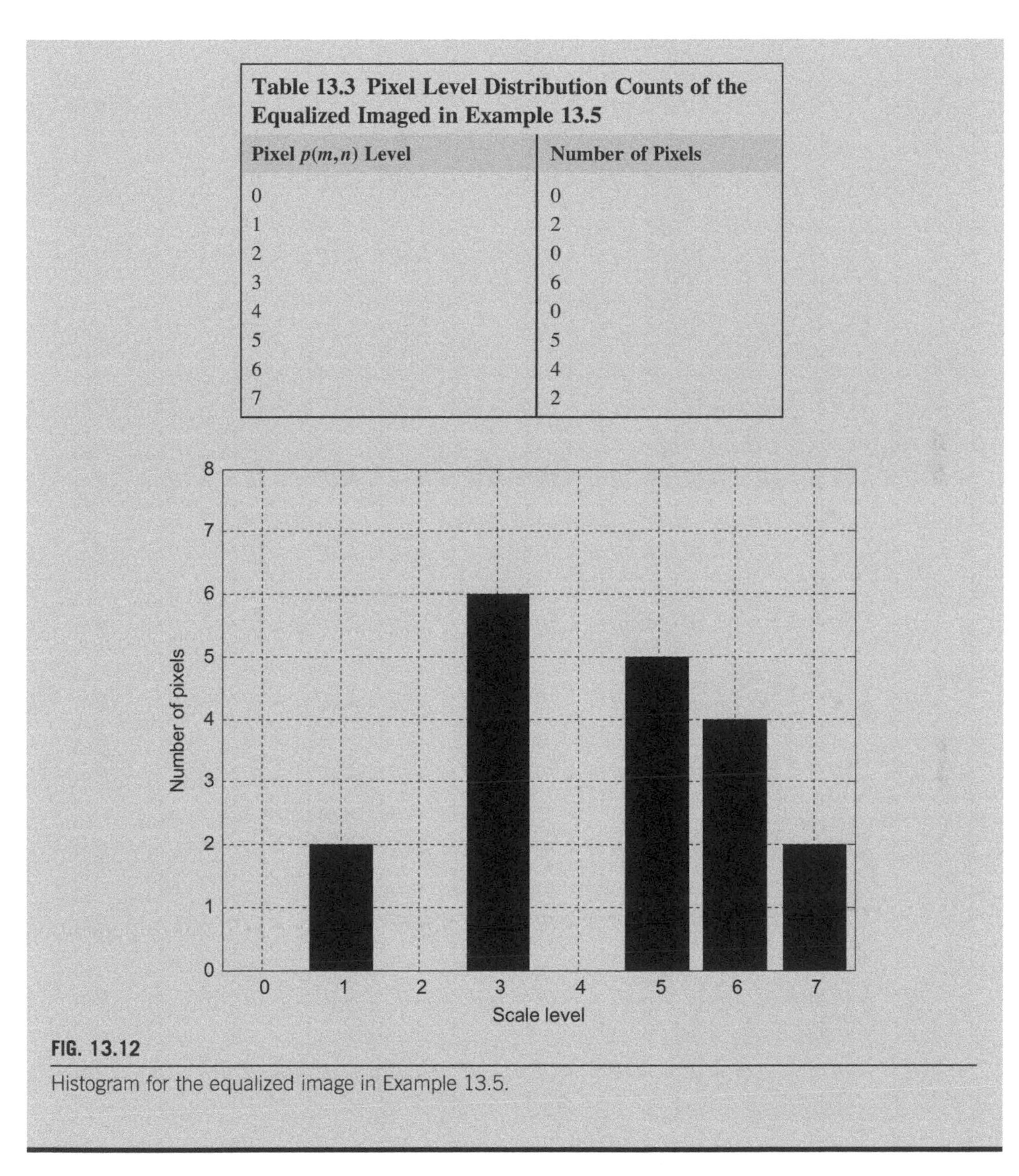

Table 13.3 Pixel Level Distribution Counts of the Equalized Imaged in Example 13.5

Pixel $p(m,n)$ Level	Number of Pixels
0	0
1	2
2	0
3	6
4	0
5	5
6	4
7	2

FIG. 13.12

Histogram for the equalized image in Example 13.5.

Next, we apply the image histogram equalization to enhance a biomedical image of a human neck in Fig. 13.13A, while Fig. 13.13B shows the original image histogram. We see that there are many pixel counts residing at the lower scales in the histogram. Hence, the image looks rather dark and may be underexposed.

Fig. 13.14A and B show the equalized grayscale image using the histogram method and its histogram, respectively. As shown in the histogram, the equalized pixels reside more on the larger scale, and hence the equalized image has improved contrast.

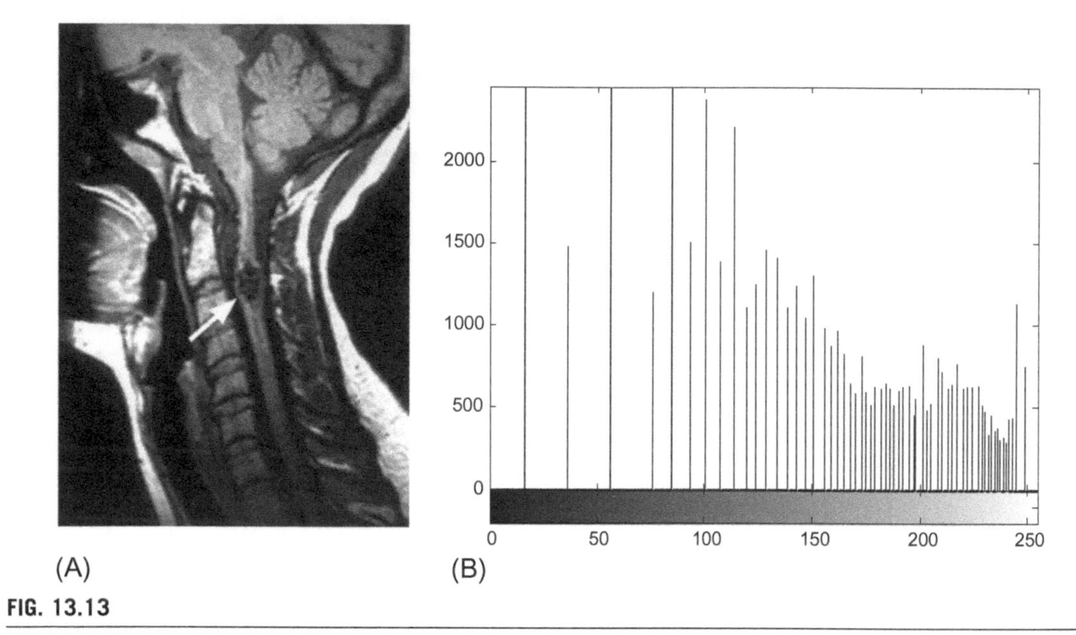

FIG. 13.13

(A) Original grayscale image, (B) Histogram for the original grayscale image.

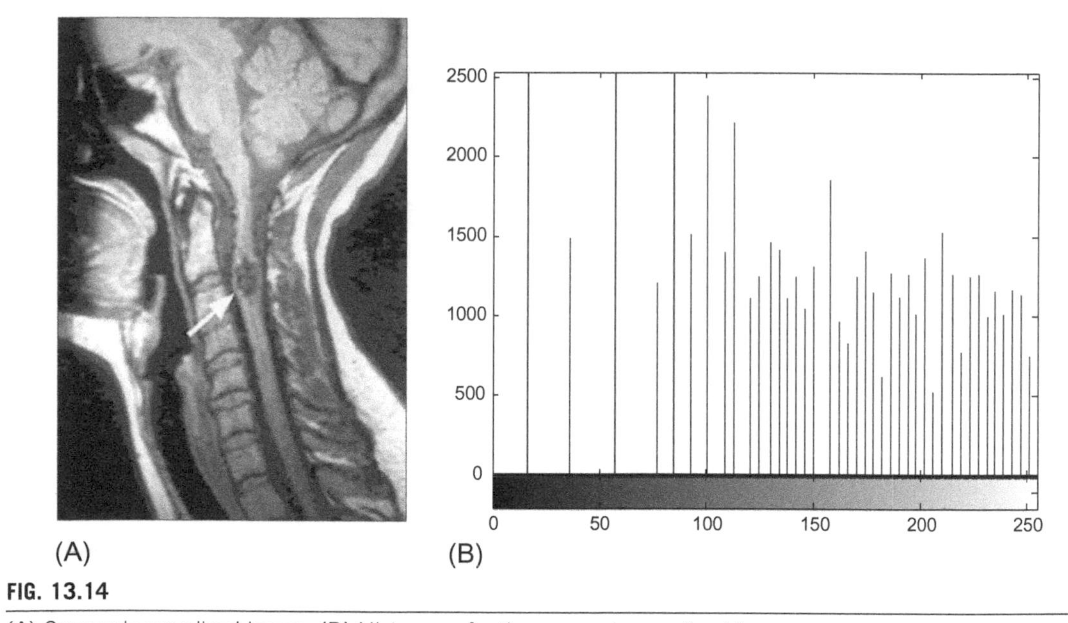

FIG. 13.14

(A) Grayscale equalized image. (B) Histogram for the grayscale equalized image.

13.2.2 24-BIT COLOR IMAGE EQUALIZATION

For equalizing the RGB image, we first transform RGB values to YIQ values since Y channel contains most of the signal energy, about 93%. Then Y channel is equalized just like the grayscale equalization to enhance the luminance. We leave the I and Q channels as they are, since these contain color information only and we do not equalize them. Next, we can repack the equalized Y channel back to the YIQ format. Finally, the YIQ values are transformed back to the RGB values for display. Fig. 13.15 shows the procedure.

Fig. 13.16A shows an original RGB color outdoors scene that is underexposed. Fig. 13.16B shows the equalized RGB image using the method for equalizing Y channel only. We can verify significant improvement with the equalized image showing much detailed information.

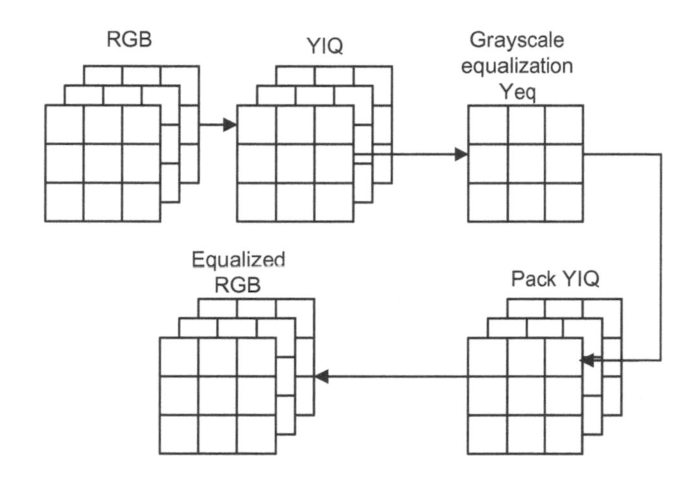

FIG. 13.15

Color image equalization.

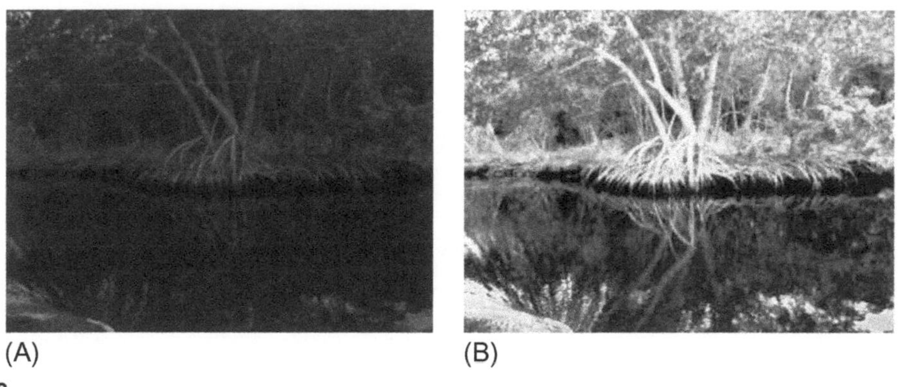

FIG. 13.16

(A) Original RGB color image (B) Equalized RGB color image (For interpretation of the references to colour in this figure legend, the reader is referred to the web version of this article.)

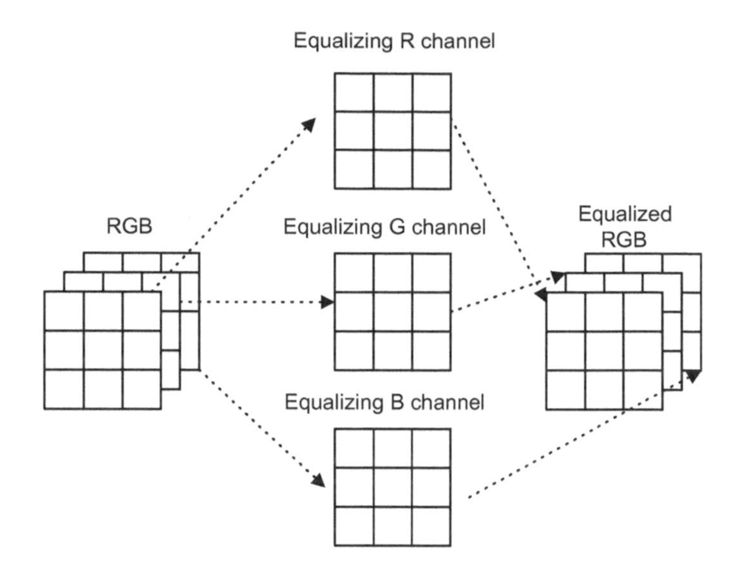

FIG. 13.17

Equalizing RGB channels.

We can also using the histogram equalization method to equalize each of the RGB channels, or their possible combinations. Fig. 13.17 illustrates such a procedure.

Some color effects can be observed. Equalization of the R channel only would make the image looks more red since the red pixel values are stretched out to the full range. Similar observations can be made for equalizing the G channel, or the B channel only. The equalized images for the R, G, and B channels, respectively, are shown in Fig. 13.18. The image from equalizing the R, G, and B channels simultaneously is shown in the upper left corner, which offers improved image contrast.

13.2.3 8-BIT INDEXED COLOR IMAGE EQUALIZATION

Equalization of the 8-bit color indexed image is more complicated. This is because the pixel value is the index for color map entries, and there are three RGB color components for each entry. We expect that after equalization, the index for each pixel will not change from its location on the color map table. Instead, the RGB components in the color map are equalized and changed. The procedure is described in the following and is shown in Fig. 13.19.

Step 1: The RGB color map is converted to the YIQ color map. Note that there are only 256 color table entries. Since the image contains the index values, which point to locations on the color table containing RGB components, it is natural to convert the RGB color table to the YIQ color table.

Step 2: The grayscale image is generated using the Y channel value, so that grayscale equalization can be performed.

Step 3: Grayscale equalization is executed.

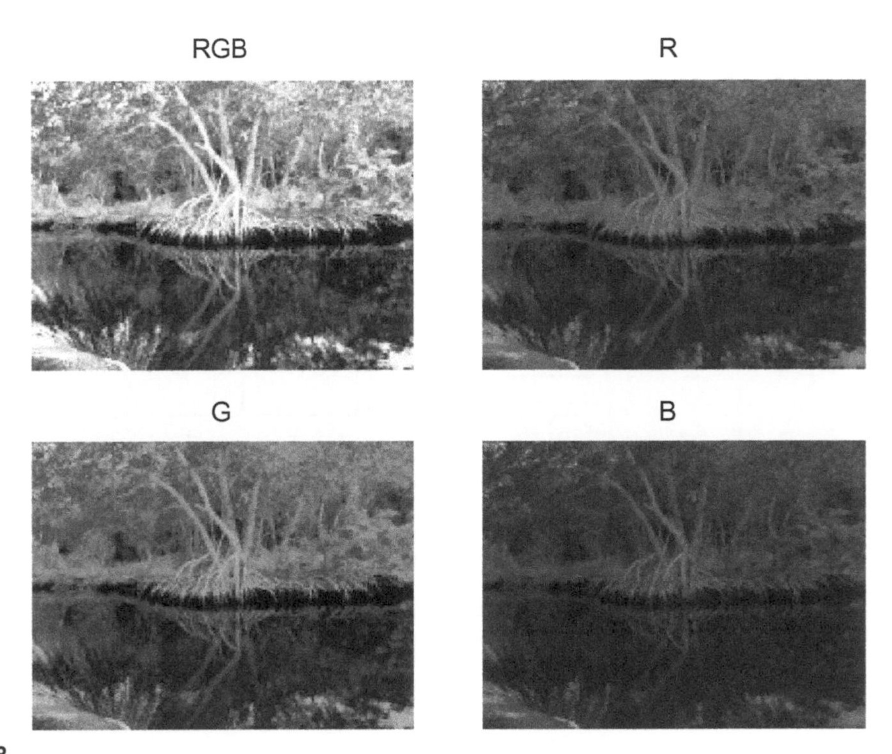

FIG. 13.18

Equalization effects for RGB channels. (For interpretation of the references to colour in this figure legend, the reader is referred to the web version of this article.)

Step 4: The equalized 256 Y values are divided by their corresponding old Y values to obtain the relative luminance scale factors.

Step 5: Finally, the RGB values are each scaled in the old RGB color table with the corresponding relative luminance scale factor and are normalized as new RGB channels in color table in the correct range. Then the new RGB color map is the output.

Note that original index values are not changed; only the color map content is.

Using the previous outdoors picture for the condition of underexposure, Fig. 13.20 shows the equalized indexed color image. We see that the equalized image displays much more detail.

13.2.4 MATLAB FUNCTIONS FOR EQUALIZATION

Fig. 13.21 lists MATLAB functions for performing equalization for the different image formats. The MATLAB functions are explained as follows:

histeq = grayscale histogram equalization, or 8-bit indexed color histogram equalization

imhis t = histogram display

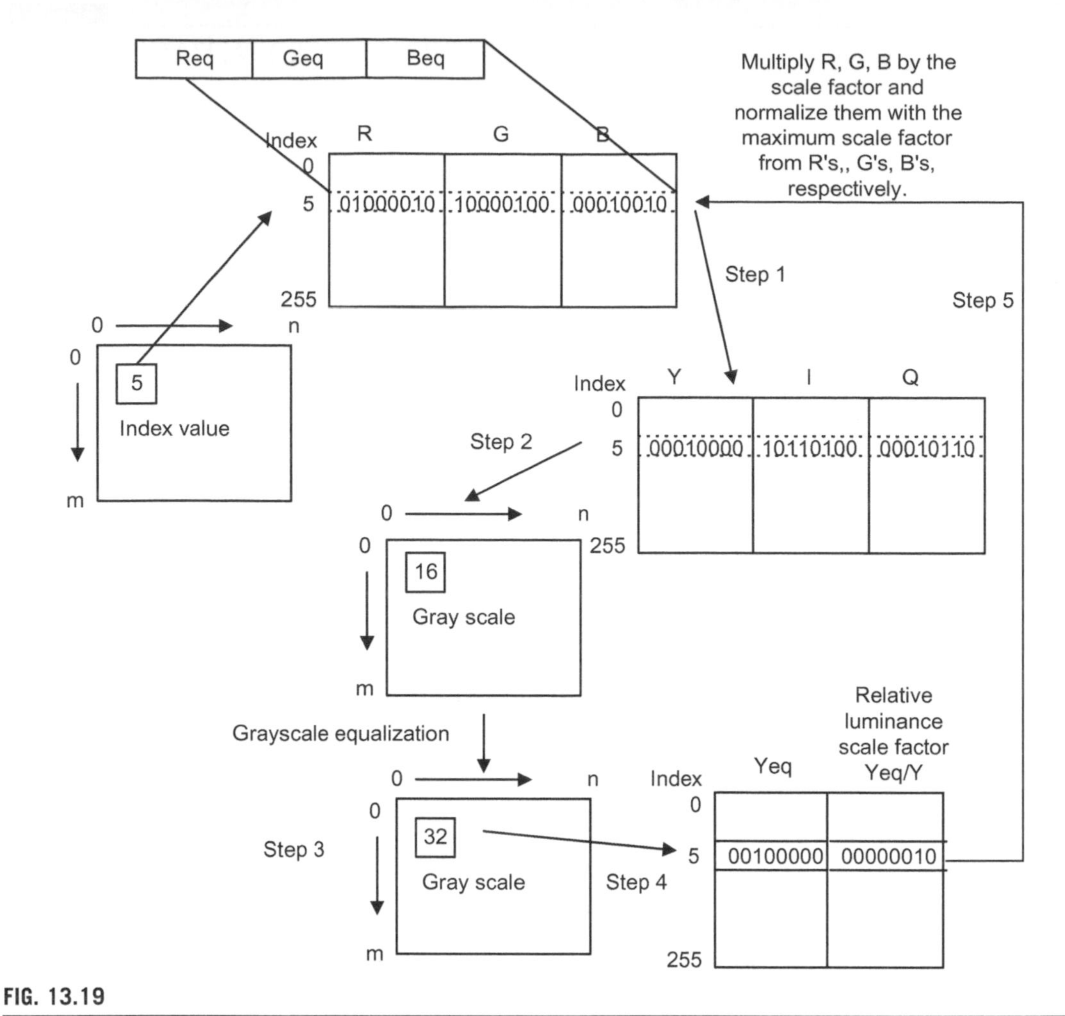

FIG. 13.19

Equalization of the 8-bit indexed color image.

FIG. 13.20

Equalized indexed the 8-bit color image. (For interpretation of the references to colour in this figure legend, the reader is referred to the web version of this article.)

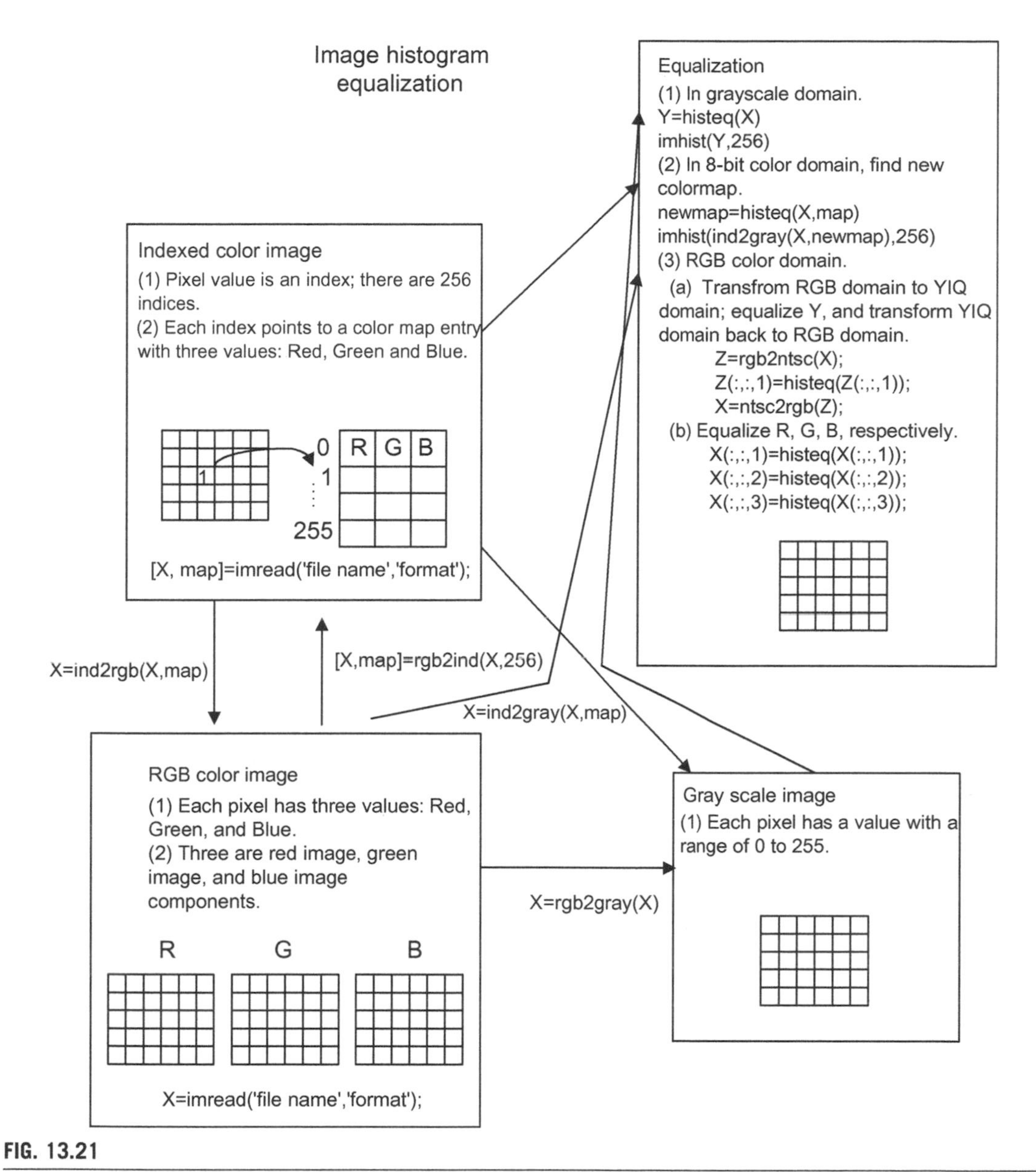

FIG. 13.21

MATLAB functions for image equalization.

rgb2ntsc = 24-bit RGB color image to 24-bit YIQ color image conversion

ntsc2rgb = 24-bit YIQ color image to 24-bit RGB color image conversion

Examples using the MATLAB functions for image format conversion and equalization are given in Program 13.1.

Program 13.1. Examples of image format conversion and equalization.

```
disp('Read the rgb image');
XX=imread('trees','JPEG');            % Provided by the instructor
figure, imshow(XX);
title('24-bit color')
disp('the grayscale image and histogram');
Y=rgb2gray(XX);                         %RGB to grayscale conversion
figure, subplot(1,2,1);imshow(Y);
title('original');subplot(1.2.2);imhist(Y, 256);

disp('Equalization in grayscale domain');
Y=histeq(Y);
figure, subplot(1,2,1); imshow(Y);
title('EQ in grayscale-domain'); subplot(1,2,2); imhist(Y, 256);

disp('Equalization of Y channel for RGB color image');
figure
subplot(1,2,1); imshow(XX);
title('EQ in RGB color');
subplot(1,2,2); imhist(rgb2gray(XX),256);

Z1=rgb2ntsc(XX);                     % Conversion from RGB to YIQ
Z1(:,:,1)=histeq(Z1(:,:,1));         %e Equalizing Y channel
ZZ=ntsc2rgb(Z1);                     %Conversion from YIQ to RGB

figure
subplot(1,2,1); imshow(ZZ);
title('EQ for Y channel for RGB color image');
subplot(1,2,2); imhist(im2uint8(rgb2gray(ZZ)),256);

ZZZ=XX;
ZZZ(:,:,1)=histeq(ZZZ(:,:,1)); %Equalizing R channel
ZZZ(:,:,2)=histeq(ZZZ(:,:,2)); %Equalizing G channel
ZZZ(:,:,3)=histeq(ZZZ(:,:,3)); %Equalizing B channel
figure
subplot(1,2,1); imshow(ZZZ);
title('EQ for RGB channels');
subplot(1,2,2); imhist(im2uint8(rgb2gray(ZZZ)),256);

disp('Equalization in 8-bit indexd color');
[Xind, map]=rgb2ind(XX, 256); % RGB to 8-bit index image conversion
newmap=histeq(Xind,map);
figure
subplot(1,2,1); imshow(Xind,newmap);
title('EQ in 8-bit indexed color');
subplot(1,2,2); imhist(ind2gray(Xind,newmap),256);
```

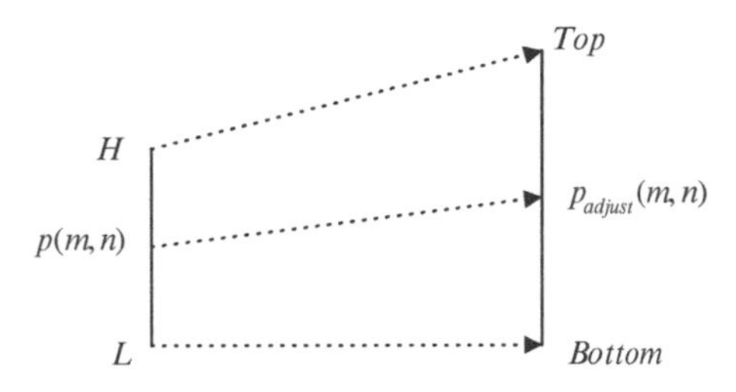

FIG. 13.22

Linear level adjustment.

13.3 IMAGE LEVEL ADJUSTMENT AND CONTRAST

Image level adjustment can be used to linearly stretch pixel level in an image to increase contrast and shift the pixel level to change viewing effects. Image level adjustment is also a requirement for modifying results from image filtering or other operations to an appropriate range for display. We study this technique in the following subsections.

13.3.1 LINEAR LEVEL ADJUSTMENT

Sometimes, if the pixel range in an image is small, we can adjust the image pixel level to make use of a full pixel range. Hence, contrast of the image is enhanced. Fig. 13.22 illustrates linear level adjustment.

The linear level adjustment is given by the following formula:

$$p_{adjust}(m,n) = Bottom + \frac{p(mn) - \mathrm{L}}{\mathrm{H} - \mathrm{L}} \times (\mathrm{Top} - \mathrm{Bottom}), \tag{13.6}$$

where $p(m,n)$ = original image pixel

$p_{adjust}(m,n)$ = desired image pixel

H = maximum pixel level in the original image

L = minimum pixel level in the original image

Top = maximum pixel level in the desired image

Bottom = minimum pixel level in the desired image

Besides adjusting image level to a full range, we can also apply the method to shift the image pixel levels up or down.

EXAMPLE 13.6

Given the following image (matrix filled with integers) with a grayscale ranging from 0 to 7, that is, with each pixel encoded in 3 bits,

$$\begin{bmatrix} 3 & 4 & 4 & 5 \\ 5 & 3 & 3 & 3 \\ 4 & 4 & 4 & 5 \\ 3 & 5 & 3 & 4 \end{bmatrix},$$

Continued

EXAMPLE 13.6—CONT'D

(a) Perform level adjustment to full range.
(b) Shift the level to the range from 3 to 7.
(c) Shift the level to the range from 0 to 3.

Solution:

(a) From the given image, we set the following for level adjustment to the full range:

$$H=5,\ L=3,\ Top=2^3-1=7,\ Bottom=0.$$

Applying Eq. (13.6) yields the second column in Table 13.4.

(b) For the shift-up operation, it follows that

$$H=5,\ L=3,\ Top=7,\ Bottom=3.$$

(c) For the shift-down operation, we set

$$H=5,\ L=3,\ Top=3, Bottom=0.$$

The results for (b) and (c) are listed in the third and fourth column, respectively, of Table 13.4. According to Table 13.4, we have three images:

$$\begin{bmatrix} 0 & 4 & 4 & 7 \\ 7 & 0 & 0 & 0 \\ 4 & 4 & 4 & 7 \\ 0 & 7 & 0 & 4 \end{bmatrix} \begin{bmatrix} 3 & 5 & 5 & 7 \\ 7 & 3 & 3 & 3 \\ 5 & 5 & 5 & 7 \\ 3 & 7 & 3 & 5 \end{bmatrix} \begin{bmatrix} 0 & 2 & 2 & 3 \\ 3 & 0 & 0 & 0 \\ 2 & 2 & 2 & 3 \\ 0 & 3 & 0 & 2 \end{bmatrix}$$

Table 13.4 Image Adjustment Results in Example 13.6

Pixel $p(m,n)$ Level	Full Range	Range [3 7]	Range [0–3]
3	0	3	0
4	4	5	2
5	7	7	3

Next, applying the level adjustment for the neck image of Fig. 13.13A, we get results as shown in Fig. 13.23: the original image, the full range stretched image, the level shift-up image, and the level shift-down image. As we can see, the stretching operation increases image contrast while the shift-up operation lightens the image and the shift-down operation darkens the image.

13.3.2 ADJUSTING THE LEVEL FOR DISPLAY

When two 8-bit images are added together or undergo other mathematical operations, the sum of two pixel values could be as low as 0 and as high as 510. We can apply the linear adjustment to scale the range back to 0–255 for display. The following is the addition of two 8-bit images, which yields a sum that is out of the 8-bit range:

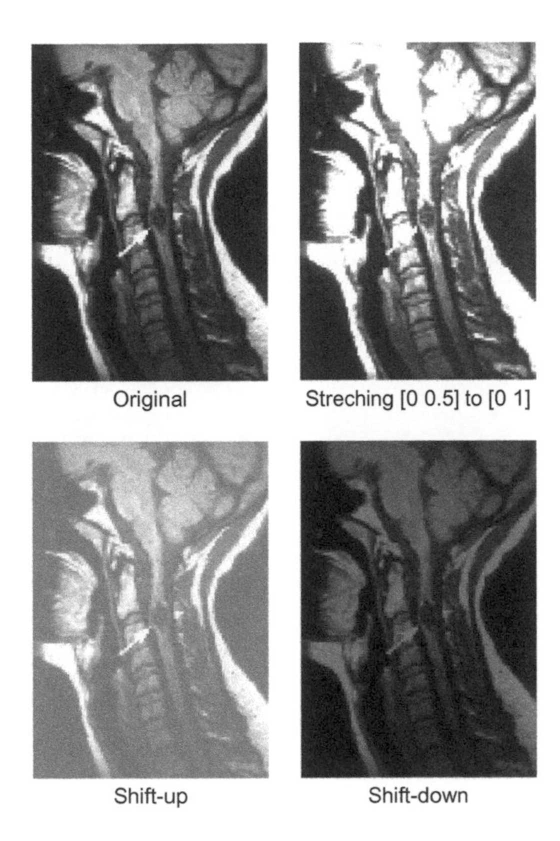

FIG. 13.23

Image level adjustment.

$$\begin{bmatrix} 30 & 25 & 5 & 170 \\ 70 & 210 & 250 & 30 \\ 225 & 125 & 50 & 70 \\ 28 & 100 & 30 & 50 \end{bmatrix} + \begin{bmatrix} 30 & 255 & 50 & 70 \\ 70 & 3 & 30 & 30 \\ 50 & 200 & 50 & 70 \\ 30 & 70 & 30 & 50 \end{bmatrix} = \begin{bmatrix} 60 & 280 & 55 & 240 \\ 140 & 213 & 280 & 60 \\ 275 & 325 & 100 & 140 \\ 58 & 179 & 60 & 100 \end{bmatrix}$$

To scale the combined image, modify Eq. (13.6) as

$$p_{scaled}(m,n) = \frac{p(mn) - \text{Minimum}}{\text{Maximum} - \text{Minimum}} \times (\text{Maximum scale level}). \tag{13.7}$$

Note that in the image to be scaled,

$$\text{Minimum} = 325$$

$$\text{Minimum} = 55$$

$$\text{Maixmum scale level} = 255,$$

we have after scaling:

$$\begin{bmatrix} 5 & 213 & 0 & 175 \\ 80 & 149 & 213 & 5 \\ 208 & 255 & 43 & 80 \\ 3 & 109 & 5 & 43 \end{bmatrix}.$$

13.3.3 MATLAB FUNCTIONS FOR IMAGE LEVEL ADJUSTMENT

Fig. 13.24 lists applications of the MTALAB level adjustment function, which is defined as:

J = imajust (I, [bottom level, top level], [adjusted bottom, adjusted top], gamma)

I = input intensity image

J = output intensity image

gamma = 1 (linear interpolation function as we discussed in the text)

0 < gamma < 1 lightens image; and gamma > 1 darkens image.

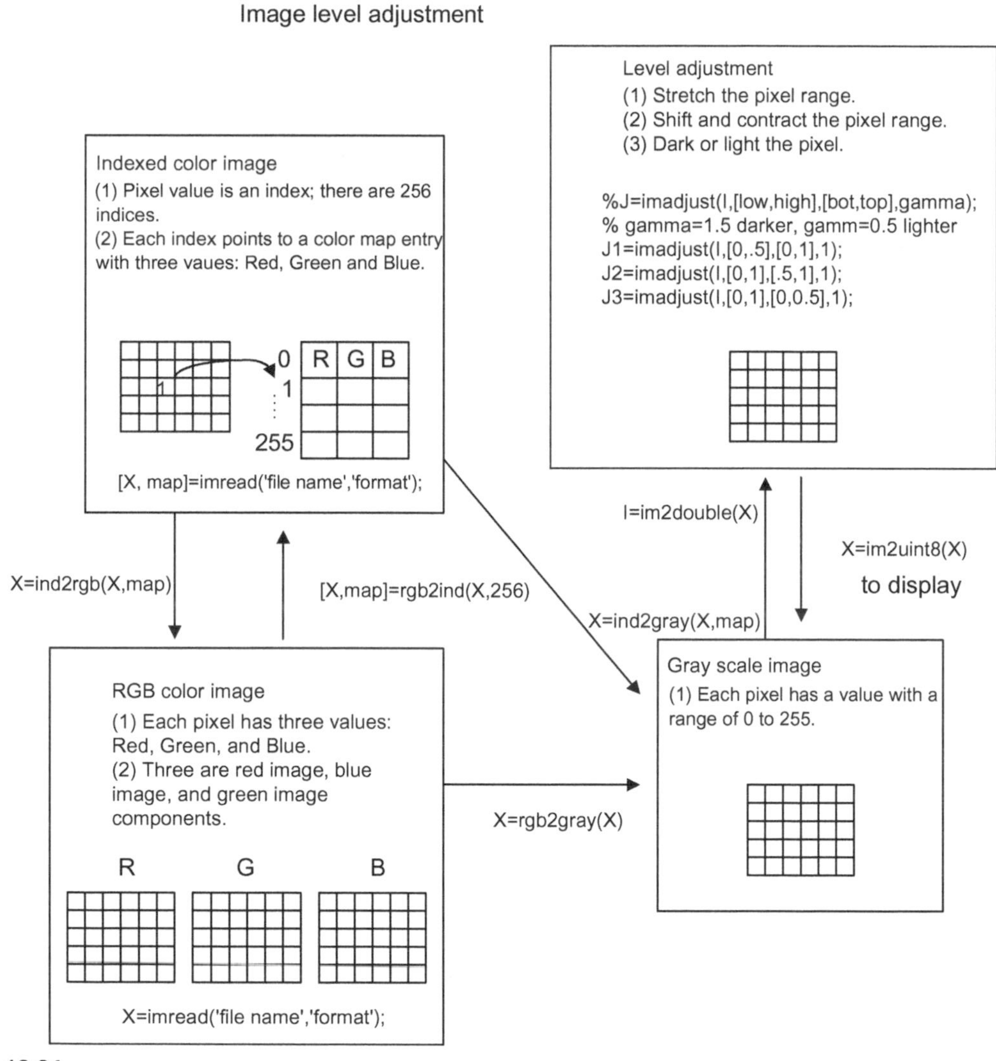

Image level adjustment
Level adjustment
(1) Stretch the pixel range.
(2) Shift and contract the pixel range.
(3) Dark or light the pixel.
%J=imadjust(I,[low,high],[bot,top],gamma);
% gamma=1.5 darker, gamm=0.5 lighter
J1=imadjust(I,[0,.5],[0,1],1);
J2=imadjust(I,[0,1],[.5,1],1);
J3=imadjust(I,[0,1],[0,0.5],1);
Indexed color image
(1) Pixel value is an index; there are 256 indices.
(2) Each index points to a color map entry with three vaues: Red, Green and Blue.
0
1
255
R
G
B
[X, map]=imread('file name','format');
I=im2double(X)
X=im2uint8(X)
to display
X=ind2rgb(X,map)
[X,map]=rgb2ind(X,256)
X=ind2gray(X,map)
Gray scale image
(1) Each pixel has a value with a range of 0 to 255.
RGB color image
(1) Each pixel has three values: Red, Green, and Blue.
(2) Three are red image, blue image, and green image components.
R
G
B
X=imread('file name','format');
X=rgb2gray(X)

FIG. 13.24

MATLAB functions for image level adjustment.

13.4 IMAGE FILTERING ENHANCEMENT

As with one-dimensional (1D) digital signal processing, we can design a digital image filter such as lowpass, highpass, bandpass, and notch to process the image to obtain the desired effect. In this section, we discuss the most common ones: lowpass filters to remove the noise, median filters to remove impulse noise, and edge-detection filter to gain the boundaries of objects in images. More advanced treatment of this subject can be explored in the well-known text by Gonzalez and Wintz (1987).

13.4.1 LOWPASS NOISE FILTERING

One of the simplest lowpass filter is the average filter. The noisy image is filtered using the average convolution kernel with a size 3×3 block, 4×4 block, 8×8 block, and so on, in which the elements in the block have the same filter coefficients. The 3×3, 4×4, and 8×8 average kernels are as follows:

3×3 average kernel:

$$\frac{1}{9}\begin{bmatrix} 1 & 1 & 1 \\ 1 & 1 & 1 \\ 1 & 1 & 1 \end{bmatrix} \tag{13.8}$$

4×4 average kernel:

$$\frac{1}{16}\begin{bmatrix} 1 & 1 & 1 & 1 \\ 1 & 1 & 1 & 1 \\ 1 & 1 & 1 & 1 \\ 1 & 1 & 1 & 1 \end{bmatrix} \tag{13.9}$$

8×8 average kernel

$$\frac{1}{64}\begin{bmatrix} 1 & 1 & 1 & 1 & 1 & 1 & 1 & 1 \\ 1 & 1 & 1 & 1 & 1 & 1 & 1 & 1 \\ 1 & 1 & 1 & 1 & 1 & 1 & 1 & 1 \\ 1 & 1 & 1 & 1 & 1 & 1 & 1 & 1 \\ 1 & 1 & 1 & 1 & 1 & 1 & 1 & 1 \\ 1 & 1 & 1 & 1 & 1 & 1 & 1 & 1 \\ 1 & 1 & 1 & 1 & 1 & 1 & 1 & 1 \\ 1 & 1 & 1 & 1 & 1 & 1 & 1 & 1 \end{bmatrix} \tag{13.10}$$

Each of elements in the average kernel is 1 and the scale factor is the reciprocal of the total number of elements in the kernel. The convolution operates to modify each pixel in the image as follows. By passing the center of a convolution kernel through each pixel in the noisy image, we can sum each product of the kernel element and the corresponding image pixel value and multiply the sum by the scale factor to get the processed pixel. To understand the filter operation with the convolution kernel, let us study the following example.

EXAMPLE 13.7

Perform digital filtering on the noisy image using 2 × 2 and 3 × 3 convolutional average kernels, respectively, and compare the enhanced image with the original one given the following 8-bit grayscale original and corrupted (noisy) images.

$$4 \times 4 \,\text{original image}: \begin{bmatrix} 100 & 100 & 100 & 100 \\ 100 & 100 & 100 & 100 \\ 100 & 100 & 100 & 100 \\ 100 & 100 & 100 & 100 \end{bmatrix}$$

$$4 \times 4 \,\text{corrupted image}: \begin{bmatrix} 99 & 107 & 113 & 96 \\ 92 & 116 & 84 & 107 \\ 103 & 93 & 86 & 108 \\ 87 & 109 & 106 & 107 \end{bmatrix}$$

(a) 2 × 2 average kernel: $\frac{1}{4}\begin{bmatrix} 1 & 1 \\ 1 & 1 \end{bmatrix}$.

(b) 3 × 3 average kernel: $\frac{1}{9}\begin{bmatrix} 1 & 1 & 1 \\ 1 & 1 & 1 \\ 1 & 1 & 1 \end{bmatrix}$.

Solution:

(a) In the following diagram, we pad edges with zeros in the last row and column before processing at the point where the first kernel and the last kernel are shown in the dotted-line boxes, respectively.

99	107	113	96	0
92	116	84	107	0
103	93	86	108	0
87	109	106	107	0
0	0	0	0	0

To process the first element, we know that the first kernel covers the image elements as $\begin{bmatrix} 99 & 107 \\ 92 & 116 \end{bmatrix}$. Summing each product of the kernel element and the corresponding image pixel value, multiplying a scale factor of ¼, and rounding the result, it follows that

$$\frac{1}{4}(99 \times 1 + 107 \times 1 + 92 \times 1 + 116 \times 1) = 103.5$$

$$round(103.5) = 104.$$

In the processing of the second element, the kernel covers $\begin{bmatrix} 107 & 113 \\ 116 & 84 \end{bmatrix}$. Similarly, we have

$$\frac{1}{4}(107 \times 1 + 113 \times 1 + 116 \times 1 + 84 \times 1) = 105$$

$$round(105) = 105.$$

The process continues for the rest of image pixels. To process the last element of the first row, 96, since the kernel covers only $\begin{bmatrix} 96 & 0 \\ 107 & 0 \end{bmatrix}$, we assume that the last two elements are zeros. Then:

$$\frac{1}{4}(96 \times 1 + 107 \times 1 + 0 \times 1 + 0 \times 1) = 50.75$$

$$round(50.75) = 51.$$

Finally, we yield the following filtered image:

$$\begin{bmatrix} 104 & 105 & 100 & 51 \\ 101 & 95 & 96 & 54 \\ 98 & 98 & 102 & 54 \\ 49 & 54 & 53 & 27 \end{bmatrix}.$$

As we know, due to zero padding for boundaries, the last-row and last-column values are in error. However, for a large image, these errors at boundaries can be neglected without affecting image quality. The first 3 × 3 elements in the processed image have values that are close to those of the original image. Hence, the image is enhanced.

(b) To use 3 × 3 kernel, we usually pad zeros surround the image as shown as follows:

0	0	0	0	0	0
0	99	107	113	96	0
0	92	116	84	107	0
0	103	93	86	108	0
0	87	109	106	107	0
0	0	0	0	0	0

Similar to (a), we finally achieve the filtered image as

$$\begin{bmatrix} 46 & 68 & 69 & 44 \\ 68 & 99 & 101 & 66 \\ 67 & 97 & 102 & 66 \\ 44 & 65 & 68 & 45 \end{bmatrix}$$

Notice that after neglecting the boundaries, the enhanced 2 × 2 image is given by

$$\begin{bmatrix} 99 & 101 \\ 97 & 102 \end{bmatrix}.$$

Fig. 13.25 shows the noisy image and enhanced images using the 3 × 3, 4 × 4, 8 × 8 average lowpass filter kernels, respectively. The average kernel removes noise. However, it also blurs the image. When using a large-sized kernel, the quality of the processed image becomes unacceptable.

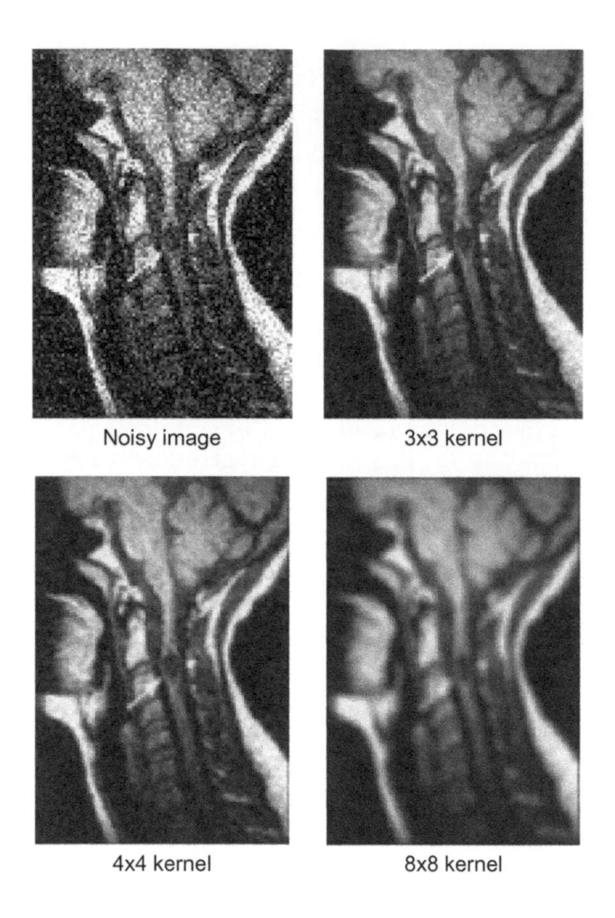

FIG. 13.25

Noise filtering using the lowpass average kernels.

The sophisticated large-size kernels are used for noise filtering. Although it is beyond the scope of the text, the Gaussian filter kernel with the standard deviation $\sigma = 0.9$, for instance, is given by the following:

$$\frac{1}{25}\begin{bmatrix} 0 & 2 & 4 & 2 & 0 \\ 2 & 15 & 27 & 15 & 2 \\ 4 & 27 & 50 & 27 & 4 \\ 2 & 15 & 27 & 15 & 2 \\ 0 & 2 & 4 & 2 & 0 \end{bmatrix}. \quad (13.11)$$

This kernel weighs the center pixel to be processed most and weighs less and less to the pixels away from the center pixel. In this way, the blurring effect can be reduced when filtering the noise. The plot of kernel values in the special domain looks like the bell shape. The steepness of shape is controlled by the standard deviation of the Gaussian distribution function, in which the larger the standard deviation, the flatter the kernel; hence, the more blurring effect will occur.

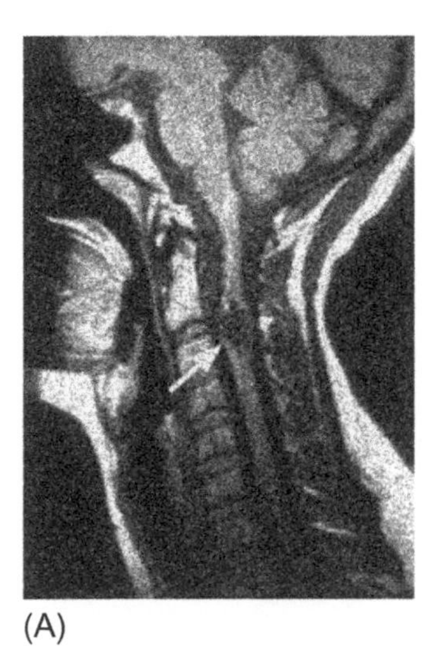
(A)

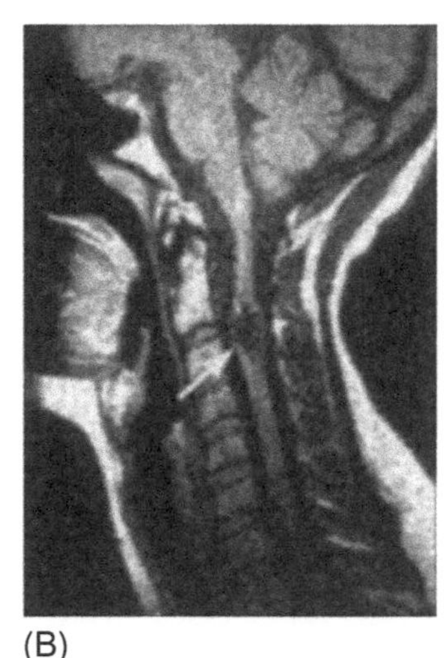
(B)

FIG. 13.26

(A) Noisy image for a human neck. (B) Enhanced image using Gaussian lowpass filter.

Fig. 13.26A shows the noisy image, while Fig. 13.6B shows the enhanced image using the 5 × 5 Gaussian filter kernel. Clearly, the majority of the noise has been filtered, while the blurring effect is significantly reduced.

13.4.2 MEDIAN FILTERING

The median filter is the one type of nonlinear filters. It is very effective at removing impulse noise, the "salt and pepper" noise, in the image. The principle of the median filter is to replace the gray level of each pixel by the median of the gray levels in a neighborhood of the pixels, instead of using the average operation. For median filtering, we specify the kernel size, list the pixel values, covered by the kernel, and determine the median level. If the kernel covers an even number of pixels, the average of two median values is used. Before beginning median filtering, zeros must be padded around the row edge and the column edge. Hence, edge distortion is introduced at image boundary. Let us look at Example 13.8.

EXAMPLE 13.8

Given a 3 × 3 median filter kernel and the following 8-bit grayscale original and corrupted (noisy) images,

$$4 \times 4 \,\text{original image}: \begin{bmatrix} 100 & 100 & 100 & 100 \\ 100 & 100 & 100 & 100 \\ 100 & 100 & 100 & 100 \\ 100 & 100 & 100 & 100 \end{bmatrix}$$

$$4 \times 4 \,\text{corrupted image by impulse noise}: \begin{bmatrix} 100 & 255 & 100 & 100 \\ 100 & 255 & 100 & 100 \\ 255 & 100 & 100 & 0 \\ 100 & 100 & 100 & 100 \end{bmatrix}$$

Continued

EXAMPLE 13.8—CONT'D

$$3 \times 3 \text{ median filter kernel}: \left[\quad\right]_{3\times 3}$$

Perform digital filtering, and compare the filtered image with the original one.

Solution:

Step 1: The 3 × 3 kernel requires zero padding 3/2 = 1 column of zeros at the left and right edges while 3/2 = 1 row of zeros at the upper and bottom edges:

0	0	0	0	0	0
0	100	255	100	100	0
0	100	255	100	100	0
0	255	100	100	0	0
0	100	100	100	100	0
0	0	0	0	0	0

Step 2: To process the first element, we cover the 3 × 3 kernel with the center pointing to the first element to be processed. The sorted data within the kernel are listed in terms of its value as.

$$0,0,0,0,0,0,100,100,255,255.$$

The median value = median (0, 0, 0, 0, 0, 100, 100, 255, 255) = 0. Zero will replace 100.

Step 3: Continue for each element until the last is replaced.

Let us see the element at the location (1,1):

0	0	0	0	0	0
0	100	255	100	100	0
0	100	255	100	100	0
0	255	100	100	0	0
0	100	100	100	100	0
0	0	0	0	0	0

The values covered by the kernel are:

$$100,100,100,100,100,100,255,255,255.$$

The median value = median (100, 100, 100, 100, 100, 100, 255, 255, 255) = 100. The final processed image is

$$\begin{bmatrix} 0 & 100 & 100 & 0 \\ 100 & 100 & 100 & 100 \\ 0 & 100 & 100 & 0 \\ 100 & 100 & 100 & 100 \end{bmatrix}.$$

Some boundary pixels are distorted due to zero padding effect. However, for a large image, the distortion can be omitted versus the overall quality of the image. The 2 × 2 middle portion matches the original image exactly. The effectiveness of the median filter is verified via this example.

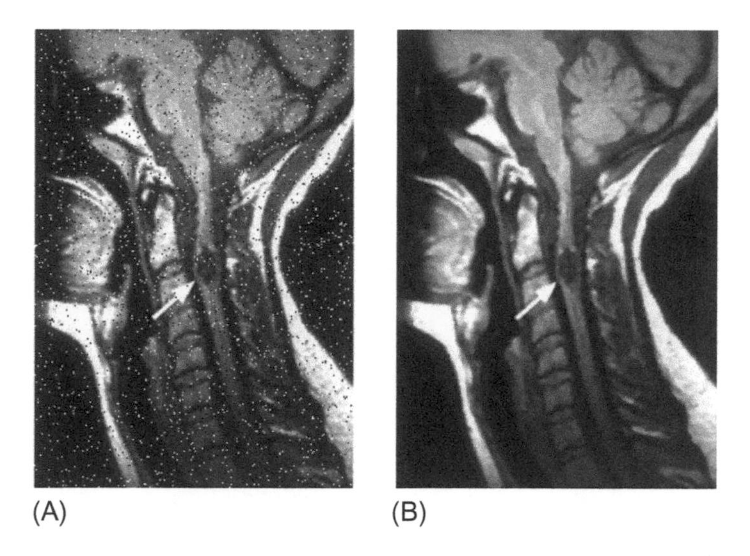

FIG. 13.27

(A) Noisy image (corrupted by "salt and pepper"noise). (B) The enhanced image using the 3×3 median filter.

The image in Fig. 13.27A is corrupted by "salt and pepper" noise. The median filter with a 3×3 kernel is used to filter the impulse noise. The enhanced image shown in Fig. 13.27B has a significant quality improvement. However, the enhanced image also seems smoothed, thus, the high-frequency information is reduced. Note that a larger size kernel is not appropriate for median filtering, because for a larger set of pixels the median value deviates from the pixel value.

13.4.3 EDGE DETECTION

In many applications, such as pattern recognition and fingerprint identification, and iris biometric identification, image edge information is required. To obtain the edge information, a differential convolution kernel is used. Of these kernels, Sobel convolution kernels are used for horizontal and vertical edge detection. They are listed as follows:

Horizontal Sobel edge detector:

$$\begin{bmatrix} -1 & -2 & -1 \\ 0 & 0 & 0 \\ 1 & 2 & 1 \end{bmatrix}. \tag{13.12}$$

The kernel subtracts the first row in the kernel from the third row to detect the horizontal difference.

Vertical Sobel edge detector:

$$\begin{bmatrix} -1 & 0 & 1 \\ -2 & 0 & 2 \\ -1 & 0 & 1 \end{bmatrix}. \tag{13.13}$$

The kernel subtracts the first column in the kernel from the third column to detect the vertical difference.

Besides Sobel edge detector, a Laplacian edge detector is devised to tackle both vertical and horizontal edges. It is listed as follows.

Laplacian edge detector:

$$\begin{bmatrix} 0 & 1 & 0 \\ 1 & -4 & 1 \\ 0 & 1 & 0 \end{bmatrix}. \tag{13.14}$$

EXAMPLE 13.9

Given the following 8-bit grayscale image,

$$5 \times 4 \text{ original image}: \begin{bmatrix} 100 & 100 & 100 & 100 \\ 110 & 110 & 110 & 110 \\ 100 & 100 & 100 & 100 \\ 100 & 100 & 100 & 100 \\ 100 & 100 & 100 & 100 \end{bmatrix},$$

Use the Sobel horizontal edge detector to detect horizontal edges.

Solution:

We pad the image with zeros before processing as follows:

$$\begin{array}{cccccc} 0 & 0 & 0 & 0 & 0 & 0 \\ 0 & 100 & 100 & 100 & 100 & 0 \\ 0 & 110 & 110 & 110 & 110 & 0 \\ 0 & 100 & 100 & 100 & 100 & 0 \\ 0 & 100 & 100 & 100 & 100 & 0 \\ 0 & 100 & 100 & 100 & 100 & 0 \\ 0 & 0 & 0 & 0 & 0 & 0 \end{array}$$

After processing using the Sobel horizontal edge detector, we have

$$\begin{bmatrix} 330 & 440 & 440 & 330 \\ 0 & 0 & 0 & 0 \\ -30 & -40 & -40 & -30 \\ 0 & 0 & 0 & 0 \\ -300 & -400 & -400 & -300 \end{bmatrix}.$$

Adjusting the scale level leads to

$$\begin{bmatrix} 222 & 255 & 255 & 222 \\ 121 & 121 & 121 & 121 \\ 112 & 109 & 109 & 112 \\ 121 & 121 & 121 & 121 \\ 30 & 0 & 0 & 30 \end{bmatrix}.$$

Disregarding the first row and first column and the last row and column, since they are at image boundaries, we identify a horizontal line of 109 in the third row.

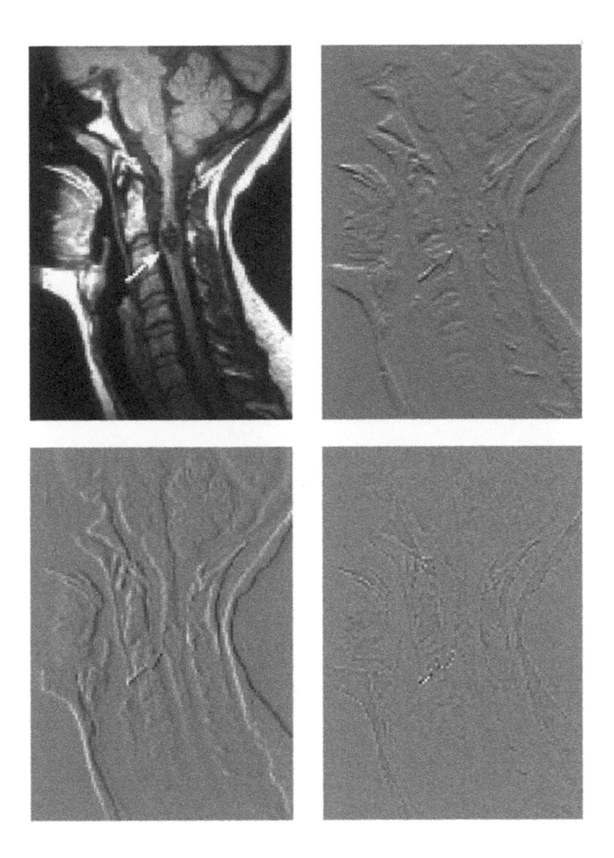

FIG. 13.28

Image edge detection. (Upper left) original image; (upper right) result from Sobel horizontal edge detector; (lower left) result from Sobel vertical edge detector; (lower right) result from Laplacian edge detector.

Fig. 13.28 shows the results from the edge detection.

Fig. 13.29 shows the edge detection for the grayscale image of the cruise ship in Fig. 13.3.

Sobel edge detection can tackle only the horizontal edge or the vertical edge, as shown in Fig. 13.29, where the edges of the image have both horizontal and vertical features. We can simply combine two horizontal and vertical edge-detected images and then rescale the resultant image in the full range. Fig. 13.29C shows that the edge-detection result is equivalent to that of the Laplacian edge detector.

Next, we apply a more sophisticated Laplacian of Gaussian filter for the edge detection, which a combined Gaussian lowpass filter and Laplacian derivative operator (highpass filter). The filter *smoothes* the image to suppress noise using the lowpass Gaussian filter, then performs Laplacian derivative operation for the edge detection, since the noisy image is very sensitive to the Laplacian derivative operation. As we discussed for the Gaussian lowpass filter, the standard deviation in the Gaussian distribution function controls degree of noise filtering before Laplacian derivative operation. A larger value of the standard deviation may blur the image; hence, some edge boundaries could be lost. Its selection should be based on the particular noisy image. The filter kernel with the standard deviation of $\sigma=0.8$ is given by

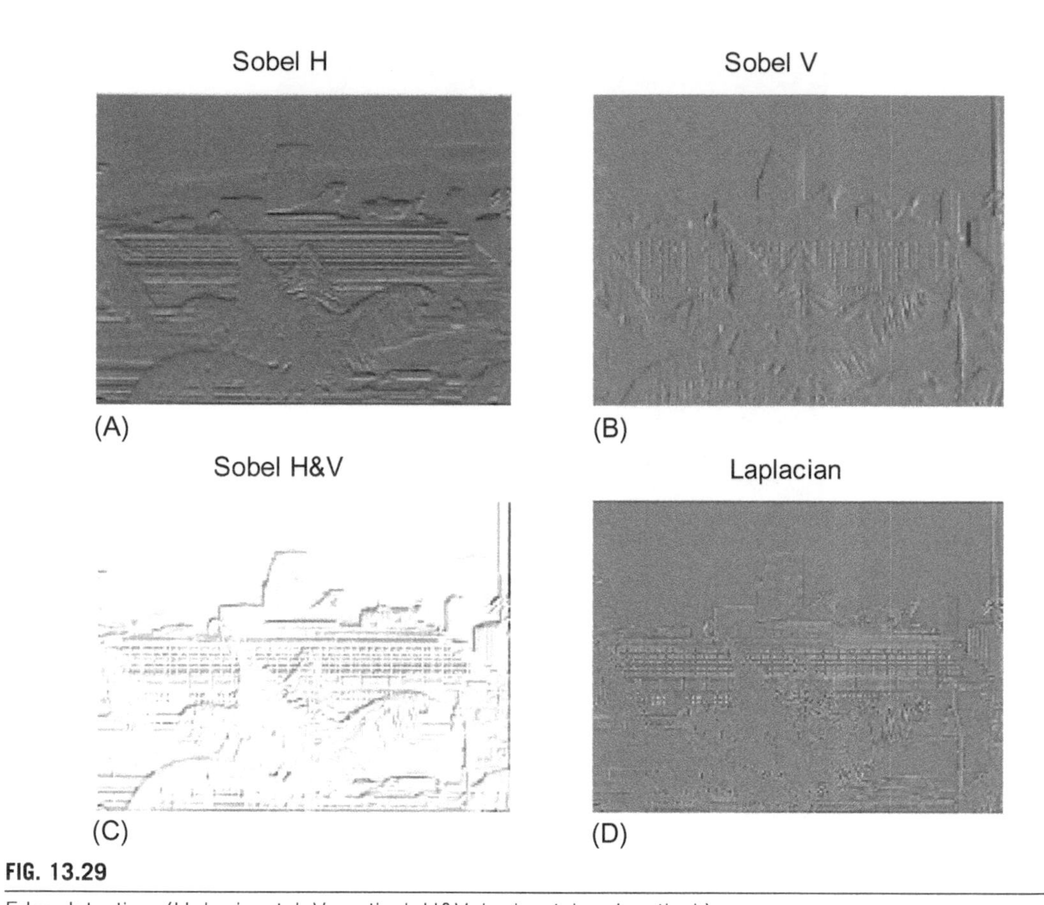

FIG. 13.29

Edge detection. (H, horizontal, V, vertical, H&V, horizontal and vertical.)

$$\begin{bmatrix} 4 & 13 & 16 & 13 & 4 \\ 13 & 9 & -25 & 9 & 13 \\ 16 & -25 & -124 & -25 & 16 \\ 13 & 9 & -25 & 9 & 13 \\ 4 & 13 & 16 & 13 & 4 \end{bmatrix}. \tag{13.15}$$

The processed edge detection using the Laplacian of Gaussian filter in Eq. (13.15) is shown in Fig. 13.30. We can further use a threshold value to convert the processed image to a black and white image, where the contours of objects can be clearly displayed.

13.4.4 MATLAB FUNCTIONS FOR IMAGE FILTERING

MATLAB image filter design and implementation functions are summarized in Fig. 13.31. MATLAB functions are explained as:

FIG. 13.30

Image edge detection using Laplacian of Gaussian filter.

```
X = image to be processed
fspecial('filter type', kernel size, parameter) = convolution kernel generation
H = fspecial('gaussian',HSIZE,SIGMA) returns a rotationally
symmetric Gaussian lowpass filter of size HSIZE with standard
deviation SIGMA (positive).
H = fspecial('log',HSIZE,SIGMA) returns a rotationally symmetric
Laplacian of Gaussian filter of size HSIZE with standard deviation SIGMA
(positive).
Y = filter2([convolution kernel], X) = two-dimensional filter using the
convolution kernel
Y = medfilt2(X, [row size, column size]) = two-dimensional median filter
```

Program 13.2 lists the sample MATLAB codes for filtering applications. Fig. 13.31 outlines the applications of the MATLAB functions.

Program 13.2. Examples of Gaussian filtering, media filtering, and Laplacian of Gaussian filtering.

```
close all;clear all; clc;
X=imread('cruise,','jpeg');        % Provided by the instructor
Y=rgb2gray(X);                     % Convert the rgb image to the grayscale image
I=im2double(Y);                    % Get the intensity image
image1_g=imnoise(I,'gaussian');    % Add random noise to the intensity image
ng=mat2gray(image1_g);             % Adjust the range
ng=im2uint8(ng);                   % 8-bit corrupted image
```

```
%Linear Filtering
K_size=5;                              % Kernel size=5x5
sigma =0.8;                          % sigma (the bigger, the smoother the image)
h=fspecial('gaussian,'K_size,sigma);   % Determine Gaussian filter coefficients
%This command will construct a Guassian filter
%of size 5x5 with a mainlobe width of 0.8.
image1_out=filter2(h,image1_g);        % Perform filetring
image1_out=mat2gray(image1_out);       % Adjust the range
image1_out=im2uint8(image1_out);       % Get the 8-bit image
subplot(1,2,1); imshow(ng),title('Noisy image');
subplot(1,2,2); imshow(image1_out);
title('5 5 Gaussian kernel');

%Median Filtering
image1_s=imnoise(I,'salt & pepper');  % Add "salt and pepper" noise to the image
mn=mat2gray(image1_s);                 % Adjust the range
mn=im2uint8(mn);                       % Get the 8-bit image

K_size=3;                                          % kernel size
image1_m=medfilt2(image1_s,[K_size, K_size]);      % Perform median filtering
image1_m=mat2gray(image1_m);                       % Adjust the range
image1_m=im2uint8(image1_m);                       % Get the 8-bit image
figure, subplot(1,2,1);imshow(mn)
title('Median noisy');
subplot(1,2,2);imshow(image1_m);
title('3 3 median kernel');

%Laplacian of Gaussian filtering
K_size =5;                             % Kernel size
sigma =0.9;                            % Sigma parameter
h=fspecial('log',K_size,alpha);       % Determine the Laplacian of Gaussian filter
image1_out=filter2(h,I);               % kernel
                                       % Perform filtering
image1_out=mat2gray(image1_out);       % Adjust the range
image1_out=im2uint8(image1_out);       % Get the 8-bit image
figure,subplot(1,2,1); imshow(Y)
title('Original');
subplot(1,2,2); imshow(image1_out);
title('Laplacian filter 5 5 kernel');
```

13.5 IMAGE PSEUDO-COLOR GENERATION AND DETECTION

We can apply certain transformations to the grayscale image so that it becomes a color image, and a wider range of pseudo-color enhancement can be obtained. In object detection, pseudo-color generation can produce the specific color for the object that is to be detected, say, red. This would significantly increase the accuracy of the identification. To do so, we choose three transformations of the grayscale level to the RGB components, as shown in Fig. 13.32.

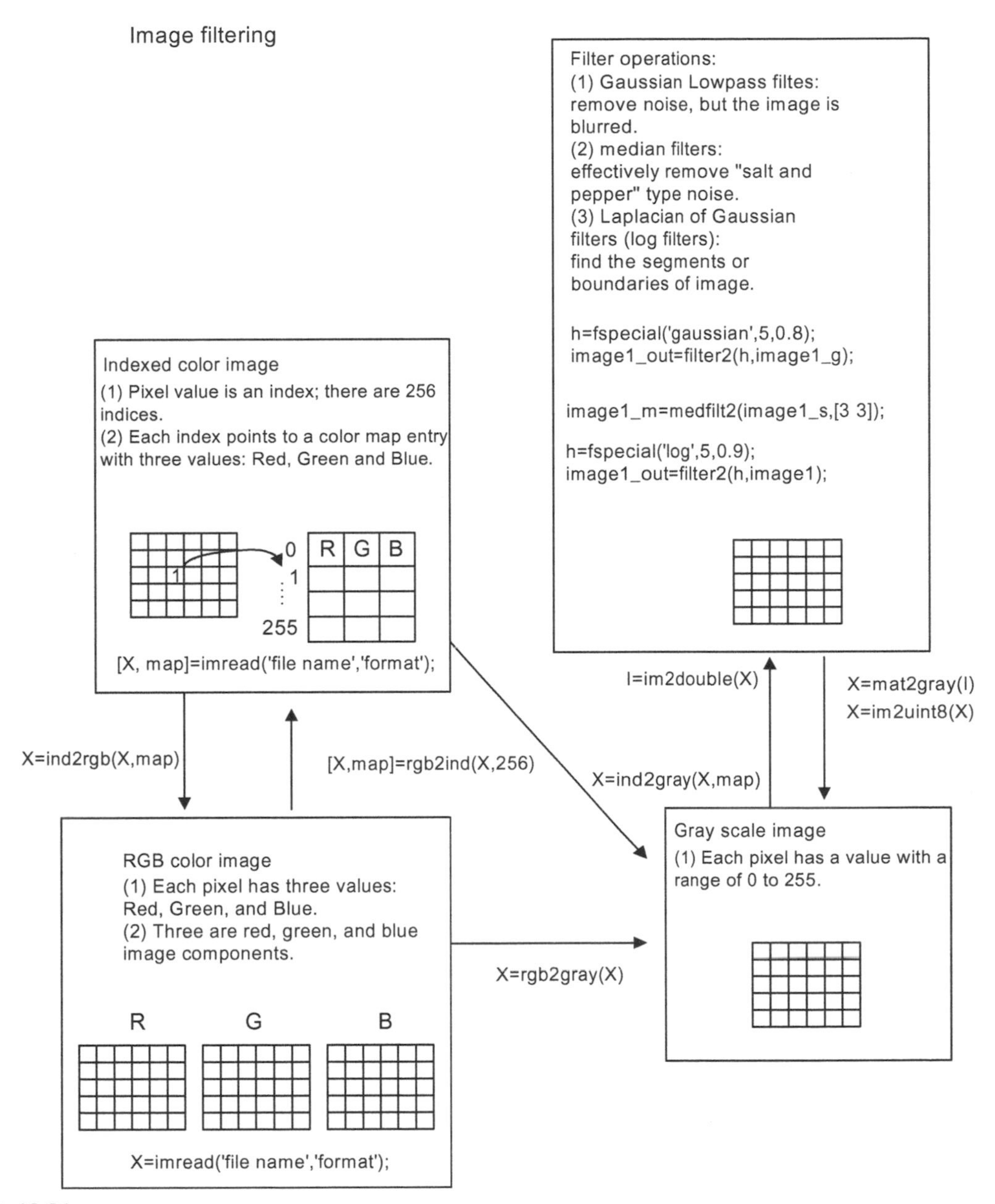
Image filtering
Filter operations:
(1) Gaussian Lowpass filtes:
remove noise, but the image is
blurred.
(2) median filters:
effectively remove "salt and
pepper" type noise.
(3) Laplacian of Gaussian
filters (log filters):
find the segments or
boundaries of image.
h=fspecial('gaussian',5,0.8);
image1_out=filter2(h,image1_g);
image1_m=medfilt2(image1_s,[3 3]);
h=fspecial('log',5,0.9);
image1_out=filter2(h,image1);
Indexed color image
(1) Pixel value is an index; there are 256
indices.
(2) Each index points to a color map entry
with three values: Red, Green and Blue.
0
1
255
R
G
B
[X, map]=imread('file name','format');
I=im2double(X)
X=mat2gray(I)
X=im2uint8(X)
X=ind2rgb(X,map)
[X,map]=rgb2ind(X,256)
X=ind2gray(X,map)
Gray scale image
(1) Each pixel has a value with a
range of 0 to 255.
RGB color image
(1) Each pixel has three values:
Red, Green, and Blue.
(2) Three are red, green, and blue
image components.
X=rgb2gray(X)
R
G
B
X=imread('file name','format');

FIG. 13.31

MATLAB functions for filter design and implementation.

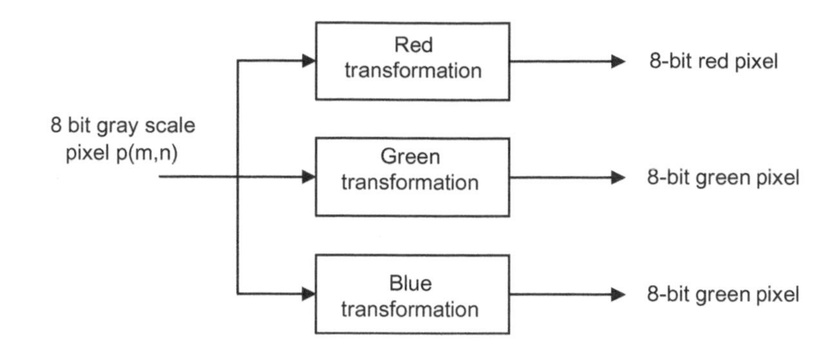

FIG. 13.32

Block diagram for transforming a grayscale pixel to a pseudo-color pixel.

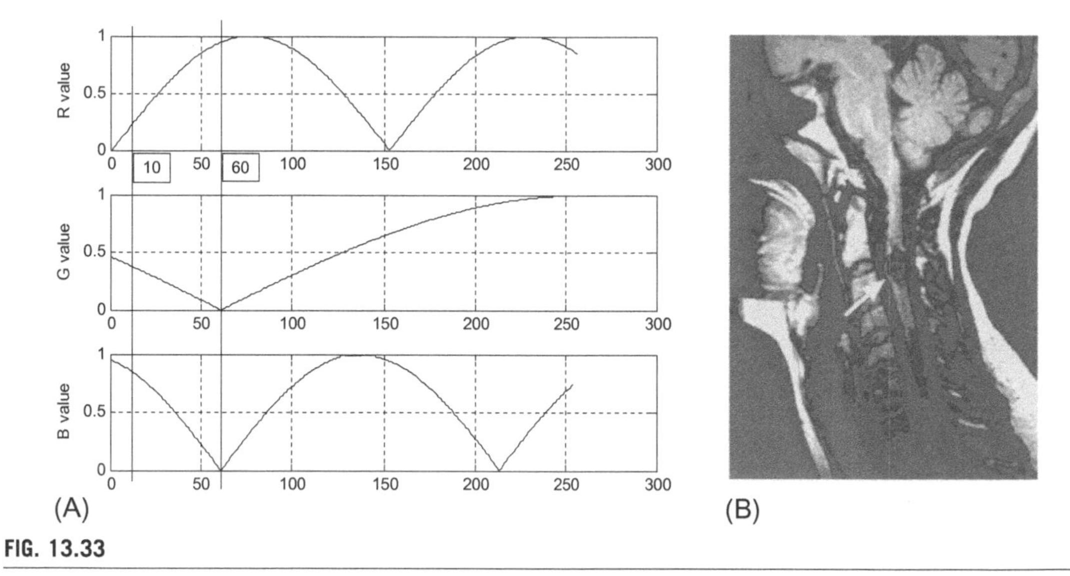

FIG. 13.33

(A) Three sine functions for grayscale transformation. (B) The pseudo-color image (For interpretation of the references to colour in this figure legend, the reader is referred to the web version of this article.)

As a simple choice, we choose three sine functions for RGB transformations, as shown in Fig. 13.33A. Changing the phase and period of one sine function can be easily done so that the grayscale pixel level of the object to be detected is aligned to the desired color with its component value as large as possible, while the other two functions transform the same grayscale level to have their color component values as small as possible. Hence, the single specified color object can be displayed in the image for identification. By carefully choosing the phase and period of each sine function, certain object(s) could be transformed to the red, green, or blue color with a favorable choice.

EXAMPLE 13.10

In the grayscale image in Fig. 13.13A, the area pointed to by the arrow has a grayscale value approximately equal to 60. The background has a pixel value approximately equal to 10. Generate the background to be as close to blue as possible, and make the area pointed to by the arrow as close to red color as possible.

Solution:

The transformation functions are chosen as shown in Fig. 13.33A, where the red value is largest at 60 and the blue and green values approach zero. At the grayscale of 10, the blue value is dominant. Fig. 13.33B shows the processed pseudo-color image.

Fig. 13.34 illustrates the pseudo-color generation procedure.

Program 13.3 lists the sample MATLAB codes for pseudo-color generation for a grayscale image.

Program 13.3. Program examples for the pseudo-color generation.

```
close all; clear all;clc
disp('Convert the grayscale image to the pseudo-color image1');
 [X, map]=imread('clipim2','gif'); % Read 8-bit index image, provided by the instructor
 Y=ind2gray(X,map);            % 8-bit color image to the grayscale conversion
% Apply pseudo-color functions using sinusoids
 C_r=304;                     % Cycle change for the red channel
 P_r=0;                       % Phase change for the red channel
 C_b=304;                     % Cycle change for the blue channel
 P_b=60;                      % Phase change for the blue channel
 C_g=804;                     % Cycle change for the green channel
 P_g=60;                      % Phase change for the green channel
 r=abs(sin(2*pi*[-P_r:255-P_r]/C_r));
b=abs(sin(2*pi*[-P_b:255-P_b]/C_b));
g=abs(sin(2*pi*[-P_g:255-P_g]/C_g));
figure, subplot(3.1,1);plot(r,'r');grid;ylabel('R value')
subplot(3,1,2);plot(g,'g');grid;ylabel('G value');
subplot(3,1,3);plot(b,'b');grid;ylabel('B value');
figure, imshow(Y);
map=[r;g;b;]';              % Construct the color map
figure, imshow(X,map);   % Display the pseudo-color image
```

13.6 IMAGE SPECTRA

In 1D signal processing such as for speech and audio, we need to examine the frequency contents, check filtering effects, and perform feature extraction. Image processing is similar. However, we need apply a two-dimensional discrete Fourier transform (2D-DFT) instead of an 1D DFT. The spectrum including the magnitude and phase is also in two dimensions. The equations of the 2D-DFT are given by:

$$X(u, v) = \sum_{m=0}^{M-1}\sum_{n=0}^{N-1} p(m, n) W_M^{um} W_N^{vn}, \tag{13.16}$$

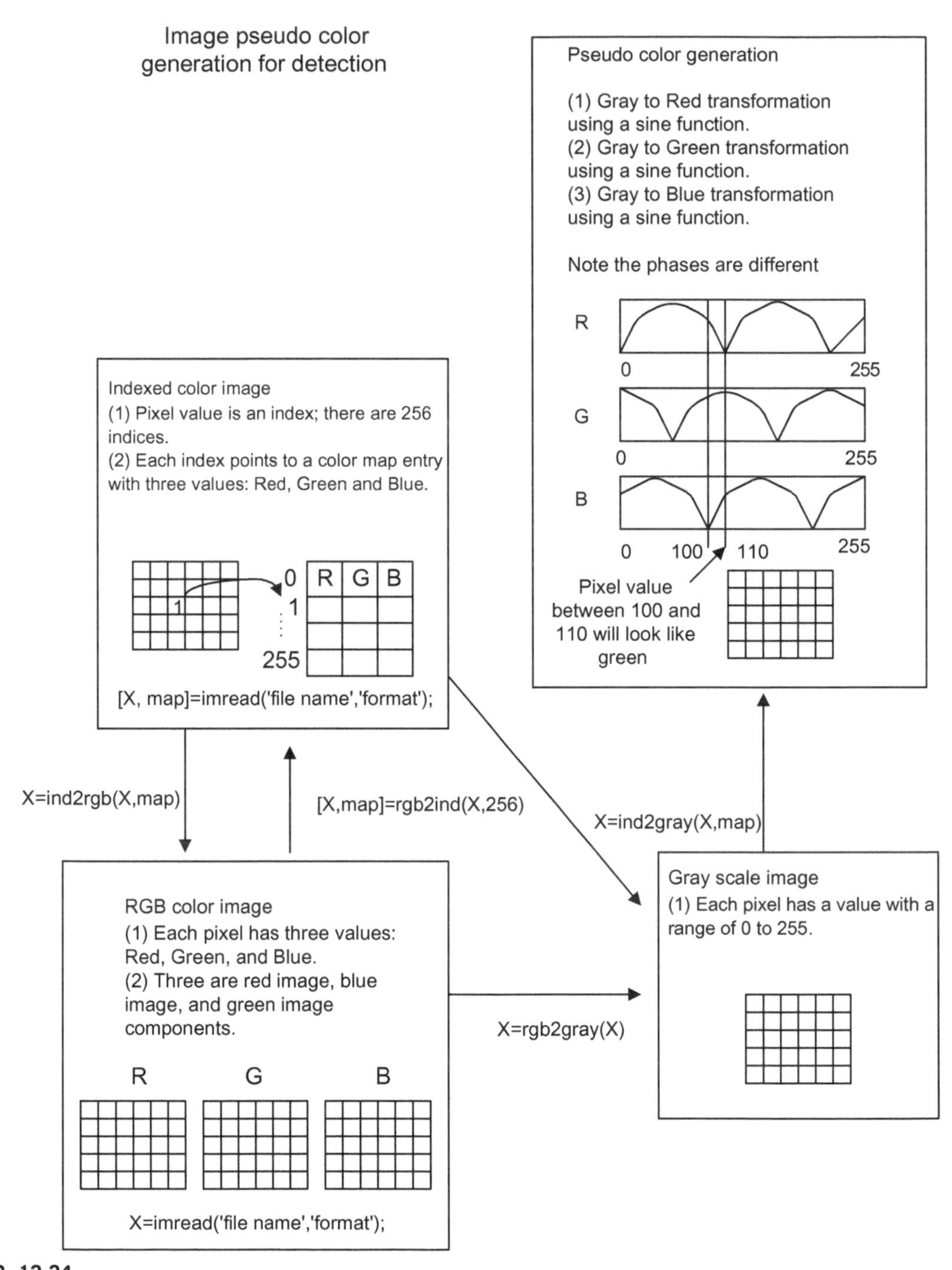

FIG. 13.34

Illustrative procedure for pseudo-color generation.

where $W_M = e^{-j\frac{2\pi}{M}}$ and $W_N = e^{-j\frac{2\pi}{N}}$.

m and n = pixel locations.

u and v = frequency indices.

Taking absolute value of the 2D-DFT coefficients $X(u,v)$ and dividing the absolute value by $(M \times N)$, we get the magnitude spectrum as

$$A(u,v) = \frac{1}{(N \times M)} |X(u,v)|. \tag{13.17}$$

Instead of going through the details of the 2D-DFT, we focus on application results via examples.

EXAMPLE 10.11

Determine the 2D-DFT coefficients and magnitude spectrum for the following 2 × 2 image.

$$\begin{bmatrix} 100 & 50 \\ 100 & -10 \end{bmatrix}.$$

Solution:

Since $M = N = 2$, applying Eq. (13.16) leads to

$$\begin{aligned} X(u,v) &= p(0,0)e^{-j\frac{2\pi u \times 0}{2}} \times e^{-j\frac{2\pi v \times 0}{2}} + p(0,1)e^{-j\frac{2\pi u \times 0}{2}} \times e^{-j\frac{2\pi v \times 1}{2}} \\ &+ p(1,0)e^{-j\frac{2\pi u \times 1}{2}} \times e^{-j\frac{2\pi v \times 0}{2}} + p(1,1)e^{-j\frac{2\pi u \times 1}{2}} \times e^{-j\frac{2\pi v \times 1}{2}}. \end{aligned}$$

For $u = 0$ and $v = 0$, we have

$$\begin{aligned} X(0,0) &= 100e^{-j0} \times e^{-j0} + 50e^{-j0} \times e^{-j0} + 100e^{-j0} \times e^{-j0} - 10e^{-j0} \times e^{-j0} \\ &= 100 + 50 + 100 - 10 = 240 \end{aligned}$$

For $u = 0$ and $v = 1$, we have

$$\begin{aligned} X(0,1) &= 100e^{-j0} \times e^{-j0} + 50e^{-j0} \times e^{-j\pi} + 100e^{-j0} \times e^{-j0} - 10e^{-j0} \times e^{-j\pi} \\ &= 100 + 50 \times (-1) + 100 - 10 \times (-1) = 160 \end{aligned}$$

Following similar operations,

$$X(1,0) = 60 \text{ and } X(1,1) = -60.$$

Thus, we have DFT coefficients as

$$X(u,v) = \begin{bmatrix} 240 & 160 \\ 60 & -60 \end{bmatrix}.$$

Using Eq. (13.17), we can calculate the magnitude spectrum as

$$A(u,v) = \begin{bmatrix} 60 & 40 \\ 15 & 15 \end{bmatrix}.$$

We can use the MABLAB function **fft2()** to verify the calculated DFT coefficients:

```
>> X=fft2([100 50;100 -10])
   X =
       240  160
       60  -60.
```

EXAMPLE 13.12

Given the following 200 × 200 grayscale image with a white rectangular (11 × 3 pixels) at its center and a black background, shown in Fig. 13.35A, we can compute the image's amplitude spectrum whose magnitudes are scaled in the range from 0 to 255. We can display the spectrum in terms of the grayscale. Fig. 13.35B shows the spectrum image.

The displayed spectrum has four quarters. The left upper quarter corresponds to the frequency components, and the other three quarters are the image counterparts. From the spectrum image, the area of the upper left corner is white and hence has the higher scale value. So, the image signal has low-frequency dominant components. The spectrum exhibits horizontal and vertical null lines (dark lines). The first vertical null line can be estimated as 200/11 = 18, while the first horizontal null line happens at 200/3 = 67. Next, let us apply the 2D spectrum to understand the image filtering effects in image enhancement.

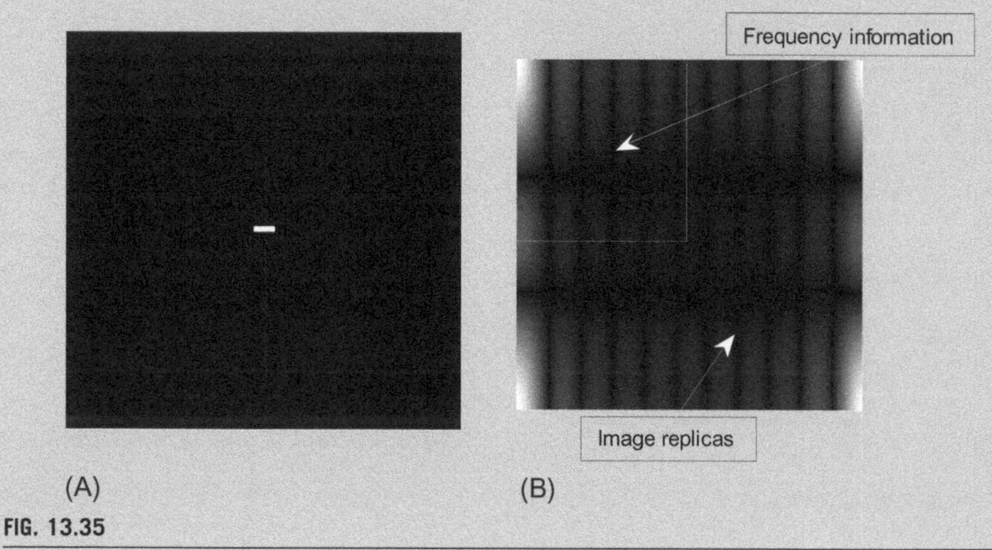

FIG. 13.35

(A) A square image. (B) Magnitude spectrum for the square image.

EXAMPLE 13.13

Fig. 13.36A is a biomedical image corrupted by random noise. Before we apply lowpass filtering, its 2D-DFT coefficients are calculated. We then compute its magnitude spectrum and scale it to the range from 0 to 255. To see noise spectral components, the spectral magnitude is further multiplied by a factor of 100. Once the spectral value is larger than 255, it is clipped to 255. The resultant spectrum is displayed in Fig. 13.36B, where we can see that noise occupies the entirety of the image.

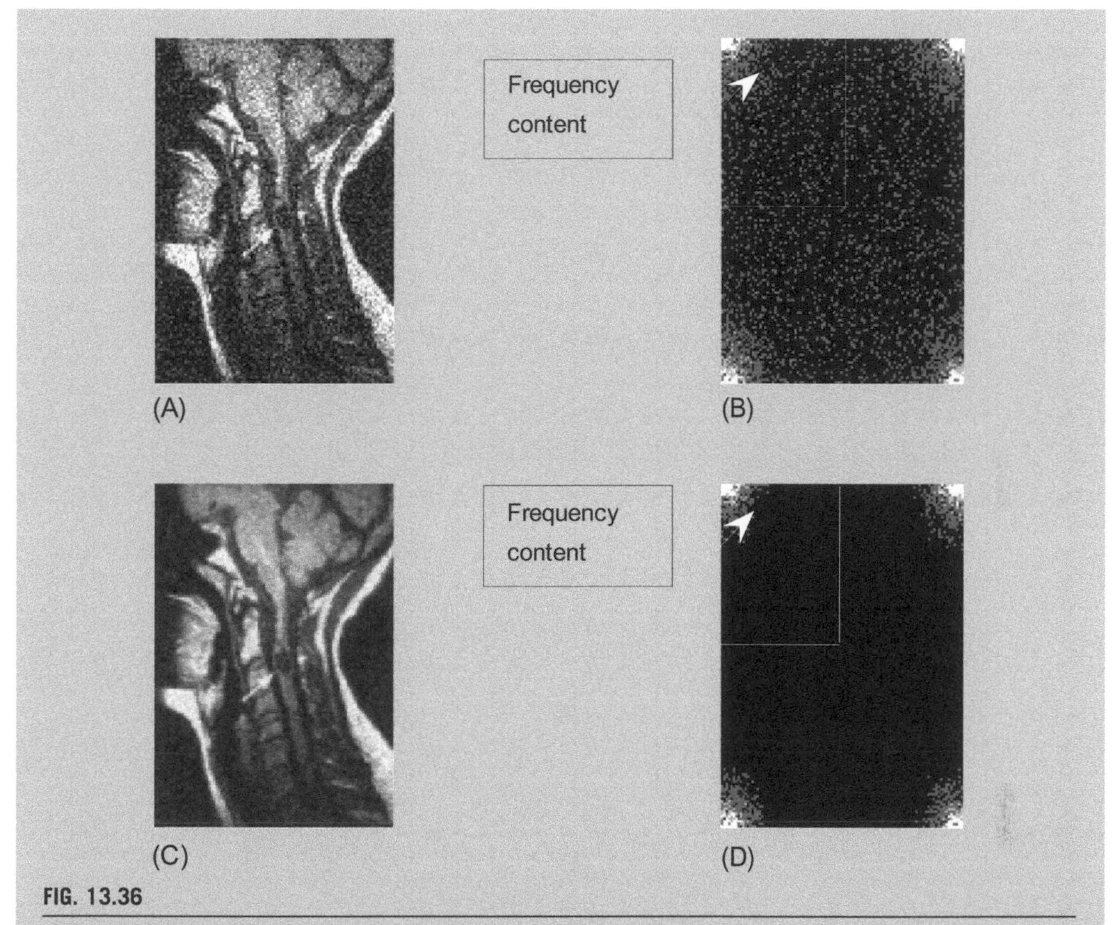

FIG. 13.36

Magnitude spectrum plots for the noisy image and the noise-filtered image. (A) The noisy image, (B) magnitude spectrum of the noisy image, (C) noise-filtered image, (D) magnitude spectrum of the noise-filtered image.

To enhance image, we apply a Gaussian lowpass filter. The enhanced image is shown in Fig. 13.36C, in which the enhancement is easily observed. Fig. 13.36D displays the spectra for the enhanced image with the same scaling process as described. As we can see, the noise is significantly reduced in comparison with Fig. 13.36B.

13.7 IMAGE COMPRESSION BY DISCRETE COSINE TRANSFORM

Image compression is necessary in our modern media systems, such as digital still and video cameras and computer systems. The purpose of compression is to reduce information storage or transmission bandwidth without losing image quality or at least without losing it significantly. Image compression can be classified as lossless compression or lossy compression. Here we focus on the lossy compression using discrete-cosine transform (DCT).

The DCT is a core compression technology used in the industry standards JPEG (Joint Photographic Experts Group) for still image compression and MPEG (Motion Picture Experts Group) for video compression, achieving compression ratio of 20:1 without noticeable quality degradation. JPEG standard image compression is used every day in real life.

The principle of the DCT is to transform the original image pixels to their DCT coefficients with the same number of the original image pixels, where the DCT coefficients have nonuniform distribution of direct current (DC) terms representing the average values, and alternate current (AC) terms representing fluctuations. The compression is achieved by applying the advantages of encoding DC terms (of a large dynamic range) with a large number of bits and low-frequency AC terms (a few, with a reduced dynamic range) with a reduced number of bits, and neglecting some high-frequency AC terms having small dynamic ranges (most of them do not affect the visual quality of the picture).

13.7.1 TWO-DIMENSIONAL DISCRETE COSINE TRANSFORM

Image compression uses two-dimensional discrete cosine transform (2D-DCT), whose transform pairs are defined as:

Forward DCT:

$$F(u,v)=\frac{2C(u)C(v)}{\sqrt{MN}}\sum_{i=0}^{M-1}\sum_{j=0}^{N-1}p(ij)\cos\left(\frac{(2i+1)u\pi}{2M}\right)\cos\left(\frac{(2j+1)v\pi}{2N}\right) \tag{13.18}$$

Inverse DCT:

$$p(i,j)=\sum_{u=0}^{M-1}\sum_{v=0}^{N-1}\frac{2C(u)C(v)}{\sqrt{MN}}F(uv)\cos\left(\frac{(2i+1)u\pi}{2M}\right)\cos\left(\frac{(2j+1)v\pi}{2N}\right) \tag{13.19}$$

where

$$C(m)=\begin{cases}\frac{\sqrt{2}}{2} & \text{if } m=0\\ 1 & \text{otherwise}\end{cases} \tag{13.20}$$

$p(i,j)$ = pixel level at the location (i,j)

$F(u,v)$ = DCT coefficient at the frequency indices (u,v).

JPEG divides an image into 8×8 image subblocks and applies DCT transform for each subblock individually. Hence, we simplify the general 2D-DCT in terms of 8×8 size. The equation for 2D 8×8 DCT is modified as:

$$F(u,v)=\frac{C(u)C(v)}{4}\sum_{i=0}^{7}\sum_{j=0}^{7}p(ij)\cos\left(\frac{(2i+1)u\pi}{16}\right)\cos\left(\frac{(2j+1)v\pi}{16}\right) \tag{13.21}$$

The inverse of 2D 8×8 DCT is expressed as:

$$p(i,j)=\sum_{u=0}^{7}\sum_{v=0}^{7}\frac{C(u)C(v)}{4}F(uv)\cos\left(\frac{(2i+1)u\pi}{16}\right)\cos\left(\frac{(2j+1)v\pi}{16}\right) \tag{13.22}$$

To become familiar with the 2D-DCT formulas, we study Example 13.14.

EXAMPLE 13.14

Determine the 2D-DCT coefficients for the following image:

$$\begin{bmatrix} 100 & 50 \\ 100 & -10 \end{bmatrix}.$$

Solution:

Applying $N=2$ and $M=2$ to Eq. (13.18) yields

$$F(u,v)=\frac{2C(u)C(v)}{\sqrt{2\times 2}}\sum_{i=0}^{1}\sum_{j=0}^{1}p(ij)\cos\left(\frac{(2i+1)u\pi}{4}\right)\cos\left(\frac{(2j+1)v\pi}{4}\right)$$

For $u=0$ and $v=0$, we achieve

$$\begin{aligned}F(0,0)&=c(0)c(0)\sum_{i=0}^{1}\sum_{j=0}^{1}p(i,j)\cos(0)\cos(0)\\&=\left(\frac{\sqrt{2}}{2}\right)^2[p(0,0)+p(0,1)+p(1,0)+p(1,1)]\\&=\frac{1}{2}\quad(100+50+100-10)=120\end{aligned}$$

For $u=0$ and $v=1$, we achieve

$$\begin{aligned}F(0,1)&=c(0)c(1)\sum_{i=0}^{1}\sum_{j=0}^{1}p(i,j)\cos(0)\cos\left(\frac{(2j+1)\pi}{4}\right)\\&=\left(\frac{\sqrt{2}}{2}\right)\times 1\times\left(p(0,0)\cos\frac{\pi}{4}+p(0,1)\cos\frac{3\pi}{4}+p(1,0)\cos\frac{\pi}{4}+p(1,1)\cos\frac{3\pi}{4}\right)\\&=\frac{\sqrt{2}}{2}\left(100\times\frac{\sqrt{2}}{2}+50\left(-\frac{\sqrt{2}}{2}\right)+100\times\frac{\sqrt{2}}{2}-10\left(-\frac{\sqrt{2}}{2}\right)\right)=80\end{aligned}$$

Similarly,

$$F(1,0)=30\,\text{and}\,F(1,1)=-30.$$

Finally, we get

$$F(u,v)=\begin{bmatrix} 120 & 80 \\ 30 & -30 \end{bmatrix}.$$

Applying the MATLAB function **dct2()** to verify the DCT coefficients as follows:

```
>> F=dct2([100 50;100 -10])
F =
120.0000 80.0000
 30.0000 -30.0000
```

EXAMPLE 13.5

Given the following DCT coefficients from a 2 × 2 image:

$$F(u,v) = \begin{bmatrix} 120 & 80 \\ 30 & -30 \end{bmatrix},$$

Determine the pixel $p(0,0)$.

Solution:

Applying Eq. (13.19) of the inverse 2D-DCT with $N=M=2$, $i=0$, and $j=0$, it follows that

$$\begin{aligned} p(0,0) &= \sum_{u=0}^{1}\sum_{v=0}^{1} c(u)c(v)F(u,v)\cos\left(\frac{u\pi}{4}\right)\cos\left(\frac{v\pi}{4}\right) \\ &= \left(\frac{\sqrt{2}}{2}\right)\times\left(\frac{\sqrt{2}}{2}\right)\times F(0,0) + \left(\frac{\sqrt{2}}{2}\right)\times F(0,1)\times\left(\frac{\sqrt{2}}{2}\right) \\ &+ \left(\frac{\sqrt{2}}{2}\right)\times F(1,0)\times\left(\frac{\sqrt{2}}{2}\right) + F(0,1)\left(\frac{\sqrt{2}}{2}\right)\times\left(\frac{\sqrt{2}}{2}\right) \\ &= \frac{1}{2}\times 120 + \frac{1}{2}\times 80 + \frac{1}{2}\times 30 + \frac{1}{2}(-30) = 100. \end{aligned}$$

We apply the MATLAB function **idct2()** to verify the inverse DCT to get the pixel values as follows:

```
>> p=idct2([120 80; 30 -30])
  p =
    100.0000 50.0000
    100.0000 -10.0000.
```

13.7.2 TWO-DIMENSIONAL JPEG GRAYSCALE IMAGE COMPRESSION EXAMPLE

To understand the JPEG image compression, we examine the 8 × 8 grayscale subblock. Table 13.5 shows a subblock of the grayscale image shown in Fig. 13.37 to be compressed.

Applying 2D-DCT leads to Table 13.6.

Table 13.5 8 × 8 Subblock

150	148	140	132	150	155	155	151
155	152	143	136	152	155	155	153
154	149	141	135	150	150	150	150
156	150	143	139	154	152	152	155
156	151	145	140	154	152	152	155
154	152	146	139	151	149	150	151
156	156	151	142	154	154	154	154
151	154	149	139	151	153	154	153

FIG. 13.37

Original image.

Table 13.6 DCT Coefficients for the Subblock Image in Table 13.5

1198	−10	26	24	−5	−16	0	12
−8	−6	3	8	0	0	0	0
0	−3	0	0	−8	0	0	0
0	0	0	0	0	0	0	0
0	−4	0	0	0	0	0	0
0	0	−1	0	0	0	0	0
−10	1	0	0	0	0	0	0
0	0	0	0	0	0	0	0

These DCT coefficients have a big DC component of 1198 but small AC component values. These coefficients are further normalized (quantized) with a quality factor Q, defined in Table 13.7.

After normalization, as shown in Table 13.8, the DC coefficient is reduced to 75, and a few small AC coefficients exist, while most are zero. We can encode and transmit only nonzero DCT coefficients and omit transmitting zeros, since they do not carry any information. They can be easily recovered by resetting coefficients to zero during decoding. By this principle we achieve data compression.

As shown in Table 13.8, most nonzero coefficients resides at the upper left corner. Hence, the order of encoding for each value is based on the zigzag path in which the order is numbered, as in Table 13.9.

According to the order, we record the nonzero DCT coefficients as a JPEG vector, shown as:

JPEG vector: **[75 –1 –1 0 –1 3 2 EOB]**

where "EOB" = the end of block coding. The JPEG vector can further be compressed by encoding the difference of DC values between subblocks, in *differential pulse code modulation* (DPCM), as discussed in Chapter 10, as well as by run-length coding of AC values and Huffman coding, which both belong to lossless compression techniques. We discuss this in the next section.

Table 13.7 The Quality Factor

16	11	10	16	24	40	51	61
12	12	14	19	26	58	60	55
14	13	16	24	40	57	69	56
14	17	22	29	51	87	80	62
18	22	37	56	68	109	103	77
24	35	55	64	81	104	113	92
49	64	78	87	103	121	120	101
72	92	95	98	112	100	103	99

Table 13.8 Normalized DCT Coefficients

75	−1	3	2	0	0	0	0
−1	−1	0	0	0	0	0	0
0	0	0	0	0	0	0	0
0	0	0	0	0	0	0	0
0	0	0	0	0	0	0	0
0	0	0	0	0	0	0	0
0	0	0	0	0	0	0	0
0	0	0	0	0	0	0	0

Table 13.9 The Order to Scan DCT Coefficients

0	1	5	6	14	15	27	28
2	4	7	13	16	26	29	42
3	8	12	17	25	30	41	43
9	11	18	24	31	40	44	53
10	19	23	32	39	45	52	54
20	22	33	38	46	51	55	60
21	34	37	47	50	56	59	61
35	36	48	49	57	58	62	63

During the decoding stage, the JPEG vector is recovered first. Then the quantized DCT coefficients are recovered according to the zigzag path. Next, the recovered DCT coefficients are multiplied by a quality factor to obtain the estimate of the original DCT coefficients. Finally, we apply the inverse DCT to achieve the recovered image subblock, which is shown in Table 13.10.

For comparison, the errors between the recovered image and original image are calculated and listed in Table 13.11.

Table 13.10 The Recovered Image Subblock

153	145	138	139	147	154	155	153
153	145	138	139	147	154	155	153
155	147	139	140	148	154	155	153
157	148	141	141	148	155	155	152
159	150	142	142	149	155	155	152
161	152	143	143	149	155	155	152
162	153	144	144	150	155	154	151
163	154	145	144	150	155	154	151

Table 13.11 The Coding Error of the Image Subblock

3	−3	−2	7	−3	−1	0	2
−2	−7	−5	3	−5	−1	0	0
1	−2	−2	5	−2	4	5	3
1	−2	−2	2	−6	3	3	−3
3	−1	−3	2	−5	3	3	−3
7	0	−3	4	−2	6	5	1
6	−3	−7	2	−4	1	0	−3
12	0	−4	5	−1	2	0	−2

The original and compressed images are displayed in Figs. 13.37 and 13.38, respectively. We do not see any noticeable difference between these two grayscale images.

13.7.3 JPEG COLOR IMAGE COMPRESSION

This section is devoted to reviewing the JPEG standard compression and examines the steps briefly. We focus on the encoder, since the decoder is just the reverse process of the encoding. The block diagram for the JPEG encoder is in Fig. 13.39.

The JPEG encoder has the following main steps:

1. Transform RGB to YIQ or YUV (U and V = chrominance components).
2. Perform DCT on blocks.
3. Perform quantization.
4. Perform zigzag ordering, DPCM, and run-length encoding.
5. Perform entropy encoding (Huffman coding)

***RGB to YIQ transformation*:**

The first transformation is of the RGB image to a YIQ or YUV image. Transformation from RGB to YIQ has previously been discussed. The principle is that in YIQ format, the luminance channel carries more signal energy, up to 93%, while the chrominance channels carry up to only 7% of signal energy. After transformation, more effort can be spent on coding the luminance channel.

FIG. 13.38

JPEG compressed image.

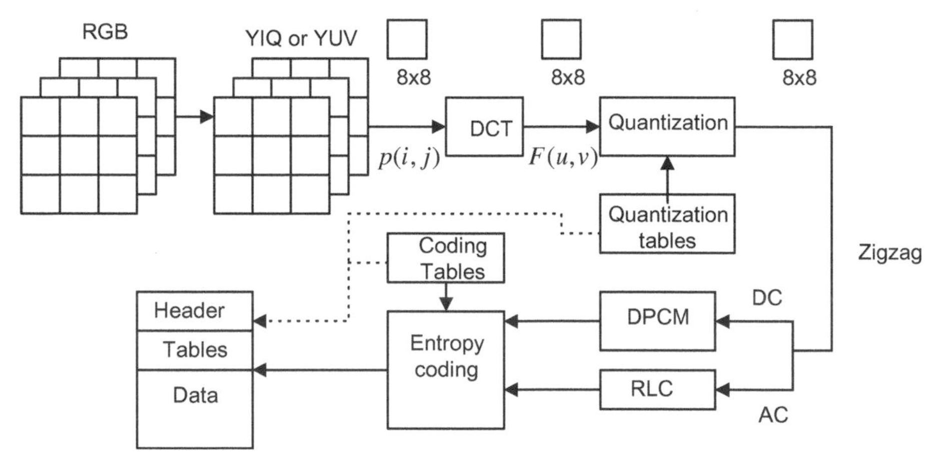

FIG. 13.39

Block diagram for JPEG encoder.

***DCT on image blocks*:**
Each image is divided into 8×8 blocks. 2D-DCT is applied to each block to obtain the 8×8 DCT coefficient block. Note that there are three blocks, Y, I, and Q.

***Quantization*:**
The quantization is operated using the 8×8 quantization matrix. Each DCT coefficient is quantized, divided by the corresponding value given in the quantization matrix. In this way, a smaller number of bits can be used for encoding the DCT coefficients. There are two different quantization tables, one for luminance as shown in Table 13.12 (which is the same as the one in the last section and listed here again for comparison) and the other one for chrominance shown in Table 13.13.

Table 13.12 The Quality Factor for Luminance

16	11	10	16	24	40	51	61
12	12	14	19	26	58	60	55
14	13	16	24	40	57	69	56
14	17	22	29	51	87	80	62
18	22	37	56	68	109	103	77
24	35	55	64	81	104	113	92
49	64	78	87	103	121	120	101
72	92	95	98	112	100	103	99

Table 13.13 The quality Factor for Chrominance

17	18	24	47	99	99	99	99
18	21	26	66	99	99	99	99
24	26	56	99	99	99	99	99
47	66	99	99	99	99	99	99
99	99	99	99	99	99	99	99
99	99	99	99	99	99	99	99
99	99	99	99	99	99	99	99
99	99	99	99	99	99	99	99

We can see that the chrominance table has numbers with larger values, so that small values of DCT coefficients will result and hence a less number of bits are required for encoding each DCT coefficient. Zigzag ordering to produce the JPEC vector is similar to the grayscale case, except that there are three JPEG vectors.

***DPCM on Direct-Current Coefficients*:**

Since each 8×8 image block has only one DC coefficient, which can be a very large number and varies slowly, we make use of DPCM for coding DC coefficients. As an example for the first five image blocks, DC coefficients are 200, 190, 180, 160, and 170. DPCM with a coding rule of $d(n) = DC(n) - DC(n-1)$ with initial condition $d(0) = DC(0)$ produces a DPCM sequence as

$$200, -10, -10, -20, 10.$$

Hence, the reduced signal range of these values is feasible for entropy coding.

***Run-length coding on Alternating-Current Coefficients*:**

The run-length method encodes the pair of

- the number zeros to skip and
- next nonzero value.

The zigzag scan of the 8×8 matrix makes up a vector of 64 values with a long runs of zeros. For example, the quantized DCT coefficients are scanned as

$$[75, -1, 0 - 1, 0, 0, -1\,3, 0, 0, 0, 2, 0, 0, \ldots, 0],$$

with one run, two runs, and three runs of zeros in the middle and 52 extra zeros toward the end. The run-length method encodes AC coefficients by producing a pair (run length, value), where the run length is the number of zeros in the run, while the value is the next nonzero coefficient. A special pair (0, 0) indicates EOB. Here is the result from a run-length encoding of AC coefficients:

$$(0,-1),(1,-1),(2,-1),(0,3),(3,2),(0,0).$$

***Lossless Entropy Coding*:**
The DC and AC coefficients are further compressed using entropy coding. JPEG allows Huffman coding and arithmetic coding. We focus on the Huffman coding here.

***Coding DC coefficients*:**
Each DPCM-coded DC coefficient is encoded by a pair of symbols (size, amplitude) with the size (4-bit code) designating the number of bits for the coefficient as shown in Table 13.14, while the amplitude is encoded by the actual bits. For the negative number of the amplitude, 1's complement is used.

For example, we can code the DPCM-coded DC coefficients 200, −10, −10, −20, 10 as

$$(8,11001000),(4,0101),(4,0101),(5,01011),(4,1010).$$

Since there needs to be 4 bits for encoding each size, we can use 45 bits in total for encoding the DC coefficients for these five subblocks.

***Coding AC coefficients*:**
The run-length AC coefficients have the format as (run length, value). The value can be further compressed using the Huffman coding method, similar to coding the DPCM-coded DC coefficients. The run length and the size are each encoded by 4 bits and packed into a byte.

Symbol 1: (run length, size)

Symbol 2: (amplitude).

The 4-bit run length can tackle only the number of runs for zeros from 1 to 15. If the run length of zeros is larger than 15, then a special code (15,0) is used for Symbol 1. Symbol 2 is the amplitude in Huffman coding as shown in Table 13.14, while the encoded Symbol 1 is kept in its format:

(run length, size, amplitude).

Let us code the following run-length code of AC coefficients:

$$(0,-1),(1,-1),(2,-1),(0,3),(3,2),(0,0).$$

Table 13.14 Huffman Coding Table

Size	Amplitude
1	−1,1
2	−3,−2,2,3
3	−7,..,−4.4, …,7
4	−15, …,−8.8, …,15
5	−31, …,−16,16, …,31
.	.
.	.
.	.
10	−1023, …,−512,512, …,1023

FIG. 13.40

JEPG compressed color image (For interpretation of the references to colour in this figure legend, the reader is referred to the web version of this article.)

We can produce a bit stream for AC coefficients as:

$$(0000, 0001, 0), (0001, 0001, 0), (0010, 0001, 0),$$
$$(0000, 0010, 11), (0011, 0010, 10), (0000, 0000).$$

There are 55 bits in total. Fig. 13.40 shows a JPEG compressed color image. The decompressed image is indistinguishable from the original image after comparison.

13.7.4 IMAGE COMPRESSION USING WAVELET TRANSFORM CODING

We can extend the 1D discrete wavelet transform (1D-DWT) discussed in chapter 12 to the 2D-DWT. The procedure is described as follows. Given an N × N image, the 1D DWT using level one is applied to each row of the image; and after all the rows are transformed, the level-1 1D DWT is applied again to each column. After the rows and columns are transformed, we achieve four first-level subbands labeled as LL, HL, LH, and HH shown in Fig. 13.41A. The same procedure repeats for the LL band only and results in the second-level subbands: LL2, HL2, LH2, and HH2 as shown in Fig. 13.41B. The process proceeds to the higher-level as desired. With the obtained wavelet transform, we can quantize coefficients to achieve compression requirement. For example, for the two-level coefficients, we can omit HL1, LH1, HH1 to simply achieve a 4:1 compression ratio. The decompression is a reverse process, that is, inversely transforms columns and then rows of the wavelet coefficients. We can apply the inverse discrete wavelet transform (IDWT) to the recovered LL band with the column and row inverse transform processes, and continue until the inverse transform at level one is completed. Let us look at an illustrative example.

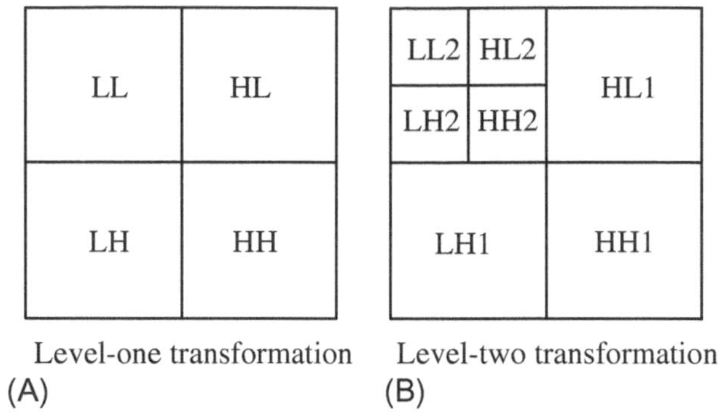

LL	HL
LH	HH

Level-one transformation
(A)

LL2	HL2	HL1
LH2	HH2	
LH1		HH1

Level-two transformation
(B)

FIG. 13.41

The two-dimensional DWT for level 1 and level 2. (A) Level-1 transformation, (B) Level-2 transformation.

EXAMPLE 13.6

Given a 4 × 4 image shown as follows:

113	135	122	109
102	116	119	124
105	148	138	122
141	102	140	132

(a) Perform 2D-DWT using the 2-tap Haar wavelet.
(b) Using the result in (a), perform 2D-IDWT using the 2-tap Haar wavelet.

Solution:

(a) The MATLAB function **dwt()** is applied to each row. Execution result for the first-row is displayed as follows:

```
>> dwt([1 1]/sqrt(2),[113 135 122 109],1)'      %Row vector coefficients
ans =176.0696 163.3417 -13.8492  9.1924
```

The completed row transform is listed as follows:

176.0696	163.3417	−13.8492	9.1924
154.1493	171.8269	−9.8995	−3.5355
178.8980	183.8478	−30.4056	11.3137
171.8269	192.3330	27.5772	5.6569

Next, the result for the first column is shown as follows:

```
>> dwt([1 1]/sqrt(2),[ 176.0696 154.1493 178.8980 171.8269 ],1)'
ans = 233.5000 248.0000 15.5000  5.0000
```

Applying each column completes level-1 transformation:

233.5000	237.0000	−17.5000	4.0000
248.0000	266.0000	−2.0000	12.0000
15.5000	−6.0000	−3.5000	9.0000
5.0000	−6.0000	−41.0000	4.0000

Now, we perform the Level-2 transformation. Applying the transform to the first row for LL1 band yields

```
>> dwt([1 1]/sqrt(2),[ 233.5000 237.0000],1)'
ans = 332.6937  -2.4749
```

After completing the row transformation, the result is given by

332.6937	−2.4749	−17.5000	4.0000
363.4529	−12.7279	−2.0000	12.0000
15.5000	−6.0000	−3.5000	9.0000
5.0000	−6.0000	−41.0000	4.0000

Similarly, the first-column MATLAB result is listed as follows:

```
>> dwt([1 1]/sqrt(2),[ 332.6937 363.4529 ],1)'
ans = 492.2500 -21.7500
```

Finally, we achieve the completed level-2 DWT as

492.2500	−10.7500	−17.5000	4.0000
−21.7500	7.2500	−2.0000	12.0000
15.5000	−6.0000	−3.5000	9.0000
5.0000	−6.0000	−41.0000	4.0000

(b) Recovering LL2 band first, the first column reconstruction is given as

```
>> idwt([1 1]/sqrt(2),[ 492.2500 -21.7500],1)'
ans = 332.6937 363.4529
```

Completing the inverse of the second column in the LL2 band gives

332.6937	−2.4749	−17.5000	4.0000
363.4529	−12.7279	−2.0000	12.0000
15.5000	−6.0000	−3.5000	9.0000
5.0000	−6.0000	−41.0000	4.0000

Now, we show the first row result for LL2 band in MATLAB as follows:

```
>> idwt([1 1]/sqrt(2),[ 332.6937 -2.4749 ],1)'
ans = 233.5000 237.0000
```

Continued

EXAMPLE 13.6—CONT'D

The recovered LL1 band is shown as follows:

$$\begin{matrix} 233.5000 & 237.0000 & -17.5000 & 4.0000 \\ 248.0000 & 266.0000 & -2.0000 & 12.0000 \\ 15.5000 & -6.0000 & -3.5000 & 9.0000 \\ 5.0000 & -6.0000 & -41.0000 & 4.0000 \end{matrix}$$

Now we are at the level-1 inverse process. For simplicity, the first column result in MATLAB and the completed results are listed as follows, respectively.

```
>> idwt([1 1]/sqrt(2),[ 233.5000 248.0000 15.5000 5.0000],1)'
ans = 176.0696 154.1493 178.8980 171.8269
```

$$\begin{matrix} 176.0696 & 163.3417 & -13.8492 & 9.1924 \\ 154.1493 & 171.8269 & -9.8995 & -3.5355 \\ 178.8980 & 183.8478 & -30.4056 & 11.3137 \\ 171.8269 & 192.3330 & 27.5772 & 5.6569 \end{matrix}$$

Finally, we perform the inverse of row transform at level one. The first row result in MATLAB is listed as follows:

```
>> idwt([1 1]/sqrt(2),[ 176.0696 163.3417 -13.8492  9.1924],1)'
ans = 113.0000 135.0000 122.0000 109.0000
```

The final inversed DWT is yielded as

$$\begin{matrix} 113.0000 & 135.0000 & 122.0000 & 109.0000 \\ 102.0000 & 116.0000 & 119.0000 & 124.0000 \\ 105.0000 & 148.0000 & 138.0000 & 122.0000 \\ 141.0000 & 102.0000 & 140.0000 & 132.0000 \end{matrix}$$

Since there is no quantization for each coefficient, we obtain a perfect reconstruction.

Fig. 13.42 shows an 8-bit grayscale image compression by applying one-level wavelet transform, in which the Daubechies wavelet filter with 16-tap is used. The wavelet coefficients (each is coded using 8-bit) are shown in Fig. 13.42A. By discarding HL, LH, and HH band coefficients, we can achieve 4:1 compression. The decoded image is displayed in Fig. 13.42C. The MATLAB program is listed in Program 13.4.

Fig. 13.43 illustrates two-level wavelet transform and compression results. By discarding HL2, LH2, HH2, HL1, LH1, and HH1 subbands, we achieve 16:1 compression. However, as shown in Fig. 13.43C, we can observe a noticeable degradation of image quality. Since the high-frequency details are discarded, the compressed image shows a significant smooth effect. In addition, there are many advanced methods to quantize and compress the wavelet coefficients. Of these compression

(A) Wavelet coefficients

(B) Original image

(C) 4:1 Compression

FIG. 13.42

(A) Wavelet coefficients; (B) original image; (C) 4:1 compression.

(A) Wavelet coefficients

(B) Original image

(C) 16:1 Compression

FIG. 13.43

(A) Wavelet coefficients; (B) original image; (C) 16:1 compression.

techniques, the embedded zerotree of the wavelet coefficient (EZW) method is the most efficient one, which can be found in Li et al. (2014).

Program 13.4. One-level wavelet transform and compression.

```
close all; clear all; clc
X=imread('cruise','JPEG');
Y=rgb2gray(X); % Convert the image into grayscale
h0=[0.054415842243144  0.312871590914520 0.675630736297712 ...
  0.585354683654425  -0.015829105256675 -0.284015542962009 ...
  0.000472484573805  0.128747426620538  -0.017369301001845 ...
  -0.044088253930837  0.013981027917411 0.008746094047413 ...
  -0.004870352993456  -0.000391740373377 0.000675449406451 ...
  -0.000117476784125];
M=length(h0);
h1(1:2:M-1)=h0(M:-2:2);h1(2:2:M)=-h0(M-1:-2:1);%Obtain QMF highpass filter
[m n]=size(Y);
% Level-1 transform
[m n]=size(Y);
for i=1:m
   W1(i,:)=dwt(h0,double(Y(i,:)),1)';
end
for i=1:n
   W1(:,i)=dwt(h0,W1(:,i),1); % Wavelet coefficients at level-1
end
%Quantization using 8-bit
wmax=double(max(max(abs(W1)))); %Scale factor
W1=round(double(W1)*2^7/wmax); %Get 8-bit data
W1=double(W1)*wmax/2^7;%Recover the wavelet
Figure (1); imshow(uint8(W1));xlabel('Wavelet coefficients');
% 8-bit Quantization
[m, n]=size(W1);
WW=zeros(m,n);
WW(1:m/2,1:n/2)=W1(1:m/2,1:n/2);
W1=WW;
%Decoding from level-1 using W1
[m, n]=size(W1);
for i=1:n
   Yd1(:,i)=idwt(h0,double(W1(:,i)),1);
end
for i=1:m
   Yd1(i,:)=idwt(h0,double(Yd1(i,:)),1)';
end
YY1=uint8(Yd1);
Figure (2),imshow(Y);xlabel('Original image');
Figure (3),imshow(YY1);xlabel('4:1 Compression');
```

Program 13.5. Two-level wavelet compression.

```
close all; clear all; clc
X=imread('cruise','JPEG');
Y=rgb2gray(X);
h0=[0.054415842243144 0.312871590914520 0.675630736297712 ...
  0.585354683654425-0.015829105256675 -0.284015542962009 ...
  0.000472484573805 0.128747426620538-0.017369301001845 ...
  -0.044088253930837 0.013981027917411 0.008746094047413 ...
  -0.004870352993456 -0.000391740373377 0.000675449406451 ...
  -0.000117476784125];
M=length(h0);
h1(1:2:M-1)=h0(M:-2:2);h1(2:2:M)=-h0(M-1:-2:1);%Obtain QMF highpass filter
[m n]=size(Y);
% Level-1 transform
[m n]=size(Y);
for i=1:m
    W1(i,:)=dwt(h0,double(Y(i,:)),1)';
end
for i=1:n
    W1(:,i)=dwt(h0,W1(:,i),1); % Wavelet coefficients at level-1 transform
end
% Level-2 transform
Y1=W1(1:m/2,1:n/2); %Obtain LL subband
[m n]=size(Y1);
for i=1:m
    W2(i,:)=dwt(h0,Y1(i,:),1)';
end
for i=1:n
W2(:,i)=dwt(h0,W2(:,i),1);
end
W22=W1; W22(1:m,1:n)=W2; % Wavelet coefficients at level-2 transform
wmax=max(max(abs(W22)));
% 8-bit Quantization
W22=round(W22*2^7/wmax);
W22=double(W22)*wmax/2^7;
Figure(1), imshow(uint8(W22));xlabel('Wavelet coefficients');
[m, n]=size(W22); WW=zeros(m,n);
WW(1:m/4,1:n/4)=W22(1:m/4,1:n/4);
W22=WW; %Discard HL2,LH2, HH2, HL1, LH1, HH1 subbands
% Decoding from Level-2 transform
[m,n]=size(W22); Wd2=W22(1:m/2,1:n/2);
% Level-2
[m n]=size(Wd2);
for i=1:n
    Wd1(:,i)=idwt(h0,double(Wd2(:,i)),1);
end
for i=1:m
    Wd1(i,:)=idwt(h0,double(Wd1(i,:))',1);
end
% Level-1
```

```
[m, n]=size(W22);Yd11=W22;
Yd11(1:m/2,1:n/2)=Wd1;
for i=1:n
Yd(:,i)=idwt(h0,Yd11(:,i),1);
end
for i=1:m
Yd(i,:)=idwt(h0,double(Yd(i,:)),1)';
end
Figure (2),imshow(Y),xlabel('Original image');
Y11=uint8(Yd); Fig. (3),imshow(Y11);xlabel('16:1 compression');
```

13.8 CREATING A VIDEO SEQUENCE BY MIXING TWO IMAGES

In this section, we introduce a method to mix two images to generate an image (video) sequence. Applications of mixing the two images may include fading in and fading out images, blending two images, or overlaying text on an image.

In mixing two images in a video sequence, a smooth transition is required from fading out one image of interest to fading in another image of interest. We want to fade out the first image and gradually fade in the second. This cross-fade scheme is implemented using the following operation:

$$\text{Mixed image} = (1-\alpha)\times \text{image}_1 + \alpha \times \text{image}_2 \qquad (13.23)$$

where α = fading in proportionally to the weight of the second image (value between 0 and 1), and $(1-\alpha)$ = fade out proportionally to the weight of the second image.

The video sequence in Fig. 13.44A consisting of six frames is generated using $\alpha=0$, $\alpha=0.2$, $\alpha=0.4$, $\alpha=0.6$, $\alpha=0.8$, $\alpha=1.0$, respectively, for two images. The equations for generating these frames are listed as follows:

$$\text{Mixed image}_1 = 1.0\times \text{image}_1 + 0.0\times \text{image}_2$$
$$\text{Mixed image}_2 = 0.8\times \text{image}_1 + 0.2\times \text{image}_2$$
$$\text{Mixed image}_3 = 0.6\times \text{image}_1 + 0.4\times \text{image}_2$$
$$\text{Mixed image}_4 = 0.4\times \text{image}_1 + 0.6\times \text{image}_2$$
$$\text{Mixed image}_5 = 0.2\times \text{image}_1 + 0.8\times \text{image}_2$$
$$\text{Mixed image}_6 = 0.0\times \text{image}_1 + 1.0\times \text{image}_2.$$

The sequence begins with the Grand Canyon image and fades in with the cruise ship image. At frame 4, the 60% of the cruise ship is faded in, and the image begins to be discernible as such. The sequence ends with the cruise ship in 100% fade-in. Fig. 13.44A displays the generated grayscale sequence. Fig. 13.44B shows the color RGB video sequence.

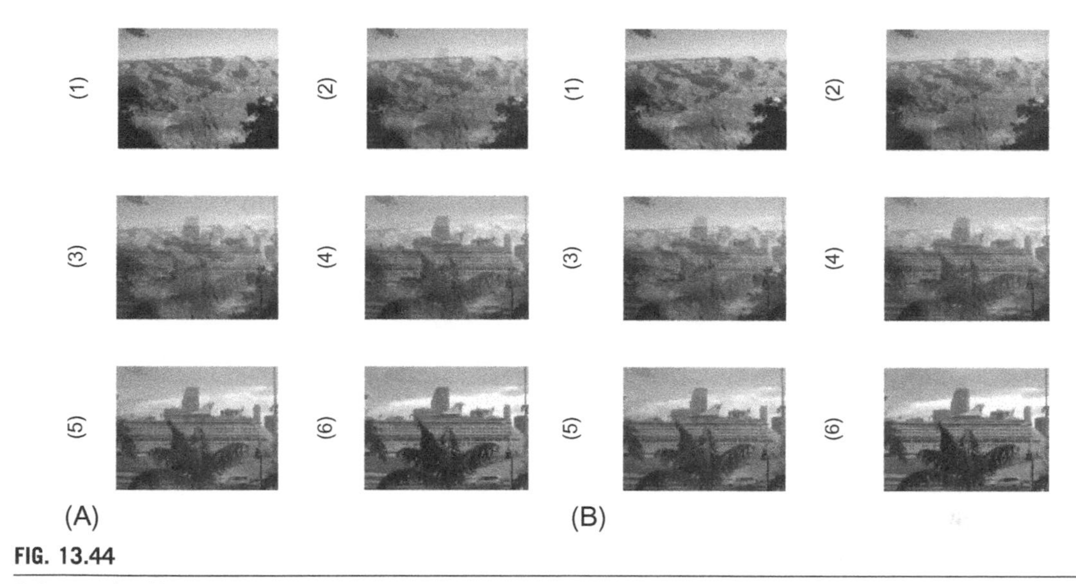

FIG. 13.44

A Grayscale video sequence.

13.9 VIDEO SIGNAL BASICS

Video signals generally can be classified as component video, composite video, and S-video.

In *component video*, three video signals—such as the RGB channels or the Y, I, and Q channels—are used. Three wires are required for connection to a camera or TV. Most computer systems use component video signals. *Composite video* has the intensity (luminance) and two-color (chrominance) components that modulate the carrier wave. This signal is used in the broadcast color TV. The standard by the US-based National Television System Committee (NTSC) combines channel signals into a chroma signal, which is modulated to a higher frequency for transmission. Connecting TVs or VCRs requires only one wire, since both video and audio are mixed into the same signal. *S-video* sends luminance and chrominance separately, since the luminance presenting black-and-white intensity contains most of the signal information for visual perception.

13.9.1 ANALOG VIDEO

In computer systems, progressive scanning traces a whole picture, called *frame* via *row wise*. A higher-resolution computer uses 72 frames per second (fps). The video is usually played at a frame rate varying from 15 to 30 frames.

In TV reception and some monitors, *interlaced scanning* is used in a cathode-ray tube display, or raster. The odd-numbered lines are traced first, and the even-numbered lines are traced next. We then get the odd-field and even-field scans per frame. The interlaced scheme is illustrated in Fig. 13.45, where the odd lines are traced, such as A to B, then C to D, and so on, ending in the middle at E. The even field begins at F in the middle of the first line of the even field and ends at G. The purpose

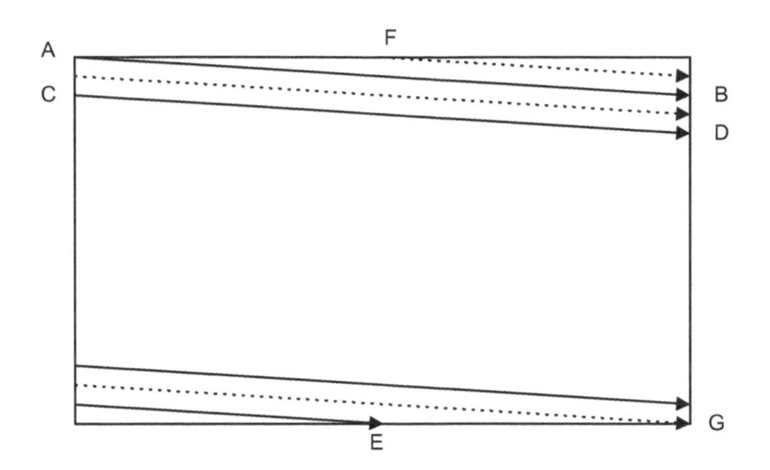

FIG. 13.45

Interlaced raster scanning.

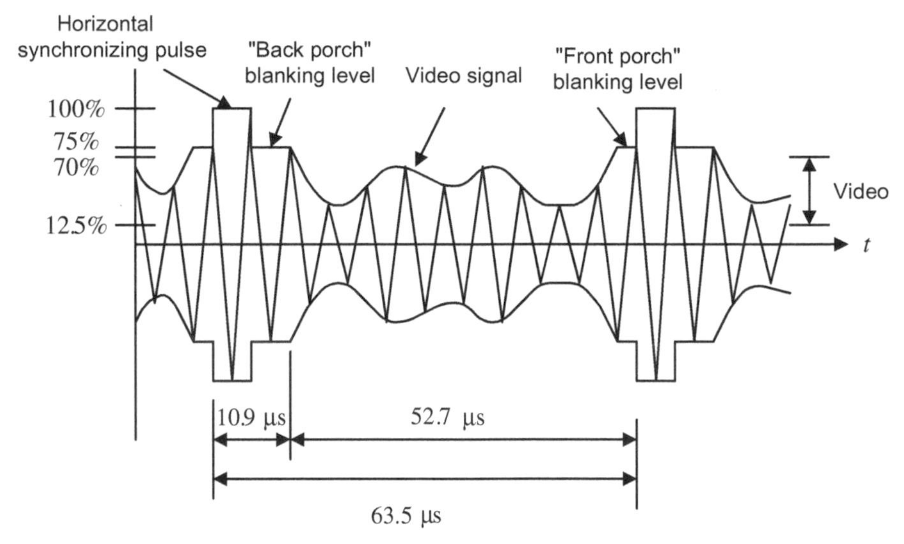

FIG. 13.46

Video-modulated waveform.

of using interlaced scanning is to transmit a full frame quickly to reduce flicker. Trace jumping from B to C is called horizontal retrace, while trace jumping from E to F or G to A is called vertical retrace.

The video signal is amplitude modulated. The modulation levels for NTSC video are shown in Fig. 13.46. In the United States, negative modulation is used, considering that less amplitudes come from a brighter scene, while more amplitudes come from a darker one. This is due to the fact that most pictures contain more white levels than black levels. With negative modulation, possible power efficiency can be achieved for transmission. The reverse process will apply for display at the receiver.

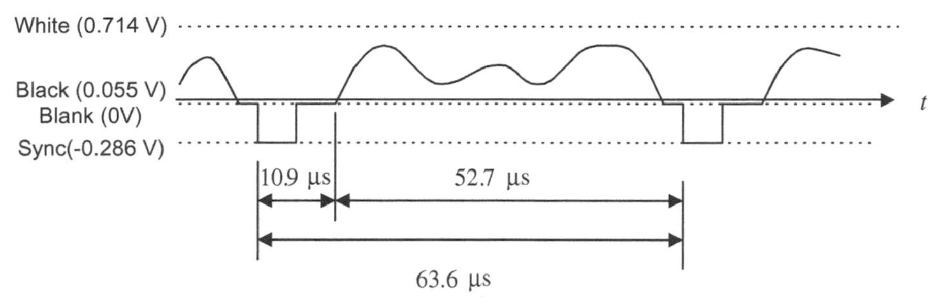

FIG. 13.47

The demodulated signal level for one NTSC scan line.

The horizontal synchronizing pulse controls the timing for the horizontal retrace. The blanking levels are used for synchronizing as well. The "back porch" (Fig. 13.46) of the blanking also contains the color subcarrier burst for the color demodulation.

The demodulated electrical signal can be seen in Fig. 13.47, where a typical electronic signal for one scan line is depicted. The white intensity has a peak value of 0.714 V, and the black has voltage level of 0.055 V, which is close to zero. The blank corresponds to 0 V, and the synchronizing pulse is at the level of −0.286 V. The time duration for synchronizing is 10.9 μs; that of the video occupies 52.7 μs; and that of one entire scan line occupies 63.6 μs. Hence, the line scan rate can be determined as 15.75 kHz.

Fig. 13.48 describes vertical synchronization. A pulse train is generated at the end of each filed. The pulse train contains six equalizing pulses, six vertical synchronizing pulses, and another six equalizing pulses at the rate of twice the size of the line scan rate (31.5 kHz), so that the timing for sweeping half the width of the field is feasible. In NTSC, the vertical retrace takes the time interval of 20 horizontal lines designated for control information at the beginning of each field. The 18 pulses of the vertical blanking occupy the time interval that is equivalent to nine lines. This leaves lines 10–20 for other uses.

A color subcarrier resides the back porch, as shown in Fig. 13.48. The eight cycles of the color subcarrier are recovered via a delayed gating circuit trigged by the horizontal sync pulse. Synchronization includes the color burst frequency and phase information. The color subcarrier is then applied to demodulate the color (chrominance).

Let us summarize NTSC video signals. The NTSC TV standard uses an aspect ratio of 4:3 (ratio of picture width to height), and 525 scan lines per frame at 30 fps. Each frame has an odd field and an even field. So there are 525/2 = 262.5 lines per field. NTSC actually uses 29.97 fps. The horizontal sweep frequency is $525 \times 29.97 = 15{,}734$ lines per second, and each line takes $1/15{,}734 = 63.6\ \mu$ sec. Horizontal retrace takes $10.9\ \mu$ sec, while the line signal takes $52.7\ \mu$ sec. For one line of image display. Vertical retrace and sync are also needed so that the first 20 lines for each field are reserved to be used. The active video lines per frame are 485. The layout of the video data, retrace, and sync data is shown in Fig. 13.49.

Blanking regions can be used for V-chip information, stereo audio channel data, and subtitles in various languages. The active line then is sampled for display. A pixel clock divides each horizontal line of video into samples. For example, vertical helical scan (VHS) uses 240 samples per line, Super VHS, 400–425 samples per line.

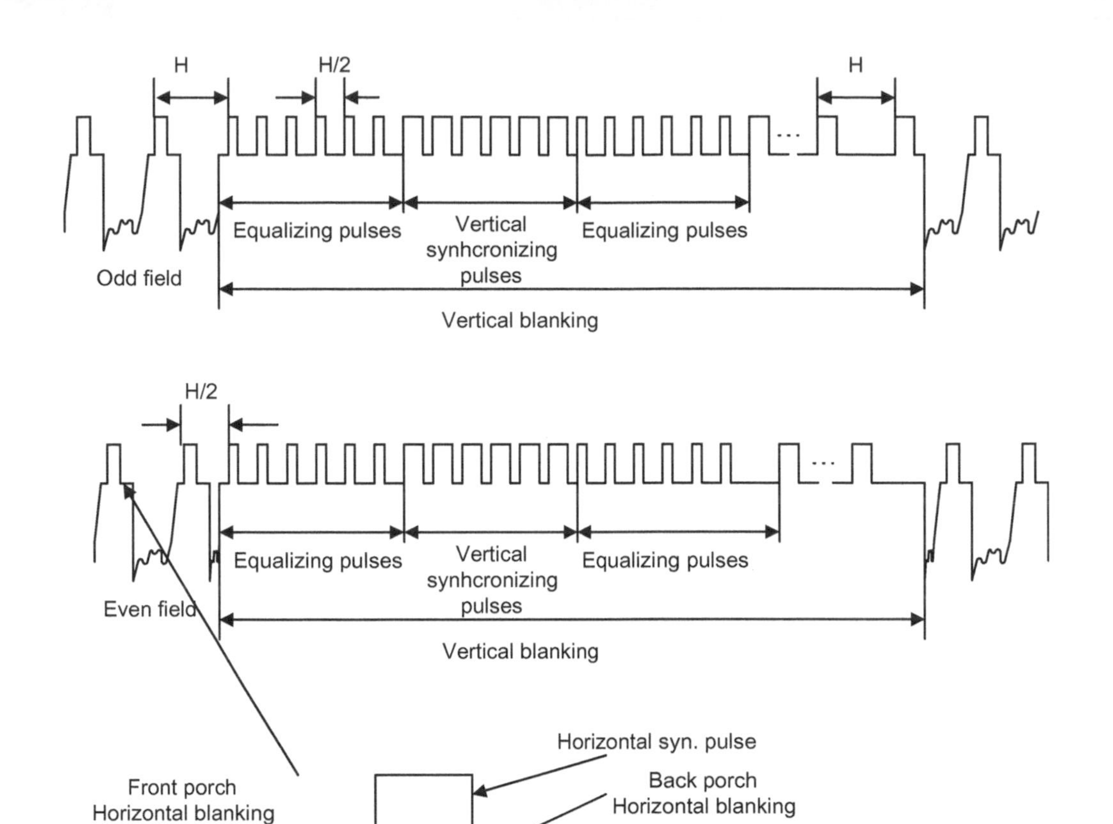

FIG. 13.48

Vertical synchronization for each field and the color subcarrier burst.

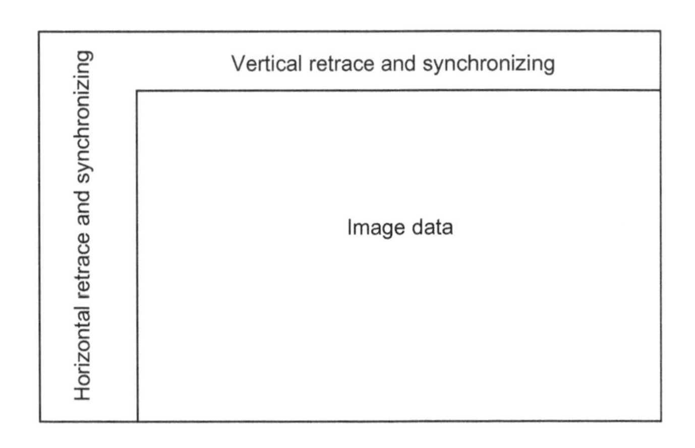

FIG. 13.49

Video data, retrace, and sync layout.

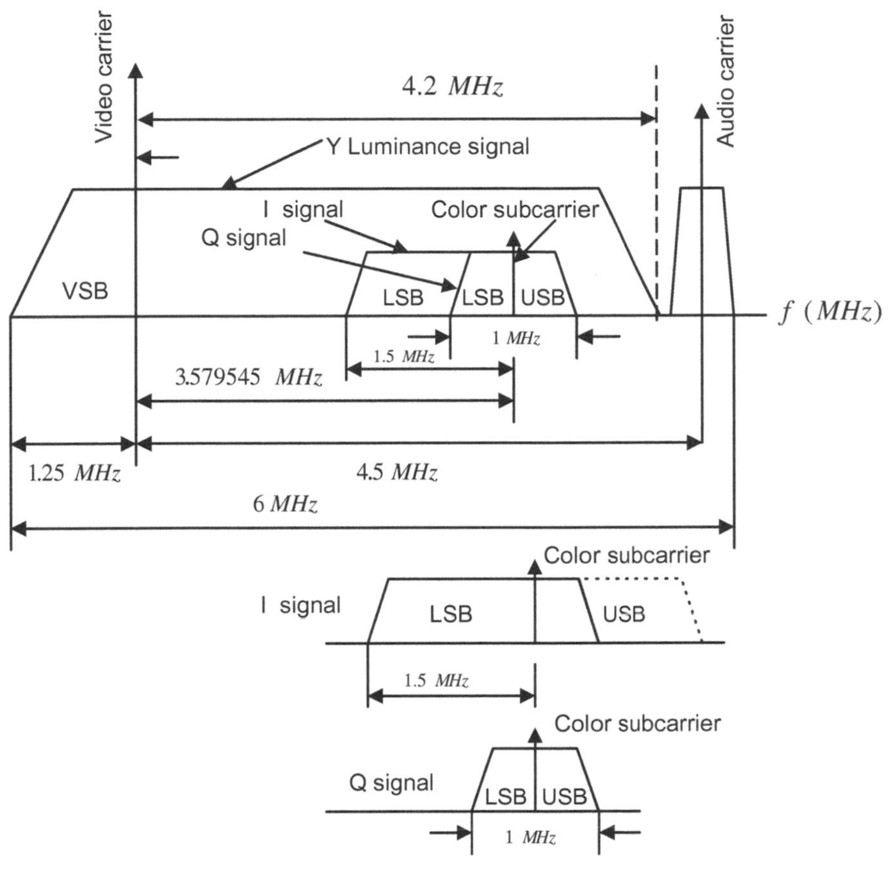

FIG. 13.50

NTSC Y, I, and Q spectra.

Fig. 13.50 shows the NTSC video signal spectra. The NTSC standard assigns a bandwidth of 4.2 MHz for luminance Y, 1.6 MHz for I, and 0.6 for Q, due to the human perception of color information. Since the human eye has higher resolution to the I color component than to the Q color component, the wider bandwidth for I is allowed.

As shown in Fig. 13.50, *vestigial sideband modulation* (VSB) is employed for the luminance, with a picture carrier of 1.25 MHz relative to the VSB left edge. The space between the picture carrier and the audio carrier is 4.5 MHz.

The audio signal containing the frequency range from 50 Hz to 15 kHz is stereo frequency modulated (FM), using a peak frequency deviation of 25 kHz. Therefore, the stereo FM audio requires a transmission bandwidth of 80 kHz, with an audio carrier located at 4.5 MHz relative to the picture carrier.

The color burst carrier is centered at 3.58 MHz above the picture carrier. The two color components I and Q undergo *quadrature amplitude modulation* (QAM) with modulated component I output, which is VSB filtered to remove the two-thirds of the upper sideband, so that all chroma signals fall within a 4.2 MHz video bandwidth. The color burst carrier of 3.58 MHz is chosen such that the chroma signal and luminance are interleaved in the frequency domain to reduce interference between them.

Generating a chroma signal with QAM gives

$$C = I\cos(2\pi f_{sc}t) + Q\sin(2\pi f_{sc}t), \tag{13.24}$$

where C = chroma component and f_{sc} = color subcarrier = 3.58 MHz. The NTSC signal is further combined into a composite signal:

$$Composite = Y + C = Y + I\cos(2\pi f_{sc}t) + Q\sin(2\pi f_{sc}t). \tag{13.25}$$

At decoding, the chroma signal is obtained by separating Y and C first. Generally, the lowpass filters located at the lower end of the channel can be used to extract Y. The comb filters may be employed to cancel interferences between the modulated luminance signal and the chroma signal (Li et al., 2014). Then we perform demodulation for I and Q as follows:

$$\begin{aligned} C \times 2\cos(2\pi f_{sc}t) &= I2\cos^2(2\pi f_{sc}t) + Q \times 2\sin(2\pi f_{sc}t)\cos(2\pi f_{sc}t) \\ &= I + I \times \cos(2 \times 2\pi f_{sc}t) + Q\sin(2 \times 2\pi f_{sc}t). \end{aligned} \tag{13.26}$$

Applying a lowpass filter yields the I component. Similar operation applying a carrier signal of 2 sin $(2\pi f_{sc}t)$ for demodulation recovers the Q component.

***PAL Video*:**

The phase alternative line (PAL) system uses 625 scan lines per frame at 25 fps, with an aspect ratio of 4:3. It is widely used in Western Europe, China, and India. PAL uses the YUV color model, with an 8-MHz channel in which Y has 5.5 MHz and U and V each have 1.8 MHz with the color subcarrier frequency of 4.43 MHz relative to the picture carrier. The U and V are the color difference signals (chroma signals) of the B-Y signal and R-Y signal, respectively. The chroma signals have alternate signs (e.g., +V and − V) in successive scan lines. Hence, the signal and its sign reversed one in the consecutive lines are averaged to cancel out the phase errors that could be displayed as the color errors.

***SECAM Video*:**

The SECAM (Séquentiel Couleur à Mémoire) system uses 625 scan lines per frame at 25 fps, with an aspect ratio of 4:3 and interlaced fields. The YUV color model is employed, and U and V signals are modulated using separate color subcarriers of 4.25 and 4.41 MHz, respectively. The U and V signals are sent on each line alternatively. In this way, the quadrature multiplexing and the possible cross coupling of color signals could be avoided by halving the color resolution in the vertical dimension.

Table 13.15 lists the frame rates, numbers of scan lines, and bandwidths for each analog broadband TV systems.

13.9.2 DIGITAL VIDEO

Digital video has become dominant over the long-standing analog method in the modern systems and devices because it offers high image quality; flexibility of storage, retrieval, and editing capabilities; digital enhancement of video images; encryption; channel noise tolerance; and multimedia system applications.

Digital video formats are developed by the Consultative Committee for International Radio (CCIR). A most important standard is CCIR-601, which became ITU-R-601, an international standard for professional video applications.

Table 13.15 Analog Broadband TV Systems

TV System	Frame Rate (fps)	Number of Scan Lines	Total Bandwidth (MHz)	Y Bandwidth (MHz)	U or I Bandwidth (MHz)	V or Q Bandwidth (MHz)
NTSC	29.97	525	6.0	4.2	1.6	0.6
PAL	25	625	8.0	5.5	1.8	1.8
SECAM	25	625	8.0	6.0	2.0	2.0

Source: Li, Z.N., Drew, M.S., Liu, J., 2014. Fundamentals of Multimedia, second ed. Springer.

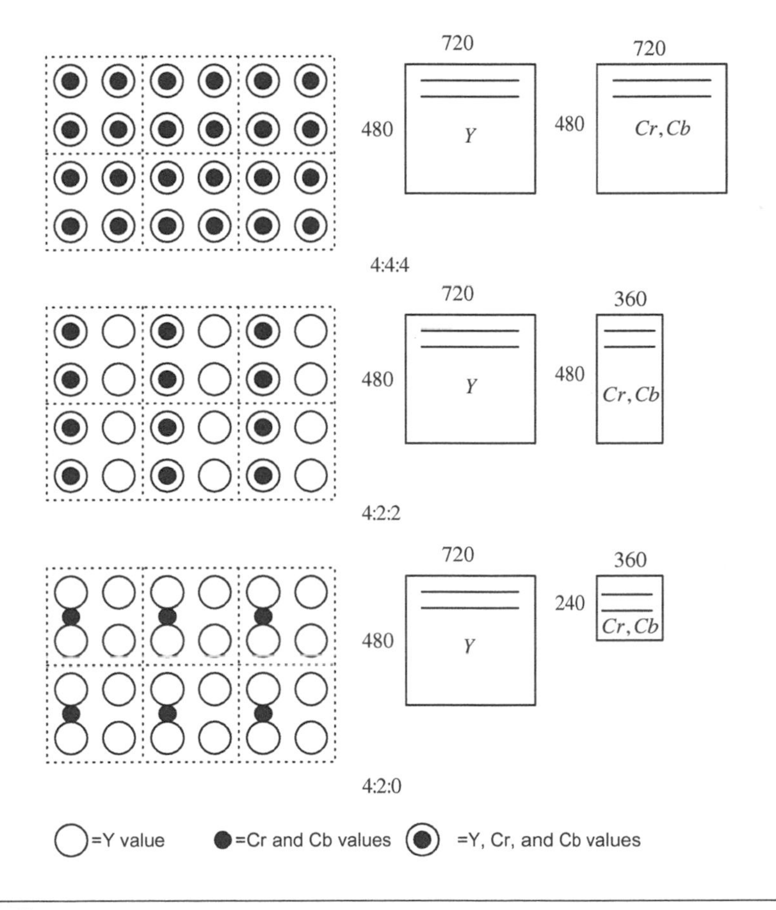

FIG. 13.51

Chroma subsampling.

In CCIR-601, chroma subsampling is carried for digital video. Each pixel is in YCbCr color space, where Y is the luminance, and Cb and Cr are the chrominance. Subsampling schemes include 4:4:4 (no chroma subsampling); 4:2:2, 4:1:1, and 4:2:0 as illustrated in Fig. 13.51.

In a 4:4:4 video format in each frame, the number of values for each chrominance component, Cb or Cr, is the same as that for luminance, Y, both horizontally and vertically. This format finds applications

Table 13.16 Digital Video Specifications

	CCIR 601 525/60 NTSC	CCR 601 625/50 PAL/SECAM	CIF	QCIF
Luminance resolution	720 × 480	720 × 576	352 × 288	176 × 144
Chrominance resolution	360 × 480	360 × 576	176 × 144	88 × 72
Color subsampling	04:02:02	04:02:02	04:02:00	04:02:00
Aspect ratio	04:03	04:03	04:03	04:03
Fields/secs	60	50	30	30
Interlaced	Yes	Yes	No	No

CCR, comparison category rating; CIF, common intermediate format; QCIF, quarter-CIF.
Source: Li, Z.N., Drew, M.S., Liu, J., 2014. Fundamentals of Multimedia, second ed., Springer.

in the computer graphics, in which the chrominance resolution is required for both horizontal and vertical dimension. The format is not widely used in video applications due to a huge storage requirement.

As shown in Fig. 13.51, for each frame in the 4:2:2 video format, the number of chrominance components for Cr or Cb is half the number of luminance components for Y. The resolution is full vertically, and the horizontal resolution is downsampled by a factor of 2. Considering the first line of six pixels, transmission occurs in the following form: (Y0, Cb0), (Y1, Cr0), (Y2,Cb2), (Y3,Cr2), (Y4,Cb4), (Y5, Cr4), and so on. Six Y values are sent for every two Cb and Cr values that are sent.

In the 4:2:0 video format, the number of values for each chrominance Cb and Cr is half the number of luminance Y for both horizontal and vertical directions. That is, the chroma is downsampled horizontally and vertically by a factor of 2. The location for both Cb and Cr is shown in Fig. 13.51. Digital video specifications are given in Table 13.16.

CIF was specified by the Comité Consultatif International Téléphonique et Télégraphique (CCITT), which is now the International telecommunications Union (ITU). CIF produces a low-bit rate video and supports a progressive scan. QCIF achieves an even lower bit rate video. Neither formats supports the interlaced scan mode.

Table 13.17 outlines the high-definition TV (HDTV) formats supported by the Advanced Television System Committee (ATSC), where "I" means interlaced scan and "P" indicates progressive scan. MPEG compressions of video and audio are employed.

13.10 MOTION ESTIMATION IN VIDEO

In this section, we study motion estimation since this technique is widely used in the MPEG video compression. A video contains a time-ordered sequence of frames. Each frame consists of image data. When the objects in an image are still, the pixel values do not change under constant lighting condition. Hence, there is no motion between the frames. However, if the objects are moving, then the pixels are moved. If we can find the motions, which are the pixel displacements, with *motion vectors*, the frame

Table 13.17 High-Definition TV (HDTV) Formats

Number of Active Pixels per Line	Number of Active Lines	Aspect Ratio	Picture Rate
1920	1080	16:09	60I 30P 24P
1280	720	16:09	60P 30P 24 P
704	480	16:9 and 4:3	60I 60P 30P 24P
640	480	4:03	60I 60P 30P 24P

Source: Li, Z.N., Drew, M.S., Liu, J., 2014. Fundamentals of Multimedia, second ed., Springer.

data can be recovered from the reference frame by copying and pasting at locations specified by the motion vector. To explore such an idea, let us look at Fig. 13.52.

As shown Fig. 13.52, the reference frame is displayed first, and the next frame is the target frame containing a moving object. The image in the target frame is divided into N × N macroblocks (20 macroblocks). A macroblock match is searched within the search window in the reference frame to find the closest match between a macroblock under consideration in the target frame and the macroblock in the reference frame. The differences between two locations (motion vectors) for the matched macroblocks are encoded.

The criteria for searching the best matching can be chosen using the mean absolute difference (MAD) between the reference frame and the target frame:

$$MAD(i,j)=\frac{1}{N^2}\sum_{k=0}^{N-1}\sum_{l=0}^{N-1}|T(m+k,n+l)-R(m+k+i,n+l+j)| \tag{13.27}$$

$$u=i, v=j\, for\, MAD(i,j)=\text{minimum}, \text{and}-p\le i,j\le p. \tag{13.28}$$

There are many search methods for finding the motion vectors, including optimal sequential or brute force searches; and suboptimal searches such as 2D-logarithmic and hierarchical searches. Here we examine the sequential search to understand the basic idea.

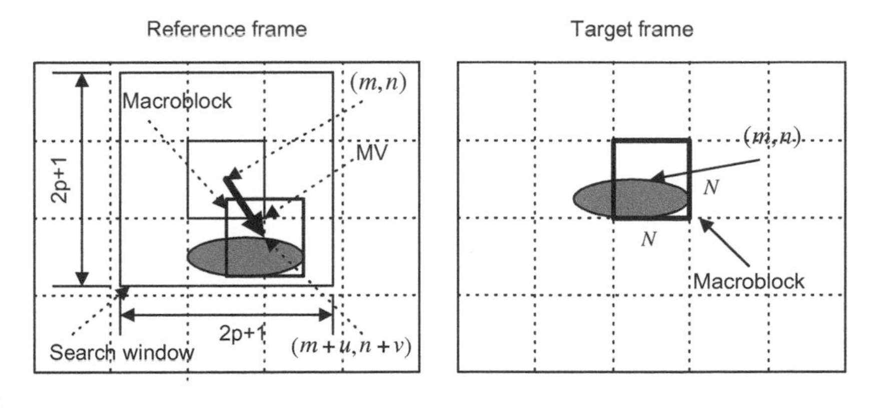

FIG. 13.52

Macroblocks and motion vectors in the reference frame and target frame.

The sequential search for finding the motion vectors employs methodical "brute force" to search the entire $(2p+1) \times (2p+1)$ search window in the reference frame. The macroblock in the target frame compares each macroblock centered at each pixel in the search window of the reference frame. Comparison using Eq. (13.27) proceeds pixel by pixel to find the best match in which the vector (i, j) produces the smallest MAD. Then the motion vector $(MV(u,v))$ is found to be $u=i$, and $v=j$. The algorithm is described as follows:

```
min_MAD=large value
for i=-p, ..., p
  for j=-p,...,p
   cur_MAD=MDA(i,j);
   if cur_MAD < min_MAD
        min_MAD= cur_MAD;
        u=i;
        v=j;
   end
 end
end
```

The sequential search provides best matching with the least MAD. However, it requires a huge amount of computations. Other suboptimal search methods can be employed to reduce the computational requirement, but with sacrifices of image quality. These topics are beyond our scope.

EXAMPLE 13.17

An 80×80 reference frame, target frame, and their difference are displayed in Fig. 13.53. A macroblock with a size of 16×16 is used, and the search window has a size of 32×32. The target frame is obtained by moving the reference frame to the right by 6 pixels and to the bottom by 4 pixels. The sequential search method is applied to find all the motion vectors. The reconstructed target frame using the motion vectors and reference image is given in Fig. 13.53.

Since $80 \times 80/(16 \times 16)=25$, there are 25 macroblocks in the target frame and 25 motion vectors in total. The motion vectors are found to be:

Horizontal direction =

$$-6\ -6\ -6\ -6\ -6\ -6\ -6\ -6\ -6\ -6\ -6\ -6\ -6$$
$$-6\ -6\ -6\ -6\ -6\ -6\ -6\ -6\ -6\ -6\ -6\ -6$$

Vertical direction =

$$-4\ -4\ -4\ -4\ -4\ -4\ -4\ -4\ -4\ -4\ -4\ -4\ -4$$
$$-4 -4 -4\ -4\ -4\ -4\ -4\ -4\ -4\ -4\ -4\ -4$$

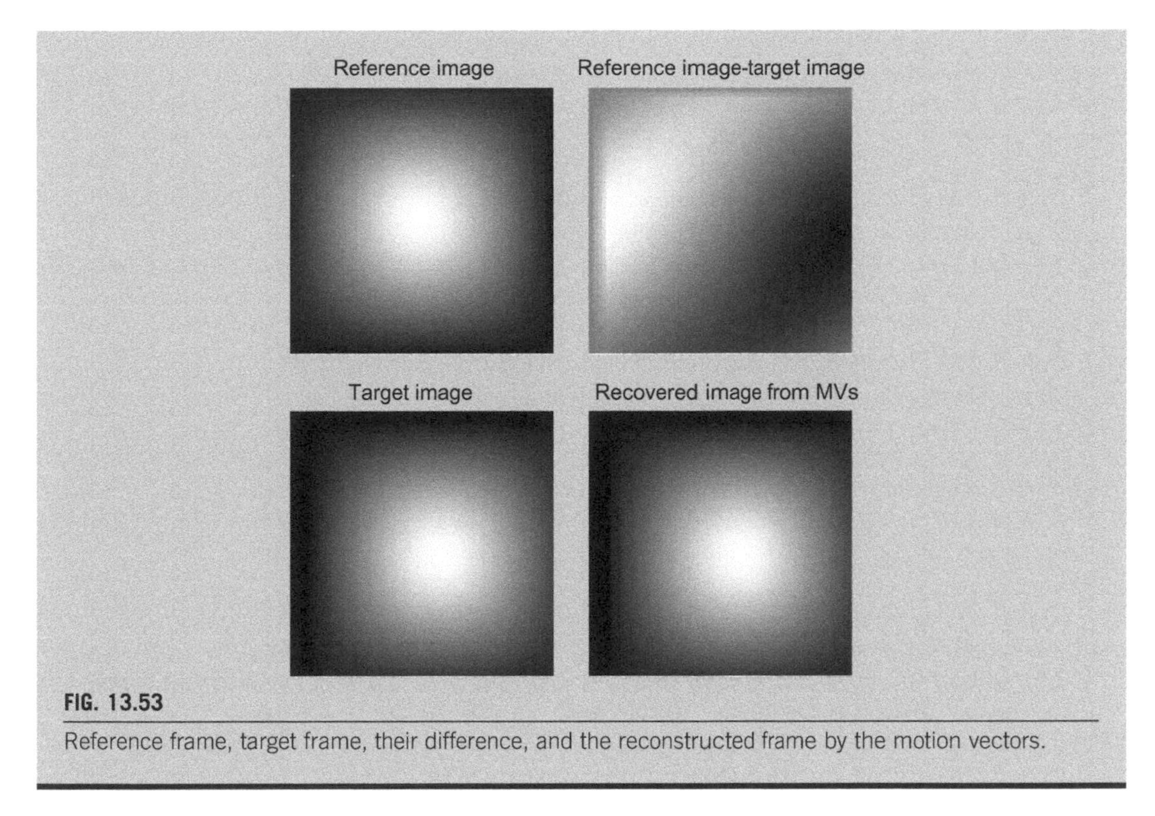

FIG. 13.53

Reference frame, target frame, their difference, and the reconstructed frame by the motion vectors.

The motion vector comprises the pixel displacements from the target frame to the reference frame. Hence, given the reference frame, directions specified in the motion vector should be switched to indicate the motion toward the target frame. As indicated by the obtained motion vectors, the target image is the version of the reference image moving to the right by 6 pixels and down by 4 pixels.

13.11 SUMMARY

1. A digital image consists of pixels. For a grayscale image, each pixel is assigned a grayscale level that presents luminance of the pixel. For an RGB color image, each pixel is assigned a red component, a green component, and a blue component. For an indexed color image, each pixel is assigned an address that is the location of the color table (map) made up of the RGB components.
2. Common image data formats are 8-bit grayscale image, 24-bit color, and 8-bit indexed color.
3. The larger the number of pixels in an image, or the larger the numbers of the RGB components, the finer is the spatial resolution in the image. Similarly, the more scale levels used for each pixel, the better the scale-level image resolution. The more pixels and more bits used for the scale levels in the image, the more storage is required.

4. The RGB color pixel can be converted to YIQ color pixels. Y component is the luminance occupying 93% of the signal energy, while the I and Q components represent the color information of the image, occupying the remainder of the energy.
5. The histogram for a grayscale image shows the number of pixels at each grayscale level. The histogram can be modified to enhance the image. Image equalization using the histogram can improve the image contrast and effectively enhances contrast for image underexposure. Color image equalization can be done only in the luminance channel or RGB channels.
6. Image enhancement techniques such as average lowpass filtering can filter out random noise in the image; however, it also blurs the image. The degree of blurring depends on the kernel size. The bigger the kernel size, the more blurring occurs.
7. Median filtering effectively remove the "salt and pepper" noise in an image.
8. The edge-detection filter with Sobel convolution, Laplacian, and Laplacian of Gaussian kernels can detect the image boundaries.
9. The grayscale image can be made into a facsimile of the color image by pseudo-color image generation, using the RGB transformation functions.
10. RGB-to-YIQ transformation is used to obtain the color image in YIQ space, or vice versa. It can also be used for color-to-grayscale conversion, that is, keep only the luminance channel after the transformation.
11. 2D spectra can be calculated and are used to examine filtering effects.
12. JPEG compression uses the 2D-DCT transform for both grayscale and color images in YIQ color space. JPEG uses different quality factors to normalize DCT coefficients for the luminance (Y) channel and the chrominance (IQ) channels.
13. The mixing two images, in which two pixels are linearly interpolated using the weights $1-\alpha$ and α, can produce video sequences that have effects such as fading in and fading out of images, blending of two images, and overlaying of text on an image.
14. Analog video uses interlaced scanning. A video frame contains odd and even fields. Analog video standards include NTSC, PAL, and SECAM.
15. Digital video carries the modulated information for each pixel in YCbCr color space, where Y is the luminance and Cb and Cr are the chrominance. Chroma subsampling creates various digital video formats. The industrial standards include CCIR601, CCR601, CIF, and QCIF.
16. The motion compensation of a video sequence produces motion vector for all the image blocks in the target video frame, which contain displacements of these image blocks relative to the reference video frame. Recovering the target frame involves simply copying each image block of the reference frame to the target frame at the location specified in the motion vector. Motion compensation is a key element in MPEG video.

13.12 PROBLEMS

13.1 Determine the memory storage requirement for each of the following images:

(a) 320×240 8-bit grayscale.

(b) 640×480 24-bit color image.

(c) 1600×1200 8-bit indexed image.

13.2 Determine the number of colors for each of the following images:

(a) 320×240 16-bit indexed image.

(b) 200×100 24-bit color image.

13.3 Given a pixel in an RGB image as follows:

$$R = 200, G = 120, B = 100,$$

convert the RGB values to the YIQ values.

13.4 Given a pixel of an image in YIQ color format as follows:

$$Y = 141, I = 46, Q = 5,$$

convert the YIQ values back to RGB values.

13.5 Given the following 2×2 RGB image,

$$R = \begin{bmatrix} 100 & 50 \\ 100 & 50 \end{bmatrix} G = \begin{bmatrix} 20 & 40 \\ 10 & 30 \end{bmatrix} B = \begin{bmatrix} 100 & 50 \\ 200 & 150 \end{bmatrix},$$

convert the image into grayscale.

13.6 Produce a histogram of the following image, which has a grayscale value ranging from 0 to 7, that is, each pixel is encoded in 3 bits.

$$\begin{bmatrix} 0 & 1 & 2 & 2 & 0 \\ 2 & 1 & 1 & 2 & 1 \\ 1 & 1 & 4 & 2 & 3 \\ 0 & 2 & 5 & 6 & 1 \end{bmatrix}$$

13.7 Given the following image with a grayscale value ranging from 0 to 7, that is, each pixel being encoded in 3 bits,

$$\begin{bmatrix} 0 & 1 & 2 & 2 & 0 \\ 2 & 1 & 1 & 2 & 1 \\ 1 & 1 & 4 & 2 & 3 \\ 0 & 2 & 5 & 6 & 1 \end{bmatrix},$$

perform equalization using the histogram in Problem 13.6, and plot the histogram for the equalized image.

13.8 Given the following image with a grayscale value ranging from 0 to 7, that is, each pixel being encoded in 3 bits,

$$\begin{bmatrix} 2 & 4 & 4 & 2 \\ 2 & 3 & 3 & 3 \\ 4 & 4 & 4 & 2 \\ 3 & 2 & 3 & 4 \end{bmatrix},$$

perform level adjustment to the full range, shift the level to the range from 3 to 7, and shift the level to the range from 0 to 3.

13.9 Given the following 8-bit grayscale original and noisy images, and 2×2 convolution average kernel,

$$4 \times 4 \text{ original image}: \begin{bmatrix} 100 & 100 & 100 & 100 \\ 100 & 100 & 100 & 100 \\ 100 & 100 & 100 & 100 \\ 100 & 100 & 100 & 100 \end{bmatrix}$$

$$4 \times 4 \text{ corrupted image}: \begin{bmatrix} 93 & 116 & 109 & 96 \\ 92 & 107 & 103 & 108 \\ 84 & 107 & 86 & 107 \\ 87 & 113 & 106 & 99 \end{bmatrix}$$

$$2 \times 2 \text{ average kernel}: \frac{1}{4}\begin{bmatrix} 1 & 1 \\ 1 & 1 \end{bmatrix},$$

perform digital filtering on the noisy image, and compare the enhanced image with the original image.

13.10 Given the following 8-bit grayscale original and noisy image, and 3×3 median filter kernel,

$$4 \times 4 \text{ original image}: \begin{bmatrix} 100 & 100 & 100 & 100 \\ 100 & 100 & 100 & 100 \\ 100 & 100 & 100 & 100 \\ 100 & 100 & 100 & 100 \end{bmatrix}$$

$$4 \times 4 \text{ corrupted image by impulse noise}: \begin{bmatrix} 100 & 255 & 100 & 100 \\ 0 & 255 & 255 & 100 \\ 100 & 0 & 100 & 0 \\ 100 & 255 & 100 & 100 \end{bmatrix}$$

$$3 \times 3 \text{ average kernel}: \begin{bmatrix} \ \\ \ \\ \ \end{bmatrix},$$

perform digital filtering, and compare the filtered image with the original image.

13.11 Given the following 8-bit 5×4 original grayscale image,

$$\begin{bmatrix} 110 & 110 & 110 & 110 \\ 110 & 100 & 100 & 110 \\ 110 & 100 & 100 & 110 \\ 110 & 110 & 110 & 110 \\ 110 & 110 & 110 & 110 \end{bmatrix},$$

apply the following edge detectors to the image:

(a) Sobel vertical edge detector.

(b) Laplacian edge detector.

and scale the resultant image pixel value to the range of 0–255.

13.12 In Example 13.10, if we switch the transformation functions between the red function and the green function, what is the expected color for the area pointed to by the arrow, and what is the expected background color?

13.13 In Example 13.10, if we switch the transformation functions between the red function and the blue function, what is the expected color for the area pointed to by the arrow, and what is the expected background color?

13.14 Given the following grayscale image $p(i,j)$:

$$\begin{bmatrix} 100 & -50 & 10 \\ 100 & 80 & 100 \\ 50 & 50 & 40 \end{bmatrix},$$

determine the 2D-DFT coefficient $X(1,2)$ and the magnitude spectrum $A(1,2)$.

13.15 Given the following grayscale image $p(i,j)$:

$$\begin{bmatrix} 10 & 100 \\ 200 & 150 \end{bmatrix},$$

determine the 2D-DFT coefficients $X(u,v)$ and magnitude $A(u,v)$.

13.16 Given the following grayscale image $p(i,j)$:

$$\begin{bmatrix} 10 & 100 \\ 200 & 150 \end{bmatrix},$$

apply the 2D-DCT to determine the DCT coefficients.

13.17 Given the following DCT coefficients $F(u,v)$:

$$\begin{bmatrix} 200 & 10 \\ 10 & 0 \end{bmatrix},$$

apply the inverse 2D-DCT to determine 2D data.

13.18 In JPEG compression, DCT DC coefficients from several blocks are 400, 390, 350, 360, and 370. Use DPCM to produce the DPCM sequence, and use the Huffman table to encode the DPCM sequence.

13.19 In JPEG compression, DCT coefficients from the an image subblock are

$$[175, -2, 0, 0, 0, 4, 0, 0, -37, 0, 0, 0, 0, -2, 0, 0, \ldots, 0].$$

(a) Generate the run-length codes for AC coefficients.

(b) Perform entropy coding for the run-length codes using the Huffman table.

13.20 Given the following grayscale image $p(i,j)$:

$$\begin{bmatrix} 10 & 100 \\ 200 & 150 \end{bmatrix},$$

apply the 2D-DWT using the Haar wavelet to determine the level-1 DWT coefficients.

13.21 Given the following level-1 IDWT coefficients $W(u,v)$ obtained using the Harr wavelet:

$$\begin{bmatrix} 200 & 10 \\ 10 & 0 \end{bmatrix},$$

apply the IDWT to determine 2D data.

13.22 Given the following grayscale image $p(i,j)$:

$$\begin{bmatrix} 100 & 150 & 60 & 80 \\ 80 & 90 & 50 & 70 \\ 110 & 120 & 100 & 80 \\ 90 & 50 & 40 & 90 \end{bmatrix},$$

apply the 2D-DWT using the Haar wavelet to determine the level-1 DWT coefficients.

13.23 Given the following level-1 IDWT coefficients $W(u,v)$ obtained using the Harr wavelet:

$$\begin{bmatrix} 250 & 50 & -30 & -20 \\ 30 & 20 & 10 & -20 \\ 10 & 20 & 0 & 0 \\ 20 & 15 & 0 & 0 \end{bmatrix},$$

apply the IDWT to determine 2D data.

13.24 Explain the difference between horizontal retrace and vertical retrace. Which one would take more time?

13.25 What is the purpose of using the interlaced scanning in the traditional NTSC TV system?

13.26 What is the bandwidth in the traditional NTSC TV broadcast system? What is the bandwidth to transmit luminance Y, and what are the required bandwidths to transmit Q and I channels, respectively?

13.27 What type of modulation is used for transmitting audio signals in the NTSC TV system?

13.28 Given the composite NTSC signal

$$Composite = Y + C = Y + I\cos(2\pi f_{sc}t) + Q\sin(2\pi f_{sc}t),$$

show demodulation for the Q channel.

13.29 Where does the color subcarrier burst reside? What is the frequency of the color subcarrier, and how many cycles does the color burst have?

13.30 Compare differences of the NTSC, PAL, and SECAM video systems in terms of the number of scan lines, frame rates, and total bandwidths required for transmission.

13.31 In the NTSC TV system, what is the horizontal line scan rate? What is the vertical synchronizing pulse rate?

13.32 Explain which the following digital video formats achieves the most data transmission efficiency:
(a) 4:4:4.
(b) 4:2:2.
(c) 4:2:0.

13.33 What is the difference between an interlaced scan and a progressive scan? Which of the following video systems use the progressive scan?
(a) CCIR-601.
(b) CIF.

13.34 In motion compensation, which of the following would require more computation? Explain.
(a) Finding the motion vector using the sequential search
(b) Recovering the target frame with the motion vectors and reference frame

13.35 Given a reference frame and target frame of size 80×80, a macroblock size of 16×16, and a search window size of 32×32, estimate the numbers of subtraction, absolute value, and additions for searching all the motion vectors using the sequential search method.

MATLAB Problems

Use MATLAB to solve Problems 13.36–13.42.

13.36 Given the image data "trees.jpg," use MATLAB functions to perform each of the following processing:
(a) Use MATLAB to read and display the image.
(b) Convert the image to the grayscale image.
(c) Perform histogram equalization for the grayscale image in (b) and display the histogram plots for both original grayscale image and equalized grayscale image.
(d) Perform histogram equalization for the color image in A and display the histogram plots of Y channel for both original color image and equalized color image.

13.37 Given the image data "cruise.jpg," perform the following linear filtering:
(a) Convert the image to grayscale image and then create the 8-bit noisy image by adding Gaussian noise using the following code:

noise_image = imnoise(I,'gaussian');
where I is the intensity image obtained from normalizing the grayscale image.

(b) Process the noisy image using the Gaussian filter with the following parameters: the convolution kernel size = 4, SIGMA = 0.8, and compare the filtered image with the noisy image.

13.38 Given the image data "cruise.jpg," perform the following filtering process:
(a) Convert the image to grayscale image and then create the 8-bit noisy image by adding "pepper and salt" noise using the following code:

noise_image = imnoise(I,'salt & pepper');
where I is the intensity image obtained from normalizing the grayscale image.

(b) Process the noisy image using the median filtering with a convolution kernel size of 4×4.

13.39 Given the image data "cruise.jpg," convert the image to the grayscale image and detect the image boundaries using Laplacian of Gaussian filtering with the following parameters:

(a) Kernel size = 4 and SIGMA = 0.9.

(b) Kernel size = 10 and SIGMA = 10.

Compare the results.

13.40 Given the image data "clipim2.gif," perform the following process:

(a) Convert the indexed image to the grayscale image.

(b) Adjust the color transformation functions (sine functions) to make the object indicated by the arrow in the image to be red and the background color to be green.

13.41 Given the image data "cruiseorg.tiff," perform JPEG compression by completing the following steps:

(a) Convert the image to the grayscale image.

(b) Write a MATLAB program for encoding: (1) dividing the image into 8×8 blocks, (2) transforming each block using the DCT, and (3) scaling and rounding DCT coefficients with the standard quality factor. Note that lossless-compressing the quantized DCT coefficients is omitted here for a simple simulation.

(c) Continue to write the MATLAB program for decoding: (1) invert the scaling process for quantized DCT coefficients, (2) perform the inverse DCT for each 8×8 image block, and (3) recover the image.

(d) Run the developed MATLAB program to examine the image quality using.

- **I.** The quality factor.
- **II.** The quality factor $\times$ 5.
- **III.** The quality factor $\times$10.

13.42 Given the image data "cruiseorg.tiff," perform wavelet-based compression by completing the following steps:

(a) Convert the image to grayscale image.

(b) Write a MATLAB program for encoding: (1) use 8-tap Daubechies filter coefficients, (2) apply the two-level DWT, and (3) perform 8-bit quantization for subbands LL2, LH2, HL2, HH2, LH1, HL1, HH1 for simulation.

(c) Write the MATLAB program for the decoding process.

(d) Run the developed MATLAB program to examine the image quality using.

- **I.** Reconstruct the image using all subband coefficients.
- **II.** Reconstruct the image using LL2 subband coefficients.
- **III.** Reconstruct the image using LL2, HL2, LH2, and HH2 subband coefficients.
- **IV.** Reconstruct the image using LL2, HL2, and LH2 subband coefficients.
- **V.** Reconstruct the image using LL2, HL2, LH2, HH2, LH1, and HL1 subband coefficients.

CHAPTER

HARDWARE AND SOFTWARE FOR DIGITAL SIGNAL PROCESSORS

14

CHAPTER OUTLINE

14.1 DIGITAL SIGNAL PROCESSOR ARCHITECTURE

Unlike microprocessors and microcontrollers, digital signal processors (DSPs) have special features that require operations such as fast Fourier transform (FFT), filtering, convolution and correlation, and real-time sample-based and block-based processing. Therefore, DSPs use different dedicated hardware architecture.

Digital Signal Processing. https://doi.org/10.1016/B978-0-12-815071-9.00014-2

We first compare the architecture of the general microprocessor with that of the DSPs. The design of general microprocessors and microcontrollers is based on the *Von Neumann architecture*, which was developed based on a research paper written by John Von Neumann and others in 1946. Von Neumann suggested that computer instructions, as we discuss, should be numerical codes instead of special wiring. Fig. 14.1 shows the Von Neumann architecture.

As shown in Fig. 14.1, a Von Neumann processor contains a single, shared memory for programs and data, a single bus for memory access, an arithmetic unit, and a program control unit. The processor proceeds in a serial fashion in terms of fetching and execution cycles. This means that the central processing unit (CPU) fetches an instruction from memory and decodes it to figure out what operation to do, and then executes the instruction. The instruction (in machine code) has two parts: the *opcode* and the *operand*. The opcode specifies what the operation is, that is, tells the CPU what to do. The operand informs the CPU what data to operate on. These instructions will modify memory, or input and output (I/O). After an instruction is completed, the cycles will resume for the next instruction. One instruction or piece of data can be retrieved at a time. Since the processor proceeds in a serial fashion, it causes most units staying in a wait state.

As noted, the Von Neumann architecture operates the cycles of fetching and execution by fetching an instruction from memory, decoding it via the program control unit, and finally executing instruction. When execution requires data movement—that is, data to be read from or written to memory—the next instruction will be fetched after the current instruction is completed. The Von Neumann-based processor has this bottleneck mainly due to the use of a single, shared memory for both program instructions and data. Increasing the speed of the bus, memory, and computational units can improve speed, but not significantly.

To accelerate the execution speed of digital signal processing, DSPs are designed based on the *Harvard architecture*, which originated from the Mark 1 relay-based computers built by IBM in 1944 at Harvard University. This computer stored its instructions on punched tape and data using relay latches. Fig. 14.2 shows today's Harvard architecture. As depicted, the DSP has two separate memory spaces. One is dedicated for the program code, while the other is employed for data. Hence, to

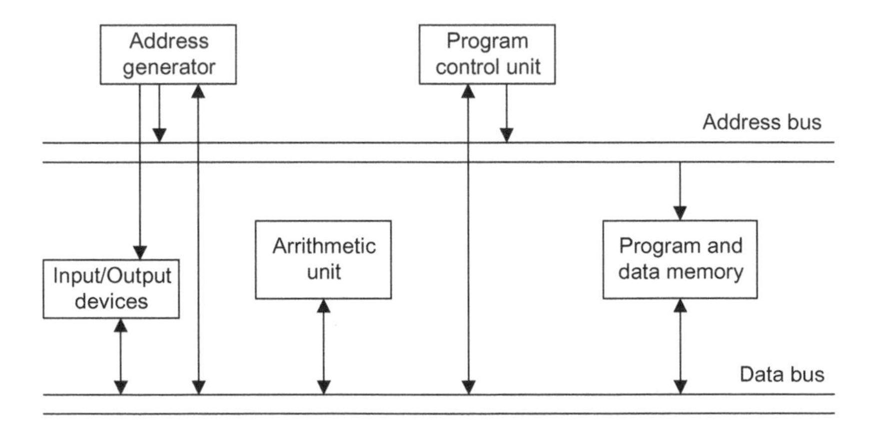

FIG. 14.1

General microprocessor based on the Von Neumann architecture.

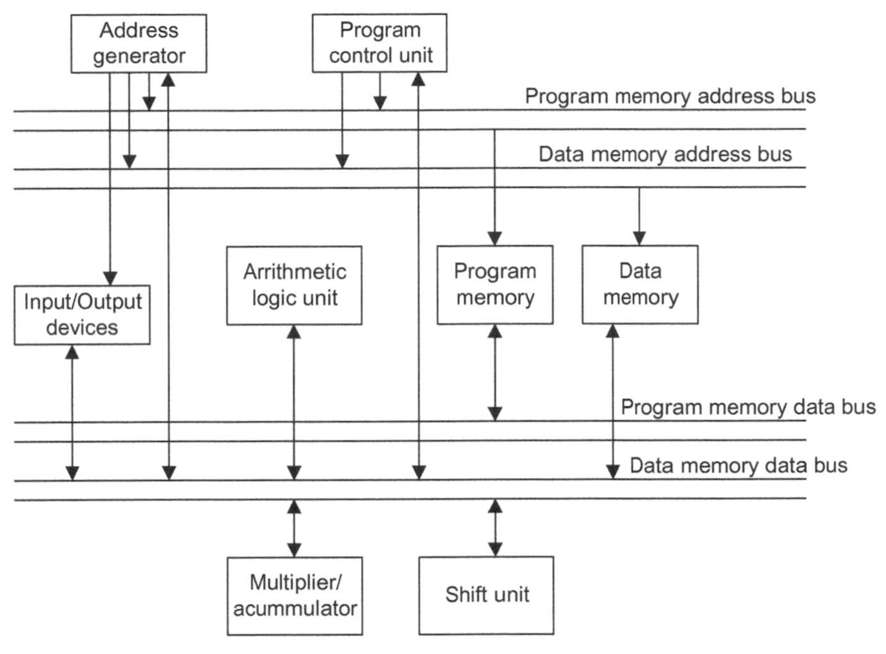

FIG. 14.2

DSPs based on the Harvard architecture.

accommodate two memory spaces, two corresponding address buses and two data buses are used. In this way, the program memory and data memory have their own connections to the program memory bus and data memory bus, respectively. This means that the Harvard processor can fetch the program instruction and data in parallel at the same time, the former via the program memory bus and the latter via the data memory bus. There is an additional unit called a *multiplier and accumulator* (MAC), which is the dedicated hardware used for the digital filtering operation. The last additional unit, the shift unit, is used for the scaling operation for the fixed-point implementation when the processor performs digital filtering.

Let us compare the executions of the two architectures. The Von Neumann architecture generally has the execution cycles described in Fig. 14.3. The fetch cycle obtains the opcode from the memory,

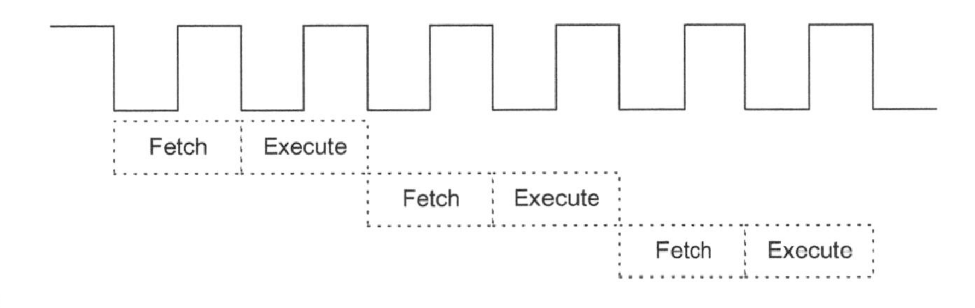

FIG. 14.3

Execution cycle based on the Von Neumann architecture.

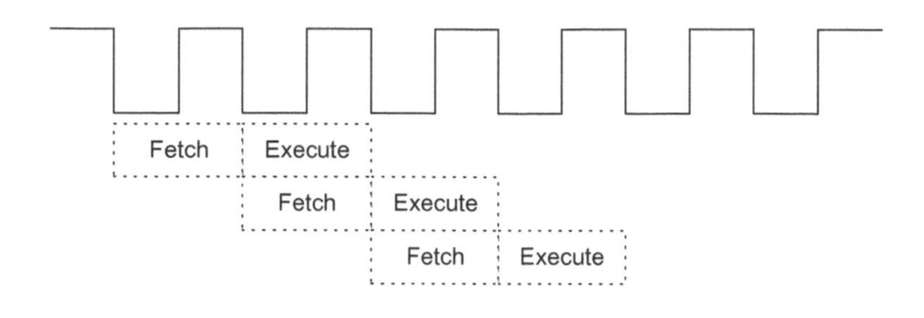

FIG. 14.4

Execution cycle based on Harvard architecture.

and the control unit will decode the instruction to determine the operation. Next is the execute cycle. Based the decoded information, execution will modify the content of the register or the memory. Once this is completed, the process will fetch the next instruction and continue. The processor operates one instruction at a time in a serial fashion.

To improve the speed of the processor operation, the Harvard architecture takes advantage of a common DSP, in which one register holds the filter coefficient while the other register holds the data to be processed, as depicted in Fig. 14.4.

As shown in Fig. 14.4, the execute and fetch cycles are overlapped. We call this the *pipelining* operation. The DSP performs one execution cycle while also fetching the next instruction to be executed. Hence, the processing speed is dramatically increased.

The Harvard architecture is preferred for all DSPs due to the requirements of most DSP algorithms, such as filtering, convolution, and FFT, which need repetitive arithmetic operations, including multiplications, additions, memory access, and heavy data flow through the CPU.

For the other applications, such as those dependent on simple microcontrollers with less of a timing requirement, the Von Neumann architecture may be a better choice, since it offers much less silica area and is thus less expansive.

14.2 DSP HARDWARE UNITS

In this section, we briefly discuss special DSP hardware units.

14.2.1 MULTIPLIER AND ACCUMULATOR

As compared with the general microprocessors based on the Von Neumann architecture, the DSP uses the MAC, a special hardware unit for enhancing the speed of digital filtering. This is dedicated hardware, and the corresponding instruction is generally referred to as the MAC operation. The basic structure of the MAC is shown in Fig. 14.5.

As shown in Fig. 14.5, in a typical hardware MAC, the multiplier has a pair of input registers, each holding the 16-bit input to the multiplier. The result of the multiplication is accumulated in the 32-bit accumulator unit. The result register holds the double-precision data from the accumulator.

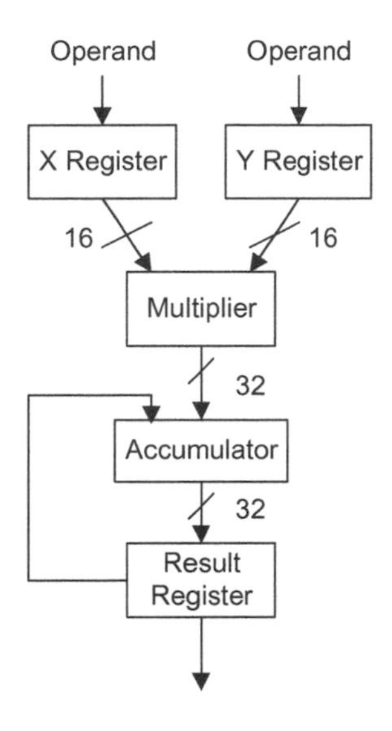

FIG. 14.5

The multiplier and accumulator (MAC) dedicated to DSP.

14.2.2 SHIFTERS

In digital filtering, to prevent overflow, a scaling operation is required. A simple scaling-down operation shifts data to the right, while a scaling-up operation shifts data to the left. Shifting data to the right is the same as dividing the data by 2 and truncating the fraction part; shifting data to the left is equivalent to multiplying the data by 2. As an example for a 3-bit data word $011_2 = 3_{10}$, shifting 011 to the right gives $001_2 = 1$, that is, $3/2 = 1.5$, and truncating 1.5 results in 1. Shifting the same number to the left, we have $110_2 = 6_{10}$, that is, $3 \times 2 = 6$. The DSP often shifts data by several bits for each data word. To speed up such operation, the special hardware shift unit is designed to accommodate the scaling operation, as depicted in Fig. 14.2.

14.2.3 ADDRESS GENERATORS

The DSP generates the addresses for each datum on the data buffer to be processed. A special hardware unit for circular buffering is used (see the address generator in Fig. 14.2). Fig. 14.6 describes the basic mechanism of circular buffering for a buffer having eight data samples.

In circular buffering, a pointer is used and always points to the newest data sample, as shown in the figure. After the next sample is obtained from analog-to-digital conversion (ADC), the data will be placed at the location of $x(n-7)$ and the oldest sample is pushed out. Thus, the location for $x(n-7)$ becomes the location for the current sample. The original location for $x(n)$ becomes a location for the past sample of $x(n-1)$. The process continues according to the mechanism just described. For each new data sample, only one location on the circular buffer needs to be updated.

The circular buffer acts like a first-in/first-out (FIFO) buffer, but each datum on the buffer does not have to be moved. Fig. 14.7 gives a simple illustration of the 2-bit circular buffer. In the figure, there is

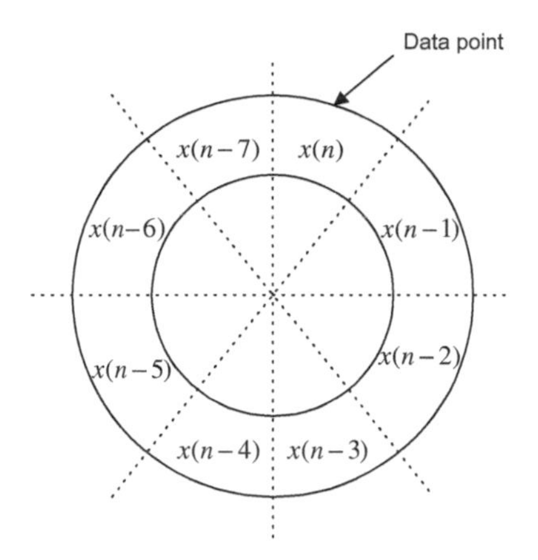

FIG. 14.6

Illustration of circular buffering.

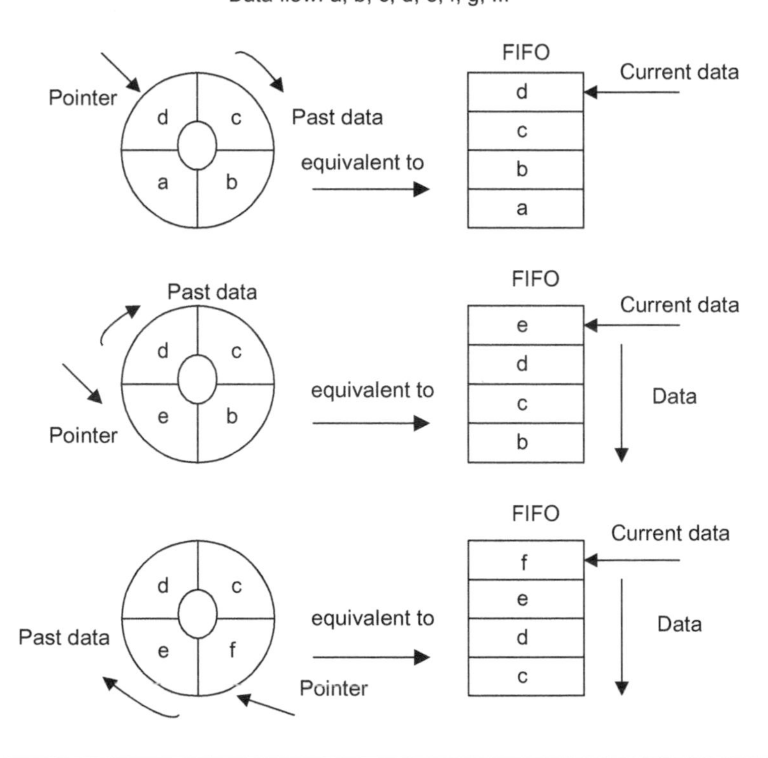

FIG. 14.7

Circular buffer and equivalent FIFO.

data flow to the ADC ($a, b, c, d, e, f, g, \ldots$) and a circular buffer initially containing a, b, c, and d. The pointer specifies the current data of d, and the equivalent FIFO buffer is shown on the right side with a current data of d at the top of the memory. When e comes in, as shown in the middle drawing in Fig. 14.7, the circular buffer will change the pointer to the next position and update old a with a new datum e. It costs the pointer only one movement to update one datum in one step. However, on the right side, the FIFO has to move each of the other data down to let in the new datum e at the top. For this FIFO, it takes four data movements. In the bottom drawing in Fig. 14.7, the incoming datum f for both the circular buffer and the FIFO buffer continues to confirm our observations.

Like finite impulse response (FIR) filtering, the data buffer size can reach several hundreds. Hence, using the circular buffer will significantly enhance the processing speed.

14.3 DSPs AND MANUFACTURES

DSPs are classified for general DSP and special DSP. The general DSP is designed and optimized for applications such as digital filtering, correlation, convolution, and FFT. In addition to these applications, the special DSP has features that are optimized for unique applications such as audio processing, compression, echo cancellation, and adaptive filtering. Here, we focus on the general DSP.

The major manufactures in the DSP industry are Texas Instruments (TI), Analog Devices, and Motorola. TI and Analog Devices offer both fixed-point DSP families and floating-point DSP families, while Motorola offers fixed-point DSP families. We concentrate on TI families, review their architectures, and study real-time implementation using the fixed- and floating-point formats.

14.4 FIXED-POINT AND FLOATING-POINT FORMATS

In order to process real-world data, we need to select an appropriate DSP, as well as a DSP algorithm or algorithms for a certain application. Whether a DSP uses a fixed- or floating-point method depends on how the processor's CPU performs arithmetic. A fixed-point DSP represents data in *2's complement integer format* and manipulates data using integer arithmetic, while a floating point processor represents number using a mantissa (fractional part) and an exponent in addition to the integer format and operates data using the floating-point arithmetic (discussed in Section 14.4.2).

Since the fixed-point DSP operates using the integer format, which represents only a very narrower dynamic range of the integer number, a problem such as the overflow of data manipulation may occur. Hence, we need to spend much more coding effort to deal with such a problem. As we see, we may use floating-point DSPs, which offer a wider dynamic range of data, so that coding becomes much easier. However, the floating-point DSP contains more hardware units to handle the integer arithmetic and the floating-point arithmetic, hence it is more expensive and slower than fixed-point processors in terms of instruction cycles. It is usually a choice for prototyping or proof-of-concept development.

When it is time to making the DSP an application-specific integrated circuit (ASIC), a chip designed for a particular application, a dedicated hand-coded fixed-point implementation can be the best choice in terms of performance and small silica area.

The formats used by the DSP implementation can be classified as fixed or floating point.

14.4.1 FIXED-POINT FORMAT

We begin with two's complement representation. Considering a 3-bit two's complement (2's complement), we can represent all the decimal numbers shown in Table 14.1.

Let us review the 2's complement number system using Table 14.1. Converting a decimal number to its 2's complement form requires the following steps:

1. Convert the magnitude in the decimal to its binary number using the required number of bits.
2. If the decimal number is positive, its binary number is its 2's complement representation; if the decimal number is negative, perform the 2's complement operation, where we negate the binary number by changing the logic 1's to logic 0's and logic 0's to logic 1's and then add a logic 1 to the data. For example, a decimal number of 3 is converted to its 3-bit 2's complement representation as 011; however, for converting a decimal number of −3, we first get a 3-bit binary number for the magnitude in decimal, that is, 011. Next, negating the binary number 011 yields the binary number 100. Finally, adding a binary logic 1 achieves the 3-bit 2's complement representation of −3, that is, 100+1=101, as shown in Table 14.1.

As we see, a 3-bit 2's complement number system has a dynamic range from −4 to 3, which is very narrow. Since the basic DSP operations include multiplications and additions, results of operation can cause overflow problems. Let us examine the multiplications in Example 14.1.

EXAMPLE 14.1

Given

(a) $2\times(-1)$

(b) $2\times(-3)$,

operate each using its 2's complement.

Solution:

(a)

```
      010
    x 001
  --------------
      0 1 0
     0 0 0
  +0 0 0
  --------------
   0 0 0 1 0
```

The 2's complement of 00010 is 11,110. Removing two extended sign bits gives 110.
The answer is 110 (−2), which is within the system.

(b)

```
      010
    x 011
  -----------
      0 1 0
    0 1 0
  +0 0 0
  -----------
    0 0 1 1 0
```

The 2's complement of 00110 is 11,010. Removing two extended sign bits achieves 010. Since the binary number 010 is 2, which is not (−6) as what we expect, overflow occurs; that is, the result of the multiplication (−6) is out of our dynamic range (−4 to 3).

Let us design a system treating all the decimal values as fractional numbers, so that we obtain the fractional binary 2's complement system shown in Table 14.2.

To become familiar with the fractional binary 2's complement system, let us convert a positive fraction number $\frac{3}{4}$ and a negative fraction number $-\frac{1}{4}$ in decimals to their 2's complements. Since

$$\frac{3}{4} = 0 \times 2^0 + 1 \times 2^{-1} + 1 \times 2^{-2},$$

its 2's complement is 011. Note that we did not mark the binary point for clarity.

Table 14.1 A 3-Bit 2's Complement Number Representation

Decimal Number	Two's Complement
3	011
2	010
1	001
0	000
−1	111
−2	110
−3	101
−4	100

Table 14.2 A 3-Bit 2's Complement System Using Fractional Representation

Decimal Number	Decimal Fraction	Two's Complement
3	3/4	0.11
2	2/4	0.10
1	1/4	0.01
0	0	0.00
−1	−1/4	1.11
−2	−2/4	1.10
−3	−3/4	1.01
−4	−4/4 = −1	1.00

Again, since

$$\frac{1}{4} = 0 \times 2^0 + 0 \times 2^{-1} + 1 \times 2^{-2},$$

its positive-number 2's complement is 001. For the negative number, applying 2's complement to the binary number 001 leads to $110 + 1 = 111$, as we see in Table 14.2.

Now let us focus on the fractional binary 2's complement system. The data are normalized to the fractional range from -1 to $1 - 2^{-2} = \frac{3}{4}$. When we carry out multiplications with two fractions, the result should be a fraction, so that multiplication overflow can be prevented. Let us verify the multiplication (010) × (101), which is the overflow case in Example 14.1. We first multiply two positive numbers:

```
      0.1 0
    x 0.1 1
  -------------
      0 1 0
    0 1  0
 +0 0  0
 --------------
   0.0 1 1 0
```

The 2's complement of $0.0110 = 1.1010$.

The answer in decimal form should be

$$1.1010 = (-1) \times (0.0110)_2 = -\left(0 \times (2)^{-1} + 1 \times (2)^{-2} + 1 \times (2)^{-3} + 0 \times (2)^{-4}\right) = -\frac{3}{8}.$$

This number is correct, as we can verify from Table 14.2, that is, $\left(\frac{2}{4} \times \left(-\frac{3}{4}\right)\right) = -\frac{3}{8}$.

If we truncate the last two least-significant bits to keep the 3-bit binary number, we have an approximated answer as

$$1.10 = (-1) \times (0.01)_2 = -\left(0 \times (2)^{-1} + 1 \times (2)^{-2}\right) = -\frac{1}{2}.$$

The truncation error occurs. The error should be bounded by $2^{-2} = \frac{1}{4}$. We can verify that

$$|-1/2 - (-3/8)| = 1/8 < 1/4.$$

With such a scheme, we can avoid the overflow due to multiplications but cannot prevent the additional overflow. In the following addition example,

```
      0.1 1
    + 0.0 1
  -------------
      1 0 0
 --------------
      1. 0 0
```

where the result 1.00 is a negative number.

Adding two positive fractional numbers yields a negative number. Hence, overflow occurs.

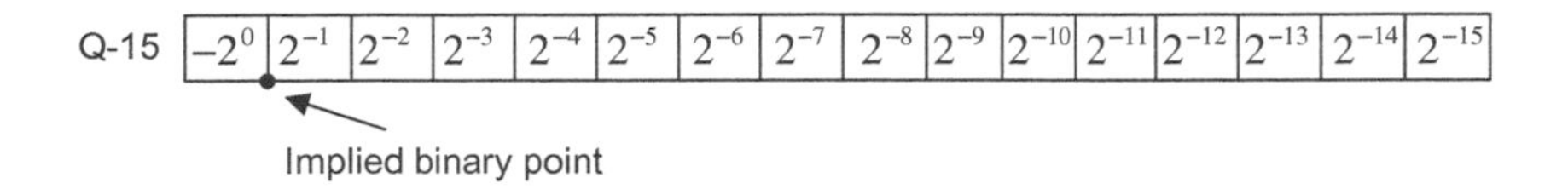

FIG. 14.8

Q-15 (fixed-point) format.

We see that this signed fractional number scheme partially solves the overflow in multiplications. This fractional number format is called the signed Q-2 format, where there are 2 magnitude bits plus one sign bit. The overflow in addition will be tackled using a scaling method discussed in the later section.

Q-format number representation is the most common one used in the fixed-point DSP implementation. It is defined in Fig. 14.8.

As indicated in Fig. 14.8, Q-15 means that the data are in a sign magnitude form in which there are 15 bits for magnitude and one bit for sign. Note that after the sign bit, the dot shown in Fig. 14.8 implies the binary point. The number is normalized to the fractional range from -1 to 1. The range is divided into 2^{16} intervals, each with a size of 2^{-15}. The most negative number is -1, while the most positive number is $1-2^{-15}$. Any result from multiplication is within the fractional range of -1 to 1. Let us study the following examples to become familiar with Q-format number representation.

EXAMPLE 14.2

Find the signed Q-15 representation for the decimal number 0.560123.

Solution:

The conversion process is illustrated using Table 14.3. For a positive fractional number, we multiply the number by 2 if the product is larger than 1, carry bit 1 as a most-significant bit (MSB), and copy the fractional part to the next line for the next multiplication by 2; if the product is <1, we carry bit 0 to MSB. The procedure continues to collect all 15 magnitude bits.

We yield the Q-15 format representation as.

$$0.100011110110010.$$

Since we only use 16 bits to represent the number, we may lose accuracy after conversion. Like quantization, a truncation error is introduced. However, this error should be less than the interval size, in this case, $2^{-15}=0.0000305017$. We verify this in Example 14.5. An alternative way of conversion is to convert a fraction, let us convert $\frac{3}{4}$ to Q-2 format. We multiply $\frac{3}{4}$ by 2^2, and then convert the truncated integer to its binary, that is,

$$(3/4)\times 2^2=3=011_2.$$

In this way, it follows that

$$(0.560123)\times 2^{15}=18354.$$

Converting 18,354 to its binary representation will achieve the same answer.

Continued

EXAMPLE 14.2—CONT'D

Table 14.3 Conversion Process of Q-15 Representation

Number	Product	Carry
0.560123×2	1.120246	1 (MSB)
0.120246×2	0.240492	0
0.240492×2	0.480984	0
0.480984×2	0.961968	0
0.961968×2	1.923936	1
0.923936×2	1.847872	1
0.847872×2	1.695744	1
0.695744×2	1.391488	1
0.391488×2	0.782976	0
0.782976×2	1.565952	1
0.565952×2	1.131904	1
0.131904×2	0.263808	0
0.263808×2	0.527616	0
0.527616×2	1.055232	1
0.055232×2	0.110464	0 (LSB)

MSB, most-significant bit; LSB, least-significant bit.

The next example illustrates the signed Q-15 representation for a negative number.

EXAMPLE 14.3

Find the signed Q-15 representation for the decimal number −0.160123.

Solution:

Converting the Q-15 format for the corresponding positive number with the same magnitude using the procedure described in Example 14.2, we have

$$0.160123 = 0.001010001111110.$$

Then after applying 2's complement, the Q-15 format becomes

$$-0.160123 = 1.110101110000010.$$

Alternative way: Since $(-0.160123) \times 2^{15} = -5246.9$, converting the truncated number −5246 to its 16-bit 2's complement yields 1.110101110000010.

EXAMPLE 14.4

Convert the Q-15 signed number 1.110101110000010 to the decimal number.

Solution:

Since the number is negative, applying the 2's complement yields

$$0.001010001111110.$$

Then the decimal number is

$$-(2^{-3}+2^{-5}+2^{-9}+2^{-10}+2^{-11}+2^{-12}+2^{-13}+2^{-14})=-0.160095.$$

EXAMPLE 14.5

Convert the Q-15 signed number 0.100011110110010 to the decimal number.

Solution:

The decimal number is

$$2^{-1}+2^{-5}+2^{-6}+2^{-7}+2^{-8}+2^{-10}+2^{-11}+2^{-14}=0.560120.$$

As we know, the truncation error in Example 14.2 is less than $2^{-15}=0.000030517$. We verify that the truncation error is bounded by

$$|0.560120-0.560123|=0.000003<0.000030517.$$

Note that the larger the number of bits used, the smaller the truncation error that may accompany it.

Examples 14.6 and 14.7 are devoted to illustrating data manipulations in the Q-15 format.

EXAMPLE 14.6

Add the two numbers in Examples 14.4 and 14.5 in Q-15 format.

Solution:

Binary addition is carried out as follows:

$$\begin{array}{r} 1.\,1\,1\,0\,1\,0\,1\,1\,1\,0\,0\,0\,0\,0\,1\,0 \\ +\quad 0.\,1\,0\,0\,0\,1\,1\,1\,1\,0\,1\,1\,0\,0\,1\,0 \\ \hline 1\,0.\,0\,1\,1\,0\,0\,1\,1\,0\,0\,1\,1\,0\,1\,0\,0 \end{array}$$

Then the result is:

$$0.011001100110100.$$

This number in the decimal form can be found to be

$$2^{-2}+2^{-3}+2^{-6}+2^{-7}+2^{-10}+2^{-11}+2^{-13}=0.400024.$$

EXAMPLE 14.7

This is a simple illustration of the fixed-point multiplication.
Determine the fixed-point multiplication of 0.25 and 0.5 in Q-3 fixed-point 2's complement format.

Solution:

Since $0.25 = 0.010$ and $0.5 = 0.100$, we carry out binary multiplication as follows:

```
        0.010
      × 0.100
     ---------
         0000
        0000
       0010
   +  0000
 --------------
    0. 001000
```

Truncating the least-significant bits to convert the result to Q-3 format, we have

$$0.010 \times 0.100 = 0.001.$$

Note that $0.001 = 2^{-3} = 0.125$. We can also verify that $0.25 \times 0.5 = 0.125$.

The Q-format number representation is a better choice than the 2's complement integer representation. But we need to be concerned with the following problems.

1. When converting a decimal number to its Q-N format, where N denotes the number of magnitude bits, we may lose accuracy due to the truncation error, which is bounded by the size of the interval, that is, 2^{-N}.
2. Addition and subtraction may cause overflow, where adding two positive numbers leads to a negative number, or adding two negative number yields a positive number; similarly, subtracting a positive number from a negative number gives a positive number, while subtracting a negative number from a positive number results in a negative number.
3. Multiplying two numbers in Q-15 format will lead to a Q-30 format, which has 31 bits in total. As in Example 14.7, the multiplication of Q-3 yields a Q-6 format, that is, 6 magnitude bits and a sign bit. In practice, it is common for a DSP to hold the multiplication result using a double word size such as the MAC operation, as shown in Fig. 14.9 for multiplying two numbers in Q-15 format. In Q-30 format, there is one sign-extended bit. We may get rid of it by shifting left by one bit to obtain Q-31 format and maintaining the Q-31 format for each MAC operation.

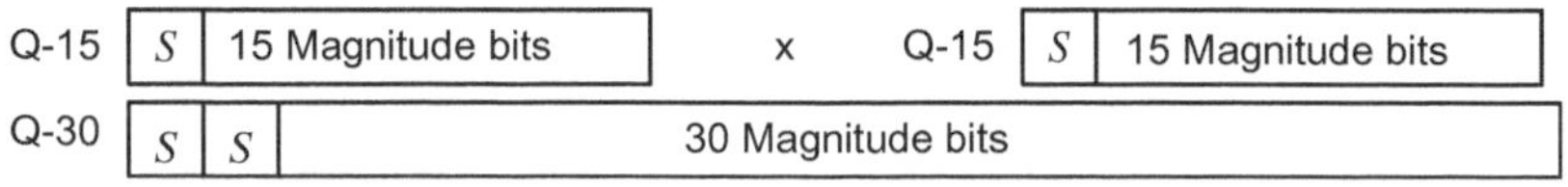

FIG. 14.9

Sign bit extended Q-30 format.

Sometimes, the number in Q-31 format needs to be converted to Q-15; for example, the 32-bit data in the accumulator needs to be sent for 16-bit digital-to-analog conversion (DAC), where the upper most-significant 16 bits in the Q-30 format must be used to maintain accuracy. We can shift the number in Q-30 to the right by 15 bits or shift the Q-31 number to the right by 16 bits. The useful result is stored in the lower 16-bit memory location. Note that after truncation, the maximum error is bounded by the interval size of 2^{-15}, which satisfies most applications. In using the Q-format in the fixed-point DSP, it is costly to maintain the accuracy of data manipulation.

4. Underflow can happen when the result of multiplication is too small to be represented in the Q-format. As an example, in a Q-2 system shown in Table 14.2, multiplying 0.01×0.01 leads to 0.0001. To keep the result in Q-2, we truncate the last two bits of 0.0001 to achieve 0.00, which is zero. Hence, underflow occurs.

14.4.2 FLOATING-POINT FORMAT

To increase the dynamic range of number representation, a floating-point format, which is similar to scientific notation, is used. The general format for the floating-point number representation is given by

$$x = M \cdot 2^E,$$

where M is the mantissa, or fractional part in Q format, and E is the exponent. The mantissa and exponent are signed numbers. If we assign 12 bits for the mantissa and 4 bits for the exponent, the format looks like Fig. 14.10.

Since the 12-bit mantissa has limits between -1 to $+1$, the dynamic range is controlled by the number of bits assigned to the exponent. The bigger the number of bits designated to the exponent, the larger the dynamic range. The number of bits for mantissa defines the interval in the normalized range; as shown in Fig. 14.10, the interval size is 2^{-11} in the normalized range, which is smaller than the Q-15. However, when more mantissa bits are used, the smaller interval size will be achieved. Using the format in Fig. 14.10, we can determine the most negative and most positive numbers as:

$$\text{most negative number} = (1.00000000000)_2 \cdot 2^{0111_2} = (-1) \times 2^7 = -128.0$$

$$\text{most positive number} = (0.11111111111)_2 \cdot 2^{0111_2} = \left(1 - 2^{-11}\right) \times 2^7 = 127.9375.$$

The smallest positive number is given by

$$\text{Smallest positive number} = (0.00000000001)_2 \cdot 2^{1000_2} = \left(2^{-11}\right) \times 2^{-8} = 2^{-19}.$$

As we can see, the exponent acts like a scale factor to increase the dynamic range of the number representation. We study the floating-point format in the following example.

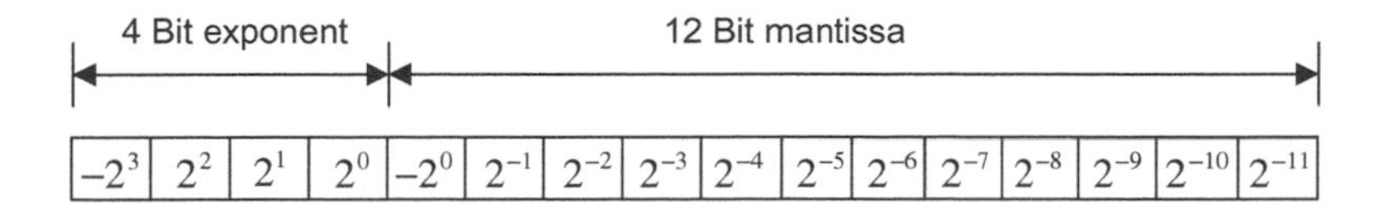

FIG. 14.10

Floating-point format.

EXAMPLE 14.8

Convert each of the following decimal numbers to the floating-point number using the format specified in Fig. 14.10.

(a) 0.1601230
(b) −20.430527

Solution:

(a) We first scale the number 0.1601230 to $0.160123/2^{-2} = 0.640492$ with an exponent of −2 (other choices could be 0 or −1) to get $0.160123 = 0.640492 \times 2^{-2}$. Using 2's complement, we have −2 = 1110. Now we convert the value 0.640492 using Q-11 format to get 010100011111. Cascading the exponent bits and the mantissa bits yields

$$1110010100011111.$$

(b) Since $-20.430527/2^5 = -0.638454$, we can convert it into the fractional part and exponent part as $-20.430527 = -0.638454 \times 2^5$.
Note that this conversion is not particularly unique, the forms $-20.430527 = -0.319227 \times 2^6$ and $-20.430527 = -0.1596135 \times 2^7$ .... are still valid choices. Let us keep what we have now. Therefore, the exponent bits should be 0101.
Converting the number 0.638454 using Q-11 format gives:

$$010100011011.$$

Using 2's complement, we obtain the representation for the decimal number −0.6438454 as

$$101011100101.$$

Cascading the exponent bits and mantissa bits, we achieve

$$0101\ 101011100101$$

The floating-point arithmetic is more complicated. We must obey the rules for manipulating two floating-point numbers. Rules for arithmetic addition are given as

$$x_1 = M_1 2^{E_1}$$

$$x_2 = M_2 2^{E_2}.$$

The floating-point sum is performed as follows:

$$x_1 + x_2 = \begin{cases} \left(M_1 + M_2 \times 2^{-(E_1 - E_2)}\right) \times 2^{E_1}, & \text{if } E_1 \geq E_2 \\ \left(M_1 \times 2^{-(E_2 - E_1)} + M_2\right) \times 2^{E_2} & \text{if } E_1 < E_2 \end{cases}.$$

As a multiplication rule, given two properly normalized floating-point numbers:

$$x_1 = M_1 2^{E_1}$$

$$x_2 = M_2 2^{E_2},$$

where $0.5 \le |M_1| < 1$ and $0.5 \le |M_2| < 1$. Then multiplication can be performed as follows:

$$x_1 \times x_2 = (M_1 \times M_2) \times 2^{E_1 + E_2} = M \times 2^E.$$

That is, the mantissas are multiplied while the exponents are added:

$$M = M_1 \times M_2$$

$$E = E_1 + E_2.$$

Examples 14.9 and 14.10 serve to illustrate manipulations.

EXAMPLE 14.9

Add two floating-point numbers achieved in Example 14.8:

$$1110\,010100011111 = 0.640136718 \times 2^{-2}$$

$$0101\;101011100101 = -0.638183593 \times 2^{5}.$$

Solution:

Before addition, we change the first number to have the same exponent as the second number, that is,

$$0101\,000000001010 = 0.00500168 \times 2^{5}.$$

Then we add two mantissa numbers.

$$\begin{array}{rl} & 0.\,0\,0\,0\,0\,0\,0\,0\,1\,0\,1\,0 \\ & 1.\,0\,1\,0\,1\,1\,1\,0\,0\,1\,0\,1 \\ + & \hline & 1.\,0\,1\,0\,1\,1\,1\,0\,1\,1\,1\,1 \end{array}$$

and we get the floating number as

$$0101\,101011101111.$$

We can verify the result by the following:

$0101\;101011101111 = -(2^{-1}+2^{-3}+2^{-7}+2^{-11}) \times 2^5 = -0.633300781 \times 2^5 = -20.265625.$

EXAMPLE 14.10

Multiply the two floating-point numbers achieved in Example 14.8:

$$1110\,010100011111 = 0.640136718 \times 2^{-2}$$

$$0101\;101011100101 = -0.638183593 \times 2^{5}.$$

Solution:

From the results in Example 14.8, we have the bit patterns for these two numbers as

$$E_1 = 1110, E_2 = 0101, M_1 = 010100011111, M_2 = 101011100101.$$

Continued

EXAMPLE 14.10—CONT'D

Adding two exponents in 2's complement form leads to

$$E = E_1 + E_2 = 1110 + 0101 = 0011,$$

which is +3, as we expected, since in the decimal domain (−2)+5=3. As previously shown in the multiplication rule, when multiplying two mantissas, we need to apply their corresponding positive values. If the sign for the final value is negative, then we convert it to its 2's complement form. In our example, $M_1 = 010100011111$ is a positive mantissa. However, $M_2 = 101011100101$ is a negative mantissa, since the MSB is 1. To perform multiplication, we use 2's complement to convert M_2 to its positive value, 010100011011, and note that the multiplication result is negative. We multiply two positive mantissas and truncate the result to 12 bits to give

$$010100011111 \times 010100011011 = 001101010100.$$

Now we need to add a negative sign to the multiplication result with the 2's complement operation. Taking the 2's complement, we have

$$M = 110010101100.$$

Hence, the product is achieved by cascading the 4-bit exponent and 12-bit mantissa as

$$0011110010111100.$$

Converting this number back to the decimal number, we can verify the result to be

$$-0.408203125 \times 2^3 = -3.265625.$$

Next, we examine overflow and underflow in the floating-point number system.

Overflow:

During operation, overflow will occur when a number is too large to be represented in the floating-point number system. Adding two mantissa numbers may lead a number larger than 1 or less than −1, and multiplying two numbers causes the addition of their two exponents so that the sum of the two exponents could overflow. Consider the following overflow cases.

Case 1. Add the following two floating-point numbers:

$$0111\,011000000000 + 0111\,010000000000$$

Note that two exponents are the same and they are the biggest positive number in 4-bit 2's complement representation. We add two positive mantissa numbers as.

$$\begin{array}{rl} & 0.\,1\,1\,0\,0\,0\,0\,0\,0\,0\,0\,0 \\ + & 0.\,1\,0\,0\,0\,0\,0\,0\,0\,0\,0\,0 \\ \hline & 1.\,0\,1\,0\,0\,0\,0\,0\,0\,0\,0\,0. \end{array}$$

The result for adding mantissa numbers is negative. Hence the overflow occurs.

Case 2: Multiply the following two numbers:

$$0111\,011000000000 \times 0111\,01100000000$$

Adding two positive exponents gives

$$0111 + 0111 = 1000 \text{ (negative, the overflow occurs.)}$$

Multiplying two mantissa numbers gives

$$0.11000000000 \times 0.1100000000 = 0.10010000000 \text{ (OK!)}.$$

Underflow:

As we discussed before, underflow will occur when a number is too small to be represented in the number system. Let us divide the following two floating-point numbers:

$$1001\,001000000000 \div 0111\,01000000000.$$

First, subtracting two exponents leads to.

1001 (negative) − 0111 (positive) = 1001 + 1001 = 0010 (positive; the underflow occurs).

Then, dividing two mantissa numbers, it follows that

$$0.01000000000 \div 0.1000000000 = 0.10000000000 \text{ (OK!)}$$

However, in this case, the expected resulting exponent is −14 in decimal, which is too small to be presented in the 4-bit 2's complement system. Hence the underflow occurs.

Now that we understand the basic principles of the floating-point formats, we can next examine two floating-point formats of the Institute of Electrical and Electronics Engineers (IEEE).

14.4.3 IEEE FLOATING-POINT FORMATS

Single-Precision Format:

IEEE floating-point formats are widely used in many modern DSPs. There are two types of IEEE floating-point formats (IEEE 754 standard). One is the IEEE single-precision format, and the other is the IEEE double-precision format. The single-precision format is described in Fig. 14.11.

The format of IEEE single-precision floating-point standard representation requires 23 fraction bits F, 8 exponent bits E, and 1 sign bit S, with a total of 32 bits for each word. F is the mantissa in 2's complement positive binary fraction represented from bit 0 to bit 22. The mantissa is within the normalized range limits between +1 and +2. The sign bit S is employed to indicate the sign of the number, where when $S = 1$ the number is negative, and when $S = 0$ the number is positive. The exponent E is in excess 127 form. The value of 127 is the offset from the 8-bit exponent range from 0 to 255, so that E-127 will have a range from −127 to +128. The formula shown in Fig. 14.11 can be

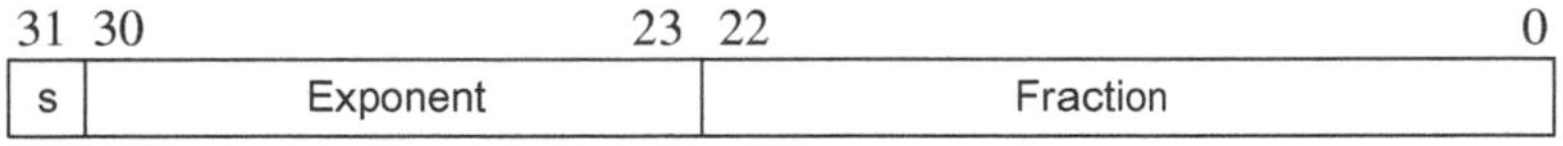

31	30 … 23	22 … 0
s	Exponent	Fraction

$$x = (-1)^s \times (1.F) \times 2^{E-127}$$

FIG. 14.11

IEEE single-precision floating-point format.

applied to convert the IEEE 754 standard (single precision) to the decimal number. The following simple examples also illustrate this conversion.

$$0\ 10000000\ 00000000000000000000000 = (-1)^0 \times (1.0_2) \times 2^{128-127} = 2.0$$
$$0\ 10000001\ 10100000000000000000000 = (-1)^0 \times (1.101_2) \times 2^{129-127} = 6.5$$
$$1\ 10000001\ 10100000000000000000000 = (-1)^1 \times (1.101_2) \times 2^{129-127} = -6.5.$$

Let us look at Example 14.11 for more explanation.

EXAMPLE 14.11

Convert the following number in the IEEE single-precision format to the decimal format:

$$110000000.010\ldots0000.$$

Solution:

From the bit pattern in Fig. 14.11, we can identify the sign bit, exponent, and fractional as:

$$s = 1,\ \ E = 2^7 = 128$$
$$1.F = 1.01_2 = (2)^0 + (2)^{-2} = 1.25.$$

Then, applying the conversion formula leads to

$$x = (-1)^1(1.25) \times 2^{128-127} = -1.25 \times 2^1 = -2.5.$$

In conclusion, the value x represented by the word can be determined based on the following rules, including all the exceptional cases:

- If $E = 255$ and F is nonzero, then $x = NaN$ ("Not a number").
- If $E = 255$, F is zero, and S is 1, then $x = -$Infinity.
- If $E = 255$, F is zero, and S is 0, then $x = +$Infinity.
- If $0 < E < 255$, then $x = (-1)^s \times (1.F) \times 2^{E-127}$, where $1.F$ represents the binary number created by prefixing F with an implicit leading 1 and a binary point.
- If $E = 0$ and F is nonzero, then $x = (-1)^s \times (0.F) \times 2^{-126}$. This is an "unnormalized" value.
- If $E = 0$, F is zero, and S is 1, then $x = -0$.
- If $E = 0$, F is zero, and S is 0, then $x = 0$.

Typical and exceptional examples are shown as follows:

$$0\ 00000000\ 00000000000000000000000 = 0$$
$$1\ 00000000\ 00000000000000000000000 = -0$$

$$0\ 11111111\ 00000000000000000000000 = \text{Infinity}$$
$$1\ 11111111\ 00000000000000000000000 = -\text{Infinity}$$

$$0\ 11111111\ 00000100000000000000000 = \text{NaN}$$
$$1\ 11111111\ 00100010001001010101010 = \text{NaN}$$

$$0\,00000001\,00000000000000000000000 = (-1)^0 \times (1.0_2) \times 2^{1-127} = 2^{-126}$$
$$0\,00000000\,10000000000000000000000 = (-1)^0 \times (0.1_2) \times 2^{0-126} = 2^{-127}$$
$$0\,00000000\,00000000000000000000001 = (-1)^0 \times (0.00000000000000000000001_2) \times 2^{0-126}$$
$$= 2^{-149} \text{ (smallest positive value)}$$

Double-Precision Format:

The IEEE double-precision format is described in Fig. 14.12.

The IEEE double-precision floating-point standard representation requires a 64-bit word, which may be numbered from 0 to 63, left to right. The first bit is the sign bit S, the next 11 bits are the exponent bits E, and the final 52 bits are the fraction bits F. The IEEE floating-point format in double-precision significantly increases the dynamic range of number representation since there are 11 exponent bits; the double-precision format also reduces the interval size in the mantissa normalized range of +1 to +2, since there are 52 mantissa bits as compare with the single-precision case of 23 bits. Applying the conversion formula shown in Fig. 14.12 is similar to the single-precision case.

EXAMPLE 14.12

Convert the following number in the IEEE double-precision format to the decimal format:

001000...0.110...0000

Solution:

Using the bit pattern in Fig. 14.12, we have

$$s = 0,\quad E = 2^9 = 512 \text{ and}$$

$$1.F = 1.11_2 = (2)^0 + (2)^{-1} + (2)^{-2} = 1.75.$$

Then, applying the double-precision formula yields

$$x = (-1)^0 (1.75) \times 2^{512-1023} = 1.75 \times 2^{-511} = 2.6104 \times 10^{-154}.$$

For purpose of completeness, rules for determining the value x represented by the double-precision word are listed as follows:

- If $E = 2047$ and F is nonzero, then $x = NaN$ ("Not a number")
- If $E = 2047$, F is zero, and S is 1, then $x = -$Inifinity

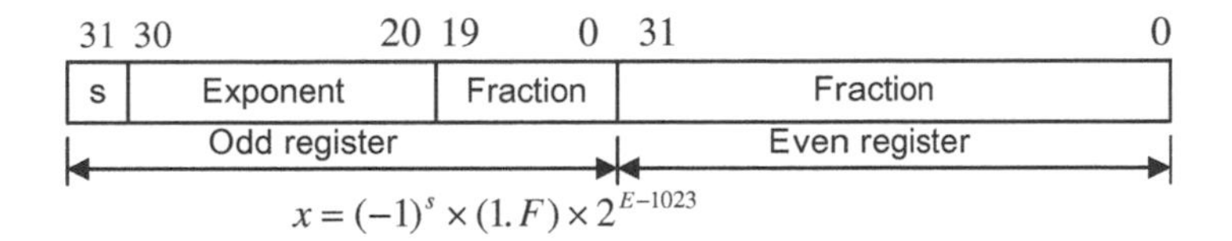

FIG. 14.12

IEEE double-precision floating-point format.

- If $E=2047$, F is zero, and S is 0, then $x=+$Inifinity
- If $0<E<2047$, then $x=(-1)^s \times (1.F) \times 2^{E-1023}$, where "1. F"is intended to represent the binary number created by prefixing F with an implicit leading 1 and a binary point
- If $E=0$ and F is nonzero, then $x=(-1)^s \times (0.F) \times 2^{-1022}$. This is an "unnormalized" value
- If $E=0$, F is zero, and S is 1, then $x=-0$
- If $E=0$, F is zero, and S is 0, then $x=0$.

14.4.4 FIXED-POINT DSPs

Analog Devices, TI, and Motorola all manufacture fixed-point DSPs. Analog Devices offers a fixed-point DSP family such as ADSP21xx. TI provides various generations of fixed-point DSPs based on historical development, architecture features, and computational performance. Some of the most common ones are TMS320C1x (first generation), TMS320C2x, TMS320C5x, and TMS320C62x. Motorola manufactures a variety of fixed-point processors, such as DSP5600x family. The new families of fixed-point DSPs are expected to continue to grow. Since they share some basic common features such as program memory and data memory with associated address buses, arithmetic logic units (ALUs), program control units, MACs, shift units, and address generators, here we focus on an overview of the TMS320C54x processor. The typical TMS320C54x fixed-point DSP architecture appears in Fig. 14.13.

The fixed-point TMS320C50 families supporting 16-bit data have on-chip program memory and data memory in various size and configuration. They include data RAM (random-access memory) and program ROM (read-only memory) used for program code, instruction, and data. Four data buses and four address buses are accommodated to work with the data memory and program memory. The program memory address bus and program memory data bus are responsible for fetching the program instruction. As shown in Fig. 14.13, the C and D data memory address buses and the C and D data memory data buses deal with fetching data from the data memory while the E data memory address bus and E data memory data bus are dedicated to move data into data memory. In addition, the E memory data bus can access the I/O devices.

Computational units consist of an ALU, an MAC, and a shift unit. For TMS320C54x families, the ALU can fetch data from the C, D, and program memory data buses and access the E memory data bus. It has two independent 40-bit accumulators, which are able to operate the 40-bit addition. The multiplier, which can fetch data from C and D memory data buses and write data via E data memory data bus, is capable of operating 17-bit $\times$ 17-bit multiplications. The 40-bit shifter has the same capability of bus access as the MAC, allowing all possible shifts for scaling and fractional arithmetic such as we have discussed for the Q-format.

The program control unit fetches instructions via the program memory data bus. Again, in order to speed up memory access, there are two address generators available: one responsible for program addresses and one for data addresses.

Advanced Harvard architecture is employed, where several instructions operate at the same time for given a given single instruction cycle. Processing performance offers 40 MIPS (million instruction sets per second). To further explore this subject, the reader is referred to Dahnoun (2000), Embree (1995), Ifeachor and Jervis (2002) and Van der Vegte (2002), as well as the web site (www.ti.com).

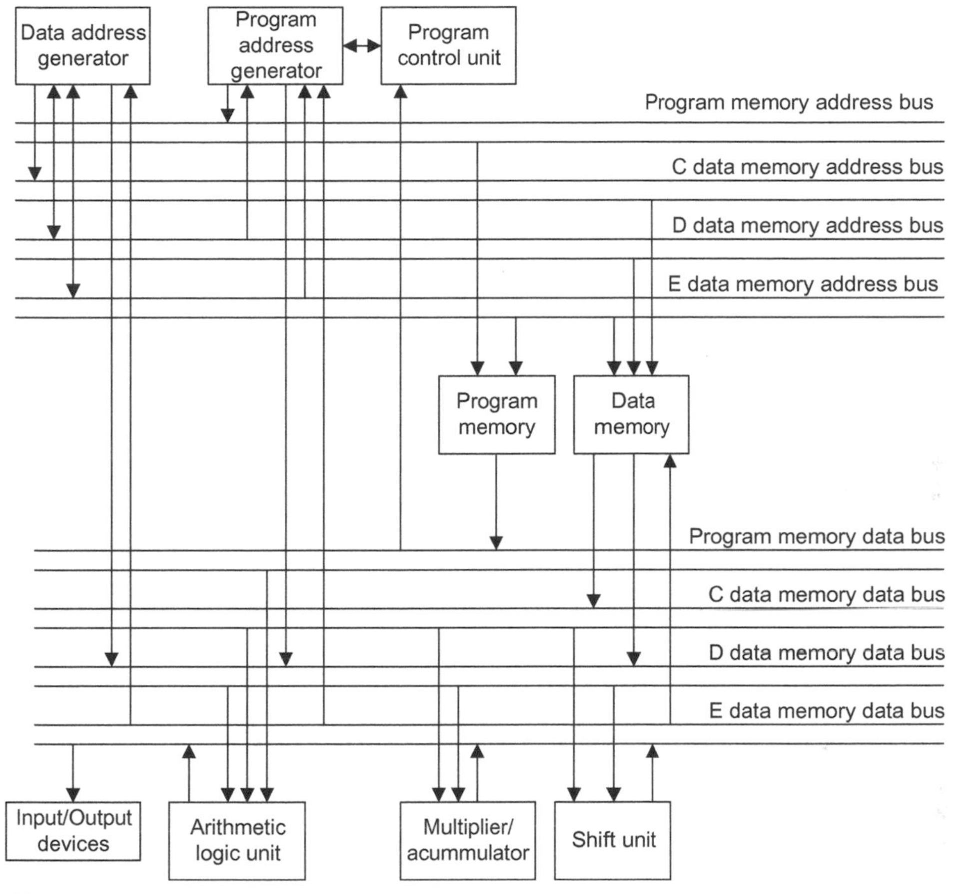

FIG. 14.13

Basic architecture of TMSC320C54x family.

14.4.5 FLOATING-POINT DSPs

Floating-point DSPs perform DSP operations using the floating-point arithmetic, as we discussed earlier. The advantages of using the floating-point processor include getting rid of finite word length effects such as overflows, round-off errors, truncation errors, and coefficient quantization error. Hence, in terms of coding, we do not need to scale input samples to avoid overflow, shift the accumulator result to fit the DAC word size, scale the filter coefficients, or apply Q-format arithmetic. The floating-point DSP with high-performance speed and calculation precision facilitates a friendly environment to develop and implement DSP algorithms.

Analog Devices provides the floating-point DSP families such as ADSP210xx and TigerSHARC. TI offers a wide range of the floating-point DSP families, in which the TMS320C3x is the first generation, followed by the TMSC320C4x and TMS320C67x families. Since the first generation of a floating-point DSP is less complicated than later generations but still has the common basic features, we overview the first-generation architecture first.

Fig. 14.14 shows the typical architecture of TI' TMS320C3x families. We discuss some key features briefly. Further detail can be found in the TMS320C3x User's Guide (Texas Instruments, 1991), the TMS320C6x CPU and Instruction Set Reference Guide (Texas Instruments, 1998), and other studies (Dahnoun, 2000; Embree, 1995; Ifeachor and Jervis, 2002; Kehtaranavaz and Simsek, 2000; Sorensen and Chen, 1997; Van der Vegte, 2002).

TMS320C3x family consists of 32-bit single-chip floating-point processors that support both integer and floating-point operations.

The processor has a large memory space and is equipped with dual-access on-chip memories. A program cache is employed to enhance the execution of commonly used codes. Similar to the fixed-point processor, it uses the Harvard architecture, where there are separate buses used for program and data so that instructions can be fetched at the same time that data are being accessed. There also exist memory buses and data buses for direct-memory access (DMA) for concurrent I/O and CPU operations, and peripheral access such as serial ports, I/O ports, memory expansion, and an external clock.

The C3x CPU contains the floating-point/integer multiplier; an ALU, which is capable of operating both integer and the floating-point arithmetic; a 32-bit barrel shifter; internal buses; a CPU register file; and dedicated auxiliary register arithmetic units (ARAUs). The multiplier operates single-cycle

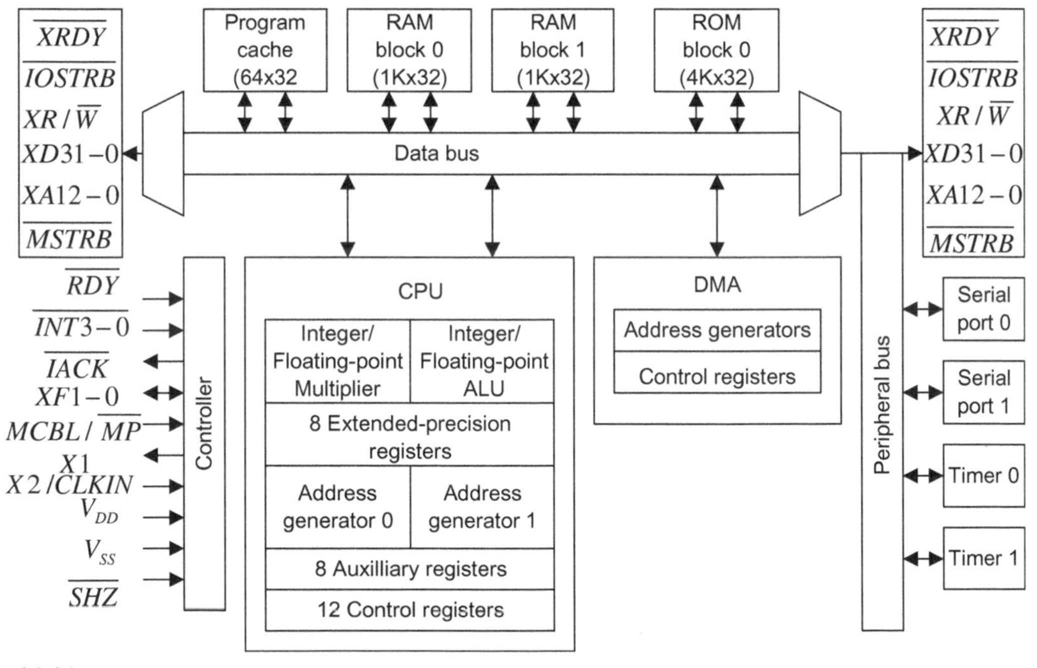

FIG. 14.14

The typical TMS320C3x floating-point DSP.

multiplications on 24-bit integers and on 32-bit floating-point values. Using parallel instructions to perform a multiplication, an ALU will cost a single cycle, which means that a multiplication and an addition are equally fast. The ARAUs support addressing modes, in which some of them are specific to DSP such as circular buffering and bit-reversal addressing (digital filtering and FFT operations). The CPU register file offers 28 registers, which can be operated on by the multiplier and ALU. The special functions of the registers include eight-extended 40-bit precision registers for maintaining the accuracy of the floating-point results. Eight auxiliary registers can be used for addressing and for integer arithmetic. These registers provide internal temporary storage of internal variables instead of external memory storage, to allow performance of arithmetic between registers. In this way, program efficiency is greatly increased.

The prominent feature of C3x is its floating-point capability, allowing operation of numbers with a very larger dynamic range. It offers implementing of the DSP algorithm without worrying about problems such as overflows and coefficient quantization. Three floating-point formats are supported. A short 16-bit floating-point format has 4 exponent bits, 1 sign bit, and 11 mantissa bits. A 32-bit single-precision format has 8 exponent bits, 1 sign bit, and 23 fraction bits. A 40-bit extended precision format contains 8 exponent bits, 1 sign bit, and 31 fraction bits. Although the formats are slightly different from the IEEE 754 standard, conversions are available between these formats.

The TMS320C30 offers high-speed performance with 60-ns single-cycle instruction execution time which is equivalent to 16.7 MIPS. For speech quality applications with an 8 kHz sampling rate, it can handle over 2000 single-cycle instructions between two samples (125 μs). With instruction enhancement such as pipeline executing each instruction in a single cycle (four cycles required from fetch to execution by the instruction itself) and a multiple interrupt structure, this high-speed processor validates implementation of real-time applications in the floating-point arithmetic.

14.5 FINITE IMPULSE RESPONSE AND INFINITE IMPULSE RESPONSE FILTER IMPLEMENTATIONS IN FIXED-POINT SYSTEMS

With knowledge of the IEEE format and of filter realization structures such as the direct-form I, direct-form II, and parallel and cascade forms (Chapter 6), we can study digital filter implementation in the fixed-point processor. In the fixed-point system, where only integer arithmetic is used, we prefer input data, filter coefficients, and processed output data to be in the Q-format. In this way, we avoid overflow due to multiplications and can prevent overflow due to addition by scaling input data. When the filter coefficients are out of the Q-format range, coefficient scaling must be taken into account to maintain the Q-format. We develop FIR filter implementation in Q-format first, and then infinite impulse response (IIR) filter implementation next. In addition, we assume that with a given input range in Q-format, the filter output is always in Q-format even if the filter passband gain is larger than 1.

First, to avoid the overflow for an adder, we can scale the input down by a scale factor S, which can be safely determined by the following equation

$$S = I_{\max} \cdot \sum_{k=0}^{\infty} |h(k)| = I_{\max} \cdot (|h(0)| + |h(1)| + |h(2)| + \cdots), \tag{14.1}$$

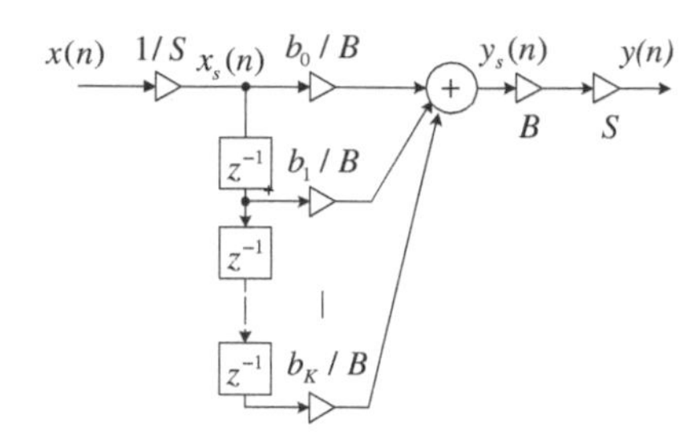

FIG. 14.15

Direct-form I implementation of the FIR filter.

where $h(k)$ is the impulse response of the adder output and $I_{\max}$ the maximum amplitude of the input in Q-format. Note that this is not an optimal factor in terms of reduced signal-to-noise ratio. However, it shall prevent the overflow. Eq. (14.1) means that the adder output can actually be expressed as a convolution output:

$$\text{adder output} = h(0)x(n) + h(1)x(n-1) + h(2)x(n-2) + \cdots. \tag{14.2}$$

Assuming the worst condition, that is, that all the inputs $x(n)$ reach a maximum value of $I_{\max}$ and all the impulse coefficients are positive, the sum of the adder gives the most conservative scale factor, as shown in Eq. (14.1). Hence, scaling down of the input by a factor of S will guarantee that the output of the adder is in Q-format.

When some of the FIR coefficients are larger than 1, which is beyond the range of Q-format representation, coefficient scaling is required. The idea is that scaling down the coefficients will make them <1, and later the filtered output will be scaled up by the same amount before it is sent to DAC. Fig. 14.15 describes the modified implementation.

In the figure, the scale factor B makes the coefficients b_k/B convertible to the Q format. The scale factors of S and B are usually chosen to be a power of 2, so the simple shift operation can be used in the coding process. Let us implement an FIR filter containing filter coefficients larger than 1 in the fixed-point implementation.

EXAMPLE 14.13

Given the FIR filter

$$y(n) = 0.9x(n) + 3x(n-1) + 0.9x(n-2),$$

with a passband gain of 4, and assuming that the input range only occupies 1/4 of the full range for a particular application, develop the DSP implementation equations in the Q-15 fixed-point system.

Solution:

The adder may cause overflow if the input data exist for $\frac{1}{4}$ of a full dynamic range. The scale factor is determined using the impulse response, which consists of the FIR filter coefficients, as discussed in Chapter 3.

$$S = \frac{1}{4}(|h(0)| + |h(1)| + |h(2)|) = \frac{1}{4}(0.9 + 3 + 0.9) = 1.2.$$

Overflow may occur. Hence, we select $S=2$ (a power of 2). We choose $B=4$ to scale all the coefficients to be <1, so the Q-15 format can be used. According to Fig. 14.15, the developed difference equations are given by

$$x_s(n) = \frac{x(n)}{2}$$

$$y_s(n) = 0.225x_s(n) + 0.75x_s(n-1) + 0.225x_s(n-2)$$

$$y(n) = 8y_s(n)$$

Next, the direct-form I implementation of the IIR filter is illustrated in Fig. 14.6.

As shown in Fig. 14.16, the purpose of a scale factor C is to scale down the original filter coefficients to the Q-format. The factor C is usually chosen to be a power of 2 for using a simple shift operation in DSP.

EXAMPLE 14.14

The following IIR filter,

$$y(n) = 2x(n) + 0.5y(n-1),$$

uses the direct-form I, and for a particular application, the maximum input is $I_{max} = 0.010....0_2 = 0.25$.

Develop the DSP implementation equations in the Q-15 fixed-point system.

Solution:

This is an IIR filter whose transfer function is

$$H(z) = \frac{2}{1-0.5z^{-1}} = \frac{2z}{z-0.5}.$$

Applying the inverse z-transform, we have the impulse response

$$h(n) = 2 \times (0.5)^n u(n).$$

To prevent overflow in the adder, we can compute the S factor with the help of the Maclaurin series or approximate Eq. (14.1) numerically. We get

$$S = 0.25 \times \left(2(0.5)^0 + 2(0.5)^1 + 2(0.5)^2 + \cdots\right) = \frac{0.25 \times 2 \times 1}{1-0.5} = 1.$$

MATLAB function **impz()** can also be applied to find the impulse response and the S factor:

```
>> h=impz(2,[1 -0.5]); % Find the impulse response
>> sf=0.25*sum(abs(h)) % Determine the sum of absolute values of h(k)
sf =1
```

Hence, we do not need to perform the input scaling. However, we need scale down all the coefficients to use the Q-15 format. A factor of $C=4$ is selected. From Fig. 14.16, we get the differences equations as

Continued

EXAMPLE 14.14—CONT'D

$$x_s(n) = x(n)$$

$$y_s(n) = 0.5_s x(n) + 0.125 y_f(n-1)$$

$$y_f(n) = 4 y_s(n)$$

$$y(n) = y_f(n).$$

We can develop these equations directly. First, we divide the original difference equation by a factor of 4 to scale down all the coefficients to be <1, that is,

$$\frac{1}{4} y_f(n) = \frac{1}{4} \times 2x_s(n) + \frac{1}{4} \times 0.5 y_f(n-1)$$

and define a scaled output

$$y_s(n) = \frac{1}{4} y_f(n).$$

Finally, substituting $y_s(n)$ to the left side of the scaled equation and rescaling up the filter output as $y_f(n) = 4y_s(n)$, we have the same results as we got before.

The fixed-point implementation for the direct-form II is more complicated. The developed direct-form II implementation of the IIR filter is illustrated in Fig. 14.17.

As shown in Fig. 14.17, two scale factors A and B are designated to scale denominator coefficients and numerator coefficients to their Q-format representations, respectively. Here S is a special factor to scale down the input sample so that the numerical overflow in the first sum in Fig. 14.17 can be prevented. The difference equations are given in Chapter 6 and listed here:

$$w(n) = x(n) - a_1 w(n-1) - a_2 w(n-2) - \cdots - a_M w(n-M)$$

$$y(n) = b_0 w(n) + b_1 w(n-1) + \cdots + b_M w(n-M).$$

The first equation is scaled down by the factor A to ensure that all the denominator coefficients are <1, that is,

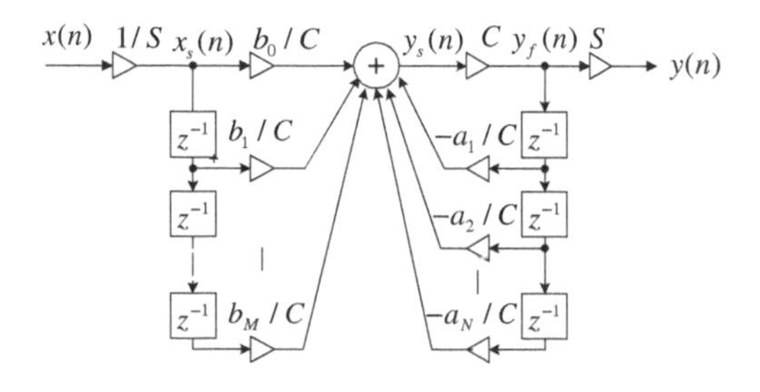

FIG. 14.16

Direct-form I implementation of the IIR filter.

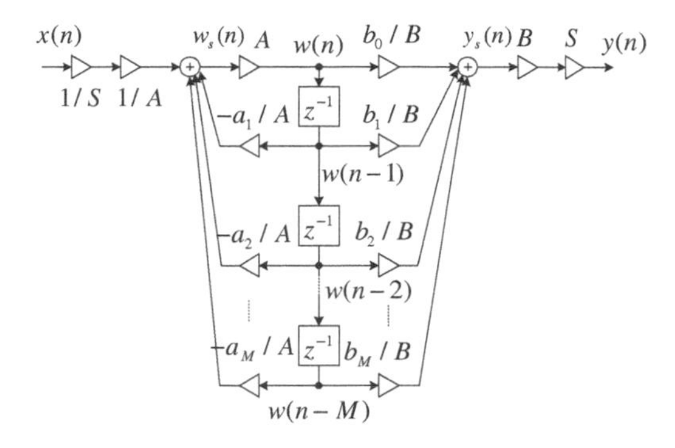

FIG. 14.17

Direct-form II implementation of the IIR filter.

$$w_s(n) = \frac{1}{A}w(n) = \frac{1}{A}x(n) - \frac{1}{A}a_1 w(n-1) - \frac{1}{A}a_2 w(n-2) - \cdots - \frac{1}{A}a_M w(n-M)$$

$$w(n) = A \times w_s(n).$$

Similarly, scaling the second equation yields

$$y_s(n) = \frac{1}{B}y(n) = \frac{1}{B}b_0 w(n) + \frac{1}{B}b_1 w(n-1) + \cdots + \frac{1}{B}b_M w(n-M)$$

and

$$y(n) = B \times y_s(n).$$

To avoid the first adder overflow (first equation), the scale factor S can safely be determined by Eq. (14.3):

$$S = I_{\max}\left(|h(0)| + |h(1)| + |h(2)| + \cdots\right), \tag{14.3}$$

where $h(k)$ is the impulse response of the filter whose transfer function is the reciprocal of the denominator polynomial, where the poles can cause a larger value to the first sum. Hence, $h(n)$ is given by

$$h(n) = Z^{-1}\left(\frac{1}{1 + a_1 z^{-1} + \cdots + az^{-M}}\right) \tag{14.4}$$

All the scale factors A, B, and S are usually chosen to be a power of 2, respectively, so that the shift operations can be used in the coding process. Example 14.15 serves as an illustration.

EXAMPLE 14.15

Given the IIR filter:

$$y(n) = 0.75x(n) + 1.49x(n-1) + 0.75x(n-2) - 1.52y(n-1) - 0.64y(n-2),$$

with a passband gain of 1 and a full range of input, use the direct-form II implementation to develop the DSP implementation equations in the Q-15 fixed-point system.

Continued

EXAMPLE 14.15—CONT'D

Solution:

The difference equations without scaling in the direct-form II implementation are given by

$$w(n) = x(n) - 1.52w(n-1) - 0.64w(n-2)$$

$$y(n) = 0.75w(n) + 1.49w(n-1) + 0.75w(n-2).$$

To prevent the overflow in the first adder, we have the reciprocal of the denominator polynomial as

$$A(z) = \frac{1}{1 + 1.52z^{-1} + 0.64z^{-2}}.$$

Using the MATLAB function leads to.

```
>> h=impz(1,[1 1.52 0.64]);
>> sf=sum(abs(h))
   sf = 10.4093
```

We choose the S factor as $S = 16$ and we choose $A = 2$ to scale down the denominator coefficients by half. Since the second adder output after scaling is

$$y_s(n) = \frac{0.75}{B}w(n) + \frac{1.49}{B}w(n-1) + \frac{0.75}{B}w(n-2),$$

we have to ensure that each coefficient is <1, as well as the sum of the absolute values

$$\frac{0.75}{B} + \frac{1.49}{B} + \frac{0.75}{B} < 1$$

to avoid the second adder overflow. Hence $B = 4$ is selected. We develop the DSP equations as

$$x_s(n) = x(n)/16$$

$$w_s(n) = 0.5x_s(n) - 0.76w(n-1) - 0.32w(n-2)$$

$$w(n) = 2w_s(n)$$

$$y_s(n) = 0.1875w(n) + 0.3725w(n-1) + 0.1875w(n-2)$$

$$y(n) = (B \times S)y_s(n) = 64y_s(n).$$

The implementation for cascading the second-order section filters can be found in Ifeachor and Jervis (2002).

A practical example is presented in the next section. Note that if a floating-point DSP is used, all the scaling concerns should be ignored, since the floating-point format offers a large dynamic range, so that overflow hardly even happens.

14.6 DIGITAL SIGNAL PROCESSING PROGRAMMING EXAMPLES

In this section, we first review the TMS320C67x DSK (DSP Starter Kit), which offers floating-point and fixed-point arithmetic. We will then investigate the real-time implementation of digital filters.

14.6.1 OVERVIEW OF TMS320C67x DSK

In this section, a TI TMS320C6713 DSK (DSP Starter Kit) shown in Fig. 14.18 is chosen for demonstration. This DSK board has an approximate size of 5×8 inches, a clock rate of 225 MHz and a 16-bit stereo codec TL V320AIC23 (AIC23) which deals with analog inputs and outputs. The onboard codec AIC23 applies sigma-delta technology for ADC and DAC functions. The codec runs using a 12 MHz system clock and the sampling rate can be selected from a range of 8–96 kHz for speech and audio processing. Other boards such as a TI TMS320C6711 DSK can also be found in the references (Kehtaranavaz and Simsek, 2000; TMS320C6x CPU and Instruction Set Reference Guide; 1999).

(A) TMS320C6713 DSK board

(B) TMS320C6713 DSK block diagram

FIG. 14.18

C6713 DSK board and block diagram. (A) TMS320C6713 DSK board. (B) TMS320C6713 DSK block diagram.

Courtesy of Texas Instruments.

The on-board daughter card connections facilitate the external units for advanced applications such as external peripheral and external memory interfaces (EMIFs). The TMS320C6713 DSK board consists of 16 MB (megabytes) of synchronous dynamic RAM (SDRAM) and 512 kB (kilobytes) of flash memory. There are four onboard connections: MIC IN for microphone input; LINE IN for line input; LINE OUT for line output; and HEADPHONE for a headphone output (multiplexed with LINE OUT). The four user DIP switches can be read from a program running within DSK board as well as they can provide with a user feedback control of interface. The four LEDs (light-emitting diodes) on the DSK board can be controlled by a running DSP program. The onboard voltage regulators provide 1.26 V for the DSP core while 3.3 V for the memory and peripherals. The USB port provides the connection between the DSK board and the host computer, where the user program is developed, compiled, and downloaded to the DSK for real-time applications using the user-friendly software called the Code Composer Studio (CCS), which we discuss this later.

In general, the TMS320C67x operates at a high clock rate of 300 MHz. Combining with high speed and multiple units operating at the same time has pushed its performance up to 2400 MIPS at 300 MHz. Using this number, the C67x can handle 0.3 MIPS between two speech samples at a sampling rate of 8 kHz and can handle over 54,000 instructions between the two audio samples with a sampling rate of 44.1 kHz. Hence, the C67x offers great flexibility for real-time applications with a high-level C language.

Fig. 14.19 shows a C67x architecture overview, while Fig. 14.20 displays a more detailed block diagram. C67x contains three main parts, which are the CPU, the memories, and the peripherals. As shown in Fig. 14.9, these three main parts are joined by an EMIF interconnected by internal buses to facilitate interface with common memory devices; DMA; a serial port; and a host-port interface (HPI).

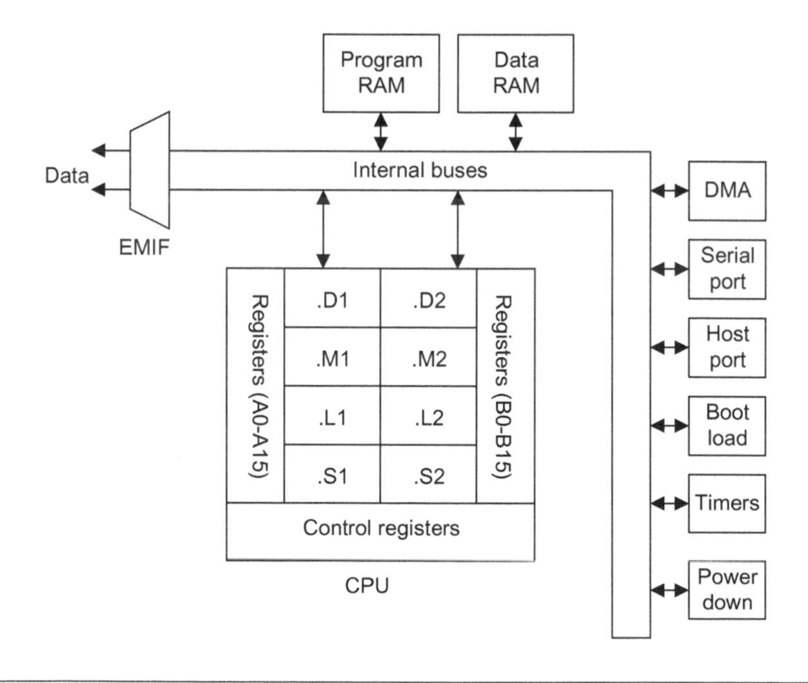

FIG. 14.19

Block diagram of TMS320C67x floating-point DSP.

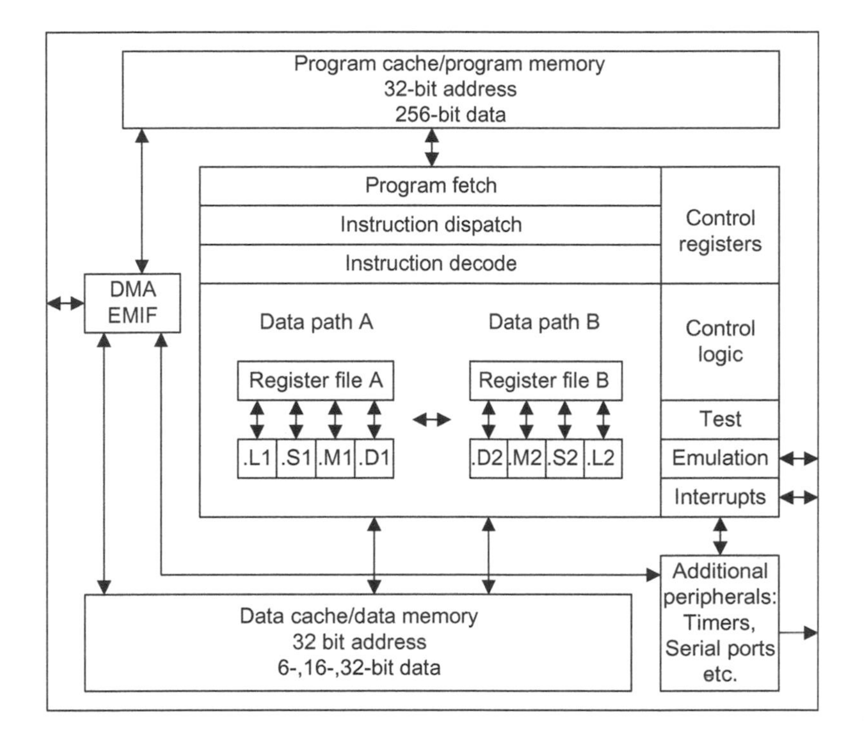

FIG. 14.20

Registers of TMS320C67x floating-point DSP.

Since this section is devoted to show DSP coding examples, C67x key features and references are briefly listed here:

(1) Architecture: The system uses Texas Instruments Veloci^{TM} architecture, which is an enhancement of the VLIW (very long instruction word architecture) (Dahnoun, 2000; Ifeachor and Jervis, 2002; Kehtaranavaz and Simsek, 2000).

(2) CPU: As shown in Fig. 14.20, the CPU has eight functional units divided into two sides A and B, each consisting of units .D, .M, .L, and .S. For each side, an .M unit is used for multiplication operation, an. L unit is used for logical and arithmetic operations, and a .D unit is used for loading/storing and arithmetic operations. Each side of the C67x CPU has sixteen 32-bit registers that the CPU must go through for interface. More detail can be found in Appendix D (Texas Instruments, 1991) as well as in Kehtaranavaz and Simsek (2000) and Texas Instruments (1998).

(3) Memory and internal buses: Memory space is divided into internal program memory, internal data memory, and internal peripheral and external memory space. The internal buses include a 32-bit program address bus, a 256-bit program data bus to carrying out eight 32-bit instructions (VLIW), two 32-bit data address buses, two 64-bit load data buses, two 64-bit store data buses, two 32-bit DMA buses, and two 32-bit DMA address buses responsible for reading and writing. There also exit a 22-bit address bus and a 32-bit data bus for accessing off-chip or external memory.

(4) Peripherals:

(a) EMIF, which provides the required timing for accessing external memory
(b) DMA, which moves data from one memory location to another without interfering with the CPU operations
(c) Multichannel buffered serial port (McBSP) with a high-speed multi-channel serial communication link
(d) HPI, which lets a host access internal memory
(e) Boot loader for loading code from off-chip memory or the HPI to internal memory
(f) Timers (two 32-bit counters)
(g) Power-down units for saving power for periods when the CPU is inactive.

The software tool for the C67x is the CCS provided by TI. It allows the user to build and debug programs from a user-friendly graphical user interface (GUI) and extends the capabilities of code development tools to include real-time analysis. Installation, tutorial, coding, and debugging information can be found in the CCS Getting Started Guide (Texas Instruments, 2001) and in Kehtaranavaz and Simsek (2000).

Particularly for the TMS320C6713 DSK with a clock rate of 225 MHz, it has capability to fetch eight 32-bit instructions every 4.4 ns (1/225 MHz). The functional block diagram is shown in Fig. 14.21. The detailed description can found in Chassaing and Reay (2008).

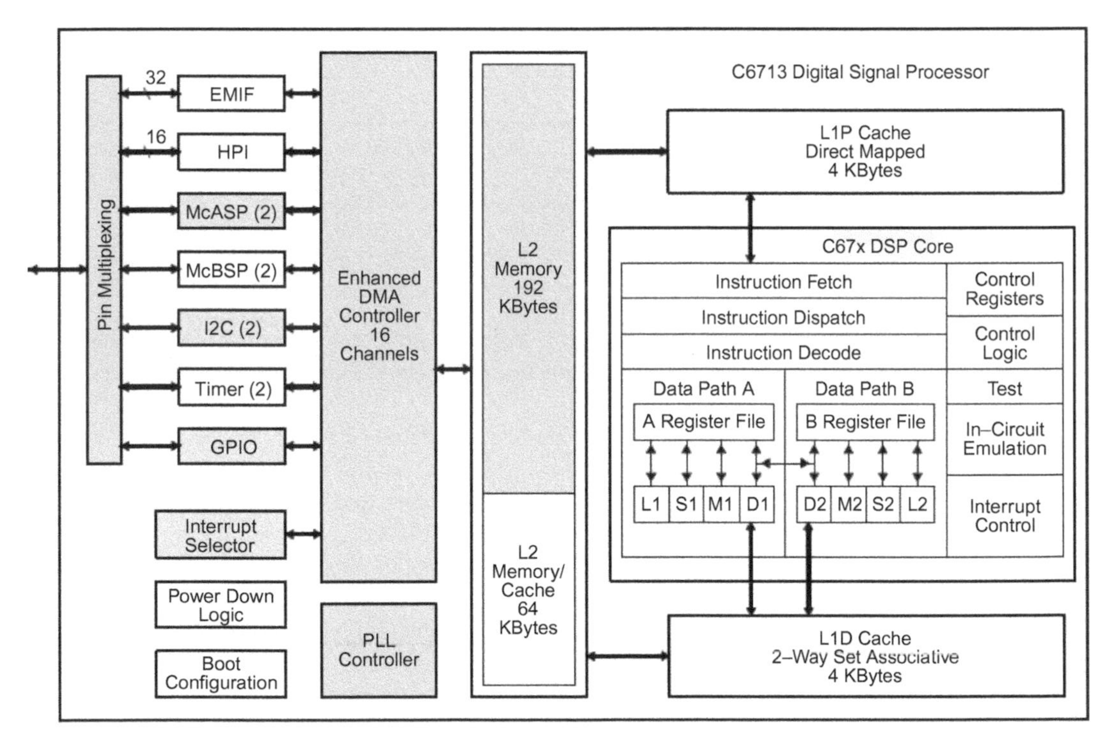

FIG. 14.21

Functional block diagram and registers of TMS320C6713.

Courtesy of Texas Instruments.

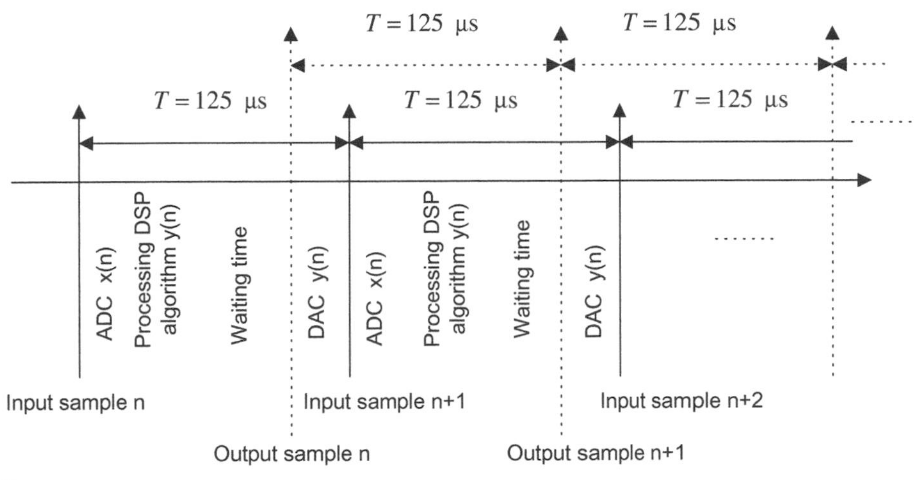

FIG. 14.22

Concept of real-time processing.

14.6.2 CONCEPT OF REAL-TIME PROCESSING

We illustrate the real-time implementation shown in Fig. 14.22, where the sampling rate is 8000 samples per second; that is, the sampling period $T = 1/f_s = 125\ \mu s$, which is the time between two samples.

As shown in Fig. 14.22, the required timing includes an input sample clock and an output sample clock. The input sample clock maintains the accuracy of sampling time for each ADC operation, while the output sampling clock keeps the accuracy of time instant for each DAC operation. The time between the input sample clock n and output sample clock n consists of the ADC operation, algorithm processing, and the wait for the next ADC operation. The numbers of instructions for ADC and DSP algorithm must be estimated and verified to ensure that all instructions have been completed before the DAC begins. Similarly, the number of instructions for DAC must be verified so that DAC instructions will be finished between the output sample clock n and the next input sample clock $n+1$. Timing usually is set up using the DSP interrupts (we will not pursue the interrupt setup here).

Next, we focus on the implementation of the DSP algorithm in the floating-point system for simplicity. A DSK setup example (Tan and Jiang, 2010) is depicted in Fig. 14.23, while a skeleton code for the verification of the input and output is depicted in Fig. 14.24.

14.6.3 LINEAR BUFFERING

During DSP such as digital filtering, past inputs, and past outputs are required to be buffered and updated for processing the next input sample. Let us first study the FIR filter implementation.

FIR Filtering:

Consider implementation for the following 3-tap FIR filter:

$$y(n) = 0.5x(n) + 0.2x(n-1) + 0.5x(n-2).$$

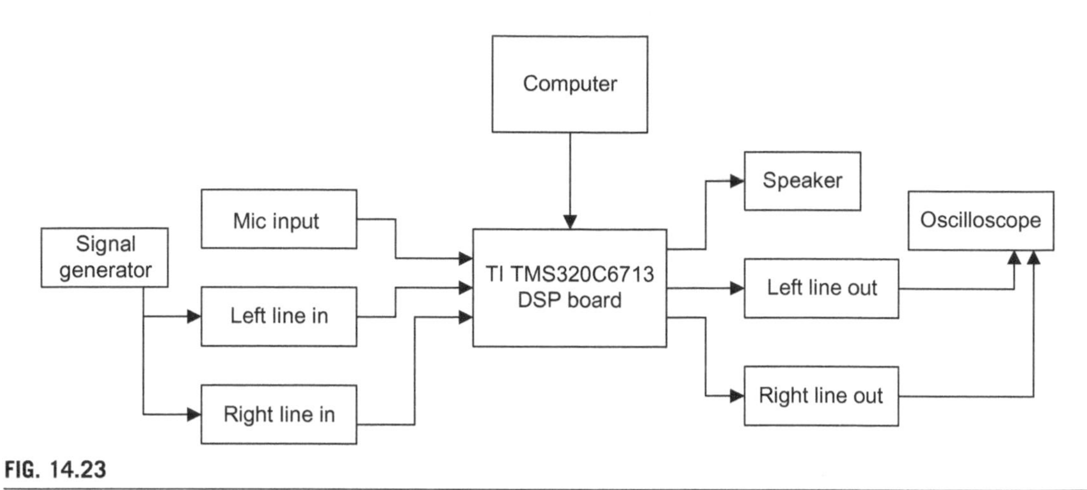

FIG. 14.23

TMS320C6713 DSK setup example.

```
float x[1]={0.0};
float y[1]={0.0};
interrupt void c_int11()
{
    float lc; /*left channel input */
    float rc; /*right channel input */
    float lcnew; /*left channel output */
    float rcnew; /*right channel output */
//Left channel and right channel inputs
   AIC23_data.combo=input_sample();
   lc=(float) (AIC23_data.channel[LEFT]);
   rc= (float) (AIC23_data.channel[RIGHT]);
// Insert DSP algorithm below
   x[0]=lc; /* Input from the left channel */
   y[0]=x[0]; /* simplest DSP equation */
// End of the DSP algorithm
    lcnew=y[0];
    rcnew=y[0];
    AIC23_data.channel[LEFT]=(short) lcnew;
    AIC23_data.channel[RIGHT]=(short) rcnew;
    output_sample(AIC23_data.combo);
}
```

FIG. 14.24

Program segment for verifying input and output.

The buffer requirements are shown in Fig. 14.25. The coefficient buffer b[3] contains three FIR coefficients, and the coefficient buffer is fixed during the process. The input buffer x[3], which holds the current and past inputs, is required to be updated. The FIFO update is adopted here with the segment of codes shown in Fig. 14.25. For each input sample, we update the input buffer using FIFO, which begins at the end of the data buffer; the oldest sampled is kicked out first from the buffer and updated with the value from the upper location. When the FIFO completes, the first memory

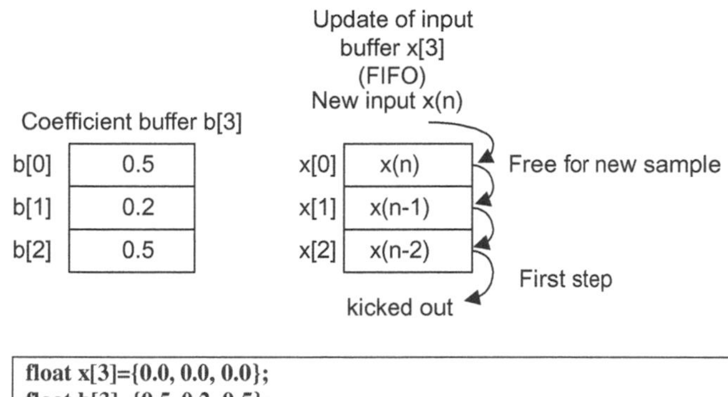


```
float x[3]={0.0, 0.0, 0.0};
float b[3]={0.5, 0.2, 0.5};
float y[1]={0.0};
interrupt void c_int11()
{
    float lc; /*left channel input */
    float rc; /*right channel input */
    float lcnew; /*left channel output */
    float rcnew; /*right channel output */
    int i;
//Left channel and right channel inputs
   AIC23_data.combo=input_sample();
   lc=(float) (AIC23_data.channel[LEFT]);
   rc= (float) (AIC23_data.channel[RIGHT]);
// Insert DSP algorithm below
   for(i=2; i>0; i--)  /* Update the input buffer x[3] */
   {  x[i]=x[i-1]; }
   x[0]= (float) lc; /*Input from the left channel */
   y[0]=0;
    for(i=0; i<3; i++)
   {  y[0]=y[0]+b[i]*x[i]; } /* FIR filtering */
// End of the DSP algorithm
    lcnew=y[0];
    rcnew=y[0];
    AIC23_data.channel[LEFT]=(short) lcnew;
    AIC23_data.channel[RIGHT]=(short) rcnew;
    output_sample(AIC23_data.combo);
}
```

FIG. 14.25

Example of FIR filtering with linear buffer update.

location x[0] will be free to be used to store the current input sample. The segment of code in Fig. 14.25 explains implementation.

Note that in the code segment, $x[0]$ holds the current input sample $x(n)$, while $b[0]$ is the corresponding coefficient; $x[1]$ and $x[2]$ hold the past input samples $x(n-1)$ and $x(n-2)$, respectively; similarly, $b[1]$ and $b[2]$ are the corresponding coefficients.

Again, note that using the array and loop structures in the code segment is for simplicity in notations and assume that the reader is not familiar with the C pointers in the C language. The concern for simplicity has to do mainly with the DSP algorithm. More coding efficiency can be achieved using the C pointers and a circular buffer. The DSP-oriented coding implementation can be found in Kehtaranavaz and Simsek (2000) and Chassaing and Reay (2008).

IIR Filtering:

Similarly, we can implement an IIR filter. It requires an input buffer, which holds the current and past inputs; an output buffer, which holds the past outputs; a numerator coefficient buffer; and a denominator coefficient buffer. Considering the following IIR filter for implementation,

$$y(n)=0.5x(n)+0.7x(n-1)-0.5x(n-2)-0.4y(n-1)+0.6y(n-2),$$

we accommodate the numerator coefficient buffer b[3], the denominator coefficient buffer a[3], the input buffer x[3], and the output buffer y[3] shown in Fig. 14.26. The buffer updates for input x[3] and output y[3] are FIFO. The implementation is illustrated in the segment of code listed in Fig. 14.26.

Again, note that in the code segment, $x[0]$ holds the current input sample, while $y[0]$ holds the current processed output, which will be sent to the DAC unit for conversion. The coefficient $a[0]$ is never modified in the code. We keep that for a purpose of notation simplicity and consistency during the programming process.

Digital Oscillation with IIR Filtering:

The principle for generating digital oscillation is described in Chapter 8, where the input to the digital filter is the impulse sequence, and the transfer function is obtained by applying the z-transform of the digital sinusoid function. Applications can be found in dual-tone multifrequency (DTMF) tone generation, digital carrier generation for communications, and so on. Hence, we can modify the implementation of IIR filtering for tone generation with the input generated internally instead of using the ADC channel.

Let us generate an 800 Hz tone with the digital amplitude of 5000. According to the section in Chapter 8 ("Applications: Generation and Detection of DTMF Tones Using the Goertzel Algorithm"), the transfer function, difference equation, and the impulse input sequence are found to be, respectively,

$$H(z)=\frac{0.587785z^{-1}}{1-1.618034z^{-1}+z^{-2}}$$

$$y(n)=0.587785x(n-1)+1.618034y(n-1)-y(n-2)$$

$$x(n)=5000\delta(n).$$

We define the numerator coefficient buffer b[2], the denominator coefficient buffer a[3], the input buffer x[2], and the output buffer y[3], shown in Fig. 14.27, which also shows the modified implementation for tone generation.

Initially, we set $x[0]=5000$. Then it will be updated with $x[0]=0$ for each current processed output sample $y[0]$.

14.6.4 SAMPLE C PROGRAMS

Floating-Point Implementation Example:

Real-time DSP implementation using a float-point processor is easy to program. The overflow problem hardly ever occurs. Therefore, we do not need to consider scaling factors, as described in the last section. The code segment shown in Fig. 14.28 demonstrates the simplicity of coding the floating-point IIR filter using the direct-form I structure.

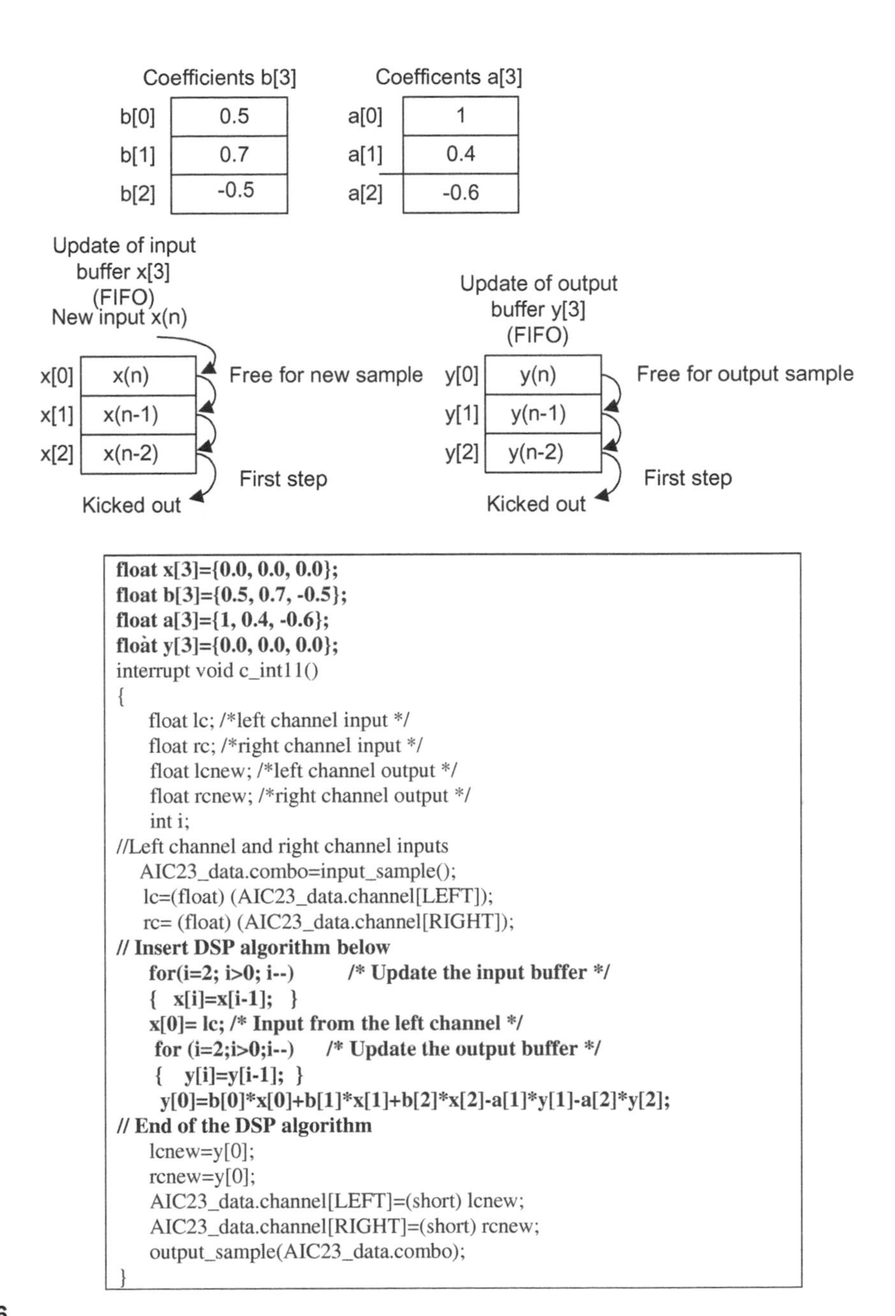


```
float x[3]={0.0, 0.0, 0.0};
float b[3]={0.5, 0.7, -0.5};
float a[3]={1, 0.4, -0.6};
float y[3]={0.0, 0.0, 0.0};
interrupt void c_int11()
{
    float lc; /*left channel input */
    float rc; /*right channel input */
    float lcnew; /*left channel output */
    float rcnew; /*right channel output */
    int i;
//Left channel and right channel inputs
   AIC23_data.combo=input_sample();
   lc=(float) (AIC23_data.channel[LEFT]);
   rc= (float) (AIC23_data.channel[RIGHT]);
// Insert DSP algorithm below
    for(i=2; i>0; i--)          /* Update the input buffer */
    {  x[i]=x[i-1];  }
    x[0]= lc; /* Input from the left channel */
     for (i=2;i>0;i--)      /* Update the output buffer */
     {   y[i]=y[i-1]; }
      y[0]=b[0]*x[0]+b[1]*x[1]+b[2]*x[2]-a[1]*y[1]-a[2]*y[2];
// End of the DSP algorithm
    lcnew=y[0];
    rcnew=y[0];
    AIC23_data.channel[LEFT]=(short) lcnew;
    AIC23_data.channel[RIGHT]=(short) rcnew;
    output_sample(AIC23_data.combo);
}
```

FIG. 14.26

Example of IIR filtering using linear buffer update.

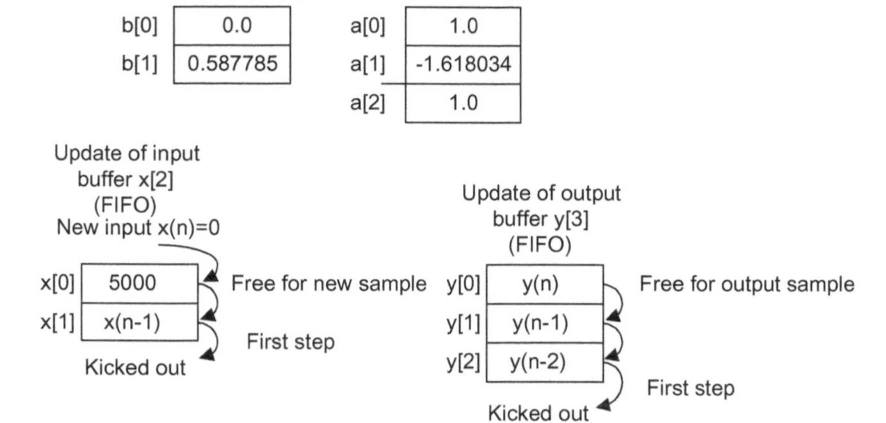


```
float x[2]={5000, 0.0}; /*initialize the impulse input */
float b[2]={0.0, 0.587785};
float a[3]={1.0, -1.618034,  1.0};
float y[3]={0.0};
interrupt void c_int11()
{    float lc; /*left channel input */
     float rc; /*right channel input */
     float lcnew; /*left channel output */
     float rcnew; /*right channel output */
     int i;
//Left channel and right channel inputs
   AIC23_data.combo=input_sample();
   lc=(float) (AIC23_data.channel[LEFT]);
   rc= (float) (AIC23_data.channel[RIGHT]);
// Insert DSP algorithm below
     y[0]=b[0]*x[0]+b[1]*x[1]+b[2]*x[2]-a[1]*y[1]-a[2]*y[2];
     for(i=1; i>0; i--)    /* Update the input buffer with zero input */
     {  x[i]=x[i-1];  }
      x[0]= 0;
      for (i=2;i>0;i--)    /* Update the output buffer */
      {   y[i]=y[i-1]; }
// End of the DSP algorithm
     lcnew=y[0];
     rcnew=y[0];
     AIC23_data.channel[LEFT]=(short) lcnew;
     AIC23_data.channel[RIGHT]=(short) rcnew;
     output_sample(AIC23_data.combo);
}
```

FIG. 14.27

Example of IIR filtering using linear buffer update and the impulse sequence input.

```
float a[5]={1.00, -2.1192, 2.6952, -1.6924, 0.6414};
float b[5]={0.0201, 0.00, -0.0402, 0.00, 0.0201};
float x[5]={0.0, 0.0, 0.0, 0.0, 0.0};
float y[5]={0.0, 0.0, 0.0, 0.0, 0.0};

interrupt void c_int11()
{
    float lc; /*left channel input */
    float rc; /*right channel input */
    float lcnew; /*left channel output */
    float rcnew; /*right channel output */
    int i;
//Left channel and right channel inputs
   AIC23_data.combo=input_sample();
   lc=(float) (AIC23_data.channel[LEFT]);
   rc= (float) (AIC23_data.channel[RIGHT]);
// Insert DSP algorithm below
    for(i=4; i>0; i--)       /* Update the input buffer */
    {
        x[i]=x[i-1];
    }
    x[0]= lc; /* Input from the left channel */
     for (i=4;i>0;i--)     /* Update the output buffer */
     {
          y[i]=y[i-1];
      }
      y[0]=b[0]*x[0]+b[1]*x[1]+b[2]*x[2]+b[3]*x[3]+b[4]*x[4]-a[1]*y[1]-a[2]*y[2]-a[3]*y[3]-a[4]*y[4];
// End of the DSP algorithm
    lcnew=y[0];
    rcnew=y[0];
    AIC23_data.channel[LEFT]=(short) lcnew;
    AIC23_data.channel[RIGHT]=(short) rcnew;
    output_sample(AIC23_data.combo);
}
```

FIG. 14.28

Sample C code for IIR filtering (floating-point implementation).

Fixed-Point Implementation Example:

When the execution time is critical, the fixed-point implementation is preferred in a floating-point processor. We implement the following IIR filter with a unit passband gain in direct-form II:

$$H(z) = \frac{0.0201 - 0.0402z^{-2} + 0.0201z^{-4}}{1 - 2.1192z^{-1} + 2.6952z^{-2} - 1.6924z^{-3} + 0.6414z^{-4}}$$

$$w(n) = x(n) + 2.1192w(n-1) - 2.6952w(n-2) + 1.6924w(n-3) - 0.6414w(n-4)$$

$$y(n) = 0.0201w(n) - 0.0402w(n-2) + 0.0201w(n-4).$$

Using MATLAB to calculate the scale factor S, it follows that.

```
» h=impz([1][1 -2.1192 2.6952 -1.6924 0.6414]);
» sf=sum(abs(h))
  sf =28.2196
```

Table 14.4 Filter Coefficients in Q-15 Format

IIR Filter	Filter Coefficients	Q-15 Format (Hex)
$-a_1$	0.5298	0x43D0
$-a_2$	−0.6738	0xA9C1
$-a_3$	0.4230	0x3628
$-a_4$	−0.16035	0xEB7A
b_0	0.0201	0x0293
b_1	0.0000	0x0000
b_2	−0.0402	0xFADB
b_3	0.0000	0x000
b_4	0.0201	0x0293

Hence we choose $S=32$. To scale the filter coefficients in the Q-15 format, we use the factors $A=4$ and $B=1$. Then the developed DSP equations are

$$x_s(n)=x(n)/32$$

$$w_s(n)=0.25x_s(n)+0.5298w_s(n-1)-0.6738w_s(n-2)+0.4231w_s(n-3)-0.16035w_s(n-4)$$

$$w(n)=4w_s(n)$$

$$y_s(n)=0.0201w(n)-0.0402w(n-2)+0.0201w(n-4)$$

$$y(n)=32y_s(n).$$

Using the method described in Section 14.5, we can convert filter coefficients into the Q-15 format; each coefficient is listed in Table 14.4.

The list of codes for the fixed-point implementation is displayed in Fig. 14.29, and some coding notations are given in Fig. 14.30.

Note that this chapter has provided only basic concepts and an introduction to real-time DSP implementation. The coding detail and real-time DSP applications will be treated in a separate DSP course, which deals with real-time implementations.

14.7 ADDITIONAL REAL-TIME DSP EXAMPLES

In this section, we examine more examples of real-time DSP implementations.

14.7.1 ADAPTIVE FILTERING USING THE TMS320C6713 DSK

The implementation for system modeling is discussed in Section 9.3 and is depicted in Fig. 14.31, where the input is fed from a function generator. The unknown system is a bandpass filter with the lower cutoff frequency of 1400 Hz and upper cutoff frequency of 1600 Hz. As shown in Fig. 14.31, the left input channel (Left Line In [LCI]) is used for the input while the left output channel (Left Line Out [LCO]) and the right output channel (Right Line Out [RCO]) are designated as the system output

```
/*float a[5]={1.00, -2.1192, 2.6952, -1.6924, 0.6414}; float b[5]={0.0201, 0.00, -0.0402, 0.00, 0.0201};*/
short a[5]={0x2000, 0x43D0, 0xA9C1, 0x3628, 0xEB7A}; /* coefficients in Q-15 format */
short b[5]={0x0293, 0x0000, 0xFADB, 0x0000, 0x0293};
int w[5]={0, 0, 0, 0, 0};
int sample;

interrupt void c_int11()
{
    float lc; /*left channel input */
    float rc; /*right channel input */
    float lcnew; /*left channel output */
    float rcnew; /*right channel output */
    int i, sum=0;
//Left channel and right channel inputs
  AIC23_data.combo=input_sample();
  lc= (float) (AIC23_data.channel[LEFT]);
  rc= (float) (AIC23_data.channel[RIGHT]);
// Insert DSP algorithm below
  sample = (int) lc; /*input sample from the left channel*/
  sample = (sample << 16); /* move to high 16 bits */
  sample = (sample>>5); /* scaled down by 32 to avoid overflow */
  for (i=4;i>0;i--)
  {
   w[i]=w[i-1];
  }
  sum= (sample >> 2); /* scaled down by 4 to use Q-15 */
  for (i=1;i<5;i++)
  {
   sum += (_mpyhl(w[i],a[i])) <<1;
  }
  sum = (sum <<2); /* scaled up by 4 */
  w[0]=sum;
  sum =0;
  for(i=0;i<5;i++)
  {
   sum += (_mpyhl(w[i],b[i]))<<1;
  }
  sum = (sum << 5); /* scaled up by 32 to get y(n) */
  sample= (sum>>16); /* move to low 16 bits */
// End of the DSP algorithm
    lcnew=sample;
    rcnew=sample;
    AIC23_data.channel[LEFT]=(short) lcnew;
    AIC23_data.channel[RIGHT]=(short) rcnew;
    output_sample(AIC23_data.combo);
}
```

FIG. 14.29

Sample C code for IIR filtering (fixed-point implementation).

short coefficient; declaration of 16 bit signed integer
int sample, result; declaration of 32 bit signed integer
MPYHL assembly instruction (signed multiply high low 16 MSB x 16 LSB)
result = (_mpyhl(sample,coefficient)) <<1;
sample must be shifted left by 16 bits to be stored in the high 16 MSB.
coefficient is the 16 bit data to be stored in the low 16 LSB.
result is shifted left by one bit to get rid of the extended sign bit, and high 16 MSB's are designated for the processed data.
Final **result** will be shifted down to right by 16 bits before DAC conversion.
sample = (result>>16);

FIG. 14.30

Some coding notations for the Q-15 fixed-point implementation.

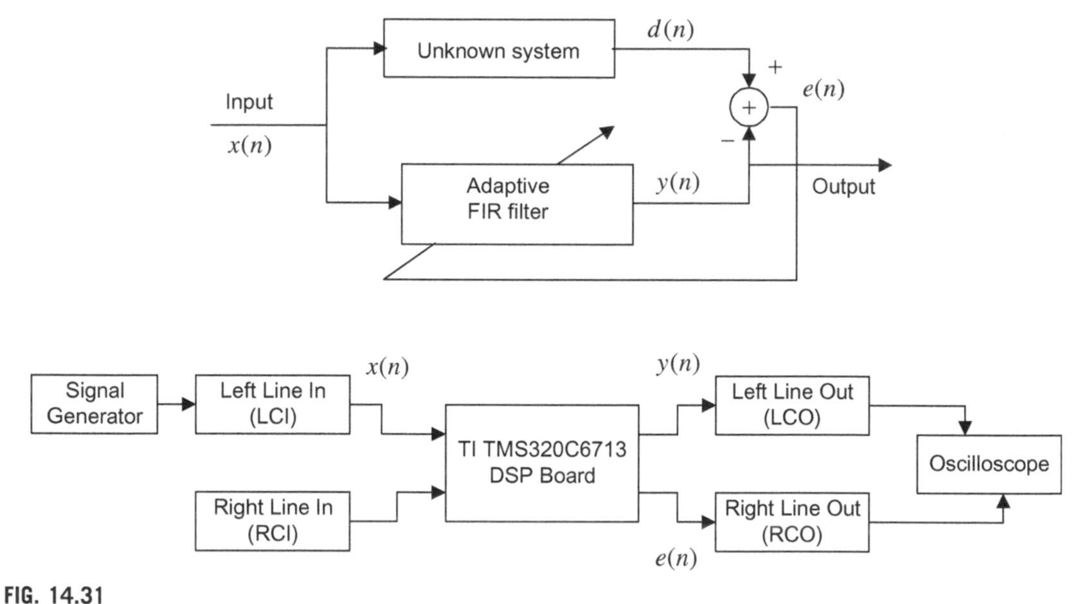

FIG. 14.31

Setup for system modeling using the LMS adaptive filter.

and error output, respectively. Note that the right input channel (Right Line In [RCI]) is not used. When the input frequency is swept from 200 to 3000 Hz, the output shows the maximum peak when the input frequency is dialed to around 1500 Hz. Hence, the adaptive filter acts like the unknown system. Fig. 14.32 lists the sample program segment.

With the advantage of the stereo input and output channels, we can conduct system modeling for an unknown analog system illustrated in Fig. 14.33, where RCI is used to feed the unknown analog system output to the DSK. The program segment is listed in Fig. 14.34.

Fig. 14.35A shows an example of an adaptive noise reduction system. Fig. 14.35B shows the details of the adaptive noise cancellation. The first DSP board is used to create the real-time corrupted signal which is obtained by mixing the mono audio source (Left Line In [LCI1]) from any audio device and the tonal noise (Right Line In [RCI1]) generated from a function generator. The output (Left line Out

```
/*Numerator coefficients */
/*for the bandpass filter (unknown system) with fL=1.4 kHz, fH=1.6 kHz*/
float b[5]={ 0.005542761540433, 0.000000000000002, -0.011085523080870,
             0.000000000000003  0.005542761540431};
/*Denominator coefficients */
/*for the bandpass filter (unknown system) with fL=1.4 kHz, fH=1.6 kHz*/
float a[5]={ 1.000000000000000, -1.450496619180500. 2.306093105231476,
            -1.297577189144526  0.800817049322883};
float x[40]={0,0,0,0,0,0,0,0,0,0,0,0,0,0,0,0,0,0,0,0,
             0,0,0,0,0,0,0,0,0,0,0,0,0,0,0,0,0,0,0,0}; /*Reference input buffer*/
float w[40]={0,0,0,0,0,0,0,0,0,0,0,0,0,0,0,0,0,0,0,0,
             0,0,0,0,0,0,0,0,0,0,0,0,0,0,0,0,0,0,0,0}; /*Adaptive filter coefficients*/
float d[5]={0.0, 0.0, 0.0, 0.0, 0.0}; /*Unknown system output */
float y[1]={0.0}; /* Adaptive filter output */
float e[1]={0.0}; /* Error signal */
float mu=0.000000000002; /*Adaptive filter convergence factor*/
interrupt void c_int11()
{
    float lc; /*left channel input */
    float rc; /*right channel input */
    float lcnew; /*left channel output */
    float rcnew; /*right channel output */
    int i;
//Left channel and right channel inputs
   AIC23_data.combo=input_sample();
   lc=(float) (AIC23_data.channel[LEFT]);
   rc= (float) (AIC23_data.channel[RIGHT]);
// Insert DSP algorithm below
   for(i=39;i>0;i--) /*Update the input buffer*/
   { x[i]=x[i-1]; }
    x[0]=lc;
   for(i=4; i>0; i--)
   { d[i]=d[i-1]; }
   d[0]=b[0]*x[0]+ b[1]*x[1]+ b[2]*x[2]+ b[3]*x[3]+ b[4]*x[4]
        -a[1]*d[1]-a[2]*d[2]- a[3]*d[3]- a[4]*d[4]; /*Unknown system output*/
// Adaptive filter
   y[0]=0;
   for(i=0;i<40; i++)
   { y[0]=y[0]+w[i]*x[i];}
    e[0]=d[0]-y[0]; /* Error output */
    for(i=0;i<40; i++)
   { w[i]=w[i]+2*mu*e[0]*x[i];} /* LMS algorithm */
// End of the DSP algorithm
    lcnew=y[0]; /* Send the tracked output */
    rcnew=e[0]; /* Send the error signal*/
    AIC23_data.channel[LEFT]=(short) lcnew;
    AIC23_data.channel[RIGHT]=(short) rcnew;
    output_sample(AIC23_data.combo);
}
```

FIG. 14.32

Program segment for system modeling.

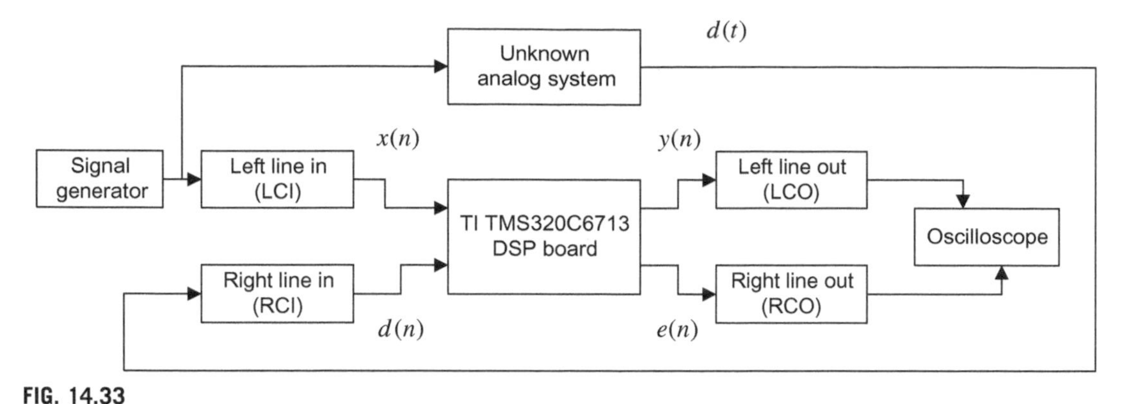

FIG. 14.33

System modeling using an LMS adaptive filter.

```
float x[40]={0,0,0,0,0,0,0,0,0,0,0,0,0,0,0,0,0,0,0,0,
             0,0,0,0,0,0,0,0,0,0,0,0,0,0,0,0,0,0,0,0}; /*Reference input buffer*/
float w[40]={0,0,0,0,0,0,0,0,0,0,0,0,0,0,0,0,0,0,0,0,
             0,0,0,0,0,0,0,0,0,0,0,0,0,0,0,0,0,0,0,0}; /*Adaptive filter coefficients*/
float d[1]={0.0}; /*Unknown system output */
float y[1]={0,0}; /* Adaptive filter output */
float e[1]={0.0}; /* Error signal */
float mu=0.000000000002; /*Adaptive filter convergence factor*/
interrupt void c_int11()
{
    float lc; /*left channel input */
    float rc; /*right channel input */
    float lcnew; /*left channel output */
    float rcnew; /*right channel output */
    int i;
//Left channel and right channel inputs
   AIC23_data.combo=input_sample();
   lc=(float) (AIC23_data.channel[LEFT]);
   rc= (float) (AIC23_data.channel[RIGHT]);
// Insert DSP algorithm below
   for(i=39;i>0;i--) /*Update the input buffer*/
   { x[i]=x[i-1]; }
   x[0]=lc;
   d[0]=rc;  /*Unknown system output*/
// Adaptive filter
   y[0]=0;
   for(i=0;i<40; i++)
   { y[0]=y[0]+w[i]*x[i];}
   e[0]=d[0]-y[0]; /* Error output */
   for(i=0;i<40; i++)
   { w[i]=w[i]+2*mu*e[0]*x[i];} /* LMS algorithm */
// End of the DSP algorithm
    lcnew=y[0]; /* Send the tracked output */
    rcnew=e[0]; /* Send the error signal*/
    AIC23_data.channel[LEFT]=(short) lcnew;
    AIC23_data.channel[RIGHT]=(short) rcnew;
    output_sample(AIC23_data.combo);
}
```

FIG. 14.34

Program segment for modeling an analog system.

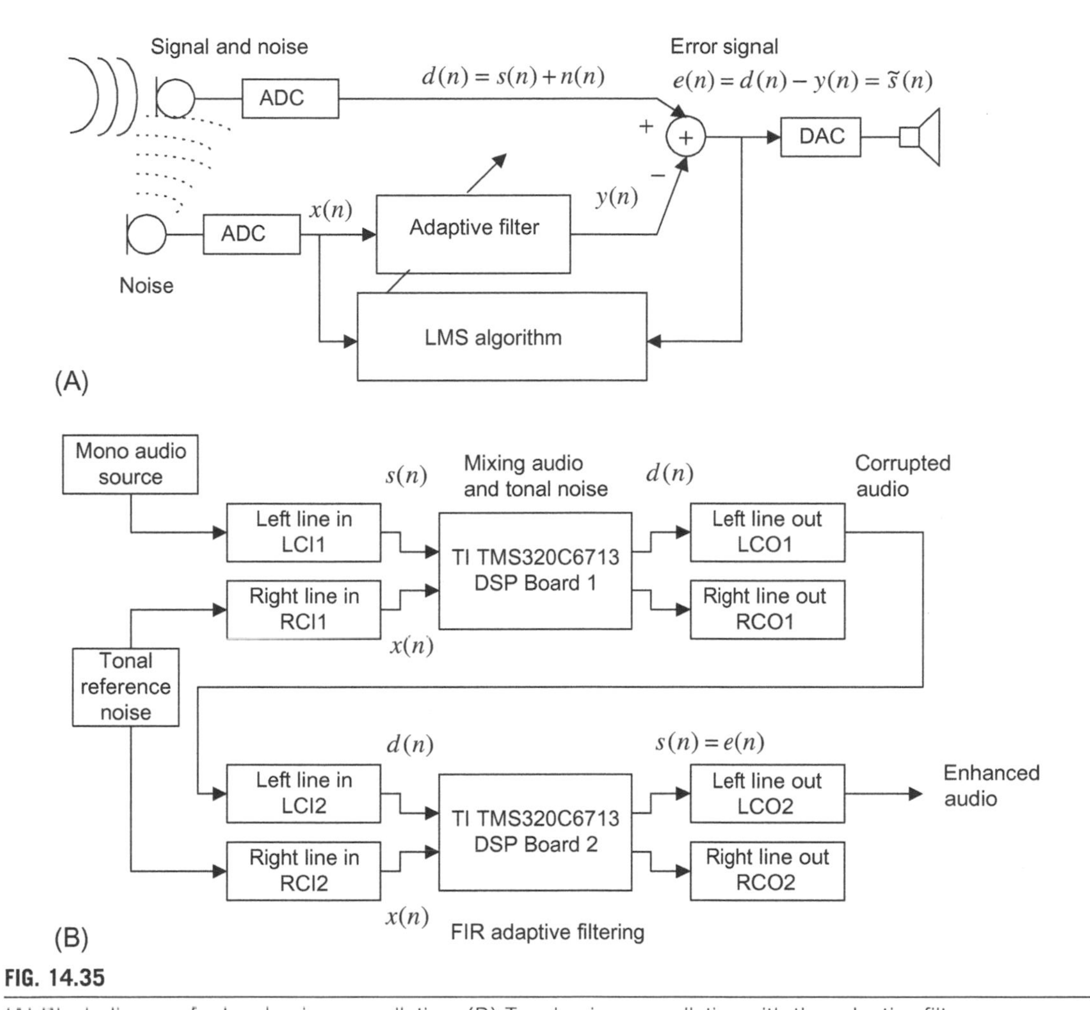

FIG. 14.35

(A) Block diagram for tonal noise cancellation. (B) Tonal noise cancellation with the adaptive filter.

[LCO1]) is the corrupted signal, which is fed to the second DSP board for noise cancellation application. The adaptive FIR filter in the second DSP board uses the reference input (Right Line In [RCI2]) to generate the output, which is used to cancel the tonal noise embedded in the corrupted signal (Left Line In [LCI2]). The output (Left Line Out [LCO2]) produces the clean mono audio signal (Jiang and Tan; 2012). Fig. 14.36 details the implementation.

Many other practical configurations can be implemented similarly.

14.7.2 SIGNAL QUANTIZATION USING THE TMS320C6713 DSK

Linear quantization (see Chapter 10) can be implemented as shown in Fig. 14.37. The program listed in Fig. 14.37 only demonstrates the left channel coding while the right channel coding can be easily extended. The program consists of both encoder and decoder. First, it converts the 16-bit 2's complement

```
float x[1]={0.0};  /* Tonal reference noise */
Float s[1]={0,0}; /* Audio signal */
float d[1]={0.0}; /* Corrupted signal*/
interrupt void c_int11()
{
    float lc; /*left channel input */
    float rc; /*right channel input */
    float lcnew; /*left channel output */
    float rcnew; /*right channel output */
    int i;
//Left channel and right channel inputs
   AIC23_data.combo=input_sample();
   lc=(float) (AIC23_data.channel[LEFT]);
   rc= (float) (AIC23_data.channel[RIGHT]);
// Insert DSP algorithm below
   s[0]=lc;
   x[0]=rc;
   d[0]=s[0]+x[0];
// End of the DSP algorithm
    lcnew=d[0]; /* Send to DAC */
    rcnew=rc; /* keep the original data */
    AIC23_data.channel[LEFT]=(short) lcnew;
    AIC23_data.channel[RIGHT]=(short) rcnew;
    output_sample(AIC23_data.combo);
}
```

(A) Program segment for DSK1 (generation of the corrupted signal).

```
float x[20]={0,0,0,0,0,0,0,0,0,0,0,0,0,0,0,0,0,0,0,0}; /*Reference input buffer*/
float w[20]={0,0,0,0,0,0,0,0,0,0,0,0,0,0,0,0,0,0,0,0}; /*Adaptive filter coefficients*/
float d[1]={0.0}; /* Corrupted signal*/
float y[1]={0,0}; /* Adaptive filter output */
float e[1]={0.0}; /* Enhanced signal */
float mu=0.000000000004; /*Adaptive filter convergence factor*/
interrupt void c_int11()
{
    float lc; /*left channel input */
    float rc; /*right channel input */
    float lcnew; /*left channel output */
    float rcnew; /*right channel output */
    int i;
//Left channel and right channel inputs
   AIC23_data.combo=input_sample();
   lc=(float) (AIC23_data.channel[LEFT]);
   rc= (float) (AIC23_data.channel[RIGHT]);
// Insert DSP algorithm below
  d[0]=lc; /*Corrupted signal*/
   for(i=19;i>0;i--) /*Update the reference noise buffer  input buffer*/
   { x[i]=x[i-1]; }
   x[0]=rc;
// Adaptive filter
  y[0]=0;
  for(i=0;i<20; i++)
  { y[0]=y[0]+w[i]*x[i];}
   e[0]=d[0]-y[0]; /* Enhanced output */
   for(i=0;i<20; i++)
  {  w[i]=w[i]+2*mu*e[0]*x[i]; }/* LMS algorithm */
// End of the DSP algorithm
    lcnew=e[0]; /* Send to DAC */
    rcnew=rc; /* keep the original data */
    AIC23_data.channel[LEFT]=(short) lcnew;
    AIC23_data.channel[RIGHT]=(short) rcnew;
    output_sample(AIC23_data.combo);
}
```

(B) Program segment for DSK 2 (LMS adaptive filter).

FIG. 14.36

Program segments for noise cancellation. (A) Program segment for DSK1 (generation of the corrupted signal). (B) Program segment for DSK 2 (LMS adaptive filter).

```
int PCMcode;
/* Sign-magnitude format: s magnitude bits, see Section 10.1,in Chapter 10 */
/* 16-bit PCM code  5-bit quantization  recovered 16-bit PCM */
// See the following example:
/*  16-bit sign-magnitude code      5-bit PCM code      16-bit recovered sign-magnitude code */
/*  sADCDEFGHIJKLMNO            sABCDE             sABCD10000000000   */
/* Note that, convert the two's complement form to the sign-magnitude form   */
/* before quantizing and at decoder site, convert                              */
/* the sign-magnitude form to the two's complement form for DAC              */
int nofbits=5; // specify the number of quantization bits
int sign[16]={0x01,0x02,0x04,0x08,0x10,0x20,0x40,0x80,
        0x100,0x200,0x400,0x800,0x1000,0x2000,0x4000,0x8000}; // add sign bit (MSB)
int mask[16]={0x0,0x01,0x03,0x07,0x0f,0x1f,0x3f,0x7f,
        0xff,0x1ff,0x3ff,0x7ff,0xfff,0x1fff,0x3fff,0x7fff};// mask for obtaining magnitude bits
int rec[16]={0x4000,0x2000,0x1000,0x0800,
            0x0400,0x0200,0x0100,0x0080,
            0x0040,0x0020,0x0010,0x0008,
            0x0004,0x0002,0x0001,0x0000};// the mid-point of the quantization interval
interrupt void c_int11()
{
    int lc; /*left channel input */
    int rc; /*right channel input */
    int lcnew; /*left channel output */
    int rcnew; /*right channel output */
    int temp, dec;
//Left channel and right channel inputs
   AIC23_data.combo=input_sample();
   lc=(int) (AIC23_data.channel[LEFT]);
   rc= (int) (AIC23_data.channel[RIGHT]);
// Insert DSP algorithm below
 /* Encoder :*/
 tmp =lc; // for the Left Line In channel
   if (tmp <0 )
   {
     tmp=-tmp;  // get magnitude bits to work with
   }
   tmp =(tmp>>(16-nofbits));
   PCMcode=tmp&mask[nofbits-1]; // get magnitude bits
  if (lc>=0)
  {
     PCMcode= PCMcode | sign[nofbits-1];  // add sign bit
  }
 /* PCM code (stored in the lower portion) */
 /* Decoder: */
    dec = PCMcode&mask[nofbits-1];  // obtain magnitude bits
    tmp= PCMcode&sign[nofbits-1]; // obtain the sign bit
    dec = (dec<<(16-nofbits)); // scale to 15 bit magnitude
    dec = dec | rec[nofbits-1]; // recovering the mid-point of the quantization interval
  lc =dec;
  if (tmp == 0x00)
  {
    lc =-dec;   // back to the two's complement form (change the sign)
  }
// End of the DSP algorithm
      lcnew=lc; /* Send to DAC */
      rcnew=lc;
      AIC23_data.channel[LEFT]=(short) lcnew;
      AIC23_data.channel[RIGHT]=(short) rcnew;
      output_sample(AIC23_data.combo);
}
```

FIG. 14.37

Encoding and decoding using the TMS320C6713 DSK.

EXAMPLE 14.16

Given a 16-bit datum: lc = −2050 (decimal), convert it to the 5-bit linear PCM code using information listed in Fig. 14.37.

Solution:

1. Encoding: We use pseudo code for illustration.

```
tmp = 2050 (decimal)=0x0802 (Hex)= 0000 1000 0000 0010 (binary)
After shifting 11 bits, tmp=0x0001 (Hex)
PCMcode =tmp&mask[5-1]=0x0001&0x000f=0x0001 (Hex) // Get magnitude bits
if lc >0 {
 PCMcode= PCMcode | sign[5-1] // add positive sign bit
} // This line is not executed since lc<0
 PCMcode=00001 (binary)
```

2. Decoding: We explain the decoding process via the pseudo code.

```
dec=PCMcode& mask[4]=0x0001& 0x000f=0x0001 (Hex)
tmp=PCMcode&sign[4]=0x0001&0x0010=0x0000 (the number is negative)
dec = dec<<11=0x0800
dec =dec | rec[4]=0x0800| 0x0400=0x0C00 (Hex)=3072 (decimal)
lc=-3072 (recovered decimal)
```

The same procedure can be followed for coding a positive decimal number.

data to the sign-magnitude format with truncated magnitude bits as required. Then the decoder converts the compressed PCM code back to the 16-bit data. The encoding and decoding can be explained in Example 14.16.

Fig. 14.38 demonstrates codes for digital μ-law encoding and decoding. It converts a 12-bit linear PCM code to an 8-bit compressed PCM code using the principles discussed in Section 10.2. Note that the program only performs left-channel coding.

14.7.3 SAMPLING RATE CONVERSION USING THE TMS320C6713 DSK

Downsampling by an integer factor of M using the TMS320C6713 is depicted in Fig. 14.39. The idea is that we set up the DSK running at the original sampling rate and update the DAC channel once for every M samples. The program (Tan and Jiang; 2008) is shown in Fig. 14.40.

Upsampling by an integer factor of L using the TMS320C6713 is depicted in Fig. 14.41. Again, we set up the DSK running at the upsampling rate f_{sL} and acquire a sample from the ADC channel once for every L samples. The program (Tan and Jiang, 2008) is listed in Fig. 14.42.

```
int ulawcode;
/* Digital mu-law definition*/
// Sign-magnitude format: s segment quantization
// s=1 for the positive value, s =0 for the negative value
// Segment defines compression
// quantization with 16 levels
// See Section 11.2, Chapter 11
/* Segment    12-bit PCM      4-bit quantization interval  */
/*  0         s0000000ABCD    s000ABCD                     */
/*  1         s0000001ABCD    s001ABCD                     */
/*  2         s000001ABCDX    s010ABCD                     */
/*  3         s00001ABCDXX    s011ABCD                     */
/*  4         s0001ABCDXXX    s100ABCD                     */
/*  5         s001ABCDXXXX    s101ABCD                     */
/*  6         s01ABCDXXXXX    s110ABCD                     */
/*  7         s1ABCDXXXXXX    s111ABCD                     */
//
/* segment    recovered 12-bit PCM       */
/*  0         s0000000ABCD               */
/*  1         s0000001ABCD               */
/*  2         s000001ABCD1               */
/*  3         s00001ABCD10               */
/*  4         s0001ABCD100               */
/*  5         s001ABCD1000               */
/*  6         s01ABCDq0000               */
/*  7         s1ABCD100000               */
/* Note that, convert the two's complement form to the sign-magnitude form   */
/* before quantizing and at decoder site, convert                            */
/* the sign-magnitude form to the two's complement form for DAC              */
interrupt void c_int11()
{
    int lc; /*left channel input */
    int rc; /*right channel input */
    int lcnew; /*left channel output */
    int rcnew; /*right channel output */
    int tmp,ulawcode, dec;
//Left channel and right channel inputs
   AIC23_data.combo=input_sample();
   lc=(int) (AIC23_data.channel[LEFT]);
   rc= (int) (AIC23_data.channel[RIGHT]);
```

FIG. 14.38

Digital μ-law encoding and decoding.

(continued)

```
// Insert DSP algorithm below
 /* Encoder :*/
 tmp =lc;
 if (tmp <0 )
 { tmp=-tmp; // Get magnitude bits to work with
 }
 tmp =(tmp>>4); // Linear scale down to 12 bits to use the u-255 law table
  if( (tmp&0x07f0)==0x0)  // Segment 0
  {  ulawcode= (tmp&0x000f); }
  if( (tmp&0x07f0)==0x0010) // Segment 1
  {   ulawcode= (tmp&0x00f);
      ulawcode= ulawcode | 0x10; }
  if( (tmp&0x07E0)==0x0020)  // Segment 2
  {   ulawcode= (tmp&0x001f)>>1;
      ulawcode = ulawcode |0x20; }
  if( (tmp&0x07c0)==0x0040)  // Segment 3
  {  ulawcode= (tmp&0x003f)>>2;
      ulawcode= ulawcode | 0x30; }
  if( (tmp&0x0780)==0x0080) // Segment 4
  {  ulawcode= (tmp&0x007f)>>3;
      ulawcode =ulawcode | 0x40;}
  if( (tmp&0x0700)==0x0100) // Segment 5
  {  ulawcode= (tmp&0x00ff)>>4;
      ulawcode = ulawcode | 0x50;}
  if( (tmp&0x0600)==0x0200)  // Segment 6
  {  ulawcode= (tmp&0x01ff)>>5;
      ulawcode = ulawcode | 0x60; }
  if( (tmp&0x0400)==0x0400) // Segment 7
  {  ulawcode= (tmp&0x03ff)>>6;
      ulawcode=ulawcode | 0x70; }
   if (lc>=0)
   {
      ulawcode= ulawcode|0x80;
   }
  /* u-law code (8 bit compressed PCM code) for transmission and storage */
  /* Decoder: */
     tmp = ulawcode&0x7f;
     tmp = (tmp>>4);
  if ( tmp == 0x0) // Segment 0
  { dec = ulawcode&0xf; }
  if ( tmp == 0x1)  // Segment 1
  { dec = ulawcode&0xf | 0x10; }
  if ( tmp == 0x2)   // Segment 2
  {  dec = ((ulawcode&0xf)<<1) | 0x20;
      dec= dec |0x01; }
  if ( tmp == 0x3)   // Segment 3
  {   dec = ((ulawcode&0xf)<<2) | 0x40;
      dec = dec|0x02; }
  if ( tmp == 0x4)   // Segment 4
  {   dec = ((ulawcode&0xf)<<3) | 0x80;
      dec = dec|0x04; }
  if ( tmp == 0x5)   // Segment 5
  {  dec = ((ulawcode&0xf)<<4) | 0x0100;
      dec = dec|0x08; }
  if ( tmp == 0x6)   // Segment 6
```

FIG. 14.38, CONT'D

```
{ dec = ((ulawcode&0xf)<<5) | 0x0200;
  dec =dec|0x10; }
if ( tmp == 0x7)  // Segment 7
{ dec = ((ulawcode&0xf)<<6) | 0x0400;
  dec = dec |0x20; }
 tmp =ulawcode & 0x80;
lc =dec;
if (tmp == 0x00)
{ lc=-dec;  // Back to 2's complement form
}
 lc= (lc<<4); // Linear scale up to 16 bits
// End of the DSP algorithm
   lcnew=lc; /* Send to DAC */
   rcnew=lc;
   AIC23_data.channel[LEFT]=(short) lcnew;
   AIC23_data.channel[RIGHT]=(short) rcnew;
   output_sample(AIC23_data.combo);
}
```

FIG. 14.38, CONT'D

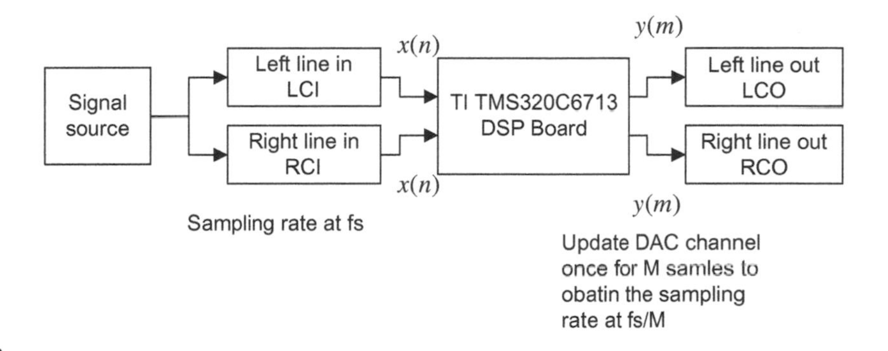

FIG. 14.39

Downsampling using the TMS320C6713.

```
int M=2;
int Mcount=0;
float x[67];
float y[1]={0.0};
interrupt void c_int11()
{
    float lc; /*Left channel input */
    float rc; /*Right channel input */
    float lcnew; /*Left channel output */
    float rcnew; /*Right channel output */
    int i,j;
    float sum;
//Left channel and right channel inputs
   AIC23_data.combo=input_sample();
   lc=(float) (AIC23_data.channel[LEFT]);
   rc= (float) (AIC23_data.channel[RIGHT]);
// Insert DSP algorithm below
   Mcount++;
   for(i=66; i>0; i--)  // Update input buffer
   { x[i]=x[i-1]; }
    x[0]=lc;  // Load new sample
    sum=0.0;
   for(i=0;i<67;i++)    // FIR filtering
   { sum=sum+x[i]*b[i]; }
   if (Mcount== M)
   {  Mcount=0;
      y[0]=sum; }  // Update DAC with processed sample (decimation)
// End of the DSP algorithm
    lcnew=y[0]; /* Send to DAC, need to apply the anti-image filter to remove the S&H effect*/
    rcnew=y[0];
    AIC23_data.channel[LEFT]=(short) lcnew;
    AIC23_data.channel[RIGHT]=(short) rcnew;
    output_sample(AIC23_data.combo);
}
```

FIG. 14.40

Downsampling implementation. (Anti-aliasing filter has 67 coefficients which stored in an array b[67].)

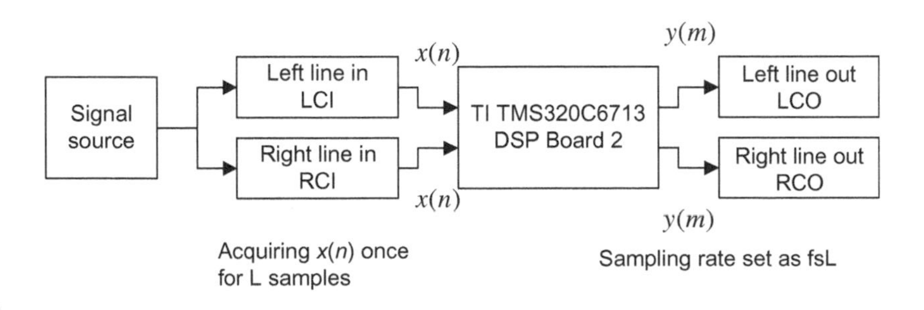

FIG. 14.41

Upsampling using the TMS320C6713.

```
int L=2;
int Lcount=0;
float x[67];
float y[1]={0.0};
interrupt void c_int11()
{
    float lc; /*Left channel input */
    float rc; /*Right channel input */
    float lcnew; /*Left channel output */
    float rcnew; /*Right channel output */
    int i,j;
//Left channel and right channel inputs
   AIC23_data.combo=input_sample();
   lc=(float) (AIC23_data.channel[LEFT]);
   rc= (float) (AIC23_data.channel[RIGHT]);
// Insert DSP algorithm below
   Lcount++;
   for(i=66; i>0; i--)  // Update input buffer with zeros
   { x[i]=x[i-1]; }
    x[0]=0;
   if (Lcount==L)
   {
     x[0]=lc;  // Load new sample for every L samples
     Lcount =0;
   }
   y[0]=0.0;
   for(i=0;i<67;i++)    // FIR filtering
   {  y[0]=y[0]+x[i]*b[i]; }
   y[0]=(float) L*y[0]
// End of the DSP algorithm
    lcnew=y[0]; /* Send to DAC */
    rcnew=y[0];
    AIC23_data.channel[LEFT]=(short) lcnew;
    AIC23_data.channel[RIGHT]=(short) rcnew;
    output_sample(AIC23_data.combo);
}
```

FIG. 14.42

Upsampling implementation. (Anti-image filter has 67 coefficients which stored in an array b[67].)

14.8 SUMMARY

1. The Von Neumann architecture consists of a single, shared memory for programs and data, a single bus for memory access, an arithmetic unit, and a program control unit. The Von Neumann processor operates fetching and execution cycles seriously.
2. The Harvard architecture has two separate memory spaces dedicated to program code and to data, respectively, two corresponding address buses, and two data buses for accessing two memory spaces. The Harvard processor offers fetching and executions in parallel.
3. The DSP special hardware units include an MAC dedicated to DSP filtering operations, a shifter unit for scaling and address generators for circular buffering.
4. The fixed-point DSP uses integer arithmetic. The data format Q-15 for the fixed-point system is preferred to avoid the overflows.

5. The floating-point processor uses the floating-point arithmetic. The standard floating-point formats include the IEEE single-precision and double-precision formats.
6. The architectures and features of fixed-point processors and floating-point processors were briefly reviewed.
7. Implementing digital filters in the fixed-point DSP system requires scaling filter coefficients so that the filters are in Q-15 format, and input scaling for adder so that overflow during the MAC operations can be avoided.
8. The floating-point processor is easy to code using the floating-point arithmetic and develop the prototype quickly. However, it is not efficient in terms of the number of instructions it has to complete compared with the fixed-point processor.
9. The fixed-point processor using fixed-point arithmetic takes much effort to code. But it offers the least number of the instructions for the CPU to execute.
10. Additional real-time DSP examples are provided, including adaptive filtering, signal quantization and coding, and sample rate conversion.

14.9 PROBLEMS

14.1 Find the signed Q-15 representation for a decimal number 0.2560123.

14.2 Find the signed Q-15 representation for a decimal number −0.2160123.

14.3 Find the signed Q-15 representation for a decimal number −0.3567921.

14.4 Find the signed Q-15 representation for a decimal number 0.4798762.

14.5 Convert the Q-15 signed number = 1.010101110100010 to a decimal number.

14.6 Convert the Q-15 signed number = 0.001000111101110 to a decimal number.

14.7 Convert the Q-15 signed number = 0.110101000100010 to a decimal number.

14.8 Convert the Q-15 signed number = 1.101000100101111 to a decimal number.

14.9 Add the following two Q-15 numbers:

$$1.101010111000001 + 0.010001111011010$$

14.10 Add the following two Q-15 numbers:

$$0.001010101000001 + 0.010101111010010$$

14.11 Add the following two Q-15 numbers:

$$1.001010101000001 + 1.010101111010010$$

14.12 Add the following two Q-15 numbers:

$$0.001010101000001 + 1.010101111010010$$

14.13 Convert each of the following decimal numbers to a floating-point number using the format specified in Fig. 14.10.

(a) 0.1101235
(b) −10.430527

14.14 Convert each of the following decimal numbers to a floating-point number using the format specified in Fig. 14.10.

(a) 2.5568921
(b) −0.678903
(c) 0.0000000
(d) −1.0000000

14.15 Add the following floating-point numbers whose formats are defined in Fig. 14.10, and determine the sum in the decimal format.

$$1101\,011100011011 + 0100\,101111100101.$$

14.16 Add the following floating-point numbers whose formats are defined in Fig. 14.10, and determine the sum in the decimal format.

$$0111\,110100011011 + 0101\,001000100101.$$

14.17 Add the following floating-point numbers whose formats are defined in Fig. 14.10, and determine the sum in the decimal format.

$$0001\,000000010011 + 0100\,001000000101.$$

14.18 Convert the following number in IEEE single precision format to a decimal format:

$$110100000.010\ldots0000.$$

14.19 Convert the following number in IEEE single precision format to a decimal format:

$$010100100.101\ldots0000.$$

14.20 Convert the following number in IEEE double precision format to a decimal format:

$$011000\ldots0.1010\ldots0000$$

14.21 Convert the following number in IEEE double precision format to a decimal format:

$$011000\ldots0.0110\ldots0000$$

14.22 Given the following FIR filter:

$$y(n) = -0.2x(n) + 0.6x(n-1) + 0.2x(n-2),$$

with a passband gain of 1 and the input being a full range, develop the DSP implementation equations in the Q-15 fixed-point system.

14.23 Given the following IIR filter,

$$y(n) = 0.6x(n) + 0.3y(n-1),$$

with a passband gain of 1 and the input being a full range, use the direct-form I method to develop the DSP implementation equations in the Q-15 fixed-point system.

14.24 Repeat Problem 14.23 using the direct-form II.

14.25 Given the following FIR filter:

$$y(n) = -0.36x(n) + 1.6x(n-1) + 0.36x(n-2),$$

with a passband gain of 2 and the input being half of range, develop the DSP implementation equations in the Q-15 fixed-point system.

14.26 Given the following IIR filter:

$$y(n) = 1.35x(n) + 0.3y(n-1),$$

with a passband gain of 2, and the input being half of range, use the direct-form I method to develop the DSP implementation equations in the Q-15 fixed-point system.

14.27 Repeat Problem 14.26 using the direct-form II.

14.28 Given the following IIR filter:

$$y(n) = 0.72x(n) + 1.42x(n-2) + 0.72x(n-2) - 1.35y(n-1) - 0.5y(n-2),$$

with a passband gain of 1 and a full range of input, use the direct-form I to develop the DSP implementation equations in the Q-15 fixed-point system.

14.29 Repeat Problem 14.28 using the direct-form II.

APPENDIX

INTRODUCTION TO THE MATLAB ENVIRONMENT

Matrix Laboratory (MATLAB) is used extensively in this book for simulations. The goal here is to help students acquire familiarity with MATLAB and build basic skills in the MATLAB environment. Hence, Appendix A serves the following objectives:

1. Learn to use the help system to study basic MATLAB command and syntax
2. Learn array indexing
3. Lean basic plotting utilities
4. Learn to write script M-files
5. Learn to write functions in MATLAB.

A.1 BASIC COMMANDS AND SYNTAX

MATLAB has an accessible help system by the help command. By issuing the MATLAB help command following the topic or function (routine), MATLAB will return the text information of the topic and show how to use the function. For example, by entering **help** on the MATLAB prompt to question a function **sum()**, we see the text information (listed partially here) to show how to use the MATLAB function **sum()**.

```
» help sum
 SUM    Sum of the elements.
For vectors, SUM(X) is the sum of the elements of X.
For matrices, SUM(X) is a row vector with the sum over
each column.
»
```

The following examples are given to demonstrate the usage:

```
» x=[ 1 2 3 1.5 -1.5 -2];        %Initialize an array
» sum(x)                         %Call MATLAB function sum
ans =
          4                      %Display the result
```

```
»
» x=[1 2 3; -1.5 1.5 2.5; 4 5 6]      %Initialize 3×3 matrix
x =                                   %Display the contents of 3×3 matrix
1.0000   2.0000   3.0000
-1.5000   1.5000   2.5000
4.0000   5.0000   6.0000

» sum(x)                              %Call MATLAB function sum
ans =
3.5000   8.5000   11.5000             %Display the results
»
```

The MATLAB can be used like a calculator to work with numbers, variables, and expressions in the command window. The following are the basic syntax examples:

```
» sin(pi/4)
ans =
0.7071
» pi*4
ans =
12.5664
```

In MATLAB, variable names can store values, vectors, and matrices. See the following examples.

```
» x=cos(pi/8)
x =
 0.9239
» y=sqrt(x)-2^2
y =
 -3.0388
» z=[1 -2 1 2]
z =
1  -2   1   2
» zz=[1 2 3; 4 5 6]
zz =
 1   2   3
 4   5   6
```

Complex numbers are natural in MATLAB. See the following examples.

```
» z=3+4i              % Complex number
z =
 3.0000 + 4.0000i
» conj(z)             % Complex conjugate of z
ans =
 3.0000 - 4.0000i
» abs(z)              % Magnitude of z
ans =
 5
» angle(z)            % Angle of z (radians)
ans =
0.9273
» real(z)             % Real part of a complex number z
ans =
 3
» imag(z)             % Imaginary part of a complex number z
ans =
 4
» exp(j*pi/4)         % Polar form of a complex number
ans =
 0.7071 + 0.7071i
```

The following shows examples of array operations. Krauss et al. (1994) and Stearns and Hush (2011) give the detailed explanation for each operation.

```
» x=[1 2; 3 4]        %Initialize 2×2 matrixes
x =
   1   2
   3   4
» y=[-4 3;-2 1]
y =
  -4   3
  -2   1
» x+y                 % Add two matrixes
ans =
  -3   5
   1   5
```

```
» x*y                 % Matrix product
ans =
   -8    5
   -20   13
» x.*y                % Array element product
ans =
   -4   6
   -6   4
» x'                  % Matrix transpose
ans =
   1   3
   2   4
» 2.^x                % Exponentiation: matrix x contains each exponent.
ans =
   2   4
   8   16
» x.^3                % Exponentiation: power of 3 for each element in matrix x
ans =
   1   8
   27  64
» y.^x                % Exponentiation: each element in y has a power
                      % specified by a corresponding element in matrix x
ans =
  -4   9
  -8   1
» x=[0 1 2 3 4 5 6]              % Initialize a row vector
x =
        0   1   2   3   4   5   6
» y=x.*x-2*x                     % Use array element product to compute a quadratic
                                 % function
y =
      0  -1   0   3   8   15   24
»
» z=[1 3]'                       % Initialize a column vector
   z =
      1
      3
   »w=x\z                        % Inverse matrix x, then multiply it by the column vector z
   w =
         1
         0
```

A.2 MATLAB ARRAY AND INDEXING

Let us look at the syntax to create an array as follows:

Basic syntax: **x = begin:step:end.**

An array x is created with the first element value of **begin**. The value increases by a value of **step** for each next element in the array and stops when the next stepped value is beyond the specified end value of **end**. In simulation, we may use this operation to create the sample indexes or array of time instants for digital signals. The **begin**, **step,** and **end** can be assigned to be integers or floating-point numbers.

The following examples are given for illustrations:

```
» n=1:3:8                        % Create a vector containing n=[1 4 7]
n =
      1   4   7
» m=9:-1:2                       % Create a vector m=[9 8 7 6 5 4 3 2 ]
m =
      9   8   7   6   5   4   3   2
» x=2:(1/4):4                    % Create x=[2 2.25 2.5 2.75 3 3.25 3.5 3.75 4]
x =
        Columns 1 through 7
        2.0000   2.2500   2.5000   2.7500   3.0000   3.2500   3.5000
        Columns 8 through 9
        3.7500   4.0000
» y=2:-0.5:-1                    % create y=[2 1.5 1 0.5 0 -0.5 -1]
y =
          2.0000   1.5000   1.0000   0.5000        0  -0.5000  -1.0000
»
```

Next, we examine creating a vector and extracting numbers in a vetcor:

```
» xx= [ 1 2 3 4 5 [5:-1:1] ]   % Create xx=[ 1 2 3 4 5  5 4 3 2 1]
xx =
         1   2   3   4   5   5   4   3   2   1
» xx(3:7)                      % Show elements from index 3 to index 7
ans =
      3   4   5   5   4
» length(xx)                   % Return of the numbet of elements in vetcor xx
ans =
        10
» yy=xx(2:2:length(xx))        % Display even indexed numbers in array xx
yy =
         2   4   5   3   1
```

A.3 PLOT UTILITIES: SUBPLOT, PLOT, STEM, AND STAIR

The following are common MATLAB plot functions for digital signal processing (DSP) simulation:

subplot opens subplot windows for plotting.

plot produces an x-y plot. It can also create multiple plots.

stem produces discrete-time signals.

stair produces staircase (sample-and-hold) signals.

The following program contains different MATLAB plot functions:

```
t=0:0.01:0.4;         %Create time vector for time instants from 0 to 0.4 seconds
xx=4.5*sin(2*pi*10*t+pi/4); % Calculate a sine function with the frequency 10 Hz
yy=4.5*cos(2*pi*5*t-pi/3);  % Calculate cos function with the frequency 5 Hz
subplot(4,1,1), plot(t,xx);grid       % Plot a sine function in window 1
subplot(4,1,2), plot(t,xx, t,yy,'-.') ;grid;   % Plot sine and cos functions in window 2
subplot(4,1,3), stem(t,xx);grid       % Plot a sine function in the discrete-time form
subplot(4,1,4), stairs(t,yy);grid     % Plot a cos function in the sample-and-hold form
xlabel('Time (sec.)');
```

Each plot is shown in Fig. A.1. Notice that dropping the semicolon at the end of the MATLAB syntax will display values on MATLAB prompt.

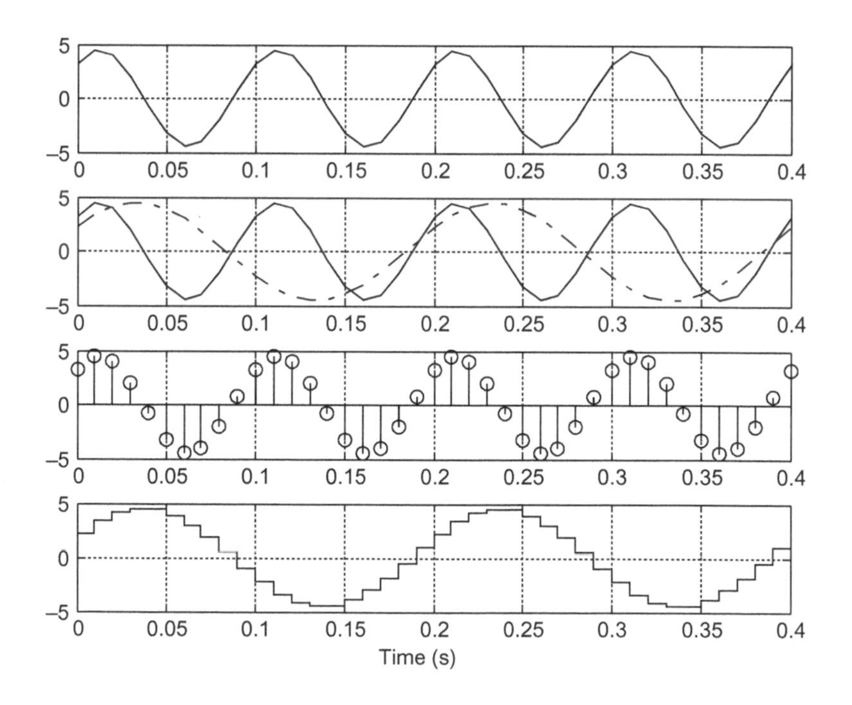

FIG. A.1

Illustration of MATLAB plot functions.

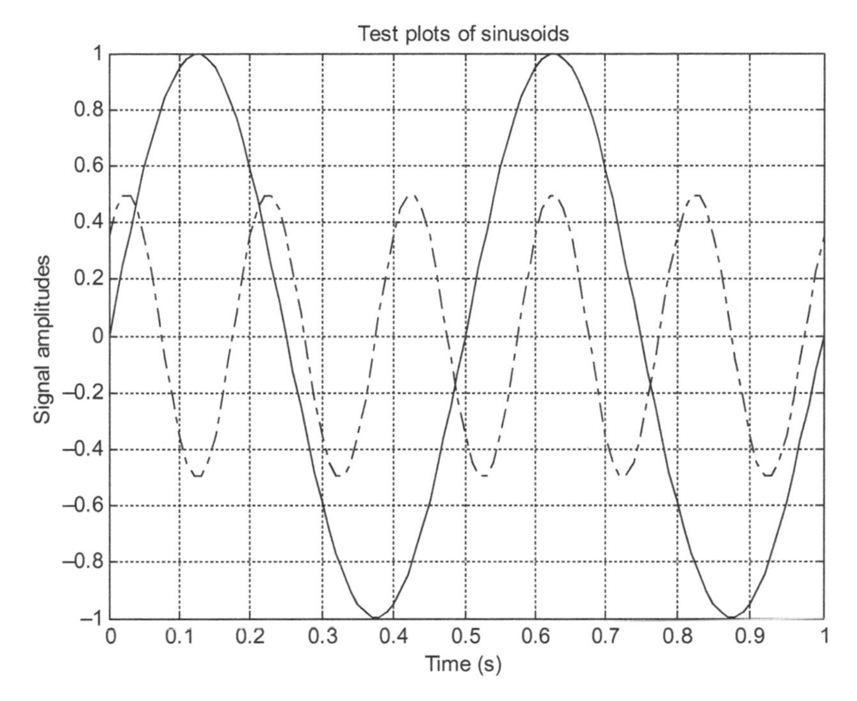

FIG. A.2

Illustration for the MATLAB script file test.m.

A.4 MATLAB SCRIPT FILES

We can create a MATLAB script file using the built-in MATLAB editor (or Windows Notepad) to write MATLAB source code. The script file is named "filename.m" and can be run by typing the file name at the MATLAB prompt and hitting the return key. The script file **test.m** is described here for illustration. Fig. A.2 depicts the plot produced by test.m.

At MATLAB prompt, run the program.

```
>>which test      % show the folder where test.m resides
```

Go to the folder that contains test.m, and run your script from MATLAB.

```
>>test            % run the test.m
>>type test       % display the contents of test.m
```

test.m

```
t=0:0.01:1;
x=sin(2*pi*2*t);
y=0.5*cos(2*pi*5*t-pi/4);
plot(t,x), grid on
```

```
title('Test plots of sinusoids')
ylabel('Signal amplitudes');
xlabel('Time (sec.)'); hold on
plot(t,y,'-.');
```

A.5 MATLAB FUNCTIONS

A MATLAB function is often used for replacing the repetitive portions of the MATLAB code. It is created using a MATLAB script file. However, the code begins with a key word **function**, followed by the function declaration, comments for the help system, and program codes. A function **sumsub.m** that computes the addition and subtraction of two numbers is listed here for illustration.

sumsub.m

```
function [sum, sub]=sumsub(x1,x2)
%sumsub: Function to add and subtract two numbers
% Usage:
% [sum, sub] = sumsub(x1,x2)
% x1 = the first number
% x2= the second number
%  sum = x1+x2;
%  sub = x1-x2
sum = x1+x2;              %Add two numbers
sub= x1-x2;               %Subtract x2 from x1
```

To use the MATLAB function, go to the folder that contains **sumsub.m**. At the MATLAB prompt, try the following:

```
»help sumsub        % Display usage information on MATLAB prompt
sumsub: Function to add and subtract two numbers
usage:
[sum, sub] = sumsub(x1,x2)
 x1 = the first number
 x2= the second number
 sum = x1+x2;
 sub = x1-x2
```

Run the function as follows:

```
» [x1, x2]=sumsub(3, 4-3i);       % Call function sumsub
» x1                              % Display the result of sum
x1 =
  7.0000 - 3.0000i
» x2                              %Display the result of subtraction
x2 =
  -1.0000 + 3.0000i
```

The MATLAB function can also be used inside an m-file. More MATLAB exercises for introduction to DSP can be explored in McClellan et al. (2015) and Stearns and Hush (2011).

APPENDIX

REVIEW OF ANALOG SIGNAL PROCESSING BASICS

B

B.1 FOURIER SERIES AND FOURIER TRANSFORM

Electronics applications require familiarity with some periodic signals such as the square wave, rectangular wave, triangular wave, sinusoid, saw tooth wave, and so on. These periodic signals can be analyzed in frequency domain with the help of the Fourier series expansion. According to Fourier theory, a periodic signal can be represented by a Fourier series that contains the sum of a series of sine and/or cosine functions (harmonics) plus a direct current (DC) term. There are three forms of Fourier series: (1) sine-cosine, (2) amplitude-phase, and (3) complex exponential. We will review each of them individually in the following text. Comprehensive treatments can be found in Ambardar (1999), Soliman and Srinath (1998), and Stanley (2003).

B.1.1 SINE-COSINE FORM

The Fourier series expansion of a periodic signal $x(t)$ with a period of T via the sine-cosine form is given by

$$x(t) = a_0 + \sum_{k=1}^{\infty} a_k \cos(k\omega_0 t) + \sum_{k=1}^{\infty} b_k \sin(k\omega_0 t) \tag{B.1}$$

whereas $\omega_0 = 2\pi/T_0$ is the fundamental angular frequency in radians per second, while the fundamental frequency in term of Hz is $f_0 = 1/T_0$. The Fourier coefficients of a_0, a_k, and b_k can be found according to the following integral equations:

$$a_0 = \frac{1}{T_0}\int_{T_0} x(t)dt \tag{B.2}$$

$$a_k = \frac{2}{T_0}\int_{T_0} x(t)\cos(k\omega_0 t)dt \tag{B.3}$$

$$b_k = \frac{2}{T_0}\int_{T_0} x(t)\sin(k\omega_0 t)dt. \tag{B.4}$$

Note that the integral is performed over one period of the signal to be expanded. From Eq. (B.1), the signal $x(t)$ consists of a DC term and sums of sine and cosine functions with their corresponding harmonic frequencies. Again, note that $k\omega_0$ is the nth harmonic frequency.

B.1.2 AMPLITUDE-PHASE FORM

From the sine-cosine form, we note that there is a sum of two terms with the same frequency. The term in the first sum is $a_k \cos(k\omega_0 t)$ and the other is $b_k \sin(k\omega_0 t)$. We can combine these two terms and modify the since-cosine form into an amplitude-phase form:

$$x(t) = A_0 + \sum_{k=1}^{\infty} A_k \cos(k\omega_0 t + \phi_k). \tag{B.5}$$

The DC term is the same as before; that is,

$$A_0 = a_0 \tag{B.6}$$

and the amplitude and phase are given by

$$A_k = \sqrt{a_k^2 + b_k^2} \tag{B.7}$$

$$\phi_k = \tan^{-1}\left(\frac{-b_k}{a_k}\right) \tag{B.8}$$

respectively. The amplitude-phase form provides very useful information for spectral analysis. With the calculated amplitude and phase for each harmonic frequency, we can create the spectral plots. One depicts a plot of the amplitude vs. its corresponding harmonic frequency, called the amplitude spectrum, while the other plot shows each phase vs. its harmonic frequency, called the phase spectrum. Note that the spectral plots are one-sided, since amplitudes and phases are plotted vs. the positive harmonic frequencies. We will illustrate these via Example B.1.

B.1.3 COMPLEX EXPONENTIAL FORM

The complex exponential form is developed based on expanding sine and cosine functions in the sine-cosine form into their exponential expressions using Euler's formula and regrouping these exponential terms. The Euler's formula is given by

$$e^{\pm jx} = \cos(x) \pm j\sin(x),$$

which can be written as two separate forms:

$$\cos(x) = \frac{e^{jx} + e^{-jx}}{2}$$

$$\sin(x) = \frac{e^{jx} - e^{-jx}}{2j}.$$

We will focus on the interpretation and application rather than its derivation. Thus the complex exponential form is expressed as

$$x(t) = \sum_{k=-\infty}^{\infty} c_k e^{jk\omega_0 t}, \tag{B.9}$$

where c_k represents the complex Fourier coefficients, which can be found from

$$c_k = \frac{1}{T_0}\int_{T_0} x(t) e^{-jk\omega_0 t} dt. \tag{B.10}$$

The relationship between the complex Fourier coefficients c_k and the coefficients of the sine-cosine form are

$$c_0 = a_0 \tag{B.11}$$

$$c_k = \frac{a_k - jb_k}{2}, \quad \text{for } k > 0. \tag{B.12}$$

Considering a real signal $x(t)$ ($x(t)$ is not a complex function) in Eq. (B.10), c_{-k} is equal to the complex conjugate of c_k, that is, $\overline{c}_k$. It follows that

$$c_{-k} = \overline{c}_k = \frac{a_k + jb_k}{2}, \quad \text{for } k > 0. \tag{B.13}$$

Since c_k is a complex value, which can be further written in the magnitude-phase from, we obtain

$$c_k = |c_k| \angle \phi_k, \tag{B.14}$$

where $|c_k|$ is the magnitude and ϕ_k is the phase of the complex Fourier coefficient. Similar to the magnitude-phase form, we can create the spectral plots for $|c_k|$ and ϕ_k. Since the frequency index k goes from $-\infty$ to ∞, the plots of resultant spectra are two sided.

EXAMPLE B.1

Consider the following square waveform $x(t)$, shown in Fig. B.1, where T_0 represents a period. Find the Fourier series expansions in terms of (a) the sine-cosine form, (b) the amplitude-phase form, and (c) the complex exponential form.

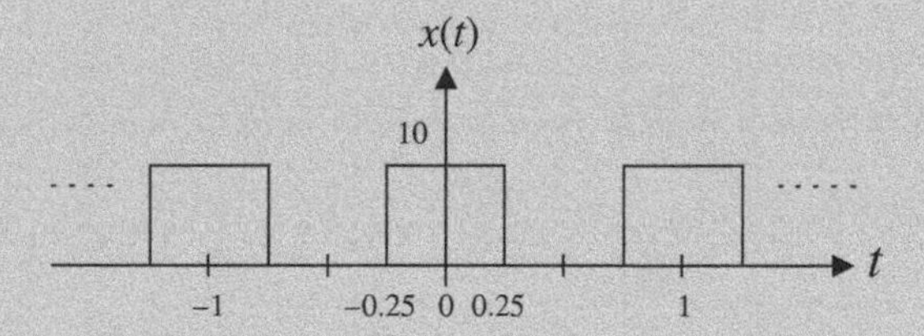

FIG. B.1

Square waveform in Example B.1.

Solution:

From Fig. B.1, we note that $T_0 = 1$ s and $A = 10$.

The fundamental frequency is.

$$f_0 = 1/T_0 = 1 \text{ Hz or } \omega_0 = 2\pi \times f_0 = 2\pi \text{ rad/s}.$$

(a) Using Eqs. (B.1) to (B.3) yield

$$a_0 = \frac{1}{T_0}\int_{-T_0/2}^{T_0/2} x(t)dt = \frac{1}{1}\int_{-0.25}^{0.25} 10dt = 5$$

$$a_k = \frac{2}{T_0}\int_{-T_0/2}^{T_0/2} x(t)\cos(k\omega_0 t)dt$$
$$= \frac{2}{1}\int_{-0.25}^{0.25} 10\cos(k2\pi t)dt$$
$$= \frac{2}{1}\frac{10 \times \sin(k2\pi t)}{k2\pi}\bigg|_{-0.25}^{0.25} = 10\frac{\sin(0.5\pi k)}{0.5\pi k}$$
$$b_k = \frac{2}{T_0}\int_{-T_0/2}^{T_0/2} x(t)\sin(k\omega_0 t)dt$$
$$= \frac{2}{1}\int_{-0.25}^{0.25} 10 \times \sin(k2\pi t)dt$$
$$= \frac{2}{1}\frac{-10\cos(k2\pi t)}{k2\pi}\bigg|_{-0.25}^{0.25} = 0$$

Thus, the Fourier series expansion in terms of the sine-cosine form is written as

$$x(t) = 5 + \sum_{k=1}^{\infty} 10\frac{\sin(0.5\pi k)}{0.5\pi k}\cos(k2\pi t)$$
$$= 5 + \frac{20}{\pi}\cos(2\pi t) - \frac{20}{3\pi}\cos(6\pi t) + \frac{4}{\pi}\cos(10\pi t) - \frac{20}{7\pi}\cos(14\pi t) + \cdots$$

(b) Making use of the relations between the sine-cosine form and the amplitude-phase form, we obtain

$$A_0 = a_0 = 5$$
$$A_k = \sqrt{a_k^2 + b_k^2} = |a_k| = 10 \times \left|\frac{\sin(0.5\pi k)}{0.5\pi k}\right|.$$

Again, noting that $-\cos(x) = \cos(x + 180^0)$, the Fourier series expansion in terms of the amplitude-phase form is

$$x(t) = 5 + \frac{20}{\pi}\cos(2\pi t) + \frac{20}{3\pi}\cos\left(6\pi t + 180^0\right) + \frac{4}{\pi}\cos(10\pi t) + \frac{20}{7\pi}\cos\left(14\pi t + 180^0\right) + \cdots$$

(c) First let us find the complex Fourier coefficients using the formula, that is,

$$c_k = \frac{1}{T_0}\int_{-T_0/2}^{T_0/2} x(t)e^{-jk\omega_0 t}dt$$
$$= \frac{1}{1}\int_{-0.25}^{0.25} Ae^{-jk2\pi t}dt$$
$$= 10 \times \frac{e^{-jk2\pi t}}{-jk2\pi}\bigg|_{-0.25}^{0.25} = 10 \times \frac{\left(e^{-j0.5\pi k} - e^{j0.5\pi k}\right)}{-jk2\pi}$$

Applying Euler's formula yields

$$c_k = 10 \times \frac{\cos 0.5\pi k - j\sin(0.5\pi k) - [\cos(0.5\pi k) + j\sin(0.5\pi k)]}{-jk2\pi} = 5\frac{\sin(0.5\pi k)}{0.5\pi k}.$$

Second, using the relationship between the sine-cosine form and the complex exponential form, it follows that

$$c_k = \frac{a_k - jb_k}{2} = \frac{a_k}{2} = 5\frac{\sin(0.5k\pi)}{(0.5k\pi)}.$$

Certainly, the result is identical to the one obtained directly from the formula. Note that c_0 cannot be evaluated directly by substituting $k=0$, since we have the indeterminate term $\frac{0}{0}$. Using L'Hospital's rule, described Appendix H, leads to

$$c_0 = \lim_{k\to 0} 5\frac{\sin(0.5k\pi)}{(0.5k\pi)} = \lim_{n\to 0} 5\frac{\frac{d(\sin(0.5k\pi))}{dk}}{\frac{d(0.5k\pi)}{dk}}$$

$$= \lim_{k\to 0} 5\frac{0.5\pi\cos(0.5k\pi)}{0.5\pi} = 5$$

Finally, the Fourier expansion in terms of the complex exponential form is shown as follows:

$$x(t) = \cdots + \frac{10}{\pi}e^{-j2\pi t} + 5 + \frac{10}{\pi}e^{j2\pi t} - \frac{10}{3\pi}e^{j6\pi t} + \frac{2}{\pi}e^{j10\pi t} - \frac{10}{7\pi}e^{j14\pi t} + \cdots$$

B.1.4 SPECTRAL PLOTS

As previously discussed, the magnitude-phase form can provide information to create a one-sided spectral plot. The amplitude spectrum is obtained by plotting A_k vs. the harmonic frequency $k\omega_0$, and the phase spectrum is obtained by plotting ϕ_k vs. $k\omega_0$, both for $k \geq 0$. Similarly, if the complex exponential form is used, the two-sided amplitude and phase spectral plots of $|c_k|$ and ϕ_k vs. $k\omega_0$ for $-\infty < k < \infty$ can be achieved, respectively. We illustrate this by the following example.

EXAMPLE B.2

Based on the solution to Example B.1, plot the one-sided amplitude spectrum and two-sided amplitude spectrum, respectively.

Solution:

Based on the solution for A_k, one-side amplitude spectrum is shown in Fig. B.2:

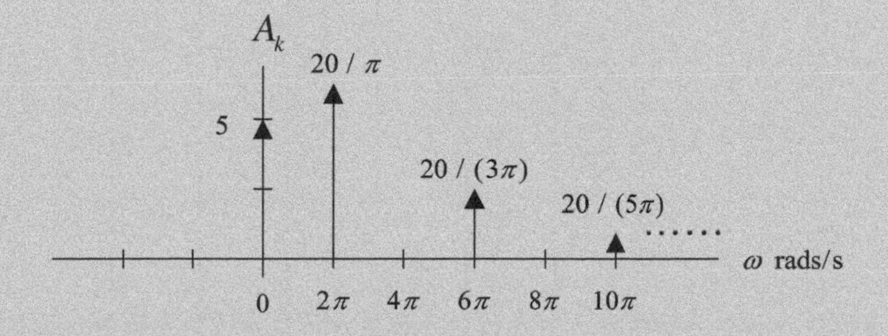

FIG. B.2

One-sided spectrum of the square waveform in Example B.2.

According to the solution of the complex exponential form, the two-sided amplitude spectrum is demonstrated as shown in Fig. B.3.

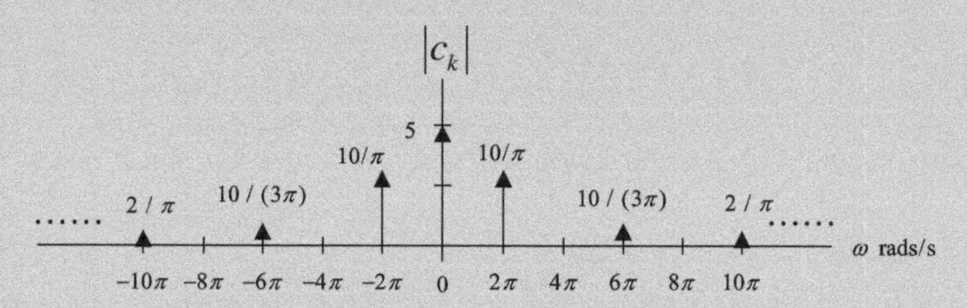

FIG. B.3

Two-sided spectrum of the square waveform in Example B.2.

A general pulse train $x(t)$ with a period T_0 seconds and a pulse width τ seconds is shown in Fig. B.4. The Fourier series expansions for sine-cosine and complex exponential forms can be derived similarly and are given as follows:

Sine-cosine form:

$$x(t) = \frac{\tau A}{T_0} + \frac{2\tau A}{T_0}\left(\frac{\sin(\omega_0\tau/2)}{(\omega_0\tau/2)}\cos(\omega_0 t) + \frac{\sin(2\omega_0\tau/2)}{(2\omega_0\tau/2)}\cos(2\omega_0 t) + \frac{\sin(3\omega_0\tau/2)}{(3\omega_0\tau/2)}\cos(3\omega_0 t) + \cdots\right) \tag{B.15}$$

Complex exponential form:

$$x(t) = \cdots + \frac{\tau A}{T_0}\frac{\sin(\omega_0\tau/2)}{(\omega_0\tau/2)}e^{-j\omega_0 t} + \frac{\tau A}{T_0} + \frac{\tau A}{T_0}\frac{\sin(\omega_0\tau/2)}{(\omega_0\tau/2)}e^{j\omega_0 t} + \frac{\tau A}{T_0}\frac{\sin(2\omega_0\tau/2)}{(2\omega_0\tau/2)}e^{j2\omega_0 t} + \cdots, \tag{B.16}$$

where $\omega_0 = 2\pi f_0 = 2\pi/T_0$ is the fundamental angle frequency of the periodic waveform.

The reader can derive the one-side amplitude spectrum A_k and the two-sided amplitude spectrum $|c_k|$. The expressions for the one-sided amplitude and two-side amplitude spectra are given by the following:

$$A_0 = \frac{\tau}{T_0}A \tag{B.17}$$

$$A_k = \frac{2\tau}{T_0}A\left|\frac{\sin(k\omega_0\tau/2)}{(k\omega_0\tau/2)}\right|, \quad \text{for } k = 1, 2, 3\ldots \tag{B.18}$$

$$|c_k| = \frac{\tau}{T_0}A\left|\frac{\sin(k\omega_0\tau/2)}{(k\omega_0\tau/2)}\right|, \quad -\infty < k < \infty. \tag{B.19}$$

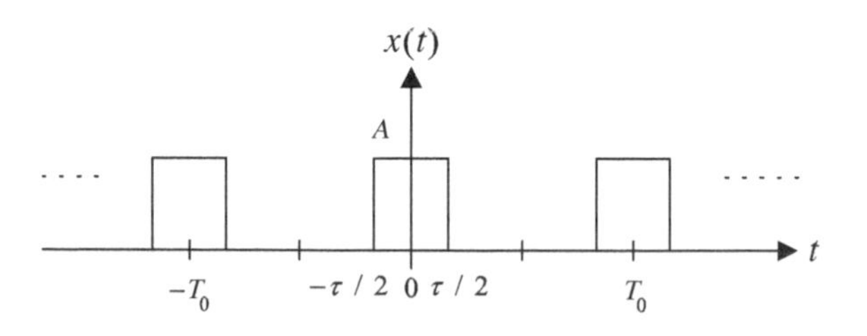

FIG. B.4

Rectangular waveform (pulse train).

EXAMPLE B.3

In Fig. B.4, if $T_0 = 1$ ms, $\tau = 0.2$ ms, and $A = 10$, use Eqs. (B.17)–(B.19) to derive the amplitude one-sided spectrum and two-sided spectrum for each of the first four harmonic frequency components.

Solution:

The fundamental frequency is

$$\omega_0 = 2\pi f_0 = 2\pi \times (1/0.001) = 2000\pi \text{rad/s}.$$

Using Eqs. (B.17), (B.18) yields the one-side spectrum as

$$A_0 = \frac{\tau}{T_0} A = \frac{0.0002}{0.001} \times 10 = 2, \text{ for } n = 0, \; n\omega_0 = 0.$$

for $k = 1$, $k\omega_0 = 2000\pi$ rad/sec:

$$A_1 = \frac{2 \times 0.0002}{0.001} \times 10 \times \left| \frac{\sin(1 \times 2000\pi \times 0.0002/2)}{(1 \times 2000\pi \times 0.0002/2)} \right| = 4 \frac{\sin(0.2\pi)}{(0.2\pi)} = 3.7420,$$

for $k = 4$, $k\omega_0 = 8000\pi$ rad/sec:

$$A_2 = \frac{2 \times 0.0002}{0.001} \times 10 \times \left| \frac{\sin(2 \times 2000\pi \times 0.0002/2)}{(2 \times 2000\pi \times 0.0002/2)} \right| = 4 \frac{\sin(0.4\pi)}{(0.4\pi)} = 3.0273,$$

for $k = 3$, $k\omega_0 = 6000\pi$ rad/sec:

$$A_3 = \frac{2 \times 0.0002}{0.001} \times 10 \times \left| \frac{\sin(3 \times 2000\pi \times 0.0002/2)}{(3 \times 2000\pi \times 0.0002/2)} \right| = 4 \frac{\sin(0.6\pi)}{(0.6\pi)} = 2.0182,$$

for $k = 4$, $k\omega_0 = 8000\pi$ rad/sec:

$$A_4 = \frac{2 \times 0.0002}{0.001} \times 10 \times \left| \frac{\sin(4 \times 2000\pi \times 0.0002/2)}{(4 \times 2000\pi \times 0.0002/2)} \right| = 4 \frac{\sin(0.8\pi)}{(0.8\pi)} = 0.9355.$$

The one-side amplitude spectrum is plotted in Fig. B.5.

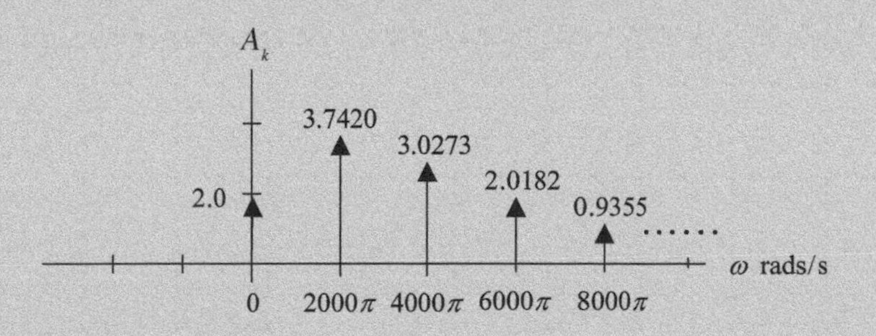

FIG. B.5

One-sided spectrum in Example B.3.

Similarly, applying Eq. (B.19) leads to

$$|c_0| = \frac{0.0002}{0.001} \times 10 \times \left| \lim_{k \to 0} \frac{\sin(k \times 2000\pi \times 0.0002/2)}{(k \times 2000\pi \times 0.0002/2)} \right| = 2 \times |1| = 2.$$

Note: we use the fact $\lim_{x \to 0} \frac{\sin(x)}{x} = 1.0$ (see L'Hospital's rule in Appendix H)

$$|c_1| = |c_{-1}| = \frac{0.0002}{0.001} \times 10 \times \left| \frac{\sin(1 \times 2000\pi \times 0.0002/2)}{(1 \times 2000\pi \times 0.0002/2)} \right| = 2 \times \left| \frac{\sin(0.2\pi)}{0.2\pi} \right| = 1.8710$$

$$|c_2| = |c_{-2}| = \frac{0.0002}{0.001} \times 10 \times \left| \frac{\sin(2 \times 2000\pi \times 0.0002/2)}{(2 \times 2000\pi \times 0.0002/2)} \right| = 2 \times \left| \frac{\sin(0.4\pi)}{0.4\pi} \right| = 1.5137$$

$$|c_3| = |c_{-3}| = \frac{0.0002}{0.001} \times 10 \times \left| \frac{\sin(3 \times 2000\pi \times 0.0002/2)}{(3 \times 2000\pi \times 0.0002/2)} \right| = 2 \times \left| \frac{\sin(0.6\pi)}{0.6\pi} \right| = 1.0091$$

$$|c_4| = |c_{-4}| = \frac{0.0002}{0.001} \times 10 \times \left| \frac{\sin(4 \times 2000\pi \times 0.0002/2)}{(4 \times 2000\pi \times 0.0002/2)} \right| = 2 \times \left| \frac{\sin(0.8\pi)}{0.8\pi} \right| = 0.4677.$$

Fig. B.6 shows the two-sided amplitude spectral plot.

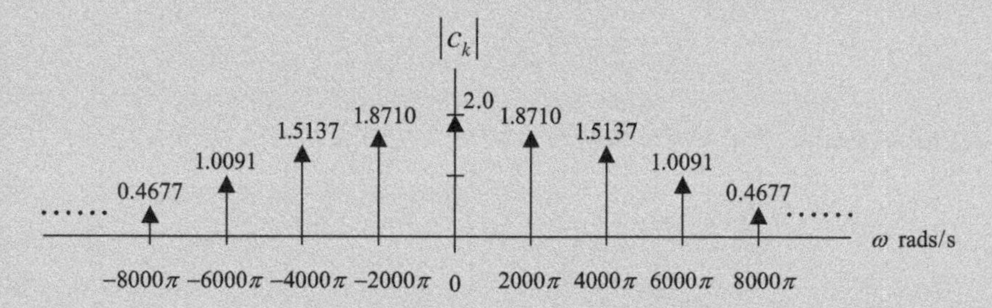

FIG. B.6

Two-sided spectrum in Example B.3.

The following example illustrates the use of table information to determine the Fourier series expansion of the periodic waveform. Table B.1 consists of the Fourier series expansions for common periodic signals in the sine-cosine form while Table B.2 shows the expansions in the complex exponential form.

EXAMPLE B.4

Given a sawtooth waveform shown in Fig. B.7, where $T_0 = 1$ ms and $A = 10$, use the formula in the table to determine the Fourier series expansion in magnitude-phase form, and determine the frequency f_3 and amplitude value of A_3 for the third harmonic. Write the Fourier series expansion also in a complex exponential form, and determine $|c_3|$ and $|c_{-3}|$ for the third harmonic.

Solution:

(a) Based on the information in Table B.1, we have

Table B.1 Fourier Series Expansions for Some Common Waveform Signals in the Sine-Cosine Form

Time domain signal $x(t)$	Fourier series expansion
Positive square wave 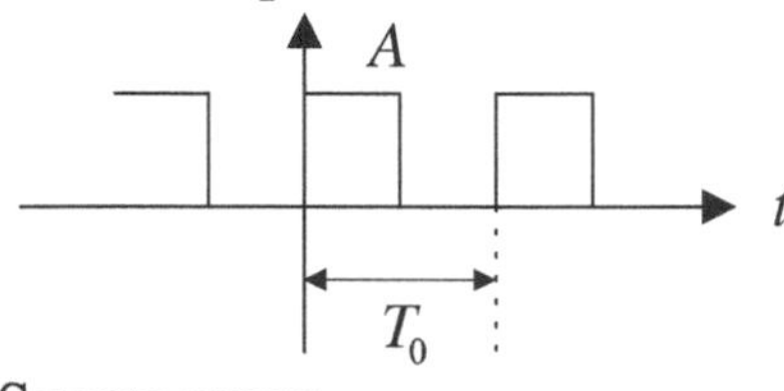	$x(t) = \frac{A}{2} + \frac{2A}{\pi}\left(\sin\omega_0 t + \frac{1}{3}\sin 3\omega_0 t + \frac{1}{5}\sin 5\omega_0 t + \frac{1}{7}\sin 7\omega_0 t + \cdots\right)$
Square wave 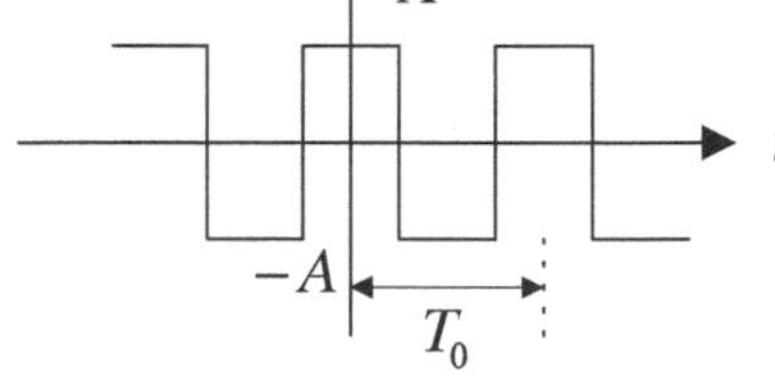	$x(t) = \frac{4A}{\pi}\left(\cos\omega_0 t - \frac{1}{3}\cos 3\omega_0 t + \frac{1}{5}\cos 5\omega_0 t - \frac{1}{7}\cos 7\omega_0 t + \cdots\right)$
Triangular wave 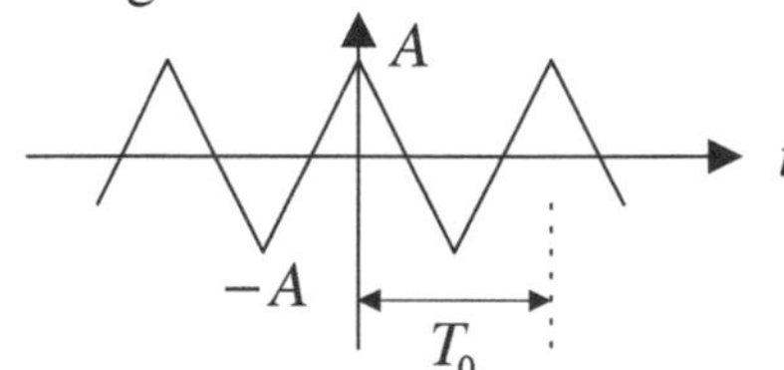	$x(t) = \frac{8A}{\pi^2}\left(\cos\omega_0 t + \frac{1}{9}\cos 3\omega_0 t + \frac{1}{25}\cos 5\omega_0 t + \frac{1}{49}\cos 7\omega_0 t + \cdots\right)$
Sawtooth wave 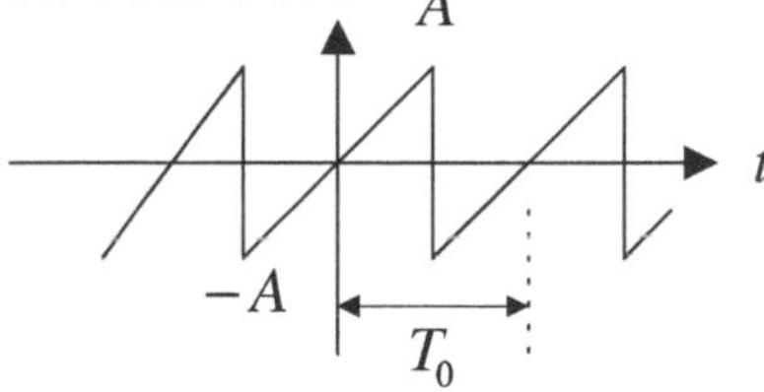	$x(t) = \frac{2A}{\pi}\left(\sin\omega_0 t - \frac{1}{2}\sin 2\omega_0 t + \frac{1}{3}\sin 3\omega_0 t - \frac{1}{4}\sin 4\omega_0 t + \cdots\right)$
Rectangular wave (Pulse train) 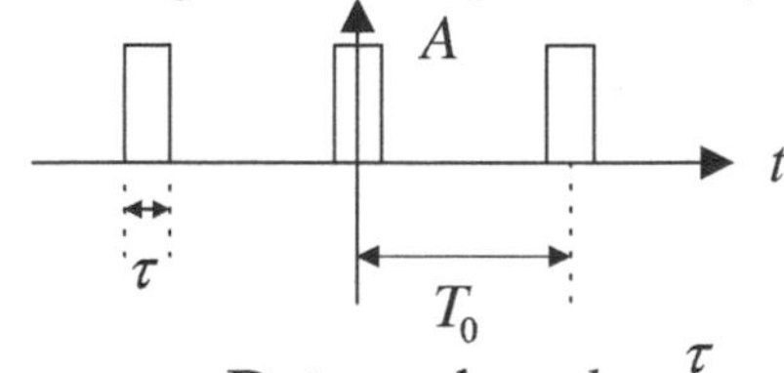 Duty cycle $= d = \frac{\tau}{T_0}$	$x(t) = Ad + 2Ad\left(\frac{\sin \pi d}{\pi d}\right)\cos\omega_0 t + 2Ad\left(\frac{\sin 2\pi d}{2\pi d}\right)\cos 2\omega_0 t + 2Ad\left(\frac{\sin 3\pi d}{3\pi d}\right)\cos 3\omega_0 t + \cdots$
Ideal impulse train 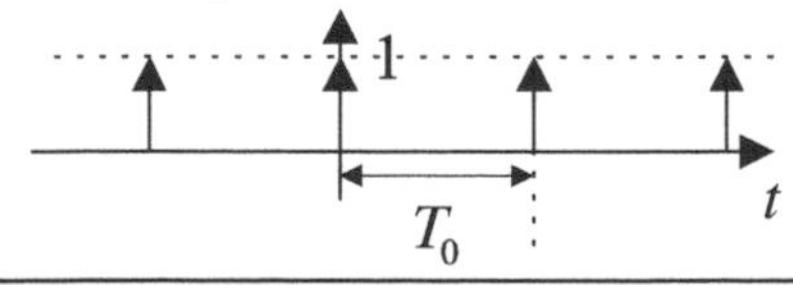	$x(t) = \frac{1}{T_0} + \frac{2}{T_0}(\cos\omega_0 t + \cos 2\omega_0 t + \cos 3\omega_0 t + \cos 4\omega_0 t + \cdots)$

Table B.2 Fourier Series Expansions for Some Common Waveform Signals in the Complex Exponential Form

Time domain signal $x(t)$	Fourier series expansion
Positive square wave 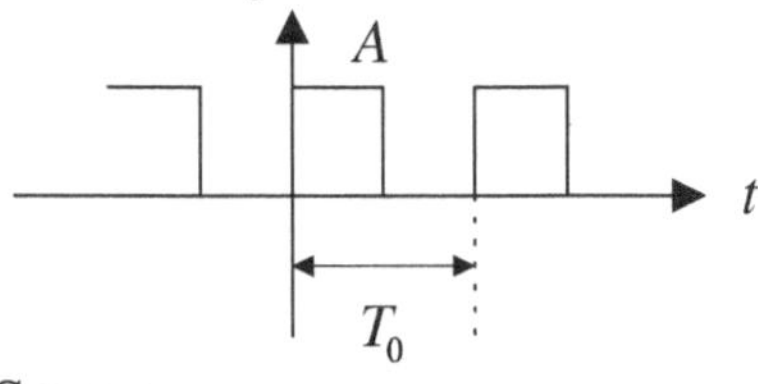	$x(t) = \cdots - \frac{A}{j3\pi}e^{-j3\omega_0 t} - \frac{A}{j\pi}e^{-j\omega_0 t} + \frac{A}{2} + \frac{A}{j\pi}e^{j\omega_0 t} + \frac{A}{j3\pi}e^{j3\omega_0 t} + \frac{A}{j5\pi}e^{j5\omega_0 t} + \cdots)$
Square wave 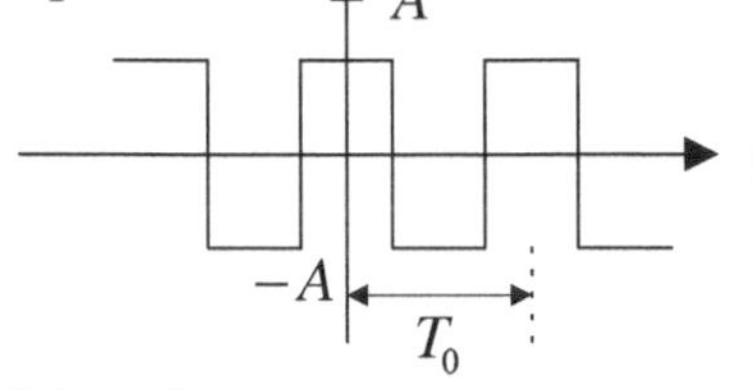	$x(t) = \frac{2A}{\pi}\left(\cdots + \frac{1}{5}e^{-j5\omega_0 t} - \frac{1}{3}e^{-j3\omega_0 t} + e^{-j\omega_0 t} + e^{j\omega_0 t} - \frac{1}{3}e^{j3\omega_0 t} + \frac{1}{5}e^{j5\omega_0 t} - \cdots\right)$
Triangular wave 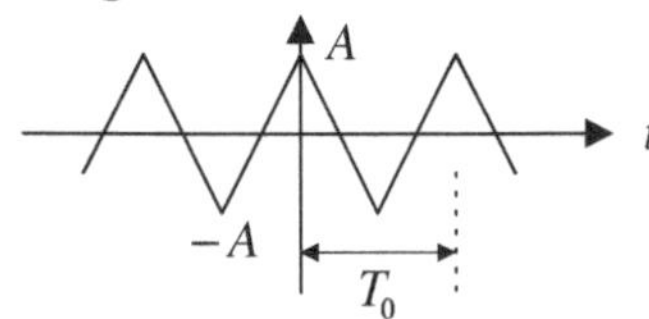	$x(t) = \frac{4A}{\pi^2}\left(\cdots + \frac{1}{25}e^{-j5\omega_0 t} + \frac{1}{9}e^{-j3\omega_0 t} + e^{-j\omega_0 t} + e^{j\omega_0 t} + \frac{1}{9}e^{j3\omega_0 t} + \frac{1}{25}e^{j5\omega_0 t} + \cdots\right)$
Sawtooth wave 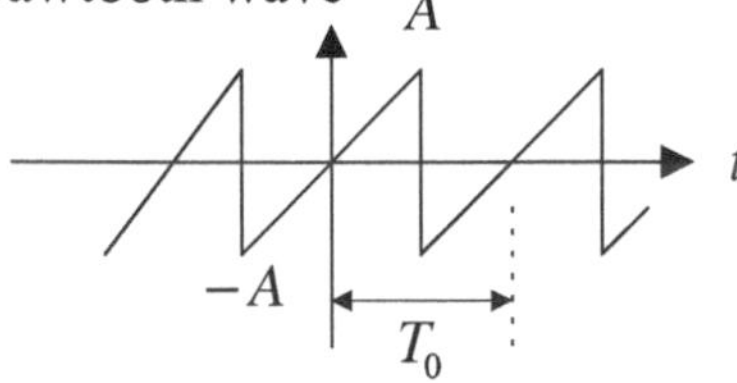	$x(t) = \frac{A}{j\pi}\left(\cdots - \frac{1}{3}e^{-j3\omega_0 t} + \frac{1}{2}e^{-j2\omega_0 t} - e^{-j\omega_0 t} + e^{j\omega_0 t} - \frac{1}{2}e^{j2\omega_0 t} + \frac{1}{3}e^{j3\omega_0 t} + \cdots\right)$
Rectangular wave (Pulse train) 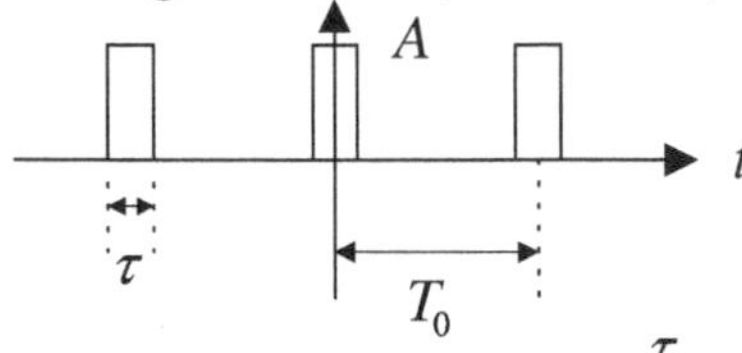 Duty cycle $= d = \frac{\tau}{T_0}$	$x(t) = \cdots + Ad\left(\frac{\sin \pi d}{\pi d}\right)e^{-j\omega_0 t} + Ad + Ad\left(\frac{\sin \pi d}{\pi d}\right)e^{j\omega_0 t} + Ad\left(\frac{\sin 2\pi d}{2\pi d}\right)e^{j2\omega_0 t} + Ad\left(\frac{\sin 3\pi d}{3\pi d}\right)e^{j3\omega_0 t} + \cdots$
Ideal impulse train 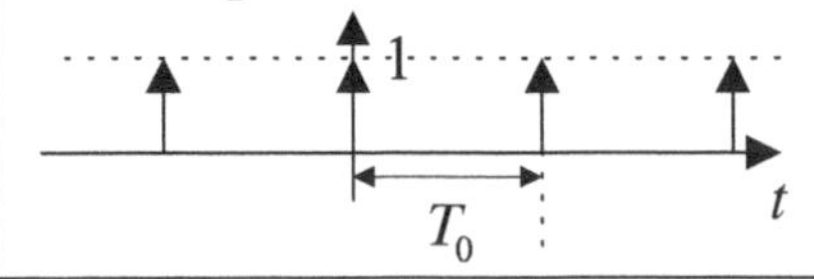	$x(t) = \frac{1}{T_0}(\cdots + e^{-j3\omega_0 t} + e^{-j2\omega_0 t} + e^{-j\omega_0 t} + 1 + e^{j\omega_0 t} + e^{j2\omega_0 t} + e^{j3\omega_0 t} + \cdots)$

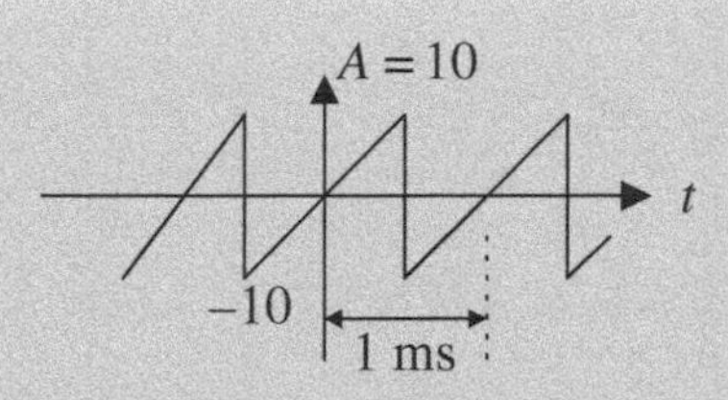

FIG. B.7

Sawtooth waveform for Example B.4.

$$x(t) = \frac{2A}{\pi}\left(\sin\omega_0 t - \frac{1}{2}\sin 2\omega_0 t + \frac{1}{3}\sin 3\omega_0 t - \frac{1}{4}\sin 4\omega_0 t + \cdots\right).$$

Since $T_0 = 1$ ms, the fundamental frequency is

$$f_0 = 1/T_0 = 1000\text{Hz, and } \omega_0 = 2\pi f_0 = 2000\pi\,\text{rad/sec}.$$

Then, the expansion is determined as

$$x(t) = \frac{2\times 10}{\pi}\left(\sin 2000\pi t - \frac{1}{2}\sin 4000\pi t + \frac{1}{3}\sin 6000\pi t - \frac{1}{4}\sin 8000\pi t + \cdots\right).$$

Using trigonometric identities:

$$\sin x = \cos\left(x - 90^0\right) \text{ and } -\sin x = \cos\left(x + 90^0\right),$$

and simple algebra, we finally obtain

$$x(t) = \frac{20}{\pi}\cos\left(2000\pi t - 90^0\right) + \frac{10}{\pi}\cos\left(4000\pi t + 90^0\right)$$
$$+ \frac{20}{3\pi}\cos\left(6000\pi t - 90^0\right) + \frac{5}{\pi}\cos\left(8000\pi t + 90^0\right) + \cdots$$

From the magnitude-phase form, we then determine f_3 and A_3 as follows:

$$f_3 = 3\times f_0 = 3000\text{Hz, and } A_3 = \frac{20}{3\pi} = 2.1221.$$

(b) From Table B.2, the complex exponential form is

$$x(t) = \frac{10}{j\pi}\left(\cdots - \frac{1}{3}e^{-j6000\pi t} + \frac{1}{2}e^{-j4000\pi t} - e^{-j2000\pi t} + e^{j2000\pi t} - \frac{1}{2}e^{j4000\pi t} + \frac{1}{3}e^{j6000\pi t} + \cdots\right).$$

From the expression, we have

$$|c_3| = \left|\frac{10}{j\pi}\times\frac{1}{3}\right| = \left|\frac{1.061}{j}\right| = 1.061 \text{ and}$$

$$|c_{-3}| = \left|-\frac{10}{j\pi}\times\frac{1}{3}\right| = \left|-\frac{1.061}{j}\right| = 1.061.$$

B.1.5 FOURIER TRANSFORM

The Fourier transform is a mathematical function that provides frequency spectral analysis for a continuous time signal. The Fourier transform pair is defined as

$$\text{Fourier transform}: \ X(\omega) = \int_{-\infty}^{\infty} x(t)e^{-j\omega t}dt \quad \text{(B.20)}$$

$$\text{Inverse Fourier transform}: \ x(t) = \frac{1}{2\pi}\int_{-\infty}^{\infty} X(\omega)e^{j\omega t}d\omega, \quad \text{(B.21)}$$

where $x(t)$ is a continuous time and $X(\omega)$ is a two-sided continuous spectrum vs. the continuous frequency variable ω, where $-\infty < \omega < \infty$. Again, the spectrum is a complex function that can further be written as

$$X(\omega) = |X(\omega)|\angle\phi(\omega), \quad \text{(B.22)}$$

where $|X(\omega)|$ is the continuous amplitude spectrum, while $\angle\phi(\omega)$ designates the continuous phase spectrum.

EXAMPLE B.5

Let $x(t)$ be a single rectangular pulse, shown in Fig. B.8, where the pulse width is $\tau = 0.5$ s. Find its Fourier transform and sketch the amplitude spectrum.

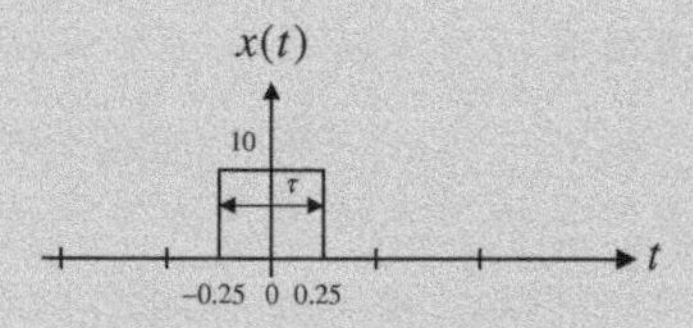

FIG. B.8

Rectangular pulse in Example B.5.

Solution:

Applying Eq. (B.21), and using the Euler's formula, we have

$$X(\omega) = \int_{-\infty}^{\infty} x(t)e^{-j\omega t}dt = \int_{-0.25}^{0.25} 10e^{-j\omega}dt$$

$$= 10\frac{e^{-j\omega t}}{-j\omega}\Bigg|_{-0.25}^{0.25} = 10 \times \frac{\left(e^{-j0.25\omega} - e^{j0.25\omega}\right)}{-j\omega}$$

$$= 10 \times \frac{\cos(0.25\omega) - j\sin(0.25\omega) - [\cos(0.25\omega) + j\sin(0.25\omega)]}{-j\omega}$$

$$= 5\frac{\sin(0.25\omega)}{0.25\omega}$$

where the amplitude spectrum is expressed as

$$|X(\omega)| = 5 \times \left|\frac{\sin(0.25\omega)}{0.25\omega}\right|.$$

Using $\omega = 2\pi f$, we can express the spectrum in terms of Hz as

$$|X(f)| = 5 \times \left| \frac{\sin(0.5\pi f)}{0.5\pi f} \right|.$$

The amplitude spectrum is shown in Fig. B.9. Note that the first null point is at $\omega = 2\pi/0.5 = 4\pi$ rad/s, and the spectrum is symmetric.

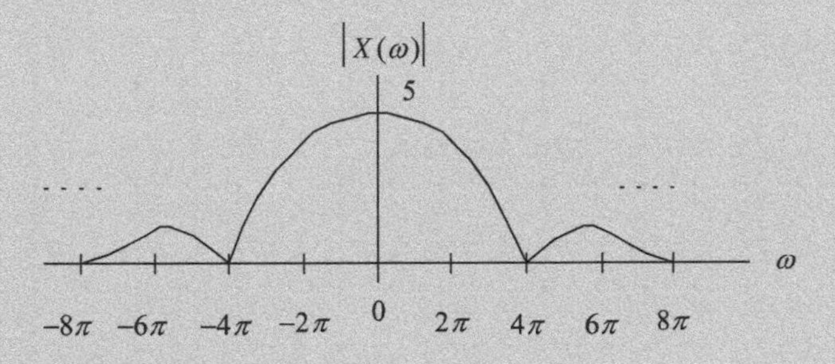

FIG. B.9

Amplitude spectrum for Example B.5.

EXAMPLE B.6

Let $x(t)$ be an exponential function given by

$$x(t) = 10e^{-2t}u(t) = \begin{cases} 10e^{-2t} & t \geq 0 \\ 0 & t < 0 \end{cases}.$$

Find its Fourier transform.

Solution:

According to the definition of the Fourier Transform,

$$X(\omega) = \int_0^{\infty} 10e^{-2t}u(t)e^{-j\omega t}dt = \int_0^{\infty} 10e^{-(2+j\omega)t}dt$$

$$= \frac{10e^{-(2+j\omega)t}}{-(2+j\omega)}\Bigg|_0^{\infty} = \frac{10}{2+j\omega}$$

$$X(\omega) = \frac{10}{\sqrt{2^2+\omega^2}} \angle -\tan^{-1}\left(\frac{\omega}{2}\right)$$

Using $\omega = 2\pi f$, we get

$$X(f) = \frac{10}{2+j2\pi f} = \frac{10}{\sqrt{2^2+(2\pi f)^2}} \angle -\tan^{-1}(\pi f).$$

The Fourier transforms for some common signals are listed in Table B.3. Some useful properties of the Fourier transform are summarized in Table B.4.

Table B.3 Fourier Transform for Some Common Signals

Time domain signal $x(t)$	Fourier spectrum $X(f)$
Rectangular pulse 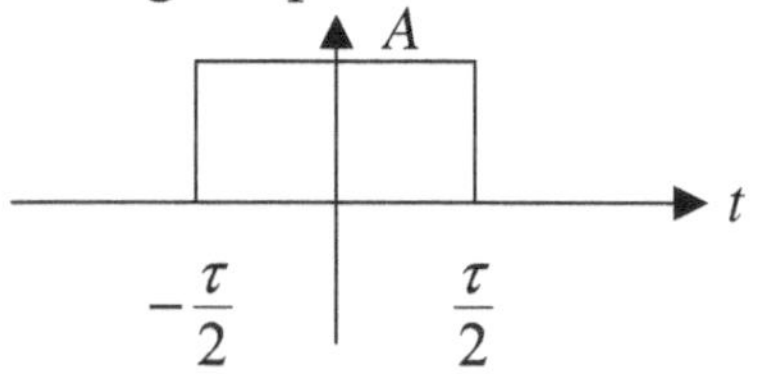	$X(f)=A\tau\frac{\sin\pi f\tau}{\pi f\tau}$
Triangular pulse 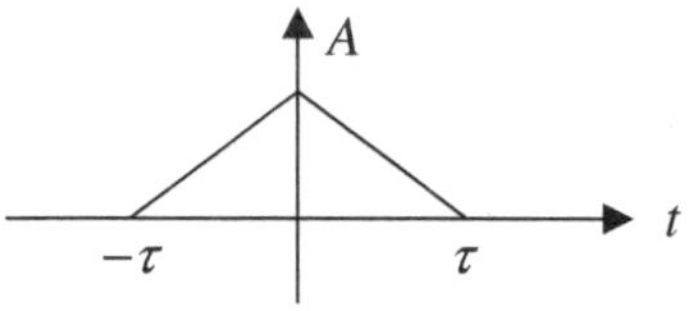	$X(f)=A\tau\left(\frac{\sin\pi f\tau}{\pi f\tau}\right)^2$
Cosine pulse 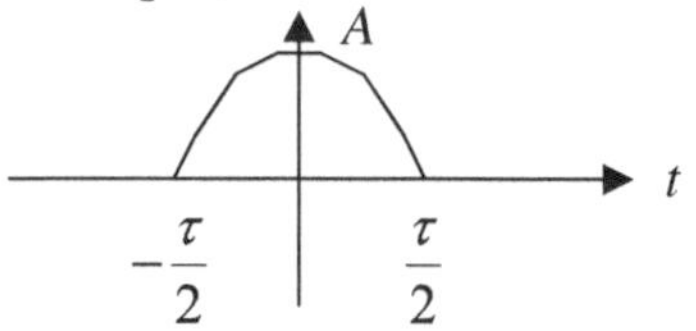	$X(f)=\frac{2A\tau}{\pi}\left(\frac{\cos\pi f\tau}{1-4f^2\tau^2}\right)$
Sawtooth pulse 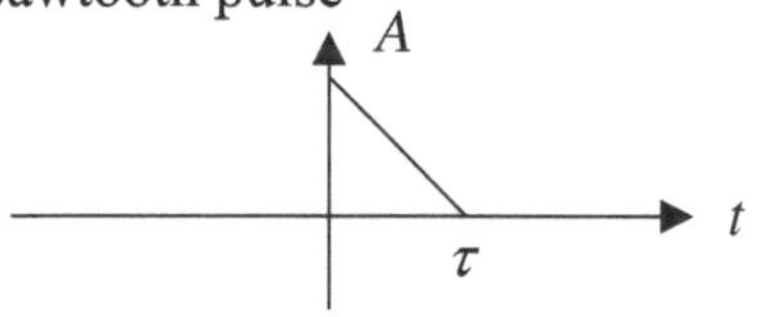	$X(f)=\frac{jA}{2\pi f}\left(\frac{\sin\pi f\tau}{\pi f\tau}e^{-j\pi f\tau}-1\right)$
Exponential function $\alpha=\frac{1}{\tau}$ 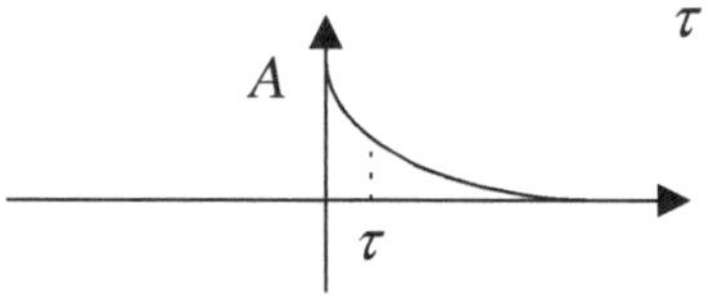	$X(f)=\frac{A}{\alpha+j2\pi f}$
Impulse function 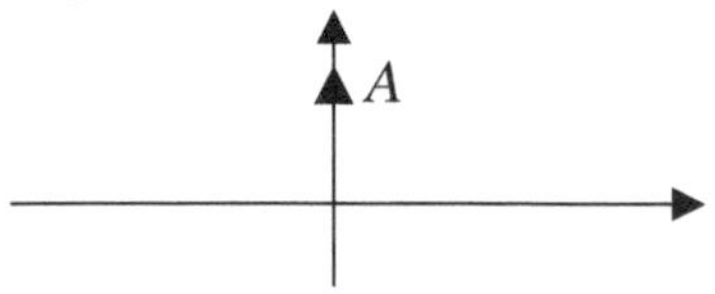	$X(f)=A$

Table B.4 Properties of Fourier Transform

Line	Time Function	Fourier Transform
1	$\alpha x_1(t)+\beta x_2(t)$	$\alpha X_1(f)+\beta X_2(f)$
2	$\frac{dx(t)}{dt}$	$j2\pi fX(f)$
3	$\int_{-\infty}^{t} x(t)dt$	$\frac{X(f)}{j2\pi f}$
4	$x(t-\tau)$	$e^{-j2\pi f\tau}X(f)$
5	$e^{j2\pi f_0 t}x(t)$	$X(f-f_0)$
6	$x(at)$	$\frac{1}{a}X\left(\frac{f}{a}\right)$

EXAMPLE B.7

Find the Fourier transforms of the following functions:
(a) $x(t)=\delta(t)$, where $\delta(t)$ is an impulse function defined by

$$\delta(t)=\begin{cases} \neq 0 & t=0 \\ 0 & \text{elsewhere} \end{cases}$$

with a property given as

$$\int_{-\infty}^{\infty} f(t)\delta(t-\tau)dt=f(\tau)$$

(b)

$$x(t)=\delta(t-\tau)$$

Solution:
(a) We first use the Fourier transform definition and then applying the delta function property,

$$X(\omega)=\int_{-\infty}^{\infty}\delta(t)e^{-j\omega t}dt=e^{-j\omega t}\big|_{t=0}=1.$$

(b) Similar to (a), we obtain

$$X(\omega)=\int_{-\infty}^{\infty}\delta(t-\tau)e^{-j\omega t}dt=e^{-j\omega t}\big|_{t=\tau}=e^{-j\omega\tau}.$$

Example B.8 shows how to use the table information to determine the Fourier transform of the non-periodic signal.

EXAMPLE B.8

Use Table B.3 to determine the Fourier transform for the following cosine pulse (Fig. B.10).

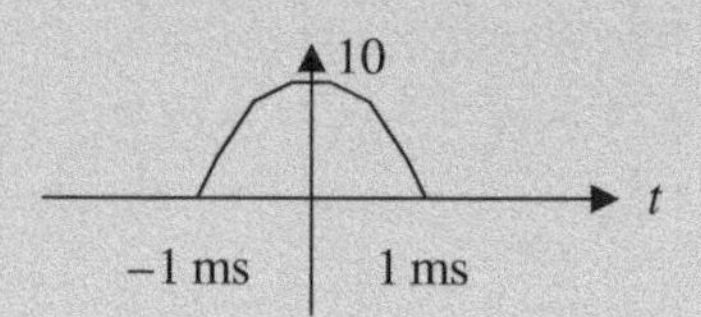

FIG. B.10

Cosine pulse in Example B.8.

Solution:

According to the graph, we can identify

$$\frac{\tau}{2} = 1ms, \text{ and } A = 1.$$

τ is given by

$$\tau = 2 \times 1ms = 0.002 \text{ seconds.}$$

Applying the formula from Table B.3 gives

$$X(f) = \frac{2 \times 10 \times 0.002}{\pi}\left(\frac{\cos \pi f 0.002}{1 - 4f^2 0.002^2}\right) = \frac{0.04}{\pi}\left(\frac{\cos 0.002\pi f}{1 - 4 \times 0.002^2 f^2}\right).$$

B.2 LAPLACE TRANSFORM

In this section, we will review Laplace transform and its applications.

B.2.1 LAPLACE TRANSFORM AND ITS TABLE

The Laplace transform plays an important role in analysis of continuous signals and systems. We define the Laplace transform pairs as

$$X(s) = L\{x(t)\} = \int_0^{\infty} x(t)e^{-st}dt \tag{B.23}$$

$$x(t) = L^{-1}\{X(s)\} = \frac{1}{2\pi j}\int_{\gamma - j\infty}^{\gamma + j\infty} X(s)e^{st}ds. \tag{B.24}$$

Note that the symbol $L\{\}$ denotes the forward Laplace operation, while the symbol $L^{-1}\{\}$ indicates the inverse Laplace operation. Some common Laplace transform pairs are listed in Table B.5.

In Example B.9, we examine the Laplace transform in light of its definition.

Table B.5 Laplace Transform Table

Line	Time Function $x(t)$	Laplace Transform $X(s)=L(x(t))$
1	$\delta(t)$	1
2	1 or $u(t)$	$\frac{1}{s}$
3	$tu(t)$	$\frac{1}{s^2}$
4	$e^{-at}u(t)$	$\frac{1}{s+a}$
5	$\sin(\omega t)u(t)$	$\frac{\omega}{s^2+\omega^2}$
6	$\cos(\omega t)u(t)$	$\frac{s}{s^2+\omega^2}$
7	$\sin(\omega t+\theta)u(t)$	$\frac{s\sin(\theta)+\omega\cos(\theta)}{s^2+\omega^2}$
8	$e^{-at}\sin(\omega t)u(t)$	$\frac{\omega}{(s+a)^2+\omega^2}$
9	$e^{-at}\cos(\omega t)u(t)$	$\frac{s+a}{(s+a)^2+\omega^2}$
10	$\left(A\cos(\omega t)+\frac{B-aA}{\omega}\sin(\omega t)\right)e^{-at}u(t)$	$\frac{As+B}{(s+a)^2+\omega^2}$
11a	$t^n u(t)$	$\frac{n!}{s^{n+1}}$
11b	$\frac{1}{(n-1)!}t^{n-1}u(t)$	$\frac{1}{s^n}$
12a	$e^{-at}t^n u(t)$	$\frac{n!}{(s+a)^{n+1}}$
12b	$\frac{1}{(n-1)!}e^{-at}t^{n-1}u(t)$	$\frac{1}{(s+a)^n}$
13	$(2\text{Real}(A)\cos(\omega t)-2\text{Imag}(A)\sin(\omega t))e^{-at}u(t)$	$\frac{A}{s+a-j\omega}+\frac{A^*}{s+a+j\omega}$
14	$\frac{dx(t)}{dt}$	$sX(s)-x(0^-)$
15	$\int_0^t x(t)dt$	$\frac{X(s)}{s}$
16	$x(t-a)u(t-a)$	$e^{-as}X(s)$
17	$e^{-at}x(t)u(t)$	$X(s+a)$

EXAMPLE B.9

Derive the Laplace transform of the unit step function.

Solution:

By the definition in Eq. (B.23),

$$X(s)=\int_0^{\infty}u(t)e^{-st}dt$$
$$=\int_0^{\infty}e^{-st}dt=\left.\frac{e^{-st}}{-s}\right|_0^{\infty}=\frac{e^{-\infty}}{-s}-\frac{e^0}{-s}=\frac{1}{s}.$$

The answer is consistent with the result listed in Table B.5.

Now we use the results in Table B.5 to find the Laplace transform of a function. Let us look at Example B.10.

EXAMPLE B.10

Perform the Laplace transform for each of the following functions.
(a) $x(t) = 5\sin(2t)u(t)$
(b) $x(t) = 5e^{-3t}\cos(2t)u(t)$.

Solution:

(a) Using line 5 in Table B.5 and noting that $\omega = 2$, the Laplace transform immediately follows that

$$X(s) = 5L\{2\sin(2t)u(t)\}$$
$$= \frac{5 \times 2}{s^2 + 2^2} = \frac{10}{s^2 + 4}.$$

(b) Applying line 9 in Table B.5 with $\omega = 2$ and $a = 3$ yields

$$X(s) = 5L\{e^{-3t}\cos(2t)u(t)\}$$
$$= \frac{5(s+3)}{(s+3)^2 + 2^2} = \frac{5(s+3)}{(s+3)^2 + 4}.$$

B.2.2 SOLVING DIFFERENTIAL EQUATIONS USING LAPLACE TRANSFORM

One of the important applications of the Laplace transform is to solve differential equations. Using the differential property in Table B.5, we can transform a differential equation from the time domain to the Laplace domain. This will change the differential equation into an algebraic equation, and we then solve the algebraic equation. Finally, the inverse Laplace operation is processed to yield the time domain solution.

EXAMPLE B.11

Solve the following differential equation using the Laplace transform:
$\frac{dy(t)}{dt} + 10y(t) = x(t)$ with an initial condition: $y(0) = 0$,
where the input $x(t) = 5u(t)$.

Solution:

Applying the Laplace transform on both sides of the differential equation and using the differential property (line 14 in Table B.5), we get

$$sY(s) - y(0) + 10Y(s) = X(s).$$

Note that

$$X(s) = L\{5u(t)\} = \frac{5}{s}.$$

Substituting the initial condition yields

$$Y(s) = \frac{5}{s(s+10)}.$$

Then we use a partial fraction expansion by writing

$$Y(s) = \frac{A}{s} + \frac{B}{s+10}$$

where

$$A = sY(s)|_{s=0} = \left.\frac{5}{s+10}\right|_{s=0} = 0.5$$

and

$$B = (s+10)Y(s)|_{s=-10} = \left.\frac{5}{s}\right|_{s=-10} = -0.5.$$

Hence,

$$Y(s) = \frac{0.5}{s} - \frac{0.5}{s+10}.$$

$$y(t) = L^{-1}\left\{\frac{0.5}{s}\right\} - L^{-1}\left\{\frac{0.5}{s+10}\right\}.$$

Finally, applying the inverse of the Laplace transform leads to using the results listed in Table B.5, we obtain the time domain solution as

$$y(t) = 0.5u(t) - 0.5e^{-10t}u(t).$$

B.2.3 TRANSFER FUNCTION

A linear analog system can be described using the Laplace transfer function. The transfer function relating the input and output of a linear system is depicted as

$$Y(s) = H(s)X(s), \tag{B.25}$$

where $X(s)$ and $Y(s)$ are the system input and response (output), respectively, in the Laplace domain, and the transfer function is defined as a ratio of the Laplace response of the system to the Laplace input under zero initial conditions given by

$$H(s) = \frac{Y(s)}{X(s)}. \tag{B.26}$$

The transfer function will allow us to study the system behavior. Considering an impulse function as the input to a linear system, that is, $x(t) = \delta(t)$, whose Laplace transform is $X(s) = 1$, we then find the system output due to the impulse function to be

$$Y(s) = H(s)X(s) = H(s). \tag{B.27}$$

Therefore, the response in the time domain $y(t)$ is called the impulse response of system and can be expressed as

$$h(t) = L^{-1}\{H(s)\}. \tag{B.28}$$

The analog impulse response can be sampled and transformed to obtain a digital filter transfer function as one of the applications. This topic is covered in Chapter 8.

EXAMPLE B.12

Consider a linear system described by the differential equation shown in Example B.11. $x(t)$ and $y(t)$ designate the system input and system output, respectively. Derive the transfer function and the impulse response of the system.

Solution:

Taking the Laplace transform on both sides of the differential equation yields

$$L\left\{\frac{dy(t)}{dt}\right\}+L\{10y(t)\}=L\{x(t)\}.$$

Applying the differential property and substituting the initial condition, we have

$$Y(s)(s+10)=X(s).$$

Thus, the transfer function is given by

$$H(s)=\frac{Y(s)}{X(s)}=\frac{1}{s+10}.$$

The impulse response can be found by taking the inverse Laplace transform as

$$h(t)=L^{-1}\left\{\frac{1}{s+10}\right\}=e^{-10t}u(t).$$

B.3 POLES, ZEROS, STABILITY, CONVOLUTION, AND SINUSOIDAL STEADY-STATE RESPONSE

This section is a review of analog system analysis.

B.3.1 POLES, ZEROS, AND STABILITY

To study the system behavior, the transfer function is written in a general form given by

$$H(s)=\frac{N(s)}{D(s)}=\frac{b_m s^m+b_{m-1}s^{m-1}+\cdots+b_0}{a_n s^n+a_{n-1}s^{n-1}+\cdots+a_0}. \tag{B.29}$$

It is a ratio of the numerator polynomial of degree of m to the denominator polynomial of degree n. The numerator polynomial is expressed as

$$N(s)=b_m s^m+b_{m-1}s^{m-1}+\cdots+b_0, \tag{B.30}$$

while the denominator polynomial is given by

$$D(s)=a_n s^n+a_{n-1}s^{n-1}+\cdots+a_0. \tag{B.31}$$

Again, the roots of $N(s)$ are called zeros, while the roots of $D(s)$ are called poles of the transfer function $H(s)$. Note that zeros and poles could be real numbers or complex numbers.

Given a system transfer function, the poles and zeros can be found. Further, a pole-zero plot can be created on the s-plane. Having the pole-zero plot, the stability of the system is determined by the following rules:

1. The linear system is stable if the rightmost pole(s) is(are) on the left-hand half plane (LHHP) on the s-plane.
2. The linear system is marginally stable if the rightmost pole(s) is(are) simple-order (first-order) on the $j\omega$ axis, including the origin on the s-plane.
3. The linear system is unstable if the rightmost pole(s) is(are) on the right-hand half plane (RHHP) of the s-plane or if the rightmost pole(s) is(are) multiple-order on the $j\omega$ axis on the s-plane.
4. Zeros do not affect the system stability.

EXAMPLE B.13

Determine whether each of the following transfer functions is stable, marginally stable, and unstable.

(a) $H(s)=\frac{s+1}{(s+1.5)(s^2+2s+5)}$

(b) $H(s)=\frac{(s+1)}{(s+2)(s^2+4)}$

(c) $H(s)=\frac{s+1}{(s-1)(s^2+2s+5)}$

Solution:

(a) A zero is found to be $s=-1$.
The poles are calculated as $s=-1.5$, $s=-1+j2$, $s=-1-j2$.
The pole-zero plot is shown in Fig. B.11(a). Since all the poles are located on the LHHP, the system is stable.

(b) A zero is found to be $s=-1$.
The poles are calculated as $s=-2$, $s=j2$, $s=-j2$.
The pole-zero plot is shown in Fig. B11(b). Since the first-order poles $s=\pm j2$ are located on the $j\omega$ axis, the system is marginally stable.

(c) A zero is found to be $s=-1$.
The poles are calculated as $s=1$, $s=-1+j2$, $s=-1-j2$.
The pole-zero plot is shown in Fig. B11(c). Since there is a pole $s=1$ located on the RHHP, the system is unstable.

B.3.2 CONVOLUTION

As we discussed before, the input and output relationship of a linear system in the Laplace domain is shown as

$$Y(s)=H(s)X(s). \tag{B.32}$$

It is apparent that in the Laplace domain, the system output is the product of the Laplace input and transfer function. But in the time domain, the system output is given as

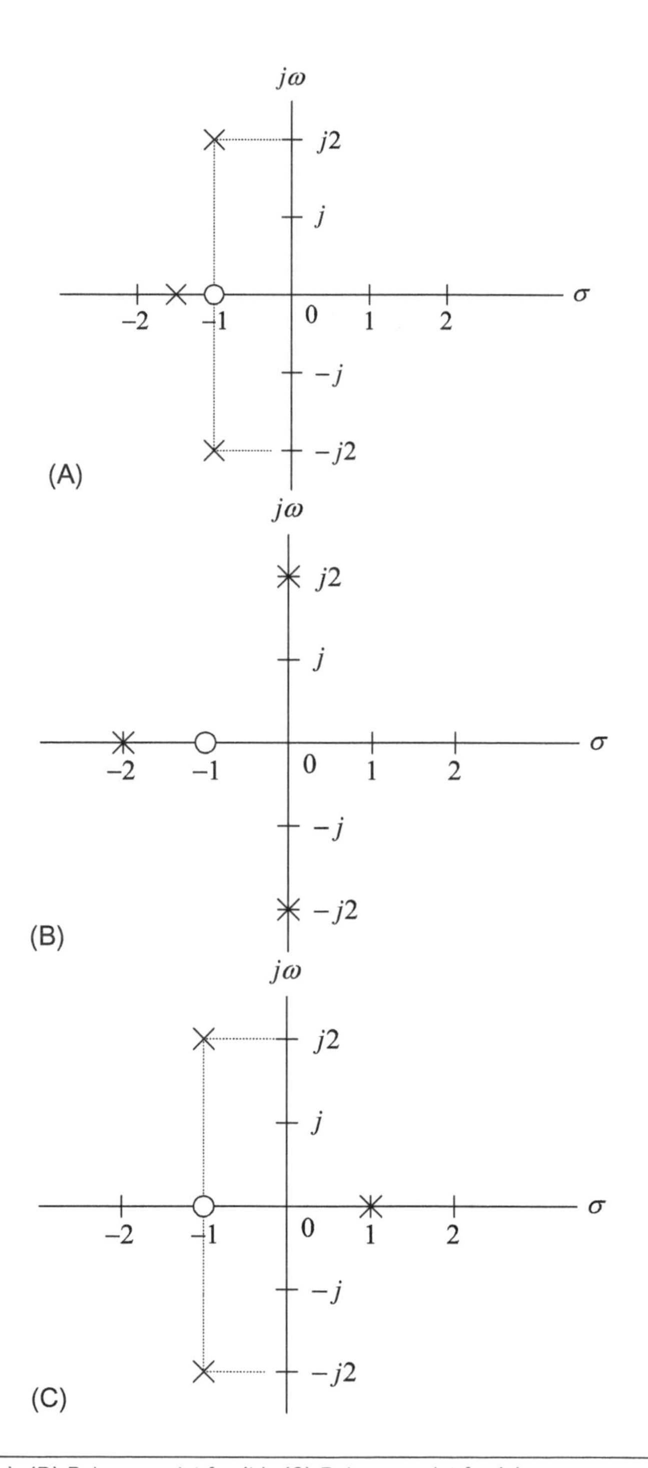

FIG. B.11

(A) Pole-zero plot for (a), (B) Pole-zero plot for (b), (C) Pole-zero plot for (c).

$$y(t) = h(t)*x(t), \tag{B.33}$$

where * denotes the linear convolution of the system impulse response $h(t)$ and the system input $x(t)$. The linear convolution is further expressed as

$$y(t) = \int_0^{\infty} h(\tau)x(t-\tau)d\tau. \tag{B.34}$$

EXAMPLE B.14

As in Examples B.11 and B.12, for a linear system, the impulse response and the input are given, respectively, by

$$h(t) = e^{-10t}u(t) \text{ and } x(t) = 5u(t).$$

Determine the system response $y(t)$ using the convolution method.

Solution:

Two signals $h(\tau)$ and $x(\tau)$ that are involved in the convolution integration are displayed in Fig. B.12. To evaluate the convolution, the time-reversed signal $x(-\tau)$ and the shifted signal $x(t-\tau)$ are also plotted for reference. Fig. B.12 shows an overlap of $h(\tau)$ and $x(t-\tau)$. According to the overlapped (shaded) area, the lower limit and the upper limit of the convolution integral are determined to be 0 and t, respectively. Hence,

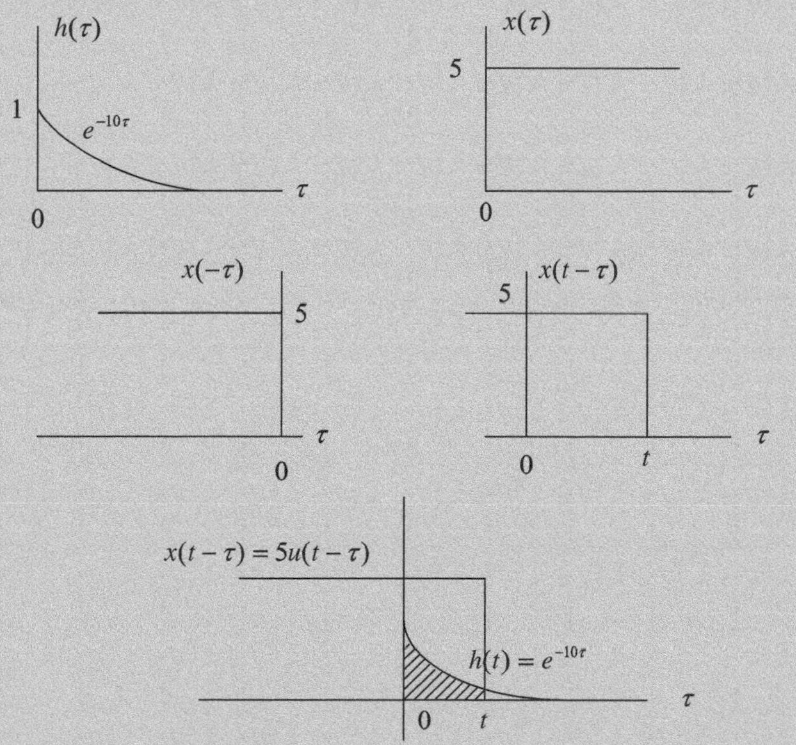

FIG. B.12

Convolution illustration for Example B.14.

$$y(t) = \int_0^t e^{-10\tau} \cdot 5d\tau = \frac{5}{-10} e^{-10\tau} \Big|_0^t$$
$$= -0.5e^{-10t} - \left(-0.5e^{-10\times 0}\right).$$

Finally, the system response is found to be

$$y(t) = 0.5u(t) - 0.5e^{-10t}u(t).$$

The solution is the same as that obtained using the Laplace transform method described in Example B.11.

B.3.3 SINUSOIDAL STEADY-STATE RESPONSE

For a linear analog system, if the input to system is a sinusoid of radian frequency ω, the steady-state response of the system will also be a sinusoid of the same frequency. Therefore, the transfer function, which provides the relationship between a sinusoidal input and a sinusoidal output, is called the steady-state transfer function. The steady-state transfer function is obtained from the Laplace transfer function by substituting $s = j\omega$, as shown in the following:

$$H(j\omega) = H(s)|_{s=j\omega}. \tag{B.35}$$

Thus we have a system relationship in a sinusoidal steady state as

$$Y(j\omega) = H(j\omega)X(j\omega). \tag{B.36}$$

Since $H(j\omega)$ is a complex function, we can write it in the phasor form:

$$H(j\omega) = A(\omega)\angle\beta(\omega), \tag{B.37}$$

where the quantity $A(\omega)$ is the amplitude response of the system defined as

$$A(\omega) = |H(j\omega)|, \tag{B.38}$$

and the phase angle $\beta(\omega)$ is the phase response of the system. The following example is presented to illustrate the application.

EXAMPLE B.15

Consider a linear system described by a differential equation shown in Example B.12, where $x(t)$ and $y(t)$ designate the system input and system output, respectively. The transfer function has been derived as

$$H(s) = \frac{10}{s+10}.$$

(a) Derive the steady-state transfer function.
(b) Derive the amplitude response and phase response.
(c) If the input is given as a sinusoid, that is, $x(t) = 5\sin(10t + 30^0)u(t)$, find the steady-state response $y_{ss}(t)$.

Solution:

1. By substituting $s = j\omega$ into the transfer function in terms of a suitable form, we get the steady-state transfer function as

$$H(j\omega) = \frac{1}{\frac{s}{10}+1} = \frac{1}{\frac{j\omega}{10}+1}.$$

2. The amplitude response and phase response are found to be

$$A(\omega) = \frac{1}{\sqrt{\left(\frac{\omega}{10}\right)^2+1}}$$

$$\beta(\omega) = \angle - \tan^{-1}\left(\frac{\omega}{10}\right).$$

3. When $\omega = 10$ rad/s, the input sinusoid can be written in terms of phasor form as

$$X(j10) = 5\angle 30^0.$$

For the amplitude and phase of the steady-state transfer function at $\omega = 10$, we have

$$A(10) = \frac{1}{\sqrt{\left(\frac{10}{10}\right)^2+1}} = 0.7071$$

$$\beta(10) = -\tan^{-1}\left(\frac{10}{10}\right) = -45^0.$$

Hence, we yield

$$H(j10) = 0.7071\angle -45^0.$$

Using Eq. (B.36), the system output in phasor form is obtained as

$$Y(j10) = H(j10)X(j10) = \left(1.4141\angle -45^0\right)\left(5\angle 30^0\right)$$

$$Y(j10) = 3.5355\angle -15^0.$$

Converting the output in phasor form back to the time domain results in the steady-state system output as

$$y_{ss}(t) = 3.5355 \sin\left(10t - 15^0\right)u(t).$$

B.4 PROBLEMS

B.1. Develop equations for the amplitude spectra, that is, A_k (one sided) and $|c_k|$ (two sided), for the pulse train $x(t)$ displayed in Fig. B.13,

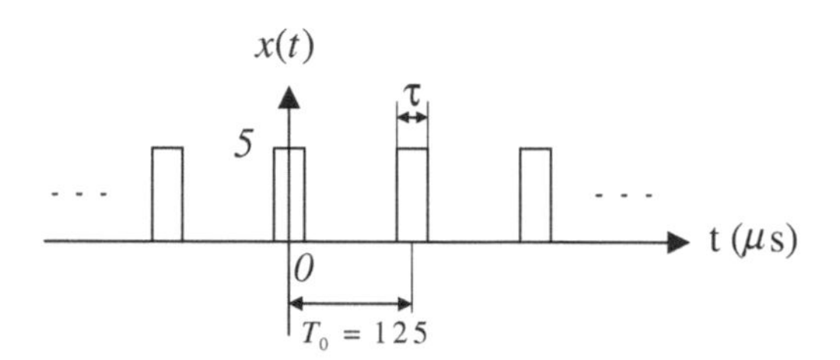

FIG. B.13

Pulse train in problem B.1.

where $\tau = 10$ s.

(a) Plot and label the one-sided amplitude spectrum up to 4 harmonic frequencies including DC.
(b) Plot and label the two-sided amplitude spectrum up to 4 harmonic frequencies including DC.

B.2. In the waveform shown in Fig. B.14, $T_0 = 1$ ms and $A = 10$. Use the formula in Table B.1 to write a Fourier series expansion in magnitude-phase form. Determine the frequency f_3 and amplitude value of A_3 for the third harmonic.

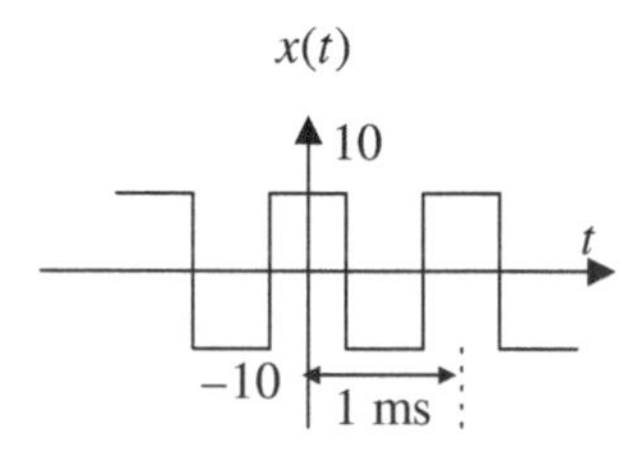

FIG. B.14

Square wave in problem B.2.

B.3. In the waveform shown in Fig. B.15, $T_0 = 1$ ms, $\tau = 0.2$ ms, and $A = 10$.

(a) Use the formula in Table B.1 to write a Fourier series expansion in magnitude-phase form.
(b) Determine the frequency f_2 and amplitude value of A_2 for the second harmonic.

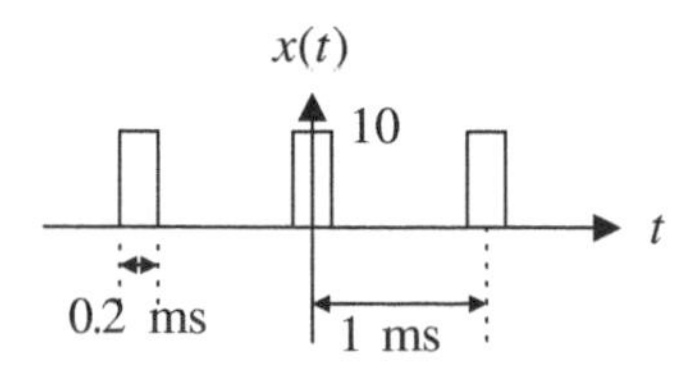

FIG. B.15

Rectangular wave in problem B.3.

B.4. Find the Fourier transform $X(\omega)$ and sketch the amplitude spectrum for the rectangular pulse $x(t)$ displayed in Fig. B.16.

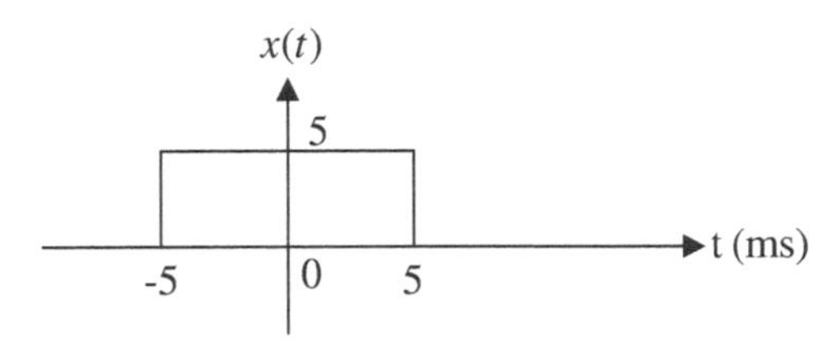

FIG. B.16

Rectangular pulse in problem B.4.

B.5. Use Table B.3 to determine the Fourier transform of the pulse in Fig. B.17.

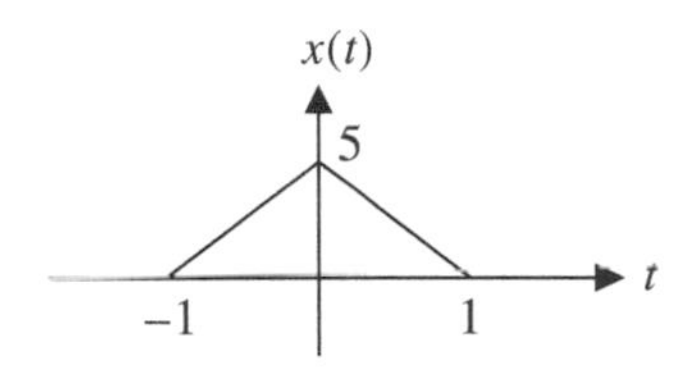

FIG. B.17

Triangular pulse in problem B.5.

B.6. Use Table B.3 to determine the Fourier transform for the pulse in Fig. B.18.

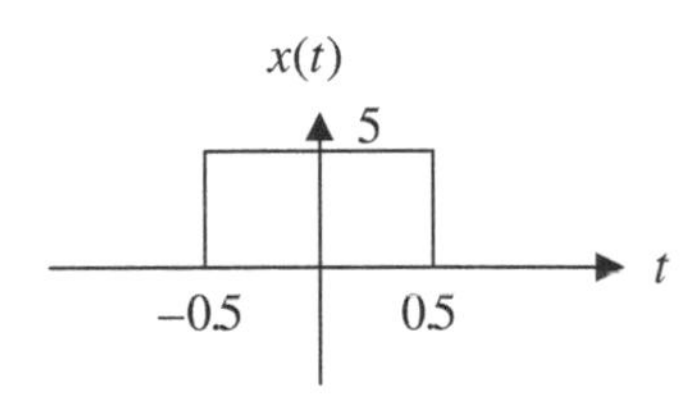

FIG. B.18

Rectangular pulse in problem B.6.

B.7. Determine the Laplace transform $X(s)$ for each of the following time domain functions using the Laplace transform pairs in Table B.5.

(a) $x(t)=10\delta(t)$
(b) $x(t)=-100tu(t)$
(c) $x(t)=10e^{-2t}u(t)$
(d) $x(t)=2u(t-5)$
(e) $x(t)=10\cos(3t)u(t)$
(f) $x(t)=10\sin(2t+45^0)u(t)$
(g) $x(t)=3e^{-2t}\cos(3t)u(t)$
(h) $x(t)=10t^5u(t)$

B.8. Determine the inverse transform of the analog signal $x(t)$ for each of the following functions using Table B.5 and partial fraction expansion.

(a) $X(s) = \frac{10}{s+2}$

(b) $X(s) = \frac{100}{(s+2)(s+3)}$

(c) $X(s) = \frac{100s}{s^2+7s+10}$

(d) $X(s) = \frac{25}{s^2+4s+29}$

B.9. Solve the following differential equation using the Laplace transform method:

$$2\frac{dx(t)}{dt} + 3x(t) = 15u(t) \text{ with } x(0) = 0.$$

(a) Determine$X(s)$.

(b) Determine the continuous signal $x(t)$ by taking the inverse Laplace transform of $X(s)$.

B.10. Solve the following differential equation using the Laplace transform method.

$$\frac{d^2x(t)}{dt^2} + 3\frac{dx(t)}{dt} + 2x(t) = 10u(t) \text{ with } x'(0) = 0 \text{ and } x(0) = 0.$$

(a) Determine$X(s)$.

(b) Determine $x(t)$ by taking the inverse Laplace transform of $X(s)$.

B.11. Determine the locations of all finite zeros and poles in the following functions. In each case, make an s-plane plot of the poles and zeros, and determine whether the given transfer function is stable, unstable, or marginally stable.

(a) $H(s) = \frac{(s-3)}{(s^2+4s+4)}$

(b) $H(s) = \frac{s(s^2+5)}{(s^2+9)(s^2+2s+4)}$

(c) $H(s) = \frac{(s^2+1)(s+1)}{s(s^2+7s-8)(s+3)(s+4)}$

B.12. Given the transfer function of a system

$$H(s) = \frac{5}{s+5},$$

and the input $x(t) = u(t)$,

(a) determine the system impulse response $h(t)$;

(b) determine the system Laplace output based on $Y(s) = H(s)X(s)$;

(c) determine the system response $y(t)$ in the time domain by taking the inverse Laplace transform of $Y(s)$.

B.13. Given the transfer function of a system

$$H(s) = \frac{5}{s+5}$$

(a) determine the steady-state transfer function;
(b) determine the amplitude response and phase response in terms of the frequency ω;
(c) determine the steady-state response of the system output $y_{ss}(t)$ in time domain using the results from (b), given an input to the system as $x(t) = 5\sin(2t)u(t)$.

B.14. Given the transfer function of a system

$$H(s) = \frac{5}{s+5},$$

and the input $x(t) = u(t)$, determine the system output $y(t)$ using the convolution method; that is, $y(t) = h(t) * x(t)$.

APPENDIX

NORMALIZED BUTTERWORTH AND CHEBYSHEV FUNCTIONS

C.1 NORMALIZED BUTTERWORTH FUNCTION

The normalized Butterworth squared magnitude function is given by

$$|P_n(\omega)|^2 = \frac{1}{1+\varepsilon^2(\omega)^{2n}}, \tag{C.1}$$

where n is the order and ε is the specified ripple on filter passband. The specified ripple in dB is expressed as $\varepsilon_{dB} = 10 \cdot \log_{10}(1+\varepsilon^2)$ dB.

To develop the transfer function $P_n(s)$, we first let $s = j\omega$ and then substitute $\omega^2 = -s^2$ into Eq. (C.1) to obtain

$$P_n(s)P_n(-s) = \frac{1}{1+\varepsilon^2(-s^2)^n}. \tag{C.2}$$

Eq. (C.2) has $2n$ poles, and $P_n(s)$ has n poles on the left-hand half plane (LHHP) on the s-plane, while $P_n(-s)$ has n poles on the right-hand half plane (RHHP) on the s-plane. Solving for poles leads to

$$(-1)^n s^{2n} = -1/\varepsilon^2. \tag{C.3}$$

If n is an odd number, Eq. (C.3) becomes

$$s^{2n} = 1/\varepsilon^2$$

and the corresponding poles are solved as

$$p_k = \varepsilon^{-1/n} e^{j\frac{2\pi k}{2n}} = \varepsilon^{-1/n}[\cos(2\pi k/2n) + j\sin(2\pi k/2n)], \tag{C.4}$$

where $k = 0, 1, \cdots, 2n-1$. Thus in the phasor form, we have

$$r = \varepsilon^{-1/n}, \text{ and } \theta_k = 2\pi k/(2n) \text{ for } k = 0, 1, \cdots, 2n-1. \tag{C.5}$$

When n is an even number, it follows that

$$s^{2n} = -1/\varepsilon^2$$

$$p_k = \varepsilon^{-1/n} e^{j\frac{2\pi k+\pi}{2n}} = \varepsilon^{-1/n}[\cos((2\pi k+\pi)/2n) + j\sin((2\pi k+\pi)/2n)], \tag{C.6}$$

where $k=0, 1, \cdots, 2n-1$. Similarly, the phasor form is given by

$$r=\varepsilon^{-1/n}, \text{ and } \theta_k=(2\pi k+\pi)/(2n) \text{ for } k=0,1,\cdots,2n-1. \tag{C.7}$$

When n is an odd number, we can identify the poles on the LHHP as

$$\begin{aligned} p_k &= -r, k=0 \text{ and} \\ p_k &= -r\cos(\theta_k)+jr\sin(\theta_k), k=1,\cdots,(n-1)/2 \end{aligned} \tag{C.8}$$

Using complex conjugate pairs, we have

$$p_k^*=-r\cos(\theta_k)-jr\sin(\theta_k).$$

Note that

$$(s-p_k)(s-p_k^*)=s^2+(2r\cos(\theta_k))s+r^2,$$

and a factor from the real pole $(s+r)$, it follows that

$$P_n(s)=\frac{K}{(s+r)\prod_{k=1}^{(n-1)/2}(s^2+(2r\cos(\theta_k))s+r^2)} \tag{C.9}$$

and $\theta_k=2\pi k/(2n)$ for $k=1, \cdots, (n-1)/2$.

Setting $P_n(0)=1$ for the unit passband gain leads to

$$K=r^n=1/\varepsilon.$$

When n is an even number, we can identify the poles on the LHHP as

$$p_k=-r\cos(\theta_k)+jr\sin(\theta_k), k=0,1,\cdots,n/2-1. \tag{C.10}$$

Using complex conjugate pairs, we have

$$p_k^*=-r\cos(\theta_k)-jr\sin(\theta_k).$$

The transfer function is given by

$$\begin{aligned} P_n(s) &= \frac{K}{\prod_{k=0}^{n/2-1}(s^2+(2r\cos(\theta_k))s+r^2)} \\ \theta_k &= (2\pi k+\pi)/(2n) \text{ for } k=0,1,\cdots,n/2-1 \end{aligned} \tag{C.11}$$

Setting $P_n(0)=1$ for the unit passband gain, we have

$$K=r^n=1/\varepsilon.$$

Let us examine the following examples.

EXAMPLE C.1

Compute the normalized Butterworth transfer function for the following specifications:
Ripple = 3 dB

$n = 2$

Solution:

$$n/2 = 1$$

$$\theta_k = (2\pi \times 0 + \pi)/(2 \times 2) = 0.25\pi$$

$$\varepsilon^2 = 10^{0.1 \times 3} - 1,$$

$$r = 1, \text{ and } K = 1.$$

Applying Eq. (C.11) leads to

$$P_2(s) = \frac{1}{s^2 + 2 \times 1 \times \cos(0.25\pi)s + 1^2} = \frac{1}{s^2 + 1.4141s + 1}.$$

EXAMPLE C.2

Compute the normalized Butterworth transfer function for the following specifications:
Ripple = 3 dB

$n = 3$

Solution:

$$(n-1)/2 = 1$$

$$\varepsilon^2 = 10^{0.1 \times 3} - 1,$$

$$r = 1, \text{ and } K = 1$$

$$\theta_k = (2\pi \times 1)/(2 \times 3) = \pi/3.$$

From Eq. (C.9), we have

$$P_3(s) = \frac{1}{(s+1)(s^2 + 2 \times 1 \times \cos(\pi/3)s + 1^2)} = \frac{1}{(s+1)(s^2 + s + 1)}.$$

For the unfactored form, we can carry out

$$P_3(s) = \frac{1}{s^3 + 2s^2 + 2s + 1}.$$

EXAMPLE C.3

Compute the normalized Butterworth transfer function for the following specifications:
Ripple $=1.5\,\text{dB}$

$n=3$

Solution:

$$(n-1)/2=1$$

$$\varepsilon^2=10^{0.1\times 1.5}-1,$$

$$r=1.1590, \text{ and } K=1.5569$$

$$\theta_k=(2\pi\times 1)/(2\times 3)=\pi/3.$$

Applying Eq. (C.9), we achieve the normalized Butterworth transfer function as

$$P_3(s)=\frac{1}{(s+1.1590)(s^2+2\times 1.1590\times\cos(\pi/3)s+1.1590^2)}$$
$$=\frac{1}{(s+1)(s^2+1.1590s+1.3433)}.$$

For the unfactored form, we can carry out

$$P_3(s)=\frac{1.5569}{s^3+2.3180s^2+2.6866s+1.5569}.$$

C.2 NORMALIZED CHEBYSHEV FUNCTION

Similar to analog Butterworth filter design, the transfer function is derived from the normalized Chebyshev function, and the result is usually listed in a table for design reference. The Chebyshev magnitude response function with an order of n and the normalized cutoff frequency $\omega=1$ radian per second is given by

$$|B_n(\omega)|=\frac{1}{\sqrt{1+\varepsilon^2 C_n^2(\omega)}}, n\geq 1, \tag{C.12}$$

where the function $C_n(\omega)$ is defined as

$$C_n(\omega)=\begin{cases}\cos(n\cos^{-1}(\omega)) & \omega\leq 1\\ \cosh(n\cosh^{-1}(\omega)) & \omega>1\end{cases}, \tag{C.13}$$

where ε is the ripple specification on the filter passband. Note that

$$\cosh^{-1}(x)=\ln\left(x+\sqrt{x^2-1}\right). \tag{C.14}$$

To develop the transfer function $B_n(s)$, we let $s=j\omega$ and substitute $\omega^2=-s^2$ into Eq. (C.12) to obtain

$$B_n(s)B_n(-s)=\frac{1}{1+\varepsilon^2 C_n^2(s/j)}. \tag{C.15}$$

The poles can be found from

$$1+\varepsilon^2 C_n^2(s/j)=0$$

or

$$C_n(s/j)=\cos\left(n\cos^{-1}(s/j)\right)=\pm j1/\varepsilon. \tag{C.16}$$

Introduce a complex variable $v=\alpha+j\beta$ such that

$$v=\alpha+j\beta=\cos^{-1}(s/j), \tag{C.17}$$

we can then write

$$s=j\cos(v). \tag{C.18}$$

Substituting Eq. (C.17) into Eq, (C.16) and using trigonometric identities, it follows that

$$\begin{aligned} C_n(s/j) &= \cos(n\cos^{-1}(s/j)) \\ &= \cos(nv)=\cos(n\alpha+jn\beta) \\ &= \cos(n\alpha)\cosh(n\beta)-j\sin(n\alpha)\sinh(n\beta)=\pm j1/\varepsilon. \end{aligned} \tag{C.19}$$

To solve Eq. (C.19), the following conditions must be satisfied:

$$\cos(n\alpha)\cosh(n\beta)=0 \tag{C.20}$$

$$-\sin(n\alpha)\sinh(n\beta)=\pm 1/\varepsilon. \tag{C.21}$$

Since $\cosh(n\beta)\geq 1$ in Eq. (C.20), we must let

$$\cos(n\alpha)=0, \tag{C.22}$$

which therefore leads to

$$\alpha_k=(2k+1)\pi/(2n), k=0,1,2,\cdots,2n-1. \tag{C.23}$$

With Eq. (C.23), we have $\sin(n\alpha_k)=\pm 1$. Then Eq. (C.21) becomes

$$\sinh(n\beta)=1/\varepsilon. \tag{C.24}$$

Solving Eq. (C.24) gives

$$\beta=\sinh^{-1}(1/\varepsilon)/n. \tag{C.25}$$

Again from Eq. (C.18),

$$\begin{aligned} &s=j\cos(v)=j[\cos(\alpha_k)\cosh(\beta)-j\sin(\alpha_k)\sinh(\beta)] \\ &\text{for } k=0,1,\cdots,2n-1 \end{aligned}. \tag{C.26}$$

The poles can be found from Eq. (C.26):

$$\begin{aligned} &p_k=\sin(\alpha_k)\sinh(\beta)+j\cos(\alpha_k)\cosh(\beta) \\ &\text{for } k=0,1,\cdots,2n-1 \end{aligned}. \tag{C.27}$$

Using Eq. (C.27), if n is an odd number, the poles on the left-hand side are solved to be

$$p_k=\sin(\alpha_k)\sinh(\beta)+j\cos(\alpha_k)\cosh(\beta) \text{ for } k=0,1,\cdots,2n-1. \tag{C.28}$$

Using complex conjugate pairs, we have

$$p_k^* = -\sin(\alpha_k)\sinh(\beta) - j\cos(\alpha_k)\cosh(\beta) \tag{C.29}$$

and a real pole

$$p_k = -\sinh(\beta), k = (n-1)/2. \tag{C.30}$$

Note that

$$(s - p_k)(s - p_k^*) = s^2 + b_k s + c_k \tag{C.31}$$

and a factor from the real pole $[s + \sinh(\beta)]$, it follows that

$$B_n(s) = \frac{K}{[s + \sinh(\beta)] \prod_{k=0}^{(n-1)/2-1} (s^2 + b_k s + c_k)}, \tag{C.32}$$

where

$$b_k = 2\sin(\alpha_k)\sinh(\beta) \tag{C.33}$$

$$c_k = [\sin(\alpha_k)\sinh(\beta)]^2 + [\cos(\alpha_k)\cosh(\beta)]^2 \tag{C.34}$$

$$\alpha_k = (2k+1)\pi/(2n) \text{ for } k = 0, 1, \cdots, (n-1)/2 - 1. \tag{C.35}$$

For the unit passband gain and the filter order as an odd number, we set $B_n(0) = 1$. Then

$$K = \sinh(\beta) \prod_{k=0}^{(n-1)/2-1} c_k \tag{C.36}$$

$$\beta = \sinh^{-1}(1/\varepsilon)/n \tag{C.37}$$

$$\sinh^{-1}(x) = \ln\left(x + \sqrt{x^2+1}\right). \tag{C.38}$$

Following the similar procedure for the even number of n, we have

$$B_n(s) = \frac{K}{\prod_{k=0}^{n/2-1} (s^2 + b_k s + c_k)} \tag{C.39}$$

$$b_k = 2\sin(\alpha_k)\sinh(\beta) \tag{C.40}$$

$$c_k = [\sin(\alpha_k)\sinh(\beta)]^2 + [\cos(\alpha_k)\cosh(\beta)]^2 \tag{C.41}$$

$$\text{where } \alpha_k = (2k+1)\pi/(2n) \text{ for } k = 0, 1, \cdots, n/2 - 1. \tag{C.42}$$

For the unit passband gain and the filter order as an even number, we require that $B_n(0) = 1/\sqrt{1+\varepsilon^2}$, so that the maximum magnitude of the ripple on passband equals 1. Then we have

$$K = \prod_{k=0}^{n/2-1} c_k / \sqrt{1+\varepsilon^2} \tag{C.43}$$

$$\beta = \sinh^{-1}(1/\varepsilon)/n \tag{C.44}$$

$$\sinh^{-1}(x) = \ln\left(x + \sqrt{x^2+1}\right). \tag{C.45}$$

Eqs. (C.32) to (C.45) are applied to compute the normalized Chebyshev transfer function. Now let us look at the following illustrative examples.

EXAMPLE C.4

Compute the normalized Chebyshev transfer function for the following specifications:
Ripple = 0.5 dB

$n=2$

Solution:

$$n/2 = 1.$$

Applying Eqs. (C.39) to (C.45), we obtain

$$\alpha_0 = (2\times 0+1)\pi/(2\times 2) = 0.25\pi$$

$$\varepsilon^2 = 10^{0.1\times 0.5} - 1 = 0.1220, 1/\varepsilon = 2.8630$$

$$\beta = \sinh^{-1}(2.8630)/n = \ln\left(2.8630 + \sqrt{2.8630^2+1}\right)/2 = 0.8871$$

$$b_0 = 2\sin(0.25\pi)\sinh(0.8871) = 1.4256$$

$$c_0 = [\sin(0.25\pi)\sinh(0.8871)]^2 + [\cos(0.25\pi)\cosh(0.8871)]^2 = 1.5162$$

$$K = 1.5162/\sqrt{1+0.1220} = 1.4314.$$

Finally, the transfer function is derived as

$$B_2(s) = \frac{1..4314}{s^2 + 1.4256s + 1.5162}.$$

EXAMPLE C.5

Compute the normalized Chebyshev transfer function for the following specifications:
Ripple = 1 dB

$n=3$

Solution:

$$(n-1)/2 = 1.$$

Applying Eqs. (C.32) to (C.38) leads to

$$\alpha_0 = (2\times 0+1)\pi/(2\times 3) = \pi/6$$

$$\varepsilon^2 = 10^{0.1\times 1} - 1 = 0.2589, 1/\varepsilon - 1.9653$$

$$\beta = \sinh^{-1}(1.9653)/n = \ln\left(1.9653 + \sqrt{1.9653^2+1}\right)/3 = 0.4760$$

$$b_0 = 2\sin(\pi/6)\sinh(0.4760) = 0.4942$$

$$c_0 = [\sin(\pi/6)\sinh(0.4760)]^2 + [\cos(\pi/6)\cosh(0.4760)]^2 = 0.9942$$

$$\sinh(\beta) = \sinh(0.4760) = 0.4942$$

$$K = 0.4942 \times 0.9942 = 0.4913.$$

We can conclude the transfer function as

$$B_3(s) = \frac{0.4913}{(s + 0.4942)(s^2 + 0.4942s + 0.9942)}.$$

Finally, the unfactored form is found to be

$$B_3(s) = \frac{0.4913}{s^3 + 0.9883s^2 + 1.2384s + 0.4913}.$$

APPENDIX

SINUSOIDAL STEADY-STATE RESPONSE OF DIGITAL FILTERS

D.1 SINUSOIDAL STEADY-STATE RESPONSE

Analysis of the sinusoidal steady-state response of the digital filters will lead to the development of the magnitude and phase responses of digital filters. Let us look at the following digital filter with a digital transfer function $H(z)$ and a complex sinusoidal input

$$x(n) = Ve^{j(\Omega n + \varphi_x)}, \tag{D.1}$$

where $\Omega = \omega T$ is the normalized digital frequency, while T is the sampling period and $y(n)$ denotes the digital output, as shown in Fig. D.1.

The z-transform output from the digital filter is then given by

$$Y(z) = H(z)X(z). \tag{D.2}$$

Since $X(z) = \frac{Ve^{j\varphi_x} z}{z - e^{j\Omega}}$, we have

$$Y(z) = \frac{Ve^{j\varphi_x} z}{z - e^{j\Omega}} H(z). \tag{D.3}$$

Based on the partial fraction expansion, $Y(z)/z$ can be expanded as the following form:

$$\frac{Y(z)}{z} = \frac{Ve^{j\varphi_x}}{z - e^{j\Omega}} H(z) = \frac{R}{z - e^{j\Omega}} + \text{sum of the rest of partial fractions}. \tag{D.4}$$

Multiplying the factor $(z - e^{j\Omega})$ on both sides of Eq. (D.4) yields

$$Ve^{j\phi_x} H(z) = R + \left(z - e^{j\Omega}\right)(\text{sum of the rest of partial fractions}). \tag{D.5}$$

Substituting $z = e^{j\Omega}$, we get the residue as

$$R = Ve^{j\phi_x} H\left(e^{j\Omega}\right).$$

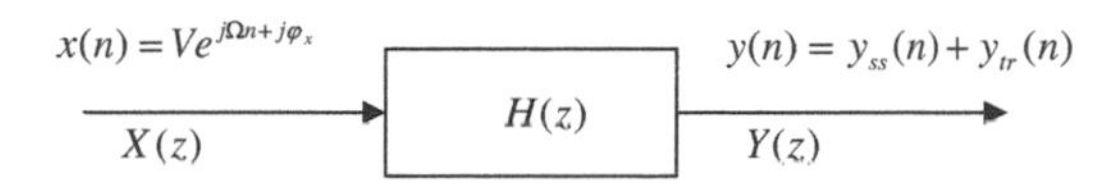

FIG. D.1

Steady-state response of the digital filter.

Then substituting $R = Ve^{j\phi_x}H(e^{j\Omega})$ back into Eq. (D.4) results in

$$\frac{Y(z)}{z} = \frac{Ve^{j\varphi_x}H(e^{j\Omega})}{z - e^{j\Omega}} + \text{sum of the rest of partial fractions,} \tag{D.6}$$

and multiplying z on both sides of Eq. (D.6) leads to

$$Y(z) = \frac{Ve^{j\varphi_x}H(e^{j\Omega})z}{z - e^{j\Omega}} + z \times \text{sum of the rest of partial fractions.} \tag{D.7}$$

Taking the inverse z-transform leads to two parts of the solution:

$$y(n) = Ve^{j\varphi_x}H\left(e^{j\Omega}\right)e^{j\Omega n} + Z^{-1}(z \times \text{sum of the rest of partial fractions}). \tag{D.8}$$

From Eq. (D.8), we have the steady-state response

$$y_{ss}(n) = Ve^{j\varphi_x}H\left(e^{j\Omega}\right)e^{j\Omega n} \tag{D.9}$$

and the transient response

$$y_{tr}(n) = Z^{-1}(z \times \text{sum of the rest of partial fractions}). \tag{D.10}$$

Note that since the digital filter is a stable system, and the locations of its poles must be inside the unit circle on the z-plane, the transient response will be settled to zero eventually. To develop filter magnitude and phase responses, we write the digital steady-state response as

$$y_{ss}(n) = V\left|H\left(e^{j\Omega}\right)\right|e^{j\Omega + j\varphi_x + \angle H\left(e^{j\Omega}\right)}. \tag{D.11}$$

Comparing Eq. (D.11) and Eq. (D.1), it follows that

$$\begin{aligned}\text{Magnitude response} &= \frac{\text{Amplitude of the steady state output}}{\text{Amplitude of the sinusoidal input}}\\ &= \frac{V\left|H\left(e^{j\Omega}\right)\right|}{V} = \left|H\left(e^{j\Omega}\right)\right|\end{aligned} \tag{D.12}$$

$$\text{Phase response} = \frac{e^{j\varphi_x + j\angle H\left(e^{j\Omega}\right)}}{e^{j\varphi_x}} = e^{j\angle H\left(e^{j\Omega}\right)} = \angle H\left(e^{j\Omega}\right). \tag{D.13}$$

Thus we conclude that

$$\text{Frequency response} = H\left(e^{j\Omega}\right) = H(z)\big|_{z=e^{j\Omega}}. \tag{D.14}$$

Since $H(e^{j\Omega}) = |H(e^{j\Omega})| \angle H(e^{j\Omega})$

$$\text{Magnitude response} = \left|H\left(e^{j\Omega}\right)\right| \tag{D.15}$$

$$\text{Phase response} = \angle H\left(e^{j\Omega}\right). \tag{D.16}$$

D.2 PROPERTIES OF THE SINUSOIDAL STEADY-STATE RESPONSE

From Euler's identity and trigonometric identity, we know that

$$\begin{aligned}e^{j(\Omega + k2\pi)} &= \cos(\Omega + k2\pi) + j\sin(\Omega + k2\pi)\\ &= \cos\Omega + j\sin\Omega = e^{j\Omega},\end{aligned} \tag{D.17}$$

where k is an integer taking values of $k=0, \pm 1, \pm 2, \cdots$. Then:

$$\text{Frequency response}: H\left(e^{j\Omega}\right) = H\left(e^{j(\Omega + k2\pi)}\right) \tag{D.18}$$

$$\text{Magnitude frequency response}: \left|H\left(e^{j\Omega}\right)\right| = \left|H\left(e^{j(\Omega + k2\pi)}\right)\right| \tag{D.19}$$

$$\text{Phase response}: \angle H\left(e^{j\Omega}\right) = \angle H\left(e^{j\Omega + 2k\pi}\right). \tag{D.20}$$

Clearly, the frequency response is periodical, with a period of 2π. Next, let us develop the symmetric properties. Since the transfer function is written as

$$H(z) = \frac{Y(z)}{X(z)} = \frac{b_0 + b_1 z^{-1} + \cdots + b_M z^{-M}}{1 + a_1 z^{-1} + \cdots + a_N z^{-N}}, \tag{D.21}$$

Substituting $z = e^{j\Omega}$ into Eq. (D.21) yields

$$H\left(e^{j\Omega}\right) = \frac{b_0 + b_1 e^{-j\Omega} + \cdots + b_M e^{-jM\Omega}}{1 + a_1 e^{-j\Omega} + \cdots + a_N e^{-jN\Omega}}. \tag{D.22}$$

Using Euler's identity, $e^{-j\Omega} = \cos\Omega - j\sin\Omega$, we have

$$H\left(e^{j\Omega}\right) = \frac{(b_0 + b_1\cos\Omega + \cdots + b_M\cos M\Omega) - j(b_1\sin\Omega + \cdots + b_M\sin M\Omega)}{(1 + a_1\cos\Omega + \cdots + a_N\cos N\Omega) - j(a_1\sin\Omega + \cdots + a_N\sin N\Omega)}. \tag{D.23}$$

Similarly,

$$H\left(e^{-j\Omega}\right) = \frac{(b_0 + b_1\cos\Omega + \cdots + b_M\cos M\Omega) + j(b_1\sin\Omega + \cdots + b_M\sin M\Omega)}{(1 + a_1\cos\Omega + \cdots + a_N\cos N\Omega) + j(a_1\sin\Omega + \cdots + a_N\sin N\Omega)}. \tag{D.24}$$

Then the magnitude response and phase response can be expressed as

$$\left|H\left(e^{j\Omega}\right)\right| = \frac{\sqrt{(b_0 + b_1\cos\Omega + \cdots + b_M\cos M\Omega)^2 + (b_1\sin\Omega + \cdots + b_M\sin M\Omega)^2}}{\sqrt{(1 + a_1\cos\Omega + \cdots + a_N\cos N\Omega)^2 + (a_1\sin\Omega + \cdots + a_N\sin N\Omega)^2}} \tag{D.25}$$

$$\angle H\left(e^{j\Omega}\right) = \tan^{-1}\left(\frac{-(b_1\sin\Omega + \cdots + b_M\sin M\Omega)}{b_0 + b_1\cos\Omega + \cdots + b_M\cos M\Omega}\right)$$

$$-\tan^{-1}\left(\frac{-(a_1\sin\Omega + \cdots + a_N\sin N\Omega)}{1 + a_1\cos\Omega + \cdots + a_N\cos N\Omega}\right) \tag{D.26}$$

Based in Eq. (D.24), we also have the magnitude and phase response for $H(e^{-j\Omega})$ as

$$\left|H\left(e^{-j\Omega}\right)\right| = \frac{\sqrt{(b_0 + b_1\cos\Omega + \cdots + b_M\cos M\Omega)^2 + (b_1\sin\Omega + \cdots + b_M\sin M\Omega)^2}}{\sqrt{(1 + a_1\cos\Omega + \cdots + a_N\cos N\Omega)^2 + (a_1\sin\Omega + \cdots + a_N\sin N\Omega)^2}} \tag{D.27}$$

$$\angle H\left(e^{-j\Omega}\right) = \tan^{-1}\left(\frac{b_1\sin\Omega + \cdots + b_M\sin M\Omega}{b_0 + b_1\cos\Omega + \cdots + b_M\cos M\Omega}\right)$$

$$-\tan^{-1}\left(\frac{a_1\sin\Omega + \cdots + a_N\sin N\Omega}{1 + a_1\cos\Omega + \cdots + a_N\cos N\Omega}\right). \tag{D.28}$$

Comparing (D.25) with (D.27) and (D.26) with (D.28), respectively, we conclude the symmetric properties as

$$\left|H\left(e^{-j\Omega}\right)\right| = \left|H\left(e^{j\Omega}\right)\right| \tag{D.29}$$

$$\angle H\left(e^{-j\Omega}\right) = -\angle H\left(e^{j\Omega}\right) \tag{D.30}$$

APPENDIX

FINITE IMPULSE RESPONSE FILTER DESIGN EQUATIONS BY FREQUENCY SAMPLING DESIGN METHOD

Recall in Section 7.5 on the "Frequency Sampling Design Method":

$$h(n) = \frac{1}{N}\sum_{k=0}^{N-1} H(k) W_N^{-kn}, \tag{E.1}$$

where $h(n)$, $0 \le n \le N-1$, is the causal impulse response that approximates the finite impulse response (FIR) filter, and $H(k)$, $0 \le k \le N-1$, represents the corresponding coefficients of the discrete Fourier transform (DFT), and $W_N = e^{-j\frac{2\pi}{N}}$. We further write DFT coefficients, $H(k)$, $0 \le k \le N-1$, into polar form:

$$H(k) = H_k e^{j\varphi_k}, \ 0 \le k \le N-1, \tag{E.2}$$

where H_k and φ_k are the kth magnitude and the phase angle, respectively. The frequency response of the FIR filter is expressed as

$$H(e^{j\Omega}) = \sum_{n=0}^{N-1} h(n) e^{-jn\Omega}. \tag{E.3}$$

Substituting (E.1) into (E.3) yields

$$H(e^{j\Omega}) = \sum_{n=0}^{N-1} \frac{1}{N} \sum_{k=0}^{N-1} H(k) W_N^{-kn} e^{-j\Omega n}. \tag{E.4}$$

Interchanging the order of the summation in Eq. (E.4) leads to

$$H(e^{j\Omega}) = \frac{1}{N}\sum_{k=0}^{N-1} H(k) \sum_{n=0}^{N-1} \left(W_N^{-k} e^{-j\Omega}\right)^n. \tag{E.5}$$

Since

$$W_N^{-k} e^{-j\Omega} = \left(e^{-j2\pi/N}\right)^{-k} e^{-j\Omega} = e^{-(j\Omega - 2\pi k/N)},$$

and using the identity

$$\sum_{n=0}^{N-1} r^n = 1 + r + r^2 + \cdots + r^{N-1} = \frac{1-r^N}{1-r},$$

we can write the second summation in Eq. (E.5) as

$$\sum_{n=0}^{N-1}\left(W_N^{-k}e^{-j\Omega}\right)^n=\frac{1-e^{-j(\Omega-2\pi k/N)N}}{1-e^{-j(\Omega-2\pi k/N)}}. \tag{E.6}$$

Using the Euler formula leads Eq. (E.6) to

$$\begin{aligned}\sum_{n=0}^{N-1}\left(W_N^{-k}e^{-j\Omega}\right)^n&=\frac{e^{-jN(\Omega-2\pi k/N)/2}\left(e^{jN(\Omega-2\pi k/N)/2}-e^{-jN(\Omega-2\pi k/N)/2}\right)/2j}{e^{-j(\Omega-2\pi k/N)/2}\left(e^{j(\Omega-2\pi k/N)/2}-e^{-j(\Omega-2\pi k/N)/2}\right)/2j}\\&=\frac{e^{-jN(\Omega-2\pi k/N)/2}\sin\left[N(\Omega-2\pi k/N)/2\right]}{e^{-j(\Omega-2\pi k/N)/2}\sin\left[(\Omega-2\pi k/N)/2\right]}.\end{aligned} \tag{E.7}$$

Substituting Eq. (E.7) into Eq. (E.5) leads to

$$H\left(e^{j\Omega}\right)=\frac{1}{N}e^{-j(N-1)\Omega/2}\sum_{k=0}^{N-1}H(k)e^{j(N-1)k\pi/N}\frac{\sin\left[N(\Omega-2\pi k/N)/2\right]}{\sin\left[(\Omega-2\pi k/N)/2)\right]}. \tag{E.8}$$

Let $\Omega=\Omega_m=\frac{2\pi m}{N}$, and substitute it into Eq. (E.8) we get

$$H\left(e^{j\Omega_m}\right)=\frac{1}{N}e^{-j(N-1)2\pi m/(2N)}\sum_{k=0}^{N-1}H(k)e^{j(N-1)k\pi/N}\frac{\sin\left[N(2\pi m/N-2\pi k/N)/2\right]}{\sin\left[(2\pi m/N-2\pi k/N)/2)\right]}. \tag{E.9}$$

Clearly, when $m\neq k$, the last term of the summation in Eq. (E.9) becomes

$$\frac{\sin\left[N(2\pi m/N-2\pi k/N)/2\right]}{\sin\left[(2\pi m/N-2\pi k/N)/2)\right]}=\frac{\sin(\pi(m-k))}{\sin(\pi(m-k)/N)}=\frac{0}{\sin(\pi(m-k)/N)}=0.$$

when $m=k$, and using L'Hospital's rule, we have

$$\frac{\sin\left[N(2\pi m/N-2\pi k/N)/2\right]}{\sin\left[(2\pi m/N-2\pi k/N)/2)\right]}=\frac{\sin(N\pi(m-k)/N)}{\sin(\pi(m-k)/N)}=\lim_{x\to 0}\frac{\sin(Nx)}{\sin(x)}=N.$$

Then Eq. (E.9) is simplified to

$$H\left(e^{j\Omega_k}\right)=\frac{1}{N}e^{-j(N-1)\pi k/N}H(k)e^{j(N-1)k\pi/N}N=H(k).$$

That is,

$$H\left(e^{j\Omega_k}\right)=H(k),\ 0\le k\le N-1, \tag{E.10}$$

where $\Omega_k=\frac{2\pi k}{N}$, corresponding to the kth DFT frequency component. The fact is that if we specify the desired frequency response, $H(\Omega_k)$, $0\le k\le N-1$, at the equally spaced sampling frequency determined by $\Omega_k=\frac{2\pi k}{N}$, they are actually the DFT coefficients; that is, $H(k)$, $0\le k\le N-1$, via Eq. (E.10). Furthermore, the inverse of the DFT using (E.10) will give the desired impulse response, $h(n)$, $0\le n\le N-1$.

To devise the design procedure, we substitute Eq. (E.2) in Eq. (E.8) to obtain

$$H\left(e^{j\Omega}\right)=\frac{1}{N}e^{-j(N-1)\Omega/2}\sum_{k=0}^{N-1}H_k e^{j\varphi_k+j(N-1)k\pi/N}\frac{\sin\left[N(\Omega-2\pi k/N)/2\right]}{\sin\left[(\Omega-2\pi k/N)/2)\right]}. \tag{E.11}$$

It is required that the frequency response of the designed FIR filter expressed in Eq. (E.11) be a linear phase. This can easily be accomplished by setting

$$\varphi_k + (N-1)k\pi/N = 0,\ 0 \le k \le N-1 \tag{E.12}$$

in Eq. (E.11) so that the summation part becomes a real value, thus resulting in the linear phase of $H(e^{j\Omega})$, since only one complex term, $e^{-j(N-1)\Omega/2}$, is left, which presents the constant time delay of the transfer function. Second, the sequence $h(n)$ must be real. To proceed, let $N=2M+1$, and due to the properties of DFT coefficients, we have

$$\overline{H}(k) = H(N-k),\ 1 \le k \le M \tag{E.13}$$

where the bar indicates complex conjugate. Note the fact that

$$\overline{W_N^{-k}} = W_N^{-(N-k)},\ 1 \le k \le M. \tag{E.14}$$

From Eq. (E.1), we write

$$h(n) = \frac{1}{N}\left(H(0) + \sum_{k=1}^{M} H(k)W_N^{-kn} + \sum_{k=M+1}^{2M} H(k)W_N^{-kn}\right). \tag{E.15}$$

Eq. (E.15) is equivalent to

$$h(n) = \frac{1}{N}\left(H(0) + \sum_{k=1}^{M} H(k)W_N^{-kn} + \sum_{k=1}^{M} H(N-k)W_N^{-(N-k)n}\right).$$

Using Eqs. (E.13) and (E.14) in the last summation term leads to

$$\begin{aligned} h(n) &= \frac{1}{N}\left(H(0) + \sum_{k=1}^{M} H(k)W_N^{-kn} + \sum_{k=1}^{M} \overline{H}(k)\overline{W_N^{-kn}}\right) \\ &= \frac{1}{2M+1}\left(H(0) + \sum_{k=1}^{M}\left(H(k)W_N^{-kn} + \overline{H}(k)\overline{W_N^{-kn}}\right)\right). \end{aligned}$$

Combining the last two summation terms, we achieve

$$h(n) = \frac{1}{2M+1}\left\{H(0) + 2\mathrm{Re}\left(\sum_{k=1}^{M} H(k)W_N^{-kn}\right)\right\},\ 0 \le n \le N-1. \tag{E.16}$$

Solving Eq. (E.12) gives

$$\varphi_k = -(N-1)k\pi/N,\ 0 \le k \le N-1. \tag{E.17}$$

Again, note that Eq. (E.13) is equivalent to

$$H_k e^{-j\varphi_k} = H_{N-k}e^{j\varphi_{N-k}},\ 1 \le k \le M. \tag{E.18}$$

Substituting (E.17) in (E.18) yields

$$H_k e^{j(N-1)k\pi/N} = H_{N-k}e^{-j(N-1)(N-k)\pi/N}, 1 \le \mathrm{k} \le M. \tag{E.19}$$

Simplification of Eq. (E.19) leads the following result:

$$H_k = H_{N-k}e^{-j(N-1)\pi} = (-1)^{N-1}H_{N-k}, 1 \le k \le M. \tag{E.20}$$

Since we constrain the filter length to be $N=2M+1$, Eq. (E.20) can be further reduced to

$$H_k=(-1)^{2M}H_{2M+1-k}=H_{2M+1-k}, 1\leq k\leq M. \tag{E.21}$$

Finally, by substituting (E.21) and (E.17) into (E.16), we obtain a very simple design equation:

$$h(n)=\frac{1}{2M+1}\left\{H_0+2\sum_{k=1}^{M}H_k\cos\left(\frac{2\pi k(n-M)}{2M+1}\right)\right\}, 0\leq n\leq 2M. \tag{E.22}$$

Thus the design procedure is simply summarized as follows: Given the filter length, $2M+1$, and the specified frequency response, H_k at $\Omega_k=\frac{2\pi k}{(2M+1)}$ for $k=0$, 1, ⋯, M, the FIR filter coefficients can be calculated via Eq. (E.22).

APPENDIX

WAVELET ANALYSIS AND SYNTHESIS EQUATIONS

F.1 BASIC PROPERTIES

The inter product of two functions is defined as

$$<x,y>=\int x(t)y(t)dt. \tag{F.1}$$

Two functions are orthogonal if

$$<x(t),x(t-k)>=\begin{cases} A & \text{for } k=0 \\ 0 & \text{for } k\neq 0 \end{cases}. \tag{F.2}$$

Two functions are orthonormal if

$$<x(t),x(t-k)>=\begin{cases} 1 & \text{for } k=0 \\ 0 & \text{for } k\neq 0 \end{cases}. \tag{F.3}$$

The signal energy is defined as

$$E=\int x^2(t)dt. \tag{F.4}$$

Many wavelet families are designed to be orthonormal:

$$E=\int \psi^2(t)dt=1 \tag{F.5}$$

$$E=\int \psi_{jk}^2(t)dt=\int \left[2^{j/2}\psi(2^jt-k)\right]^2dt=\int 2^j\psi^2(2^jt-k)dt. \tag{F.6}$$

Let $u=2^jt-k$. Then $du=2^jdt$.

Eq. (F.5) becomes

$$E=\int 2^j\psi^2(u)2^{-j}du=1. \tag{F.7}$$

Both father and mother wavelets are orthonormal at scale j:

$$\int \phi_{jk}(t)\phi_{jn}(t)dt=\begin{cases} 1 & k=n \\ 0 & \text{otherwise} \end{cases} \tag{F.8}$$

$$\int \psi_{jk}(t)\psi_{jn}(t)dt = \begin{cases} 1 & k=n \\ 0 & \text{otherwise} \end{cases}. \tag{F.9}$$

F.2 ANALYSIS EQUATIONS

When a function $f(t)$ is approximated using the scaling functions only at scale $j+1$, it can be expressed as

$$f(t) = \sum_{k=-\infty}^{\infty} c_j(k) 2^{j/2}\phi(2^j t - k).$$

Using the inner product,

$$c_j(k) = <f(t), \phi_{jk}(t)> = \int f(t) 2^{j/2}\phi(2^j t - k)dt. \tag{F.10}$$

Note that

$$\phi(t) = \sum_{n=-\infty}^{\infty} \sqrt{2}h_0(n)\phi(2t-n). \tag{F.11}$$

Substituting Eq. (F.11) into Eq. (F.10) leads to

$$c_j(k) = <f(t), \phi_{jk}(t)> = \int f(t) 2^{j/2} \sum_{n=-\infty}^{\infty} \sqrt{2}h_0(n)\phi\left[2(2^j t - k) - n\right]dt$$

$$c_j(k) = \sum_{n=-\infty}^{\infty} \int f(t) 2^{(j+1)/2} h_0(n)\phi\left(2^{(j+1)}t - 2k - n\right)dt.$$

Let $m = n+2k$. Interchange of the summation and integral leads to

$$c_j(k) = \sum_{m=-\infty}^{\infty} \int f(t) 2^{(j+1)/2} h_0(m-2k)\phi\left(2^{(j+1)}t - m\right)dt$$

$$c_j(k) = \sum_{m=-\infty}^{\infty} \left(\int f(t)\phi_{(j+1)m}(t)dt\right) h_0(m-2k). \tag{F.12}$$

Using the inner product definition for the DWT coefficient again in (F.12), we achieve

$$c_j(k) = \sum_{m=-\infty}^{\infty} <f(t), \phi_{(j+1)m}(t)> h_0(m-2k) = \sum_{m=-\infty}^{\infty} c_{j+1}(m)h_0(m-2k). \tag{F.13}$$

Similarly, note that

$$\psi(t) = \sum_{k=-\infty}^{\infty} \sqrt{2}h_1(k)\phi(2t-k).$$

Using the inner product gives

$$d_j(k) = <f(t), \psi_{jk}(t)> = \int f(t) 2^{j/2} \sum_{n=-\infty}^{\infty} \sqrt{2}h_1(n)\phi\left[2(2^j t - k) - n\right]dt$$

$$d_j(k)=\sum_{n=-\infty}^{\infty}\int f(t)2^{(j+1)/2}h_1(n)\phi\left(2^{(j+1)}t-2k-n\right)dt. \tag{F.14}$$

Let $m=n+2k$. Interchange of the summation and integral leads to

$$d_j(k)=\sum_{m=-\infty}^{\infty}\int f(t)2^{(j+1)/2}h_1(m-2k)\phi\left(2^{(j+1)}t-m\right)dt$$

$$d_j(k)=\sum_{m=-\infty}^{\infty}\left(\int f(t)\phi_{(j+1)m}(t)dt\right)h_1(m-2k). \tag{F.15}$$

Finally, applying the inner product definition for the WDT coefficient, we obtain

$$d_j(k)=\sum_{m=-\infty}^{\infty}<f(t),\phi_{(j+1)m}(t)>h_1(m-2k)=\sum_{m=-\infty}^{\infty}c_{j+1}(m)h_1(m-2k). \tag{F.16}$$

F.3 WAVELET SYNTHESIS EQUATIONS

We begin with

$$f(t)=\sum_{k=-\infty}^{\infty}c_j(k)2^{j/2}\phi(2^jt-k)+\sum_{k=-\infty}^{\infty}d_j(k)2^{j/2}\psi(2^jt-k).$$

Taking an inner product using the scaling function at scale level $j+1$ gives

$$c_{j+1}(k)=<f(t),\phi_{(j+1)k}(t)>=\sum_{m=-\infty}^{\infty}c_j(m)2^{j/2}\int\phi(2^jt-m)\phi_{(j+1)k}(t)dt$$
$$+\sum_{m=-\infty}^{\infty}d_j(m)2^{j/2}\int\psi(2^jt-m)\phi_{(j+1)k}(t)dt$$

$$c_{j+1}(k)=\sum_{m=-\infty}^{\infty}c_j(m)2^{j/2}\int\sum_{n=-\infty}^{\infty}\sqrt{2}h_0(n)\phi(2^{j+1}t-2m-n)\phi_{(j+1)k}(t)dt$$
$$+\sum_{m=-\infty}^{\infty}d_j(m)2^{j/2}\int\sum_{n=-\infty}^{\infty}\sqrt{2}h_1(n)\phi(2^{j+1}t-2m-n)\phi_{(j+1)k}(t)dt. \tag{F.17}$$

Interchange of the summation and integral, it yields

$$c_{j+1}(k)=\sum_{m=-\infty}^{\infty}\sum_{n=-\infty}^{\infty}c_j(m)h_0(n)\int 2^{(j+1)/2}\phi(2^{j+1}t-2m-n)\phi_{(j+1)k}(t)dt$$
$$+\sum_{m=-\infty}^{\infty}\sum_{n=-\infty}^{\infty}d_j(m)h_1(n)\int 2^{(j+1)/2}\phi(2^{j+1}t-2m-n)\phi_{(j+1)k}(t)dt.$$

Using the inner product, we get

$$c_{j+1}(k) = \sum_{m=-\infty}^{\infty} \sum_{n=-\infty}^{\infty} c_j(m)h_0(n) < \phi_{(j+1)(2m+n)}(t), \phi_{(j+1)k}(t) >$$
$$+ \sum_{m=-\infty}^{\infty} \sum_{n=-\infty}^{\infty} d_j(m)h_1(n) < \phi_{(j+1)(2m+n)}(t), \phi_{(j+1)k}(t) > . \tag{F.18}$$

From the wavelet orthonormal property, we have

$$< \phi_{(j+1)(2m+n)}(t), \phi_{(j+1)k}(t) >= \begin{cases} 1 & n = k - 2m \\ 0 & \text{otherwise} \end{cases}. \tag{F.19}$$

Substituting Eq. (F.19) into Eq. (F.18), we finally obtain

$$c_{j+1}(k) = \sum_{m=-\infty}^{\infty} c_j(m)h_0(k-2m) + \sum_{m=-\infty}^{\infty} d_j(m)h_1(k-2m). \tag{F.20}$$

APPENDIX

REVIEW OF DISCRETE-TIME RANDOM SIGNALS

In this Appendix, we will briefly review the important properties of the random variable, discrete-time random signal, and random process.

G.1 RANDOM VARIABLE STATISTICAL PROPERTIES

A random variable is often described using its probability distribution or its probability density function. The probability of a random variable X, which is the probability distribution function, is defined as

$$P_X(x) = \text{Probability}\,(X \le x). \tag{G.1}$$

The derivative of the probability distribution function is defined as the probability density function of a random variable X, which is given by

$$p_X(x) = \frac{\partial P_X(x)}{\partial x}. \tag{G.2}$$

Then the probability distribution function can be expressed as

$$P_X(x) = \int_{-\infty}^{x} p_X(\alpha)d\alpha. \tag{G.3}$$

The statistical properties of a random variable X such as the mean (expected value), the mean-square value $E(X^2)$, and the variance σ_X^2, are expressed as follows:

$$m_X = E(X) = \int_{-\infty}^{\infty} x p_X(x)dx \tag{G.4}$$

$$E(X^2) = \int_{-\infty}^{\infty} x^2 p_X(x)dx \tag{G.5}$$

$$\begin{aligned}\sigma_X^2 &= E\left(|X - m_X|^2\right) = E[(X - m_X)(X - m_X)^*]\\ &= \int_{-\infty}^{\infty} (x - m_X)(x - m_X)^* p_X(x)dx = E(X^2) - |m_X|^2\end{aligned} \tag{G.6}$$

where $E(\cdot)$ designates the expectation operator and "*" denotes complex conjugation; and σ_X is the standard deviation. As an example, the uniform probability density function depicted in Fig. G.1 is defined in G. 7.

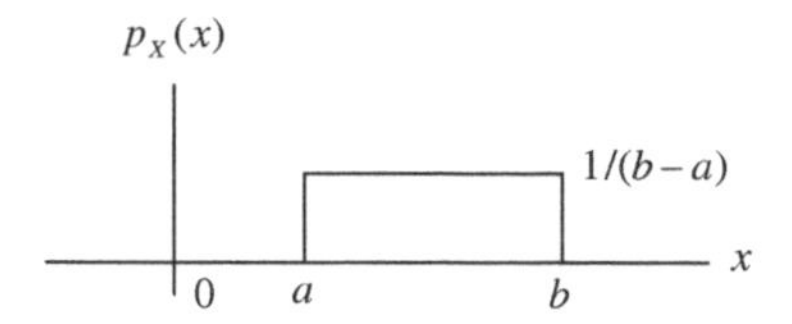

FIG. G.1

Uniform probability density function.

$$p_X(x) = \begin{cases} 1/(b-a) & a < x \le b \\ 0 & \text{elsewhere} \end{cases}. \tag{G.7}$$

The mean value, mean-square value, and variance of the uniformly distributed real random variable X can be derived in (G.8), (G.9), and (G.10), respectively.

$$m_X = \frac{1}{b-a}\int_a^b x dx = (b+a)/2 \tag{G.8}$$

$$E(X^2) = \frac{1}{b-a}\int_a^b x^2 dx = (b^2 + a^2 + ba)/3 \tag{G.9}$$

$$\sigma_X^2 = (b^2 + a^2 + ba)/3 - (b+a)^2/4 = (b-a)^2/12. \tag{G.10}$$

As another example, Fig. G.2 shows a Gaussian probability density function, which is defined in (G.11).

$$p_X(x) = \frac{1}{\sigma_X\sqrt{2\pi}} e^{-(x-m_X)^2/(2\sigma_X^2)}. \tag{G.11}$$

The mean value and mean-square value of the Gaussian distributed real random variable X can be determined by

$$E(X) = \int_{-\infty}^{\infty} \frac{x}{\sigma_X\sqrt{2\pi}} e^{-(x-m_X)^2/(2\sigma_X^2)} dx = m_Y \tag{G.12}$$

$$E(X^2) = \int_{-\infty}^{\infty} \frac{x^2}{\sigma_X\sqrt{2\pi}} e^{-(x-m_X)^2/(2\sigma_X^2)} dx = \sigma_X^2 - m_X^2. \tag{G.13}$$

For the case of two random variables X and Y, a joint probability distribution function is defined as

$$P_{XY}(x) = \text{Probability}\,(\text{X} \le x, Y \le y) \tag{G.14}$$

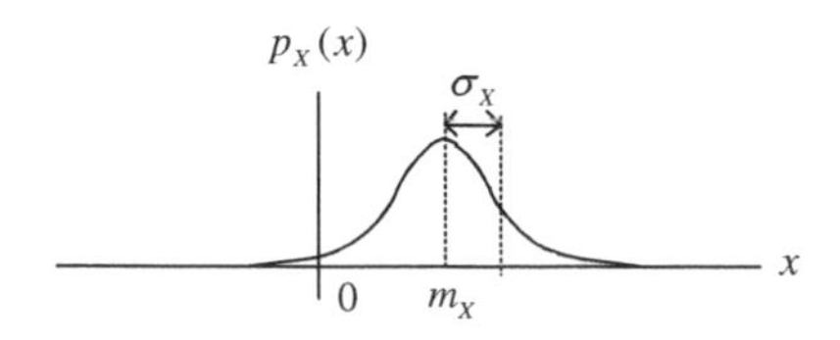

FIG. G.2

Gaussian probability density function.

After the joint distribution density function is defined as

$$p_{XY}(x,y)=\frac{\partial^2 P_X(x,y)}{\partial x \partial y}, \tag{G.15}$$

the joint distribution function can be expressed as

$$P_{XY}(x,y)=\int_{-\infty}^{x}\int_{-\infty}^{y} p_{XY}(\alpha,\beta)d\alpha d\beta. \tag{G.16}$$

The cross correlation and cross covariance of two random variables X and Y are, respectively, given by

$$\phi_{XY}=E(XY^*)=\int_{-\infty}^{\infty}\int_{-\infty}^{\infty} xyp_{XY}(x,y)dxdy \tag{G.17}$$

$$\begin{aligned}\gamma_{XY}&=E[(X-m_X)(Y-m_Y)^*]\\&=\int_{-\infty}^{\infty}\int_{-\infty}^{\infty}(x-m_X)(y-m_Y)^*p_{XY}(x,y)dxdy\\&=\phi_{XY}-m_X m_Y^*,\end{aligned} \tag{G.18}$$

where m_X and m_Y designate the mean of X and the mean of Y, respectively.

If X and Y are linear independent or uncorrelated, a relation in (G.19) holds

$$E(XY)=E(X)E(Y). \tag{G.19}$$

If X and Y are statistically independent, a relation in (G.20) holds

$$p_{XY}(x,y)=p_X(x)p_Y(y). \tag{G.20}$$

It can be seen that if X and Y are statistically independent, they are also linearly independent or uncorrelated; however, if X and Y are linearly independent (uncorrelated), they may not be statistically independent.

G.2 RANDOM SIGNAL STATISTICAL PROPERTIES

For the discrete-time signal $\{X(n)\}$ at time index n, the mean value, mean-square value, and variance are defined as follows:

$$m_{X(n)}=E[X(n)]=\int_{-\infty}^{\infty} xp_{X(n)}(x;n)dx \tag{G.21}$$

$$E\left[X^2(n)\right]=\int_{-\infty}^{\infty} x^2 p_{X(n)}(x;n)dx \tag{G.22}$$

$$\begin{aligned}\sigma^2_{X(n)}&=E\left[\left|X(n)-m_{X(n)}\right|^2\right]=E\left\{\left[X(n)-m_{X(n)}\right]\left[X(n)-m_{X(n)}\right]^*\right\}\\&=E\left[X^2(n)\right]-\left|m_{X(n)}\right|^2.\end{aligned} \tag{G.23}$$

The autocorrelation and auto-covariance are, respectively, defined by

$$\phi_{XX}(n,m)=E(X(m)X(n)^*)] \tag{G.24}$$

$$\begin{aligned}\gamma_{XX}(m,n) &= E\left[\left(X(m)-m_{X(m)}\right)\left(X(n)-m_{X(n)}\right)^*\right] \\ &= \phi_{XX}(m,n)-m_{X(m)}m_{X(n)}^*.\end{aligned} \tag{G.25}$$

Again, for the case of two discrete-time random signals $\{X(n)\}$ and $\{Y(m)\}$, the cross-correlation function is defined as

$$\phi_{XY}(m,n) = E(X(m)Y(n)^*) = \int_{-\infty}^{\infty}\int_{-\infty}^{\infty} xy^* p_{X(m)Y(n)}(x;m,y;n)dxdy. \tag{G.26}$$

And the cross-covariance function is determined by

$$\begin{aligned}\gamma_{XY}(m,n) &= E\left\{\left[X(m)-m_{X(m)}\right]\left[Y(n)-m_{Y(n)}\right]^*\right\} \\ &= \phi_{XY}(m,n)-m_{X(m)}m_{Y(n)}^*,\end{aligned} \tag{G.27}$$

where $p_{X(m)Y(n)}(x;m,y;n)$ denotes the joint probability density function of $X(m)$ and $Y(n)$. If $X(m)$ and $Y(n)$ are uncorrelated, then $\gamma_{XY}(m,n)=0$.

G.3 WIDE-SENSE STATIONARY RANDOM SIGNALS

In the wide-sense stationary (WSS) random process, the statistical properties are either independent of time or of time origin. The mean $E[X(n)]$ has the same constant value m_X for all the values of time index n while the auto-correlation and auto-covariance functions only depend on the difference $l=(m-n)$ of time indices m and n, that is,

$$m_X = E[X(n)] \tag{G.28}$$

$$\phi_{XX}(m,n) = \phi_{XX}(l+n,n) = \phi_{XX}(l) = E[X(m)X(n)^*)] = E[X(n+l)X(n)^*)] \tag{G.29}$$

$$\begin{aligned}\gamma_{XX}(m,n) &= \gamma_{XX}(n+l,n) = \gamma_{XX}(l) = E\{[X(m)-m_X][X(n)-m_X]^*\} \\ &= E\{[X(n+l)-m_X][X(n)-m_X]^*\} = \phi_{XY}(l)-|m_X|^2.\end{aligned} \tag{G.30}$$

The mean-square value and variance of a WSS random process $\{X(n)\}$ are expressed as

$$E\left[X^2(n)\right] = \phi_{XX}(0) \tag{G.31}$$

$$\sigma_X^2 = \gamma_{XX}(0) = \phi_{XY}(0)-|m_X|^2. \tag{G.32}$$

The cross-correlation and cross-covariance functions between two WSS random processes $\{X(n)\}$ and $\{Y(n)\}$ are defined as

$$\phi_{XY}(l) = E[X(n+l)Y(n)^*)] \tag{G.33}$$

$$\gamma_{XY}(l) = E\{[X(n+l)-m_X][Y(n)-m_Y]^*\} = \phi_{XY}(l)-m_X m_Y^*. \tag{G.34}$$

The symmetry properties give the following relations:

$$\phi_{XX}(-l) = \phi_{XX}^*(l) \tag{G.35}$$

$$\gamma_{XX}(-l) = \gamma_{XX}^*(l) \tag{G.36}$$

$$\phi_{XY}(-l) = \phi_{YX}^*(l) \tag{G.37}$$

$$\gamma_{XY}(-l) = \gamma_{YX}^*(l). \tag{G.38}$$

G.4 ERGODIC SIGNALS

For an ergodic signal, the time averages of its properties is a limit when the length of the realization goes to infinity equal to its ensemble averages, which are derived from the expectation operator. For an ergodic signal, we can compute the mean value, variance, and auto-covariance using following formulas:

$$m_X = \lim_{N\to\infty} \frac{1}{2N+1} \sum_{n=-N}^{N} x(n) \tag{G.39}$$

$$\sigma_X^2 = \lim_{N\to\infty} \frac{1}{2N+1} \sum_{n=-N}^{N} [x(n) - m_X]^2 \tag{G.40}$$

$$\gamma_{XX}(l) = \lim_{N\to\infty} \frac{1}{2N+1} \sum_{n=-N}^{N} [(x(n+l) - m_X)(x(n) - m_X)]. \tag{G.41}$$

Practically, the causal sequence is used, that is, $x(n)=0$ for $n<0$. The following equations are adopted instead.

$$\hat{m}_X = \lim_{N\to\infty} \frac{1}{N+1} \sum_{n=0}^{N} x(n) \tag{G.42}$$

$$\hat{\sigma}_X^2 = \lim_{N\to\infty} \frac{1}{N+1} \sum_{n=0}^{N} [x(n) - m_X]^2 \tag{G.43}$$

$$\hat{\gamma}_{XX}(l) = \lim_{N\to\infty} \frac{1}{N+1} \sum_{n=0}^{N} [(x(n+l) - m_X)(x(n) - m_X)]. \tag{G.44}$$

As an illustrative example, we consider a random signal given by

$$x(n) = \sin(n\Omega + \theta)$$

The probability density function of phase θ is uniformly distributed and defined as

$$p_\theta(\theta) = \begin{cases} 1/(2\pi) & 0 < \theta \le 2\pi \\ 0 & \text{elsewhere} \end{cases}.$$

According to the ensemble averages, we derive

$$m_x = E[x(n)] = E[\sin(n\Omega + \theta)] = 0$$

$$\begin{aligned} \sigma_x^2 &= E[x^2(n)] = E[\sin^2(n\Omega + \theta)] = E[1/2 - (1/2)\cos(2n\Omega + 2\theta)] \\ &= E(1/2) - (1/2)E[\cos(2n\Omega + 2\theta)] = 1/2 \end{aligned}$$

$$\begin{aligned} r_{xx}(l) &= E\{\sin[(n+l)\Omega + \theta]\sin(n\Omega + \theta)\} = (1/2)E\{\cos l\Omega - \cos[(2n+l)\Omega + 2\theta]\} \\ &= E[(\cos l\Omega)]/2 - E\{\cos[(2n+l)\Omega + 2\theta]\}/2 = \cos l\Omega/2. \end{aligned}$$

We can show that the random signal is ergodic in mean, variance, and auto-correlation function, respectively. Using (G.42)–(G.44), we can see that

$$\hat{m}_X = \lim_{N\to\infty} \frac{1}{2N+1} \sum_{n=-N}^{N} \sin(n\Omega + \theta_0) \to 0$$

that is, it is ergodic in mean;

$$\hat{\sigma}_X^2 = \lim_{N\to\infty}\frac{1}{2N+1}\sum_{n=-N}^{N}\sin^2(n\Omega+\theta_0)$$
$$= \lim_{N\to\infty}\frac{1}{2N+1}\sum_{n=-N}^{N}(1/2-\cos(2n\Omega+2\theta_0)/2] \to 1/2.$$

This indicates that the signal is ergodic in variance. Finally, we see

$$\hat{\gamma}_{XX}(l) = \lim_{N\to\infty}\frac{1}{2N+1}\sum_{n=-N}^{N}[\sin[(n+l)\Omega+\theta_0]\sin(n\Omega+\theta_0)$$
$$= \lim_{N\to\infty}\frac{1}{2N+1}\sum_{n=-N}^{N}(1/2)\{\cos(l\Omega)-\cos[(2n+l)\Omega+2\theta_0)]\} \to \cos(l\Omega)/2.$$

We can further conclude that the signal is ergodic in autocorrelation function.

G.5 STATISTICAL PROPERTIES OF LINEAR SYSTEM OUTPUT SIGNAL

To proceed, we assume that the linear system is causal, that is, $h(k)=0$ for $k<0$, where $h(k)$ is the impulse response of the causal time-invariant linear system, that is,

$$y(n) = \sum_{k=0}^{\infty} h(k)x(n-k) \tag{G.45}$$

where $x(n)$ and $y(n)$ are the system input and output, respectively. Note that we only tackle the causal time-invariant system here. For the derivation of the general case for $h(k)\neq 0, -\infty<k<\infty$, the results can be found in the textbook by Mitra (2006). For the causal time-invariant system, it can be verified that the mean value of the system output process is determined by

$$m_y = E[y(n)] = E\left\{\sum_{k=0}^{\infty} h(k)x(n-k)\right\} = \sum_{k=0}^{\infty} h(k)E[x(n-k)] = m_x\sum_{k=0}^{\infty} h(k). \tag{G.46}$$

The auto-correlation function can be derived as follows:

$$\begin{aligned}\phi_{yy}(n+l,n) &= E[y(n+l)y(n)] = E\left\{\left(\sum_{k=0}^{\infty} h(k)x(n+l-k)\right)\left(\sum_{i=0}^{\infty} h(i)x(n-i)]\right)\right\}\\ &= \sum_{k=0}^{\infty}\sum_{i=0}^{\infty} h(k)h(i)E[x(n+l-k)x(n-i)]\\ &= \sum_{i=0}^{\infty}\sum_{k=0}^{\infty} h(k)h(i)\phi_{xx}(n+l-k,n-i).\end{aligned} \tag{G.47}$$

Considering the process is a WSS, (G.47) can be written as

$$\phi_{yy}(l) = \phi_{yy}(n+l,n) = \sum_{i=0}^{\infty}\sum_{k=0}^{\infty} h(k)h(i)\phi_{xx}(l+i-k). \tag{G.48}$$

By expanding (G.48), it is verified that

$$\phi_{yy}(l) = \sum_{j=-\infty}^{\infty} \sum_{k=0}^{\infty} h(k)h(k+|j|)\phi_{xx}(l-j). \tag{G.49}$$

The cross-correlation function between the system output and input sequences is given by

$$\begin{aligned}\phi_{yx}(l) &= \phi_{yx}(n+l,n) = E[y(n+l)x(n)] = E\left\{\left(\sum_{k=-\infty}^{\infty} h(k)x(n+l-k)\right)x(n)\right\} \\ &= \sum_{k=-\infty}^{\infty} h(k)E[x(n+l-k)x(n)] = \sum_{k=-\infty}^{\infty} h(k)\phi_{xx}(l-k).\end{aligned} \tag{G.50}$$

G.6 Z-TRANSFORM DOMAIN REPRESENTATION OF STATISTICAL PROPERTIES

Taking the two-side z-transform of the auto-correlation, it follows that

$$\begin{aligned}\Phi_{yy}(z) &= \sum_{l=-\infty}^{\infty} \phi_{yy}(l)z^{-l} = \sum_{l=-\infty}^{\infty} \sum_{j=-\infty}^{\infty} \sum_{k=0}^{\infty} h(k)h(k+|j|)\phi_{xx}(l-j)z^{-l} \\ &= \sum_{j=-\infty}^{\infty} \sum_{k=0}^{\infty} h(k)h(k+|j|)z^{-j}\Phi_{xx}(z).\end{aligned} \tag{G.51}$$

For the causal linear system, that is, $h(k)=0$ for $k<0$, it can be verified that

$$\begin{aligned}H(z)H(z^{-1}) &= [h(0)+h(1)z^{-1}+h(2)z^{-2}+\cdots][h(0)+h(1)z^{1}+h(2)z^{2}+\cdots] \\ &= \sum_{j=-\infty}^{\infty} \sum_{k=0}^{\infty} h(k)h(k+|j|)z^{-j}.\end{aligned} \tag{G.52}$$

Therefore,

$$\Phi_{yy}(z) = H(z)H(z^{-1})\Phi xx(z). \tag{G.53}$$

It is straightforward to apply the two-side z-transform of the cross-correlation to obtain the following:

$$\Phi_{yx}(z) = H(z)\Phi xx(z). \tag{G.54}$$

We can also obtain each power spectrum by substituting $z=e^{j\Omega}$ to (G.53) and (G.54), respectively, that is,

$$\Phi_{yy}(e^{j\Omega}) = H(e^{j\Omega})H(e^{-j\Omega})\Phi xx(e^{j\Omega}) = |H(e^{j\Omega})|^2\Phi xx(e^{j\Omega}) \tag{G.55}$$

$$\Phi_{yx}(e^{j\Omega}) = H(e^{j\Omega})\Phi xx(e^{j\Omega}). \tag{G.56}$$

Using the inversion formula in Chapter 5, we obtain the following relations:

$$\phi_{yy}(l) = \frac{1}{2\pi j}\oint_C H(z)H(z^{-1})\Phi_{xx}(z)z^{l-1}dz \tag{G.57}$$

$$\phi_{yy}(0) = \frac{1}{2\pi j}\oint_C H(z)H(z^{-1})\Phi_{xx}(z)\frac{dz}{z} \tag{G.58}$$

$$\sigma_y^2 = \frac{1}{2\pi j}\oint_C H(z)H(z^{-1})\Phi_{xx}(z)\frac{dz}{z} - m_y^2. \tag{G.59}$$

Specifically, for a zero-mean white noise, $X(n+l)$ and $X(n)$ are uncorrelated; and $m_{X(n)}=0$. The auto-correlation function for a zero-mean white noise can be written as

$$\phi_{xx}(l) = E[X(n+l)X(n)] = \begin{cases} \sigma_X^2 & l=0 \\ 0 & l \neq 0 \end{cases}. \tag{G.60}$$

That is,

$$\phi_{xx}(l) = \sigma_X^2 \delta(l). \tag{G.61}$$

Thus, the z-transform of the auto-correlation is given by

$$\Phi_{xx}(z) = \sigma_x^2. \tag{G.62}$$

Now, using the zero-mean white noise as a time-invariant system input, the variance of the system output can then be determined by

$$\sigma_y^2 = \frac{\sigma_x^2}{2\pi j}\oint_C H(z)H(z^{-1})\frac{dz}{z} \tag{G.63}$$

APPENDIX

SOME USEFUL MATHEMATICAL FORMULAS

Form of a complex number:

$$\text{Rectangular form}: a + jb,\ \text{where}\ j = \sqrt{-1} \tag{H.1}$$

$$\text{Polar form}: Ae^{j\theta} \tag{H.2}$$

$$\text{Euler formula}: e^{\pm jx} = \cos x \pm j\sin x \tag{H.3}$$

Conversion from the polar form to the rectangular form:

$$Ae^{j\theta} = A\cos\theta + jA\sin\theta = a + jb, \tag{H.4}$$

where $a = A\cos\theta$ and $b = A\sin\theta$.

Conversion from the rectangular form to the polar form:

$$a + jb = Ae^{j\theta}, \tag{H.5}$$

where $A = \sqrt{a^2 + b^2}$.

We usually specify the principal value of the angle such that $-180^0 < \theta \leq 180^0$. The angle value can be determined as

$$\theta = \tan^{-1}\left(\frac{b}{a}\right)\ \text{if}\ a \geq 0$$

(that is, the complex number is in the first or fourth quadrant in the rectangular coordinate system);

$$\theta = 180^0 + \tan^{-1}\left(\frac{b}{a}\right)\ \text{if}\ a < 0\ \text{and}\ b \geq 0$$

(that is, the complex number is in the second quadrant in the rectangular coordinate system); and

$$\theta = -180^0 + \tan^{-1}\left(\frac{b}{a}\right)\ \text{if}\ a < 0\ \text{and}\ b \leq 0$$

(that is, the complex number is in the third quadrant in the rectangular coordinate system). Note that

$$\theta\ \text{radian} = \frac{\theta\ \text{degree}}{180^0} \times \pi$$

$$\theta\ \text{degree} = \frac{\theta\ \text{radian}}{\pi} \times 180^{\circ}.$$

Complex numbers:

$$e^{\pm j\pi/2} = \pm j \tag{H.6}$$

$$e^{\pm j2n\pi} = 1 \tag{H.7}$$

$$e^{\pm j(2n+1)\pi} = -1 \tag{H.8}$$

Complex conjugate of $a+jb$:

$$(a+jb)^* = conj(a+jb) = a-jb \tag{H.9}$$

Complex conjugate of $Ae^{j\theta}$:

$$\left(Ae^{j\theta}\right)^* = conj\left(Ae^{j\theta}\right) = Ae^{-j\theta} \tag{H.10}$$

Complex number addition and subtraction:

$$(a_1+jb_1)\pm(a_2+jb_2) = (a_1\pm a_2)+j(b_1\pm b_2) \tag{H.11}$$

Complex number multiplication:
  Rectangular form:

$$(a_1+jb_1)\times(a_2+jb_2) = a_1a_2 - b_1b_2 + j(a_1b_2+a_2b_1) \tag{H.12}$$

$$(a+jb)\cdot conj(a+jb) = (a+jb)(a-jb) = a^2+b^2 \tag{H.13}$$

Polar form:

$$A_1e^{j\theta_1}A_2e^{j\theta_2} = A_1A_2e^{j(\theta_1+\theta_2)} \tag{H.14}$$

Complex number division:
  Rectangular form:

$$\begin{aligned}\frac{a_1+jb_1}{a_2+jb_2} &= \frac{(a_1+jb_1)(a_2-jb_2)}{(a_2+jb_2)(a_2-jb_2)}\\ &= \frac{(a_1a_2+b_1b_2)+j(a_2b_1-a_1b_2)}{(a_2)^2+(b_2)^2}\end{aligned} \tag{H.15}$$

Polar form:

$$\frac{A_1e^{j\theta_1}}{A_2e^{j\theta_2}} = \left(\frac{A_1}{A_2}\right)e^{j(\theta_1-\theta_2)} \tag{H.16}$$

Trigonometric identities:

$$\sin x = \frac{e^{jx}-e^{-jx}}{2j} \tag{H.17}$$

$$\cos x = \frac{e^{jx}+e^{-jx}}{2} \tag{H.18}$$

$$\sin\left(x\pm 90^0\right) = \pm\cos x \tag{H.19}$$

$$\cos\left(x\pm 90^0\right) = \mp\sin x \tag{H.20}$$

$$\sin x\cos x = \frac{1}{2}\sin 2x \tag{H.21}$$

$$\sin^2 x + \cos^2 x = 1 \tag{H.22}$$

$$\sin^2 x = \frac{1}{2}(1 - \cos 2x) \tag{E.23}$$

$$\cos^2 x = \frac{1}{2}(1 + \cos 2x) \tag{H.24}$$

$$\sin(x \pm y) = \sin x \cos y \pm \cos x \sin y \tag{H.25}$$

$$\cos(x \pm y) = \cos x \cos y \mp \sin x \sin y \tag{H.26}$$

$$\sin x \cos y = \frac{1}{2}(\sin(x+y) + \sin(x-y)) \tag{H.27}$$

$$\sin x \sin y = \frac{1}{2}(\cos(x-y) - \cos(x+y)) \tag{H.28}$$

$$\cos x \cos y = \frac{1}{2}(\cos(x-y) + \cos(x+y)) \tag{H.29}$$

Series of exponentials:

$$\sum_{k=0}^{N-1} a^k = \frac{1-a^N}{1-a}, a \neq 1 \tag{H.30}$$

$$\sum_{k=0}^{\infty} a^k = \frac{1}{1-a}, |a| < 1 \tag{H.31}$$

$$\sum_{k=0}^{\infty} k a^k = \frac{1}{(1-a)^2}, |a| < 1 \tag{H.32}$$

$$\sum_{k=0}^{N-1} e^{\left(j\frac{2\pi nk}{N}\right)} = \begin{cases} 0 & 1 \leq n \leq N-1 \\ N & n = 0, N \end{cases} \tag{H.33}$$

L'Hospital's rule:

If $\lim_{x \to a} \frac{f(x)}{g(x)}$ results in the undetermined form $\frac{0}{0}$ or $\frac{\infty}{\infty}$, then

$$\lim_{x \to a} \frac{f(x)}{g(x)} = \lim_{x \to a} \frac{f'(x)}{g'(x)} \tag{H.34}$$

where $f'(x) = \frac{df(x)}{dx}$ and $g'(x) = \frac{dg(x)}{dx}$.

Solution of the quadratic equation:

For a quadratic equation expressed as

$$ax^2 + bx + c = 0, \tag{H.35}$$

the solution is given by

$$x = \frac{-b \pm \sqrt{b^2 - 4ac}}{2a}. \tag{H.36}$$

Solution of the simultaneous equations:

Simultaneous linear equations are listed as follows:

$$\begin{aligned} a_{11}x_1 + a_{12}x_2 + \cdots + a_{1n}x_n &= b_1 \\ a_{21}x_1 + a_{22}x_2 + \cdots + a_{2n}x_n &= b_2 \\ \cdots \\ a_{n1}x_1 + a_{n2}x_2 + \cdots + a_{nn}x_n &= b_n \end{aligned} \tag{H.37}$$

Solution is given by Cramer's rule, that is,

$$x_1 = \frac{D_1}{D},\ x_2 = \frac{D_2}{D}, \cdots,\ x_n = \frac{D_n}{D}. \tag{H.38}$$

where D, D_1, D_2, …, D_n are the $n \times n$ determinants. Each is defined as follows:

$$D = \begin{vmatrix} a_{11} & a_{12} & \cdots & a_{1n} \\ a_{21} & a_{22} & \cdots & a_{2n} \\ \vdots & \vdots & \ddots & \vdots \\ a_{n1} & a_{n2} & \cdots & a_{nn} \end{vmatrix} \tag{H.39}$$

$$D_1 = \begin{vmatrix} b_1 & a_{12} & \cdots & a_{1n} \\ b_2 & a_{22} & \cdots & a_{2n} \\ \vdots & \vdots & \ddots & \vdots \\ b_n & a_{n2} & \cdots & a_{nn} \end{vmatrix} \tag{H.40}$$

$$D_2 = \begin{vmatrix} a_{11} & b_1 & \cdots & a_{1n} \\ a_{21} & b_2 & \cdots & a_{2n} \\ \vdots & \vdots & \ddots & \vdots \\ a_{n1} & b_n & \cdots & a_{nn} \end{vmatrix} \tag{H.41}$$

$$\cdots$$

$$D_n = \begin{vmatrix} a_{11} & a_{12} & \cdots & b_1 \\ a_{21} & a_{22} & \cdots & b_2 \\ \vdots & \vdots & \ddots & \vdots \\ a_{n1} & a_{n2} & \cdots & b_n \end{vmatrix} \tag{H.42}$$

$$D = (-1)^{1+1} a_{11} M_{11} + (-1)^{1+2} a_{12} M_{12} + \cdots (-1)^{1+n} a_{1n} M_{1n}, \tag{H.43}$$

where M_{ij} is an $(n-1) \times (n-1)$ determinant obtained from D by crossing out the ith row and jth column. D can also be expanded by any row or column. As an example by using the second column,

$$D = (-1)^{1+2} a_{12} M_{12} + (-1)^{2+2} a_{22} M_{12} + \cdots (-1)^{n+2} a_{n2} M_{n2}. \tag{H.44}$$

2×2 determinant:

$$D = \begin{vmatrix} a_{11} & a_{12} \\ a_{21} & a_{22} \end{vmatrix} = a_{11}a_{22} - a_{12}a_{21}. \tag{H.45}$$

3×3 determinant:

$$\begin{aligned} D &= \begin{vmatrix} a_{11} & a_{12} & a_{13} \\ a_{21} & a_{22} & a_{23} \\ a_{31} & a_{32} & a_{33} \end{vmatrix} \\ &= (-1)^{1+1} a_{11} \begin{vmatrix} a_{22} & a_{23} \\ a_{32} & a_{33} \end{vmatrix} + (-1)^{1+2} a_{12} \begin{vmatrix} a_{21} & a_{23} \\ a_{31} & a_{33} \end{vmatrix} + (-1)^{1+3} a_{13} \begin{vmatrix} a_{21} & a_{22} \\ a_{31} & a_{32} \end{vmatrix} \\ &= a_{11}(a_{22}a_{33} - a_{23}a_{32}) - a_{12}(a_{21}a_{33} - a_{23}a_{31}) + a_{13}(a_{21}a_{32} - a_{22}a_{31}). \end{aligned} \tag{H.46}$$

Solution for two simultaneous linear equations:

$$\begin{aligned} ax + by &= e \\ cx + dy &= f \end{aligned} \tag{H.47}$$

The solution is given by

$$x = \frac{D_1}{D} = \frac{\begin{vmatrix} e & b \\ f & d \end{vmatrix}}{\begin{vmatrix} a & b \\ c & d \end{vmatrix}} = \frac{ed - bf}{ad - bc}, \tag{H.48}$$

$$y = \frac{D_2}{D} = \frac{\begin{vmatrix} a & e \\ c & f \end{vmatrix}}{\begin{vmatrix} a & b \\ c & d \end{vmatrix}} = \frac{af - ec}{ad - bc}. \tag{H.49}$$

Answers to Selected Problems

Chapter 2

2.1 Hint:

(b)

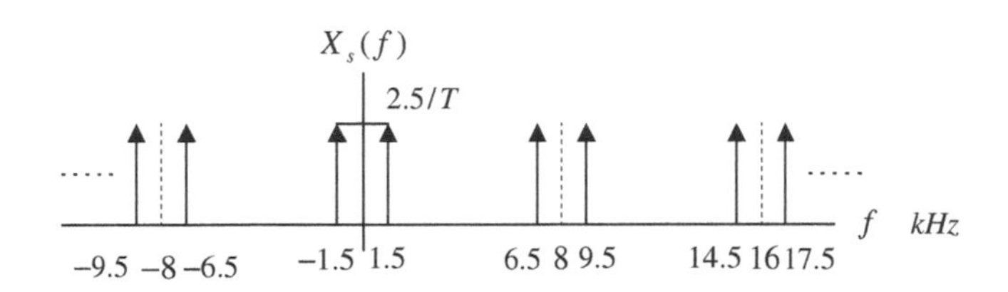

2.2 Hint:

(a)

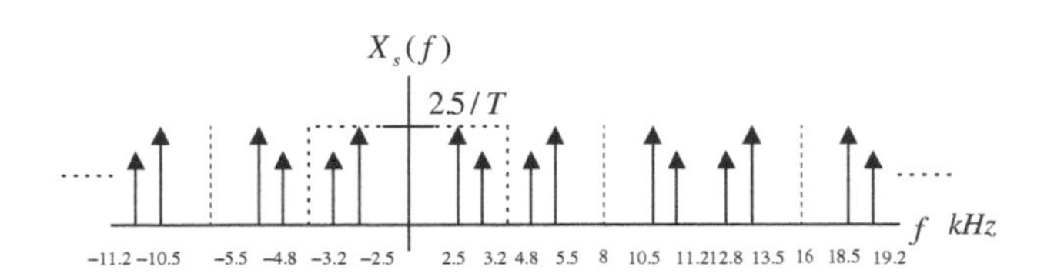

2.5

Hint:

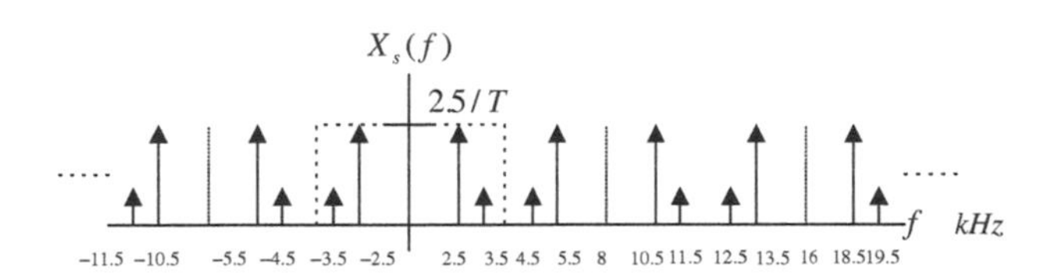

(c) The aliasing frequency = 3.5 kHz.

2.9

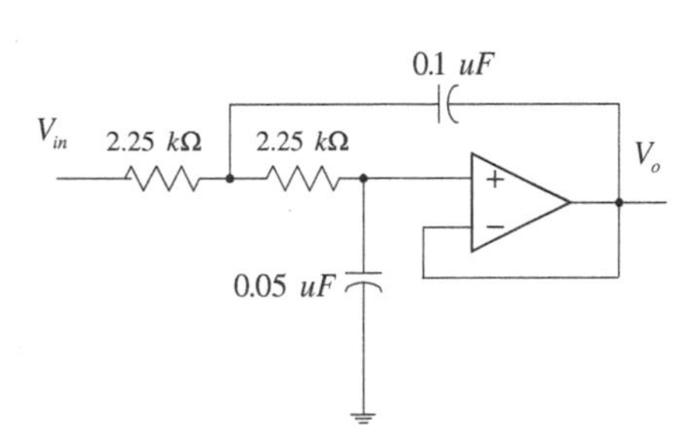

2.10 %aliasing noise level = 8.39%

2.13

(a) % aliasing noise level = 57.44%.

(b) % aliasing noise level = 20.55%.

2.17
(a) %distortion = 24.32%.
(b) %distortion = 5.68%.

2.18 $f_c = 4686$ Hz.

2.21
b1b0 = 01.

2.22 $V_0 = 1.25$ V.

2.25
(a) $L = 2^4 = 16$ levels, (b) $\Delta = \frac{x_{\max} - x_{\min}}{L} = \frac{5}{16} = 0.3125$.
(c) $x_q = 3.125$, (d) binary code = 1010, (e) $e_q = -0.075$.

2.27
(a) $L = 2^3 = 8$ levels, (b) $\Delta = \frac{x_{\max} - x_{\min}}{L} = \frac{5}{8} = 0.625$.
(c) $x_q = -2.5 + 2 \times 0.625 = -1.25$, (d) binary code = 010, (e) $e_q = -0.05$.

2.29
(a) $L = 2^6 = 64$ levels, (b) $\Delta = \frac{x_{\max} - x_{\min}}{L} = \frac{20}{64} = 0.3125$,
(c) $SNR_{dB} = 1.76 + 6.02 \times 6 = 37.88$ dB.

Chapter 3

3.1

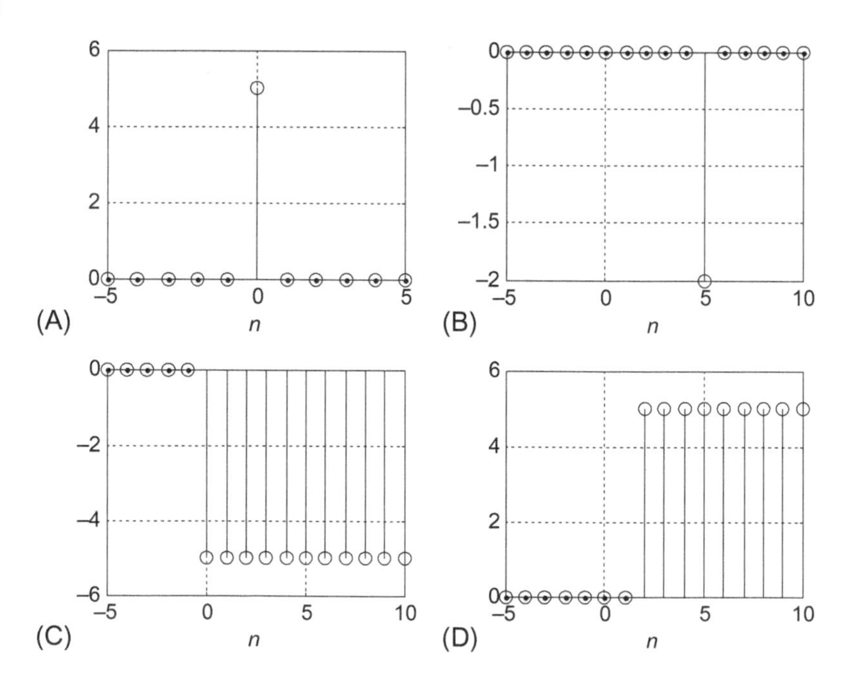

3.2 Hint:

(a)

n	0	1	2	3	4	5	6	7
$x(n)$	1.000	0.5000	0.2500	0.1250	0.0625	0.0313	0.0156	0.0078

(d)

n	0	1	2	3	4	5	6	7
$x(n)$	0.0000	1.1588	1.6531	1.7065	1.5064	1.1865	0.8463	0.5400

3.5

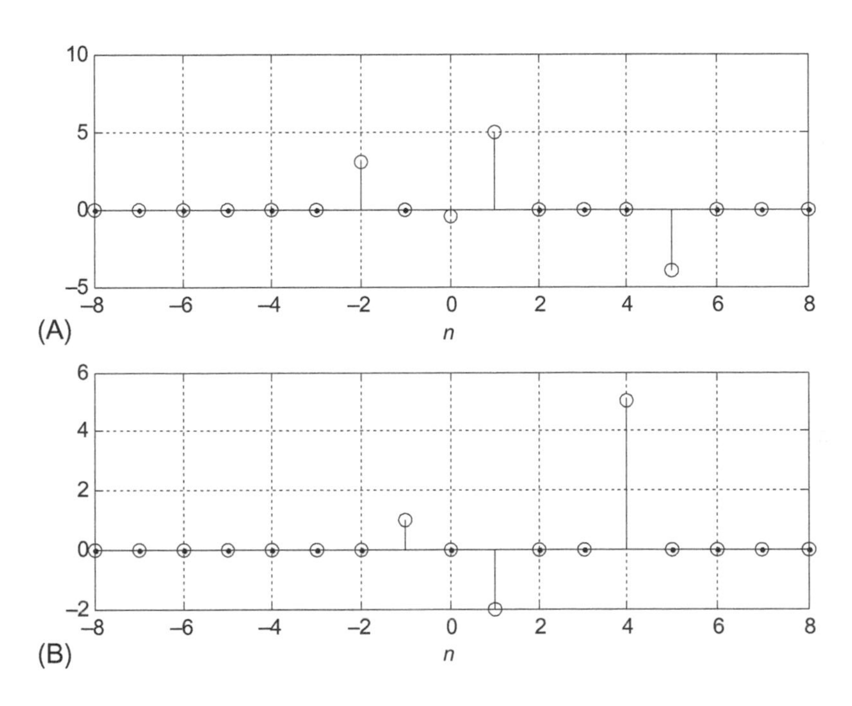

3.6

(a) $x(n)=3\delta(n)+\delta(n-1)+2\delta(n-2)+\delta(n-3)+\delta(n-5)$

(b) $x(n)=\delta(n-1)-\delta(n-2)+\delta(n-4)-\delta(n-5)$

3.9

(a) $x(n)=e^{-0.5n}u(n)=(0.6065)^n u(n)$

(b) $x(n)=5\sin(0.2\pi n)u(n)$

(c) $x(n)=10\cos(0.4\pi n+\pi/6)u(n)$

(d) $x(n)=10e^{-n}\sin(0.15\pi n)u(n)=10(0.3679)^n\sin(0.15\pi n)u(n)$

3.10

(a) nonlinear system, (b) linear system, (c) nonlinear system.

3.13
(a) time invariant.

3.15
(a) causal system, (b) noncausal system, (c) causal system.

3.16
(a) $h(n) = 0.5\delta(n) - 0.5\delta(n-2)$
(b) $h(n) = (0.75)^n$; $n \geq 0$
(c) $h(n) = 1.25\delta(n) - 1.25(-0.8)^n$; $n \geq 0$

3.19
(a) $h(n) = 5\delta(n-10)$, (b) $h(n) = \delta(n) + 0.5\delta(n-1)$.

3.20

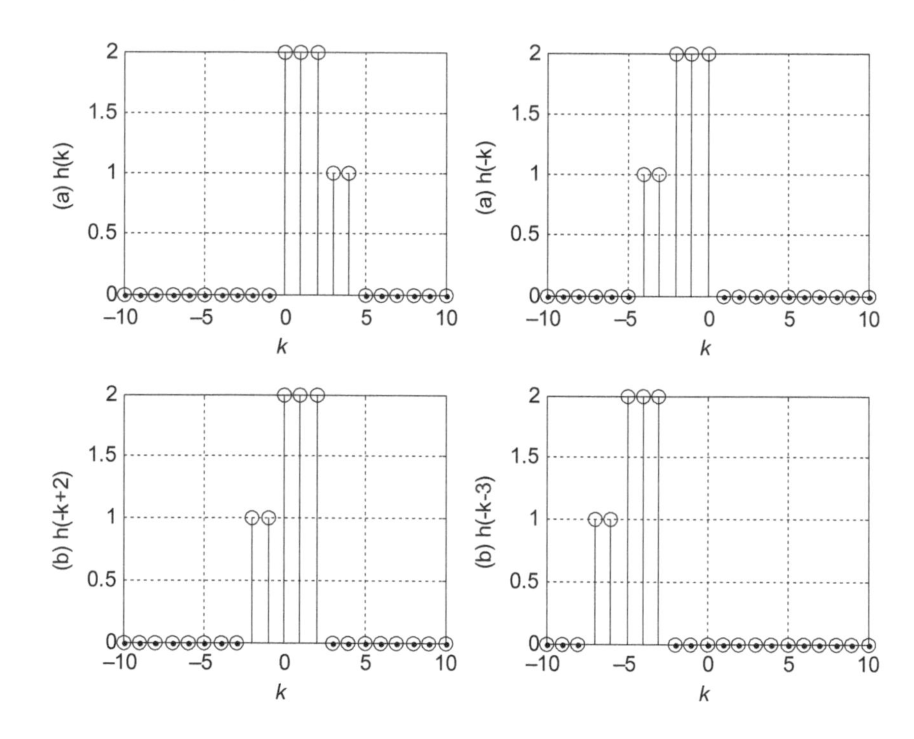

3.22
$y(0) = 4, y(1) = 6, y(2) = 8, y(3) = 6, y(4) = 5, y(5) = 2, y(6) = 1,$
$y(n) = 0$ for $n \geq 7$

3.24
$y(0) = 0, y(1) = 1, y(2) = 2, y(3) = 1, y(4) = 0$
$y(n) = 0$ for $n \geq 4$

3.25
Since $h(n) = 0.5\delta(n) + 100\delta(n-2) - 20\delta(n-10)$
and $S = 0.5 + 100 + 20 = 120.5$ = finite number, the system is stable.

3.28
(a) $h(n) = (0.75)^n u(n), S = \sum_{k=0}^{\infty}(0.75)^k = 1/(1-0.75) = 4 = \text{finite, the system is stable.}$
(b) $h(n) = (2)^n u(n), S = \sum_{k=0}^{\infty}(2)^k = 1+2+2^2+\cdots = \infty = \text{infinite, the system is unstable.}$

Chapter 4

4.1
$X(0) = 1, X(1) = 2-j, X(2) = -1, X(3) = 2+j$

4.5
$x(0) = 4, x(1) = 3, x(2) = 2, x(3) = 1$

4.6
$X(0) = 10, X(1) = 3.5-4.3301j, X(2) = 2.5-0.8660j, X(3) = 2, X(4) = 2.5+0.8660j, X(5) = 3.5+4.3301j$

4.9
$\overline{x}(0) = 4, \overline{x}(4) = 0$

4.10
$\Delta f = 2.5\text{Hz}$ and $f_{\max} = 10\text{kHz}$

4.13
$N = 4096, \Delta f = 0.488, \text{Hz}$

4.15
(a) w = [0.0800 0.2532 0.6424 0.9544 0.9544 0.6424 0.2532 0.0800].
(b) w = [0.0 0.1883 0.6113 0.9505 0.9505 0.6113 0.1883 0.0].

4.16
(a) xw = [0 0.4000 0 -0.8000 0 0].
(b) xw = [0 0.3979 0 -0.9121 0 0.0800].
(c) xw = [0 0.3455 0 -0.9045 0 0].

4.19
(a) $A_0 = 0.1667$, $A_1 = 0.3727$, $A_2 = 0.5$, $A_3 = 0.3727$.
$\varphi_0 = 0^0$, $\varphi_1 = 154.43^0$, $\varphi_2 = 0^0$, $\varphi_3 = -154.43^0$.
$P_0 = 0.0278$, $P_1 = 0.1389$, $P_2 = 0.25$, $P_3 = 0.1389$.

(b) $A_0 = 0.2925$, $A_1 = 0.3717$, $A_2 = 0.6375$, $A_3 = 0.3717$.
$\varphi_0 = 0^0$, $\varphi_1 = 145.13^0$, $\varphi_2 = 0^0$, $\varphi_3 = -145.13^0$.
$P_0 = 0.0856$, $P_1 = 0.1382$, $P_2 = 0.4064$, $P_3 = 0.1382$.

(c) $A_0 = 0.1875$, $A_1 = 0.4193$, $A_2 = 0.5625$, $A_3 = 0.4193$.
$\varphi_0 = 0^0$, $\varphi_1 = 153.43^0$, $\varphi_2 = 0^0$, $\varphi_3 = -153.43^0$.
$P_0 = 0.0352$, $P_1 = 0.1758$, $P_2 = 0.3164$, $P_3 = 0.1758$.

4.21 $X(0) = 10$, $X(1) = 2-2j$, $X(2) = 2$, $X(3) = 2+2j$, 4 complex multiplications.

4.22 $x(0) = 4$, $x(1) = 3$, $x(2) = 2$, $x(3) = 1$, 4 complex multiplications.

4.25 $X(0) = 10$, $X(1) = 2-2j$, $X(2) = 2$, $X(3) = 2+2j$, 4 complex multiplications.

4.26 $x(0) = 4$, $x(1) = 3$, $x(2) = 2$, $x(3) = 1$, 4 complex multiplications.

Chapter 5

5.1

(a) $X(z)=\frac{4z}{z-1}$, (b) $X(z)=\frac{z}{z+0.7}$.

(c) $X(z)=\frac{4z}{z-e^{-2}}=\frac{4z}{z-0.1353}$,

(d) $X(z)=\frac{4z[z-0.8\times\cos(0.1\pi)]}{z^2-[2\times0.8z\cos(0.1\pi)]+0.8^2}=\frac{4z(z-0.7608)}{z^2-1.5217z+0.64}$

(e) $X(z)=\frac{4e^{-3}\sin(0.1\pi)z}{z^2-2e^{-3}z\cos(0.1\pi)+e^{-6}}=\frac{0.06154z}{z^2-0.0947z+0.00248}$

5.2

(a) $X(z)=\frac{z}{z-1}+\frac{z}{z-0.5}$

(b) $X(z)=\frac{z^{-4}z\left[z-e^{-3}\cos(0.1\pi)\right]}{z^2-[2e^{-3}\cos(0.1\pi)]z+e^{-6}}=\frac{z^{-3}(z-0.0474)}{z^2-0.0948z+0.0025}$

5.3

(c) $X(z)=\frac{5z^{-2}}{z-e^{-2}}$, (e) $X(z)=\frac{4e^{-3}\sin(0.2\pi)}{z^2-2e^{-3}\cos(0.2\pi)z+e^{-6}}=\frac{0.1171}{z^2-0.0806z+0.0025}$.

5.5

(a) $X(z)=15z^{-3}-6z^{-5}$

(b) $x(n)=15\delta(n-3)-6\delta(n-5)$

5.9

(a) $X(z)=-25+\frac{5z}{z-0.4}+\frac{20z}{z+0.1}$, $x(n)=-25\delta(n)+5(0.4)^n u(n)+20(-0.1)^n u(n)$.

(b) $X(z)=\frac{1.6667z}{z-0.2}-\frac{1.6667z}{z+0.4}$, $x(n)=1.6667(0.2)^n u(n)-1.6667(-0.4)^n u(n)$.

(c) $X(z)=\frac{1.3514z}{z+0.2}+\frac{Az}{z-P}+\frac{A^*z}{z-P^*}$

where $P=0.5+0.5j=0.707\angle 45^0$, and $A=1.1625\angle-125.54^0$

$x(n)=1.3514(-0.2)^n u(n)+2.325(0.707)^n\cos(45^0\times n-125.54^0)$.

(d) $X(z)=\frac{4.4z}{z-0.6}+\frac{-0.4z}{z-0.1}+\frac{-1.2z}{(z-0.1)^2}$,

$x(n)=4.4(0.6)^n u(n)-0.4(0.1)^n u(n)-12n(0.1)^n u(n)$.

5.10

$Y(z)=\frac{-4.3333z}{z-0.5}+\frac{5.333z}{z-0.8}$, $y(n)=-4.3333(0.5)^n u(n)+5.3333(0.8)^n u(n)$.

5.13

$Y(z)=\frac{9.84z}{z-0.2}+\frac{-29.46z}{z-0.3}+\frac{20z}{z-0.4}$

$y(n)=9.84(0.2)^n u(n)-29.46(0.3)^n u(n)+20(0.4)^n u(n)$

5.14

(a) $Y(z)=\frac{-4z}{z-0.2}+\frac{5z}{z-0.5}$, $y(n)=-4(0.2)^n u(n)+5(0.5)^n u(n)$.

(b) $Y(z)=\frac{5z}{z-1}+\frac{-5z}{z-0.5}+\frac{z}{z-0.2}$,

$y(n)=5u(n)-5(0.5)^n u(n)+(0.2)^n u(n)$

5.17

(a) $Y(z)=\frac{Az}{z-P}+\frac{A^*z}{z-P^*}$, $P=0.2+0.5j=0.5385\angle 68.20^0$, $A=0.8602\angle -54.46^0$

$y(n)=1.7204(0.5382)^n\cos\left(n\times 68.20^0-54.46^0\right)$

(b) $Y(z)=\frac{1.6854z}{z-1}+\frac{Az}{z-P}+\frac{A^*z}{z-P^*}$, where $P=0.2+0.5j=0.5385\angle 68.20^0$, $A=0.4910\angle -136.25^0$

$y(n)=1.6845u(n)+0.982(0.5382)^n\cos\left(n\times 68.20^0-136.25^0\right)$

Chapter 6

6.1

(a) $y(0)=0.5$, $y(1)=0.25$, $y(2)=0.125$, $y(3)=0.0625$, $y(4)=0.03125$.
(b) $y(0)=1$, $y(1)=0$, $y(2)=0.25$, $y(3)=0$, $y(4)=0.0625$.

6.3

(a) $y(0)=-2$, $y(1)=2.3750$, $y(2)=-1.0312$, $y(3)=0.7266$, $y(4)=-0.2910$.
(b) $y(0)=0$, $y(1)=1$, $y(2)=-0.2500$, $y(3)=0.3152$, $y(4)=-0.0781$.

6.4

(a) $H(z)=0.5+0.5z^{-1}$, (b) $y(n)=2\delta(n)+2\delta(n-1)$, $y(n)=-5\delta(n)+10u(n)$.

6.5

(a) $H(z)=\frac{1}{1+0.5z^{-1}}$, (b) $y(n)=(-0.5)^n u(n)$, $y(n)=0.6667u(n)+0.3333(-0.5)^n u(n)$.

6.9

$H(z)=1-0.3z^{-1}+0.28z^{-2}$, $A(z)=1$, $N(z)=1-0.3z^{-1}+0.28z^{-2}$.

6.12

(a) $y(n)=x(n)-0.25x(n-2)-1.1y(n-1)-0.18y(n-2)$
(b) $y(n)=x(n-1)-0.1x(n-2)+0.3x(n-3)$

6.13

(b) $H(z)=\frac{(z+0.4)(z-0.4)}{(z+0.2)(z+0.5)}$

6.15

(a) zero: $z=0.5$, poles: $z=-0.25$ ($|z|=0.25$), $z=-0.5\pm 0.7416j$ ($|z|=0.8944$), stable.
(b) zeros: $z=\pm 0.5j$, poles: $z=0.5$ ($|z|=0.5$), $z=-2\pm 1.7321j$ ($|z|=2.6458$), unstable.
(c) zero: $z=-0.95$, poles: $z=0.2$ ($|z|=0.2$), $z=-0.7071\pm 0.7071j$ ($|z|=1$), marginally stable.
(d) zeros: $z=-0.5$, $z=-0.5$, poles: $z=1$ ($|z|=1$), $z=-1$, $z=-1$ ($|z|=1$), $z=0.36$ ($|z|=0.36$), unstable.

6.17

$H(z)=0.5z^{-1}+0.5z^{-2}, H(e^{j\Omega})=0.5e^{-j\Omega}+0.5e^{-j2\Omega}$

$\left|H(e^{j\Omega})\right|=0.5\sqrt{(1+\cos\Omega)^2+(\sin\Omega)^2}, \angle H(e^{j\Omega})=\tan^{-1}\left(\frac{-\sin\Omega-\sin 2\Omega}{\cos\Omega+\cos 2\Omega}\right)$

6.19

$$H(z)=\frac{1}{1+0.5z^{-2}}, H(e^{j\Omega})=\frac{1}{1+0.5e^{-j2\Omega}}$$

$$|H(e^{j\Omega})|=\frac{1}{\sqrt{(1+0.5\cos 2\Omega)^2+(0.5\sin 2\Omega)^2}}, \angle H(e^{j\Omega})=-\tan^{-1}\left(\frac{-0.5\sin 2\Omega}{1+0.5\cos 2\Omega}\right)$$

6.21

(a) $H(z)=0.5+0.5z^{-1}$, $H(e^{j\Omega})=0.5+0.5e^{-j\Omega}$.

$$|H(e^{j\Omega})|=\sqrt{(0.5+0.5\cos\Omega)^2+(0.5\sin\Omega)^2}, \angle H(e^{j\Omega})=\tan^{-1}\left(\frac{-0.5\sin\Omega}{0.5+0.5\cos\Omega}\right).$$

(b) $H(z)=0.5-0.5z^{-1}$, $H(e^{j\Omega})=0.5-0.5e^{-j\Omega}$.

$$|H(e^{j\Omega})|=\sqrt{(0.5-0.5\cos\Omega)^2+(0.5\sin\Omega)^2}, \angle H(e^{j\Omega})=\tan^{-1}\left(\frac{0.5\sin\Omega}{0.5-0.5\cos\Omega}\right).$$

(c) $H(z)=0.5+0.5z^{-2}$, $H(e^{j\Omega})=0.5+0.5e^{-j2\Omega}$.

$$|H(e^{j\Omega})|=\sqrt{(0.5+0.5\cos 2\Omega)^2+(0.5\sin 2\Omega)^2}, \angle H(e^{j\Omega})=\tan^{-1}\left(\frac{-0.5\sin 2\Omega}{0.5+0.5\cos 2\Omega}\right).$$

(d) $H(z)=0.5-0.5z^{-2}$, $H(e^{j\Omega})=0.5-0.5e^{-j2\Omega}$.

$$|H(e^{j\Omega})|=\sqrt{(0.5-0.5\cos 2\Omega)^2+(0.5\sin 2\Omega)^2}, \angle H(e^{j\Omega})=\tan^{-1}\left(\frac{0.5\sin 2\Omega}{0.5-0.5\cos 2\Omega}\right).$$

6.23

$$H(z)=\frac{0.5}{1+0.7z^{-1}+0.1z^{-2}}, y(n)=0.5556u(n)-0.111(-0.2)^n u(n)+0.5556(-0.5)^n u(n).$$

6.25

(a) $y(n)=x(n)-0.9x(n-1)-0.1x(n-2)-0.3y(n-1)+0.04y(n-2)$

(b) $w(n)=x(n)-0.3w(n-1)+0.04w(n-2)$

$y(n)=w(n)-0.9w(n-1)-0.1w(n-2)$

(c) Hint: $H(z)=\dfrac{(z-1)(z+0.1)}{(z+0.4)(z-0.1)}$

$w_1(n)=x(n)-0.4w_1(n-1)$

$y_1(n)=w_1(n)-w_1(n-1)$

$w_2(n)=y_1(n)+0.1w_2(n-1)$

$y(n)=w_2(n)+0.1w_2(n-1)$

(d) Hint: $H(z)=2.5+\dfrac{2.1z}{z+0.4}-\dfrac{3.6z}{z-0.1}$

$y_1(n)=2.5x(n)$

$w_2(n)=x(n)-0.4w_2(n-1)$

$y_2(n)=2.1w_2(n)$

$w_3(n)=x(n)+0.1w_3(n-1)$

$y_3(n)=-3.6w_3(n)$

$y(n)=y_1(n)+y_2(n)+y_3(n)$

6.26 (a) $y(n) = x(n) - 0.5x(n-1)$
(b) $y(n) = x(n) - 0.7x(n-1)$
(c) $y(n) = x(n) - 0.9x(n-1)$
Filter in (c) emphasizes high frequencies most.

Chapter 7
7.1
(a) $H(z) = 0.2941 + 0.3750z^{-1} + 0.2941z^{-2}$, (b) $H(z) = 0.0235 + 0.3750z^{-1} + 0.0235z^{-2}$.

7.3
(a) $H(z) = 0.1514 + 0.1871z^{-1} + 0.2000z^{-2} + 0.1871z^{-3} + 0.1514z^{-4}$
(b) $H(z) = 0.0121 + 0.1010z^{-1} + 0.2000z^{-2} + 0.1010z^{-3} + 0.0121z^{-4}$

7.5
(a) $H(z) = -0.0444 + 0.0117z^{-1} + 0.0500z^{-2} + 0.0117z^{-3} - 0.0444z^{-4}$
(b) $H(z) = -0.0035 + 0.0063z^{-1} + 0.0500z^{-2} + 0.0063z^{-3} - 0.0035z^{-4}$

7.7
(a) Hanning window.
(b) filter length = 63.
(c) cutoff frequency = 1000 Hz.

7.9
(a) Hamming window.
(b) filter length = 45.
(c) lower cutoff frequency = 1500 Hz, upper cutoff frequency = 2300 Hz.

7.11 Hint:
(a) $y(n) = 0.25x(n) - 0.5x(n-1) + 0.25x(n-2)$.
(b) $y(n) = 0.25[x(n) + x(n-2)] - 0.5x(n-1)$.

7.13
$N = 3$, $\Omega_c = 3\pi/10$, $\Omega_0 = 0$, $H_0 = 1$, $\Omega_1 = 2\pi/3$, $H_1 = 0$

$H(z) = 0.3333 + 0.3333z^{-1} + 0.3333z^{-2}$

7.15
$H(z) = -0.1236 + 0.3236z^{-1} + 0.6z^{-2} + 0.3236z^{-3} - 0.1236z^{-4}$

7.17
$H(z) = 0.1718 - 0.2574z^{-1} - 0.0636z^{-2} + 0.2857z^{-3} - 0.0636z^{-4} - 0.2574z^{-5} + 0.1781z^{-6}$

7.19 $W_p = 1$, $W_s = 12$.

7.21 $W_p = 1$, $W_s = 122$.

7.23
Hamming window, filter length = 33, lower cutoff frequency = 3500 Hz.

7.24
Hamming window, filter length = 53,
lower cutoff frequency = 1250 Hz, upper cutoff frequency = 2250 Hz.

7.25

Lowpass filter: Hamming window, filter length = 91, cutoff frequency = 2000 Hz.
Highpass filter: Hamming window, filter length = 91, cutoff frequency = 2000 Hz.

Chapter 8

8.1 $H(z) = \dfrac{0.3333 + 0.3333z^{-1}}{1 - 0.3333z^{-1}}$

$y(n) = 0.3333x(n) + 0.3333x(n-1) + 0.3333y(n-1)$

8.3

(a) $H(z) = \dfrac{0.6625 - 0.6625z^{-1}}{1 - 0.3249z^{-1}}$

$y(n) = 0.6225x(n) - 0.6225x(n-1) + 0.3249y(n-1)$

8.5

(a) $H(z) = \dfrac{0.2113 - 0.2113z^{-2}}{1 - 0.8165z^{-1} + 0.5774z^{-2}}$

$y(n) = 0.2113x(n) - 0.2113x(n-2) + 08165y(n-1) - 0.5774y(n-2)$

8.7

(a) $H(z) = \dfrac{0.1867 + 0.3734z^{-1} + 0.1867z^{-2}}{1 - 0.4629z^{-1} + 0.2097z^{-2}}$

$y(n) = 0.1867x(n) + 0.3734x(n-1) + 0.1867x(n-2) + 0.4629y(n-1) - 0.2097y(n-2)$

8.9

(a) $H(z) = \dfrac{0.0730 - 0.0730z^{-2}}{1 + 0.8541z^{-2}}$

$y(n) = 0.0730x(n) - 0.0730x(n-2) - 0.8541y(n-2)$

8.11

(a) $H(z) = \dfrac{0.5677 + 0.5677z^{-1}}{1 + 0.1354z^{-1}}$

$y(n) = 0.5677x(n) + 0.5677x(n-1) - 0.1354y(n-2)$

8.13

(a) $H(z) = \dfrac{0.1321 - 0.3964z^{-1} + 0.3964z^{-2} - 0.1321z^{-3}}{1 + 0.3432z^{-1} + 0.6044z^{-2} + 0.2041z^{-3}}$

$$y(n) = 0.1321x(n) - 0.3964x(n-1) + 0.3964x(n-2) - 0.1321x(n-3)$$
$$-0.3432y(n-1) - 0.6044y(n-2) - 0.2041y(n-3)$$

8.15

(a) $H(z) = \dfrac{0.9609 + 0.7354z^{-1} + 0.9609z^{-2}}{1 + 0.7354z^{-1} + 0.9217z^{-2}}$

$$y(n) = 0.9609x(n) + 0.7354x(n-1) + 0.9609x(n-2)$$
$$-0.7354y(n-1) - 0.9217y(n-2)$$

8.17

(a) $H(z) = \dfrac{0.0242 + 0.0968z^{-1} + 0.1452z^{-2} + 0.0968z^{-3} + 0.0242z^{-4}}{1 - 1.5895z^{-1} + 1.6690z^{-2} - 0.9190z^{-3} + 0.2497z^{-4}}$

$$y(n) = 0.0242x(n) + 0.0968x(n-1) + 0.1452x(n-2) + 0.0968x(n-3) + 0.0242x(n-4)$$
$$+ 1.5895y(n-1) - 1.6690y(n-2) + 0.9190y(n-3) - 0.2497y(n-4)$$

8.19 $H(z) = \dfrac{1}{1 - 0.3679z^{-1}}$

(a) $H(z) = \dfrac{1}{1 - 0.3679z^{-1}}$

$y(n) = x(n) + 0.3679y(n-1)$

8.21

(a) $H(z) = \frac{0.1 - 0.09781z^{-1}}{1 - 1.6293z^{-1} + 0.6703z^{-2}}$

$y(n) = 0.1x(n) - 0.0978x(n-1) + 1.6293y(n-1) - 0.6703y(n-2)$

8.23

$H(z) = \frac{0.9320 - 1.3180z^{-1} + 0.9320z^{-2}}{1 - 1.3032z^{-1} + 0.8492}$

$y(n) = 0.9320x(n) - 1.3180x(n-1) + 0.9329x(n-2) + 1.3032y(n-1) - 0.8492y(n-2)$

8.25

$H(z) = \frac{0.9215 + 0.9215z^{-1}}{1 + 0.8429z^{-1}}$

$y(n) = 0.9215x(n) + 0.9215x(n-1) - 0.8429y(n-1)$

8.27

$H(z) = \frac{0.9607 - 0.9607z^{-1}}{1 - 0.9215z^{-1}}$

$y(n) = 0.9607x(n) - 0.9607x(n-1) + 0.9215y(n-1)$

8.29

(a)

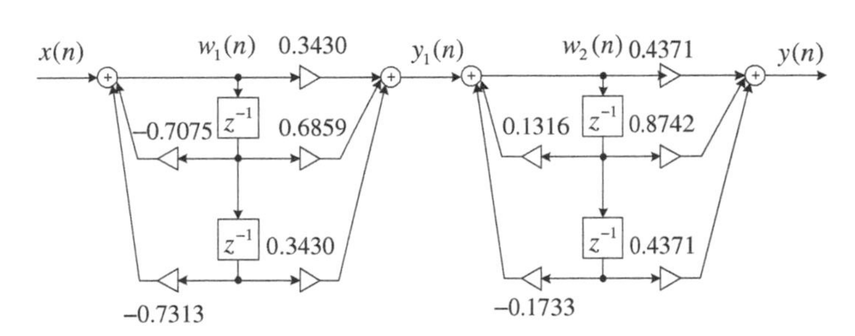

(b) for section 1: $w_1(n) = x(n) - 0.7075w_1(n-1) - 0.7313w_1(n-2)$

$y1(n) = 0.3430w_1(n) + 0.6859w_1(n-1) + 0.3430w_1(n-2)$

for section 2: $w_2(n) = y_1(n) + 0.1316w_2(n-1) - 0.1733w_2(n-2)$

$y_2(n) = 0.4371w_2(n) + 0.8742w_2(n-1) + 0.4371w_2(n-2)$

8.30 $H(z) = \frac{0.9511z^{-1}}{1.0000 - 0.6180z^{-1} + z^{-2}}$, $y(n) = 0.9511x(n-1) + 0.618y(n-1) - y(n-2)$

8.32 (a) $H_{852}(z) = \frac{0.6203z^{-1}}{1 - 1.5687z^{-1} + z^{-2}}$, $H_{1477}(z) = \frac{0.9168z^{-1}}{1 - 0.7986z^{-1} + z^{-2}}$

(b) $y_{852}(n) = 0.6203x(n-1) + 1.5678y_{852}(n-1) - y_{852}(n-2)$

$y1477(n) = 0.9168x(n-1) + 0.7986y_{1477}(n-1) - y_{1477}(n-2)$

$y9(n) = y_{1477}(n) + y_{852}(n)$

8.34 $X(0) = 2$, $|X(0)|^2 = 4$, $A_0 = 0.5$ (single side).

$X(1) = 1 - j3$, $|X(1)|^2 = 10$, $A_1 = 1.5811$ (single side).

8.36 $A_0=2.5$, $A_2=0.5$.

8.39 Chebyshev notch filter 1: order =2

$$H(z)=\frac{0.9915-1.9042z^{-1}+0.9915z^{-2}}{1.0000-1.9042z^{-1}+0.9830z^{-2}}$$

Chebyshev notch filter 2: order =2

$$H(z)=\frac{0.9917-1.3117z^{-1}+0.9917z^{-2}}{1.0000-1.3117z^{-1}+0.9835z^{-2}}$$

8.41 filter order=4;

$$H(z)=\frac{0.1103+0.4412z^{-1}+0.6618z^{-2}+0.4412z^{-3}+0.1103z^{-4}}{1.0000+0.1509z^{-1}+0.8041z^{-2}-0.1619z^{-3}+0.1872z^{-4}}$$

8.43 filter order=4;

$$H(z)=\frac{0.0300-0.0599z^{-2}+0.0300z^{-4}}{1.0000-0.6871z^{-1}+1.5741z^{-2}-0.5177z^{-3}+0.5741z^{-4}}$$

8.45 $H(z)=\dfrac{0.5878z^{-1}}{1-1.6180z^{-1}+z^{2}}$

$y(n)=0.5878x(n-1)+1.6180y(n-1)-y(n-2)$

8.47 $X(0)=1$; $|X(0)|^2=1$; $A_0=0.25$ (single side).
$X(1)=1-j2$; $|X(1)|^2=5$; $A_1=1.12$ (single side)

Chapter 9

9.1 $w^*=2$, and $J_{\min}=10$.

9.3 $w^*=-5$, and $J_{\min}=50$.

9.5 $w^*\approx w_3=1.984$, and $J_{\min}=10.0026$.

9.7 $w^*\approx w_3=-4.992$, and $J_{\min}=5.0001$.

9.9

(a) $y(n)=w_0x(n)+w_1x(n-1)$

$e(n)=d(n)-y(n)$

$w_0=w_0+0.2\times e(n)x(n)$

$w_1=w_1+0.2\times e(n)x(n-1)$

(b) for $n=0$

$y(0)=0$

$e(0)=3$

$w_0=1.8$

$w_1=1$

for $n=1$

$y(1)=1.2$

$e(1)=-3.2$

$w_0=2.44$

$w_1=-0.92$

for $n=2$

$y(2)=5.8$

$e(2)=-4.8$

$w_0=0.52$

$w_1=0.04$

9.13 (a) $n(n)=0.5\cdot x(n-5)$

(b) $xx(n)=5000\cdot\delta(n)$, $yy(n)=0.7071xx(n-1)+1.4141yy(n-1)-yy(n-2)$

(c) $d(n)=yy(n)-n(n)$

(d) for $k=0, \cdots, 24$, $w_k=0$

$$y(n)=\sum_{k=0}^{24} w_k x(n-k)$$

$e(n)=d(n)-y(n)$

for $k=0, \cdots, 24$

$w_k=w_k+2\mu e(n)x(n-k)$

9.15

(a) $w_0=w_1=w_2=0$, $\mu=0.1$

$y(n)=w_0x(n)+w_1x(n-1)+w_2x(n-2)$

$e(n)=d(n)-y(n)$

$w_0=w_0+0.2e(n)x(n)$

$w_1=w_1+0.2e(n)x(n-1)$

$w_2=w_2+0.2e(n)x(n-2)$

(b) for $n=0$: $y(0)=0$, $e(0)=0$, $w_0=0$, $w_1=0$, $w_2=0$

for $n=1$: $y(1)=0$, $e(1)=2$, $w_0=0.4$, $w_1=0.4$, $w_2=0$.

for $n=2$: $y(2)=0$, $e(2)=-1$, $w_0=0.6$, $w_1=0.2$, $w_2=-0.2$

9.17

(a) $w_0=w_1=0$, $\mu=0.1$

$x(n)=d(n-3)$

$y(n) = w_0 x(n) + w_1 x(n-1)$

$e(n) = d(n) - y(n)$

$w_0 = w_0 + 0.2e(n)x(n)$

$w_1 = w_1 + 0.2e(n)x(n-1)$

(b) for $n=0$: $x(0)=0$, $y(0)=0$, $e(0)=-1$, $w_0=0$, $w_1=0$

for $n=1$: $x(1)=0$, $y(1)=0$, $e(1)=1$, $w_0=0$, $w_1=0$.

for $n=2$: $x(2)=0$, $y(2)=0$, $e(2)=-1$, $w_0=0$, $w_1=0$.

for $n=3$: $x(3)=-1$, $y(3)=0$, $e(3)=1$, $w_0=-0.2$, $w_1=0$.

for $n=4$: $x(4)=1$, $y(4)=-0.2$, $e(4)=-0.8$, $w_0=-0.36$, $w_1=0.16$.

9.18
(a) 30 coefficients

9.20
for $k=0, \cdots, 19$, $w_k=0$

$$y(n) = \sum_{k=0}^{19} w_k x(n-k)$$

$e(n) = d(n) - y(n)$

for $k=0, \cdots, 19$

$w_k = w_k + 2\mu e(n)x(n-k)$

Chapter 10
10.1 (a) $\Delta=0.714$.

(b) for $x=1.6$ V, binary code $=110$, $x_q=1.428$ V, and $e_q=-0.172$ V.

for $x=-0.2$ V, binary code $=000$, $x_q=0$ V, and $e_q=0.2$ V.

10.5 for $x=1.6$ V, binary code $=111$, $x_q=1.132$ V, and $e_q=-0.468$V.

for $x=-0.2$ V, binary code $=010$, $x_q=-0.224$ V, and $e_q=-0.024$V.

10.9 (a) 0 0 0 1 0 1 0 1.

(b) 1 1 1 0 0 1 1 1.

10.10
(a) 0 0 0 0 0 0 0 0 0 1 1 1, (b) 1 0 1 1 0 0 1 1 0 0 0 0.

10.13
010

001

010

10.14
for binary code =110, $\hat{x}(0) = \tilde{x}(0) + d_q(0) = 0 + 5 = 5$
for binary code =100, $\hat{x}(1) = \tilde{x}(1) + d_q(1) = 5 + 0 = 5$
for binary code =110, $\hat{x}(2) = \tilde{x}(2) + d_q(2) = 5 + 2 = 7$

10.17
(a) 1:1.
(b) 2:1.
(c) 4:1.

10.18
(a) 128 KBPS.
(b) 64 KBPS
(c) 32 KBPS.

10.21
(a) 12 channels.
(b) 24 channels
(c) 48 channels

10.22
$X_{DCT}(0) = 54, X_{DCT}(1) = 0.5412, X_{DCT}(2) = -4, X_{DCT}(3) = -1.3066$

10.25
$X_{DCT}(1) = 33.9730, X_{DCT}(3) = -10.4308, X_{DCT}(5) = 1.2001, X_{DCT}(7) = -1.6102$

10.26
(a) inverse DCT: 10.0845 6.3973 13.6027 −2.0845.
(b) recovered inverse DCT: 11.3910 8.9385 15.0615 −3.3910
(c) quantization error: 1.3066 2.5412 1.4588 −1.3066

10.29
(a) −9.0711 −0.5858
−13.3137 −0.0000
−7.8995 0.5858
(b) 3, 4, 5, 4

Chapter 11
11.3
(a)

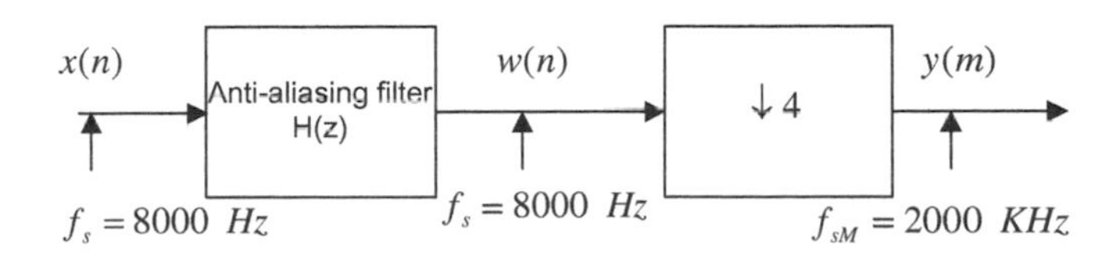

(b) Hamming window, $N = 133$, $f_c = 900$ Hz.

11.4

(a)

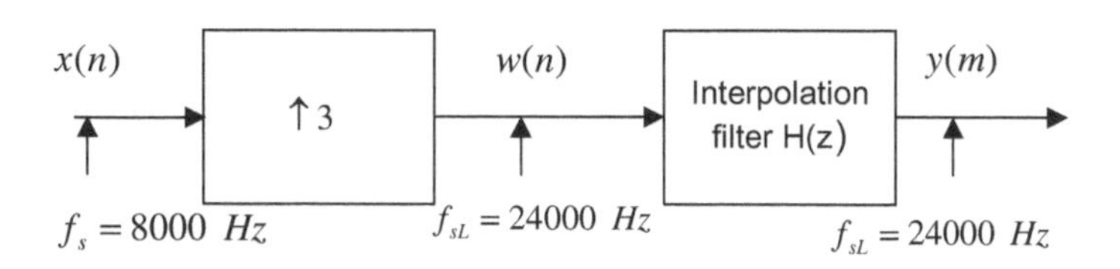

(b) Hamming window, $N = 133$, $f_c = 3700$ Hz.

11.7

(a)

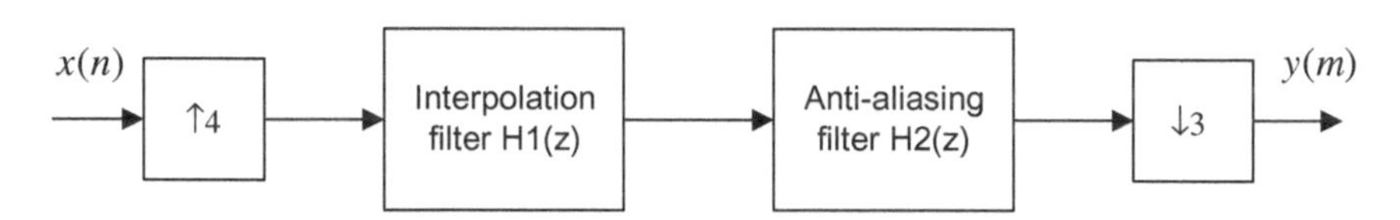

(b) Combined filter $H(z)$: Hamming window, $N = 133$, $f_c = 2700$ KHz.

11.8

(a)

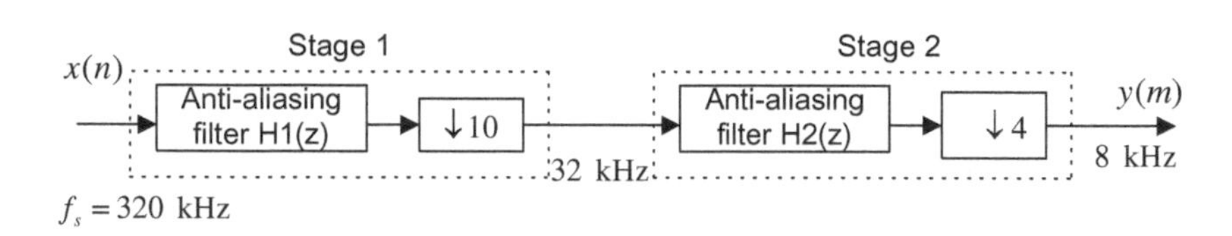

(b) $M_1 = 10$, and $M_2 = 4$.
(c) Filter specification for $H_1(z)$: Hamming window, choose $N = 43$, $f_c = 15{,}700$ Hz.
(d) Filter specification for $H_2(z)$: Hamming window, $N = 177$, $f_c = 3700$ Hz.

11.9

(a)

(b)

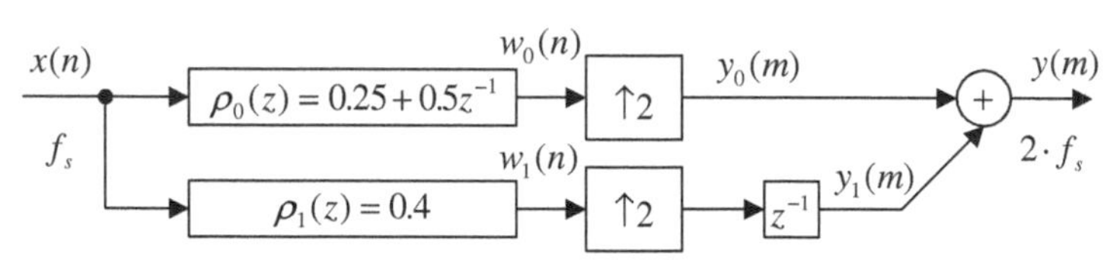

11.10

(a)

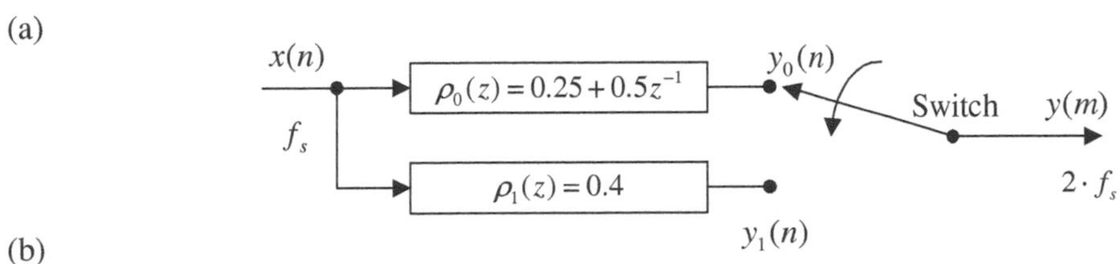

(b)

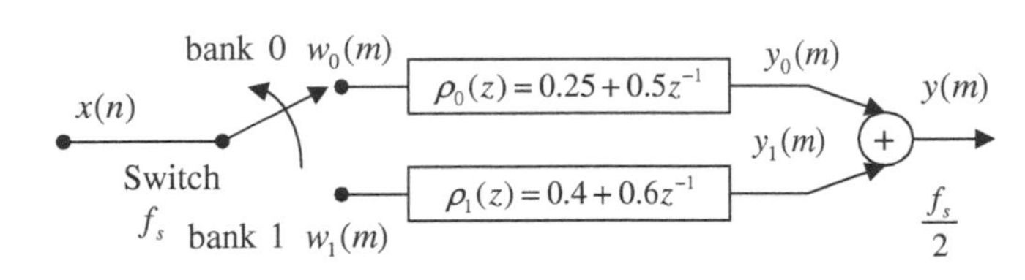

11.15 (a) $f_s = 2f_{\max}2^{2(n-m)} = 2 \times 15 \times 2^{2\times(16-12)} = 7680$ kHz.

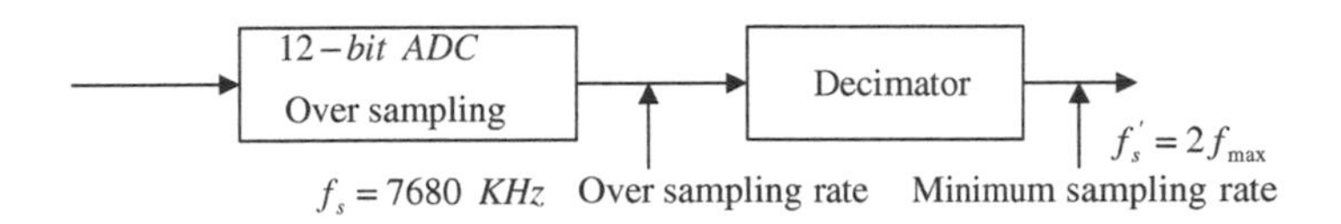

11.17

(a)

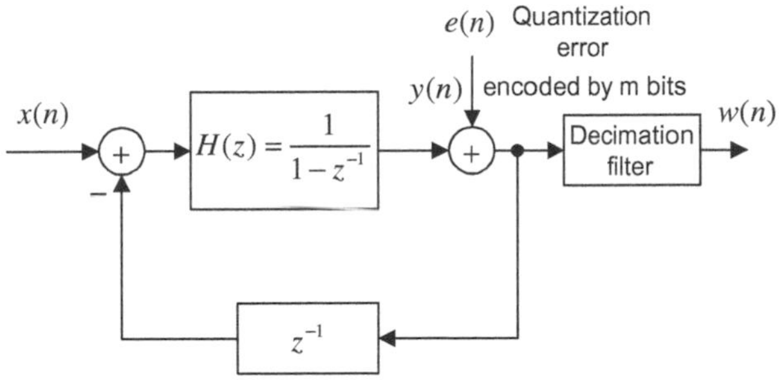

(b) $n = 1 + 1.5 \times \log_2\left(\frac{128}{2\times4}\right) - 0.86 \approx 6$ bits

11.18

(a)

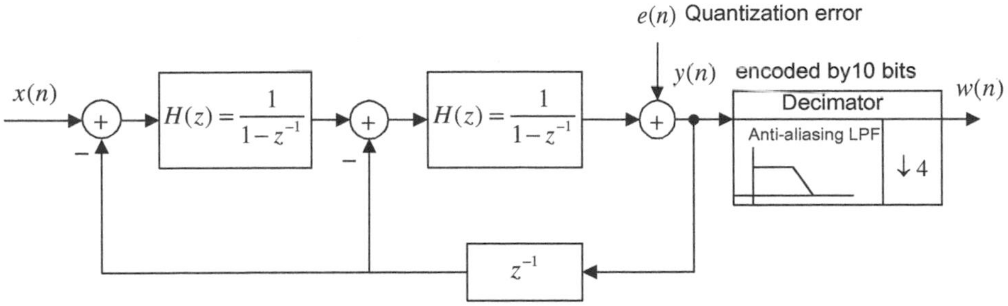

(b)

$$n = m + 2.5 \times \log_2\left(\frac{f_s}{2f_{\max}}\right) - 2.14 = 10 + 2.5 \times \log_2\left(\frac{160}{2 \times 20}\right) - 2.14 = 12.86 \approx 13\text{ bits}$$

11.21

(a) $f_c/B = 6$ is an even number, which is the case 1, we select $f_s = 10$ kHz

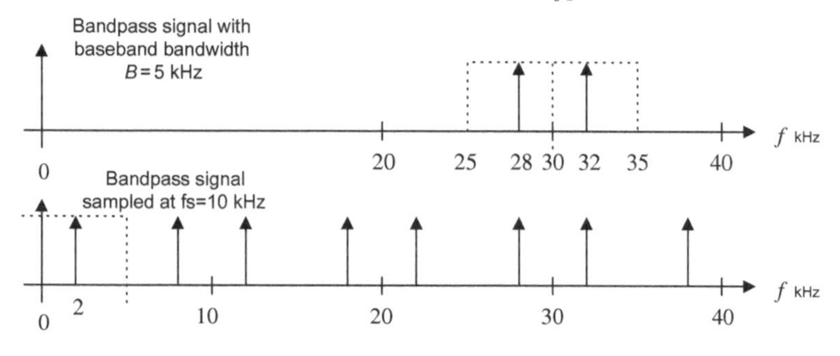

(b) Since $f_c/B = 5$ is an odd number, we select $f_s = 10$ kHz.

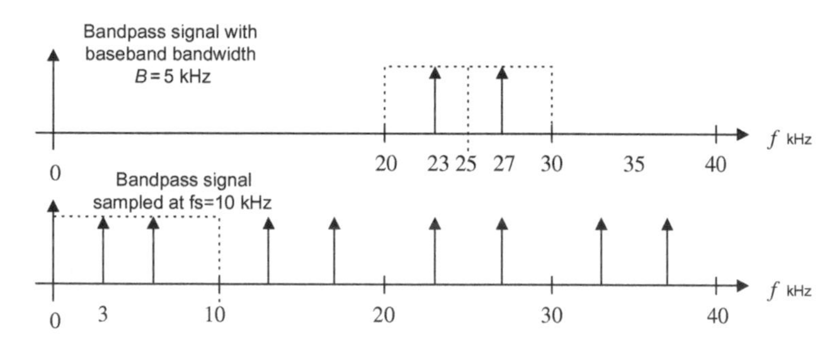

(c) Now, $f_c/B = 6.6$which is a non-integer. We extend the bandwidth $\overline{B} = 5.5$ kHz, so $f_c/\overline{B} = 6$ and $f_s = 2\overline{B} = 11$kHz.

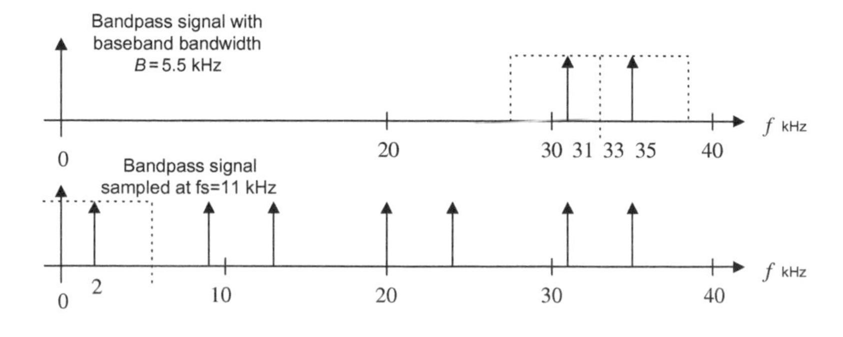

Chapter 12

12.1

(a) $B = f_{sM}/2 = f_s/(2M) f_c = 2(f_s/(2M)) = 2B,\ f_c/B = 2 = \text{even}.$

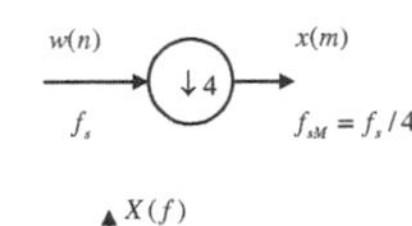

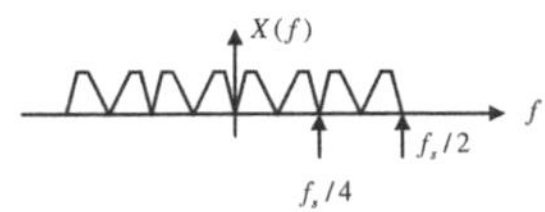

(b) $B = f_{sM}/2 = f_s/(2M) f_c = f_s/(2M) = B,\ f_c/B = 1 = \text{odd}.$

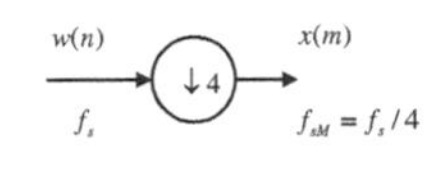

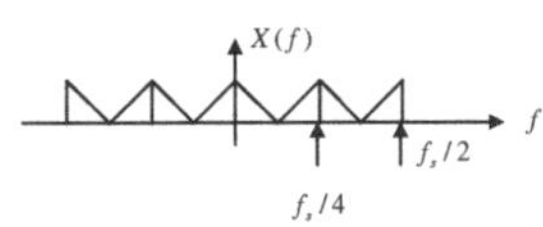

12.3

From Eq. (13.7), $\overline{Y}(z) = \frac{1}{2}(W(z) + W(e^{-j\pi}z))$, $\overline{Y}(e^{j\Omega}) = \frac{1}{2}\left(W(e^{j\Omega}) + W\left(e^{j(\Omega-\pi)}\right)\right)$, $W(e^{j(\Omega-\pi)})$ is the shifted version of $W(e^{j\Omega})$ by $f_s/2$.

(a)

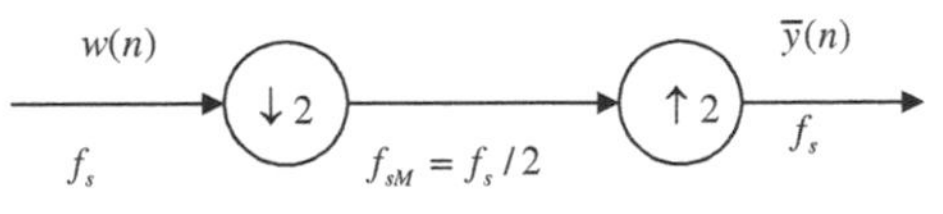

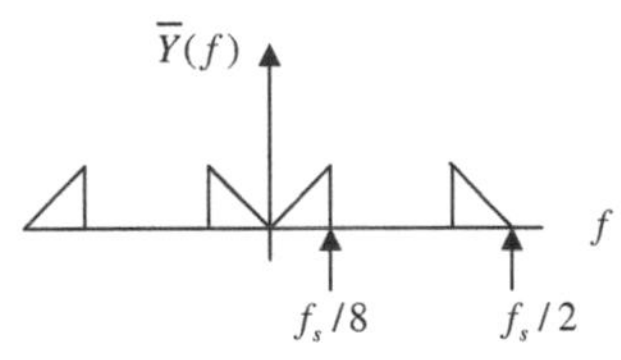

(b)

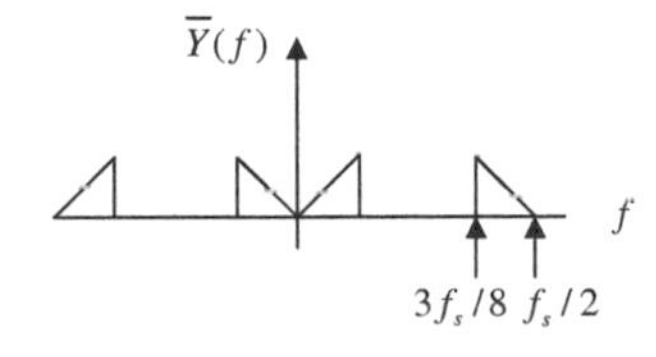

12.5

$H_1(z) = -\frac{1}{\sqrt{2}} + \frac{1}{\sqrt{2}}z^{-1} G_0(z) = \frac{1}{\sqrt{2}} + \frac{1}{\sqrt{2}}z^{-1}, G_1(z) = \frac{1}{\sqrt{2}} - \frac{1}{\sqrt{2}}z^{-1}$

12.7

$H_1(z) = 0.129 + 0.224z^{-1} - 0.837z^{-2} + 0.483z^{-3}$
$G_0(z) = -0.129 + 0.224z^{-1} + 0.837z^{-2} + 0.483z^{-3}$
$G_1(z) = 0.483 - 0.837z^{-1} + 0.224z^{-2} + 0.129z^{-3}$

12.9

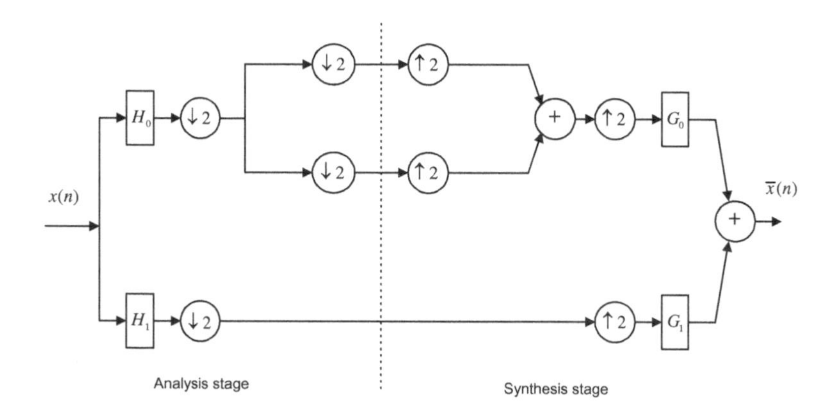

12.11

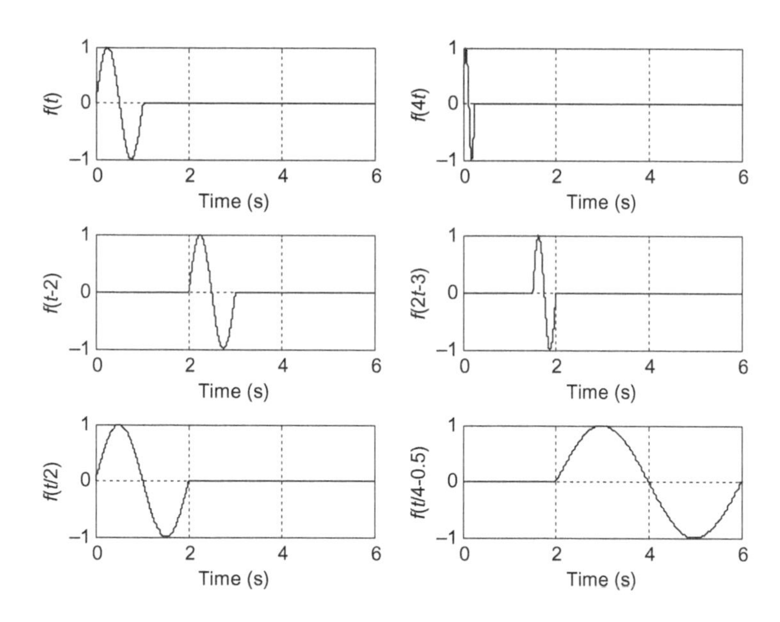

12.13

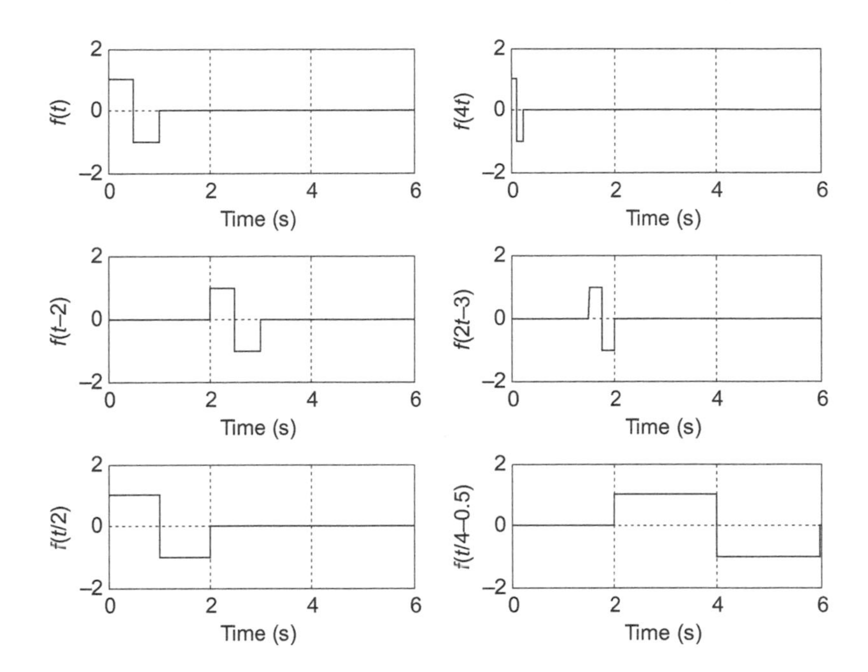

12.15

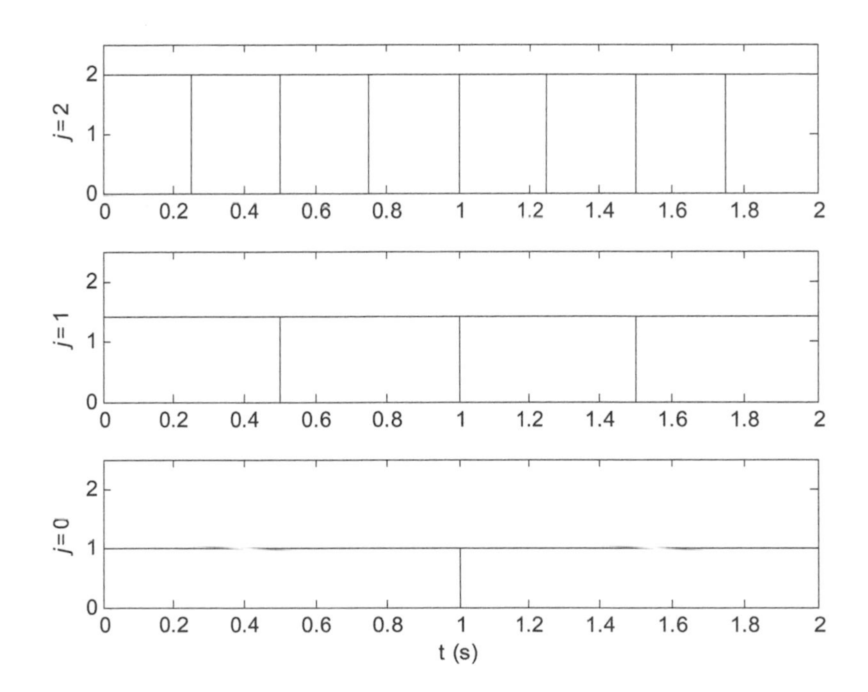

12.17
(a) $f(t) = 4\phi(2t) - 2\phi(2t-1)$
(b) $f(t) = \phi(t) + 3\psi(t)$

12.19
(a) $f(t) = (2/\pi)\phi(2t) - (2/\pi)\phi(2t-1)$
(b) $f(t) = (2/\pi)\psi(t)$

12.21

(a)
$$\sum_{k=-\infty}^{\infty} \sqrt{2}h_0(k)\phi(4t-k) = \sqrt{2}h_0(0)\phi(4t) + \sqrt{2}h_0(1)\phi(4t-1)$$
$$= \sqrt{2}\times 0.707\phi(4t) + \sqrt{2}\times 0.707\phi(4t-1) = \phi(4t) + \phi(4t-1) = \phi(2t)$$

(b)
$$\sum_{k=-\infty}^{\infty} \sqrt{2}h_1(k)\phi(4t-k) = \sqrt{2}h_1(0)\phi(4t) + \sqrt{2}h_1(1)\phi(4t-1)$$
$$= \sqrt{2}\times 0.707\phi(4t) + \sqrt{2}(-0.707)\phi(4t-1) = \phi(4t) - \phi(4t-1) = \psi(2t)$$

12.23 $w(k)$= [5.5000 0.5000 7.0711 2.1213].

12.25 $c(k)$= [2.2929 3.7071 2.4142 -0.4142].

12.27 $c(k)$=[2.1213 3.5355 2.8284 0].

Chapter 13

13.1 (a) 76.8 kbytes, (b) 921.6 kbytes, (c) 1920.768 kbytes.

13.3 $Y = 142$, $I = 54$, $Q = 11$.

13.5 $\begin{bmatrix} 53 & 44 \\ 59 & 50 \end{bmatrix}$

13.7 $\begin{bmatrix} 1 & 4 & 6 & 6 & 1 \\ 6 & 4 & 4 & 6 & 4 \\ 4 & 4 & 6 & 6 & 6 \\ 1 & 6 & 7 & 7 & 4 \end{bmatrix}$

13.9 $\begin{bmatrix} 102 & 109 & 104 & 51 \\ 98 & 101 & 101 & 54 \\ 98 & 103 & 100 & 51 \\ 50 & 55 & 51 & 25 \end{bmatrix}$

13.10 $\begin{bmatrix} 0 & 100 & 100 & 0 \\ 0 & 100 & 100 & 100 \\ 0 & 100 & 100 & 100 \\ 0 & 100 & 0 & 0 \end{bmatrix}$

13.11
(a)
Vertical Sobel detector: $\begin{bmatrix} -1 & 0 & 1 \\ -2 & 0 & 2 \\ -1 & 0 & 1 \end{bmatrix}$, Processed image $\begin{bmatrix} 225 & 125 & 130 & 33 \\ 249 & 119 & 136 & 6 \\ 249 & 119 & 136 & 6 \\ 255 & 125 & 130 & 0 \\ 255 & 128 & 128 & 30 \end{bmatrix}$.

(b) Laplacian edge detector: $\begin{bmatrix} 0 & 1 & 0 \\ 1 & -4 & 1 \\ 0 & 1 & 0 \end{bmatrix}$, Processed image: $\begin{bmatrix} 0 & 106 & 106 & 0 \\ 106 & 255 & 255 & 106 \\ 106 & 255 & 255 & 106 \\ 117 & 223 & 223 & 117 \\ 0 & 117 & 117 & 0 \end{bmatrix}$.

13.13 Blue color is dominant in the area pointed by the arrow; red color is dominant in the background.

13.15 $X(u,v) = \begin{bmatrix} 460 & -40 \\ -240 & -140 \end{bmatrix}$ and $A(u,v) = \begin{bmatrix} 115 & 10 \\ 60 & 35 \end{bmatrix}$.

13.16
Forward DCT: $F(u,v) = \begin{bmatrix} 230 & -20 \\ -120 & -70 \end{bmatrix}$.

13.17 Inverse DCT: $p(i,j) = \begin{bmatrix} 110 & 100 \\ 100 & 90 \end{bmatrix}$.

13.19
(a) (0,-2) (3,4), (2,-3), (0,7), (4, −2), (0,0).
(b) (0000, 0010, 01), (0011, 0011, 100), (0010, 0010, 00),
(0000, 0011, 111), (0100, 0010, 01), (0000, 0000)

13.20
$w = 230.0000\ -20.0000;$
$-120.0000\ -70.0000.$

13.21
$f = 110.0000\ 100.0000;$
$100.0000\ 90.0000.$

13.23
$f = 115.0000\ 145.0000\ 25.0000\ 45.0000.$
$105.0000\ 135.0000\ 5.0000\ 25.0000.$
$30.0000\ 20.0000\ 7.5000\ 27.5000.$
$10.0000\ -0.0000\ -7.5000\ 12.5000.$

13.28 hint:

$$\begin{aligned} Composite \times 2\sin(2\pi f_{sc}t) &= Y \times 2\sin(2\pi f_{sc}t) + I\cos(2\pi f_{sc}t) \times 2\sin(2\pi f_{sc}t) + Q \times 2\sin^2(2\pi f_{sc}t) \\ &= Y \times 2\sin(2\pi f_{sc}t) + I\sin(2 \times 2\pi f_{sc}t) + Q - Q\cos(2 \times 2\pi f_{sc}t) \end{aligned}$$

Then apply lowpass filtering.

13.35
$\frac{80 \times 80}{16 \times 16}(16^2 \times 32^2 \times 3) = 19.661 \times 10^6$ operations

Chapter 14
14.1
0.2560123 (decimal) = 0. 0 1 0 0 0 0 0 1 1 0 0 0 1 0 1 (Q-15).

14.2

$-0.2160123 \times 2^{15} = -7078_{10} = 1110010001011010$

-0.2160123 (decimal) $=$ 1. 1 1 0 0 1 0 0 0 1 0 1 1 0 1 0 (Q-15).

14.5

0. 1 0 1 0 1 0 0 0 1 0 1 1 1 1 0 $=0.6591186$.
1. 0 1 0 1 0 1 1 1 0 1 0 0 0 1 0 (Q-15) $=-0.6591186$.

14.6

0. 0 0 1 0 0 0 1 1 1 1 0 1 1 1 0 (Q-15) $=0.1400756$.

14.9

1. 1 0 1 0 1 0 1 1 1 0 0 0 0 0 1 + 0. 0 1 0 0 0 1 1 1 1 0 1 1 0 1 0.
= 1. 1 1 1 1 0 0 1 1 0 0 1 1 0 1 1.

14.13

(a) 1101 011100001100.
(b) 0100 101,011,001,001.

14.15

1101 011100011011 (floating) $= 0.8881835 \times 2^{-3}$(decimal).
0100 101,111,100,101 (floating) $= -0.5131835 \times 2^{4}$ (decimal).
0.8881835×2^{-3}(decimal)$= 0.0069389 \times 2^{4}$ (decimal) $=$ 0100 000000001110 (floating).

0100 101,111,100,101 (floating) + 0100 000000001110 (floating).
= 0100 101,111,110,011 (floating) $= -8.1016$ (decimal).

14.18

$(-1)^1 \times 1.025 \times 2^{160-127} = -8.8047 \times 10^9$

14.20

$(-1)^0 \times 1.625 \times 2^{1536-1023} = 4.3575 \times 10^{154}$

14.25

$B=2, S=2$

$x_s(n) = \dfrac{x(n)}{2}$

$y_s(n) = -0.18x_s(n) + 0.8x_s(n-1) + 0.18x_s(n-2)$

$y(n) = 4y_s(n)$

14.26

$S=1, C=2$

$x_s(n) = x(n), y_s(n) = 0.75x_s(n) + 0.15y_f(n-1), y_f(n) = 2y_s(n), y(n) = y_f(n)$

14.29

$S=8, A=2, B=4$

$x_s(n) = x(n)/8, w_s(n) = 0.5x_s(n) - 0.675w(n-1) - 0.25w(n-2), w(n) = 2w_s(n)$

$y_s(n) = 0.18w(n) + 0.355w(n-1) + 0.18w(n-2), y(n) = 32y_s(n)$

Appendix B

B.1 $A_0=0.4$, $A_1=0.7916$, $A_2=0.7667$, $A_3=0.7263$, $A_4=0.6719$
$|c_0|=0.4$, $|c_1|=|c_{-1}|=0.3958$,
$|c_2|=|c_{-2}|=0.3834$, $|c_3|=|c_{-3}|=0.3632$, $|c_4|=|c_{-4}|=0.3359$.

B.3
$x(t)=2+3.7420\times\cos(2000\pi t)+3.0273\times\cos(4000\pi t)$
$+2.0182\times\cos(6000\pi t)+0.9355\times\cos(8000\pi t)+\cdots$

$f_2=2000$ Hz, $A_2=3.0273$

B.5 $X(f)=5\left(\dfrac{\sin\pi f}{\pi f}\right)^2$

B.7 (a) $X(s)=10$, (b) $X(s)=-100/s^2$, (c) $X(s)=\dfrac{10}{s+2}$

(d) $X(s)=\dfrac{2e^{-5s}}{s}$, (e) $X(s)=\dfrac{10s}{s^2+9}$, (f) $X(s)=\dfrac{14.14+7.07s}{s^2+9}$

(g) $X(s)=\dfrac{3(s+2)}{(s+2)^2+9}$, (h) $X(s)=\dfrac{12000}{s^6}$

B.9
(a) $X(s)=\frac{7.5}{s(s+1.5)}$, (b) $x(t)=5u(t)-5e^{-1.5t}u(t)$.

B.11
(a) zero: $s=3$, poles: $s=-2$, $s=-2$, stable.
(b) zeros: $s=0$, $s=\pm2.236j$, poles: $s=\pm3j$, $s=-1\pm1.732j$, marginally stable.
(c) zeros: $s=\pm j$, $s=-1$, poles: $s=0$, $s=-3$, $s=-4$, $s=-8$, $s=1$, unstable.

B.13
(a) $H(j\omega)=\dfrac{1}{\dfrac{j\omega}{5}+1}$, (b) $A(\omega)=\dfrac{1}{\sqrt{1+\left(\dfrac{\omega}{5}\right)^2}}\beta(\omega)=\angle-\tan\left(\frac{\omega}{5}\right)$.

(c) $Y(j2)=4.6424\angle-21.80^0$ that is, $y_{ss}(t)=4.6424\sin(2t-21.80^0)u(t)$.

References

Ahmed, N., Natarajan, T., 1983. Discrete-Time Signals and Systems. Prentice Hall, Englewood Cliffs, NJ.
Akansu, A.N., Haddad, R.A., 2001. Multiresolution Signal Decomposition: Transforms, Subbands, and Wavelets, second ed. Academic Press, Inc.
Alkin, O., 1993. Digital Signal Processing—A Laboratory Approach Using PC-DSP. Prentice-Hall, Englewood Cliffs, NJ.
Ambardar, A., 1999. Analog and Digital Signal Processing, second ed. Brooks/Cole Publishing Company, ITP.
Brandenburg, K., 1997. Overview of MPEG Audio: Current and Future Standards for Low-Bit-Rate Audio Coding. J. Audio Eng. Soc. 45(1/2).
Carr, J.J., Brown, J.M., 2001. Introduction to Biomedical Equipment Technology, fourth ed. Prentice-Hall, Upper Saddle River, NJ.
Chassaing, R., Reay, D., 2008. Digital Signal Processing and Applications with the TMS320C6713 and TMS320C6416 DSK, second ed. John Wiley & Sons.
Chen, W., 1986. Passive and Active Filters—Theory and Implementations. John Wiley & Sons.
Dahnoun, N., 2000. Digital Signal Processing Implementation Using the TMS320C6000TM DSP Platform. Prentice-Hall.
Deller, J.R., Hansen, J.H.L., Proakis, J.G., 1999. Discrete-Time Processing of Speech Signals. Wiley-IEEE Press.
El-Sharkawy, M., 1996. Digital Signal Processing Applications with Motorola's DSP56002 Processor. Prentice-Hall, Upper Saddle River, NJ.
Embree, P.M., 1995. C Algorithms for Real-Time DSP. Prentice-Hall, Upper Saddle River, NJ.
Farhang-Boroujeny, B., 2013. Adaptive Filters: Theory and Applications, second ed. John Wiley.
Gonzalez, R.C., Wintz, P., 1987. Digital Image Processing, second ed. Addison-Wesley Publishing Company.
Grover, D., Deller, J.R., 1998. Digital Signal Processing and the Microcontroller. Prentice-Hall, Upper Saddle River, NJ.
Haykin, S., 2014. Adaptive Filter Theory, fifth ed. Prentice-Hall, Englewood Cliffs, NJ.
Ifeachor, E.C., Jervis, B.W., 2002. Digital Signal Processing: A Practical Approach, second ed. Prentice Hall, Upper Saddle River, NJ.
Jiang, J., Tan, L., 2012. In: Teaching Adaptive Filters and Applications in Electrical and Computer Engineering Technology Program.2011 Proceedings of the American Society for Engineering Education, San Antonio, Texas, June.
Kehtaranavaz, N., Simsek, B., 2000. C6x-Based Digital Signal Processing. Prentice Hall, Upper Saddle River, NJ. 07458.
Krauss, T.P., Shure, L., Little, J.N., 1994. Signal Processing TOOLBOX for Use with MATLAB. The Math Works Inc.
Kuo, S.M., Morgan, D.R., 1996. Active Noise Control Systems: Algorithms and DSP Implementations. John Wiley & Sons.
Li, Z.N., Drew, M.S., Liu, J., 2014. Fundamentals of Multimedia, second ed. Springer.
Lipshiz, S.P., Wannamaker, R.A., Vanderkooy, J., 1992. Quantization and dither: a theoretical survey. J. Audio Eng. Soc. 40 (5), 355–375.
Lynn, P.A., Fuerst, W., 1999. Introductory Digital + Processing with Computer Applications, second ed. John Wiley & Sons.
Maher, R.C., 1992. On the nature of granulation noise in uniform quantization systems. J. Audio Eng. Soc. 40 (1/2), 12–20.

Maxim Integrated Products, 2018. MAX1402 +5V, 18-Bit, Low-Power, Multichannel, Oversampling (Sigma-Delta) ADC. Retrieved August 4, 2018 from https://www.maximintegrated.com/en/products/analog/data-converters/analog-to-digital-converters/MAX1402.html.

McClellan, J.H., Oppenheim, A.V., Shafer, R.W., Burrus, C.S., Parks, T.W., Schuessler, H., 1998. Computer Based Exercises for Signal Processing Using MATLAB. Prentice-Hall, Upper Saddle River, NJ.

McClellan, J.H., Shafer, R.W., Yoder, M.A., 2015. DSP First—A Multimedia Approach, second ed. Prentice-Hall, Upper Saddle River, NJ.

Mitra, S.K., 2006. Digital Signal Processing: A Computer-Based Approach, third ed. McGraw-Hill.

Nelatury, S.R., 2007. Additional correction to the impulse invariance method for the design of IIR digital filters. Digital Signal Process. 17, 530–540. Elsevier.

Nelson, M., 1992. The Data Compression Book. M&T Publishing, Inc.

Oppenheim, A.V., Shafer, R.W., 1975. Discrete-Time Signal Processing. Prentice-Hall, Englewood Cliffs, NJ.

Oppenheim, A.V., Shafer, R.W., Buck, J.R., 1998. Discrete-Time Signal Processing, second ed. Prentice-Hall, Upper Saddle River, NJ.

Pan, D., 1995. A tutorial on MPEG/audio compression. IEEE Multimedia 2 (2), 60–74.

Phillips, C.L., Harbor, R.D., 2000. Feedback Control Systems, fifth ed. Prentice-Hall, Englewood Cliffs, NJ.

Phillips, C.L., Nagle, H.T., 1995. Digital Control System Analysis and Design, third ed. Prentice-Hall, Englewood Cliffs, NJ.

Porat, B., 1997. A Course in Digital Signal Processing. John Wiley & Sons.

Princen, J., and Bradley, A. B., "Analysis/synthesis filter bank design based on time domain aliasing cancellation," IEEE Trans. Acoust. Speech Signal Process., Vol. ASSP 34, No. 5, 1986.

Proakis, J.G., Manolakis, D.G., 2007. Digital Signal Processing: Principles, Algorithms, and Applications, fourth ed. Prentice-Hall, Upper Saddle River, NJ.

Rabbani, M., Jones, P.W., 1991. Digital Image Compression Techniques. The Society of Photo-Optical Instrumentation Engineers (SPIE), Bellingham, Washington.

Rabiner, L.R., Schafer, R.W., 1978. Digital Processing of Speech Signals. Prentice-Hall, Englewood Cliffs, NJ.

Randall, R.B., 2011. Vibration-Based Conditioning Monitoring: Industrial, Aerospace, and Automotive Applications. John Wiley & Sons.

Roddy, D., Coolen, J., 1997. Electronic Communications, fourth ed. Prentice Hall.

Sayood, K., 2012. Introduction to Data Compression, fourth ed. Morgan Kaufmann.

Smith, M. J. T., amd Barnwell, T. P., "A procedure for designing exact reconstruction filter banks for tree-structured sub-band coders," Proc. IEEE ICASSP, pp. 27.1.1-27.1.4, 1984.

Soliman, S.S., Srinath, M.D., 1998. Continuous and Discrete Signals and Systems, second ed. Prentice-Hall, Upper Saddle River.

Sorensen, H.V., Chen, J.P., 1997. A Digital Signal Processing Laboratory Using TMS320C30. Prentice-Hall, Upper Saddle River.

Stanley, W.D., 2003. Network Analysis with Applications, fourth ed. Prentice-Hall, Upper Saddle River, NJ.

Stearns, S.D., David, R.A., 1996. Signal Processing Algorithms in MATLAB. Prentice-Hall, Upper Saddle River, NJ.

Stearns, S.D., Hush, D.R., 1990. Digital Signal Analysis, second ed. Prentice-Hall, Englewood Cliff, NJ.

Stearns, S.D., Hush, D.R., 2011. Digital Signal Processing with Examples in MATLAB. CRC Press LLC, second ed.

Tan, L., Jiang, J., 2008. A simple DSP project for teaching real-time signal rate conversions. Technol. Interf. J. 9 (1). Fall.

Tan, L., Jiang, J., 2010. In: Improving Digital Signal Processing Course with Real-time Processing Experiences for Electrical and Computer Engineering Technology Students.2010 Proceedings of the American Society for Engineering Education, Louisville, Kentucky, June.

Tan, L., Jiang, J., 2012. Novel adaptive IIR notch filters for frequency estimation and tracking. Chapter 20, In: Streamlining Digital Signal Processing: A Tricks of the Trade Guidebook. IEEE Press/Wiley & Sons, pp. 197–205.

Tan, L., Wang, L., 2011. Oversampling technique for obtaining higher-order derivative of low-frequency signals. IEEE Trans. Instrum. Meas. 60 (11), 3677–3684.

Tan, L., Jiang, J., Wang, L., 2012. Pole radius varying IIR notch filter with transient suppression. IEEE Trans. Instrum. Meas. 61 (6), 1684–1691.

Tan, L., Jiang, J., Wang, L., April 2014. Multirate processing technique for obtaining integer and fractional order derivatives of low-frequency signals. IEEE Trans. Instrum. Meas. 63 (4), 904–912.

Texas Instruments, 1991. TMS320C3x User's Guide. Texas Instruments, Dallas, Texas.

Texas Instruments, 1998. TMS320C6x CPU and Instruction Set Reference Guide. Literature ID# SPRU 189C, Texas Instruments, Dallas, Texas.

Texas Instruments, 2001. Code Composer Studio: Getting Started Guide. Texas Instruments, Dallas, Texas.

Tomasi, W., 2004. Electronic Communications Systems: Fundamentals Through Advanced, fifth ed. Prentice Hall, Upper Saddle River, NJ.

Vaidyanathan, P.P., 1993. Multirate Systems and Filter Banks. Prentice Hall, Upper Saddle River, NJ.

Vaidyanathan, P. P., 1990. "Multirate digital filters. Filter banks, polyphase networks, and applications: a tutorial," Proc. IEEE, Vol. 78, pp. 56–93.

Vegte, J., 2002. Fundamentals of Digital Signal Processing. Prentice Hall, Upper Saddle River, NJ.

Vetterli, M., Kovacevic, J., 1995. Wavelets and Subband Coding. Prentice Hall PTR, Englewood Cliffs, NJ.

Webster, J.G., 2009. Medical Instrumentation: Application and Design, fourth ed. John Wiley & Sons, Inc.

Widrow, B., Stearns, S.D., 1985. Adaptive Signal Processing. Prentice-Hall, Englewood Cliff, NJ.

Yost, W.A., 2000. Fundamentals of Hearing: An Introduction, fourth ed. Academics Press.

Index

Note: Page numbers followed by *f* indicate figures, *t* indicate tables, and *b* indicate boxes.

A

J

K

L

M

N

R

S

T

U

V

W

Y

Z